55th Annual Conference of the Environmental Design Research Association 2024 (EDRA55)

Human-Centric Environments

Portland, Oregon, USA
19-22 June 2024

Editors:

Ihab Elzeyadi
Sharmin Kader

ISBN: 979-8-3313-2813-9
DOI: https://doi.org/10.52202/082078

Printed from e-media with permission by:

Curran Associates, Inc.
57 Morehouse Lane
Red Hook, NY 12571

Some format issues inherent in the e-media version may also appear in this print version.

Printed with permission by Curran Associates, Inc. (2025)

For permission requests, please contact The Environmental Design Research Association (EDRA) at the address below.

The Environmental Design Research Association (EDRA)

2010 Pear St.
#1060 Pinole, CA
94564-9998

www.edra.org

Additional copies of this publication are available from:

Curran Associates, Inc.
57 Morehouse Lane
Red Hook, NY 12571 USA
Phone: 845-758-0400
Fax: 845-758-2633
Email: curran@proceedings.com
Web: www.proceedings.com

HUMAN-CENTRIC ENVIRONMENTS

CONFERENCE PROCEEDINGS

55th Annual Conference of the
Environmental Design Research Association

CO-EDITORS
Ihab Elzeyadi, Ph.D. and Sharmin Kader, Ph.D.

PRODUCTION EDITOR
Sharmin Kader, Ph.D.

ASSOCIATE EDITORS
Debarati Majumdar Narayan, Ph.D.
Elif Tural, Ph.D.
Alexandre H. Champagne, Ph.D.
Lisa Lim, Ph.D.
Marzieh Rezaei, Ph.D.
Deni Ruggeri, Ph.D.
Bo Zhang, Ph.D.

edra55
PORTLAND, OREGON

PO Box 43023 Washington, D.C., 20010, USA

Web: www.edra.org
Email: proceedings@edra.org
ISBN 979-8-9855428-4-4

THANK YOU TO OUR SPONSORS!

PLATINUM SPONSORS

UNIVERSITY OF OREGON | College of Design

UNIVERSITY OF OREGON | School of Architecture & Environment

BRONZE SPONSOR

HACIENDA
COMMUNITY DEVELOPMENT CORP.

GOLD SPONSOR

DLRGROUP
ARCHITECTURE ENGINEERING PLANNING INTERIORS

SILVER SPONSOR

MOVING BOUNDARIES
COLLABORATIVE 2024

PLATINUM & SILVER SPONSOR

LEVER

EXHIBITOR,

BALL STATE UNIVERSITY
R. Wayne Estopinal College of Architecture and Planning

WITH SUPPORT FROM

TRAVEL PORTLAND

TABLE OF CONTENTS

TABLE OF CONTENTS

PREFACE

In 1969, EDRA promoted an innovative design discipline, both humanistic and scientific, that aims to create a cultural shift in the way "we approach design for humans and by humans while sustaining our environment." As the EDRA55 logo visualizes, we planned this conference to celebrate the 55th birthday of EDRA's mission and focus on how we can create human-centric environments as an outcome of a design process in all shapes, forms, and scales.

It is an immense pleasure and honor to be part of the organization of the EDRA 55 conference. This compiled proceeding documents the length and breadth of submissions and activities of this event, which focused on the design of environments and settings that celebrate the integration of human and environmental well-being at all scales of cities, ecologies, neighborhoods, cultural landscapes, buildings, interior spaces, and products. The conference theme asked the question: how can design bring focus to the human dimension, hence the fostering of human-centric environments? The overwhelming response from the conference attendees and submissions provided a multitude of answers that considered human needs, wants, satisfaction, comfort, and well-being, at multiple scales, and in relationship to other species and the global environment in their critical investigation of the relationship between people and their built and natural environments.

Human-centric environments can be seen as both a Panacea and a Pariah. It provides the power of explanation and forecasting human interactions and forces of design in the built and natural environment towards a mutual understanding of co-existence within our planet in the face of climate change. Over the years, EDRA's annual meetings have addressed societal and environmental challenges, disclosed evidence-based solutions at multiple scales, and gathered researchers, practitioners, and policymakers from multiple disciplines. EDRA 55-PDX continued this tradition and was a wonderful testimonial that this spirit is alive in transdisciplinary discussions, for the sustaining interest of environmental design fields to enrich human experiences and people's interactions with their environment. The conference explored human-centric environments and design in six tracks and scales: Nature and Urban Ecology, Sustainable Building, and Infrastructure, Resilient Design for Climate Change, Design for Health and Well-being, Spatial Justice, Diversity, and Inclusion by Design, and Transdisciplinary Design and Collaboration/Open Track.

EDRA55 delegates were greeted by the city of Portland, Oregon, a strategic setting that embodied the conference theme. With its pioneering vision of a walkable and sustainable city that is focused on a human-centric framework of development and urban renewal, Portland offered a wonderful backdrop for our conference with its progressive sustainable urbanism projects, award-winning parks, sustainable and green livable communities, and net-zero energy buildings. All these settings are connected by alternative modes of transportation, equitable models of housing affordability, neighborhood revitalizations, and historic, unparalleled cultural landscapes.

EDRA55 was a great success in terms of scholarly publications. The proceedings include 24 peer-reviewed full papers, 395 abstracts presentations, 105 posters, 35 Symposia, 15 Workshops, 8 Intensive sessions, 14 Researcher/Practitioner Storytelling,18 Graduate Students Workshops, 8 Media Shorts , and 12 Work-in-Progress submissions . The event saw over 600 registered participants. The acceptance rate for the peer-reviewed papers was also highly competitive, at approximately 77%.

I want to take this opportunity to acknowledge the tremendous work of my colleagues and members of the EDRA55-PDX local organizing committee, Philip Speranza and Chandra Robinson, for their collaboration and service to make this conference a success. In addition, I thank Nasrin Golshany for her help in organizing the Confex peer-review process and session programming. It was wonderful to kickstart EDRA 55 with a special Juneteenth Celebration and the Juneteenth Keynote of Wilson Smith, Nike Design Intelligence Manager of Creative Outreach. The conference opening dual Keynotes were delivered by Farshid Moussavi and Tatiana Bilbao. Bill Leddy delivered the second-day Keynote, and the EDRA Plenary was led by EDRA founder, Henri Sanoff. JohnPaul Jones delivered the closing keynote. It was amazing to have such wonderful and engaging world-class keynote speakers and thought leaders in one event and it was a great honor that they all accepted my invitation to join us at EDRA 55.

My sincere gratitude goes to the EDRA Board for entrusting me with the opportunity to organize and plan this conference and for all their help and support along the way. This compilation of the proceedings would have never been possible without the dedication of EDRA Board members, Sharmin Kader, Elif Tural, and Debarati "Mimi" Majumdar Narayan, who collaborated on the process of editing, organizing, and compiling all the pieces to document this monumental task. I also want to acknowledge the support of our double platinum sponsors for the conference, the College of Design, University of Oregon. Most importantly, I want to thank all EDRA 55 participants for their thoughtful submissions, peer reviews, presentations, enthusiasm, discussions, and for investing time and resources to travel and engage with us for this conference. I hope this proceeding brings back memories of the wonderful times we spent together from June 19-22, 2024, in Portland, Oregon.

Until we meet again, Happy Trails to You!

Ihab Elzeyadi, Ph.D.
Chair, EDRA 55 Local Organizing Committee
Professor of Architecture,
University of Oregon - USA,

On behalf of the EDRA 55 proceedings committee:

Sharmin Kader, Ph.D.
Past Chair, EDRA Board
Assistant Professor, Ball State University - USA

Elif Tural, Ph.D.
Chair, EDRA Board
Associate Professor at Virginia Tech - USA

Debarati "Mimi" Majumdar Narayan, Ph.D.
Chair Elect, EDRA Board
Chair, Program Committee of the EDRA Board
Adjunct Faculty, University of Maryland - USA

A FEW MOMENTS FROM EDRA55

Event: Ihab Elzeyadi, Chair of the EDRA 55 Local Organizing Committee, opening the Juneteenth keynote.

Event: After the Keynote Speech by Henry Sanoff, Professor Emeritus of Architecture, North Carolina State University From Left to Right, Ihab Elzeyadi, Henry Sanoff, Lorraine Maxwell, and Debarati Majumdar Narayan

Event: Knowledge-Network Lunch Meeting of the "Research in Professional Practice"

Event: Tim Ganey (Design Leader) from DLR Group speaking during the Graduate Student Mentorship Workshop at the DLR Group Portland office

Event: Wilson Smith providing his Juneteenth keynote Speech, "Empathy Driven Design."

Event: Participants in the Graduate Student Mentorship Workshop.

Event: Ihab Elzeyadi giving a speech with Elif Tural (Chair of the EDRA Board) during the Welcoming Session.

A FEW MOMENTS FROM EDRA55

Event: EDRA Board Members Dinner with the Founder of EDRA, Henry Sanoff. Pictured from left to right: Debarati "Mimi" Majumdar Narayan (Chair-Elect), Marzieh "Melika" Rezaei (Assistant Executive Director), Celen Pasalar (Board Member), Lisa Lim (Board Member), Lorraine E. Maxwell (Emeritus Board Member), Henry Sanoff (Guest), Julie Stevens (Board Member), Hannah Arnett (Board Member), Sharmin Kader (Past-Chair), Elif Tural (Chair of the Board), Alina Osnaga (Board Member), and B. Sanborn, Ph.D. (Board Member).

Event: Bill Leddy providing his Keynote Speech, "Empathic Design: Celebrating the Rich Diversity of the Human Condition."

Event: Knowledge-Network Lunch Meeting of the "Children and Youth Environments"

Event: After the award ceremony, the EDRA55 Banquet on Friday, June 21, 2024. Pictured from left to right: Philip Speranza (EDRA 55 Local Committee and Associate Professor of the University of Oregon), Ihab Elzeyadi (Chair of the EDRA 55 Local Organizing Committee and Professor of the University of Oregon), and Elif Tural (Chair of the EDRA Board and Associate Professor of Virginia Tech).

Event: Juneteenth Reception & Celebration on Wednesday, June 19, 2025. From Left to Right, Debarati "Mimi" Majumdar Narayan (Chair-Elect of the EDRA Board and Chair of the EDRA55 Program Committee), Elif Tural (Chair of the EDRA Board), Henry Sanoff (Founder of EDRA), and Sharmin Kader (Past Chair of the EDRA Board).

Event: After the dinner and award ceremony, the EDRA55 Banquet on Friday, June 21, 2024.

CONFERENCE TRACKS

EDRA55-PDX CELEBRATES HUMAN-CENTRIC DESIGN IN ALL FORMS OF ENVIRONMENTS AND AT DIFFERENT SCALES:

TRACK 1:

Nature and Urban Ecology

This track will examine how we generate knowledge to address the great challenges of urban ecological settings in cities. Examples of topics include the uses of public and semi-public spaces, sustainable urbanism, adaptation to climate change, cultural landscapes, preservation of tangible and intangible heritage, waste, and pollution management (soil, air, noise, water), natural resource depletion, loss of biodiversity, loss of cultural diversity, and urban sprawl. How do we design human-centric environments dedicated to urban ecology and the preservation of cultural landscapes to successfully manage these challenges both locally and globally?

TRACK 2:

Sustainable Buildings and Infrastructure

This track will focus on human-centric design and evaluation of sustainable building and infrastructure. How do we recognize and plan for the needs of diverse populations shaped by people's unique perspectives, life experiences, socio-demographics, cultural influences, identities, geography, and preferences, to name a few. What impacts did our buildings, urban centers, and indoor environments have on vulnerable populations including children, ethnic minorities, young adults, the elderly, people with mental health disorders, and people facing physical spatial abilities? How do we design sustainable urban centers that provide healthy levels of density, livability, and affordability, especially regarding dignified housing for all citizens? How would the design of net-zero energy, water, and waste in buildings provide a blueprint for better ecological design and sustainable neighborhoods?

TRACK 3:

Resilient Design for Climate Change

This track will focus on the numerous ways planning and design professionals, researchers, and educators define resiliency and the scale of design actions to adapt and reduce the impacts of climate change on people. How does the concept of resilient design, its approaches, methods, and tools be used to design human-centric environments? What is the theoretical foundation for research and practice focused on resilient design? What are the emerging design tools (data, simulation, virtual) and how are they being deployed in design projects? What are the ways researchers are studying the built and natural environment in all scales to provide sustainability and resilience in the face of climate change?

CONFERENCE TRACKS

EDRA55-PDX CELEBRATES HUMAN-CENTRIC DESIGN IN ALL FORMS OF ENVIRONMENTS AND AT DIFFERENT SCALES:

TRACK 4:

Design for Health and Wellbeing

This track will focus on research and practice models to promote human-centric spaces that are designed to promote health and well-being for all people, communities, transspecies, ecosystems, and living systems. This track will uncover and question the methods, conversations, and outcomes for the research, design, and policy debates under the umbrella of healthy human-centric environments in everyday spaces as well as healthcare facilities.

TRACK 5:

Spatial Justice, Equity, and Inclusion by Design

This track focuses on celebrating environmental design work, scholarship, and advocacy at the intersections of gender, race, ethnicity, indigeneity, sexuality, and economic inequality that is enriched by and enriching of their lived experience. This track will highlight how professionals and academics are crossing boundaries to address emerging issues that compromise spatial justice, equity, and inclusion by design and engage in critical conversations about effective approaches, new perspectives on human-centric engagement, and empowerment of marginalized and under-served communities.

TRACK 6:

Transdisciplinary Design and Collaboration/Open Track

This track examines the means and practices for sustaining productive collaborations and partnerships to promote human-centric design at different scales for all types of environments. What are the benefits and challenges of transdisciplinary collaborations? What critiques emerge when blurring and/or borrowing from multiple disciplines? How can we investigate human-centric design within transspecies and transdisciplinary frameworks? What methodologies are being embraced by environmental design researchers and practitioners to innovate and stimulate research collaborations and community engagement? What new methods, ways of working, ethics or insights can help advance the practice of human-centric design? What other topics and focus can we bring to the discussion of human-centric design that we have not considered? What open topics can help us re-think and re-question our conference theme and tracks?

KEYNOTE ADDRESS & PLENARY SESSIONS

EDRA55 Portland featured keynotes and plenaries from world renowned speakers.

WILSON SMITH

EDRA55 Juneteenth Keynote: Empathy Driven Design

Wilson W. Smith III is the Nike Design Intelligence Manager of Creative Outreach. During his 40-year tenure at Nike, Smith worked on projects that run the gamut of the design profession—retail, graphics, architecture, footwear, experience, brand, and apparel—and designed and developed product for some of the greatest athletes in Nike's history, including Michael Jordan, Serena Williams, Andre Agassi, Rodger Federer. In addition, Wilson is a Professor in Product Design at U of O, where his studio focus is Adaptive Design Enabling Athletes with Disabilities. He's also co-taught Intro to Product Design in the UO Sports Product Management graduate program.

Wilson is the recipient of 2018 Ellis F. Lawrence Medal from the College of Design for distinguished alumnus. Graduating in 1980 with an architectural degree from U of O, Wilson initially worked for Skidmore, Owings and Merrill out of Portland. He began his Nike career in 1983, assisting Tinker Hatfield in Corporate Architecture. Transitioning into footwear design, his focus became athlete inspired products in Tennis, Basketball, Cross Training, and Brand Jordan, where he designed Air Jordan XVI & XVII. Black Enterprise Magazine named Wilson one of America's Top Black Designers.

What truly defines Wilson is his incredible passion around coaching, teaching and giving back. Wilson's commitment to advancing the next generation, and the power of design to positively impact the world has created a presence and legacy that expands beyond his product designs.

In 2019, he was honored with the Jan Ernst Matzeliger Award as the first Black designer in the athletic footwear industry, and with Nike's inaugural Design Achievement Award. For Fall '22, Nike Classics created the "Thank You, Wilson" shoe story of his retro '96 Air More Uptempo. Wilson is also a worship leader at SouthLake Church, West Linn, Oregon.

FARSHID MOUSSAVI

EDRA55 Thursday Keynote 1: Housing as a Sustainable Common

Farshid Moussavi is an architect, founder of Farshid Moussavi Architecture and Professor in Practice of Architecture at Harvard University Graduate School of Design. She was previously co-founder of Foreign Office Architects (FOA). Educated at Harvard's Graduate School of Design, University College London and Dundee University, Moussavi has a longstanding involvement with academic institutions worldwide.

She was head of the Institute of Architecture at the Academy of Fine Arts in Vienna and has held teaching posts at the Architectural Association in London, the Berlage Institute in Rotterdam, the Hoger Architectuur Instituut Sint-Lucas in Gent, and in the United States at Harvard, UCLA, Princeton and Columbia University.

Moussavi has served on key design and architectural advisory panels and international design juries including for the British Council, the Mayor of London's "Design for London" advisory group, the London Development Agency, the RIBA Gold and Presidential Medals and the Stirling Prize for Architecture.

She has previously served as a trustee of the Whitechapel Gallery in London and the London Architecture Foundation, and a member of the Steering Committee of the Aga Khan Award for Architecture. Currently, she is a trustee of the Norman Foster Foundation London and New Architecture Writers (NAW) which focuses on black and minority ethnic emerging writers who are under-represented across design journalism and curation.

She is also one of the London Mayor's 42 Design Advocates. Moussavi was elected a Royal Academician in 2015 and Professor of Architecture at the RA Schools in 2017. In 2018, she was appointed Officer of the Order of the British Empire (OBE) in the Queen's Birthday Honours for services to architecture. She has published three books, 'The Function of Ornament', 'The Function of Form'and 'The Function of Style', based on her research and teaching at Harvard. Her fourth and most recent book is titled Architecture and Micropolitics, Four Buildings 2011-2022, Farshid Moussavi Architecture (2022).

TATIANA BILBAO

EDRA55 Thursday Keynote 2: Architecture as a Collective Act

Tatiana Bilbao began her eponymous studio in 2004 with the aim of integrating social values, collaboration and sensitive design approaches to architectural work. The work of the office intersects with research allowing to design for diverse circumstances and in reconstruction or crisis scenarios.

Prior to founding her firm, Bilbao was an Advisor in the Ministry of Development and Housing of the Government of the Federal District of Mexico City, during this period she was part of the General Development Directorate of the Advisory Council for Urban Development in the City.

Bilbao holds a recurring teaching position at Yale University School of Architecture and has taught at Harvard University GSD, Columbia University GSAPP, Rice University, University of Andrés Bello in Chile, and Peter Behrens School of Arts at HS Dusseldorf in Germany.

Her work has been published in The New York Times, A + U, Domus, Arquitectura Viva, El Croquis, El País, among others. Also, Bilbao has been recognized with the Kunstpreis Berlin in 2012, was named in 2010 as an Emerging Voice by the Architecture League of New York, the Global Award for Sustainable Architecture Prize by the LOCUS Foundation in 2014, the Marcus Prize Award 2019, Tau Sigma Delta Gold Medal of 2020, The Honorary Fellow of The Royal Architectural Institute of Canada (RAIC) in 2021, Richard Neutra Award, the AW Architect in 2022 and most recently The Miller Prize in 2023.

BILL LEDDY

EDRA55 Friday Keynote: Empathic Design: Celebrating the Rich Diversity of the Human Condition

William Leddy, FAIA, is a Founding Principal of San Francisco-based LEDDY MAYTUM STACY Architects (LMSA), the 2017 recipient of the national American Institute of Architects Firm Award. For over three decades he has been a national leader in the design of environments that promote inclusion and social justice while advancing urgent climate action.

LMSA has received over 175 regional, national and international design awards and has been recognized by numerous organizations including the American Institute of Architects, the French Institute of Architects, the Norwegian Association of Architects, the U.S. Department of Energy, the San Francisco Museum of Modern Art, and the National Building Museum. LMSA is one of only two firms in the nation to have received twelve or more national AIA Committee on the Environment Top Ten Green Project awards – the Institute's most prestigious recognition of integrated design excellence.

Leddy has lectured widely and served as visiting professor at the Southern California Institute of Architecture and the California College of the Arts, as the Howard A. Friedman Visiting Professor at the University of California, Berkeley, and the Pietro Belluschi Distinguished Visiting Professor at the University of Oregon. Leddy is also a long-time advocate for expanding our shared values of design excellence to embrace the intersection of inclusion, diversity, and rapid decarbonization in the built environment.

His 2010 design for the Ed Roberts Campus in Berkeley, California - an international center for the Disabled Rights Movement - is widely recognized as a landmark of advanced Universal Design at a civic scale.

As 2013 Chair of the national AIA Committee on the Environment Advisory Group, Leddy initiated the national AIA COTE Top Ten for Students competition to help prepare young designers to address a rapidly changing climate. From 2019 to 2023, Leddy served as the AIA California Vice President of Climate Action, working to accelerate the decarbonization of the built environment in California and beyond. Under his leadership, California passed first-in-the-nation legislation requiring continuing education in zero net carbon design for architecture licensure, as well as the first state-wide embodied carbon emissions building code.

LMSA's new book - "Practice with Purpose: A Guide to Mission Driven Design" - was published in 2023.

JOHNPAUL JONES

EDRA55 Saturday Keynote: A Conversation with the Land / I Don't Believe in Magic

Johnpaul Jones has a distinguished 52-year career as an architect and founding partner of Jones & Jones.

Earning his Bachelor of Architecture from the University of Oregon in 1967, his design philosophy emerged from his Choctaw-Cherokee ancestors, which connects his work to the natural world, animal world, spirit world, and human world.

Mr. Jones' designs have won widespread acclaim for their reverence for the earth, for paying deep respect to regional Indigenous architectural traditions and native landscapes, and for heightening understanding of Indigenous People and their diverse Native cultures of America.

Johnpaul has led the design of numerous cultural centers and museums with tribes spanning the North American continent, culminating in his 12-year engagement as overall lead design consultant for the Smithsonian's National Museum of the American Indian on the Mall in Washington, D.C. A Fellow in the American Institute of Architects, his designs have won a stream of local and national awards.

His awards include the 2005 Distinguished Service Award from the University of Oregon (his alma mater), the AIA Seattle Medal (2006), the Executive Excellence Award from the American Indian Science and Engineering Society (2006), the Pietro Belluschi Distinguished Professorship from the University of Oregon (2011), the Island Treasure Award from the Bainbridge Island Art and Humanities Council (2013), the Washington State Governor's Heritage Award (2014), and the National Humanities Medal from the National Endowment for the Humanities (2014) conferred by President Barack Obama.

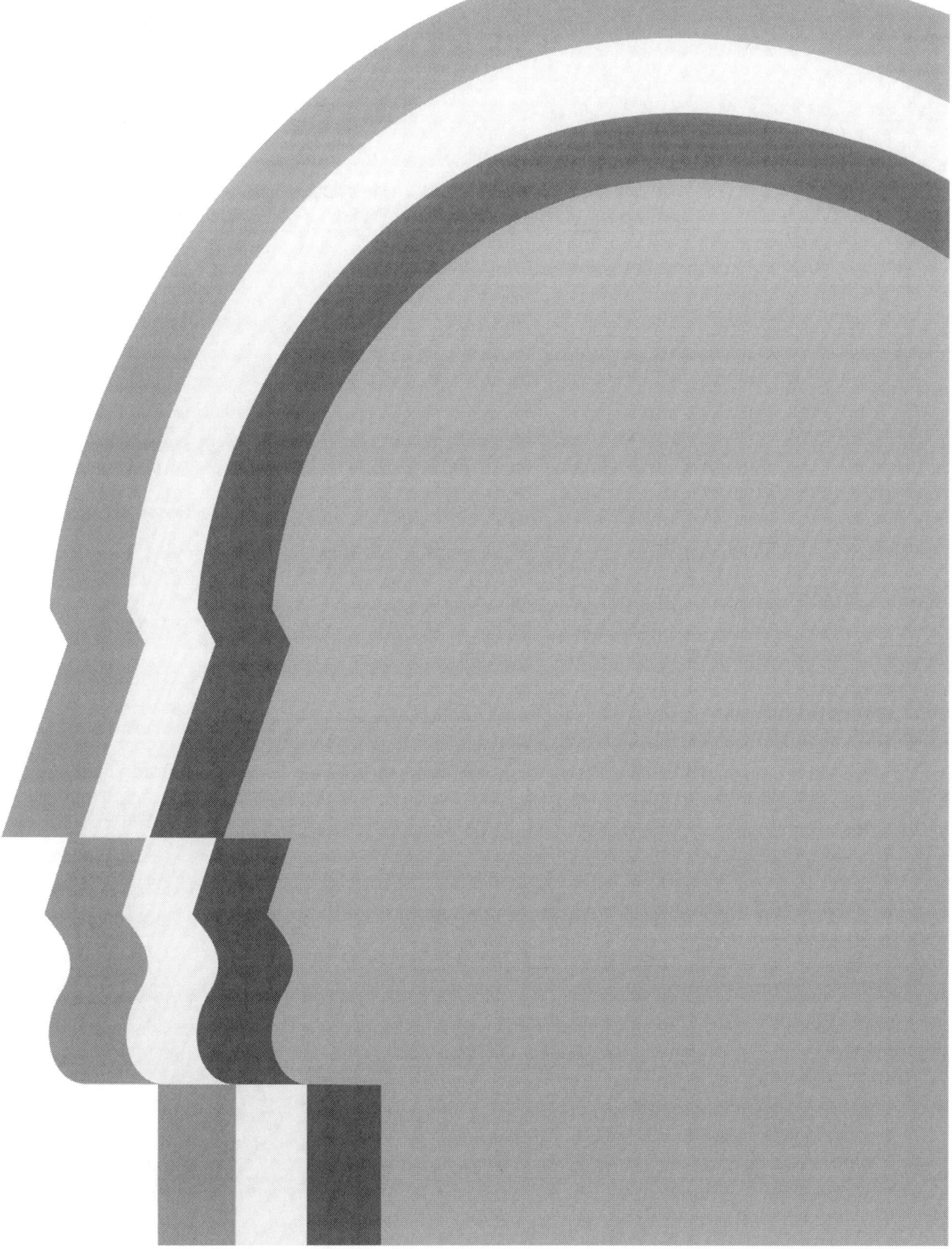

MOBILE SESSIONS

SUSTAINING INFILL

Mobile Session One

ABSTRACT

Portland is a city rich with infill opportunities and experiments across a range of scales to support the health and social life of its communities. This session will tour three recent mixed-use infill projects in Downtown and Northeast Portland that foreground sustainable performance and innovative material use: the PAE Living Building (ZGF Architects), Albina Yard (LEVER Architecture), and Mississippi Workshop (Waechter Architecture).

Here, participants will explore what it means to design and operate the world's first developer-driven Living Building, and to experience first-hand the tectonic, environmental, and spatial affordances of cross-laminated timber construction and the "forest to frame" processes that support this work.

The tour will begin Downtown and will end in the Northeast, with transportation via the MAX Yellow Line. Additional projects will be highlighted along the way, including a recent co-housing project by Beebe Skidmore Architects. At the conclusion of the tour, there will be an opportunity for a social gathering over coffee in the courtyard of the Mississippi building.

Session Hosts

Mark Fretz, Assistant Professor
Department of Architecture,
University of Oregon &
Interim Co-Director, Institute for
Health in the Built Environment

Justin Fowler, Director,
Portland Architecture Program,
University of Oregon

SUSTAINABLE URBANISM IN PORTLAND

Mobile Session Two

ABSTRACT

Portland has been a laboratory for sustainable urban design and has a wealth of examples addressing active transportation, transit oriented design, stormwater management, ecological development, equitable design, and health-focused design.

This walking tour will take participants from the South Waterfront project, through the Park Blocks and city center, and up to the Pearl District, discussing notable, sustainability-oriented design approaches throughout.

This session will also introduce participants to the Sustainable Urban Design Framework developed by Prof. Larco.

This framework organizes sustainability related outcomes in relation to different scales of urban design interventions and can be used as a design tool, an evaluation tool, or as a tool to assist with stakeholder engagement. This framework is the focus of the Sustainable Urban Design Handbook, recently published by Routledge.

Session Hosts

Nico Larco, AIA, Professor of Urban Architecture, University of Oregon & Director, Urbanism Next
Co-Director Sustainable Cities Initiative

UNDERSTANDING PLACE: NATIVE AMERICAN STUDENT & COMMUNITY CENTER

Mobile Session Three

ABSTRACT

This Mobile Session will tour the Native American Student and Community Center (NASCC) at Portland State University. NASCC's mission is to provide a location where Indigenous students connect to other students, faculty, staff, and community members in an inclusive and supportive environment.

The Center welcomes the greater Indigenous Community to PSU's campus, provide opportunities for shared learning and understanding. The Portland area is home to one of the largest urban Indigenous American populations in the nation, 40,000 strong. NASCC, which recently celebrated its 20th Anniversary, is in large part due to the efforts of PSU Indigenous students who proposed, lobbied, fought, and fundraised for this site to support Indigenous students and the surrounding Indigenous communities.

A repurposed traffic triangle located just north of the I-405 entrance, the site's design transformed a "found space" into an important community hub. Participants will be able to see and experience the range of interior spaces, exterior gathering spaces, and a garden of Indigenous plants.

The Mobile Session will include design professionals, artists, managers, former Indigenous students, and current Indigenous students who created this visionary project and continue to maintain its vibrancy.

Session Hosts

Lynn Paxson, Professor of Architecture, Iowa State University

EXPLORING TRAUMA-INFORMED DESIGN AT PARROTT CREEK

Mobile Session Four

ABSTRACT

Parrott Creek, an 80-acre natural site, is an alternative to incarceration for adjudicated youth in Oregon - one that seeks to create a trauma-informed environment and help break unhealthy cycles.

It is also home to the Cultural Ecology Program, an Indigenous land stewardship project, the overriding goal of which is to engage with and center Indigenous leadership, values, and partnerships to restore and steward the natural area for community and cultural use. Engaging the youth in their residential program and the wider community, this project works to address trauma held at the individual, community, and environmental levels.

This mobile session will span across the disciplines of landscape architecture, architecture, and ecology to explore trauma-informed design as a methodology for communities and sites in need and expand one's understanding of restorative justice.

Session Hosts

Ameyalli Mañon-Ferguson, Cultural Ecology Program Manager, Parrott Creek Child & Family Services

Elisandra Garcia, Director of Engagement / Project Designer, Eldorado

UNDERSTANDING PORTLAND'S RESPONSE TO HOUSING DEPRIVATION

Mobile Session Five

ABSTRACT

This mobile walking tour will provide an overview of Portland's plan to end housing deprivation and visit four exemplary Permanent Supportive Housing (PSH) projects in Portland's Pearl District and Old Town.

PSH delivers affordable long-term housing with integrated support services to people who have experienced chronic housing deprivation and has been associated with improved health outcomes, lower emergency service and healthcare use, and increased perceived levels of freedom, autonomy, choice, and control among residents.

The number of PSH dwellings in the United States has nearly doubled between 2007 and 2020 and these efforts have demonstrated remarkable success. For example, 99% of the 532 tenants who moved into Portland PSH in 2020 still resided in stable housing one year later. In addition to supportive housing and funding policies, successful PSH requires research and collaboration across design, housing, social service, and healthcare sectors with attention to human-centered and health-promoting design.

Session hosts have active and published research on PSH and wellbeing and will share their knowledge between building visits.

Session Hosts

Christina Bollo, Asst. Prof. of Architecture (Housing Specialization), University of Oregon

Kimberly Rollings, Institute for Social Research, University of Michigan

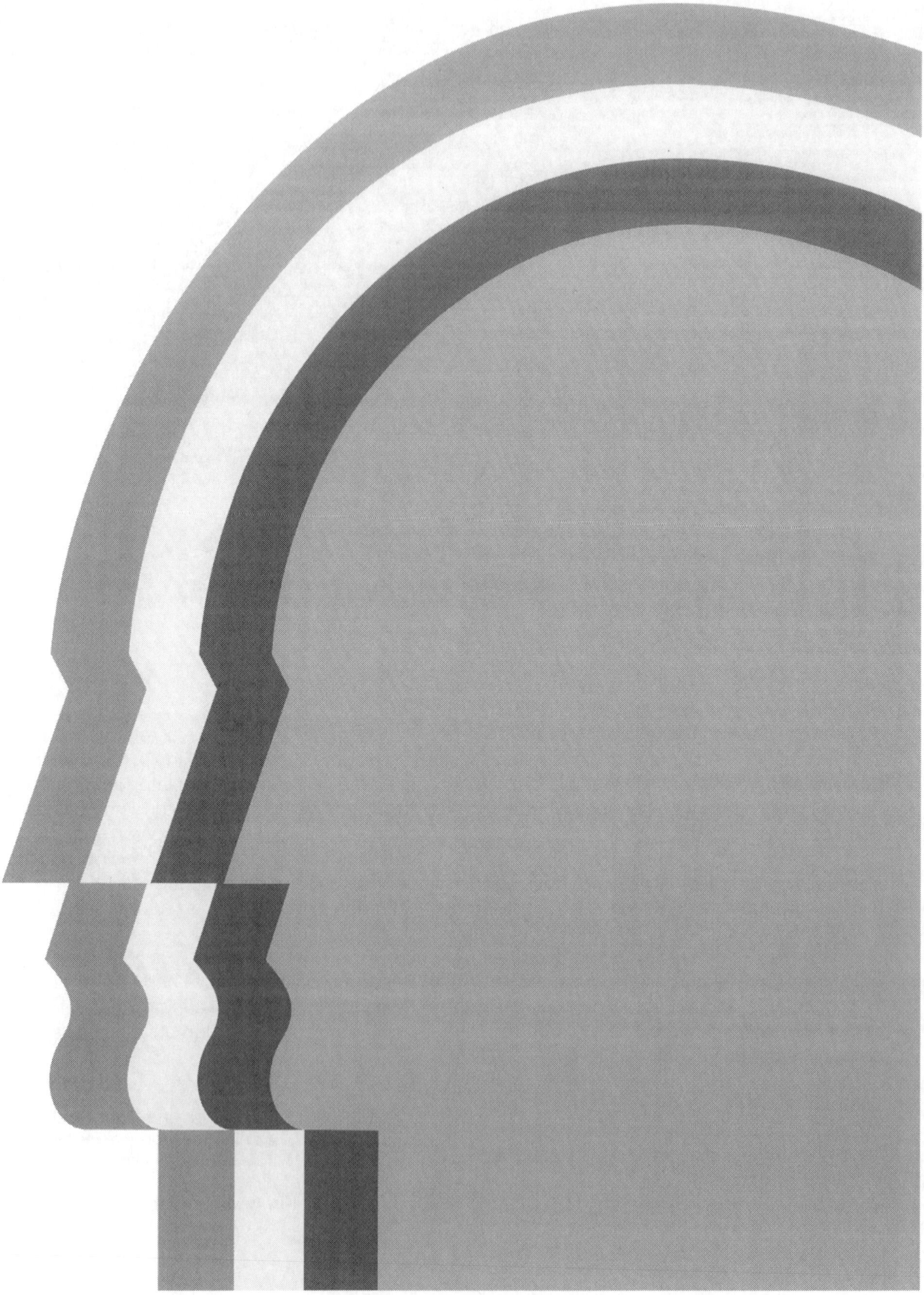

AWARDS & RECOGNITION

Ceremonies

Great Places Awards
Paper & Poster Awards
2025 Michael Brill Grant Recipient
Certificate of Research Excellence (CORE)

GREAT PLACES AWARDS

The 26th Annual Ceremony

EDRA's Great Places Awards seek to recognize work that combines expertise in design, research, and practice, and contributes to the creation of dynamic, humane places that engage our attention and imagination.

Partnership: EDRA is a proud partner with Public for Public Spaces (PPS) for the Great Places Awards.

Categories

- Place Art
- Place Book
- Place Design
- Place Planning
- Place Research

2024 GPA In Numbers

- 10 Jurors
- 31 Submissions

United States, Peru, Mexico, Venezuela, China, Canada, Korea

GREAT
PLACES
AWARDS

FINAL JUROR LIST

Place Art Jurors

Stan Carroll, AIA
Stan Carroll Public Art Works

Christina Hite, FASLA
Dix Hite + Partners

Place Book Jurors

Ashraf Salama, Ph.D.
Department Head/Professor of Architecture & Built Environment, University of Northumbria

Mohammad Gharipour, Ph.D.
Chair/Professor, Architecture
University of Maryland

Place Design Jurors

Ervin Garip, Ph.D.
Professor, Department of Interior Architecture Istanbul Technical University

George Hallowell, Ph.D.
Professional Faculty, School of Architecture North Carolina State University

FINAL JUROR LIST

Place Planning Jurors

Ellen Dunham Jones
Professor, School of Architecture Georgia Institute of Technology

Jason Sowell
Associate Professor, Department of Architecture University at Buffalo

Place Research Jurors

Lun Ge, Ph.D.
Beijing University of Civil Engineering and Architecture

Jeri Brittin, Ph.D.
Social and Behavioral Sciences Director, HDR

PLACE ART WINNER

A Symphony in Nature: The Umbrella Pavilion Unfolds

Authors: Zhongzong Zeng, Yingying Huang, Xinran Zheng, and Yongquan Chen

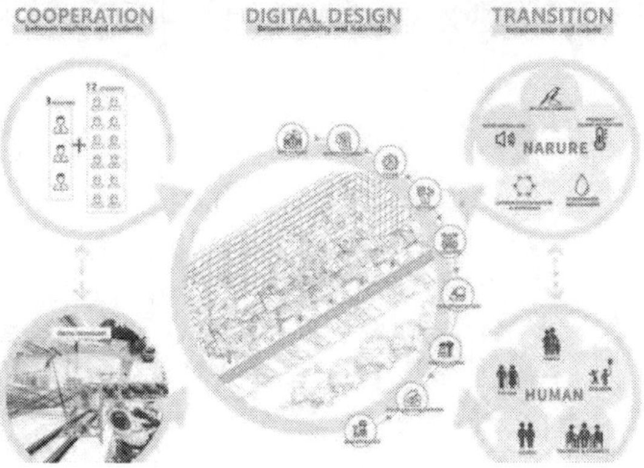

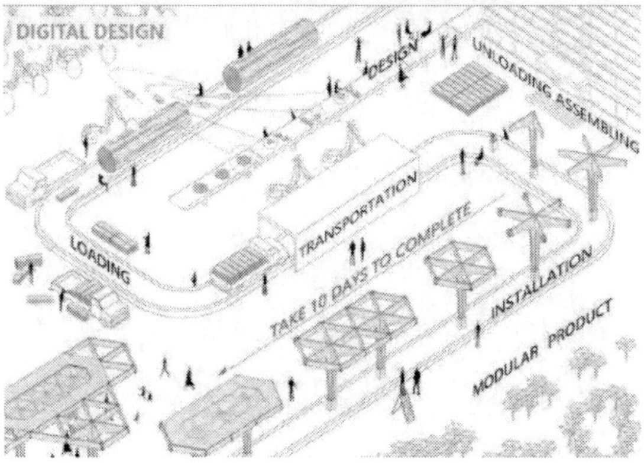

JUROR'S COMMENT:

"As an artistic prototype and tool for placemaking, this installation sets a high bar. From a holistic sustainable intervention for streetscapes, allowing for carbon sequestration, infiltration, enclosure and shade, while stunning in it's form, the project exhibits forward thinking artistic vision.

The level of community engagement is left a bit to the imagination; it's unclear how that was integrated in the process. The creative process is well documented, however.

The umbrella pavilion is a beautiful structure, with implications for future practice. The role this piece plays in transforming an existing fast passed circulation space into a vibrant and slower paced social space is significant.

The formal dialog the piece inspires between the existing lines of evenly spaced trees and the adjacent architectural entry points weaves two disparate geometrical ideas together masterfully. The detailing is precise and was well conducted. Well Executed."

PLACE ART HONORABLE MENTION

Handmade Urbanism: DIT Do-It-Together Popup Rooftop Skatepark
Authors: Yoon Zoosun

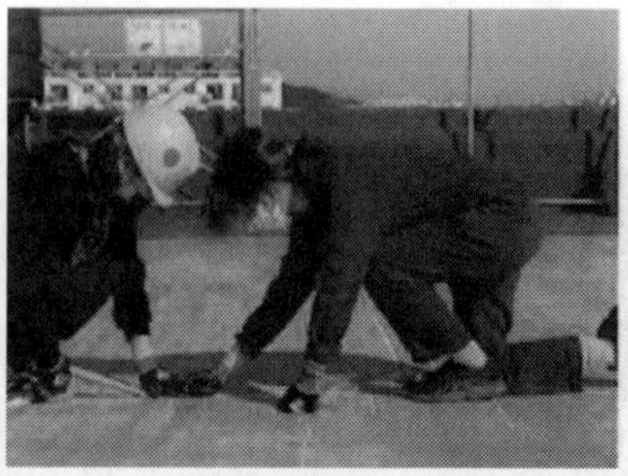

JUROR'S COMMENT:

"This project stands out as an example for community engagement, centering skatepark craft and artistry as a tool for placemaking, engagement, and social change. The result is a beautiful, dynamic skatepark that facilitates community creativity and action. This project brought a tear to our eyes by the apparent selfless community spirit and meaningful impact. Proliferation of these types of projects across the globe would go a great distance toward making this a more prideful and non-digital socially engaged world. What got created here was not the apparent skate park but a much larger aura of connection, community pride, and friendships."

PLACE BOOK WINNER

Landscape Architecture for Sea Level Rise: Innovative Global Solutions
Authors: Galen Newman and Zixu Qiao

JUROR'S COMMENT:

"This book assesses national and global strategies to address a critical and timely issue -- flooding and sea level rise. Brilliantly structured, it identifies structural, non-structural, and hybrid approaches to heighten flood resilience.

Utilizing a consistent investigation, the book offers design-based tactics for protecting people and the environment. Involving a significant number of contributors, this book underscores the need for departing beyond traditional engineering solutions towards more nature-based approaches.

Receiving praise from academics, researchers, practitioners and professional organizations, this book is an exceptional resource for urban management professionals, planners, landscape architects in fighting climate change through innovative design responses.

The book scores very high in the four categories of evaluation including context, process, outcomes, and impact. The Jury views the book as a holistic examination of pioneering strategies for addressing flooding and sea level rise in diverse contexts across the United States and globally.

Offering instrumental and perceptive insights into a mix of engineering and nature-based solutions, it offers great lessons for urban managers, planners, and designers to combat climate change through thoughtful design."

PLACE BOOK HONORABLE MENTION

Public Space: Notes on Why It Matters, What We Should Know, and How to Realize Its Potential
Author: Vikas Mehta

JUROR'S COMMENT:

"This book is an admirable endeavor that expands urban design discourse on public space. Structured in a non-linear fashion, the book includes three core sections—paradoxes, possibilities, and propositions—preceded by an initial in-depth discussion and reflection on why care for public space and what its fundamentals are.

The format of the book makes it accessible to a broader audience as it is well animated with interesting illustrations that invite engagement. Believing in the immense capacity of public space and amid diverse social challenges, the book calls for the need to address equity and inclusivity in the context of crises including social injustice, climate change, and pandemics.

Scoring considerably high in the four categories of assessment, the jury views the book as a comprehensive exploration of public space by drawing on established urban design discourse and presenting advanced ideas in an accessible style.
In essence, the book contributes to the contemporary debates on access and inclusivity and the positive social impact of public space on cities and communities."

PLACE DESIGN WINNER

Howard Park Renovation

Authors: Alliance Architects; The Lakota Group; M/E Design Services; Stantec; Abonmarche; Tango Design; Paul Fujawa Engineers; Ambient Energy

JUROR'S COMMENT:

The Howard Park Renovation project is an inspiring transformation of an 1870s historic park that had fallen into neglect and disrepair over the last decade. The project is an example of enhancing the emotional bond to the park with a meaningful connection both to existing users and to the wider community of South Bend, Indiana. Because of its history as the first public park in the city and its insightful design and human-centered experience, it has the potential to enhance the life quality of a wide range of users now and into the future.

Through numerous hours of community meetings and stakeholder interviews, revered elements of the park were identified for renovation, and new activities and dimensions were designed to create a reinvigorated park that is both four-season and operable both day and night. New project elements include parkland and playgrounds, new water features, several new flexible support buildings, and carefully designed activity areas to allow varied uses from rock concerts to farmer's markers with the purpose of inclusivity and flexibility.

The new functions and activity spaces added to the project, along with significant reinvention of the original components of the park will likely broaden the constituency for the park and represent a very dynamic and human-centered place as well as providing a catalyst for new business development in the surrounding East Bank neighborhood. The most apparent indication of the park design responding to user needs is the reported 300,000 first annual users and activity programming that is flexible and inclusive enough to allow both night and day activities throughout all four seasons."

PLACE DESIGN HONORABLE MENTION

The Carousel for Companionship

Authors: Joseph Altshuler and Zack Morrison

JUROR'S COMMENT:

"The Carousel for Companionship project creates an impressive graphically driven social interaction space. The project provides a model for urban 'in-between' spaces, reinvigorating an underutilized urban lot using community volunteer labor and a limited budget.

The design is meant to tie the existing lot to adjoining streets and neighborhoods using large-scale graphics, shapes and colors derived from nearby well-known buildings, and a new user-friendly carousel.

The project brings together the fields of graphic design, landscape design and urban design within a social framework to create a lively urban area based on human-environment interaction. It offers a strong urban context and unique experience, strengthening the concept of place attachment and creating a bond between the users and the city.

The design serves as a community stage for performances, urban room for personal experiences, and un-programmed activities. The project also reveals a debate at the interface of permanence and temporality of architectural installations."

PLACE PLANNING WINNER

Olympia Armory Creative Campus Concept Plan
Authors: Yuansi Cai; Lesley Bain

JUROR'S COMMENT:

"The Creative Campus Concept Plan retrofits an entire block containing former Armory buildings into a multidisciplinary arts hub.

It is impressive both for its multi-faceted, deep-listening, community engagement process as well its comprehensive integration of the physical design, operational model, and business plan. Impactful steps towards implementation provide evidence of the plan's widespread local and national support.

These include a $1 million Heritage Capital grant, a $75,000 Our Town grant from the National Endowment for the Arts, and a successful city- wide vote to raise approximately $2.5 million/year for access to arts, science, cultural and heritage programming for City residents."

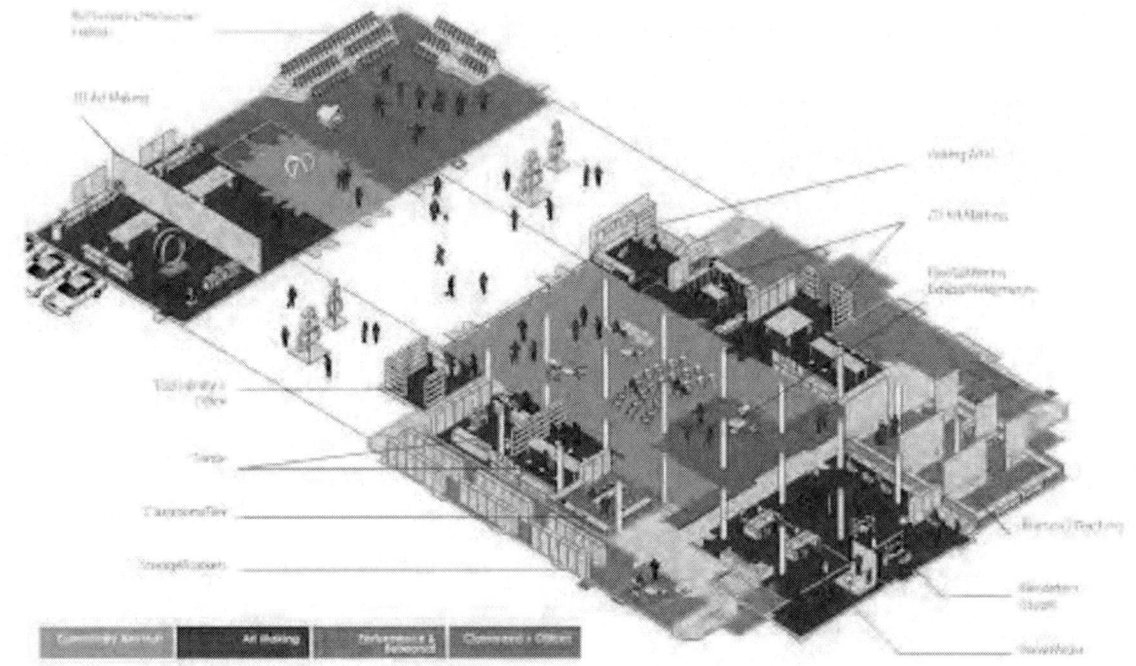

PLACE PLANNING HONORABLE MENTION

OUSD The Center Farm: Seeding Urban Resilience through Collaborative Design

Authors: Patricia Algara, Martha Clifford, MaFe Gonzalez, Kira Maritano, Sarka Volejnikova, Kat Romo, and Michelle Oppen

JUROR'S COMMENT:

"The Center Farm Landscape Plan seeks to increase access to healthy food and urban nature in a food-insecure neighborhood. Undertaken for the Oakland Unified School District and Trust for Public Land, the project used extensive, trans-disciplinary workshops to integrate the diverse knowledge and expectations from educators, nutritionists, urban ag experts, students, and residents...

The project clearly outlines the planning process and assessment by different actors who have a unique perspective on the project's design, implementation, and long term management; these actors ranged from experts in education and urban ag to the students and community members are the project's ultimate end users.

The planning process integrated these viewpoints at different stages of the design process to balance criteria necessary for operating an urban farm with user needs and desires."

PLACE RESEARCH WINNER

Making Participants the Photographers: Meaning of Home As Seen through the Eyes of Rural Residents in China

Author: Wei Zhao

JUROR'S COMMENT:

"The submission, including a written paper as additional material, describes a two-phased qualitative study using participant photovoice and interviews, as well as informal observation and other relevant secondary data sources for context...

The findings are not particularly surprising, but are vitally important as the Chinese government pushes its "Socialist Countryside" agenda on rural residents, and similarly rehouses other groups including the elderly.

...the findings are important in reinforcing and furthering knowledge in the field of place research. Overall, this is a very good study and interesting and well-written paper..."

PLACE RESEARCH HONORABLE MENTION

Environmental Affordance for Social Support and Community in a New State-of-the-Science Rehabilitation Hospital

Authors: Francesqca Jimenez, Abigail Heithoff, Carolyn Frey, Amber Walter, Devin Bowers, Sandra Tkacz, Allen Buie, Melissa Moseley, Hamed Aali, Amol Karmarkar, Richard Kunz

JUROR'S COMMENT:

"The research identifies a critical---and growing---concern within health care: how does the design of healthcare facilities improve/ speed recovery/ healing, foster patient wellness, support patients' families, and facilitate staff needs in the provision of care?

The submission identifies a set of frameworks and methods (focus groups, interviews, mockups, among others) that informed the architecture's design and underscored a post-occupancy evaluation of the building's impact on patient experience."

PAPER & POSTER AWARDS

BEST PAPER

Authors:

Ms. Shreya Jaiswal, LEED AP, D2D Green Design
Martha Bohm, University at Buffalo/ School of Architecture
Korydon Smith, University of Buffalo
Dr. Edward Steinfeld, University at Buffalo

Paper Title:

Sensory Engagement And Spatial Harmony: Unveiling The Role Of Fire In Multisensory Design

HONORABLE MENTION BEST PAPER

Author:

Sharmin Kader, Ph.D., Asso. AIA, Ball State University College of Architecture and Planning

Paper Title:

A Retrospective Review of END-of-Life Care Environments from EDRA1 to EDRA55

BEST STUDENT PAPER

Authors:

Yasaman Gholami, Ph.D. Student, School of Architecture, Georgia Institute of Technology
Raha Rastegar, Ph.D. Candidate, School of Architecture, Georgia Institute of Technology
Herminia Machry, Ph.D., EDAC, Assistant Professor, School of Architecture & Design, University of Kansas
Elahn Little, Research Scientist, SimTigrate Design Lab, Georgia Institute of Technology
Leandro Miletto Tonetto, Ph.D., Associate Professor, School of Industrial Design, Georgia Institute of Technology

Paper Title:

Could Social Space Design Boost Social Interactions Among Older Adults with Mild Cognitive Impairment?

HONORABLE MENTION BEST STUDENT PAPER

Authors:

Ms. Jianjiao Liu, University of Melbourne
Raymond Green, University of Melbourne.

Paper Title:

Nature Nurtures: Investigating Factors Impacting Children's Outdoor Activities and Nature Exposure in Beijing, China

BEST STUDENT POSTER DESIGN

Recipient:

Anna Ellis, Florida State University

Poster Title:

Designing for Wellbeing: A Trauma-Informed Approach to Academic Social Spaces

BEST PAPER AWARD

Sensory Engagement And Spatial Harmony: Unveiling The Role Of Fire In Multisensory Design

The research delves into the broader theme of biophilia, place attachment, stress reduction, and overall well-being, exploring the potential of fire as a biophilic design element to contribute to these aspects consequentially. Fire is a biophilic element overlooked in previous stress-restoration research primarily focused on natural materials, vegetation, daylighting, and other biophilic elements. The work is informed by how the historical relocation of fire and its consequential effects on sensory experiences lead to stress and frustration within the built environment.

Employing qualitative research methods, namely grounded theory and semi-structured interviews conducted in an indoor environment with a fireplace, viz., The One World café at the University at Buffalo (case study location). The study comprehensively examines spatial, emotional, sensory, and cultural dimensions of sensory engagement with the fireplace. The research methods allow the researcher to delve deeper into the multisensory experiences of the users from past (memories) to the present. The experience of twenty-seven participants from semi-structured interviews reveal a connection between the diminished presence of fire and a decline in sensory and social experiences, influencing the degree of attachment to a space or element in a space.

The evaluation of the fireplace at One World Café case study highlights the fireplace's limitations in terms of scale and sensory boundaries, emphasizing the influence of proximity and direct sensory engagement on individuals' responses to fire. In conclusion, the multisensory engagement provided by fire intensifies the perceived authenticity of the experience and strengthens the connection to nature. It also implied that fire instigates impactful and memorable experiences that enhance individuals' connection to a specific place. Incorporating sensory overlap strategies, designers can enrich environmental design, prioritizing factors like olfactory stimulation and cultural associations. Additionally, the research advocates further investigating the potential of other biophilic elements and multisensory encounters, aiming to find creative design-oriented solutions.

Authors:
Ms. Shreya Jaiswal, LEED AP, D2D Green Design
Martha Bohm, University at Buffalo/ School of Architecture
Korydon Smith, University of Buffalo
Dr. Edward Steinfeld, University at Buffalo

HONORABLE MENTION BEST PAPER AWARD

A Retrospective Review of END-of-Life Care Environments from EDRA1 to EDRA55

Hospice is a place to provide end-of-life care to individuals certified as "terminal" and to their families (Government Printing Office, 2000 in Swenson, 2009, p. 9). The environmental design research on end-of-life care environment is significant because the place experience has an imperative impact on the patient's "quality of life" (Cohen et al., 2001) and the possibility of a "good death" as defined by Tong and his colleagues (2003).

For over 55 years, EDRA has been promoting theory, research, and design related to environmental gerontology. A good amount of research was conducted on environments for older adults (e.g., nursing homes) considering the 'Third Stage' or 'Fourth Stage' of life. Some were published and presented at EDRA conferences, but very few research studies have focused on the 'Last Stage' of life, the end-of-life environments (e.g., hospice). Additionally, the COVID-19 pandemic has created many unprecedented scenarios and challenges for hospice facilities (Rogers et al. 2021). It is significant to develop an understanding of the cutting-edge knowledge on the hospice or end-of-life care facility environments, having a retrospective review would be significant to develop future research and design directions.

This paper has been written considering a literature review that includes the following search approaches: a) database (e.g., PsycINFO, Google Scholar, ProQuest Dissertations & Thesis), b) EDRA01-EDRA55 proceedings, c) books, reports, and guidelines published in last 20 years, d) white papers, webinars, pod-cast, or online panel discussions.

In the last fifty-five years of EDRA, a significant amount of development has been made: identification of therapeutic goals, development of research-based design guidelines, reference books on hospice environments, development of the environmental assessment tool for hospice settings considering US healthcare and building design system, and considering Canadian healthcare system for home-less hospice care populations.

Additionally, a few pieces of literature have come out after the pandemic. This article has discussed the key findings of all this literature and identifies the future research directions of end-of-life care environments, which carries significance to enhance our ability to design more thoughtful and efficacious in the era of an increasingly aging population when there is a growth of constructing new hospice facilities.

Author:
Sharmin Kader, Ph.D., Asso. AIA, Ball State University College of Architecture and Planning

BEST STUDENT PAPER

Could Social Space Design Boost Social Interactions Among Older Adults with Mild Cognitive Impairment?

The aging population, particularly those living with Mild Cognitive Impairment (MCI), confronts a tendency for social isolation. Promoting social interactions is recognized as a key element and potential protective factor against both social isolation and cognitive decline. However, individuals with MCI encounter difficulties in engaging socially. Previous studies recognize the impact of built environments on social interactions. However, there is a gap in understanding how the design of social spaces influences individuals with MCI in non-residential settings, particularly from an objective perspective.

This study investigated spatial utilization patterns of individuals with MCI during social interactions in a non-residential therapeutic setting, focusing on understanding social behaviors in relation to two seating layout characteristics—having seating close and facing each other and having available seats for everyone — proposed by MCI healthcare professionals through the literature. Field observations were conducted at the Cognitive Empowerment Center—a therapeutic facility for MCI-focused therapeutic activities in Atlanta, GA.

Using a behavioral mapping tool, 1310 social interactions were recorded involving members of the therapeutic program, their care partners, and staff. The analysis identified social interactions by types, group sizes, locations, participants' roles, and postures, using ANOVA and the Bonferroni test.

The results showed that seating availability and positioning impacted the number and group size of social interactions observed in each space for individuals with MCI and their care partners. No significant association was found involving interaction types. Study findings highlight how seating availability and layout can promote socialization for older adults with MCI, given their needs for physical and auditory comfort.

Authors:
Yasaman Gholami, Ph.D. Student, School of Architecture, Georgia Institute of Technology
Raha Rastegar, Ph.D. Candidate, School of Architecture, Georgia Institute of Technology
Herminia Machry, Ph.D., EDAC, Assistant Professor, School of Architecture & Design, University of Kansas
Elahn Little, Research Scientist, SimTigrate Design Lab, Georgia Institute of Technology
Leandro Miletto Tonetto, Ph.D., Associate Professor, School of Industrial Design, Georgia Institute of Technology

HONORABLE MENTION STUDENT PAPER

Nature Nurtures: Investigating Factors Impacting Children's Outdoor Activities and Nature Exposure in Beijing, China

Children growing up in cities today are increasingly disconnected from nature, potentially compromising their quality of life. This research investigated factors influencing children's engagement in outdoor activities and nature exposure. We conducted in-depth interviews with 81 parent-child dyads living in Beijing, China.

Open-ended data was content analyzed, which revealed that most parents perceive their children's outdoor activities and nature exposure to be suboptimal. The influencing factors extend beyond the child's inherent personality and include constraints related to three main external factors: 1) time and companionship, 2) accessibility to natural environmental infrastructure and services, and 3) accidental factors. Time constraints mainly stem from academic pressures, compressing children's discretionary time, which indirectly relates to a lack of peer companionship.

Parental lifestyle, beliefs and work commitment were also identified as being influential, given respondents' age and reliance on parental guidance. The use of electronic devices further encroaches on children's time outdoors. Accessibility to infrastructure and services is primarily concerned with access to natural (green) spaces in daily life, particularly in residential areas, and safety concerns. Organizational factors, such as community-organized events organized by non-profit and for-profit organizations, also played a role. Government policies, including the "double reduction" policy, restrictions on electronic device usage, and initiatives encouraging child-friendly cities were discussed.

Accidental factors, including weather conditions and health issues (e.g., COVID-19), were likewise influential. Implications for landscape architects and urban planners were underscored, emphasizing the need to provide opportunities to facilitate children's outdoor activities and nature exposure, and minimize barriers to this end in the city.

Authors:
Ms. Jianjiao Liu, University of Melbourne
Raymond Green, University of Melbourne,

BEST STUDENT POSTER DESIGN AWARD

Designing for Wellbeing: A Trauma-Informed Approach to Academic Social Spaces

Trauma can be minor, short-lived, or long-term and can have any combination of physical, emotional, psychological, or cognitive effects on individuals. Generation Z ages 13 to 25 are facing higher rates of anxiety, depression, and distress than any other generation, resulting in an increased demand for mental health services (American Psychological Association, 2022). This prevalence of trauma-related challenges emphasizes the impact of trauma as a significant public health concern.

In addition to affecting one's life experiences, trauma influences how one experiences the environment (SAMHSA, 2014). Specifically, the design of the built environment can impact users' perceptions of those spaces and affect their wellbeing. Because of the pervasive effects that trauma can have on people, users who have experienced trauma may not always feel comfortable or motivated to engage with public spaces. This research study concerns trauma-informed design approaches and their potential to provide spaces that address the unique needs of students who have experienced trauma. Such an approach may effectively mitigate stress levels and, in turn, positively impact users by enhancing their overall wellbeing.

Previous research suggests that trauma-informed design approaches in homeless shelters and supportive housing may have a positive effect. However, there is a gap in inquiry on how a trauma-informed design framework can be applied to other building typologies, such as academic environments, for the betterment of trauma-experienced students. University administrations, architects, and designers seek to design academic spaces that would better meet student needs; however, existing university campus facilities may not support the needs of students who have experienced trauma (American Psychological Association, 2022).

This Master of Fine Arts thesis research study uses a mixed methods approach to answer the primary research question, "What architectural qualities influence the academic social space preferences of college students who have experienced trauma?" Surveys, including image-based questions to gauge architectural preferences and reflection questions to learn more about user experiences, were distributed online to currently enrolled college students who have experienced trauma, yielding a total of 100 completed surveys. Ten follow-up interviews took place to gain a deeper understanding of participants' choices. Participants took into consideration these built environment qualities based on the researcher's informed expertise and literature review: scale and proportion, access to views, curvilinearity and rectilinearity, biophilic elements, lighting, color, and materiality. Participants were further asked how these design qualities affect their emotional wellbeing.

This study intends to assist designers in making meaningful design decisions that will have a positive impact on college students who have experienced trauma and mental health issues. Data analysis is currently in progress. However, early findings suggest that certain spatial qualities may positively contribute to students' sense of wellbeing, including features such as higher ceilings, access to views, curvilinear elements, biophilic elements, layered lighting, prospect and refuge, lighter colors, and natural materials. Final results will be shared in the presentation if selected.

References:

American Psychological Association. (2022, November 15). Increased need for mental health care strains capacity. https://www.apa.org/news/press/releases/2022/11/mental-health-care-strains

Substance Abuse and Mental Health Services Administration. (2014). Trauma-Informed Care in Behavioral Health Services. https://store.samhsa.gov/sites/default/files/d7/priv/sma14-4816.pdf

MICHAEL BRILL GRANT
FOR URBAN COMMUNICATION, DESIGN, AND HUMAN BEHAVIOR

ABOUT THE AWARD

- Established in 2015
- Supports innovative research that seeks to enhance the quality of community life by examining human communicative behavior in the context of technologies and physical environments
- Honors Mike Brill whose life's work encompassed these fields
- The Urban Communication Foundation was founded by EDRA member Gary Gumpert in 2005
- Award is funded by the Urban Communication Foundation and by the Environmental Design Research Association.
  Chair: Gary Gumpert, PhD
  Jury Coordinator 2024: Peter Hecht, PhD
  Award amount: $ 3,000

THE MICHAEL BRILL GRANT COMMITTEE

Gary Gumpert, Ph.D.
Susan Drucker, J.D.
Susan Weideman, Ph.D.
Sheila Gobes-Ryan, Ph.D.
Richard Wener, Ph.D.
Peter Hecht, Ph.D.

2024 RECIPIENT

Yasaman Gholami, B.S., M.L.A.
PhD Student in Architecture, Georgia Institute of Technology

Project Title: Being Together in Computer-mediated Communication: Exploring the Role of the Built Environment

The project is at the intersection of social relationships, digitally mediated communications and the influences of physical space.

CERTIFICATE OF RESEARCH EXCELLENCE AWARDS (CORE)

MERIT CERTIFICATES & CORE CERTIFICATES

ABOUT THE AWARD:

EDRA Certificate of Research Excellence (EDRA CORE) recognizes and celebrates rigorous and innovative practice-based environmental design research studies.

CORE serves as a bridge between the design industry and the larger environmental design research community.

MERIT WINNERS:

Blueprint for Mitigating Nurse Burnout: A Social Listening Perspective

Designing Cancer Infusion Centers: Results from a Pre and Post-Occupancy Evaluation

Universal Platform for Critical Access Hospitals: Impact on Operations

Optimizing Hosptial Arrival Experiences: An Integrated Research Protocol for Post-Occupancy Evaluation

Full Circle: Post-Occupancy Evaluation of a Consolidated Transplant Clinic on Patient Experience and Staff Efficiencies

edra core
improving environments through research excellence

CORE WINNERS:

Architecture Matters: Exploring Audiences' Holistic Theatrical Experience through Social Media

Cognitative Commons: Connecting a Library's Evolution and Student Achievement

Addressing Multi-Billion Dollar Challenge: Advancing Knowledge of How High-Quality School Environments can Positively Affect Educational Outcomes

Net Positive Education: Measuring the Impact of Two High Performance Schools

Free Address, Work from Anywhere, and Everything in between: A Post-Occupancy Evolution of an Office Offering Work Flexibility through Design and Policy

The Hybrid Report: Designing for the New Hybrid Work Era

KPI-Driven Design Tool to Inform Post-Pandemic Workplace Design

A Firmwide Exploratory Mixed Methods Research on Post-Occupancy Evolution in Architecture Practice

Getting it Right: A Network-Based Approach to Site Selection for Collaboration

Quiet Carrels and Collaborative Hubs: Campus Library as Third Place

Regreening Suburbia: An Analysis of Urban Greening Approaches in U.S. Sprawl Retrofitting Projects

Ameliorating Cold Stress: Examining the Effects of a Severe Winter Storm on Residents of Subsidized Housing Neighborhoods and Modeling a Design Intervention

FINAL JUROR LIST

Nooshin Ahmadi, M.Arch
Lecturer, Cornell Human Ecology, Cornell University

Jacob Culbert, Ph.D.
Design Anthropologist, HGA Architects & Engineers

Shan Jiang, Ph.D.
Director of Research, GBBN

Saleh Kalantari, Ph.D.
Associate Professor, Director of DAIL, Cornell Human Ecology, Cornell university

Galen Newman, Ph.D.
Professor and Department Head, School of Architecture, Texas A&M University

Widya Ramadhani, Ph.D, EDAC.
Design Researcher, Perkins Eastman

FINAL JUROR LIST

B Sanborn, AIA Allied Member, EDAC
Global Design Research Leader, Principal DLR Group

Brian Schermer, Ph.D., AIA
Associate Professor, University of Wisconsin-Milwaukee, Principal at Workshop Architects

Nicholas Watkins, Ph.D.
Director of Research, Ware Malcomb

Bo Zhang, Ph.D.
Associate Professor, Oklahoma State University

BLUEPRINT FOR MITIGATING NURSE BURNOUT:
A SOCIAL LISTENING PERSPECTIVE

Submitted By: HKS, CADRE and MillerKnoll

Study Team: Rutali Joshi, Deborah Wingler, Edward Hunt, Michelle Ossmann

DESIGNING CANCER INFUSION CENTERS: RESULTS FROM A PRE AND POST-OCCUPANCY EVALUATION

Submitted By: HGA and Elizabeth Healthcare

Study Team: Mahshid Jalalian, Nancy Doyle, Lou Ann Bunker-Hellmich, Abbie Ochsner, and Kara Freihoefer

UNIVERSAL PLATFORM FOR CRITICAL ACCESS HOSPITALS: IMPACT ON OPERATIONS

Submitted By: HGA and Centracare

Study Team: Terri Zborowsky, Erin LaBrec, Shawn Sullivan, and Centracare

OPTIMIZING HOSPITAL ARRIVAL EXPERIENCES: AN INTEGRATED RESEARCH PROTOCOL FOR POST-OCCUPANCY EVALUATION

Submitted By: GBBN and The University of Kansas

Study Team: Shan Jiang, Hui Cai, Monalipa Dash, Kirsten Miller, Zahra Daneshvar, and Jackson Feiknopf

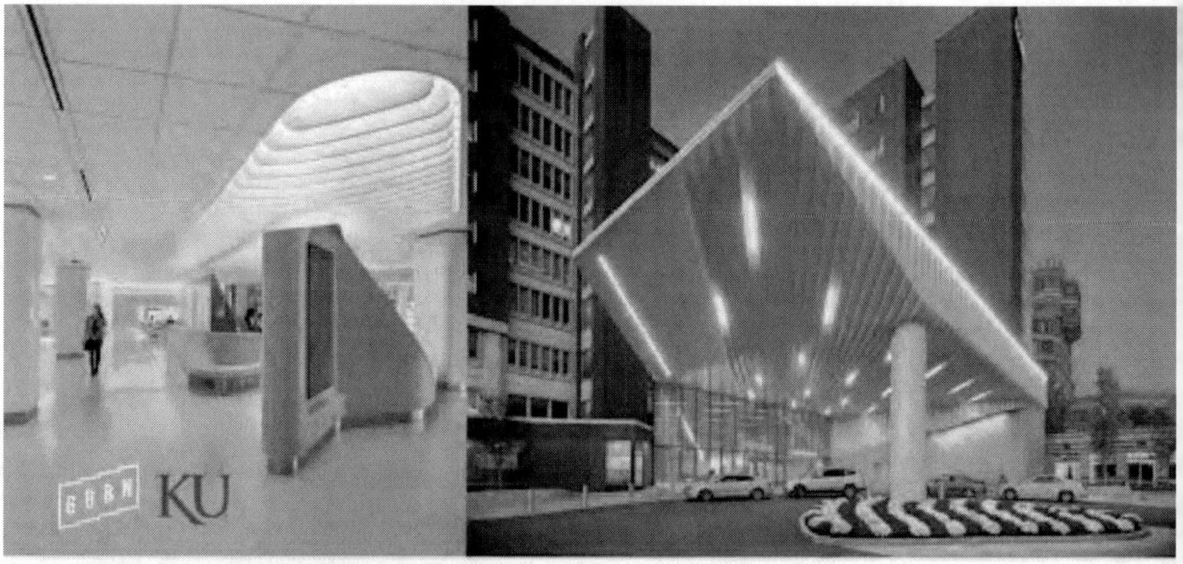

**OPTIMIZING HOSPITAL ARRIVAL EXPERIENCES:
AN INTEGRATED RESEARCH PROTOCOL FOR POST-OCCUPANCY EVALUATION**

FULL CIRCLE: POST-OCCUPANCY EVALUATION OF A CONSOLIDATED TRANSPLANT CLINIC ON PATIENT EXPERIENCE AND STAFF EFFICIENCIES

Submitted By: Flad UW Health

Study Team: Kara Freihoefer, Elizabeth Strutz, Preeti Dalvie, Erin Frush, Laura Serebin, Sinh Dinh, and Karron Conger

Success Outcome

- Patients walk **40-60% less in the new consolidated model**
- Average entry sequence is **30 minutes when no pre-work** and **60 minutes with pre-work**; meeting predicted assumptions
- Electronic tracking boards are **highly integrated into staff workflows**; 80% of visits to exam rooms involves interaction with e-boards
- Since opening, clinic volumes increased **1-2% monthly**

ARCHITECTURE MATTERS: EXPLORING AUDIENCES' HOLISTIC THEATRICAL EXPERIENCE THROUGH SOCIAL MEDIA

Submitted By: GBBN

Study Team: Shan Jiang, Marcene Kinney, Christopher Yost and Julia Clements

COGNITIVE COMMONS: CONNECTING A LIBRARY'S EVOLUTION AND STUDENT ACHIEVEMENT

Submitted By: Multi Studio and Free State High School

Study Team: Michael Ralph, Brittany Coudriet and Marusa Hegeman

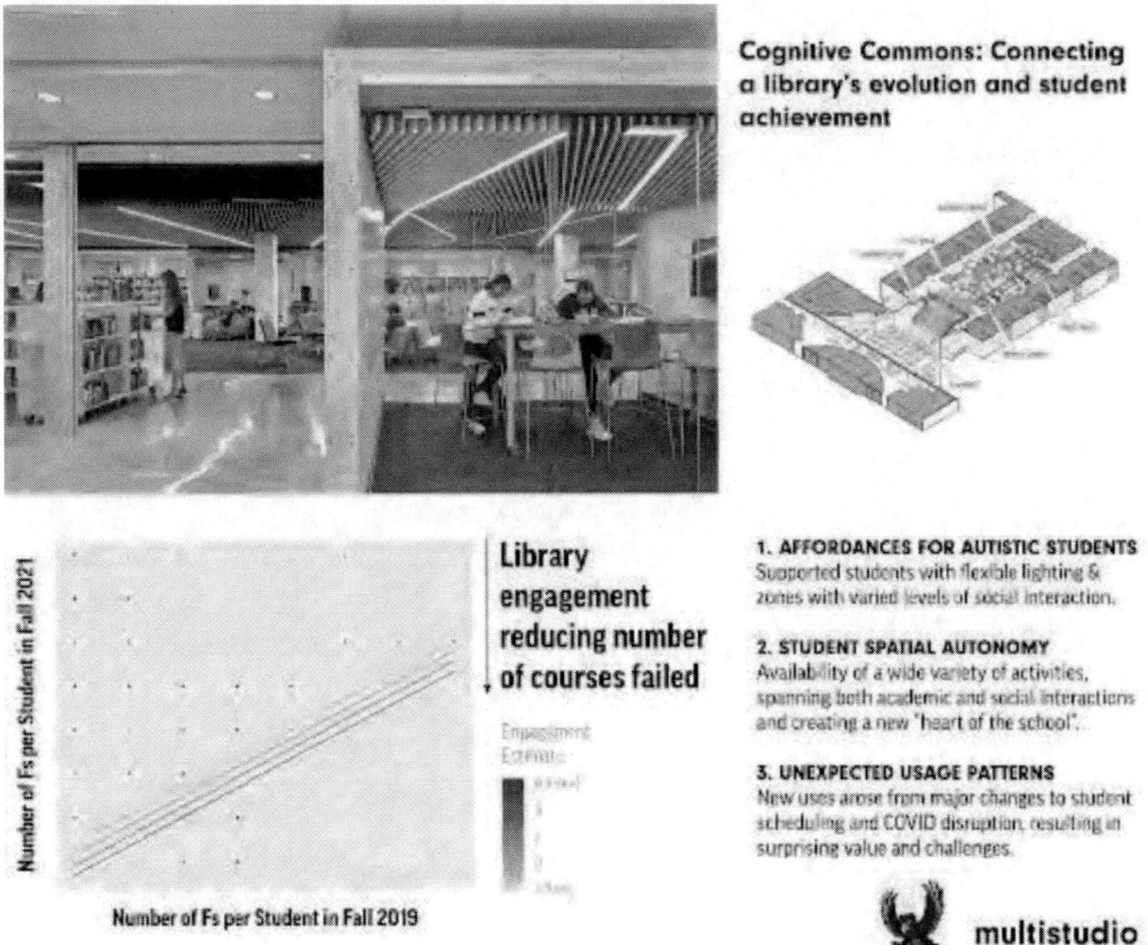

ADDRESSING MULTI-BILLION DOLLAR CHALLENGE: ADVANCING KNOWLEDGE OF HOW HIGH-QUALITY SCHOOL ENVIRONMENTS CAN POSITIVELY AFFECT EDUCATIONAL OUTCOMES

Submitted By: Perkins Eastman, Drexel University and Invontics

Study Team: Emily Chmielewski, Heather Jauregui, Sean O'Donnell, Karen Gioconda, Bruce Levine, and Lance Kruse

NET POSITIVE EDUCATION: MEASURING THE IMPACT OF TWO HIGH PERFORMANCE SCHOOLS

Submitted By: Perkins Eastman

Study Team: Widya Ramadhani, Emily Chmielewski, Heather Jauregui, and Sean O'Donnell

Net Positive Education™:
Measuring the Impact of Two-High Performance Schools

PERKINS— EASTMAN
Human by Design

FREE ADDRESS, WORK FROM ANYWHERE, AND EVERYTHING IN BETWEEN: A POST-OCCUPANCY EVOLUTION OF AN OFFICE OFFERING WORK FLEXIBILITY THROUGH DESIGN AND POLICY

Submitted By: Perkins Eastman

Study Team: Hanna Negami, Widya Ramadhani, Rebecca Milne, Heather Jauregui, Jeff Young, Jennifer Askey, Connor Glass, and Jennica Deely

THE HYBRID REPORT: DESIGNING FOR THE NEW HYBRID WORK ERA

Submitted By: Corgan

Study Team: Melissa Hoelting, Mahdi Afkhamiaghda, Samantha Flores and Emily Strain

KPI-DRIVEN DESIGN TOOL TO INFORM POST-PANDEMIC WORKPLACE DESIGN

Submitted By: Arcadis

Study Team: Helia Taheri, Shreya Somani, Amy Keller Frye and Kyujin (Q) Song

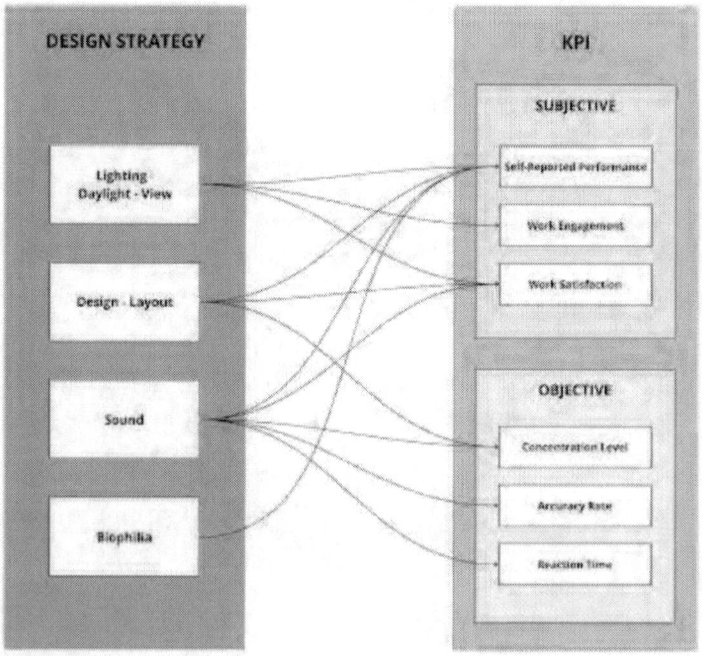

A FIRMWIDE EXPLORATORY MIXED METHODS RESEARCH ON POST-OCCUPANCY EVOLUTION IN ARCHITECTURE PRACTICE

Submitted By: Arcadis

Study Team: Helia Taheri, John Naranjo, Rachel Bruce and Wei Xia

ARCADIS

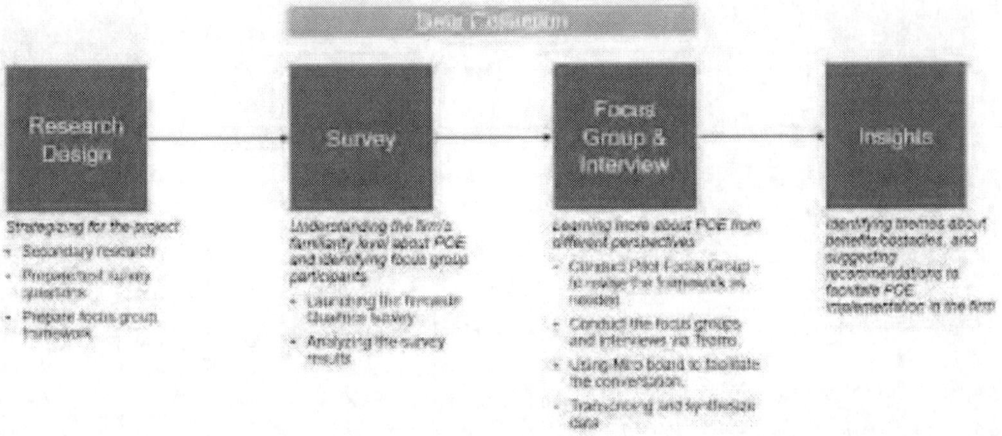

GETTING IT RIGHT: A NETWORK-BASED APPROACH TO SITE SELECTION FOR COLLABORATION

Submitted By: Cannon Design

Study Team: Felix Kabo

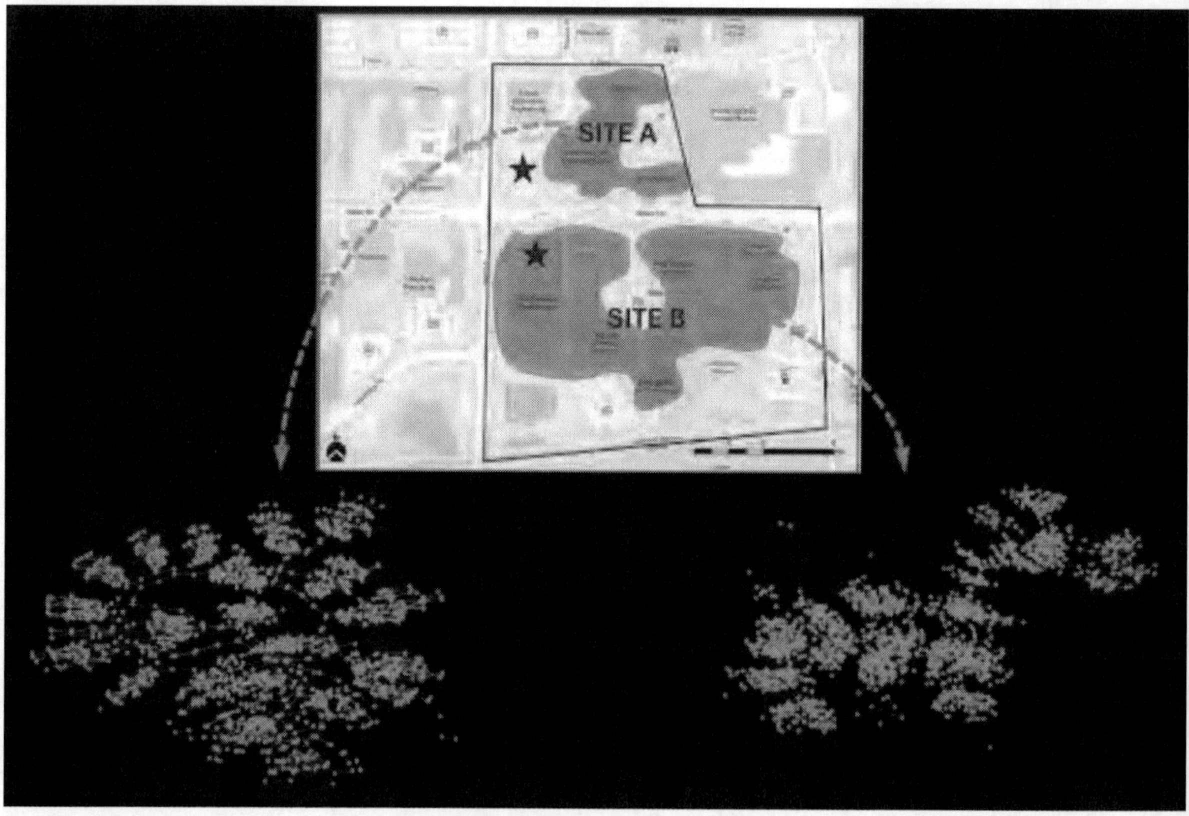

QUIET CARRELS AND COLLABORATIVE HUBS: CAMPUS LIBRARY AS THIRD PLACE

Submitted By: University of Florida and
University of San Carlos - Talamban Campus

Study Team: Margaret Portillo, Jason Meneely, Laura Spears,
Adrian Perez Del Monte, Laura I. Spears, Valrie Minson, Jean Bossart
and Sara Gonzales

REGREENING SUBURBIA: AN ANALYSIS OF URBAN GREENING APPROACHES IN U.S. SPRAWL RETROFITTING PROJECTS

Submitted By: University of British Columbia

Study Team: Keunhyun Park

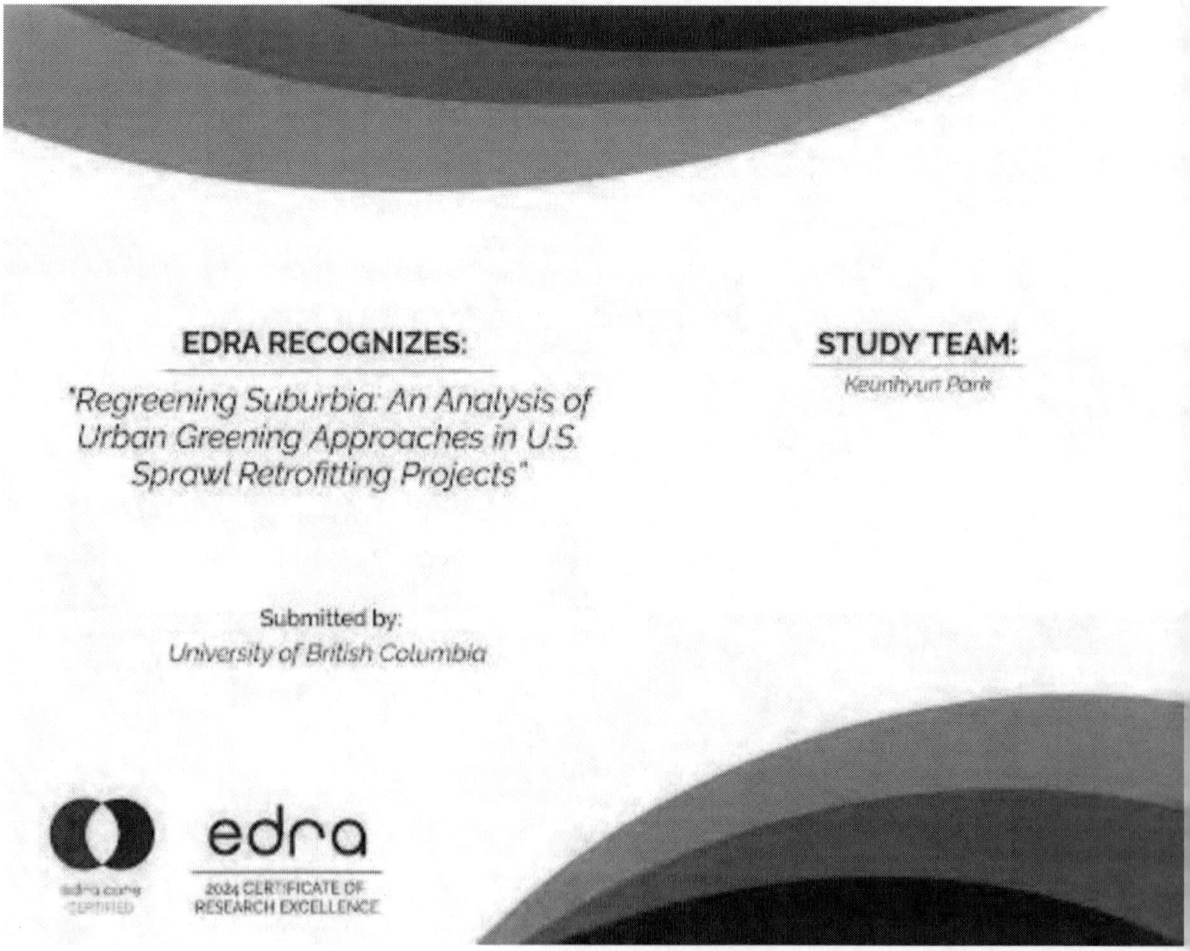

AMELIORATING COLD STRESS: EXAMINING THE EFFECTS OF A SEVERE WINTER STORM ON RESIDENTS OF SUBSIDIZED HOUSING NEIGHBORHOODS AND MODELING A DESIGN INTERVENTION

Submitted By: Texas A&M University

Study Team: Dongying Li, Xiaoyu Li, Yue Zhang and Robert Brown

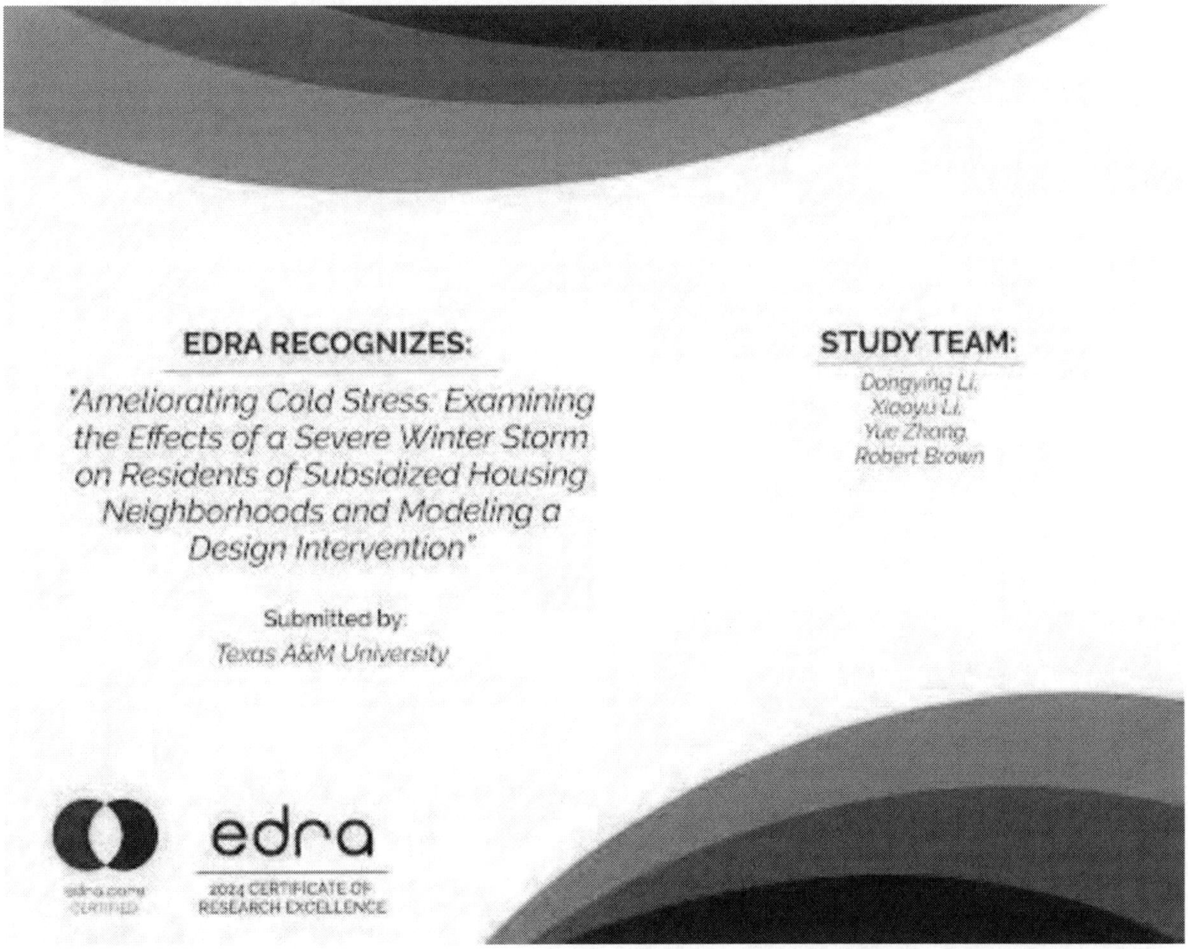

REFEREED FULL PAPERS

IMPACT OF OBJECTIVE WALKABILITY ON SENSE OF COMMUNITY AT DEIR GHBAR, AMMAN, JORDAN: A GIS METHOD OF ANALYSIS

Track 5: Spatial Justice, Equity, and Inclusion by Design
Knowledge Network: International Housing Research, Residential Environments
Methods: Mixed Methods, Survey

Al-Homoud (German Jordanian University); Sarah Mousa Alzghoul (German Jordanian University)

1 ABSTRACT

Amman is experiencing phenomenal population growth, involving multicultural residents. Contemporary neighborhoods in Amman do not seem to be socially active, appearing to lack a sense of community. This social problem is the result of car-oriented physical planning, the neighborhoods lacking safe walkability and outdoor activities. The purpose of this study was to link together walkability, the attributes of the built environment, and sense of community for the residents of the Deir Ghbar neighborhood using geographic information system (GIS) maps. A relationship was hypothesized between neighborhood walkability and sense of community. GIS spatial statistics tools were used to spatially test the hypothesis. The spatial effect of objective walkability on sense of community was calculated and mapped using GIS spatial regression analysis. Objective walkability was found to have a positive effect on sense of community through three of its sub-components: street connectivity, retail density, and residential density. Land-use mix produced no significant effect. GIS spatial analysis model revealed that only objective residential density had a significant spatial effect on sense of community in the correlation model. Residential density is associated with residential land classification type. Our findings invite planners to create a greater spatial diversity of residential density to avoid spatial socioeconomic polarization. Research outcomes can inform local community planners and decision-makers in Jordan to create new strategy to evaluate walkability of neighborhoods using GIS.

1.1 KEYWORDS

Walkability; GIS; Urban; Sustainable; Networks; Jordan; Sense of community; Amman

1.2 INTRODUCTION

In 2015, Amman's population has doubled, reaching 4,077,450. As of 2020, the Amman Governorate populations reached 4,536,500 inhabitants. By the end of 2025 the population of the city is estimated to grow 2.2% annually, reaching 6.4 million (UN Habitat, 2022). The capital city's population represents 42% of Jordan's total population, according to the Jordanian Department of Statistics (DoS) in 2015. These demographic changes are associated with massive migration from the surrounding countries. This rapid growth in population, in addition to the diverse cultural backgrounds, has changed the social structure of the city.

Neighborhoods have become characterized by their lack of social capital and sense of belonging. This problem has been exacerbated by a dependence on cars. Based on 2017 census data, the growth rate in cars per person can be estimated at 8.3%. As a result, residential areas now suffer from drive-through traffic. In addition, the lack of pedestrian infrastructure has made walking difficult, even for short distances, with the Greater Amman Municipality (GAM) having done very little to improve walkability in the city. Some actions have been taken in the city towards this end, such as sidewalk regulation number 19/2014, which focused on regulating the construction of sidewalks to specific standards and keeping the sidewalks only for pedestrian use. In addition, the GAM initiated the Transport and Mobility Master Plan 2010, aimed at improving public transport efficiency and discouraging the dependence on cars. However, there has been an absence of a comprehensive framework in spatial planning to achieve such walkability. A comprehensive strategy is thus needed.

Several studies have focused on the importance of walkability in improving the sense of community (French et al., 2013; Wilkerson et al., 2012). According to Chavis and Wandersman (1990), establishing a sense of community can work as a system for stimulating healthy communities and individuals. A sense of community can turn people who live next door from complete strangers into neighbors who are partners in the local community and capable of sharing common activities, interests, and perspectives (Farahani, 2016).

In this study, we evaluated the attributes that affect walkability, as well as the impact of walkability on the sense of community. The study area was the Deir Ghbar district, which is a contemporary residential area not more than 15 years old. At the national level, no previous studies have addressed this topic.

https://doi.org/10.52202/082078-0001

2 RESEARCH OBJECTIVES

To achieve our aim, we had the following objectives:
1) measure objective walkability,
2) measure the sense of community, and
3) visualize the resulting spatial maps using geographic information system (GIS) tools.

3 LITERATURE REVIEW

3.1 Residential urban accessibility

Accessibility measures explain the effect of land use and transport planning on the functioning of communities. There are four main components to accessibility: 1) a land-use indicator measured by density, quality, and spatial distribution; 2) a transportation indicator measured by the time, cost, and effort needed to reach the desired land-use activity; 3) a temporal indicator of the various constraints and opportunities at play during different times of the day; and 4) an understanding of a person's needs, abilities, and opportunities (Geurs and Van Wee, 2004). Pedestrian-friendly planning is characterized by compactness, high density, street pattern design, and land-use mix, which affect neighborhood accessibility, walkability, and livability (Krizek, 2003; Tasi, 2014). Accessibility may vary based on the spatial distribution of retail shops and services relative to population density (Handy and Clifton, 2001). Pedestrian-friendly planning and walkability affect the sense of community and livability in residential areas (Tsai, 2014).

3.2 Objective walkability

The objective walkability measure depends on the features of the built environment (Mayne et al., 2013). A residential area that has high street connectivity, high residential density, and a high land-use mix could encourage high walkability (D'Haese et al., 2014). Other features used in various walkability metrics include: street design, specifically street density; the availability and affordability of sidewalks; accessibility; street connectivity and directness; safety; and easy road-crossing; the absence of high-speed traffic; land-use density; pedestrian-traffic separation and buffering; land-use mix; street planting and landscaping; security; and a sense of place at the local level (Cervero et al., 2009; Lo, 2009). However, the most frequently used parameters in objective walkability indices in the literature can be summarized into the following four measures (Agampatian, 2014; Dygryn et al., 2010; Frank et al., 2010; French et al., 2013; Jun and Hur, 2015; Leslie et al., 2007; Wood et al., 2010):

1. Land-use mix: This measure indicates the heterogeneity of the land use in a specific area. A higher proportion of non-residential uses to residential uses means a higher proportion of people willing to walk to various destinations. Land-use mix can be measured by (a) the ratio of non-residential to residential zone use, and (b) the ratio of non-residential uses to unit area.

2. Street connectivity: This measure represents the degree of connectivity in a network of roads. A well-connected network offers a wider variety of alternative routes to reach a destination, among which people can choose the shortest and easiest route. This in turn encourages people to walk and increases walkability. One of the methods used to quantify this measure is by computing the number of intersections. The most frequently used measure to indicate street connectivity is the number of street centerline intersections (a true intersection with three or more branching streets) per unit area.

3. Retail density: This measure indicates the density of commercial uses per unit area. The higher the density, the greater the number of retail stores, which in turn positively affects walkability. Various measures are used to indicate this parameter, the most frequent being (a) the retail floor area ratio--the ratio of commercial building square footage to land square footage, and (b) the number of retail locations in each area.

4. Residential density: This can be measured by population density or household density. Household density is considered the most appropriate measure for residential areas because it indicates the residential density, whereas the population density can indicate populations of non-residential use. Household density can be calculated for each census block.

3.3 Walkability indices and GIS

Features of the built environment affecting walkability can be quantified and measured using GIS based on existing data, thus facilitating the quantification and visualization of walkability, while GIS-based spatial patterns can also be visualized (Leslie et al., 2007). Walkability indices measuring built features in the neighborhood are widely used and have recently gained in popularity (Manaugh and El-Geneidy, 2011). The variables most frequently used in this method are population density, land-use mix, access to recreational facilities, street patterns, sidewalk coverage, vehicular traffic, crime, and a composite variable/index based on selected measures from previous studies (Brownson et al., 2009). Three measures of the built environment were combined to produce the walkability index: 1) net residential

https://doi.org/10.52202/082078-0001

density; 2) street connectivity; and 3) land-use mix--the uniformity of the spatial distribution of residential, commercial, and office land uses (Leslie et al., 2007). Other scholars have used the same measures, with one addition--commercial density (Dygryn et al., 2010).

Leslie et al. (2007) developed a walkability index measured using dwelling density, connectivity, land-use attributes, and net retail area. The measures were quantified as follows: 1) dwelling density--a point-shaped dwelling layer was prepared by counting the number of dwellings in each Census Collection Districts (CCD). Then, the overall residential dwelling density is calculated by dividing the dwelling counts by the area of the CCD where they were located. To unify the measures, the results in each CCD were ranked on a scale from 1 to 10; 2) connectivity--measured by the number of street connections at each intersection of the centerline of a street. The density is measured by the number of intersections in each square kilometer for each CCD, the results being ranked on a scale from 1 to 10; (c) land use, divided into two affective variables--mixed land use (heterogeneity) and non-mixed land use (homogeneity). An entropy score for each CCD is created by using the entropy index formula. The resulting score is either 0 (representing single land use) or 1 (representing mixed land use). This score is converted into a standard score (1-10); and (d) gross retail areas (GRAs)--measured by field survey and divided by the total parcel area (P) to obtain the net retail area (NRA) for each parcel that consists of retail uses NRA = GRA/P.

3.4 Sense of community and walkability

A sense of community refers to the sense of belonging, of identity, and reinforcement. The concept represents a situation of familiar neighbors living next door who have shared experiences, activities, and history (Farahani and Lozanovska, 2014; Pretty et al., 2006). Farahani (2016) concluded that creating a strong sense of community and attachment to the local community is vital to the quality of life in neighborhoods. Certain elements can add valuable characteristics to the neighborhood environment, such as safety and security, residential satisfaction, community identity, local organizations and civic participation, and health and well-being. Pretty et al. (2006) found that sense of community and social capital affect health and well-being at the community level. On one hand, a sense of community can protect against the physical and psychological symptoms of illness. On the other hand, it can have some negative impacts, such as the exclusion of newly arrived minority groups and migrants or of "others" who are different. Sense of community has been defined as the term used by residents, politicians, and social scientists to represent the association between a person and their social structure (Chavis and Wandersman, 1990). According to McMillan and Chavis (1986), the definition of sense of community consists of four main parts: 1) membership, or a sense of belonging; 2) influence, or a sense of mattering; 3) reinforcement integration, or the fulfillment of needs; and 4) shared emotional connections, which include shared history, places, a common time, and homogeneous experiences.

The built environment possesses features that can influence sense of community in four main ways: 1) the availability of public places, such as parks, plazas, and commercial streets; 2) pedestrian accessibility, walkability, and a human-scale built form; 3) mixed land use; and 4) greenery. Social life and interaction can be encouraged by the qualities of the built environment that promote walking (Farahani and Lozanovska, 2014). Building a walkable neighborhood is a way to increase sense of community because walkability increases social capital and familiarity between neighbors. Recent studies have begun to focus on the attributes of the built environment that influence walkability (French et al., 2013).

4. METHODS

A correlational design using quantitative research data was used to study the impact of walkability (independent variable) on sense of community (dependent variable). The hypothesis of the study was: objective walkability affects sense of community. The dependent variable "sense of community" was defined as: 1) feeling at home in the neighborhood; 2) the neighborhood sticking together; and 3) having relationships with neighbors. Sense of community was measured on a 5-point Likert scale.
The independent variable "objective walkability" was defined by an index and measured by two methods-a GIS ordinary least squares (OLS) tool and SPSS tools. The index attributes and measures were as follows:

1. Land-use mix was defined by the heterogeneity of the land uses and was operationally measured using the following formula: land-use mix = area of non-residential use/area of census data block (CDB) in acres. Using GIS, land-use mix was measured as a proportion, on a scale of 0-1. The results were collapsed into five levels using GIS--the higher the number, the greater the land-use mix.
2. Street connectivity was defined by the density of street intersections in each area, and was operationally measured using the following formula: connectivity = accumulated number of street branches in each node connected to or within the CDB/area of the CDB in acres. Connectivity was measured using GIS. The results were an open true number, which was collapsed into five levels-the higher number of intersections per unit area, the greater the connectivity.

https://doi.org/10.52202/082078-0001

3. Retail density was defined by the amount of area designated for commercial use (commercial density). The measure was operationally defined by the following formula: retail density = number of retail uses in a CDB/area of the CDB in acres. The retail density was measured using GIS as a proportion on a scale of 0-1. The results were collapsed into five levels using GIS-the higher the number, the greater the retail density.
4. Residential density was defined by the number of residential households per unit area and operationally defined by the following formula: residential density = number of households in each CDB/area of the CDB in acres. Residential density was measured using GIS, and the results were an open true number. As above, the measure was collapsed into five levels, where the higher the number, the greater the residential density.

4.1 Research setting

The Deir Ghbar neighborhood is a modern residential area located on the western side of Amman. It is one of the wealthiest neighborhoods in Amman, where the cost of living is relatively high. Deir Ghbar has wide streets, although its walkability is low and pedestrian facilities are unavailable. The streets have discontinuous and narrow sidewalks, which are almost entirely occupied by trees. The area suffers from through traffic, and the residents depend on neighboring districts for their commercial activities. The total area of Deir Ghbar is 1.9 km2, divided between 0.6 km2 of streets and 1.3 km2 of land parcels. Deir Ghbar has three types of housing (A,B, &C).

The most dominant housing types in Deir Ghbar are A and B (87%), both considered upper-income housing in Amman because the price of land is high. Note that Deir Ghbar has the largest land area and the lowest floor area ratio compared to the other districts.

4.2 Study population and sampling

The total study population was 17,450 inhabitants, with 9,112 males and 8,338 females (DoS, 2015). The unit of analysis was the head of the household. The study area contained a total of 4,679 households. The sampling frame was the households distributed over the CDBs. There was a total of 50 CDBs in the area. Based on GIS counts, the number of households in these blocks ranged from 62 to 446. Figure 3 shows the official division of the Deir Ghbar CDBs with the number of households in each indicated.
Sample size: The sample size was calculated based on the formula used by Krejcie and Morgan (1970). The outcome was a sample size of about 380 households from all the Deir Ghbar CDBs.
Sampling technique: Random proportional cluster sampling was employed using the CDBs to reach the target sample size of 380, with eight subjects being drawn from each CDB (380/50 = ~8).
Sampling procedure: The minimum number of households in any CDB was 62. We counted every fifth unit until eight subjects were obtained for each CDB. The counting proceeded along each street for the units on the same side of the street, in both horizontal and vertical directions. For apartment buildings, the researcher entered the building and counted the units starting from the right-hand side until the fifth unit was reached. If the subject refused to participate in, the researcher moved to the next household until eight subjects were reached in each CDB.

4.3 Data collection

Quantitative data from field surveys was used to investigate objective walkability and sense of community. The surveys were conducted using face-to-face interviews. Instrument A was a structured questionnaire. The subjects received a letter of introduction and were assured that the data would be confidential. Instrument B consisted of spatial maps obtained from the GAM and the DoS, which defined the study boundaries and study area CDBs to explore walkability using GIS. These maps were also used to create maps for exploring the spatial distribution of walkability and sense of community along the CDBs. The obtained maps included: 1) a land-use map to explore land-use mix; 2) a street centerline map to explore street connectivity; 3) a retail location map to explore retail density; and 4) a map of the buildings with building height for exploring residential density.

5 ANALYSIS AND DISCUSSION

5.1 Digital data sources and GIS

Digital data sources in the form of a GIS shapefile were used to develop an objective walkability index. Geographic vector data, which were all spatially referenced according to the official coordination system used in Jordan, were employed to calculate the parameters of the walkability index. The data included land-use mix, street connectivity, retail density, and residential density. The collected data was mapped and analyzed using GIS software.

5.2 SPSS projection on GIS maps

The collected data were mapped and analyzed using GIS software. To assess correlations quantitatively and visually between the variables, a GIS-based spatial regression analysis model was developed and mapped. The relationships between the variables were spatially measured for each CDB using a GIS OLS spatial regression analysis, which was applied using a spatial statistics toolbox in GIS software. The SPSS data was projected onto GIS mapping using a spatial joint tool in GIS, with the joint features being the CDBs of Deir Ghbar.

https://doi.org/10.52202/082078-0001

5.3 Dependent variable - sense of community

Data were obtained from the survey questionnaire and computed for each CDB by summing the results of the questionnaire for all the subjects in each block and dividing the results by the number of subjects. The results were mapped using GIS to facilitate studying the spatial distribution of sense of community (see Figure 1).

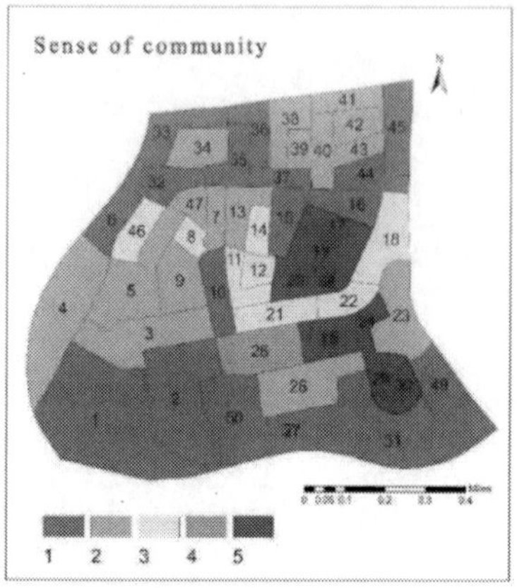

Figure 1. Spatial distribution of sense of community. (2023). Image by the authors.

The results showed that the lowest sense-of-community values were recorded around the periphery of the neighborhood (blue and green blocks in Figure 1). The highest values were recorded at the center of the neighborhood (red blocks in Figure 1), with the highest value recorded in Type C housing.

5.4 Independent variable - objective walkability

Analysis of the objective walkability attributes was performed using GIS software as follows:

Land-use mix. This was calculated using GIS with the formula: land-use mix - area of non-residential uses/area of CDB in acres. The total land-use mix values in the area were 1.17. A total of 16 CDBs in the study area contained non-residential land use, with the remaining 34 being completely residential, with a zero land-use mix value. The results were expressed as a ratio ranging from 0 to 1. The highest recorded land-use mix value was 0.2 in CDB 41, the lowest was 0.0063 in CDB 1. Figure 2 shows the spatial distribution of land-use mix values across the CDBs in the study area. This process followed Leslie et al.'s (2007) calculation method.

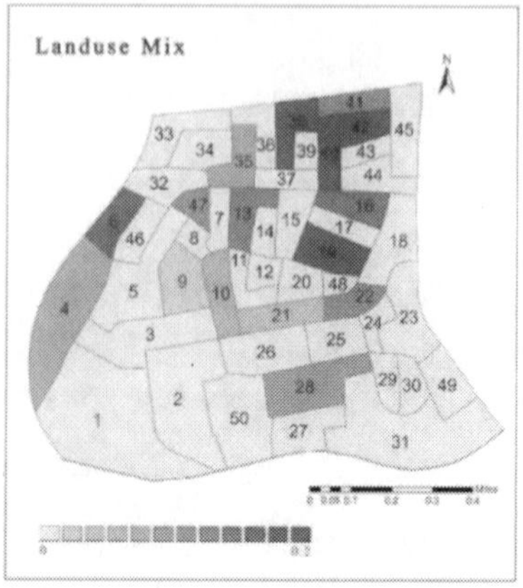

Figure 2. Spatial distribution of land-use mix. (2023). Image by the authors.

Street connectivity. This was calculated using GIS and the formula: connectivity - accumulated number of street branches in each node connected to or within the CDB/area of the CDB in acres). The highest recorded connectivity value was 8 in CDB 8, the lowest being 1 in CDB 50. Figure 3 shows the spatial distribution of connectivity values across the CDBs in the study area. This process followed calculations of Leslie et al. (2007).

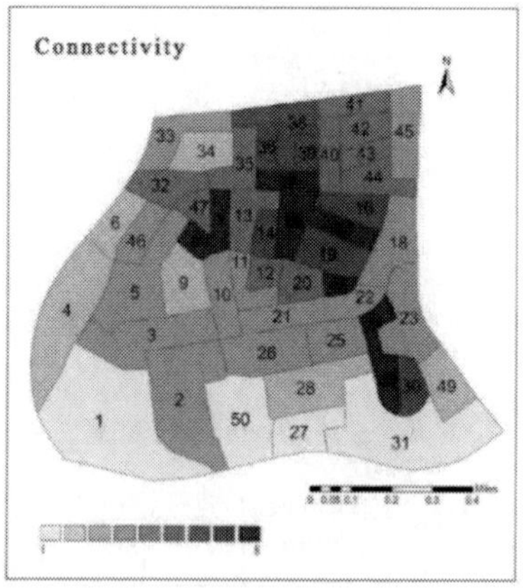

Figure 3. Spatial distribution of connectivity. (2023). Image by the authors.

Retail density. This was calculated using GIS using the formula: retail density - number of retail uses in a CDB/area of the CDB in acres. The total retail density value in the area was 1.46. A total of 10 CDBs from the study area contained retail land use, with the remaining 40 had no retail land use. The highest retail density value was 0.25 in CDB 47, the lowest was 0.02 in CDB 1. Figure 4 (left) shows the spatial distribution of retail density values across CDBs in the study area. This process followed that of Leslie et al. (2005).

https://doi.org/10.52202/082078-0001

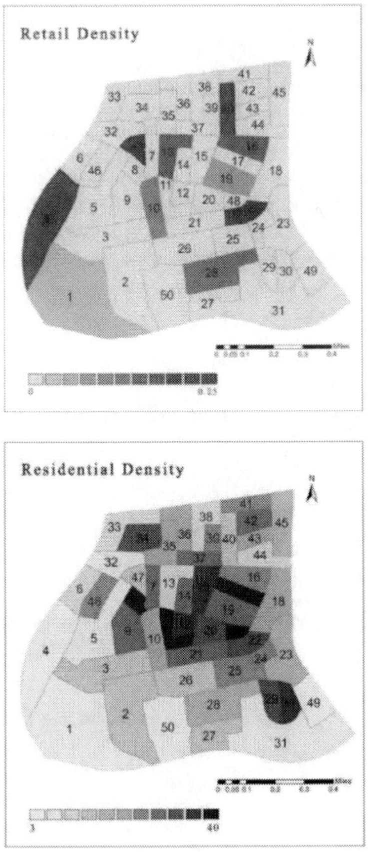

Figure 4: Top: Spatial distribution of retail density. Bottom: Spatial distribution of residential density values. (2023). Image by the authors.

Residential density. This was calculated using GIS and the formula: residential density = number of households in each CDB/area of the CDB in acres. The highest residential density value was 40 in CDB 48, the lowest was 3 in CDB 1. Figure 4 (right) shows the spatial distribution of residential density values among the CDBs in the study area. This method followed Leslie et al.'s (2007) calculation process. To obtain a standard measure, the results were collapsed into 10 levels (from 1 to 10).

Total objective walkability. This was calculated for each CDB by summing the values for land-use mix, street connectivity, retail density, and residential density using the geometric overlay tool in GIS. The lowest objective walkability was 4.03 for CDB 1, the highest was 48 for CDB 48. Figure 5 shows the spatial distribution of the objective walkability values among the CDBs in the study area.

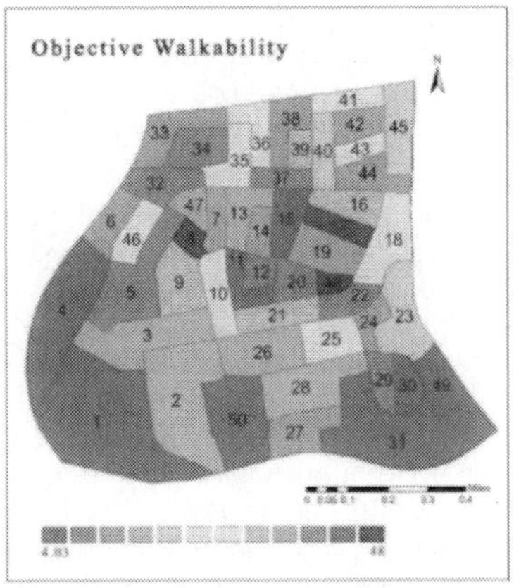

Figure 5. Spatial distribution of total objective walkability. (2023). Image by the authors.

5.5 Spatial modeling hypothesis relationships using GIS spatial statistics

The quantitative data for sense of community and objective walkability obtained from the field surveys were mapped by GIS software using the joint spatial tool. The CDBs were considered as a joint field to relate the data from each block with the spatial location of the CDBs. Sense of community was mapped on a scale of 1–5. Objective walkability was collapsed to a scale of 1–5 to unify the scales of all variables and facilitate comparison. The analysis included spatial regression and hotspot analyses.

Spatial regression analysis for sense of community by objective walkability-spatial relationship: A GIS OLS tool was used to evaluate the relationships between two or more variables. In addition, an OLS linear regression was employed to spatially test the stated hypothesis that objective walkability and its four components affect sense of community.
Overall model significance. To assess model significance, a joint F-statistic and measures of overall model statistical significance were carried out. The model was found to be statistically significant ($p = 0.001$), as indicated in Table 1.

https://doi.org/10.52202/082078-0001

Table 1. Results for overall model significance.

Model significance			
Joint F-statistic [e]:	50.09	Prob(>F), (4,45) degrees of freedom:	0.001*

Objective land-use mix by sense of community: There was no significant spatial effect on sense of community. In this model, land-use mix had little impact because it contributed positively, with a 0.001-unit change in sense of community for every 1-unit change in objective land-use mix (t = 0.01, p = 0.99, a = 0.001) (Table 2).

Objective street connectivity by sense of community: There was no significant spatial effect on sense of community in this model. Street connectivity had little impact because it positively contributed, with a 0.26-unit change in sense of community for every 1-unit change in objective street connectivity (t = 1.10, p = 0.25, a = 0.26) (Table 2).

Objective retail density by sense of community: There was no significant spatial effect on sense of community in this model, which showed little impact. Retail density contributed positively, with a 0.002-unit change in sense of community for every 1-unit change in objective retail density (t = 0.01, p = 0.99, a = 0.002) (Table 2).

Objective residential density by sense of community: There was a significant spatial effect on sense of community in this model. Residential density positively contributed, with a 0.43-unit change in sense of community for every 1-unit change in objective residential density (t = 2.13, p = 0.02, a = 0.43) (Table 2).

Table 2. OLS GIS regression results for each explanatory variable.

Variable	Mean	Std. deviation	Coefficient (a)	Probability (p-value	t-statistic	Std. error
Sense of Community	**2.34**	**1.38**	—	—	—	—
Objective Walkability Components						
Land-Use Mix	0.4	0.94	0.001	0.99	0.01	0.23
Street Connectivity	0.56	1.13	0.26	0.25	1.10	0.23
Retail Density	0.3	0.81	0.002	0.99	0.01	0.29
Residential Density	0.1	1.3	0.43	0.02	2.13	0.20

Objective residential density and objective land-use mix were the only subcomponent of objective walkability that recorded a significant spatial impact on sense of community. Deir Ghbar residents had a high sense of community when they had a high objective residential density and a high objective land-use mix.

The analysis provided focuses on the use of digital data sources and GIS to develop an objective walkability index and assess its impact on the sense of community in a specific area. The study utilized various parameters such as land-use mix, street connectivity, retail density, and residential density to calculate the walkability index. GIS software was used to map and analyze the collected data.

The GIS-based spatial regression analysis model was used to examine the correlations between variables. By projecting SPSS data onto GIS maps, they were able to assess the relationships visually and quantitatively between the variables.

The study also examined the spatial distribution of the sense of community across the neighborhood using GIS mapping.

The results showed that the sense of community was lower on the periphery of the neighborhood and higher at the center, with Type C housing recording the highest value. Objective walkability attributes were analyzed using GIS software, including land-use mix, street connectivity, retail density, and residential density. Each parameter was calculated using specific formulas, and the results were expressed as values or ratios. The spatial distribution of these attributes across the study area was mapped for visualization and analysis.

Furthermore, the study employed spatial modeling and GIS spatial statistics to explore the relationships between sense of community and objective walkability. An OLS linear regression model was used to test the hypothesis that objective walkability and its components affect the sense of community. The overall model was found to be statistically significant.

https://doi.org/10.52202/082078-0001

However, when examining the individual components, it was observed that only objective residential density had a significant spatial effect on sense of community. This means that higher residential density corresponded to a higher sense of community. Other components such as land-use mix, street connectivity, and retail density did not show a significant spatial impact.

In summary, this analysis provides valuable insights into the use of digital data sources and GIS to develop a walkability index and assess its relationship with the sense of community. The findings suggest that residential density plays a vital role in fostering a stronger sense of community in the study area.

6. CONCLUSIONS

GIS shapefiles and digital data sources were used to develop an objective walkability index based on parameters such as land-use mix, street connectivity, retail density, and residential density. In addition, spatial mapping and analysis using GIS software facilitated the examination of the spatial distribution of variables, such as sense of community and objective walkability, within the study area.

On the other hand, sense of community was found to be highest in the center of the neighborhood, particularly in Type C housing. Housing Type C recorded the highest values for sense of community, with Type A showing lower values. This finding is justified because the Type C residences were characterized by higher densities, whereas the Type A residences were characterized by lower densities. The lowest values for sense of community occurred around the periphery and on the edges of the neighborhood, whereas the highest values were recorded in the center. Sense-of-community values were low along highways and arterial roads, where speed and the volume of traffic were likely to be higher.

Further, objective walkability attributes were analyzed using GIS software, revealing variations in land-use mix, street connectivity, retail density, and residential density across different community development blocks (CDBs). Among objective walkability components, only objective residential density and objective land-use mix significantly impacted the sense of community. The lowest values were recorded along the southern edge of the neighborhood, while the highest were near the center. Objective walkability values were higher where there was a higher density of buildings and a more complex road network.

Recommendations, implications, and design guidelines provide valuable insights for urban planners and policymakers aimed at creating more walkable, connected, and socially cohesive neighborhoods as follows:

6.1 Recommendations

General recommendations include the following:

- Enhancing land-use mix, street connectivity, retail density, and residential density could contribute to improving walkability and fostering a sense of community in the study area.
- Based on the findings, focusing on improving residential density and land-use mix could have the most significant impact on enhancing the sense of community.
- Implementing policies or interventions that encourage mixed-use development, pedestrian-friendly street networks, and increased residential density could help create more walkable and socially connected neighborhoods.
- Considering the positive influence of high residential density and land-use mix on the sense of community, urban planning interventions should focus on promoting mixed-use developments and increasing the density of residential areas.
- Enhancing street connectivity could positively contribute to walkability, but it did not show a significant effect on the sense of community. Nonetheless, improving street connectivity remains valuable for overall neighborhood development.

Further, the findings of this study provide valuable insights and recommendations for local municipality planners and authorities in Amman to improve walkability and enhance community connectivity. Here are some of the key recommendations:

1. Utilize GIS mapping and spatial statistics to assess built-environment features, their impact on community cohesion, and identify locations requiring physical interventions.
2. Formulate distinct spatial policies pertaining to land-use management to regulate land-use mix, retail density, residential density, and street connectivity at the neighborhood level. These policies should align with pedestrian-oriented planning principles.
3. Implement pedestrian-oriented planning and design criteria within neighborhoods to reduce reliance on cars and improve walkability.
4. Adopt traffic planning and design policies that discourage through traffic and enhance safety within neighborhoods.
5. Evaluate the road hierarchy and street connectivity within neighborhoods to identify areas for improvement and optimize traffic flow.

https://doi.org/10.52202/082078-0001

6. Establish guidelines for neighborhood road design to limit traffic and ensure high connectivity of local roads.
7. Enact legislation mandating the construction of affordable, safe, and obstacle-free sidewalks within neighborhoods.
8. Initiate a pilot project to rehabilitate sidewalks at the neighborhood level, with Deir Ghbar identified as a potential site for this endeavor.
9. Implement legislation regarding the planting of trees and placement of street furniture along sidewalks in residential neighborhoods.
10. Incorporate design elements in sidewalk construction that enhance accessibility for individuals with mobility limitations and accommodate shopping carts and baby carriages.
11. Develop a comprehensive maintenance strategy for sidewalks to ensure their long-term usability.
12. Harness the advantages derived from i increasing residential density and promoting spatial diversity.
13. Address socioeconomic polarization in Amman resulting from the association between housing prices/land value and residential density. Stakeholders should carefully consider the necessity of introducing diverse and flexible pricing in the housing market.

By implementing these recommendations, local authorities can work towards enhancing walkability, promoting community cohesion, and improving the quality of life for residents in Amman.

6.2 Implications

- The study highlights the importance of considering both physical and social factors in urban planning and design. Walkability and the sense of community are intertwined and should be addressed together.
- The findings can inform urban planners, policymakers, and developers in promoting sustainable and livable neighborhoods by integrating walkability-enhancing features and fostering community cohesion.
- Recognizing the spatial variations in walkability and the sense of community can guide targeted interventions and resource allocation to areas with the greatest need.
- The spatial distribution maps highlight areas with low sense of community values around the neighborhood's periphery, indicating the need for targeted community-building initiatives in those regions.
- Areas with higher residential density and land-use mix should be recognized as potential hotspots for fostering a sense of community.

6.3 Design Guidelines

- When designing neighborhoods, emphasis should be placed on creating diverse land-use patterns within proximity to residential areas to enhance the sense of community.
- Careful consideration should be given to the spatial organization of neighborhoods, with particular attention to ensuring a balance between centralized and distributed community spaces to maximize social interactions.
- Enhancing residential density through appropriate zoning regulations and efficient land use allocation can contribute to stronger community bonds.
- Promote mixed-use development by integrating residential, commercial, and recreational spaces within proximity, encouraging active street life and reducing dependency on private vehicles.
- Develop a well-connected street network that prioritizes pedestrian and cyclist safety, providing direct pathways between key destinations and fostering social interactions.
- Design neighborhoods with a range of housing options and densities to accommodate diverse populations, ensuring a balance between privacy and opportunities for social interactions.
- Foster neighborhood amenities, such as parks, community centers, and gathering spaces, serve as focal points for social activities and strengthen the sense of community.

Overall, the integration of GIS and spatial analysis provides valuable insights into the relationship between walkability and the sense of community, enabling evidence-based decision-making in urban planning and design. By prioritizing walkability and community-building, cities can create more livable, sustainable, and socially vibrant environments.

https://doi.org/10.52202/082078-0001

7. REFERENCES

Agampatian, R. (2014). Using GIS to measure walkability: A Case study in New York City. Master Thesis. School of Architecture and the Built Environment, KTH, Stockholm, Sweden.

Brownson, R. C., Hoehner, Ch. M., Day, K., Forsyth, A. & Sallis, J. (2009). Measuring the Built Environment for Physical Activity: State of the Science. American journal of preventive medicine. 36 (4). 99-123.

Cervero, R., Sarmiento, O.L., Jacoby, E., Gomez, L. F., & Neiman, A. (2009). Influences of Built Environments on Walking and Cycling: Lessons from Bogotá. International Journal of Sustainable Transportation. 3 (4). 203-226. https://doi.org/10.1080/15568310802178314

Chavis, D. M., & Wandersman, A. (1990). Sense of Community in the Urban Environment: A Catalyst for Participation and Community Development. American Journal of Community Psychology. 18, pp 55-81.

D'Haese, S., Van Dyck, D., Bourdeaudhuij,I,, Deforche, B., & Greet, C. (2014). The association between objective walkability, neighborhood socioeconomic status, and physical activity in Belgian children. International Journal of Behavioral Nutrition and Physical Activity. 11 (104). https://doi.org/10.1186/s12966-014-0104-1

Dygryn, J., Mitáš, J. & Stelzer, J. (2010). The influence of built environment on walkability using geographic information system. Journal of Human Kinetics. 24, 93-99. https://doi.org/10.2478/v10078-010-0025-2

Farahani, L., & Lozanovska, M. (2014). A framework for exploring the sense of community and social life in residential environments. International Journal of Architectural Research. 8 (3). DOI:10.26687/archnet-ijar.v8i3.412

Farahani, L. (2016). The value of the sense of community and neighboring. Housing, Theory and Society. 33 (3), (357-376) https://doi.org/10.1080/14036096.2016.1155480

French, S., Wood, L., Foster, S. A., Giles-Corti, B., Frank, L., & Learnihan, V. (2013). Sense of community and its association with the neighborhood built environment. Environment and Behavior. 46 (6), pp677-697. https://doi.org/10.1177/0013916512469098

Frank, L. D., & Devlin, A., Johnstone, Sh., & Loon, J. V. (2010). Neighborhood Design, Travel, and Health in Metro Vancouver: Using a Walkability Index. Executive summary. Active Transportation Collaboratory, UBC. https://act-trans.ubc.ca/files/2011/06/WalkReport_ExecSum_Oct2010_HighRes.pdf

Geurs, K. T. & Van Wee, B. (2004). Accessibility Evaluation of land-use and transport strategies: review and research Directions. Journal of Transport Geography. 12, pp 127-140.

Handy, S. & Clifton, K. (2001). Evaluating neighborhood accessibility: Possibilities and practicalities. Journal of Transportation and Statistics. 4 (2).

Jun, H. & Hur, M. (2015). The relationship between walkability and neighborhood social environment: The importance of physical and perceived walkability. Applied Geography. ^2, pp 115-124. https://doi.org/10.1016/j.apgeog.2015.04.014.

Krejcie, R.V., & Morgan, D.W., (1970). Determining Sample Size for Research Activities. Educational and Psychological Measurement. 30, 607-610. https://doi.org/10.1177/001316447003000308

Krizek, K.J. (2003). Operationalizing neighborhood accessibility for land use-Travel behavior research and regional modeling. Journal of Planning Education and Research. 22 (3), pp 270-287. https://doi.org/10.1177/0739456X02250315

Leslie E, Coffee N, Frank L, Owen N, Bauman A, Hugo G. (2007). Walkability of local communities: using geographic information systems to objectively assess relevant environmental attributes. Health Place. 13 (1), pp111-22. doi: 10.1016/j.healthplace.2005.11.001

Lo, R. H. (2009). Walkability: What is it? Journal of Urbanism: International Research on Placemaking and Urban Sustainability. 2 (2) Pp145-166. https://doi.org/10.1080/17549170903092867

Manaugh, K, & El-Geneidy, A. (2011). Validating walkability indices: How do different households respond to the walkability of their neighborhood? Transportation Research Part D: Transport and Environment. 16 (4), pp309-315. https://doi.org/10.1016/j.trd.2011.01.009.

Mayne, D.J., Morgan, G.G., Willmore, A., Rose, N., Jalaludin, B., Bambrickand, H., & Bauman, A. (2013). An objective index of walkability for research and planning in the Sydney Metropolitan Region of New South Wales, Australia: an ecological study. International Journal of Health Geographics. 12 (61). https://doi.org/10.1186/1476-072X-12-61

https://doi.org/10.52202/082078-0001

McMillan, D.W, & Chavis. D.M. (1986). Sense of community: A definition and theory. Journal of Community Psychology. 14(1), 6–23. https://doi.org/10.1002/1520-6629(198601)14:1<6::AID-JCOP2290140103>3.0.CO;2-I

Pretty, G., Bishop, B., Fisher, A., & Sonn, Ch.. (2006). Psychological sense of community and its relevance to well-being and everyday life in Australia. The Australian Community Psychologist. The Australian Psychological Society Ltd. Melbourne.

Tsai, Te-I. (2014). Neighborhood accessibility as a measure for defining sustainable urban form and assessment tools. Journal of Habitat Engineering and Design. 6 (1), 119-135.

UN Habitat. (2022). Urban Planning and Infrastructure in Migration Contexts: AMMAN SPATIAL PROFILE Jordan. March 2022. https://unhabitat.org/sites/default/files/2022/04/220411-final_amman_profile.pdf

Wilkerson, A., Carlson, N.E., Yen, I.H., & Michael, Y. (2012). Neighborhood physical features and relationships with neighbors: Does positive physical environment increase neighborliness? Environment and Behavior. 44 (5), 595-615. https://doi.org/10.1177/0013916511402058

Wood, L., Frank, L.D., & Giles-Corti, B. (2010). Sense of community and its relationship with walking and neighborhood design. Social Science & Medicine. 70(9), 1381-1390. https://doi.org/10.1016/j.socscimed.2010.01.021.

https://doi.org/10.52202/082078-0001

ASSESSING THE EFFECTS OF VIEWS OF NATURE AND SCENTS ON WORKPLACE PERFORMANCE AND PRODUCTIVITY MEASURES

Track 4: Design for Health and Wellbeing
Knowledge Network: Environmental Design Research Education, Work Environments
Methods: Experimental/Quasi-Experimental Research, Survey
Sarayu Chandramouli (Texas A&M University); Suma Katabattuni (Rekovar Inc.,);
Marco A Palma (Texas A&M University); Jay E Maddock (Texas A&M University)

1 ABSTRACT
In the pursuit of "Design for Health and Wellbeing", this study combines architectural concepts from a public health perspective to investigate the dynamic relationship between the built environment and human cognitive wellbeing. Rooted in the Attention Restoration Theory (ART), Stress Reduction Theory (SRT), and the emerging concept of "smellscape," our research explores the combined effects of visual and olfactory elements of nature within controlled office environments. The experiment is based on the well-evidenced fact that time spent in nature improves mental, physical, and psychological wellbeing combined with the idea that nature in office spaces improves productivity and performance.

Although nature is a multisensory experience, the literature emphasizes much on nature views. This paper presents a subsect of a study that explores nature views, scents, and the combined effects of views and scents. The incorporation of smellscape theory adds a novel dimension to this investigation, recognizing the significant role of olfactory stimuli in shaping individuals' perceptions and experiences of space. This extended abstract focuses on the combination of views and scents in office workspaces. The hypotheses posit that the integration of views of nature and natural fragrances, as part of the smellscape, yields a synergistic enhancement in cognitive performance required for work performance. Employing a between-subjects 2x2 experimental study design, the experiments were carried out through O-Tree, an open-sourced coding platform, and iMotions 9.3, a neurophysiological eye-tracking software to develop an online environment (Chen et al., 2016). The meticulously designed approach delves into uncharted territory, examining how the fusion of sensory nature experiences and smellscape influences stress levels, memory retention, attention, abstract intelligence, risk-taking tendencies, and attitudes toward cheating. The sample consisted of 128 students to resemble the young working adult population, recruited from the Texas A&M University System, who were asked to come to the lab during the span of the study.

The results of this paper indicate a significant difference in task performance and positive emotional dimensions in participants with increased nature content, more so with multiple stimuli. The eye tracking data shows increased positive emotions, and positive facial expressions in condition 1 (views) and condition 3 (both). The eye-tracking data also suggests increased attention and engagement across all three conditions followed by surprise and frustration. The exception to this positive state was prominent for the Stress test where, despite the condition participants were in, their answers indicated no happiness, and contempt was identified multiple times as a leading expression.

By selecting cognitive tasks mirroring real-world office tasks, the study not only deepens the understanding of the intricate relationship between nature, smellscape, and cognition but also provides actionable insights for architects, designers, planners aiming to craft spaces that holistically support their occupants. On a smaller scale, the findings may also encourage more people to try and incorporate nature into their own personal work areas, for their comfort. This study contributes to understanding how nature views and scents influence cognitive wellbeing in office settings, offering insights for architects designing spaces to support occupants' holistic wellbeing and productivity, and users to incorporate

Keywords: *Biophilia, Cognition, Healing Environments, Health Outcomes, Nature connectedness, Nature Scents, Nature Views, Offices.*

2. INTRODUCTION
One of the great challenges of our time is to bring the beneficial experience of nature into the design of contemporary buildings, landscapes, communities, and cities." – Stephen R. Kellert (Kellert, 2018).
The World Health Organization (WHO) states that whether people are healthy or not is determined in large part by their circumstances and environment. To a great extent, factors such as where we live, the state of our environment, genetics, our income and education level, and our relationships with friends and family all have considerable impacts on health (Braveman & Gottlieb, 2014). Every space around us

https://doi.org/10.52202/082078-0002

that we use, either the built or natural environment helps to orient ourselves, by providing us with a strong connection to the surrounding space, and this sense of embodiment is how we come to understand and relate to the world. In the field of urban design, this connection that all people share with their surroundings comes from the overarching idea called "placemaking", a concept that inspires people to collectively reinvent, and re-imagine spaces at the heart of each community (Laenen et al., 2020).

One underlying premise behind placemaking can be traced to the innate connection that human beings have with their natural environment. The definition of nature can be multi-faceted, and extensive, but for this paper, it extends to the rural countryside, comprising farmland, forests, hills, mountains, lakes, rivers, and associated trails. Nature also includes blue spaces, including beaches, dunes, and activities that occur in the open ocean. In biophilic design and environments, the concept of nature is often central (Frumkin & Fox, 2011). It emphasizes the integration of natural elements, materials, patterns, and processes into the built environment to create spaces that foster human well-being, reduce stress, and enhance overall quality of life. Several studies have proposed that nature embodies certain qualities that human beings have an inherent tendency to interact with. Starting in the 17th century, nature began to be an integral part of the healthy design and planning of the newly industrializing world, with large parks starting to burgeon across big cities and smaller towns alike (Stigsdotter et al., 2011).

The surge in interest in creating spaces and places that support health and wellbeing in recent ages is a renaissance in public health and design thinking. Even controlling demographic characteristics and socioeconomic status, studies have shown that the less greenspace an individual is exposed to, the higher the risks of morbidity and mortality. The health outcomes that are linked to nature are numerous ranging from mental health issues (depression, anxiety, stress, migraines, ADHD), to physical health issues (diabetes, cardiovascular issues, cancer, obesity, respiratory illnesses, etc.) (Cecily et al., 2006; Hartig et al., 2014). Laura Prestwood discusses how every place, whether built or physical, provides human beings with a connection to their surroundings, and this sense of embodiment is how people understand and relate to the world (Prestwood, 2010). Since people spend a major part of their day inside buildings, the consensus among experts and researchers is that the quality of the building can either induce or alleviate stress.

Public Health has been an important concern so far since at least 1926 when public and community health was declared a fundamental human right (Tulchinsky & Varavikova, 2014). The World Health Organization (WHO) later added disease control to improve physical and mental health (Frumkin, 2001). The American Psychological Association (APA) conducted a series of research on nature and health, starting with "Nurture vs. Nature", a set of studies within psychology to find a common ground between the two concepts, and identified areas of nature that would benefit the human mind. Following this came the "Nurtured by Nature" which was set to explore nature's connectedness, mental wellbeing, and cognition.

This series explores nature exposure for office employees, during their general workday, from their workspaces, as opposed to time spent in nature generally. Although wholesome exposure to nature is the most beneficial, most people cannot take time out of their workday to do so. Incorporating stimuli of nature into the built environment (office spaces) is the second best way to enable for partial exposure to nature (Weir, 2020). Second, public health research encompasses exploring the nature experience (relationships shared between man and nature) on different levels based on the socio-ecological model (McLeroy et al., 1988). This includes personal levels (individuals), and societal levels (population and community health) (Hartig et al., 2014). Among the many works that discuss the positive effects of nature, the book "Cognitive Architecture: Designing for How We Respond to the Built Environment by Sussman and Hollander discusses that people are generally healthier in an environment that is supportive, aesthetic, and has a mixed land-use with minimal repetition (Sussman & Hollander, 2014). When it concerns the presence of nature within office spaces, works by Ulrich, and Grinde & Patil have demonstrated strong positive effects on self-esteem and performance including decision-making and impulsivity (Grinde & Patil, 2009; Ulrich et al., 1991). One study conducted in the Netherlands showed a quick upward elevation in moods after a 5-minute exposure to nature, which impacted performance (Brown et al., 2013). Three studies reported that exposure to nature can be invigorating and improves attention and both short- and long-term memory span (Hartig et al., 2003; Marc et al., 2008; Pilotti et al., 2015). Two other studies note an improvement in communication skills, and social interactions which in turn affect team dynamics and collaboration (Bratman et al., 2012; Herzog et al., 1997).

https://doi.org/10.52202/082078-0002

3 LITERATURE REVIEW

Acknowledging health as a multidimensional concept, encompassing physical, mental, emotional, social, and intellectual aspects, is crucial for a holistic understanding. This perspective recognizes well-being as an intricate interplay of various factors rather than merely the absence of diseases. Our exploration focuses on mental, emotional, social, and intellectual dimensions, understanding that improvements in these domains inherently enhance physical health through factors like increased exercise and an overall healthier lifestyle. Extensive literature reviews have established the profound impact of nature views on health outcomes across these dimensions (Bratman et al., 2012; Hartig et al., 2003; Lee et al., 2011; Mackay & Neill, 2010; Taylor et al., 2002). However, a notable gap persists in our understanding of the positive effects of nature's scents on well-being. The literature review, focusing on studies published in English between 2010 and 2023, specifically explores the connection between views, scents, and various facets of well-being, serving as a foundation for our subsequent analyses.

3.1 Biophilia and Biophilic Hypothesis

The hypothesis that humans have an inherent inclination to affiliate with nature is referred to as biophilia (Gillis & Gatersleben, 2015). In 1984, biologist Edward O. Wilson (dubbed the Father of Biodiversity) explained that the time spent in nature evokes positive emotions, and called this connection, Biophilia (a term coined by Erich Fromm in 1964). He explained that these affiliations are the evolutionary results of repeated contact with the natural environment and an organic dependency on life-like processes, that are now essential for survival and thereby a universal primary need (Wilson, 1986). In the 1990s, a group of scholars that included Stephen Kellert started a field of research focusing on an emerging hypothesis "Biophilic Hypothesis" to expand the addition of nature, into cities, and buildings alike to make it a component of human lifestyles (Kellert & Wilson, 1995). Over the last two decades, research on natural contact has increased to expand the evidence base linking positive relationships and outcomes. Nature contact has since expanded to include not just gardens and greenspaces but also oceans, beaches, mountains, and other terrestrial and aquatic ecosystems. The realization that an unpleasant environment can cause stress, anxiety, discomfort, and reduce productivity has helped in making major strides in opening doors for research regarding the health benefits of natural environments. Furthermore, evidence of nature's ability to soothe pain and improve moods has caused significant curiosity among researchers and professionals from multiple disciplines. In recent times, this theory has been implemented to create spaces with access to nature to enhance human experience and health (emotional, physical, psychological, and mental). Two major explanatory theories in the field of environmental psychology account for the restorative power of nature: The Attention Restoration Theory (ART) (Kaplan, 1992), and the Stress Reduction Theory (SRT) by Ulrich (Ulrich, 2002).

ART suggests that mental fatigue or exhaustion and drop in concentration can be restored with time spent in nature, or simply by looking at it. This is much different than simply taking a break (without nature), as ART posits that nature captures our attention effortlessly and enjoyably, allowing for mental recovery. Taking a break in a non-natural environment may still involve mental effort or distractions, making it less effective for attention restoration. The lack of breaks and the longevity of indoor time are very prevalent in the current world, a trend that results in directed attention fatigue. There is something about nature itself that allows for cognitive processing to increase, resulting in increasing attention (focus). Kaplan and Kaplan proposed that people experience four phases on the way to restoration: a clearer head, alleviation of mental fatigue, soft fascination, and reflection and restoration (Kaplan & Kaplan, 1989). Building on this, Ulrich found that even pictures of nature increased feelings of affection, friendliness, and playfulness in children. With this being said about ART, the Kaplans stipulate that for nature to be restorative, it must possess four qualities:

1. An ability to generate "awe" in people (Fascination) – The greater the "awe" of a place, the easier it gets to give directed attention a rest and let involuntary attention take its place.
2. The ability of the space to enable an individual to be away from the place and yet be able to relax and let their minds wander away from daily stress (being away).
3. The detail behind each component in the natural environment – the feeling of being able to absorb the effects of the elements in the space (extension).
4. Characteristics present in an environment that meets the expectations of an individual (preferences, and goals of that individual - compatibility) (Hartig et al., 2014; Pearson & Craig, 2014).

Stress Reduction Theory (SRT) was built with a similar framework as ART and can be thought of as another facet of it. It was formulated by Roger Ulrich, and it is a widely used framework for understanding the relationship between stress and work (a person-environment fit) (Kaplan, 1992). According to Ulrich, this mismatch can be caused by a variety of factors, including physical, social, and psychological factors. Like ART, SRT also suggests that exposure to nature can lead to physiological stress recovery,

https://doi.org/10.52202/082078-0002

improved mood, and reduced symptoms of anxiety and depression. Within the field of environmental health and design, SRT holds great significance as the physical spaces around people and the built environment that live/work in can impact their wellbeing and stress levels (Ulrich et al., 1991). For example, designers can create spaces that enhance personal resources, such as access to natural light, views of nature, and opportunities for social interaction (Evans & McCoy, 1998; Ulrich, 2002; Ulrich et al., 1991). These design elements can help reduce stress and improve well-being by creating an environment that is more supportive of individuals' needs.

Because of the present-day urban population growth, the value of nature has only become more integral for many people, who spend hours every day indoors without any exposure to the outdoors. In response, people are turning to nature to escape from the urban environment and to improve their health and well-being. Public health professionals and planners have now started investing in the idea of equitable planning to ensure every neighborhood has fair access to green spaces, a topic that is also huge amongst researchers, and not discussed in this paper. In our day-to-day lives now, green spaces come in the form of parks, community gardens, and urban forests – they provide a way for people to connect with nature and enjoy its benefits. But if nature only exists as an independent entity, then for the people inside office buildings, spending time in nature becomes an ordeal as they would have to practically extract themselves from their work to interact with nature. To circumvent this issue, cities are incorporating more green spaces into their design and promoting access to nature for all residents. Public health campaigns are also emphasizing the importance of spending time in nature and encouraging people to take advantage of the benefits that it can provide. Overall, nature has become more important in urban areas because it offers a way to counteract the negative effects of urbanization and pollution on health and well-being. But this does not imply that sole exposure or connectedness to nature is sufficient for the elimination of health conditions, nor is it purely a solution for long-term wellbeing. Furthermore, to accrue all possible benefits related to exposure of nature, a strong physiological connection is needed between people and their natural health, although short-term or indirect exposures can benefit multiple facets of health (Martin et al., 2020).

3.2 Overview of Nature Scents

A substantial number of studies addressing biophilia focus on views (visual stimulation), although nature can be experienced through other senses. One of the other predominant senses of nature explores the scents in the immediate environment and the resulting implications of them. But unfortunately, even with this being the case, many studies consider anthropogenic smells as the source of nuisance, while some others explore the impact of "unpleasant" smells on well-being thereby creating a gap in knowledge on the connection between nature, scents, and well-being (Bentley et al., 2023; Van der Stoep et al., 2021).

The use of smells to promote health and well-being has occurred throughout history, and such evidence can be used to inform the design of Salutogenic sensory gardens, which can be particularly beneficial for those with visual or hearing impairments. The role of natural environments for good health and well-being was particularly apparent during the COVID-19 pandemic, and the lack of smell impaired people's quality of life and emotional well-being. There is a significant body of evidence demonstrating links between the olfactory system in general and metabolic health, mood, cognition, and neurological disorders (Green et al., 2023; Johnson et al., 2019). Views and scents associated with nature elicit distinct responses in individuals. Views of nature are restorative to the visual senses, providing a range of colors and textures, which have significant impacts on mental health, productivity, and performance. On the other hand, scents trigger emotional dimensions based on personal associations, memories, and cultural meanings, playing a crucial role in human health and well-being. The olfactory system is directly connected to the limbic system, which governs emotions, memories, and learning.

Discussion of olfactory receptors within nature backgrounds explores the holistic odor experience of a place such as a garden or a park. In 1990, researcher Douglas Porteus described the term "Smellscapes" as the totality of the olfactory landscape in a specific environment (Henshaw, 2013). Along with Henshaw, many other researchers explored topics relating to smellscapes like the scents of emotions or olfactory stimuli, affective dimensions in odor perceptions, and more. Natural odors emanating from flowers, trees, and plants have been shown to increase positive feelings of joy, mood, nostalgia/memory, and energy.

https://doi.org/10.52202/082078-0002

The most important scent (correlated with positive human experience) is the "earthy smell" given off by the soil after contact with rain which is caused by dead microbes at the level of 5 ppt. This contact produces the chemical odorant called "Geosmin", causing human beings to smell "petrichor" (Neff, 2018).[1]

Recent theoretical frameworks, such as the one presented by Marselle and colleagues (2021), suggest that specific traits, like scents, can exert their influence on various dimensions of well-being through diverse pathways, including the restoration of cognition. Rachel Herz, a psychologist who is most recognized for her work on the psychology of scents has published numerous articles on the impact of scents on wellbeing. In 1995, she explained how odor-evoked memories can influence attention and emotional responses, highlighting their roles in shaping cognitive and affective processes (Herz & Cupchik, 1995).

4. RESEARCH OBJECTIVES

This paper includes a multisensory approach with nature views and scents, to probe if the combination of the stimuli influences the performance of the sample, based on the findings of the literature. The research was carefully carried out to manipulate the environment to emulate an office, and the scents to be strictly nature-based. This study investigates the direct effects of views of nature and scents on each (and combined) outcome metrics such as attention retention, short-term memory skills and recall abilities, abstract reasoning & intelligence, risk aversion / taking attitudes, and cheating attitudes. Originally, this study measured individual effects of views and scents of nature and then the additive effects of the two senses.

For this paper, the research questions are as follows:

Do nature views and scents together in office environments enhance:

- Attention,
- Memory and recall abilities,
- Abstract reasoning in participants

after taking a 5-minute break also with the two stimuli?

The general hypotheses are:

H0: There is no significant improvement in the above outcome attributes in participants working on routine office games in a biophilic office space with both nature views and scents as opposed to a workspace with just either one of the stimuli.

Do nature views and scents together in office spaces impact (reduce) decision-making skills, particularly:

- Risk-taking/ risk aversion behaviors
- Attitudes towards cheating in levels

in participants after taking a 5-minute break also with nature views?

H0: There is no significant difference in the risk-taking attitudes and cheating attitudes in participants in office environments with nature views and scents compared to those conducted in office environments with either one stimulus (views only or scents only).

This paper explores three conditions from the original study that had one additional control group.

1 For the purposes of this study, owing to Geosmin or petrichor not being available as a fragrance in the market, a fragrance called "A walk in the woods" was chosen. This fragrance creates a wholesome smellscapes of a woodland – forest area, comprising of musk, amber, citrus, vetiver and cedarwood with a top note of "bergamot".

https://doi.org/10.52202/082078-0002

5. METHODS

This study was conducted collaboratively between the School of Public Health and the Human Behavior Lab, with assistance from the Department of Computer Science & Engineering at Texas A&M University. Utilizing eight computer stations positioned in the laboratory with scenic views through the windows (figure 1), participants were recruited anonymously through an online recruitment platform usually used by the Human Behavior Lab and organized in groups of eight per session. Due to the between-subjects design, each group of participants was randomly assigned to a single condition. Approval for the study was obtained from the Institutional Review Board (IRB) of the Texas A&M University System (Approval #: IRB2021-1370).

The experiment employed O-Tree software, an open-source Python-based platform designed by Computer Science experts (Chen et al., 2016), to create and implement six computerized games. These games were operated using iMotions 9.3, a neurophysiological integration software that recorded eye-tracking and facial expressions. The study sessions were homogeneous, with all participants in one session assigned to and tested for one condition. The experiments began in late 2022 and ended in May 2023. Participants engaged in six games in 90 minutes, including five breaks of five minutes each. The first game intentionally induced stress to generalize performance outcomes.
An exit survey, administered via O-Tree, followed the completion of the games, and total incentives were calculated.

Apart from the stress-inducing game, participants played five additional games, each aligned with literature-based measurement criteria for specific outcomes. Games were derived or adapted from previous lab studies or current literature. The final games included Letter Identification, Word List Learning, Raven's Progressive Matrices, Holt & Laury, Hidden Lies Tests (Axelrod, 2001; Ferreira, 2019; Holt & Laury, 2005; Kunda et al., 2013; Oechssler & Sofianos, 2019; Zhang et al., 2019). The details and images about the games are provided in the section "Gaming Protocols". The final games were programmed into the Human Behavior Laboratory systems, incorporating measurement metrics such as time, accuracy, speed, number of questions completed, tendencies to cheat, and risk-taking attitudes.

Nature scents[2] were released using "scent-wave" diffusers, employing cold air diffusion technology to disperse fragrances evenly. Scents included cedar, vetiver, golden amber, fruity citrus, and musk. Break rooms during these sessions were infused with scents. Intensity control prevented adverse effects, and sessions concluded with the same exit survey.

2 Cold air diffusion technology is a method of dispersing essential oils into the air without using heat. This technology preserves the therapeutic properties of oils, ensures consistent particle size for uniform distribution, allows controlled and quiet operation.

https://doi.org/10.52202/082078-0002

Figure 1. View of Nature from Lab, Texas A&M Research Parkway (Photographs)

5.1 Materials and Sample

The study unfolded at the Human Behavior Lab situated at 1500 Research Pkwy, surrounded by the expansive Research Park. This paper is one part of a study that had a total of 256 participants. For this paper, a meticulous recruitment process recruited 192 students from the university body, after ensuring their eligibility. Participants, grouped in sessions of 64 (n=192), spent 90 minutes during the evenings to mitigate any morningness-eveningness variability (Adan & Almirall, 1991).

The study also ensured uniformity by excluding individuals not meeting criteria, such as those under 18, legally blind, with eye corrective surgery, photosensitive epilepsy, cardiac pacemakers, or implantable cardioverter-defibrillators. Participants were made to acknowledge the criteria before signing up for the study. In addition, exclusions extended to individuals with allergies/sensitivities to scents. This attention to detail aimed to ensure a robust dataset and control for potential confounding factors, ultimately contributing to the validity of the study's findings. In each session, participants received detailed instructions on the study's expectations, including eye tracking and calibration procedures. Rigorous guidelines were set to maintain consistency in seating positions, preventing data loss in facial tracking and eye tracking. Participants with faulty calibration during the session were excluded from emotive tracking analysis, while their task performance data remained valuable for analysis.

5.2 Development of Online Environments

Experiments were conducted using O-tree, a web-based behavioral experiment platform, within the controlled confines of the Human Behavior Lab. The centralized control mainframe facilitated this process, allowing participants to interact with experimental games on individualized systems via a dedicated IP address configured by the lab (e.g., 127.111.987.100/otree_experimentalview/).

Notably, sessions falling under Condition 2 (Views) occurred exclusively within the server session designated as 127.011.987.100/otree_ **experimentalview/**. Post-session, data was retrieved completely for each session from the mainframe. An uninterrupted server operation was crucial for this and couldn't be terminated during the study. However, server termination was a prerequisite for launching other experiments on the same infrastructure.

O-tree coding featured a built-in progress monitor, tracking participant engagement and payment statuses. Payment status tracking, integrated into O-tree by the research team, automated incentive calculations ($15 for attendance and up to $10 based on performance). The graphical interface provided real-time updates with a 15-second refresh rate. The setup within O-tree consists of three components (Chen et al., 2016):

- Rooms and Sessions coded on the PyCharm Terminal, modeled after Python's structural framework,
- A mainframe/ server computer (either a cloud server or a local system, in this case, a local system) from which the games are deployed, and monitored (figure 2-4), using the following structure:

https://doi.org/10.52202/082078-0002

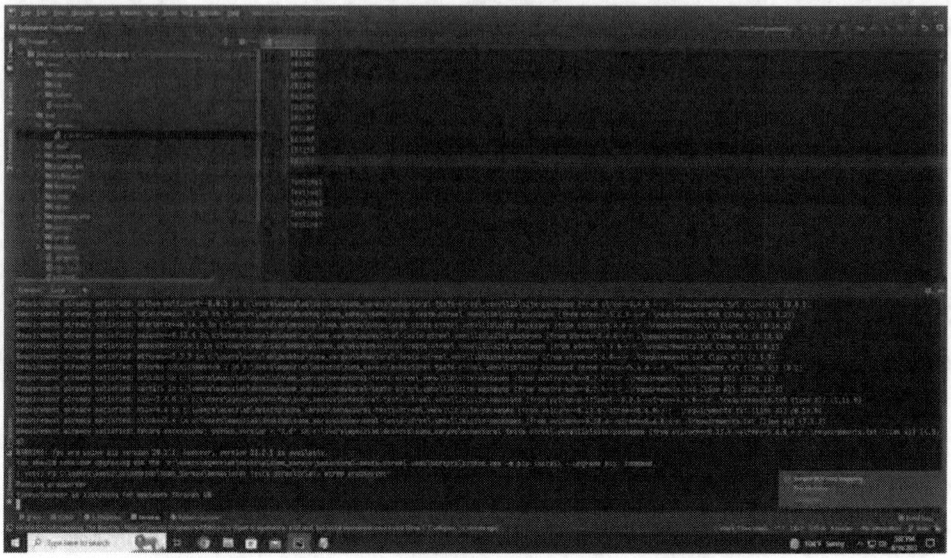

Figure 2. PyCharm Terminal (03/10/2022). Human Behavior Lab; Texas A&M University

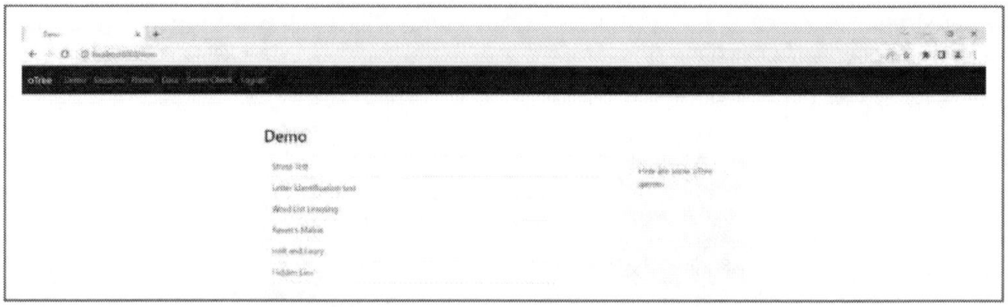

Figure 3. Control Panel: O-tree Server (3/10/22). Human Behavior Lab; Texas A&M University

```
"cd C: \users\ageczlab\Desktop\ suma_sarayu\Behavioral-Tests-OTree"
"python -m venv.venv"
".\. venv\ scripts\ activate.ps1
(.venv) PS C:\ Users\ageczlab\Desktop\ suma_sarayu\Behavioral-Tests-
OTree>
cd .\trial\"
"pip install -r. \requirements.txt"
OTree prodserver
```

Figure 4. Code Scheme: PyCharm Deployment (3/10/22). Human Behavior Lab; Texas A&M University

https://doi.org/10.52202/082078-0002

For analyzing eye movements and facial expressions, sessions were run via iMotions 9.3, a biometric research software platform, equipped with Tobii X2-60 eye trackers. The iMotions Facial Expression Analysis Module integrated Affectiva's AFFDEX for recognizing various emotions (figure 5).

This seamless integration facilitated the synchronization of facial emotions with stimuli within the software. The platform recorded eye movements, andresearchers manually noted peak points during breaks for analysis.

Upon entering the lab, participants were instructed to sit facing the computer for eye calibration. The iMotions platform tracked eye positions and gazes throughout sessions, providing comprehensive analysis and visualization tools. Data could be exported for further examination.

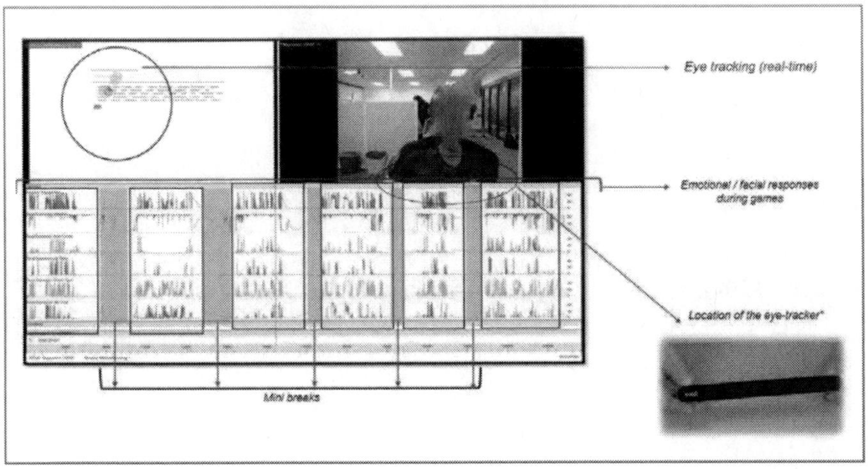

Figure 5. iMotions Full Analysis metrics screen of Respondent (6/23/2022). Human Behavior Lab; Texas A&M University

5.3 Gaming Protocols

1. Stress Test: Adapted from the "Trier Social Stress Test" (Ferreira, 2019), participants engaged in an arithmetic task to induce stress. They sequentially subtracted a random two-digit odd number from a random four-digit number for twenty rounds, digitally.
2. Letter Identification (Attention Skills): Participants completed a letter identification task adapted from previous research (Zhang et al., 2019). They counted occurrences of a designated letter within a paragraph of random letters within a five-minute timeframe.
3. Word List Learning (Memory Skills and Recall): Derived from the Wechsler Adult Intelligence Scale (WAIS) (Axelrod, 2001), participants memorized and recalled words presented in lists. Rounds involved recalling words viewed for 15 seconds over four repetitions.
4. Raven's Test (Abstract Reasoning and Intelligence): Based on the Raven's Progressive Matrices (RPM), participants solved matrix reasoning questions within seven minutes. Questions varied in difficulty, with participants rewarded for accuracy.
5. Holt and Laury Risk Assessment (Risk Aversion): Participants made decisions on paired lottery choices, reflecting attitudes toward risk. The probabilities of outcomes would change over 10 rounds (Holt & Laury, 2005).
6. Hidden Lies (Cheating Game): Participants engaged in a task measuring cheating tendencies. They transferred tokens to a virtual partner (computer), with opportunities for dishonest behavior (Oechssler & Sofianos, 2019).

5.4 Exit Survey

Once all the games were completed by all participants, one was called to pick a game for the payoffs. If game 1 (mental math) was chosen, the participants were paid the sum of all total correct answers. For the other games, the payoffs were calculated as detailed in each respective section above. Once the game was chosen, and rounds, if necessary, the participants were asked to continue with the last part of the session: the exit survey.

https://doi.org/10.52202/082078-0002

The survey included questions relating to demographics, class schedules, morningness-eveningness items (circadian rhythm variability) (Adan & Almirall, 1991), perceived stress levels, technological proficiency, and the time they spend using computers. To understand if the two senses of nature (view and scent) influenced performances, self-reporting measures of noticing outdoor/ indoor ambiance were also added.

6 RESULTS

6.1 Performance Analysis

The study examined the impact of nature views and scents on performance improvements through a thorough statistical analysis. Employing a one-way Analysis of Variance (ANOVA), the three conditions were compared for performance differences. All tables were produced using IBM SPSS v.29, with each of the outcome variables being the dependent variable. In the case of a significant ANOVA result, post hoc Tukey's HSD was employed to identify differing conditions. Interpretation relied on examining p-values, with a significance level set at $p < 0.05$. This approach aimed to provide insights into how nature views, scents, or their combination influence performance improvements in a controlled office environment

Table 1. Demographic Frequency Analysis

GENDER	FREQUENCY	CUMULATIVE %
Male	89	46.4
Female	102	53.1
Other	1	0.5

AGE	FREQUENCY	CUMULATIVE %
18 - 24	159	82.8
25 - 34	32	16.7
35 - 44	1	0.5

https://doi.org/10.52202/082078-0002

Table 1. Demographic Frequency Analysis Continued

RACE	FREQUENCY	CUMULATIVE %
White (Non-Hispanic)	47	24.5
Black (Non-Hispanic)	8	4.2
Asian (Non-Hispanic)	71	37
Mexican American	23	12
Other Hispanic	11	5.7
Others	29	15.1
Don't want to say	3	1.6

EDUCATION	FREQUENCY	CUMULATIVE %
Bachelor's	102	53.1
Master's	65	33.9
Ph.D./Doctorate	18	9.4
Other	7	3.6

YEAR OF EDUCATION	FREQUENCY	CUMULATIVE %
Freshman	21	10.9
Sophomore	20	10.4
Junior	30	15.6
Senior	38	19.8
Graduate/Professional	83	43.2

CLASS HOURS	FREQUENCY	CUMULATIVE %
0 - 1 hours	80	41.7
1 - 3 hours	63	32.8
3 - 5 hours	39	20.3
5 or more hours	10	5.2

https://doi.org/10.52202/082078-0002

Table 1. Demographic Frequency Analysis Continued

HOURS WORKED	FREQUENCY	CUMULATIVE %
0 - 1 hours	113	58.9
1 - 3 hours	31	16.1
3 - 5 hours	27	14.1
5 or more hours	21	10.9

WHAT TIME SLEEP IS NEEDED	FREQUENCY	CUMULATIVE %
8-9 pm	11	5.7
9 - 10:15 pm	31	16.1
10:15 - 12:30 pm	77	40.1
12:30 - 1:45 am	40	20.8
1:45 - 3am	28	14.6
3 am or later	5	2.6

HOURS OF SLEEP LAST NIGHT	FREQUENCY	CUMULATIVE %
0 - 2 hours	9	4.7
5 - 8 hours	154	80.2
8 - 10 hours	28	14.6
10 hours or more	1	0.5

PEAK PRODUCTIVITY TIME	FREQUENCY	CUMULATIVE %
8 - 10 am	44	22.9
11 am - 1 pm	97	50.5
3 - 5 pm	32	46.7
7 - 9 pm	19	9.9

https://doi.org/10.52202/082078-0002

Table 1. Demographic Frequency Analysis Continued

IF YOU WENT TO BED AT 11PM, WHAT LEVEL OF TIREDNESS WOULD YOU BE?	FREQUENCY	CUMULATIVE %
Not tired at all	57	29.7
A little tired	77	40.1
Fairly tired	44	22.9
Very tired	14	7.3

"FEELING BEST" TIME PERIOD	FREQUENCY	CUMULATIVE %
12 - 4 am	7	3.6
4 - 7 am	9	4.7
7 - 9 am	24	12.5
9 am - 4 pm	94	49
4 - 10 pm	44	22.9
10 pm - 12 am	14	7.3

MORNING PERSON VS. EVENING PERSON	FREQUENCY	CUMULATIVE %
Morning	35	18.2
Prefer moring over evening	56	29.2
Prefer evening over morning	62	32.3
Evening	39	20.3

STRESS LEVELS: PRE-STRESSOR	FREQUENCY	CUMULATIVE %
Not stressed at all	99	51.6
A litte stressed	63	32.8
Fairly stressed	24	12.5
A lot stressed	6	3.1

https://doi.org/10.52202/082078-0002

Table 1. Demographic Frequency Analysis Continued

STRESS LEVELS: POST-STRESSOR	FREQUENCY	CUMULATIVE %
No changes	33	17.2
A little less stressed	40	20.8
A lot less stressed	16	8.3
A little more stressed	67	34.9
A lot more stressed	36	18.8

HOW DID THE MINI BREAKS AFFECT YOUR STRESS LEVELS?	FREQUENCY	CUMULATIVE %
Not at all	75	39.1
A little change	80	41.7
Fair amount	31	16.1
A lot	6	3.1

TIME SPENT IN NATURE	FREQUENCY	CUMULATIVE %
Not at all	10	5.2
Less than 30 minutes	20	10.4
30 minutes - 1 hour	37	19.3
1 - 2 hours	47	24.5
2 - 3 hours	31	16.1
3 - 4 hours	19	9.9
4 - 5 hours	16	8.3
5 - 7 hours	8	4.2
7 hours or more	4	2.1

https://doi.org/10.52202/082078-0002

Table 1. Demographic Frequency Analysis Continued

DID YOU NOTICE YOUR OUTDOOR SURROUNDINGS?	FREQUENCY	CUMULATIVE %
Yes	166	86.5
No	26	13.5

DID YOU NOTICE YOUR INDOOR CONDITIONS?	FREQUENCY	CUMULATIVE %
Yes	147	76.6
No	45	23.4

HOURS IN FRONT OF A SCREEN - OUTSIDE OF WORK	FREQUENCY	CUMULATIVE %
0 - 1 hour	4	2.1
1 - 2 hours	25	13
2 - 4 hours	59	30.7
4 - 6 hours	57	29.7
6 hours or more	47	24.5

COMPUTER PROFICIENCY	FREQUENCY	CUMULATIVE %
Excellent	50	26
Very good	87	45.3
Fair	51	26.6
A lot	4	2.1

https://doi.org/10.52202/082078-0002

ANOVA RESULTS		SUM OF SQUARES	DF	MEAN SQUARE	F	SIG.
Stress Test (1): Mental Math	Between Groups	.010	2	.005	.203	.817
	Within Groups	4.859	189	.026	-	-
	Total	4.870	191	-	-	-
Attention Test (2): Letter ID	Between Groups	1.885	2	.943	3.308	.039
	Within Groups	53.859	189	.285	-	-
	Total	55.745	191	-	-	-
Memory Recall (3): Word List Learning	Between Groups	9.292	2	4.64	8.494	<.001
	Within Groups	103.375	189	.547	-	-
	Total	112.667	191	-	-	-
Abstract Reasoning (4): Raven's Advanced Matrix	Between Groups	3.375	2	1.688	.922	.399
	Within Groups	345.875	189	1.830	-	-
	Total	349.250	191	-	-	-
Risk Aversion (5): Holt & Laury	Between Groups	.323	2	.161	3.333	.038
	Within Groups	9.156	189	.048	-	-
	Total	9.479	191	-	-	-
Cheating Behavior (6): Hidden lies Test	Between Groups	11.142	2	5.571	3.419	.035
	Within Groups	306.324	189	1.629	-	-
	Total	317.466	191	-	-	-

Table 2. Performance Tests (One way ANOVA) between groups/conditions

As seen in the table above, four of the six games played showed significance while the stress test and Raven's Matrix did not. Since our null hypothesis was framed for six individual games and outcomes, depending on our results we rejected the hypothesis on a case-by-case basis. The null hypothesis for these four games can be individually rejected for tasks relating to attention, memory recall, risk aversion and cheating behaviors.

To know the effect sizes and the direction of the change, a further analysis was conducted using Tukey's post-hoc. The Tukey criterion (T) is defined as:

T = q (c,n-c) √MSE / Ni, where
q (c,n-c) = Range Distribution, based on c and n-c, df
c= Conditions
n= sample size
MSE = Mean Sq. Error
N = sample size of the treatment group; the Tukey's comparison is provided in Table 3.

Table 3. Tukey's Honestly Significant Difference Test

ANOVA RESULTS	(I) CONDITION CASE	(J) CONDITION CASE	MEAN DIFF (I-J)	SIG.	CI (LL; UL)
Attention Test (2): Letter ID	Views Only	Scents Only	.234*	0.037	0.01; 0.46
		Both	0.172	.165	-0.05; 0.39
	Scents Only	Both	-.062	.786	-0.29; 0.16
Memory Recall (3): Word List Learning	Views Only	Scents Only	.344	0.02	0.03; 0.65
		Both	.531*	<.001	0.22; 0.84
	Scents Only	Both	.188	.325	-0.12; 0.50
Risk Aversion (5): Holt & Laury	Views Only	Scents Only	-.016	.915	-0.11; 0.08
		Both	-.094	.041	-0.19; 0.00
	Scents Only	Both	-.078	.113	-0.17; 0.01
Cheating Behavior (6): Hidden lies Test	Views Only	Scents Only	.580	.030	-0.05; 1.12
		Both	.190	.680	-0.35; 0.72
	Scents Only	Both	-.391	.196	-0.92; 0.14

https://doi.org/10.52202/082078-0002

From Table 3 above, the differences were noted between mostly with the views condition. Letter ID shows that participants who were assigned to the views-only condition outperformed the other two groups. The p-values, group means and confidence levels ($p<0.05$; MD = 0.234; {0.01, 0.46}) show an improvement between views and scents-only groups. With memory recall, participants displayed better performances in the views only against the other groups. The findings for Game 5 (risk aversion) should be interpreted differently. In the views condition, participants underperformed group 3 (both); meaning that they took lower risks. Some argue that risk-taking is an important factor in work environments, while others maintain a stance on being conservative. Perhaps, this means that those participants were conservative in their risk-taking, while people with both views and scents overindulged. Although the significance in this case was weak ($p = 0.041$), more investigation aimed specifically at risky behaviors can shed deeper, and more insightful evidence regarding the nature of the relationship between risk aversion behaviors and nature stimuli (views, scents, , and even other multiple stimuli).

A secondary analysis was conducted using the questions from the exit survey based on participants' habits in their regular lives, specifically how long they spent in nature in a typical week. A total of 147 participants (out of 192) were used for this analysis based on their answers (time spent >=4 hours). The findings obtained from the secondary analysis yielded very similar results to the one shown above, although game 6 (cheating) yielded an insignificant p value. The tables are not shown here for the subset analysis, owing to the premise of this paper, and the lack of varying results.

6.2 Eye Tracking Analysis

The initial examination of eye-tracking data using iMotions used an eye tracker attached to the monitors. Once turned on, they use infrared signals to detect, and assess participants' eye movements and facial expressions. They focused on key elements such as fixation points, gaze clusters, and various metrics related to visual attention. Three metrics were used to analyze eye tracking: Blink Detection, Fixation Metrics (through AOIs), and Gaze Duration. But before any of these metrics could be properly analyzed, the data was annotated to exclude all portions of each participant in the times they left the lab (null data: Zero data points during the mini-break). We placed live markers after data collection to help us quickly split the data. We also used triggers sent through the Application Programming Interface (API) and added manual markers while checking the data.

Once annotations were copied onto all participants, station-wise data was exported into CSV files, and manually filtered for the null data. While this was being done, Areas of Interest were calculated to measure gaze fixations on the screen. They play a pivotal role in extracting insightful metrics from eye-tracking data. These metrics (manually determined and helped compare where participants were looking and fixating between the two groups) offer a comprehensive understanding of respondent behavior during stimulus engagement. The analysis was done using automatic R notebook algorithms. The AOI maps (Figure 6, Table 4) correlated with this evidence, with participants seemingly looking at the peripheral edges (towards windows) of the monitor less during the scents-only condition. Since the eye-tracker itself could not detect eye tracking outside the area of the monitor, this could be a vital factor when looking at the TFD data, as positive distractions warranted participants deflecting their gaze outside of the monitor. The AOIs were manually set to cover the central portion of the screen (area 1), and the perimeters (areas 2 + 3). These areas were set so any eye movements with area 1 were considered as engaging with their monitors, and the gaze that fell outside of this area (area 2+3) was towards the windows in the back. Table 4 shows a summary of time and percentages across conditions for primary fixation data. Total fixation duration data showed increased gaze duration in the scents-only condition as opposed to the two other conditions, as also seen in the cumulative areas of interest. The total fixation duration is a tested metric to assess "engagement" in tasks and is measured in Msec units. Figure 7 shows scents, followed by "both" conditions to be above the views only condition. This is probably because the scents only contained no views, the participants kept their gaze on the monitor for a more prolonged period. The distance of blink detection was also set to 50" from the screen/eye tracker, meaning that if the distance between the trackers and their eyes exceeded this specified distance, then the tracking information would not be recorded. Although this did not automatically exclude the null points, it helped annotate. The blink detection algorithm also considers parameters that determine the sensitivity of the eye-tracking. These include:

- To adjust the sensitivity of the blink detection, three parameters can be modified:
- Minimum duration [ms] between two blinks, Minimum blink duration (default: 20 milliseconds), Maximum blink duration (default: 20 milliseconds).

For our study, parameters were changed to 30, 20, and 20 for enhanced specificity in blink rate analysis after discussions with the iMotions team. Heatmaps were initially generated but since the trackers were not positioned to record gaze movements outside, they were later excluded as the major AOI regions were completely centered on the screens.

https://doi.org/10.52202/082078-0002

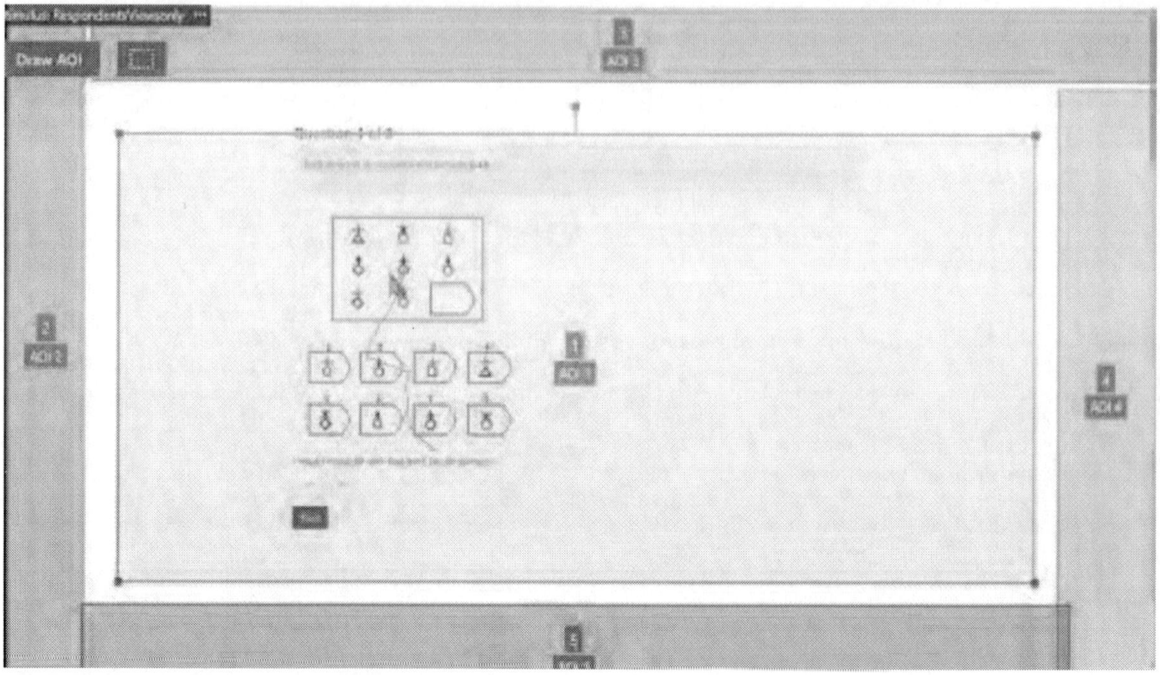

Figure 6. Example of Manual AOIs: Yellow: Area 1; Pink & Green: Areas 2+3 (iMotions 9.3)

AREAS OF INTEREST - SUMMARY TABLE (3 CONDITIONS)						
AOI Metrics	Views Only		Scents Only*		Both	
	Area 1	Area 2 + 3	Area 1	Area 2 + 3	Area 1	Area 2 + 3
Avg. Total AOI(ms)	3315063.50 (55 mins)		3430725 (57 mins)		3789171 (63 mins)	
AOI Duration (ms)	2631506	683557.50	3430725.5	615018.5	2562917.50	812751
AOI %	78.3%	20.57%	72%	13.05%	67%	21.4%
FIXATION BASED METRICS						
Fixation Count	1291	27	1008	22	1259	31
Revisit Count	13	9	10	4*	13	8
Dwell Time (mins)	8.1	2.20	9.7	1.02	8.8	3.25

Table 4. Areas of Interest for the three conditions (I: Views; II: Scents only; III: Both)

The study analyzes fixation-based metrics to derive insights. Total AOIs represent the session's duration excluding null data. AOI duration denotes the average time participants spent looking at different AOI areas, with AOI % calculated accordingly. Fixation count indicates uninterrupted attention in AOIs, while revisit count measures revisits after distraction, reflecting engagement levels. Dwell Time, converted from milliseconds to minutes, represents the total viewing duration. Throughout the study, the views section showed lower fixations and dwell time compared to scents but similar engagement to the "both" group.

https://doi.org/10.52202/082078-0002

6.3 Facial Expression Analysis

Facial expressions were assessed using the Affectiva AFFDEX algorithm, which detects specific areas on participants' faces (action units) and links these to emotions based on the EMFACS system (Ekman & Friesen, 1978). The Affdex system categorizes seven emotions—anger, sadness, disgust, fear, joy, surprise, and contempt—along with measures for valence and engagement (iMotions, Denmark; Affectiva, USA). The emotions of anger, sadness, disgust, fear, joy, and surprise were considered in the analysis, excluding contempt due to its limited reliability (iMotions, Denmark). Expressions with a magnitude below a margin of 50 m/s were coded as 0 and those that fell above were 1s. All emotions were analyzed to detect the most frequently displayed by participants in each condition for each game. Synchronously, on average the most frequent emotion displayed was "surprise", followed by "engagement/ attention" and then "frustration". Some of the other positive expressions include happiness, and valence while a major percentage of participants across all conditions displayed frustration, anger, and sadness with the stress test. Data was plotted individually for engagement, joy, surprise, anger, and frustration on a scale of 0 – 1000 m/s for each condition (not shown in this paper).

Figure 7. Avg. Total Fixation Duration – Condition (12/28/22). Human Behavior Lab; Texas A&M University

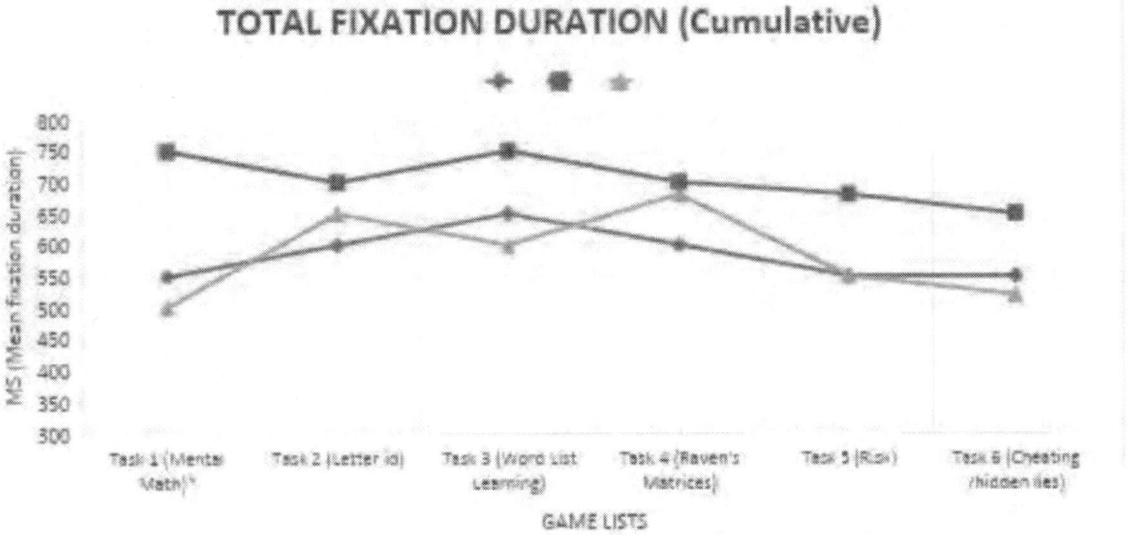

Table 5. Readings. Total Fixations over session – Condition Wise (3/12/2023). Human Behavior Lab; Texas A&M University

CONDITIONS	VIEWS ONLY	SCENTS ONLY	BOTH
Task 1 (Mental Math)	550	750	500
Task 2 (Letter id)	600	700	650
Task 3 (Word List Learning)	650	750	600
Task 4 (Raven's Matrices)	600	700	680
Task 5 (Risk)	550	680	550
Task 6 (Cheating /hidden lies)	550	650	520

https://doi.org/10.52202/082078-0002

7. CONCLUSIONS

The study revealed noteworthy disparities between individuals exposed to views of nature, scents, and those in offices with multiple stimuli. While stress tests and Raven's matrices did not exhibit significant distinctions, other cognitive tasks unveiled compelling findings. Participants with access to nature views demonstrated markedly superior performance in terms of accuracy, speed, and overall task completion. Additionally, they exhibited a greater frequency of positive emotions throughout the tasks. Eye-tracking data corroborated these observations, indicating heightened attention and engagement among individuals in nature-exposed conditions, as evidenced by consistently prolonged Total Fixation Duration.

Notably, Tukey's test highlighted the salience of the views condition, showing a statistically significant divergence from other conditions. This underscores the substantive impact of views of nature on cognitive performance and emotional states. Moreover, parallel results observed in a scent-focused test reinforce the pivotal role of natural elements in indoor environments.

These findings underscore the imperative of integrating nature into workspaces, elucidating its potential to enhance both performance and well-being. Further research is warranted to elucidate the long-term implications and refine practical applications in professional settings.

8. DISCUSSION

These findings carry significant implications for the design of office spaces and environments, extending beyond the immediate scope of this study. Integrating nature views into workplaces emerges as a promising strategy to enhance cognitive performance, emotional well-being, and job satisfaction among workers. The observed improvements in accuracy, speed, and emotional experiences during tasks underscore the potential for views of nature and scents to create a more conducive and productive work environment. Notably, the inclusion of performance analysis of computer tasks and eye-tracking dimensions provides quantitative insights into human responses to biophilic environments. The data in this paper, when analyzed together helps complete the study with results suggesting that even though incorporating multiple stimuli of nature adds benefits to workplace performance and wellbeing, views have been shown once again to be an invaluable primary stimulus to add. This underscores the importance of positive distractions in maintaining cognitive engagement and productivity over time. The practical implications of this research extend beyond the immediate implementation of nature views. Integrating scents into indoor areas with minimal investment can yield significant effectiveness in enhancing well-being, provided precautions are taken to maintain a balanced sensory environment. Furthermore, this study displays the importance of considering both biophilic design features and the characteristics of indoor spaces in shaping physiological and cognitive responses. This highlights the need for further research exploring diverse design patterns and indoor settings to identify optimal biophilic elements for stress reduction and cognitive improvement. Ultimately, the scientific evidence presented in this paper lays the groundwork for developing design guidelines that promote tailored biophilic strategies, benefiting a wide range of indoor spaces and occupants.

Moreover, the ease with which natural elements can be incorporated into office environments is noteworthy. Users themselves can often incorporate nature elements after the initial design phase, although integrating greenspaces and smellscapes during conception would serve the purpose more effectively and align better with the objectives of this study. Even simplistic elements, devoid of excessive complexity, can significantly contribute to the overall well-being and productivity of office occupants.

9. Directions for Future Research

This study opens avenues for exploring various ways to integrate nature into workspaces, extending beyond the specific focus on views and scents. Investigating different combinations of natural elements holds promise for uncovering additional additive effects on cognitive performance and well-being. Furthermore, future research can broaden our understanding of biophilia by exploring aspects such as natural sounds, water features, and haptic biophilic features, which were not addressed in this study. Leveraging advanced technologies like shimmer for measuring sweat levels (EDA) and incorporating physiological measures such as Heart Rate Variability (HRV) and fMRI could provide deeper insights into the immediate health impacts of biophilic settings. This nuanced approach to physiological assessment aims to capture a broader spectrum of responses to natural stimuli, contributing to a more comprehensive understanding of biophilic effects. Additionally, extending the duration of "mini-breaks" in future studies could provide an opportunity to explore the long-term effects of sustained nature exposure through longitudinal research. Such studies could provide valuable insights into the enduring benefits of biophilic design on cognitive performance and stress reduction, informing both research endeavors and practical applications in diverse indoor settings.

https://doi.org/10.52202/082078-0002

10. REFERENCES

Adan, A., & Almirall, H. (1991). Horne & Östberg morningness-eveningness questionnaire: A reduced scale. Personality and Individual differences, 12(3), 241-253.

Axelrod, B. N. (2001). Administration duration for the wechsler adult intelligence scale-III and wechsler memory scale-III. Archives of Clinical Neuropsychology, 16(3), 293-301.

Bentley, P. R., Fisher, J. C., Dallimer, M., Fish, R. D., Austen, G. E., Irvine, K. N., & Davies, Z. G. (2023). Nature, smells, and human wellbeing. Ambio, 52(1), 1-14. https://doi.org/10.1007/s13280-022-01760-w

Bratman, G. N., Hamilton, J. P., & Daily, G. C. (2012). The impacts of nature experience on human cognitive function and mental health. Annals of the New York Academy of Sciences, 1249(1), 118-136.

Braveman, P., & Gottlieb, L. (2014). The social determinants of health: it's time to consider the causes of the causes. Public health reports (Washington, D.C. : 1974), 129 Suppl 2(Suppl 2), 19-31. https://doi.org/10.1177/00333549141291s206

Brown, D. K., Barton, J. L., & Gladwell, V. F. (2013). Viewing nature scenes positively affects recovery of autonomic function following acute-mental stress. Environ Sci Technol, 47(11), 5562-5569. https://doi.org/10.1021/es305019p

Cecily, M., Mardie, T., Anita, P., Peter, B., Lawrence St, L., & Leger, L. S. (2006). Healthy nature healthy people: 'Contact with nature' as an upstream health promotion intervention for populations. Health promotion international. https://doi.org/10.1093/heapro/dai032

Chen, D. L., Schonger, M., & Wickens, C. (2016). oTree—An open-source platform for laboratory, online, and field experiments. Journal of Behavioral and Experimental Finance, 9, 88-97. https://doi.org/https://doi.org/10.1016/j.jbef.2015.12.001

Ekman, P., & Friesen, W. V. (1978). Facial action coding system. Environmental Psychology & Nonverbal Behavior.

Evans, G. W., & McCoy, J. M. (1998). When buildings don't work: The role of architecture in human health. Journal of environmental psychology, 18(1), 85-94.

Ferreira, S. O. (2019). Emotional activation in human beings. Psicologia USP, 30, e180176-e180176.

Frumkin, H. (2001). Beyond toxicity: human health and the natural environment. Am J Prev Med, 20(3), 234-240. https://doi.org/10.1016/s0749-3797(00)00317-2

Frumkin, H., & Fox, J. (2011). Contact with nature. Making healthy places: Designing and building for health, well-being, and sustainability, 229-243.

Gillis, K., & Gatersleben, B. (2015). A review of psychological literature on the health and wellbeing benefits of biophilic design. Buildings, 5(3), 948-963.

Green, J. D., Reid, C. A., Kneuer, M. A., & Hedgebeth, M. V. (2023). The proust effect: Scents, food, and nostalgia. Current Opinion in Psychology, 101562.

Grinde, B., & Patil, G. G. (2009). Biophilia: does visual contact with nature impact on health and well-being? International journal of environmental research and public health, 6(9), 2332-2343. https://doi.org/10.3390/ijerph6092332

Hartig, T., Evans, G. W., Jamner, L. D., Davis, D. S., & Gärling, T. (2003). Tracking restoration in natural and urban field settings. Journal of environmental psychology, 23(2), 109-123.

Hartig, T., Mitchell, R., De Vries, S., & Frumkin, H. (2014). Nature and health. Annual review of public health, 35, 207-228.

Henshaw, V. (2013). Urban smellscapes: Understanding and designing city smell environments. Routledge.

Herz, R. S., & Cupchik, G. C. (1995). The emotional distinctiveness of odor-evoked memories. Chem Senses, 20(5), 517-528. https://doi.org/10.1093/chemse/20.5.517

Herzog, T. R., Black, A. M., Fountaine, K. A., & Knotts, D. J. (1997). Reflection and attentional recovery as distinctive benefits of restorative environments. Journal of environmental psychology, 17(2), 165-170.

Holt, C. A., & Laury, S. K. (2005). Risk aversion and incentive effects: New data without order effects. American economic review, 95(3), 902-912.

Johnson, M. B., Kingston, R., Utell, M. J., Wells, J., Singal, M., Troy, W. R., Horenziak, S., Dalton, P., Ahmed, F. K., & Herz, R. S. (2019). Exploring the science, safety, and benefits of air care products: perspectives from the inaugural air care summit. Inhalation Toxicology, 31(1), 12-24.

Kaplan, R. (1992). The psychological benefits of nearby nature. In. Timber Press.

https://doi.org/10.52202/082078-0002

Kaplan, R., & Kaplan, S. (1989). The experience of nature: A psychological perspective. Cambridge university press.

Kellert, S. R. (2018). Nature by design: The practice of biophilic design. yale university press.

Kellert, S. R., & Wilson, E. O. (1995). The biophilia hypothesis. Island press.

Kunda, M., McGreggor, K., & Goel, A. K. (2013). A computational model for solving problems from the Raven's Progressive Matrices intelligence test using iconic visual representations. Cognitive Systems Research, 22-23, 47-66. https://doi.org/https://doi.org/10.1016/j.cogsys.2012.08.001

Laenen, A., Kolgen, S., & Patel, S. (2020). Placemaking and (Digital) Storytelling.

Lee, J., Park, B.-J., Tsunetsugu, Y., Ohira, T., Kagawa, T., & Miyazaki, Y. (2011). Effect of forest bathing on physiological and psychological responses in young Japanese male subjects. Public health, 125(2), 93-100.

Mackay, G. J., & Neill, J. T. (2010). The effect of "green exercise" on state anxiety and the role of exercise duration, intensity, and greenness: A quasi-experimental study. Psychology of sport and exercise, 11(3), 238-245.

Marc, G. B., John, J., & Stephen, K. (2008). The Cognitive Benefits of Interacting With Nature. Psychological science. https://doi.org/10.1111/j.1467-9280.2008.02225.x

Martin, L., White, M. P., Hunt, A., Richardson, M., Pahl, S., & Burt, J. (2020). Nature contact, nature connectedness and associations with health, wellbeing and pro-environmental behaviours. Journal of environmental psychology, 68, 101389. https://doi.org/https://doi.org/10.1016/j.jenvp.2020.101389

McLeroy, K. R., Bibeau, D., Steckler, A., & Glanz, K. (1988). An ecological perspective on health promotion programs. Health education quarterly, 15(4), 351-377.

Neff, E. P. (2018). Stop and smell the geosmin. Lab Animal, 47(10), 270-270.

Oechssler, J., & Sofianos, A. (2019). The Binary Lottery Procedure does not induce risk neutrality in the Holt-Laury and Eckel-Grossman tasks.

Pearson, D. G., & Craig, T. (2014). The great outdoors? Exploring the mental health benefits of natural environments [Opinion]. Frontiers in psychology, 5. https://doi.org/10.3389/fpsyg.2014.01178

Pilotti, M., Klein, E., Golem, D., Piepenbrink, E., & Kaplan, K. (2015). Is viewing a nature video after work restorative? Effects on blood pressure, task performance, and long-term memory. Environment and Behavior, 47(9), 947-969.

Prestwood, L. E. (2010). Architectural design factors of domestic violence shelters that affect outcomes for female domestic violence victims: A naturalistic inquiry to establish grounded theory for future research. Texas A&M University.

Stigsdotter, U. K., Palsdottir, A. M., Burls, A., Chermaz, A., Ferrini, F., & Grahn, P. (2011). Nature-based therapeutic interventions. In Forests, trees and human health (pp. 309-342). Springer.

Sussman, A., & Hollander, J. (2014). Cognitive Architecture: Designing for How We Respond to the Built Environment. https://doi.org/10.4324/9781315856964

Taylor, A. F., Kuo, F. E., & Sullivan, W. C. (2002). Views of nature and self-discipline: Evidence from inner city children. Journal of environmental psychology, 22(1-2), 49-63.

Tulchinsky, T. H., & Varavikova, E. A. (2014). Chapter 1 - A History of Public Health. In T. H. Tulchinsky & E. A. Varavikova (Eds.), The New Public Health (Third Edition) (pp. 1-42). Academic Press. https://doi.org/https://doi.org/10.1016/B978-0-12-415766-8.00001-X
Ulrich, R. (2002). Health Benefits of Gardens in Hospitals.

Ulrich, R. S., Simons, R. F., Losito, B. D., Fiorito, E., Miles, M. A., & Zelson, M. (1991). Stress recovery during exposure to natural and urban environments. Journal of environmental psychology, 11(3), 201-230.

Van der Stoep, N., Van der Smagt, M. J., Notaro, C., Spock, Z., & Naber, M. (2021). The additive nature of the human multisensory evoked pupil response. Sci Rep, 11(1), 707. https://doi.org/10.1038/s41598-020-80286-1

Weir, K. (2020). Nurtured by nature. Monitor on Psychology, 51(3), 50-56.

Wilson, E. O. (1986). Biophilia. Harvard university press.

Zhang, Y., Xu, Z., & Palma, M. A. (2019). Conveniently dependent or naively overconfident? An experimental study on the reaction to external help. PLoS One, 14(5), e0216617. https://doi.org/10.1371/journal.pone.0216617

https://doi.org/10.52202/082078-0002

THE IMPACT OF INTEGRATING NATURAL ELEMENTS ON HEALTHCARE DESIGN: A BRIEF REVIEW

Track 4: Design for Health and Wellbeing
Knowledge Network: Health, Interior Design
Methods: Case Study, Meta-Analysis
Daphne A.U. Chilaka (Texas Tech University); Ladan Khalvati (Texas Tech University); Kristi Gaines (Texas Tech University)

1 ABSTRACT

In healthcare design, the integration of natural elements offers promising potential for improving patient outcomes and well-being. This review examined the impact of natural elements on patients' health, staff well-being, and the overall atmosphere of healthcare facilities. A comprehensive search was conducted on databases such as Web of Science, SCOPUS, and Google Scholar, utilizing keywords like "architecture," "healthcare design," "biophilic design," "hospital design," and "natural elements." Out of 70 articles, 29 research articles were selected based on inclusion criteria relevant to integrating natural elements like greenery, natural light, water features, and biophilic design principles in healthcare environments. Thematic analysis was conducted on the selected studies to identify common themes, trends, and significant findings concerning the impact of integrating natural elements in healthcare design. The results indicated that hospital environments with sterile conditions can cause stress among individuals, while the incorporation of elements like greenery and natural light creates a sense of tranquility. Studies showed that nature exposure leads to reduced stress, improved recovery, and shorter hospital stays, with patients reporting higher satisfaction, engagement, and treatment compliance. Additionally, integrating natural elements may lead to healthcare providers experiencing reduced burnout and improved job satisfaction, with increased cost savings and resource allocation in healthcare facilities. The evidence suggests that while challenges may exist, embracing nature's qualities can create more supportive and therapeutic healthcare spaces, contributing to better patient outcomes and care quality. Further research may be needed to explore the relationship between biophilic design interventions and measurable health outcomes in healthcare settings.

1.1 Keywords
Architecture, healthcare design, biophilic design, hospital design, natural element

2 INTRODUCTION

In recent years, the significance of the physical environment in healthcare settings has garnered increasing attention among researchers, architects, and healthcare professionals (Bernhardt et al., 2022). The realization that the design of healthcare facilities can profoundly impact patient outcomes and well-being has led to a paradigm shift in the approach to healthcare design. Moving away from sterile and clinical spaces, a new design philosophy emphasizing the integration of natural elements has emerged, offering a more biophilic and patient-centric approach (Kadwell, 2019). This paper presents a brief review of the impact of integrating natural elements in healthcare design, focusing on its effects on patient health, staff well-being, and the overall ambiance of healthcare facilities.

Healthcare environments that emphasize functionality and efficiency, often fail to consider the emotional and psychological needs of patients. The clinical nature of these spaces can evoke feelings of anxiety, stress, and isolation, thereby potentially hindering the healing process (Howze et al., 2004). In contrast, incorporating natural elements, such as greenery, natural light, water features, and biophilic design principles, seeks to bridge the gap between the built environment and nature's inherent healing qualities. By integrating elements that mimic the tranquility of natural settings, healthcare designers aim to create spaces that promote a sense of calm, comfort, and emotional well-being for patients (Kellert, 2018).

Numerous studies have investigated the effects of nature-inspired design in healthcare settings, revealing a plethora of benefits for patients. Exposure to natural elements has been linked to reduced stress levels, improved mood, and enhanced overall well-being. Patients recovering in biophilic environments have shown faster recovery rates, shorter hospital stays, and increased satisfaction with their healthcare experience (Scates et al., 2020). Moreover, interactions with nature have been found to improve pain management, decrease the need for pain medication, and boost immune function, thus positively impacting the healing process. Beyond its effects on patients, the integration of natural elements in healthcare design also extends to the well-being of healthcare staff. Nurses, doctors, and other healthcare professionals often work in high-stress environments, and the incorporation of natural elements can help alleviate job-related stress and

https://doi.org/10.52202/082078-0003

contribute to higher job satisfaction (Creasy, 2012). A supportive and pleasant work environment fosters a more compassionate and empathetic healthcare team, ultimately benefiting patient care. Additionally, the impact of integrating natural elements goes beyond individual patients and healthcare professionals. Healthcare facilities that embrace biophilic design principles have the potential to positively influence the overall healthcare system (Moslehian et al., 2023). By promoting patient healing and satisfaction, these environments may lead to reduced readmission rates and improved resource allocation, thus enhancing operational efficiency and potentially reducing healthcare costs. However, despite the numerous benefits of integrating natural elements in healthcare design, challenges exist in implementing these principles universally (Rich et al., 2021). Limited space, budget constraints, and regulatory requirements often pose obstacles to the full realization of biophilic designs. Nevertheless, the growing body of research supporting the positive impact of nature-inspired design continues to inspire architects, designers, and healthcare stakeholders to explore innovative solutions (Sadick et al., 2023).

This paper seeks to provide a concise review of the evidence surrounding the impact of integrating natural elements in healthcare design. By exploring the effects on patient health, staff well-being, and the overall ambiance of healthcare facilities, we aim to underscore the importance of creating healing environments that embrace nature's therapeutic qualities. As we delve into the existing literature, we hope to inspire further research, collaboration, and innovation in healthcare design, ultimately fostering spaces that prioritize patient-centered care and contribute to improved healthcare outcomes)

3 METHODS

This study employed a literature review methodology to explore the impact of integrating natural elements in healthcare design. The objective is to investigate the effects of nature-inspired design on patient health, staff well-being, and the overall ambiance of healthcare facilities. A comprehensive search was conducted on databases such as Web of Science, SCOPUS, and Google Scholar, utilizing keywords like "architecture", "healthcare design", "biophilic design", "hospital design", and "natural elements". These databases were chosen for Interior, environmental design, and health-related research. Of the 70 articles collected, 29 research articles were selected based on inclusion criteria relevant to integrating natural elements like greenery, natural light, water features, and biophilic design principles in healthcare.

Figure 1. PRISMA Flow Chart

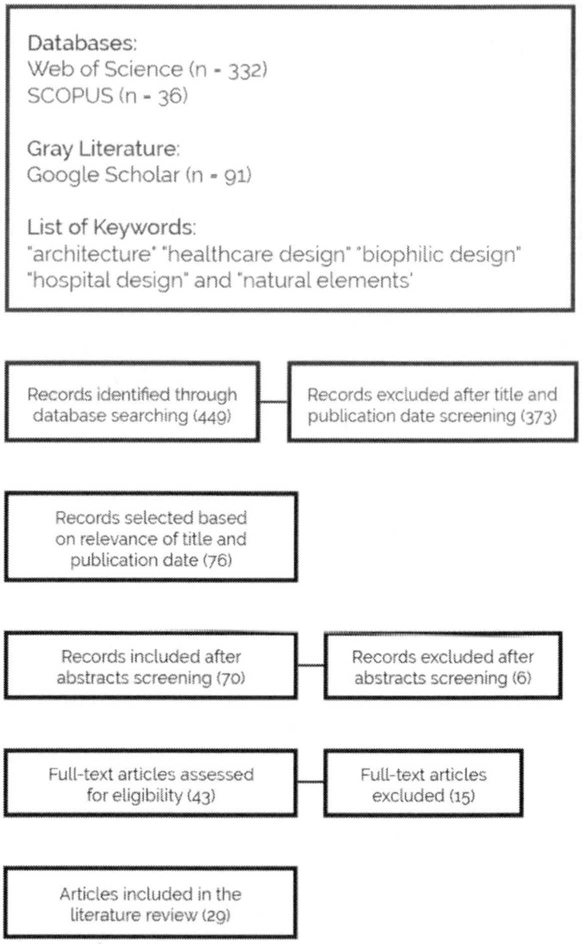

As shown in Figure 1, peer-reviewed articles were filtered from these databases, and duplicates were removed. Published peer-reviewed articles from the years 2004 to 2024 were selected based on screening of the title and publication date, for a broad range of accurate and relevant findings. Abstracts of the articles identified in the search and subsequently of the selected full-text texts were screened for relevance to the research question using PRISMA (Preferred Reporting Items for Systematic Reviews and Meta-Analyses). Finally, 29 articles were selected to be included in the literature review. Through this systematized collection of scholarly articles thematic analysis to identify common themes, trends, and significant findings concerning the impact of integrating natural elements in healthcare design was conducted.

https://doi.org/10.52202/082078-0003

3 RESULTS

This section explores various studies conducted by independent researchers and experts, highlighting the impact of integrating natural elements in healthcare design. The incorporation of natural elements into healthcare design has been the subject of extensive research, that reveals detailed insights into how nature promotes and influences positive health outcomes for both patients and healthcare staff.

3.1 Stress Reduction and Well-being

Patient Outcomes and Recovery. Integrating natural elements into healthcare environments has been consistently associated with stress reduction and improved well-being for patients. Studies by Park et al. (2018) and Dijkstra et al. (2008) demonstrate the stress-reducing effects of natural light and indoor plants, respectively on patients' outcomes and recovery. Park et al. (2018) utilized extensive medical data to demonstrate the effectiveness of natural light in hospitals. The analysis included 38,788 patients with beds near the window and 46,233 patients with beds near the door. The results showed that patients situated near windows experienced shorter hospital stays, indicating the potential value of natural light in healthcare design. Additionally, research by Dijkstra et al. (2008) explored the stress-reducing impacts of indoor plants, highlighting their potential to enhance patient well-being. Participants were presented with scenarios of hospitalization and exposed to photos of medical facilities using or excluding indoor plants. Those exposed to rooms with indoor plants reported lower stress levels, and mediation analysis confirmed that the stress reduction occurred via the room's inherent beauty. The study highlights the potential of indoor plants to enhance patients' well-being and suggests that making the environment more attractive can contribute to patients' recovery.

Focus on Patient Sleep and Recovery. The impact of natural elements on patient sleep and recovery is another significant theme in healthcare design research. Hadi et al. (2019) examined the effects of bright light exposure on patient sleep and recommended effective lighting interventions to promote better sleep quality. With a specific focus on patient sleep, Hadi et al. (2019) examined the impact of bright light exposure on patient sleep in healthcare environments. Their analysis of thirty-four articles revealed that different lighting conditions could enhance various sleep characteristics. Patients' sleep results can be improved by exposing them to brief bursts of strong light in the morning, moderate levels of light in the evening, and low illumination levels during the day. The results highlight how crucial lighting systems are in healthcare facilities. Meanwhile, the inconsistent nature of study designs and findings calls for additional research on illumination interventions. Future research should be more explicit on the spectrum characteristics of light sources to enhance patient sleep. The review suggests effective lighting conditions for implementation in healthcare settings to promote better patient sleep. Additionally, designers should consider non visual responses to light and the spectrum and intensity of lighting over patient beds to enhance patient well-being. The study emphasized the importance of lighting systems in healthcare facilities and called for additional research on illumination interventions, specifically highlighting the need for clarity on the spectrum characteristics of light sources.

Health Benefits for Healthcare Staff. The impact of natural elements extends beyond patient care to benefit healthcare staff as well. A study by Karanikola et al. (2020) emphasizes the positive impact of exposure to the natural environment on the well-being of healthcare workers, highlighting the importance of incorporating green spaces and promoting interaction with nature. In the study, the Hospital of Alexandroupolis was evaluated to assess the interior and exterior design of the healthcare environment. The primary objective of the research was to reduce stress levels among the staff, particularly nursing employees. The results indicate that there is a shortage of suitable green areas, and the ones that are currently available only receive mediocre satisfaction. The study emphasizes the need for brief escapes to support stress relief processes and improve working conditions. Exposure to the natural environment, both directly and indirectly, through windows and outdoor spaces has a positive impact on the well-being of healthcare workers. While some improvements have been made in the outdoor environment, further enhancements are still necessary. The study suggests that incorporating solutions like short breaks, providing personal spaces with views, and encouraging interaction with plants and animals can help improve the hospital's work and healing environment. The study also highlights the importance of addressing the behavior of patients' visitors to support the well-being of healthcare staff. Improvements in existing infrastructures and proper maintenance are essential for a more satisfactory healthcare environment. Job satisfaction varies among different categories of hospital employees, with nursing and other staff experiencing more stress due to direct patient contact. The study recommends specific measures to alleviate stress and improve working conditions, particularly for this group of staff members.

https://doi.org/10.52202/082078-0003

Staff Well-being and Performance. Several studies highlight the importance of natural elements in supporting the well-being and performance of healthcare staff (Karanikola et al. 2020; Kamali & Abbas, 2012). While much of the research focuses on the benefits of natural elements for patients, there is also growing recognition of their positive impact on staff well-being. Studies by Karanikola et al. (2020) and Kamali & Abbas (2012) highlight the importance of incorporating green spaces, natural light, and comfortable break areas to reduce stress levels among healthcare workers and improve job satisfaction. Kamali & Abbas (2012) delved into the influence of lighting design on nursing services in a Malaysian healthcare center, employing a comprehensive methodology that included site observations, interviews, and surveys with 120 nurses. Their findings emphasized the age-dependent sensitivity of nurses to lighting conditions, revealing that older nurses faced challenges in certain tasks under natural lighting compared to their younger counterparts. Considering the implications of these findings, the study recommended optimizing task assignments based on nurses' age to enhance performance. Moreover, it advocated for improvements in lighting design and equipment in areas where difficulties were identified. The importance of these findings extends beyond the initial study, calling for further research in diverse hospital settings to validate and refine the suggested lighting optimizations

Promoting Health and Well-being is a recurring outcome of incorporating natural elements into healthcare environments, as highlighted in several studies. Overall, these studies emphasize the stress-reducing properties of nature's components in medical environments and their positive impact on patient's and staff's well-being. Future research could explore other environmental stimuli and replicate the findings in real healthcare settings with actual patients.

3.2 User Preferences and Satisfaction

Preference for Natural Settings across various studies is shown in a consistent choice for natural settings among both patients and healthcare staff. Barnhart et al. (1998) conducted a study to determine the outdoor behavior preferences of patients. The study comprised an equal percentage of patients and staff (50% each), who provided their ratings for preferred behaviors. The data collection was done through an unbiased computer survey. The study found that participants preferred natural settings for passive behaviors, such as sitting, watching, and talking, and private settings for active behaviors, such as walking and socializing. The differences between staff and patient preferences for settings and behaviors were minimal. These findings revealed that nature was the most preferred location for staff and patients, especially when combined with specific behaviors. Thus, underscoring the innate human connection to nature and the importance of integrating natural elements into healthcare design to create supportive environments.

Preference for Natural Elements is another theme noted in several studies. In a similar study by Nyrud et al. (2014), wood as a design element, was found to be preferred in healthcare spaces. The study highlighted the preference for rooms with a moderate amount of wood, emphasizing the importance of careful design considerations. Nyrud et al. (2014) undertook research to determine people's preference for natural building materials, specifically wood, in the patient rooms of a hospital in Norway. The researchers sent an online survey to the hospital staff, including physicians, nurses, and other staff. The participants were asked to evaluate digitally altered images of patient rooms with different amounts of wood on surfaces. The study revealed that both the staff and the patients preferred an intermediate degree of wood in the rooms. Less preferred were rooms with no wood or with too much wood. The study did not find any significant differences in preferences between different professional groups. The research highlights the importance of designing hospital interiors carefully, considering user preferences, and incorporating biophilic design principles. Although the study did not explore the effects of these preferences on clinical or health outcomes, it suggests that using natural elements in hospital settings has practical significance in improving the work environment and potentially enhancing services provided to patients. Careful attention to user preferences regarding the amount of wood used significantly impacts the overall atmosphere. Thus, underlining the need for meticulous design considerations in integrating natural elements like wood into healthcare spaces.

3.3 Biophilic Design Elements

Creating a healing environment and enhancing the patient experience are key goals of incorporating natural design elements into healthcare environments. The concept of biophilic design aims to integrate nature into built environments and has emerged as a key approach to healthcare design. Several studies emphasize the importance of integrating biophilic design principles in healthcare design.

Promotion of Healing and Recovery. Incorporating natural elements process in healthcare facilities is closely associated with promoting healing enhancing

https://doi.org/10.52202/082078-0003

comfort and facilitating the recovery process of patients. In 2016, Nejati et al. conducted a study that used visual simulation techniques to investigate the restorative potential of nature-based design elements in the break rooms of hospital staff. Indoor plants, nature-inspired artwork, window views, and direct access to the outdoors via a balcony were among the various design measures evaluated by nurses for their potential to promote healing. The study found that the perceived restorative effects of the break rooms gradually improved with larger levels of natural content, with a direct connection to nature being ranked as the most helpful for stress alleviation. The incorporation of various natural elements into break areas piqued the interest of nurses, who suggested that doing so could improve the well-being of healthcare staff. In an earlier study by Zhao et al. (2022), the significance of biophilic design in medical facilities was investigated, particularly in the context of COVID-19, which boosted the requirement for sustainability and health. The research explores the biophilic design in various countries, with some lacking biophilic design while others have integration without high standardization and recognition. The article emphasizes the importance of re-establishing the relationship between structures and ecosystems to improve people's health. The study proposes appropriate design methods based on the health system, relevant natural characteristics, and health benefits of biophilic design in hospital spaces. The research suggests optimizing the interaction between natural elements and people, increasing emotional connections in spaces, and realizing systematic compounding with diverse designs. Biophilic design is seen as a way to make medical centers sustainable, and its application can contribute to the health of both clients and medical personnel. The study highlights the importance of carefully selecting natural factors to avoid potential risks and enhance the positive impact of biophilic design. Overall, biophilic design can play a crucial role in improving the health and well-being of individuals in hospital environments.

Impact on Psychological Well-being. Research consistently demonstrates the positive impact of natural elements on psychological well-being in healthcare settings. Avinç & Selçuk (2021) highlight the potential of biophilic design to transform healthcare facilities into healing environments by reconnecting humans with nature. The importance of access to outdoor spaces has been emphasized during the COVID-19 pandemic for people's psychological well-being. The study explores how biophilic design principles can be applied to healthcare facilities. The findings suggest that healthcare facilities can be transformed into healing environments through the successful implementation of biophilic design concepts. The conclusion highlights that biophilic design strives to reconnect humans with nature, promoting healing and well-being. By integrating natural elements such as natural patterns and geometries, hospitals can create healthier environments for patients, staff, and visitors, leading to the transition to an eco-centric age. A 2018 study by Totaforti investigated the importance of nature in promoting the health and well-being of individuals. The study discusses biophilic design, which aims to integrate nature into buildings, and its potential benefits in healthcare spaces. The article emphasizes the need for a cultural shift in designing spaces that reconnect individuals with nature, which can enhance their psychophysical well-being. The humanization of healthcare spaces and contact with nature can have a positive impact on patient's emotional well-being, reducing stress and pain. However, further research is necessary to understand the impact of the human-nature relationship on health and to adopt a new vision of medicine and healthcare environments. The article suggests that biophilic design should not be limited to incorporating plants into the built environment but should consider a more holistic approach based on understanding the human-nature relationship.

3.4 Strategies for Integrating Natural Elements

The Importance of Nature Exposure, both direct and indirect, is identified as a critical factor in promoting well-being in healthcare environments. Several studies reviewed, emphasize the need for green spaces and outdoor areas in healthcare facilities to support stress relief processes and improve working conditions for healthcare staff (Jung et al., 2023; Tekin et al., 2023; Karanikola et al., 2020; Park et al., 2018; Nejati et al., 2016; Hobday & Dancer 2013; Dijkstra et al. 2008, Ulrich, 2008)). These studies highlight the therapeutic benefits of natural elements in enhancing the overall healthcare experience.

Natural Light and Air Circulation. Studies investigate the impact of natural light and air circulation on health outcomes in healthcare settings. Park et al. (2018) demonstrate the effectiveness of natural light in hospitals, with patients near windows experiencing shorter hospital stays. Building on this theme, Hobday & Dancer (2013) wanted to understand the role of sunshine and air circulation in preventing infections in healthcare settings. Their review of historical and recent data suggested that modern healthcare designs, prioritizing comfort, may inadvertently contribute to pathogen persistence. The findings imply that protection is provided via air circulation from airborne pathogen transmission, but there is limited evidence for the role of sunlight. As antimicrobial resistance increases, further research is needed to clear the inherent dangers of airborne germs and the possible advantages of increased sunshine and clean air exposure. The study proposed

https://doi.org/10.52202/082078-0003

a reevaluation of lighting strategies to better protect against bioterrorism threats, with an emphasis on increased exposure to sunlight and fresh air.

Natural Building Materials. Integrating natural elements into healthcare design is also aligned with broader goals of environmental sustainability. Researchers, such as Avinç & Selçuk (2021), Hobday & Dancer (2013), and Nyrud et al. (2012, 2010), emphasize the potential of biophilic design to create sustainable healthcare facilities that minimize environmental impact and optimize resource efficiency. This reflects a shift towards eco-centric approaches to healthcare architecture that prioritize both human well-being and environmental stewardship. In the study done by Nyrud et al. (2010), the authors shifted the focus to the psychological impact of using wood as a building material. Their study, part of a larger project exploring the health benefits of wood in indoor settings, highlighted the preference for rooms with a moderate amount of wood. The study involved evaluating the psychological impact of different wood interiors in a hospital room, with hospital employees as the expert group and patients as the control group. Participants were asked to rate rooms with varying amounts of wood on walls, floor, and ceiling. The results showed that rooms using wood at a moderate level were most preferred, and rated as pleasant, natural, calming, and secure. Rooms with no wood and rooms with wood on all surfaces were the least preferred. The study indicates that there are restrictions on the amount of wood that is favored in interior settings. The study emphasizes the importance of understanding preferences as they can guide designers in creating environments that promote health and well-being. Integrating natural elements like wood into the built environment can offer people greater access to nature, which is essential as urbanization reduces natural exposure. Further research within the project will explore the potential health benefits of natural elements in hospitals and other built settings, offering valuable insights for planners and environmental designers aiming to improve indoor environments. In an earlier study conducted by Nyrud et al. in 2012, the effects of using wood in medical centers, specifically on indoor air quality, were evaluated. The study looked at three room layouts, including regular patient rooms without wood panels, and rooms with birch wood and oak wood wall panels. The results showed no significant differences in volatile organic compound (VOC) ventilations. As a result, it can be concluded that wooden paneled walls in hospital facilities have a negligible impact on indoor air quality. This finding suggests that it may be challenging to influence patients' perceptions of the atmosphere inside the hospital. Natural light and indoor plants have also been extensively studied.

Role of Technology. Advanced technologies, including virtual reality (VR), have been employed to delve deeper into the psychological impact of biophilic design. Studies by Jung et al. (2023) and Tekin et al. (2023) demonstrate the positive effects of biophilic elements on reducing stress, enhancing emotional well-being, and improving overall comfort in medical facilities. Jung et al. (2023) utilized VR and brainwave analysis to explore the effects of biophilic design elements on the psychological well-being and comfort of hospital room occupants. The researchers used electroencephalogram (EEG) and VR methods to examine brain and emotional responses. They tested three types of biophilic designs: a control room, a room with a digital wall, and a room with both plant walls and a digital wall. The study found that incorporating biophilic elements into the design of medical facilities can reduce stress, induce positive emotional changes, and alleviate anxiety. The researchers suggest that using biophilic design in healthcare can improve both physical and psychological well-being. The use of VR can offer a deeper understanding of the effects of biophilic design on human experiences. However, the study notes certain limitations, such as the need to measure both state and trait anxiety, and the consideration of different machine learning algorithms. Nonetheless, the research encourages the adoption of biophilic design principles to enhance well-being and provides a roadmap for improving environmental research methodologies. Virtual reality (VR) has been utilized to explore the psychological impact of biophilic design, revealing its potential to reduce stress and anxiety among hospital room occupants (Jung et al., 2023).

Optimizing Design for Specific Needs: Several studies highlight the importance of tailoring healthcare design to specific user needs and contexts. Kamali & Abbas (2012) emphasize the age-dependent sensitivity of nurses to lighting conditions, suggesting optimizations based on age to enhance performance. Additionally, the significance of user involvement in the design process is emphasized in various studies. Bengtsson & Grahn (2014) introduced the Quality Evaluation Tool (QET), which emphasizes the importance of involving users in creating purposeful outdoor spaces in healthcare facilities. The Quality Evaluation Tool (QET) is based on theory triangulation, integrating hypotheses and data from studies on how people's health and well-being are affected by the outdoors. The QET consists of 19 environmental qualities, including six for comfort for access to the great outdoors and the nearby wildlife. It also includes concepts for involving users in the creation of designs and basic design principles. The discussion relates to the QET and the specific healthcare setting being investigated. The tool's flexibility allows it to be useful in various settings, considering the needs and preferences of

https://doi.org/10.52202/082078-0003

different user groups and supporting the healing process. The QET is designed to be continuously improved based on empirical experience and more evidence. The article emphasizes the importance of involving users in the design process and tailoring the QET to specific environmental contexts to create purposeful and supportive outdoor environments in healthcare settings. Gaminiesfahani et al. (2020) conducted a scoping review of the research on the effects of architectural and landscape characteristics on the health and well-being of children in pediatric healthcare buildings. The study identified nine design qualities that affect healing and well-being, including accessibility to the outdoors, art, music, natural light, less crowded spaces, and quieter surroundings. The research underscores the importance of considering the experiences of sick children and their families while in the hospital and calls for a more comprehensive understanding of design variables to provide holistic and context-specific design recommendations. The inclusion of garden/green spaces is highlighted as particularly beneficial in children's healthcare environments.

Standardized Evidence-Based Design (ED) Frameworks. Several studies propose standardized design frameworks to guide the incorporation of natural elements into healthcare environments. For instance, Tekin et al. (2023) developed a standardized biophilic design framework based on the experiences of different user groups, emphasizing emotional, mental, and spiritual health alongside physical treatment. The framework identifies key parameters that are important for different groups, such as access to nature, natural light, thermal comfort, privacy, and quietness. The research highlights the need to consider emotional, mental, and spiritual health in addition to physical treatment in healthcare settings. It offers guidelines for future research and design direction and stresses the importance of context-specific design frameworks that are based on user needs. The study suggests that a rigorous design structure should be compared to both qualitative source data and objective data from science to provide comprehensive recommendations. Similarly, researchers like Nyrud et al. (2010, 2014) and Park et al. (2018) conducted empirical studies to evaluate user preferences and the impact of natural elements on patient outcomes. This trend underscores the need for designers to adopt a systematic approach to healthcare design that integrates empirical evidence, user feedback, and best practices in healthcare design.

Interdisciplinary Perspectives. Many studies advocate for interdisciplinary perspectives and holistic design approaches to better understand the relationship between humans and nature in healthcare environments. Seymour (2016) developed a model that combines perspectives from evolutionary psychology, environmentalism, evolutionary biology, and social economics. This inclusive approach aims to comprehensively understand the relationship between humans and nature and its impact on health. By integrating theoretical concepts from diverse fields, this model provides a holistic understanding of the intricate connections between human health and the environment, guiding nuanced healthcare design strategies. In the research article, Seymour (2016) delved into the evolving interaction between humans and nature and its potential effects on people's health, drawing from diverse interdisciplinary perspectives. To more fully characterize the relationship between humans and nature, it covers current discussions and literature in evolutionary psychology, environmentalism, evolutionary biology, and social economics. Based on the definition provided by the World Health Organization, the idea of "health" is also investigated, and its interplay with the human-nature relationship is examined through a developing conceptual model. The article acknowledges the existing gaps and limitations in research on the impact of nature on different aspects of health (physical, mental, and social) and advocates for a clearer comprehension of the difficulties present in these interlinkages. In conclusion, these studies collectively underscore the significance of integrating natural elements, including lighting design, building materials, and nature exposure, in healthcare settings. The findings not only provide insights into optimizing healthcare environments for both providers and patients but also call for ongoing research to refine and expand our understanding of these intricate relationships.

4 DISCUSSION

Integrating natural elements in healthcare design offers significant potential for enhancing patient outcomes and well-being. By moving away from sterile-like hospital spaces and embracing biophilic principles, healthcare facilities can create environments that promote tranquility, reduce stress, and support the healing process. The evidence from numerous studies highlights the positive impact of nature-inspired design on patients' recovery, satisfaction, and overall health. Additionally, healthcare providers may experience reduced burnout and improved job satisfaction in these supportive environments. The incorporation of natural elements in healthcare design can also extend to cost savings and improved resource allocation, as enhanced patient healing and satisfaction may lead to reduced readmission rates and operational efficiencies.

https://doi.org/10.52202/082078-0003

4.1 Limitations of the Research

Despite the growing evidence supporting the benefits of integrating natural elements in healthcare design, many studies identify challenges and opportunities for further research and improvement. Hobday & Dancer (2013) highlight the need for reevaluating lighting strategies to better protect against pathogen transmission, calling for additional research on the role of sunshine and air circulation in preventing infections. Similarly, Hadi et al. (2019) call for additional research on illumination interventions to enhance patient sleep. Seymour (2016) also advocates for a clearer understanding of the human-nature relationship and its impact on health. These findings underscore the ongoing efforts to refine design strategies and promote the widespread implementation of biophilic design principles in healthcare settings.

While these challenges exist in integrating natural elements in healthcare design universally, the growing body of research continues to inspire innovation in healthcare architecture and design. To maximize the benefits of integrating natural elements, healthcare designers and stakeholders must consider user preferences, context-specific needs, and the therapeutic qualities of different natural elements. By prioritizing patient-centered care and creating healing environments that reconnect individuals with nature, we can contribute to improved healthcare outcomes and better quality of care. As we move forward, the collaboration between researchers, architects, healthcare professionals, and policymakers will be crucial in further exploring the potential of nature-inspired design in healthcare settings. By embracing biophilic principles and creating purposeful and supportive environments, we can revolutionize the healthcare experience, foster well-being, and promote sustainable, patient-centric healthcare facilities.

5 CONCLUSIONS

In healthcare design, the integration of natural elements has emerged as a promising approach, embracing biophilic principles that prioritize patients' well-being. Aseptic healthcare spaces often induce stress, while incorporating natural elements like greenery and natural light can create tranquil environments. Research shows that exposure to nature-inspired design reduces stress levels, improves mood, and enhances overall well-being. Patients recovering in biophilic spaces tend to experience faster recovery rates, shorter hospital stays, and increased satisfaction with their healthcare experience. Moreover, healthcare providers may benefit from reduced burnout and improved job satisfaction in such settings. Embracing nature's therapeutic qualities has the potential to not only enhance patient outcomes but also lead to cost savings and improved resource allocation. Although challenges exist in implementing biophilic design universally, the growing body of evidence continues to inspire architects, designers, and healthcare stakeholders to explore innovative solutions. By prioritizing patient-centered care and creating healing environments that incorporate natural elements, we can contribute to better patient outcomes and overall care quality.

5.1 Opportunities for Future Research

While a majority of the research focuses on the psychological and emotional benefits of natural elements in healthcare design, there is also growing interest in their potential impact on clinical outcomes. Studies by Park et al. (2018) and Hadi et al. (2019) suggest that factors such as natural light exposure and access to outdoor spaces may influence patient recovery rates and sleep quality. This indicates the need for further research to explore the relationship between biophilic design interventions and measurable health outcomes in healthcare settings.

6 REFERENCES

Avinç, G. M., & Selçuk, S. A. (2021) An Evaluation OF Biophilic Design Parameters in Hospital Buildings. PACE-2021 International Congress on the Phenomenological Aspects of Civil Engineering.

Barnhart, S. K., Perkins, N. H., & Fitzsimonds, J. (1998). Behavior and outdoor setting preferences at a psychiatric hospital. Landscape and Urban Planning, 42(2-4), 147-156.

Bengtsson, A., & Grahn, P. (2014). Outdoor environments in healthcare settings: A quality evaluation tool for use in designing healthcare gardens. Urban forestry & urban greening, 13(4), 878-891.

Bernhardt, J., Lipson-Smith, R., Davis, A., White, M., Zeeman, H., Pitt, N., Shannon, M., Crotty, M., Churilov, L., & Elf, M. (2022). Why hospital design matters: A narrative review of built environments research relevant to stroke care. International Journal of Stroke, 17(4), 370-377.

Creasy, T. M. (2012). The wellness clinic: A new approach to healthcare design.

Dijkstra, K., Pieterse, M. E., & Pruyn, A. (2008). Stress-reducing effects of indoor plants in the built healthcare environment: The mediating role of perceived attractiveness. Preventive medicine, 47(3), 279-283.

https://doi.org/10.52202/082078-0003

Gaminiesfahani, H., Lozanovska, M., & Tucker, R. (2020). A scoping review of the impact on children of the built environment design characteristics of healing spaces. HERD: Health Environments Research & Design Journal, 13(4), 98-114.

Hadi, K., Du Bose, J. R., & Choi, Y.-S. (2019). The effect of light on sleep and sleep-related physiological factors among patients in healthcare facilities: A systematic review. HERD: Health Environments Research & Design Journal, 12(4), 116-141.

Hobday, R., & Dancer, S. (2013). Roles of sunlight and natural ventilation for controlling infection: Historical and current perspectives. Journal of Hospital Infection, 84(4), 271-282.

Howze, E., Baldwin, G., & Kegler, M. (2004). Environmental health promotion: Bridging traditional environmental health and health promotion. Health Education & Behavior, 31(4), 429-440.

Jung, D., Kim, D. I., & Kim, N. (2023). Bringing nature into hospital architecture: Machine learning-based EEG analysis of the biophilia effect in virtual reality. Journal of Environmental Psychology, 89, 102033.

Kadwell, M. E. (2019). Intergenerational healing: Redefining the healthcare environment.

Kamali, N. J., & Abbas, M. Y. (2012). Healing environment: enhancing nurses' performance through proper lighting design. Procedia-Social and Behavioral Sciences, 35, 205-212.

Karanikola, P., Andrea, V., Tampakis, S., & Tsolakidou, A. (2020). Indoor and outdoor design in healthcare environments: The employees' views in the general university hospital of Alexandroupolis, Greece. Environments, 7(8), 61.

Kellert, S. R. (2018). Nature by design: The practice of biophilic design. Yale University Press.

Moslehian, A. S., Roös, P. B., Gaekwad, J. S., & Van Galen, L. (2023). Potential risks and beneficial impacts of using indoor plants in the biophilic design of healthcare facilities: A scoping review. Building and Environment, 110057.

Nejati, A., Rodiek, S., & Shepley, M. (2016). Using visual simulation to evaluate restorative qualities of access to nature in hospital staff break areas. Landscape and Urban Planning, 148, 132-138.

Nyrud, A., Bysheim, K., & Bringslimark, T. (2010). Health benefits from wood interiors in a hospital room. Proceedings of the International Convention of Society of Wood Science and Technology and United Nations Economic Commission for Europe, Geneva,

Nyrud, A. Q., Bringslimark, T., & Bysheim, K. (2014). Benefits from wood interior in a hospital room: A preference study. Architectural Science Review, 57(2), 125-131.

Nyrud, A. Q., Bringslimark, T., & Englund, F. (2012). Wood use in a hospital environment: VOC emissions and air quality. European Journal of Wood and Wood Products, 70(4), 541.

Park, M. Y., Chai, C.-G., Lee, H.-K., Moon, H., & Noh, J. S. (2018). The effects of natural daylight on length of hospital stay. Environmental health insights, 12, 1178630218812817.

Rich, R. K., Jimenez, F. E., Puumala, S. E., DePaola, S., Harper, K., Roy, L., & Brittin, J. (2021). From fable to reality at Parkland Hospital: The impact of evidence-based design strategies on patient safety, healing, and satisfaction in an adult inpatient environment. HERD: Health Environments Research & Design Journal, 14(3), 65-82.

Sadick, A.-M., Kamardeen, I., & Vu, X. P. (2023). Challenges for implementing biophilic strategies in Australian building design. Journal of Building Engineering, 74, 106849.

Scates, D., Dickinson, J. I., Sullivan, K., Cline, H., & Balaraman, R. (2020). Using nature-inspired virtual reality as a distraction to reduce stress and pain among cancer patients. Environment and Behavior, 52(8), 895-918.

Seymour, V. (2016). The human–nature relationship and its impact on health: A critical review. Frontiers in public health, 260.

Tekin, B. H., Corcoran, R., & Gutiérrez, R. U. (2023). A systematic review and conceptual framework of biophilic design parameters in clinical environments. HERD: Health Environments Research & Design Journal, 16(1), 233-250.

Totaforti, S. (2018). Applying the benefits of biophilic theory to hospital design. City, Territory and Architecture, 5, 1-9.

Ulrich, Roger. (2008). Biophilic theory and research for healthcare design. Biophilic Design: The Theory, Science, and Practice of Bringing Buildings to Life. 1, 87-106.

Zhao, Y., Zhan, Q., & Xu, T. (2022). Biophilic design as an important bridge for sustainable interaction between humans and the environment: Based on practice in Chinese healthcare space. Computational and Mathematical Methods in Medicine, 2022.

https://doi.org/10.52202/082078-0003

USER-CENTRIC RENOVATION DESIGN STRATEGIES OF TRADITIONAL OFFICES FOR SUPPORTING HYBRID WORK: ENGAGING EMPLOYEES THROUGH RESEARCH

Track 4: Design for Health and Wellbeing
Knowledge Network: Participation, Work Environments
Methods: Case Study, Mixed Methods
Suncica Milosevic (University of Utah); Ajla Aksamija (University of Utah)

ABSTRACT

This study investigated renovation strategies of office spaces to post-pandemic, hybrid working environments with an emphasis on user inclusion in the holistic design process – from pre-design inquiry, design synthesis and development, and final design validation. The goal was to determine renovation strategies that would adapt the existing office spaces to the new working modes and space needs with flexibility for future adaptability and growth, where the strategies would be based on the specific culture of the organization.

The study was conducted through a combination of research methods. Qualitative methods included archival research of the latest construction documents, empirical documentation using photographs of the current space conditions, observations of the current use of these office spaces, and thematic analysis of an anonymous general survey and focus group interviews with different teams and hierarchies within the organization. Quantitative methods included quantitative analysis of the survey and focus group interview results, computational modeling of 3D BIM models for the two case study office spaces, reiterative visualization of the proposed design strategies, and review and validation of egress and accessibility compliance. This resulted in several design options for each of the two case studies, with each being presented to the decision-makers of the organization and jointly evaluated against advantages, disadvantages, and budgetary considerations. This revision and validation process led to the development and approval of the final design outcomes.

The results of this study underscore the importance of a user-integrated design process, through which significant improvements can be achieved to provide well-functioning spaces and more comfortable and inclusive work environments. With a previously published study that elaborates on the full spectrum of research methods, this paper highlights the importance of user integration in the holistic design process and elaborates on the different design options that led to the final design.

Keywords

hybrid office design strategies, hybrid office renovation, user-inclusive workplace environments, post-pandemic office design, employee-engaged office design

1 INTRODUCTION

1.2 Background on hybrid work

While remote and hybrid work modes existed well before the onset of the COVID-19 pandemic, the pandemic has accelerated and escalated their prominence on a global scale and fully transformed space-use patterns and office cultures (Appel-Meulenbroek et al. 2022). Specifically, the rise of these work modes began in the early 2000s across many corporate offices due to the advancements in technology that allowed for increased work efficiencies and time management related to work travel and time outside of the office (Appel-Meulenbroek et al. 2022 and Torten et al. 2016). In addition, the numerous benefits of remote work, such as increased employee autonomy, flexibility with time management, and increased employee satisfaction in ever-increasing demands of work schedules, together with economic advantages to both employers and employees, had made these work modes critical to employee attraction and retention for many U.S. organizations (Geertgens 2022, Nediari et al. 2020, Nguyen 2018, Torten et al. 2016). Thus, for organizations that were already offering these work-mode flexibilities, the urgent shift to working fully remotely or to work from home (WFH) at the initial phase of the pandemic was much easier to accommodate, than more traditionally structured work organizations. However, the already evolved technologies, their further development and proliferation, and shifting work-culture trends were relatively quickly followed across all non-essential work sectors.

With a large share of employees worldwide working remotely or from home, predominantly industry-driven and publicly disseminated studies on the benefits and drawbacks of working remotely and on adaptive office strategies were gaining much attention. For example, a study by a global

https://doi.org/10.52202/082078-0004

architectural design firm, Perkins and Will (Eitland 2020), highlighted numerous benefits for both the organization and the employees for remote work. This study also emphasized that long-term separation from the office posed adverse effects on the mental health and well-being of employees, such as a reduced sense of connectedness and belonging, challenges in communication and collaboration, and a reduced sense of well-being and productivity associated with increased work hours, burnout, and difficulty of maintaining work-life balance. Their research results concluded with a projection for a more even distribution of time spent between remote work and office work in the post-pandemic future (Eitland 2020). Similar findings were concluded in a recent study by Babapour Chafi et al. (2022), where similar benefits and drawbacks of remote work were noted, and where an explicit notion of hybrid work mode was projected as an ideal balance between remote and office work modes. This balanced work mode, now more commonly referred to as hybrid work, has been increasingly implemented as an official or a more permanent work-mode model since it can balance the changing work culture and negative effects of remote work if the office spaces and integrated digital technologies are carefully designed and adapted to the diverse and evolving needs of the users or employees (Babapour Chafi et al. 2022 and Barriga et al. 2021).

Regarding office space design, the early 2000s also introduced the open office plan workspaces as opposed to the traditional and space-confining cubicles. Non-traditional types of spaces in the workplaces started to be introduced, such as wellness rooms, fitness rooms, breakrooms, and other "amenity" spaces to encourage socialization and workplace collaboration (Geertgens 2022, Nediari et al. 2020, and Nguyen 2018). However, with the onset of the COVID-19 pandemic, social distancing was an immediate challenge to resolve in these open-plan environments, and initial studies were focused on the pandemic-period challenges of social distancing and revisiting how users should occupy and circulate through office space to avoid risks of infection. Meanwhile, in the post-pandemic work culture shift of increased remote and hybrid work, while increased personal space and overall hygiene are prioritized to ensure the health and well-being of employees (Hogarty 2020a, Hogarty 2020b, SCB 2020), social distancing is not such a critical factor because many corporate offices are uncomfortably empty and uninviting. This leaves many open-ended questions such as how to maximize and optimize existing office space, how to integrate non-permanent, evolving trend functions and zones, and how to re-enliven the social atmosphere and office culture.

One such open-ended scenario was questioned by the leadership team of two hybrid work-mode, but largely vacant and uninviting office spaces, within an academic institution of the University of Utah, that presented the two case study offices for this research. The goal of this study was to investigate renovation considerations and design strategies for a post-pandemic, hybrid office environment, as the existing physical space was insufficient for future growth and was non-functional for this novel work mode structure. Importantly, the office spaces lacked the vibrant social atmosphere that was present before the COVID-19 pandemic, and the purpose was to investigate renovation design strategies that would support the new work culture and create an inviting, vibrant workspace.

2 LITERATURE REVIEW

A previously published complementary study by the authors of this article (Aksamija and Milosevic, 2023), elaborates on (a) the history of hybrid work and the effects of the COVID-19 pandemic on working environments, (b) the many challenges and opportunities of hybrid work, and (c) on the current hybrid office design trends and strategies. It highlights the existing research gaps in hybrid office transitions, where studies for adapting traditionally structured office spaces for hybrid work are extremely limited, and it stresses that user inclusion in the holistic design process is detrimental to the success of such adaptations where careful attention must be paid to the unique and individualized office cultures apart from referencing the current and evolving office design trends. This article, however, elaborates on the user inclusion process that was a core element of this study and highlights the multiple design options that were part of the iterative and user-inclusive decision-making design process.

2.2 Rising emphasis on the importance of user inclusion in the design process

A very recent literature review study was conducted by Hettithanthri, Hansen, and Munasinghe (2023) on peer-reviewed publications focusing on the architectural design processes. Their results raise an alarm about the persisting gap in user integration or participation in the design process, which is detrimental to the success of any design project, as the design outcomes must align with the actual needs of the users. This core value of designing for people, or users, is emphasized across academia, but – unfortunately, often lost in translation in conventional practice where time is key to completing projects; where designs rely on established experience by the architect or designer; and where the client, who may not also be the user(s), typically has a predetermined outcome or vision in mind. Moreover, the early design process, such as pre-schematic and schematic phases, is reduced to as few weeks as possible, often

https://doi.org/10.52202/082078-0004

eliminating the possibility of user integration into the design or decision-making. Some progressive firms are increasingly integrating community research and user input into their design process, whether through direct contact with the users, such as Perkins and Will and DLR Group who conduct internal community planning research and user/community group meetings for their public transit and education facilities (Perkins&Will 2022, DLR Group 2023), or through indirect contact with the users, such as architectural departments within large academic institutions like the University of Chicago Medical Campus, that rely on internally available academic research on user/community needs for their healthcare facilities (UCM 2022). However, these are a few examples and still atypical processes that seem to be integrated within large corporations that can afford the time and resources for this level of user-input integration and user participation in the overall design process.

On a broad scale, these persisting gaps have been gaining increasing attention and evolving into entire human-centric movements and even parallel or branching professions to architectural practice, bearing names such as community-engaged design, public interest design, inclusive design, democratic design, etc. (Ferduous and Bell 2021, Abendroth and Bell 2018, Zallio and Clarkson 2021), where user inclusion and space equity and accessibility are at the forefront of the process, rather than the physical object outcome or product. They call to the importance beyond typical accessibility and physical needs of space environments to understanding and "empathizing" with the users regarding how they behave, socialize, use, and access space. A specific article in the Ferduous and Bell (2021) book, titled The Empathetic Designer: Emotional Intelligence in the Design Studio by Orlowski (2021) argues that future professionals within the architectural discipline will need to expand their roles to engage in empathy and pursue direct engagement with their specific clients and/or communities for which they are designing. This call for previously ignored responsibilities is reflected in the literature above by Hettithanthri, Hansen, and Munasinghe (2023), which finds limitations to this very "empathy" in academic training, referring to it as a critical skill for real-world practice in place of the traditional linearity in design, which may or may not include users only at the mere beginning of the project conceptualization, rather than a holistic and reiterative design process.

When it comes to specifics of interior design, a novel publication was just released by Hicks (2023) on community-engaged interior design which outlines recommended frameworks that ensure user inclusion and participation through a holistic design process, from pre-schematic design to construction completion and occupancy reviews, that can be applied in both academia and practice. Hicks (2023) outlines the importance of numerous qualitative research methods. These include qualitative methods such as archival analysis of construction documents to assess existing design conditions, empirical analysis of current state photographs, sketches, and observations of ways in which the spaces are currently occupied and utilized, and community- or user-engaged surveys, interviews, and programming or adjacency exercises to plan arrangement of space functions according to user needs, which are more challenging for renovations with set existing spaces to work with. The author also emphasizes the importance and associated challenges of synthesizing, or "translating" all of the gathered research data into a "hierarchy" of design strategies that are physically and economically feasible, as well as the importance of presenting these results to the community or users in a manner which is catered to the spoken and visual language of the audience and which leaves room for improvement. While the purpose of our study did not proceed beyond design development into the design implementation and construction stages, many of its recommended steps for inclusive design development were implemented, including the verification of the final design for egress and occupancy code requirements and accessibility so that if the final design outcomes were to be implemented, the organization would be prepared for permitting with their local agencies.

Thus, while existing literature on user inclusion may also be limited and not widely implemented in architectural practice, these recent publications are demonstrating that there is an increased awareness of the importance of user inclusion in the holistic design process, whether it be an individual, a group of individuals, or an entire community, and that integration of these skills in the academic training of design professionals is one critical path to ensuring their translation in practice and real-world application.

2.3 Summary of current design trends and strategies for hybrid offices

Recent research suggests that future office spaces need to prioritize holistic themes of flexibility for both employees and physical space, as well as the health and well-being of users, the community, and the environment (Cheung 2021, Geertgens 2022, IWFM 2022, K2 Space 2021a and 2021b, Rousell 2021). Flexibility for employees may include comfortable and adjustable furniture, personal touches such as the use of art and color, and availability and easy access to resources and technology, flexible partitions and furniture arrangements for individualized and group work, and the integration of nontraditional office functions such as private call booths, collaboration hubs, social lounges, maker

https://doi.org/10.52202/082078-0004

spaces, fitness spaces, meditation spaces, dining spaces, outdoor spaces. (Cheung 2021, IWFM 2022, K2 Space 2021a and 2021b). To support the health and well-being of employees, some recommended strategies may include branding, color, and signage for a sense of culture and belonging, spaces that maximize access to natural light and outdoor views, spaces that foster face-to-face connections, spaces that accommodate introverted and extroverted work habits, private and public spaces for wellness and recharge, and inclusion of biophilic design elements both for décor and open space zoning. To support the health and well-being of the community and natural environment, office renovations should strive to meet sustainability certification standards such as LEED, BREEAM, and WELL, consider up-cycling existing furniture and implement materials with high recycled content, optimize interior design for natural light and ventilation, implement automated control systems, and include additional facilities that encourage active lifestyles and sustainable modes of commute, such as fitness areas, showers, locker rooms, and bicycle storage (Cheung 2021).

Some additional strategies for office renovations to accommodate hybrid work modes include accommodation and balance of both independent work and teamwork spaces, minimizing auditory and visual distractions with furniture types and placement, providing private conversation/individual concentration areas, offering various sizes of group meeting spaces for social interaction and collaboration, equipping spaces with technology for remote and in-office collaboration, implementing technologies showing schedules and occupancy, ensuring ample circulation and personal space for comfort and privacy, and maximizing access to outdoor views and natural daylight while integrating biophilic elements for well-being (Appel-Meulenbroek et al. 2022, Babapour Chafi et al. 2022, Cheung 2021, Fayard 2021, Geertgens 2022, Howell 2022, IWFM 2022, Knight 2022, Nediari et al. 2021, Rousell 2021, Smite et al. 2023, Steelcase 2021).

However, while the existing research indicates that hybrid mode transitions will continue to emphasize the flexibility of space and integration of strategies that will support and positively contribute to employees' sense of well-being (Geertgens 2022, IWFM 2022, K2 Space 2021a and 2021b, and Rousell 2021), it also emphasizes that no one-size-fits-all method exists for transforming traditional offices into hybrid work environments and that the individual office cultures must be considered to ensure comfortable and inclusive adaptations for all members (Appel-Meulenbroek et al. 2022, Fayard et al. 2021, Capuano 2022, Nakai et al. 2022).

For these reasons, simply relying on current office design trends and recommendations poses risks of a negative impact on office culture. The emphasis on user input and user participation in renovation strategies for hybrid work mode office spaces is critical in this context, as any implemented adaptations or changes directly influence the culture and social cohesion of an organization.

3 RESEARCH OBJECTIVES

Given the few existing studies for adapting traditionally structured office spaces for hybrid work, this study focused on the following primary objectives:

1. To investigate existing conditions and users' perception of their office spaces to the investigated offices.
2. To determine current and projected future work patterns and physical space needs for the investigated case study office spaces.
3. To determine employee (or user) perceptions of work culture and physical space needs for a hybrid working environment.
4. To provide space planning strategies and recommendations for hybrid office space design that meets current and projected future needs, and that aligns with the specific office culture.

At the same time, the overarching goal of this research was to emphasize the importance of user input and user inclusion in the design process. This was conducted by evaluating the physical conditions of the existing office spaces, investigating the employees' working patterns and office culture, synthesizing those results against current industry trends and recommendations, presenting several design options which were further refined through user input, and ultimately developing final design renovation strategies that would meet both the current and the projected future needs while supporting a hybrid work-mode structure, and that would align with the organization's work patterns and culture.

4 METHODS

The study implemented both qualitative and quantitative research methods for architectural research (Aksamija 2021), which align with the recently published methods by Hicks (2023) on community-engaged interior design, as described in Section 2. Qualitative methods included archival and empirical research of the existing conditions, as well as user input through a general online survey and seven focus group interviews. The survey included 71 participants and consisted of three different categories of questions: (1) 24 questions regarding working patterns and characteristics, (2) 23 questions regarding the workplace environment, and (3) 4

https://doi.org/10.52202/082078-0004

questions regarding participants' demographics. The seven focus groups included 49 participants and were organized to interview teams that work across all three work modes (fully remote, hybrid, and fully in-office) and different office structure hierarchies among general staff, management, and leadership. The quantitative methods included quantitative analysis of survey and interview results, computational modeling and visualization of the existing and proposed design strategies, and review and validation of the final design's egress and accessibility compliance.

The key goals of the focus group interviews were to identify both challenges and opportunities with the current arrangement of interior space programming and adjacencies, including specifics of moveable and fixed furniture systems, to understand how users of these specific organization structures conduct their work, socialize, and interact with each other and outside clients. The aforementioned complementary study by the authors (Aksamija and Milosevic 2023) illustrates the general survey and focus group interview structure, the specifics of survey and interview questions, as well as the detailed summaries of those results. This article will refer to those results and focus on the overarching reiterative process of several design options that illustrate how the ultimate, final designs were derived at each of the two case study offices.

5 RESULTS AND DISCUSSION

5.1 Case studies overview

While part of the same department of an academic institution, the two case study offices are located on different parts of the university campus and are hosted in different typologies of buildings, with Office A located in an administrative and academic instruction building and Office B located at the university's main library building. This administrative unit oversees online and digital learning for the university, including all online academic programs. At the time of the study, this unit employed approximately 120 employees and was projected to double its workforce shortly. It had already implemented a formal hybrid work structure, where the majority of employees work either fully remote or hybrid, and a small percentage fully in-office.

Figure 1 depicts the existing conditions for the two case-study office spaces, Office A and Office B, highlighting some key limitations in their office space layouts regarding space circulation, space adjacencies and distribution, and quality and access to natural light and outdoor views. In both offices, individual workstations are located near high-traffic areas and exposed to significant noise and privacy issues. However, unlike Office B, Office A lacks a formal reception area, which further increases foot traffic near individual workstations and the break room. Additionally, at Office A, the proximity of general staff workstations to the fully-glazed, private management offices affects personal comfort and privacy. Figure 1 also highlights that the distribution of individual and teamwork spaces is unbalanced, as both offices mainly consist of individual workstations or private offices with limited team spaces. This is especially critical for Office B, which does not have any small or large group areas beyond two informal areas of the office with a single table and two chairs, one at the main reception and one by the north window wall area. Such a setup that hinders opportunities for collaborative work may not be suitable for hybrid work models, as there is a lack of spaces for individual, private meetings as well as a lack of team spaces. Figure 1 further emphasizes differences in daylight access and access to outdoor views. In Office A, the layout limits the general staff's access to natural light due to the placement of executive offices along the perimeter of the office plan, while some workstations also suffer from glare issues due to their specific orientation relative to the windows. Meanwhile, Office B faces challenges with overall limitations of daylight access due to its north-facing windows and the unusual, triangular floor plan, where only a few of the private offices and workstations enjoy outdoor views and direct exposure to daylight.

5.2 Summary of the general survey and focus group interviews

These observations were supported by the outcomes of the general survey and focus group interviews, where the predominant issues were the quality of artificial light and access to views and natural light, lack of privacy, and exposure to noise and distractions, in addition to the expressed need for more comfortable, ergonomic furniture. Regarding space distribution, the specific types of spaces that are currently unavailable, but were requested include private call booths or private face-to-face meeting spaces, individual, high-focus workstation zones, small team collaboration zones, a client greeting area or casual socialization and gathering space, a designated eating area or pantry (specific to Office B only), and a quiet, resting area. These results gave insight that reconfiguration of interior layouts is necessary to reduce disruptive circulation patterns, to offer a more balanced proportion of individual and team spaces, and to propose spatial layouts that ensure more equitable access to daylight and outdoor views, in addition to ensuring that furniture and computer screen orientation reduces and prevents glare, and provides increased sense of privacy from predominant adjacencies and circulation patterns. More detailed breakdowns, including figures and tables, of the qualitative and quantitative general survey results are available

https://doi.org/10.52202/082078-0004

in the complementary study by the authors (Aksamija and Milosevic 2023).

These general survey results set a precedent for the focus group interviews, so that a more in-depth insight may be gained on the previously expressed issues at the two respective office spaces. Table 1 summarizes the resulting general patterns of the focus group interviews common to both case study offices, and is followed by a list of additional, case study-specific patterns that needed to be taken into account for the adaptive interior designs. The third column indicates which specific focus group interviews of the total seven exhibited these patterns, and the fourth column indicates the number of participants that expressed these perspectives as well as the total percentage of participants, which gave insight for hierarchical prioritization of patterns that need to be addressed with the proposed design options.

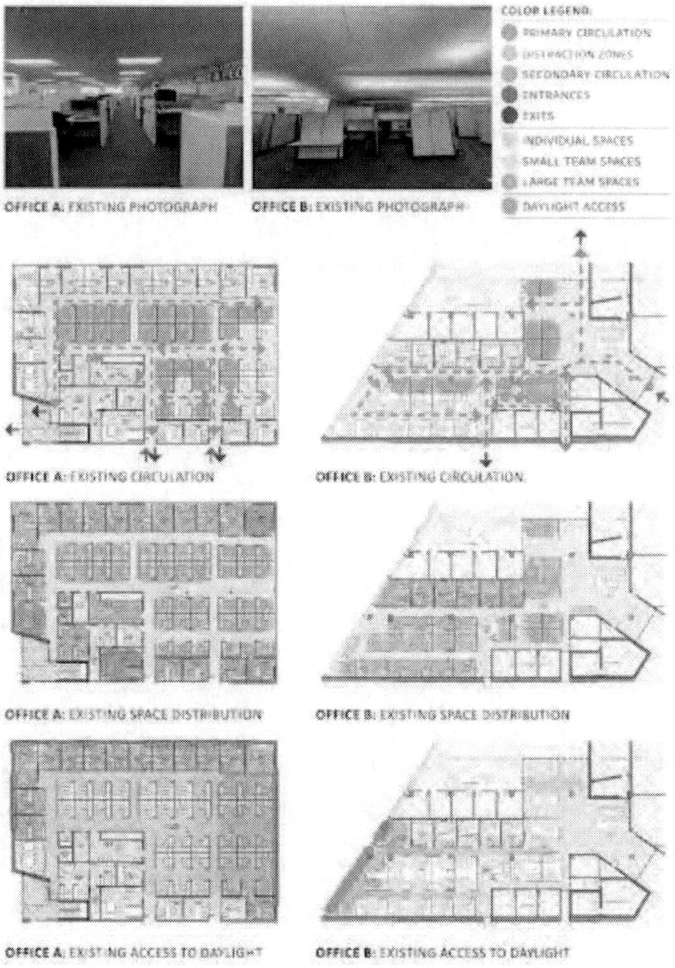

Figure 1. Existing photographs and conditions of the case study offices, Office A and Office B, illustrating current circulation patterns, space adjacencies and distribution, and accessibility to daylight and outdoor views. Photos and diagrams by the authors.

https://doi.org/10.52202/082078-0004

#	PATTERN	DESCRIPTION	FOCUS GROUP #	(N) AND %N; (N=49)
1	True hybrid work-mode	• Desire to work from the office 1-3 days a week for a fully remote group to alleviate a sense of isolation and feel integrated within the work organization. • Similarly, the desire to work from home 1-3 days a week for the fully in-office group to also benefit from the occasional flexibility and comfort of working from home.	1 - remote 3 - in-office 4 - mixed-mode	n = 22 %N = 45%
2	Shared workstations and private offices	• Openness for shared individual workstations on an alternating or scheduled pattern among general staff. • Similarly, there is openness for shared private, executive offices on alternating schedules among leadership staff, and for multifunctionality of those more private and daylit rooms. • Suggested display of schedules.	1 - remote 2 - hybrid 4 - mixed-mode 5 - hybrid 6 - hybrid 7 - mixed mode	n = 44 %N = 90%
3	Removal of "cubicle farms" and area zoning by teams with color and "home-like" décor	• Generally negative attitude toward cubicles, or large areas of open desk stations out of concerns for privacy and noise distractions. • Preference for designated permanent team zones to alleviate a sense of non-belonging and temporary presence at the office. • Ideally, team zones will be cognizant of inter-team dynamics and follow practical adjacencies. • Also, team zones should allow for more personal and welcoming touches of color and décor to express team identities.	1 - remote 4 - mixed-mode 7 - mixed mode	n = 20 %N = 41%
4	Private call "pods", team "touch-base" spaces, and "flex" spaces.	• Need for private call spaces, and spaces for quick, in-person, face-to-face conversations with clients or colleagues that are not disruptive to the cohort due to large volumes of virtual calls and visits. • Similarly, there is a need for small and large team meeting spaces, that are not necessarily formal like a conference room, but flexible to serve as team "touch-base" or "huddle" spaces, mixed-mode meeting spaces, and spaces for social gathering.	1 - remote 2 - hybrid 6 - hybrid 7 - mixed mode	n = 31 %N = 63%

https://doi.org/10.52202/082078-0004

#	PATTERN	DESCRIPTION	FOCUS GROUP #	(N) AND %N; (N=49)
5	Daylighting + views	• Need for reoriented workstations to reduce glare. • Similarly, there is a need for a redistribution of workstations to optimize access and equity of daylight and views to the outdoors. • Need for improved distribution, quality, and color of artificial lighting, as some workstations have very little access to light and others are burdened by too much, cool-tone light. • Suggested desk lamps and shading devices or screens.	1 - remote 2 - hybrid 4 - mixed-mode 5 - hybrid 6 - hybrid	n = 41 %N = 84%
6	Privacy and reduction of noise	• Need for reoriented workstations to accommodate more visual privacy for monitors. • Need for reduction of foot traffic and noise from colleagues who either have visitors or are on virtual calls. • Sometimes there are five or more persons on the same virtual call while seated in their individual, adjacent cubicles, which can be very disruptive.	1 - remote 2 - hybrid 4 - mixed-mode 5 - hybrid 6 - hybrid	n = 41 %N = 84%
7	Shared equipment and supplies storage, and reduction of personal storage	• Need for designated equipment and supplies storage to alleviate visual clutter and improve the organization of the office. • Need for shared team storage. • Additionally, there is a need for designated printing, copying, and faxing equipment, and kitchen appliances to reduce the burden on power from individually owned equipment. • Expressed openness to reduce general personal storage, much of which is unused. • In return, suggested personal lockers, mailboxes, and personal, lockable, and moveable storage to accommodate a hybrid, flexible schedule and reduce the need to carry all personal items home.	1 - remote 2 - hybrid 4 - mixed-mode 5 - hybrid 6 - hybrid	n = 41 %N = 84%

https://doi.org/10.52202/082078-0004

#	PATTERN	DESCRIPTION	FOCUS GROUP #	(N) AND %N; (N=49)
8	Improved accessibility and wayfinding, and more comfortable furniture	• Need for accessible, automated entries at both offices with improved wayfinding signage for visiting clients. • More open space for circulation and between individual workstations to accommodate social distancing. • Need for equitable distribution of standing desks and a need for seating space for visitors and clients. • Similarly, there is a need for the reduction of tall cubicles that limit visibility and wayfinding for those in a wheelchair or of short stature.	1 - remote 2 - hybrid 5 - hybrid 6 - hybrid 7 - mixed mode	n = 36 %N = 73%
* Total number of participants = 49 * Focus groups 2, 5, 6, and 7 already work hybrid work-mode schedules and were thus not included in pattern #1.				

The additional and case study-specific requests for Office A included a formal reception space and an open space for large staff meetings. For Office B they included a formal break room or pantry area with kitchen appliances (ideally a full kitchen with plumbing fixtures), and an informal, flexible lounge or meeting space. These results reiterated the areas of concern that resulted from archival and empirical analysis, as well as the general survey. In total, three interior design options were developed for Office A and five for Office B, each of which was presented to the leadership and management teams of the organization for discussion and review.

Figures 2 and 3 illustrate three design options each, highlighting the ideal design option (which integrated the majority of user input and recommendations and considered the addition of plumbing fixtures and partition adjustments), followed by the intermediate and final design options (which did not add any plumbing fixtures or shift any interior partitions.

https://doi.org/10.52202/082078-0004

Figure 2. Design options for Office A. Diagrams by the authors.

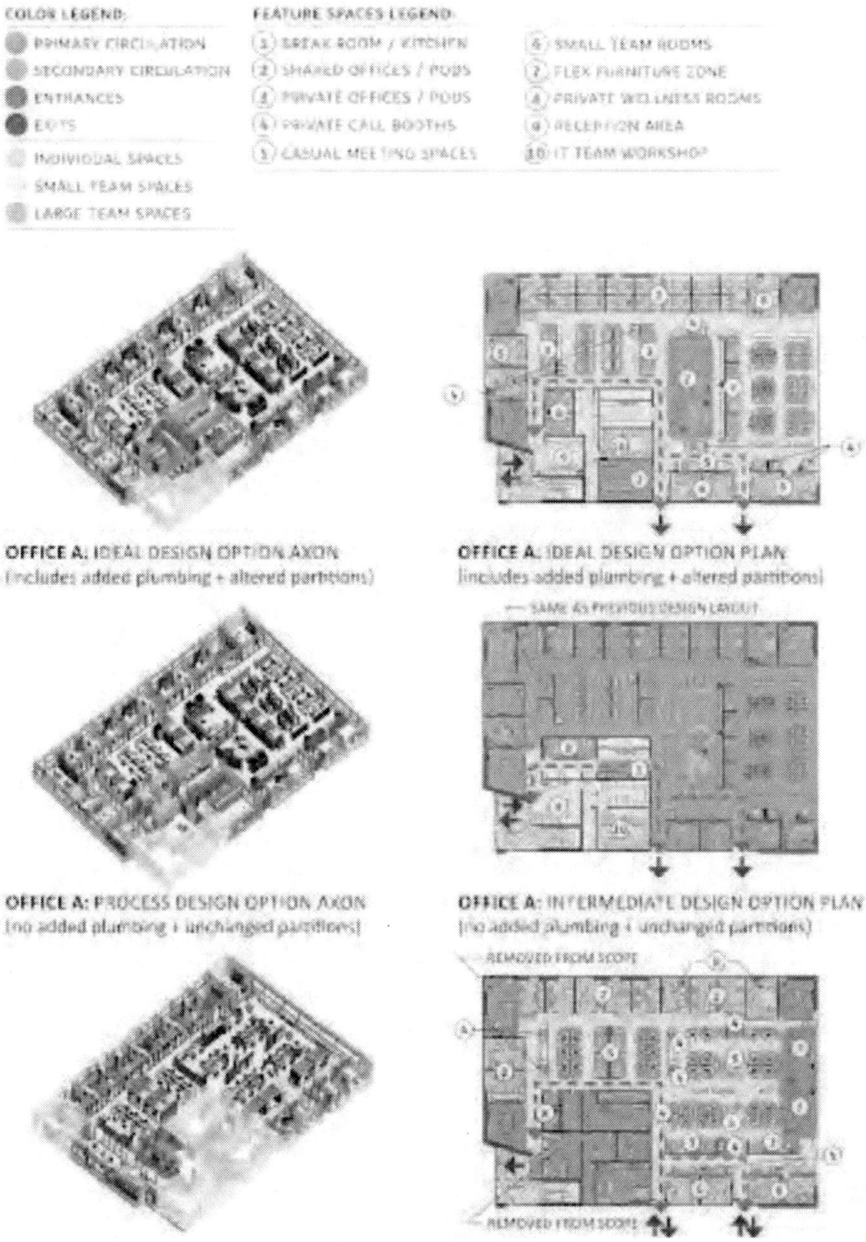

https://doi.org/10.52202/082078-0004

Figure 3. Design options for Office B. Diagrams by the authors.

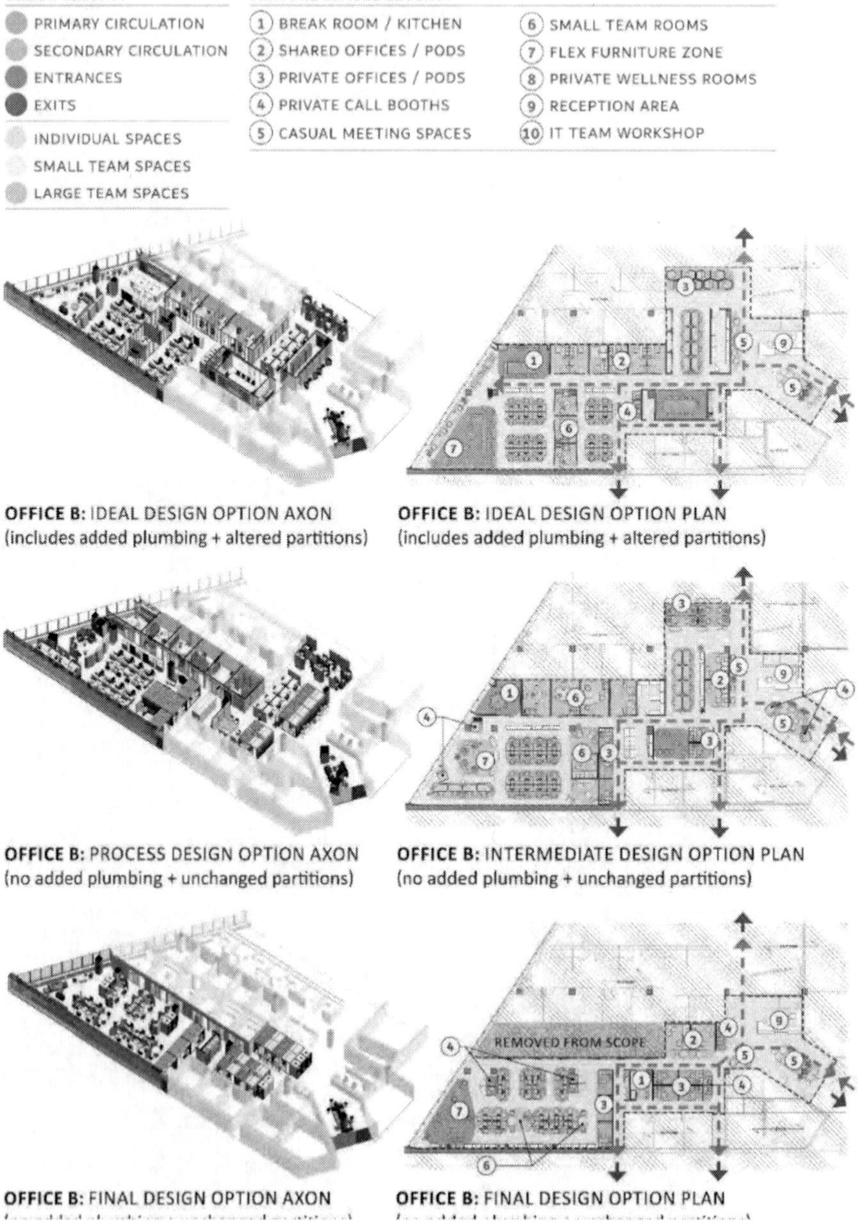

https://doi.org/10.52202/082078-0004

5.3 Discussion of the distinctive design options

While the study sought to accommodate every aspect of these focus group patterns, including the additional case study-specific items, some desirable strategies were beyond the scope of this study, such as addressing the quality of artificial lighting, quality and distribution of ambient temperature, and quantity of electrical outlets. However, most user input patterns were possible to address and integrate into the final design.

The general findings of the focus group interviews reiterated the similar areas of concern that resulted from archival and empirical analysis, as well as the general survey. However, they helped provide more depth and insight into how different teams work both individually, as a group, and with other teams of the organization – which was immensely helpful in understanding the specific needs for quantities and locations of individualized workstations and team areas. Additionally, they provided insight into the nuanced areas of concern, such as fears of uncertainty regarding having a designated area for work in the office, how sharing of some larger offices would be scheduled, whether one would be able to keep some personal belongings at the office, etc. There was also expressed fear of losing the sense of belonging and the tight-knit work culture if "everything" is temporary and constantly changing. These examples showed that the majority of the expressed concerns revolved around the logistics of a hybrid work mode and a sense of certainty and belonging within a changing work culture. Beyond this level of insight that influenced the individual case study design options, the focus group interviews helped the research team gain trust with the collective organization and provided them a voice in decision-making regarding their working environment. The study made sure to include even the fully remote teams in these interviews for these very purposes of inclusion, even if they may not choose to or need to visit and/or use the office spaces. Something interesting in our results was the insight that even the fully-remote teams of this organization would like to have an option of working from the office sometimes, which reiterated the importance of this process to ensure that they are welcomed and accommodated in all proposed design options.

Looking at the results of Table 2, **Pattern 2 (Shared workstations and private offices)** had the highest discussion rate of **90%**, which was not expected but reflected an overwhelmingly positive attitude toward sharing of individual workstations, including private executive offices, for more equitable access to office resources. Scheduled sharing of the private, executive offices was recognized for its opportunities of accessibility to spaces that offer more privacy for high-focus tasks or sensitive information conversations and meetings, as well as for their accessibility to more daylight and outdoor views. The second highest discussion rate was reflected in **Patterns 5, 6, and 7 (Daylighting and views, Privacy and reduction of noise, Shared equipment and supplies storage, and reduction of personal storage)** at **84%**, which emphasizes the importance of addressing access to daylight and views, privacy and noise disruptions, and redistribution of storage space. The third highest discussion rate was reflected in **Pattern 8 (Improved accessibility and wayfinding, and more comfortable furniture)** at **73%**, which emphasized the importance of ensuring that all types of spaces at each case study office offered an accessible space of equal quality as well as ensuring that circulation to all office spaces was accessible. Additionally, this informed that it was important to specify comfortable furniture and ensure that furniture sizes and placement do not obstruct views for those who may be of shorter stature or in a wheelchair. This concern was specifically brought up in one of the focus group meetings, where it was expressed that the current, high-wall, and narrowly-spaced cubicles make circulation through these offices uncomfortable and disorienting. While these percentages helped develop an overall hierarchy, authors deemed each theme as equally important and sought to implement each of the eight themes, or patterns, in all iterations of the proposed design options.

Upon presenting several design options, as illustrated in Figures 2 and 3, due to budgetary reasons, the leadership team of the organization expressed a strong preference for further optimization of the design options where no plumbing and no partitions were added or altered. Additionally, some limitations were introduced, such as maintaining the original private, executive offices – specifically, the corner executive office at Office A and the full row of private offices at Office B (as hatched in gray and noted as "*removed from scope*"

https://doi.org/10.52202/082078-0004

in the final design option diagrams of Figures 2 and 3). This eliminated an opportunity to make those oversized rooms into shared, team spaces. Additional limitations included maintaining the core spaces at Office A per the existing configuration, which includes the already existing break room and kitchen, printing and supplies area, and workstations for the AV/IT team (also hatched in gray and noted as "removed from scope" in the final design option diagram of Figure 2). This request meant that a formal reception area, which originally used to be in this part of the floor plan, could not be re-introduced and that the circulation path could not be opened up for easier wayfinding and a proper entrance zone at this office. Similarly, regarding Office B, given the inability to introduce plumbing fixtures, such as a kitchen sink or a dishwasher, the participants' requests for a formal kitchen and eating area could not be met. Thus, some of the specific needs at each case study office could not be resolved, and those design options were eliminated.

However, the presentation of several design options to the collective organization allowed for transparency on these impactful decisions. Tables 2 and 3 provide itemized summaries of the requested revisions at the two respective offices.

https://doi.org/10.52202/082078-0004

#	REVISION	DESCRIPTION	LOCATION ON PLAN
1	Removed scope	• Remove the north row of spaces that include the added formal break room/kitchen and shared offices or small team spaces, and revert to the original function and layout of private, executive offices.	• Northwest row spaces that include space type #1 and spaces type #2 or #6.
2	Removed scope	• Remove the north cluster of individual workstations from the scope of work, as that area may need to be shared with another collective. • However, keep the shared, 2-person office pods in this area of the plan if possible, and some casual seating with private call booths near the reception.	• North spaces type #3 removed, and spaces type #2, #5, and #4 to be near reception.
3	Resize the flex furniture zone and provide several individual workstation zones	• Keep the southwest corner recommendation for a large, flexible furniture zone but resize it to smaller so that more clusters of individual workstations can be accommodated. • Further, segment the currently shown individual workstation zones and integrate open-plan breakaway areas instead of small team pods.	• Southwest corner space type #7 and adjacent clusters of workstations with team areas space type #6.
4	Flip orientation of individual office pods to face teams	• Recommendation for private office pods along the predominant foot-traffic zones was well received. • However, would prefer that the glazed portion of these private office pods be flipped to face the open-plan clusters of individual workstations.	Centrally located space type #3
5	Remove large team space but may integrate a dry pantry with some casual seating in the same zone	• Given the sheer number of privately owned electrical appliances in this office, such as personal coffee machines, mini refrigerators, printers, copiers, etc., and the inability to budget a full kitchen/break room, the proposed centrally located large team space may integrate a small dry pantry with a refrigerator and a coffee machine with counter top seating.	• Centrally located large team space to be space type #1.

5.4 Discussion of the final design

Using the specific feedback, as described in Tables 2 and 3, the final design options were developed and are shown in Figures 4 and 5. Significant improvements were made, and these final design options were once more reviewed and approved by the organization's leadership and management teams. As shown in these final figures, at both case-study locations, improved circulation, privacy, and reduction of noise distractions were achieved with the rearrangement of the furniture layouts in such ways as to create barriers against those distraction areas.

Referencing the Table 1 summary of user input themes or patterns, these examples include the following: (a) private, sound-isolated office pods **(Pattern 4)**, (b) marker board screens at open desk stations **(Pattern 6)**, (c) moveable, sound absorbing partitions **(Pattern 4)**, (d) personal storage locker rows **(Pattern 7)**, and (e) colorful finishes and plants **(Pattern 3)**.

https://doi.org/10.52202/082078-0004

Figure 4. Office A final design plan diagrams and interior views that illustrate original and new design strategies. Diagrams and illustrations by the authors.

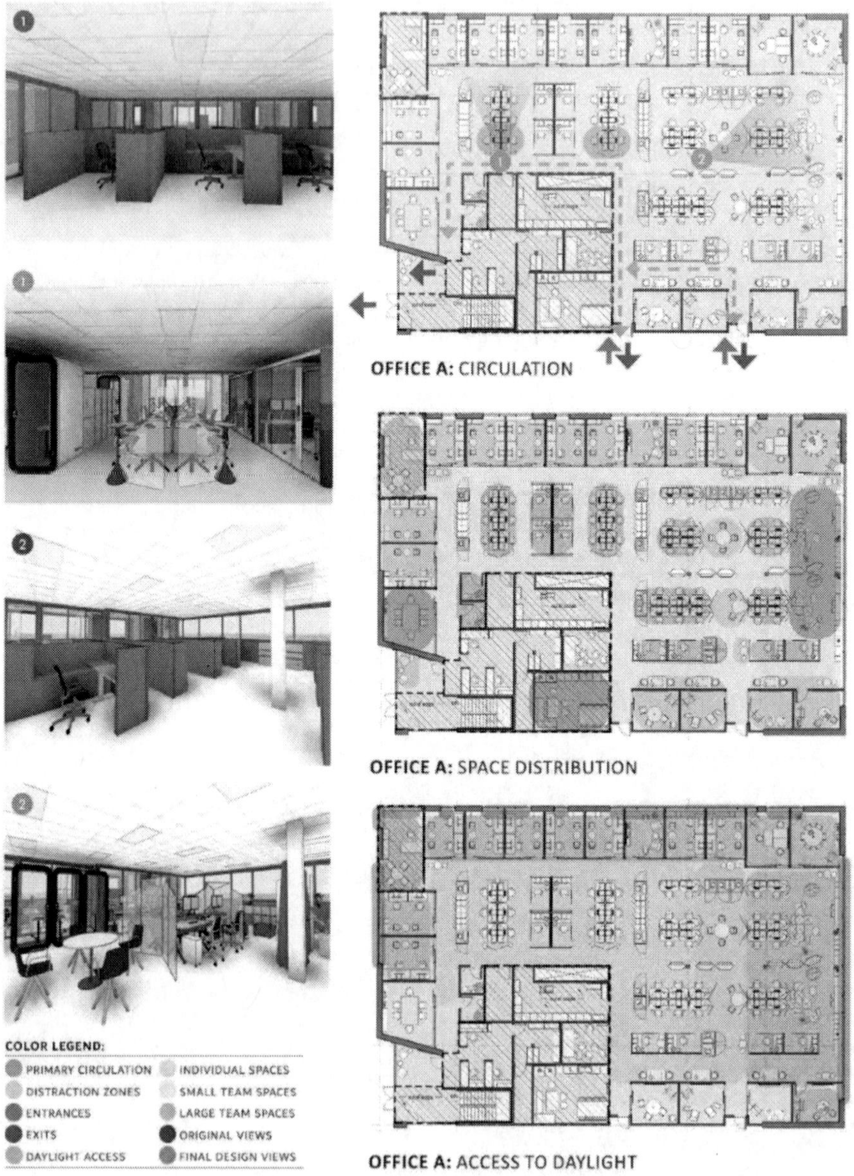

https://doi.org/10.52202/082078-0004

Figure 5. Office B final design plan diagrams and interior views that illustrate original and new design strategies. Diagrams and illustrations by the authors.

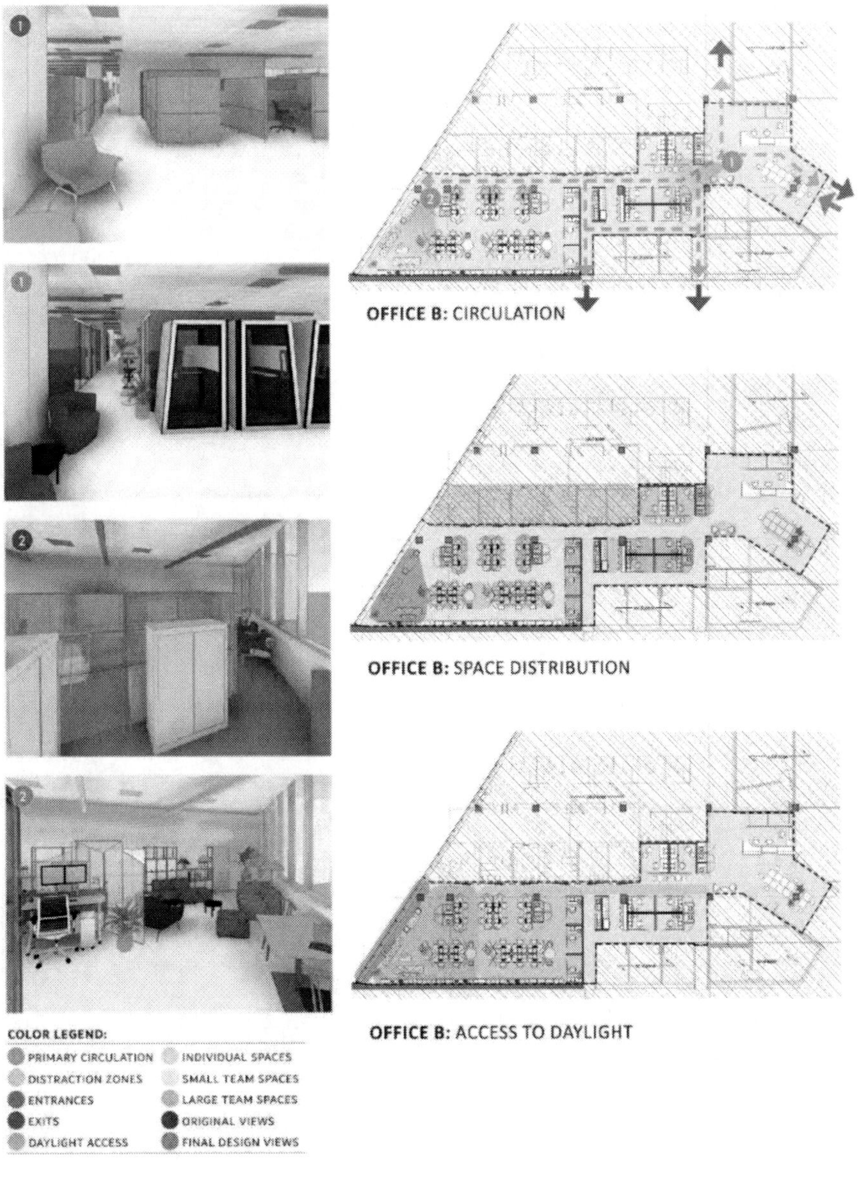

https://doi.org/10.52202/082078-0004

Workstations have been zoned by teams and clustered in groups, along with independent workstations that are not assigned to individuals but rather assigned to a specific team. This allows shared spaces for hybrid, scheduled work mode. Privacy concerns and circulation were also considered in the design, where movable screens can help adjust for privacy needs, reduce glare, and allow for reconfiguration. Private call pods have been included in the design, as well as team storage, lockers, and personal movable storage units.

Significant improvements were made to introduce more team spaces and areas compared to the original design while providing a higher quantity of individual workstations. Both offices gained flexible zones that can accommodate casual gatherings for large and small groups or serve as casual workspaces. These "flex" zones have been intentionally placed adjacent to windows to provide an area for all with access to direct daylight and outdoor views. It is important to note that these reorganizations of space at both locations were verified against the latest permit-approved egress code plans to ensure that occupancy calculations and egress path distances are not changed nor affected by the new proposed layouts or occupancy counts. Additionally, circulation spaces and furniture specifications that meet accessibility guidelines were implemented for all types of spaces to ensure a more welcoming and inclusive working environment.

6 CONCLUSIONS

The changing work practices, where remote and hybrid work modes are gaining popularity among various organizations, require different approaches for the design of office spaces. Traditional offices that solely rely on individual workstations may not be functional or desirable for these changing work modes. Offices are increasingly becoming places for collaboration, meetings, and occasional independent work, and as such must provide a balanced proportion of individual workstations, meeting and collaborative spaces, socialization, and spaces for supporting employees' well-being.

This study investigated renovation strategies for traditional office spaces, where two different offices were analyzed. These offices are utilized by an academic unit responsible for overseeing online programs and digital technologies for online learning. Through qualitative and quantitative research methods, the study investigated current conditions, users' perspectives and reflections, and renovation strategies for adapting these office spaces to hybrid work mode. Qualitative methods included archival and empirical research, where construction documents were analyzed, observations of the current use of these office spaces, general surveys, and focus groups with different teams within the organization were conducted. Quantitative methods included numeric analysis of the survey results and focus groups, computational modeling, and reiterative visualization of the proposed design strategies. Several design options were proposed, and each option was presented to the teams and decision-makers, and evaluated against advantages, disadvantages, and budgetary considerations. This led to the final design outcome, which would provide shared and independent workstations zoned by teams, shared collaborative spaces, private and sound-isolated office pods, personal storage, and moveable, sound-absorbing partitions. The final design improved circulation, privacy, and reduction of noise, and created a welcoming environment. The study shows that a user-integrated design process, where employees are included in the design decision-making process, is vital for providing well-functioning spaces and more comfortable and inclusive work environments.

7 REFERENCES

Abendroth, L. M., & Bell, B. (2018). Public interest design practice guidebook: SEED methodology, case studies, and critical issues. Routledge.

Aksamija, A. (2021). Research methods for the architectural profession. Routledge.

Aksamija, A., and Milosevic, S. (2023). Post-pandemic office spaces: Considerations and design strategies for hybrid work environments. Inquiry – The ARCC Journal for Architectural Research (ENQ), 20 (1), 41-64. https://doi.org/10.17831/enqarcc.v20i1.1192

Appel-Meulenbroek, R., Kemperman, A., Van De Water, A., Weijs-Perrée, M., & Verhaegh, J. (2022). How to attract employees back to the office? A stated choice study on hybrid working preferences. Journal of Environmental Psychology, 81. https://doi.org/10.1016/j.jenvp.2022.101784

Babapour Chafi, M., Hultberg, A., & Bozic Yams, N. (2022). Post-pandemic office work: Perceived challenges and opportunities for a sustainable work environment. Sustainability, 14(1), 294. https://doi.org/10.3390/su14010294

Barriga M., Raúl, H., Campoverde Aguirre, R., Coello-Montecel, D., Ochoa Pacheco, P., & Paredes-Aguirre, M. I. (2021). The influence of work-family conflict on burnout during the COVID-19 pandemic: The effect of teleworking overload. International Journal of Environmental Research and Public Health, 18(19), 10302.

https://doi.org/10.52202/082078-0004

Capuano, A. (2022). Post-pandemic workplace design and the plight of employees with invisible disabilities: Is Australian labour law and anti-discrimination legislation equipped to address new and emerging workplace inequalities?. University of New South Wales Law Journal 45(2), 873-913.

Cheung, F. (2021). Designing holistic well-being in the age of hybrid working. Architecture/Design CTBUH (Council on Tall Buildings and Urban Habitat) Journal, 2021(4), 36-43.

DLR Group. (2023). Austin Independent School District Long-range plan: Long-range planning rooted in equity. DLR Group. https://www.dlrgroup.com/work/austin-independent-school-district-long-range-plan/. Accessed: December 2, 2023.

Eitland, E. (2020). At home: What happened when we shifted to WFH (and what we can learn). Perkins and Will COVID Insights and Perspectives, Perkins and Will. https://perkinswill.com/news/at-home-what-happened-when-we-shifted-to-wfh-and-what-we-can-learn/. Accessed: February 5, 2023.

Geertgens, E. (2022). The future of [work] space – what comes next?. Facility Management Journal (FMJ), 32(1), 22-28. http://fmj.ifma.org/publication/?m=30261&i=733359&p=22&ver=html5

Fayard, A. L., Weeks, J., & Khan, M. (2021). Designing the hybrid office: From workplace to "culture space". Harvard Business Review 99(2), 114. https://hbr.org/2021/03/designing-the-hybrid-office

Ferdous, F., & Bell, B. (2021). All-inclusive engagement in architecture: Towards the future of social change. Routledge.

Hettithanthri, U., Hansen, P., & Munasinghe, H. (2023). Exploring the architectural design process assisted in conventional design studio: a systematic literature review. International Journal of Technology and Design Education, 33(5), 1835–1859. https://doi.org/10.1007/s10798-022-09792-9

Hicks. (2023). Community-engaged interior design: An illustrated guide. Routledge. https://doi-org.ezproxy.lib.utah.edu/10.4324/9781003366102

Hogarty, S. (2020a). Tips from experts for social distancing at work. WeWork Ideas. https://www.wework.com/ideas/professional-development/business-solutions/tips-from-experts-for-social-distancing-at-work. Accessed Oct 12, 2023.

Hogarty, S. (2020b). How to practice social distancing at work. WeWork Ideas. https://www.wework.com/ideas/professional-development/business-solutions/how-to-practice-social-distancing-at-work. Accessed Oct 12, 2023.

Howell, T. (2022). Coworking spaces offer a post-pandemic office alternative. MIT Sloan Management Review, 63(2), 1-3.

IWFM. (2022). Roundtable report – The future of office design. Institute of Workplace and Facilities Management (IWFM) in partnership with MillerKnoll. https://www.iwfm.org.uk/resource/the-future-of-office-design.html. Accessed: February 5, 2023.

Knight, C., Olaru, D., Lee, J. A., & Parker, S. K. (2022). The loneliness of the hybrid worker, MIT Sloan Management Review, 63(4), 10-12.

K2 Space (a). (2021). Hybrid office design. K2 Space. https://k2space.co.uk/knowledge/hybrid-office-design/#1. Accessed February 24, 2022.

K2 Space (b). (2021). Office design trends 2023. K2 Space. https://k2space.co.uk/knowledge/office-design-trends-2023/. Accessed February 24, 2022.

Nakai, T., Tsuji, T., Tsuda, H., Sotodate, T., Namba, Y., Uenishi, T., Iwasaki, K., Kokubo, K., & Tomita, H. (2022). Working conditions, work productivity, quality of life, and depressive symptoms in undiagnosed adults with and without attention-deficit/hyperactivity disorder (ADHD) symptoms during the COVID-19 pandemic. Neuropsychiatric Disease and Treatment 18, 1561-1572.

Nediari, A., Roesli, C. & Simanjuntak, P.M. (2021). Preparing post COVID-19 pandemic office design as the new concept of sustainability design. IOP Conference Series: Earth and Environmental Science, 729(1), 12095. https://iopscience.iop.org/article/10.1088/1755-1315/729/1/012095

Nguyen, T. (2018). Spatial and motivational programming in office environments: Comparison of co-working to other types of commercial spaces. Perkins & Will Research Journal (PWRJ), 10(1), 54-84.

Orlowski, E. M. (2021). The emphathetic designer: Emotional intelligence in the design studio. In F. Ferdous, & B. Bell (Eds.), All-inclusive engagement in architecture: Towards the future of social change (pp. 38-52). Routledge.

Perkins&Will. (2022). Three cities, three ways urban design can create social equity. Perkins and Will. https://perkinswill.com/insights/three-cities-three-ways-urban-design-can-create-social-equity/. Accessed: December 2, 2023.

https://doi.org/10.52202/082078-0004

Rousell, F. (2021). How to design a hybrid workspace – evolving the modern office, Jolie Journal by Jolie Studio Research. https://www.joliestudio.co.uk/journal/how-to-design-a-hybrid-workspace. Accessed February 5, 2023.

SCB (2020). The post COVID-19 workplace. Solomon Cordwell Buenz. https://www.scb.com/2020/05/the-post-covid-19-workplace/. Accessed Feb 20, 2023.

Smite, D., Moe, N. B., Hildrum, J., Gonzalez-Huerta, J., & Mendez, D. (2023). Work-from-home is here to stay: Call for flexibility in post-pandemic work policies. The Journal of Systems and Software, 195, 111552. https://doi.org/10.1016/j.jss.2022.111552

Steelcase. (2021). Collaboration in the hybrid workplace – A guide for designing spaces to support in-office and remote collaboration. Steelcase. https://www.steelcase.com/hybrid-collaboration-guide/. Accessed February 5, 2023.

Torten, R., Reaiche, C., & Caraballo, E.L. (2016). Teleworking in the new millennium. The Journal of Developing Areas, 50(5), 317-26. https://www.jstor.org/stable/26415595

UCM. (2022). At the forefront of health equity (2022 Community benefit report highlights). University of Chicago Medicine (UCM). https://community.uchicagomedicine.org/2022/assets/Community-Benefit-Report-Brochure-2022.pdf

Zallio, & Clarkson, P. J. (2021). Inclusion, diversity, equity and accessibility in the built environment: A study of architectural design practice. Building and Environment, 206, 108352. https://doi.org/10.1016/j.buildenv.2021.108352

https://doi.org/10.52202/082078-0004

RESEARCH ON CITIZEN'S EVALUATION OF URBAN PUBLIC SERVICE IN CURITIBA

Track 5: Spatial Justice, Equity, and Inclusion by Design
Knowledge Network: Communication, Residential Environments
Methods: Survey
Keiro Hattori (Ryukoku University)

ABSTRACT

Curitiba has been known for its innovative urban planning policy and has been praised for its success in urban planning worldwide. However, its innovativeness seems to peak in the early 1990s, when Jaime Lerner was the mayor for the third term. Curitiba is still perceived as the international brand in the city planning circle to this day, however, some factors were starting to undermine its planning and governance model in Curitiba. How do people in Curitiba perceive today's planning policy? Do they still appreciate it, or do they despair of its current planning policy? To answer these questions, a questionnaire survey was conducted among the citizens of Curitiba. The objective of this research is to clarify how people in Curitiba evaluate the current planning policy and to comprehend the level of its erosion if there are any. The questionnaire asked several questions to understand what the citizens of Curitiba appreciated and what they did not appreciate in terms of its urban policy. It also tried to grasp how people saw the recent change in urban conditions, whether positive or negative. The study also analyzed the differences among respondents' attributes. The research found that citizens were quite satisfied and recognized the improvement in public services regarding urban design, however, it also revealed that they were not satisfied and recognized the retrogression of the management of public services.

Keywords
Curitiba, Questionnaire Survey, Evaluation of Public Services

1 INTRODUCTION

1.1 Objective of the Research

Curitiba, the capital of the state of Parana and its largest city, has been known for its innovative urban planning policy and has been praised for its success in urban planning worldwide (Macedo, 2004, Santos, 2011, Marchiori, 2014). However, its innovativeness seems to peak in the early 1990s, when Jaime Lerner was the mayor for the third term. Curitiba is still perceived as the international brand in the city planning circle to this day, however, some skepticism has been articulated by several critics. Irazábal (2005) indicated that some factors were starting to undermine its planning and governance model in Curitiba. Mártinez pointed out the flaw of Cruritiba's sustainability and its reality did not live up to its reputation (Mártinez, 2016).

However, their assumptions were based on observation, participant observation, structured and semi-structured interviews, and media content analysis, but not from actual citizens of Curitiba, who are the main and most important clients for city planning. How do people in Curitiba perceive today's planning policy? Do they still appreciate it, or do they despair of its current planning policy? To answer this question, a questionnaire survey was conducted among the citizens of Curitiba. The objective of this research is to clarify how people in Curitiba evaluate the current planning policy and to comprehend the level of its erosion if there are any.

1.2 Methodology

The questionnaire survey was conducted in August of 2016. The questionnaire was distributed by two approaches. One approach was distributing them at several communal meetings. The questionnaire that was printed on paper was distributed and collected at the meetings. The other approach was through the Internet. The questionnaire was posted on the internet and the students in Catholic Panama University were asked to participate in the questionnaire with help from its professors.
The questionnaire asked several questions to understand what the citizens of Curitiba appreciated and what they did not appreciate in terms of the urban policy of Curitiba. It also tried to grasp how people saw the recent change in urban conditions, whether positive or negative.

2 RESULT OF QUESTIONNAIRE SURVEY

2.1 Profile of Respondents

Overall, there were 394 responses. 329 responses were by paper and 65 by the Internet. Of all the respondents, 234 were female and 157 were male. In terms of age structure, it was fairly evenly distributed. 47% of the respondents were born in Curitiba. 27% of the respondents were born out of Parana State.

2.2 Respondents' evaluation

Respondents were asked to grade nine public services into five categories: very good; good; average; bad; and very bad. Overall, people have a positive evaluation of Curitiba's urban policy. 49.9%

https://doi.org/10.52202/082078-0005

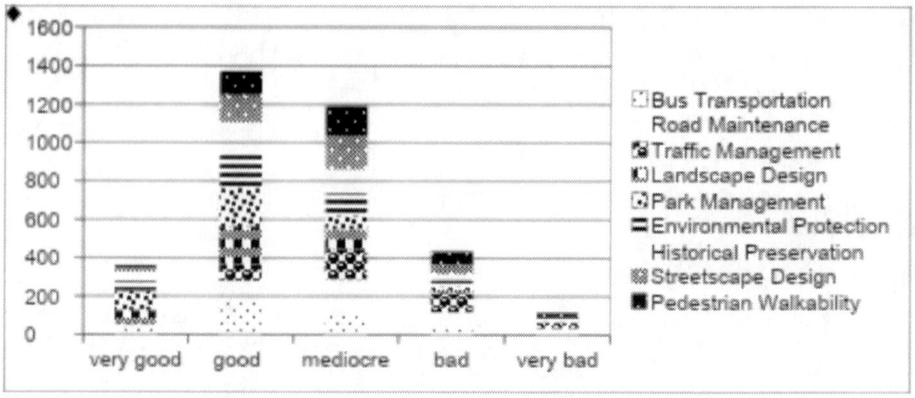

Figure 1. Overall evaluation of Curitiba's public service

Table 1 depicts how each public service is evaluated. "Park Management" is the highest evaluated public service with 76.5% of respondents evaluating it as "very good" or "good" and only 5.2% evaluating it as "bad" or "very bad." Landscape Design is the second most highly evaluated public service.

"Traffic Management" is the lowest evaluated public service with 33.2% of respondents evaluating it as "bad" or "very bad." "Road Maintenance" is the second lowly evaluated public service. Overall, respondents recognize that the city of Curitiba has been conducting fairly well in the field of environmental and landscape design, but poorly in the field of traffic issues including the walkability of pedestrians.

Table 1. Evaluation of each public service.

OVERALL EVALUATION	PUBLIC SERVICE	THE RATIO OF "VERY GOOD AND GOOD"	THE RATIO OF "VERY BAD AND BAD"
Highly evaluated	Park Management Landscape Design	76.5% 69.2%	5.15% 4.91%
Overall good evaluation	Environmental Protection Bus Service Historic Preservation	60.1% 59.5% 55.3%	10.3% 11.0% 11.9%
Relatively good evaluation	Streetscape Design	41.2%	13.6%
Split of opinions	Pedestrian Walkability Road Maintenance Traffic Management	35.5% 26.0% 24.5%	24.1% 28.9% 33.2%

of respondents evaluated the overall policy as "very good" or "good," compared to 15.8% of respondents who evaluated "bad" or "very bad." (Figure 1).

2.3 How they evaluate the change

The respondents were asked how they evaluated the change in the nine public services since they first came to Curitiba. Figure 2 shows respondents' evaluation of how public services have changed since they first came to Curitiba. The majority of respondents answered there has been no change or a little bit of improvement in terms of public services.

More than 50% of respondents replied that "park management" has been improving. The services that many people perceived as getting worse are "Traffic Management" and "Road Maintenance." In terms of "Bus Service," their opinion splits. 45% of respondents answered, "it has improved," whereas 30% answered, "it has receded."

Table 2 categorized the evaluation of public services to four categories. Out of nine services, respondents evaluated the change of three services quite positively (Park Management, Environmental protection, and Landscape Design), two services improved somewhat (Historic Preservation, Streetscape Design), and "traffic management" receded. Three services had a split of opinions (Bus Service, Pedestrian Walkability, and Road Maintenance).

https://doi.org/10.52202/082078-0005

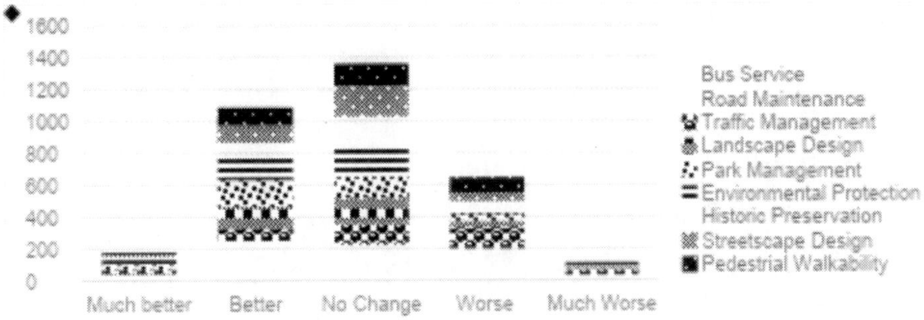

Figure 2. Overall evaluation of Curitiba's public service

Table 2. The evaluation of the change of public services into four categories

OVERALL EVALUATION	PUBLIC SERVICE	THE RATIO OF "VERY GOOD AND GOOD"	THE RATIO OF "VERY BAD AND BAD"
Improved significantly	Park Management Landscape Design Environmental Protection	52.2% 43.7% 42.2%	12.7% 13.1% 10.6%
Improved Somewhat	Historic Preservation Streetscape Design	33.2% 31.2%	14.6% 16.9%
Split of Opinions	Bus Service Pedestrian Walkability Road Maintenance	46.1% 32.9% 30.2%	31.0% 30.8% 32.4%
Deteriorated somewhat	Traffic Management	24.0%	40.5%

3 ANALYSIS

This chapter intends to find if there is any difference in respondent's attributes regarding their evaluation of the nine public services.

3.1 By Gender

Table 3 shows the P-value on the nine public service evaluation by gender. The topics that are evaluated differently by gender are "Streetscape Design" and "Traffic Management." The two public services are delved deeper into their differences.

For the evaluation of "Traffic Management," male respondents have both negative and positive opinions regarding this matter than female respondents. It can be assumed that male respondents are more opinionated than female respondents regarding traffic management. For the evaluation of "Streetscape Design" by gender, male respondents have a much more negative opinion regarding this matter than female respondents.

Table 3. P-value by gender

PUBLIC SERVICE	P-VALUE
Bus Transportation	0.157
Road Maintenance	0.712
Traffic Management	0.018
Landscape Design	0.123
Park Management	0.260
Environmental Protection	0.338
Historical Preservation	0.664
Streetscape Design	0.003
Pedestrian Walkability	0.838

Table 4. P-value by structure

PUBLIC SERVICE	P-VALUE
Bus Transportation	0.003
Road Maintenance	0.000
Traffic Management	0.002
Landscape Design	0.074
Park Management	0.010
Environmental Protection	0.230
Historical Preservation	0.077
Streetscape Design	0.312
Pedestrian Walkability	0.000

https://doi.org/10.52202/082078-0005

3.2 By Age Structure

Table 4 shows the P-value on the nine public services by age structure. There are differences between Age Structure for "Bus Transportation," "Road Maintenance," "Traffic Management," "Park Management," and "Pedestrian Walkability." Overall, public policies regarding management issues see the disparity among opinions by age structure. Urban design policies seem to have similar opinions among all ages. The following sections will dissect the difference in age structure for public services that have different opinions by age structure.

The survey results indicate that older generation (older than 60 years old) tend to evaluate "bus transportation" service higher than the younger generation. The most severe evaluation came from 40 to 59. The survey found that nearly 50% of respondents whose ages are 40 to 50 years old have answered that "road maintenance" is "bad" or "very bad", followed by the age group of "50 to 60 years old." 16% of respondents whose ages are 50 to 60 years old answered that "road maintenance" is "very bad." There are few public services that "70 or older" have a harsh opinion.

However, why such a disparity among age groups? There was also a significant statistical difference between these two factors comparing relation between age groups and car ownership. It indicates that certain age groups have higher car ownership than others. If you calculate the P-value of car ownership by each age group for the evaluation of "road maintenance", there seems to be not much of a difference except for ages between 20 to 40 (Table 5).

Figure 3 shows the evaluation of "road maintenance" by age group with car and without a car. As these figures indicate, respondents under age 50 tends to evaluate "road maintenance" generously when they do not own a car, but this tends to turn around when respondents are older than 50 years old.

Table 5. P-value of car ownership for the evaluation of "road maintenance by age group

AGE GROUP	P-VALUE
20 or younger	0.968
20-30	0.020
30-40	0.052
40-50	0.047
50-60	0.150
60 years or older	0.106

When respondents are older than 50 years old, respondents who do not own a car tend to have more severe opinions than those who own a car.

Regarding the evaluation of the "Traffic Management," "50 to 59 years old" has the highest percentage of "very bad" respondents. Overall, this generation and "40 to 49 years old" has 40% or more negative responses ("bad" or "very bad"). Only "70 or older" has some positive responses for "traffic management." Respondents who are older than 60 years old tend to have a positive view of traffic management.

Regarding the evaluation of the "Park Management," 39 years old and younger has a much more positive view than others. 40 to 59 years old has a relatively negative opinion.

Regarding the evaluation of the "pedestrian walkability," "40 to 49 years old" have a negative opinion. Nearly 20% of "50 to 59 years old" respondents have answered that they are "very bad." "70 or older," "60 to 69 years old," and "under 20" have a positive opinion.

Regarding the evaluation of the "Traffic Management," "50 to 59 years old" has the highest percentage of "very bad" respondents. Overall, this generation and "40 to 49 years old" has 40% or more negative responses ("bad" or "very bad"). Only "70 or older" has some positive responses for "traffic management." Respondents who are older than 60 years old tend to have a positive view of traffic management. Regarding the evaluation of the "Park Management," 39 years old and younger has a much more positive view than others. 40 to 59 years old has a relatively negative opinion.

Regarding the evaluation of the "pedestrian walkability," "40 to 49 years old" have a negative opinion. Nearly 20% of "50 to 59 years old" respondents have answered that they are "very bad." "70 or older," "60 to 69 years old," and "under 20" have a positive opinion.

3.3 By Respondent's Birthplace

Table 6 shows the P-value on the nine public services by respondent's birthplace. There are no differences between their opinion toward public services depending on Birthplace besides Historical Preservation.

Regarding the "historical preservation," respondents who were born in Curitiba are much more opinionated than respondents who were born outside of Curitiba.

https://doi.org/10.52202/082078-0005

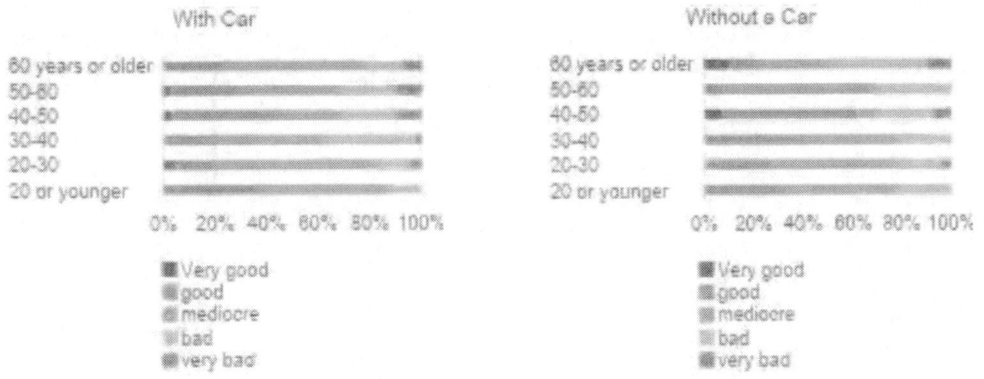

Figure 3. The evaluation of "road maintenance" by age group with car and without a car.

3.4 Where Respondents Live

Table 7 shows the P-value on the nine public services where respondents live. Besides "landscape design" and "environmental protection," the evaluation of public policies differs depending on where respondents live. This indicates that public service differs by area and there may be discrimination in terms of geography. If so, it should be modified.

In terms of the evaluation of "Bus Transportation" on where respondents live, respondents living in Cajuru district stand out in the sense that they have the most negative view regarding bus transportation service. Respondents living in Matriz, and CIC seem to be content with the service.

In terms of the evaluation of "Road Maintenance" where respondents live, respondents living in the southern suburban neighborhoods of Pinheirinho, Boqueirao, and Barrio Novo have a relatively positive opinion regarding "road maintenance" than the other parts of the city.

Regarding the evaluation of "Traffic Management" on where respondents live, a certain number of respondents living in Cajuru, and Santa Felicidade have a quite negative opinion regarding traffic management. 14% of Respondents living in both neighborhoods evaluated the city's traffic management as being "very bad." In contrast, respondents living in the suburban neighborhoods of Pinheirinho, Boqueirao, and Bairro Novo have a more positive view of traffic management than the rest of the neighborhood.

Regarding the evaluation of "Park Management" on where respondents live, three neighborhoods (Boa Vista, Santa Felicidade, and Matriz) have a relatively positive opinion. On the contrary, respondents living in CIC seem not to be impressed with the service as much as respondents living in the other area. Regarding the evaluation of "historical preservation," there is a difference of opinion depending on where respondents live. Respondents living in Santa Felicidade and Matriz, where a lot of historical buildings are located, have both positive and negative opinions. Respondents living in Boa Vista have an overall good evaluation.

Regarding the evaluation of "streetscape design," there is a difference of opinion depending on where people live. Respondents living in Boa Vista have a quite good evaluation, followed by Santa Felicidade and Matriz.

Table 6. P-value by respondent's birthplace

PUBLIC SERVICE	P-VALUE
Bus Transportation	0.485
Road Maintenance	0.088
Traffic Management	0.785
Landscape Design	0.277
Park Management	0.130
Environmental Protection	0.213
Historical Preservation	0.068
Streetscape Design	0.804
Pedestrian Walkability	0.677

Table 7. P-value by respondent's live

ADDRESS	P-VALUE
Bus Transportation	0.033
Road Maintenance	0.009
Traffic Management	0.006
Landscape Design	0.062
Park Management	0.000
Environmental Protection	0.108
Historical Preservation	0.002
Streetscape Design	0.003
Pedestrian Walkability	0.001

https://doi.org/10.52202/082078-0005

Regarding shows the evaluation of "pedestrian activity," there is a difference of opinion depending on where people live. Respondents living in Boa Vista have a quite good evaluation, followed by Santa Felicidade and Matriz. Respondents living in Cajuru and Portão have a relatively negative opinion in terms of pedestrian walkability.

3.5 Summary of the Analysis

Table 8 indicates the P-value of each nine public services by respondent's attributes. The strongest disparity of the evaluation is manifested in where respondents live. This indicates that there is a disparity among regions in providing public services, especially in park management.

There is also a disparity among age differences, especially in road maintenance and pedestrian walkability. Overall, 40 to 49-year-olds are more critical of public services than other age groups. There is only a small difference in gender and there is little difference in birthplace. From a public service standpoint, there is no difference in respondents' attribute in landscape design and environmental protection, but strong difference in road maintenance, park management, and pedestrian walkability.

4 CONCLUSION

Curitiba was a post child of good planning making good cities in the 1990s. Jaime Lerner, the mayor of Curitiba from 1989 to 1992 became a prominent figure in the urban planning circle which led him to become the president of the International Union of Architects in 2002.

However, it has been more than 20 years since Jaime Lerner has withdrawn from the political arena. A lot has changed regarding Curitiba's policies since then. The author has conducted a questionnaire survey of the residents of Curitiba to understand how the residents evaluate the city's nine urban policies since the departure of Jaime Lerner.

Overall, the public services have been evaluated highly. However, "traffic management," "road maintenance," and "pedestrian-walkability" have lower evaluation than "park management," "landscape design" and "environmental protection." There is an issue regarding the "management" of the public services. These management services were also perceived to be retrogressing rather than improving. "Traffic Management," "Road Maintenance," and "Pedestrian Walkability" should be improved to better the situation.

By analyzing the respondent's attributes, the research found that there is a strong disparity in the evaluation of "road maintenance," "traffic management," "park management," "streetscape design," and "pedestrian walkability." The strongest disparity of the evaluation is manifested in where respondents live. This indicates that there is a disparity among regions in providing public services and should be improved. The author hopes that the data provided in this paper may contribute to making the public policies of Curitiba better in some ways.

5 REFERENCES

Irazábal, C. (2005). City making and urban governance in the Americas. Routledge.

Macedo, J. (2004). Curitiba. Cities, 21(6), 537-549. https://doi.org/10.1016/j.cities.2004.08.008

Marchiori, R. (2014). Ex-capital ecológica, Curitiba precisa reinventar sua marca, Gazeta Do Povo (09/05/2014).

Santos, E. (2011). Curitiba, Brazil – pioneering in developing bus rapid transit and urban planning solutions. LAP Lambert Academic Publishing.

Table 7. P-value by respondent's live

PUBLIC SERVICE	GENDER	AGE	BIRTHPLACE	ADDRESS
Bus Transportation	0.157	0.003	0.485	0.033
Road Maintenance	0.712	0.000	0.088	0.009
Traffic Management	0.018	0.002	0.785	0.006
Landscape Design	0.123	0.074	0.277	0.062
Park Management	0.260	0.010	0.130	0.000
Environmental Protection	0.338	0.230	0.213	0.108
Historical Preservation	0.664	0.077	0.068	0.002
Streetscape Design	0.003	0.312	0.804	0.003
Pedestrian Walkability	0.838	0.000	0.677	0.001

https://doi.org/10.52202/082078-0005

NON-GOVERNMENTAL ACTORS' ENGAGEMENT IN MICRO-REGENERATION: CASE STUDIES FROM GUANGZHOU, CHINA

Track 6: Transdisciplinary Design and Collaboration/Open Track
Knowledge Network: Communication, Participation
Methods: Case Study
Bo Peng (University of Washington)

1 INTRODUCTION

In the past decade, urban development in China has witnessed a gradual shift from traditional urban expansion towards the optimization and reconstruction of existing urban spaces. Experience substantiates that large-scale demolition and reconstruction cannot efficaciously address the complex challenges posed by the scarcity of spatial resources, the intricate juxtaposition of public and private interests, the irreversible damage to existing urban fabric, and the threats to human well-being. The current impetus towards the restoration of ecological and urban environments has further accentuated the imperative to facilitate sustainable urban development through systematic, organic, and progressive urban regeneration strategies. This also extends to the overarching objective of improving the quality of the living environment, promoting social and economic development, and preserving existing urban culture and fabric, which has become one of the most significant concerns of the contemporary era.

Micro-regeneration is one of the urban regeneration approaches characterized by small-scale interventions of community public spaces, with low costs, within a relatively short time. (Ma & Ying, 2016) It enables the improvement of citizens' living environment through underutilized public spaces while preserving the integrity of existing urban fabrics and minimizing disruptions to people's everyday lives. Meanwhile, these micro-regeneration projects can function as a catalyst and exemplar, providing valuable insights for subsequent or other urban regeneration projects.

Nonetheless, the planning and implementation of micro-regeneration projects still encounter various challenges. Given the social and political background of China's urban development, inadequate civic engagement during the decision-making processes leads to a tendency for discrepancies or conflicts between governmental plans and citizens' opinions. More importantly, most urban regeneration projects often involve the redistribution of limited resources, which poses great challenges to coordinating a large diversity of demands. In this sense, the micro-regeneration of old neighborhoods is essentially a socio-spatial activity centering on humanistic needs, which requires civic engagement to coordinate various actors to ensure the effectiveness and justice of the projects.

Under these circumstances, this paper seeks to explore how non-governmental actors engage in the planning processes of micro-regeneration projects by discussing the identities of actively engaged non-governmental actors, the actions taken by these actors, and the resulting impacts. The two selected cases emphasize the critical need to have an inclusive communication and decision-making platform to incorporate diverse social forces into urban regeneration, thereby ensuring the human-centric orientation of the projects.

1.1 Civic Engagement in China's Urban Regeneration Progress

The rapid progress of urban urbanization and urban transformation in many Asian countries have resulted in the need to regenerate the old neighborhoods (Ye, 2011). The current issues of these sites include the degradation of existing building structures, lack of public spaces, facilities, and infrastructures, inadequate safety management, and the absence of community cultures (Cai et al., 2017). As China has prioritized enhancing sustainability and livability within its urban development strategies since 2010, micro-regeneration of old neighborhoods also has become part of China's urban development agenda (Li et al., 2014; Zhao et al., 2019). Different from traditional large-scale urban redevelopment projects, micro-regeneration of old neighborhoods is characterized by low cost and short project cycle, primarily targeting areas with poor living environment (Zhao et al., 2019).

Micro-regeneration of old neighborhoods emphasizes the engagement of diverse stakeholders and the preservation of existing urban fabric and community cultures, while introducing new structures and facilities to support and enhance the communities' living environment, economic and cultural vitality (Cai et al., 2017). Consequently, considering micro-regeneration of old neighborhoods as the governance of communities' everyday spaces necessitates more

https://doi.org/10.52202/082078-0006

inclusive engagement of actors (Wang, 2018). The sustainable development of these neighborhoods is realized through the coordination of diverse interests and needs from different stakeholders. While the government and relevant administrative departments use urban regeneration as a means to address urban issues and promote economic development, non-governmental actors, particularly citizens, play essential roles in influencing the decision-making of the government and the private sector through their preferences and behaviors (Bromley et al., 2005; Verdini, 2015). At the same time, a comprehensive institutional framework with relevant laws, regulations, and policies is essential to protect and coordinate the public interests and address the concerns of diverse actors (Rossi, 2004).

Effectively incorporating diverse actors into the decision-making processes of public affairs, particularly in the context of urban development, is a valuable research topic in China. Mills (1959) considered civic engagement in public affairs has the potential to improve the decision-making mechanism of the traditional bureaucracy system. On that note, Davidoff (1962) argued that professionals should not only listen to but also convey the opinions of relevant stakeholders, rather than making decisions directly on behalf of the public. This stance serves as a bedrock for civic engagement in the realm of urban planning and design. His concept of "Advocacy Planning" aims to encourage citizens to participate in urban development processes, as a way to ensure their rights of the city (Davidoff, 1965). While the rise of post-modernist thinking in the 1970s manifested in bottom-up resistance toward top-down policies (Healey, 2003), democratic pluralism views civic engagement as a form of direct democracy, which has been widely accepted in the Global North (Nelissen, 1982).

In addressing various public affairs and urban challenges, a single organization, or actor, is often insufficient, requiring collaboration and planning among multiple groups of actors (Watson, 2014). The decisions of how to manage and reshape spaces in declined neighborhoods through civic engagement can enhance the understanding, of both governmental and non-governmental stakeholders, of the environmental and social potentials of existing urban spaces (Kim et al., 2020). However, due to the constraints caused by the previous planned economy and conventional technocrat thinking in urban development, many urban regeneration projects remain top-down in nature, with power centralized in governmental actors (Li et al., 2014). Even though China's Urban and Planning Law addresses civic engagement in Article 26, it still lacks detailed instructions or necessary support for its implementation, potentially leading to citizens' symbolic participation (Arnstein, 1969).

1.2 The Involvement of Non-governmental Actors

Given that micro-regeneration is not only a practice of spatial planning but also involves complicated relationships of property rights, the coordination of diverse interests, and the governance of public affairs, the inclusive engagement of diverse actors is crucial for its implementation and effectiveness (Zhang & Li, 2016). As there is a tendency to combine the top-down approaches with the bottom-up initiatives in civic engagement in China's urban development (Kim et al., 2020), non-governmental actors have emerged as important forces in micro-regeneration projects.

There are various definitions of Non-governmental Organizations (NGOs), shaped by different cultures or socio-economic conditions. These definitions often characterized NGOs as non-governmental, non-profit, voluntary in nature, and dedicated to civic-minded issues [citation]. Nevertheless, discussing civic engagement in China requires a more inclusive consideration of the actors involved. Building on the idea of NGOs, this paper uses the term "non-governmental actors" to include other self-governing civic actors that operate independently from the government and market forces. Thereby, the involvement of non-governmental actors in the decision-making of public affairs becomes essential to lay the groundwork for civic engagement, voluntary actions, and the transition to social-governed civil society (Deng, 2004).

According to their functions and roles, non-governmental actors can be classified into grassroots actors and grassroots support actors (Wang, 1999). Grassroots actors are grounded in local communities, concerning the interests and development of their community members, with relatively high mobility. On the other hand, grassroots support actors often take the form of regional or national network alliances, composed mostly of professionals. These supportive actors aim to take collective actions towards common goals, defending the interests of specific grassroots groups by offering financial, professional, organizational, planning, or coordination services.

The independent and civic nature of the non-governmental actors empowers them to exert positive impacts on overseeing the operations of the governmental entities and safeguarding the public interests (Wang, 2013). Civic actions, particularly the organized collective actions, taken by non-governmental actors play a crucial role in protecting public interests and social justice. Such a goal is achieved by urging the government to enhance transparency of their decisions and decision-making processes, consider the opinions of diverse stakeholders, and provide more opportunities for

https://doi.org/10.52202/082078-0006

civic engagement. Though the rise of civil society in China has provided favorable conditions for civic actions, the impact of non-governmental actors is constrained by limited funding and capacity as well as the lack of comprehensive participatory mechanisms, influenced by the longstanding planned economic structure and centralized power in the government (Wang, 2013). While previous research on micro-regeneration tends to focus on design strategies (Liu et al., 2016), there is an increasing emphasis on understanding the roles and implications of non-governmental actors in shaping the urban environment.

2 RESEARCH OBJECTIVES

- This research seeks to address the following objectives:
- To understand the significance of non-governmental actors' engagement in two distinct modes of civic engagement, characterized by a more confrontational approach in the Yong Qing Fang project and a more collaborative one in the Pan Tang Wu Yue project.
- To explore the roles of non-governmental actors that are actively engaged in the two micro-regeneration projects.
- To examine the actions taken by non-governmental actors during the planning and decision-making processes of micro-regeneration projects, and discuss factors that may influence their actions.
- To analyze the implications of these actions and the effectiveness of different civic engagement approaches used in the two micro-regeneration projects.

3 METHODS

3.1 Case Study Approach

With the increasing attention to civic engagement in urban regeneration projects, this research adopts case studies as the main approach to examine the roles of non-governmental actors and the ways they were engaged in the micro-regeneration projects. The two selected projects of this research are Yong Qing Fang and Pan Tang Wu Yue in Guangzhou, China, representing pioneering efforts in involving non-governmental actors in the planning and decision-making processes. However, the spectrum of actors involved and the ways they were involved both had implications in their respective design and planning strategies, which can offer valuable references for addressing practical challenges for other micro-regeneration projects.

Building on the research objectives in the previous session, this paper tries to answer the following specific questions: (1) What were the factors that influenced which non-governmental actors were actively engaged in the projects? (2) what were the actions that these non-governmental actors took? And (3) how might these actions serve as forms of civic engagement? In order to answer these questions, this paper draws selectively on the Power-Interest Grid (Mitchell et al., 1997; Ackermann & Eden, 2011) a method commonly used in stakeholder analysis. In the case of micro-regeneration projects, "interests" usually include concerns about the outcomes of a project or a decision, including economic interests, community interests, professional interests/responsibilities, and self-realization; "power," on the other hand, often refers to the processes or effect that influence other actors' thoughts, behaviors, or perceptions, such as knowledge, resources, social skills, and social status. These factors are crucial in shaping the actions taken by the actors and the influence these actions wielded.

3.2 The Yong Qing Fang Project: From Unilateral Decision-making to Public Agendas

Enning Road, located in Liwan District, Guangzhou, is a historical street with the longest and most well-preserved arcade integrating the Cantonese culture of folk opera. The traditional brick-wood structures in the Enning Road area suffered from poor lighting, ventilation, and sanitation, resulting in a high concentration of buildings classified as highly dilapidated (Yang, 2007). Consequently, the initial Liwan District Enning Road Reconstruction Pilot Plan proposed large-scale reconstruction of the whole area. However, due to the absence of civic engagement in the planning processes, such a plan ignored the existing community and its cultural significance faced substantial opposition from the public, including citizens, scholars, and various other non-governmental stakeholders. The final plan went through several major revisions to accommodate the evolving interests and demands of the public (Lin et al., 2012).

Located in one of the most livable and vibrant blocks with several historical and cultural heritage sites along Enning Road, the "Yong Qing District Dilapidated (Old) Houses Renovation and Revitalization Project" (hereafter as the Yong Qing Fang) represents the government's attempt to resolve the impasse of the Enning Road project. The micro-regeneration approach is employed to adapt the design strategies to the emerging and changing demands of the public. It is also the first time that developers were involved in an urban regeneration project in a historical preservation area. The persistent conflicts among the residents, developers, and government, as well as the divergent concerns regarding historical preservation and urban redevelopment, had posed considerable challenges in achieving consensus for the development and regeneration of the Enning Road area. Based on the

https://doi.org/10.52202/082078-0006

relationships of various actors involved in the project, the regeneration process of the Enning Road area can be delineated into the following four key stages:

- The initiation of the government-led demolition and reconstruction (2006-2008): The upcoming Asian Games in 2010 prompted the redevelopment of the Enning Road area, given the significant location in the old urban center. In order to align with the timeline of the Asian Games, the government initiated the demolition immediately after identifying the historical structures for preservation and outlining the envisioned functions and spatial configuration. However, this decision was made without adequate consultation with the local community. Such hasty arrangements triggered discontent among local residents, who voiced concerns not only about the rapid demolition but also about the compensation offered in exchange for the requisition of their properties. Despite the residents' spontaneous protests against the demolition, it was the media's coverage of the Enning Road redevelopment project that compelled the government to reconsider its demolition plans. During this phase, the project remained predominantly under the government's determination, leaving the local community to find themselves in a passive and disadvantaged position, with only the preliminary involvement of the news media.

- The awakening of public awareness in historic preservation and the active involvement of diverse actors (2009-2011): The local community leveraged the influence of the news media to express their concerns regarding the project, with a particular focus on issues such as demolition, relocation, and historic preservation. Meanwhile, they also expressed a collective desire to take proactive actions to improve the environment of their neighborhood. Seizing the opportunity of the City Image Construction for Asian Games symposium, the local community effectively captured the attention of the city mayor, Wan Qingliang, seeking essential governmental support. Such exposure also resonated with the interests of other concerned citizens, particularly college students, scholars, and social activists, leading to the establishment of two self-organized civic organizations, which then wielded substantial influence on promoting the preservation of the cultural heritage and the significance of civic engagement in the Enning Road project (Huang, 2017). Throughout this period, an alliance of voices from local residents, news media, concerned citizens, and certain scholars coalesced into a public discourse that propelled the Enning Road event forward.

- The postponement of the reconstruction plan, and the decline in public attention (2012-2014): The official inclusion of the Enning Road area into the 23rd historic preservation zone in Guangzhou brought forth constraints regarding the low Floor Area Ratio and a concurrent increase in the land's market value. These factors collectively contributed to a diminished interest among developers to invest in this regeneration project. As the government shifted its focus towards the design of the Guangzhou Cantonese Opera Museum and the planning of the Lizhiwan Bay project, other demolition activities within the Enning Road area were suspended. Consequently, social attention on the Enning Road area had waned, leading to a gradual disbandment of civic groups due to the limited availability of their volunteer members.

- The controversy of incorporating developers into historic preservation via micro-regeneration (2015-2022): The introduction of the concept of "micro-regeneration" in the publication of Urban Renewal Approaches of Guangzhou Municipality in 2016, paved the way for the Yongqing District Dilapidated (Old) Houses Renovation and Revitalization Project. The Yong Qing Fang project served as a pilot project to experiment with micro-regeneration in the old city center, as well as the city's first attempt to engage developers in urban regeneration projects in historically significant areas. The Yong Qing Fang project adopts the Build-Operate-Transfer (BOT) development mode, in which the government collaborates with real estate developers to undertake urban regeneration projects. The government allocates 15-year land use rights to the developers, while the developers are responsible for the investment, construction, and subsequent operation of the project, as well as managing both the profits and losses incurred over this 15-year period. Such an approach reignited public discussion once again. Drawing on past experience, the government established a Co-creation Committee for this second phase of the Yong Qing Fang project, to facilitate direct communication among a diverse array of stakeholders, including members of the People's Congress representatives in Liwan district, members of the district's CPCC, community planners, representatives of residents, businesses, media, and professionals (Enning Road Establishes Guangzhou's First Historic District Renewal 'Co-Creation' Committee, 2018).

3.3 The Pan Tang Wu Yue Project: From Consultation to Co-creation and Co-preservation of Heritage

Pan Tang Wu Yue, a 4-hectare urban village populated by over 400 households, is another project located within one of the historical preservation

https://doi.org/10.52202/082078-0006

zones in Liwan District. It is one of the few remaining original rural settlements in Guangzhou that has retained its spatial layout, urban fabric and landscape patterns dating back to the Qing dynasty, marked by the presence of multiple traditional clan families. Under the influence of rapid urbanization, rural settlements have gradually turned into what is commonly known as "urban villages" ("Cheng Zhong Cun") in Chinese cities. Moreover, Pan Tang Wu Yue is an urban village with rich cultural traditions and a unique kinship-oriented community culture, in which the community's affairs are deliberated and determined collectively by representatives from selected families.

Similar to the Yong Qing Fang project, the initial urban regeneration plan of the Pan Tang Wu Yue featured large-scale demolition and reconstruction. A large portion of the houses had been requisitioned by the state, leading to extended periods of vacancy, which emphasized the pressing need for maintenance and renovation. With the issue of Urban Renewal Approaches of Guangzhou Municipality in 2016, Pan Tang Wu Yue was also designated as one of the pilot projects of micro-regeneration, aiming to preserve the historical urban fabric and incrementally revitalize the neighborhood. Notably, it is the first micro-regeneration project of a historical district that relies solely on governmental funding. Thereby, to avoid opposition as encountered in the Yong Qing Fang project, the Pan Tang Wu Yue project introduced community planners into the planning processes, marking the city's first attempt to facilitate civic engagement from the bottom-up (Xu et al., 2017). Based on the level of civic engagement, the process of the Pan Tang Wu Yue project can be delineated into the following four stages:

- Government-led spatial interventions (2013-2016): In contrast to conventional property-led urban regeneration projects, the Pan Tang Wu Yue project is government-led, driven by the overarching objectives of preserving the neighborhood's historical features. The early phase of the Pan Tang Wu Yue project primarily focused on improving the streetscape and renovating dilapidated structures and infrastructures. Thus, a large portion of the project's planning and implementation were undertaken by governmental departments, in collaboration with design professionals and construction teams. While there were over 80 households that were reluctant to relocate, their participation in the project was limited to the preliminary survey of public opinion.
- The introduction of community planners and face-to-face conversations (2017): Two architects from the leading design team of the project, the Urban Elephant Architects Design Studio, drawing on their educational background in Taiwan and extensive research on Guangzhou's Xiguan culture, assumed the role of community planners in this project. Conducting comprehensive fieldwork, they interacted with residents across all age groups to learn more about the local culture and foster connections with them at the same time. Meanwhile, they facilitated both formal and informal conversations, including "tea parties", community meetings, and message boards, among residents and relevant governmental officials and professionals, to address and discuss concerns from various stakeholders. However, due to the residents' lack of confidence in their influence in the decision-making processes, those who participated exhibited a pessimistic manner about the project.
- An exploration to discover local culture and foster trust (2017-2018): In addition to communication in the form of meetings, the community planners also hosted various types of activities to help both the residents and the professionals explore community culture. The "Exhibition of Pan Tang Wu Yue's History, Present, and Future" and the "Remembering My Home Together: The Oral History of Pan Tang Wu Yue" are the two most well-known events. These activities served as bridges, fostering trust between the community and the government in a more humanistic manner. They played a crucial role in helping the community understand the overall planning and objectives of the project, while lowering the barrier for them to express their concerns. At the same time, the success of these activities also attracted attention and participation from a broader scope of civic actors and social media, including the "Back Home" initiative group ("Fan Wu Qi" in Cantonese) and a WeChat social media platform called the "Traditional History and Culture of Pantang, Guangzhou" (Xu et al., 2017), further expanding the collaborative network for the Pan Tang Wu Yue project (Rui et al., 2023).
- Co-creation and civic engagement in practice (2018-2020): The participatory approaches have received recognition and support from the government, facilitating the collaboration of the community planners with other relevant departments to establish an intermediate entity, the Pan Tang Wu Yue Co-creation Committee, aiming to coordinate resources and strategies across various departments and sectors, cultivating a culture of collaboration and collective decision-making. As one of the professional teams involved in the project, the

https://doi.org/10.52202/082078-0006

Architecture Research and Design Institute of South China University of Technology utilized its academic influence to initiate a collaborative design workshop with several universities, bringing additional external support with fresh perspectives into the project. The designs of the public space surrounding the Sanguanmiao and the new gateway were collectively developed by the residents and the professionals based on the preliminary fieldwork from the design workshop.

4 RESULTS

4.1 Objectives of the Projects and Governance Strategies

Yong Qing Fang and Pan Tang Wu Yue projects were both launched against the backdrop of the 2007 Asian Game, undergoing a transformative evolution in urban planning – initially characterized by large-scale demolition and reconstruction, followed by a shift towards preservation, and now emphasizing regeneration. The Yong Qing Fang project received more attention and controversy compared to the Pan Tang Wu Yue project, owing to its larger demolition area. Concurrently, Guangzhou witnessed a noteworthy transition in its urban redevelopment policy at that time, transitioning from a singular focus on property-led redevelopment to an overarching commitment to comprehensive regeneration. The unique historical locations of these two projects also prompted the initial experiment of micro-regeneration, with corresponding protocols for historical preservation established subsequently. The absence of well-defined policies led to a lack of clear instructions or guidance, necessitating the projects to progress incrementally. Nevertheless, it is precisely because of the experiment of incremental and small-scale experiments in these two projects that enabled professionals to distill valuable experiences for the micro-regeneration approaches.

Yong Qing Fang. The original intent of the Yong Qing Fang project was to renovate the dilapidated houses on Enning Road as a pilot project for the old city reconstruction. The project hoped to transform the neighborhood into a commercial destination like Xintiandi in Shanghai, with a complete demolition of existing structures in mind. However, as the Enning Road event progressed, the Yong Qing Fang project opted to preserve the majority of its extant buildings while repurposing previously vacant or abandoned areas into public usable spaces. The district has maintained its commercial orientation, dominated by restaurants and retail businesses, and supplemented by Airbnb accommodations and exhibitions. The initial vision of establishing the maker and office spaces had to be revised due to limited business interests. With the growing influx of tourists into Yong Qing Fang, some local residents who are living around the commercial area have also ventured into various businesses.

Pan Tang Wu Yue. Due to the neighborhood's rich traditional culture, privately-owned buildings have retained their residential functions, with renovated facades. Structures that were requisitioned by the state have been transformed for commercial activities or into pocket parks based on the evaluation of the sites. Nonetheless, Pan Tang Wu Yue Front Street and Pan Tang Wu Yue New Street have preserved their mixed-used residential character. Meanwhile, because of the neighborhood's proximity to the children's clothing market on the 8th Zhongshan Road, the Pan Tang Wu Yue New Street has organically developed a series of convenience businesses to cater to the clothing market's needs, while some houses have been leased to nearby merchants or converted into warehouse purposes.

Such differences in project objectives led to distinct governance strategies. In the Yong Qing Fang case, the BOT model entails a partnership between the government and the private capital. Within the core commercial area, the government and the developer group oversee the district's planning and design, while the local residents mainly maintain control over their own property spaces. On the other hand, Pan Tang Wu Yue, as a government-led and government-sponsored project aimed at improving the community's public spaces and services, introduces community planners to facilitate residents' interest in participating in the project, thereby exploring the approaches and mechanisms for civic engagement.

4.2 Active Non-governmental Actors and Their Actions

The different objectives and strategies of the two projects also significantly influenced the actors who were actively involved in the projects and the interests they sought to pursue. In the Yong Qing Fang project, in which the initial plan involved demolishing a neighborhood with historical heritage, many non-governmental actors proactively took part in the project to express opposition to this plan. Their objectives were primarily focused on preserving the historical value of the heritage and the economic value of privately owned properties. In the Pan Tang Wu Yue project, in which there is a strong community network in the neighborhood and the objective was to improve the environment for the existing community, non-governmental actors were intentionally engaged to foster collaborative relationships in order to coordinate diverse interests centered around community spaces and development.

Non-governmental actors that have been actively engaged in the decision-making and planning processes of the Yong Qing Fang project were the local residents, the news media, the Enning Road

https://doi.org/10.52202/082078-0006

Folk Focus Group, the Enning Road Academic Concern Group, professionals hired by the government, and the Co-creation Committee (see Table 1).

Given that the primary objectives of the Yong Qing Fang project were oriented towards enhancing the city's image and promoting economic development of the district, the initial plan to demolish the whole area and the later plan centered on the creative economy didn't thoroughly communicate with the residents. Local residents' concerns of their living environment and property rights didn't receive adequate attention from the initial redevelopment plan. Although possessing rich knowledge and connections to the cultural and historical fabric of the neighborhood, the residents found themselves constrained by limited resources and influence over the top-down planning agenda. The lack of proper communication platform and detailed guidelines for civic engagement compelled them to adopt a more confrontational approach in articulating their opposition to the project, such as through protesting and submitting petitions (Chen et al., 2008; Chen & He, 2010; Chen et al., 2011).

Under such a circumstance, the news media, such as the New Express and Yang Cheng Evening News, played an extremely significant role throughout the project, which made the Enning Road event a milestone of civic engagement in Guangzhou's urban regeneration progress. Although the news media followed the Yong Qing Fang project for the controversy of the event rather than immediate interests, its credibility helps promote the local event into a public discussion by reaching out to the governmental officials at the higher level. The issue of culture preservation was uncovered by the media, arguing that the regeneration of the Yong Qing Fang area was not only about the living environment of the local residents, but also closely connected with the urban memory and cultural identity of Guangzhou city. With the influence of the media, the residents were able to address their concerns about the demolition and reconstruction, which were closely related to personal interests, but at the same time, connect their own interests with a broader topic of cultural preservation and property rights. The New Express published a series of consecutive reports under the title "the Last Enning Road", with a central theme commemorating the culture heritage of the district. The relevant topics were continuously circulated and strengthened by the media among the public, engendering widespread societal interest and attracting potential participants.

Two self-organized civic groups, the Enning Road Folk Focus Group and the Enning Road Academic Concern Group, were also substantial forces in transforming the preservation of the Enning Road area into a public discourse (Huang, 2017). The Enning Road Folk Focus Group started from a small-group discussion on Douban, a Chinese social network website, in 2010. It was initiated by a college student who is also a local resident in Guangzhou. As the internet facilitated the spread of the topic, this group evolved into one of the most potent advocates for the preservation of Enning Road. The group worked closely with the community, and conducted interviews and surveys to convey residents' voices, while establishing rapport with them. Two noteworthy achievements based on their comprehensive investigations include the Social Assessment Report on the Enning Road Regeneration and Reconstruction Project that delineates the projects' impacts on the economic and social life of the residents and the Opinion Report on the Enning Road Regeneration and Reconstruction Plan that presents references based on opinions of the residents and experts regarding the planning schedule, public interests, public participation and etc. Another civic group, the Enning Road Academic Concern Group, emerged as an offshoot of the Folk Focus Group, focusing on raising awareness of the cultural heritage of the Enning Road area, through organizing academic forums and compiling community magazines and newspaper. These two civic groups not only promoted the preservation of the Enning Road area, but also raised citizens' awareness regarding historic preservation of local culture in general.

Relying primarily on governmental funding, the Pan Tang Wu Yue project directed its efforts towards improving the neighborhood's environment and infrastructure while preserving the historic heritage and culture of the local community. Thereby, the major non-governmental actors actively involved in the Pan Tang Wu Yue project included community planners, the local residents, and the Co-creation Committee, with support from other professional teams and civic groups that prioritize heritage preservation and community interests over economic interests (see Table 2).

Serving as community planners, the two architects affiliated with the Urban Elephant Architects Design Studio played critical roles as intermediaries between the government and the residents, facilitating communication and collaboration. Beyond offering technical and intellectual support to the formal planning, they also incorporated community building approaches into the project as a way to nurture mutual trust and share local knowledge. While the conventional public participation methods like community meetings did not achieve desired outcomes to solicit constructive opinions from the

https://doi.org/10.52202/082078-0006

Table 1 The Power and Interests of Major Non-governmental Actors in the Yong Qing Fang Project (2023). Edited by the author.

NON-GOVERNMENTAL ACTORS	INTERESTS	POWER
• Local residents	• Community interests, economic interests.	• Knowledge, social skills
• News media	• Professional interests/responsibilities	• Resources, social skills
• The Enning Road Folk Focus Group	• Community interests, self-realization	• Knowledge, social skills
• The Enning Road Academic Concern Group	• Community interests, self-realization	• Knowledge, social skills
• Professionals hired by the government	• Community interests, professional interests/responsibilities	• Knowledge, social skills
• Co-creation Committee	• Community interests, professional interests/responsibilities, self-realization	• Knowledge, social skills, social status

community, the community planners organized a variety of community activities to inform and engage the ordinary public, including the "Exhibition of Pan Tang Wu Yue's History, Present, and Future" and the "Remembering My Home Together: The Oral History of Pan Tang Wu Yue", effectively reducing the barriers to participate in public processes. The community planner in the Pan Tang Wu Yue project have established a closer connection with the local community than in other conventional urban regeneration projects.

Drawing on their professional interests, the two community planners also started the "Back Home Initiative" ("Fan Nguk Kei" in Cantonese) and established a corresponding non-profit organization in 2014 (Xu et al., 2017). This initiative is dedicated to fostering dialogue among diverse actors, particularly aimed at bridging the gap between the governmental/professional actors and the ordinary public. The "Back Home Initiative" group leverages social media and public exhibitions as mediums to facilitate discussion concerning the preservation and development of Pan Tang Wu Yue area. The group also sees the the Pan Tang Wu Yue project as an opportunity to establish a communication platform featuring professional knowledge and civic engagement that can be applied to a broader context.

Although being the most immediate stakeholders of the project, the residents' engagement was primarily facilitated and actualized by the community planners. In addition to the basic requirements for improved public spaces and services, the residents of Pan Tang Wu Yue were also motivated by their attachment to the neighborhood and commitment to preserve the community's traditional culture. Thus, while the government-assigned missions for the community planners was to implement the micro-regeneration, most activities held by the community planners aimed at encouraging the residents think about how the design can be better integrated with their everyday lives and cultural activities. Through this process of consensus building, the community planners were able to enroll more actors who shared an interest in preserving the community's culture. The network expanded from current residents to those who had relocated but maintained a connection with the neighborhood, as well as other volunteers with interests in the culture.

These activities further prompted the development of a WeChat social media platform, the "Guangzhou Pantang History and Traditional Culture", managed by a group of volunteers who are either residents or are closely connected to the community. The group utilizes the social media platform to advocate for the traditional culture of the Pantang community, offering a space for discussion and sharing its history, culture, related activities, and also the progress of micro-regeneration in the neighborhood. Actively engaged in the oral history activity and the design of Sanguanmiao, the group posted information to recruit participants for these activities. The influence of social media enables the group to effectively frame the Pan Tang Wu Yue project as an example of civic engagement and co-creation.

A Co-creation Committee was established in both projects, as a a formalized platform to bring diverse actors together to engage in discussions on the planning of the neighborhood. The committee is comprised of representatives from the local residents, community planners, the media, professionals, and governmental officials. Although most of the operational responsibilities were still undertaken by community planners or other professional actors, the establishment of the Co-creation Committee signified the provision of a secure platform for local residents to communicate with other key decision-makers. Additionally, the committee empowers its members to undertake proactive actions; for instance, the community planners leverage the committee as a conduit not only for initiating and coordinating design reviews but also for soliciting financial support from the government for community initiatives. However, it is

https://doi.org/10.52202/082078-0006

Table 2 The Power and Interests of Major Non-governmental Actors in the Pan Tang Wu Yue Project (2023). Edited by the author.

NON-GOVERNMENTAL ACTORS	INTERESTS	POWER
• Community planners	• Community interests, professional interests/responsibilities, self-realization	• Knowledge, social skills, social status
• Local residents	• Community interests, self-realization	• Knowledge, social skills
• Professionals hired by the government	• Community interests, professional interests/responsibilities	• Knowledge, social skills
• Co-creation Committee	• Community interests, professional interests/responsibilities, self-realization	• Knowledge, social skills, social status
• "Back Home Initiative" Group	• Community interests, professional interests/responsibilities, self-realization	• Knowledge, social skills
• "Guangzhou Pantang History and Traditional Culture" Group	• Community interests, self-realization	• Knowledge, social skills

noteworthy that the Co-creation committees do not seem to effectively encourage civic engagement, as the predominant form of action is community meetings and the meetings cannot address all the concerns due to time constraints.

For most non-governmental actors, their concerns revolve around community interests and professional interests/responsibilities. However, compared to the Pan Tang Wu Yue project in which the residents share common goals of preserving their community culture while improving their living environment, residents in the Yong Qing Fang project expressed more individual concerns about rights of their own properties or the financial compensation associated to the project. In a similar vein, served as third-party actors, the community planners in the Pan Tang Wu Yue project aim to achieve neighborhood regeneration by facilitating communication and building consensus, while the goal of news media in the Yong Qing Fang project was to make a significant impact with their reports, making residents' actual concerns part of the news narratives rather than defending their interests. The objectives among actors are critical in determining the effectiveness of collective actions in urban regeneration projects. In the Yong Qing Fang project, considerable efforts were invested in reconciling the conflicted interests, resulting in several postponements of the implementation. On the contrary, in the Pan Tang Wu Yue project, the actors with sharing objectives were able to and willing to coordinate and collaborate, maximizing the impact of their efforts in implementing the project.

4.3 Forms of Civic Engagement

In a similar vein, the main interests and powers that these non-governmental actors have also determined the corresponding civic engagement approaches, and their interactions with other actors involved in the project (Table 3).

In the Yong Qing Fang case, the influence of non-governmental actors on the project manifests through individual actors announcing their opinions through a network alliance established though the publicization of their interests and concerns. The collective demands voiced by residents initially came from their interests to optimize benefits from the demolition and regeneration project, which directly reflected in the increase in compensation packages and the sustained presence of "nail households" resistant to relocate. However, the impact of these demands on governmental decisions transcends their immediate objectives. Given the recurrent media coverage and scholarly discussions surrounding this contestation, the Yong Qing Fang project has evolved from an urban regeneration project of the physical environment into a public discourse. This discourse encompasses multifaceted issues related to urban development, including property rights, the preservation of traditional culture, citizens' entitlement to information and participation, and other topics related to the idea of publicity.
In the Enning Road event, the appeal to protect personal interests transcended the boundaries of private and public domains, capturing widespread attention from those who were not directly related to the project. Such a network alliance of discourses expanded the scope of this grassroots effort, giving them opportunities to engage in communication and negotiation with the authorities.

Meanwhile, the involvement of the two academic focus groups prompted a shift in the project's emphasis and discourse, transitioning from demolition and reconstruction to enhancing the urban landscape and preserving historical and cultural elements. Nevertheless, the developer-driven micro-regeneration, oriented towards fostering a creative economy, also encountered resistance from the local residents. Due to the

https://doi.org/10.52202/082078-0006

Table 3 The Comparison of Civic Engagement in the Yong Qing Fang and Pan Tang Wu Yue projects (2023). Edited by the author.

	YONG QING FANG PROJECT	PAN TANG WU YUE PROJECT
Key actors	Opposition groups	Knowledge, social skills, social status
Forms of collaboration	Networked alliance of discourses	Knowledge, social skills
Concerns	Opposition of the demolition; defend residents' property rights	Knowledge, social skills
Approaches	Confrontational activities	Knowledge, social skills, social status
Foundation	Pressure of societal impact; property rights	Knowledge, social skills
Activities	Individual complaints; collective petitions	Knowledge, social skills
Results	The postponement of the demolition; revision of project and urban regeneration policies; establishment of the Co-creation Committee	Knowledge, social skills

lack of established civic engagement procedures, community members continued to face challenges in effectively conveying their opinions to the authorities or the planning team through institutional channels. Reflecting on the previous round of confrontational activities, the government has adjusted the civic engagement mechanism. The first Co-creation Committee, established by the Liwan District Government, emerged as a platform for facilitating face-to-face communication among diverse actors. This Co-creation Committee system has then developed into a critical approach in collaborative planning and placemaking in Guangzhou, including the Pan Tang Wu Yue project.

Civic engagement in the Yong Qing Fang project is primarily characterized by the externalization and publicization of confrontational activities rooted in the interests of the local community. The opposing actors employed various strategies, such as establishing a network alliance, increasing social impacts, and protests, to articulate their dissent from the bottom up. These efforts seek to transcend the constraints of the existing institutional framework, advocating for adjustments and revision in the demolition policies, the design proposals, and the opinion feedback mechanism. Notably, the formation of oppositional alliances also gradually transformed the dynamic of the governance and decision-making processes. In this mode of civic engagement, the legitimacy of these bottom-up actions lies in the impact of societal discourse and the intrinsic property rights of the local residents.

In the case of Pan Tang Wu Yue, community planners emerge as critical non-governmental actors to lead collaborative activities grounded in placemaking. They played a crucial role in facilitating connections among local residents and between the residents and the authorities through various forms of activities, thereby empowering the community to articulate its opinions. This process gradually encourages a collaborative partnership between planners and the residents, underscoring not only the critical role of community planners as mediators, but also the transformative shift in the traditional dynamics of governance and civic engagement in urban regeneration projects. This is evident in the two-way directions of collaboration modes: external support that served as the institutional foundation to empower the community, and connection and collaboration within the community to propel the project.

Here, collaboration was fostered through organizing and participating in community activities and establishing shared goals related to the project among residents. Since 2016, the community planners have organized more than ten community activities. In the early stages, these activities mainly sought to cultivate trust with the residents, including events like tea parties, photography exhibitions, and conventional community meetings to acquaint residents with the project. Later, as familiarity between the community planners and the local residents deepened, the focus shifted towards more sophisticated activities, for instance, the collaborative development of a restoration design for a cultural relic in the community using a 1:1 model, among residents, several NGOs, and experts in historic preservation. These activities facilitated the formation of partnerships between community planners and the residents, leading to collective decision-making and management of community affairs, which also bolstered social cohesion and fostered collective identities within the community. Furthermore, the recognition and endorsement of such effective

https://doi.org/10.52202/082078-0006

civic engagement efforts by governmental bodies and the media have served to attract younger generations back to the community, which in turn, has augmented residents' overall enthusiasm and capacities for participating in the project.

The community planners in the Pan Tang Wu Yue projects assumed the role of intermediaries to connect the internal and external forces of the community. Their collaboration with the local residents, realized through various forms including community activities, social platforms, and knowledge sharing, makes up for the inadequate attention to the local communities in the existing urban regeneration framework. Concurrently, by sharing their expertise and networking with other professional resources, they are able to bridge the gaps in governmental functions, especially regarding professional knowledge in specific fields and diversification of financial support. In this sense, the efficacy of collaboration in the Pan Tang Wu Yue project is embedded in the community planner system itself, while rooted in individual's rights to participate in public affairs.

5 CONCLUSIONS

This paper reflects on the governance actions of non-governmental actors in micro-regeneration projects. Through examining actions taken by non-governmental actors in two projects with different backgrounds and objectives, it explores the implications of different non-governmental actors and project objectives on the modes of civic engagement, particularly those at the grassroots level, and their impacts on the projects. In the Yong Qing Fang project, the community augmented the impact of their active opposition by establishing a network alliance with the media and academic groups; while in the Pan Tang Wu Yue project, community planners served as mediators to facilitate connections among various actors, especially between the government and the community. Despite operating in distinct ways, both cases underscore the significant roles of non-governmental intermediaries in promoting civic engagement. On one hand, by collaborating with communities in various types of activities, these intermediaries foster the will or the capacity of the grassroots to participate in urban regeneration projects; on the other hand, they find it more advantageous to utilize their professional knowledge and skills to establish trust and secure support from government authorities, which in turn, help them empower the communities and carry out collaborative activities.

The different modes of civic engagement and involved actors observed in the Yong Qing Fang project and the Pan Tang Wu Yue project reveal a paradigmatic shift in urban regeneration strategies, transitioning from unilateral decision-making to multi-actor collaboration, a more human-centric approach of placemaking. Based on the case studies, two elements are crucial to facilitate such a shift. First is whether there is a platform to enable diverse actors to share information, knowledge and opinions as well as engage in conversation throughout the planning process, thereby reaching consensus to take effective actions. Secondly, the introduction of intermediaries, such as community planners, can bridge the gap and foster collaborative relationships among actors with diverse interests and influences on the projects. While this paper mainly explores the implications of non-governmental actors' actions in civic engagement using empirical results, there is room for further discussion and research regarding strategies and models of collaborative placemaking in China, especially from the perspective of social governance and state-society relationship.

6 REFERENCES

Ackermann, F., & Eden, C. (2011). Strategic management of stakeholders: theory and practice. Long Range Planning, 44(3), 179–196. https://doi.org/10.1016/j.lrp.2010.08.001.

Arnstein, S. R. (1969). A ladder of citizen participation. Journal of the American Institute of Planners, 35(4), 216–224. https://doi.org/10.1080/01944366908977225.

Bromley, R. D. F., Tallon, A. R., & Thomas, C. J. (2005). City centre regeneration through residential development: Contributing to sustainability. Urban Studies, 42(13), 2407–2429. https://doi.org/10.1080/00420980500379537.

Cai, Y., Yang, X., & Li, D. (2017). Research on the content and countermeasures of "micro-regeneration" of old urban residential areas. Urban Development Research, 24(4).

Chen, W., Gao, G., Guo, X., Wu, S., & Ao, C. (2008, May 12). Homeowners in Guangzhou collectively submitted a petition to the National People's Congress to suggest revisions of the demolition regulations. New Express. https://news.sina.com.cn/c/2008-05-12/012513861324s.shtml.

Chen, W., & He, S. (2010, February 5). 183 Households collectively opposed against the Enning Road Plan. New Express. https://news.sina.cn/sa/2010-02-05/detail-ikmxzfmi9722937.d.html.

Chen, W., He, S., & Lin, S. (2011, June 29). Residents of Enning Road published a letter to support self-help redevelopment. New Express. http://news.sina.com.cn/c/2011-06-29/034922723118.shtml.

https://doi.org/10.52202/082078-0006

Davidoff, P. (1965). Advocacy and pluralism in planning. Journal of the American Institute of Planners. 31(4), 331–338. https://doi.org/10.1080/01944366508978l8.

Davidoff, P., & Reiner, T. A. (1962). A choice theory of planning. Journal of the American Institute of Planners, 28(2), 103–115. https://doi.org/10.1080/01944366208979427.

Deng, G. (2004). The new environment for the development of non-governmental organizations in China. Society, 10.

Enning Road establishes Guangzhou's first historic district renewal "Co-creation" Committee. (2018, September 7). Southern Metropolis Daily. https://www.sohu.com/a/252552986_161795

Healey, P. (2003). Collaborative planning in perspective. Planning Theory, 2(2), 101–123. https://doi.org/10.1177/14730952030022002.

Huang, D. (2017). How do people get engaged in civic participation? A case study of citizen activism in rebuilding Enning Road, Guangzhou. Chinese Journal of Sociology, 3(2), 237–267. https://doi.org/10.1177/2057150X17702089

Kim, G., Newman, G., & Jiang, B. (2020). Urban regeneration: Community engagement process for vacant land in declining cities. Cities, 102, 102730. https://doi.org/10.1016/j.cities.2020.102730

Li, L. H., Lin, J., Li, X., & Wu, F. (2014). Redevelopment of urban village in China – A step towards an effective urban policy? A case study of Liede village in Guangzhou. Habitat International, 43, 299–308. https://doi.org/10.1016/j.habitatint.2014.03.009

Lin, D., Zhou, K., & Wang, S. (2012). Discussing old city renewal and public participation through the "Enning Road Event." Diversity and Inclusion - Proceedings of the 2012 China Urban Planning Annual Conference. 2012 China Urban Planning Annual Conference.

Liu, Y., Wu, F., Liu, Y., & Li, Z. (2016). Changing neighbourhood cohesion under the impact of urban redevelopment: A case study of Guangzhou, China. Urban Geography, 38(2), 266–290. https://doi.org/10.1080/02723638.2016.1152842

Ma, H., & Ying, K. (2016). Micro-regeneration of community public space exploring approaches to community building in the context of organic urban regeneration in Shanghai. Time+Architecture, 4.

Mills, C. W. (1959). The power elite. Oxford University Press.

Mitchell, R. K., Agle, B. R., & Wood, D. J. (1997). Toward a theory of stakeholder identification and salience: Defining the principle of who and what really counts. The Academy of Management Review, 22(4), 853–886. https://doi.org/10.2307/259247.

Nelissen, N. J. M. (1982). Urban renewal participation experiments: Heralds of a new local democracy? Council of European Municipalities.

Rossi, U. (2004). The multiplex city: The process of urban change in the historic centre of Naples. European Urban and Regional Studies, 11(2), 156–169. https://doi.org/10.1177/0969776404041421.

Rui, G., Li, R., & Xing, Y. (2023). Research on participatory planning and design strategy in conservation area joint development based on memory place representation: A case study of Puntong Village micro-regeneration in Guangzhou. New Architecture, 1.

Verdini, G. (2015). Is the incipient Chinese civil society playing a role in regenerating historic urban areas? Evidence from Nanjing, Suzhou and Shanghai. Habitat International, 50, 366–372. https://doi.org/10.1016/j.habitatint.2015.09.008

Wang, C. (2018). Community micro-renewal mechanism directed by good governance. Planners, 34(2).

Wang, S. (1999). Pluralism and Unification: An International Comparison of the Third Sector. Zhejiang People's Press.

Wang, Z. (2013). NGO development and reflections in the field of urban planning in China. Shanghai Urban Planning Review, 6.

Watson, V. (2014). Co-production and collaboration in planning – The difference. Planning Theory & Practice, 15(1), 62–76. https://doi.org/10.1080/14649357.2013.866266

Xu, H., Li, P., & Li, R. (2017). Reading, planning, communicating, imaging and designing: A record of an urban micro-regeneration in Puntoon Wuyue Village. Architectural Journal, 1.

Yang, X. (2007). Discovering the values of historical district and constructing an industry of folk cultures——The preservation and utilization of Arcade Street on Enning Road, Guangzhou. Urban Development Studies, 5.

https://doi.org/10.52202/082078-0006

Ye, L. (2011). Urban regeneration in China: Policy, development, and issues. Local Economy: The Journal of the Local Economy Policy Unit, 26(5), 337–347. https://doi.org/10.1177/0269094211409117

Zhang, C., & Li, X. (2016). Urban redevelopment as multi-scalar planning and contestation: The case of Enning Road project in Guangzhou, China. Habitat International, 56, 157–165. https://doi.org/10.1016/j.habitatint.2016.05.008

Zhao, Y., Zang, B., & Zhu, X. (2019). Latest developments in urban renewal research at domestic and abroad. Urban Development Studies, 26(10).

https://doi.org/10.52202/082078-0006

BUILDING CODES AND HOUSING RESILIENCE IN CALIFORNIA: DISPLACED HOMEOWNERS' PERCEPTIONS ON SUSTAINABLE REBUILDING

Track 3: Resilient Design for Climate Change
Knowledge Network: Cultural Aspects of Design, Sustainable Planning, Design & Behavior
Methods: Case Study, Interviews
Mariana Garcia Junqueira (Kansas State University - APDesign)

ABSTRACT

Traditionally, the federal public land management policy has largely addressed wildfires as unwanted (i.e., to be prevented or suppressed). It was only in the early 1970s that federal policy began to recognize wildland fire as a historical, ecological factor (Pyne, 1982) that could not be fully eliminated. This new ecological understanding of wildfires gained traction in 1985, and became a nationally recognized issue. Being prone to recurring wildfires, California has constantly implemented policies that require or incentivize home hardening to seek greater housing resilience. However, property owners do not always correctly perceive risks. Especially while navigating disaster displacement, homeowners' rationale for rebuilding more resilient homes can represent an additional challenge to their traumatic experiences. Building codes ensure that minimum, acceptable standards are used in designing, constructing, and maintaining housing. Understanding how code change impacts the rebuilding process from the homeowner perspective is imperative in informing new directions for more appropriate, human-centered sustainable rebuilding. By focusing on wildfire-responsive building codes, we ask, "How have changing building codes impacted displaced homeowners' rebuilding experiences?" A historical analysis characterized how building codes changed throughout time for increased wildfire resilience in California and qualitative interviews with individuals displaced in 2018 revealed that participants experienced added emotional, social, and financial tolls related to building codes. These results pointed to two anemic areas (i.e., education and context-based innovation), leading to a systematic selection and analysis of successful documented experiences worldwide. Recommendations incorporate sustainable building practices that consider the displaced homeowner's perspective, supporting social-ecological health and vitality.

Keywords

Disaster displacement, building codes, home hardening, homeowner perception, post-disaster sustainable building practices.

1 INTRODUCTION

General building code changes are constantly driven by technological advancements and an increased body of knowledge on specific areas that support sustainable development goals (e.g., improvement of urban dwellers' health and well-being or urbanization's environmental impact). In addition, lessons learned from the devastating impacts of natural disasters can promote planning and policy change in various ways.Unlike recent times, most fires historically spread during conditions that were less than high to extreme severity. However, as one of the consequences of climate change, current wildfires during severe conditions have led to extreme, uncontrollable fire conditions. It is suggested that wildfires are inevitable, however, fire disasters are not.

This paper indicates that Wildland-Urban (WU) fire disasters are principally a home ignition problem (where an effective reduction is possible), and not a problem of controlling extreme wildfires (where control is not possible). Further, because WU fire destruction is a home/structure ignition problem, an appropriate problem definition must relate the "wildland" ignition factors to the "urban" ignition due to a structure's flammable materials and design.

Therefore, building codes have been constantly updated to promote Home Hardening, which aims to increase housing resilience and reduce the susceptibility of a home being lost in a WU fire disaster. Among U.S. states, California has gone the furthest in mandating the take-up of wildfire resilience investments by property owners. However, the application of these codes varies throughout the state, and is still very much dependent on homeowners' proper perception of risks.

2 RESEARCH OBJECTIVES

Understanding how code change impacts the process of rebuilding from the perspective of the displaced homeowner is imperative in informing new directions for more appropriate, human-centered, sustainable rebuilding. By focusing on wildfire-responsive building codes, we ask, "How have changing building codes impacted displaced homeowners' rebuilding experiences?"

https://doi.org/10.52202/082078-0007

3 METHODS

3.1 Research Design

A qualitative interpretivist framework was best suited for this study. Interpretivist researchers view experience and perspective as valuable sources of knowledge and seek to uncover the interpretations of reality that people have based on their understanding of the world. This study has thus defined the person-in-environment as the unit of analysis, and obtaining "as complete a picture as possible" of the phenomenon was needed to understand how these people see the world, the way they perceive constraints, and how they choose to act (Mazumdar & Geis, 2001, p. 262).

A Naturalistic Field Research (NFR) method was carried out because it neither focuses solely on the human/social dimension (as in the Social Sciences) nor strictly on the physical aspects (which is often the case in Architecture, Planning, and Design). In this way, this method allowed for the study of both human/social and physical dimensions in their symbiotic relations (i.e., a socio-physical phenomenon) in context. Data collection involved multiple sources of information aimed at achieving an in-depth understanding of the phenomenon to allow for triangulation or confirmation (Mazumdar & Geis, 2001; Yin, 1994). Data collection techniques used in this research are described next.

3.2 Data Collection

Archival data, government statistics, and documents were analyzed to corroborate objective information (e.g., dates, meteorological data, and names of organizations involved in specific events mentioned by survivors). This also included a code survey, in which relevant codes, regulations, and requirements (e.g., California regulations and requirements, local building codes, etc.) were analyzed to track code changes in response to fire disasters, among other updates.

In addition, significant data came from interviews conducted in 2019. These include twenty-five survivors of nineteen households displaced by the Woolsey Fire. Some interviewees were interviewed more than once. Interviews were conducted by phone or in person in appointments that each lasted between 1 and 4 hours. These interviews were audio-recorded and later transcribed for analysis. Eligible participants were individuals (18 years of age or older) from any of the communities affected by the Woolsey Fire who suffered a total loss of their homes. Recruitment started through social media (i.e., Facebook). By identifying virtual groups dedicated to fire survivors, we involved looking for keywords included in their names (e.g., Woolsey Fire, California wildfires, etc.), and reaching out to each group's administrator, who invited eligible participants to contact me (i.e., process of referral).

In-depth, semi-structured interviews were conducted mostly in the participants' temporary housing. Questions were brief and simple, so that the different dimensions introduced in the answers could be pursued. Introducing questions asked to all survivors included their experience of the disaster and subsequent displacement, as well as their relocation and rebuilding plans. Such opening questions often yielded spontaneous, rich descriptions where the participants themselves provided the highlights of what they experienced, i.e., participants select the main dimensions of the phenomenon under investigation (Kvale, 1996). In this way, relevant questions arose from interactions during each interview, and each interview had the potential to generate new questions to be included in a subsequent interview.

Opting for an in-depth approach to interviews was informed by the type of knowledge sought and was necessarily linked to the research question and underlying philosophical assumptions of this study. In-depth interviews are especially effective for collecting information to answer certain types of research questions, particularly when the interest is in "understanding the perceptions of participants or learning how participants come to attach certain meanings to phenomena and events..." (Taylor & Bogdan, 1998, p. 98). Therefore, this type of interview was the most appropriate to seek a deeper understanding of people's experiences (e.g., feelings and thoughts), and what these meant to them. Furthermore, specifying questions framed in a more operational manner (e.g., What did you think at the time? How did you feel about that?") was also utilized to comprehend participants' emotional reactions to certain narrated events. This allowed for learning from the participants as much as they were willing to share, as the goal was to get past the general accounts and reach for answers that reflected many layers of depth.

3.3 Data Analysis

In NFR, analyses will emerge from the interaction between collected data and focusing decisions. The purpose of this process is to achieve interpretations that arise from the data (Lofland & Lofland, 1995). Coding initiated the process of sorting, organizing, and categorizing data. Of critical importance was not to simply label parts of the data, but rather, to develop categories and, subsequently, themes that emerged from the array of data. Simultaneously, memos were written to elaborate on the coding and categories process, allowing for continual reflection and ultimately achieving "an interrelated set of memos that form a coherent analysis" (Lofland & Lofland, 1995, p. 193). As a result, the relevant literature and the findings in this study are organized

https://doi.org/10.52202/082078-0007

in two empirical sections: The Wildland-Urban (WU) Fire Problem in California, and Displaced Homeowners' Perceptions on Sustainable Rebuilding.

4 RESULTS

4.1 The Wildland-Urban (WU) Fire Problem in California

This first section focuses on characterizing the problem at hand by presenting a conceptual and historical overview of fire management approaches adopted in the United States throughout time, and fire survivors' conceptualization of the fire problem in California. This is followed by an overview of relevant codes, regulations, and requirements that articulate how building codes have changed to respond to failed responses to fire disasters and its significance in improving housing resilience, especially amid homeowners' often inadequate perceptions of risk.

The fire destruction of urban, suburban, and exurban development during wildfires has been commonly referred to as the wildland-urban interface (WUI) fire problem (Cohen, 2010; 2019) at least since 1974 (Butler, 1974). Traditionally, the federal public land management policy has largely addressed wildfires as unwanted (i.e., to be prevented or to be suppressed at the smallest area possible). It was only in the early 1970s that federal policy began to recognize wildland fire as a historical, ecological factor (Cohen, 2010; Pyne, 1982) that could not be fully eliminated.

This new ecological understanding of wildfires gained traction in 1985, after a period of significant destruction of property and lives (Cohen, 2010), and home destruction related to wildfires became a nationally recognized issue, motivating fire protection officials from federal and state agencies to initiate a collaborative effort with the National Fire Protection Association (NFPA). This initial fire management response resulted in the 1986 Wildfire Strikes Home conference (Cohen, 2010), and since then, wildland fire management policy has recognized the WUI fire problem as a principal issue in several documents including the National Fire Plan (2000), the Federal Wildland Fire Management Policy (2001), the 10-Year Comprehensive Strategy (2001), and the Healthy Forests Restoration Act (2003). These initiatives consolidated programs such as the Firewise and Fire Adapted Communities, which focus on fire exclusion and suppression.

However, there is no historical evidence or current management trend to suggest that all wildfires can be excluded, and if not excluded, controlled with an initial suppression response (Cohen, 2019). For instance, examinations of WUI fire disasters over one hundred years in the U.S. showed that fire suppression has successfully controlled around 95 percent of wildfires upon discovery and initial attack during the conditions when most historical fires spread. Meanwhile, wildfires that produced extreme fire behavior conditions – characterized by high wind speeds, low relative humidity and continuous flammable vegetation producing rapid fire growth rates, high fire intensities and showers of burning embers (firebrands) across wide areas downwind from the wildfire – are among the 5 percent of wildfires where initial attack has failed (Cohen, 2010). When attempted wildfire exclusion and suppression fail, the third step in the reactive fire management approach is to resort to structure protection (Cohen, 2010). This process is described in Figure 1.

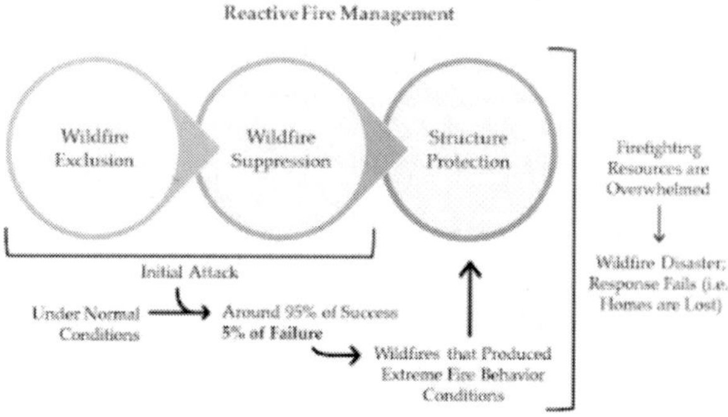

Figure 1. The Reactive Fire Management Approach. Diagram by the author based on based on Cohen (2010; 2019).

When the initial attack fails, and a wildfire occurrence with extreme wildfire conditions, structure protection is not guaranteed. In extreme fire conditions, firefighting resources are overwhelmed quickly, and the response fails, characterizing a fire disaster, in which homes are often lost to partial or total fire damage. Due to climate change, extreme fire behavior conditions are becoming more common, and it is reasonable to assume the inevitability of wildfire occurrence with extreme wildfire conditions (Baylis & Boomhower, 2021a; Williams, 2013), resulting in more frequent and more intense fire disasters.

Given the inevitability of extreme wildfire conditions, the inevitability of WUI fire disasters seems assured if the emphasis continues to be on a reactive management of wildland and structure firefighting approach. Fortunately, scholarship has advanced enough to say that WUI fires can occur without a WUI

https://doi.org/10.52202/082078-0007

fire disaster, even in extreme wildfire conditions. This happens because although this type of community destruction is initiated by wildfires, structure ignitions are not determined by geographic location; rather, the fire destruction is determined by the ignition conditions of a structure in relation to its immediate surroundings. If homes are made ignition-resistant previously discussed (Cohen, 2010; 2019) indicate that required heat transfer for a sustained home ignition is determined by the conditions of a structure in relation to a structure's immediate surroundings.

However, the global (e.g., climate change) and local variables (e.g., a structure and its surroundings)

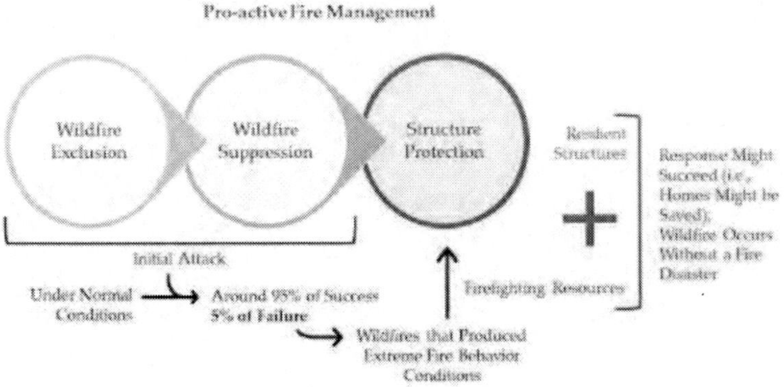

and most homes do not ignite, fire protection resources are not overwhelmed, the available structure protection is effective for ignitions that do occur, and WUI fire disasters are prevented (Cohen, 2019).

Hence, extreme wildfires can occur without WUI fire disaster if the emphasis shifts to a pro-active management of wildland and structure fire-fighting approach. This process can have a different outcome because of changes (i.e., Resilient Structures) that affect the third stage (i.e., Structure Protection) of the fire management process, as diagrammed in Figure 2

Figure 2. The Pro-active Fire Management Approach. Diagram by the author based on based on Cohen (2010; 2019).

This indicates that WU fire disasters are principally a home ignition problem (where an effective reduction is possible), and not a problem of controlling extreme wildfires (where control is not possible). In this way, community destruction during wildfires is more appropriately referred to as simply wildland-urban (WU) fire (Cohen, 2010). Because WU fire destruction is a home/structure ignition problem, an appropriate problem definition must relate the "wildland" ignition factors of firebrands and flames to the "urban" ignition due to a structure's flammable materials and design (as shown in Figure 2). In essence, the WU fire correct definition is a statement relating the heat transfer of burning objects (flames and firebrands) to a structure's ignition. The patterns of destruction herein described are not always known, especially by those most impacted by the WU fire disasters: the displaced survivors. As previously described, climate change impacts extreme fire behavior conditions, making inevitable the occurrence of wildfires with extreme wildfire conditions (Baylis & Boomhower, 2021a; Williams, 2013).

Accordingly, "Climate Change" and "Global Warming" were terms constantly used interchangeably by participants to assign blame for the latest fire disasters in two opposite ways. First, most participants blamed the environmental phenomena for changing long-standing, "sustainable" patterns (e.g., of local weather, water scarcity, and normal fire seasons) that led to more disastrous fires. However, some mentioned the phenomena while assigning blame elsewhere. Connie, for instance, said that government institutions and the scientific community justify the disastrous fires based on a hoax. In other words, she described the above-mentioned phenomena as a fabricated explanation to deflect blame for their failed responses to disasters. This reinforces what researchers who defend the enforcement of renewed building codes versus the voluntary take up of recommendations have reasoned, that the enforcement of housing resilience through code change can be justified by the common misperception of risk by homeowners (Baylis & Boomhower, 2021b; Turner & Vaughan, n.d.).

In the U.S, building codes usually refer to model codes developed in a national forum, taking the views of all interested parties – as well as state and

https://doi.org/10.52202/082078-0007

local codes -- into consideration. Codes are designed to protect buildings and the people and property inside them from extreme events such as WU fires. They also ensure structural integrity; electrical, plumbing, and mechanical system safety, as well as accessibility and practical and achievable levels of energy efficiency (Turner & Vaughan, n.d.).

Architects and engineers must ensure their building designs meet or exceed the minimum, legally mandated code requirement for a particular jurisdiction. The local building department reviews plans submitted to them before construction and issues permits before the project can proceed. Inspectors verify the project, as constructed, complies with the code. Building industry stakeholders affected by building codes include contractors and subcontractors, manufacturers of building products and materials, insurance companies, facility managers, building owners and tenants, and building inspectors. However, there seems to still be a lack of communication of the potential benefits (e.g., reducing WU fire disaster risks) to those most impacted code change at the residential level.

4.2 Fire-Responsive Code Change

As previously discussed, the patterns of fire destruction in the U.S. (Cohen, 2010; 2019) indicate that required heat transfer for a sustained home ignition is determined by the conditions of a structure in relation to a structure's immediate surroundings. This area, a structure/home and its immediate surroundings, is called the Home Ignition Zone (HIZ) (Cohen, 2001; 2010; Cohen & Stratton, 2008). Further research into how structures can ignite during extreme wildfire conditions has enabled the identification and assessment of ignition vulnerabilities with guidance for effectively increasing home ignition resistance during extreme wildfire conditions (How to Prepare Your Home for Wildfires, 2022).

Learning from failed responses, in this case, calls for code change to increase housing resilience by regulating the HIZ at various levels. In this way, land use planning can reduce wildfire risk by helping communities grow and develop with wildfire in mind. Land use planning tools such as zoning, plans, regulations, and building codes can influence the design, layout, and placement of homes built in wildfire-prone areas (Wildfire Risk to Communities, 2022).

Land use planning measures are diverse and can be customized to fit the needs of the neighborhood, community, or county. For example, land use planning regulations can require defensible space in the home ignition zone, adequate water supply, road widths and access, and home hardening materials for homes built in high wildfire hazard areas. Land use planning tools can also restrict development in areas of highest hazard. These and other tools are described in Table 1.

Focusing on building codes and standards allows for the advancement of the HIZ, by introducing important concepts such as Home Hardening, which is an important component of the HIZ. Also referred to as Defensible Space, the HIZ is an area 100-200 feet from the foundation and includes vegetation, the home itself, and other structures or attachments like decks, furniture, fences, and outbuildings (Wildfire Risk to Communities, 2022).

Homes lost to wildfire are most often ignited by embers or small, low-intensity fires. Ember ignition can occur when embers enter the building through vents or an open window. Once inside, embers can ignite furnishings or other combustible materials stored there. Ember ignition can also occur when embers accumulate and ignite combustible parts of the building, such as a wood shake roof, combustible decking, or debris accumulated on a roof or in a gutter. Vegetation or other nearby combustible materials can be ignited by embers, causing a spot fire and subjecting a portion of the siding, windows, or doors to fire (Cohen, 2010; 2019; Wildfire Risk to Communities, 2022). Therefore, building codes have been constantly updated to promote Home Hardening, which aims to help reduce the susceptibility of a home to a WU fire disaster. In the Home Hardening process, several components of the home should be addressed, including the roof, gutters and eaves, vents, siding, windows and doors, decks, fences, and other attachments.

Among U.S. states, California has gone the furthest in mandating the take-up of wildfire resilience investments by property owners. However, the application of these codes varies throughout the state (Baylis & Boomhower, 2021). In areas where CAL FIRE provides firefighting services (State Responsibility Area – SRA), the state directly determines building standards. Within incorporated cities and other areas with their own fire departments (Local Responsibility Area – LRA), local governments have historically had greater control over code requirements (Baylis & Boomhower, 2021).

The first of these was the Bates Bill of 1992 (Assembly Bill 337), which is further described, along with the subsequent code changes, in Table 2.

https://doi.org/10.52202/082078-0007

Land Use Planning Tools with Potential to Reduce WU Fire Disaster Risk

Watershed Management Plans	Reduces wildfire through fuel treatments, protecting vital water resources.
Land Preservation Tools	Encourages agricultural lands to buffer development from wildfires.
Building Codes	Requires ignition-resistant construction materials for new developments and retrofits.
Forest Management Projects	Reduces fuels within the wildland-urban interface (WUI).
Landscaping Regulations	Requires property owners to manage hazardous vegetation and maintain their properties.
Steep Slope Ordinances	Restricts development within high wildfire-risk areas.
Land Use and Development Codes	Incentivizes developers to plan open space and recreational trails, creating fuel breaks.
Subdivision Design Standards	Requires risk reduction features, such as minimum road widths, secondary access, and adequate water supply.
Local Government Support	Supports fire adapted communities through good land use planning.

Table 1. Land Use Planning Tools with Potential to Reduce WU Fire Disaster Risk. Table by the author based on Wildfire Risk to Communities.

Fire-responsive Code Change in California

The Bates Bill, 1992	It encouraged stronger building standards in LRA areas by requiring CAL FIRE to produce maps of recommended Very High Fire Hazard Severity Zones (VHFHSZ). Local governments could then choose to adopt these recommended hazard maps and their accompanying building standards.
Assembly Bill 3819, 1994	It increased requirements for ignition-resistant roofs in all SRA areas and in the subset of LRA areas where local governments had adopted the VHFHSZs. • In 1995, it required Class B roofs on newly-constructed or reroofed homes in regulated areas. • In 1997, the requirement increased to Class A roofs in high-hazard areas.
Assembly Bill 423, 1999	It simplified enforcement of the new roofing codes by outlawing the use of unrated roofing materials throughout the state.
Chapter 7A standards of the California Building Code, 2008	Applied to all homes built in 2008 or later in SRA areas and in LRA areas where the VHFHSZ designations had been accepted, to many dimensions: • Roofs must be rated class A or B, • Eaves and exterior siding must be fire resistant, • Vents must be covered by a fine wire mesh to resist ember intrusion, • Windows and doors must resist fire for at least 20 minutes, • Decks and other building appendages must be built of noncombustible materials, and • Additional requirements for defensible space.

Table 2. Fire-Responsive Code Change in California. Table by the author based on Baylis & Boomhower (2021).

https://doi.org/10.52202/082078-0007

A recent, comprehensive empirical evaluation of California's strict wildfire building codes, by analyzing a new dataset of property-level data on U.S. homes destroyed by wildfires up to 2020 (Baylis & Boomhower, 2021). First, this study found a remarkable impact of code change in housing resilience to WU fire disasters by comparing three types of jurisdictions: SRA areas, where compliance with California building codes was mandatory, LRA areas that were ever recommended by CAL FIRE as VHFHSZ areas (regardless of whether local governments accepted the designation), and the areas without fire-responsive building codes (which included LRA areas in California that were never recommended for consideration as VHFHSZ, as well as areas in areas of other states without any state or local fire-responsive building codes.

The empirical strategy isolates the effect of building code changes using a fixed effects design that compares outcomes for pre- and post-code homes on the same residential street. This approach narrows the comparison to homes experiencing essentially identical wildfire exposures. For instance, one analysis reported that about 35% of homes exposed to fire that were built prior to the mid-1990s were destroyed. The destruction decreased to about 20% for homes built after the mid-1990s. This sharp improvement in resilience corresponds in time to the post-Oakland Firestorm building code reforms. Evidence also showed that homes built before about 1980 may be less likely to be destroyed than homes built just prior to the roof requirements, because they are more likely to have been re-roofed at least once after the mid-1990s and complied with the new requirements for ignition-resistant materials at roof replacement (Baylis & Boomhower, 2021).

Among other things, the study also looked at the spillover benefits of code-induced mitigation to neighboring homes. In other words, the extent to which each resilient home improves overall community resilience. In sum, the results show that compared to reliance on voluntary action alone, California's fire-responsive building codes reduced average structure loss risk during a WU fire disaster by 16 percentage points, or about a 40% reduction. They also reduced the risk to a close neighbor's home by about 2 percentage points or 6%. These striking results imply materially different levels of resilience in communities with and without such codes. "Moreover, the spatial externalities provide a classic rationale for public policy intervention even if homeowners were fully informed and rational about wildfire risk" (Baylis & Boomhower, 2021, p. 33).

Having documented these large resilience benefits, the next section turns to homeowners' perceptions on sustainable rebuilding, which included dealing with an array of fire-responsive codes.

4.3 Displaced Homeowners' Perceptions on Sustainable Rebuilding

The Woolsey Fire was a precise illustration of a WU fire disaster, according to the historical and current approaches to the WU fire problem in California, as described earlier. This section describes the way in which the events unfolded while the Woolsey Fire reached the homes of participants in this study (i.e., the circumstances of their forced displacement), after the failed response that prioritized a reactive wildland and structure fire-fighting approach. These circumstances shaped fire survivors' subsequent experiences of displacement, relocation, and rebuilding after the fire.

The disaster killed three people, destroyed 1,643 structures, and burned approximately 96,949 acres (Top 20 Most Destructive California Wildfires, 2019). Although not the largest nor the deadliest wildfire in Califor-nia's history, the Woolsey Fire was so damaging in terms of home environments that it currently ranks seventh among the most destructive wildfires in California's history and the fourth most destructive fire in the last decade (Top 20 Most Destructive California Wildfires, 2019), after destroying at least 300 homes, and immediately displacing their occupants.

Previous studies on displacement suggest that, in general, more reliable data is available when analyzing internal displacement than when looking at cross-border migration trends, "...because vulnerability differs significantly across disaster-affected communities" (Mearns & Norton, p. 26). Thus, the somewhat homogeneous make-up of the population in the area affected by the Woolsey Fire validates them as a group of people with similar circumstances. For instance, the median household income in the area affected does not vary greatly, with the majority of households affected earning between $100,000 and $150,000, there is a 75 percent rate of homeownership, in comparison to renting and other arrangements, the population is predominantly white and the me-dian age it between 45 to 54 years (The U.S. Census Bureau GIS mapping files by Census Tract, 2019; 2020).

According to the After-Action Review of the Woolsey Fire Incident (AAR), commissioned by the Los Angeles County to evaluate the causes and their response to the event, meteorologists and fire technical staff saw evidence of increasing fire weather conditions on November 6, 2018 (Citygate Associates, 2019). Per existing protocols, fire agencies upstaffed predetermined resources, and CalFire

https://doi.org/10.52202/082078-0007

relocated resources across the State. On November 8, the relative humidity dropped from 100% at 7:00 AM to 7% at 11:00 AM, and although the weather forecast called for moderate temperatures in the mid-70s, winds blowing from the east became sustained, hot, and dry. At 2:30 PM, these winds were sustained at 23 miles per hour (mph) with gusts up to 37 mph. Meteorologically, "the conditions for a large-scale fire incident were perfect" (Citygate Associates, 2019, p. 19).

Historically, two distinct fire seasons existed in California, depending on geography. These regional differences had always allowed for statewide firefighting resources to be redeployed to Southern California, to stand ready for regional threats once Northern California fires had ceased. Although the Woolsey Fire erupted late in the traditional fire season of Southern California, Northern California was still experiencing extreme fire weather conditions. Approximately eight hours before the ignition of the Woolsey Fire, the Camp Fire (the deadliest in California history) ignited in Butte County just east of Paradise. Significant to the Woolsey Fire is that over 1,000 fire personnel were assigned to the Camp Fire, pulling from the statewide pool of firefighting resources (Citygate Associates, 2019).

Seven hours later, the major Hill Fire ignited in the Hill Canyon area of Santa Rosa Valley, only 3.5 miles east of the City of Camarillo. The Ventura County Fire Department deployed a full brush fire ground and aerial unit response to the Hill Fire. Initial fire attack resources were outflanked, and the fire quickly jumped Highway 101 within approximately 15 minutes from the time of dispatch, running approximately three miles in 15 minutes. Crews were quickly overpowered, and direct operations were rendered ineffective by strong winds, steep topography, and flammable, drought-stricken fuels in the path of the fire (Citygate Associates, 2019). Based on these initial conditions, the Hill Fire was believed to have the potential to reach the coast by nightfall.

Meanwhile, at approximately 2:24 PM, the Woolsey Fire was first reported at the Ventura County Emergency Communications Center as a request for a "smoke check," resulting in the initial deployment of the Rocketdyne Fire Department and one Ventura County fire engine. Initial ignition occurred in the Woolsey Canyon of the Santa Susana Mountain range on the old Santa Susana Field Laboratory (SSFL) property in Chatsworth. At 2:33 PM, before fire engines arrived on the scene, the event was upgraded to a full wildland fire unit response, which also triggered Los Angeles County Fire and the Los Angeles City Fire to respond per the established Santa Susana Mutual Threat Zone (MTZ) agreement. The first two ground units and a helicopter arrived on the scene between 2:45 PM and 2:51 PM. All reported a wind-driven fire growing from five to approximately 30 acres in six minutes. The initial wind direction pushed the fire in a southerly direction on the first day. The fire then crossed the 101 Freeway and headed into the Santa Monica Mountains, in alignment with the geographic canyon features, and thus, conducive to a rapid-fire run toward the Pacific Ocean (Citygate Associates, 2019).

The Santa Susana MTZ agreement calls for a massive initial joint response from three fire departments: Ventura County Fire, Los Angeles County Fire, and Los Angeles City Fire. However, due to the two other competing fires in the state, local and eventually regional and statewide mutual aid systems were unable to respond to the many incident resource requests. Through the mutual aid request system records, the AAR emphasized that resources were repeatedly requested for the Woolsey Fire, but those requests went unfulfilled in the first 21.5 hours of the incident. The report concluded that during the first two days of the Woolsey Fire, 53 percent of the fire engine mutual aid requests were unfilled, which totals 874 units not sent to the site (Citygate Associates, 2019). During these 48 hours, the fire burned 96 percent of its final footprint.

The large differences in wildfire acres burned in California are primarily due to variable weather conditions and situations in which fires occur in remote locations that are difficult to access (California's Forests and Rangelands: 2010 Assessment, 2019). In other words, the size of a fire is influenced by the presence of strong winds, the dryness of vegetation due to lack of rainfall, and the ease of accessibility to firefighters. In this case, the extensive drought of previous years, the lack of availability of personnel, and the difficult terrain in which the fire gained momentum played an important role in the disastrous outcome. Additionally, the high fire frequency in the Santa Monica Mountains, especially in the corridor that goes from the Santa Susana Pass in the Simi Hills down to the coast through Malibu Canyon, led to very little "old growth" chaparral and a complex vegetation pattern created by overlapping fires. These non-native, younger species burned easier, faster, and were certainly factors in the quick spreading of the wildfire. Therefore, the fire immediately presented unprecedented complexities, that led to one survivor describing their experience as "a surreal disaster response" (Ralph, July 12, 2019, Interview).

By noon on Friday (November 9), the fire had burned all the way to the Pacific Ocean in Malibu. Most of the fire growth occurred after the fire crossed Highway 101 and required approximately 6.5 hours to reach the ocean. At times the agencies re-sponding to the fire were "overwhelmed in the first hours by this incident's speed and weight of impact, exposing

https://doi.org/10.52202/082078-0007

some limitations between the agencies and systems as they meshed into a single, wide-area regional response team in less than 24 hours" (Citygate Associates, 2019, p.5).

The expected fire behavior and rate of spread far exceeded the past experiences of emergency responders, policymakers, and the community members in the Santa Monica Mountains fire corridor areas with canyon or drainage generally running to the ocean. Experienced firefighters recounted witnessing fire behavior unlike anything they have witnessed in the past (Citygate Associates, 2019), and this surprise factor is also consistent with most of the survivors' recounted experiences.

Patricia, for instance, remembers that her children called her that night to say that their old neighborhood was on fire, which was only three miles away from where she had been living. To her surprise, flying embers were the cause of houses burning in their old neighborhood. On this point, Patricia remarked: "You would think that just the houses up against the mountains would be in danger, not the ones in the middle of the community. But the ones in the middle of the community were burning too". Patricia also attributed her late evacuation to the disbelief of such unexpected fire behavior:

We didn't really think the fire was going to come, [that] it was going to burn your house down, because for the past 50 years, the fires would go straight to the canyons, to the beach, and they would pass right by us... We are... like in a mini valley. So, they just went all around us every time (Patricia, July 12, 2019, Interview).

Only when Patricia saw the flames coming over the crest did she realize that "it didn't look good". Other participants expressed similar surprise on how the fire behaved that night in comparison to many other fires of previous decades they had experienced while living in the area. Cecilia, for example, recalls coming back from a camping trip on that Thursday morning and hearing about the Hill Fire. However, she only heard about the rapid start and spread of the Woolsey Fire in the evening, at about 7:00 PM:

So, I started paying attention to that one... I started slowly packing some things up, and I told my son, you know, "It's getting a little closer," and he said, "Don't worry," because we have evacuated about 5... 7 times... And he goes, "We always come back, it'll be fine." It would have to go through... all this [pointing the nearby grocery store and school] to get to us, but we didn't take into consideration [the fire] coming from the back to us (Cecilia, August 6, 2019, Interview).

Thus, not only did the fire spread quickly, but it also shifted its usual course. According to another survivor named Linda, as the fire advanced, she monitored her surroundings from her house that sat upon the top of a crest. Linda's neighbor, half a mile down the hill, could not "see north," so she kept calling to check in on the fire progress. Linda recalls not seeing anything and repeatedly replying: "It still has to go down three more ravines before it gets here... just don't worry about it... it's nowhere near us". However, at about 9:15 the next morning, a Monrovia Sheriff's Department patrol car came up the street with sirens, as she recounts:

He wasn't telling us a mandatory evacuation, he was screaming, "Evacuate!" I'm looking to the north, north-south, and there's nothing up there... So, I walk outside... I had to walk up a little... the fire was there. So, the previous 10, 15 years that I've been doing that, 58 years I've been dealing with [fires], all the fires have come from the north-south. This fire came from the west. So, by the time I got to the driveway, it was too late... I ran back to my house, grabbed the animals... Put [them] in the car and drove down my canyon... And by the time I got [a] that point of less than a mile away, I watched my house blow up. It was that fast (Linda, August 19, 2019, Interview).

Thus, expectations by community members regarding the fire agencies' capabilities to safely order evacuations and to confront a fast-moving wildfire were harshly reset by the many surprises presented by the Woolsey Fire.

The Woolsey Fire was so surprising that Patricia and Ralph never received an official evacuation order. To warn them, a neighbor called at about 2:30 AM. Patricia and Ralph had a unique perspective as the fire approached, as she was the president of their homeowners' association, and Ralph, the chairman of the emergency preparedness committee. Their own evacuation decisions were thus tempered by the absence of official evacuation orders and firefighting response:

We looked at each other, and we said, "Are we going to start putting things in the cars? Or... should we go around? You're the president. I'm the chair of the emergency preparedness committee. Should we just go around to all the homes, that 214 other homes, and evacuate the elderly and the handicapped because everybody is sleeping?" (Ralph, July 12, 2019, Interview).

Throughout the night, while evacuating neighbors and tending to his wife, who tried to pack some clothes and documents, Ralph's frustration became directed at the lack of firefighting presence. As he fought back the tears eight months later, he described one of his biggest disappointments that night:

https://doi.org/10.52202/082078-0007

We responded to a 93-year-old man's house because we saw a fire truck come ahead of the ambulance. In this country, when you have an ambulance going through any zone, if a fire truck can accompany, they do, but the fire truck, in this case, was ahead of the ambulance by about five minutes. We responded to that man's house with the fire engine, the fire truck, and we asked them, "When will backup be here?" And they said, "Soon," but backup never showed up before the fire got there. The ambulance came, they picked up the 93-year-old man, his caretaker had called, and he left in an ambulance, and I saw the ambulance go out with the fire truck behind it (Ralph, July 12, 2019, Interview).

Especially in the Malibu coastal area, residents expected a more robust firefighting response. Still, as described in the AAR, many factors prevented that from happening (Citygate Associates, 2019).

Throughout this study, participants who lost their homes in the Malibu coastal area expressed frustration regarding inadequate fire response. Cassie, who lives about a mile and a half up from the Pacific Coast Highway, explained why she thinks her area burned so terribly: "It was because it came over in the Santa Monica Mountains, where there were no structures. No air drops. No effing fireman, no water... I'm so angry. My house and 200 of my... neighbors did not have to burn".

Furthermore, a decisive event that was continually mentioned by almost every participant in this study: We evacuated and we went down to Zuma Beach parking lot. Where they had 10 or 12 fire trucks lined up sitting there because some head honcho told them, for their orders, they were not to put out structure fires that were there, to only save lives. Excuse me, we all evacuated. So what life are you going to save? (Cassie, December 17, 2019, Interview).

Linda also mentioned the Zuma beach incident, but from a different perspective: "At one point I had my brother's T-shirt on that said, 'Fire Department' and said 'Captain' on it. I had to turn it inside out. My neighbors were screaming at me" (Linda, August 19, 2019, Interview).

In this context, the ARR also addressed the Zuma Beach incident:

Due to the Unable to Fill (UTF) requests and the firestorm's impact on roads through the mountains, units in Malibu consisted only of locally assigned units and a very small quantity of mutual aid. This is why the public observed that the normally heavy fire unit response was not occurring... On Friday evening, there were still nowhere near the units requested. As a result, the public observation was, of course, valid; active structural fire-fighting in the coastal plain in the early hours of the fire was not possible (Citygate Associates, 2019, p. 29).

They concluded by recommending "Public education to increase awareness regarding public impact when a devastating fire incident shifts from direct control methods to Fire Front Following and Life Safety First tactics" (Citygate Associates, 2019, p. 8). Thus, whether they were long-standing responsibilities or new orders given during the event, the fact remains that such protocols were not known by the community, leaving many members feeling unheard and abandoned.

The community has a perception that public agencies can continuously protect them, but as an incident and the size of the Woolsey Fire shows, this is not always pos-sible. Hence, the urgency to abandon reactive wildland and structure firefighting approaches to WU fires and focus policy and planning efforts on more proactive strategies, such as the effective reduction of home/structure ignition through new standards of building safety.

Furthermore, in massive, dynamic wildfires, some life and structure loss are tragic but expected possibilities; however, what occurred in less than 24 hours was not anticipated by any prior plan or preparedness exercise. This only reinforces the need to learn from failures and promote fire-responsive change in housing codes, regulations, and requirements.

So far, this section has described the way in which the events unfolded while the Woolsey Fire reached the homes of participants in this study (i.e., the circumstances of their forced displacement), after the failed response that prioritized a reactive wildland and structure firefighting approach. These circumstances shaped fire survivors' subsequent experiences of planning their relocation and rebuilding after the fire.

4.4 Displaced Homeowners' Perceptions on Sustainable Rebuilding

All participants reported similar traumatic experiences of the fire, and post-disaster psychological consequences ranging from frustration to anger and trauma. The psychological implications of such experiences are thoroughly discussed elsewhere (Junqueira, 2020). Of importance to this discussion is that decisions regarding relocation and rebuilding were made amid such challenging conditions. First, this section will discuss participants' relocation experiences, characterized by multiple moves and trade-offs to remain local.

A common reason provided by participants to explain their pattern of serial relocation (multiple moves) was the fact that during the first few weeks after the

https://doi.org/10.52202/082078-0007

fire, they had little knowledge about their options. By the time they were able to turn their attention to begin planning their relocation, fewer solutions were attainable.

In a state that is experiencing a widespread housing shortage—ranking 49th among the 50 United States in housing units per resident (The U.S. Census Bureau GIS mapping files by Census Tract, 2019; 2020), the abrupt loss of homes to disasters worsens the existing imbalance between supply and demand. One of the reasons that exacerbated housing market struggles seems to be that efforts from the Woolsey Fire overlapped with those of people displaced by other fires (e.g., Thomas, 2017; Hill, 2018; and Getty, 2019). Thus, people displaced by distinct fires that hit contiguous areas of Southern California in a two-year interval were often navigating the housing market together.

A considerable group of participants reported relocating as opportunities arose. For instance, Cecilia evacuated with her son Hudson and their dog to a friend's house in Pacific Palisades. Within twelve hours, they heard that their home was gone. According to Cecilia, she reported the incident to her home insurance company the next day and received no guidance: "I have never had to use insurance in that way. So, I had no idea how it worked. And he didn't tell me anything. So, I reported it. That's all I did" (Cecilia, August 6, 2019, Interview). After living on her friend's couches for two weeks, Cecilia moved two more times after different arrangements were made by her church leaders. After one month, she was able to start looking for an apartment closer to her original neighborhood, after receiving information about how to afford temporary housing.

Cecilia's trajectory illustrates the relocation pattern found among participants in this study and is representative of what happened after the Woolsey Fire in Southern California. This relocation pattern was characterized by consecutive moves with stays that became longer towards the end of the process. Another characteristic seen in Cecilia's journey, also found in most participants' relocation patterns, was the desire to stay local, i.e., to return to the area of their pre-fire residence. This desire heavily influenced participants' decision-making processes during relocation and rebuilding planning.

In some instances, push drivers (e.g., being a fire-prone area or the expensive housing market) were downplayed by interviewees, and appropriate relocation was delayed and even avoided, reasoned by one's emotional investment was deposited into a place or a group (e.g., the original area of residence). Therefore, temporary, local relocations were predominant in the aftermath of the Woolsey Fire. Further, the desire to remain local was often a reason to compromise on the size or quality of the temporary housing, i.e., a decision-making process that involved trade-offs.

Challenges to permanent relocation were similar to those encountered while in temporary housing: expensive housing market, downsizing, and other trade-offs. A different mindset was found, however, when discussing permanent rebuilding plans. One neighborhood that originally housed many participants in this study is used to illustrate some of these challenges, identified as significant roadblocks that burdened survivors' processes of recovery.

A central issue was the fact that rebuilding after the fire was taking much longer than it should, in many cases. Justine, one of the most vocal survivors in her community, signaled part of the problem, more than one year after the fire, as she rhetorically asked: "So what's wrong with this picture? Why aren't people rebuilding their houses?" (December 17, 2019, Interview). What Justine called a "post-fire chaos" included, on average, an additional six months to have a burned property cleaned and cleared through state-wide programs, which added a first layer of uncertainty and delay to the process of rebuilding.

In many cases, survivors depleted their rebuilding resources long before they were allowed to start the rebuilding process. Some had their alternative living expenses (ALE) benefits expiring within six to 12 months from the day of the fire and had to transition to paying rent out-of-pocket, some were simultaneously paying multiple bills on the burned-down house and the temporary housing, some were not insured, underinsured, or were never even approved for ALE benefits through FEMA due to their lifestyle (e.g., autonomous worker or not legally married partners).

Moreover, a rapidly changing housing market and urbanistic legislations hindered some survivors from planning for rebuilding in their pre-fire lots, mainly in two ways. First, the one participant in this study who was a long-time renter (of more than three decades) was communicated by her landlord that he would not be rebuilding a house similar to the one she lived in, but rather, he would take the opportunity to build a multi-family property for renting. Even among homeowners, being "priced-out" of their own lots was also a concern, especially for residents of unincorporated areas in the Santa Monica Mountains.

For instance, homeowners displaced by this one neighborhood expressed concern for being in a one-of-a-kind collection of individual homes owned by co-operative members due to a concession from July of 1986, in which residents purchased the park by forming a non-profit mutual benefit corporation that continues to this day. Now, displaced shareholders

https://doi.org/10.52202/082078-0007

participating in this study expressed fear of being forced out of such a unique arrangement.

Second, urbanistic legislations that materialized in the last few decades are not "catching up" to burned-out homeowners. Many participants expressed dissatisfaction with the fact that they can no longer rebuild the exact same house that was lost. This has caused some hesitation in deciding to rebuild, not only because of the new house would have a different size and/or design, but also because survivors are dragging what they foresee as a stressful process of navigating new building codes and processes

An additional roadblock was the fact that FEMA identified a floodplain area cutting through the center of the neighborhood, as shown in Figure 3, which requires action per the Floodplain Management Executive Order 11988, of 1977. The objective of this executive order is to require agencies to avoid the long and short-term adverse impacts associated with the occupancy and modification of floodplains and to avoid direct and indirect support of floodplain development wherever there is a practicable alternative.

Homes that were standing (after surviving the fire) could remain in place. However, homes that burned down in that area could not be rebuilt.

The weather and development patterns in California leave watersheds and property prone to severe post-fire flood damage. This well-documented series of events is referred to as the fire/flood sequence, therefore, the complete impacts of wildfire can only be seen over a period of several years (Levy & Korkosz, 1997). Though naturally occurring in the Santa Monica Mountains, debris flows are aggravated by any disturbance of slopes, soils, or vegetation, including roads, housing pads, fire lines, and fires.

The well-being of the area residents is closely reliant upon the overall condition of the watershed. Although most residents mentioned the terrible flood that followed the fire (acknowledging the water management issue), their overall perception was that government and environmental agencies were imposing yet another roadblock to their rebuilding.

Another roadblock was identified by those who advanced in the rebuilding planning process. Code change was commonly seen as an inconvenient

Figure 3: Floodplain identified by FEMA.

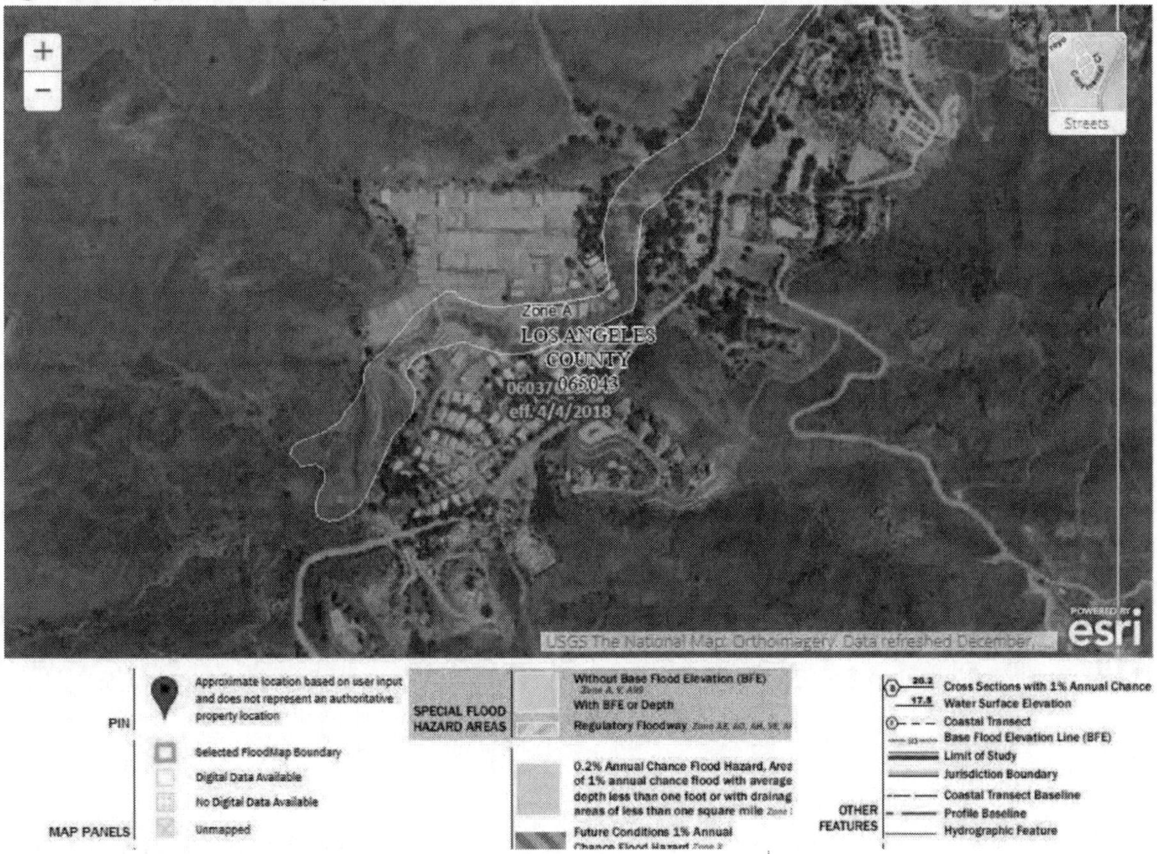

https://doi.org/10.52202/082078-0007

obstacle to rebuilding. The most common changes were building setbacks, which would culminate in rebuilding a house considerably smaller than the original (pre-fire) home.

Another barrier to rebuilding a home similar to the original was put in place by the local homeowners' association in charge of regulating building aesthetic characteristics in the neighborhood. According to participants, this regulation became much stricter, including a restricted color palette (only white, grey, and beige) for new housing facades.

At some point during the year following the fire, some participants decided to move on and remake their homes elsewhere. For instance, Rita, although underinsured, was able to buy a new house in a different area in Southern California. There is an additional waiting time while her house is built, which contributes to her being the maximum number of moves among participants in this study (i.e., twelve in August and twenty by November 2019). Her decision-making process in rebuilding also involved compromise. In essence, Rita is downsizing and leaving her neighborhood of decades but hoping to still settle in Southern California sooner rather than later, since her original neighborhood encountered so many roadblocks that delayed the possibility of rebuilding on the same lot.

5 CONCLUSIONS

The results described above revealed that participants experienced added emotional, social, and financial tolls related to changes in building codes. These results pointed to two anemic areas (i.e., education and context-based innovation), leading to a systematic selection and analysis of successful documented experiences worldwide.

Code changes enable innovation in the building sector and permit more innovative products and processes to gain market share. As new building requirements are reviewed through the model code revision process (currently every three years in the U.S.), innovative stakeholders make their cases for inclusion. These innovations add value to buildings and improve their impact on health and safety.

Recommendations should incorporate sustainable building practices that consider the displaced homeowner's perspective, supporting social-ecological health and vitality. In this sense, how code changes are communicated to displaced homeowners is an area in need of further study.

7 REFERENCES

Baylis, P., & Boomhower, J. (2021a). Mandated vs. voluntary adaptation to natural disasters: The case of U.S. wildfires. NBER Working Paper 29621.

Baylis, P., & Boomhower, J. (2021b). Building codes and community resilience to natural disasters. Working Paper.

Butler, C. P. (1974). The urban/wildland fire interface. Washington State University Papers, 74(15).

California's Forests and Rangelands: 2010 Assessmenthttp://frap.fire.ca.gov/data/assessment2010/pdfs/california_forest_assessment_nov22.pdf

Citygate Associates. (2019). After Action Review of the Woolsey Fire Incident. County of Los Angeles.

Cohen, J. D. (2001). Wildland-Urban fire: A different approach. In Proceedings of the Firefighter Safety Summit, Missoula, MT & Fairfax, VA, 6-8 Nov.

Cohen, J. D., & Stratton, R. (2008). Home destruction examination: Grass Valley Fire, Lake Arrowhead, California. Tech. Paper R5-TP-026b. USDA Forest Service – Pacific Southwest Region.

Cohen, J. D. (2019). An analysis of wildland-urban fire with implications for preventing structure ignition. White Paper.

Cohen, J. D. (2020). The wildland-urban interface fire problem. Fremontia, 38, pp. 16-22.
How to Prepare Your Home for Wildfires. https://www.nfpa.org/-/media/Files/Firewise/Fact-sheets

Junqueira, M. (2020). Living in transition: Displacement and temporary living arrangements after the Woolsey Fire in Southern California. In Proceedings of the Environmental Design Research Association (EDRA) 51 Annual Conference, Tempe, AZ, US, 4-7 Apr. 2020.

Kvale, S. (1996). InterViews: An Introduction to Qualitative Research Interviewing. Sage Publications.

Levy, L., & Korkosz, J. (1997). Water resources management plan: Santa Monica Mountains National Recreation Area. The National Park Service.

Lofland, J., & Lofland, L. H. (1995). Analyzing social settings: A guide to qualitative observation and analysis, 3rd ed. Wadsworth Publishing Company.

https://doi.org/10.52202/082078-0007

Mazumdar, S., & Geis, G. R. (2001). Case study method for research on disability. In B. N. Altman, & S. N. Barnartt (Eds.), Research in Social Science and Disability: Exploring theories and expanding methodologies (pp. 255-275). Elsevier.

Mearns, R., & Norton, A. (2010). Social dimensions of climate change: Equity and vulnerability in a warming world. The World Bank.

Pyne, S. (1982). Fire in America: A cultural history of wildland and rural fire. Princeton University Press.

Taylor, S. J., & Bogdan, R. (1998). Introduction to qualitative research methods: A guidebook and resource, 3rd ed. John Wiley & Sons.

The U.S. Census Bureau GIS mapping files by Census Tracthttps://www.census.gov/geographies/mapping-files.html
T
op 20 Most Destructive California Wildfires. https://www.fire.ca.gov/media/5511/top20_destruction

Turner, E., & Vaughan, J. (n.d.) The Value and Impact of Building Codes.

Wildfire Risk to Communities. https://wildfirerisk.org/reduce-risk/home-ignition-zone/#research

Williams J. (2013). Exploring the onset of high-impact mega-fires through a forest land management prism. Forest Ecology and Management, 294, pp. 4-10.

Yin, R. K. (1994). Case Study Research: Design and Methods. Sage Publications.

https://doi.org/10.52202/082078-0007

PAVED PARADISE: A VIEW FROM OREGON

Track 2: Sustainable Buildings and Infrastructure
Knowledge Network: Cities & Globalization, Sustainable Planning, Design & Behavior
Methods: Case Study, Interpretive-Historical Research
James Tice (University of Oregon)

Introduction

Parking in the American city is an omnipresent problem that should be better known because it affects our everyday lives in important ways, shapes our cities, and usually not for the better. A deeper understanding of its history and concomitant problems may just help improve the quality of life in the city for all, allowing us to counteract its worst effects. Parking has emerged as a topic that has finally attracted the attention it deserves, not just in urban design circles but among others who care deeply about civic life and the future of livable, sustainable, and beautiful cities (Grabar).

Donald Shoup: The High Cost of Free Parking

In his 2017 publication, The High Cost of Free Parking, Donald Shoup, identifies parking as one of the most devastating aspects of contemporary American culture (Shoup, 2017). He points out, in his 700-page book, that the automobile and the need to accommodate it through storage—parking—has so overtaken our way of life and affected our environment to such an extent that it is almost impossible to imagine a world in which it does not exist. He asks why is parking free—requiring acres of public land at enormous cost in resources and quality of the environment—while housing is so expensive? Parking, especially off-street parking, has proven to be costly, environmentally wasteful, and one of the biggest obstacles to affordable housing. Shoup (2017) goes on to say, and this is perhaps the most astonishing fact in his book, that in the world of planning professionals, community leaders, and well-intentioned citizens this scourge on our way of life has, by and large, has not been insufficiently recognized much less seriously studied. Its impact it seems, has been symmetrical: the more profound the problem, the more it has been ignored. A notable exception is Henry Grabar's (2023) Paved Paradise: How Parking Explains the World. I have relied primarily on both texts to support my argument below.

Electric vehicles

No doubt the growing acceptance and consequent proliferation of electric vehicles will have important benefits for the health of our environment by reducing carbon emissions, but there is no guarantee that this emergent pattern will seriously reduce the number of vehicles on the road or their need to be 'housed' somewhere. The problem of parking will likely remain.

Forbidden Urban Jewels

My epiphany about the deleterious effects of parking came in 1974. I became aware of a wonderful housing type in Los Angeles. Courtyard Housing in Los Angeles (Polyzoides et al.) presented the history and analyzed a beautiful and often unassuming housing type quite common in LA but found in other parts of the country as well: courtyard housing. This desirable housing type found fertile soil in Los Angeles in the 20s as an affordable solution to the large influx of Mid-Westerners who were drawn to the sun and promise of the good life in Southern California.

A typical example of courtyard housing consists of six to twelve living units assembled around a garden court. The small scale and modest accommodations disguised their profound impact on the city in the 20s where they provided housing for people of modest means to 'fit' into neighborhoods otherwise dominated by the single-family house. The 'low courts' found a unique and well-developed architectural expression in many parts of the city like East Los Angeles, where today they still provide homes for a largely immigrant population which benefits from the low-cost affordability they provide. Meanwhile, the West Hollywood 'high courts', which are near the film studios, were home to movie stars like Rudolf Valentino and Clark Gable. Even here, modest accommodations of twelve to fifteen units surrounding one or more courtyards were typical and well-adjusted to a standard-sized lot or multiple thereof. Despite being well-loved by their past and current occupants, I was shocked to learn that these buildings were all 'illegal'. Why? Because of parking! Building codes in Los Angeles by mid-century specified exorbitant parking standards that could never be met within the limits of these lovely, sometimes exquisitely designed courtyards. There were simply not enough parking spaces on ANY of these sites to meet updated codes. This was the case even though many potential occupants did not need or even want more parking. But the city would not allow even modest deviations to its codes and consequently, these jewels could no longer be realized. Instead, the 'dingbat' type emerged: a three or four-story structure raised above street level with a large parking area on the ground floor and a continuous curb cut that obliterated the sidewalk and any 'curb' appeal (Grabar). The effect on the street landscape was a wasteland of hollowed-out carports.

https://doi.org/10.52202/082078-0008

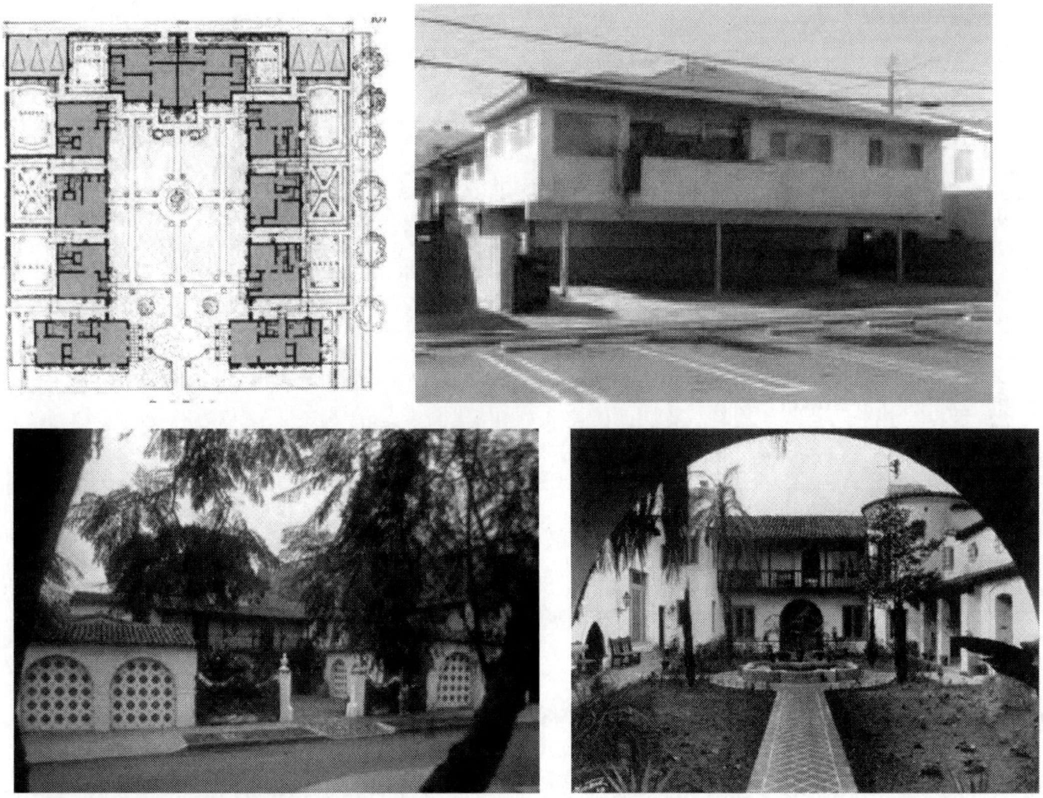

Figure 1. Courtyard housing in Los Angeles, above, replaced by 'dingbat" apartments with parking on grade. (Photo: Julius Shulman and others)

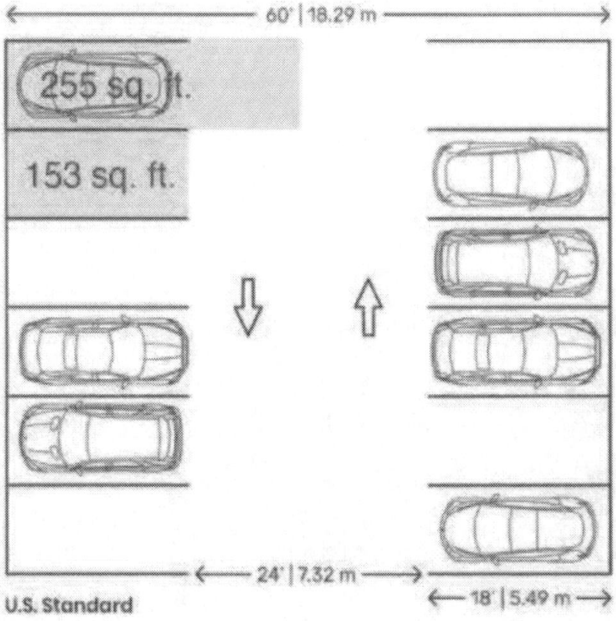

Figure 2. 12 parking spaces = 3,060 sq. ft. (Image from Dimensions.com)

https://doi.org/10.52202/082078-0008

Parking 101

A typical parking stall is 8.5ft wide by 18 ft long. With added space for access and backing up, the space requirement for one car is 255 sq ft (Shoup estimates that 300 sq. ft. is more realistic). Parking requirements for a one-bedroom or studio apartment can equal the size of the apartment and cost 20K or more for construction.

Donald Shoup's growing cadre of followers, calling themselves "Shoupistas", informs us that there are 290.8 million cars in the U.S. as of 2022, or eight parking spaces for every car (taking up about as much land as the state of West Virginia). For all its promiscuity, the American automobile spends 95% of its life in a solitary parking space. There are many negative side effects of parking and parking lots. Environmentally, they contribute to the heat island effect in the city; contribute to carbon emissions by using asphalt, and can result in acres of nonpermeable surfaces that cause harmful stormwater runoff. Aesthetically parking is a blight on the urban landscape. Has anyone ever seen a beautiful parking lot or parking garage? Vast parking lots are a scourge to livable, walkable neighborhoods and cities. Surface parking in cities has taken over acres of land, that could be used for affordable housing, parks, civic structures, offices, and retail the latter could add to the city's tax base and thus pay for services that now go begging. As noted, affordable housing is one of the most important and urgent needs in the U.S. and has been—until recently—held hostage to unreasonable parking standards, required by municipal codes. In some urban areas in Phoenix, this can add a hefty $40,000 (Grabar) to the construction cost of a single housing unit. This practice has had the effect of driving up the cost of housing for those who can least afford it, thus making it impossible for millions to live in affordable dwellings.

Whether occupants at all stages of their lives need vehicles or whether renters or homeowners need or even want the cars required by code is a primary question. Codes have notoriously treated every situation the same, 'one size fits all', without recognizing the diverse needs of the occupants and special circumstances based on location and context.

Urban Deserts: Parking Dominates the Urban Landscape

Beginning in the 1950s mid-sized cities bent over backward to accommodate parking in hopes of bolstering their local economies in large part to compete with new shopping malls on their periphery which offered ample, 'free' parking. To respond to this threat, the strategy was to require minimum standards of off-street parking of all new construction so that downtown could compete with the new shopping centers. Standards were created arbitrarily and uniformly without concern for the unique conditions of the site, diverse demographics, or other contextual considerations, like adjacency to transit lines or denser urban settings where walking or mass transit is preferred to the automobile.

Parking Code 'Alchemy'

The history of municipal parking codes in the U.S. since the mid-century is revealing (Shoup) (Grabar). It helps explain the failure of our current parking laws which have been irrational, self-defeating, and sometimes downright bizarre. Parking codes evolved through various planning policies that have codified every aspect of parking into a pseudo-science that Shoup compares to the practice of bloodletting. Although well-intentioned, and purported to be rational and 'scientific', the codes read more like a medieval alchemist's recipe. Here are some random examples taken from Shoup and Grabar:

- 1.8 spaces per hospital bed
- 1 space for every four beds in a nursing home
- 1 space for every four seats in a theatre
- 1 space for every six seats in a stadium, church, chapel, mosque, synagogue, or temple
- 3 spaces for every thousand gallons in a swimming pool
- 1 space for every pool table in the pool hall
- 5.4 spaces for each bowling lane in a bowling alley
- 1 space for every three holes in a golf course

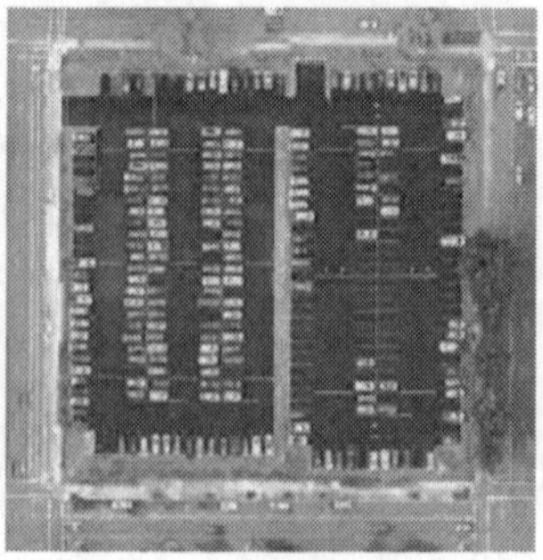

Figure 4. Southdale Mall, Victor Gruen, Edina, MN, 1956 (Photo: Guy Gilette)

https://doi.org/10.52202/082078-0008

Victor Gruen: Mall Master

Shopping malls as the antithesis of downtown emerged in part because of the promise of free parking and easy access. Corporate America could use the benefits of mass merchandising to effectively drive out the competition on Main Street, with its heterogeneous mix of small retail, mom-and-pop stores, and walkable neighborhoods. Variety, if not quality, was sacrificed for theoretically, at least, less expensive goods in controlled environments with, yes, free parking, within a vast asphalt enclosure insulated from the vicissitudes of city life and harsh environments: crime and inclement weather. The emigre architect Victor Gruen was the first to promote the mega shopping center in places like the one in Southdale, Edina, MN which opened in 1956 (Hardwick). It featured 5,200 parking spaces. Gruen set the standard for other malls across the U.S. including Valley River Center just north of Eugene which opened in 1969. As a regional draw, it is the largest shopping center between Portland and Sacramento. Its parking embraces The Urban Land Institute of 1982 recommendation (ULI) that parking be determined by the anticipated surge that occurs on Black Friday, the day after Thanksgiving. The one-day capacity that requires a superabundance of parking space sets the standard for the other 99% of the year (Tachieva, 2011).

Conceptual Framework: Urban Design Principles and the Figure/Ground Plan

Some fundamental ideas about urban design and cartographic representation can make ideas about city form and related parking impact more apparent. I cannot stress enough how important it is to represent the form and space of the city in a manner that makes it visually compelling and comprehensible (Graves, 2009). To use the medical analogy, to effect a cure, the patient must first be subject to an accurate diagnosis. The figure/ground drawing is, in fact, a very effective urban diagnostic tool.

The Good City: Urban Design Principles

Léon Krier (1980), proposes useful visual guides for urban designers. The one showing how the res publica and res economica can weave together into one whole, civitas, is especially telling as it presupposes a diverse, multi-faceted realm consisting of civic institutions, private residences, and commercial activities and public space. Another diagram by Krier examines the city block which he emphasizes is quite literally the 'building block' of the traditional city. He shows two configurations, one anti-urban and the other urban. The latter provides a model for reinforcing the edges of each block, thus defining the street: the public space in most cities. Jane Jacobs (1961) was the first to recognize the importance of the well-integrated and defined street as the social backbone of the traditional city. Portland's Pioneer Square, like the defined street, is an excellent example of a contained, public space of the type proposed by Krier. This approach to urban design where urban space is paramount employing 'form-based codes', or more generically, 'rules of form', is increasingly replacing the land-use planning and functional zoning that prevailed post WWII (Talen, 2011).

The 1748 figure/ground map of Rome by Giambattista Nolli (Tice et al.) is a good example of a clear, concise graphic image of the city and its formal and spatial qualities. Most of us have used Google Maps or various online mapping programs. If one wishes to examine the city for its physical and spatial properties these maps can often be misleading because they typically blur the city's most salient physical characteristics: its solids and voids. If we examine two views of Rome's Piazza Navona in the Nolli plan, we can see clearly that a very large space was 'carved out' of the dense urban fabric of Rome, the black. The solid and white void, are immediately evident. Note that this is the essential quality of space for the piazza for conceptually it is a large

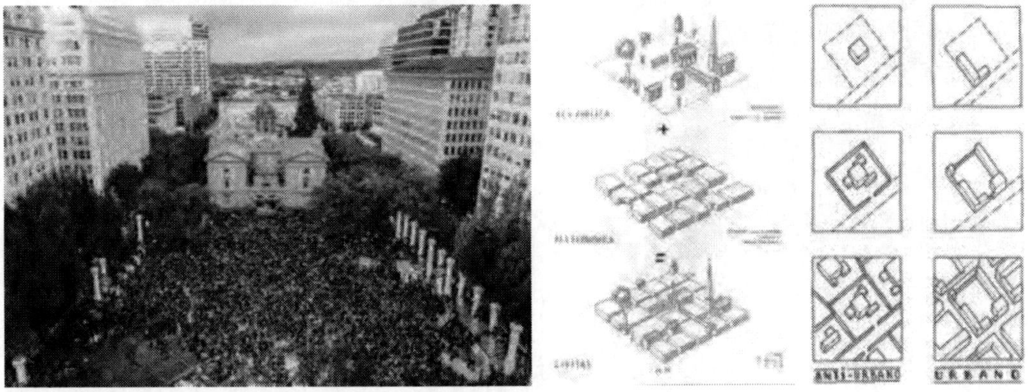

Figure 5. Diagrams of the Good City and Urban Blocks by Léon Krier & Pioneer Courthouse Square, Portland, OR (Photo: Joshua Blake)

https://doi.org/10.52202/082078-0008

outdoor room. When we look at the corresponding section of Open Street Map at the same location and scale, that quality dissolves into a graphic haze. Overlapping and competing values of light and dark, traffic patterns that emphasize the automobile and not the pedestrian, faint building outlines, and distracting text all conspire to obliterate any spatial or formal sense of the city's reality. It makes it impossible for us to imagine the piazza's social and cultural functions as a receptacle providing for all manner of human activities and celebrations, as it is one of the great living rooms of Rome. Likewise, a comparison of Eugene's downtown using the Nolli map method shows the spatial and formal qualities of the city directly. The related Google Map at the same scale, while providing information about traffic patterns and land use, fails to make clear its physical and spatial character. As I will argue below, the form and space of the city can promote, or discourage, the social life of its inhabitants. This figure/ground method can also facilitate comparisons with other cities especially when rendered at the same scale (Jenkins, 2008). The figure/ground plans below show their comparative building density, that is, the percentage of unbuilt voids to built solids. At 50 percent density, Eugene displays almost two to three times the amount of open space compared to other well-known city centers.

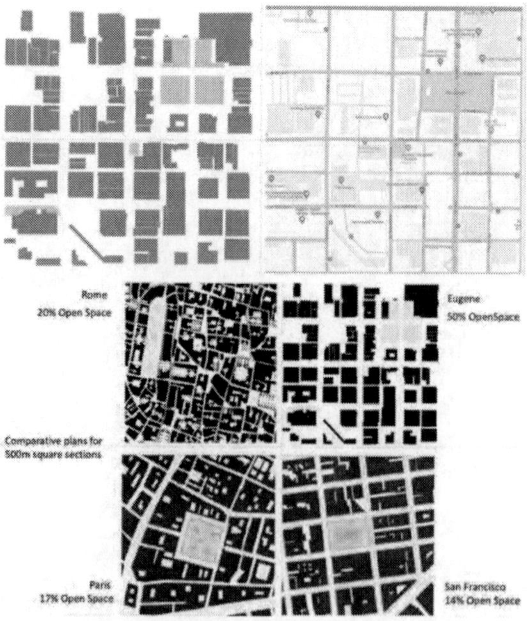

Figure 7. Above: Eugene 2023 F/G Map vs. Google Map covering the same area.
Below: Comparative Urban Densities at the same scale (Paris and San Francisco drawings by E. Jenkins).

The following maps of Eugene use the figure/ground technique, with buildings as black and open space as white, to make clear the physical, spatial dimensions that determine the form of the city and render its unique character. And although it is a two-dimensional plan, giving only a hint at its three-dimensional reality (for example it cannot show the change in building heights near the center of the city as opposed to its perimeter), it is a good place to start. The figure/ground gives a kind of foothold on the complexity of the urban fabric and its morphology allowing one to begin to analyze the city's deep structure (Rowe, Koetter).

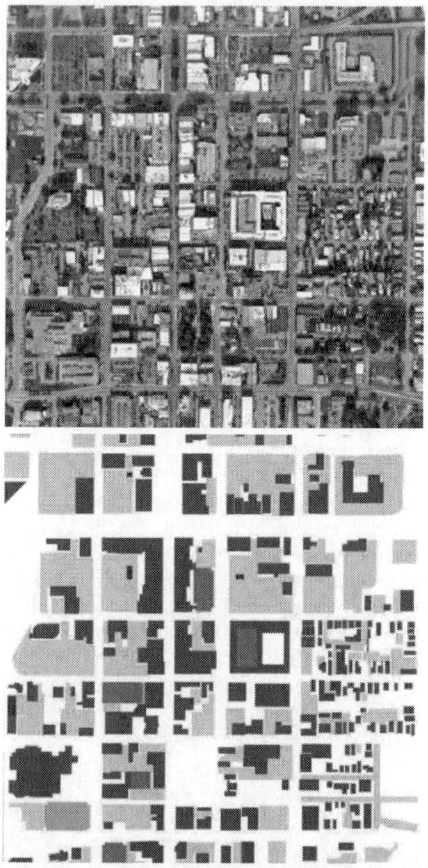

Figure 8. One Quarter Square Mile of Pensacola, FL, Satellite image 2021, (image: Google Earth) Figure/Ground with Parking in salmon, parking garages in maroon. Off-street parking occupies 57.6 Acres.

https://doi.org/10.52202/082078-0008

The Pensacola Syndrome

"The Pensacola Syndrome" describes the widespread malaise in America after WWII resulting from the demolition of buildings in the center city, many of historic value, to provide off-street parking to compete with the perceived convenience afforded by parking at large shopping malls (Grabar). In the process, many cities have eliminated the very reason for people to live, enjoy, and cherish their city in the first place. This cultural lobotomy has typically resulted in a barren city landscape characterized by cultural amnesia dominated by half-empty parking lots. The effects have been felt across America, and Eugene, OR is no exception. It has created a numbing, sterile urban landscape and contributed to social, economic, and environmental ills. In the words of Gertrude Stein, there is simply "no there, there".

Incorporated in 1862 on land occupied by the Kalapuya people', the city of Eugene f was founded by its namesake, Eugene Skinner. It is the county seat of Lane County and is located about 110 miles south of Portland at the southern tip of the Willamette Valley and 60 miles east of the Pacific coast. As of 2020, Eugene had a population of 175,000 making it the second-largest city in Oregon after Portland. Its economic base traditionally relied on the timber industry and agriculture but today the largest employer is the University of Oregon with 19,500 students. The state's flagship institution of higher learning, located to the east of downtown, the UO was founded in 1876; from the beginning, it was charged with a liberal arts curriculum as its major focus, as opposed to the engineering curricula at

A Sketch Portrait of Eugene, OR

Figure 9A. Eugene, OR, View from Skinner's Butte looking south along the Willamette Street axis in yellow. Spencer's Butte is shown as the highest point on the horizon. (Photo: Creative Commons)

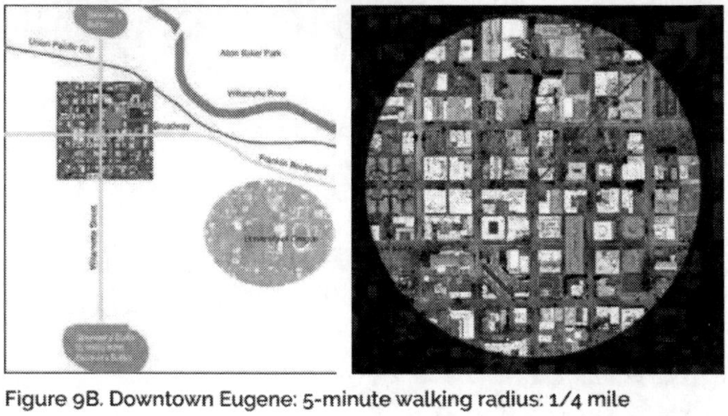

Figure 9B. Downtown Eugene: 5-minute walking radius: 1/4 mile

I would like to bring this topic home to Eugene, OR (Wikipedia). Specifically, I would like to trace the role of parking and its effect on the city from the mid-20th century to the present day focusing on the central core of the city. As we shall see, nearly one-third of the city's area is occupied by surface parking today. The portion of Eugene under examination is the traditional city core, roughly 1/2 mile by 1/2 mile or 1/4 square mile, covering an area that approximates the time needed to transverse its boundaries (it takes 5 minutes to walk the distance of 1/2 mile at a leisurely pace).

Oregon State in Corvallis. Eugene has a unique cultural identity as home to the Oregon Bach Festival, and an equally strong identity as 'Track Town USA' as it hosts important international track events.

The location of the city was set by two important topographic factors: the Willamette River, prone to flooding, and Skinner's butte a basalt outcropping that bears the name of its founder. The city developed on the south flank of the butte as the river was naturally diverted to the north edge. Farther to the south, Spencer's Butte, describes a north-south axis or cardo which coincides with Willamette Street, the city's historic Main Street. The east-west axis or

https://doi.org/10.52202/082078-0008

decumanus is defined by Broadway which links the city to the university campus to the east, and the Whiteaker neighborhood to the west. Willamette and Broadway intersect at the city's 'forum', Kesey Square. The historic Park Blocks and county courthouse occur in the northeast quadrant of the city, defining the historic heart of the city and the original public market which functions as such to this day.

Eugene is a microcosm of the mid-sized American city, reflecting larger patterns of urban development throughout the United States since the mid-fifties. As I begin to examine the physical impact of parking on the city's form, its "physique", I am assuming that its physical well-being is relevant to its moral and spiritual well-being as well. Like the ancient Romans who posited that a healthy body and healthy mind are inseparable, I believe that the fate of a city is inextricably tied to its form.

Comparison of Eugene's Urban Structure and Off-Street Parking

The transformation of Eugene's urban structure over the last 80 years unfolds with alarming consistency: less built fabric defining streets and its historic park blocks and more and more land area devoted to parking which has dramatically metastasized throughout the center since mid-century. Since then the city's fabric has been eroded by a complex set of forces: increasing dependency on the automobile and concomitant invasive traffic patterns and parking requirements, ill-conceived Federal Urban Renewal policies which encouraged the demolition of 'sub-standard' buildings in downtown, along with architectural and planning professions that were increasingly anti-urban in their approach to design within the city.

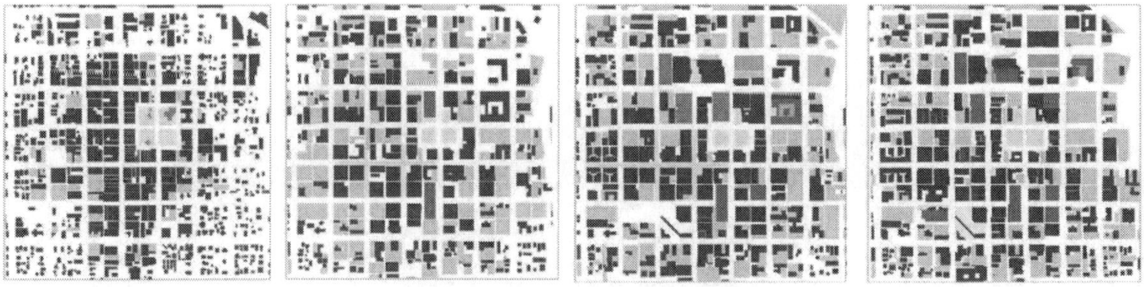

1944 1974 2000

Figure 10. Figure/Ground Plans showing off-street parking in salmon color; parking structures in maroon.

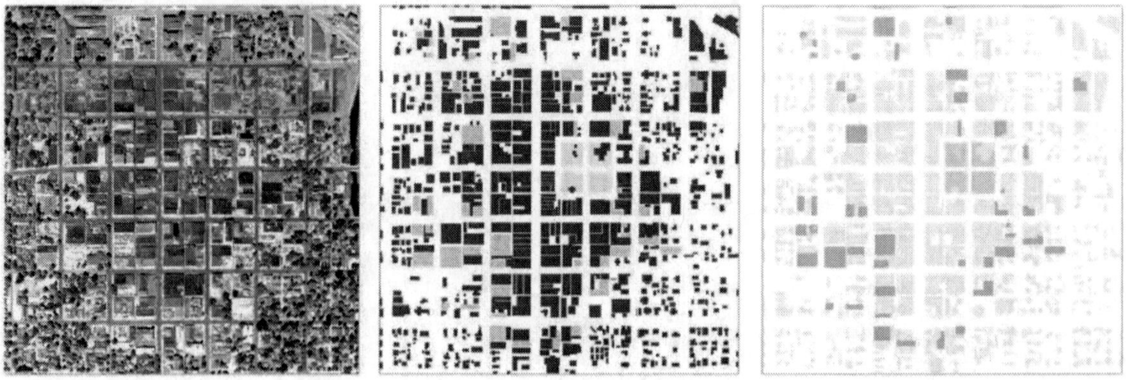

Aerial Photo **Figure/Ground with Parking** **Parking Index: 4.4, Acres: 4.8**

Figure 11. 1944 Analysis of City Plans with Parking

https://doi.org/10.52202/082078-0008

1944: Intact

The 1944 aerial photo and figure/ground with off-street parking indicated in color, shows a compact downtown with a strong gestalt describing continuous streets and well-defined Park Blocks. Located in the northeast quadrant of the blocks, the Lane County Courthouse is rendered as a free-standing object that emphasizes its honorific role as a civic monument and symbol of justice and the rule of law. It is the most memorable architecture in the city's center. A few parking lots play a subsidiary role and are generally located on the periphery of the city. The city's growth patterns display a slight north-south development along the Willamette Street axis although, overall, the center conforms to a roughly concentric model. The east-west axis, and especially the development to the east, has been held in check by three conditions: the periodic flooding of the river, the Union Pacific rail lines, and large canning operations along its river's edge which were serviced by the railroad and a mill race diverted from the river.

Figure 12. 1974 Analysis of City Plans with Parking

Aerial Photo | **Figure/Ground with Parking** | **Parking Index: 47.0, Acres: 51.2**

1974: Eroded

The aerial photo and figure/ground with off-street parking in color show that an astonishing 30 percent of the city fabric has been demolished. New off-street parking fills virtually all the sites destroyed, and two large parking structures have been added. Beyond creating off-street parking to compete with the strip malls and Valley River Center, Eugene was a (willing?) victim of the Federal Government's practice of removing 'outdated' structures through its urban renewal program, now recognized as one of the most devastating urban policies in the post-WWII era (Jacobs, J.). Perhaps the practice is best summed up for Eugene, not as 'urban renewal', but as 'urban removal'. The city's experiment with a pedestrian-only street along Broadway is shown in green. It was a valiant, but ultimately failed, attempt to make the downtown more attractive to retail users and shoppers. A new city hall, with a large parking lot at the basement level, is within the city center district. Although the southern quadrants of the Park Blocks remain intact, overall, the Park Blocks were seriously compromised. In a one-two punch, the historic courthouse was demolished only to be replaced by a sprawling new facility that encroached on the northeast quadrant. Meanwhile, the northwest quadrant was nearly obliterated by a two-story parking structure.

https://doi.org/10.52202/082078-0008

Figure 13. 2000 Analysis of City Plans with Parking

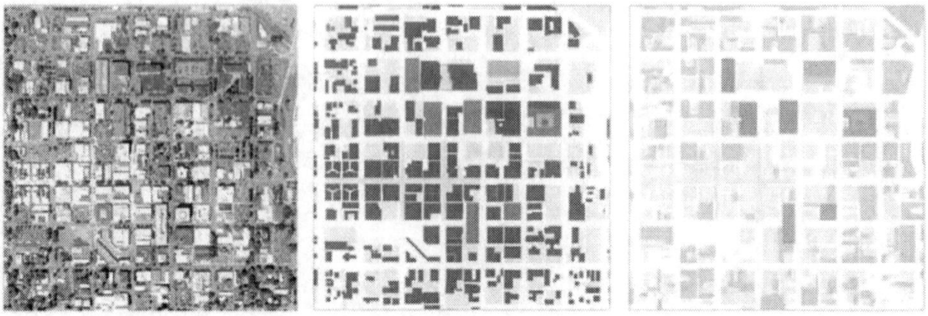

Aerial Photo | Figure/Ground with Parking | Parking Index: 47.0, Acres: 51.2

2000: Modest Improvements

The aerial photo and figure/ground with off-street parking show a modest increase in the built fabric with a concomitant decrease in off-street parking. There is an emphasis on Broadway, the east-west axis giving this major street a measure of definition and meaning to the urban core. Its crossing with Willamette Street is defined by a modest plaza, now christened Kesey Square. A new performing arts center, The Hult, occupies the northern end of Willamette Street and features a dedicated five-story parking structure adjacent while a new bus transit hub to the south, also close to Willamette, suggests a counter-measure to the dependency on the automobile.

Figure 14. 2023 Analysis of City Plans with Parking

Aerial Photo | Figure/Ground with Parking | Parking Index: 47.0, Acres: 51.2

2023: Current State

The Google Earth satellite image shows the city in its current state. Some gaps in the urban fabric evident after mid-century have been filled. Importantly the city has a new public library in its downtown, with public parking in the basement. However, the city still lacks a cohesive urban structure and concomitant viable commercial, residential, and civic functions. The intrusive parking structure in the Park Blocks was finally removed and replaced now by a market building and plaza with limited surface parking, perhaps eventually to be built upon by a new city hall. The former city hall, however, was demolished in 2015. The entire city block it occupied is now a vast surface parking lot. The city's parking index has improved slightly but parking is still omnipresent. A 'report card' follows at the end of this essay summing up Eugene's status.

https://doi.org/10.52202/082078-0008

Figure 15. Transformation of Eugene's 19th century Ideal Park Blocks (1944–2023)

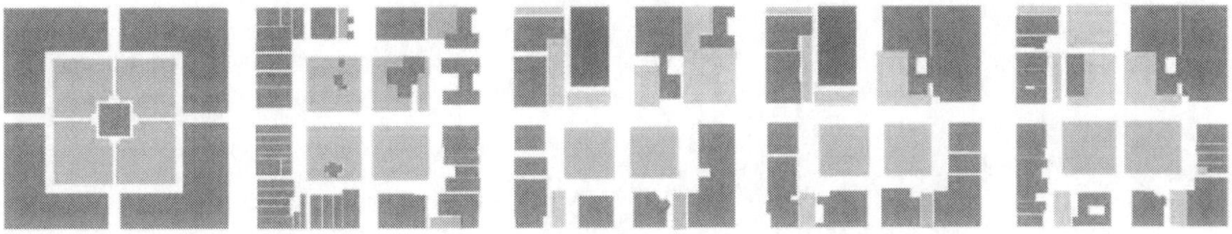

Park Blocks and the Erosion of the Ideal

The erosion of Eugene's Park Blocks demonstrates some of the negative effects brought about by the anti-urban forces that promoted parking at any cost. The idealized 'four square' park blocks as shown in the 1944 map with the Lane County Courthouse as a free-standing object have been systematically compromised. This ideal is a well-known urban type reminiscent of the courthouse square and its prominent courthouse celebrated by Colin Rowe and John Hejduk in their essay "Lockhart Texas" (Rowe, Hejduk). In 1959 the handsome Richardsonian courthouse was demolished to make way for a new county building of indifferent architectural and urban value. The damage done to the integrity of the Park Blocks from an urban design point of view was exacerbated by the large sprawling county building which encroached upon its domain on the northeast corner. This undercut the original intention of an identifiable 'four square' park and consequently denied the pedestrian promenade that previously encircled all four blocks. In 1974, as the map detail from that year shows, a two-story parking structure in the northwest quadrant, absorbed most of that block, effectively canceling its urban role. Further demolitions and building setbacks with additional automobile parking nibbled at the edges of the parks so that by 2000 the blocks were seriously compromised. Their role as an ordered, outdoor room for celebrations and civic pride had evaporated, and parking was at least partially to blame. By 2023, there were concerted efforts to restore the urban significance of the Park Blocks some of which were partially successful. The two-story parking structure in the northwest quadrant was removed making room for a public plaza and market building but the fate of its northern portion is still in the balance. Overall, the Park Blocks have been subjected to anti-urban forces since mid-century, parking being one of the most invasive, from which it is now struggling to recover.

Figure 16. Park Blocks, Lane County Courthouse, 1898–1959 (destroyed) Willamette Street Parade, ca. 1945 (Photos: Dotson and Anon.)

The historic preservation movement provides an important commentary on Eugene's Park Blocks and its downtown streets. It suggests that a change in attitude is necessary if the city's old mistakes are not to be repeated. Jacqueline Kennedy Onassis has given voice to this issue:

Is it not cruel to let our cities die by degrees, stripped of all her proud monuments, until there is nothing left of all her history and beauty to inspire our children? If they are not inspired by her past, where will they find the strength to fight for her future? Americans care about their past, but for short-term gain, they ignore it and tear down everything that matters. Maybe ... this is the time to take a stand to reverse the tide so that we won't all end up in a uniform world of steel and glass boxes (Peterson, Littenberg).

Report Card for Eugene, OR: Parking and the City as of 2023

Eugene's current state shows a city recovering from some of the most debilitating aspects of post-war development, but it still has a long way to go to recover some of the best urban qualities that were present in its pre-WII state. Off-street surface parking remains one of the most conspicuous impediments to the realization of a healthy downtown.

https://doi.org/10.52202/082078-0008

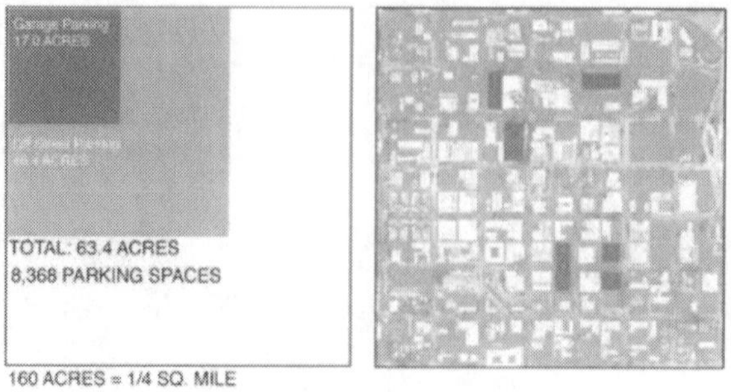

Figure 17. Parking Report Card: Eugene 2023

- Off-street parking makes up 29% of the land area of the central city or 46.4 acres
- Garage parking totals 17.0 acres
- Total space devoted to parking - 63.4 acres
- TOTAL NUMBER OF PARKING SPACES AT 300 SQ. FT. EACH - 8,368 PARKING SPACES

A guestimate based on visual inspection of current satellite images shows that up to 60% of the total surface parking space in Eugene goes unused during a typical weekday.

Parking Reform

Shoup discusses several strategies to realign off-street parking requirements with actual needs. Among them are: time sharing (tradeoffs) so that the same parking spaces can be used for daytime use (office employees) with nighttime users (apartment dwellers); reductions based on transit zones; and variable demands based on demographics; and fees for individuals who work in office buildings or occupants of apartments to be waived for those who do not wish to have the cost of parking attached to their salary or rent (Shoup). He also advocates digital signs posting real-time availability for paid curb-side parking reducing cruising time and inconvenience, a method now in effect in Bend, OR, and other cities such as San Francisco.

Figure 18. Downtown Los Gatos, CA. showing parking garages located mid-block while curb-side parking occurs on Santa Cruz Avenue (left in the image). (Image: Google Earth)

Putting Parking in its Place

From an urban design point of view, the city of Los Gatos, CA provides an approach to parking that mitigates some of its most negative effects by putting parking in its place (Stouhi). Not only quantity but placement of parking in the city's fabric can be crucial. The strategy adopted is a simple variation of Krier's diagram above where the city block is reinforced at the perimeter by buildings while block interiors are occupied by long-term parking lots (two split-level stories) which also happen to be free, suggesting the municipality's willingness to pay for this amenity. The continuity of the exterior pedestrian street is maintained (along with short-term curb-side parking). Intermittent passages through the street wall provide links between the pedestrian realm of the street and the interior block parking. The lesson is simple: all parking need not be banned categorically; by placing it strategically within the urban fabric, so that it no longer takes pride of place, its negative effects can be mitigated through intelligent design.

The Future of Parking in the City

In 2016, Portland, Oregon, waived parking requirements for affordable housing developments, showing how zoning changes could make urban housing more available and affordable. In 2017, Buffalo, (Hess, Rehler) New York, became the first U.S. city to stop requiring development projects to include at least a minimum amount of parking, codes that had been on the books since 1953. Other cities followed, including Hartford, Connecticut, and Santa Monica, California. Many cities are now considering reforms, and a bill pending before the California Legislature would remove minimums for new buildings near public transportation across the Golden State.

In May 2021, Minneapolis struck down minimum parking requirements for new development as part of its climate and greenhouse gas emission goals. From San Diego and Salt Lake City to Raleigh, North Carolina, and Richmond, Virginia, cities are considering similar changes. In the future, U.S. cities could look quite different, designed for pedestrians rather than parked cars.

The following was posted by the Eugene City Council on November 17, 2023 (Eugene City):
Eugene City Council aligns parking requirements with new state climate rules, following the provisions of the Climate-Friendly and Equitable Communities (CFEC) rules. After substantial public involvement, on November 13 the Eugene City Council adopted changes to the land use code to eliminate off-street parking requirements citywide. Additionally, these changes include other required parking reforms such as requiring tree canopy

https://doi.org/10.52202/082078-0008

or solar panels to cover large parking lots and establishing parking maximums for some uses where they did not yet exist.

In response to a 2020 state mandate setting goals to reduce pollution, Oregon's Land Conservation and Development Commission adopted the Climate-Friendly and Equitable Communities (CFEC) rules. Eugene and Springfield, among other metropolitan areas across the state, are required to change housing and transportation planning systems to encourage more climate-friendly development and reduce emissions from transportation.

These late developments are tremendously helpful initiatives for the health of the city that just may be able to put parking in its place and give cities across the country, including Eugene, a fighting chance to become the place where we all would like to live, work, and take pride.

References

"Eugene City Council aligns parking requirements with new state climate rules", (2023). https://www.eugene-or.gov/civicalerts.aspx?aid=6309]

Grabar, Henry, (2023). Paved Paradise: How Parking Explains the World, Penguin Press, New York, 2023.

Graves, Charles P. Jr., (2009). The Genealogy of Cities, Kent State University Press, Kent (OH).

Hardwick, M. Jeffrey, (2004). Mall maker: Victor Gruen, architect of an American dream, University of Pennsylvania Press, Philadelphia.

Hess, Daniel Baldwin and Rehler, (2023). Jeffrey, "Minus Minimums: Development Response to the Removal of Minimum Parking Requirements in Buffalo (NY)", Journal of the American Planning Association, https://www.tandfonline.com/doi/full/10.1080/01944363.2020.1864225

Jacobs, Allan B., (1993). Great Streets, MIT Press, Cambridge (MA).

Jacobs, Jane, (1961). The Death and Life of Great American Cities, Random House, New York.

Jenkins, Erik J, (2008). To Scale: One Hundred Urban Plans, Routledge, Abingdon (Eng) and New York (NY).
Krier, Léon and Culot, Maurice, (1980). Leon Krier: drawings, 1967–1980, Ed. des Archives d'architecture moderne, Bruxelles.

Peterson, Steven and Littenberg, Barbara, (2020). Space & anti-space: the fabric of place, city, and architecture, Oro Editions, Novato, CA.

Polyzoides, S., Sherwood, R. and Tice, J., (1992. 2nd ed). Courtyard Housing in Los Angeles, Princeton Architectural Press, New York.

Rowe, Colin and Hejduk John, (1957). "Lockhart, Texas", Architectural Record, 121 (3).

Rowe, Colin and Koetter, Fred, (1978). Collage City, MIT Press, Cambridge, MA.

Shoup, Donald, (2017). The High Cost of Free Parking, Routledge, Taylor & Francis Group, London and New York.

Stouhi, Dima, (2023). "From empty grids to interactive playgrounds", Arch Daly, https://www.archdaily.com/960672/from-empty-grids-to-interactive-playgrounds-parking-lots-and-their-evolving-identities

Talen, Emily, (2011). City Rules: How Regulations Affect Urban Form, Island Press, Washington, D.C.

Tachieva, Galina, (2010). Sprawl repair manual, Island Press, Washington.

Tice, James, et al. (n.d.). The Interactive Nolli Map Website 2.0, https://web.stanford.edu/group/spatialhistory/nolli//index.html

Urban Land Institute, (1982). Parking Requirements for Shopping Centers, Washington, D.C.

Wikipedia, (n.d.). Eugene, Oregon, https://en.wikipedia.org/wiki/Eugene,_Oregon

Note: All diagrams are by the author except for Figure 2, "Parking Standards", Figure 5, "The Good City and the Urban Block" by Léon Krier, and Figure 7, "Comparative Figure/Grounds" (partial plans of Paris and San Francisco by Erik Jenkins). Aerial photographs are courtesy of the University of Oregon Map Library. Google Earth maps are found online at [https://earth.google.com/web/].

https://doi.org/10.52202/082078-0008

MODERN SLAVERY: A FRAMEWORK FOR MITIGATING FORCED LABOR IN THE BUILDING MATERIALS SUPPLY CHAIN

Track 2: Spatial Justice, Equity, and Inclusion by Design
Knowledge Network: Researchers in Professional Practice, Sustainable Planning, Design & Behavior
Methods: Case Study, Interviews
Joon Ta (MSR Design)

ABSTRACT

Despite the common belief that slavery belongs to the past, it persists in contemporary society, manifesting in diverse forms. The COVID-19 pandemic, economic shockwaves, widespread job loss, and an upsurge in poverty, have left millions of workers worldwide in situations of heightened vulnerability and their fundamental principles and rights at work at greater risk. Modern slavery is a global phenomenon occurring in every country and industry, with the 2021 Global Estimates revealing that 50 million people in situations of modern slavery on any given day, either forced to work against their will or in a marriage that they were forced into (International Labour Organization, 2022). The current understanding of forced labor within the built environment predominantly concentrates on materials and countries deemed to be high-risk. High-risk materials are more likely to include forced labor in their making, growing, or extraction. High-risk countries are more likely to use forced labor practices to manufacture a product or perform construction activities. There exists a pressing need to expand knowledge on this topic. The research aims to raise awareness of the issue, proposing a framework and offering resources as an advocacy tool to empower design professionals in recognizing and mitigating the risk of forced labor within the supply chains of specified building materials.

KEYWORDS

modern slavery, forced labor, materials supply chain transparency, rubber flooring

1 INTRODUCTION

The research is a collaborative effort between a university and a leading architecture firm as part of a Master's program at the university. The graduate student is a primary researcher with the support of academic advisors and firm mentors. The topic was formed based on the mutual interest of the student and the firm as they understand the importance of social equity in the built environment and the gap in the literature related to the topic of forced labor in the building materials supply chain. The research is conducted from June 2023 to December 2023.

The research questions include:

1. How can design professionals mitigate the risk of having forced labor in the supply chain of building materials?

- What types of work are most likely to involve forced labor in the supply chain of building materials?
- At which stage of the life cycle does forced labor most likely happen in the supply chain of building materials?
- What questions can design professionals ask manufacturers to collect information on forced labor in the materials supply chain? What information should they look for in the documentation provided by manufacturers and what does the information mean?

To answer these questions, it is important to understand the forced labor definition and the material supply chain and establish connections between the two elements to gain insight into the present status of forced labor within the building materials supply chain.

1.1 Forced Labor Definition

Since forced labor and modern slavery are not defined by law, the research uses the definition of the International Labour Organization (ILO). According to the ILO Forced Labour Convention, 1930 (No. 29), forced or compulsory labor is "all work or service which is exacted from any person under the threat of a penalty and for which the person has not offered himself or herself voluntarily" (International Labour Organization, 1930). This definition consists of three elements:

1. Work or service refers to all types of work occurring in any activity, industry, or sector including in the informal economy.
2. Menace of any penalty refers to a wide range of penalties used to compel someone to work.
3. Involuntariness: The terms "offered voluntarily" refer to the free and informed consent of a worker to take a job and his or her freedom to leave at

https://doi.org/10.52202/082078-0009

any time. This is not the case for example when an employer or recruiter makes false promises so that a worker takes a job he or she would not otherwise have accepted (International Labour Organization, n.d.).

Forced labor is a form of modern slavery that covers a set of specific legal concepts including forced labor, concepts linked to forced labor (i.e., debt bondage, slavery, slavery-like practices, and human trafficking), and forced marriage (International Labour Organization, 2022) (Figure 1). Forced labor is further divided into privately-imposed forced labor and state-imposed forced labor. As reported by ILO and Walk Free Foundation (2017), some means of coercion of forced labor exploitation include:

- Physical violence
- Sexual violence
- Threats of violence
- Threats against family
- Locked in work or living quarters
- Kept drunk/drugged
- Punished through deprivation of food, sleep, etc.
- Punished through fine/financial penalty
- Threats of legal action
- Withheld passport or other documents
- Had to repay debt
- Withheld wages
- Too far from home and nowhere to go
- Other penalty

State-imposed forced labor includes the following means of coercion according to ILO and Walk Free Foundation (2017):

Abuse of conscription: Any work or service exacted from conscripts that is not purely military, such as work of general interest, or the use of conscripts for economic development purposes.

Obligation to perform work beyond normal civic obligations: Any work or civic obligation to participate in public works or in civil/civic services that go beyond normal civic obligations, for instance, the requisitioning of persons to perform public work, mass mobilization of children, students, residents, civil servants, and any individual for participation in government events; forced mobilization of citizens at the benefit of private actors.

- Abuse of the obligation to participate in minor communal services: Work imposed on members of a community which is not minor in scale, not in the direct interest of the community, and has not benefited from prior consultation of the members of the said community on the need for such works.
- Compulsory prison labor of prisoners in remand, or administrative detention: Mandatory labor of prisoners in remand or administrative detention.
- Compulsory prison labor exacted for the benefit of private individuals, companies, or associations: Mandatory labor of prisoners in privatized prisons or prisoners in public prisons placed at the disposal of private entities inside or outside the prison premises. Amounts to forced labor when prisoners have not given their free, formal, and informed consent to work and when conditions of work do not approximate those of a free labor relationship.
- Compulsory prison labor exacted from persons under certain circumstances: Compulsory prison labor exacted from persons:
- As a punishment for holding or expressing political views or views ideologically opposed to the established political, social, or economic system.
- For labor discipline.
- As a punishment for having participated in strikes.
- Forced recruitment of children by governments or militia groups: Forced or compulsory recruitment of children under 18 for use in armed conflict, whether by military forces, paramilitary, or rebel groups.
- Note: While forced recruitment of children by governments or militia groups is part of the typology used for measurement, ultimately this form was not able to be measured.
- Compulsory labor for the purpose of economic development: Compulsory labor or services exacted as a method of mobilizing and using labor for purposes of economic development.
- Development work
- Prison labor

Privately-imposed forced labor and some state-imposed forced labor terms do not have further descriptions.

Article 2(2) of Convention No. 29 (International Labour Organization, 1930) describes several limited exceptions to the "forced labor" definition. According to this provision of the Convention, the following five situations do not constitute forced labor:

- Work exacted under compulsory military service for the necessity of national defense provided that the work imposed on conscripts is of purely military character.
- Normal civic obligations of a fully self-governing country, such as compulsory jury service, or the duty to assist a person in danger.
- Prison labor as a consequence of a conviction in a court of law, provided it is carried out under the supervision and control of a public authority and the convicted person is not hired to or placed at the disposal of private individuals, companies, or associations.
- Work exacted in cases of emergency, such as

https://doi.org/10.52202/082078-0009

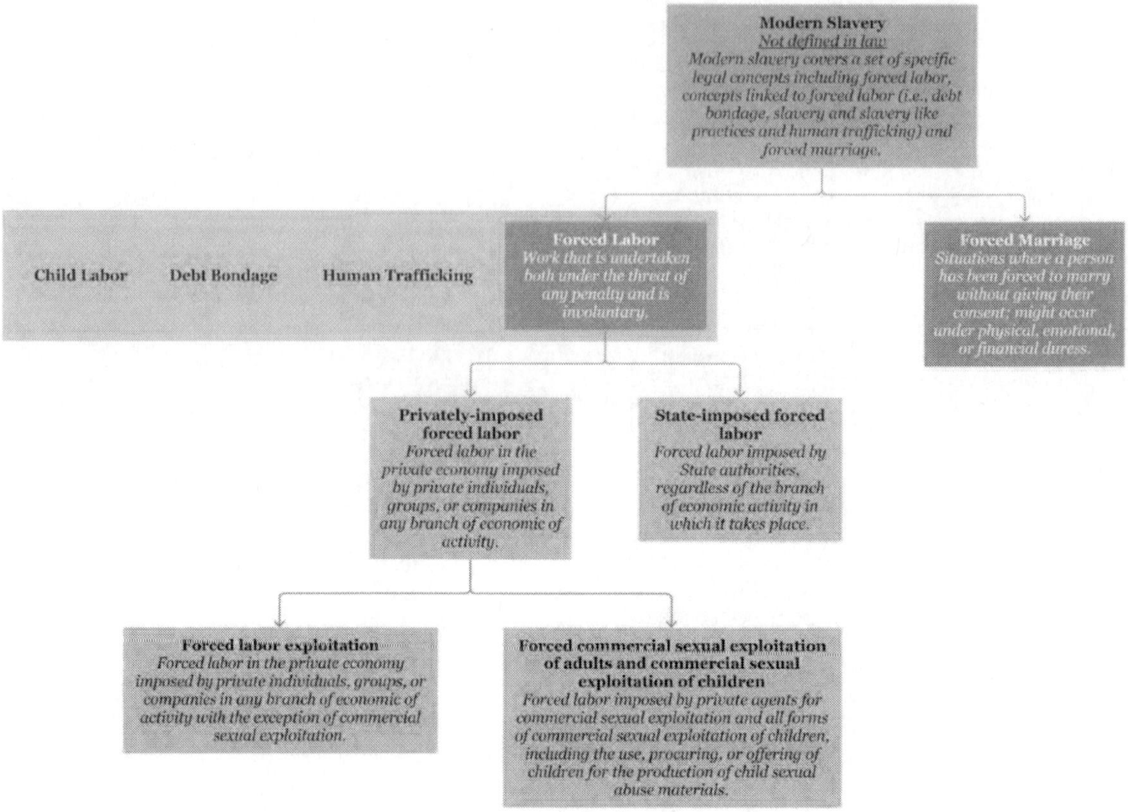

war, calamity, or threatened calamity (e.g. fire, flood, famine, earthquake) or any circumstance that would endanger the lives or well-being of the whole or part of the population.

- Minor communal services performed by the members of a community in the direct interest of that community, provided that the community or its direct representatives are consulted regarding the need for such services.

1.2 Material Supply Chain

According to the American Institute of Architects (AIA) and Carbon Leadership Forum (CLF), the life cycle stages for building products include (Figure 2):

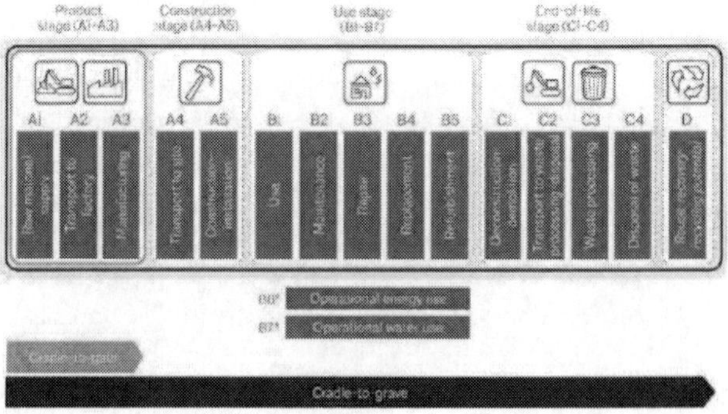

https://doi.org/10.52202/082078-0009

Figure 2: Life Cycle Stages for Building Products

Product stage (A1-A3)

- A1 – Raw material supply
- A2 – Transport to factory
- A3 – Manufacturing
- Construction stage (A4-A5)
- A4 – Transport to site
- A5 – Construction-installation
- Use stage (B1-B7)
- B1 – Use
- B2 – Maintenance
- B3 – Repair
- B4 – Replacement
- B5 – Refurbishment
- End-of-life stage (C1-C4)
- C1 – Deconstruction/demolition
- C2 – Transport to waste processing/disposal
- C3 – Waste processing
- C4 – Disposal of waste
- D – Reuse-recovery-recycling potential

Forced labor can happen in any of a building material's life cycle, however, the high-risk activities often happen during the product stage (A1-A3) where the supply chain structure can get complex with global suppliers. Some of the high-risk activities include raw materials extraction, including mining, extraction process with hazardous conditions, carrying heavy loads, transporting, working underground, etc. (Design for Freedom, n.d.-b). Global suppliers are hard to manage because of the different laws and regulations in each country. Moreover, suppliers often possess just enough power to resist top-down approaches to corporate social responsibility by further "outsourcing" their workforce in favor of temporary, contract labor who receive less legal protection (Parella, 2019), thus making the supply chain even more complex and difficult to manage.

Forced Labor in the Building Materials Supply Chain

The manufacturing sector accounts for nearly one-fifth of all forced labor exploitation of adults, about 3 million people (International Labour Organization, 2022). Manufacturing involves the transformation of raw materials from agriculture, forestry, fishing, and mining or quarrying, as well as the transformation of other manufacturing products into new products. Most forced labor cases occur in production in the lower tiers of domestic or global supply chains (International Labour Organization, 2022).

One of the leading resources to address forced labor in the built environment is provided by Design for Freedom. Design for Freedom was founded by Grace Farms Foundation in 2020 to reimagine architecture by raising awareness and inspiring responses to disrupt forced labor in the building materials supply chain (Design for Freedom, n.d.-a). It is important to note that Design for Freedom defined the scope of their data as forced labor, child labor, and human trafficking, which are not hundred percent aligned with ILO's definition. Thus, the statistics between the two sources may not be comparable.

According to Design for Freedom (n.d.), high-risk materials include:

- Bricks
- Copper
- Glass
- Minerals (Mica, Gypsum, Silica)
- Polysilicon | Solar Panels
- Precursors
- Rubber
- Steel and Iron
- Stone
- Textiles
- Timber

These high-risk materials are further linked to high-risk countries, raw materials, and applications & end products (Figure 3). The data from Design for Freedom was visualized by the researcher using Tableau Public to be used as a tool that helps understand what high-risk materials are, where they are from, and what applications and products include them.

2 RESEARCH OBJECTIVES

The research aims to achieve two primary objectives. Firstly, to raise awareness of forced labor within the building materials supply chain. Secondly, to propose a Forced Labor In The Building Materials Supply Chain Mitigation Framework with resources, serving as an advocacy tool to empower design professionals in recognizing and effectively mitigating the risks of having forced labor within the supply chain of building materials. The research aims to contribute to a more informed and proactive industry approach towards ethical and responsible sourcing practices.

https://doi.org/10.52202/082078-0009

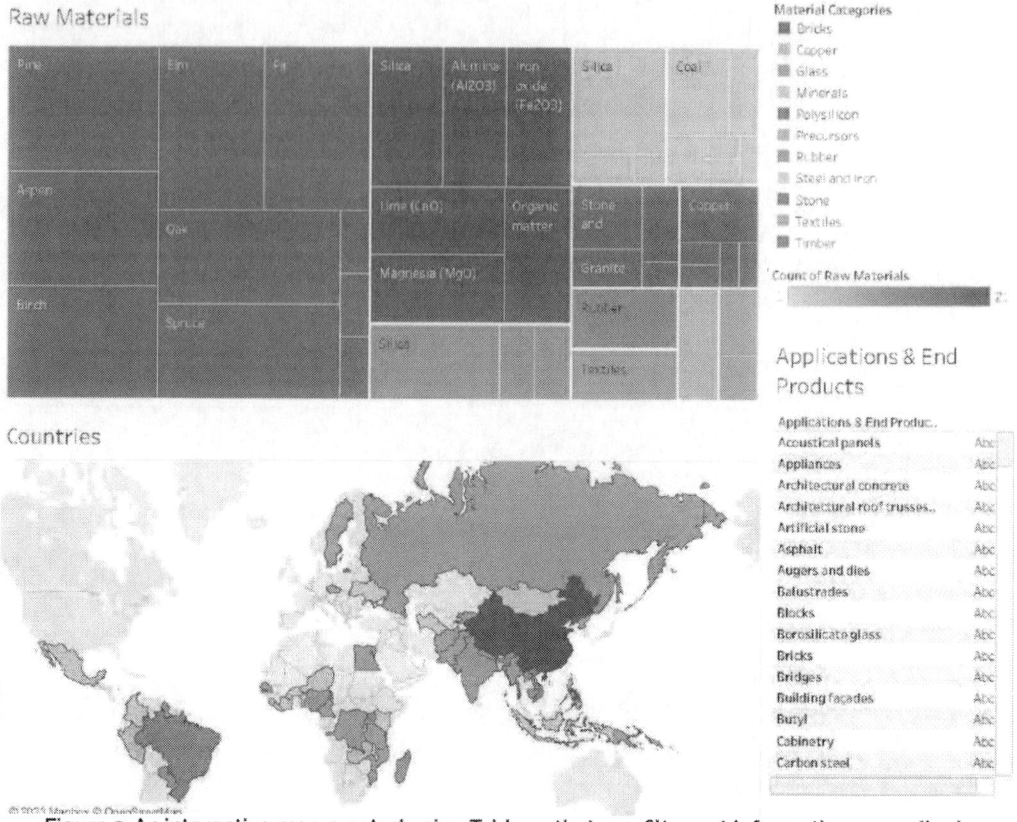

Figure 3: An interactive map created using Tableau that can filter out information accordingly.

3 METHODS

3.1 Literature Review

Literature review includes reviewing global estimates and trends of modern slavery and high-risk goods to be aware of modern slavery issues as a whole, and existing principles and standards, frameworks, certifications, and resources on forced labor in the built environment as well as in the materials supply chain in general to understand the current practice related to forced labor. One important finding is that each organization has its version of forced labor or modern slavery definition since modern slavery is not defined by law. Thus, it is important to note the scope of forced labor covered by each organization when using their information.

Legal issues are closely related to forced labor. Since the supply chain has a complex and global nature, understanding labor laws and international laws is necessary. The literature reveals complex issues of the corporate social responsibility (CSR) model, audit methods, international frameworks, and acts. In the U.S., California is the only state that introduced the Transparency in Supply Chains Act, which requires covered corporations to disclose their efforts to ensure that their supply chains are free from slavery and human trafficking, including information about their practices concerning verification, audits, certifications, internal accountability standards and procedures, and training (Parella, 2019).

There is a need to improve human rights compliance as there is no serious enforcement mechanism for nondisclosure, and existing disclosure and due diligence do not establish liability for human rights abuses even when a business enterprise causes or contributes to those abuses; liability attaches only for failure to follow certain processes established under national law (Brudney, 2023). One way to improve compliance is external reputational mechanisms (Parella, 2019).

However, the problem with reputational markets for social compliance is that the actors who supply the reputational rewards or sanctions-such as consumers-often cannot tell the difference between a good compliance program and public relations "greenwashing." This confusion creates the risk that "good firms" are under-compensated and "bad firms" are overcompensated by the reputational markets, thereby reducing the likelihood that the former will actually invest in a meaningful compliance program given that the costs invested in a compliance

https://doi.org/10.52202/082078-0009

program are unrecognized and unrewarded in the reputational markets (Parella, 2019).

While there are efforts to address forced labor at the corporate level, there are weaknesses in the supply chain where suppliers resist top-down approaches to corporate social responsibility by further "outsourcing" their workforce in favor of temporary, contract labor who receive less legal protection; or when on-site inspections do occur, suppliers counteract this threat by orchestrating the process to provide a false image of compliance (Parella, 2014). There needs to be more research on forced labor in the built environment since the issue has yet to be widely explored. This research aims to use two case studies of flooring manufacturers as a way to understand and raise awareness of forced labor in the building materials supply chain and to create a framework and provide resources to help designer professionals minimize the risk of having forced labor in the supply chain of specified building materials.

3.2 Case Study

The research involves case studies including interviews with two rubber flooring manufacturers. The goal is to raise awareness, create a framework, and provide resources for design professionals to navigate and minimize the risk of having forced labor in the supply chain of specified building materials. Rubber flooring manufacturers were chosen because rubber is a high-risk material according to Design for Freedom.

Manufacturer 01 was chosen because they had well-developed policies and documentation that address human rights issues including a Code of Ethics, Responsible Sourcing Code of Conduct, Human Rights Statement, Whistleblowing Procedure (Ethics Hotline), Corporate Social & Environmental Responsibility Report, supply chain risk mapping, Environmental Product Declaration (EPD), Cradle to Cradle certification, Living Product Challenge certification.

Manufacturer 02 was chosen because they are a pilot in sustainability and are known to have good sustainability practices. They also have a Code of Conduct, Ethics Hotline, sustainability report, supply chain risk mapping, Cradle to Cradle certification, and Programme for the Endorsement of Forest Certification (PEFC).

Overall, these two manufacturers are good starting case studies to understand the current leading practices in addressing forced labor in the building materials supply chain and set the standard for the framework and what design professionals can expect from manufacturers regarding forced labor issues. Interview questions were developed based on unanswered questions after the literature review. Manufacturer 01 interview was via a 30-minute Zoom meeting with Sustainability Product Manager. The connection was made through the local sales rep. The Zoom meeting was recorded for reference. Manufacturer 02 interview was via email with the Vice President of Global Sustainability. They are both aware of the research effort and the intent of the interviews.

Below is the list of questions for the two manufacturers. They were modified for the research report to keep the manufacturers anonymous.

Manufacturer 01 Interview Questions

1. Can you provide the full list of 19 countries where Manufacturer 01's production sites are located?
2. Do these include suppliers' factories?
3. Can you share the list of suppliers for Manufacturer 01's rubber flooring and where they are located?
4. Where is the natural rubber from?
5. Does Manufacturer 01 Code of Conduct apply to all suppliers or just Tier 1 suppliers?
6. If audit only applies to Tier 1 suppliers, what are some strategies to minimize the risk of modern slavery in Tier 2+ suppliers?
7. Does Manufacturer 01 monitor suppliers' compliance with the Code of Conduct?
8. What are Manufacturer 01's strategies for auditing to make sure the information is reliable? In the past, there were incidents where child labor was found in the supply chain of large retail companies even after the supplier's factory had been inspected and certified by the Worldwide Responsible Apparel Production Group (WRAP) (Parella, 2014).
9. How does Manufacturer 01 assess partners to ensure they comply with international standards relating to human rights at work (mentioned in Manufacturer 01 Code of Ethics)?
10. In Manufacturer 01's Human Rights Statement, it was mentioned that there are training sessions tailored to groups of employees more exposed to risks (fair competition, anti-corruption, and ethics), notably teams in France, Mexico, and Russia. How do you identify which countries are more at risk?
11. Can you explain more about the process of working with EcoVadis on Labor & Human Rights criteria?
12. How does the information on Labor & Human Rights provided by EcoVadis impact Manufacturer 01's practice? How does Manufacturer 01 use the information?
13. Do you subscribe to EcoVadis membership? Does the information get updated annually?

https://doi.org/10.52202/082078-0009

Manufacturer 02 Interview Questions

1. Does Manufacturer 02 have any policies/ code of conduct/code of ethics/statements or certifications that address modern slavery in the supply chain?
2. If yes, do they only apply to Tier 1 suppliers or all suppliers involved in the supply chain?
3. Does Interface have whistleblowing systems?
4. Can you share the list of suppliers for Manufacturer 02's rubber flooring and where they are located?
5. What has Manufacturer 02 been doing to minimize the risk of having modern slavery in the supply chain? (CSR risk mapping, audit, etc.)

4 RESULTS

With the information gathered from the literature review and interviews, supply chain diagrams were created for Manufacturer 01 and Manufacturer 02 (Figures 4 & 5). Manufacturer 01 is located in the U.S. and Manufacturer 02 is located in Germany. Manufacturer 01 opted not to reveal the origin of their natural rubber, whereas Manufacturer 02 openly communicated that their natural rubber is sourced from Vietnam, a country considered high-risk for rubber production. However, it is worth noting that despite this risk, Manufacturer 02 was able to provide a Programme for the Endorsement of Forest Certification (PEFC), emphasizing their commitment to sustainable forest management. Neither of the manufacturers disclosed the details of their supplier lists.

In general, both manufacturers exhibit similar supply chain structures and policies across the entire supply chain. This includes having a Code of Conduct or Code of Ethics for their respective operations, a Code of Conduct Agreement with Tier 1 suppliers that involves audits with a specified notice period, an Ethics Hotline for both manufacturers and Tier 1 suppliers and risk mapping for the entire supply chain conducted either by an external entity like EcoVadis or by an internal Procurement Team.

A notable observation is that there exists a common gap in both supply chains, wherein Tier 2 and higher suppliers operate without being directly influenced by the policies and statements of the manufacturers.

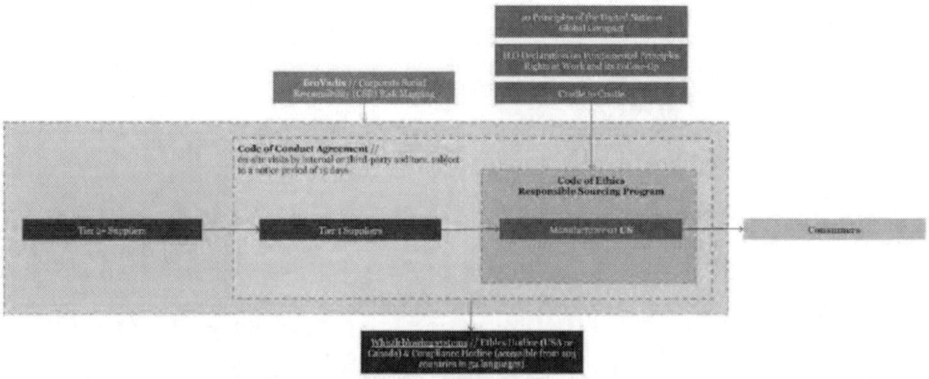

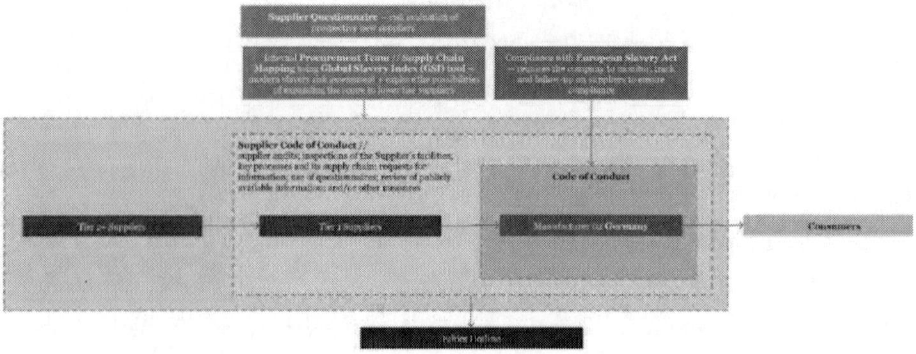

https://doi.org/10.52202/082078-0009

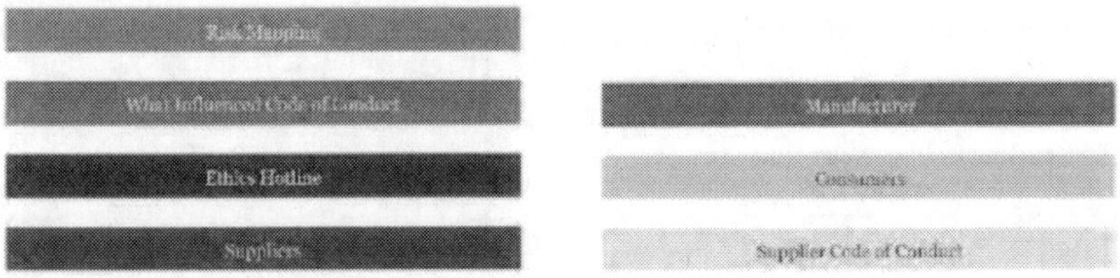

5 THE FRAMEWORK

Learning from the literature review and lessons from case studies, the Forced Labor in the Building Materials Supply Chain Mitigation Framework was developed to best suit designers' and architects' workflow. The Framework includes four steps:
Step One: Identify High-Risk Materials
Step Two: Gather Public Information
Step Three: Request Non-Transparent Information
Step Four: Further Education

5.1 Step One: Identify High-Risk Materials

The data of high-risk materials and countries from the Design for Freedom Toolkit was visualized using Tableau Public to create an interactive tool that helps identify high-risk materials, countries, and applications & end products quickly and easily. The map can be accessed here: https://bit.ly/highriskidentificationmap

5.2 Step Two: Gather Public Information

Table 1 illustrates the list of documentation entailing labor practices information that can be found on manufacturers' websites.

NAME OF DOCUMENTATION	LABOR PRACTICES INFORMATION TO LOOK FOR IN THE DOCUMENTATION
Corporate Social Responsibility (CSR) Report or Sustainability Report	Statements on suppliers' social responsibility, labor, human rights, and societal issues.
Code of Ethics	Suppliers-related information, human rights statement, health and safety statement, whistleblowing system
Responsible Sourcing or Code of Conduct for Suppliers	Information on modern slavery, forced labor, child labor, human trafficking, health, and safety.
Human Rights Statement	Information on modern slavery, forced labor, child labor, human trafficking, health, and safety.
Whistleblowing Procedure	Is it available in different languages? Is it open to all suppliers or only Tier 1 suppliers?
Health Product Declaration (HPD)	Are there chemicals of concern? (e.g. PVC's vinyl chloride monomer building block is a known human carcinogen, which is hazardous for workers.)

https://doi.org/10.52202/082078-0009

NAME OF DOCUMENTATION	LABOR PRACTICES INFORMATION TO LOOK FOR IN THE DOCUMENTATION
Environmental Product Declaration (EPD)	Some EPDs include locations of material suppliers. Check if they are on the high-risk countries list using the Map from Step One.
Living Product Challenge Certification	Did the product meet the requirements for Ethical Supply Chain and Just Organizations?
Cradle to Cradle Certification	Which level of Social Fairness did the product achieve? Refer to the "Certifications & Organizations Addressing Forced Labor" table for details on what each level covers.
Just Label	Supply chain transparency
Declare Label	Red-List Free materials
Responsible Sourcing Certification	ResponsibleSteel Forest Stewardship Council - FSC Programme for the Endorsement of Forest Certification - PEFC etc

5.3 Step Three: Request Non-Transparent Information

If transparency information cannot be found on manufacturers' websites or additional information needs to be requested, the following email template can be used. The email template was adapted from the Design for Freedom toolkit and is meant to start the conversation and collect the most necessary information. For more in-depth questions, refer to Design for Freedom's supplier questionnaire in the Design for Freedom Toolkit.

Dear [Manufacturer],

Our organization has committed to working toward more equitable building materials supply chains. We seek material transparency in the products we specify on our projects. Forced labor is a global phenomenon occurring in every country and industry, with the 2021 Global Estimates revealing that 28 million individuals endure such conditions daily. We strive to prioritize products and manufacturers that support human rights and equitable practices. We invite you to be a collaborator in this important work. We appreciate your commitment to this effort.

Do you have any policies, code of conduct, code of ethics, or statements that address forced labor in the supply chain?

• What grievance and whistle-blowing mechanisms do you have in place for employees?
Please describe.

Please list the country(ies) where your organization manufactures this product.

Do you have Corporate Social Responsibility (CSR) risk mapping? Please describe.

• Do you have an audit program and/or other due diligence program for your suppliers? If yes, do they only apply to Tier 1 suppliers or all suppliers involved in the supply chain?

• Do you monitor suppliers' compliance?

5.4 Step Four: Further Education

One of the most important steps is to continue educating ourselves and team members on the importance of the topic. Below are some additional resources for further education.

1. ***ILO Global Estimates of Modern Slavery***
   https://www.ilo.org/global/topics/forcedlabour/lang--en/index.htm
2. ***50forfreedom***
   https://50forfreedom.org/
3. ***Walk Free***
   https://www.walkfree.org/
4. ***Design for Freedom Toolkit***
   https://www.designforfreedom.org/home/design-for-freedom-toolkit/
5. ***U.S. Living Wage***
   https://livingwage.mit.edu/
6. ***U.K. Living Wage***
   https://www.livingwage.org.uk/
7. ***Global Living Wage***
   https://globallivingwage.org/

https://doi.org/10.52202/082078-0009

6 CONCLUSIONS

In conclusion, human rights transparency in the building materials supply chain is in its early phase, underscoring the need for proactive inquiry and a demand for greater transparency. The Forced Labor in the Building Materials Supply Chain Mitigation Framework serves as a valuable tool to initiate discussions with manufacturers. For more comprehensive inquiries, the Design for Freedom Toolkit provides additional resources. In instances where anticipated transparency information is not provided, it is advisable to escalate the inquiry to senior leaders within the manufacturer or have leaders from your organization seek the required information.

Firms and design professionals are encouraged to take the following steps to enhance their commitment to ethical practices in the building materials supply chain. Firstly, it is essential to familiarize themselves with third-party organizations that specialize in forced labor practices. These entities can provide valuable expertise and guidance in navigating the complexities of ensuring human rights transparency. Secondly, initiate conversations with manufacturers even in cases where they may decline to provide the requested information. Demonstrating a demand for transparency concerning labor and human rights will progressively establish industry standards and expectations for manufacturers in providing such information. Thirdly, it is crucial to evaluate the alignment of the firm's ESG (Environmental, Social, and Governance) goals with the ongoing initiative. Ensuring that ethical considerations are integrated into broader sustainability goals enhances the overall impact of the initiative. Additionally, in the selection and specification of materials, prioritize those with lower forced labor risks. Incorporate social equity as a key factor in the evaluation and selection process, fostering a commitment to responsible sourcing. Lastly, consider taking a proactive stance by becoming a signatory for the AIA Materials Pledge and the Interior Design Pledge for Positive Impact. This commitment further solidifies a dedication to responsible practices and contributes to a collective industry effort to address and mitigate forced labor issues in the supply chain.

7 DISCUSSION

When evaluating risks related to human rights or forced labor, businesses should prioritize the well-being of individuals rather than solely focusing on the impact on the business. Although these concerns may often align, ensuring the protection of victims, rather than solely safeguarding the business's reputation, should be a key consideration in any response to incidents of forced labor.
In cases where harm is attributable to a supplier or third party, the immediate reaction may be to terminate the business relationship. However, this is seldom the optimal approach, as it may leave victims vulnerable to recurring harm. Instead, companies should collaborate with third parties to enhance their practices, safeguard workers, and assess the necessity of remediation, either by themselves or by encouraging other involved parties, including business partners, to address any harm caused.
A comprehensive approach involves implementing a corrective action plan that encompasses both short-term and long-term measures to prevent future harm, alongside providing remediation to victims. Remedial actions may include facilitating access to health, legal, or psychosocial services, supporting repatriation, providing financial compensation, and taking preventive measures against potential future harm.

If a supplier remains uncooperative, companies should explore ways to leverage their influence to encourage the supplier to take corrective action. If, despite efforts, cooperation remains elusive, there may be instances where terminating the business relationship is necessary. In such cases, the primary focus should be on safeguarding workers to prevent any adverse effects resulting from the breakdown in the supplier relationship. Engaging third parties, such as civil society organizations or law enforcement, can be instrumental in providing support wherever possible during this process (Walk Free, 2022).

8 REFERENCES

Brudney, J. (2023). Hiding in plain sight: An ILO convention on labor standards in global supply chains. Chicago Journal of International Law.

Cradle to Cradle Products Innovation Institute. (n.d.). Cradle to Cradle Certified, Version 4.0 - Cradle to Cradle Products Innovation Institute. C2ccertified.org. https://c2ccertified.org/the-standard/version-4-0

Design for Freedom. (n.d.-a). About. Design for Freedom. https://www.designforfreedom.org/about/

Design for Freedom. (n.d.-b). Design for Freedom | Toolkit. Design for Freedom. Retrieved November 2, 2023, from https://www.designforfreedom.org/home/design-for-freedom-toolkit/

End Slavery Now. (n.d.). Combating Child Labor in the Afghan Carpet Sector. Endslaverynow.org. Retrieved December 6, 2023, from https://www.endslaverynow.org/blog/articles/combating-child-labor-in-the-afghan-carpet-sector

EPD. (n.d.). The EPD | EPD International. Www.environdec.com. https://www.environdec.com/all-about-epds/the-epd

https://doi.org/10.52202/082078-0009

International Labour Organization and Walk Free Foundation. (2017, September 19). Methodology of the global estimates of modern slavery: Forced labour and forced marriage. Www.ilo.org. https://www.ilo.org/global/topics/forced-labour/publications/WCMS_586127/lang--en/index.htm

International Labour Organization. (1930). C029 - Forced Labour Convention, 1930 (No. 29). https://www.ilo.org/dyn/normlex/en/f?p=NORMLEXPUB:12100:0::NO::P12100_INSTRUMENT_ID,P12100_HISTORICAL:312174,Y

International Labour Organization. (2016). ILO standard on forced labour: The new protocol and recommendation at a glance. https://www.ilo.org/wcmsp5/groups/public/@ed_norm/@declaration/documents/publication/wcms_508317.pdf

International Labour Organization. (2022, September 12). Global estimates of modern slavery: Forced labour and forced marriage. Www.ilo.org. https://www.ilo.org/global/topics/forced-labour/publications/WCMS_854733/lang--en/index.htm
International Living Future Institute. (2016, November 4). Living product challenge | Living-Future.org. International Living Future Institute. https://living-future.org/lpc/

Kristine, L. (n.d.-a). Brothers carrying stone - Nepal. Lisa Kristine. Retrieved December 6, 2023, from https://lisakristine.com/portfolio-items/brothers-carrying-stone-nepal/?portfolioCats=108#

Kristine, L. (n.d.-b). Stacking - Nepal. Lisa Kristine. Retrieved December 6, 2023, from https://lisakristine.com/portfolio-items/stacking-nepal/

Lewis, M., Huang, M., Stephanie, S., & Simonen, K. (2021, October). AIA-CLF embodied carbon toolkit for architects Part II: Measuring embodied carbon. Carbon Leadership Forum. https://carbonleadershipforum.org/toolkit-2-measuring/

Parella, K. (2014). Outsourcing corporate accountability. Washington Law Review, 89(3), 747-818.

Parella, K. (2019). Improving human rights compliance in supply chain. Notre Dame Law Review, 95(2), 727-794.

PEFC. (n.d.). What is PEFC? Pefc.org. https://pefc.org/discover-pefc/what-is-pefc

Walk Free. (2022). Modern slavery response & remedy framework. https://www.walkfree.org/reports/modern-slavery-response-remedy-framework/

https://doi.org/10.52202/082078-0009

NATURE-BASED SOLUTIONS: SPATIAL RECOGNITION AND OPTIMIZATION OF LARGE-SCALE DRAINAGE SYSTEMS IN PLAIN LAKE CITIES

Track 3: Resilient design for climate change
Knowledge Network: Movement in Designed Environments: Nature & Ecology, Sustainable Planning, Design & Behavior
Methods: Case Study, Design Methods
Huaiyu Xue (Southeast University); Xueyuan Wang (Southeast University); Yuning Cheng (Southeast University)

1 ABSTRACT

Situated in Hubei Province, China, Wuhan is a rapidly developing plain city renowned as the "city of hundreds of lakes." It boasts an extensive water network, crucial for regulating and storing water. However, the city has encountered challenges such as frequent waterlogging and imbalanced water storage due to rapid urbanization. To address this issue, the present study draws inspiration from the concept of "Nature-Based Solutions" (NBS). Focusing on Wuhan's primary drainage system, this research explores the identification and establishment of storage and drainage areas, along with comprehensive planning strategies at the city level. These endeavors align with a sustainable urban stormwater system cycle, harmonizing artificial interventions with natural stormwater processes. This approach ensures equilibrium between the drainage network and natural forces within the system. The findings of this study are as follows: Firstly, by employing the SCS-CN model and surface equal volume filling method, areas capable of accommodating excess surface runoff were identified. This identification was based on simulations of the urban stormwater process, incorporating topographical factors and stormwater recurrence intervals. Secondly, considering construction land data, actual submergence areas, and waterlogging incidents, the large-scale drainage system was organized with a focus on controlling surface elevation. Lastly, the concept of the "storage and drainage function area of the large-scale drainage system" was introduced. This approach involved a hierarchical distribution for optimizing the layout of urban land use in Wuhan. It encompassed various zones, including water storage and drainage areas, enhanced self-drainage regions, waterlogging reduction areas, and low-intensity development areas.

1.1 Keywords

Large-scale drainage system; Nature-Based solutions; hydrologic process simulation; SCS-CN model; waterlogging risk assessment

2 INTRODUCTION

Since the beginning of China's reform and opening-up, Wuhan has entered a period of rapid development in its physical economy. However, within the urban built-up area, there have been issues of green spaces and lakes being encroached upon and the hardening of riverbank buffer zones. The urban landscape has gradually shifted from soft green spaces to hard paved surfaces, leading to weakened infiltration and increased runoff, resulting in the accumulation of water in low-lying areas during heavy rainfall.

Distinguishing itself from flood risks in rivers and mountainous regions, urban rainwater flooding in plains occurs when the intensity of rainfall exceeds the combined capacity of urban spatial storage and drainage systems, leading to water accumulation within the city. Unlike other factors that restrict the development of plain cities, the development of urban areas with high construction density intertwines a dense network of production and living, which makes the urban system sensitive and fragile due to the interdependencies among various subsystems. Therefore, when cities face heavy rainstorms, the resulting losses increase in tandem with urban development.

In order to deal with a series of eco-environmental problems brought about by urban development, a new urban stormwater management method based on the traditional concept of grey infrastructure was proposed (Barbosa et al., 2012; Roy et al., 2008) The stormwater regulation and storage system, as a spatial means of sustainable response to waterlogging (Che et al., 2014), appears in various forms under different development backgrounds, examples include sustainable drainage systems(Hoang & Fenner, 2016; Zhou, 2014), green rainwater infrastructure(Cao et al., 2021) and low-impact development facilities(Jia et al., 2012). The concept of a large-scale drainage system has emerged as a sustainable spatial solution to combat urban waterlogging. Xie Yingxia proposed the construction of a large-scale drainage system to address urban waterlogging disasters in China (Xie, 2013). The theoretical basis originates from the United

https://doi.org/10.52202/082078-0010

Kingdom: small-scale drainage systems typically consist of rainwater pipe networks constructed under low recurrence interval standards to reduce the impact of high-frequency, low-intensity rainfall events (Che et al., 2013). Large-scale drainage systems are based on higher recurrence intervals and various spatial configurations, designed to discharge runoff generated by extreme rainfall events beyond the capacity of small-scale drainage systems.(Li et al., 2017; Cheng et al., 2019) Hydrodynamic models from hydrology, such as SWMM(Chen et al., 2022), InfoWorks CS(Koudelak & West, 2008), MOUSE, HEC-HMS, and SUSTAIN, have been widely used to study urban waterlogging and flood control(Hu et al., 2021). In recent years, these models have been integrated with ARCGIS, RS, and GPS technologies to construct urban rainwater flood process models at multiple scales.

"Nature-based Solution" (NBS) does not advocate a return to the original state of nature or the overlaying of emerging artificial technologies to form ecological management strategies. Instead, it seeks to utilize technological advancements to enhance our understanding of natural systems and harness their inherent capabilities.(Kabisch et al., 2016; Keesstra et al., 2018) It considers the combination of natural forces and artificial technologies as the core components of the proposed solution, particularly in the context of plain cities and their waterlogging issues. The aim is to alleviate pressure on artificial rainwater pipe networks and provide sustainable solutions for urban waterlogging.

Operating through a "storage and discharge" mechanism on the surface, the proposed system heavily relies on the potential energy generated by the natural vertical space within cities. The hydrological processes, land cover types, elevation, and blue-green spaces at the surface largely determine the layout of functional zones within the large-scale drainage system. Therefore, it is crucial to analyze the spatial occurrence and processes of urban waterlogging.

The research focuses on establishing the link between spatial patterns and rainfall-flooding processes through hydrological process simulations. By considering the reciprocal feedback between patterns and processes, a spatial pattern is constructed. Through simulation-based risk assessments of urban flood inundation, factors such as simulated flood extent and water depth are analyzed in conjunction with current land use, remote sensing data, monitoring of waterlogged areas, and other relevant factors. Based on this analysis, functional zones within the large-scale drainage system are designed, optimizing the planning and control strategies for Wuhan's urban green space system and water network. Moreover, solutions are proposed for the identified risks associated with the current construction sites and public facilities in Wuhan, aiming to enhance the city's spatial storage capacity and alleviate urban waterlogging issues at the level of spatial planning.

3 METHODS AND MATERIALS

3.1 Construction Method of the Large-scale Drainage System in Plain Lake City

The planning for artificial drainage to cope with urban waterlogging in most cities in China typically includes the determination of institutional frameworks, zoning, relevant formulas, parameters, and the arrangement of pipe networks and pumping stations. Such planning and design processes are more suitable for cities where land connectivity and drainage are not limited by surrounding rivers and lakes. In plain and water network-based areas like Wuhan, which serves as a representative, the unique natural conditions of internal river and lake systems and the specific drainage space require the construction of a large-scale drainage system that aligns with the water network foundation and responds to the vertical space of the city.

One of the peculiarities of Wuhan is that during heavy rainfall, the water level of rivers and channels exceeds the land elevation. To prevent river water from flowing back into the built-up areas, Wuhan requires a critically integrated hydrological network, regional storage areas, and localized depressions for water retention. By reserving low-lying spaces for water storage in this manner, rainwater can be discharged through natural flow. However, the irreversible nature of land development and the economic constraints of high-density urbanization make it impossible to solely rely on self-drainage for spatial planning. Thus, it is crucial to identify key areas based on the actual rainfall and flooding processes and form a complete large-scale drainage system by combining self-drainage, storage and discharge, and forced drainage methods.

The large-scale drainage system in a plain and lake-based city includes, at a minimum, the internal lake and river channel systems, storage facilities, and artificially modified multi-functional drainage channels. The pattern matching of rainfall and flooding processes in the natural water bodies of the city will directly affect the layout of the overall storage and discharge zones of the drainage system. In order to achieve self-drainage and storage through urban river and lake systems, the planning and design should adhere to high recurrence interval storm standards, calculate the required storage and discharge capacity, organize the space for rainfall and flooding processes in a rational manner, and incorporate vertical adjustments into spatial planning .

https://doi.org/10.52202/082078-0010

3.2 Identification Method of the Functional Area of Storage and Drainage of the Large-scale Drainage System Based on Rainfall-Flood Process

The study utilizes the SCS-CN model and the ARCGIS platform to calculate the surface runoff volume under different rainfall return periods. This is achieved by subtracting the estimated network flow, vegetation and structure interception, and the infiltration of different land cover types from the rainfall amount. Additionally, the study incorporates the simulation of rainfall-runoff processes based on surface elevation to obtain the spatial distribution and height of flood-prone areas under various return periods.

The SCS-CN model is a method developed by the United States Soil Conservation Service to simulate peak flow and surface runoff volume in response to extreme precipitation events within small watersheds(Shi et al., 2009). In this research, multiple small watershed units are delineated based on the flow direction and existing land use conditions. For each unit, the SCS model is employed to calculate the total runoff volume during different intense rainfall return periods. By approximating the cumulative height through volume-based methods, the spatial pattern of flood runoff areas is determined. Finally, the interpretation results of inundation maps, identification of risk-buffer zones based on inundation points, as well as the analysis of green infrastructure and land use status, are overlaid to conduct a comprehensive analysis of the rainfall-runoff pattern. The resulting functional zones for the stormwater drainage system are illustrated in Figure 1.

3.3 Study Area and Data

The city of Plainwater is formed by sediment carried by rivers, with a dense network of waterways and numerous lakes and reservoirs. The water area accounts for a high proportion, and there are often small and large rivers scattered throughout the city. Wuhan is located in the middle reaches of the Yangtze River and the eastern part of the Jianghan Plain. It belongs to a residual hill-type river-lake alluvial plain. As a typical plain lake city, it has the largest lake area in China, with a water area of 803.17 square kilometers at normal water levels. The urbanization rate has reached 80.49%, but the expansion of urban construction land has caused a serious problem of uncontrolled water storage. In July 2016, continuous rainfall for about a week resulted in widespread flooding in the city, causing systematic disasters in multiple areas, with some communities remaining flooded for up to a week even after the rain stopped. The point lakes and linear rivers within the scope of Wuhan City are the core of urban water storage. The vertical spatial organization of the internal lake pattern and water network will directly determine the city's natural runoff capacity. They are ideal research subjects for applying large-scale drainage systems to enhance the city's flood control and water storage capabilities.

For this study, we used the GDEMV2 10-meter resolution digital elevation model of Wuhan City from 2013. High-resolution images of Wuhan City were obtained through the "LocaSpaceViewer 3D Digital Earth" platform. Land use data was derived from the interpretation of Landsat 30-meter remote sensing images for the entire Wuhan City area. Based on the LUCC classification system, land use types were catego-rized into six primary types, including cropland, woodland, grassland, water area, construction land, and unused land, as well as 25 secondary types, including forested land, shrubland, sparse forest land, other woodland, and high, medium, and low coverage grassland. The inundation points in Wuhan's construction land were obtained from the Wuhan Water Affairs Bureau. According to the "Design Standards for Drainage and Flood Control Systems in Wuhan City (2013)", the values for 24-hour rainfall intensity (Table 1) and flood control standards (Table 2) were determined.

The surface runoff coefficient represents the ratio of the total amount of runoff to the total amount of precipitation in the same period and represents the ratio of precipitation to runoff. Combining with the Outdoor Drainage Design Code GB50014-2006, the values of the surface runoff coefficient under different land-use types were determined (Table 3).

4 RESULTS

4.1 Simulation Results of Rain Flood Process and Identification of Storage and Discharge Areas

The surface runoff coefficient represents the ratio of the total runoff volume to the total precipitation volume during a specific time period. It indicates the proportion of rainfall that is converted into runoff and reflects various factors such as soil type, land use patterns, and soil moisture conditions. In this study, the surface runoff coefficients for different land use types were adopted from the "Outdoor Drainage Design Code GB50014-2006". During the rainfall-runoff simulation process, it was crucial to determine a hydrological process scale that is compatible with the urban spatial scale, ensuring the integrity of all small watershed units within the Wuhan city area. As a result, the study area was divided into 15 small watersheds. By combining the surface runoff coefficients with land use data, the surface runoff, inundation volume, water storage range, and cumulative elevation under different return periods were calculated for each small watershed unit in Wuhan. (Figure 1) Furthermore, by overlaying the remotely sensed interpreted flood extent and buffering the risk zones around inundation points, the flood pattern for each small watershed unit was determined. Finally, all the information was integrated to identify the main functional zones of the stormwater drainage system in Wuhan.

4.2 Analysis of Green Space Structure and Storage and Drainage Capacity in Wuhan

Analysis of the Green Space Structure within the Storage and Drainage Zones of Wuhan's Stormwater Drainage System under Different Return Periods (Figure 2): Within the Level 1 storage and drainage zone, the green spaces annually accumulate a significant amount of rainfall, primarily concentrated in the northern part of the city. Green spaces located near lakes, ponds, or rivers serve as the initial recipients of water overflow during the rainy season. In conjunction with the existing river and lake beds, they form new catchment basins that collectively receive surface runoff from surrounding areas or upstream watersheds. The lowest-lying green spaces within the watershed units form small depressions upon which rainfall collects after initial interception and infiltration, resulting in the accumulation of runoff at the lowest point.

The storage and drainage zones of higher return periods extend outward to a certain distance, with a noticeable increase in the southern part of the city, indicating an expansion of the drainage capacity in response to intensified rainfall. Over extended periods, these areas can function normally, but during lower-frequency events, there is a high risk of waterlogging.

The green spaces within the secondary storage and drainage zones correspond to the surface spaces capable of collecting rainwater during low-frequency, high-risk extreme weather events. They fall within the deployment range of the surface stormwater drainage system. Green spaces outside the storage and drainage zones have the capacity for infiltrating and retaining rainwater, as their elevation is higher than the drainage areas within the watershed. In plain urban areas, these green spaces play a runoff-generating role when the soil reaches saturation. Green spaces located along the runoff lines and buffer zones possess the ability to direct flow. Currently, cultivated land is the primary component of green space within the storage and drainage scope (Table 1). On the one hand, due to production needs, farmland itself has a large area, wide distribution, and a certain rainwater collection function; therefore, farmland within the storage and drainage scope with high-frequency recurring periods should pay attention to flood prevention construction. On the other hand, there is a significant amount of land invasion in the city for pond reclamation, and this type of soft space has a high water retention capacity. For this type of land, a strategy of returning farmland to ponds should be adopted.

The forest and grasslands within the built-up area's storage and drainage scope are mostly park green spaces that contain lakes and ponds. During rainfall, rainwater collected here is absorbed by vegetation and soil, and excess runoff is mainly directed to lakes and ponds. This not only meets the irrigation needs of park green spaces but also plays an essential role in rainwater harvesting and utilization as an integral part of urban hydrological circulation.

4.3 Analysis of Water System Structure and Storage and Drainage Capacity in Wuhan

An analysis was conducted on the distribution of water systems and land use structure within the storage and drainage areas of Wuhan City under different heavy rainfall return periods (Figure 3, Table 2). The water systems within the storage and drainage areas can be classified into two categories. Firstly, rivers, canals, and lakes serve as natural depressions where all surface runoff ultimately converges. Secondly, wetlands such as swamps and tidal flats play a role in storing runoff, as well as purifying water quality. Due to its function of receiving upstream water and runoff from surrounding areas, the water level of the Yangtze River during flood seasons is higher than that of the surrounding developed land, and it is artificially dammed, thus, it cannot serve as a runoff channel.

Lakes and ponds within the storage and drainage areas have the highest capacity for storing rainwater within the city. They are capable of storing the maximum amount of rainwater with minimal land occupation. Moreover, their high rainwater collection capacity offers the potential for establishing a natural ecological replenishment system. Therefore, it is crucial to ensure unobstructed pathways for surrounding runoff to flow into the lakes. Most lakes and ponds are situated within the storage and drainage areas, especially under higher return periods of heavy rainfall. Compared to forests, grasslands, and dry fields, lakes and ponds generally maintain a dynamic equilibrium of infiltration saturation, which weakens their infiltration capacity during rainfall events. Additionally, lakes, with their clearly defined basins, can store more rainwater within the same land area compared to dispersed wetlands and tidal flats. Therefore, enhancing land management and control within and around lakes, as well as pre-flood season scheduling, can effectively address urban waterlogging. Although water systems within the storage and drainage areas corresponding to a 100-year return period have lower frequencies, they play a critical role in mitigating urban waterlogging during high-intensity rainfall events. Therefore, they should have emergency flood control functions and the potential for flexible land use conversion. A few lakes located outside the storage and drainage areas, due to topographic and elevation factors, have limited

https://doi.org/10.52202/082078-0010

capacity to collect runoff within small watershed units. Consequently, they are unable to serve as significant runoff channels to collect rainwater over a large area. As a result, they can be considered as terminal discharge points for strong outflows.

5 DISCUSSIONS

The urban ecological planning emphasizes the development of the urban ecological functional zoning, and includes flood disasters as one of the factors for ecological sensitivity evaluation. The physical space of the planning is mostly composed of urban green spaces and water systems. It emphasizes achieving the goals of improving the urban water environment and optimizing water resources allocation through water ecological restoration and ecological landscape construction. The planning proposes the division of urban water environmental functional zones and control units. Therefore, the study analyzes the functional zoning of the two material carriers, green spaces and water systems, in the urban space. Based on the comprehensive consideration of the different level distribution of simulated storage and discharge ranges, regional torrential rain intensity and waterlogged points, and the construction status of Wuhan city, the research area is divided into four categories: storage and discharge core area, self-discharge reinforcement area, construction control area, and local optimization area. Suggestions and control strategies for planning applications are provided for each category of areas.

5.1 How to Formulate the Control Strategy for the Storage and Drainage of the Green Space System and Water System?

Green spaces such as parks can better intercept, retain, and purify rainwater, playing a critical role in flood control and storage during the rainy season, while also providing various ecological services for the city such as open spaces and educational areas. The core area for water storage and discharge includes urban green spaces within the simulated storage and discharge range, protecting existing green spaces that play a crucial role in regulating water flow. Urban development, town construction, and industrial layout are prohibited in this area, while new green spaces should be added at the point where runoff enters the storage and discharge area. This area needs to have greater flexibility to cope with unexpected heavy rainfalls and must face extreme weather conditions. The construction control area includes the simulated construction land within the storage and discharge range.
The construction intensity should be reasonably controlled, and combined with LID facilities, it should have hydrological functions that meet the regulation process without changing the nature of the land, ensuring that the runoff path is not obstructed. The self-discharge enhancement area includes most of the green spaces within the city's scope, and because of its elevation, it should enhance the self-discharge function of the runoff area, impeding rainwater and reducing the production of runoff. It is also the largest permeable area for rainwater in the city, mainly consisting of farmland, forests, and grasslands, forming the overall ecological resource foundation of the city. The local optimization area involves complex factors, such as the formation of flooded spots and flooded areas in different storm intensities within the built-up area of the city, including urban rainwater pipes overflowing due to exceeding load standards. Through the addition of street-side green spaces, sunken greenbelts, and rain gardens, local optimization can be achieved by guiding the water flow path and turning points through design direction, guiding the rainwater runoff to storage and discharge or self-discharge areas through a combination of gray and green facilities (Figure 4).

On the other hand, the function of the water system in the storage and drainage area is to regulate runoff and store rainwater. Urban lakes, as important rainwater regulation spaces, should control their ability to store water, water quality, and quantity, and functional zones. The differences in the amount of rainwater that each lake can regulate should be determined, and functional zones should be established, along with the development of rainwater treatment plans for different regulation needs. The core storage and drainage area includes simulated rivers, lakes, and ponds within the range of the storage and drainage area. During rainfall, it represents the lowest point within the small watershed unit, and serves as the terminal for surface runoff collection. The existing water body area should be preserved, and urban development and rural construction should be strictly prohibited. At the same time, the slope of the shoreline should be reduced, and soft shorelines should replace hard ones. For wetlands such as beaches and marshes, research should be strengthened on soil water retention capacity, and based on the water retention and infiltration capabilities of different wetland soils, the functions of each wetland in the urban hydrological process and corresponding planning measures should be determined. The construction control zone includes built-up areas within the simulated storage and drainage area, and a large number of original lakes and ponds that have been encroached upon by construction land. This area should be appropriately returned to the lake by returning land and fields, while controlling construction intensity and scientifically designing vertical standards. The self-discharge strengthening zone includes other water system spaces within the city's jurisdiction. Within a certain range of hydrological processes, this area corresponds to the confluence function, but there are still water systems with lower elevations in a larger range. Therefore, although it cannot become the core water storage space, it can still receive a certain

https://doi.org/10.52202/082078-0010

amount of rainwater and collect runoff from smaller units. At the same time, this area can serve as a space for scheduling and strong discharge, collecting water in various ways, and sharing the pressure of other areas' drainage. (Figure 4) The local optimization zone is the area where the simulated runoff route is located within the storage and drainage range, corresponding to the runoff process in the hydrological process. The core runoff line takes on the main runoff function, while secondary runoff lines can better penetrate into the built-up area to collect surface runoff rainwater. Some runoff lines that pass through urban construction land cannot flow normally due to obstacles from urban buildings and interference from underground pipelines. Therefore, in suitable areas of construction land, the original stone edge ditch drainage system can be replaced with a roadside open ditch drainage system. Some runoff lines that have more natural plots can use surface grass ditch spaces to form a runoff system and connect with the drainage area.

5.2 How to Reduce the Risk of Waterlogging of Public Facilities in the Completed Area of Wuhan?

The built-up areas within the drainage area are faced with the risk of internal waterlogging. Once a hundred-year rainstorm occurs, which lasts for a long time with great intensity, the standard of the urban underground drainage pipe network is difficult to cope with such a large volume of water. This often leads to the paralysis of critical infrastructure such as the power and transportation systems. The scope of impact is extensive, requiring a comprehensive drainage system based on natural solutions, as well as a well-developed flood prevention and emergency management system. For the selected public buildings within the drainage area, including government buildings, historical sites, university campuses, bus stops, hospitals, parking lots, gas stations, and railway stations, an analysis of the risk of waterlogging and recommendations are provided under different rainfall intensities (Figure 5).

For the drainage area within a higher recurrence period range, in cases where land use cannot be changed, it is necessary to combine the surrounding green space system with concentrated transformation of Low Impact Development (LID) facilities. This targeted approach enhances rainwater discharge and runoff adjustment functions. At the same time, emergency flood prevention plans and disaster avoidance strategies should be implemented. This area not only has strong drainage capacity, relieving the hydraulic pressure of surrounding areas, but can also be utilized for the construction of flood-resistant public facilities that match its landscape benefits and drainage advantages. Examples of such facilities include urban pedestrian walkways and educational demonstration sites, while avoiding the placement of important functions such as gas stations and transportation hubs.

For public facilities within the drainage area with a lower recurrence period, the construction intensity within the drainage area should be controlled first. Surface runoff should be directed to nearby green water systems. Simultaneously, the internal topography and pipe network should be inspected to ensure unobstructed surface runoff routes through vertical design and pipe network dredging. Public service facilities located near waterlogging points and along drainage paths are advised to construct nearby green spaces or artificial storage facilities. By prioritizing specific units and implementing joint defense measures among multiple units, overall drainage capacity can be improved, ensuring water ecological safety.

6 CONCLUSIONS

The research aims to fully utilize the advantageous lake resources in Wuhan City by incorporating elevation topography and substrate coverage with grey infrastructure. This vertical organization aims to control the spatial arrangement of water-green areas to match the storage and drainage processes. The planning should optimize the vertical water system for waterlogging, establish low-lying storage areas, and enhance connectivity through facilities like grass ditches and bioretention zones. These efforts collectively form a comprehensive drainage system, resembling a plain-lake urban model.

Within the framework of the NBS concept, this research emphasizes optimizing the layout of the large-scale drainage system. It seeks to strike an ideal balance between grey and green infrastructure, expanding urban development from a two-dimensional plane to a three-dimensional realm. From a macro perspective, it involves a hybrid grey-green system for controlling urban rainfall and flood processes within a heterogeneous spatial context. On a meso level, conflicts between buildings, surrounding land, and transportation infrastructure need to be addressed, along with temporal contradictions between short-term and long-term plans, as well as local and overall urban areas. In the process of developing solutions, simulating rainfall and flood processes serves as a tool for planners to better understand nature, but the actual effectiveness lies in the storage and drainage spaces themselves, rather than artificial technologies. Future research should focus on refining the construction of the large-scale drainage system at different scales and improving the practical approaches to integrating the drainage system with urban planning and development.

https://doi.org/10.52202/082078-0010

7 REFERENCES

Barbosa, A. E., Fernandes, J. N., & David, L. M. (2012). Key issues for sustainable urban stormwater management. Water Research, 46(20), 6787–6798. https://doi.org/10.1016/j.watres.2012.05.029

Cao, X., Huang, N., Chen, R., Jiang, Y., & Shi, Y. (2021). Evaluating rainwater storage capacity of green infrastructure for urban flood management. Urban Water Journal, 18(9), 673–680. https://doi.org/10.1080/1573062X.2021.1925704

Che Wu, Yang Zheng, Zhao Yang, & Li Junqi. (2013) . Prevention and control of urban waterlogging and analysis of drainage system in China. China water supply and drainage, 29(16) , 13-19. https://doi.org/10.19853/j.zgjsps.1000-4602.2013.16.004

Chee Wu, Yan Pan, Zhao Yang, & Tian F. (2014) . Development and analysis of modern international stormwater management system. China water supply and drainage, 30(18) , 45-51. https://doi.org/10.19853/j.zgjsps.1000-4602.2014.18.010

Chen, W., Wu, H., Kimball, J. S., Alfieri, L., Nanding, N., Li, X., Jiang, L., Wu, W., Tao, Y., Zhao, S., & Zhong, W. (2022). A Coupled River BasinUrban Hydrological Model (DRIVEUrban) for Real Time Urban Flood Modeling. Water Resources Research, 58(11), e2021WR031709. https://doi.org/10.1029/2021WR031709

Cheng Xiaowen, Ling Yunfei, Jia Lingyu, & You Xueyi. (2019) . Planning method and case practice of urban large drainage system. Water supply and drainage, 55(s 1) , 60-63. https://doi.org/10.13789/j.cnki.wwe1964.2019.s1.15

Hoang, L., & Fenner, R. A. (2016). System interactions of stormwater management using sustainable urban drainage systems and green infrastructure. Urban Water Journal, 13(7), 739–758. https://doi.org/10.1080/1573062X.2015.1036083

Hu, C., Xia, J., She, D., Song, Z., Zhang, Y., & Hong, S. (2021). A new urban hydrological model considering various land covers for flood simulation. Journal of Hydrology, 603, 126833. https://doi.org/10.1016/j.jhydrol.2021.126833

Jia, H., Lu, Y., Yu, S. L., & Chen, Y. (2012). Planning of LID–BMPs for urban runoff control: The case of Beijing Olympic Village. Separation and Purification Technology, 84, 112–119. https://doi.org/10.1016/j.seppur.2011.04.026

Kabisch, N., Frantzeskaki, N., Pauleit, S., Naumann, S., Davis, M., Artmann, M., Haase, D., Knapp, S., Korn, H., Stadler, J., Zaunberger, K., & Bonn, A. (2016). Nature-based solutions to climate change mitigation and adaptation in urban areas: Perspectives on indicators, knowledge gaps, barriers, and opportunities for action. Ecology and Society, 21(2), art39. https://doi.org/10.5751/ES-08373-210239

Keesstra, S., Nunes, J., Novara, A., Finger, D., Avelar, D., Kalantari, Z., & Cerdà, A. (2018). The superior effect of nature based solutions in land management for enhancing ecosystem services. Science of The Total Environment, 610–611, 997–1009. https://doi.org/10.1016/j.scitotenv.2017.08.077

Koudelak, P., & West, S. (2008). Sewerage network modelling in Latvia, use of InfoWorks CS and Storm Water Management Model 5 in Liepaja city. Water and Environment Journal, 22(2), 81–87. https://doi.org/10.1111/j.1747-6593.2007.00079.x

Li Junqi, Wang Yaotang, Wang Wenliang, & Yang Zheng. (2017) . Method and case of planning and design of urban road for large drainage system. Water supply and drainage, 53(4) , 18-24. https://doi.org/10.13789/j.cnki.wwe1964.2017.0101

Roy, A. H., Wenger, S. J., Fletcher, T. D., Walsh, C. J., Ladson, A. R., Shuster, W. D., Thurston, H. W., & Brown, R. R. (2008). Impediments and Solutions to Sustainable, Watershed-Scale Urban Stormwater Management: Lessons from Australia and the United States. Environmental Management, 42(2), 344–359. https://doi.org/10.1007/s00267-008-9119-1

Shi, Z.-H., Chen, L.-D., Fang, N.-F., Qin, D.-F., & Cai, C.-F. (2009). Research on the SCS-CN initial abstraction ratio using rainfall-runoff event analysis in the Three Gorges Area, China. CATENA, 77(1), 1–7. https://doi.org/10.1016/j.catena.2008.11.006

Xie Yingxia. (2013) . The development trend of drainage planning is viewed from the frequent occurrence of urban waterlogging. Urban planning, 37(2) , 45-50.

Zhou, Q. (2014). A Review of Sustainable Urban Drainage Systems Considering the Climate Change and Urbanization Impacts. Water, 6(4), 976–992. https://doi.org/10.3390/w6040976

https://doi.org/10.52202/082078-0010

META-BIM TOOL FOR COLLABORATIVE FACILITY PROGRAMMING AND DESIGN OF HEALTH CARE ENVIRONMENT

Track 4: Design for Health and Wellbeing
Knowledge Network: Health, POE/Programming
Methods: Mixed Methods

Akikazu Kato (Mie University)

1 ABSTRACT

The challenge of using Building Information Modeling (BIM) systems is becoming an important issue in the building industries of the USA and Japan, as well as other countries. BIM enables front-loading decision-making in programming and design phases, meaning more is discussed and decided in the basic design stage, rather than in the succeeding detailed design stage. This shift in workload increases liability issues for the architects and engineers, while the process enhances the quality of each phase in the project's development. The use of BIM provides a holistic picture for the general contractors as compared to the conventional construction process in which various information is scattered throughout architectural, structural, mechanical, and electrical drawings and specification sheets. Thus, the system will increase productivity in the construction phase. When the information needed in the management and operation of the building is suitably extracted before the actual completion of the building, which might be called the virtual handover stage, the use of BIM can support efficiency in the facility management stage. Lastly, BIM enables simulations by use of evidence uncovered through research, benefitting the building performance assessment stage. This paper provides a number of case studies, including the development of 3D visual drawings and knowledge attained from building performance evaluation (BPE) studies, which were incorporated into simulations using an application programming interface (API) in BIM systems. These attempts led to the development of a complete design interface, and a BPE feedback loop in the early stages of programming and design was realized. With the above achievements in mind, the evolution of Meta-BIM will be discussed.

1.1 Keywords

Building Information Modeling, Facility Programming and Design, Meta-BIM, Building Performance Evaluation, Health Care Facilities

1 INTRODUCTION

Building Information Modeling (BIM) is one of the most promising developments in the architecture, engineering, and construction (AEC) industries (Eastman, 2011). Fundamentally transforming modes of design practice and standards of building design, delivery, and operation (Kensek & Noble, 2014). Also, BIM will benefit the process of building performance evaluation.

The practical use of BIM is now evolving, with more participants than previously involved. Clients/users, programmers/designers, contractors, and facility managers seek innovative processes during design and construction in order to reexamine workflow, increase productivity, and solidify industry leadership among competitors.

The challenge of using BIM systems is becoming an important issue in the building industries of the USA and Japan, as well as other countries. BIM enables front-loading decision-making in programming and design phases, meaning more is discussed and decided in the basic design stage, rather than in the succeeding detailed design stage. This shift in workload increases liability issues for the architects and engineers, while the process enhances the quality of each phase in the project's development. The use of BIM provides a holistic picture for the general contractors as compared to the conventional construction process in which various information is scattered throughout architectural, structural, mechanical, and electrical drawings and specification sheets. Thus, the system will increase productivity in the construction phase. When the information needed in the management and operation of the building is suitably extracted before the actual completion of the building, which might be called the virtual handover stage, the use of BIM can support efficiency in the facility management stage. Lastly, BIM enables simulations by use of evidence uncovered through research, benefitting the Building Performance Evaluation (BPE). The process model of BPE takes into consideration performance evaluation over the entire life cycle of a building. The six phases and feedback loops are strategic planning – effectiveness review, programming – program review, design – design review, construction – commissioning, occupancy – Post-Occupancy Evaluation (POE), and re-use/recycling – market/need analysis. (Preiser, Hardy, & Schramm, 2018).

https://doi.org/10.52202/082078-0011

2 RESEARCH OBJECTIVES

This paper aims to clarify the current status of BIM usage mainly in the USA and Japan, and to compare the differences and similarities in the cross-cultural context so that the most useful methodology can be found. This methodology can then be applied to guide students in higher education who will be the next leaders and practitioners in the coming years. And, by using the case studies of a children's hospital redevelopment project and hospital inpatient quarter design projects in Japan, this paper explains the advantages and discusses the development of Meta-BIM as a major player in the current innovations regarding the building delivery cycle. The term 'Meta-BIM' was initially used in a chapter of the 2018 book (Kato et al., 2018). The recent six years have brought enough experience and achievements to discuss the meta or the multi-dimensional use of BIM.

3 LITERATURE ON BIM

In American academia, the "BIM Handbook" (Eastman et al., 2008) was published as an informative, seminal book describing the state-of-the-art technology of BIM systems in 2008 and was updated in 2011 and 2018. The book provided powerful examples of BIM use for building owners, managers, designers, engineers, and contractors. "Building Information Modeling: BIM in Current and Future Practice" by Kensek and Noble stretched the boundaries of BIM, and challenges the profession to examine its full potential. It considers, for example, how BIM could support design thinking and reasoning, support simulation, and provide insights into the profession and the direction it is heading (Kensek & Noble, 2014). In professional practice, a similar publication on integrating BIM and sustainability appeared in 2008, that was "Green BIM: Successful Sustainable Design with Building Information Modeling" (Krygiel & Nies 2008). The American Institute of Architects (AIA) upholds that these revolutionary movements are intended to create environmentally friendly design through a streamlined process. An industry leader, Kimon Onuma proclaims to have used BIM as early as 1993, leading BIM-storm from 2006. BIM-storm is a cloud-based charrette on architectural design projects joining multiple owners, users, and consulting members. In Japan, an introductory publication "Innovating Industry: BIM Construction Revolution" (Yamanashi, 2009) was published in 2009. The book claimed that the IT revolution led by BIM will impact business and the economy more significantly than the Industrial Revolution.
A number of publications on operating manuals for BIM have been published in the USA and Japan, including those by Kensek (2014) and Ieiri (2014). Additionally, guidebooks were published for facility managers, including those by Teicholz (2013) and the JFMA BIM-FM Research Council (2015).

4 IMPORTANT FEATURES OF BIM

When BIM is used in project programming and design, all of the information regarding the building is linked automatically. When a designer changes a geometric shape in a building plan drawing, related information in elevations, sections, detailed working drawings, information in finishing schedule charts, and/or structural, mechanical, and electrical drawings, all relevant pieces of building information are changed automatically. It could be said that the geometric shape is simply one portion of information regarding the building.

BIM systems also give accurate figures of used building materials, like volume of concrete, total area of wallpaper, number of door knobs, and others. It detects potential incidents, for example, an air-conditioning duct colliding with a structural beam, and warns designers and supervisors to make necessary adjustments.

5 BIM FOR HOSPITAL PROJECTS IN LITERATURE

In the case study of Maryland General Hospital: MGH in Baltimore, Maryland, an expansion project of approximately 9,600 square meters (103,000 square feet) was carried out, reaching completion in March 2010. It was connected to an existing structure built in the 1950s and included 8 new operating suites, 4 specialty rooms, an 18-bed intensive-care unit (ICU), a pharmacy, and a laboratory (Eastman et al., 2011). MGH, founded in 1881, is part of the University of Maryland Medical System. On June 6, 2013, Maryland General Hospital was renamed "University of Maryland, Midtown Campus"

Throughout the development of this project, information from different sources was integrated using BIM to form a system consisting of a centralized database and integrated facilities management software. The system also uses bar-coding for equipment and tablet PCs for on-site data handling. This made data available at all times, helping to eliminate waste from facilities management, optimize and increase the lifecycle of equipment, increase efficiency in preventive maintenance, and provide accurate and electronic as-built documents.

6 VISUAL SIMULATION OF PATIENT ROOMS

The author was involved in the redevelopment project of a 200-bed children's hospital in Ohbu City, Aichi Prefecture, Japan—the Aichi Children's Health and Medical Center—in collaboration with Yasui Architects & Engineers. The facility, completed in September 2015, added 7,200 square meters (77,500 square feet) and includes a tertiary emergency department, a new operating department, a Pediatric Intensive Care Unit (PICU), and support facilities. Because there was the existing main building, it was found difficult

https://doi.org/10.52202/082078-0011

to incorporate a full-scale BIM. BIM was used only in a new building block. Autodesk Revit Architecture software was used to develop 3D drawings of major functional areas of the new addition, with renovations to be made to portions of the existing structure. During the construction period, 3D drawings of rooms for visual simulation were developed, positioning medical equipment and furnishings such as patient beds and vital sign monitors. These drawings were made to facilitate a better understanding of the final working drawings. Figure 1 is a 3D drawing of the PICU. Because normally the provision of medical equipment and furniture is not included in the construction contract, the contractor tends not to be responsible for the final view. Thus, the staff found developed images helpful to attain a better understanding of the new environment. Also, this practice helped the staff to discuss earlier room operation methods. Moreover, an FM system development project was carried out using similar BIM data, and the case of MGH outlined above has been quite informative because of the similarities between the two projects. (Seki, Kato, & et al., 2016). In another patient room renovation project, a series of interior perspective drawings were developed using the Revit system. The target hospital, Hekinan Municipal Hospital, was newly built in 1986, and the author was working at the design firm Kume Architects & Engineers by then. The developed drawings for the recent study varied from a design of a cozy hotel room, to a design using plenty of wood surfaces expecting a relaxing effect (Figure 2), and to a design to inspire patients using primary colors resembling Rietveld design of Schroeder House interior. The variation of simulated images provoked fruitful thoughts among hospital staff and activated a lively discussion (Yamaguchi & Kato, 2020).

7 BIM PROJECTS FOR HOSPITAL WARD DESIGN

One of the fundamental issues in hospital ward design is the balance of patient privacy versus nursing efficiency. The key performance index for patient privacy was considered to be the ratio of private patient rooms in contrast to multiple beds. Even in Japan with the national health insurance system, some hospitals are now provided with 100% private rooms. As an example in this paper, Gero Hospital is such with total private rooms. The index for nursing efficiency is the walking distance of nurses. Early in the 1980s, the author was involved in a time-motion study of nurses' activities in hospital wards and simulations of nurses' walking distances were carried out to evaluate the efficiency of ward floor plan types (Taniguchi, Kato, & et al., 1984). In more recent studies, similar time-motion studies were carried out in Hekikan Municipal Hospital and Gero Prefectural Hospital mentioned above.

As shown in Figure 3, the ward floor type of Hekinan was a single-double corridor type with dispersed nursing corners to enable the closer distance between the nursing base and patient rooms to realize the shorter walking distance of nurses. A total of nearly 90 beds were placed on one floor forming one nursing administrative unit, and a fairly large-sized nursing base was placed in the floor center combined with two nursing corners in each wing consisting of a total of three nursing units in the initial design concept. The number of beds among the three nursing units varied from 13 beds, including care units for most seriously ill patients, to those units with 36 to 40 beds in the wings. However, at the time of the survey, nursing units consisted of two units, West Ward with 50 beds and East Ward with 40 beds. This arrangement was causing quite a long walking distance in the West Ward.

Figure 1: 3D Drawing of Pediatric Intensive Care Unit (2016) by the author and Seki, S.

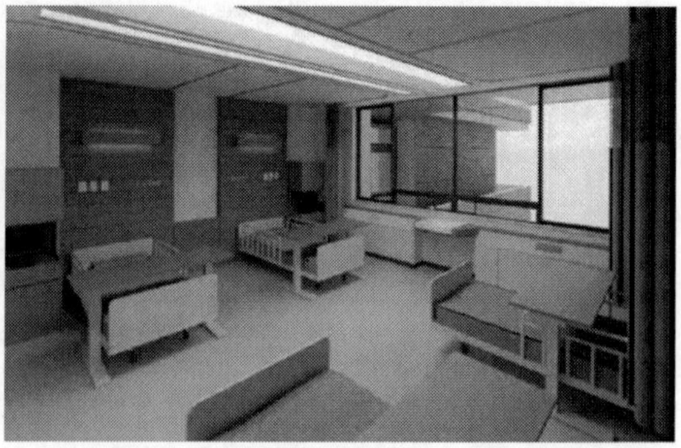

Figure 2: Four Bed Room Design Using Plenty of Wood Surfaces (2020) by the author and Yamaguchi, K.

https://doi.org/10.52202/082078-0011

The floor plan type of the Gero Prefectural Hospital was a circular floor type with unified single rooms where five single rooms were placed facing a common unit hall. (Figure 4) Each of the West and East Ward wings consisted of 38 bed rooms with the nursing base in the center. When single rooms are aligned along the corridor, the length of the corridor tends to be longer than having multiple bed rooms. However, because single rooms were clustered as units of five, the corridor length of Gero tended to be shorter, meaning shorter walking distance for the nurses.

Two types of numerical operations research were carried out. One was the hill climbing algorithm to find the most probable shortest distance path between the nursing base and the patient room. This gave the shortest sum of nurses' walking distance to evaluate the efficiency of ward floor type (Takeuchi, Kato, & et al., 2020). Another was the genetic algorithm to find the most suitable placements of patients according to the seriousness of their illness. Because nurses tend to visit more seriously ill patients, their placements closer to the nursing base tend to be crucial for the total walking distance. The most optimized patient placement is shown in Figure 4. However, in reality, condensing most seriously ill patients and/or empty bed rooms in certain parts is not practical and probable. To realize the most optimal situation, patients need to be displaced quite often. Thus, the most inappropriate placement was simulated and the comparisons of two extremes were carried out to attain the efficiency of ward floor types. Figure 5 shows one of the result drawings, one with the most inappropriate patient placement resulting in an increase of less than 6% over the most optimal placement. The increasing percentage marked about 30% in the Hekinan simulation and it was concluded the circular floor type of Gero has the higher efficiency. The data on the walking behaviors of four nurses was used in the simulation (Kato, A., Yamaguchi, K., & et al., 2021). The two numerical simulation methods were programmed using Python language and incorporated into the Revit BIM system using Dynamo as an Application Programming Interface (API), a supporting software so that various simulations concerning nurses' walking distance were carried out during the programming and/or design stages of hospital ward plans.

8 META-BIM DISCUSSION

In the course of the evolution of humans and architecture, the initial breakthrough might have been the use of geometry in Ancient Egypt, like the triangulation drawings, or triangular surveying, to verify one's land holdings in the fertile flood plain along the Nile and to calculate the volume of stone in

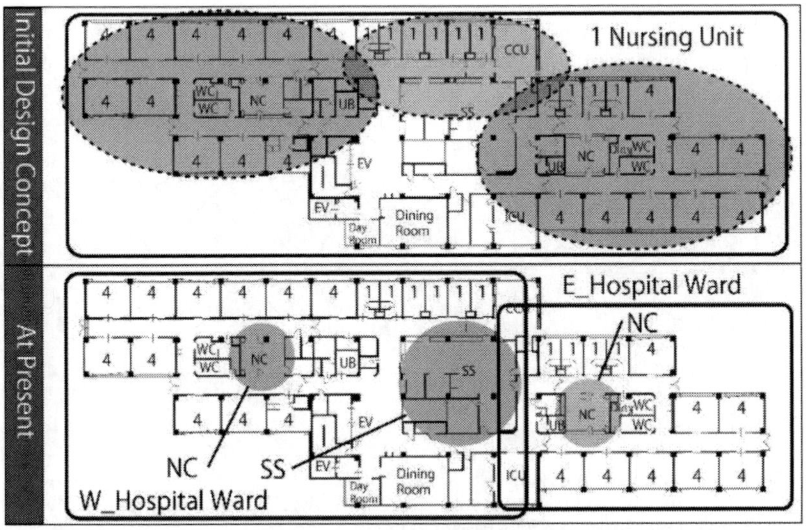

Figure 3: Comparison of Ward Composition at Hekinan Municipal Hospital (2019) by the author and Takeuchi, K.

https://doi.org/10.52202/082078-0011

building a giant pyramid. The second might have been in the Renaissance, the use of perspective drawings first described in 1435 by Leon Battista Alberti in his Della Pittura (Bertol, 1997). The use of perspective drawing on a wall and/or ceiling created the extended spatial image perception as in the School of Athens (Scuola di Atene) drawn by Raphael in the Vatican between 1509 and 1511, certainly the first Virtual Reality (VR). The third epoch was the use of computers starting in the 1960's (Kalay et al., 2014). It should be noted that in the same era, the Environmental Design Research Association (EDRA) was founded, and Post-Occupancy Evaluation (POE) became a mainstream field of specialization, which later was extended to the Building Performance Evaluation (BPE) (Preiser & Hardy, 2015). Thus, the current use of BIM is staged as the fourth major innovation.

Early BPE projects of hospital facilities started with the behavioral tracking of pathways of building occupants, including one carried out in a hospital ward (Nuffield, 1955) and another in the simulation of the Yale Traffic Index (Thompson & Goldin, 1975). Inspired by these studies, the author carried out a time-motion study of nurses' activities in hospital wards and conducted simulations of nurses' walking distances to evaluate ward floor plan types as mentioned previously. These POE/BPE studies are now accumulated to create a knowledge base regarding the programming, design, and management of hospitals and healthcare facilities.

Starting from the latter half of the 1990s, with the start of Windows'95, more efficient personal computers, and internet access, the studies on computer usage in facility programming and design became easier. Fuzzy Inference and Genetic Algorithms in workplace studies were carried out using the open environment of AutoCAD software by evaluating the performance of floor plan types and creating a suitable floor layout (Kato, 1994; Kato et al., 2001). Application Programming Interface (API) is provided in current BIM software like Autodesk Revit, which has a similar interface as AutoCAD where an external program can be loaded and performed on the BIM display (Takeuchi, Kato, & et al., 2020; Kato, Yamaguchi, & et al., 2021).

As mentioned in case studies by combining the knowledge base of BPE studies with simulations using API in BIM systems, a complete design interface can establish a BPE feedback loop in the early stages of programming and design. As the automated development of geometric forms continues, the next phase of evolution will lead to the development of Meta-BIMs. Moreover, by using the accreditation evaluations of Leadership in Energy and Environmental Design (LEED), Sustainable Development Goals (SDGs), and/or Joint Commission International (JCI) standards, more knowledge can be acquired for the simulations resulting in more profound Evidence-based Design (EBD). Finally, Meta-BIM should be incorporated with the generative Artificial Intelligence: AI. The systems such as ChatGPT and/or Bing are found quite useful.

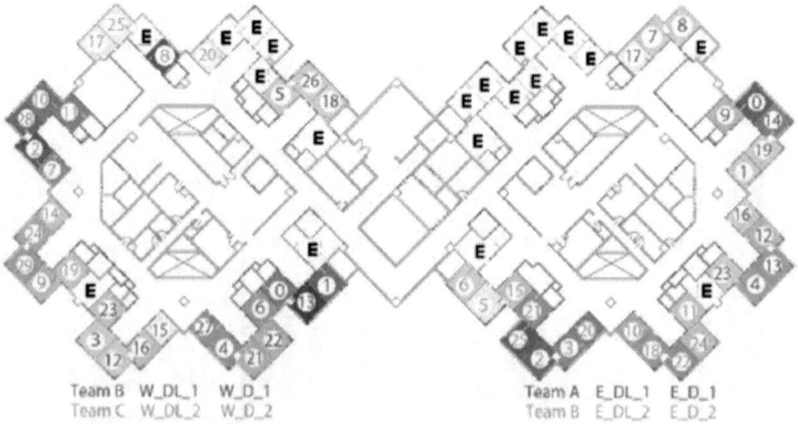

Figure 4: Ward Plan of Gero Prefectural Hospital showing Optimized Patient Placement, Shaded rooms are for more seriously ill patients. E for empty room. (2020) by the author and Yamaguchi, K.

https://doi.org/10.52202/082078-0011

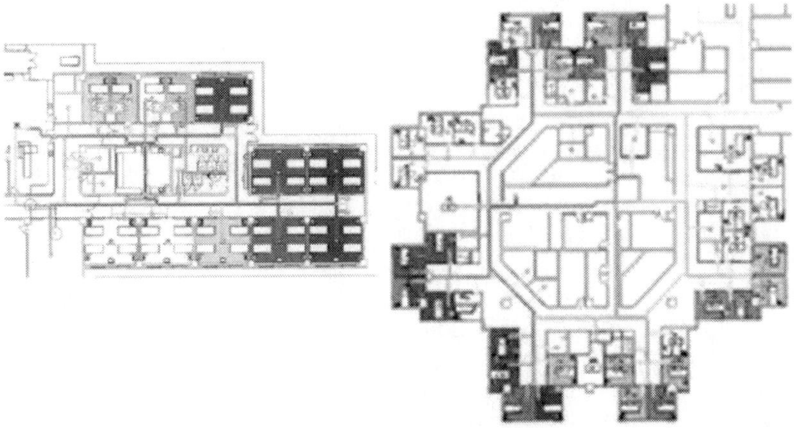

Figure 5: Comparison of Nurses' Walking Distance in Two Hospital Wards. The difference of two extremes was about 30% at Hekinan (left) and less than 6% at Gero (right). (2020) by the author and Yamaguchi, K.

And, the image generation system such as Microsoft Image Creator as well. However, to be used for the programming and design practice, at least the source of reference should be clarified. Or, the issue of copyright will never be resolved.

Lastly, this paper has shown that the use of Meta-BIM might bring extra merits in carrying out simulations within the BIM display. Starting from the 1960's, the exercise of POE/BPE brought about a volume of knowledge regarding facilities in use. The development of PCs has realized the use of numerical simulations and other AI technology, and computer programming using API enables knowledge simulations on a BIM display. Thus, it is hoped that Meta-BIM will extend the skills and minds of practitioners and stakeholders for the better usage of related buildings in their sustainable life cycle.

9 REFERENCES

Bertol, D. (1996). Designing Digital Space: An Architect's Guide to Virtual Reality. John Wiley & Sons.

Eastman, C. et.al. (2008). BIM Handbook: A Guide to Building Information Modeling for Owners, Managers, Designers, Engineers, and Contractors, John Wiley & Sons, first edition.

Eastman, C. et.al. (2011). BIM Handbook: A Guide to Building Information Modeling for Owners, Managers, Designers, Engineers, and Contractors, John Wiley & Sons, second edition.

Ieiri, R. (2014). BIM in This Much (This Much Series), Shuwa System Publishing. [In Japanese]
JFMA BIM-FM Research Council (2015). BIM Utilization Guidebook for Facility Manager, JFMA. [In Japanese]

Kalay, Y. E., Schaumann, D., Seung, W. H., and Simeone, D. (2014). Beyond BIM: Next-Generation Building Information Modeling to Support Form, Function, and Use of Buildings, Chapter 24, In Kensek, K. M., and Noble, D. E. (2014).

Kato, A. (1994). A Study on Development of Floor Zoning Support System, Facility Management Tools Using Fuzzy Inference, Strategies & Technologies for Maintenance & Modernization of Building, CIB W70 Tokyo Symposium, Vol.1, pp.231-236.

Kato, A.(2019). Teaching BIM and BPE Using Problem-Based Learning (PBL) at Mie University, Japan, Proceedings of the 50th Annual Conference of the Environmental Design Research Association, edra.

Kato, A., Le Roux, P. C., and Kitakami, Y. (2001). Application of Genetic Algorithms in Office Building Configurations, Reproduction Model Simulation in a Japanese Local Government Office, Proceedings of EDRA32 and paper handout, Environmental Design Research Association, Edinburgh, U.K.

Kato, A., Mori, S., and Taniguchi, G. (2018). Multi-Tasking Architectural Computer Programs: BIM. In: Preiser, W.F.E., Hardy, A., Wilhelm, J. J. (Eds.). Adaptive Architecture, Changing Parameters and Practice, Routledge, pp.227-232

Kato, A., Takeuchi, T. Fujita, Y.(2019). Consideration on Educational Environment of the BIM Using PBL in the Department of Architecture at Mie University, Social Equity by Design, Designing Connections through Community, Proceedings of the 49th Annual Conference of the Environmental Design Research Association, ISBN 978-1-387-97114-5, 2018, pp.63-70

https://doi.org/10.52202/082078-0011

Kato, A., Yamaguchi, K., Takeuchi, T.(2020). Building Performance Evaluation in Hospital Ward Planning by Nursing Travel Distance Simulation Using BIM, Proceedings of 51th Annual Conference of Environmental Design Research Association, edra51Tempe, Arizona, pp.40-49

Kensek, K. M. (2014). Building Information Modeling, Pocket Architecture Technical Design Series, Routledge.

Kensek, K. M. and Noble, D. E. (2014). Building Information Modeling: BIM in Current and Future Practice, John Wiley & Sons.

Krygiel E. and Nies B. (2008). Green BIM: Successful Sustainable Design with Building Information Modeling, Sybex

McDonnell, M. J., & Hahs, A. K. (2009). Comparative ecology of cities and towns: Past, present and future. In M. J. McDonnell, A. K. Hahs, & J. H. Breuste (Eds.), Ecology of cities and towns - A comparative approach (pp. 71-89). Cambridge.

Nuffield Provincial Hospitals Trust (1955). Studies in the Functions and Design of Hospitals, Oxford University Press.

Preiser, W. F. E., and Hardy, A.E. (2015). Historical Review of Building Performance Evaluation. In: Preiser, W.F.E., Davis, A., Salama, A., and Hardy, A. (eds)(2015) Architecture Beyond Criticism: Expert Judgment and Performance Evaluation. London: Routledge.

Preiser, W. F. E., Hardy, A.E., and Schramm, U. Ed. (2018). Building Performance Evaluation, From Delivery Process to Life Cycle Phases, Second Edition, Springer

Sabol, L. and Kendra, S. (2015). Putting BIM in Practice at the Smithonian Institution, IFMA's World Workplace 2015, Denver, Colorado, October 7-9, 2015.

Sasaki, D. and Yasui, I. (2015). Attack Using BIM, Nikkei Architecture Magazine, August 25, 2015 issue. [In Japanese]

Seki, S., Kato, A., et.al. (2016). Development Studies on Facility Management System Using BIM for Children's Hospital, Utiliza-tion of BIM in Renovation and Extension Project for Aichi Children's Health and Medical Center, Research Paper Report of Tokai Branch Conference, Architectural Institute of Japan, Vol.54, No.402, pp.317-320. [In Japanese]

Taniguchi, G. et.al. (1984). A Study on Nursing Activities in Hospital Wards, Transactions of the Architectural Institute of Japan, No.344, October 1984, pp.116-125. [In Japanese]

Teichholz, P. (2013). BIM for Facility Managers, IFMA Foundation, John Wiley & Sons.

Thompson, J. D., and Goldin, G. (1975). The Hospital: A Social and Architectural History, Yale University Press.

Vogel, B., Molich, T., & Klar, N. (2009). Der Wildkatzenwegeplan: Ein strategisches Instrument des Naturschutz (The Wildcat Infrastructure Plan - A strategic instrument of nature conservation). Naturschutz und Landschaftsplanung, 41, 333-340.

Yamaguchi, K., Kato, A. (2020). Study on Evaluation of Hospital Patient Room Design Using Semantic Differential Method, Research Paper Report of Tokai Branch Conference, Architectural Institute of Japan, Vol.58, No.445, pp.425-428. [In Japanese]

Yamashina, K. (2009). Innovating the Industry: BIM Construction Revolution, Nihon Jitsugyo Publishing. [In Japanese]

https://doi.org/10.52202/082078-0011

ARCHITECTURAL NEGOTIATIONS IN MONKEY-HUMAN COEXISTENCE: BEYOND HUMAN-CENTRIC URBAN STRATEGIES

Track 1: Nature and Urban Ecology
Knowledge Network: Environmental & Architectural Phenomenology
Methods: Observation

Pitchawut Virutamawongse (Chiang Mai University)

1 ABSTRACT

This study investigates the adaptive architectural modifications within Lopburi, Thailand, a city known for its significant monkey population that influences both the socio-cultural fabric and the urban environment. The research delineates the interactions between humans and monkeys, particularly focusing on the architectural strategies employed by residents to prevent monkey encroachments. Through a detailed analysis of Lopburi's old town, the paper identifies various defensive elements like mesh grilles, corrugated metal, and monkey deterrent wiring, examining their distribution and effectiveness. It reveals a complex 'urban negotiation', challenging the conventional human-centric approach to urban design. The study highlights the juxtaposition between the necessity of protecting property and the desire for open, harmonious urban spaces. It proposes a nuanced understanding of coexistence that goes beyond mere conflict, considering the potential for integrated strategies that respect both human and animal needs in urban ecosystems. This research contributes to the discourse on biodiversity in urban spaces, advocating for a shift in architectural thought to encompass multi-species urban planning.

1.1 Keywords

Urban Ecology, Architectural Adaptation, Multi-Species Coexistence, Monkey Deterrent Strategies, Lopburi.

2 INTRODUCTION

Cities, environments created to satisfy human needs, are sometimes built over natural resources that were once home to various living organisms. Urbanization occasionally forces these organisms to leave their natural habitats for new, challenging environments. Some animals return to urban areas and adapt their behaviors to benefit from the resources there. Lopburi, Thailand, is one such location relevant to this phenomenon. Located in Thailand's central region, Lopburi is steeped in history, evident in its numerous historical structures and architectural sites. The province is well-known as a tourist destination for those interested in history and wishing to experience a city cohabitating with a large population of monkeys (Cripps & Olarn, 2020). Several places in Lopburi are connected to the story of the Ramayana, one of which is specifically related to 'Hanuman,' a pivotal monkey character in the tale. Consequently, the story of Hanuman and the monkeys is undeniably intertwined with Buddhism, Thailand's predominant religion (Intrawut, 2002). This religious connection significantly impacts the relationship between humans and monkeys in the urban environment of Lopburi.

In 2023, the overall monkey population in Lopburi reached 10,000, with about 2,000 residing in the old town, well-known for the cohabitation of humans and monkeys (Thairath, 2023). The monkeys play a significant role in various aspects, such as tourism, the local economy, and most importantly, as the representative image of the province. Their presence has attracted visitors, making Lopburi a popular tourist spot, which in turn has boosted the local economy and created job opportunities. However, monkeys can also cause various problems, such as property damage, and sometimes they attack people to obtain what they need. Consequently, the existence of monkeys in Lopburi has both positive and negative effects on the urban environment and its inhabitants. The issues and opportunities presented by the monkeys have been discussed repeatedly to date (The Citizen ThaiPBS, 2023). Various strategies have been implemented, yet they have not been successful, primarily due to methodologies approached from a human-centric perspective. Therefore, to fully understand the situation and create a suitable environment in Lopburi, it is crucial to learn from our mistakes in neglecting their intelligence, adaptability, behavior, and capabilities in urban settings.

3 RESEARCH OBJECTIVES

Urban Animal and Adaptivity in Urban Environment

The phenomenon of animals emerging in urban environments is a global occurrence. In some cases, animals significantly benefit from urban contexts. For instance, in Tokyo, a city vibrant day and night, crows have discovered an overwhelming source of food from garbage in commercial and residential areas (Morishita et al., 2003), enjoying urban greenery in the midst of the concrete jungle. Conversely, in other scenarios, urban animals exploit the environment

https://doi.org/10.52202/082078-0012

under specific conditions or times. Coyotes in urban settings, for instance, tend to hide during the day to avoid human conflict, emerging at night to forage for food (Gehrt, 2005). Furthermore, studies have indicated that urban animals often exhibit superior innovative skills, being faster learners, bolder, and better at problem-solving compared to their rural counterparts (Audet et al., 2015).

However, urban animal adaptation can also create issues. In Tokyo, crows searching for food in the morning often tear garbage bags apart, causing early morning street messiness (Kurosawa et al., 2003). Similarly, urban raccoons have learned to open garbage bins, causing daily problems for residents (Jeong et al., 2016). Understanding their behavior and sensory capabilities beyond human capacity could enhance our quality of life in remarkable ways. For example, the presence of water monitor lizards in Bangkok indicates the health of the surrounding environment (Mahaprom et al., 2015; Matichon Online, 2018). Additionally, the daily behavior of crows seeking food after sunrise can help identify improperly managed garbage spots in the city (Virutamawongse & Judhinaputra, 2016).

The situation with monkeys in Lopburi is similar to these urban animal cases. Originally forest inhabitants, they were displaced by urban development but gradually returned, adapting to the urban dynamics. Buddhism, the dominant religion in the area, influences the local mindset, leading people to feed and be kind to the monkeys. As a result, the city and its residents provide ample food and living spaces for the monkeys, making it too convenient for them to leave. Using their climbing abilities and observational skills, monkeys access areas unreachable to humans and find food even in difficult times. This behavior sometimes leads to safety concerns, as interactions with humans can result in injuries from bites or scratches and property invasions (Anothai et al, 2020).

To prevent these conflicts, human shelters, including houses and public buildings, have been redesigned to deter monkey invasions. However, these architectural strategies have compromised functionality and quality of life and have not fully succeeded in mitigating monkey-related problems. Therefore, this research will investigate and analyze the redesigned architecture to understand the relationship between monkey behavior and capabilities and the strategies employed in architectural redesign.

4 METHODS

4.1 Study Context

The selected site is located in the old town of Lopburi, as shown in Figure 1. The shrine 'Phra Prang Sam Yod' is a historic architectural site popular with tourists. Surrounding Phra Prang Sam Yod is a commercial area, now largely vacant due to ongoing conflicts caused by monkeys. Despite this, several historically significant places remain, such as the Malai Rama Theatre to the north of Phra Prang Sam Yod, which is now a dwelling for a group of monkeys. Additionally, to the west, there is the still-active Manora market and the 'Sarn Pra Karn' shrine on the circle island, which is sacred to locals. Within a 200-meter radius of Phra Prang Sam Yod, there are five territories of monkey groups, housing a total of 2,000 monkeys. The area focusing on architectural adaptations to prevent monkey invasions is divided into two zones.

Figure 1: The layout of the old town of Lopburi with the focused study areas highlighted.

https://doi.org/10.52202/082078-0012

Zone A, on the east side of Phra Prang Sam Yod, includes commercial spaces on the ground floor, with hotels and housing units above. Zone B, to the south of Phra Prang Sam Yod, is mostly deserted; however, the remaining residents use the ground floor for storage.

4.2 Analyzing Facade Modification and Local Approaches to Deter Monkey Incursions

The architecture in Zones A and B was built during the Modern era. Generally, most buildings feature architectural elements such as cantilevered slabs or vertical fins, which inadvertently aid monkeys in climbing and accessing levels above the ground. Consequently, these architectural features, combined with public utilities like electrical wires present in Lopburi's old town, essentially turn the area into a playground for monkeys. As a result, monkey movements can cause damage to habitats, commercial areas, and public utilities. Therefore, as a countermeasure, locals have begun to adapt their living methods by adding elements to their buildings to prevent monkeys from gaining access to their properties, damaging windows, or climbing to upper floors. This research will analyze the types of additional elements introduced to prevent issues caused by monkeys, compare the proportion of these elements to the remaining areas, and examine the proportion of areas that have been damaged by monkeys where no protective elements are in place.

5 RESULTS

5.1 Architectural Modifications for Monkey Deterrence

According to the gathered information, four types of additional architectural elements have been implemented to prevent damage caused by monkeys, as illustrated in Figure 2. Corrugated metal is one method used to stop monkeys from climbing and reaching higher levels of buildings. Building owners attach corrugated metal almost vertically and cover the entire width of the units, severing any connection with the exterior context. This approach is primarily seen in Zone B, near the road's edge. Another common method is the installation of monkey deterrent wiring above the ground floor, observable along the roadside buildings in Zone A. This simple method demonstrates local ingenuity in finding a straightforward solution to the monkey problem. The materials include PVC pipes used as posts, wrapped with wire, and connected continuously from one end of a building unit to the other. This approach is cost-effective, and allows for ventilation and light, but may not be durable in the long term. Another tactic, which also offers an economic opportunity, is to convert the entire surface of a unit into an advertisement board. This is observed in a specific area in Zone A and effectively seals the entire surface, making it difficult for monkeys to climb. Lastly, a popular and effective approach is to attach mesh grills to the building's surface. Although monkeys can climb the mesh, they are unable to access or damage anything inside the protected area.

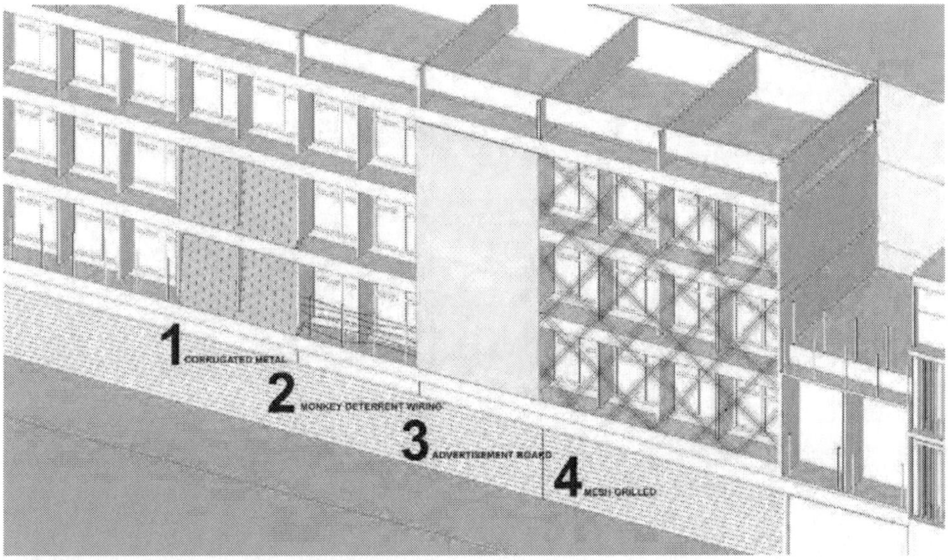

Figure 2: Types of additional architectural elements implemented to deter monkeys.

https://doi.org/10.52202/082078-0012

4.3 Analysis of Architectural Interventions for Monkey Deterrence in Urban Buildings

In this section, the entire frontage of the buildings along the road, including the architectural modifications, will be analyzed to understand the proportion of the actual buildings and the areas that have been modified to prevent monkey-related property damage. Figure 3 presents information on Zone A, showing several buildings starting from the left: a bank, a hotel, a row of commercial buildings, a clinic, and at the far end, a street food establishment with residential space above. The buildings at both extremities largely exist without any monkey deterrent elements, due to facades primarily composed of concrete and glass. Conversely, the hotel and commercial areas have installed monkey deterrent wiring to protect the structures from monkeys. The difference between the two is that the hotel utilizes only monkey deterrent wiring, while the commercial area employs a mix of methods to prevent monkeys from accessing the building. In total, there are 48 roadside-facing units without any additional elements to deter monkeys. Of these, only 15 units are without any damage or have not yet been damaged by monkeys, which accounts for only 31.25% overall.

In Zone B, as depicted in Figure 4, nearly all buildings have implemented architectural elements to prevent damage from monkeys, with the exception of the central building, which is abandoned. In this area, methods such as using advertisement boards and monkey deterrent wiring are not observed. Instead, the majority of the buildings are almost entirely covered by mesh grilles or corrugated metal. The building at the far right end is one example where the owner has adapted to the presence of monkeys. Notably, the main entrance of this building is equipped with a mesh grille door to control monkey access intermittently. Moreover, only 8 out of the 17 units facing the street have been damaged by monkeys, which represents 47.05%. This suggests that there is a possibility that the monkeys no longer find it necessary to damage the abandoned building to gain entry.

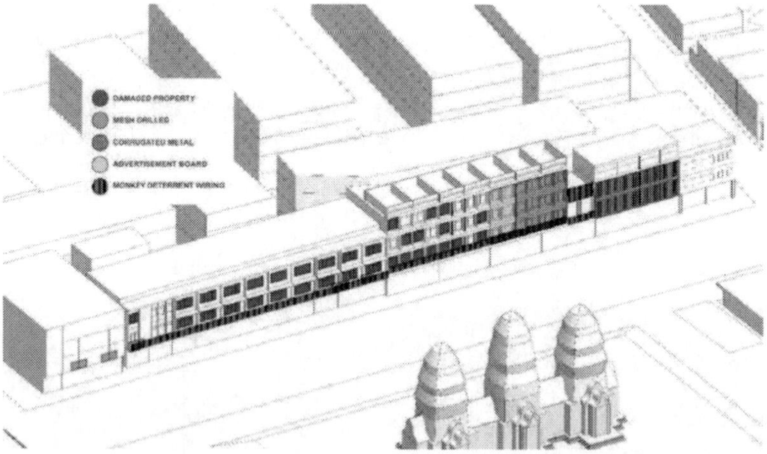

Figure 3: Architectural Adaptations in Zone A for Monkey Deterrence

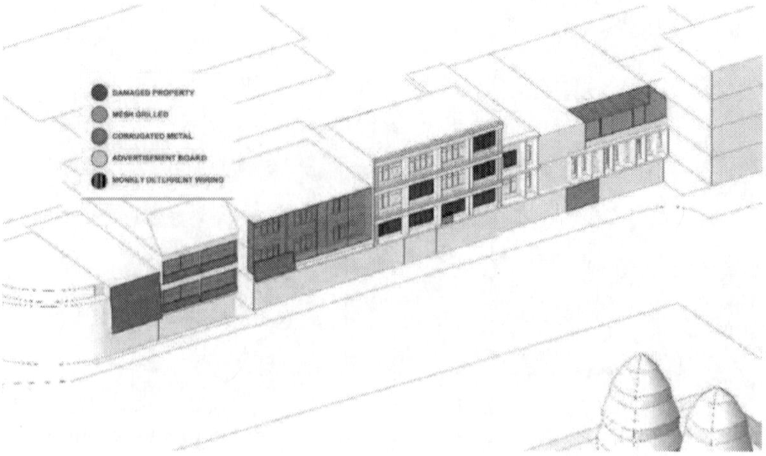

Figure 4: Architectural Adaptations in Zone B for Monkey Deterrence

https://doi.org/10.52202/082078-0012

6 CONCLUSIONS

In conclusion, there are various methods employing additional architectural elements to prevent monkey encroachment in Lopburi's old town, uniquely tailored to this specific context. Monkeys have adapted to the urban environment, seeking food and shelter, and performing whatever actions necessary, which sometimes cause difficulties for people and damage to property. Conversely, humans have observed and contended with the monkeys' adaptability, which has been disruptive to their daily lives. Nevertheless, the locals have learned, observed, and adapted in search of a harmonious approach to coexist with the monkeys within the urban landscape. This has involved integrating additional elements into building facades, sacrificing some benefits such as natural light, air, and urban vistas for survival in a shared urban environment. Thus, this unique urban context displays a potent sense of negotiation, where a human-centric approach does not function as effectively as it might elsewhere.

7 DISCUSSION

The research findings shed light on the resurgence of monkeys in their former habitats from which they were displaced by urban development, prompting a reevaluation of our coexistence with nature, not just in this research site study or urban settings but across all environments.

This study has focused on the architectural adaptations that have emerged as survival mechanisms in a context shared by humans and monkeys. It casts doubt on the efficiency of a bottom-up approach to development. Individual building owners grapple with the unique challenges posed by their non-human neighbors, employing various methods they believe will be effective. However, this research suggests that isolated efforts fall short of the collective achievement observed in monkey communities. The sporadic implementation of monkey deterrents leaves unprotected areas vulnerable, and the monkeys, with their high intelligence, climbing skills, and adaptability, inevitably exploit these weaknesses. This contrasts sharply with the often directionless nature of human top-down development, which fails to address the situation holistically.

The study also highlights the trade-offs inherent in modifying architecture to counter wildlife challenges. While barriers like mesh grilles and corrugated metal sheets protect against monkeys, they also detract from the quality of urban life by obstructing natural light and airflow. Additionally, the persistent noise from monkey activity remains an unresolved issue. The discussion extends to the potential overreliance on deterrents, suggesting that their complete success may not equate to a holistic triumph. True success in urban design should also encompass safe and harmonious interactions in communal spaces, which current measures do not fully address.
Moreover, the monkeys' use of urban structures for navigation suggests a pressing need to re-envision city planning with cohabitation in mind. Creating safe corridors for wildlife mobility could mitigate the risks and damages currently experienced.

Lastly, recognizing the intelligence and capabilities of monkeys should inspire a shift in perspective, from viewing them as nuisances to considering them as integral to the urban ecosystem. The ultimate goal is to discover ways in which their presence can enhance the urban experience for all.

8 REFERENCES

Anothai, J. Sarunya, H., & Thanapoom, R. (2020) Incidences and characteristics of monkey-related injuries among locals and tourists in Mueng District, Lopburi Province, 2013 - 2017. Chulalongkorn Medical Journal, 64(1), 029-037.

Audet, J.-N., Ducatez, S., & Lefebvre, L. (2015). The town bird and the country bird: Problem solving and immunocompetence vary with urbanization. Behavioral Ecology, 27(2), 637–644. https://doi.org/10.1093/beheco/arv201

Cripps, K., & Olarn, K. (2020, October 14). Ancient Thai city grapples with surging monkey population. CNN. https://edition.cnn.com/travel/article/lopburi-monkey-temple-thailand-intl-hnk/index.html

Intrawut, P. (2002). () [Legend of Lopburi (Lavo)]. Damrong Journal of the Faculty of Archaeology Silpakorn University, 1(1), 1-28

Jeong, Y., Jantz, L. M., & Smith, J. (2016). Investigation into seasonal scavenging patterns of raccoons on human decomposition. Journal of Forensic Science, 61(2), 467-471.

Kurosawa, R., Kanai, Y., Matsuda, M., & Okuyama, M. (2003). Conflict between humans and crows in greater Tokyo - garbage management as possible solution. Global Environmental Research, 7(2), 139-147

Mahaprom, R., & Kulabtong, S. (2018). Observation of feeding habit of the Asian water monitor, Varanus Salvator (Laurenti, 1768) (Squamata Varanidae) on a Asian toad, Duttaphrynus melanostictus (Schneider, 1799) (Anura Bufonidae) in Thailand. Biodiversity Journal, 9(3), 213–216. https://doi.org/10.31396/biodiv.jour.2018.9.3.213.216

Matichon Online. (2018, April 17). [Population of Asian water monitors increased due to few numbers of their preys]. Matichon Online. https://www.matichon.co.th/local/quality-life/news_920361

https://doi.org/10.52202/082078-0012

Morishita, E., Itao, K., Sasaki, K., & Higuchi, H. (2003). Movements of crows in urban areas, based on PHS tracking. Global Environmental Research, 7(2), 181-192.

Sol, D., Griffin, A.S., Bartomeus, I., & Boyce H. (2011). Exploring or avoiding novel food resources? The novelty conflict in an invasive bird. PLoS ONE 6(5): e19535. https://doi.org/10.1371/journal.pone.0019535

Thairath. (2023, December 23). [Updating Monkey Populations: The First Step to Solving the Issues] Thairath. https://www.thairath.co.th/scoop/theissue/2746255

The Citizen ThaiPBS. (2023, December 30). : " " [The future of Human City and Monkey]. The Citizen ThaiPBS. https://thecitizen.plus/node/92753

Virutamawongse, P., & Judhinaputra, G. A.(2016). Sentinel crows, Sustainable improvement of urban spatial quality by cybernetic interventions on both natural and artificial environments [Master's thesis, The University of Tokyo].

https://doi.org/10.52202/082078-0012

COULD SOCIAL SPACE DESIGN BOOST SOCIAL INTERACTIONS AMONG OLDER ADULTS WITH MILD COGNITIVE IMPAIRMENT?

Track 4: Design for Health and Wellbeing
Knowledge Network: Environment-Gerontology, Health
Methods: Case Study, Observation

Yasaman Gholami (SimTigrate Design Lab, Georgia Institute of Technology); Raha Rastegar (SimTigrate Design Lab, Georgia Institute of Technology); Herminia Machry (Georgia Institute of Technology); Elahn Little (Research Scientist, SimTigrate Design Lab, Georgia Institute of Technology); Leandro Miletto Tonetto (Associate Professor, School of Industrial Design, Georgia Institute of Technology

1 ABSTRACT

The aging population, particularly those living with Mild Cognitive Impairment (MCI), confronts a tendency for social isolation. Promoting social interactions is recognized as a key element and potential protective factor against both social isolation and cognitive decline. However, individuals with MCI encounter difficulties in engaging socially. Previous studies recognize the impact of built environments on social interactions. However, there is a gap in understanding how the design of social spaces influences individuals with MCI in non-residential settings, particularly from an objective perspective.

This study investigated spatial utilization patterns of individuals with MCI during social interactions in a non-residential therapeutic setting, focusing on understanding social behaviors in relation to two seating layout characteristics— having seating close and facing each other and having available seats for everyone —proposed by MCI healthcare professionals through the literature. Field observations were conducted at the Cognitive Empowerment Center—a therapeutic facility for MCI-focused therapeutic activities in Atlanta, GA. Using a behavioral mapping tool, 1310 social interactions were recorded involving members of the therapeutic program, their care partners, and staff. The analysis identified social interactions by types, group sizes, locations, participants' roles, and postures, using ANOVA and the Bonferroni test.

The results showed that seating availability and positioning impacted the number and group size of social interactions observed in each space for individuals with MCI and their care partners. No significant association was found involving interaction types. Study findings highlight how seating availability and layout can promote socialization for older adults with MCI, given their needs for physical and auditory comfort.

1.1 Keywords

Social Interaction, Cognitive Impairment, Environmental Design, Human-Centered Design, Therapeutic Setting

2 INTRODUCTION

Social isolation and loneliness are critical health issues, affecting nearly a quarter of adults aged 65 years and older (Donovan & Blazer, 2020). This problem becomes even more significant for older adults living with Mild Cognitive Impairment (MCI) (Farhang et al., 2022), a diagnosis for early stages of cognitive decline that may lead to dementia (Koepsell and Monsell, 2012). Social interactions have been identified as an effective strategy in addressing social isolation and are considered potential protective factors against cognitive decline and incident dementia (Lydon et al., 2022). Studies suggest that meaningful social interaction may even lead to better cognitive functioning and increased cognitive reserve (Lydon et al., 2022). However, individuals with MCI often face challenges in engaging socially, presenting a unique obstacle to the proposed strategy.
In this context, the built environment has long been recognized as a crucial factor influencing communication and social interactions (Pfeiffer, 1980), with recent studies indicating that the built environment can enhance the likelihood and quality of social interactions, irrespective of users' cognitive abilities (Sullivan & Chang, 2011). For individuals facing cognitive challenges, specific studies demonstrate that the spatial layout and interior arrangement can support social interactions for those living with dementia in residential care facilities (Ferdous & Moore, 2015).

Consideration should be given to the fact that older adults living with MCI often reside independently at home, engaging in social interactions in settings that are less structured than nursing homes or assisted living facilities (Zhu et al., 2023). Only recently, specialized therapeutic facilities for MCI been designed (Aflatoony et al., 2020). One is the Atlanta-based Cognitive Empowerment Program (CEP). These

https://doi.org/10.52202/082078-0013

facilities offer opportunities for individuals living with MCI to interact socially within evidence-based design therapeutic spaces (Mynatt et al., 2022).

To better understand how social behaviors are taking place in this type of non-residential therapeutic setting, this study investigated the spatial utilization patterns of individuals with MCI and their care partners during social interactions, looking at specific aspects of the interior design of social spaces that can enhance their social interactions.

2.1 Social Engagement and Mild Cognitive Impairment (MCI)

The aging population faces an increasing challenge in the realm of social isolation and social loneliness, especially for individuals living with Mild Cognitive Impairment (MCI). MCI represents an intermediate stage between the expected cognitive changes associated with normal aging and the more advanced cognitive decline that could potentially evolve into dementia (Koepsell & Monsell, 2012).This stage presents a complex and pressing challenge. Individuals with MCI, typically aged over 65, encounter a range of cognitive challenges, including memory deficits, language impairments, practical judgment difficulties, decision-making constraints, and decreased responsiveness to stimuli (Eshkoor et al., 2015).

Studies have shown that active social interactions are not only crucial for preventing social isolation and loneliness but also promising in preserving cognitive function (Hartel & Carstensen, 2006; Sanaei et al., 2023) and potentially mitigating cognitive decline (Bourassa et al., 2017; Lydon et al., 2022). However, older adults facing cognitive challenges often struggle to maintain social ties and networks, which would otherwise serve as protective factors against cognitive decline (Zhu et al., 2023). As cognitive decline progresses, individuals with MCI lose confidence to engage with others, leading to a shrinking of social networks and fewer opportunities for interaction outside their homes (Yates, Clare, & Woods, 2017). This challenge is intensified by a higher incidence of anxiety and depression among individuals with MCI, further complicating the interplay between psychological and cognitive factors (Ma, 2020). Consequently, individuals experiencing cognitive decline often tend to withdraw from social activities (Guarnera et al., 2023).

2.2 Built Environment Features Impacting Social Interactions

Studies are showing that the built environment has the potential to facilitate social interactions by increasing the likelihood of unplanned interactions (Dempsey, 2008). Studies on older adults with advanced cognitive impairments, such as dementia and Alzheimer's, suggest that environmental designs facilitating physical and visual connections to other spaces, as well as designs that control the level of surrounding stimuli and distractions, can positively impact social interactions (Day et al., 2000; Dempsey, 2008).

In this area, five general design criteria—viewability, variety, privacy, and a sense of homelikeness—have been developed and suggested for designing assisted living facilities that support social engagement for individuals living with Dementia (Lee & Rodiek, 2013). These design criteria are recommended as they impact the likelihood of encounters among older adults in seating areas, as well as their awareness of the presence of others, referred to as co-presence (Lee & Rodiek, 2013). Another more recent work on memory-care residential facilities identified nineteen environmental affordances that foster social interaction among people living with Dementia (Ferdous, 2020).

Reviewing the literature on the role of the built environment in social interactions for people living with cognitive impairment reveals a lack of studies in the context of MCI and non-residential therapeutic facilities. One (da Rosa et al., 2023) recent study identified five environmental design priorities that contribute to social interaction for people living with MCI in social spaces, based on surveys responded to by healthcare professionals with expertise in MCI (da Rosa et al., 2023). Three environmental priorities were related to interior-exterior features—visual access to natural environments, natural lighting in living and dining spaces, access to outdoor environments—, while two applied to interior spaces and seating layout—having seating close and facing each other, and having available seats for everyone in social areas—(da Rosa et al., 2023).

The current study aimed to investigate how these two seating attributes contribute to patterns of social interactions for individuals living with MCI in a non-residential therapeutic facility using the case study of CEP in Atlanta, GA.

3 RESEARCH OBJECTIVES

The primary goal of this research was to contribute to the current understanding of environmental design criteria aimed at fostering social interactions for individuals living with MCI in non-residential therapeutic settings. In pursuit of this objective, the current study aims to investigate the spatial utilization patterns of individuals with MCI for social interaction. This investigation considers the two seating layout characteristics proposed by da Rosa and colleagues (2023): having seating close and facing each other and having available seats for everyone. To be more precise, the study seeks to address how social spaces are used for social interactions by individuals living

https://doi.org/10.52202/082078-0013

with MCI, with a focus on these two attributes. For this research, the CEP building area in Atlanta, GA, was chosen as the study setting due to being used solely and frequently by individuals living with MCI, their care partners (family members or friends), and staff (e.g., therapeutic instructors and clinicians).

4 METHODS

A cross-sectional observational study was conducted in March and April 2022, spanning eight days, to map how specific spaces at CEP, intentionally designed for socialization, were being used by program members, care partners, and staff during unscheduled social interactions between therapeutic activities. Considering the environmental characteristics of those specific spaces based on two prioritized interior design attributes—having seating close and facing each other and ensuring seats are available to everyone in social areas—the observed spaces were assessed. These characteristics were then evaluated based on observed behaviors in these spaces.
The study obtained IRB approval from the Georgia Institute of Technology.

4.1 Participants

During the first four days of observations, the social behaviors of two groups of CEP Members (all previously diagnosed with MCI), referred to as Group 1 and Group 2, were followed. For the subsequent four days, groups 3 and 4 were observed during the same daytime period while they were socially engaged. Groups 1 and 2, consisted of eleven members with MoCA scores ranging from 15 to 25 and eleven care partners. The MoCA (Montreal Cognitive Assessment) is a widely used screening tool designed to assess various cognitive domains and identify potential cognitive impairment, including MCI. A MoCA score typically ranges from 0 to 30, and the interpretation of scores may vary. Generally, a score between 18 and 25 is indicative of MCI (Wood et al., 2020). Groups 1 and 2 were observed while being acquainted with each other for about 10 weeks.
Groups 3 and 4, consisted of nine members with MoCA scores ranging from 19 to 25 and nine care partners. These two groups were observed while being acquainted with each other for about 28 weeks. On each day of observation, 14 to 15 staff members were present in the CEP.

4.2 Study Setting

The CEP facility is dedicated to individuals living with MCI to use along with their care partners (friend or family member) for approximately eight hours per week in scheduled therapeutic activities, such as yoga, tai chi, nutrition, and occupational therapy classes. Their participation in these activities spans over the course of six months to one year.
The CEP facility design was developed based on design principles formulated through a process that included a literature review, case studies, and co-design charrettes involving designers, researchers, individuals living with MCI, and their care partners (Aflatoony et al., 2020).

Five social spaces were designed in the CEP to support social interactions: Sitting Area 1, Sitting Area 2, Tech Bar, Dining Area, and the Kitchen. As presented in Figure 1 (CEP floorplan), Sitting Area 1 is located at the entrance zone adjacent to the reception area. This space is purposefully designed as a waiting area, primarily intended for use upon arrival or during intervals between scheduled activities. As depicted in Figure 1, Sitting Area 2 closely resembles Seating Area 1. The Tech Bar, primarily designed to provide members with a dedicated space for addressing their technical issues with a technician, is in an alcove area near Seating Area 2 and the reception. The Dining Area is designed to accommodate all cohorts of members during lunchtime, and it is in a corner adjacent to various

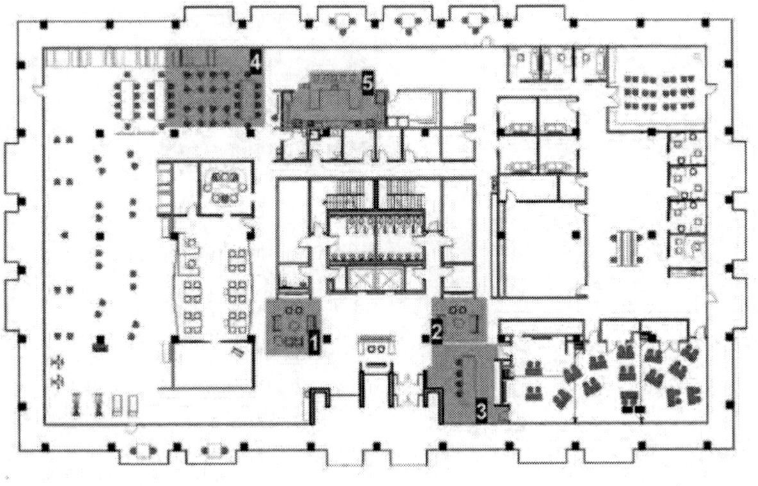

Figure 1. CEP social spaces shown in the center floor plan, 1: Seating Area 1, 2: Seating Area 2, 3: Tech Bar, 4: Dining Area, Kitchen

https://doi.org/10.52202/082078-0013

Table 1. The characteristics of the CEP social spaces regarding their seating layout attributes.

	Sitting Area 1	Sitting Area 2	Tech Bar	Dining Area	Kitchen
Picture					
Design Attribute 1: **Seating close and facing each other**	Users can select their seating position—next to each other, across from each other (with distances of 2.6 m and 2.8 m), or in a corner arrangement.	Users can select their seating position—next to each other, across from each other (with distances of 2.1 m), or in a corner arrangement.	All seats in this zone face a bookshelf, limiting seating options to sitting next to each other.	Users can choose their seating position—next to each other, across from each other, or in a corner arrangement. Options include small tables (1.25m between chairs), long tables (1.25m between chairs, 3.75m at ends), and booths (0.95m between chairs).	All seats face away from the corridor towards the kitchen cabinets, limiting seating options to sitting next to each other.
Design Attribute 2: **Available seats for everyone in social areas**	Accommodates up to eight individuals with two 2-seater couches, two individual sofas, and two individual accent chairs.	Accommodates up to six individuals with two 2-seater couches, and two individual accent chairs.	Accommodates up to five individuals with five bar stools.	Accommodates up to sixty individuals with a variety of seating options, including booths and chairs.	Accommodates up to six individuals with the sole option of bar stools.

activities. Lastly, the Kitchen is positioned in an alcove space branching from a corridor connecting the dining room to the counseling rooms. The Kitchen is designed as a venue for conducting cooking classes for members.

All these social spaces at the CEP building were assessed based on the two design attributes prioritized by MCI experts in (da Rosa et al., 2023). Table 1 presents the characteristics of these spaces based on those attributes.

4.3 Metrics and Tools

After conducting a descriptive assessment of the environmental characteristics of social spaces within the CEP, a tablet-based behavioral mapping tool was developed by the authors. The tool was designed to document and categorize interactions by their types (momentary: lasting ≤ 5 seconds or conversational: lasting > 5 seconds), group sizes, postures (seated, standing or walking), and specific locations within five designated observation zones depicted in Figure 1.

The CEP floor plan was subdivided into multiple zones to enable simultaneous observation of all target spaces with multiple observers, ensuring consistency in the time of observation. Then, different notations were defined for these subdivided zones through the tool for recording interactions and distinguishing their categorization. A screenshot of the tool used for the observation of seating area 1 is shown in Figure 2. The criteria included the recording of interactions involving either members or care partners, excluding interactions between staff only.

Notably, the program schedule played a pivotal role in shaping the movement patterns and presence of members and care partners in specific zones. For example, when members and care partners were scheduled for classes near seating area 1 and seating area 2, their break times were primarily spent in these areas between classes, rather than opting for the kitchen or other social spaces. To achieve a more comprehensive observation for a better understanding of social behaviors in space, consideration was also given to observing adjacent spaces to social spaces, as illustrated in Figure 3. However, the interpretation of results was conducted by sticking to the research questions.

https://doi.org/10.52202/082078-0013

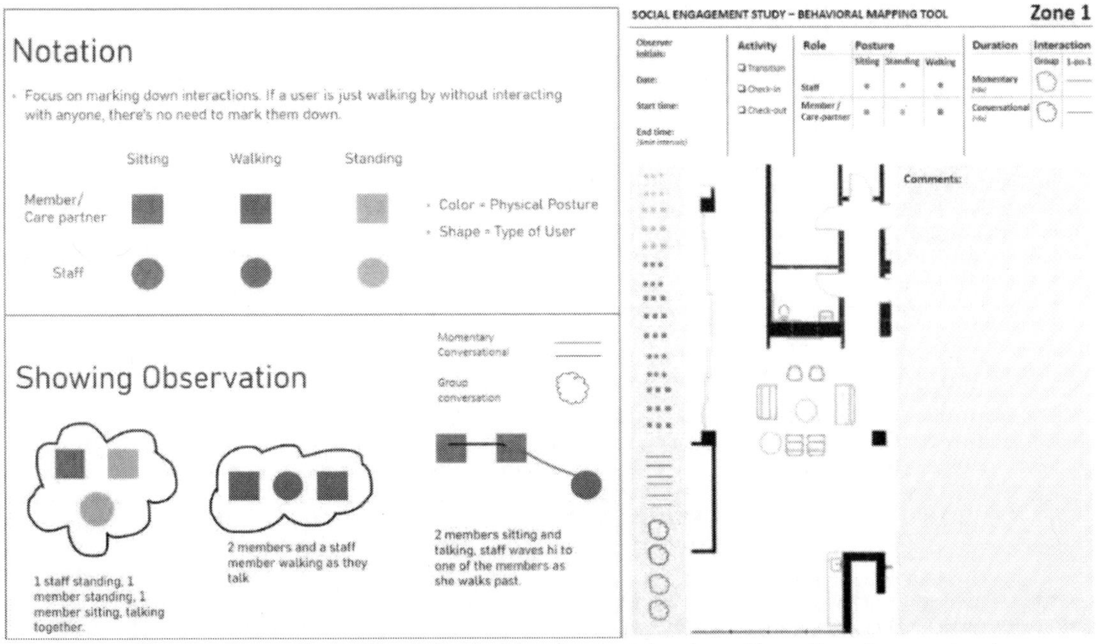

Figure 2. Screenshot of the Behavior Mapping tool developed by authors to record and categorize social interactions at Cognitive Empowerment Center's social spaces

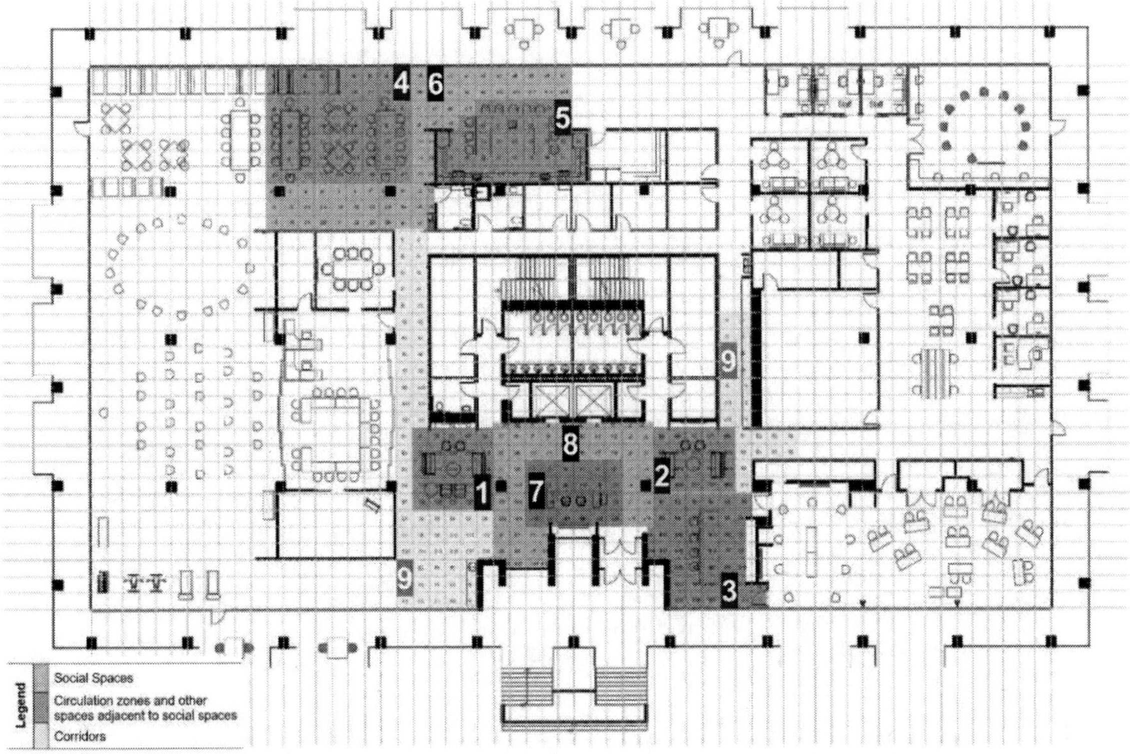

Figure 3. Observed Zones: 1) Seating Area 1, 2) Seating Area 2, 3) Tech Bar, 4) Dining Area, 5) Kitchen, 6) Circulation zone adjacent to Kitchen and Dining Area, 7) Reception [mostly used during check-in times], 8) Circulation zone adjacent to Seating Area 1&2 and the reception area, 9) Others: corridors

https://doi.org/10.52202/082078-0013

4.4 Procedure

Following the design of the behavior mapping tool, as described in the metric and tool subsection, a three-day pilot observation was conducted to revise and finalize both the tool and instructions, to enhance usability of the tool. Once the tool was finalized, seven researchers were trained to use the tool with an iPad and an Apple Pencil to record interactions. On observation days, the researchers visited the CEP building, selected zones to start the observation, starting at 9:30 am (coinciding with member and care partner check-in) and concluding each day at noon.

Social interactions were observed during 15-minutes intervals between scheduled activities (as arrival check-ins, transitions between therapeutic activities). Each researcher was assigned to one specific zone, where they recorded information about all interactions observed in one observation page, filling out a new page every 5-minutes interval. In each page, the researcher recorded their name, the time and date of their observation, and drew the location of the interaction along with its respective notations.

At the end of each observation day, all data were transferred to an Excel spreadsheet, including the date, time interval, number of individuals involved in each interaction, participants' roles (members and care partners versus staff who were identifiable by name tags), their posture, type of interaction, and location based on the grid number in the tool. The data were then analyzed using the method described in the following section.

4.5 Data Analysis

After the data were transferred to the Excel spreadsheet, statistical analyses were conducted using SPSS. The average percentage of individuals involved in each interaction, relative to the total number of individuals involved in all interactions within that specific zone, was calculated using Mean. The objective of calculating and analyzing the percentage of people was to account for variations in the number of individuals present in each space. In other words, instead of considering the actual number of people interacting, the focus was on the percentage of individuals involved in each interaction, referred to as 'group sizes' for simplicity and coherence. For instance, in Seating Area 1, if there were 10 individuals involved in interactions, and the average number of individuals involved in each interaction was 3, then the mean percentage for group size would be 30%. This indicates how large interactions in that specific zone were, considering the total number of individuals present and involved in interactions in Seating Area 1.

Also, multiple comparisons between means were performed using ANOVA (Analysis of Variance) and the Bonferroni test.

Additionally, the distribution of momentary and conversational interactions was examined using the chi-squared test. A t-test was also employed to determine whether significant differences existed between the observations made during the first four days compared to the second four days. These statistical tests allowed for the exploration of any notable patterns or variations in the social interactions observed during the study period. The patterns of interactions were then analyzed, with a focus on their relationship to the environmental features present within the observed zones.

Since the aim of this study was to map how social interactions took place in particular spaces and relate them to the design features of each one, the analysis was concentrated solely on single-zone interactions. By single-zone interactions, it was meant that all the individuals involved were located in just one of the observation zones depicted in Figure 3. So, if, in an interaction involving 5 individuals, 4 were located in the Kitchen, and the other one was located in the Circulation zone adjacent to the Kitchen, this interaction was no longer considered a single-zone interaction. However, If all five parties were located either in the Kitchen or the Circulation zone adjacent to the Kitchen, it was considered a single-zone interaction.

5 RESULTS

Of the 1310 interactions observed, 795 occurred in single zones (78.5%) – as identified in Figure 3 – and 218 involved at least one side of the interaction in another zone (21.5%). Also, there were no meaningful differences between the results of groups 1 and 2 (first four days) versus groups 3 and 4 (second four days). So, the presented results are based on all 795 interactions in single zones that were observed during the eight days.

5.1 Types of Interactions

The distribution of momentary and conversational interactions was equivalent in most zones. In two of them, the momentary interactions were overrepresented compared to the others: at the Reception ($X2\ (1{,}795) - 4.818$, $p < .05$) and at the Seating area #2 ($X2\ (1{,}795) - 4.262$, $p < .05$). In all the other zones, there were no significant differences observed, as presented in Table 2.

https://doi.org/10.52202/082078-0013

Table 2. Frequency of momentary and conversational interactions

Area	Measures	Conversational	Momentary	Total	X^2
Seating area 1	Count	117	25	142	X^2 (1,795) = .343, p > .05
	Expected Count	119.3	22.7	142	
	Adjusted Residual	-0.6	0.6		
Seating area 2	Count	60	19	79	X^2 (1,795) = 4.262, **p < .05**
	Expected Count	66.4	12.6	79	
	Adjusted Residual	-2.1	2.1		
Tech Bar	Count	39	3	42	X^2 (1,795) = 2.577, p > .05
	Expected Count	35.3	6.7	42	
	Adjusted Residual	1.6	-1.6		
Dining area	Count	169	37	206	X^2 (1,795) = .817, p > .05
	Expected Count	173.1	32.9	206	
	Adjusted Residual	-0.9	0.9		
Kitchen+ Others*	Count	23	6	29	X^2 (1,795) = .498, p > .05
	Expected Count	24.4	4.6	29	
	Adjusted Residual	-0.7	0.7		
Circulation zone adjacent to the Dining areas & Kitchen	Count	64	10	74	X^2 (1,795) = .368, p > .05
	Expected Count	62.2	11.8	74	
	Adjusted Residual	0.6	-0.6		
Reception	Count	153	18	171	X^2(1,795) = 4.818, **p < .05**
	Expected Count	143.7	27.3	171	
	Adjusted Residual	2.2	-2.2		
Circulation zone adjacent to Seating areas and the reception	Count	43	9	52	X^2 (1,795) = .074, p > .05
	Expected Count	43.7	8.3	52	
	Adjusted Residual	-0.3	0.3		

*Out of 29 recorded interactions, 11 occurred in the Kitchen, and the remaining took place in zone 9, highlighted in yellow in Figure 3.

4.4 Group Size of Interactions

Group size, which refers to the percentage of individuals involved in each interaction relative to the total number of individuals involved in all interactions within a specific zone, was calculated. Table 3 presents the total number of interactions observed in each zone, along with the standard deviation and mean for the group size and the roles of individuals involved in interactions.

As shown in the table, Seating Area 1, Seating Area 2, and the Dining Area exhibit larger group size interactions. This implies that regardless of the number of people present in these areas, the percentage of participation in group interactions was higher compared to other zones.

Additionally, the results of the ANOVA test indicate significant differences between zones ($F (7,787) = 19.082$, $p < .001$). Focusing on the target social spaces, observed interactions in the Dining area involved a larger group size compared to the Tech bar ($p < .001$), and in Seating Areas #1 and #2, observed interactions involved larger group sizes than those in the tech bar ($p < .001$).

Table 3. Group size of interactions

Area	Members & Care Partners			Staff			Members, Care Partners, Staff		
	N	Mean Group Size (%)	SD Group Size (%)	N	Mean Group Size (%)	SD Group Size (%)	N**	Mean Group Size (%)	SD Group Size (%)
Seating area 1	142	**18.57**	10.80	14	8.10	3.86	142	**9.27**	5.11
Seating area 2	79	**18.61**	11.78	27	7.16	2.57	79	**10.39**	4.97
Tech Bar	42	10.08	4.98	33	8.48	4.17	42	8.34	3.08
Dining area	206	**20.52**	13.67	11	6.67	0.00	206	**10.08**	6.44
Kitchen+Others*	29	14.96	6.91	9	6.67	0.00	29	8.17	2.87
Circulation zone adjacent to the Dining areas & Kitchen	74	19.64	9.93	9	7.41	2.22	74	9.95	4.75
Reception	171	10.16	5.25	152	6.84	1.07	171	8.02	2.59
Circulation zone adjacent to Seating areas and the reception	52	14.49	6.43	17	6.67	0.00	52	8.08	3.07
Total	795	16.52	10.99	272	7.13	2.13	795	9.22	4.85

N stands for Number and SD stands for Standard Deviation.

*Out of 29 recorded interactions, 11 occurred in the Kitchen, and the remaining took place in zone 9, highlighted in yellow in Figure 3.

** Since the dataset includes only interactions that involved at least one member or care partner, the number of interactions listed in this column matches the count in the column that presents data exclusively for members and care partners. However, when the involvement of staff members is also considered, the number of individuals engaged in interactions changes. This leads to different mean and standard deviation (SD) values in the column that represents interactions involving members, care partners, and staff.

4.4 Group Size and posture during interactions

Table 4 shows the means and standard deviations of the group size of members and care partners while sitting, standing, and walking during social interactions in each zone. As presented in the table, Seating Area 1, Seating Area 2, and the Dining Area exhibit larger group sizes in seated interactions. However, considering standing and walking postures, the group size was larger in the circulation area around the Dining and Kitchen than in the circulation around the Seating areas or in other corridors.

The results of the ANOVA test indicated significant differences in group size while sitting in some zones ($F(3,391) = 4.712$, $p < .01$). Focusing on the social spaces, the Dining, Seating #1, and Seating #2 areas involved a larger group size compared to the Tech Bar ($p < .001$, $p < .05$, and $p < .05$, respectively).

Table 4. Group Size and Posture of members and care partners during the interactions

Area	Sitting			Standing			Walking		
	N	Mean Group Size (%)	SD Group Size (%)	N	Mean Group Size (%)	SD Group Size (%)	N	Mean Group Size (%)	SD Group Size (%)
Seating area #1	134	**17.44**	10.87	17	10.73	5.17	14	8.40	2.97
Seating area #2	73	**17.71**	12.04	12	8.72	4.59	10	8.55	3.65
Tech Bar	29	10.05	5.24	8	7.74	2.27	6	11.70	6.06
Dining area	159	**19.56**	14.96	65	12.25	6.05	28	12.17	8.02
Kitchen + Others	0	0	0	16	15.06	8.70	14	13.78	4.82
Circulation zone adjacent to the Dining areas & Kitchen	0	0	0	23	**18.68**	10.48	60	**17.18**	8.82
Reception	0	0	0	148	10.25	5.62	26	9.02	3.75
Circulation zone adjacent to Seating areas and the reception	0	0	0	17	15.80	8.97	35	13.85	4.79
Total	395	17.80	12.80	306	11.77	6.94	193	13.25	7.28

N stands for Number and SD stands for Standard Deviation.

https://doi.org/10.52202/082078-0013

5 DISCUSSION AND CONCLUSIONS

It is critical to promote social interactions for older adults living with MCI, especially in facilities used by these individuals as a group. Thinking of how the design of social areas in these types of settings have the power to impact social interactions for this populations, this study shows that specific seating characteristics have a relevant role in the way people with MCI choose to interact with each other, with their care partners and with staff.

The study considered two key attributes: 1) seating close and facing each other, and 2) available seats for everyone in social areas. Within the CEP social spaces, five different areas were examined. The Dining Area accommodates 100% of members and care partners with seats close to each other and facing. Seating Area 1 accommodates 40% of members and care partners with an arrangement of seats close to each other and facing. Seating Area 2 offers seating for about 30% of members and care partners, with a similar seating arrangement to Seating Area 1. The Kitchen provides seats for about 30% of members and care partners, arranged next to each other. The Tech Bar has seats available for 25% of members and care partners, with a similar arrangement to the Kitchen. As the results showed, when considering spaces with seats arranged facing each other and close, the dining area, with the most available seats, had the highest number of interactions and a significantly larger group size of interactions. Not only does the dining area facilitate a large number of interactions and a substantial group size, but the circulation around the dining area and kitchen also significantly supports larger standing and walking group interactions.

Regarding the two seating areas, even though they are both located close to each other and on either side of the reception and waiting area, seating area 1 had almost twice as many interactions as seating area 2, with just a 10% higher availability of seats. However, concerning group size, they had almost the same percentage.

When considering the kitchen area and TechBar, where the seats do not face each other, the number of interactions and group size were considerably less than in other spaces, even though the kitchen has the same number of seats available as seating area 2.

According to the results of the study, the two attributes of seating positions and seating availability play a considerable role in supporting the number and group size of social interactions, both in seated interactions and in general interactions (regardless of posture) for individuals living with MCI and their care partners in non-residential social spaces as healthcare experts suggested.

To conclude, the results of this study confirm that the two environmental features— having seating close and facing each other and having available seats for everyone —impact the number of social interactions as well as the group size of interactions for people living with MCI in social therapeutic settings. However, the study does not show a significant association between these two design features and the type of interactions.

Regarding the availability of seats, the results emphasize the importance of seats being available for everyone in social spaces. As shown in the current studied space, social spaces with similarities in the likelihood of the presence of users within (due to their location within the larger space) and with similarities in seating position, like Seating Area 1 and Seating Area 2, exhibited different numbers of social interactions due to variations in the number of available seats. Sufficient seating for individuals during social interactions contributes to physical comfort, making the availability of seats a crucial factor when designing for older adults. It's important to consider that the significance of physical comfort could be even greater when designing for individuals with Mild Cognitive Impairment (MCI) who may struggle with balance issues and may only feel compelled to maintain more meaningful conversations while seated.

Considering the seating arrangement, among the various seating spaces observed, those providing face-to-face interaction with proximity supported more and larger social interactions, while those positioned next to each other (not face-to-face) were used for smaller social interactions. Social spaces with the same number of seats available, and with similarities in the likelihood of the presence of users within (due to their nearness), such as Seating Area 2 and Tech Bar with two different seating positions, exhibited considerable differences in creating both the number and group size of interactions.

The current study has limitations that could point to future research directions. Primarily, the interactions of members (individuals living with MCI) were not distinguished from those of their care partners. Additionally, the observed spaces exhibit variations beyond their seat availability and arrangement, which may contribute to differences in social interaction behavior not accounted for in the current study. Furthermore, as CEP is a structured program, the program schedule played a pivotal role in shaping the movement patterns and presence of members and care partners in specific zones, potentially affecting the study's variables. Although differences in MoCA scores did not show any significant differences in the social behaviors of the observed groups, future studies may consider accounting for the varying levels of MCI with a larger number of participants.

https://doi.org/10.52202/082078-0013

6 REFERENCES

Aflatoony, L., Tonetto, L. M., Campiglia, G. C., & DuBose, J. R. (2020). Photo narrative: Co-designing a built environment with seniors with mild cognitive impairments.

Bourassa, K. J., Memel, M., Woolverton, C., & Sbarra, D. A. (2017). Social participation predicts cognitive functioning in aging adults over time: comparisons with physical health, depression, and physical activity. Aging & Mental Health, 21(2), 133–146.

da Rosa, V., Herminia, M., & Tonetto, L. M. (2023). Designing therapeutic and social spaces for older adults facing Mild Cognitive Impairment: Priorities in spatial and furniture layout.

Day, K., Carreon, D., & Stump, C. (2000). The therapeutic design of environments for people with dementia: a review of the empirical research. The Gerontologist, 40(4), 397–416.

Dempsey, N. (2008). Does quality of the built environment affect social cohesion? Proceedings of the Institution of Civil Engineers-Urban Design and Planning, 161(3), 105–114.

Donovan, N. J., & Blazer, D. (2020). Social isolation and loneliness in older adults: review and commentary of a national academies report. The American Journal of Geriatric Psychiatry, 28(12), 1233–1244.

Eshkoor, S. A., Hamid, T. A., Mun, C. Y., & Ng, C. K. (2015). Mild cognitive impairment and its management in older people. Clinical Interventions in Aging, 687–693.

Farhang, M., Miranda-Castillo, C., Behrens, M. I., Castillo, E., Mosquera Amar, S., & Rojas, G. (2022). Impact of social isolation and coping strategies in older adults with mild cognitive impairment during the covid-19 pandemic: A qualitative study. Aging & Mental Health, 26(7), 1395–1416.

Ferdous, F. (2020). Positive social interaction by spatial design: a systematic review of empirical literature in memory care facilities for people experiencing dementia. Journal of Aging and Health, 32(9), 949–961.

Ferdous, F., & Moore, K. D. (2015). Field observations into the environmental soul: Spatial configuration and social life for people experiencing dementia. American Journal of Alzheimer's Disease & Other Dementias®, 30(2), 209–218.

Guarnera, J., Yuen, E., & Macpherson, H. (2023). The Impact of Loneliness and Social Isolation on Cognitive Aging: A Narrative Review. Journal of Alzheimer's Disease Reports, Preprint, 1–16.

Hartel, C. R., & Carstensen, L. L. (2006). When I'm 64. National Academies Press.

Koepsell, T. D., & Monsell, S. E. (2012). Reversion from mild cognitive impairment to normal or near-normal cognition: risk factors and prognosis. Neurology, 79(15), 1591–1598.

Lee, M.-A., & Rodiek, S. D. (2013). Residents' preference for spatial features in sitting areas at assisted living facilities-focused on direct or indirect social interaction for older adults. International Journal of Human Ecology, 14(1), 87–102.

Lydon, E. A., Nguyen, L. T., Nie, Q., Rogers, W. A., & Mudar, R. A. (2022). An integrative framework to guide social engagement interventions and Technology Design for Persons with Mild Cognitive Impairment. Frontiers in Public Health, 9, 750340.

Ma, L. (2020). Depression, anxiety, and apathy in mild cognitive impairment: current perspectives. Frontiers in Aging Neuroscience, 9.

Mynatt, E. D., Vickers, K., LaForce, S., Farmer, S. K., Johnson, J. M., Doiron, M., Ramesh, A., Fain, W. B., Zubatiy, T., & Rodriguez, A. D. (2022). Pivoting an MCI Empowerment Program to Online Engagement. Proceedings of the ACM on Human-Computer Interaction, 6(GROUP), 1–26.

Pfeiffer, T. S. (1980). Behaviour and interaction in built space. Built Environment (1978-), 35–50.

Sanaei, M., Machacek, M., Gilbert, S. B., Wu, P., & Oliver, J. (2023). Comparing Perceptions of Performance Across Virtual Reality, Video Conferencing, and Face-to-Face Collaborations. 2023 IEEE International Conference on Systems, Man, and Cybernetics (SMC), 4556–4561.

Sullivan, W. C., & Chang, C.-Y. (2011). Mental health and the built environment. Making Healthy Places: Designing and Building for Health, Well-Being, and Sustainability, 106–116.

Wood, J. L., Weintraub, S., Coventry, C., Xu, J., Zhang, H., Rogalski, E., Mesulam, M.-M., & Gefen, T. (2020). Montreal Cognitive Assessment (MoCA) performance and domain-specific index scores in amnestic versus aphasic dementia. Journal of the International Neuropsychological Society, 26(9), 927–931.

Zhu, D., Al Mahmud, A., & Liu, W. (2023). Social connections and participation among people with mild cognitive impairment: barriers and recommendations. Frontiers in Psychiatry, 14.

https://doi.org/10.52202/082078-0013

DEVELOPING AN ANALYTICAL METHOD FOR WALKABILITY MEASUREMENT TOOLS

Track 5: Spatial Justice, Equity, and Inclusion by Design
Knowledge Network: Active Living by Design, Movement in Designed Environments: Nature & Ecology
Methods: Mixed Methods
Eda Paykoç Özçelik (Yasar University, Faculty of Architecture); Tonguç Akış (IZTECH)

1 ABSTRACT

As the most basic human action, walking is equitable and inclusive. To define how walking-friendly an area is, the notion of walkability has been used and framed in various ways. Because this notion is studied in different disciplines, there is neither an accurate definition nor a definite criteria. Since the 1990s, walkability has been measured with hard-copy walkability measurement tools (WMTs) and, more recently, by web-based walkability measurement tools. In the literature, WMTs are generally used to measure the walkability level in selected cases or areas. However, few studies have examined the underlying processes of these WMTs. Accordingly, the aim of this paper is to define and categorize the foundations and main ideas of the underlying principles of walkability measurement tools through mapping, clustering, tabling, and analyzing. The method was developed based on two specific charts: a Comprehensive Data Chart and a Content Analysis Chart. The Comprehensive Data Chart defines and categorizes the properties of WMTs while the Content Analysis Chart critically compares their content. With the help of two walkability measurement tools, SPACES and WalkScore, two WMTs are compared to better illustrate the proposed method. Thus, this study provides a new method to analyze WMTs in more detail.

1.1 Keywords

Walkability; Walkability Measurement Tools; Analytical Method Development; Content Analysis

2 INTRODUCTION

As the basic and oldest human action, walking is an interesting topic in the area of urban research. Walking can impact the development of the built environment, and the significant progress in technology made the focus even more essential. To walk is a subjective action, because individuals have different physical and perceptual abilities and limitations, but also because it can be performed with or without goals. In any case, the environment's physical and social characteristics directly affect people's willingness and tendency to walk. Among these characteristics, there are factors that can encourage walking, as both a mode of travel or an activity. In this context, the environment is defined as "walking-friendly" or "walkable." To encourage walking, researchers with distinct perspectives from various disciplines have studied walkability to reach a wider audience. Yet, there are still missing points and uncertainties about how walkability is covered, defined and measured. For example, the term walkability lacks a common definition accepted by all disciplines and an agreed set of criteria. Besides, in the global literature, measures rely on using different walkability measurement tools (WMTs).

The main aim of using WMTs is to measure and analyze walkability, on the assumption that it is a quantifiable and measurable concept, although data is also collected regarding other everyday life issues. The present study does not consider whether walking is a quantifiable concept. Rather, it focuses on developing a critical method for understanding and analyzing the underlying principles of different WMTs with two selected representative WMTs - SPACES and WalkScore.

3 RESEARCH OBJECTIVES

Although there are several types of walkability studies in the literature, there is a lack of research into the theoretical framework of WMTs. This paper aims to fill this gap in the literature by developing a scientific method for critically understanding the design process and underlying ideas of WMTs (Author 1, 2019). For representing this developed method, two WMTs are selected as representatives.

The study addresses two main research questions. The first question concerns what the general basis of WMTs are, including their practical and theoretical formations (how their developers define them), each tool's engines, algorithms, and data sources, the tool's main focus, its evaluation criteria, how data is collected and analyzed, scoring procedures, and presentation of results. The second question concerns what the tools cover and what they miss in their content.

4 LITERATURE REVIEW

4.1. Historical Development of Walkability

Many people like walking because it is safe, practical, comfortable, and cheap compared to other activities (Aghaabbasi, Moeinaddini, Shah, Shekari, and Kermani, 2018; Bereitschaft, 2018). As a recreational or a transportation mode, walking is affected by

https://doi.org/10.52202/082078-0014

the qualities of the built environment (Zuniga-Teran et al., 2017). It is also influenced by the inclusion of technology into everyday lives, e.g. due to the enlargement of cities or to the increased use of automobiles. Nowadays, even if an area is described as highly walking-friendly, many people prefer automated transportation over walking because of time constraints. In the context of technological progress and capitalist society, the "liveability of the urban environment" has become a hot topic. Academics and researchers have considered "walkability" being one of the underlying concepts (Ghadimkhani, 2011). Theoreticians, such as Jane Jacobs, Kevin Lynch, and Dan Burden, were engaged in a discourse on New Urbanism, the Howard's Garden City, and the Drummond's Chicago Plans (Pak and Verbeke, N.D.; Azmi and Karim, 2012; Azmi, Karim, and Ahmad, 2013; Biernecka-Lievestro, 2014; Fitzsimons, 2013).

In the academic literature, walkability first appeared in transportation and planning, particularly the American Highway Capacity Manual (HCM) developed by the Institute of Transportation Engineers' Journal in 2000, in an article entitled "US Federal Highway Administrations' Pedestrian Facilities Users' Guide" (Do, 2002) and the City of Portland Pedestrian Master Plan of 1998 (Fitzsimons, 2013). The notion then spread in other disciplines, e.g. public health, medicine, psychology, gerontology, endocrinology, philosophy, sociology and architecture (Choi, 2012; 2013; 2014; Fitzsimons, 2013; Barros, 2014). By developing different perspectives on the same notion, these disciplines have contributed to a better understanding and more detailed analysis of walkability in all its behavioral, physical and perceptual aspects (De Cambra, 2012; Moudon and Lee (2003); Fitzsimons (2013)). Given its multidisciplinary appeal, walkability is a vast topic discussed in manifold articles, papers, projects, reports, theses, and dissertations.

4.2. Walkability in the Literature

A distinct clustering has emerged across these various studies into four broad themes (Table 1) (Andrews, Hall, Evans, and Colls, 2012). However, given the rapidly increasing number of studies on walkability, a more detailed re-clustering is needed. Researchers have been focusing on different aspects of walkability and used their own set of criteria, making difficult the conversation on walkability among disciplines. Besides, walkability is also characterized by a set of subjective issues in addition to objective qualities.

Table 1 *Four Themes of Walkability Studies (Source: authors)*

#	Themes	Content
1	**Measuring walkability and its potential facilitators**	demographic and social variables related to the social composition of places, including age, ethnicity, socio-economic status, and the contextual factors of places, including availability, land use designations, sidewalks, and crossings
2	**Walkability and diverse quantifiable outcomes**	walking rates, distances and health status
3	**Perceptions of walkability**	familiarity and safety
4	**Evaluations of creating walkable environments**	

Although the definitions and contents of WMTs vary, their benefits are widely accepted, particularly their contribution to public health (Ghadimkhani, 2011), their economic-related everyday aspects (Battista and Manaugh, 2017; Lefebvre, Ropars, Morency, Singleton, and Clifton, 2017), social advantages, assistance in making effective planning decisions (Ghadimkhani, 2011; Nilles and Kaparis, 2018), as well as environmental sustainability and energy efficiency, perceptions and level of creativity (Bereitschaft, 2018). According to the most current literature, the promotion of walkability requires innovative intelligent urban properties and smart technological interventions while taking sustainability into consideration. Through these solutions, walkability has become of the most important factors in smart city definitions (Table 2). Through these elements of a smart city, walkability can be promoted in a more innovative way (Kamel, 2013).

https://doi.org/10.52202/082078-0014

Table 2 *Four Clusters of Walkability in Smart Cities (Source: authors)*

#	Clusters
1	**smart innovations, inconspicuous and safe walkability, including Intelligent pedestrian traffic lights and intelligent street lighting (efficient and intuitive urban ground lighting and presence-based dynamic post lighting)**
2	**smart innovations in convivial walkability, such as augmented reality, time-scope, digital play tables, intelligent musical streets, smart information panels and digital columns**
3	**intelligent wind towers, intelligent sheds, bus stop shelters, public toilets, retractable street furniture, vending machines and pavements**
4	**smart innovations in consistent and sustainable walkability, intelligent photovoltaic (PV) sheds, solar trash receptacles and smart irrigation systems**

4.3. Measuring Walkability

Given the various broad definitions and scope, several media and methods are used to measure walkability. It can be measured with hard-copy (paper-based) tools, which are mainly used to analyze the selected case area. Researchers, academics, and institutes considered relevant authorities use hard-copy WMTs to assess the relationships between person, environment, and the act of walking in order to evaluate walkability in a case study area objectively and quantitatively (Maghelal and Capp, 2011). Hard-copy tools include audit tools, checklists, inventories, level-of-service scales, surveys, instruments, indices and surveys (De Cambra, 2012; Maghelal and Capp, 2011). Objective and subjective variables are used together when the case study area in between a street scale and a neighborhood scale (Biernecka-Lievestro, 2014; De Cambra, 2012).

With the integration of technology, hard-copy WMTs have evolved into web-based tools. The most common methods currently in use are surveys, direct observations, unobtrusive indicators, GIS technologies, motion detectors, and self-report measures (De Cambra, 2012; Choi, 2012; Meester, Dyck, Bourdeaudhuij, Deforche, and Cardon, 2013; Duncan, Aldstadt, Whalen, Melly, and Gortmaker, 2011). Due to an increasing demand for digital methods in the measure of the urban environment livability, new methods or tools have appeared, such as Remote Sensing, Google Street View, and Microsoft Visual Earth. These latter options are quicker and more practical, and also more secure for collecting data (Dobesova and Krikva, 2012; Shatu and Yiğitcanlar, 2018).

5 METHODS

This study was conducted in several steps. The first was a detailed literature review on walkability and WMTs conducted by comprehensively searching for academic sources that included the keyword 'WMT' in their literature review, such as journal papers, conference papers, academic reports and these. In addition, an online search was conducted for WMTs used by private-sector professionals. The aim of this literature review was to make a literature review about walkability as well as listing the WMTs in detail (Author 1, 2019).

In the second step, WMTs were selected as the most comprehensive representatives for analysis: one hard-copy WMT (Systematic Pedestrian and Cycling Environmental Scan - SPACES) and one web-based WMT (WalkScore). These two WMTs are analyzed in detail by searching the following general and academic databases: Google, Google Scholar, Science Direct, EBSCOhost, Elsevier, and Taylor and Francis. In addition, some printed sources and webpages were reviewed.

After collecting the data, an analytical filter was developed in step three for clustering the information and identifying the main underlying principles of the WMTs. Using this critical filter, the information about WMTs could be clustered under several main headings to compare the tools. During this process, two main charts were created. The first chart, the Comprehensive Data Chart, presented the WMTs' main characteristics (Table 3). The main purpose of the chart was to show details like the tool developer, target user profile, methodological details, question types, and presence of supplementary items. The second chart, the Content Analysis Chart, presented the questions and criteria for each tool in detail in order to show which topics were covered or missing (Table 4). The Comprehensive Data Chart was generated using concrete data from the reviewed sources whereas the Content Analysis Chart was created by applying a critical lens.

https://doi.org/10.52202/082078-0014

Table 3 *Comprehensive Data Chart (Source: authors)*

COMPREHENSIVE DATA CHART	DEVELOPER	AIM & INTEREST	TARGET PROFILE	METHODOLOGY	QUESTION FORMATIONS	PRESENCE OF GUIDES
HARD-COPY						
WEB-BASED						

Table 4 *Content Analysis Chart (Source: authors)*

CONTENT ANALYSIS CHART	#1			#2		#3	
	PERCEPTIONS	USERS	ENVIRONMENTAL	PHYSICAL	FUNCTION	NETWORK	POLICY
HARD-COPY							
WEB-BASED							

5.1 Chart I: Comprehensive Data Chart

Following the comprehensive literature review, the Comprehensive Data Chart was created by analyzing, filtering and clustering the collected data. The information for the chart was taken from all sources without any interference. The sources for the chart were the tool itself, publications about the tool (academic articles, web articles), publications by the tool developer(s) (academic articles, web articles), websites (the tool itself, the tool developer's, other related sources) and other publications (tool developer's thesis and tool developer's guides). Based on the data, the following six main groups were created to comprehensively represent each tool's general characteristics: tool developer, tool developer's aim and interest, tool target profile, tool methodology, tool question formation, and tool guides.

These six main groups also had sub-groups, as follows. "Developer" included tool type and details about who developed the tool. The type of developer could be a firm (private or state), individual researchers, a research team from the same or different organizations or research interests, a government facility, or research project outcome, as the direct output of a project. The tool name and tool type were also specified.

The "aim and interest" group identified the interest area and general aim of the tool using quotes about the tool or the tool developer itself from the developer's webpages, interviews, or speeches. The interest area refers to the developer's main research field while the general aim of the tool developer includes specific quotes about the tool's target. "Target profile" included information about the user, target location, presence of adapted versions and adaptation of criteria scales. The target user profile had five categories (adults, children, parents, elderly and disabled). Target location, referring to the main geographic area that the tool targeted, was needed for tools that lacked a specific user profile. Presence of an adapted version identified whether the tool's content and methodology were adapted for different locations. Adaptation of criteria scales recorded whether the scales were 'local', meaning adapted to geography, or 'global', meaning that the criteria were the same for all locations.

"Methodology" included information about data collection, in-processing and evaluation. Data collection concerns data collection methods and their details. The data collection methods could include numeric data, direct sampling, open-source, crowdsourcing, software, prioritization, observation, questionnaires, GIS data, self-report and on-street. This category also recorded if a specific program was used or specific data was collected from a source, such as the crime rate. Usage of numeric data was recorded as 'yes' or 'no' while the number of team members recorded how many people were involved in data collection. Data collection also included Auditor's privacy (questions asked to the auditor like name, surname, ID, email, address) and maps used (printed or online).

The In-process phase of the methodology included criteria weighting, recorded as same or different, engines and specific algorithms being used, which included the name and other details, the method of in-process scoring, such as ordinal, Likert, nominal, multiple-choice, yes/no, checking, review and algorithmic scoring.

https://doi.org/10.52202/082078-0014

information for the third phase of the methodology, tools' and tool developers' definitions of scoring, included the developer's quotes related to quantification as part of evaluation. Final scoring methods like ordinal, Likert, ranging and review are also a part of this phase. For a more detailed analysis, this phase also recorded if there was a range in the final score and visualization of the final score.

"Question formation" included information about the types of questions used in the tool, specifically open-ended versus close-ended. The former includes two-option responses, one best answer, rating scales, ordered choice, 'other, please specify', items in series, paired comparisons, matching, 'check all that apply', listing and ranking.

Finally, "presence of guides" recorded if the tool had a user guide / manual and/or a scoring guide as plus (+ existing) or minus (- not existing).

5.2 Chart II: Content Analysis Chart

The Content Analysis Chart was prepared by following five steps to identify each tool's scope regarding the issues covered and note any missing points. First, the content of each tool (criteria/questions) was directly copied and pasted from the tools into a single file unchanged. Second, the tool content of each tool was clustered under the following 24 headings in alphabetical order (Table 5). Third, the content was converted into a single consistent phrase format because the form of the content in each tool was different. Fourth, sub-groups were created for each of the 24 main groups. Finally, the chart was made more readable by assigning a symbol for each group, preceded by digits to represent each sub-groups.

Table 5 *Content Analysis Chart - 24 headings (Source: authors)*

Accessibility	Facilities and Amenities	Pavement/Sidewalk	Speed
Aesthetics	Finding the Way and Navigation	Pedestrians	Street Characteristics
Buildings	Land Use and Density	Policy	Topographic Characteristics
Connectivity	Lighting	Public Space	Traffic
Crossings	Neighborhood Characteristics	Public Transportation	Walking and Cycling
Disability	Obstructions	Safety	Walking Paths

The 24 groups were themselves grouped under the following seven themes: Perception/Abstract for more subjective, perceptual aspects; Environment-Based for more spatial items; Physical for more objective, concrete aspects of the immediate environment; Function for the functional characteristics of buildings and neighborhoods; Network for traffic and transportation-related dimensions; User for the person using the tool; and Policy for walkability-related regulations and laws. To represent these charts, one hard-copy walkability measurement tool – Systematic Pedestrian and Cycling Environmental Scan (SPACES) – and one web-based walkability measurement tool – WalkScore – were selected.

As one of the first auditing tools for measuring and analyzing the relationships between the built environment and physical activity (both walking and cycling) (Transportation Research Board Institute of Medicine of the National Academies, 2005), SPACES was originally designed for use in Australia (Clifton, Smith, and Rodriguez, 2007). It was developed by a group of researchers from the University of Western Australia in 2000 based on a previous study (Calgary Regional Partnership, N.D.; MIDSS Measurement Instrument Database for the Social Sciences, N.D.). The tool's observational infrastructure comprises 37 items in the following four main groups (A Guide to: Systematic Pedestrian and Cycling Environment Scan (SPACES), N.D.; Calgary Regional Partnership, N.D.; Romanow et al., 2013):

https://doi.org/10.52202/082078-0014

1. Types of buildings and features
2. General assessment of paths that could be used for walking and/or cycling (e.g., type, location, material, slope, condition)
3. General street assessment (e.g., number of lanes on a road, kerb type, traffic control devices, crossing facilities, presence of street lights, presence of destinations, car and bicycle parking facilities)
4. Opportunities for natural surveillance, garden maintenance, cleanliness and views, and perceived attractiveness or difficulty for walking and cycling

During data collection, a trained researcher observes the area while walking, focusing on the street segment level (Calgary Regional Partnership, N.D.; A Guide to: Systematic Pedestrian and Cycling Environment Scan (SPACES), N.D.). Street maps are also provided to aid the audit while there is a manual for grading to guide scaling and reduce subjectivity and variability between different observers (A Guide to: Systematic Pedestrian and Cycling Environment Scan (SPACES), N.D.). GIS systems are used to provide information that cannot be observed, such as traffic volume and speed, street and intersection design, locations of local destinations like parks, shops and public transport points (Calgary Regional Partnership, N.D.). SPACES was also adapted for application in New Zealand (Oliver et al., 2015; A Guide to: Systematic Pedestrian and Cycling Environment Scan (SPACES), N.D.; Badland, Opit, Witten, Keams, and Mavoa, 2010).

WalkScore is one of the most well-known and widely used web-based WMTs in the literature (Barnes, Winters, Ste-Marie, McKay and Ashe, 2016). It was developed from a project by the Washington-based Front Seat Management Company in 2002 (Arribas-Bel, 2014; Duncan, Aldstadt, Whalen, Melly, and Gortmaker, 2011). It mainly evaluates and supports walkability and pedestrian access to facilities (Hirsch, Moore, Evenson, Rodriguez, and Diez Roux, 2013) by focusing on land use mix, density and street-grid density with analog parameters (Gilderbloom, Riggs and Meares, 2015). The tool grades different types of amenities based on their distance under the following five main categories: education (schools and libraries), recreational areas (parks and greenery areas), retail (bookstore, grocery stores, and clothing shops), food (restaurants, coffee shops and bars), and entertainment and leisure (cinemas and theaters) (Brown et al., 2013; El-Geneidy, Lierop, and Wasfi, 2016; Lin et al., 2016; Kok and Jennen, 2012). The evaluation criteria are based on seven items (WalkScore, 2019):

1. A center (walkable neighborhoods have a center, whether a main street or a public space)
2. People (enough people for businesses to flourish and for public transit to run frequently)
3. Mixed-income/mixed-use (affordable housing located near businesses)
4. Parks and public spaces (plenty of public places to gather and play)
5. Pedestrian design (buildings are close to the street, parking lots relegated to the back),
6. Schools and workplaces (close enough that most residents can walk from their homes),
7. Complete streets (streets designed for cyclists, pedestrians and transit)

The tool uses various data sources to get difference kinds of information, including Google AJAX Search Application Program Interface (API), Localize and Census, Open Street Map, Education.com, and Transit Agencies (Duncan, Aldstadt, Whalen, Melly, and Gortmaker, 2011; El-Geneidy, Lierop, and Wasfi, 2016; Langlois, Wasfi, Ross, and El-Geneidy, 2016; Xu and Wang, 2015). The tool uses a specific evaluation algorithm, Google Search Algorithm, to measure the walkability of each area (Xu and Wang, 2015) and calculate the distance-walkability relation (Brewster, Hurtado, Olson, and Yen, 2009). WalkScore can be easily used by manually entering the name or geographic coordinates of the location (Harvey, Aultman-Hall, Hurley and Troy, 2015).

Amenities at different distances get different grades while each category is weighted equally. For the final score, each location is graded between 0-100 (Lin et al., 2016; Walk Score, 2019). There are three different final grades: Walk Score, Bike Score and Transit Score (Chandra et al., 2016).

WalkScore was originally developed for real estate purposes (Carr, Dunsinger and Marcus, 2010) as indicated by their vision statement: "Beds: 3 Baths: 2 WalkScore:84". By being used by real estate brokers and realtors, the walkability value of space is capitalized (Arribas-Bel, 2014). WalkScore is available throughout the USA and several places in Canada, Australia, New Zealand, the UK and Ireland (Barnes, Winters, Ste-Marie, McKay and Ashe, 2016; Duncan, 2013).

6 RESULTS

The features of SPACES and WalkScore are presented in the Content Analysis Chart (Table 6) and Comprehensive Data Chart (Table 7).

https://doi.org/10.52202/082078-0014

Table 6 *Results of Content Analysis Chart (Source: authors)*

CONTENT ANALYSIS CHART		SPACES	WALK SCORE
PERCEPTIONS	ACCESSIBILITY	-	0
	CONNECTIVITY	α	α
	WAYFINDING & NAVIGATION	✜	-
	SAFETY	•	-
USER	DISABILITY	-	-
	PEDESTRIAN	-	¶ / 3¶
ENVIRONMENTAL	AESTHETICS	◉ / 2◉ / 4◉ / 5◉	-
	BUILDINGS	4■	-
	NEIGHBORHOOD CHARACTERISTICS	-	-
	PUBLIC SPACES	-	✖ / 2✖
	STREET CHARACTERISTICS	-	-
	TOPOGRAPHIC QUALITIES	2 Θ	-
PHYSICAL	PEDESTRIAN CROSSINGS	◆ / 2◆	-
	WAYFINDING & NAVIGATION	✜	-
	LIGHTING	✪ / 2✪	-
	OBSTRUCTIONS	⌘ / 3⌘	-
	PAVEMENTS / SIDEWALK	2⌥	-
	WALKING PATHS	2 ψ	-
FUNCTION	FACILITIES & AMENITIES	✚ / 2✚	3✚
	LAND USE & DENSITY	★ / 2★	★
NETWORK	PUBLIC TRANSPORTATION	-	-
	SPEED	-	-
	TRAFFIC	2 Δ / 5 Δ / 7 Δ	-
	WALKING & CYCLING	2 Ω / 3 Ω	-
POLICY	POLICY	-	-

https://doi.org/10.52202/082078-0014

Table 7 *Results of Comprehensive Data Chart (Source: authors)*

COMPREHENSIVE DATA CHART			TOOL	
			SPACES	WALK SCORE
DEVELOPER	TYPE		Research project outcome (Calgary Regional Partnership, N.D. & A GUIDE TO: Systematic Pedestrian & Cycling Environment Scan (SPACES), N.D. - website)	Firm (Arribas-Bel, 2012 & Duncan, Aldstadt, Whalen, Melly & Gortmaker, 2011 - academic articles)
	DETAILS		Terri J. Pikora, Billie Giles-Corti, Matthew W. Knuman, Fiona C. Bull, Konrad Jamrozik & Rob J. Donovan (Calgary Regional PartnershIP, N.D. & A Guide To: Systematic Pedestrian & Cycling Environment ScaN (SPACES), N.D. - website)	Front Seat Management Company LLC. (Arribas-Bel, 2012 & Duncan, Aldstadt, Whalen, Melly & Gortmaker, 2011 - academic articles)
AIM & INTEREST	INTEREST AREA		Health promotion, urban futures, physical activity (Calgary Regional Partnership, N.D. & A Guide To: Systematic Pedestrian & Cycling Environment Scan (SPACES), N.D. - website)	Real estate (Carr, Dunsinger & Marcus, 2010 - academic article & developer's website - https://www.walkscore.com/about.shtml)
	GENERAL AIM		Measuring physical activity-related aspects of the built environment (Pikora, Bull, Jamrozik, Knuiman Gles-Corti & Donovan, N.D.; Lee & Talen, 2014 & Pikora, Giles-Corti, Knuman, Bull, Jamrozik & Donovan, 2006 - academic articles) // evaluating the physical environment in local neighborhoods for activity of walking & cycling (Premier's Council for Active Living New South Wales & New Centre for Physical Activity & Health, 2007 - website)	"Our vision is for every property listing to read: beds: 3 baths: 2 walk score: 84. we want to make it easy for people to evaluate walkability & transportation when choosing where to live" (developer's website - ttps://www.walkscore.com/about.shtml) // EVALUATING & supporting walkability & pedestrian access to transportation (Hirsch, Moore, Evenson, Rodriguez & Diex Roux, 2013 - academic article) // promoting walkable neighborhoods (Arribas-Bel, 2014 - academic article)
TARGET PROFILE	USERS	ADULT CHILDREN PARENTS ELDERLY DISABLED	Australia (Clifton, Smith & Rodriguez, 2007 - academic article)	USA, Australia, New Zealand, England, Ireland, several addresses in Canada (Barnes, Winters, Ste-Marie, Mckay & Ashe, in press & Duncan, 2013 - academic article)
	TECHNOLOGICAL & LOCATION-BASED ADAPTATION		New Zealand (Oliver et al., 2015 - academic article, A Guide To: Systematic Pedestrian & Cycling Environment Scan (SPACES), N.D. - website & Badland, Opit, Witten, Keams & Mavoa, 2010 - academic article)	No information

https://doi.org/10.52202/082078-0014

METHODOLOGY	DATA COLLECTION	METHODS	Observational + GIS data (Calgary Regional Partnership, N.D - website)	Numeric data from several data sources (Duncan, Aldstadt, Whalen, Melly & Gortmaker, 2011; El-Geneidy, Lierop & Wasfi, 2016; Langlois, Wasfi, Ross & El-Geneidy, 2016 & Xu, Wen & Wang, 2015 - academic articles)
		DETAILS	GIS data: traffic volume & speed, street & intersection design, location of local destinations (e.g., parks, shops, & public transport points) (Calgary Regional Partnership, N.D - website)	Google Ajax Search Application Program Interface (API) - Localeze & Census - Open Street Map - Education.Com - transit agencies (over 200) (Duncan, Aldstadt, Whalen, Melly & Gortmaker, 2011; El-Geneidy, Lierop & Wasfi, 2016; Langlois, Wasfi, Ross & El-Geneidy, 2016 & Xu, Wen & Wang, 2015 - academic articles)
		USAGE OF NUMERIC DATA	Yes (A Guide To: Systematic Pedestrian & Cycling Environment Scan (SPACES), N.D. - website)	Yes (Duncan, Aldstadt, Whalen, Melly & Gortmaker, 2011, El-Geneidy, Lierop & Wasfi, 2016, Langlois, Wasfi, Ross & El-Geneidy, 2016 & Xu, Wen & Wang, 2015 - academic articles)
		NUMBER OF TEAM MEMBERS	Individual (A Guide To: Systematic Pedestrian & Cycling Environment Scan (SPACES), N.D. - website)	-
		AUDITOR PRIVACY	Author ID (Tool)	No information required
		MAPS BEING USED	Printout map (Tool)	Updateable online map (Google) (Tool)
	IN-PROCESS	CRITERIA WEIGHT	No information	No information
		ENGINES & ALGORITHMS	No information	Special distance-decay algorithm (Brewster, Hutrado, Olson & Yen, 2009, Brown et al., 2013 & Cole, Dunn, Hunter, Owen & Sugiyama, 2015 - academic articles)
		IN-PROCESS SCORING	Nominal (checking)	Algorithmic scoring
	EVALUATION	TOOLS' DEFINITIONS OF SCORING	Checking (Tool)	Algorithmic scoring (Chudyk, Winters, Moniruzzaman, Ashe, Gould & Mckay, 2015 & Cole, Dunn, Hunter, Owen & Sugiyama, 2015 - academic articles)
		DEVELOPERS' DEFINITIONS OF SCORING	Number, average, coding, assessment, tick (Tool)	Get scores (Tool's website - main page - www.walkscore.com)
		FINAL SCORING	-	Evaluate (Tool's website - main page - www.walkscore.com)
			No information	Ordinal (ranging)
		RANGING THE FINAL SCORE	No information	From 0 to 100 (Tool developer's website - www.walkscore.com)
		VISUALIZATION	No information	Yes (only numeric) (Tool developer's website - www.walkscore.com)
QUESTION FORMATIONS	OPEN-ENDED		-	-
	CLOSE-ENDED	TWO-OPTION RESPONSES	+	
		ONE BEST ANSWER	+	
		RATING SCALE	+	
		ORDERED CHOICE	-	

https://doi.org/10.52202/082078-0014

		"OTHER, PLEASE SPECIFY"	-	
		ITEMS IN SERIES	+	
		PAIRED COMPARISONS	-	
		MATCHING	-	
		CHECK ALL THAT APPLY	+	
		LISTS	-	
		RANKING	-	
PRESENCE OF GUIDES	USER GUIDE / MANUAL		+	-
	SCORING GUIDE		-	+

The Content Analysis Chart shows that WalkScore's content disregards all the physical qualities of the space, its network qualities and related policies. In contrast, SPACES does not cover users and policy settings. The greatest difference between SPACES and WalkScore is that SPACES covers all aspects of the physical qualities of the space whereas WalkScore ignores them all.

The Comprehensive Analysis Chart shows that WalkScore was developed by a private firm. Due to its developer, its main interest area is real estate and its economic relationship to walkability. In contrast, SPACES was developed as a research project main interested in health-related aspects of walkability. Due to its online features, WalkScore can be used in several countries whereas the printout characteristics of SPACES limits it to a single area. WalkScore only uses databases. Its users (walkers) are not a data source. In contrast, SPACES combines both ready-made data sets and user data. During data evaluation, WalkScore represents all steps in detail whereas there is not much information about SPACES. Regarding question formation, both tools avoid open-ended question types, although these are the most appropriate forms for personal comments. WalkScore only uses ratings whereas SPACES include several types. Finally, WalkScore lacks a user manual but has a scoring guide whereas SPACES has a user guide but no scoring guide.

7 CONCLUSIONS & DISCUSSION

This study primarily aimed to develop a method to identify the targets and background ideas of two WMTs. Because WMTs analyze walkability based on numerical scores, they imply that walking is a quantifiable concept. However, this paper does not consider whether walkability can be measured quantitatively. Instead, the analysis showed that the main aim of WMTs is not really walking, but other issues like economics. They also consider the act of walking as an individual activity rather than a collective movement.

This study also aimed to contribute to the current literature of walkability in several ways. Firstly, a critical method was developed to cover both web-based and hard-copy walkability measurement tools in detail. Although there are several types of research about walkability, the present study is important because it applied a more comprehensive comparative analysis to WMTs.

Secondly, the study critically analyzed two walkability measurement tools in the literature. These studies all consider walkability individually whereas this study covered the broader field of walkability while accepting all possible tools that are used. Moreover, by deploying the criteria for comparison, it considered two WMTs as cases. Applying this critical lens revealed each WMT's theoretical framework and conceptual details.

Third, the method developed for this study could be applied to both hard-copy and web-based WMTs. These are more open to revision to delineate the historical development of WMTs.

Future studies can apply this method to other WMTs beyond the scope of our paper. The methods can also be applied to newly developed or in-process tools as a means for future WMTs to analyze themselves in terms of their missing or more focused parts, or the parameters that they should include and pay attention to. Content analysis and comprehensive analysis methods can also be validated through a real case study to develop a new tool. Finally, this method can also be applied to tools that also or only cover bikeability.

https://doi.org/10.52202/082078-0014

8 REFERENCES

A Guide to: Systematic Pedestrian and Cycling Environment Scan (SPACES) (N.D.). [pdf]. Retrieved 10/25/17 from http://weightmanagement.hiirc.org.nz/assets/sm/Resource527/attachments/mmowi5risr/SPACES%20guide.pdf?download=true

Aghaabbasi, M., Moeinaddini, M., Shah, M. Z., Shekari, Z. A., and Kermani, M. A. (February, 2018). Evaluating the capability of walkability audit tools for assessing sidewalks. Sustainable Cities and Society, 37, 475-484. https://doi.org/10.1016/j.scs.2017.12.001

Andrews, G. J., Hall, E., Evans, B., and Colls, R. (December, 2012). Moving beyond walkability: on the potential of health geography. Social Science and Medicine, 75(11), 1925-1932. https://doi.org/10.1016/j.socscimed.2012.08.013

Arribas-Bel, D. (May, 2014). Accidental, open and everywhere: emerging data sources for understanding the cities. Applied Geography, 49, 45-53. https://doi.org/10.1016/j.apgeog.2013.09.012

Azmi, D. I., and Karim, H. A. (2012). Implication of walkability towards promoting sustainable urban neighborhood. Procedia- Social and Behavioral Sciences, 50, 204-213. https://doi.org/10.1016/j.sbspro.2012.08.028

Azmi, D. I., Karim, H. A., and Ahmad, P. (2013). Comparative study of neighborhood walkability to facilities between two precincts of Putrajaya. Procedia- Social and Behavioral Sciences, 105, 513-524. https://doi.org/10.1016/j.sbspro.2013.11.055

Badland, H. M., Opit, S., Witten, K., Kearns, R. A., and Mavoa, S. (December, 2010). Can virtual streetscape audits reliably replace physical streetscape audits?. Journal of Urban Health: Bulletin of New York Academy of Medicine, 87(6), 1007-1016. https://doi.org/10.1007/s11524-010-9505-x

Barnes, R., Winters, M., Ste-Marie, N., McKay, H., and Ashe, M. C. (December, 2016). Age and retirement status differences in associations between the built environment and active travel behavior. Journal of Transport and Health, 3(4), 513-522. https://doi.org/10.1016/j.jth.2016.03.003

Barros, A. P. (July, 2014). The Systemic Focus on Walkability and Urban Form. Paper presented at 21st International Seminar on Urban Form, Porto. [pdf]. Retrieved 4/28/16 from https://www.researchgate.net/publication/269572405_The_systemic_focus_on_walkability_and_urban_form

Battista, G. A., and Manaugh, K. (January, 2017). Using embodied videos of walking interviews in walkability assessment. Transportation Research Record: Journal of the Transportation Research Board, 2661(1), 12–18. https://doi.org/10.3141/2661-02

Bereitschaft, B. (August, 2018). Exploring perceptions of creativity and walkability in Omaha, NE. City, Culture and Society. https://doi.org/10.1016/j.ccs.2018.08.002

Biernecka-Lievestro, J. (2014). Analyzing the association between walkability and regional economic vitality (Published Doctoral Dissertation). George Mason University [pdf]. Retrieved 11/12/16 from http://citeseerx.ist.psu.edu/viewdoc/

download?doi=10.1.1.924.4982andrep=rep1andtype=pdf

Brewster, M., Hurtado, D., Olson, D., and Yen, J. (2009). Walkscore.com: a new methodology to explore associations between neighborhood resources, race and health. Paper Presented at 137st APHA Annual Meeting and Exposition 2009, Philadelphia, Pennsylvania. [pdf]. Retrieved 11/12/16 from https://www.researchgate.net/publication/266900527_Walkscorecom_A_New_Methodology_to_Explore_Associations_Between_Neighborhood_Resources_Race_and_Health

Brown, S. C., Pantin, H., Lombard, J., Toro, M., Huang, S., Plater-Zyberk, E., Perrino, T., Perez-Gomez, G., Barrera-Allen, L., and Szapocznik, J. (August, 2013). Walk Score: associations with purposive walking in recent cuban immigrants. American Journal of Preventive Medicine, 45(2), 202-206. https://doi.org/10.1016/j.amepre.2013.03.021

Calgary Regional Partnership. (N.D.). Bikeability Audit Tool [pdf]. Retrieved 10/25/17 from http://greenfield.calgaryregion.ca/tools/greenfield_process_bikeabilityAudit.pdf

Carr, L. J., Dunsiger, S. I., and Marcus, B. H. (November, 2010). Walk Score as a global estimate of neighborhood walkability. American Journal of Preventive Medicine, 39(5), 460-463. https://doi.org/10.1016/j.amepre.2010.07.007

Chandra, S., Braughton, M., Galicia, L. D., Sanchez, A., Medina, M., and Aldrete, R. (2016). A multi-modal transportation score to evaluate infrastructure supply-demand for commuters. Procedia Engineering, 145, 304-311. https://doi.org/10.1016/j.proeng.2016.04.078

Choi, E. (2012). Walkability as an urban design problem (Published Bachelor Thesis). KTH Royal Institute of Technology. [pdf]. Retrieved 9/12/16 from https://www.divaportal.org/smash/get/diva2:551283/FULLTEXT01.pdf

Choi, E. (2013). Understanding walkability: dealing with the complexity behind pedestrian behavior. In Proceedings of the 9th International Space

https://doi.org/10.52202/082078-0014

Syntax Symposium, pp. 4-14. [pdf]. Retrieved 9/12/16 from https://pdfs.semanticscholar.org/64f0/6a8fb7427f65139574012ed22523dd2391ea.pdf?_ga=2.183792166.1858781259.1549801968-1146575004.1549801968

Choi, E. (2014). Walkability and the complexity of walking behavior. ITU AIZ, 11(2), 87-99. [pdf]. Retrieved 11/14/16 from http://www.az.itu.edu.tr/azvol11no2web/8-choi-1102.pdf

Clifton, K. J., Smith, A. D. L., and Rodriguez, D. (2007). The development and testing of an audit for the pedestrian environment. Landscape and Urban Planning, 80(1-2), 95-100. https://doi.org/10.1016/j.landurbplan.2006.06.008

De Cambra, P. J. M. (2012). Pedestrian accessibility and attractiveness indicators for walkability assessment Manitoba (Published Master Thesis). [PDF document]. Retrieved 11/14/16 from https://fenix.tecnico.ulisboa.pt/downloadFile/395144992898/Dissertacao.pdf

Do, A. (2002). Walking the safety walk. Public Roads, 66(2), 2–5. Retrieved 12/17/21 from https://highways.dot.gov/public-roads/septemberoctober-2002/walking-safety-walk

Dobesova, Z., and Krikva, T. (March 21, 2012). Walkability index in the urban planning: a case study in Olomouc City. Advances in Spatial Planning. 179-196. https://doi.org/10.5772/36587

Duncan, D. T. (August, 2013). What's your Walk Score? web-based neighborhood walkability assessment for health promotion and disease prevention. American Journal of Preventive Medicine, 45(2), 244-245. https://doi.org/10.1016/j.amepre.2013.04.008

Duncan, D. T., Aldstadt, J., Whalen, J., Melly, S. J., and Gortmaker, S. L. (2011). Validation of Walk Score for estimating neighborhood walkability: an analysis of four US metropolitan areas. International Journal of Environmental Research and Public Health, 8(11), 4160-4179. https://doi.org/10.3390/ijerph81141603

El-Geneidy, A., Leirop, D. V., and Wasfi, R. (October, 2016). Do people value bicycle sharing? a multilevel longitudinal analysis capturing the impact of bicycle sharing on residential sales in Montreal, Canada. Transport Policy, 51, 174-181. https://doi.org/10.1016/j.tranpol.2016.01.009

Fitzsimons, L. (2013). A Multidisciplinary Examination of Walkability: It's Concept, Assessment and Applicability (Published Doctoral Dissertation). Dublin City University [PDF document]. Retrieved 11/14/16 from http://doras.dcu.ie/19387/2/Lorraine_Fitzsimons_D'Arcy_PhD_Vol_2.pdf

Ghadimkhani, P. (2011). Increasing walkability in public city centers: the case of Tunali Hilmi Street, Ankara (Published Master Thesis). Middle East Technical University. [pdf]. Retrieved 4/28/16 from https://etd.lib.metu.edu.tr/upload/12613235/index.pdf

Gilderbloom, J. I., Riggs, W. W., and Meares, W. L. (February, 2015). Does walkability matter? an examination of walkability's impact on housing values, foreclosures and crime. Cities, 42, 13-24. https://doi.org/10.1016/j.cities.2014.08.001

Harvey, C., Aultman-Hall, L., Hurley, S. E., and Troy, A. (October, 2015). Effects of skeletal streetscape design on perceived safety. Landscape and Urban Planning, 142, 18-28. https://doi.org/10.1016/j.landurbplan.2015.05.007

Hirsch, J. A., Moore, K. A., Evenson, K. R., Rodriguez, D. A., and Diez Roux, A. V. (August, 2013). Walk score and transit score and walking in the multi-ethnic study of atherosclerosis. American Journal of Preventive Medicine, 45(2), 158-166. https://doi.org/10.1016/j.amepre.2013.03.018

Kamel, M. A. E. (2013). Encouraging walkability in GCC cities: smart urban solutions. Smart and Sustainable Built Environment, 2(3), 288-310. https://doi.org/10.1108/sasbe-03-2013-0015

Kok, N., and Jennen, M. (July, 2012). The impact of energy labels and accessibility on office rents. Energy Policy, 46, 489-497. https://doi.org/10.1016/j.enpol.2012.04.015

Langlois, M., Wasfi, R. A., Ross, N. A., and El-Geneidy, A. M. (2016). Can transit- oriented developments help achieve the recommended weekly level of physical activity?. Journal of Transport and Health, 3(2), 181-190. https://doi.org/10.1016/j.jth.2016.02.006

Lefebvre-Ropars, G., Morency, C., Singleton, P. A., and Clifton, K. J. (December, 2017). Spatial transferability assessment of a composite walkability index: the Pedestrian Index of the Environment (PIE). Transportation Research Part D: Transport and Environment, 57, 378-391. https://doi.org/10.1016/j.trd.2017.08.018

Lin, T., Xia, J., Robinson, T. P., Olaru, D., Smith, B., Taplin, J., and Cao, B. (2016). Enhanced huff model for estimating park and ride (pnr) catchment areas in Perth, WA. Journal of Transport Geography, 54, 336-348. https://doi.org/10.1016/j.jtrangeo.2016.06.011

Maghelal, P. K., and Capp, C. J. (2011). Walkability: a review of existing pedestrian indices. URISA Journal, 23(2), 5-19. [pdf]. Retrieved 10/24/17 https://www.researchgate.net/publication/279588344_'Walkability_A_Review_of_

https://doi.org/10.52202/082078-0014

Existing_Pedestrian_Indices

Meester, F. D., Dyck, D. V., Bourdeaudhuij, I. D., Deforche, B., and Cardon, G. (March, 2013). Do psychosocial factors moderate the association between neighborhood walkability and adolescents' physical activity?. Social Science and Medicine, 81, 1-9. https://doi.org/10.1016/j.socscimed.2013.01.013

MIDSS – Measurement Instrument Database for the Social Sciences (N.D.). Systematic Pedestrian and Cycling Environment Scan (SPACES) Instrument. Retrieved 10/25/17 from http://www.midss.org/content/systematic-pedestrian-and-cycling-environmental-scan-spaces-instrument

Nilles, M., and Kaparias, I. (December, 2018). Investigating the relation of highway design standards with network-level walkability: the case study of Luxembourg. International Journal of Transportation Science and Technology, 7(4), 254-263. https://doi.org/10.1016/j.ijtst.2018.08.001

Oliver, M., Witten, K., Blakely, T., Parker, K., Badland, H., Schofield, G., Ivory, V., Pearce, J., Mavoa, S., Hicnkson, E., Sweetsur, P., and Kearns, R. (2015). Neighborhood built environment associations with body size in adults: mediating effects of activity and sedentariness in a cross-sectional study of New Zealand adults. BMC Public Health, 15(1), 956-966. https://doi.org/10.1186/s12889-015-2292-2

Pak, B., and Verbeke, J. (N.D.). Walkability as performance indicator for urban spaces. Paper presented at ECCADE [PDF document]. Retrieved 4/28/16 from http://papers.cumincad.org/data/works/att/ecaade2013_178.content.pdf

Romanow, N. T. R., Couperthwaite, A. B., McCormack, G. R., Nettel-Aguirre, A., Rowe, B. H., and Hagel, B. E. (2013). Assessing inter-rater agreement of environmental audit data in a matched case-control study on bicycling injuries. Injury Prevention, 19(5), 336-341. https://doi.org/10.1136/injuryprev-2012-040500

Shatu, F., and Yigitcanlar, T. (June, 2018). Development and validity of a virtual street walkability audit tool for pedestrian route choice analysis – SWATCH. Journal of Transport Geography, 70, 148-160. https://doi.org/10.1016/j.jtrangeo.2018.06.004

Transportation Research Board Institute of Medicine of the National Academies (2005). Does the built environment influence physical activity?: examining the evidence [PDF document]. Retrieved 10/25/17 from http://onlinepubs.trb.org/onlinepubs/sr/sr282.pdf

Walk Score (2019). Walk Score Main Page. Retrieved 10/15/16 from https://www.walkscore.com/

Walk Score (2019). Walk Score Methodology. Retrieved 2/17/19 from https://www.walkscore.com/methodology.shtml

Xu, Y., and Wang, F. (July, 2015). Built environment and obesity by urbanicity in the U.S. Health and Place, 34, 19-29. https://doi.org/10.1016/j.healthplace.2015.03.010

Zuniga-Teran, A. A., Orr, B. J., Gimblett, R. H., Chalfoun, N. V., Marsh, S. E., Guertin, D. P., and Going, S. B. (2017). Designing healthy communities: testing the walkability model. Frontiers of Architectural Research, 6(1), 63-73. https://doi.org/10.1016/j.foar.2016.11.005

https://doi.org/10.52202/082078-0014

SENSORY ENGAGEMENT AND SPATIAL HARMONY: UNVEILING THE ROLE OF FIRE IN MULTISENSORY DESIGN

Track 4: Design for Health and Wellbeing
Knowledge Network: Environmental & Architectural Phenomenology, Health
Methods: Grounded Theory, Interviews
Shreya Jaiswal (D2D Green Design); Martha Bohm (University at Buffalo / School of Architecture); Korydon Smith (University of Buffalo); Edward Steinfeld (University at Buffalo)

1 ABSTRACT

The research delves into the broader theme of biophilia, place attachment, stress reduction, and overall well-being, exploring the potential of fire as a biophilic design element to contribute to these aspects consequently. Fire is a biophilic element overlooked in previous stress-restoration research, primarily focused on natural materials, vegetation, daylighting, and other biophilic elements. The work is informed by how the historical relocation of fire and its consequential effects on sensory experiences lead to stress and frustration within the built environment.

Employing qualitative research methods, namely grounded theory and semi-structured interviews conducted in an indoor environment with a fireplace, viz., the One World café at the University at Buffalo (case study location). The study comprehensively examines spatial, emotional, sensory, and cultural dimensions of sensory engagement with the fireplace. The research methods allow the researcher to delve deeper into the multisensory experiences of the users from the past (memories) to the present. The experience of twenty-seven participants from semi-structured interviews reveals a connection between the diminished presence of fire and a decline in sensory and social experiences, influencing the degree of attachment to a space or element in a space.

The evaluation of the fireplace at the One World Café case study highlights the fireplace's limitations in terms of scale and sensory boundaries, emphasizing the influence of proximity and direct sensory engagement on individuals' responses to fire. In conclusion, the multisensory engagement provided by fire intensifies the perceived authenticity of the experience and strengthens the connection to nature. It also implied that fire instigates impactful and memorable experiences that enhance individuals' connection to a specific place. Incorporating sensory overlap strategies, designers can enrich environmental design, prioritizing factors like olfactory stimulation and cultural associations. Additionally, the research advocates further investigating the potential of other biophilic elements and multisensory encounters, aiming to find creative design-oriented solutions.

1.1 Keywords

Fire, Built Environment, Biophilia, Multisensory Engagement, Stress Reduction

2 INTRODUCTION

Driven by the growing understanding of our innate connection to nature - Biophilia, this research delves into the potential of incorporating fire, a multisensory element, into the built environment to reduce stress and frustration. Recognizing the gaps in existing literature around fire's potential as a biophilic tool within architectural spaces (Yao, Zhang, & Gong, 2021), this study investigates the impact of its diminished presence on stress and frustration levels. The built environment has a direct influence on occupant mood (Wang, et al., 2021), and thus, incorporating biophilic principles like natural elements, including fire, holds promise as a strategy for stress reduction in architectural spaces.

The main aim of using WMTs is to measure and analyze walkability, on the assumption that it is a quantifiable and measurable concept, although data is also collected regarding other everyday life issues. The present study does not consider whether walking is a quantifiable concept. Rather, it focuses on developing a critical method for understanding and analyzing the underlying principles of different WMTs with two selected representative WMTs – SPACES and WalkScore.

2.1 History of Fire in the Built Environment

The historical significance of fire in human evolution and social development is well-documented. Early societies, as described by Warren (1914), were brought together by the discovery of fire, leading to the formation of deliberative assemblies and social interactions. In earlier times, the central hearth was a focal point for warmth, cooking, and protection. Heschong emphasizes the symbolic importance of fire, associating it with protection, community, and home (Heschong, 1978).

In the built environment, the evolution of fireplaces, explored in 'Environmental Issues for Architecture' (Smith, 2011), traces the transition from open-hearth systems of Penhurst Castle to more sophisticated

https://doi.org/10.52202/082078-0015

designs like the chimney. Innovations like the Chipping Campden fireplace and Savot's Convection fireplace highlight unique regional adaptations. However, as technology advanced, the multisensory aspects of fire diminished, muting smell, auditory, and thermal experiences. The development of the Pennsylvanian Fireplace marked an efficient breakthrough, and central heating systems became prominent during the Industrial Revolution. The hearth saw a shift from the hall's center to the basement of homes. This gradual shift was a move towards the sanitization of our homes. As safety in the built space increased, our ability to hear, smell, see, and get warmth from the hearth [multisensory experience] decreased.

The progression from a central hearth to hidden heating systems has resulted in a loss of the sensory richness that once played a vital role in human civilization. In the modern era, where artificial or digital fireplaces are promoted from a fire-safety point-of-view, modern fireplaces have come at the expense of extinguishing a fundamental aspect of our daily lives - multisensory engagement.

2.2 Multisensory Elements
Elements such as fire, air, and water, which simultaneously engage multiple senses, are termed multisensory elements (Spence & Gallace, 2011). The dominant perception method revolves around sensory reception, with extensive research comparing the impact of various sensory elements. While most studies focus on visual stimuli and stress reduction, the prevailing sense in modern life, there is a lack of exploration into the combined impact of all senses, termed "the multisensory experience." Studies by (Windhorst & Williams, 2015) and (Song, Ikei, & Miyazaki, 2019) indicate a preference for multisensory experiences over single-sense inputs.

Multiple researchers have categorized senses to characterize the depth of the experience. This study will use the framework by (Erwine, 2016), which is derived from the work of (Tuan, 1990), (Pallasmaa, 2012) and (Heschong, 1978). (Erwine, 2016) categorizes senses as visual, somatic, olfactory, acoustic, cultural, and thermal. She mentions that in the quest for standardized comfort, the modern focus on visual-centric design neglects the richness of multisensory experiences. It highlights that different senses provide access to various aspects of a space, emphasizing the importance of a multisensory approach to creating memorable places. Erwine delineates that each sense possesses volumetric boundaries, varying in rigidity, permanence, and texture. In the context of a bonfire, the visual boundary, extending far and encompassing activities and light, is the largest. Following the visual boundary is the auditory boundary, enveloping voices, laughter, and the fire's hiss. Subsequently, the olfactory and somatic boundary, though smaller, offers a more intimate experience involving the scent of fire and food and tactile sensations. The boundaries of olfactory, somatic, and auditory senses exhibit a gradient, gradually decreasing, while the visual boundary is more defined. The innermost circle

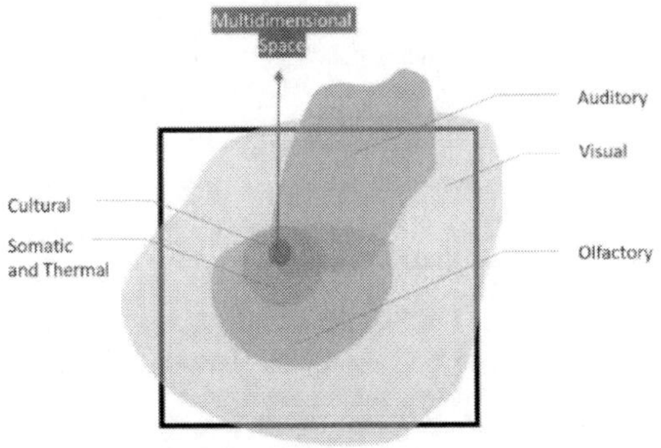

Figure 1 – Bonfire with its multisensory boundaries (2023). Figure by the authors.

comprises cultural and thermal senses, the former directly linked to perception and the latter associated with immediate environmental temperature. This paper will concentrate on the convergence of these senses, referred to as the "multidimensional space," while exploring the volumetric boundaries and how they affect a user's experience beyond the conventional three dimensions of tectonic space.

2.3 Biophilic Elements
Edward O. Wilson describes biophilic elements as the "innate tendency to focus on life and life-like processes," with emotional ties to life persisting through human evolution (Kellert & Wilson, 1993). The multisensory nature of biophilic elements has been recognized for its stress-reducing effects, re-establishing a connection to nature within built environments. Biophilic elements encompass natural features introduced into such spaces, including plants, nature views, air, fire, curvilinear shapes, and biomorphism. As noted by (Zhong, Schröder, & Bekkering, 2022) the evolving concept of Biophilia underscores the significant relationship between human well-being and interaction with nature. This connection has proven to positively impact restoration and reduce psychological stresses, promoting calmness, relaxation, and rejuvenation (Ulrich, et al., 1991)

Biophilia's prominence is reflected in building rating systems like WELL, Living Building Challenge, and the 9 Foundations of a Healthy Building. Many researchers have proposed various categorizations of biophilic elements. In this study, the categorization by Kellert (Kellert S. R., 2008) and Browning and Ryan (Browning & Ryan, 2020) was adopted due to its alignment with research objectives, mainly

https://doi.org/10.52202/082078-0015

focusing on multisensory elements. Kellert proposed the organization of biophilic elements into – direct experience of nature (light, water, air, etc.), indirect experience of nature (images, materials, texture, etc.), and experience of space and place (prospect, refuge, organized complexity, etc.). Fire fell in the direct experience of nature category in the organization by Kellert. While Browning and Ryan took a more objective approach and categorized the biophilic elements as nature in the space (visual/non-visual connection to nature, sensory stimuli, etc.), nature analogous (biomimicry, material connection, etc.), and Nature of space (prospect, refuge, mystery, etc.)

2.4 Stress Reduction Theory

Ulrich introduced Stress Reduction Theory (SRT), proposing that landscapes with vegetation, water, modest depth, and complexity have been beneficial for human survival across generations (Ulrich, et al., 1991). Such environments, including lakes, mountains, green spaces, and parks, are preferred for stress reduction (Kuo & Sullivan, 2001). Joye & De Block (2011) demonstrated that sensory-rich environments significantly reduced stress, incorporating elements like plants, water, and bird sounds.

Incorporating biophilic elements such as natural light, plants, and water in the built environment has been linked to stress reduction (Kellert & Wilson, 1993). Studies using visual stimulation and sound aromatherapy to replicate nature experiences have also shown effectiveness in stress recovery (Alvarsson, Wiens, & Nilsson, 2010), (Song, Ikei, & Miyazaki, 2019). Therefore, adopting biophilic and multisensory strategies, such as incorporating fire in the built environment, can be considered effective in reducing stress.

Due to the geographical relevance (Buffalo, New York) and the chosen criteria, fire emerged as a compelling biophilic element. Its inherent multisensory qualities, encompassing visual fascination, warmth, sounds, and distinct smell, make it highly suitable for exploration within the research framework.

3 RESEARCH OBJECTIVES

The central question that this paper aims to answer is "By eliminating fire from the built environment, have we suppressed its potential to contribute to place attachment?"

The absence of fire raises inquiries about its potential impact on place attachment within the built environment. Fire can establish a distinctive ambiance through its crackling sound, dancing flames, warm glow, and the aroma of burning wood, evoking feelings of comfort, bonding, and familiarity. These sensory experiences significantly influence emotional states, shaping our overall perception of a place. Eliminating fire from the built environment may inadvertently impede opportunities for social engagement, hindering the formation of social bonds and potentially compromising the well-being of occupants.

To answer the central question, the following themes were explored –

1. What emotions does fire evoke and why? Can fire reduce stress?With its cultural significance and historical associations, fire elicits diverse emotions and offers a multisensory engagement. The exploration of fire as a stress-reducing element delves into individuals' emotional well-being experiences and dependence on fire.
2. How do people experience fire, in its presence and as a memory?Analyzing the diverse ways individuals experience fire, both in its immediate presence and as a recollection, provides valuable insights into design strategies that amplify its influence and generate lasting impressions.
3. How do we perceive fire in a spatial context?Exploring the impact of the location of fire within a built environment, examining whether spatial information can be derived from its placement, and if there are preferences regarding the type of fire people engage with.
4. What senses are dominant in the experience of fire?It seeks to understand whether a singular sense can adequately convey the full impact of fire or if a multisensory experience is essential for its comprehensive effect.

4 METHODS

Qualitative methodology (Figure 2) – grounded theory with a constructivist paradigm utilizes open-ended questions to engender a theory through an inductive process. It explores twenty-seven participants' direct and memory-based encounters with fire through think-aloud method and semi-structured interviews as proposed by Groat & Wang; and Hatch (Groat & Wang, 2013) (Hatch, 2023).

https://doi.org/10.52202/082078-0015

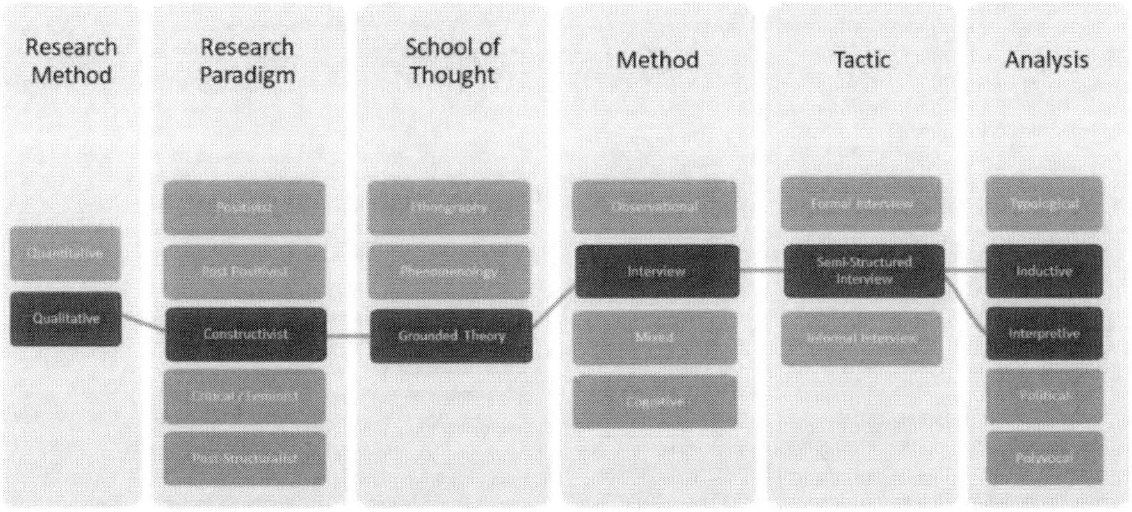

Figure 2 – Research Methodology (2023). Figure by the authors.

The study additionally maps the participants' spatial positioning relative to the fireplace and assesses their reactions to memory and present-based inquiries. It examines the differences in participants' responses corresponding to their positions within the space, reflecting upon place attachment.

The constructivist paradigm aligns seamlessly with the specific research as existing theories fall short in addressing the focal point of the study, making the constructivist paradigm suitable for the co-construction of a theory through an iterative process. For instance, addressing questions like "How does fire make us feel, and why?" involves collaborative efforts with respondents to gather insights and perspectives. The aim is to develop a theory informed by participant experiences and interpretations, ensuring it emerges from the collected data, considering contextual nuances and diverse viewpoints.

Grounded theory, employed as a methodological tool, facilitates an iterative process encompassing restructuring, analysis, interpretation, and revisitation of interviews. This iterative approach serves to enrich the dataset and rectify any inherent shortcomings. It proves instrumental in unraveling the multifaceted dimensions of individuals' experiences with fire, both in the present and as recollected memories, capturing the complexity inherent in the spatial context.

To enhance data richness, the study integrates semi-structured interviews, offering flexibility within the conversation to allow participants to guide the direction while adhering to the predetermined question framework. The "think-aloud method" further enhances the study by prompting participants to articulate their thoughts and experiences in real time during spatial encounters, yielding valuable insights into the spatial aspects of their experiences. The interviews persisted until saturation, ensuring a comprehensive understanding of adequate cases.

Data analysis, undertaken through a combination of interpretive and inductive methods, proves pivotal in extracting meaningful themes and patterns directly from the data without imposing preconceived notions. Initial categorizations from the inductive analysis serve as the foundational basis for developing a interview guide. In tandem with the inductive approach, interpretive analysis deepens the understanding and interpretation of the data, involving a thorough exploration of the meaning, context, and significance of participants' responses and experiences. This interpretive lens also facilitates spatial mapping, allowing the researcher to connect participants' responses with their perceptions of sensory stimuli within the spatial context.

4.1 Pilot

In the preliminary investigation, a university café, One World Cafe, was chosen with a central gas-based fireplace enclosed with glass as the focal point. Interviews were conducted with randomized selection of six individuals and a group of three users who actively selected this café, ensuring a naturalistic setting aligned with their preference for a natural environment. The pilot was conducted as a preliminary learning phase preceding the launch of a comprehensive interview. Using a semi-structured interview and think-aloud method, the researcher gained a nuanced understanding of user experience by asking about sensory perceptions, emotional responses, cultural influences, and spatial references elicited by fire. The iterative nature of the methodology allowed for flexible adjustments in the interview guide to refine and improve the quality

https://doi.org/10.52202/082078-0015

of the questions asked. It aimed to elicit in-depth insights into participants' perceptions, emotions, and experiences concerning the café's fireplace.

Upon scrutinizing the seven test cases, it became evident that an additional domain, termed "characteristics of fire," was necessary alongside the initially identified four domains, i.e., sensory perceptions, emotional responses, cultural influences, and spatial references. Fire-based sources were primarily perceived as a means of warmth rather than joy, with fireplaces eliciting emotional rather than spatial responses for most participants. During the observation phase, it was noted that seats near the fireplace were popular and frequently occupied, though availability varied with the time of day.

Participants better recalled fire-related memories when prompted with references to the cafe's fireplace. Discussions predominantly revolved around the visual and thermal aspects, with less emphasis on auditory or olfactory elements unless specifically addressed. Some participants naturally shared fire-related memories during conversations about the cafe's fireplace, facilitating a seamless transition to related questions. Conversely, others required specific prompts to recollect such memories. The recurrent theme among participants was the association of fire with memories of family gatherings, barbeques, and bonfires. While some initially deemed the fireplace as the sole source of fire-related memories, they became more open to sharing diverse experiences upon the reminder that any form of fire could be discussed.

After the pilot interviews, modifications were made to the interview guide to address perceived limitations in sensory element exploration and incorporate inquiries regarding fire's role in stress reduction. Furthermore, participants were surveyed on their opinions regarding the potential installation of additional fireplaces on campus, including preferred locations. Upon analyzing the data derived from the revised interview guide, a recurrent theme surfaced in participants' responses regarding fireplaces. A notable differentiation emerged between "natural" and "artificial" fireplaces, prompting additional inquiries to elucidate these distinctions and categorize the experiences referenced during interviews.

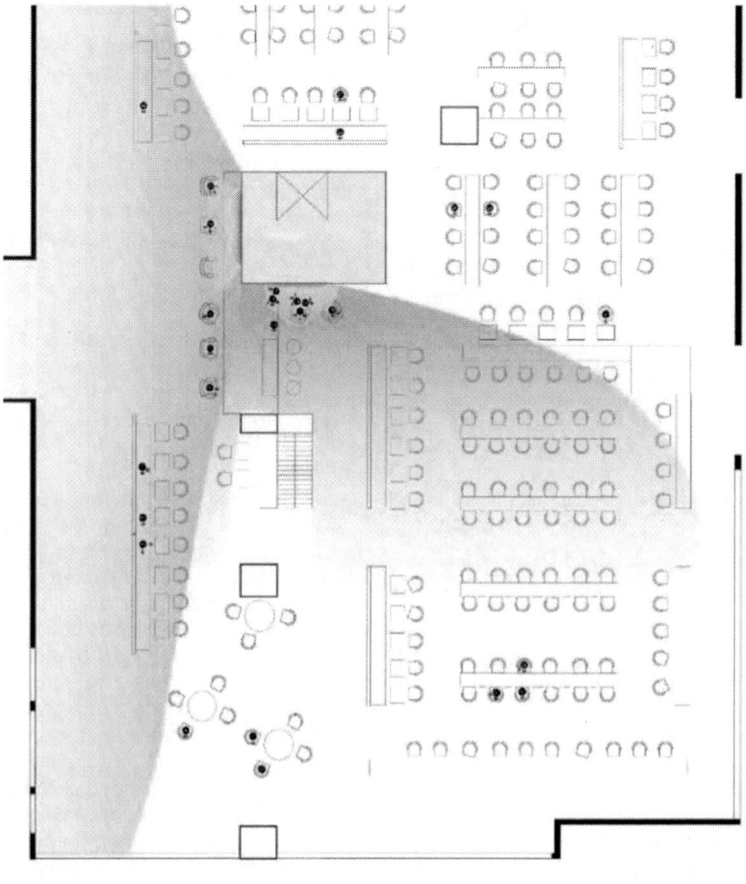

Figure 3 – Planar sensory mapping of One World Cafe (2023). Figure by the authors.

https://doi.org/10.52202/082078-0015

During interviews at One World Café, the fireplace assumed a central role, and the spatial mapping of sensory boundaries was undertaken with the utilization of self as the measuring instrument. The resulting visual representation depicted the visual, thermal, auditory, and olfactory boundaries, along with participants' positions within the space, as illustrated in Figure 03.

Analysis of the spatial mapping yielded noteworthy insights. Firstly, participants close to the fireplace provided more detailed descriptions and memories of the fireplace and fire-related experiences than those at a distance. Secondly, those nearer to the fireplace tended to harbor a more positive outlook towards fire than their counterparts farther away. Thirdly, participants positioned at a distance or facing away from the fireplace encountered challenges recalling their experiences with fire, attributable to its limited sensory zone within the spatial context. Lastly, some participants reported experiencing warmth even when situated beyond the sensory boundaries of the fireplace.

The interview transcripts were coded hierarchically to analyze the participants' responses. The structure had three levels: theme, category, and sub-category. Each group had its theme, category, and sub-category. They are described in Table 1.

Table 1. Coding Hierarchy (2023). Table by the authors.

Code	Theme	Category	Sub-Categories
	Sensory experience	At Present	Visual Olfactory Auditory Somatic Thermal Cultural
		Memory-based	Visual Olfactory Auditory Somatic Thermal Cultural
	Spatial References		
	Stress Reduction Emotions		
	Type of Fire	Fireplace	"Natural" "Artificial"
		Others	

The questionnaire for the interview commenced by orienting participants within the space, prompting them to situate themselves and become fully present. Following this, questions addressed their awareness of the fireplace, initial impressions, and any perceived effects on their mental or physical state. Participants were then asked to engage their senses to observe the fireplace, identifying the sensory perceptions associated with it. Subsequently, inquiries delved into personal experiences, memories, and cultural significance related to fireplaces or firepits, exploring the earliest memories and the factors that contribute to the memorability of these experiences, including cultural, emotional, and memory-related aspects. Moving on to sensory qualities, participants described the sensory attributes of fire based on both memory and present experiences, expressing how it makes them feel and whether they have ever utilized fire in stressful situations for relaxation. The interview concluded with open-ended questions regarding the fireplace's characteristics, its location, and potential improvements.

https://doi.org/10.52202/082078-0015

Some of the interviews are as follows -

P1 – "As for me, I come from the city, but I have fond memories of visiting a cabin with my cousin. It was a lot of fun, and I have a deep appreciation for nature. Nowadays, it seems like everyone is always on their laptops or iPads, and there's not much human interaction. But when you're out in nature, gathered around a fireplace, there's a sense of connection and warmth that's hard to replicate. The fireplace at the cabin was small and open, and we would arrange our beds around it, cook food, and just have a good time. There's warmth that comes with that connection, and it's something that can't be replaced by anything else."

P14 – "I love candles. I have to light a candle, and it relaxes me...if I have a lot of work that I was doing, or like just a really stressful day, then I have like a stressful week candle that I light. I always like have a candle like next to a Christmas tree for warm, cozy Christmas vibes. I like how it smells. But also seeing the flame go up and the crackling that's also like quite comforting..."

P5 – "...barbeques were my favorite thing. Smelling nice food not the gasoline smell, when you are outside, it smells natural. It is comforting and nostalgic..."

5 RESULTS

5.1 Synthesis

Table 2. Type of Responses to Fire as a Memory and a Live Stimulus (2023). Table by the authors.

Visual	Poetic, Patterned, artistic, abstract, dramatic, dancing	Flickering, sizzling	Sparkly	Aesthetic, attractive, ambient	Scary, see smoke	hurts eyes, annoying, glare	wood browns and blacks	Small
	Bright	Light Source	Dynamic, dancing, flowy	Consistent	Dim	distracting		
Thermal	Warm, cozy, heat				Pricking			
Olfactory	Sweet, scent	Burnt, chard	Outdoorsy, woody	Likeness	Suffocating, smokey	Gasoline	Unscented	
	Food like, stove fire	Homey	Festive		Dislike			
Auditory	Quiet	Ambient, soothing	Crackle	Hiss, swoosh				

https://doi.org/10.52202/082078-0015

Emotional Cultural	Calm, relax, content	Deep thoughts, focus, mesmerize	Cozy, curl up	Love the fire, spiritual	distracting	anxious	double edged sword, neither bad or good	not drawn to it
	Exciting, happy, dramatic, fascinate	family, nostalgic, bonding, community	Safe, convenient	renewing, positive, inspiring, creative	hurtful, uncontrolled	wasteful		Neutral
	Reminds of holidays	burn the leaves, smoke	Incense, sage	family trips, vacations				
	Positive				Negative		Neutral	

Fire, pivotal for human survival since its discovery, has evolved from a survival necessity to a recreational element. Table 2 categorizes participants' responses to fire as positive, negative, or neutral.

In Figure 4, despite instances of negative experiences, participants' overall encounters with fire remained predominantly positive or neutral, averting negative outcomes. Notably, a higher number of participants expressed negative responses to olfactory experiences compared to other negative encounters, underscoring the substantial influence of olfactory stimuli on individual experiences. Auditory and cultural experiences consistently evoked positive or neutral responses, potentially attributed to the low intensity of auditory stimuli, which tends not to interfere negatively, remaining a positive stimulus. Consequently, fire establishes an inevitable link with individual identity, traditions, and religious practices, contributing to its positive connotation. This emotional attachment contributes to the formation of place attachment, as individuals develop connections with environments where they have positive experiences, such as those involving fire.

An additional noteworthy finding is that while only a limited number of participants experienced the intersection of sensory zones, a majority reported stress reduction. This observation implies that stress reduction may not necessitate the engagement of all senses.

It brings to light the question, "How do we experience a fire in memory and as a live experience? Is there a difference and a dominant sense in forming experience?"

https://doi.org/10.52202/082078-0015

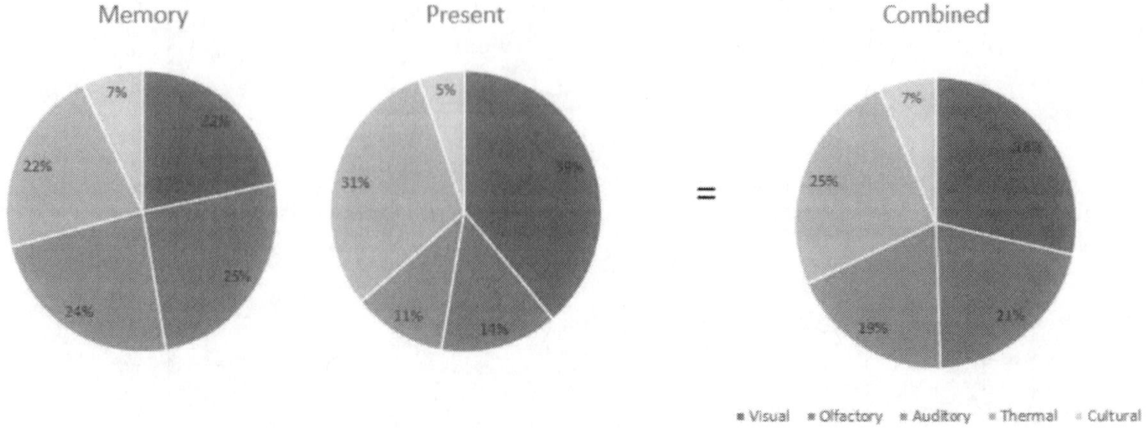

Figure 4 – Fire as a Memory and Live Experience (2023). Figure by the authors.

To comprehend the dual facets of fire experiences—both as a memory and at the present—data from these experiences were categorized into visual, olfactory, auditory, thermal, and cultural senses and analyzed proportionally. The analysis delineated that in memory-based experiences, the olfactory sense dominated, closely followed by the auditory sense, while visual and thermal senses constituted significant proportions, albeit smaller than the olfactory and auditory senses. The cultural factor represented the smallest percentage.

Conversely, in live experiences, the visual sense took precedence, followed by the thermal sense, collectively forming the majority of the sensory experience. Olfactory and auditory senses comprised a quarter of the total share, with the cultural sense being the least dominant.

When amalgamating both experiences, visual stimuli remained the dominant sense, trailed by thermal and olfactory senses, with the auditory sense following and the cultural sense emerging as the least prevalent. The analysis of impact of senses on one's experience underscores fire's multisensory nature, wherein all elements, except the cultural sense, possess the potential to influence an experience. The prevalence of the visual sense is attributed to visually dominant built environments. The sensory engagement can contribute to the formation of strong emotional bonds with places where such experiences occur, enhancing place attachment.

Repetition of certain terminology during the interview, such as "cozy," "warm," "relaxed," "calm," "natural," and "crackling," suggests stress reduction effects of fireplace, although further analytical scrutiny is required. Explicit attributions of fire as a stress reduction element were recurrent in interview excerpts, emphasizing its potential to foster a meaningful space, albeit not necessarily requiring the engagement of all senses. Environments that incorporate natural features like fireplaces can evoke feelings of comfort, belonging, and attachment, thereby strengthening individuals' connections to those places.

Participants cited various forms of fire in the interview, with campfires and candles comprising over 50% of responses, followed by fireplaces and firepits. A notable categorization emerged during discussions, distinguishing fire as "natural" and "artificial." This classification revealed a prevalent preference for "natural" fireplaces due to their perceived ability to connect individuals to nature, engage all senses, and the intrinsic process of starting the fire.

Associations with "artificial" fireplaces predominantly centered on safety, with a consensus among participants that these options are secure, making them suitable for unsupervised areas like universities and libraries. Meanwhile, "natural" fireplaces were preferred for homes due to the engagement of multiple senses, such as smell and sound, and the perceived authenticity, enjoyment, and gratification associated with making a fire. A "natural" fireplace's warmth and intensity were considered superior to an "artificial" one.

5.2 Major Findings

The exploration of fire's inherent potential and its effective utilization is a crucial component in designing buildings that aim to provide more than just visual appeal but a truly enriching and fulfilling experience. While prioritizing safety measures in our environment is essential, it has inadvertently resulted in the loss of certain elements that connect us with

https://doi.org/10.52202/082078-0015

nature and evoke a sense of loss. The question arises is how we can reconcile this loss and find a harmonious balance between safety and restoring these essential connections with nature.

1. **Sensory overlap is essential for the creation of meaningful and enriching experiences.**
   Sensory overlap is pivotal in crafting meaningful experiences within a built environment, as evidenced in the case of One World Café. Individuals positioned where multiple sensory boundaries intersect reported heightened engagement, fostering a greater likelihood of return visits. This convergence of sensory stimuli contributes significantly to forming favorable spatial memories. This suggests that the richness of sensory experiences can deepen individuals' attachment to a place.

   A notable proportion (one in four participants) responded adversely to olfactory stimuli, underscoring the substantial impact of smell on overall environmental experiences. Whether pleasant or unpleasant, odors are crucial in shaping individuals' perceptions and emotional responses within a given space.

   These findings highlight the intricate nature of sensory experiences and their profound influence on individuals' perceptions and emotions. Designers, armed with this understanding, can implement strategies prioritizing sensory overlap. Factors such as olfactory stimulation and cultural associations can be considered, thereby enriching environmental design to create more meaningful and engaging spaces.

2. **People are drawn to the full sensory experience of fire and associate it with nature.**
   The association with primal elements endows fire with an authentic and natural essence, invoking a sense of Biophilia. This innate affinity for nature compels individuals to actively seek and engage with fire, establishing a meaningful connection with the natural world. The nexus between fire and nature is further bolstered by the multisensory engagement it affords. Study participants expressed that the appeal of fire extends beyond its visual allure, encompassing the auditory resonance of crackling sounds, the warmth emanating from the flames, the distinctive scent of burning wood, and the tactile interaction with the fire itself. This full sensory engagement enhances the perceived authenticity of the experience, thereby reinforcing the profound connection to nature. It further suggests that incorporating natural elements into built environments can enhance place attachment by tapping into individuals' innate connections with nature.

3. **One World Café's fireplace was found ineffective due to its small scale and tight sensory boundaries.** The limited impact and influence of the fireplace at One World Café at the University at Buffalo can be attributed to its small scale and constrained sensory boundaries. The proximate nature of these boundaries restricts the spatial reach of sensory engagement, thereby diminishing the overall effectiveness of the fireplace as a focal point within the café. Participants seated closer to the fire displayed heightened interest and relevance, whereas those positioned at a distance exhibited disinterest or perceived lesser significance in the experience. The physical distance from the fire appears to influence the level of interest and the degree of sensory connection experienced by participants.

4. In the case of the small-scale fireplace at One World Café, the limited space and sensory boundaries restricted the number of individuals who could simultaneously benefit from the fire-related experience. The scarcity of shared resources diminished the communal aspect of the fireplace, which is often crucial in fostering a sense of connection and shared experience among individuals within a space.

6 CONCLUSION

In conclusion, this research delves into the historical evolution of fire, particularly the relocation of the hearth, resulting in diminished multisensory experiences and elevated stress levels. The study explores the biophilia concept, assessing fire's potential for stress reduction. Employing qualitative methodologies, including grounded theory and semi-structured interviews, it formulates practical insights for design stakeholders across dimensions. The relocation and diminished presence of fire in the built environment underscore the need for optimizing sensory immersion to foster transformative experiences and profound place attachment.

The study highlights fire's significant role as a biophilic and multisensory element, offering practical implications for emotionally fulfilling environments. Sensory overlap is crucial in creating meaningful experiences, as observed at One World Café, emphasizing the impact of smell. People are naturally drawn to the full sensory experience of fire, revealing a profound connection with nature. However, the ineffectiveness of One World Café's fireplace is attributed to its small scale and tight sensory boundaries, limiting communal aspects.

The research methodology's flexibility, utilizing the think-aloud method and spatial mapping, addressed challenges in collecting spatial data, providing valuable insights into participant interactions with fire. Spatial mapping played a crucial role in unearthing

https://doi.org/10.52202/082078-0015

implicit data, contributing to a nuanced analysis of the relationship between participants, the fireplace, and the surrounding spatial context.

7 DISCUSSION

7.1 Applicability for Design

The study underscores the significance of fire in human evolution and its vital role in promoting psychological and physiological well-being. Recognizing this importance, designers have an opportunity to reintegrate fire into the built environment and harness its potential for creating positive experiences. The theories derived from this research offer valuable insights that can directly inform design practices –

1. How can a synergy between energy, safety, environment, and fire be formed to bring back fire into our spaces?

2. Balancing the intricate interplay among energy, safety, environment, and fire in design raises a pivotal inquiry. Addressing concerns encompassing fire hazards, energy efficiency, and environmental impacts while concurrently emphasizing the promotion of well-being and superior experiential quality necessitates a meticulous and holistic design approach.

3. How can designs benefit from a multifaceted approach than designing visually dominant spaces?
   Designing spaces beyond visual dominance requires a shift in perspective to embrace the full range of sensory experiences the fire provides. By recognizing the importance of haptic and kinesthetic senses, designers can create environments that engage users on a deeper level. The olfactory sense also offers a powerful opportunity to enhance spatial experiences. By carefully curating olfactory inputs and integrating them into the design, spaces can evoke specific moods, create associations, and trigger emotional responses. Prioritizing the design of sensory-rich spaces, we can foster deeper connections between individuals and their surroundings, resulting in more meaningful and memorable experiences.

4. How can sensory roles be divided amongst various elements in space to replicate the effects of fire?
   Designers may strategically assign warmth, movement, visual effects, and scent to approximate the ambiance of a fireplace in a multisensory environment. For instance, radiators can emit warmth akin to a fire, HVAC systems can simulate gentle air movement, digital screens can display realistic fire visuals, and fragrance technologies can diffuse the scent of burning wood. It will answer whether artificial sensory engagement can elicit the desired emotional response.

5. How can multisensory elements of fire used to create a sense of community, safety, and stress reduction?
   The multisensory aspects of fire, including its warmth, visual allure, soothing crackling sounds, and distinct aroma, can evoke a range of positive emotions and contribute to a sense of well-being. Can these be incorporated in communal spaces, such as cafes, libraries, and lounges, to allow designers to craft environments that foster relaxation, social interaction, and a sense of belonging?

6. How can fire be altered to suit space typologies?

   Participants expressed their preferences for specific fireplaces in different settings during the interviews. These preferences varied depending on the function and ambiance desired for each space. For instance, participants mentioned a preference for artificial fireplaces in libraries. This preference could be attributed to the need for a controlled and safe environment in libraries, where open flames may pose a fire hazard. On the other hand, participants expressed a preference for natural fireplaces in their homes which could be attributed to the desire for an authentic and organic experience connecting them to nature and creating a comforting atmosphere within their living spaces.

Designers can adapt fire elements by understanding these preferences and considering different space typologies' specific requirements and desired experiences. This could involve using artificial fireplaces, natural fireplaces, varying sizes, or alternative forms of fire representation, such as candles, to create the desired atmosphere and ambiance in each space.

7.2 Methodological Observations

The research methodology employed in this thesis embraced a flexible approach that allowed for adjustments and modifications throughout the process.

Collecting spatial data was challenging, as it did not come naturally to the participants. In this study, a focused effort was made to collect spatial data pertaining to participants' interactions with fire. The think-aloud method was implemented, prompting participants to articulate their spatial interactions and perceptions in the presence of fire. Questions such as "Can you describe the space as you move around it?" and "Can you describe the mental image?"

https://doi.org/10.52202/082078-0015

were employed. This method facilitated participant reflection on spatial experiences, offering valuable insights into the spatial dimensions of their encounters with fire. The approach significantly enhanced the understanding of how individuals engage with fire in their physical surroundings, enriching the research topic's comprehensive analysis.

Spatial mapping also played a crucial role in the research by unearthing implicit data and establishing connections between participants' responses and physical locations within the space. This approach enabled the exploration of spatial dynamics, highlighting the significance of participants' positioning within the space w.r.t. fireplace and its influence on their perceptions and data-rich responses. Integrating spatial mapping techniques contributed to a nuanced analysis of the research findings, offering deeper insights into the multifaceted relationship between participants, the fireplace, and the surrounding spatial context.

7.3 Further Research Questions

The impact of fire on individuals can vary depending on the specific space in which it is experienced, and demographics may also play a role in shaping these effects.

1. How does fire's effect on people change from space to space? Does demographics play a role?

   Diversifying the study to various locations, such as cafes, restaurants, or residential areas, is crucial for a more comprehensive understanding. Examining different settings with unique characteristics and demographics will provide broader perspectives on how fire impacts individuals in diverse contexts.

2. What physiological levels of stress reduction are seen in participants with varied demographics?

   To comprehensively understand stress reduction across diverse demographics, quantifying stress reduction associated with various sensory engagements is crucial. Researchers can systematically analyze physiological responses such as heart rate variability, cortisol levels, and skin conductance to provide precise data for future studies and evidence-based design practices. Considering safety concerns with fire, exploring "artificial" elements as proxies for multisensory experiences warrants future studies to compare their effectiveness against real fire in stress reduction.

3. How can water, air, or other biophilic elements benefit from the multisensory design approach?

   Researchers can identify the elements that evoke the most substantial positive reaction by systematically examining individuals' physiological and psychological responses to various biophilic stimuli. This knowledge can then be translated into practical design recommendations and guidelines for creating captivating and health-promoting environments.

7.4 Limitations

This research has certain limitations. Firstly, the study at a university café involved a restricted age group (18 to 30 years), potentially excluding experiences and perspectives from individuals beyond this demographic. The study, focusing on fire-related emotions, sensory engagement, cultural significance, and spatial arrangements, mitigates the impact of this age restriction, as it is not heavily reliant on demographics.

Secondly, the research was confined to a specific location, centering on a single café with a particular gas-based fireplace, potentially limiting the generalizability of findings to diverse settings or alternative fireplace types. Weather and geographical considerations are pertinent, as the observed preference for the fireplace could be influenced by the café's colder climate and winter season data collection. However, participants' memories were not exclusively tied to winter climates.

Additionally, the measurement of sensory boundaries in the spatial diagram at the One World Café relied on personal approximation, recognizing the subjectivity and individual variability inherent in sensory boundaries.

8 ACKNOWLEDGMENT

I wish to express my sincere gratitude for the support and guidance I received during this research, which served as my thesis project at M.Arch, University at Buffalo. My heartfelt thanks go to my thesis guides, Martha Bohm, Korydon Smith, and Edward Steinfield. I am also appreciative of the financial assistance provided by the University at Buffalo through a thesis grant.

Furthermore, I extend my thanks to EDRA for the valuable opportunity to present my thesis before peers and professionals in the field. The positive response and constructive feedback from the audience have been inspiring, motivating me to submit this paper.

https://doi.org/10.52202/082078-0015

9 REFERENCES

Alvarsson, J. J., Wiens, S., & Nilsson, M. E. (2010). Stress recovery during exposure to nature sound and environmental noise. International journal of environmental research and public health, 7(3), 1036-1046.

Joye, Y., & De Block, A. (2011). Nature and environment as spaces for stress reduction: The case of student nurses in a simulated shiftwork environment. Landscape and Urban Planning, 102(3), 210-216. Kellert, S., & Wilson, E. (1993). The Biophilia Hypothesis. Bulletin of Science, Technology & Society, 15(1), 52-53.

Klepeis, Nelson, N. &., Ott, W. &., Robinson, W. &., & John. (2001). The National Human Activity Pattern Survey (NHAPS): A Resource for Assessing Exposure to Environmental Pollutants.

Kuo, F. E., & Sullivan, W. C. (2001). Environment and crime in the inner city: Does vegetation reduce crime?. . Environment and behavior, 33(3), 343-367.

Song, C., Ikei, H., & Miyazaki, Y. (2019). Physiological effects of forest-related visual, olfactory, and combined stimuli on humans: An additive combined effect. Urban Forestry & Urban Greening, 44, 126437.

Spence, C., & Gallace, A. (2011). Multisensory design: Reaching out to touch the consumer. Psychology & Marketing, 28(3), 267-308.

Ulrich, R. S., Simons, R. F., Losito, B. D., Fiorito, E., Miles, M. A., & Zelson, M. (1991). Stress recovery during exposure to natural and urban environments. Journal of environmental psychology, 11(3), 201-230.

Wang, L., Zhou, Y., Wang, F., Ding, L., Love, P. E., & Li, S. (2021). The influence of the built environment on people's mental health: an empirical classification of causal factors. Sustainable Cities and Society, 74, 103185.

Windhorst, E., & Williams, A. (2015). "It's like a different world": Natural places, post-secondary students, and mental health. Health & Place, 34, 241-250.

Yao, W., Zhang, X., & Gong, Q. (2021). The effect of exposure to the natural environment on stress reduction: A meta-analysis. Urban forestry & urban greening, 57, 126932.

Zhong, W., Schröder, T., & Bekkering, J. (2022). Biophilic design in architecture and its contributions to health, well-being, and sustainability: A critical review. Frontiers of Architectural Research, 11(1), 114-141.

Browning, W. D., & Ryan, C. O. (2020). Nature Inside: A biophilic design guide (1st ed.). RIBA Publishing.

Erwine, B. (2016). Creating sensory spaces: The architecture of the invisible. Taylor & Francis.

Groat, L. N., & Wang, D. (2013). Architectural Research Methods. John Wiley & Sons.

Hatch, J. A. (2023). Doing qualitative research in education settings. State University of New York press.

Heschong, L. (1978). Thermal delight in architecture. (Doctoral dissertation, Massachusetts Institute of Technology).

Kellert, S. R. (2008). Biophilia. In Encyclopedia of Ecology (pp. 462-466).

Pallasmaa, J. (2012). The Eyes of the Skin: Architecture and the Senses. Wiley.

Smith, D. L. (2011). Environmental Issues for Architecture. Wiley.

Tuan, Y. (1990). Topophilia: A Study of Environmental Perceptions, Attitudes, and Values. Columbia University Press.

Warren, H. (1914). The Ten Books on Architecture. United Kingdom: Harvard University Press.

Wilson, E. O. (1986). Biophilia-the human bond with other species. Havard University Press

https://doi.org/10.52202/082078-0015

EXAMINING CHANGING IDENTITY IN ADAPTIVE REUSE: THE CASE OF MÜZE GAZHANE IN ISTANBUL

Track 2: Sustainable Buildings and Infrastructure
Knowledge Network: Sustainable Planning, Design & Behavior, Historic Preservation
Methods: Case Study, Interviews
Ipek Aksel (Yeditepe University); Cagri Imamoglu (Bilkent University)

1 ABSTRACT

This study explores the identity evolution of Müze Gazhane, a museum repurposed from a historic gasworks in Istanbul. It investigates the substantial impact of its evolving identity on visitors' meaning-making processes and satisfaction levels. Employing a comprehensive two-tiered methodology, the research combines questionnaire and semi-structured interviews to offer a nuanced understanding of the museum's impact. Findings suggest that Müze Gazhane may transcend its role as a conventional museum, becoming an integral part of daily life, especially for the residents of its Hasanpaşa neighborhood. The historical significance of Müze Gazhane emerges as a crucial factor in visitor preferences. Participants of the study reported finding meaning in repeat and frequent visits to the museum and acknowledged its transformation journey. Thus, the study highlighted the museum's transformative role within the community, suggesting an improvement in the life satisfaction of the residents of its neighborhood. Many visitors reported recognizing the diverse potentials of the museum: Its historical backdrop seemed to elevate their experiences, with heightened satisfaction when they learned about the new aspects of the history of the heritage area during their visits. The present research stands as a pioneering endeavor in Istanbul, uniquely evaluating (place) identity, meaning-making, and satisfaction levels within a museum context, with Müze Gazhane's changing identity adding a distinctive dimension. The study advocates for continued exploration of the societal contributions of museums repurposed from industrial heritage buildings, emphasizing their positive impacts on social cohesion and visitor experiences.

1.1 Keywords
Identity; Meaning-making; Visitors' Satisfaction; Industrial Heritage; Museum

2 INTRODUCTION

Driven by the growing understanding of our innate Museums have undergone a notable transformation from their historical role as venues for the exhibition of private collections to becoming more inclusive, interactive, and communal spaces catering to a diverse audience. This evolution is characterized by a shift towards increased openness and accessibility to the public, transcending the traditional function of museums solely as repositories for artifacts (Packer, 2008). Presently, museums strive to accommodate various individual needs, such as entertainment, education, socialization, and gathering, reflecting a departure from their exclusive focus on the past. Accordingly, contemporary museums are emancipating themselves from the confines of being mere repositories of history. They are progressively assuming roles as hubs for communal engagement, active participation, and knowledge transfer, operating at the organizational, structural, and managerial levels within their communities (He, Li, Cheng, & Wu, 2019). This transformative paradigm is often referred to as the "new museum," a term popularized by Peter Vergo in the 1950s. The distinguishing feature of new museums lies in their socially oriented approach, emphasizing communication with society and seeking to fortify the nexus between the city and its inhabitants. Strategies implemented by new museums aim to engage diverse demographic groups inclusively (Arabacıoğlu & Ergin, 2016). Concurrently, the restoration of heritage buildings within the city center of Istanbul has gained prominence in recent times. While certain areas are under protection, there is a discernible trend of repurposing structures in these locales into museums (Aykaç, 2020). The extension of this approach, transforming heritage buildings into new museums throughout the city, is pivotal not only for its potential contribution to tourism but also for fostering stronger bonds between residents and the urban landscape. Each museum that replaces a heritage building has the potential to exhibit distinctive characteristics shaped by its historical and social context. Consequently, a nuanced examination of a specific museum case becomes imperative to comprehend its identity and assess the interplay of key factors influencing both the museum and visitor experiences, providing a comprehensive understanding of their relationship.

3 RESEARCH OBJECTIVES

In the present study, we aimed to examine the evolving identity of a museum that has been adapted from an industrial heritage site and its influence on visitor experiences. Our primary objective was to

https://doi.org/10.52202/082078-0016

contribute to a nuanced understanding of those museums as instrumental entities with the potential to develop connections between residents and their neighborhoods. While numerous studies have scrutinized visitor experiences through the lenses of meaning-making processes and satisfaction levels, there is a discernible gap in research focusing specifically on the experiences of visitors in repurposed museums derived from heritage buildings.

A comprehensive exploration that takes into account both the historical and current identities of these buildings, evaluated in conjunction with visitors' experiences, may yield valuable insights.
In line with our objectives, we evaluated visitor experiences by concentrating on their meaning-making processes and satisfaction levels within a specific case, carefully considering the unique identity of the museum in question. Thus, the central research question guiding this inquiry was: What is the influence of the changing identity of Müze Gazhane on visitors' meaning-making and satisfaction levels? We tackled this question by employing a two-tiered approach encompassing both quantitative and qualitative methodologies.

4 METHOD

4.1 Site

The selection of Müze Gazhane for our case study was predicated upon its distinctive identity as a repurposed industrial heritage site within Istanbul (Figure 1). Notably, the site has undergone a transformative trajectory, originally functioning as a gasworks (Hasanpaşa Gasworks) that played a pivotal role in illuminating the city. Subsequently, it evolved into a cultural complex and museum, continuing to enrich the community through artistic and cultural events. Of significance is the grassroots restoration effort initiated by individuals residing in the neighborhood, some of whom remain actively engaged at the organizational level of the museum. Situated in the heart of the Hasanpaşa neighborhood in the Kadıköy district of Istanbul, the freely accessible museum attracts a diverse audience, making it a focal point for local residents. Despite gaining popularity and witnessing an increasing influx of visitors, its recent public debut underscores the need for a comprehensive investigation into visitor experiences, considering their meaning-making processes and satisfaction levels while accounting for the museum's evolving identity.

Figure 1 Müze Gazhane Square (Retrieved on 1/8/2024: https://muze-gazhane.istanbul/hakkinda/)

https://doi.org/10.52202/082078-0016

4.2 Participants, Measures, and Procedure

The study was conducted in Müze Gazhane between April and October of 2022 by employing a two-tiered approach, as mentioned above. The first phase of the study involved the collection of quantitative data via a questionnaire administered to sixty seven participants (50 women, 17 men, major age range 18-35) utilizing the quota sampling method. The questionnaire consisted of three parts: the first encompassed demographic data, visit frequency, whether participants visited alone or with others/ company, familiarity with the Hasanpaşa gasworks, awareness of the museum's renovation process, and overall meaning-making and satisfaction levels post-visit. In the second and third parts, participants were asked to respond to 30 questions involving the perceived identity of Müze Gazhane, their meaning-making processes, and satisfaction levels by using a 5-point Likert scale. To assess the perceived changing identity of Müze Gazhane, we asked questions such as "Are you familiar with Müze Gazhane's history and its renovation process?"; "Are you happy with the transformation of Müze Gazhane?"; and "Do you think it is important to visit Müze Gazhane because it is a cultural heritage area?". Similarly, to understand the meaning-making and satisfaction levels of visitors, we asked both direct and indirect questions, for example, "Do you agree that the changing identity of Müze Gazhane makes this place more meaningful?"; "Are you satisfy with the facilities offered in Müze Gazhane?"; and "Do you agree that witnessing the changing identity of Müze Gazhane helped you discover new things about yourself?". Quantitative analyses were performed using Statistical Package for Social Sciences (SPSS) software version 19 (IBM Corp, 2012).

Concurrently, the qualitative phase involved semi-structured interviews with twelve participants (7 female, 5 male, mean age 39), selected voluntarily through convenience sampling. Ten main interview questions were predetermined to initiate meaningful conversations regarding perceived changing identity, meaning-making, and satisfaction level. Changing identity-related questions such as "Are you familiar with the renovation process of Müze Gazhane?"; "Is this changing identity a reason for you to visit Müze Gazhane?". Meaning-making and satisfaction level-related questions: "Does this changing identity make Müze Gazhane more meaningful for you?"; "Did your positive emotions increase towards your neighborhood/region with the opening of Müze Gazhane?"; "Are you happy with the diverse range of activities and experiences that Müze Gazhane offers?". Additional questions emerged organically during the interviews. Responses were both noted and voice-recorded, with participant consent. Qualitative data from the second phase were categorized and coded based on predefined themes and subsequently subjected to an inter-rater content analysis conducted by two researchers to ensure the reliability of the results. Some themes and sub-themes were found to be mutually inclusive, reflecting the nuanced and interconnected nature of participants' evaluations.

5 RESULTS

5.1 Questionnaire

Initial analyses involved a descriptive examination of demographic characteristics, visit frequency, participants' familiarity, and locality. Reliability tests employing Cronbach's Alpha were conducted to ensure questionnaire consistency (= .84 showing a strong reliability of the questionnaire). Mean scores were calculated, and diverse statistical tests, including one-sample t-tests, independent sample t-tests, and chi-square tests for independence (x2), were employed to scrutinize the influence of Müze Gazhane's changing identity on visitors' meaning-making and satisfaction levels.

The demographic overview revealed that more than half of the questionnaire participants were residents of the Anatolian Side of Istanbul (where the museum is located), with approximately one-third residing in the (Hasanpaşa) neighborhood. Nearly half of the respondents visited Müze Gazhane at least once a week. Most participants attended the museum accompanied by friends, partners, or family. Over half of the participants possessed prior knowledge of Hasanpaşa Gasworks, and a similar proportion was acquainted with Müze Gazhane's renovation process. A majority expressed a willingness to share their experiences and indicated a likelihood of revisiting the museum.

To gauge the influence of Müze Gazhane's changing identity on visitors' meaning-making and satisfaction levels, participants were queried about the importance of Müze Gazhane as a cultural heritage area. The majority asserted the significance of visiting Müze Gazhane for its cultural heritage status. Following an introduction to Müze Gazhane's changing identity, participants were prompted to assess the impact of this evolution on their decision to visit. Over half of the participants affirmed that this transformation of Müze Gazhane from a heritage area influenced their visitation choice.

Spearman's rho test results showed that the changing identity of Müze Gazhane, visitors' meaning-making, and satisfaction levels were significantly correlated. Participants who reported that the changing identity of the museum was important for them also seemed to find their visit more meaningful compared to those who did not (r (65) - .47; p<.01). Similarly, those who reported the importance of Müze Gazhane's changing

https://doi.org/10.52202/082078-0016

identity exhibited higher satisfaction levels (r (65) = .42; p<.01). Participants who reported that their museum visit was meaningful also seemed to have higher satisfaction levels (r (65) = .59; p<.01).

5.2 Interview

In the qualitative part, answers were subjected to inter-rater thematic analysis. To ensure the reliability of the analysis between two raters, Cohen's Kappa was run. There was a substantial agreement between evaluations of two raters (= .65). The thematic analysis identified three key variables: (place) identity, visitors' meaning-making, and satisfaction levels, encompassing distinct sub-themes. For instance, expressions emphasizing the transformative impact of Müze Gazhane's renovation process were incorporated under both themes and various sub-themes, such as its evolving identity through renovation and its contemporary impact on the visitor experience (Figure 2).

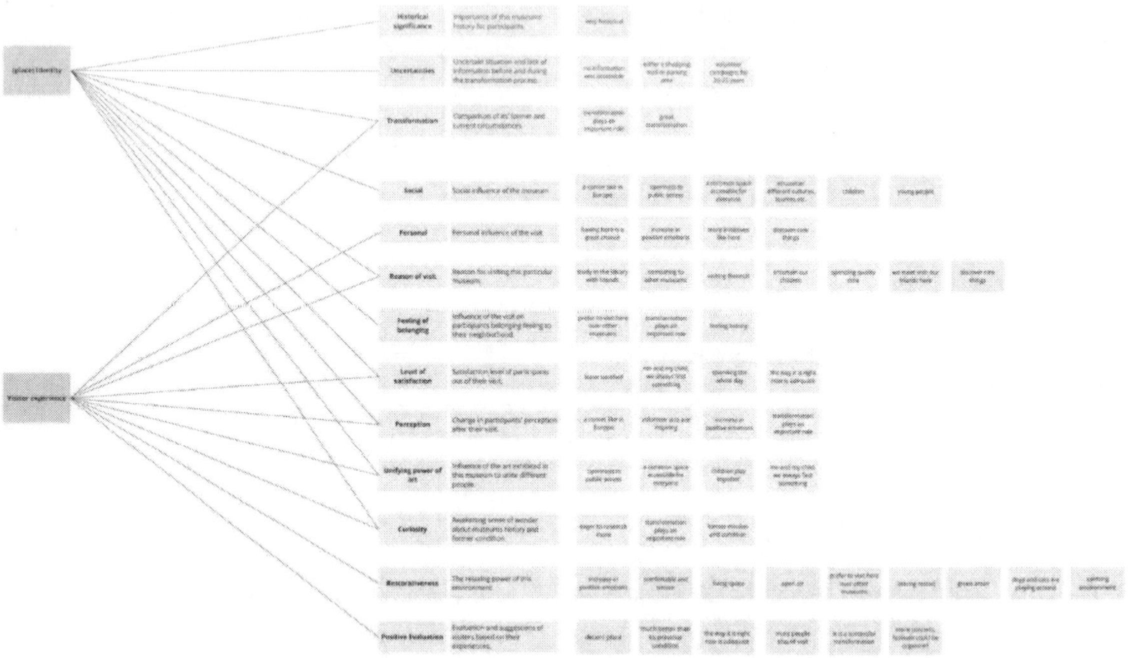

Figure 2 *Summary of Qualitative Analysis*

6 CONCLUSIONS

These findings underscore the transformative potential of Müze Gazhane in positively influencing the lives and experiences of its visitors. We hope our present study may contribute to increasing awareness regarding the repurposed utilization of the existing stock of heritage buildings. Furthermore, it may stimulate further investigations regarding the influence of such museums, with the ultimate goal of enhancing visitor experiences and formulating guidelines that contribute to the strengthening of the relationships of residents with their neighborhoods and cities.

7 DISCUSSION

The transformation of Müze Gazhane from its initial role as a gasworks area marked a significant shift in its identity, thereby influencing the experiences of its visitors. The primary objective was to discern the perceived significance of Müze Gazhane's identity in the eyes of its visitors. Initially, participants were tasked with assessing the importance of Müze Gazhane as an industrial heritage. A consensus emerged, with nearly all participants acknowledging the crucial cultural heritage status of Müze Gazhane in their answers to the questionnaire and interview questions. Subsequently, participants were apprised of the comprehensive renovation process that culminated in a substantial alteration of the gasworks area's identity. One of the participants mentioned, "I studied at the ITU Faculty of Architecture, so I was aware of the transformation process of Müze Gazhane since my graduate studies, and I had the chance to follow up on the news" (42 y.o. female). Participants upon gaining this awareness, affirmed that the altered identity rendered Müze Gazhane

https://doi.org/10.52202/082078-0016

more meaningful to visit. "When I came here for the first time, I was not familiar with its history. However, I feel more interested now, and I would like to research more about it" (23 y.o. male). They expressed a belief in its significance as a heritage area deserving preservation and maintenance, attributing additional value to the evolving identity of Müze Gazhane. For example, a 44-year-old male participant mentioned: "Even though it escaped from former municipality governors' notice, Müze Gazhane is an important place that opened the doors for the public regarding integrating the new with the old."

The qualitative analysis sheds light on Müze Gazhane's historical significance, portraying it not merely as a sporadic destination for occasional visits but rather as an integral part of visitors' daily routines. The majority of visitors emerged as regular patrons, integrating Müze Gazhane into their everyday lives. "I bring my dog here for a walk every day, also meet the same faces, it gives a nice community feeling" (56 y.o. female). Their visits extend beyond a traditional museum experience, encompassing activities such as spending quality time with family and friends, making use of open-air facilities, and engaging in scholarly pursuits within the library. An 18-year-old female student shares his experience: "I come here to study for my exams. This is a very nice park, full of green areas, and it also has a library. I did not have time to visit the exhibitions yet, though. Especially open-air areas are my favorite. Spending time on the grass makes me feel better."

Interestingly, the act of visiting the museum emerged as a background motive, with visitors pointing to various reasons regarding their preference to visit Müze Gazhane. Notably, the presence of accessible green spaces played a pivotal role, offering an open environment for everyone. Additionally, the availability of diverse museums and workshop areas contributed to the appeal. Furthermore, ancillary amenities such as cafes, restaurants, and a 24-hour operational library were cited as influential factors that enhance the overall attraction of Müze Gazhane. This multifaceted appeal underscores Müze Gazhane's role as a dynamic and multifunctional cultural space, extending beyond its traditional museum identity to cater to the diverse needs and preferences of its visitors.

The responses pertaining to the identity, meaning-making, and satisfaction levels of visitors reveal that their engagement with Müze Gazhane transcends the mere exploration of museum content. For example, a 42-year-old female participant mentioned: "When you visit other museums in İstanbul, you go to see the exhibitions and then leave the museum; it doesn't offer you a broader experience than that. However, in Müze Gazhane, it's not only about coming here to visit exhibitions and museums but more about spending quality time with friends and family, sometimes by exchanging new things, sometimes dining together. Simply enjoy your whole day, starting from the morning till evening." Visitors not only have the opportunity to acquire new knowledge related to the museum, but also discover novel aspects of themselves, contributing to the heightened meaningfulness and satisfaction of their visit. In this regard, a 52-year-old female participant emphasized how her visits to Müze Gazhane changed her relationship with the neighborhood: "After the renovation of Müze Gazhane, it influenced my perception of my neighborhood. I find it meaningful to visit here every day." As the related literature suggests (e.g., Doering, 1999; Falk, 2006; Morrissey, 2002; Paris, 1997), museum experiences can serve as platforms for both learning and self-discovery.

In the broader context, the principal criterion for assessing the value of museums lies in their positive impact on individuals' lives. A meaningful museum experience is derived from a visit that proves beneficial, resulting in positive changes (Driver, Brown & Peterson, 1991; Packer, 2008; Prentice, 1996). A beneficial visit is characterized as a "change that is viewed to be an advantageous improvement in the condition or a gain to an individual, a group or society" (Driver, Brown & Peterson, 1991, p. 4). The renovation of Müze Gazhane has led to perceived positive changes in the daily lives of a majority of its visitors, culminating in an enhancement of their overall quality of life. This can also be found in the words of one of the participants: "Istanbul is ours, so it is important to support consistent projects that embrace this city's heritage. Municipalities' only goal should not be building roadways and parks or supplying clean water and gas. They should also carry out projects which nourish society from a sociocultural perspective" (42 y.o. male participant). "When I stop by Müze Gazhane a couple of times a week, I feel relaxed. Before this place was opened, there was no such a place to go around" (42 y.o female participant). Consequently, there exists a positive correlation between Müze Gazhane's changing identity and the meaning-making and satisfaction levels reported by its visitors.

8 REFERENCES

Arabacıoğlu, B. & Ergin, G. (2016). The "new" museum comprehension: "inclusive museum." Proceedings of the International Conference on New Trends in Architecture and Interior Design, pp. 84-89.

Aykaç, P. (2020). Museumization in the urban context: The transformation of Sultanahmet District into a museum area (published in Turkish). Türkiye Bilimler Akademisi Kültür Envanteri Dergisi, 0(21), 145–163. https://doi.org/10.22520/tubaked.2020.21.008

https://doi.org/10.52202/082078-0016

Doering, Z. D. (1999). Strangers, guests, or clients? Visitor experiences in museums. Curator: The Museum Journal, 42(2), 74–87. https://doi.org/10.1111/j.2151-6952.1999.tb01132.x

Driver, B. L., P.J. Brown, and G. L. Peterson. (1991). Benefits of Leisure. Venture Publishing, Inc.

Falk, J. H. (2006). An identitycentered approach to understanding museum learning. Curator: The Museum Journal, 49(2), 151–166. https://doi.org/10.1111/j.2151-6952.2006.tb00209.x

He, H., Li, Z., Cheng, X., & Wu, J. (2019). Gamified participatory museum experience for future museums. Lecture Notes in Computer Science. Springer Science+Business Media. https://doi.org/10.1007/978-3-030-23538-3_15

Morrissey, K. (2002). Pathways among objects and museum visitors. Perspectives on Object-Centered Learning in Museums (pp. 285-299). Lawrence Erlbaum Associates.

Müze Gazhane Square [Online image]. Retrieved January 8, 2024. Müze Gazhane. https://muzegazhane.istanbul/hakkinda/

Packer, J. (2008). Beyond learning: Exploring visitors' perceptions of the value and benefits of museum experiences. Curator: The Museum Journal, 51. 10.1111/j.2151-6952.2008.tb00293.x.

Paris, S. (Ed.). (1997b). Understanding the Visitor Experience: Theory and Practice, Part 1. Washington: Journal of Museum Education.

Prentice, R. (1996). Managing implosion: The facilitation of insight through the provision of context. Museum Management and Curatorship, 15(2), 169–185. https://doi.org/10.1080/09647779609515478

https://doi.org/10.52202/082078-0016

EXAMINING THE IMPACT OF PERCEIVED VISIBILITY ON EMPLOYEE PERFORMANCE IN OPEN PLAN WORKPLACES

Track 4: Design for Health and Wellbeing
Knowledge Network: Interior Design
Methods: Meta-Analysis, Survey
Mingming Zhao (The Chinese University of Hong Kong)

1 ABSTRACT

The shift to hybrid work patterns in the post-pandemic era underscores the imperative for a delicate balance between collaboration and privacy for concentrated work, which is essential for sustaining high work engagement and productivity in open-plan workplace settings. However, the impact of architectural designs on employee performance, especially concerning visibility - visual exposure and openness experienced by office workers, has not been fully understood. The main objective of this paper is to review and summarize how visibility influences the dual aspects of workplace productivity: concentration on individual tasks and collaboration on group tasks. Methods. This paper adopts a combination of thematic analysis and a narrative synthesis approach, with 10 peer-reviewed research papers published in scholarly articles, industry publications, and empirical research between 2003 and 2023. The analysis highlights that visual openness in workplaces promotes collaboration but can paradoxically reduce physical movement and affect spontaneous interactions. In terms of productivity, while excessive visual exposure in open-plan offices diminishes job satisfaction, a balanced approach enhances productivity, especially among younger employees. The review identifies a lack of consensus on the optimal degree of visibility for enhancing productivity. There is also a scarcity of studies differentiating the impact of visibility on various types of work tasks. The significance of this research lies in its provision of a thorough understanding of how visibility impacts employee performance in open-plan workspace settings. It advocates for a balanced approach to workspace architecture, emphasizing adaptable environments that accommodate both collaborative and focused work requirements.

1.1 Keywords

Visibility, open plan workplaces, work performance, concentration, collaboration

2 INTRODUCTION

2.1 Open Plan and Activity-Based Workplaces (ABWs)

The transformation of workplace design is closely influenced by the evolution of technology and economic development in history. Since the 1980s, with the trend of "going west" for many innovative high-tech enterprises in America, Silicon Valley became the source of a utopian beacon heralding a new era in office transformation (Duffy, 2008). Driven by the idea of the "spontaneous encounter" and emphasis on employee enjoyment, the College-style Offices designed as all-inclusive ecosystems cover expansive square feet and provide various amenities to employees, which is meant to be a self-contained universe. Concurrently, Non-territorial Offices and Activity-based Workplaces (ABWs) have been increasingly experimented (Appel-Meulenbroek et al., 2021). By the 1990s, with the dominant work culture of innovation, collaboration, and knowledge sharing within organizations, the focus had shifted towards creating workspaces reminiscent of familial or cafe atmospheres which were characterized by open layouts and shared spaces punctuated with private meeting areas. It revolves around the idea that the workplace should be designed to support the specific activities that employees engage in, rather than assigning each worker a permanent desk or office. Eismann et al. (2022) proposed models of ABW including core functional spaces and supportive functional spaces both using a desk-sharing system. The design of an ABW is based on providing a range of settings suited for different tasks, such as collaboration spaces, quiet zones, meeting rooms, and lounges. The technology giant Microsoft has successfully implemented an ABW approach in several of its global locations. In their Netherlands headquarters, Microsoft has done away with assigned seating, instead providing a variety of workspaces to cater to different activities, such as concentration zones, collaborative areas, and relaxation spaces (Searer, 2012).

Co-working spaces typify the community-based aspects of activity-based workplaces. One of the leading examples of community-based ABWs is the global co-working giant, WeWork. Founded in 2010, WeWork has evolved from a single New York-based co-working space to a global community reaching over 100 countries. Their signature open-concept, high-design spaces aim to foster a sense of community, collaboration, and innovation among its diverse users ranging from freelancers to established enterprises (Di Risio, 2021). In many ways, ABWs are

https://doi.org/10.52202/082078-0017

the antitheses of the traditional hierarchical office paradigm, which offers positive and energizing places to work contributing to higher job satisfaction and improved performance (Eismann et al., 2022). Tabak (2009) tried to make a taxonomy of office work activities. He distinguished three different ways to differentiate activities: (1) nature of the activity – social, physiological, or job-related; (2) individual or group activities; and (3) planned or unplanned activities. ABWs generally have open-plan layouts, intended to increase the interaction between employees from different departments through serendipitous meetings, plus many places for formal and informal

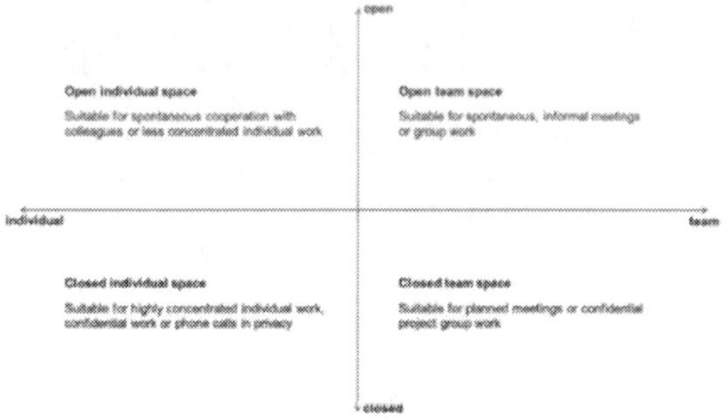

Figure 1. Taxonomy of core functional spaces (Eismann et al., 2022).

meetings. Small concentration spaces and the possibility to work from home are also offered because openness and communication can be a distraction for those employees trying to concentrate. For the taxonomy of core functional spaces, it was concluded that two basic dimensions of workspace design directly affect productive work: individual-team and closed-open as shown in Figure 1.

2.2 The Workplace Design in the Post-Pandemic Era

In the post-pandemic era, the role of open-plan workplaces has become more prominent by providing new opportunities. The COVID-19 pandemic has fundamentally shifted perceptions of the significance of remote work, propelling "hybrid work" models to the forefront in the post-pandemic era (Tostevin, 2020). Many companies are now embracing a hybrid model where employees work part-time from home and part-time in the office. It brings implications for workplace design with greater flexibility and adaptability, more interactions, and enhanced well-being promotion. With fewer resources dedicated to individual workstations, ABWs inherently support this flexibility by allowing workers to choose a workspace that best suits their current activity, whether that is engaged collaboration, individual concentration, or relaxation, and therefore promoting the well-being and mental health of employees (McLaurin, J. P., 2023). Dropbox is navigating the post-pandemic work scenario by adopting a remote-first approach while still valuing the significance of in-person collaboration through its Dropbox Studios. These spaces are not designed for everyday work but rather for team collaboration, group activities, and company events. Although the design of post-pandemic workplaces is widely believed to be more open team space oriented with thoughtful individual spaces for the hybrid work model, designing office spaces that effectively cater to this new norm remains a debate. The workplace design today requires satisfying much more complex functional needs for both collaborative and concentrated work, emphasizing engaged experience to enhance employee performance (McLaurin, J. P., 2023).

2.3 The Interplay Between Workspace Design and Employee Performance

Winston Churchill's aphorism "We shape our buildings and thereafter they shape us," has become famous in spatial psychology, architecture, and planning. Many studies have examined the interrelationships between individuals and their physical surroundings from a psychological perspective, and how the natural and built environments change human experience and behavior. In 1936, influenced by Gestalt psychology, famed psychologist Kurt Lewin developed an influential equation in his Field Theory: "B = ƒ(P, E)" (Lewin, 2015). It stated that human behavior (B) is a function of interaction between the person (P) and the environment (E) which comprises "the totality of facts which determine the behavior of an individual at a certain moment" (Lewin, 2015). In this model "P" includes the entirety of the person (internal factors) including their past, their present, their expectations of the future, their personality, their capabilities, their motivations, their desires, and so on. "E" includes all aspects of the person's environment at the time of any behavior (external factors) including their physical environment, but also their social environment and contexts. Hence, an individual can behave in very different ways if their environment changes. In addition, an individual can behave in very different ways in the same situation as they change as a person over time. Similarly, Stallworth and Kleiner (1996) highlighted that the "person-environment fit" hypothesis is particularly valuable in this context. The field of environmental psychology provides valuable insights to further investigate the complex relationship between the workspace environments and office workers' performance and satisfaction. Several research evidence suggests that work environments design affects multiple aspects of employee and organizational performance, including productivity (Kaushik et al., 2021), interaction and collaboration (Hua et al., 2011), communication (Zamani & Gum, 2019), and employee mood (Vimalanathan & Ramesh Babu, 2014). There is a positive link between PBWE and the psychological aspect of work concentration, comfort, job satisfaction, and emotional well-being (Grant et al., 2019).

https://doi.org/10.52202/082078-0017

Based on Maslow's hierarchy of needs, Grant et al. (2019) developed an adapted hierarchy of needs model (Figure. 2) for the Positive Built Workplace Environment (PBWE). At the bottom level, there are physiological needs that are supported by the built workplace environment: a comfortable space with appropriate lighting, air quality, view, thermal, and acoustic design. At the next level, there are safety needs. These include a physically safe, injury-free, and bullying-free environment, fair pay and remuneration, reasonable benefits, properly defined role descriptions, and an appropriate management structure. Safety in this context also incorporates the feeling of predictability and consistency (Walumbwa & Schaubroeck, 2009). People in the building need to feel that the services provided by the building management and the building itself are reliable and robust. At the next level (level 3) there are belongingness needs. The actual physical space should be designed to facilitate positive interactions and connections between individuals and groups. Such design should occur at the micro level as represented in well-designed workplaces, and at the macro level as represented by the design of the actual building and its precincts. Multiple theories have attempted to explain the reciprocal relationship between spatial properties and human intuitive responses. The architectural theory proposed by Joedicke in 1985 suggests that the most basic quality of architectural space, its spaciousness, is an important constituent of its experience. Appleton's (1975) habitat and prospect–refuge theories are the best-known explanations for environmental preference in architecture, which suggest preference patterns for certain configurations combining enclosure and openness. Geometry (implicating vertex density and roundness) and order (comprising also legibility, clarity, and coherence) are also suggested as important basic structural qualities of architecture affecting both experience and behavior (Weber, 1997). In this review, two visual-spatial qualities of visibility including openness and exposure will be primarily discussed.

2.4 Visibility in Open Plan Workplaces

In contemporary office design, the concept of visibility in open-plan workplaces has garnered considerable attention for its multifaceted impact on the work environment. Visibility relates to the ability to see (visual openness) and be seen (visual exposure) within a space, which is a direct function of spatial design. In an open-plan workspace, high levels of visibility are usually observed. This increased visibility entails a heightened exposure to environmental stimuli, yielding mixed outcomes. On the beneficial side, it includes aspects such as natural light and views of nature, enhancing the workspace ambiance. Elevated visibility is also theorized to facilitate greater interaction and collaboration among employees, potentially fostering a more dynamic and cohesive organizational culture (Zamani & Gum, 2019). However, this same attribute can paradoxically lead to reduced levels of personal privacy and increased susceptibility to environmental distractions, which can adversely affect individual performance and well-being. Additionally, Zamani & Gum (2019) note

Figure 2. Hierarchical of Needs for the PBWE (Grant et al., 2019).

https://doi.org/10.52202/082078-0017

that the accompanying lack of visual barriers can lead to sensory overload. Therefore, the design of open-plan offices necessitates a balanced approach, harmonizing the need for visibility with the imperative of creating a supportive and productive work environment.

Regarding important psychosocial characteristics, privacy and having the ability to work distraction-free is one of the strongest predictors of perceived individual productivity (Leaman & Bordass, 1999). People need a certain feeling of anonymity and do not like the feeling of always being exposed, which is closely related to the level of enclosure (Clements-Croome, 2006). Flexible workplace layouts and open workstation set-ups are popularly asserted to be able to positively affect workers' job satisfaction as they promote efficient collaboration. Quiet workspaces in open-plan offices also provide better adaptation and improve the work environment (Haapakangas et al., 2019). Veitch et al. (2007) claimed that a higher density might decrease workers' satisfaction. Dense workstation layout significantly increased the risk of dissatisfaction with privacy and acoustics. Moreover, it was evidenced that higher satisfaction with a workplace from those with higher partitions, also implying better privacy, led to higher satisfaction (Yildirim et al., 2007). It has been argued that an environment which offers some sense of enclosure, while still allowing a degree of visual engagement with adjacent environments, is best able to satisfy basic human emotional needs (Appleton, 1975). In contrast, a strong or overwhelming sense of exposure or enclosure allegedly leads to feelings of anxiety (Lidwell et al., 2003).

2.5 Research Questions, Objectives, and Significance

Accordingly, the study of visibility in workplace design emerges as a crucial factor for enhancing employee performance, presenting broad application potential and the possibility to redefine contemporary workplace design paradigms. While the importance of visibility is increasingly recognized in architectural research, comprehensive reviews that summarize its impact in open-plan workplace environments are scarce. Most existing studies and literature reviews concentrate on lighting, layout, workstation arrangement, and spatial syntax for enhancing productivity (J. Miller et al., 2022; Nguyen et al., 2022; Patel et al., 2021; Wright & Thompson, 2019). However, the impact of architectural designs on employee performance, especially concerning visibility - the degree of visual exposure and openness experienced by office workers, has not been fully understood. This gap underscores the need for more in-depth research and application of visibility-focused design strategies, which could significantly improve employee productivity, well-being, and overall workplace efficiency.

The primary objective of this paper is to review and summarize how visibility (visual openness and visual exposure) influences the dual aspects of workplace productivity: concentration on individual tasks and collaboration on group tasks. It also aims to explore the effective design factors that can balance the need for visibility and how it can ultimately be applied in the post-pandemic workplace design to optimize worker performance and well-being. The following research questions are intended to be addressed:

- Q1. How does visual openness affect collaboration and concentration among workers in open-plan workplaces?
- Q2. How does visual exposure affect their collaboration and concentration among workers in open-plan workplaces?
- Q3. What design factors can effectively balance the need for visibility to optimize worker performance?

To answer the questions and explore the impact of visibility designs on employee performance - concentration on individual tasks and collaboration on group tasks, this paper reviews previous workplace environment research papers, particularly those related to visual openness and visual exposure. The significance of this research extends beyond a basic analysis, filling a critical gap in the study of environmental design by thoroughly exploring the role of visibility. It not only illuminates how visibility impacts employee performance in open-plan workspace design but also holds the potential to inform future design practices. By delving into this often-overlooked aspect, the study paves the way for developing comprehensive design guidelines that can enhance workplace efficiency, collaboration, and overall employee well-being.

3 MATERIALS AND METHODS

This paper adopts a combination of thematic analysis and a narrative synthesis approach. The thematic analysis allows for the systematic categorization of literature into distinct themes, facilitating an organized and focused examination of specific aspects of the subject, such as "Effects of Visibility on Collaboration," and "Effects of Visibility on Collaboration." This approach helps in identifying patterns, trends, and gaps across different studies. A narrative approach then is used to synthesize the findings within and across the identified themes. It involves summarizing and explaining the results of the studies, discussing how they relate to each other, and concluding the overall state of research in this area. This synthesis not only elucidates the current state of research but also highlights areas that warrant further exploration, thereby guiding future studies.

3.1 Database

To conduct a comprehensive literature review, this paper employed a wide range of literature search strings related to open-plan workplaces,

https://doi.org/10.52202/082078-0017

environmental design, visibility, work performance, and well-being. Different terminological variants were taken into account. The following strings are utilized in all database searches: TS (title, abstract, and author keywords) = (("visibility" OR "visual exposure" OR "visual openness" OR "visual privacy") AND ("Workplace" OR "office" OR "workspace") AND ("emotion" OR "affective" OR "productivity" OR "health" OR "wellness" OR "well-being") AND (open workplace OR open plan workplace OR open plan office)). A total of 35 papers were retrieved from the following two databases: Web of Science and Scopus. The databases were chosen based on two considerations: (1) They contain a wide range of relevant articles covering at least one of the subject areas of architecture and design; and (2) They are widely recognized in academia for archiving high-quality peer-reviewed research. The literature search was conducted on Nov 7, 2023, and was based on abstracts, keywords, and titles of research papers indexed during the decade 2003 - 2023.

3.2 Criteria for the Study Selection

Inclusion and exclusion criteria in Table 1 are applied in the screening and selection phases. In the general screening, five articles are removed using paper type, language, and validity as the main screening criteria, as studies must be peer-reviewed publications in English with full-text access. Twenty eligible articles are further screened in the step of title and abstract screening, mainly for the context and topic. First, the context had to be open-plan workplace settings that are not closed and traditional cubicle-only workspaces. Second, they had to be in the field of architectural design, studying the impact of visibility on employee performance. Third, the study must focus on visual openness or visual exposure but no other design elements only such as acoustic or ergonomics. Although it may lead to not all relevant aspects of environmental elements affecting perceived concentration and collaboration being retrieved, it helps to ensure that the studies included in this review are highly relevant, focused, and valid. Finally, a total of 10 English-language research papers published in peer-reviewed scholarly articles, industry publications, and empirical research between 2003 and 2023 were selected for analysis.

Table 1. Eligibility Criteria of Paper Selection.

Criteria	Inclusion	Exclusion
Context	Open plan Workplaces	Classroom; Hospital; Closed and traditional cubicle workspaces
Topic	Visibility (visual openness and exposure)	Acoustics; Thermal Comfort; Lighting; Ergonomics
Year	2003 - 2023	/
Language	English	/
Paper type	Peer-reviewed publications	Review papers; Conference proceedings; Dissertation; Not academic articles (blog, magazine, newspaper, etc.); Manuscripts and personal communications
Validity	Full-text access	No full-text access; Repeated experiments

https://doi.org/10.52202/082078-0017

4 RESULTS

4.1 Data Extraction and Synthesis

The following Table 2 illustrates a review spreadsheet of visibility in open plan workplaces research:

Table 2. Thematic Literature Review Spreadsheet.

#	Year	Authors	Type of Research Design	Environmental Setting	Questions/Objectives	Findings/Key Takeaways
1	2005	Lee and Brand	Survey	Five different organizations	The effects of distractions, flexible use of workspace and personal control over the work environment on perceived job performance, job satisfaction, group cohesiveness, and inclinations to work alone or in an enclosed space and their interrelationships.	More personal control over the physical workspace (e.g., adjustment) and easy access to meeting places led to higher perceived group cohesiveness and job satisfaction. Contrary to expectation, the results indicated that distractions may have little influence on self-rated performance. Providing office workers with more control over their workplace may address individual, interpersonal, and group needs for flexibility and, in turn, may contribute to group cohesiveness.
2	2012	Stryker et al.	Field study	Two R&D sites of a large Midwestern U. S. pharmaceutical company	The relationship between the design of the physical workplace and F2F communications by focusing on the technical side of sociotechnical theory (STT), which centers on task efficiency, and views the physical design and layout of the workplace.	The visibility of the work environment and the amount of collaboration opportunity, defined as formal and informal space available for meetings and collaboration, are related to F2F communication. It suggests that in high-visibility locations, it is work environment visibility and not type of workstation—open, low-walled workstation versus closed office—that is the salient and determining feature in promoting F2F communication. Job performance for workers in high-technology jobs is improved in work environments with fewer physical barriers.
3	2018	Weziak-Bialowolska et al.	Survey	456 employees located in 20 regional office locations within the same architectural firm	The relationship between perceptions of three characteristics of the open-plan office (acoustical privacy, visual privacy, and office density), and the impact they yield on employees' judgment as well as affect-driven behaviors.	Working in an open-plan office limits the experience of privacy and intensifies the perception of intrusion among employees of an architectural company, mostly architects and designers. Additionally, employees' perception of lack of privacy and high office density negatively affect job satisfaction, work engagement, and internal work relations as well as increases the number of limited ability days. Interestingly, the lack of privacy and high office density seem to positively affect expressive personal relations among coworkers and job performance.

https://doi.org/10.52202/082078-0017

4	2018	Yunus and Ernawati	Mixed method: census, factor analysis, and hierarchical regression analysis	An educational institution in office design changes from a combi, cellular-like office to a more open, non-territorial office.	The relationship between office redesign and employee productivity; The impact of privacy on work productivity across different generations.	Collaboration and privacy exert a positive influence on work productivity. The relationship between privacy and work productivity is stronger for Generation Y than for senior employees, namely, the Baby Boomers and Generation X. Practical implications For Gen Y employees who prefer freedom, mobility, and flexibility to personalization in their workplace, the open office could be a better solution for organizations that aim for both work productivity and efficiency.
5	2019	Zamani and Gum	Mixed methods: surveys, systematic observations, space syntax	A case organization piloting an activity-based flexible offices (AFO)	How environmental features and psychological or job needs impact observed and perceived satisfaction, communication, collaboration, and perceived productivity.	Collaboration instances were higher in AWS, especially more visible and accessible open areas, supporting higher impromptu interactions and enhanced perceptions of the productivity of team members and cross-team members. Privacy requirements are linked to a greater demand for enclosed AWS. A fit between the AFO environment and needs impacted workplace satisfaction, communication, collaboration, and perceived productivity.
6	2020	Bentinck et al.	Interview before and after relocation into the new building	A new building of a Dutch University Institute	How well a new building of a Dutch University Institute (DI) supported interaction and perceived privacy among faculty members.	The transparent and centrally organized floor plan supported face-to-face communication but generated a lack of privacy for faculty members. Lack of visual privacy and the sense of being controlled by others were related to the hierarchical position of teachers in this Higher Education Institute (HEI) between students and the dean, which caused tension and diminished their well-being.
7	2021	Sailer et al.	Mixed method: Survey and spatial qualities analysis (isovist and visual field)	UK headquarters of a global technology company	The effect a workplace layout (such as its openness) has on perceptions of staff regarding teamwork, focused work, and perceived productivity.	Staff are less likely to rate their workplace environment favorably when they have higher numbers of desks within their field of vision; and when they are facing away from the room with a relatively larger area behind their back compared to the area surrounding them. Aspects of teamwork that are negatively affected include sharing information with others, as well as team identity and cohesion. Focused work (concentration) and working productively are impacted even more so with the largest effect sizes throughout. These findings highlight the relevance of investigating detailed spatial qualities of micro-locations in workplace layouts.
8	2022	Koohsari et al.	Nationwide online survey (n = 2,265)	12 personal controls in open-plan workplaces	The associations of perceived workplace layout design with workers' active and sedentary behavior in office workers in Japan.	The visibility and proximity of co-workers were positively associated with any work-related sedentary time in shared and open-plan offices. The relevance of workplace layout design to participation and the duration of workers' active and sedentary time in the context of Japan.
9	2022	Pollard et al.	Daily movement location data of 22 office workers over 4 weeks	A 1,220㎡-office workplace	The relationships between specific spatial and environmental attributes and office workers' movement behavior.	Increased visibility of colleagues had significant negative associations with several measures of daily movement behavior, including the total area utilized each day.
10	2022	Jeon et al.	Survey	Thirty-four subjects experience four open-plan offices (OPOs) with four different visual layouts.	The effects of the environment on indoor soundscape perception for designing pleasant OPOs.	The same OPO soundscape, window area ratio, and floor area showed positive correlations with visual privacy and visual satisfaction. Task performance exhibited contradictory relationships with the visual parameters. Wide, open workspaces yielded high work satisfaction, whereas narrow, confined workspaces yielded high task performance. The relationship between work satisfaction and task performance must be considered for a balanced OPO design, providing practical guidance for designing pleasant OPO environments.

https://doi.org/10.52202/082078-0017

4.2 Theme and Timeline Analysis

The thematic literature review from 2003 to 2023 reveals an evolving understanding of the role of visibility in workplace design. Over the twenty years, the literature has evolved from a general understanding of visual openness in workplace spaces to a more complex analysis of open-plan and specific spatial designs. Before 2018, studies primarily concentrated on visual openness in workplace design, as seen in the works of Lee and Brand (2005) and Stryker et al. (2012). These studies highlighted how visual openness in personal workspaces and general work environments facilitated interactions and collaboration. Starting with Zamani and Gum in 2019, there's a noticeable shift. The focus expands from mere visual openness to also include visual exposure. This shift reflects a deeper understanding of how different degrees of visual accessibility in workplaces, such as in activity-based flexible offices, impact employee interactions. In recent years, Bentinck et al. (2020), Pollard et al.(2022), and Koohsari et al. (2022) further delve into this concept. They investigate the implications of visual exposure in more dynamic settings, such as transparent, central floor plans and increased colleague visibility. These studies signify effective approaches toward fostering collaboration through strategic visual design. The evolution from a focus on visual openness to a nuanced understanding of visual exposure marks a significant trend in workplace design research. Early studies viewed openness as a broad concept that generally enhanced collaboration and interaction. However, recent studies, particularly since 2018, have started to differentiate between openness and exposure, recognizing that while openness can foster interaction, uncontrolled visual exposure can impact concentration and productivity.

Figure 3. Literature Publication Timeline.

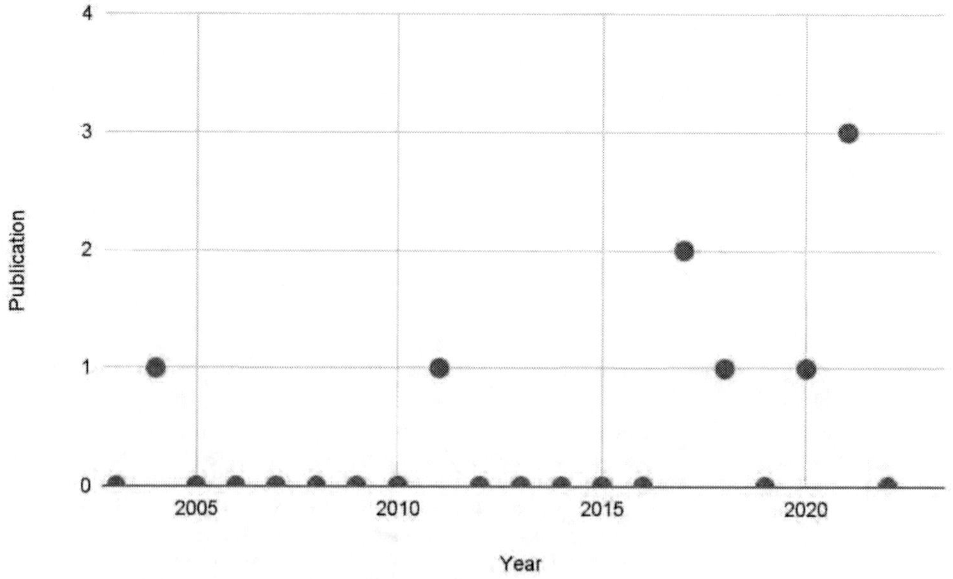

https://doi.org/10.52202/082078-0017

The selected literature is organized into two themes: Theme 1 - Visibility and Collaboration/Interaction, and Theme 2 - Visibility and Concentration/Productivity, which offers a comprehensive overview of how visibility in workplace design influences both collaboration and concentration. Theme 1 encapsulates studies that explore the relationship between visual aspects of workplace design and their impact on employee interaction and collaboration. Theme 2 focuses on how visual elements, specifically the balance between openness and exposure, affect concentration and productivity in the workplace. The following Table 3 illustrates a list of thematic literature reviews:

Table 3. Thematic Literature Review Spreadsheet.

	Year	Authors	Visibility	Context
Theme 1: Visibility and Collaboration/Interaction	2005	Lee and Brand	Visual Openness	Personal workspace control
	2012	Stryker et al.	Visual Openness	General work environment
	2019	Zamani and Gum	Visual Openness	Activity-based flexible offices
	2020	Bentinck et al.	Visual Openness and Exposure	Transparent, central floor plan
	2022	Pollard et al.	Visual Openness and Exposure	Increased visibility of colleagues
	2022	Koohsari et al.	Visual Openness and Exposure	Worker visibility and proximity
Theme 2: Visibility and Concentration/Productivity	2005	Lee and Brand	Visual Openness	Personal workspace control
	2018	Weziak-Bialowolska et al.	Visual Exposure	Open-plan offices
	2018	Yunus and Ernawati	Visual Openness and Exposure	Privacy designs
	2021	Sailer et al.	Visual Openness and Exposure	More desks in vision and backs to the room
	2022	Jeon et al.	Visual Openness and Exposure	Narrower spaces

4.3 Visibility and Collaboration/Interaction

The literature review under this theme spans a period from 2005 to 2022, presenting a diverse range of studies that examine the role of visibility in workplace design. Lee and Brand (2005) and Stryker et al. (2012) laid the groundwork by focusing on visual openness. Lee and Brand (2005) emphasized personal workspace control, linking visual openness to increased job satisfaction and group cohesiveness. Stryker et al. (2012) demonstrated that visual openness fosters face-to-face communication. These early studies underscored the positive aspects of visual openness but lacked depth in exploring potential downsides, such as privacy concerns or overstimulation. Zamani and Gum (2019) and Bentinck et al. (2020) moved beyond the initial positive outlook, introducing a more balanced view. Zamani and Gum's (2019) study in activity-based flexible offices showed that while visual openness promotes collaboration, it also affects satisfaction and productivity, which indicates a need for a balance between open and enclosed spaces. Bentinck et al.'s (2020) work further highlighted this balance by illustrating how a transparent and central floor plan promotes

https://doi.org/10.52202/082078-0017

interaction but reduces privacy as well. These studies marked a shift towards recognizing the complexity of designing workspaces that facilitate interaction without compromising individual needs. In 2022, the studies by Pollard et al. and Koohsari et al. introduced a new dimension to the discussion: the impact of visibility on movement behavior. Both studies found that increased visibility and exposure could lead to reduced movement and increased sedentary time. This perspective is significant as it connects physical workplace design with broader health implications which is a facet previously underexplored in earlier research. The following Table 4 illustrates a thematic review of visibility and collaboration/interaction:

Table 4. Theme 1 Literature Review Spreadsheet.

Year	Authors	Visibility	Context	Relationship	Impacts
2005	Lee and Brand	Visual Openness	Personal workspace control	+	Address individual needs and group cohesiveness.
2012	Stryker et al.	Visual Openness	General work environment	+	Foster face-to-face communication.
2019	Zamani and Gum	Visual Openness	Activity-based flexible offices	+	Promote collaboration; Satisfaction and perceived productivity are impacted by the fit between environment and needs.
2020	Bentinck et al.	Visual Openness and Exposure	Transparent, central floor plan	+	Promote interaction but reduce privacy.
2022	Pollard et al.	Visual Openness and Exposure	Increased visibility of colleagues	-	Reduce daily movement.
2022	Koohsari et al.	Visual Openness and Exposure	Worker visibility and proximity	-	Reduce movement and Increase sedentary time.

The thematic review of literature under theme 1 concerning high-visibility workplaces and their impact on collaboration and communication reveals a diverse and somewhat paradoxical landscape. An important element emerging from these studies, particularly highlighted by Lee and Brand (2005), is the role of personal control over the workspace. It suggests that visual openness, when combined with a degree of personal control, can be particularly effective in fostering a collaborative atmosphere. The consensus in most studies is that visual openness in the workplace supports and enhances collaboration. This increased visibility facilitates easier and more frequent face-to-face interactions among colleagues. However, this relationship is not straightforward and entails several complexities and contradictory points worth exploring. One is that visual openness can lead to reduced physical movement within the workplace. This finding raises questions about its potential impact on collaboration. While open environments are designed to encourage interaction, the actual physical behavior of employees might not align with this intention. This paradox highlights the complexity of designing work environments.

In summary, while high-visibility workplaces are generally conducive to collaboration and communication, the relationship is not as linear as it might seem. The interplay between visual openness, personal control, and movement within the workspace is complex. Designing an effective workplace requires a balanced approach that considers these multifaceted interactions. It calls for an environment that not only promotes visibility and accessibility but also respects individual preferences and behaviors, ensuring that openness does not inadvertently hinder movement and spontaneous collaboration.

https://doi.org/10.52202/082078-0017

4.4 Visibility and Concentration/Productivity

The literature under Theme 2: Visibility and Concentration/Productivity, presents a complex picture of how visibility within the workplace influences concentration and productivity. This theme encompasses studies from 2005 to 2022, each offering distinct insights into the relationship between workplace design, particularly in terms of visual exposure, and employee concentration and productivity. Lee and Brand (2005) found that visual openness not only positively enhances group cohesiveness but also addresses individual needs. Weziak-Bialowolska et al. (2018) delve into the concept of visual exposure in open-plan offices. Their findings indicate a negative impact: reduced privacy leads to lower job satisfaction and work engagement, suggesting that excessive visual exposure can be counterproductive. In the same year, Yunus and Ernawati (2018) is unique in highlighting the age of office workers. They found that less visual exposure can boost productivity, particularly among Generation Y employees, indicating a generational difference in how visual aspects are perceived and impact productivity. By 2021, the study by Sailer et al. adds another dimension to the discussion. It shows that an excess of visual openness and exposure, such as having more desks in one's field of vision, negatively affects teamwork and productivity. This suggests that too much visual exposure can be overwhelming and distracting, hindering the ability to focus and work productively. Jeon et al. (2022) contribute further by examining the impact of visual openness in narrower spaces with both less visual openness and exposure. They report a positive outcome which can increase task performance in such environments. This study hints at the potential benefit of controlled visual exposure in specific spatial contexts. The following Table 5 illustrates a thematic review of visibility and concentration/productivity:

Table 5. Theme 2 Literature Review Spreadsheet.

Year	Authors	Visibility	Context	Relationship	Impacts
2005	Lee and Brand	Visual Openness	Personal workspace control	+	Address individual needs and group cohesiveness.
2018	Weziak-Bialowolska et al.	Visual Exposure	Open-plan offices	-	Reduce privacy, negatively impacting job satisfaction and work engagement.
2018	Yunus and Ernawati	Visual Openness and Exposure	Privacy designs	+	Boost productivity, with stronger privacy-productivity links for Generation Y.
2021	Sailer et al.	Visual Openness and Exposure	More desks in vision and backs to the room	-	Affect teamwork and productivity.
2022	Jeon et al.	Visual Openness and Exposure	Narrower spaces	+	Boost task performance.

https://doi.org/10.52202/082078-0017

The literature presents a dichotomy in how visual openness and exposure affect concentration and productivity. On one hand, certain levels of visual openness can enhance productivity and cater to individual preferences. On the other hand, excessive visual exposure can lead to distractions, reduced privacy, and decreased job satisfaction, ultimately impacting productivity negatively which is particularly among younger generations. The thematic review under theme 2 underscores the importance of striking a balance in workplace design. While visual openness can be beneficial up to a point, excessive exposure can have adverse effects on employee well-being and productivity. The challenge lies in designing workspaces that provide the right amount of visual openness to boost productivity without compromising privacy and concentration. Future workplace designs might benefit from customizable visual environments that cater to diverse employee needs and preferences, acknowledging that one size does not fit all when it comes to visual aspects of workspace design.

5 DISCUSSION

The thematic reviews present a comprehensive literature exploration of the interplay between workplace visibility and two critical aspects of organizational performance: collaboration/interaction and concentration/productivity. The selected literature is organized into two themes: Theme 1 - Visibility and Collaboration/Interaction, and Theme 2 - Visibility and Concentration/Productivity. The studies under Theme 1 consistently highlight the pivotal role of visual openness in fostering workplace collaboration and interaction. However, the studies by Pollard et al. and Koohsari et al. (2022) introduce a complex paradox: while openness facilitates interaction, it can simultaneously lead to reduced physical movement, potentially impacting spontaneous collaboration. This paradox underscores a critical insight which is that workplace design cannot solely focus on enhancing visibility without considering its broader behavioral implications. The reduction in movement noted in later studies suggests that while employees may appreciate the ease of interaction that openness provides, they might also settle into a more static routine, possibly due to constant visibility or a sense of exposure. Therefore, designing workspaces that promote collaboration requires a careful balance between visual openness and creating an environment that encourages active, dynamic interaction. Theme 2 delves into the relationship between visibility and concentration/productivity. The potential downsides of excessive visual exposure in open-plan offices are emphasized, such as reduced privacy and negative impacts on job satisfaction and work engagement. The evolution of research in this area demonstrates a growing recognition of the need for a balanced approach to visibility in workplace design. While certain levels of openness can enhance focus and productivity, as suggested by Yunus and Ernawati (2018) and Jeon et al. (2022), there is a threshold beyond which visibility becomes counterproductive. This balance is especially pertinent in the context of varying individual preferences and the diverse nature of tasks within the modern workplace.
Drawing from both themes, it becomes evident that the design of workspaces must be approached with an understanding of the complex and often contrasting effects of visibility. Workplaces should not only aim to facilitate ease of interaction and collaboration but also respect individual needs for privacy and concentration. This dual objective calls for innovative design solutions that can adapt to the diverse requirements of modern workforces. Future research should continue to explore these dynamics, perhaps delving deeper into how individual differences, such as personality traits or job roles, interact with workplace visibility. Additionally, as the nature of work continues to evolve, especially with the rise of hybrid and remote work models, understanding how these findings translate into a variety of work settings becomes increasingly important.

6 CONCLUSIONS

In conclusion, the body of research analyzed in Themes 1 and 2 offers valuable insights into the intricate relationship between workplace visibility and key organizational outcomes including collaboration, interaction, concentration, and productivity. These studies collectively highlight the importance of balancing visual openness with the need for privacy and controlled exposure. This insight is crucial for designing workspaces that cater to the diverse needs of the modern workforce. The findings lay a groundwork for future research in the area and offer practical guidance for architects, designers, and organizational leaders in creating more effective and adaptive work environments. Insights from this research can inform policy-making and strategic decisions around workspace planning, particularly in an era where work dynamics are rapidly evolving due to technological advancements and the rise of remote and hybrid work models.

The research exhibits certain limitations. Firstly, only two databases and ten papers are selected in this paper, which may limit the generalizability of the findings. Secondly, there is a relative lack of emphasis on individual differences such as personality, work style preferences, or job roles, as well as other confounding factors that can significantly influence how employees interact with their workspace. Thirdly, many of the studies rely on surveys and self-reported data, which can be subject to biases and may not accurately capture actual behavior or performance. Lastly, most studies are cross-sectional, providing a snapshot at a single point in time. Longitudinal

https://doi.org/10.52202/082078-0017

studies could offer more comprehensive insights into how changes in workplace design impact employee behavior and productivity over time. Several gaps found in the literature provide opportunities for my future research: It identifies a lack of consensus on the optimal degree of visibility for enhancing productivity. There is also a limited exploration of how different types of tasks (creative vs. analytical) are affected by visibility and to develop guidelines for optimal workspace design in the post-pandemic era. As technology becomes increasingly integrated into work environments, its role in moderating the impact of workplace visibility on collaboration and productivity remains underexplored.

7 REFERENCES

Appel-Meulenbroek, R., Groenen, P., & Janssen, I. (2011). An end-user's perspective on activity-based office concepts. Journal of Corporate Real Estate, 13(2), 122–135. https://doi.org/10.1108/14630011111136830

Appleton, J. (1975). The experience of landscape. Landscape Research, 1(10), 15–16. https://doi.org/10.1080/01426397508705780

Bernstein, E. S., & Turban, S. (2018). The impact of the 'open' workspace on human collaboration. Philosophical Transactions of the Royal Society B: Biological Sciences, 373(1753), 20170239. https://doi.org/10.1098/rstb.2017.0239

Bentinck, S. A., van Oel, C. J., & van Dorst, M. J. (2020). Perception of privacy in a university building: The transparency paradox. Frontiers of Architectural Research, 9(3), 579-587. https://doi.org/10.1016/j.foar.2020.03.004

Clements-Croome, D. (Ed.). (2006). Creating the productive workplace (2nd ed). Taylor & Francis. Di Risio, A. (2021). The Largest Coworking Companies in the US: 2021. Coworking Resources. https://www.coworkingresources.org/blog/largest-coworking-companies

Duffy, F. (2008). Work and the city. Black Dog.

Eismann, T. T., Pakos, O., Rücker, M., Meinel, M., Maier, L., & Voigt, K.-I. (2022). Understanding the Mechanisms of Activity-based Workspaces: A Case Study. Environment and Behavior, 54(1), 170–210. https://doi.org/10.1177/00139165211008240

Grant, A. M., O'Connor, S. A., & Studholme, I. (2019). Towards a Positive Psychology of Buildings and Workplace Community:the Positive Built Workplace Environment. International Journal of Applied Positive Psychology, 4(1–2), 67–89. https://doi.org/10.1007/s41042-019-00019-2

Haapakangas, A., Hallman, D. M., Mathiassen, S. E., & Jahncke, H. (2019). The effects of moving into an activity-based office on communication, social relations and work demands – A controlled intervention with repeated follow-up. Journal of Environmental Psychology, 66, 101341. https://doi.org/10.1016/j.jenvp.2019.101341

Hua, Y., Loftness, V., Heerwagen, J. H., & Powell, K. M. (2011). Relationship Between Workplace Spatial Settings and Occupant-Perceived Support for Collaboration. Environment and Behavior, 43(6), 807–826. https://doi.org/10.1177/0013916510364465

Jeon, J. Y., Jo, H. I., Santika, B. B., & Lee, H. (2022). Crossed effects of audio-visual environment on indoor soundscape perception for pleasant open-plan office environments. Building and Environment, 207, 108512. https://doi.org/10.1016/j.buildenv.2021.108512

Kaushik, A., Arif, M., Ebohon, O. J., Arsalan, H., Rana, M. Q., & Obi, L. (2021). Effect of indoor environmental quality on visual comfort and productivity in office buildings. Journal of Engineering, Design and Technology. https://doi.org/10.1108/JEDT-09-2021-0474

Koohsari, M. J., McCormack, G. R., Nakaya, T., Shibata, A., Ishii, K., Lin, C. Y., & Oka, K. (2022). Perceived workplace layout design and work-related physical activity and sitting time. Building and Environment, 211, 108739. https://doi.org/10.1016/j.buildenv.2021.108739

Lee, S. Y., & Brand, J. L. (2005). Effects of control over office workspace on perceptions of the work environment and work outcomes. Journal of environmental psychology, 25(3), 323-333. https://doi.org/10.1016/j.jenvp.2005.08.001

Lewin, K. (2015). Principles of Topological Psychology. MARTINO FINE Books. Lidwell, W., Holden, K., & Butler, J. (2003). Universal principles of design. Rockport.

McLaurin, J. P. (2023, July 12). 5 Trends Driving the New Post-Pandemic Workplace. Gensler. https://www.gensler.com/blog/5-trends-driving-the-new-post-pandemic-workplace

Pollard, B., Engelen, L., Held, F., Van Buskirk, J., Spinney, R., & de Dear, R. (2022). Associations between spatial attributes, IEQ exposures and occupant movement behaviour in an open-plan office. Building and Environment, 212, 108812. https://doi.org/10.1016/j.buildenv.2022.108812

https://doi.org/10.52202/082078-0017

Sailer, K., Koutsolampros, P., & Pachilova, R. (2021). Differential perceptions of teamwork, focused work and perceived productivity as an effect of desk characteristics within a workplace layout. PloS one, 16(4), e0250058. https://doi.org/10.1371/journal.pone.0250058

Searer, S. (2012, January 27). Microsoft Headquarters—Amsterdam. Office Snapshots. https://officesnapshots.com/2012/01/27/microsofts-amsterdam-headquarters/

Stallworth, O. E., & Kleiner, B. H. (1996). Recent developments in office design. Facilities, 14(1/2), 34–42. https://doi.org/10.1108/02632779610108512

Stryker, J. B., Santoro, M. D., & Farris, G. F. (2012). Creating Collaboration Opportunity: Designing the Physical Workplace to Promote High-Tech Team Communication. IEEE Transactions on Engineering Management, 59(4), 609-620. https://doi.org/10.1109/TEM.2011.2170995

Tabak, V. (2009). User simulation of space utilisation: system for office building usage simulation. Technische Universiteit Eindhoven. https://doi.org/10.6100/IR640457

Tostevin, P. (2020, November 9). Office demand in a post-pandemic world. Savills Impacts. https://www.savills.com/impacts/market-trends/how-are-the-office-markets-faring-in-10-key-cities-arou nd-the-world.html

Veitch, J. A., Charles, K. E., Farley, K. M. J., & Newsham, G. R. (2007). A model of satisfaction with open-plan office conditions: COPE field findings. Journal of Environmental Psychology, 27(3), 177–189. https://doi.org/10.1016/j.jenvp.2007.04.002

Vimalanathan, K., & Ramesh Babu, T. (2014). The effect of indoor office environment on the work performance, health and well-being of office workers. Journal of Environmental Health Science and Engineering, 12(1), 113. https://doi.org/10.1186/s40201-014-0113-7

Walumbwa, F. O., & Schaubroeck, J. (2009). Leader personality traits and employee voice behavior: Mediating roles of ethical leadership and work group psychological safety. Journal of Applied Psychology, 94(5), 1275–1286. https://doi.org/10.1037/a0015848

Weber, R. (1997). On the aesthetics of architecture: A psychological approach to the structure and the order of perceived architectural space (Reprint). Avebury.

Węziak-Białowolska, D., Dong, Z., & McNeely, E. (2018). Turning the mirror on the architects: A study of the open-plan office and work behaviors at an architectural company. Frontiers in Psychology, 9, 2178. https://doi.org/10.3389/fpsyg.2018.02178

Yildirim, K., Akalin-Baskaya, A., & Celebi, M. (2007). The effects of window proximity, partition height, and gender on perceptions of open-plan offices. Journal of Environmental Psychology, 27(2), 154–165. https://doi.org/10.1016/j.jenvp.2007.01.004

Yunus, E. N., & Ernawati, E. (2018). Productivity paradox? The impact of office redesign on employee productivity. International Journal of Productivity and Performance Management, 67(9), 1918-1939. https://doi.org/10.1108/IJPPM-12-2017-0350

Zamani, Z., & Gum, D. (2019). Activity-based flexible office: Exploring the fit between physical environment qualities and user needs impacting satisfaction, communication, collaboration and productivity. Journal of Corporate Real Estate, 21(3), 234-253. https://doi.org/10.1108/JCRE-08-2018-0028

https://doi.org/10.52202/082078-0017

NATURE NURTURES: INVESTIGATING FACTORS IMPACTING CHILDREN'S OUTDOOR ACTIVITIES AND NATURE EXPOSURE IN BEIJING, CHINA

Track 4: Design for Health and Wellbeing
Knowledge Network: Children, Youth & Environment
Methods: Case Study, Interviews
Jianjiao Liu (University of Melbourne); Raymond Green (The University of Melbourne)

1 ABSTRACT

Children growing up in cities today are increasingly disconnected from nature, potentially compromising their quality of life. This research investigated factors influencing children's engagement in outdoor activities and nature exposure. We conducted in-depth interviews with 81 parent-child dyads living in Beijing, China. Open-ended data was content analyzed, which revealed that most parents perceive their children's outdoor activities and nature exposure to be suboptimal. The influencing factors extend beyond the child's inherent personality and include constraints related to three main external factors: 1) time and companionship, 2) accessibility to natural environmental infrastructure and services, and 3) accidental factors. Time constraints mainly stem from academic pressures, compressing children's discretionary time, which indirectly relates to a lack of peer companionship. Parental lifestyle, beliefs and work commitment were also identified as being influential, given respondents' age and reliance on parental guidance. The use of electronic devices further encroaches on children's time outdoors. Accessibility to infrastructure and services is primarily concerned with access to natural (green) spaces in daily life, particularly in residential areas, and safety concerns. Organizational factors, such as community-organized events organized by non-profit and for-profit organizations, also played a role. Government policies, including the "double reduction" policy, restrictions on electronic device usage, and initiatives encouraging child-friendly cities were discussed. Accidental factors, including weather conditions and health issues (e.g., COVID-19), were likewise influential. Implications for landscape architects and urban planners were underscored, emphasizing the need to provide opportunities to facilitate children's outdoor activities and nature exposure, and minimize barriers to this end in the city.

1.1 Keywords

Children; Contact with nature; Green spaces; Outdoor activities; Influencing factors

2 INTRODUCTION

Nowadays, children are faced with a variety of challenges attributed to the rapid urbanization, technology innovation and changes in lifestyles. Notably, children's independent mobility has been significantly curtailed by higher-density housing and expanded transportation (H. E. Woolley & Griffin, 2015). A prevailing trend is seen in the increased duration that children spend indoors, limiting their engagement in outdoor, nature-based activities. This growing inclination towards engagement in indoor activities has contributed to a noticeable disconnection with the natural world, resulting in a decline in their overall quality of life – a phenomenon that has been termed "Nature-Deficit Disorder" (Louv, 2008).

Over the last two decades, researchers have provided empirical evidence to support the notion that contact with nature can support children's overall physiological and psychological health and well-being (Tim Gill, 2014). Various studies have shown that environments rich in natural elements have the qualities to meet children's needs for stimulating and varied play environments, thus improving their motor fitness through playing and exploring in nature (Fjørtoft & Sageie, 2000). In addition to promoting physical well-being, studies have delved into the positive impact of nature experiences on enhancing psychological well-being and cognitive development of children, such as their emotional functioning, attention restoration and alleviation of stress (Bagot et al., 2015; Liu & Green, 2023; Taylor et al., 2002). These benefits are not confined to the general population of children but also extend to those grappling with psychological challenges such as Attention Deficit Disorder (Taylor & Kuo, 2009). Engaging in outdoor play within nature-based environments has been found to empower children to actively communicate with their peers, learn essential social skills, enhance their creativity, independent mobility, confidence, and develop good work ethics (Laaksoharju et al., 2012; Raney et al., 2019).

https://doi.org/10.52202/082078-0018

China has undergone a rapid process of urbanization, and children growing up in cities are, as a consequence, now experiencing diminished opportunities to have contact with nature. The shrinking and fragmentation of urban green space systems have physically restricted children from having adequate access to nature. Children also miss out on certain chances to experience nature due to the increasing use of electronic devices, heavy academic pressures in the context of the Chinese education system, and their parents' demanding work commitments and an inclination towards excessive protection (from crime, stranger danger, and traffic-related problems).

The good news is that in recent years, there is a growing recognition in China of the crucial importance and contribution of exposure to nature for children's development, and the Chinese government is developing a series of policies to improve the current situation with the aim of improving the lives of children (Wang et al., 2018). In March 2021, "the construction of child-friendly cities" was officially written into China's 14th Five-Year Plan, which put forward specific requirements for strengthening the construction of out-of-school activity venues, and improving public service facilities for children, to ensure their rights to survival, development, protection, and participation (The People's Republic of China, 2021). Child-friendly spaces in communities, schools, and public places are an essential part of the construction of child-friendly cities because of the wide distribution, rich variety, short renovation period, and strong public perceptibility.

Since 2021, Chinese authorities have implemented the "Double Reduction" policy, outlined in the guideline "Opinions on Further Reducing the Burden of Homework and Off-School Tutoring on Students Undergoing Compulsory Education" (Ministry of Education of the People's Republic of China, 2021) . This policy aims to alleviate the excessive academic burdens on school-aged students, and enhance school services to cater to diverse student needs. Simultaneously, the policy prohibits extracurricular tutoring institutions from organizing subject-based training during weekends, national holidays, and school vacations. National pilot cities, such as Beijing, Shanghai, and Guangzhou, are paving the way for the policy's nationwide implementation.

China is also making efforts to prevent children from becoming addicted to online games. The second revised version of the Law of the People's Republic of China on the Protection of Minors took effect on June 1st, 2021, prohibiting Internet products and service providers from providing products and services to children that induce addiction induce addiction to minors. Providers of online gaming, streaming, audio/video, and social media need to implement functions such as time limits, authorization and consumption controls for children. The National Press and Publication Administration of China (2021) has further imposed strict requirements and restrictions on real-name registration and login, as well as the hours of providing online game services to children, with all online game enterprises only allowed to provide services to minors for one hour from 20:00 to 21:00 daily on Fridays, Saturdays, Sundays and holidays. The study reported here aimed to understand the multifaceted factors influencing opportunities for children's engagement in outdoor activities and contact with nature in Beijing, China. We also investigated the effectiveness of governmental initiatives aimed at fostering such objectives. Through a comprehensive discussion of these factors and their interrelationships, the study endeavored to provide evidence-based recommendations and strategies to facilitate the reconnection of children with nature in the city. The implications for policymakers, educators, landscape architects and urban planners are underscored, highlighting the importance of creating opportunities for facilitating children's outdoor activities and exposure to nature while simultaneously minimizing barriers within the urban environment.

3 METHODOLOGY

3.1 Research area and population

This study was conducted in four core administrative districts in Beijing, namely Haidian, Chaoyang, Dongcheng, and Xicheng based on the degree of urbanization, level of economic development, and population density. We involved children living or attending school in these four districts, as well as their parents as our participants. Children in middle childhood, typically ranging from 8 to 12 years old, were the target respondents. Within this age range, children exhibit more independence and a natural curiosity about their environment, and have generally established familiarity with their immediate outdoor surroundings.

3.2 Data collection and analysis

We recruited participants through social media advertisements, hardcopy posters, and by utilizing a snowballing approach through our personal contacts. This initially involved firstly speaking to children's parents or guardians to obtain their consent before reaching out to the children to ask about their willingness to participate. The study protocol received approval from the Human Research Ethics Committee of the University of Melbourne (Reference No. 2021-22465-24141-3).

We initially collected demographic information from parent-child dyads, including details about the children's age, gender, parents' age, average household income, education level, etc. We then conducted in-depth interviews with parents and children, focusing on various aspects of the children's

https://doi.org/10.52202/082078-0018

outdoor activities and their interactions with nature (including where, when, what, how long, with whom), and the parents' opinions with regard to the effects of government policies, along with other influencing factors. To ensure confidentiality, each child-parent dyad was assigned a random number. Child participants were labelled as C, and parent participants were labelled as P, followed by their assigned numbers.

We employed the NVivo software for content analysis of open-ended data obtained from interviews. Open data coding was utilized to explore influencing factors related to children's outdoor activities and nature exposure. Subsequently, we revisited the interview content to further refine the coding process and identify underlying mechanisms.

4 RESULTS

Parents generally believe that their children did not have adequate opportunities to engage in outdoor activities and to have contact with nature. The time for outdoor experiences of children could be broadly categorized into school time, after-school hours, and weekends. During school hours, activities are confined to physical education classes, morning exercises, and recesses. Most schools prohibit outdoor activities during recesses to minimize the risk of accidents such as falls and potential injury. The nature of schools and their geographical constraints make it difficult to provide children with diverse, unstructured outdoor experiences, contrasting with their parents previous school experiences who, as one parent mentioned (P[62]) said, they typically "rushed out of classrooms as soon as the bell ring ... and this is actually a release of the state of emotion and entire body. I find it quite beneficial, but nowadays ... ".

The situation slightly improves after school and on weekends, when children might have the opportunity to engage in activities in nearby residential areas, parks, and even suburbs. These activities span from swinging on swings to fishing and hiking. However, these activities were influenced by various factors. Some parents mentioned the impact of children's personalities. For instance, some children preferred to have quiet time at home rather than engaging in outdoor activities, though such cases were not common. The majority mentioned external factors, such as parental guidance, available infrastructure, with academic pressure being the most common. We categorize these external influencing factors into the following three aspects – time and companionship, accessibility to natural environmental infrastructure and services, and accidental factors.

4.1 Time and companionship

3.1.1 Academic pressure

Academic pressure was the most frequently mentioned factor affecting the scheduling of children's outdoor activities and opportunities for nature experience. Despite the abundance of high-quality educational resources in Beijing, their uneven distribution intensifies competition due to the substantial population. The Chinese education system emphasizes exams and grades, and students face frequent assessments, such as the entrance examinations for middle school and college. Current primary school students also face a heavier workload in subjects compared to previous generations, let alone the demands and competitive pressure coming from schools, teachers, and peers. However, children's academic pressures likely stem from greater parental expectations, with many extracurricular courses being arranged for them. In addition to academic tutoring, many children are also involved in extracurricular activities such as sports and arts (e.g., playing the piano, dancing), further restricting their schedules and time to be outdoors.

As previously mentioned, the Chinese government has implemented the "Double Reduction" policy, and the interviews suggested that parents feel this policy has achieved some success, particularly in reducing pressure from school and placing restrictions on extracurricular classes, thereby allowing for more discretionary time. For example, P[06] and P[18], respectively, said that:

"Before the implementation of the 'Double Reduction' policy, she [her child] had to return home first to complete her homework. Now she can play for a while after school."

"After the 'Double Reduction' policy, children might have a bit more relaxed schedule ... It does have an impact on outdoor activities. Previously, after school at 3:30 pm, teachers might still give lectures and assign homework. Now, this is not allowed, and has turned into playtime, such as handicraft classes, physical exercises, etc. The 'Double Reduction' policy has increased opportunities for physical and extracurricular activities."

However, there are still many practical problems, mainly manifesting in higher demands on parents' time and efforts. For example, P[23] and P[23], respectively, said that:

"The kid's academic pressure has reduced, but for me, the pressure is still heavy because, after the policy, the kids come home early, have less homework, which means more parental supervision needed. After they come home, I have to pay constant attention to them because I'm worried about other unforeseen events."

"The biggest impact on us is that both my spouse and I feel an increased burden. The work

https://doi.org/10.52202/082078-0018

commitments and frequent business trips limit our time to dedicate to her study. Although she is in primary school, there are many things that are different from what we learned back then ... Before the policy, we outsourced these things to professionals, but, we can't find professional people to do these things now. In fact, the demands for us have become higher, requiring greater time investment and continuous learning – a considerable challenge. The shift has brought about increased responsibilities and challenges for dual-income families like us."

Many parents mentioned that the most fundamental reason is that the university admission policy with respect to selection and examination has not changed. Education is a long-term process, and the implementation of a policy may not be able to change parental mindsets in a short time. As P[88] mentioned: "Most parents have achieved their current status through education and examinations, so they hope their children can secure their future through education. They fear their children falling behind. However, parents are also quite confused because, under the influence of policies, they are uncertain about what the future holds for their children. There is a lot of anxiety among families in our surroundings".

Therefore, the pressure from parents has not decreased, and these anxious parents still require homework and tutoring for their children through various channels. For example, P[35] suggested that:

"After the 'Double Reduction', children have even less outdoor play time. There's less school homework, but after those extracurricular institutions were closed, we switched to one-on-one tutoring and online classes. She used to attend extracurricular classes on Saturdays and Sundays, but it's no longer possible, so we can only move them to Monday to Friday evenings, which reduces her play time during weekdays. She also has sports and art classes on weekends. So overall, the time for outdoor activities may increase in school, but not after school."

Therefore, for some children, their academic pressures have not lightened but merely shifted from weekends to weekdays, and the available time for outdoor activities or contact with nature has not increased. It became evident from the interviews that the most crucial aspect is firstly changing the parents' mindset, however, achieving this is not an overnight task and would require further policy support to alleviate parental anxiety concerning their children's future.

3.1.2 Parental guidance

The outdoor activities of children in this age group are largely guided by their parents, mainly out of safety concerns. Parents frequently mentioned that they generally are very willing to support their children's outdoor activities and contact with nature. However, the actual implementation is influenced by the parents' personalities, lifestyles, beliefs, and time constraints. If parents are willing to engage in outdoor activities or have a fondness for nature, they are more likely to actively take their children outdoors and involve them in such activities. One mother P[45], who is a nature enthusiast, stated that:

"I love animals and plants, so when I take my kid out, I try to let her observe some animals and plants. For example, yesterday when we were playing outside, we saw many spider webs – large and small, intricately woven. We stood there and watched for a long time, studying and observing them. A few days ago, we went on a trip, and the hotel was next to a resort with lots of plants. At night, I took her with a flashlight provided by the hotel, one for each of us, into the woods to find cicadas. There were also many hedgehogs in the small woods, and I would let her observe them. Even though we got bitten by bugs, I thought it's worth it. We saw these things in books, on iPads, or on TikTok, but when it comes to what they are really like, how fast they crawl, and how they forage, she wouldn't know unless she sees it for herself."

This mother consciously guides her child to connect with nature, observe the natural world, and cultivates her child's interest in nature. She also inspires the child to think about nature and ask questions:

"Behind the resort, there is a beach with no stones or shells, very beautiful, and it's a popular place for social media check-ins. However, I don't think it's useful because I feel like the beach has been artificially developed. Even though the sand is very fine, the child can't find what she wants. I deliberately drove to many different beaches for her to observe – some with rocks, some with small stones, and some with mud. She found it very exciting and kept asking me questions like 'what is this?', 'why is there so much mud?', and we analyzed it together ... I have also followed many bloggers focusing on animal or science on TikTok, and she might come across their works while browsing, and she watches these people's content too."

On the contrary, if parents are not interested or unwilling to go outdoors, their children's opportunities to connect with nature may be reduced. For instance, P[27] mentioned that:

"Those parents who are outgoing, sociable, and enjoy making friends, may organize outings with other families who have children of similar ages. However, I am a homebody, so compared to other parents or children, we do relatively less outdoor activities."

In addition to the lifestyle, beliefs, and actions of parents, their time management is also crucial, particularly in dual-income households. Many parents express their willingness to take their children

https://doi.org/10.52202/082078-0018

outdoors but found it challenging due to work commitments. Three parents mentioned challenges related to having a second child, especially when caring for a younger child that demands constant attention. Balancing the demands of children requires parents to allocate their time more judiciously. Planning and executing outdoor activities demand extra effort and coordination, which, in turn, may lead to time constraints and impact the time originally designated for outdoor pursuits. Some families have grandparents assisting in childcare, but this may also pose limitations, such as what P[23] mentioned: "Grandparents may help look after the child, but if they play with the child for a day, they will be exhausted and they will be tired from taking care of the children, which will inevitably hinder the child's play time."

3.1.3 Peer influence

Based on the interviews, common interest and companionship were the two ways that peers influence children's outdoor activities and contact with nature.

Common interest. Having a common interest in outdoor activities shared among friends can motivate children to explore nature together, which make the outdoor experience more enjoyable and fulfilling. These positive experiences, together with shared interests, were identified as the most consistent sources of motivation for continuous engagement in nature-based outdoor activities. Among the participants, there were two children who shared a passion for birdwatching. They joined the volunteer organization "Friends of Nature" and visited different parks to observe and record various bird species almost every weekend. It is impressive that they are able to identify dozens of common bird species and describe their physical characteristics in detail. Some outdoor activities also involve teamwork, and shared interests among friends can lead to collaborative efforts, such as in building a treehouse, which can enhance the overall experience. As P[81] mentioned: "The group of people you often play with, do they like to go into nature, or are your playmates willing to involve you in going together? This is very important. Recently, there was an occasion when some moms of her male classmates asked if we wanted to go hiking together during the Qingming Festival ... The children went to a secret spot in the park that they are building together. It's a treehouse constructed with wooden sticks and willow branches for the first layer. Whenever someone has time, they would gather there to build and expand, creating a secret base. Now, it has reached the second layer."

Even if children have not yet developed a common interest, they frequently emulate the behaviour of their peers. This can be out of curiosity, or part of socialization. Just as playing the same games or buying similar toys, these experiences provide them with common topics for communication. If a child's friends express enthusiasm for nature-related activities, the child is likely to show interest in similar activities. For example, P[22] said:

"Maybe I will take my child camping recently because she heard from other classmates that camping was very interesting, and she really wants to go."

When interacting with peers, children have the opportunity to learn from one another, and there might be instances where they are driven by a friendly competitive spirit. The desire to keep up with or surpass their peers is evident, and this dynamic extends to outdoor activities, as highlighted in the experiences shared by C[81], a 10-year-old girl:

"When we went hiking, they didn't like walking on paved paths; instead, they chose to climb through wild trails, and I followed along. Sometimes, when I struggled to climb over rocks, they offered to help, but I managed to jump over on my own at the last second."

This peer influence is not all positive. If friends like to stay at home, play video games or use electronic devices, then they may also develop such tendencies. For example, P[48] said that:

"When we were out on a trip with other families, we wanted the children to get out of the car and to see and explore the grassland, but it was evident that they weren't very interested. They seemed happiest when sitting in the car playing video games than enjoying the scenery along the way."

Peer companionship. Children may simply need the company of their peers. They often enjoy certain experiences more when they can share it with friends. This may have nothing to do with the environment they are in or the activities they are engaged. Whether it is hiking, biking, or simply walking in a park, the activities can become more attractive when friends are involved. This is reflected in what P[33] and P[79] respectively said:

"Sometimes when parents take their children to outdoor activities, the children may not be very interested, but if they have peers as companions, they would be much more motivated".

"She plays in the neighborhood, mainly with a few close friends. The equipment for fitness or entertainment purposes in the neighborhood are not really appealing to her. They may just play the games they like. However, sometimes they can't meet each other if they have extracurricular tutoring courses."

P[73] shared an interesting but somewhat sad story, that although there was a very scenic park nearby, her child preferred to play in an extremely small space in the neighborhood just because there are kids there to play with:

https://doi.org/10.52202/082078-0018

"My child is a little fat and doesn't really enjoy structured exercises. He prefers simple activities like running around with other kids. Even something as basic as kicking stones with friends can keep him engaged for a long time ... He would rather go back to the confined space in our neighborhood because there are other children there. Although the Youth Lake Park is larger with a great natural environment, he wouldn't choose it because there are no familiar playmates. He found it less interesting when there were no children to play with, so he chose to go home and play video games."

In fact, this desire for companionship may extend beyond peers and parents. P[22] shared an interesting story about a popular dad in her neighborhood:
"There is a dad in our neighborhood who is a sports enthusiast, and all the children like him very much. Many older boys even wait for him to come home from work every day. He genuinely enjoys the company of children and forms a great bond with them. In this process, you can observe that children are influenced by him. They tend to enjoy spending time outdoors engaging in games or fitness activities. My child initially preferred staying at home, reading, and didn't know what to do outdoors alone. However, he can motivate her, organize activities, and arrange games, such as playing hide and seek ... I believe that having such a group or the presence of an organized leader within the group can have a greater impact on children's outdoor activities. While a child might find interesting places on their own, many times, they don't know what to play. In such situations, having a playmate or someone more experienced to guide them makes the experience more enjoyable for them."

However, the lack of peer companionship has increasingly become a serious problem, as reported by many parents, directly influencing children's willingness to engage in outdoor activities. Parents also mentioned reasons such as a lack of age-appropriate partners around their residence and misaligned schedules among peers. The lack of peers around the residence is mainly a distance issue and may primarily affect outdoor activities after school. However, a mismatch of timing is a much bigger issue that children are currently facing. Primarily, this is because children have less discretionary time, with most of it taken up by extracurricular academic courses or organized sports programs; coupled with parents' tight working schedules, leaving children with limited time and opportunity to be with their peers outside of school. As P[55] and P[85] respectively said:

"Another reason why children are less active outside is that they don't have companions to play with. It may be a little difficult for children of different ages to blend in. He used to really like to go out and play because he had a lot of friends, but now many of them have moved out, and others might not go out because of their studies. The children in the neighborhood may not be the same age as him, and he can't keep up with them. That's why he's gradually gotten less inclined to go out."
"As he's getting older and higher in grade, children his age are playing outdoors less frequently. It's challenging for him to find peers of the same age to play with."

3.1.4 Extra screen time

The escalating use of electronic devices has become a serious concern, occupying the already limited discretionary time of children. This situation could be partly explained by the inherent attractiveness of electronic products, such as immersive video games and engaging social media content. The lack of peer companionship, as mentioned earlier, can also divert children's attention from outdoor activities to electronic devices. For some parents who require extra working hours, like P[09]:

"Sadly, I allow my child to spend a relatively long time on the screen especially when I need to work, because having something occupying her attention allows me to focus on my work."

The trend has been further exacerbated by home-based learning during the COVID pandemic. Nearly all school classes were conducted online during the lockdowns, leading to a sharp increase in children's daily screen time, let alone the fact that many extracurricular activities had also been transitioned to online platforms. Many parents reported a noticeable decline in their children's eyesight, possibly a consequence of a sedentary lifestyle, along with adversely impacting their self-discipline. As P[52] expressed:

"Children are not very good at self-control. Some children are truly getting addicted, and they secretly play in their rooms and end up skipping online classes."

As previously mentioned, the Chinese government is making efforts to prevent children from becoming addicted to online games, streaming, audio/video content, and social media, with time limits, authorization and consumption controls implemented. Some parents acknowledged that these measures have somewhat controlled their children's dependence on electronic devices, but not as significantly as was anticipated. Children can find ways to bypass these regulations, such as using their parents' accounts to log in, and there are currently no clear regulations addressing content on social media. One parent, P[55], expressed concerns about the negative impact of the internet on her child's reading skills, cognitive abilities, and development, stating:

"Children can't understand certain things at their age, and it's not always clear-cut right or

https://doi.org/10.52202/082078-0018

wrong. Some online influencers are more concerned about click rates, so what they create might mislead kids, and sometimes explaining to our children has limited effect. Moreover, it affects their interest in reading physical books … Even though we have many books at home, children can't get into reading them. They feel bored. Short videos on screens have sound, quick images, many funny commentary tracks, and the option for increased playback speed. This elevates their attention to a high threshold. When we try to redirect their focus to simple text, it becomes challenging. The child may insist on having someone next to them, requiring constant company or negotiation – asking for a favor in exchange for reading. Consequently, parents end up spending a significant amount of money and effort on enrolling them in various activities to divert their attention from the screen. During this summer vacation, we spent a lot on basketball classes, art lessons, and camp programs in an attempt to reduce his online obsession, all of which are quite expensive."

The majority of parents stated that the time and manner in which their children use electronic devices still heavily relied on parental supervision, including setting specific rules for children's screen time, or trying to shift their obsession through other activities, though this can potentially increase the financial burden on parents.

4.2 Accessibility to natural environmental infrastructure and services

3.2.1 Available natural (green) spaces

Opportunities for outdoor nature experiences for children growing up in cities are primarily concentrated in various community-level, city-level, and national-level parks, with occasional access to suburban areas. Beijing boasts a rich array of national parks with natural and historical significance, such as the Summer Palace, the National Botanical Garden, and the Olympic Forest Park. These parks offer beautiful scenery and vast expanses; the suburbs of Beijing are mountainous, all of which provide numerous opportunities for children to engage in outdoor nature experiences. As P[33] said:

"Beijing is surrounded by mountainous areas. Within just an hour's drive, and sometimes even half an hour, we can reach the mountains, providing convenient access to nature. Additionally, there are some well-maintained parks within the city area that offer good environments. I believe these resources are quite beneficial."

However, some parents highlighted the lack of accessible natural spaces, with a focus on the accessibility. The high population density in urban areas contributes to significant transportation challenges, prompting parents to carefully plan routes and timing, especially during weekends when most parents and children have available time. In contrast, community green spaces are generally considered more accessible, though these spaces are often few in number, small in size, or of low quality, with many followed a kit-fence-carpet pattern (consisting of plastic or metal recreational equipment, rubberized flooring, and fencing) (H. Woolley, 2007). For instance, P[38] and P[73] respectively mentioned that:

"Within a five-minute walk from our home, there is no decent park where children can play. This is because the high building density and lack of reserved space."

"Our residence isn't exactly a typical community; it's just two rather isolated buildings. There's a tiny space for recreation, with a tiny wooden pavilion. Apparently, a compassionate parent, finding it quite dull, generously donated a basketball hoop. However, there's only about three steps from the outside iron fence to this basketball hoop. If we wanted to play badminton, we'd likely obstruct the pedestrian path, annoying passersby. Moreover, our shuttlecocks would either end up stuck in the trees or tangled in the nearby bushes."

Many parents express concerns about the safety of children's activity areas within residential areas, primarily focusing on traffic issues and worries about potential injuries or encounters with strangers. For instance, P[81] mentioned:

"I'm particularly worried about her crossing roads or getting hit while playing in the skateboard park; I think it's a serious issue. I might feel somewhat reassured when there is an adult present, or when she plays in our enclosed community. However, if she wants to play in the park across the street, even if it involves crossing just one road, I still have concerns. Additionally, regarding trust in people, I educate my child to be sincere and friendly towards others, but at the same time, she needs to maintain a certain level of vigilance. Overall, safety is our top priority when considering these factors."

The Chinese government is actively promoting the construction of child-friendly cities to create suitable environments and services for the growth and development of children. In terms of landscape architecture and urban planning, the focus is on making urban public spaces child-friendly (such as communities, schools, hospitals, and parks), optimizing the planning and construction of sidewalks and bike lanes to ensure the safety of children, providing children with environments rich in natural and ecological features as well as recreational facilities. Some parents have noticed these changes, primarily due to the construction of many "micro spaces" – community parks or pocket parks around residential areas, as commented on by P[09]:

"Looking at a longer timeframe, such as

https://doi.org/10.52202/082078-0018

five years, we can observe changes in Beijing. A noticeable improvement is the cleanliness and utility of roadside parks. While walking or accompanying children to extracurricular courses, there are always places where children can play for a while."

However, providing suitable places for children to engage in outdoor activities and connect with nature remains a formidable challenge. Some parents mentioned certain behavioral guidelines in urban parks widely endorsed by families, schools, and society, such as avoiding trampling on lawns, picking flowers, and breaking tree branches. Many vegetated areas are enclosed by low fences, discouraging children from entering. It is partly because of the application of insecticides in certain green spaces, which may pose potential health risks to children. Another reason lies in large population base, and the cost of maintaining vegetated landscapes of high quality. This may be rooted in a collectivist Chinese cultural background, where nature is meant for everyone to appreciate. This, to some extent, limits the ways in which children can closely interact with nature, though these guidelines can be somewhat ambiguous in suburban parks and rural areas, leading to educational and moral dilemmas. As P[55] said:

"In the park, activities such as climbing trees or picking flowers are not encouraged and is usually restricted by park officials. It is not likely to have a close encounter; everything is within the fences – only for distant views. Simple touching of plants is allowed, but running, playing, and rolling on the lawn are definitely out of the question, let alone climbing trees. This is the case in city parks. Therefore, sometimes we choose to go to suburban areas. As parents, we often find ourselves in a dilemma. For example, when we see our child picking flowers, should we correct them? Should we criticize their behavior as wrong? Sometimes, he asks me, 'Mom, didn't you tell me not to pick flowers? Why can others do it? What exactly can I do and what not?' At this age, he might experience some internal conflicts ... Sometimes, park officials criticize the way parents educate their children if their children don't act as expected, so we feel double pressure. I remember when we went hiking, we might take a shortcut and step on the grass if nobody was watching, but as soon as someone intervened, we quickly came down. It feels like we are doing something wrong, because we are taking advantage of loopholes. This feeling is unpleasant, as if we are teaching our children to exploit loopholes, leading to conflicts in terms of awareness of rules."

Over the past three years and during the COVID pandemic, camping has become increasingly popular. Many parks have also designated areas specifically for camping to cater to this changing demand. However, there is still a long way to go.

3.2.2 Organized events

Just like the popular dad mentioned earlier, children also enjoy collective activities. Opportunities for children to participate in organized activities are mainly concentrated in schools and various for-profit and non-profit organizations, and nature education institutions. For instance, in schools, children's outdoor activities often focus on teacher-led physical exercise, such as running, rope-jumping, and various ball games, with their contact with nature often passive through visual means. Some children mentioned planting and picking activities organized by schools, and some schools even allocate areas for planting within the schoolyards, allowing children to participate in cultivating, caring for, and harvesting of plants. As C[14] (a 10-year-old boy) mentioned, a picking festival organized by teachers:

"There are persimmon trees in the school, and the teachers organize a picking festival every year. They put the ladders in place and let us pick the persimmons ourselves, large persimmons. Two representatives from each class are sent to pick them."

It is worth noting that schools organize annual spring and autumn excursions, taking children to both indoor and outdoor places like botanical gardens, zoos, the Summer Palace, or museums. Children expressed deep impressions of these collective activities, but due to the pandemic, these school-organized activities, in recent years, have been suspended. As C[35] (a 11-year-old girl) said:

"The last time was at the botanical garden, and we saw dandelions. Inside the greenhouse, the teacher explained things like tropical rainforests, what could be touched, what was poisonous, and should not be touched. Then we could explore freely on our own ... I enjoyed spring excursions very much, but due to the pandemic, those experiences are no longer possible."

Organized outdoor activities beyond the school mainly include physical sports, such as football, basketball, and volleyball, organized by for-profit organizations. Participating in these sports activities promotes physical health, but it is also related to entrance examination for middle school – children are evaluated not only on academic achievements but also on their performance in physical education exams, such as running (800-meter for girls and 1000-meter for boys), basketball, and volleyball. Some also mentioned nature-based activities, such as river rafting and camping activities held by profit-driven organizations. For example, P[15] shared an experience of his child participating in a three-day summer camp:

"Even though it's been a long time since they participated in an organized summer camp, lasting about three days, their impression of it remains exceptionally vivid. They still remember how the

https://doi.org/10.52202/082078-0018

teacher guided them in kayaking, teaching them how to paddle, tie knots with ropes, climb mountains, and identify plants."

Some parents noted that the fees for taking part in these nature experiential activities are often too high to afford, and these organizations are not yet mature in their development, with varying quality. They found it challenging to discern the good from the bad, leading to concerns about safety considering the child's age and self-management capabilities, which is why many parents hesitate to involve their children in these activities. In comparison, nature education activities can be organized by both for-profit and non-profit entities. For instance, there is a volunteer organization called "Friend of Nature" mentioned by some parents and children. This organization has many branches, each with specific themes. Some focus on bird observation, others on plant identification, and there are those centered around monitoring river pollution or conducting parent-child activities. These branch groups organize various nature-themed activities almost every week, providing the public with opportunities to connect with and understand nature. Additionally, as P[86] mentioned, they can find families with similar parenting beliefs within these organizations, and children can also make friends who share similar interests. However, due to the nonprofit nature of these organizations, their activities have participant limitations, and their visibility and promotional methods may not be sufficient. As a result, many families are not familiar with them.

4.3 Accidental factors

3.3.1 Climate and weather

In addition to the above mentioned influencing factors, some parents also pointed out the impact of climate and weather. For instance, P[15] remarked:

"Children generally only play in the residential area during seasons other than winter. In winter, there are no outdoor activities once they come back from school."

In hot summer weather, children also avoid outdoor activities, "further reducing the time spent outdoors", as P[30] pointed out. School activities are also influenced by the weather, as P[73] mentioned:

"Many schools do not allow children to go to the outdoor playground during recess. They spend almost the entire day in the classroom, except for physical education classes and exercise sessions, but they are not allowed to go outside if there is severe haze or rainy weather."

It is reasonable that favorable weather conditions can support outdoor activities and adventure, while inclement weather may pose safety concerns. However, some climate and weather conditions can be transformed into opportunities, such as rainy days which might inspire children to splash in puddles and discover the sensory delights of raindrops, and children can and should be allowed to experience and explore such conditions. As C[30] (a 10-year-old girl) stated:

"I usually don't go out in winter unless it's snowing. I like snow because it's fun; I can build snowmen."

3.3.2 The COVID-19 pandemic

Unlike climate and weather, the outbreak of the COVID-19 pandemic was basically unpredictable and had a substantial influence. Many parents mentioned adverse impacts on their children's physical well-being, particularly in terms of vision impairment and weight gain, due to a more sedentary lifestyle resulting from reduced time outdoors and increased screen exposure. Moreover, some children reported experiencing psychological impacts.

During the pandemic, there was a notable increase in the utilization of green spaces, primarily community parks, within the neighborhood. These spaces emerged as crucial hubs for outdoor activities, taking on heightened importance as many public parks remained closed. The increased reliance on local residential parks underscored their importance for individuals seeking a respite from indoor confinement. Beyond the immediate residential areas, children missed out on broader opportunities, such as school-sponsored excursions, inter-provincial or cross-border travels, denying them culturally, environmentally, and socially enriching experiences.

Nevertheless, for certain families, the transition to remote learning and working from home offered increased flexibility in scheduling and the chance to spend more time outdoors in close proximity to their homes. Exploring local environments and culture, as well as new outdoor nature experiences like camping, had become increasingly popular. As P[33] stated:

"Since we couldn't allocate time to visit distant tourist destinations, activities like hiking in nearby mountains and exploring the suburbs have become more frequent. Considering the current situation with the pandemic and policies in Beijing, outdoor camping has become particularly popular, in turn leading to a rise in ways to explore local natural attractions".

5 DISCUSSION

The above findings indicate that children's outdoor nature experiences are influenced by various factors. Apart from accidental factors that are somewhat unpredictable, the remaining two factors – time and companionship, accessibility to natural environmental infrastructure and services, are interrelated. For instance, children's involvement in extracurricular

https://doi.org/10.52202/082078-0018

classes can diminish their free, unstructured time and restrict opportunities for socializing with peers, potentially steering them towards increased use of electronic devices. Moreover, the absence of easily accessible green spaces might require parental supervision due to safety considerations, meaning children's outdoor activities significantly hinge on parental scheduling, which exacerbates the constraints on their discretionary time and chances of forming steady play companionships. Following the examination of the outcomes, we present two directions to tackle the challenges arising from these three factors, which relies on collective support from diverse stakeholders, encompassing efforts from both top-down and bottom-up approaches.

The first direction lies in an effort to provide children with greater access and more green spaces in which they can engage in outdoor nature experiences. The Chinese government has launched several initiatives toward achieving this goal, including the "child-friendly city" initiative mentioned above. Notably, the task of making green spaces accessible, safe, attractive, and enjoyable necessitates landscape designers and urban planners to assume vital roles in this endeavor. Accessible natural environments within urban settings, particularly community green spaces, are frequently associated with reduced commuting distances and time, which helps to ease the parents' concerns about their children's safety and offer more flexibility in scheduling time for both parents and children to visit these places. This, in turn, provides children with greater room for independent mobility. Such easily accessible green spaces can also act as a protective buffer against various accidental factors.

Due to safety considerations and other factors, many play spaces for children in residential areas are enclosed by fences. Woolley (2007) developed a Kit-Fence-Carpet (KFC) style for such spaces, Kit of fixed equipment, Fence to keep children in, and Carpet of rubber surfacing, which is quite common in urban settings but there is doubt about whether this pattern is genuinely the most preferred and beneficial for children. Subsequent studies have consistently validated children's inclination towards natural environments, a phenomenon elucidated by various theories, such as the theory of affordance and the biophilia hypothesis (Heft, 1988; Wang et al., 2018; Wilson, 1993). The former highlights how nature can offer support for a variety of children's activities, such as the presence of loose parts – movable natural elements which support creative and imaginative play, and provide children with opportunities to interact with and comprehend their environment (H. Woolley & Lowe, 2013). The latter articulates the innate connection humans have with nature, or biophilia, emphasizing its role as soft fascination and its influence on children's psychological restoration, even if experienced solely through visual means (Kaplan, 1995; Ulrich, 1984).

It needs to be mindful to incorporate more natural elements into the open spaces. It is worth noting that the elements of nature that are included should not merely serve as a visual backdrop enclosed within fences, akin to a display in a showcase. Instead, designated spaces should be reserved for children to have direct encounters with elements of the natural world, such as trees and other forms of vegetation, while also meeting both park regulations and reflecting collective cultural preferences. Stephen Kellert and Elizabeth Calabrese developed a framework for biophilic design which constitutes three kinds of experience of nature, namely the direct experience of nature (e.g., plants, animals, water), the indirect experience of nature (natural replacements, such as photos and videos of nature), and the experience of space and place (Kellert & Calabrese, 2001).

In addition, these environmental settings can serve as venues for organized events and natural educational programs. During our interviews with parents, a common concern emerged regarding the quality of organized nature-based events and the relatively high fees associated with these activities. It is essential to have government-level policy guidance and financial allocations, community-level support and services, as well as active collaboration with volunteer organizations. This could further provide equal opportunities for children to actively engage in nature-based outdoor activities and also for them to find companions, which is likely to lead to a reduction in reliance on electronic devices and similar distractions.

Beyond overcoming of physical barriers to children's outdoor activities and nature exposure, a shift in mindset, particularly among parents, is needed. While the majority of parents acknowledge that outdoor nature-based activities are good for children's development, there exists a lack of comprehensive understanding of these benefits. Some parents primarily associate outdoor activities with physical health, opting to schedule structured sports programs for their children. Consequently, this practice does not allow the children ample free time to actively and creatively engage with nature. Moreover, many parents hold high expectations for their children's academic achievements. Although the government has implemented certain policies, such as the "Double Reduction" policy as previously discussed, the effectiveness of these measures may be limited if parental attitudes do not undergo a significant transformation. It is understandable that parents' concerns are deeply rooted in a desire for their children to have a prosperous future, especially

https://doi.org/10.52202/082078-0018

in a highly competitive environment. Addressing these concerns requires corresponding policy guidance from the government (e.g., gradually shifting away from exam-centric selection systems and support of employment), awareness campaigns and educational programs targeting parents. However, it takes time for such a change in mindset to take place.

6 CONCLUSIONS

This study identified multifaceted factors influencing opportunities for children's engagement in outdoor activities and contact with nature through in-depth interviews with 81 parent-child dyads living in Beijing, China. We also investigated the effectiveness of governmental initiatives aimed at fostering such connections, such as the "Double Reduction" policy and the "child-friendly city" initiative. The influencing factors extend beyond the child's inherent personality and include three main external factors – time and companionship, accessibility to natural environmental infrastructure and services, and accidental factors. We present two directions to tackle the challenges of providing greater opportunities for children to access nature in the city arising from these three factors. These include provision of more accessible green spaces (especially within communities) and a shift in (parental) mindsets, which relies on collective support from diverse stakeholders of policymakers, educators, landscape architects and urban planners, employing efforts from both top-down and bottom-up approaches.

7 REFERENCES

Bagot, K. L., Allen, F. C. L., & Toukhsati, S. (2015). Perceived restorativeness of children's school playground environments: Nature, playground features and play period experiences. Journal of Environmental Psychology, 41, 1–9. https://doi.org/10.1016/j.jenvp.2014.11.005

Fjørtoft, I., & Sageie, J. (2000). The natural environment as a playground for children. Landscape description and analyses of a natural playscape. Landscape and Urban Planning, 48(1–2), 83–97. https://doi.org/10.1016/S0169-2046(00)00045-1

Heft, H. (1988). Affordances of children's environments: A functional approach to environmental description. Children's Environments Quarterly, 5, 29–37.

Kaplan, S. (1995). The restorative benefits of nature: Toward an integrative framework. Journal of Environmental Psychology, 15(3), 169–182. https://doi.org/10.1016/0272-4944(95)90001-2

Kellert, S. R., & Calabrese, E. F. (2001). The Practice of Biophilic Design (Issue October).

Laaksoharju, T., Rappe, E., & Kaivola, T. (2012). Garden affordances for social learning, play, and for building nature-child relationship. Urban Forestry and Urban Greening, 11(2), 195–203. https://doi.org/10.1016/j.ufug.2012.01.003

Liu, J., & Green, R. J. (2023). The effect of exposure to nature on children's psychological well-being: A systematic review of the literature. Urban Forestry and Urban Greening, 81(August 2022), 127846. https://doi.org/10.1016/j.ufug.2023.127846

Louv, R. (2008). Last Child in the Woods: Saving Our Children From Nature-Deficit Disorder (Updated and Expanded). Algonquin Books of Chapel Hill.

Ministry of Education of the people's republic of China (2021). Ease the burdens of excessive homework and off campus turoring for students undergoing compulsory education. (n.d.). Retrieved August 20, 2021 from Http://Www.Gov.Cn/Xinwen/2021-07/24/Content_5627132.Html.

National Press and Publication Administration of China (2021). Notice on Further Strict Management to Effectively Prevent Minors from Being Addicted to Online Games. Retrived August 30,2021 from http://www.nppa.gov.cn/nppa/contents/279/98792.shtml. (n.d.).

Raney, M. A., Hendry, C. F., & Yee, S. A. (2019). Physical Activity and Social Behaviors of Urban Children in Green Playgrounds. American Journal of Preventive Medicine, 56(4), 522–529. https://doi.org/10.1016/j.amepre.2018.11.004

Taylor, A. F., & Kuo, F. E. (2009). Children With Attention Deficits Concentrate Better After Walk in the Park. Journal of Attention Disorders, 12(5), 402–409.

Taylor, A. F., Kuo, F. E., & Sullivan, W. C. (2001). Coping with ADD: the surprising connection to green play settings. Environment and Behavior, 33(1), 54–77.

Taylor, A. F., Kuo, F. E., & Sullivan, W. C. (2002). Views of nature and self-discipline: Evidence from inner city children. Journal of Environmental Psychology, 22, 49–63. https://doi.org/10.1006/jevp.2001.0241

The people's republic of China (2021).The 14th Five-Year Plan for National Economic and Social Development and the Long-Range Objectives Through the Year 2035. (n.d.). Retrieved August 20, 2021 from Http://Www.Gov.Cn/Xinwen/2021-03/13/Content_5592681.Html.

Tim Gill. (2014). The Benefits of Children's Engagement

https://doi.org/10.52202/082078-0018

with Nature: A Systematic Literature Review. Children, Youth and Environments, 24(2), 10. https://doi.org/10.7721/chilyoutenvi.24.2.0010

Ulrich, R. S. (1984). View through a window may influence recovery from surgery. Science, 224(4647), 420–421. https://doi.org/10.1126/science.6143402
Wang, X., Woolley, H., Tang, Y., Liu, H. yi, & Luo, Y. (2018). Young children's and adults' perceptions of natural play spaces: A case study of Chengdu, southwestern China. Cities, 72(August 2017), 173–180. https://doi.org/10.1016/j.cities.2017.08.011

Wilson, E. O. (1993). The Biophilia Hypothesis. In E. O. W. Stephen R. Kellert (Ed.), Biophilia and the conservation ethic.

Woolley, H. (2007). Where do the children play? How policies can influence practice. Proceedings of the Institution of Civil Engineers: Municipal Engineer, 160(2), 89–95. https://doi.org/10.1680/muen.2007.160.2.89

Woolley, H. E., & Griffin, E. (2015). Decreasing experiences of home range, outdoor spaces, activities and companions: changes across three generations in Sheffield in north England. Children's Geographies, 13(6), 677–691. https://doi.org/10.1080/14733285.2014.952186

Woolley, H., & Lowe, A. (2013). Exploring the Relationship between Design Approach and Play Value of Outdoor Play Spaces. Landscape Research, 38(1), 53–74. https://doi.org/10.1080/01426397.2011.640432

https://doi.org/10.52202/082078-0018

TOWARDS HUMAN-CENTERED HISTORIC PRESERVATION THROUGH THE MAKING OF PLACE: A SYSTEMATIC REVIEW

Track 4: Design for Health and Wellbeing
Knowledge Network: Active Living by Design, Health
Methods: Case Study, Statistical/Modeling Analysis
Dr. Chunyao Liu, Ph.D, IDEC, LEED GA, and Ms. Huabing Yu, LEED GA, Arizona State University, Tempe, AZ

1 ABSTRACT
Since the enactment of the National Historic Preservation Act in 1966, historic preservation practice in the United States has been significantly shaped by governmental regulations and financial incentives. A proliferation of market-driven preservation projects emerged as tools for economic development, adhering to official guidelines that predominantly emphasize the preservation of physical elements. This emphasis, however, has led contemporary preservation practices to often overlook the sentimental aspects that involve individuals' emotional bonding, everyday experiences, and perceptions of historic places.

The sentimental aspect within the built environment is encapsulated in well-established concepts related to "place" within the field of environmental psychology, including place identity (PI), place attachment (PA), placemaking (PM), and sense of place (SoP). This systematic literature review explores the intersection of two professions — environmental psychology and historic preservation. It examines 116 manuscripts related to both place-related concepts (PA, PI, PM, and SoP) and historic preservation within the context of built environment published from 1980 till 2023, followed by an in-depth analysis of 30 selected manuscripts that investigate the relationships between place-related concepts and historic preservation practice. The findings of this paper identify topical research areas, research gaps, frequently used methodologies, and most importantly, the intricate relationships between "place" and "historic preservation." Finally, this paper underscores the potential for interdisciplinary research bridging environmental psychology and historic preservation, offering vital insights for architects, designers, and preservationists to contribute to more human-centered historic preservation projects through the making of place.

1.1 Keywords
Place identity; Place attachment; Historic preservation; Sense of place; Literature review

2 INTRODUCTION
Historic preservation, a discipline rooted in the intrinsic human desire for security and connection to roots (Powell, 1999), has undergone a transformative journey from its early beginnings in the mid-19th century to its current state both in theory and practice (Minner, 2016). Initially driven by patriotic sentiments, preservation efforts in the United States, exemplified by iconic house museums like George Washington's House in Mount Vernon, were intimately tied to the sentimental aspects of preserving historical places (Tyler et al, 2009). However, with the institutionalization of the National Historic Preservation Act of 1966 and the establishment of the National Register of Historic Places, a shift occurred. Preservation projects, once driven by emotional ties and individual experiences, have increasingly adopted top-down approaches, emphasizing physical characteristics over sentimental connections. Moreover, market-driven initiatives, spurred by legislative acts like the Tax Reform Acts of 1976 and 1986, further propelled historic preservation into the realm of economic development (Grevstad-Nordbrock 2015; Liu, 2016). As Farrar argues, "far too often, preservation is only pursued as a path to economic development, as cities and towns have increasingly realized the potential for marketing their histories" (2011, p. 728). As a result, the psychological dimensions of historic preservation, encapsulating individuals' emotional bonds, daily experiences, and perceptions of historical places, gradually faded into the background.

Preservation scholars' concerns about people's psychological needs and emotional attachment to historic preservation could be traced back to 1996 when the Municipal Art Society published an influential report titled "History Happened Here: A Plan for Saving New York City's Historically and Culturally Significant Sites." This report suggests that historic preservation should appreciate place attachment and focus on cultural values and social history rather than solely focusing on their architectural design (Minner, 2016; Kaufman, 1996, 2009). However, after two decades, there is still a dearth of literature that focuses on place-related concepts (such as PA, PI, and SoP) in historic preservation. Therefore, it is of great significance for preservationists to turn to other academic fields for insight and to incorporate the knowledge of environmental psychology.

https://doi.org/10.52202/082078-0019

In the field of environmental psychology, Place Attachment (PA), Place Identity (PI), and Sense of Place (SoP) are three related concepts that describe the emotion in person–place bonding, frequently studied in this field (Scannell & Gifford, 2010). Place Attachment (PA) is defined as a cognitive-emotional bond that individuals develop towards places (Gifford, 2014). Place Identity (PI) refers to the concept of 'placeness' (Boaç, 2009) and is described as a "potpourri of memories, conceptions, interpretations, ideas, and related feelings about specific physical settings, as well as types of settings" (Proshansky et al., 1983, p. 60). Sense of Place (SoP) is a complex connection between humans and the physical and/or social character of a place (Scannell & Gifford, 2010). The scholarly exploration of individuals' relationships with places dates back to 1974 (Lewicka, 2008). Over the past 30 years, environmental psychologists have expanded on this area of inquiry, fostering a rich understanding of place constructs (McMillan & Chavis, 1986; Manzo & Perkins, 2006; Alawadi, 2017). While there is still some disagreement about precisely defining individuals' bonds with places, concepts such as 'place attachment,' 'place identity,' and 'sense of place' (Hernández et al., 2007) remain fundamental for preservationists. Preservation projects, to truly resonate with users, must prioritize the creation of "place" over "space," acknowledging that "space" is considered an independent entity, while "place" is intimately connected to users' personal experiences and meanings (Buffel et al., 2014).

This systematic literature review seeks to reassess and underscore the importance of psychological aspects within the realm of historic preservation. Through an analysis of literature concerning both place-related concepts (PA, PI, PM, and SoP) and historic preservation within the context of the built environment, this paper aims to address three fundamental questions: 1) How do preservationists, researchers, and practitioners in the field of historic preservation and related disciplines emphasize place identity, place attachment, sense of place, and other place-related constructs in their research? 2) What relationships exist between historic preservation and these place-related constructs? 3) What are the research gaps apparent in this literature?

By scrutinizing definitions, theories, and pertinent case studies, this systematic review evaluates the existing body of literature and discusses the potential for interdisciplinary exploration between environmental psychology and historic preservation. It contends that through this interdisciplinary collaboration that integrates environmental psychology into historic preservation, architects, designers, and preservationists can contribute to a more human-centered approach to historic preservation projects. This approach extends beyond the preservation of physical elements, emphasizing the safeguarding of a profound sense of place and regional identity. Such an approach becomes increasingly pertinent in the face of evolving societal dynamics, including rapid urbanization, commercialization, and globalization.

The structure of this literature review is organized as follows: Part 2 presents the literature selection protocol and screening methodology. Part 3 systematically classifies and describes the chosen literature, delving into various topical areas. It analyzes the diverse research methods employed and examines the scales and characteristics of the historic preservation cases studied in the reviewed manuscripts. Part 4 elaborates on the intricate relationship between place-related constructs and historic preservation. Finally, Part 5 encapsulates the research conclusion, summarizing the key insights derived from the literature review.

3 RESEARCH METHODOLOGY

The research methodology follows the literature review guidelines known as PRISMA (Preferred Reporting Items for Systematic Reviews and Meta-Analyses) (Moher et al., 2009). As the earliest literature that mentioned the concept of 'place' concerning historic preservation dates back to 1984, and to be more comprehensive, this review paper examines literature published within the timeframe between 1984 and 2023. This time frame is also when financial incentives started to play a vital role in historic preservation projects in the United States, calling for more attention to humans' connections to historic places.

Based on the research questions mentioned in the Introduction, which focus on the intersection of the concept of 'place' in environmental psychology and historic preservation, this review paper uses two sets of terms for searching. The first set includes place-related terms from the field of environmental psychology, including 'place identity,' 'place attachment,' 'sense of place,' 'place-making,' and other place-related terms like 'place dependence,' 'spirit of place,' and 'notion of place.' The second set includes 'historic preservation,' 'historic building,' 'adaptive reuse,' and 'heritage,' as these terms are related to issues and approaches in historic preservation as well as interventions with historic buildings.

Qualified literature for selection includes peer-reviewed journal articles and conference proceedings published in the English language. The databases used for searching are Google Scholar, Scopus, Web of Science, JSTOR, and Elsevier. The scope of this paper includes research papers that have implications for the built environment, such as architecture, interior design, and urban design. A total of 1068 related records were identified during the initial search.

https://doi.org/10.52202/082078-0019

Table 1. The screening process for conducting a systematic literature review.

Screening Process	Record Data
Initial count:	1068
Screening Process Step 1: Read title, abstract, and keywords	
Exclusion Criterion: 1) Research papers not related to the built environment. For example, forest preservation, marine management, education, etc.	
2) HP and Place-related concepts merely appear in the research background, rather than being the central research focus. For example, the application of digital technology in historic environments, social media participation, and the tourism economy, etc.	
3) Manuscripts that are not peer-reviewed journal articles or conference proceedings. For example, non-peer-reviewed research reports and book chapters.	
Number of articles after screening 1:	148
Screening Process Step 2: Collect full text.	
Exclusion Criterion: 1) Articles are not accessible.	
Number of articles after screening 2:	116
Screening Process Step 3: Read the full text.	
Exclusion Criterion: 1) The main text does not relate to HP or Place-related concepts.	
2) The main text does not provide an analysis of the relationship between HP and Place-related concepts, which cannot answer the research questions of this paper.	
3) The main text does not align with the scope of this paper. The research results are not helpful to historic preservation researchers, architects, or designers. For example, papers on tourism or archaeology.	
The number of articles after screening 3:	30
The 30 papers serve as the basis for the content analysis.	

After the initial search, this literature review adopted a screening process with specific exclusion criteria to eliminate literature that does not align with the research scope. It also excludes repetitive, inaccessible, and unpublished literature. Table 1 summarizes the screening process, exclusion criteria, and the number of identified records.

After a detailed review of abstracts and accessibility checks, 116 papers were selected after the first two screening steps. After a thorough reading of the full text and the application of screening and exclusion criteria, this paper finally selected 30 publications for a more detailed review. Table 2 lists the journals included in this review.

4 RESEARCH TRENDS AND SUBSTANTIVE AREA

4.1 Research trends and topical areas

Within the 116 selected publications screened after the first two steps of the screening process, Figure 1 presents the number of studies published between 1998 and 2023 per year, illustrating a consistent upward trend over the past decade. This literature review reveals a significant increase in selected literature focusing on place-related concepts and historic preservation. This surge indicates a growing scholarly interest in issues related to historic

https://doi.org/10.52202/082078-0019

preservation concerning individuals' perceptions of place (e.g., PA, PI, & SoP). Additionally, it underscores the emergence of interdisciplinary research bridging the areas of environmental psychology and historic preservation as a topical and trending research direction.

Figure 2 illustrates the network visualization of keyword occurrences in the selected 116 publications, revealing the topical areas and their corresponding keywords in the realm of historic preservation (HP) and place-related concepts. Notably, the analysis reveals that the central theme in the research body is the "sense of place." The term "place attachment" emerges as a crucial and closely associated keyword with "sense of place." The two keywords, "sense of place" and "place attachment," are also closely related to historic preservation, emphasizing the significance of emotional and psychological connections people establish with historic places.

Among the 30 selected manuscripts for further examination, twelve (1, 3, 6, 12, 13, 14, 18, 22, 26, 27, 29) delve into the concept of place attachment; seven manuscripts (2, 10, 13, 14, 16, 22, 30) explore place identity; sixteen manuscripts (4, 5, 7, 8, 10, 11, 13, 15, 17, 19, 20, 21, 22, 23, 24, 28) study the concept of sense of place; and four articles (9, 21, 22, 25) contribute to the topic of placemaking. This diverse array of thematic emphases within the selected manuscripts collectively contributes to a comprehensive understanding of how individuals connect with and perceive the historical environment.

Table 2. A list of reviewed literature.

No.	Author Year		Type of HP	Location of cases	Conclusion/Summary	Research Method
1	Alawadi, K., 2017	PA	Community Preservation	Sha'biyat Al Defaa' and Sha'biyat Al Shorta, Dubai	The author criticizes the top-down approach and suggests individuals' feelings and bonds to social settings can be used as engines for HP.	Qualitative
2	Armada, B. J, 2010	PI	Adaptive Use	National Civil Rights Museum, Tennessee, US	Through HP of King's assassination site, the place identity was transformed, and community members forcefully articulated the importance of place to collective identity.	Qualitative
3	Bonney, C. G, 2015	PA	Adaptive Use	Savannah Historic District, US	PA can ease tensions between heritage, preservation, and planning caused by the current HP Guidelines, and it identified architectural features that can strengthen PA through community participation.	Qualitative
4	Burnham, 1998	SoP	Urban Preservation	Angkor, Cambodia; South Pass, US; Pompeii, Italy; Dubrovnik, Croatia; Jaisalmer, India; Fort Apache, US	Immovable heritage conservation requires new approaches that leverage public appreciation of sites' SoP to secure their future.	Qualitative
5	Cantillon & Baker, 2022	SoP	Community Preservation	Kingston and Arthur's Vale Historic Area, Norfolk Island	The local impact of transnational heritage policies and discourses that seek to define heritage values and manage heritage sites without appropriate engagement with communities and their SoP.	Mixed method
6	Cordina et al., 2022	PA	Cultural Heritage Conservation	Kandovan, Iran, and Cappadocia, Turkey	This study also reinforces the importance of PA and participation as mediating factors between cultural motivation and conservation commitment for both developed and developing sites.	Mixed method
7	Dameria et al., 2020	SoP	Cultural Heritage Conservation	N/A	This research is about seeking, discovering, and compiling principles that can be used to explain PI, PA, and place dependence in heritage contexts.	Mixed method
8	Dameria et al., 2022	SoP	Urban Preservation	Semarang Old Town, Indonesia	Building owners can develop strong attitudes in all three dimensions of SoP and strong intentions in both dimensions of sustainable heritage behavior, but not all attitudes and intentions have strong relationships.	Mixed method

https://doi.org/10.52202/082078-0019

9	Ertan, T., & Egercioglu, Y, 2015	PM	Urban Preservation/ Regeneration	Malacca, Malaysia	The author concludes that the environmental perception of the community can act as a vital tool to achieve a quality urban preservation project. Design principles and elements are suggested.	Qualitative
10	Farrar, M. E, 2011	PI, SOP	General Historic Preservation	American cities	Current HP, as one mode of PM, can have depoliticizing effects by conjuring up peculiar, selective, or even wholly imaginary pasts. SOP support PI.	Literature review
11	Gray, 2022	SoP	Preservation of Historic Building(s)	Charleston's Dock Street Theatre, US	the "restored" theater called on Charleston's historic SoP to serve both contemporary and future generations, becoming a tangible form of political and cultural agency	Qualitative
12	Huang, 2018	PA	Community Preservation	Kucapungane (Rukai) people in Taiwan	Research suggests that PA and place memories can be used to prepare for the next challenge, giving heritage protection a new meaning in disaster preparedness.	Mixed method
13	Hussein et al., 2020	PA, PI, SoP	Urban Landscape Conservation	N/A	Historic Urban Landscapes have an important role in maintaining cultural memory and, consequently, psychosocia well-being.	Qualitative
14	Jiang et al., 2022	PI & PA	Urban Preservation	Suzhou, China	From a historical urban landscape perspective, urban landscape changes caused by the development of urban forms have implications for PI and PA.	Mixed method
15	Masuwan & Lertwattanaruk, 2020	SoP	Preservation of Historic Building(s)	Phuket, Thailand	FBC has the potential to enhance the character and vitality of historic areas by improving façade design *and retaining* SoP and community pride.	Mixed method
16	Parkinson et al., 2017	PI	Urban Preservation	Three small towns, Irish	Memories convey the symbolism and significance of specific places, and the public uses these memories to construct their sense of PI.	Qualitative
17	Raqena et al., 2020	SoP	Community Preservation	Kampung Morten, Melaka	The relevance of SoP helps to raise community awareness of the potential threats of modernization and helps to raise awareness of the conservation value of bult heritage sites.	Mixed method
18	Rojak, 2015	PA	Historic Preservation of Interiors	Brewpub in Greensboro, North Carolina, US	By studying the relationship between the physical built environment & PA, the author concludes that the benefits of HP can extend beyond tax incentives to include positive social outcomes. The historical character of the interior environment supports PA	Mixed method

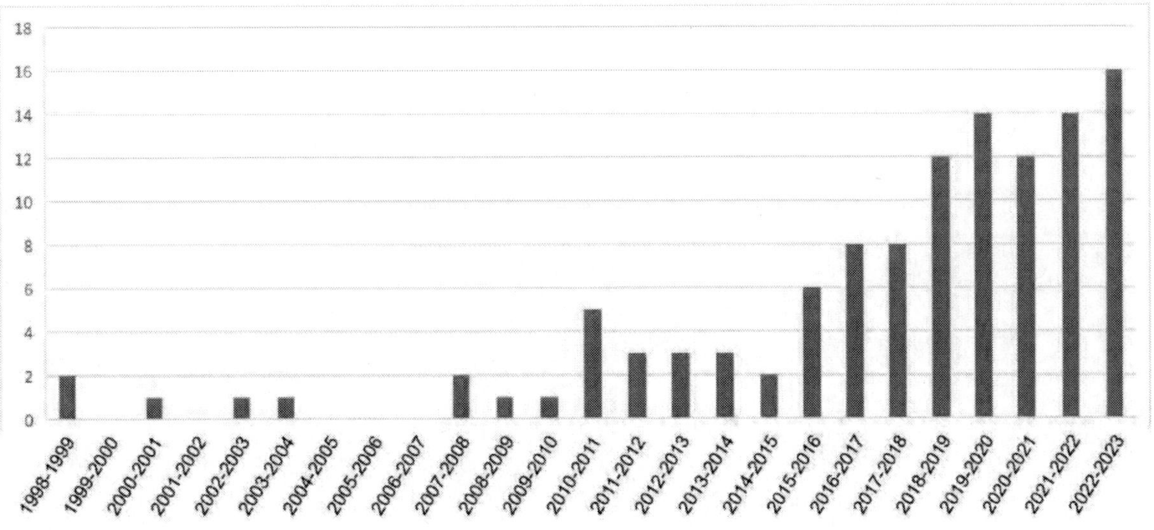

Figure 1. Number of selected publications per year after the first two steps of the screening process. The total number of publications is 116.

https://doi.org/10.52202/082078-0019

19	Rowntree, L, 1981	SOP	Urban Preservation	Salzburg, Austria	Through HP of landscape, Salzburg has created and preserved a unique SOP over the last 100 years. Codified landscape tastes become a conservative element of culture.	Qualitative
20	Salah Ouf, 2001	SoP	Urban Preservation	Gold Market, Dubai	SoP is created within a small urban core to reinforce the strong essence of the heritage and attract further conservation efforts.	Qualitative
21	Samir et al., 2019	SoP, PM	Urban Landscape Conservation	N/A	Urban heritage sites are established by the PM approach which is critical to the connection between people and place and achieves SoP.	Mixed method
22	Shamsuddin, S., & Ujang, N, 2008	PA, PI, PM, SOP	General Historic Preservation	Traditional streets in Malaysia	Physical elements and activities are essential in creating SOP. The author argues that historical significance in HP can evoke PA, and PA enhances PI.	Mixed method
23	Silva, 2015	SoP	Urban Preservation	Bhaktapur, Nepal	Conceptualize conservation and development activities by understanding the core and risk dimensions of SoP.	Qualitative
24	Spennemann, 2022	SoP	Community preservation	N/A	HP atmosphere creates SoP that supports individual mental health and community resilience in the face of adversity or fundamental change.	Qualitative
25	Thirumaran et al., 2023	PM	Preservation of Historic Building(s)	Three luxury hotels in Singapore	This can have a significant impact on PM when the historicity of heritage hotel exhibitions is linked to local social narratives when presented to potential visitors.	Mixed method
26	Wahyudie et al., 2021	PA	Community preservation	N/A	A study of PA theory supporting the protection of historic religious built environments.	Literature Review
27	Walker, A. J., & Ryan, R. L, 2008	PA	Community preservation & landscape	Monmouth, Maine, US	A positive correlation exists between rural residents' level of support for conservation planning (HP) and their level of attachment to the landscape (PA). PA can be a strong motivation for residents' support for land protection strategies.	Mixed method
28	Wang, 2012	SoP	Urban preservation	Pingyao, China	Authenticity is revealed through urban conservation efforts that preserve the physical environment of the city and maintain SoP within the city through an evolutionary process.	Mixed method
29	Wells, 2017	PA	Urban preservation	Charleston and I'On, US	Find a balance between HP and modern development to maximize people's emotional PA.	Mixed method
30	Zhou et al., 2023	PI	Urban preservation	Suzhou, China	A historical heritage that focuses on self-discovery and active conservation through citizen participation will help people understand the origins of their culture and why it should be preserved, thereby promoting PI and pride.	Mixed method

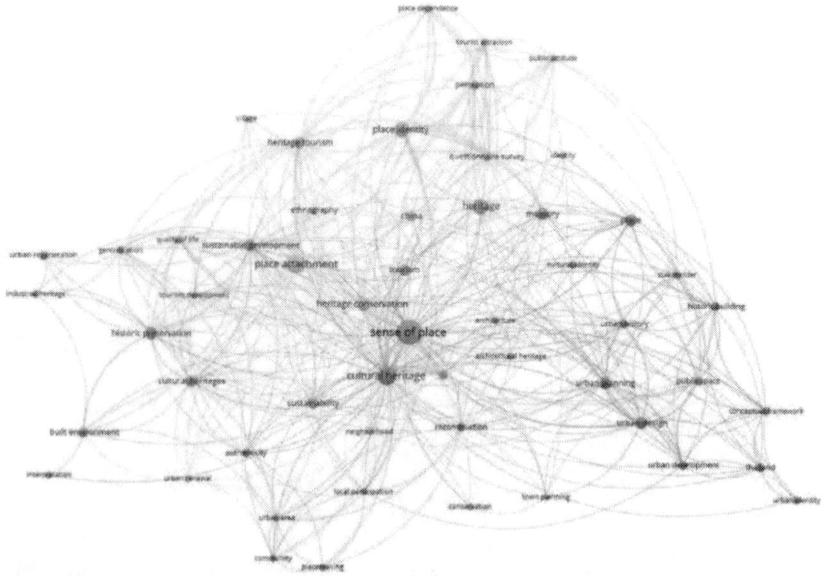

Figure 2. Network visualization showing keyword occurrences in the selected 116 publications.

https://doi.org/10.52202/082078-0019

Regarding academic disciplines, "urban planning" and "urban design" are the most relevant fields connecting with the intersection of place-related concepts and historic preservation. Other research areas that are tightly linked to the selected literature include sustainability, urban regeneration, heritage tourism, and architecture. Furthermore, the keyword "cultural heritage" stands out as a key category within the broader domain of historic preservation, underscoring the importance of preserving cultural aspects in historic sites. This analysis highlights the multifaceted nature of research in historic preservation and the sense of place, encompassing an individual's emotional connections, cultural significance, and environmental psychological factors alongside the preservation of physical historic elements.

4.2 An analysis of research methods

This review paper organizes research methodologies into four categories: qualitative research (1, 2, 3, 4, 9, 11, 13, 16, 19, 20, 23, 24), quantitative research (N/A), mixed-method research (5, 6, 7, 8, 12, 14, 15, 17, 18, 21, 22, 25, 27, 28, 29, 30), and literature review (10, 26) articles. The distribution of these methodologies among the selected literature is presented in Table 2. Notably, a predominant trend emerges, with the majority of the reviewed articles opting for mixed-method approaches. Out of the 30 qualified articles, 16 employ mixed-method research. Twelve of out 30 articles used qualitative research methods. For an in-depth examination, Table 3 provides a comprehensive synthesis of detailed information, showcasing the specific types of research methods employed by the respective articles.

In the reviewed literature, case study research emerges as the predominant method, featured in 24 out of the 30 selected articles. Within the 24 manuscripts, nine of them (1, 4, 5, 6, 9, 16, 22, 25, 29) study multiple-case studies as comparable cases, while the remaining 15 (2, 3, 8, 11, 12, 14, 15, 17, 18, 19, 20, 23, 27, 28, 30) conduct single-case studies. Case study research serves as a popular and feasible research method for historic preservation research, which is often intricate, exploratory, based on a case-by-case situation, and not statistically generalizable (Yin, 2018). The unique characteristics of each historical site are considered in historic preservation research, and the chosen cases in these studies span the globe, reflecting the diverse nature of the field. Section 3.3 expands on the analysis of cases.

Within the case study framework, archival research, surveys (or questionnaires), and interviews stand out as the most used tools for gathering data. Archival or historical research is commonly used in the field of historic preservation, offering a valuable lens into the past. Surveys and interviews are commonly associated with the environmental psychology literature. The integration of methods brings a human psychological dimension into the study of the history and preservation of the built environment.

One noteworthy aspect is the popular incorporation of visual media within the selected literature, particularly through methods like photo elicitation. Four research papers (3, 18, 27, 29) have relied on photo-elicitation - a method that helps to understand "the subjective experience of an informant in a particular environment" (Wells & Baldwin, 2012, p. 388). The popularity of using visual media to enhance investigations highlights the significance of photos in capturing individuals' intricate perceptions and emotions.

4.3 An analysis of historic preservation cases

The analysis of 24 papers employing case study research reveals interesting patterns regarding the scale and location of the chosen cases, shown in Table 4. The cases on architecture and interiors, combined, account for only 6 papers, reflecting a relatively lower focus on these specific scales. Conversely, the majority of the literature, comprising 17 papers, focuses on urban environments. This suggests a pronounced emphasis on placemaking within urban settings and a lack of research in placemaking at intimate scales, such as architecture and interiors. This research gap highlights a potential area for future exploration and investigation.

Furthermore, the distribution across different types of locations provides additional insights. A significant portion of the cases, encompassing 15 papers, is situated in Asia, indicating a substantial research focus on this region. The prominence of Asian cases may suggest unique challenges or opportunities related to historic preservation and people's perceptions in Asian contexts. Notably, these 15 papers in Asia span different countries, highlighting the diversity of contexts within the region. Following this, 7 cases are located in North America, with all of them situated within the United States. Europe contributes 5 cases, showcasing a comparatively smaller but still considerable representation. The prominence of cases in the United States might be influenced by the screening protocol requiring English language content, but it also suggests the possibility of a more concentrated scholarly focus or heightened tensions between people's perceptions and historic preservation in this region. However, despite these influences, the selected literature encompasses a diverse array of regions, suggesting a global interest in, or attention to, the intersection of historic preservation and people's perceptions.

https://doi.org/10.52202/082078-0019

Table 3. The different types of research methods.

Code	Total Amount	No.
Case study	24	1, 2, 3, 4, 5, 6, 8, 9, 11, 12, 14, 15, 16, 17, 18, 19, 20, 22, 23, 25, 27, 28, 29, 30
Archival or historical research	11	2, 3, 4, 9, 11, 12, 14, 15, 16, 19, 20
Questionnaire or survey	11	6, 8, 12, 13, 17, 18, 21, 27, 28, 29, 30
Interview	9	1,3, 5, 15, 17, 18, 22, 23, 29
Photo elicitation	4	3, 18, 27, 29
Observation	2	1, 22
Textual analysis	1	25

Table 4. Case Study Scale, Location, and Place-HP Relationship.

Scale of Place	Amount	No.	Region	Location of Cases	Place-HP Relationship
Architecture	3	2	America	National Civil Rights Museum, Tennessee, US	HP fosters the transformation of PI
		3	America	Savannah Historic District, US	PA supports HP
		18	America	Brewpub in Greensboro, North Carolina, US	HP supports PA; HP supports PI
Interior Architecture	3	11	America	Charleston's Dock Street Theatre, US	HP supports SoP
		17	Asia	Kampung Morten, Melaka	SoP supports HP
		25	Asia	Three luxury hotels in Singapore	HP supports PM
District or Communities	5	1	Asia	Sha'biyat Al Defaa' & Sha'biyat Al Shorta, Dubai	PA motivates HP
		5	Europe	Kingston & Arthur's Vale Historic Area, Norfolk Island	HP & SoP have an ambiguous relationship
		12	Asia	Old settlement of Kucapungane in Taiwan	HP supports PA and PM
		20	Asia	Gold Market, Dubai	SoP supports HP
		22	Asia	Traditional streets in Malaysia	HP evokes PA; PA enhances PI
Rural	1	27	America	Monmouth, Maine, US	PA supports HP.
Urban	17	4	Asia	Angkor, Cambodia	HP and SoP mutual support
			America	South Pass, US	
			Europe	Pompeii, Italy	
			Europe	Dubrovnik, Croatia	
			Asia	Jaisalmer, India	
			America	Fort Apache, US	
		6	Asia	Kandovan, Iran, and Cappadocia, Turkey	PA supports HP
		8	Asia	Semarang Old Town, Indonesia	HP and SoP mutually support
		9	Asia	Malacca, Malaysia	PM supports HP.
		14	Asia	Suzhou, China	HP supports PI and PA
		15	Asia	Phuket, Thailand	SoP supports HP
		16	Europe	Three small towns, Irish	HP supports PI
		19	Europe	Salzburg, Austria	HP supports SOP
		23	Asia	Bhaktapur, Nepal	SoP supports HP
		28	Asia	Pingyao, China	HP supports SoP
		29	America	Charleston and I'On, US	HP supports PA
		30	Asia	Suzhou, China	HP supports PI

https://doi.org/10.52202/082078-0019

5 COMPLEX RELATIONSHIP BETWEEN PLACE AND HISTORIC PRESERVATION

5.1 Place-HP relationship: the context

The primary focus of this paper revolves around the exploration of the nuanced relationship between place-related concepts and historic preservation. Within the 116 papers selected during the initial screening process (Step 1 & Step 2), many scholars have traditionally assumed a mutually beneficial connection between historic preservation and place-related concepts. Some scholars believe that preserving historical sites fosters a sense of place or place identity, and some other scholars suggest that people's attachment or a historic place can spur historic preservation activities. For example, in the abstract of a research paper in the field of urban heritage conservation, the authors suggest that "[u]rban heritage anchors the identity and culture of cities and is a key element in the development of a sense of place" (Liu et al., 2022, p.1). Another research paper on sustainable urban development says that the historic buildings' "inherent value" and "community's memory" contribute to the reinforcement of its "sense of place" (Lewin & Goodman, 2013, p.22). Another journal article on urban regeneration and industrial heritage suggests that industrial heritage serves as "a tool for improving the city's image" and place identity, which is a "catalyst for urban regeneration" (Cizler, 2012, p.223).

However, while numerous research papers, as exemplified above, touch upon the relationship between place-related concepts and historic preservation in their abstracts as contextual information, a significant portion of them do not investigate or explore this connection. Does the attachment to a place consistently contribute to the preservation of historic structures or districts? Or, does historic preservation invariably enhance the sense of place in historic neighborhoods? The answers to these questions necessitate further investigation, as the 116 selected research papers, although providing contextual information, fall short of comprehensive exploration. The somewhat taken-for-granted perspective and the lack of in-depth consideration regarding the Place-HP relationship highlight the need for a more thorough examination. The subsequent section delves into this intricate relationship by meticulously scrutinizing the 30 manuscripts selected for this literature review.

As a note, this paper does not investigate the relationships, similarities, or differences between various place-related constructs. In the field of environmental psychology, numerous studies have explored the relationships and overlaps among Place Attachment (PA), Place Identity (PI), Place Meaning (PM), and Sense of Place (SOP), suggesting a potential topic for another research paper. As Droseltis & Vignoles, (2010)Droseltis & Vignoles, (2010) argue, the relationship between PI and PA has sparked controversy. Different perspectives exist, including the view that PA and PI are distinct constructs, PA subsumes PI, or both PA and PI are subsumed by another construct such as SOP. For instance, Shamsuddin & Ujang, (2008) assert that PA significantly contributes to SOP, while Cheshmehzangi & Heat, (2012) propose that PI creates SOP. Despite these discussions, Droseltis & Vignoles, (2010) highlight the challenge of the 'blurring of the semantic differentiation between emotional attachment and place identity,' rendering these terms almost interchangeable in practice. Therefore, as the primary focus of this paper is on people's perceptions and emotions toward a place'shistoric preservation, this literature review does not prioritize the investigation of the distinctions or relationships between these place-related concepts.

5.2 Place-HP relationship: a thorough investigation

This part further delves into the Place-HP relationship by exploring the selected 30 manuscripts. The relationship between HP and place-related terms is presented in Table 5 and described in the following diagram (Figure 3). In Figure 3, the upper portion illustrates the impacts of place-related concepts on historic preservation, while the lower section shows the impacts of historic preservation on place-related concepts. Table 4 provides a detailed breakdown, utilizing the symbol "" to denote an impact. As a side note, the strength of these relationships cannot be measured or compared within this review paper.

https://doi.org/10.52202/082078-0019

Table 5. Categories and explanations of the Place-HP relationship.

Place & HP	N	Summary of Place-HP relationship	Impact	In-text Segments
SoP → HP	4	Conservation of immovable heritage capitalizes on public appreciation of the site's SoP.	Positive	"The field (conservation of immovable heritage) needs new methodologies that harness public appreciation of a site's "sense of place" to guarantee its future." (p. 149).
	7	Predictors of human behavior in HP can be studied when SoP is considered in the heritage context.	Positive	"…seeing the sense of place as an attitude can be effective in the context of heritage when it wants to investigate its role as a predictor of human behavior that supports heritage conservation." (p. 145).
	11	Urban's historical SoP serves to inherit culture in HP.	Positive	"the "restored" theater called on Charleston's historic sense of place to serve both contemporary and future generations, becoming a tangible form of political and cultural agency…" (p. 63).
	15	Preservation of SoP enhances vitality and pride in historic areas.	Positive	"FBCs (Form-Based Codes) have the potential to enhance the character and vibrancy of the historic area by improving façade design and preserving the sense of place and community pride." (p. 1).
	17	SoP helps to raise awareness of the HP of heritage sites' value.	Positive	"…further suggesting the relevancy of the 'sense of place' within Kampung Morten. This could help to create awareness within the community on potential threats of modernization and help increase the level of appreciation towards the value of conservation for architectural heritage sites like Kampung Morten, Melaka." (p. 35).
	20	SoP attracts further HP efforts.	Positive	"[T]he creation of a sense of place within a small urban nucleus to anchor a strong essence of heritage and attract further conservation efforts."
PM→HP	9	PM improves urban preservation (HP) quality.	Positive	"Through a placemaking phase, local residents can be involved … environmental perception of the community can indeed act as a vital tool to achieve a quality urban preservation project" (p. 59).
	21	The PM approach to urban heritage sites builds connections between people and places to obtain SoP	Positive	"... dealing with urban heritage sites by building on the placemaking approach which counts for the connection between people and the place. ..., hence, achieving a sense of place" (p. 322).
PA→HP	1	PA is important enough to justify expanded HP.	Positive	"... attachment to place and community are important enough to justify expanding historic preservation to include preservation of social bonds and cohesion" (p. 2993).

https://doi.org/10.52202/082078-0019

	3	Enriching PA results in greater effort from residents for HP.	Positive	"Integrating strategies that enhance attachment among residents into the design regulatory process can ameliorate some of the foundational conflicts between heritage, preservation, and planning" (p. 199).
	6	Mediating factors between PA and HP commitment are significant	Positive	"This study also reinforces the importance of place attachment and participation as mediating factors between cultural motivation and conservation commitment for both developed and developing sites" (p. 334).
	12	PA gives HP a new meaning in disaster preparedness.	Positive	"...people's attachment to and knowledge of places and memories of disasters can be used to prepare for the next challenge; this allows heritage preservation to take on the new meaning of disaster preparation" (p. 6).
	26	PA supports the HP of a religious built environment.	Positive	"a study of place attachment theory in supporting the preservation of the historical religious built environment" (p. 1).
	27	PA serves as a predictor to support the HP of landscape	Positive	"This study renders new evidence for landscape planners that attachment to the land can be a strong motivation for building residents' support for land protection strategies" (p. 141).
HP→SOP	10	HP doesn't support SOP as it focuses less on memory than accuracy and marketing.	Not Positive	"Yet it is evident that historic preservation is not an ideal solution to our diminishing sense of place. Instead, it often does more to insulate us from place memory than to cultivate it, for several reasons" (p. 278)
	13	HP of urban landscapes is important to place memory.	Positive	"... HULs (as a repository of memories) encompass have an important role in maintaining cultural memory and, consequently, psychosocial well-being" (p. 2).
	19	Through codified HP of landscape, a unique SOP is preserved and created.	Positive	"Salzburg has taken major steps to create and preserve a unique sense of place over the last one hundred years. ... Landscape tastes, once codified, become a conservative element of culture and change little through time" (p. 74).
	23	Conceptualizing HP in terms of the core and risk dimensions of the SoP is significant.	Positive	"... conservation and development activities could be conceived through an appreciation of both the core and the risk dimensions of the sense of place."
	24	The historical atmosphere created the SoP.	Positive	"[I]t is this nostalgia, comfort and inclusiveness that creates the sense of place and thus supports individual mental health and community resilience in the face of adversity or fundamental change."
	28	HP maintains the SoP within the city.	Positive	"... the nature of authenticity is revealed through the urban conservation effort of retaining the physical setting of the city and maintaining of the sense of place within that city through the evolutionary process."

https://doi.org/10.52202/082078-0019

HP→PA	18	HP of the character of the interior environment supports PA.	Positive	"Preserving the historically built environment and carrying the aesthetic into the building interior could be the important factors for human-place bonding in the brewpub setting" (p. 48).
	22	Historical significance in HP evokes PA, especially in long-term users.	Positive	"The historical significance of the traditional streets as the earliest shopping streets in Kuala Lumpur evoked a lot of meanings to the users thus influencing their degree of attachment" (p. 409).
	29	To maximize PA, a balance between HP and the new environment is needed.	Positive	"... is necessary to establish the desirable balance point between decay and patina to maximize people's emotional attachment to place" (p. 465).
HP ⇌ SoP	5	SoP and HP influence each other but have a vague relationship.	Ambigu ous	"... the case study of the Paradise Hotel highlights the local effects of transnational heritage policies and discourses that seek to define heritage value and manage heritage sites without appropriate engagement with communities and their sense of place" (p. 110).
	8	SoP is positively related to HP behavior.	Positive	"... that in an urban heritage setting, the building owner's sense of place dimensions, which tended to be strong, were positively related to the sustainable heritage behavior dimensions, which also tended to be strong. … each sense of place dimension may be positively associated with the conservation behavior" (p. 36).
HP→PI	2	HP influences and fosters PI.	Positive	"By rescuing King's assassination site from destruction, community members preserved a precious part of history and forcefully articulated the importance of place to collective identity" (p. 912).
	14	Changes in the urban historical landscape affect PI and PA	Positive	"… the impact of changes in the urban landscape caused by the development of urban form from the HUL (the historic urban landscape) perspective on local identity and place attachment …"
	16	Collected memories are conveyed through HP to build public PI.	Positive	"…the case study material discussed in this paper suggests that attitudes have shifted. This change is substantially underpinned by local collected memory, which informs the symbolism and meanings attached to particular places, and which the public uses to construct its sense of place identity..." (p. 516).
	30	Citizen participation in active HP, thereby promoting PI.	Positive	"... "historical heritage" is seen from afar to one that is focused on self-discovery and active preservation through citizen involvement. This view will help people understand the origins of their culture and why it should be preserved, thus promoting a place of identity and pride" (p. 2).
HP→PM	25	Heritage hotels can have an impact on PM by conveying their historicity to visitors.	Positive	"We posit a significant impact on place-making when heritage hotels exude historicity and are connected to the local social narrative when presented to prospective visitors" (p. 249).

https://doi.org/10.52202/082078-0019

Within the 30 selected manuscripts, scholars explore the relationships between historic preservation and different types of place-related terms in different contexts. 14 articles (1, 3, 4, 6, 7, 9, 11, 12, 15, 17, 20, 21, 26, and 27) describe the impacts of place-related concepts on historic preservation, and 14 articles (2, 10, 13, 14, 16, 18, 19, 22, 23, 24, 25, 28, 29, and 30) describe the impacts from the other way, looking at how historic preservation affects place-related concepts. Two articles (5 and 8) talk about mutual relationships between the two (Figure 3).

Remarkably, the impacts outlined in the 14 articles describing the influence of place-related concepts on historic preservation are uniformly positive. Many scholars emphasize the importance of local residents and communities in historic preservation, as exemplified in articles 1, 3, 9, and 27. They suggest that place-related terms, such as place attachment or place-making, contribute to the involvement of residents, and the support of residents has positive impacts on historic preservation. The positive impact of place-related terms on historic preservation is highlighted by their role in enhancing, fostering, and supporting effective historic preservation practices. These positive impacts include improving the quality of preservation (9), easing the tension between preservation and urban development (1, 3), and motivating people to support historic preservation (1, 27).

In examining articles detailing the impacts of historic preservation on place-related concepts, a consistent pattern emerges, with most suggesting a positive influence, except for article number 10. Despite initial appearances of contradiction, these results exhibit underlying consistency, which is explained as follows. In literature 2, 18, 19, and 22, place-related terms are positively influenced by historic preservation, which includes the conservation of essential characteristics such as interior elements (18), historic streets (22), historic features at memorial sites (2), and the historic sense of place in the city (19). These examples illustrate a proper approach to historic preservation, where meticulous preservation of historical features can evoke people's memories and enhance people's connection to the place.

Contrastingly, literature number 10 asserts that historic preservation, particularly in the United States, tends to prioritize accuracy and economic development over preserving people's memories (Farrar, 2011). Here, the author criticizes an improper approach to historic preservation, suggesting that this mode does not support the sense of place. Therefore, the seemingly conflicting results in these articles reveal a consistent theme – the effectiveness of historic preservation in enhancing place-related concepts depends on the method employed. Some articles highlight the positive influence when historic preservation aligns with the preservation of meaningful historical features, while others critique instances where preservation prioritizes economic considerations over preserving the intangible aspects that contribute to a sense of place. This finding underscores the importance of considering research prerequisites and different kinds of interventions in the context of historic preservation.

While the qualified articles collectively acknowledge the significance of place-related concepts within the framework of the historic built environment, there arises a notable concern regarding these positive relationships. In literature number 2, a unique case is presented where historic preservation efforts successfully transformed the original place identity associated with the shamefulness of an assassination site into a new identity linked with independence and pride (Armada, 2010). This raises a crucial issue: people's perceptions of place might be very vulnerable and can be easily exploited or controlled through activities like historic preservation. The preservation of historic properties has the potential to manipulate or control these perceptions, and if preservation is driven by false or inappropriate purposes, it can easily hinder or reshape a place's identity in an undesirable way.

It is noteworthy that literature number 2 stands out as the only paper addressing the preservation of a historic building as a museum. Consequently, the findings from this particular study might not be directly comparable to other papers that predominantly focus on everyday buildings, residential areas, and urban spaces intricately tied to people's daily lives. This distinction emphasizes the need for nuanced consideration when generalizing the implications of preservation efforts across diverse types of historic sites.

Interestingly, among the 30 selected articles, four specifically criticize the prevailing practices in historic preservation. These critiques are found in literature numbers 1, 3, 10, and 29. These papers highlight issues with the existing governmental regulations and guidelines governing historic preservation. The criticisms encompass the top-down nature of these regulations (1), their economic-driven orientation (1 and 10), their doctrinaire approach (29), and their exclusive focus on the physical environment while neglecting users' perceptions of historic places (3 and 10). This collective critique points to a pressing need for the revision of current historic preservation regulations and guidelines, suggesting an imperative for reform in the field.

https://doi.org/10.52202/082078-0019

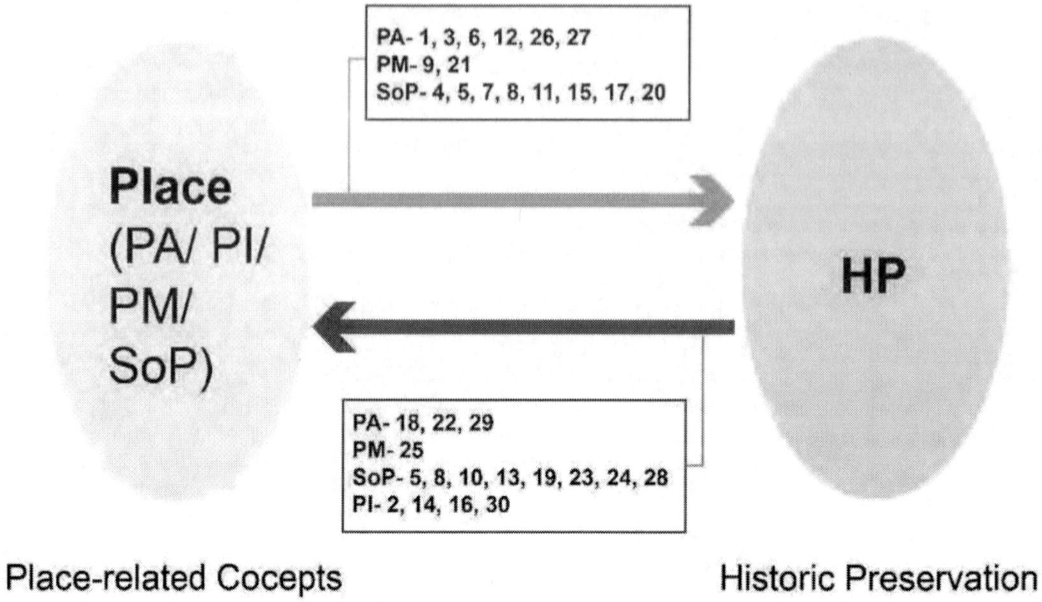

6 CONCLUSION

This paper conducts a systematic literature review analyzing 30 manuscripts that explore the intersection of place-related concepts (PA, PI, PM, and SoP) and historic preservation. Through an examination of topical research areas, methodologies, and the intricate relationships between "place" and "historic preservation," the review uncovers research gaps, proposes potential directions, and contemplates interdisciplinary opportunities at the intersection of environmental psychology and historic preservation.

This literature review identifies a heightened concern among scholars regarding the exploration of the intricate relationships between historic preservation and people's perceptions of place. The preservation of the community and urban preservation receive more attention than the preservation of the interior or rural environment. Mundane buildings and urban spaces overshadow historic landmarks or museums, showing an emphasis on placemaking in everyday life.

Regarding methodology, the review highlights a current trend of incorporating research methods from environmental psychology into historic preservation. A clear preference for a mixed-method approach is evident, with case studies being the most frequently employed method. Within these case studies, the utilization of interviews, surveys, and photo elicitation is prominent, reflecting the nuanced and exploratory nature of research in historic preservation. Significantly, the absence of papers solely relying on quantitative methods underscores the nuanced and unmeasurable nature of historical and cultural aspects in historic preservation.

The literature review reveals a substantial research gap in the study of place-related concepts and the preservation of the interior environment. Given the profound emotional connections people share with historic interiors, this unexplored area presents a promising avenue for future research. Moreover, the integration of place-related concepts into preservation regulations and guidelines emerges as an often-criticized yet underexplored area. Criticisms of existing regulatory systems within several papers provide recommendations for future research to address current shortcomings.

The intricate connections between place-related concepts and historic preservation emphasize the necessity for more nuanced research in the future. The findings indicate that historic preservation practices significantly impact people's perceptions of place, both positively and as indicated in some critiques, potentially negatively. Acknowledging the intricacies and potential drawbacks inherent in this relationship is essential for future research endeavors that guide effective preservation practices.

Looking forward, establishing connections between environmental psychology and historic preservation through interdisciplinary research emerges as a pertinent and achievable avenue. The identified research gaps and trends offer a practical guide for researchers navigating the complexities of this interdisciplinary field, enhancing our understanding of

https://doi.org/10.52202/082078-0019

the intricate relationships between place and historic preservation. This is particularly significant in the context of contemporary challenges such as rapid urbanization, commercialization, and globalization. The exploration of these opportunities and the alignment of research efforts with identified gaps and trends can empower scholars to contribute meaningfully to advancing our comprehension of the interplay between place and the preservation of cultural heritage.

7 REFERENCES

Alawadi, K. (2016). Place attachment as a motivation for community preservation: The demise of an old, bustling, Dubai Community. Urban Studies, 54(13), 2973–2997. https://doi.org/10.1177/0042098016664690

Armada, B. J. (2008). Place politics: Material transformation and community identity at the National Civil Rights Museum. Journal of Black Studies, 40(5), 897–914. https://doi.org/10.1177/0021934708321848

Boğaç, C. (2009). Place attachment in a foreign settlement. Journal of Environmental Psychology, 29(2), 267–278. https://doi.org/10.1016/j.jenvp.2009.01.001

Bonney, C. G. (2015). Beyond Aesthetics: Fostering Place Attachment through the Design Regulatory Process (dissertation).

Burnham, B. (1998). Architectural heritage: The paradox of its current state of risk. International Journal of Cultural Property, 7(1), 149–165. https://doi.org/10.1017/s0940739198770146

Cantillon, Z., & Baker, S. (2022). Ketch Yorlye daun paradise: Sense of place, heritage and belonging in Norfolk Island's Kingston and Arthur's Vale Historic Area. Thesis Eleven, 172(1), 93–113. https://doi.org/10.1177/07255136221133185

Cheshmehzangi, A., & Heat, T. (2012). Urban identities: Influences on socio-environmental values and spatial inter-relations. Procedia - Social and Behavioral Sciences, 36, 253–264. https://doi.org/10.1016/j.sbspro.2012.03.028

Cizler, J. (2012). Urban regeneration effects on industrial heritage and local community – case study: Leeds, UK. Sociologija i Prostor, 223–236. https://doi.org/10.5673/sip.50.2.5

Cordina, R., Gannon, M., Taheri, B., Okumus, F., & Lochrie, S. (2021). Committed to conservation: Tourism in developed and developing contexts. International Journal of Tourism Research, 24(2), 323–336. https://doi.org/10.1002/jtr.2504

Dameria, C., Akbar, R., Indradjati, P. N., & Tjokropandojo, D. S. (2020). A conceptual framework for understanding sense of place dimensions in the heritage context. Journal of Regional and City Planning, 31(2), 139–163. https://doi.org/10.5614/jpwk.2020.31.2.3

Dameria, C., Akbar, R., Indradjati, P. N., & Tjokropandojo, D. S. (2022). The relationship between residents' sense of place and sustainable heritage behaviour in Semarang Old Town, Indonesia. International Review for Spatial Planning and Sustainable Development, 10(1), 24–42. https://doi.org/10.14246/irspsd.10.1_24

Droseltis, O., & Vignoles, V. L. (2010). Towards an integrative model of place identification: Dimensionality and predictors of intrapersonal-level place preferences. Journal of Environmental Psychology, 30(1), 23–34. https://doi.org/10.1016/j.jenvp.2009.05.006

Ertan, T., & Eğercioğlu, Y. (2015). An integrated approach to urban preservation a suggestion to combine two methods of urban regeneration in order to achieve a further developed Urban Historic Center Preservation. 3rd Annual International Conference on Architecture and Civil Engineering (ACE 2015). https://doi.org/10.5176/2301-394x_ace15.128

Farrar, M. E. (2010). Amnesia, nostalgia, and the politics of place memory. Political Research Quarterly, 64(4), 723–735. https://doi.org/10.1177/1065912910373553

Gray, S. (2022). "Restoring" Charleston's Dock Street Theatre. The Public Historian, 44(3), 58–91. https://doi.org/10.1525/tph.2022.44.3.58

Hernández, B., Carmen Hidalgo, M., Salazar-Laplace, M. E., & Hess, S. (2007). Place attachment and place identity in natives and non-natives. Journal of Environmental Psychology, 27(4), 310–319. https://doi.org/10.1016/j.jenvp.2007.06.003

Huang, S.-M. (2018). Heritage and postdisaster recovery: indigenous community resilience. Natural Hazards Review, 19(4). https://doi.org/10.1061/(asce)nh.1527-6996.0000308

Hussein, F., Stephens, J., & Tiwari, R. (2020). Towards psychosocial well-being in historic urban landscapes: The contribution of cultural memory. Urban Science, 4(4), 59. https://doi.org/10.3390/urbansci4040059

Jiang, J., Zhou, T., Han, Y., & Ikebe, K. (2022). Urban heritage conservation and modern urban development from the perspective of the historic urban landscape approach: A case study of suzhou. Land, 11(8), 1251. https://doi.org/10.3390/land11081251

https://doi.org/10.52202/082078-0019

Lewin, S. S., & Goodman, C. (2013). Transformative renewal and urban sustainability. Journal of Green Building, 8(4), 17–38. https://doi.org/10.3992/jgb.8.4.17

Liu, C. (2016). Financial incentives and the adaptive reuse of historic interiors: Three case studies from Iowa (thesis). Iowa State University Digital Repository.

Liu, Y., Jin, X., & Dupre, K. (2022). Engaging stakeholders in contested urban heritage planning and management. Cities, 122, 103521. https://doi.org/10.1016/j.cities.2021.103521

Manzo, L. C., & Perkins, D. D. (2006). Finding common ground: The importance of place attachment to community participation and planning. Journal of Planning Literature, 20(4), 335–350. https://doi.org/10.1177/0885412205286160

Masuwan, K., & Lertwattanaruk, P. (2020). Incorporating form-based codes into the design-based approach to historic building conservation in Phuket, Thailand. Sustainability, 12(9), 3859. https://doi.org/10.3390/su12093859

McMillan, D. W., & Chavis, D. M. (1986). Sense of community: A definition and theory. Journal of Community Psychology, 14(1), 6–23. https://doi.org/10.1002/1520-6629(198601)14:1<6::aid-jcop2290140103>3.0.co;2-i

Minner, J. (2016). Revealing synergies, tensions, and silences between preservation and planning. Journal of the American Planning Association, 82(2), 72–87. https://doi.org/10.1080/01944363.2016.1147976

Mohd Rahil, N. R., Abdul Ghani, M. Z., & Sarkom, Y. (2020). Architectural heritage values and sense of place of Kampung Morten, Melaka. Planning Malaysia, 18. https://doi.org/10.21837/pm.v18i14.816

Moher, D., Liberati, A., Tetzlaff, J., & Altman, D. G. (2009). Preferred reporting items for systematic reviews and meta-analyses: The Prisma statement. PLoS Medicine, 6(7). https://doi.org/10.1371/journal.pmed.1000097

Parkinson, A., Scott, M., & Redmond, D. (2016). Revalorizing colonial era architecture and townscape legacies: Memory, identity and place-making in Irish towns. Journal of Urban Design, 22(4), 502–519. https://doi.org/10.1080/13574809.2015.1106917

Rojak, D., & Cole, L. B. (2016). Place attachment and the historic brewpub: A case study in Greensboro, North Carolina. Journal of Interior Design, 41(1), 33–50. https://doi.org/10.1111/joid.12066

Rowntree, L. (1981). Creating a sense of place. Journal of Urban History, 8(1), 61–76. https://doi.org/10.1177/009614428100800103

Salah Ouf, A. M. (2001). Authenticity and the sense of place in urban design. Journal of Urban Design, 6(1), 73–86. https://doi.org/10.1080/13574800120032914

Samir, H., Samargandi, S., & Mohammed, M. F. (2019). Placemaking as an approach to foster cultural tourism in heritage sites. WIT Transactions on The Built Environment. https://doi.org/10.2495/str190281

Scannell, L., & Gifford, R. (2010). Defining place attachment: A tripartite organizing framework. Journal of Environmental Psychology, 30(1), 1–10. https://doi.org/10.1016/j.jenvp.2009.09.006

Shamsuddin, S., & Ujang, N. (2008). Making places: The role of attachment in creating the sense of place for traditional streets in Malaysia. Habitat International, 32(3), 399–409. https://doi.org/10.1016/j.habitatint.2008.01.004

Silva, K. D. (2015). The spirit of place of Bhaktapur, Nepal. International Journal of Heritage Studies, 21(8), 820–841. https://doi.org/10.1080/13527258.2015.1028962

Spennemann, D. H. (2022). The nexus between cultural heritage management and the mental health of urban communities. Land, 11(2), 304. https://doi.org/10.3390/land11020304

Thirumaran, K., Mohammadi, Z., Azzali, S., Eijdenberg, E. L., & Donough-Tan, G. (2023). Transformed landscapes, tourist sentiments: The place making narrative of a luxury heritage hotel in Singapore. Journal of Heritage Tourism, 18(2), 243–264. https://doi.org/10.1080/1743873x.2023.2169151

Wahyudie, P., Antariksa, A., Wulandari, L. D., & Santosa, H. (2021). Place attachment in supporting the preservation of religious historical built environment. IOP Conference Series: Earth and Environmental Science, 737(1), 012035. https://doi.org/10.1088/1755-1315/737/1/012035

Walker, A. J., & Ryan, R. L. (2008). Place attachment and landscape preservation in rural New England: A maine case study. Landscape and Urban Planning, 86(2), 141–152. https://doi.org/10.1016/j.landurbplan.2008.02.001

Wang, S. (2010). In search of authenticity in historic cities in transformation: The case of Pingyao, China. SSRN Electronic Journal. https://doi.org/10.2139/ssrn.1716539

https://doi.org/10.52202/082078-0019

Wells, J. C. (2017). How are old places different from new places? A psychological investigation of the correlation between patina, spontaneous fantasies, and place attachment. International Journal of Heritage Studies, 23(5), 445–469. https://doi.org/10.1080/13527258.2017.1286607

Wells, J. C., & Baldwin, E. D. (2012). Historic preservation, significance, and age value: A comparative phenomenology of historic Charleston and the nearby new-urbanist community of I'on. Journal of Environmental Psychology, 32(4), 384–400. https://doi.org/10.1016/j.jenvp.2012.06.002

Yin, R. K. (2018). Case study research and applications: Design and methods. Sage.

Zhou, T., Zang, T., Jiang, J., Yang, X., & Ikebe, K. (2023). Analysis of the influencing factors of social participation awareness on Urban Heritage Conservation: The example of Suzhou, China. Sustainability, 15(3), 2713. https://doi.org/10.3390/su15032713

https://doi.org/10.52202/082078-0019

IS WHAT YOU THINK WHAT YOU FEEL? HOW PHYSIOLOGICAL RESPONSES DEMONSTRATE THE PERCEPTION MEASUREMENTS' LIMITATIONS

Track 4: Design for Health and Wellbeing
Knowledge Network: Health, Interior Design
Methods: Experimental/Quasi-Experimental Research, Mixed Methods
Hossein Saedi (NC State University)

1 ABSTRACT

Human interaction with nature triggers separate processes in the human body. These interactions have proven to have different impacts on individuals. However, many researchers have solely focused on perception to measure such impacts. What has been far less examined is how the body reacts during those interactions and if the physiological and perceptual reactions are aligned. Therefore, an exploratory experimental research has been conducted to examine whether or not the presence of natural greenery in indoor environments induces restorative physiological responses in individuals. Also, are individuals' perceptions of the restorative qualities of these environments aligned with their physiological responses?

One hundred and eighty two residents of a high-rise residential building were randomly assigned to walk through one of the two versions of their building's lobby. The two versions were modeled in a virtual reality (VR). One version included natural greenery, and the other version did not. The participants' eye movement patterns were recorded via VR goggles as indicators of physiological responses. At the end of the experiment, participants filled out a Perceived Restorative Scale questionnaire to measure the perceptual impacts. The results revealed a significant difference in the eye movement pattern between the two groups. Only those who walked through the lobby with greenery demonstrated validated patterns related to restorative environments. However, the members of both groups did not perceive the lobby as restorative. The outcomes of this research suggest that measuring human perception as the indicator of the restorative impact of an environment might not be enough and may lead to premature conclusions.

1.1 Keywords

Fully Immersive Virtual Reality Environment (FIVRE); Physiological Responses; Perceived Restorativeness; Greenery; Built Environment.

2 INTRODUCTION

The intrinsic connection between humans and natural environments is well-documented, offering a multitude of physical and psychological benefits (White et al., 2021; Keniger et al., 2013; Bratman, Hamilton, & Daily, 2012; Frumkin, 2001; Baycan & Nijkamp, 2009). Nevertheless, as urbanization accelerates and green spaces dwindle (Wang et al., 2019; Hamedani Golshan & Saedi, 2017 (a); Nowak et al., 2008; Barati et al., 2013; Grimm et al., 2008), the need to maximize the restorative potential of even brief interactions with vegetation within urban settings becomes increasingly crucial.

Past research has explored various modes of interaction with nature, such as 2D images, nature videos, and 360° photographs. However, these approaches often fall short of replicating the multi-sensory richness of real-world natural experiences (Schulz, 2015; Skerrett, 2000; Stamps, 1999; Kaplan & Kaplan, 1989).

In response to these limitations, our study harnesses the fully immersive virtual reality environment (FIVRE) to create an experiential bridge between individuals and nature. Recent research has demonstrated that FIVRE can evoke physiological and psychological responses similar to real-world experiences (Jiang et al., 2021; Yin et al., 2019; Bailenson, 2018; Pascoe et al., 2017; Kuliga et al., 2015; Riva et al., 2007). By utilizing FIVRE, this research aims to investigate how natural greenery within indoor environments influences individuals' perception of the restorative quality of a space and their physiological responses to that space. Our study delves into the restorative role of vegetation by closely examining participants' eye movement patterns within a FIVRE. Through gaze tracking in FIVRE, this study unveils whether the presence of natural greenery induces perceptual and physiological responses that align with the qualities attributed to restorative environments. In doing so, this research contributes to the understanding of how even micro-interactions with indoor greenery have the potential to enhance individuals' health and well-being by inducing positive perceptions of their surroundings and positive physiological responses in them.

https://doi.org/10.52202/082078-0020

In this high-rise residential building lobby setting, researchers focus on micro-interactions defined as spontaneous and informal encounters with natural green elements lasting less than a minute. This research seeks to explore the sensory and perceptual impact of these interactions on individuals' well-being. Our fundamental question is whether micro-interaction with natural green elements, thoughtfully integrated into the design of a high-rise residential building lobby, enhances participants' perception of the restorative quality of that space and results in physiological responses associated with restorative indoor environments.

The outcomes of this research offer insights into the potential of indoor greenery to stimulate sensory and perceptual responses, enriching our understanding of how nature can be harnessed to enhance the quality of indoor spaces in densely populated urban environments.

3 RESEARCH OBJECTIVES

This paper endeavors to assess the restorative efficacy of incorporating natural greenery into indoor built environments. The study delves into the nuanced impacts of micro-interactions with such greenery on various facets of individuals' psycho-physiological health, namely perception and sensation also referred to as physiological responses. Additionally, the research seeks to explain potential shortcomings associated with relying solely on human perception as the indicator of an environment's restorative quality. In doing so, alternative measures are proposed to complement perception, providing a more comprehensive approach to evaluating the restorative quality of an environment.

4 RESEARCH METHODOLOGY

This quasi-experimental exploratory study aims to assess the influence of micro-visual interaction with vegetation within a residential building lobby on participants' perceptions of the restorative quality of the lobby and physiological responses. The research methodology employed quantitative data collection methods, including eye and gaze tracking technology and a perceptual questionnaire. In line with the convention in experimental research, efforts were made to establish causal relationships between independent and dependent variables (Thyer, 2012). In this research, the independent variable under scrutiny is the presence of vegetation, while the two primary dependent variables are participants' perceptions of the lobby's restorative qualities and their eye movement patterns.

To maintain experimental consistency among all participants, a fully immersive virtual reality environment (FIVRE) was employed. Nevertheless, certain mediating variables, beyond the researchers' control, could potentially influence outcomes, such as participants' workload, sleep deprivation, or personal family matters. Additionally, participants' individual backgrounds and prior experiences with vegetation, as well as various design aspects of the vegetation used, such as color, height, and materials, could conceivably impact their perceptions of the lobby's restorative qualities.

Given the complexity of the study's variables, it is essential to note that causation cannot be solely attributed to the mere presence or absence of green elements. Numerous unmeasured factors may contribute to the observed relationships. Nonetheless, the data on individuals' gaze locations and the duration of their fixations provide valuable insights into which elements play the most significant role in shaping their overall experience.

4.1 Context

The research context, as visualized in Figure 1, was a fully immersive virtual reality environment (FIVRE) replicating a residential complex lobby. Within this FIVRE, participants encountered two distinct renditions of the same lobby: one featuring natural green elements in its design and the other devoid of such elements.

Employing VR technology for this research affords the advantage of timely and economical design modifications of the study context. Furthermore, VR technology utilization expands the possibilities for multiple experimental iterations (Saedi & Boone, 2018; Saedi et al., 2021). Also, the incorporation of virtual reality technology provided the researchers with enhanced flexibility in adjusting the research location while mitigating potential accessibility challenges commonly encountered with physical spaces (Saedi & Rice, 2019). Furthermore, existing research indicates that a virtual reality environment can engender a rejuvenating effect akin to that of a real-world physical setting, with the principal distinction being that the revitalizing effect derived from a virtual reality experience may be somewhat less potent than that of a physical environment. Consequently, opting for VR was deemed a more manageable and, potentially, more stringent assessment method for these reasons.

4.2 Participants

A total of 182 occupants residing in a high-rise residential complex willingly volunteered for full participation in the experiment, as depicted in Figure 2. Each participant was exposed to just one of the two study context variations. In terms of gender distribution, 63% were male, and 37% were female, with an average age of thirty-five. Of these participants, 164 successfully completed the entire experiment. Notably, seven participants failed to respond to more than half of the PRS questions, rendering their data ineligible for the final analysis.

https://doi.org/10.52202/082078-0020

Figure 1. Snapshots of two versions of residential building lobby at study sites (Top: Toward the building entrance. Bottom: Toward the building elevators.)

Consequently, the dataset was ultimately subjected to analysis comprised of information collected from the 164 compliant participants.

4.3 Experiment Process

Upon arriving at the residential complex, each participant was invited to a separate room to practice and get comfortable using VR equipment. This space did not have natural ventilation and exterior windows, relying solely on artificial lighting and consistent air conditioning. These controlled conditions were maintained to remove potential variables like varying room temperatures or lighting. Figure 2 summarizes the experiment's steps and timelines, followed by all participants.

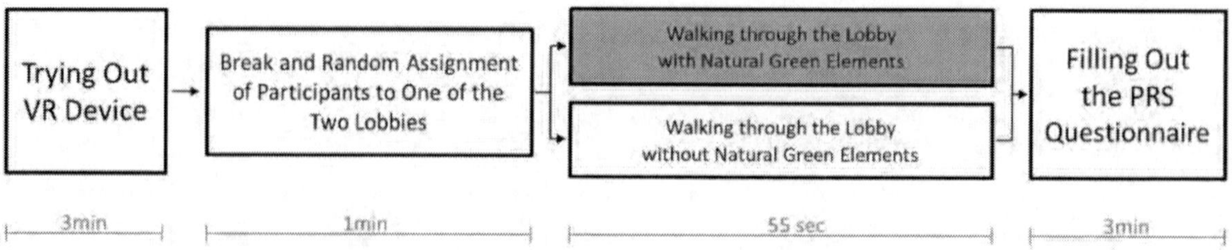

Figure 2 The Procedure that Participants Completed to Successfully Participate in the Experiment

https://doi.org/10.52202/082078-0020

After the practice round, researchers randomly assigned each participant to explore one of two FIVRE versions of the building lobby. Researchers guided participants to a conference room measuring 12x8 meters, depicted in Figure 3. The room was stripped of furniture, leaving only 12 chairs on two sides for participants to rest if they felt any discomfort related to VR technology usage, like nausea (LaViola, 2000). The 10x5 meter area was marked on the floor, outlining the space available during the VR experience. Four SteamVR (SV) base stations tracked participants' movements as they put on VR headsets and ventured through the VR environment. As they moved in the room, their virtual positions shifted in relation to their starting point, tracked by the embedded GPS in the VR headsets.

The study utilized the HTC VIVE Pro-Eye headset. Participants began their immersive journey into the VR context by pressing buttons on their VR controllers. To complete the task and record their experience, they repeated this action when reaching the elevator. Participants were instructed to walk through the lobby from the building's entrance to the elevators at their regular pace, avoiding any rushing or pauses and refraining from interacting with the virtual surroundings, including the vegetation and lobby itself.

4.4 Measures

Researchers used two measures in this study. The first measure was the eye-movement pattern used to measure the physiological responses to the two study contexts. The second measure was the Perceived Restorative Scale (PRS), which was utilized to determine how the participants perceived the restorative characteristics of the two lobbies.
Eye movement patterns have been suggested in numerous studies to have the capacity to serve as criteria for measuring individuals' physiological response to an environment (Holmqvist et al., 2011; Bradley et al., 2008; Loughland et al., 2002; van Dam & van Ee, 2018). Eye tracking provides various patterns regarding individuals' engagement with the surrounding world. Eye tracking allows researchers to study different measures related to the eye movement pattern, such as the fixation number, fixation duration, and eye travel distance. Each of these patterns has its own specific meaning and significance. For instance, a higher number of fixations on a particular part of an image indicates difficulty in comprehending the information present in that area (Wilson & Ehmke, 2007). Multiple research studies have focused on evaluating, elucidating, and interpreting these patterns. Studies such as the study by Berto et al. in 2008 focused on the eye movement pattern in environments with restorative qualities. These studies could identify eye movement patterns that were significantly different between the environments considered restorative and the environments that are not considered restorative (Saedi & Rice, 2022; Franěk, Šefara, Petružálek, Cabal, & Myška, 2018; Nordh, Hagerhall, & Holmqvist, 2013). Those studies suggested that the fixation number and eye travel distance are significantly less in the restorative environments (Wei et al., 2019; Kim, Kim, & Kang, 2019; Galyean, Murphy, & Kontar, 2016). The studies also suggested that the fixation duration is significantly longer in restorative environments (Ohly et al., 2016; Lee et al., 2014; Bowler et al., 2010). These findings also align with the fascination characteristic of a restorative environment put forward by the Attention Restoration Theory (Kaplan & Kaplan, 1989; Herzog, Black, Fountaine, and Knotts, 1997; Joye et al., 2013).

Besides eye-tracking that focuses on eye movement regardless of what eyes are looking at, FIVREs afford researchers to conduct gaze tracking. Gaze tracking involves electronically identifying the focal point of an individual's gaze, known as the fixation location. Additionally, it can track and record the movement of this fixation point, termed fixation duration (Vicente et al., 2015). The gaze-tracking system utilizes the center of the image rendered in front of the participants' eyes as the reference for the gaze's center. This allows the software to document the trajectory of the gaze during the entire virtual reality (VR) experience, also referred to as eye travel distance (Vicente et al., 2015). The data acquired through gaze tracking was employed to create heat maps, illustrating the location and duration of visual attention for every element within the two FIVREs (Duchowski, 2003; Rayner, 2009; Li et al., 2020). Gaze tracking was facilitated using the HTC Vive Pro Eye virtual reality goggles. Subsequently, the data derived from the VR experience was exported to iMotions 8.1 to generate the heat maps.

https://doi.org/10.52202/082078-0020

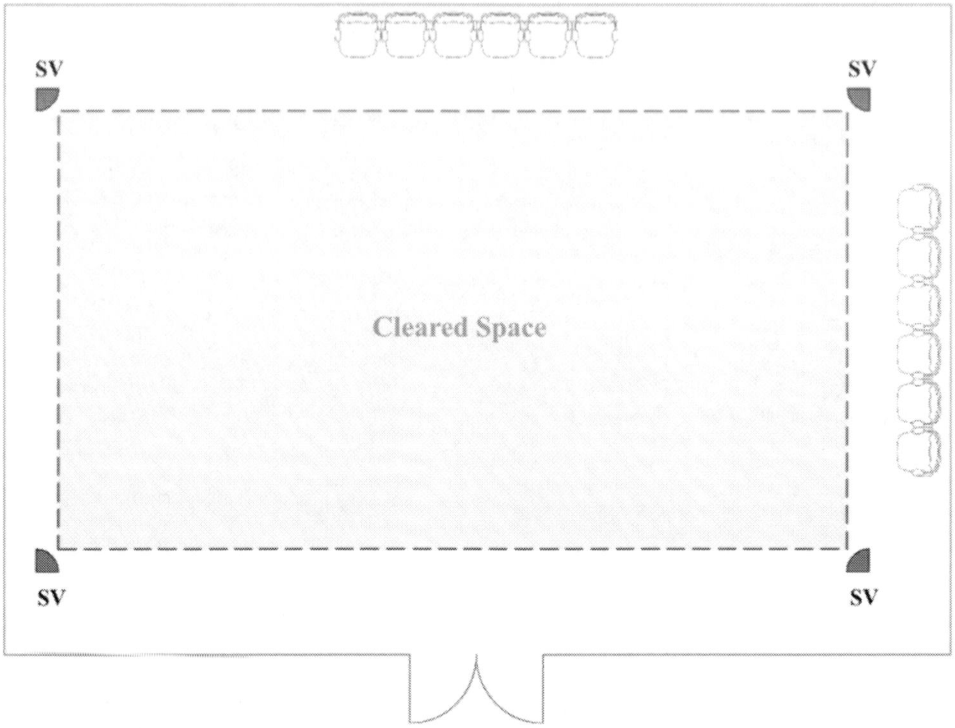

Figure 3 The Layout of the Building's Conference Room, the Cleared Space, and the Four SteamVRs (SV) Base Stations

Figure 4 Snapshots of the Two Cumulative Heatmaps (Left: Lobby with no Vegetation; Right: Lobby with Vegetation)

https://doi.org/10.52202/082078-0020

For each participant, a gaze-tracking point cloud was generated, and subsequently, a cumulative point cloud was created for each version of the study context. These point clouds were then normalized based on the overall fixation location and fixation duration of both participant groups combined. The result of this normalization process was utilized to produce a single cumulative heatmap for each FIVRE. The visual outputs were represented using a color spectrum ranging from violet to red in a rainbow pattern, where violet denoted the shortest fixation duration, and red indicated the longest fixation duration (Figure 4).

It is worth mentioning that the results of eye-tracking and gaze-tracking in terms of fixation number, fixation duration, and eye travel distance are identical. The difference is that with gaze tracking, researchers can generate heatmaps overlaid on the environment that the participants experienced.

The Perceived Restorative Scale is a questionnaire designed to analyze individuals' impressions of the four aspects considered by Attention Restoration Theory for environments with restorative capabilities, namely Extent, Being Away, Fascination, and Compatibility. The Perceived Restorative Scale (Hartig et al., 1997) was the basis for assessing the participants' perception of the lobby's restorative ability. Researchers omitted a few questions from the original twenty-six PRS questionnaire (Hartig et al., 1997) as they applied to outdoor spaces, while this study took place in an indoor space. The result was a questionnaire consisting of eleven questions with Cronbach's of 0.82. Researchers evaluated participants' responses on a 6-point Likert Scale with number one indicating "1= Not at all" and number six representing "6= Very much so."

5 RESULT

5.1 Eye and gaze -tracking

In this study, the fixation number of individuals from two groups was examined. In this evaluation, the fixation number for each group was recorded separately using virtual reality goggles and then calculated by the iMotion Software version 8.1. Additionally, the mean and standard deviation of the fixation number for participants in the study were calculated based on the lobby they experienced, whether it had natural green elements or lacked them (Lobby with Natural Green Elements: $M= 36.15$, $SD= 6.49$; Lobby without Natural Green Elements: $M= 39.15$, $SD= 5.23$).

Subsequently, a one-way analysis of variance (ANOVA) was used to compare the mean results calculated for the two groups. The output of this analysis indicated that the design of the lobby significantly influenced the fixation number in individuals ($F (1, 164) = 19.38$, $p< 0.001$, $2 = 0.25$).

The multiple post-hoc analyses test with Bonferroni correction ($p=0.05/2$) indicated that the eye fixation number in experiencing the lobby with natural green elements was significantly lower compared to the lobby without natural green elements. Furthermore, analyzing the correlation between the mean scores of the four characteristics of a restorative environment and the mean number of eye fixations in the participants demonstrated a significant inverse correlation. This means that as participants' perception of the environment's restorative capacity increased, their eye fixation number decreased (Table 1).

The fixation duration mean and standard deviation were calculated separately for all participants involved in the study (Lobby with Natural Green Elements: $M = 0.41$, $SD = 0.09$; Lobby without Natural Green Elements: $M = 0.37$, $SD = 0.06$). A one-way repeated measures ANOVA demonstrated that the design of the building lobby had a significant impact on the fixation duration mean ($F1.84, 108.70 = 9.82$, $p < 0.001$, $2 = 0.14$).

Since Mauchly's Test of Sphericity was violated, the Huynh-Feldt correction was applied (= 0.92). The post hoc Tukey multiple comparison test with Bonferroni correction ($p=0.05/2$) indicated a significant difference in the mean duration of eye fixations among the participants in the two groups. The results also revealed a significant correlation between the four characteristics of the restorative environments and the fixation duration. This result suggests that the more restorative the participants perceived the environment, the longer their eye fixation duration.

https://doi.org/10.52202/082078-0020

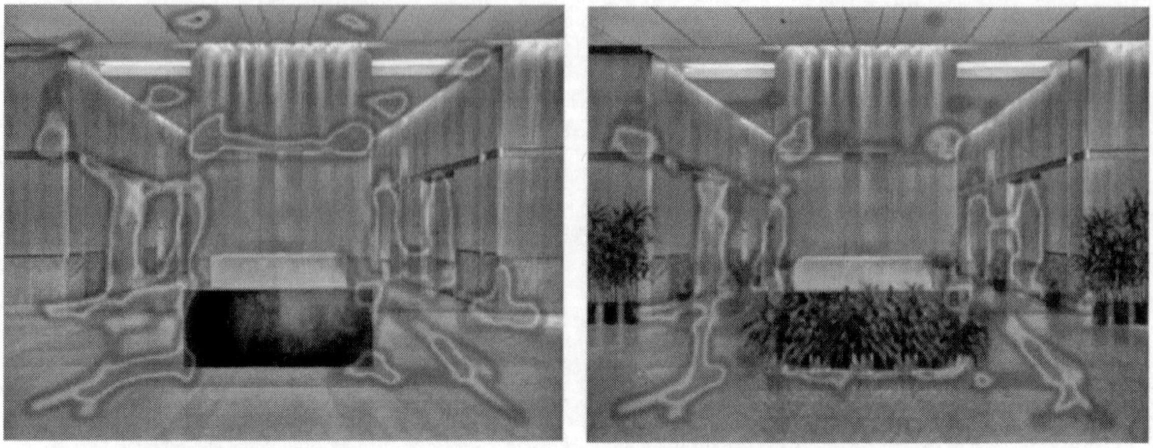

Figure 5 Snapshots of the Two Cumulative Heatmaps for the Study Site, Vegetation Data Excluded(The Point Cloud Associated with the Vegetation was Excluded.)

The participants' eye travel distances were calculated separately for each participant in the study, and the average was measured for the two versions of the lobby. A one-way ANOVA with repeated measures demonstrated that the natural green elements utilized in the lobby design significantly impacted the eye travel distance ($F (1,164) = 4.23$, $p < 0.05$, $h2 = 0.07$). The mean and standard deviation of the data also indicated greater eye travel distance for the lobby without natural elements and a lesser eye travel distance for the lobby with natural elements (Lobby with Natural Green Elements: M = 7284.39, SD = 1303.90; Lobby without Natural Green Elements: M = 6883.14, SD = 1540.85).

The multiple post-hoc analyses test with Bonferroni correction ($p=0.05/2$) indicated that the eye travel distance was significantly higher for the participants who experienced the lobby without natural green elements. Among the four characteristics of the restorative environments, only the mean score of Extent showed a significant inverse correlation with the mean eye travel distance. This inverse correlation implied that a higher perception of the Extent of the environment was associated with shorter eye travel distances (Table 1).

Moreover, the heatmaps generated (Figure 4) illustrated that in the lobby featuring the natural green elements, participants predominantly focused their visual attention on areas with those elements. Conversely, in the lobby lacking greenery, the visual attention clusters primarily centered around the pathways leading to the elevator, the elevator doors, the unit numbers on the walls, and the physical corners within the space.

The eye-tracking results revealed a significant difference in fixation number, fixation duration, and eye-travel distance among participants in the lobby featuring vegetation compared to those in the alternate group. Notably, the sole distinction between the two lobby versions was the presence or absence of vegetation. Consequently, it is plausible to attribute the differences observed in the eye movement pattern results to this varying factor.

However, an alternative argument could suggest that participants in the lobby with vegetation might inherently possess a preference for vegetation, influencing their attention levels. To address this concern, the data within the point cloud related to visual attention directed towards vegetation was isolated from the rest of the dataset. The new dataset was then employed to produce two additional heatmaps (Figure 5) corresponding to the two different versions of the lobby, utilizing the same methodology as detailed in section 4.4.

Figure 5 illustrates that there is no significant difference in the fixation location and fixation duration throughout the VR experience. Without vegetation, all participants in both groups concentrated the majority of their visual attention on comparable elements or areas within the lobby. These results provided researchers with a more robust basis for comparing the heatmaps. Additionally, data collected from the group exposed to the lobby with vegetation uncovered a notable inverse correlation between the duration of visual contact with nature and lower eye travel distance.

Researchers conducted a correlational analysis to validate that the considerable visual attention directed toward vegetation was linked to the significantly lower fixation number, eye travel distance, and significantly higher fixation duration and not merely influenced by the larger volume occupied by vegetation in the space. The same point cloud utilized for generating Figure 5

https://doi.org/10.52202/082078-0020

was employed to investigate if there was a correlation with other elements in the lobby that garnered substantial visual attention from participants, excluding vegetation. The analysis revealed no significant correlation. This result held for the correlational analysis between elements in the lobby without vegetation and individuals' fixation number, eye travel distance, and fixation duration.

5.2 The Perceived Restorative Scale

The data collected from the PRS had a standard distribution and did not include any outliers (≥2.85SD). An independent t-test was used to compare the PRS scale findings between the two groups. The PRS Data demonstrated no significant difference in the perception of the two groups' participants of the two different lobbies' restorative quality introduced by Kaplan and Kaplan in 1989. According to ART, researchers expected to observe that participants who walked through the lobby with vegetation found the lobby significantly more restorative compared to the participants in the other group. The independent t-test result demonstrated that the lobby with natural green elements (M=3.46, S.E.=0.10) and the lobby with no natural green elements (M= 3.13, S.E.= 0.12) were not perceived as significantly different among participants in terms of their restorative quality (t (164) = -3.48, p= 0.061).

Table 1 The Correlation Between the Mean Score of the Restorative Characteristic of an Environment Suggested by the Attention Restoration Theory (ART) and the Mean Eye Movement Patterns

Eye Movement Pattern	Restorative Environments Characteristics			
	Exten t	Fascination	Compatibility	Being Away
Fixation Number	-0.54*	-0.36*	-0.49*	-0.51*
Fixation Duration	0.46*	0.33*	0.45*	0.47*
Eye Travel Distance	-0.43*	-0.19	-0.28	-0.32

Correlations marked by an asterisk are significant at $p < 0.05$.

6 CONCLUSIONS

It has been revealed through the data that micro-visual interaction with natural green elements in a residential building has the potential to have a significant positive impact on the physiological responses of those who walk through that lobby.

Many studies have used perceptual impacts to evaluate the effect of interacting with natural green elements on people's attention levels. Limited studies have utilized physiological responses to measure that effect. Both methods are validated methods used in research related to attention restoration. What seems to be neglected is that encounters with an environment impact peoples' perceptions and physiological responses simultaneously. So, measuring one impact without measuring the other may not provide a comprehensive perspective on the consequences and eliminate the benefits gained from the triangulation of results.

The outcomes of this research demonstrated that physically walking through a VR representation of the lobby with vegetation had no significant impact on the participants' perception of the lobby's restorative characteristics, as measured by the (PSR) questionnaire. On the other hand, participants' physiological responses, as measured by their eye movement patterns, were significantly different and aligned with the validated eye movement patterns in restorative environments after experiencing the lobby with vegetation. These unparalleled results reinforce the idea that different criteria, perception of the space's restorative quality, and the physiological responses after experiencing the space, that researchers utilize to measure an environment's impact on people's attention restoration do not necessarily support each other. Therefore, these two measures should be considered competing factors, and how the restorative quality of an environment is measured may need to be reevaluated.

These results also underscore the necessity of a more holistic approach that brings together all the factors that influence an environment's restorative effect on individuals' attention restoration and reinforces the importance of using multiple measurement methods that allow for triangulation.

https://doi.org/10.52202/082078-0020

The outcome of this research suggests that adding vegetation to the design of an interior space has the potential to change how individuals visually interact with that environment. A change in visual interaction may be a major contributing factor to the restorative effect of designed environments with vegetation. A shift in visual fixation location and fixation duration to green elements significantly impacts the level of restorative effect. Individuals who looked at green elements less often and for less duration demonstrated significantly fewer restorative benefits. These results indicate that the specific characteristics of green elements and how they have the potential to attract and sustain attention may be very important. In addition, the results show that individuals can benefit from the positive impacts of interacting in a VR environment with elements resembling natural vegetation, and this benefit can occur even if the interaction is of short duration.

7 DISCUSSION

The results of this study can be categorized into two conflicting discoveries. The initial observation indicated that moving through the lobby featuring incorporated greenery had a noteworthy influence on the participants' physiological responses, as assessed through their eye movement patterns.

The second discovery indicated that experiencing the vegetated lobby did not lead to a significant enhancement in participants' positive perception of the space concerning the four characteristics of a space examined in accordance with Attention Restoration Theory, which assesses environments with restorative potential. These results were at odds with prior research findings that proposed visual engagement with nature positively affects individuals' perceptions as evaluated using these four restorative attributes of the environment (Abbott et al., 2016; Warm et al., 2008; Kaplan & Kaplan, 1989).

Past research in the field of attention restoration has consistently shown that having visual exposure to natural surroundings, whether for just a few minutes or extended periods spanning hours to days, can notably improve an individual's level of attention (Berto, 2005; Bratman et al., 2015). This present study extends the existing body of research by proposing that the beneficial effects associated with visual engagement with intricate natural settings can also be achieved through micro-interactions with a limited amount of vegetation, lasting for less than a minute.

The researchers employed a fully immersive virtual reality environment (FIVRE) to minimize the potential influence of confounding variables such as noise, room temperature, and lighting on the study's outcomes (Saedi & Rice, 2021). Previous research has demonstrated that VR environments can elicit physiological and psychological responses comparable to those generated by real-world environments. Furthermore, it was anticipated that, under similar conditions, the impact of interacting with vegetation in physical environments would be more pronounced than in VR environments (Bailenson, 2018; Banfi et al., 2018).

One concern associated with the use of VR in this experiment was that VR environments, while capable of producing similar responses, might have a lesser impact on participants, potentially failing to reveal significant differences. Nevertheless, the VR research environment did yield noteworthy variations in physiological responses. It could be contended that if this research were conducted in a physical environment, the observed differences might be even more conspicuous.

While the primary focus of this study was not on the specific attributes of the virtual experience itself, its findings have broadened the repertoire of tools employed to investigate the impacts of nature on attention restoration. This expansion encompasses moving beyond conventional methods involving still images (Laumann et al., 2003), videos (Wang et al., 2016; Pilotti et al., 2015), and 360-degree images (Chung et al., 2018; Moreno et al., 2018) to encompass the fully immersive virtual reality environments (FIVRE).

There are several facets that this study did not thoroughly investigate, offering potential avenues for future research. Initially, this study concentrated on examining the potential influence of visual interaction with indoor vegetation on an individual's attention level. What was left unexplored, however, is the duration of the significant effect on participants over time.

Secondly, it's important to note that vegetation can possess various attributes, including color and height, each of which could potentially affect an individual's attention. In the present study, researchers did not concentrate on a particular type of vegetation or its specific characteristics, but instead employed the options available in Lumion. Now that we have identified the significant impact of interacting with vegetation on attention levels, future studies can delve into the role of vegetation and its design characteristics in the restoration process. These proposed studies have the potential to provide more direct insights into design and planning recommendations.

Lastly, this study assessed the sensational and perceptual effects of interacting with indoor vegetation on participants' attention levels. However, it's essential to acknowledge that there could be additional psychological processes that warrant consideration when evaluating the overall impact.

https://doi.org/10.52202/082078-0020

8 REFERENCES

Abbott, L. C., Taff, D., Newman, P., Benfield, J. A., & Mowen, A. J. (2016). The influence of natural sounds on attention restoration. Journal of Park and Recreation Administration, 34(3), n/a. Retrieved from https://proxying.lib.ncsu.edu/index.php/login?url=https://search.proquest.com/docview/2006815277?accountid=12725

Bailenson, J. (2018). Experience on demand: What virtual reality is, how it works, and what it can do. W. W. Norton & Company.

Banfi C, Kemény F, Gangl M, Schulte-Körne G, Moll K, Landerl K (2018). Visual attention span performance in German-speaking children with differential reading and spelling profiles: No evidence of group differences. PLOS ONE, 13(6): e0198903.

Baycan, T., & Nijkamp, P. (2009). The role of information in the choice of reducing noise pollution: A multinomial hybrid choice model. Transportation, 36(6), 685-704.

Berto, R. (2005). Exposure to restorative environments helps restore attentional capacity. Journal of Environmental Psychology. https://doi.org/10.1016/j.jenvp.2005.07.001

Berto, R., Massaccesi, S., & Pasini, M. (2008). Do eye movements measured across high and low fascination photographs differ? Addressing Kaplan's fascination hypothesis. Journal of Environmental Psychology, 28(2), 185-191. https://doi.org/10.1016/j.jenvp.2007.11.004

Bowler, D. E., Buyung-Ali, L. M., Knight, T. M., & Pullin, A. S. (2010). Restoration in nature: A systematic review of the restorative benefits of interacting with nature in outdoor recreational settings. Environmental Science & Technology, 44(10), 3947-3955. https://doi.org/10.1021/es903183r

Bradley, M. M., Miccoli, L., Escrig, M. A., & Lang, P. J. (2008). Eye movement measures of emotion during a visual search task with emotional faces. Journal of Psychophysiology, 22(3), 148-161. https://doi.org/10.1027/0269-8803.22.3.148

Bratman, G. N., Hamilton, J. P., & Daily, G. C. (2012). The Impacts of Nature Experience on Human Cognitive Function and Mental Health. Annals of the New York Academy of Sciences, 1249(1), 118-136.

Bratman, G. N., Hamilton, J. P., Hahn, K. S., Daily, G. C., & Gross, J. J. (2015). Nature experience reduces rumination and subgenual prefrontal cortex activation. Proc Natl Acad Sci USA, 112(28), 8567. https://doi.org/10.1073/pnas.1510459112

Chung, K., Lee, D., & Park, J. Y. (2018). Involuntary attention restoration during exposure to mobile-based 360° virtual nature in healthy adults with different levels of restorative experience: Event-related potential study. J Med Internet Res, 20(11), e11152. https://doi.org/10.2196/11152.

Duchowski, A. T. (2003). Eye tracking methodology: Theory and practice. Springer Verlag.

Franěk, M., Šefara, D., Petružálek, J., Cabal, J., & Myška, K. (2018). Differences in eye movements while viewing images with various levels of restorativeness. Journal of Environmental Psychology, 57, 10-16. https://doi.org/10.1016/j.jenvp.2018.05.001

Frumkin, H. (2001). Beyond toxicity: Human health and the natural environment. American Journal of Preventive Medicine, 20(3), 234-240.

Galyean, T., Murphy, S., & Kontar, F. (2016). The Restorative effect of the urban environment on ocular microtremor. Journal of Environmental and Public Health, 2737028. https://doi.org/10.1155/2016/2737028

Grimm, N. B., Faeth, S. H., Golubiewski, N. E., Redman, C. L., Wu, J., Bai, X., & Briggs, J. M. (2008). Global Change and the Ecology of Cities. Science, 319(5864), 756-760.

Hamedani Golshan, H. & Saedi, H. (a) (2017). Nanotechnology and green building; The effect of nanotechnologies on optimal use of environmental resources in building, Payam-e-Mohandes Magazine, No.69, 18-27.

Hartig, T., Korpela, K., Evans, G., & Garling, T. (1997). A measure of restorative quality in environments. Housing, Theory and Society, 14(4).

Holmqvist, K., Nyström, M., Andersson, R., Dewhurst, R., Jarodzka, H., & van de Weijer, J. (2011). Eye tracking: A comprehensive guide to methods and measures. Oxford University Press.

Jiang, Q., Bao, W., Liu, Q., & Jiang, Z. (2021). A review of immersive virtual reality technology and its application in the medical field. Frontiers in Psychiatry, 12, 597460.

J

oye, Y., Bolderdijk, J. W., & Steg, L. (2013). Vertical greenery: A field experiment in restorative effects of urban green features on mood. Journal of Environmental Psychology, 36, 76-84.

Kaplan, R., & Kaplan, S. (1989). The Experience of Nature: A Psychological Perspective. Cambridge University Press.

https://doi.org/10.52202/082078-0020

Keniger, L. E., Gaston, K. J., Irvine, K. N., & Fuller, R. A. (2013). What are the benefits of interacting with nature? International Journal of Environmental Research and Public Health, 10(3), 913-935.

Kim, S., Kim, M. H., & Kang, M. J. (2019). Influence of visual scene on fixations during walking in a real-world environment. Frontiers in Psychology, 10, 2517. https://doi.org/10.3389/fpsyg.2019.02517

Kuliga, S.F., Thrash, T., Dalton, R.C., Holscher, C. (2015). Virtual reality as an empirical research tool exploring user experience in a real building and a corresponding virtual model. Comput Environ Urban, 54, 363–375.

Laumann, K., Gärling, T., & Stormark, K. M. (2003). Selective attention and heart rate responses to natural and urban environments. Journal of Environmental Psychology, 23(2), 125-134.

Lee, I., Choi, H., Bang, K. S., Kim, S., Song, M., & Lee, B. (2014). An experimental study on the physiological and psychological effects of forest bathing (Shinrin-Yoku) program on stress recovery: A study on middle-aged male and female office workers. Environmental Health and Preventive Medicine, 19(4), 260-265. https://doi.org/10.1007/s12199-014-0407-1

Li, G., Anguera, J. A., Javed, S. V., Khan, M. A., Wang, G., & Gazzaley, A. (2020). Enhanced attention using head-mounted virtual reality. Journal of Cognitive Neuroscience, 32(8), 1438-1454.

Loughland, C., Williams, L. M., & Gordon, E. (2002). Eye movement as a biomarker of schizophrenia: Using an integrated eye movement score. Psychiatry Research, 113(2), 155-167. https://doi.org/ 10.1016/S0925-4927(01)00076-4

Moreno, A., Baker, S., Varey, K., & Hinze-Pifer, R. (2018). Bringing attention restoration theory to the classroom: A tablet app using nature videos to replenish effortful cognition. Trends in Neuroscience and Education, 12, 7-21. https://doi.org/10.1016/j.tine.2018.07.002

Nordh, H., Hagerhall, C. M., & Holmqvist, K. (2013). Tracking restorative components: Patterns in eye movements as a consequence of a restorative rating task. Landscape Research, 38(1), 101-116. https://doi.org/10.1080/01426397.2012.691468

Nowak, G. J., Gellin, B. G., & MacDonald, N. E. (2008). The evolution of influenza vaccination strategies: The scripted use of vaccines in the U.S. & Canada. Expert Review of Vaccines, 7(4), 417-427.

Ohly, H., White, M. P., Wheeler, B. W., Bethel, A., Ukoumunne, O. C., Nikolaou, V., & Garside, R. (2016). Attention restoration theory: A systematic review of the attention restoration potential of exposure to natural environments. Journal of Toxicology and Environmental Health, Part B, 19(7), 305-343. https://doi.org/10.1080/10937404.2016.1196155

Pascoe, M. C., Thompson, D. R., Castle, D. J., Jenkins, Z. M., & Ski, C. F. (2017). Psychophysiological mechanisms of yoga for psychological health. Frontiers in Psychiatry, 8, 210.

Pilotti, M., Klein, E., Golem, D., Piepenbrink, E., & Kaplan, K. (2015). Is viewing a nature video after work restorative? Effects on blood pressure, task performance, and long-term memory. Environment and Behavior, 47(9), 947-969. https://10.1177/0013916514533187.

Rayner, K. (2009). Eye movements and attention in reading, scene perception, and visual search. The Quarterly Journal of Experimental Psychology, 62(8), 1457e1506.

Riva, G., Wiederhold, B. K., & Mantovani, F. (2007). Neuroscience of virtual reality: From virtual exposure to embodied medicine. Cyberpsychology & Behavior, 10(2), 45-56.

Saedi, H. & Boone, K.M (2018). Virtual reality and student understanding of the third dimension in landscape architecture, 2018 Council of Educators in Landscape Architecture Conference Proceedings.

Saedi, H. & Rice, A. (2021). Towards a More Just Design: The Potential Mental Health Benefits of Vegetation, 2021 Environmental Design Research Association conference proceedings.

Saedi, H. & Rice, Art. (2019). Minor change for major impact: How can minor changes in the daily urban environment impact an aspect of psycho-physiological health? Prometheus, No.4, 96-99.

Saedi, H., & Einifar, A., & Barati, N. (2021). The impact of micro interaction with natural green elements through virtual reality on attention restoration in a high-rise residential building's lobby. Journal of Architecture and Urban Planning, 13(30), 5-21. https://www.sid.ir/en/journal/ViewPaper.aspx?id=821171t

Saedi, H., & Rice, A. (2022). A deeper understanding of the impact on the restorative quality of green environments as related to the location and duration of visual interaction. Journal of Digital Landscape Architecture, 412-424.

Schulz, K. (2015). Handbook of Eye Movement Research. Springer.

https://doi.org/10.52202/082078-0020

Skerrett, D. (2000). The Healthy Eye. Wiley.

Stamps, A. (1999). Mystery of the Eye and the Shadow of Blindness. PressVision.

Thyer, B. A. (2012). Quasi-experimental research designs. Oxford University Press.
van Dam, L. C. J., & van Ee, R. (2018). Pupil dilation reflects perceptual selection and predicts subsequent stability in perceptual rivalry. Proceedings of the National Academy of Sciences, 115(28), E6595-E6601. https://doi.org/10.1073/pnas.1721495115

Vicente, F., Galván, R., & García-Magariño, I. (2015). Using eye movement patterns as implicit relevance feedback for information retrieval. Expert Systems with Applications, 42(4), 1755-1766.

Wang, D., Mann, S., & Liu, P. (2019). Using virtual reality to estimate the impact of environments on travelers' mental workload. Transportation Research, Part A: Policy and Practice, 130, 504-520.

Wang, X., S. Rodiek, C. Wu, Y. Chen, and Y. Li. (2016). Stress recovery and restorative effects of viewing different urban park scenes in Shanghai, China. Urban forestry & urban greening, 15:112–22.

Warm, J. S., Parasuraman, R., & Matthews, G. (2008). Vigilance requires hard mental work and is stressful. Hum Factors, 50(3), 433-441. https://doi.org/10.1518/001872008X312152

Wei, W., Yan, Y., Jia, W., & Shi, J. (2019). Eye tracking to investigate the impact of environmental restorativeness on driver's visual attention and detection of roadside advertisements. Transportation Research, Part F: Traffic Psychology and Behaviour, 65, 344-356. https://doi.org/10.1016/j.trf.2019.08.003

White, M. P., Alcock, I., Wheeler, B. W., & Depledge, M. H. (2021). Would you be happier living in a greener urban area? A fixed-effects analysis of panel data. Psychological Science, 32(11), 1733-1741.

Yin, J., Zhang, X., & Rong, J. (2019). Immersive virtual reality in education: A literature review. Computers & Education, 142, 103647.

https://doi.org/10.52202/082078-0020

DATA OF A CUCKOO'S NEST: FINE-GRAINED DATA AND PHENOMENA IN THE KESEY FARM PROJECT MASTER PLAN AND ARCHIVE GALLERY DESIGN

Track 1: Nature and Urban Ecology
Knowledge Network: Cyberspace & Digital Environments, Sustainable Planning, Design & Behavior
Methods: Design Methods, Practice-based Research

Philip J Speranza (University of Oregon)

1 ABSTRACT

Farming landscapes are often chosen as artistic retreats outside the city. They offer a sense of escape, a connection with nature via subsistence (Thoreau), healthy environmental qualities of air (Sunyer, 2017), light, heat, humidity, sound, and smell, and freedom from neighboring eyes or political order of law. Some historical retreats carry a magical feeling of memories of past use. They can attract pilgrimages from around the world. When a historic retreat requires preservation and architectural design is needed, how can data-driven design methods be used to measure and reconcile a magic of the place? These bucolic locations can now be quantitatively measured using geolocational data from sensors, simulation, and drone photography. These design capabilities provide a new way to understand, capture, and add to the magical feeling of a place. This paper will discuss the investigation of a data-driven design process that links natural and cultural phenomena using an array of sensor-based kits for master planning and architectural design of the historic farm of 1962 One Flew Over the Cuckoo's Nest author Ken Kesey for a contemporary artist-in-residence program led by his granddaughter Kate Smith-Kesey and a future archive and gallery building. Theory about place and multi-sensory phenomenological design (Pallasma, 1996) will be explained within the context of current technological methods of data collection, analysis, and design. Visioning and programming of the non-profit will be described. The site planning design process will explain analysis techniques for sensor and simulation methods. Design methods will then be shared that link quantitative data with qualitative phenomenological design. How can the cultural purpose of a place adapt over time, and what are the key scales of time and space to change the land use from agricultural farming to a community center? What are local and global implications knowing this was the place of Ken Kesey's contribution to new global perspectives on mental health and the use of mood-altering states in creative processes? Conclusions will focus on learning outcomes for the architect, the Kesey family, non-profit stakeholders, and student interns measuring the fine-grained urbanism qualities in time and space.

1.1 Keywords

Data-Driven Design, Phenomenology, Historic Retreat, Simulation, Cultural Adaptation

2 THEORETICAL CONTEXTUALIZATION

Environmental design processes begin with understanding the time-based experience, observation, and recording of a place. Ideas of place attachment (Norberg Schulz 1980), attachment (Latour, 2008), place attachment (Manzo & Perkins 2006) are engaged with methods of data collection around film (Whyte, 1980), drawing patterns (Appleyard 1980) and non-physical use programming of events (Tschumi, 1994). Architects such as Steven Holl speaks to the 'anchoring' of a project in the sensory experience (1995) of a site and apriori pre-theoretical ground, referencing Heideggers's philosophical importance of lived experience. These ideas were of relevance at the place of Ken Kesey's home, writing studio, and communal gatherings.

A conceptual design question for a contemporary Ken Kesey Farm Project artist-in-residence campus and archive/ gallery considered the equal importance of ideas as perception of the outside world and conception of the internal mind (Kant 2002). How do visiting creative artists and visiting researchers understand both the past uses of the place and methods of mind-altering conceptualization offered by Kesey? What is the importance of individual and collective experience of three periods: 1) the Merry Pranksters, departure point of the Further bus tours, Grateful Dead concerts including the Springfield Creamery concert Sunshine Daydream (Oregonlive 2022) and other 1960's hippy culture documented by Allen Ginsberg; 2) the post-hippy era of creative retreat for Gary Fisher's invention of the mountain bike, Paul Newman, Bill Murray and other; and 3) and the contemporary use in recent years of young artists and visiting pilgrims from around the world? How can Kesey's granddaughter, the Kesey family, and the recently formed 501(c) non-profit pass along the spirit of past experience in the current age of land use NIMBYism, building code health and safety regulation, and arts administration management? The aim of this paper is to share an inquiry to relate data-driven, site-acquired sensory data with an architectural position based on phenomenology.

https://doi.org/10.52202/082078-0021

Tadao Ando's photographic framing of moving light on benches, Holl's conceptual anchoring of light and sensory experience, RCR Arquitecte's engagement of weather atmosphere, and Solid Objectives idea of aura all embrace phenomenology in the architectural approach. Other environmental designers today have a more technical anchoring using computational sensors and relationships to human comfort - Philippe Rahm for thermodynamics of the body, Battle Roig for sensors in park design, Anne Lacaton and Philippe Vassal for passive cooling in affordable housing, and Sarah Williams geospatial demographic research Million Dollar Blocks (2020). Current technologies empower designers to extend their perception, analysis, visual communication, and design proposals for a place. A set of critical research questions thusly engages the scales of space for the current Kesey Project community center and the scales of time for its use, and this project ultimately challenges a land use definition of community and the continued legacy and value of this place to society.

3 METHOD: MEASURING FINE-GRAINED PLACE, SITE, AND BUILDING

The link between phenomenology, data-driven analysis, and design process will be shared 1) first looking at sensor driven site analysis, and then design for the 2) master planning of the Kesey Farm for the overall artist-in-residence program, and 3) architectural design of an archive for preserving and sharing materials for individual researchers and gallery space for exhibiting artist-in-residence and other work around collective gatherings and intellectual discourse of the arts locally, in Oregon and globally. The design process was pro-bono work done by the architecture and urban design practice of the author, handed over to a land use planning consultant for submission to county land use approval subsequently denied. A new approach around heritage and value beyond local community use is being investigated.

3.1. Site analysis

Kesey Farm, Pleasant Hill, Oregon. On-site visits occurred to the location in Pleasant Hill, Oregon, approximately fifteen minutes outside Eugene. The first site visits were provided by the Kesey family of both the rural farm grounds and the original barn house with their testimony of life there and answered questions. The house's ground floor is adorned with paintings from the 1960s using psychedelic colors - tied-eye shirts were first done here - by visitors of every surface including walls, floors, and ceilings. The kitchen and living offer a view north to dairy grazing fields. A narrow stair led to one of Ken Kesey's writing studios facing west and then to a large barn space on the second floor with folklore stories of transient guest inhabitants, LSD parties, and a point when open flame candles led Ken to ask people to leave for the safety of his family and others. The visit followed with a passage to a rear addition with a later writing studio facing south with solar overhangs overlooking the family's dairy grazing fields and distant hills.

The site visits then explored parts of the land including a summer pond and tree-covered marshall for dry but shaded in the summer. The adjacent open field has a fence-enclosed family cemetery for Ken and his son Jed who died in the high school wrestling team bus accident. The walk closer to the road revealed perceptions of privacy from passersby and noise from passing vehicles. First visits occurred in the spring but continued in autumn, winter, and a Music in the Landscape event with a concert pianist but then each of approximately one hundred guests was given headsets to either be with the piano or wander on their own into the landscape. The result of these in-person sensory visits was the need for passive cooling ventilation and shading in summer, sound filtration from the road, and ideas of individual vs collective experience.

Off-site Climate data analysis, visualization, and simulation. To understand the perception of locational climate data, a comparative study with other Temperate Mediterranean CsB Koppen Climate Zones was done with Palo Alto, Rotterdam and Tokyo. Ken wrote Cuckoo's Nest while at Stanford in Palo Alto as an experimental subject of government-sponsored LSD testing. Other locations were of recent family travel or comparison.

- Humidity peaked in summer at 30% in Eugene and almost 100% in Tokyo with NOAA data characterized as "muggy."
- Rainfall (Daily chance of precipitation) in the summer was almost zero in Eugene and relatively steady but peaking at 8 inches in late summer in Japan and in northern Europe.
- Temperature average annual peaks were 106-108 degrees in July and August in Eugene, and still in the 80's in July and August in Palo Alto.
- Temperature average high to low in July and August were 82.2 to 51.7 degrees F in Eugene and 75 to 59 degrees F in Palo Alto, demonstrating greater swings in Eugene.
- Results show that the summer in the Eugene area is hotter, almost no rainfall, higher peak temperatures and slightly higher ranges of daylight/nighttime temperatures.

Metabolism was studied with an eye toward human comfort, heat and humidity relationships, and metabolism, wind rose analysis was done by season revealing summer breezes from the north at 10-12 MPH. A timeline diagram was then established for key uses on site: making meals, farm work, maintenance, and gathering events. The timelines showed key farm work primarily in the spring through fall, events focused in the summer months and maintenance

https://doi.org/10.52202/082078-0021

focused in the summer.

An ecological collage was done using the above data relating specific seasonal periods, max and min average temperatures, and the time-based perception of fowl organized by mammals, birds, and insects, and fauna organized by shrubs, vines, grasses, and trees. The results visualize patterns of key ecological overlap with high arts programming periods like the summer including high periods for birds and insects more than mammals, and more vegetation filtering opportunities in the spring before summer dryness and heat suppress vegetation growth.

On-Site Arduino Sensor Data Acquisition. Ten custom-built and scripted Ardunio-based sensor kits with DHT-22 temperature and humidity sensors, sound and light sensors were placed around the site based on perceived experiential locations by the Kesey family including the summertime marsh tree grove, locations by the noise of the street, around the house proper and at possibly locations of the proposed archive gallery building near the entry road and facing west. The sensors were placed for two-to-three-day data collection periods for battery life between April 28th and May 9th, 2019. The data was visually analyzed at peak time periods 1) morning 8-10 am, middays of 12 pm-2 pm, and afternoons 4-6 pm. This period of the year fits well with visioning for programming activities for the artist-in-residence program around moderate temperature and peak vegetation filtering opportunities. Off-site solar radiance simulation was also used to be compared with on-site sensor-based solar radiance analysis. Midway through the data collection, it was inferred that a grazing cow pushed the sensor along the pond location underwater and data was scrubbed accordingly. The potential effect of livestock on sensor locations was noted.
The results of this can be seen in the figure below.
Temperature and Radiance Analysis:

- Radiance simulation shows opportunities for comfortable lower temperatures behind lower trees west of the site and taller trees in the east fry marsh. This is confirmed with sensor data in marsh but lack of resolution of sensors placed at the base of the west trees did not show that. Still, cooler areas were proven under cover of the east marsh and at the pond location possibly from evaporative cooling.
- Humidity levels were lowest at the vegetation garden location south of the barn house, and highest toward the cemetery location.
- Sound levels were high at the road location but were highest in unobstructed areas some distance from the road but near the bus barn, possibly amplified by reverberation. High levels were also measured midday in the east marsh area, and afternoon by the pond.

Figure 1. Kesey family, Kesey and the Merry Pranksters Further bus, Allen Ginsburg Bill Murry, artist in residence programming by Kate Smith-Kesey.

https://doi.org/10.52202/082078-0021

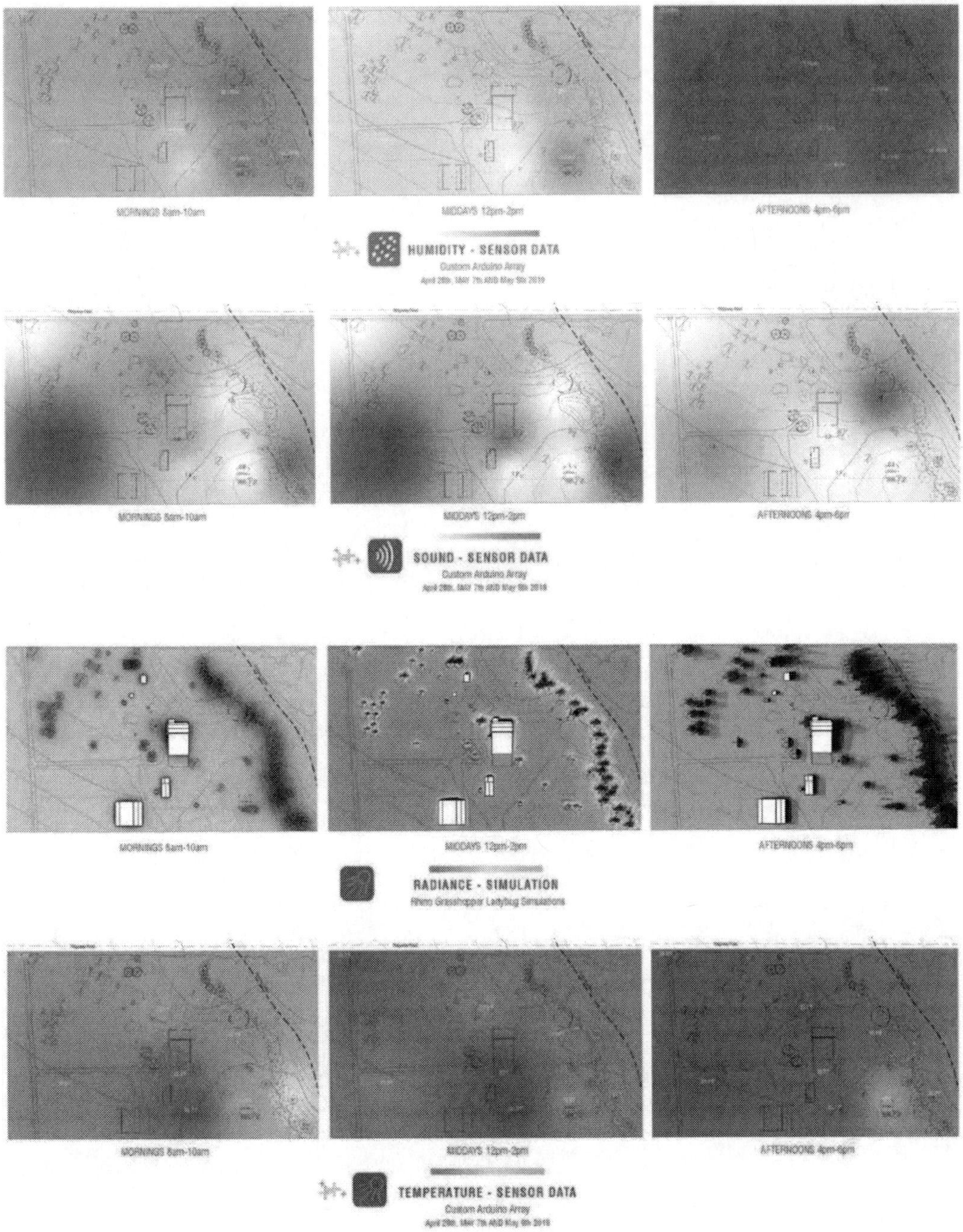

Figure 2. On-site Sensor Data for One Test Period. Diagram by the authors.

- A radiance simulation was also done across the four seasons of the year with visualized patterns of peak heated areas as expected in the open areas throughout the year but noticed in the autumn and winter as desirable and led to a small-scale difference of coniferous versus deciduous trees, and particular species of ash that lost their leaves early in the fall. Trees provided little shelter summer and spring likely from high sun azimuth. Areas north of building structures were shaded.

The design process documented Lane County Land Use Code requirements for the parcel zoning of Exclusive Farm Use. Any new construction would require a land use review. Allowable uses for non-farm units include a community center operated by a non-profit, and thus the Kesey Farm Project 503(C) was filed federally alongside the design study. A planning consultant explained the County and watch group's concerns for effective parking accommodation, traffic on the adjacent road, calendar of events and large gatherings, noise, and sanitary facilities. An interview and visit occurred with the local University of Oregon Natural History Museum archive preservation specialist and their facility. The interview presented important planning criteria for both the master plan and archive facilities to consider the following: fire, vermin, earthquake, vandalism, theft, humidity, heat, and airflow. Both the land use process and the preservation criteria added perspective to environmental qualities of air, heat, humidity, sound, and light begun off-site and on-site.

3.2 Data-driven Design Process, Master Plan of Artist-in-Residence Programming

Kesey family testimony about the artist's individual and collection experience of the site, integrated with human-scale data led to a design process around human comfort with both social and natural phenomena in mind. A human comfort generative design diagram was organized around atmosphere conditions from cool to warm, filtering capabilities of vegetation from ground cover to canopy, and metabolic activity of people from resting/sleeping to intensive exercise (left, figure 4). Atmosphere considered wine, rain, moisture, and heat in parallel with perceptual conditions of sound and visibility. Vegetation filtration was organized with input from a landscape architectural designer including grasses seen as moving in the wind, shrubs such as blueberry, huckleberry, and red flowering currant, vines, and trees at the creek such as cottonwood, ash, and oak, in a proposed orchard such as apple, plum, pear, and hazelnut and around the house such as port orchard cedar and red maple. These were aligned left to right, from cool to warm, aligned with metabolic activities from resting to moderate activities such as standing to paint, draw, and creating in general, to walking and working, then running dancing, and hiking, and lastly high-intensity activities such as cycling, skiing, and climbing, all capable in the larger 'site' of Oregon within an hour from the Kesey Farm location. The aim of this diagram before formal site organization was to establish an organized awareness of atmospheric phenomena, strategies to filter them with vegetation, and aligned with human activities on the site.

Three design strategies emerged after several sketching studies were done considering the overall site plan and various moments of human programming experience. Some of those design studies were based on existing conditions such as immediately adjacent to the historic barn house and others were invented such as conditions at and around the proposed archive gallery to the west along the entrance drive. The design strategies include: 1) Fragmented space with spaces with no hard edges, spaces that bleed and encourage chance encounters. This would suggest spaces of disconnection, artists' studios, spaces for hyper communal space and fire puts, south patio, showers and the historic barn attic. 2) South Orientation considering view to Oregon pastoral context as experienced historically by Kesey and others; 3) Atmospheric filters primarily by season including winter radiation from the south and setting west sun and night flush cooling, winter need for light and deep sun infiltration, spring damp and cold and fall as relatively dry, with the year-round consideration for sound. For each of these conditions, a small thumbnail of a unit of human experience was graphically generated.

The human comfort diagram and the design strategies were then used to inform a master plan site plan and 3D design. The site plan includes locations of key historic and proposed structures, connective pedestrian paths and vehicular drives, and specific vegetation planting. The drawing uses icons to help communicate design making from the diagramming process to the master plan proposal. A composite rendering was then generated that helps visualize informed process of the design with superimposed data from a point in time. The aim of this drawing is less to communicate a final master plan but to highlight the 1) data of atmospheric conditions, 2) the vegetation as part of a shaped Oregon landscape and 3) the magical moments of creative human activities hoped for at the Farm.

https://doi.org/10.52202/082078-0021

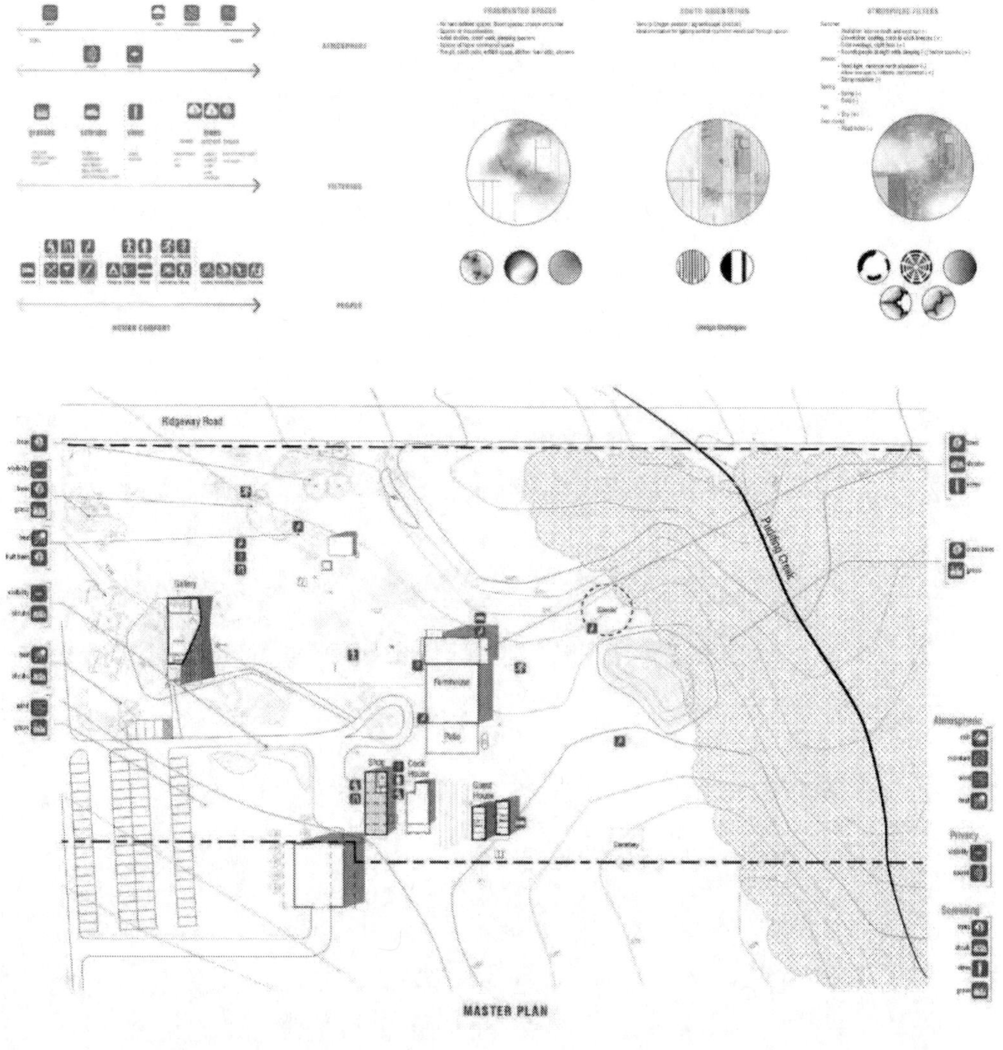

Figure 3. Human Comfort diagrams, design strategies diagrams, and master plan. Diagram by the authors.

https://doi.org/10.52202/082078-0021

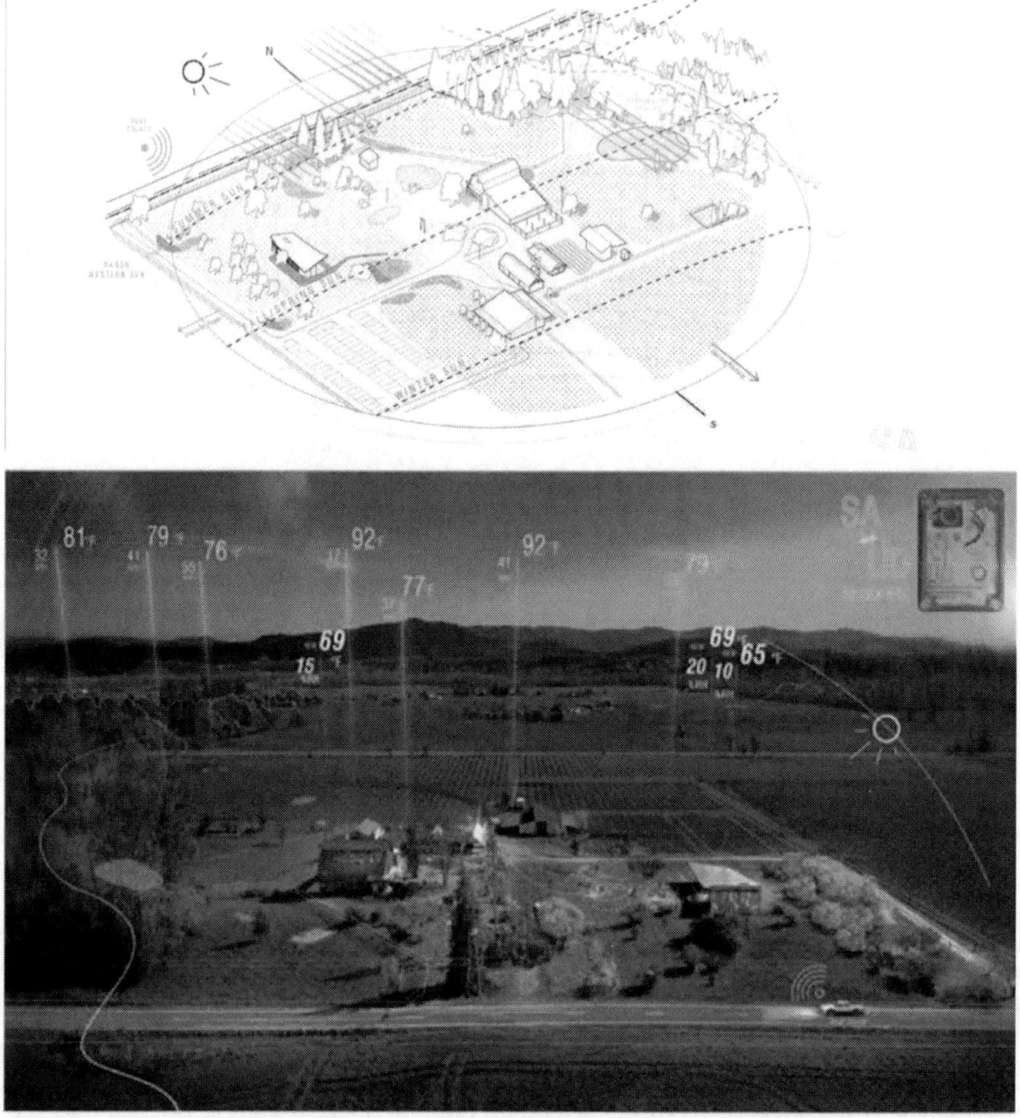

Figure 4. 3D diagram of environmental qualities and master plan aerial perspective with sensor data informing trees, shrubs and grass species to filter light, sound and humidity. Diagram by the authors.

https://doi.org/10.52202/082078-0021

3.3 Data-driven Design Process, Architecture of Archive Gallery Building, Knowledge Learning

Design of the archive gallery building began with meetings with the Kesey Family to see and discuss the collection of material and a visit and discussion with the Museum of Natural Science to discuss the technical needs such as protection from pests, bacteria, fire, theft, seismic damage and water damages are minimized, as well as space storage efficiency. In parallel a visit to the historic barn house led to research into heavy timber Dutch barn construction especially regarding the interior hay loft space with exposed structure as a place of hippy celebration sheltering from the Oregon weather winter and summer. The study of other arts foundations such as the Judd Foundation in NYC and Marfa, TX, Summer Springs, Zittel Residency, and Steven Myron Holl Foundation. The programming organization of the archive gallery also included administrative space and a porch for interior and exterior gathering. A series of parti, or spatial organization digital 3D modeling studies of assembly, bar, and square were done to understand radiance analysis for summer and winter to achieve strategies of summer passive cooling and winter passing heating.

This tested organization of spaces was done alongside an apriori (Holl) study of how visitors might interact with the Ken Kesey collection of written material and articles such as clothing.
The interior use was explained as a Public Use and Private use. The four primary public uses were:
1) public porch and gathering, 2) rotating gallery display, 3) archival display, and 4) central reading table. The private use focused on preserving each piece in the collection. The archive is a workspace to receive, decontaminate, provide support, wrap, and sometimes box all princesses before placing them in non-visible or visible storage. A conceptual diagram of spiraling knowledge was made to explain the experience of the visitor and their interaction with the archive material of Kesey, communicating both the depth of knowledge and the intimacy of the individual experience perspective on Kesey by the visitor was important. It also represented the idea of Kesey not only as an individual but as a larger collective movement, part of the hippy movement but over the decades of his life, the surrounding of friends, visitors, and family resulted in a collective experience of shared knowledge, shared experience, storytelling, travel represented in the Further bus tours and the territorial idea of the experiencing the American landscape.

While drawings of the plan, 3D digital massing, site context, and physical models were done, the design process was done collaboratively with the Kesey Family through a series of feelings-rendered drawings aimed more at exploring the sensory experience of the space by individuals and collective gatherings, rather than a technical explanation of architecture or construction alone. The feeling renderings were the result of a process of considering the environmental qualities of phenomena of the data gathering of light, heat, sound and humidity gathered in the master planning analysis work. As the data collection occurred by the student workers alongside their human experience of the historic site while listening to stories from the family and seeing pictures of the hippy culture, the design process aimed to simultaneously consider both the quantitative hard science of the data but also the visceral, human experiences of art, love, mind exploration, joy and knowledge that spiritually engage people. The renderings included ideas of being connected, in nature, together and revelation. These images were not aimed to render a final design but talk to a connection of human experience, the interior environment and the structural building that would shelter them.

https://doi.org/10.52202/082078-0021

Figure 5. Interior study of archive reading table, barn study for construction type, knowledge spiral diagram, and exhibition and archive space. Diagram by the authors.

https://doi.org/10.52202/082078-0021

4 KESEY FAMILY AND STUDENT DESIGNER EXPERIENCE

The masterplan and archive gallery data-driven design process was done by a small studio including current university undergraduate and graduate students as well as full-time staff. The students had taken computational urbanism course work including the ability to geospatially acquire data at the geographically small, human-scale as well as application of this method on other professional projects. At least one of the students had a strong educational and professional experience in art, providing a knowledge bridge to the Artist-in-residence program. The students, working hourly in the office for the Kesey Farm Project, engaged in site visits, client meetings, data acquisition, and other on-site and in-person interactions with family and artist stakeholders. At least one student work continued on to a media course around art, inclusion, and displacement for the Black community in Portland's Albina neighborhood and is currently working to extend their capstone university project to a real-world project in Portland.

Participation and feedback on the master plan and archive gallery were elicited on two occasions - a public event called Music in the Landscape and a fund-raising dinner for the Kesey Farm Project. The fund-raising dinner provided an opportunity for Kate Kesey-Smith to present the design work, a short outdoor slide presentation, printed posters, pamphlets for people to take home, and a book for comments. Feedback was provided about location, use, and master plan vegetation planning. The Music in the Landscape event featured over a hundred visitors listening to an outdoor piano concert while each visitor had wireless headsets to physically roam into the landscape of the farm creating a play of individual and collective experience at the control of the visitor. These two events of the fund-raising dinner and Music in the Landscape were observed by the designers as ways that people might interact with the masterplan exterior space activities and the archive gallery indoor and exterior porch spaces.

Submission of the master plan and archive gallery was made to the Lane County Land Use Department for proposed additional construction and use permits. The agricultural zoning has an allowable use for community center buildings which previously would include places of worship and secular community centers. The submission was prepared by a retired county planner using limited material from the designers' materials but primarily included the site plan. The review process was marked by public input and County officials forwarding to the Kesey family protesting letters from LandWatch Lane County citing parking requirements in metric systems and outside legal requirements from the County. Questions were raised about traffic, parking, and noise but were all responded to carefully. The proposed land use was denied citing that the use would not be locally oriented by land use requirement. This process provided an interesting learning experience for the designers, the non-profit, and Kesey Family that the local county land use would not acknowledge the value of cultural contribution, and while national and international visitors may occur in the future somehow that could diminish the value of both the Kesey Farm artist-in-residence use and preservation value of the archive gallery, and the interaction between local and non-local learning. Meanwhile, the designer was in Barcelona that subsequent summer and witnessed an exhibit including video and other material of the Kesey Farm and Ken Kesey at the Museum of Contemporary Art of Barcelona MACBA demonstrating the important line of Kesey between the Beatniks of William Burrough and Jack Kerouac, and the psychedelic lineage after Kesey and the Merry Pranksters including the work of the Beatles and subsequent counter-culture of race by Jean-Michel Basquiat and gender in the art world by Pauline Boty.

5 CHALLENGES / LIMITATIONS AND NEXT STEPS

The primary challenge of connecting phenomena-oriented design with data-driven design processes is the connection of the spiritual human experience with quantitative analysis. The ability of the designer to process and then translate between the qualitative and the quantitative in both the analysis and then design synthesis process is central. In the end, the design process begins with and relies on clear and consistent communication with the owner stakeholders. Challenges were met when the design meetings did not always include all family members, with Kate Kesey-Smith not present at many of the early meetings. Email and other non-present communication were done for this project in 2019 but the Zoom and other distant sharing informed by COVID would have likely provided more inclusion. Sharing and receiving feedback at the fundraising event and Music in the Landscape could have included more interaction and input and may have elicited a critical champion in the community to help address the problem that emerged between the local and non-local value cited by the Lane County Land Planning & Zoning review.

A serious challenge during that Count land use review was the positioning of the project as a quiet, local use versus one that would serve the local community but would be a global heritage site with careful consideration for potential environmental qualities such as sound and traffic, qualities carefully studied in the data-driven design process. That challenge to engage a local government that forwarded private input from LandWatch NIMBY group rather than

https://doi.org/10.52202/082078-0021

citing legal requirements could have been affronted differently either legally or via political input from a higher State authority. Such challenges political and legal were limitations beyond the expertise of the environmental design team. Another related challenge was the bottom-up artist-in-residence program initiated in the spirit of Ken Kesey by granddaughter Kate Smith-Kesey, as a challenge to the land-use authorities when the family's aim of the master planning and archive gallery process was alignment with land use requirements and building requirements of health and safety. The next steps for the project might include embracing the issues of County land use from the bottom up or looking for a champion in the arts and culture community of Oregon from the top down to acknowledge the cultural value of such a community center to an otherwise limited cultural exhibit space of Eugene-Springfield though it does demonstrate increased travelers around athletics including the 2022 World Track and Field Championships, summer Bach Festival, and other University athletic and cultural programming.

6 CONCLUSIONS

Data-based inquiry in the context of a master plan offers a new way to see and experience the site in a way that is congruous with the digital design processes of today. The sensor-based analysis was a way to communicate and draw human experiences that were previously emergent from the lived experience of Kesey and that counterculture on the farm. The contemporary condition and or that of this new project was that of owners, professionals, fundraising sponsors of arts management, and a rigorous county land use process. Proving the 'magic' was necessary for the sponsorship and regulatory processes today. Furthermore, the science of decay from moisture, risk of seismic and fire peril, and other forces on the archived materials was not originally considered. Thus the analysis using data-driven processes was able to convince a careful approach at the master plan scale to use vegetation for filtering phenomena and key strategies at the scale of the building to bring in light for archive reading and exhibition purposes.

Communication phenomena require some mode of visual communication that includes time. Creating a series of diagrams both for studying sequences analysis periods and of design thinking evolution was needed not only for the designers but to explain the work to important stakeholders. Ultimately this may have been a missed opportunity in the land use submission not to include graphic timelines and other data images of use to communicate the local and global community use and the infrequent event planning impact of sound and traffic on neighbors. The most powerful conclusion is the challenge of adaptive reuse, or continued intentional use, over changing context over generations. Kesey's use of the space adapted over his lifetime, and his granddaughter's artist-in-residence took on similar aims. However, perhaps given the spotlight of Ken Kesey's use, Kate Smith-Kesey's adapted use was scrutinized through the lens of contemporary land use, building safety, and the need for formal arts funding processes. The scale of distance that the farm is now a destination for Kesey pilgrims from around the globe is vastly different from fifty years ago.

7 REFERENCES

Holl, S. (1995). Anchoring: Selected projects 1975-1991. Princeton Architectural Press.

Kant, I., interpretation by Kitcher, P. (2002). Kant's three critiques: Critique of pure reason/critique of practical reason/critique of judgment. Hackett Publishing Company.

Kesey, K. (1962). One flew over the Cuckoo's nest. Signet.

Latour, B. & Yaneva, A. (2008). Give me a gun and I will make all buildings move: An ant's view of architecture." In R. Geiser (ed.), Explorations in Architecture. Birkhäuser.

Manzo, L. C., & Perkins, D. D. (2006). Finding common ground: The importance of place attachment to community participation and planning. Journal of Planning Literature. https://doi.org/10.1177/0885412205286160.

Norberg-Schulz, C. (1980) Genius Loci: Towards a Phenomenology of Architecture. Rizzoli.

Pallasmaa, J. (1996). The geometry of feeling: a look at the phenomenology of architecture. In K. Nesbitt (Ed.), Theorizing a new agenda for architecture, an anthology of architectural theory (pp. 447-453). Princeton Architectural Press.

Perry, D. (2022). Grateful Dead's 1972 Oregon concert saved Kesey family dairy, became counterculture legend. Oregonlive. https://www.oregonlive.com/history/2022/08/grateful-deads-1972-oregon-concert-saved-kesey-family-dairy-became-counterculture-legend.html.

Tschumi, B. (1994). The Manhattan transcripts. Wiley.

Whyte, W. H. (1980). The social life of small urban spaces. Conservation Foundation.

Williams, S. (2020). Data action: Using data for public good. The MIT Press.

https://doi.org/10.52202/082078-0021

SPATIAL JUSTICE IN RESOURCE MANAGEMENT: ENGAGING COMMUNITIES FOR EQUITABLE OUTCOMES IN THE FOOD-ENERGY-WATER NEXUS OF PHOENIX AMA

Track 5: Spatial Justice, Equity, and Inclusion by Design
Knowledge Network: Cities & Globalization, Participation
Methods: Focus Groups, Interviews
Marzieh Rezaei Ghaleh (Arizona State University); Dave White (Arizona State University)

1 ABSTRACT
This research investigates the intersection of sustainability and justice within the Phoenix Active Management Area's (AMA) Food-Energy-Water (FEW) nexus, an area confronting critical water scarcity and heat challenges. Engaging directly with Phoenix AMA's struggles, this study situates its investigation within the broader discourse on spatial justice and the management of FEW resources. Utilizing participatory stakeholder engagement and anticipatory modeling, we explore Phoenix AMA's unique vulnerabilities, including its governance efforts toward sustainable water usage, food, and energy production. Through this localized lens, our interdisciplinary approach examines the intricacies of FEW systems, highlighting the role of socio-ecological-technological changes and uncertainties. By integrating diverse stakeholder perspectives—from city officials to community organizations—into our model development and evaluation, the study reveals the critical importance of collaborative governance and inclusive participation for advancing spatial justice in resource management. This engagement proved pivotal in refining the model's applicability and enhancing its utility for policymaking and resource allocation in Phoenix AMA. The insights garnered provide actionable strategies for addressing Phoenix AMA's specific sustainability challenges, offering a valuable framework for other urban areas grappling with similar FEW nexus complexities. Our findings advocate for a nuanced understanding of spatial justice in environmental planning, underscoring the necessity of aligning sustainable resource management with equity and inclusion.

1.1 Keywords
Spatial Justice, Inclusive Engagement, Food-Energy-Water Nexus, Participatory Modeling, Phoenix AMA

2 INTRODUCTION

In the realm of sustainable development, the intricate interplay between food, energy, and water systems—collectively known as the Food-Energy-Water (FEW) nexus—has emerged as a critical area of study. This nexus represents a complex web of interactions increasingly recognized as pivotal to addressing global challenges such as climate change, resource scarcity, and urbanization (Schulterbrandt Gragg et al., 2018; Zhang et al., 2019). Within this context, spatial justice becomes a crucial lens for understanding and addressing the distribution of resources and opportunities (Allouche et al., 2019; Cansino-Loeza et al., 2022).

Spatial justice, as a key component of the FEW nexus—particularly in urban settings like the Phoenix Active Management Area (AMA)—emphasizes the equitable distribution of resources and services across different geographical areas. It seeks to rectify the imbalances that lead to disparities in access to food, energy, and water resources (White et al., 2017). In urban environments, these disparities are often stark, with some areas suffering from resource scarcity while others enjoy abundance. This uneven distribution leads to significant inequities in quality of life, health outcomes, and economic opportunities. The Phoenix AMA, characterized by its water stress (Bolin et al., 2010) and diverse policy landscape (Jacobs & Holway, 2004), provides a compelling case study for examining these interactions. As a microcosm of broader global trends, the challenges faced by the Phoenix AMA in balancing the FEW nexus offer valuable insights into sustainable urban resource management under conditions of uncertainty and change (K. Larson et al., 2015; Zhang et al., 2019). The spatial distribution of resources in this area critically influences access and equity, making the concept of spatial justice particularly relevant.

In this research, a multidisciplinary group of researchers at Arizona State University developed an anticipatory modeling framework that integrates participatory stakeholder engagement for informed decision-making within the FEW nexus (White et al., 2017). By uniting stakeholders from diverse sectors and governance levels, this study emphasizes the stakeholder engagement process, focusing on the insights gained and lessons learned, with particular emphasis on spatial justice. Stakeholder engagement is not merely a consultative process but a fundamental component of model development, ensuring that the framework is grounded in the realities of those it affects and serves.

https://doi.org/10.52202/082078-0022

Key to this approach is the recognition that sustainable management of the FEW nexus requires not only technological solutions but also an understanding of socio-economic and political dimensions. This paper, therefore, explores how interdisciplinary collaboration and stakeholder participation can enhance the resilience and sustainability of FEW systems. The overarching goal is to provide a replicable model for similar regions facing interconnected FEW system challenges, thereby contributing to the broader discourse on sustainable resource management, with a specific focus on spatial justice, equity, and inclusion by design.

2.1 Literature Review

The Food-Energy-Water (FEW) nexus presents a complex set of interdependencies crucial to sustainable development, drawing on a rich body of literature that advocates for an integrated approach (Botai et al., 2021; Olawuyi, 2020). Foundational studies by Bazilian et al. (2011) and Rasul & Sharma (2016) underscore the necessity of viewing these life-sustaining resources not in isolation but as a cohesive system, a perspective that reveals both synergies and conflicts within this nexus. This holistic view is vital for devising sustainable and efficient resource management strategies, particularly in urban settings like the Phoenix Active Management Area (AMA).

A significant and growing area within this literature is the concept of spatial justice in environmental governance. Spatial justice refers to the fair and equitable distribution of environmental benefits and burdens, including access to resources such as food, energy, and water (G. Walker, 2009). Walker (2012) and Banerjee & Steinberg (2015) examine how spatial disparities in resource access can exacerbate social inequalities, highlighting the role of geography in shaping environmental justice. Studies such as those by Walker & Bulkeley (2006) demonstrate that uneven access to resources is not just a matter of physical distribution but also involves power dynamics, institutional practices, and policy decisions. These disparities can lead to inequitable outcomes, with marginalized communities often bearing the brunt of resource scarcity and environmental degradation.

The literature also emphasizes the importance of participatory stakeholder engagement in addressing spatial justice concerns. Works by Reed et al. (2009) and Ansell & Gash (2007) highlight the need for inclusive decision-making processes that involve a diverse array of stakeholders. This engagement is crucial for ensuring that the perspectives and needs of all community members—including those from marginalized groups—are considered in the governance of the FEW nexus (Benson et al., 2015; Melloni et al., 2020).

Recent studies have expanded traditional participatory methods to explicitly address equity and inclusion, particularly in urban settings. Voinov et al. (2016) and Colloff et al. (2019) explored various engagement methods, including workshops and scenario development, which can be particularly effective in bringing together different community segments. These participatory approaches are instrumental in identifying and addressing spatial injustices, ensuring that all voices are heard and that the benefits and burdens of environmental decisions are shared more equitably.

Governing the interdependent systems of food, energy, and water presents unique challenges, as identified by Larson et al. (2015) and Pahl-Wostl (2019). Their work emphasizes the need for integrated policy approaches to effectively manage these complex dynamics. Traditional governance frameworks, which are often siloed and sector-specific, are inadequate for addressing the intricacies of the FEW nexus (Howarth & Monasterolo, 2017). The literature advocates for a shift toward more integrative, collaborative governance models that can navigate the complexities of the nexus (J. L. Jones & White, 2021; Urbinatti et al., 2020).

The evolution of modeling approaches in resource management, discussed by Basco-Carrera et al. (2018) and Kotir et al. (2017), reflects a transition to more adaptive and inclusive strategies. These anticipatory models, which accommodate changing environmental conditions and stakeholder inputs, are crucial for responsive and effective resource management. The integration of stakeholder knowledge, as demonstrated by Dilling & Lemos (2011) and Butler & Adamowski (2015), ensures that models are not only scientifically sound but also practically relevant and reflective of diverse community needs.

The review identifies a gap in empirical studies assessing the efficacy of participatory models in nexus governance. Further research is needed to explore how these models can be adapted to different socio-political contexts, ensuring their effectiveness across diverse settings. Additionally, the integration of emerging technologies in anticipatory modeling, such as artificial intelligence and machine learning, represents a promising area for future research, potentially enhancing governance models' adaptability and precision, especially in addressing environmental challenges and uncertainties.

https://doi.org/10.52202/082078-0022

In summary, the literature underscores the importance of integrating spatial justice, equity, and participatory engagement in the governance of the FEW nexus. This integrated approach is crucial for addressing the challenges of resource management in urban settings and ensuring sustainable, equitable outcomes for all community members.

2.2 Theoretical Framework

The Engage INFEWS Conceptual Framework, refined by Kliskey et al. (2021), presents a holistic model for incorporating community and stakeholder input into the study of Food-Energy-Water Systems (FEWS). This framework integrates a range of models and best practices to effectively embed diverse perspectives into FEWS research. Its structure consists of ten interrelated elements spanning engagement, technical, and monitoring aspects.

At the core of the framework is situational awareness, fostering an environment of mutual respect and shared values among diverse stakeholders. It emphasizes early and ongoing engagement, collaborative power dynamics, and co-ownership of processes. Integral to the framework are regular and transparent communication, the fusion of stakeholder insights with scientific methods, and a reflective approach to monitoring. This framework is unique in its dual focus on process and outcome evaluations, targeting scientific impact, participant empowerment, and broader societal benefits, thereby ensuring thorough community involvement in FEWS research.

In its adapted form, the framework places a stronger emphasis on spatial justice and equity, highlighting inclusive stakeholder engagement as a key component. This adaptation prioritizes understanding and addressing spatial disparities within FEWS, promotes equitable resource distribution, and broadens stakeholder representation to include marginalized groups. It aims to empower communities through localized initiatives and co-ownership, ensuring that local knowledge significantly influences FEWS research. Moreover, it evaluates the effectiveness of engagement processes and outcomes through a spatial justice lens, focusing on reducing disparities and enhancing equitable resource access.

The revised Engage INFEWS Framework thus serves as a comprehensive guide for integrating stakeholder and community engagement in FEWS research. With a reinforced commitment to spatial justice and equity, it ensures that sustainable, equitable, and effective project outcomes are achieved.

2.3 Study Context

The Phoenix Active Management Area (AMA), located in central Arizona, serves as a vital case study for examining the interplay of the Food-Energy-Water (FEW) nexus. As the fifth most populous city in the United States, Phoenix presents a unique urban landscape characterized by diverse socio-economic demographics and distinct environmental challenges, making it an ideal microcosm for exploring spatial justice within FEW systems (Bolin et al., 2010; Chow et al., 2012).

The city's demographic diversity—ranging from affluent urban districts to economically disadvantaged neighborhoods—highlights disparities in access to essential resources such as food, energy, and water (Gober & Trapido-Lurie, 2006). This socio-economic divide underscores the significance of spatial justice, particularly in understanding and addressing the uneven distribution of resources and infrastructure that perpetuate inequality. The Phoenix AMA's rapid population growth further exacerbates challenges in sustainable urban planning and resource management (Bolin et al., 2010; Larson, Wiek, et al., 2013).

The demand for food, energy, and water is intricately linked to urban development patterns, often resulting in disproportionate impacts on marginalized communities. For instance, areas with limited access to green spaces and local food sources experience heightened food insecurity, while disparities in energy infrastructure contribute to inequitable access to affordable and reliable energy (Jones et al., 2023; Maxim & Grubert, 2022; Siegner et al., 2018). Water scarcity, a pressing concern in this semi-arid region, is further intensified by climate change, affecting both the quantity and quality of water available to different communities (Larson, White, et al., 2013).

This complex socio-environmental setting provides a fertile ground for exploring how participatory models in the FEW nexus can address spatial justice issues. It highlights the importance of including a broad spectrum of voices, particularly those from marginalized and underrepresented communities, in decision-making processes. By understanding the specific needs and challenges these communities face, policymakers can develop more equitable and sustainable policies and interventions.

Thus, the Phoenix AMA not only serves as a crucial context for studying the FEW nexus but also as a platform for advancing spatial justice through inclusive stakeholder engagement and interdisciplinary research.

The Phoenix Active Management Area (AMA) encompasses a complex and interconnected web of Food-Energy-Water (FEW) components, which are critical for the region's sustenance and growth. Water, in particular, is a scarce and invaluable resource in this semi-arid environment. The Salt River, Verde River, and groundwater reservoirs serve as primary water

https://doi.org/10.52202/082078-0022

sources, but their availability is subject to climatic variations and population demands (Sampson et al., 2011). Energy is predominantly derived from fossil fuels and electricity generation, though there is a growing emphasis on renewable sources (Mounir et al., 2019).

The region's food supply chain, which includes agriculture, distribution, and consumption, faces challenges related to water-intensive farming practices and ensuring equitable access to fresh, nutritious food in urban and rural areas (Guan, 2022). The distribution and accessibility of FEW resources in the Phoenix AMA exhibit notable disparities, significantly impacting different communities (Eakin et al., 2016; Gober et al., 2010; Jacobs & Holway, 2004). Water availability varies across the region, with some communities relying heavily on groundwater pumping, leading to aquifer depletion and land subsidence, while others have more sustainable access to water sources (Bolin et al., 2010). This imbalance in water resources affects urban and rural areas alike, influencing agriculture, drinking water quality, and ecosystem health.

In terms of energy, marginalized neighborhoods often bear the brunt of environmental pollution and energy infrastructure development (Biswas et al., 2022). Moreover, disparities in clean and affordable energy access contribute to energy poverty, particularly among low-income households.

Access to fresh and healthy food is another critical concern in the Phoenix AMA. Food deserts, where nutritious food options are scarce, are prevalent in certain urban areas, disproportionately affecting low-income communities (Caldwell, 2023). These disparities in food access contribute to health inequalities, exacerbating diet-related diseases and food insecurity.

The distribution and accessibility of FEW resources in the Phoenix AMA have far-reaching consequences for different communities. Spatial planning and policy interventions are essential tools for addressing these disparities, promoting spatial justice, and fostering a more equitable and sustainable FEW nexus in the region.

3 RESEARCH OBJECTIVES

Our research examines the dynamics of stakeholder engagement within the Phoenix Active Management Area (AMA)'s Food-Energy-Water (FEW) nexus, setting forth objectives that focus on evaluating and synthesizing the impact of participatory engagement on spatial justice and resource management.
A primary goal of this study is to assess how stakeholder workshops within the Phoenix AMA have facilitated meaningful engagement, highlighted and addressed spatial disparities, and influenced decision-making within the FEW nexus. This objective centers on analyzing participant feedback and insights, examining their contributions to shaping equitable and sustainable management strategies. The evaluation aims to understand the workshops' role in fostering inclusive dialogue, identifying key areas of success or shortcomings, and determining their overall impact on the development of responsive management strategies within the FEW nexus.
In addition to evaluating the effectiveness of these workshops, our study seeks to synthesize insights from these engagement efforts to understand the broader role of stakeholder participation in advancing spatial justice within the FEW nexus. This includes a detailed review of both achievements and challenges encountered during the engagement process, with particular emphasis on how these experiences have influenced efforts toward equitable resource distribution. By analyzing the outcomes of these participatory efforts, we aim to clarify how such approaches can mitigate spatial disparities and enhance the sustainability of resource management practices in urban settings like the Phoenix AMA. Ultimately, these objectives are designed to provide a comprehensive analysis of stakeholder engagement within the FEW nexus, shedding light on its effectiveness in promoting equitable and sustainable resource management. Through this focused lens, we seek to contribute to a deeper understanding of the critical role of participatory engagement in shaping inclusive and responsive resource governance frameworks tailored to the diverse needs of the Phoenix AMA community.

4 METHODS

This study, firmly anchored in the conceptual framework proposed by Kliskey et al. (2021), takes a practical approach to integrating community and stakeholder engagement within Food-Energy-Water Systems (FEWS), with a particular focus on the Phoenix Active Management Area (AMA). The primary aim is to move beyond theoretical understanding and apply this knowledge in practice.

At the core of this approach is the development of a participatory, anticipatory modeling framework designed to facilitate social learning and promote integrated decision-making across various sectors. This framework plays a critical role in addressing environmental complexities and is explicitly designed to prioritize spatial justice, equity, and inclusive participation. By employing this methodology, the research highlights the importance of understanding and managing FEW systems in a way that addresses disparities in resource access and distribution.
It ensures that voices from previously excluded community segments are actively heard and incorporated into decision-making processes.
The study aligns with its objectives by evaluating the

https://doi.org/10.52202/082078-0022

effectiveness of stakeholder engagement workshops, analyzing how their outcomes are incorporated into policy and practice, and examining the role of stakeholder engagement in advancing spatial justice within the FEW nexus.

4.1 Participants Selection

To ensure broad and inclusive stakeholder engagement, this study employed a nonprobability sampling strategy, which was instrumental in capturing a diverse range of perspectives, particularly from individuals with extensive professional experience in the Food-Energy-Water (FEW) systems of the Phoenix Active Management Area (AMA). Stakeholders were initially identified through multiple sources, including a university research center database, a climate change-focused conference, and a workshop on the water-energy nexus (J. L. Jones & White, 2021; White et al., 2017). Recognizing the need for broader representation, the participant pool was expanded to include communities historically excluded from resource management discussions. This deliberate inclusion was essential to ensure that the study captured diverse experiences and perspectives, particularly from groups most affected by inequities in resource allocation within the FEW nexus.

4.2 Data Collection and Analysis Procedures

Data collection for this study, spanning from 2017 to 2022, consisted of a series of meticulously designed workshops complemented by individual interviews (Jones & White, 2021; White et al., 2017; Jones & White, 2022). Over this period, fifteen focus groups and fourteen semi-structured interviews were conducted, engaging stakeholders in in-depth discussions about the Food-Energy-Water (FEW) nexus within the Phoenix Active Management Area (AMA).
The first two workshops were held at the ASU Decision Theater in Tempe, an immersive environment designed for collaborative decision-making. This facility, equipped with advanced technology such as a 270-degree three-dimensional theater space and sophisticated video conferencing tools, provided an interactive setting that facilitated dynamic discussions. As circumstances evolved, later workshops transitioned to virtual platforms like Zoom while maintaining the participatory and interactive nature of these sessions.

Each workshop began with a concise ten-minute introduction outlining the project objectives and expectations, followed by a detailed thirty-minute presentation from the model developers. These presentations covered the historical development of the model, the simulation methodologies applied, and guidance on navigating various scenario simulations. The sessions served as catalysts for robust discussions, where participants engaged deeply with the material, asked clarifying questions, and shared insights based on their professional expertise and experiences.

As the study progressed, the focus of the workshops evolved. Initially, discussions centered on understanding governance dynamics within the FEW systems of the Phoenix AMA and the roles of various stakeholders. Later sessions shifted toward a critical evaluation of the integrated model developed during the research. Stakeholders provided constructive feedback on the model's credibility, salience, legitimacy, and practical applicability, which was essential for its continued refinement. These workshops functioned as platforms for mutual learning and knowledge co-creation, where participants shared decision-making experiences that significantly contributed to the iterative enhancement of the modeling framework.

Through these collaborative efforts, the workshops not only deepened the collective understanding of the FEW nexus but also played a crucial role in the iterative improvement and validation of the model. This process ensured its relevance and applicability in real-world contexts within the Phoenix AMA, reinforcing the importance of stakeholder-driven decision-making in sustainable resource management.

5 RESULTS

The workshops conducted between 2017 and 2022 on the Food-Energy-Water (FEW) nexus in the Phoenix Active Management Area (AMA) produced significant findings aligned with the study's research objectives. The involvement of a diverse group of stakeholders contributed to a multifaceted understanding of the FEW nexus, which is essential for developing sustainable management strategies.

A key outcome of these discussions was the recognition of the role of economic factors in the FEW nexus. Stakeholders emphasized the necessity of substantial investment in renewable energy infrastructure and examined the economic implications of land use, particularly regarding trade-offs between land value and agricultural production. The economic impact of importing food to compensate for reduced local agriculture in the AMA emerged as a critical concern. Discussions also highlighted the economic and cultural significance of high-water-use crops, such as alfalfa for dairy production, demonstrating the intricate connections between water use, agriculture, and the local economy.

Technological advancements, particularly in renewable energy and electric vehicles, were identified as transformative elements within the FEW nexus. Stakeholders explored how FEW systems must adapt to evolving energy profiles, including the

https://doi.org/10.52202/082078-0022

transition to renewable sources such as solar energy. The impact of technological developments on the power grid and their implications for water and energy use were also discussed. The integration of electric vehicles posed both challenges and opportunities, affecting energy demand management and grid stability.

Policy-related discussions played a central role in the workshops, focusing on shifts toward decarbonization and the potential impacts of federal policies on land use and agricultural practices. Stakeholders emphasized the need for policies that balance environmental sustainability, economic viability, and social acceptability. Discussions on legislative initiatives, such as the Carbon Act and measures to transition from fossil fuels to the electrical sector, were particularly noteworthy.

The role of agriculture in the FEW nexus was another key focus, particularly in optimizing water and energy use without compromising crop yields or economic viability. Stakeholders acknowledged the importance of considering agricultural practices within a broader socio-economic framework. The discussions included policy strategies to promote efficient irrigation methods and sustainable crop choices, ensuring long-term agricultural resilience.

The development and refinement of the integrated model, a central objective of the research, were also significant outcomes of these workshops. Participants underscored the need to balance model complexity with usability, leading to the development of effective visualization tools, such as Sankey diagrams.

However, there was a consistent demand for simpler visualization aids to enhance communication with non-expert stakeholders. The iterative engagement process played a crucial role in refining the model's relevance, credibility, and practical applicability, ensuring that it accurately represented the diverse perspectives of the Phoenix AMA community. These insights form the foundation for adaptable and flexible management strategies capable of addressing uncertainties such as climate change, economic fluctuations, and technological advancements. The findings underscore the necessity of adopting holistic and integrated resource management approaches that incorporate diverse stakeholder perspectives, particularly those from underrepresented communities. This approach is critical for ensuring sustainable resource management and policymaking within the FEW nexus.

The study's results highlight the need for a collaborative governance model that bridges scientific research with stakeholder engagement. By integrating expert knowledge with lived experiences, decision-making processes can become more informed, inclusive, and responsive to the community's diverse needs. The emphasis on participatory engagement reaffirms the importance of spatial justice in resource management, demonstrating how inclusive decision-making can drive equitable and sustainable outcomes in complex socio-environmental systems.

6 DISCUSSION

The workshops conducted from 2017 to 2022 on the Food-Energy-Water (FEW) nexus within the Phoenix Active Management Area (AMA), framed by the Engage INFEWS Conceptual Framework (Kliskey et al., 2021), provided valuable insights into the intricate dynamics of these interconnected systems. By engaging a diverse array of stakeholders, these sessions unraveled the complexities of the FEW nexus and generated knowledge essential for developing sustainable management strategies. This engagement fostered a comprehensive understanding that aligns with the study's objectives and theoretical foundations.

A key finding from these discussions was the critical role of economic factors in the FEW nexus, emphasizing the need for substantial investments in renewable energy infrastructure and a nuanced understanding of land use impacts. This insight resonates with the Engage INFEWS framework's emphasis on integrating diverse perspectives to promote spatial justice and equity. Stakeholders deliberated on land and crop values and the economic implications of food importation on local agriculture, demonstrating the framework's capacity to capture a wide range of viewpoints and facilitate shared decision-making.

Technological advancements, particularly in renewable energy and electric vehicles, emerged as transformative forces in the FEW nexus. Discussions reinforced the framework's focus on situational awareness and the need for FEW systems to adapt to changing energy landscapes. These conversations supported the study's objective of evaluating how stakeholder workshops contribute to understanding technological advancements and shaping adaptive management strategies.

Policy discussions during the workshops centered on the transition toward decarbonization and the potential effects of federal policies on land use and agriculture. These discussions aligned with the Engage INFEWS framework's call for policies that balance environmental sustainability, economic feasibility, and social acceptability. By integrating stakeholder perspectives with scientific methodologies, the framework enhances the applicability of models and policy recommendations to real-world governance challenges.

https://doi.org/10.52202/082078-0022

The role of agriculture in the FEW nexus was another significant focus, particularly in optimizing water and energy use while maintaining economic viability. Discussions on efficient irrigation practices and sustainable cropping options highlighted the need for a broader socio-economic perspective, reinforcing the framework's commitment to spatial justice. The integration of agricultural considerations within a holistic policy and planning framework further emphasized the necessity of stakeholder participation in shaping equitable resource distribution.

The iterative development and refinement of the integrated model emerged as a critical outcome of these workshops. This process underscored the importance of balancing model complexity with accessibility, a central tenet of the Engage INFEWS framework. Stakeholders consistently advocated for clearer visualization tools to improve communication with non-expert participants, reinforcing the framework's commitment to transparency and stakeholder-driven decision-making.

The insights generated from these workshops lay the foundation for management strategies that are adaptable and flexible, capable of responding to uncertainties such as climate change, economic fluctuations, and technological innovations. They highlight the necessity of a holistic, integrated approach to resource management that considers diverse stakeholder perspectives, particularly those from underrepresented communities. This approach aligns with the Engage INFEWS framework's emphasis on participatory governance and equity-driven decision-making.

In conclusion, the workshops affirmed the effectiveness of the Engage INFEWS Conceptual Framework in guiding stakeholder engagement within the FEW nexus. The collaboration facilitated by this framework enriched the understanding of FEW interdependencies and underscored the importance of inclusive, evidence-based decision-making. The insights gained from these workshops will be invaluable for informing future policies, technological advancements, and sustainable resource management practices in the Phoenix AMA and beyond. These findings mark a significant step toward achieving spatial justice and sustainability within the FEW nexus.

7 CONCLUSIONS

The workshops conducted between 2017 and 2022 in the Phoenix Active Management Area (AMA), focused on the Food-Energy-Water (FEW) nexus, have yielded critical insights that align closely with both the Engage INFEWS Conceptual Framework (Kliskey et al., 2021) and the study's research objectives. This study concludes by emphasizing these key findings and their broader implications for sustainable resource management and policy development. A major takeaway from the stakeholder engagements was the significant role economic factors play in the FEW nexus, underscoring the need for increased investments in renewable energy infrastructure and a deeper understanding of land use decisions. These discussions highlighted the necessity of integrating economic considerations into frameworks for spatial justice and equitable resource distribution. Conversations on the cultural and economic importance of water-intensive crops, along with the trade-offs between land and crop values, contributed to a more inclusive approach to resource management.

Technological advancements, particularly in renewable energy and electric vehicles, were recognized as transformative elements within the FEW nexus. The ability of FEW systems to adapt to these changes reinforces the need for adaptive management strategies that respond to evolving energy landscapes. This finding aligns with the Engage INFEWS framework's emphasis on situational awareness and integrating scientific analysis with stakeholder perspectives to develop resilient governance approaches.

Policy discussions reinforced the importance of designing strategies that balance environmental sustainability, economic feasibility, and social equity. The need for decarbonization and the potential impacts of federal policies on land use and agricultural practices were central themes, highlighting the Engage INFEWS framework's call for inclusive policymaking that integrates environmental, economic, and social considerations. These insights provide a foundation for policies that prioritize spatial justice while ensuring sustainable resource use. The iterative refinement of the integrated model was another key contribution of this research, demonstrating the importance of stakeholder engagement in enhancing the model's relevance and usability. Stakeholders emphasized the need to balance model complexity with accessibility, leading to the development of intuitive visualization tools that align with the framework's commitment to transparency and stakeholder empowerment.

https://doi.org/10.52202/082078-0022

This study makes a significant contribution to the discourse on sustainable resource management within the FEW nexus by demonstrating the effectiveness of the Engage INFEWS Conceptual Framework in aligning diverse stakeholder perspectives. The findings not only apply to the Phoenix AMA but also offer valuable insights for other regions facing similar resource management challenges. This research underscores the necessity of adopting holistic, adaptive, and inclusive approaches to managing FEW systems, ensuring that resource governance is both sustainable and equitable.

8 REFERENCES

Allouche, J., Middleton, C., & Gyawali, D. (2019). The water-food-energy nexus: Power, politics and justice. Routledge.

Ansell, C., & Gash, A. (2007). Collaborative Governance in Theory and Practice. Journal of Public Administration Research and Theory, 18(4), 543–571. https://doi.org/10.1093/jopart/mum032

Banerjee, D., & Steinberg, S. L. (2015). Exploring spatial and cultural discourses in environmental justice movements: A study of two communities. Journal of Rural Studies, 39, 41–50. https://doi.org/10.1016/j.jrurstud.2015.03.005

Basco-Carrera, L., Meijers, E., Sarısoy, H. D., Şanli, N. O., Coşkun, S., Oliemans, W., Van Beek, E., Karaaslan, Y., & Jonoski, A. (2018). An adapted companion modelling approach for enhancing multi-stakeholder cooperation in complex river basins. International Journal of Sustainable Development & World Ecology, 25(8), 747–764. https://doi.org/10.1080/13504509.2018.1445668

Bazilian, M., Rogner, H., Howells, M., Hermann, S., Arent, D., Gielen, D., Steduto, P., Mueller, A., Komor, P., Tol, R. S. J., & Yumkella, K. K. (2011). Considering the energy, water and food nexus: Towards an integrated modelling approach. Energy Policy, 39(12), 7896–7906. https://doi.org/10.1016/j.enpol.2011.09.039

Benson, D., Gain, A. K., & Rouillard, J. J. (2015). Water governance in a comparative perspective: From IWRM to a'nexus' approach? Water Alternatives, 8(1), 756–773.

Biswas, S., Echevarria, A., Irshad, N., Rivera-Matos, Y., Richter, J., Chhetri, N., Parmentier, M. J., & Miller, C. A. (2022). Ending the Energy-Poverty Nexus: An Ethical Imperative for Just Transitions. Science and Engineering Ethics, 28(4), 36. https://doi.org/10.1007/s11948-022-00383-4

Bolin, B., Seetharam, M., & Pompeii, B. (2010). Water resources, climate change, and urban vulnerability: A case study of Phoenix, Arizona. Local Environment, 15(3), 261–279. https://doi.org/10.1080/13549830903575604

Botai, J. O., Botai, C. M., Ncongwane, K. P., Mpandeli, S., Nhamo, L., Masinde, M., Adeola, A. M., Mengistu, M. G., Tazvinga, H., Murambadoro, M. D., Lottering, S., Motochi, I., Hayombe, P., Zwane, N. N., Wamiti, E. K., & Mabhaudhi, T. (2021). A Review of the Water–Energy–Food Nexus Research in Africa. Sustainability, 13(4), 1762. https://doi.org/10.3390/su13041762

Butler, C., & Adamowski, J. (2015). Empowering marginalized communities in water resources management: Addressing inequitable practices in Participatory Model Building. Journal of Environmental Management, 153, 153–162. https://doi.org/10.1016/j.jenvman.2015.02.010

Caldwell, R. (2023). Understanding Limited Food Access in Mohave County, Arizona: Influencing Factors, Current Realities, and Potential Solutions [PhD Thesis]. Northern Arizona University.

Cansino-Loeza, B., Munguía-López, A. D. C., & Ponce-Ortega, J. M. (2022). A water-energy-food security nexus framework based on optimal resource allocation. Environmental Science & Policy, 133, 1–16. https://doi.org/10.1016/j.envsci.2022.03.006

Chow, W. T. L., Chuang, W.-C., & Gober, P. (2012). Vulnerability to Extreme Heat in Metropolitan Phoenix: Spatial, Temporal, and Demographic Dimensions. The Professional Geographer, 64(2), 286–302. https://doi.org/10.1080/00330124.2011.600225

Colloff, M. J., Doody, T. M., Overton, I. C., Dalton, J., & Welling, R. (2019). Re-framing the decision context over trade-offs among ecosystem services and wellbeing in a major river basin where water resources are highly contested. Sustainability Science, 14(3), 713–731. https://doi.org/10.1007/s11625-018-0630-x

Dilling, L., & Lemos, M. C. (2011). Creating usable science: Opportunities and constraints for climate knowledge use and their implications for science policy. Global Environmental Change, 21(2), 680–689. https://doi.org/10.1016/j.gloenvcha.2010.11.006

Eakin, H., York, A., Aggarwal, R., Waters, S., Welch, J., Rubiños, C., Smith-Heisters, S., Bausch, C., & Anderies, J. M. (2016). Cognitive and institutional influences on farmers' adaptive capacity: Insights into barriers and opportunities for transformative change in central Arizona. Regional Environmental Change, 16(3), 801–814. https://doi.org/10.1007/s10113-015-0789-y

Gober, P., Kirkwood, C. W., Balling, R. C., Ellis, A.

https://doi.org/10.52202/082078-0022

W., & Deitrick, S. (2010). Water Planning Under Climatic Uncertainty in Phoenix: Why We Need a New Paradigm. Annals of the Association of American Geographers, 100(2), 356–372. https://doi.org/10.1080/00045601003595420

Gober, P., & Trapido-Lurie, B. (2006). Metropolitan Phoenix: Place making and community building in the desert. University of Pennsylvania Press.

Guan, X. (2022). Defining a Roadmap Towards a More Sustainable Food-Energy-Water (FEW) Nexus in the Phoenix Metropolitan Region Through Integrated Modeling [PhD Thesis]. Arizona State University.
Howarth, C., & Monasterolo, I. (2017). Opportunities for knowledge co-production across the energy-food-water nexus: Making interdisciplinary approaches work for better climate decision making. Environmental Science & Policy, 75, 103–110. https://doi.org/10.1016/j.envsci.2017.05.019

Jacobs, K. L., & Holway, J. M. (2004). Managing for sustainability in an arid climate: Lessons learned from 20 years of groundwater management in Arizona, USA. Hydrogeology Journal, 12(1), 52–65. https://doi.org/10.1007/s10040-003-0308-y

Jones, A., Nock, D., Samaras, C., Qiu, Y. (Lucy), & Xing, B. (2023). Climate change impacts on future residential electricity consumption and energy burden: A case study in Phoenix, Arizona. Energy Policy, 183, 113811. https://doi.org/10.1016/j.enpol.2023.113811

Jones, J. L., & White, D. D. (2021). A social network analysis of collaborative governance for the food-energy-water nexus in Phoenix, AZ, USA. Journal of Environmental Studies and Sciences, 11(4), 671–681. https://doi.org/10.1007/s13412-021-00676-3

Jones, J. L., & White, D. D. (2022). Understanding barriers to collaborative governance for the food-energy-water nexus: The case of Phoenix, Arizona. Environmental Science & Policy, 127, 111–119. https://doi.org/10.1016/j.envsci.2021.10.025

Kliskey, A., Williams, P., Griffith, D. L., Dale, V. H., Schelly, C., Marshall, A.-M., Gagnon, V. S., Eaton, W. M., & Floress, K. (2021). Thinking Big and Thinking Small: A Conceptual Framework for Best Practices in Community and Stakeholder Engagement in Food, Energy, and Water Systems. Sustainability, 13(4), 2160. https://doi.org/10.3390/su13042160

Kotir, J. H., Brown, G., Marshall, N., & Johnstone, R. (2017). Systemic feedback modelling for sustainable water resources management and agricultural development: An application of participatory modelling approach in the Volta River Basin. Environmental Modelling & Software, 88, 106–118. https://doi.org/10.1016/j.envsoft.2016.11.015

Larson, K. L., White, D., Gober, P., Kirkwood, C., Smith, V., Nelson, M., Redman, C., & Wittlinger, S. (2013). Advancing Science in Support of Water Policy and Urban Climate Change Adaptation at Arizona State University's Decision Center for a Desert City: A Synthesis of Interdisciplinary Research on Climate, Water, and Decision-Making Under Uncertainty. Decision Center for a Desert City, Arizona State University: Tempe, AZ, USA.
Larson, K. L., Wiek, A., & Withycombe Keeler, L. (2013). A comprehensive sustainability appraisal of water governance in Phoenix, AZ. Journal of Environmental Management, 116, 58–71. https://doi.org/10.1016/j.jenvman.2012.11.016

Larson, K., White, D., Gober, P., & Wutich, A. (2015). Decision-Making under Uncertainty for Water Sustainability and Urban Climate Change Adaptation. Sustainability, 7(11), 14761–14784. https://doi.org/10.3390/su71114761

Maxim, A., & Grubert, E. (2022). Anticipating Climate-Related Changes to Residential Energy Burden in the United States: Advance Planning for Equity and Resilience. Environmental Justice, 15(3), 139–148. https://doi.org/10.1089/env.2021.0056

Melloni, G., Turetta, A., Bonatti, M., & Sieber, S. (2020). A Stakeholder Analysis for a Water-Energy-Food Nexus Evaluation in an Atlantic Forest Area: Implications for an Integrated Assessment and a Participatory Approach. Water, 12(7), 1977. https://doi.org/10.3390/w12071977

Mounir, A., Mascaro, G., & White, D. D. (2019). A metropolitan scale analysis of the impacts of future electricity mix alternatives on the water-energy nexus. Applied Energy, 256, 113870. https://doi.org/10.1016/j.apenergy.2019.113870

Olawuyi, D. (2020). Sustainable development and the water-energy-food nexus: Legal challenges and emerging solutions. Environmental Science & Policy, 103, 1–9. https://doi.org/10.1016/j.envsci.2019.10.009

Pahl-Wostl, C. (2019). Governance of the water-energy-food security nexus: A multi-level coordination challenge. Environmental Science & Policy, 92,

https://doi.org/10.52202/082078-0022

356–367. https://doi.org/10.1016/j.envsci.2017.07.017

Rasul, G., & Sharma, B. (2016). The nexus approach to water–energy–food security: An option for adaptation to climate change. Climate Policy, 16(6), 682–702. https://doi.org/10.1080/14693062.2015.1029865

Reed, M. S., Graves, A., Dandy, N., Posthumus, H., Hubacek, K., Morris, J., Prell, C., Quinn, C. H., & Stringer, L. C. (2009). Who's in and why? A typology of stakeholder analysis methods for natural resource management. Journal of Environmental Management, 90(5), 1933–1949. https://doi.org/10.1016/j.jenvman.2009.01.001

Sampson, D. A., Escobar, V., Tschudi, M. K., Lant, T., & Gober, P. (2011). A provider-based water planning and management model – WaterSim 4.0 – For the Phoenix Metropolitan Area. Journal of Environmental Management, 92(10), 2596–2610. https://doi.org/10.1016/j.jenvman.2011.05.032

Schulterbrandt Gragg, R., Anandhi, A., Jiru, M., & Usher, K. M. (2018). A Conceptualization of the Urban Food-Energy-Water Nexus Sustainability Paradigm: Modeling From Theory to Practice. Frontiers in Environmental Science, 6, 133. https://doi.org/10.3389/fenvs.2018.00133

Siegner, A., Sowerwine, J., & Acey, C. (2018). Does Urban Agriculture Improve Food Security? Examining the Nexus of Food Access and Distribution of Urban Produced Foods in the United States: A Systematic Review. Sustainability, 10(9), 2988. https://doi.org/10.3390/su10092988

Urbinatti, A. M., Benites-Lazaro, L. L., Carvalho, C. M. D., & Giatti, L. L. (2020). The conceptual basis of water-energy-food nexus governance: Systematic literature review using network and discourse analysis. Journal of Integrative Environmental Sciences, 17(2), 21–43. https://doi.org/10.1080/1943815X.2020.1749086

Voinov, A., Kolagani, N., McCall, M. K., Glynn, P. D., Kragt, M. E., Ostermann, F. O., Pierce, S. A., & Ramu, P. (2016). Modelling with stakeholders – Next generation. Environmental Modelling & Software, 77, 196–220. https://doi.org/10.1016/j.envsoft.2015.11.016

Walker, G. (2009). Beyond Distribution and Proximity: Exploring the Multiple Spatialities of Environmental Justice. Antipode, 41(4), 614–636. https://doi.org/10.1111/j.1467-8330.2009.00691.x

Walker, G. P. (2012). Environmental justice: Concepts, evidence and politics. Routledge.

Walker, G. P., & Bulkeley, H. (2006). Geographies of environmental justice. 35(5), 655–659.

White, D., Jones, J., Maciejewski, R., Aggarwal, R., & Mascaro, G. (2017). Stakeholder Analysis for the Food-Energy-Water Nexus in Phoenix, Arizona: Implications for Nexus Governance. Sustainability, 9(12), 2204. https://doi.org/10.3390/su9122204

Zhang, P., Zhang, L., Chang, Y., Xu, M., Hao, Y., Liang, S., Liu, G., Yang, Z., & Wang, C. (2019). Food-energy-water (FEW) nexus for urban sustainability: A comprehensive review. Resources, Conservation and Recycling, 142, 215–224. https://doi.org/10.1016/j.resconrec.2018.11.018

https://doi.org/10.52202/082078-0022

CHILD'S PLAY: A CASE STUDY OF AN EVIDENCE-BASED HEALING SPACE FOR CHILDREN IN A REFUGEE CAMP

Track 6: Transdisciplinary Design and Collaboration/Open Track
Knowledge Network: Children, Youth & Environment
Methods: Case Study
Rachel Ni Dhugain (None); Renee Desneige Christensen (Extensions)

1 ABSTRACT
The purpose of this case study was to identify the impact of applying an evidence-based approach that incorporates neuroscience in designing therapeutic environments for children in refugee camps. This case study included a transdisciplinary literature review to establish a preliminary framework for an outdoor play space designed for children at a refugee camp in Cyprus. The project was selected to evaluate the design-based strategies as an essential modality for trauma treatment for children. This project achieves the ambitious aim of translating concepts from neuroscience into trauma-informed design strategies. The outcome of this case study supports the development of a theoretical framework for the design of children's play spaces as an important factor in trauma-based interventions in refugee camps.

1.1 Keywords
trauma-informed design, human-centric perspective, evidence-based design

2 INTRODUCTION
Exposure to mass conflict and displacement can have a devastating impact on mental health with up to one-third exposed experiencing long-term detrimental impacts to their physical, social, emotional or spiritual well-being (Steel, et al., 2009). Children, in particular, are at an increased risk of trauma and potential outcomes such as difficulty in concentrating, hyperactivity, attention-seeking, anxiety, depression and Post-traumatic Stress Disorder (PTSD) (Cantani, 2018; Osofsky, 1995; Cooley-Quille et al.,1995; Martinez & Richters, 1993; Bronstein & Montgomery, 2011; Fazel & Stein, 2002; Hadfield, Ostrowski & Ungar, 2017). Even after escaping the direct exposure to conflict and violence, refugee children continue to experience mental health and wellbeing challenges (Browder, 2018; Brown, Miller & Mitchell, 2006; Fazel & Stein, 2002; Graham, Minhas & Paxton, 2016; Kaplan, et al., 2016; Potochnick, 2018).

It is estimated that 43.3 million children have been displaced as a consequence of conflict and violence as of the end of 2022 (UNICEF, 2023a). Refugee camps are temporary facilities built to provide immediate protection and assistance to people who have been forced to flee their homes due to war, persecution or violence. The average length of time that refugees spend in camps varies depending on the crisis. In many cases refugees may spend years and even decades living in refugee camps and it is common to have entire generations growing up in the camps (UNHCR, 2023a).
While camps are not established to provide permanent solutions, they provide the most basic needs such as food, water, shelter, medical treatment. Organizations like UNICEF work with local and international partners to provide other services for the refugee camp occupants. For example, protection services for children. However, the spaces in which the services are delivered are often in makeshift spaces such as industrial tents. These spaces rarely support therapeutic interventions which can impact.

2.1 PURPOSE OF STUDY
Translating research outcomes into design strategies can be a challenging endeavor for architects and designers. While some design firms have invested in research teams, few understand the need to leverage emerging evidence from science and technology. Further, design pedagogy is often rooted in schools of thought that reflect historical precedents and are difficult to dislodge. Frameworks have the potential to significantly contribute to the advancement of environmental design by providing researchers and practitioners with a theoretical foundation that can be easily translated into design strategies. However, there is currently a lack of a robust theoretical framework for designing trauma-informed spaces for children that includes findings from neuroscience. The development of a robust evidence-based framework can guide future innovative design for children's play spaces as a therapeutic intervention tool. The purpose of this case study is to evaluate the design of a children's play space as a therapeutic intervention tool for trauma therapy through rich description and an analysis guided by a focus on research studies in neuroscience. The outcome of this case study will inform the development of a multidisciplinary framework for an evidence-based, trauma-informed design framework for designing therapeutic environments for children.

https://doi.org/10.52202/082078-0023

3 LITERATURE REVIEW

3.1 Trauma

Trauma results from an "event, series of events, or a set of circumstances that is experienced by an individual as physically or emotionally harmful or life-threatening and that has lasting adverse effects on the individual's functioning and mental, physical, social, emotional, or spiritual well-being." (SAMHSA, Trauma and Justice Strategic Initiative, 2012, p.2). Living under conditions of mass trauma, such as those experienced by refugee children, has been shown to disrupt children's basic need for safety (Burgin, 2022) and imprint in their minds a deep conviction that the world is not a safe place. Typically, exposure to trauma is associated with behavioral outcomes such as increased chaos, loss of control or the lack of predictability.

Trauma begins to live and work through the body - the physical sensations and emotions that may not be consciously associated with memories of past trauma (Van der Kolk, 2014). Trauma frequently creates disorganization in low-brain areas that regulate life support functions, regions of the brain that lack a rational or logical framework of thought but instead produce somatic symptomatology (Van der Kolk, 2014). Without the balanced low-brain areas, higher-brain areas in the cortex cannot function properly. Neurological adaptations take place within the hypothalamic pituitary adrenal (HPA) axis (the stress response system) and regulatory connections between the medial prefrontal cortex and the amygdala, which result in the physical disconnection experienced by trauma survivors. (Cohodes, 2021; Grasser, 2023).

3.2 Movement-oriented trauma therapy

Movement can be a specific pathway that stimulates brain plasticity and therefore promotes healing, learning capacities, and a sense of well-being (Perry, 2014). Moving the body engages underlying perceptions, sensations, and actions of higher-order aspects of cognition such as memory, language, and concept formation and has been found to be an effective approach in trauma therapy (Berrol, 1992; Levine & Land, 2015; Lee et al., 2023; Sanchez-Alonso & Aslin, 2022; Tomaszewski et al. 2023. While general physical activity has been found to be effective in the treatment of trauma (Rosenbaum et al., 2015), dance therapy emerged as a holistic approach grounded in the percept that human reception, processing and response inextricably link the mind and body (Marsden, 1986; Schilder, 1950; Stuss, 1988; Kolb & Whishaw, 1985; Sweigard, 1974; Paillard, 1986; Delong, 1970).

Play, especially social play, serves an important developmental role by modifying the neural mechanisms (Bell, Pellis & Kolb 2010; Himmler, Pellis & Kolb 2013), impacting social skills (Byrd & Briner, 1999; van den Berg et al. 1999) the regulation of emotions (da Silva et al. 1996; von Frijtag et al. 2002), and executive functions (Baarendse et al. 2013). Data on several primate species (e.g., Kalcher-Sommersguter et al. 2011; Kempes et al., 2008), including humans (Lindsey & Colwell, 2013; Pellegrini, 1995), are consistent with these findings. For example, studies have found that rats which were isolated and deprived from social play during a critical developmental phase developed cognitive deficits including impairments in decision making and impulse control (Baarendse, 2013; Einon,1978). Another study by Kyle, Burghardt & Cooper (2019) found that "social play deprivation disrupts neural circuits regulating aggression...stress responsivity (and) suggests that juvenile social play functions to promote coping with stress and appropriate social behavior in adulthood." Similarly, Cooper (2023) discovered that dendritic morphology in mPFC neurons were altered and flexible, context-dependent behavior was impaired within adulthood due to play deprivation.

Activities such as pouring water, playing in the sand or dirt, or swinging, engages the widespread emotional, motor, sensory and cognitive systems in predictable, controllable and moderate ways (Perry, 2014). Additionally, swings have a significant role in therapeutic play for children. The swing's repetitive motion supports the development of body awareness and balance. The movement creates a stimulation of the proprioceptive and the vestibular systems which not only helps children's coordination but improves their awareness of their body in space and time. Additionally, swinging has been shown to have a natural relaxing function. The gentle rocking motion of swinging has been shown to stimulate the raphe system in the brain which keeps serotonin levels balanced thereby impacting anxiety and stress. Play activities are most helpful when repeated at heart rate of 60-80 bpm and within a sensory modality that the child finds pleasurable. These findings contribute to the development of body-oriented, play-based interventions that create a context for healing, self-care and self-expression in the treatment of child trauma.

3.3 The built environment as an intervention tool

Even prior to birth, our first relationship is with the physical environment. Before social encounters or cognitive meaning-making our bodies have immediate responses to the space we are in. The responses occur in our biological systems long before cognitive processes. When the sensory input is received the body immediately dispatches a physiological response. Recent studies have found that exposure to certain architectural forms, such as room proportions (Cruz-Garza et al., 2021; Shemesh et al., 2021), wall curvature (Vartanian et al. 2015;

https://doi.org/10.52202/082078-0023

Shemesh et al., 2021) and window arrangement and size (Fich et al., 2014; Cruz-Garza et al., 2021) evoke physiological reactions. In addition, integration of biophilic elements into spatial designs have been found to elicit stress-reducing responses and enhance human health and wellbeing (Salingaros, 2015; Aspinal et al., 2015; Joung et al., 2015; Neal et al., 2017; Coburn et al., 2019; Salingaros & Sussman, 2020; Yin et al., 2020). Further, biomorphic designs and fractal patterns have also been shown to reduce stress responses (Ryab et al., 2014; Hagerhall et al., 2008; Chiang, Li & Jane, 2017; Bailey et al., 2018). For example, Salingaros and Masden (2008) found that natural forms activate opioid receptors which are responsible for the pain reduction and suppress the release of stress-related endocrine hormones. Beneficial responses are also evidenced by higher frontal lobe alpha waves (Aspinal et al., 2015; Joung et al., 2015; Neale et al., 2017) and lesser frontal lobe activation (Joung et al., 2015; Park et al., 2017), indicating a more relaxed state.

Architecture can heighten or calm the body's response to perceived stressors by modulating environmental stimuli and atmospheres (Pallassma, 2012). The design of these spaces hold the potential to foster an intimate and safe relationship with a space so that healing can take place. However, in sensory-depriving refugee camps, most children cannot find nurturing stimuli to engage with.

4 CASE STUDY IN CYPRUS

4.1 Temple of the Golden Trees

Beyond the use of sustainable materials and passive climate principles, the design of the Temple of the Golden Trees incorporates aesthetic qualities that impact the senses to engage the children in re-embodying themselves through physical experiences. By attending to a well-ordered spatial complexity in the design, users can choose their level of social, physical, and sensory engagement.

The refined details, soft edges, and materials, and scale create a space which supports four existential embodied needs in relation to the environment: Biophilia; Refuge & Intimacy; Exploration & Play; and Human Resonance. The spatial and temporal articulation between the interior and exterior provides multiple options so that whether extroverted or introverted, younger or older, in a larger group or alone, the child can establish safety in temporary territories while easing into the reconnection between physical and mental by controlling their own gradual immersion into the open space within. Multiple pathways and exit routes support gentle interactions at the margins, offering opportunities to see who is using the space before deciding whether to connect with or strategically avoid one another (Figure 1 and Figure 2).

Figure 1. Prospect and refuge

Figure 2. Prospect and refuge

The openings on all sides provide passive daylight while the semitransparent overhead canvas provides protection from rain and direct exposure to the sun. The indigenous plants incorporated throughout the space create an ambiance of gentle relaxation and provide cues to the tangible sense of place and time.

The compression and expansion created by the overhead canvas provides a spatial field offering kinetic options while maintaining a human scale. Once inside, the space 'stations' act as identity anchors to allow the child to experience their body in relation to space and time. The child can see themselves in the environment and to be part of something bigger than themselves, combat isolation with a sense of belonging and feel valued after being marginalized.

The sincere use of materials throughout expresses that it is made of what it appears to be made. There is no deception or tokenized representations and the stalwart design disregards the complicated nuances of other

https://doi.org/10.52202/082078-0023

people's practices. The materials convey their capabilities in creating a stronger connection with acoustic, olfactory, and tactile experience. Running across the floor barefoot resonates within the body, providing direct feedback of the interaction between the body and the environment. The body is fully present, interacting with the environment, creating a supportive relationship between the two.

The variety in materiality, colors, shapes and textures stimulates the visual sense without overwhelming it. Wooden columns with a gentle curvature evoke an abstracted representation of a tree trunk and branches which provide a safe and playful opportunity to challenge the body and respond to environmental demands. The wood, ropes, woven materials and silks provide a complexity in materiality which stimulates multiple sensory systems. The hammock's soft fabric folds around the user providing a cocoon-like sensory refuge where one can comfortably control their level of interaction with others (Figure 3).

The woven basket swing provides a pleasurable repetitive somatic experience (Figure 2). Additionally, the community swings provide an opportunity to engage with others in cooperative play. Each participant plays a role and they must coordinate their movements (Figure 4a, 4b, 4c, and 4d). The cadence is negotiated by all participants and a harmonious movement is achieved when all participants seek regulation of their movements in relationship to each other.

The climbing rope and nets provide yet another opportunity to challenge the body's responsiveness to environmental demands and bring a heightened sense of awareness. The two-tiered spaces provide a different somatic experience - the body above or below. Here, the children manipulate their experience with space and time, controlling each as they climb on top of and underneath the nets.

The open space allows for spontaneous play, which supports a sense of agency (Figure 1). The grassy area provides yet another opportunity to challenge the body and respond to environmental demands. Various body postures provide contact with the ground and allow a direct connection between the body's sensations and the environment, awakening a sense of awareness of the boundaries of each (Figure 5a, 5b, 5c, and 5d).

Figure 3. Hammocks

Figure 4(a). Community swing

https://doi.org/10.52202/082078-0023

Figure 4(b). Community swing

Figure 4(c). Community swing

Figure 4(d). Community swing

https://doi.org/10.52202/082078-0023

Figure 5a. Grass area

Figure 5c. Grass area

Figure 5b. Grass area

Figure 5d. Grass area

https://doi.org/10.52202/082078-0023

5 EVALUATION, DISCUSSION AND FUTURE RESEARCH

As is the case with all mammals, humans use interactive play within the environment to develop a sense of self, cultivate relationships with others and connect the past, present, and future - all crucial in healing from trauma. However, In sensory-depriving refugee camps, most children cannot find nurturing stimuli to engage with. The Office for the New Earth (ONE) seamlessly weaved together a broad range of cross-disciplinary concepts into a rich contextual and cohesive design in the therapeutic space to stimulate children's sense of dignity and imagination by applying extra attention to refined details, soft edges and materials, and scale. The design of the Temple of GoldenTrees achieves the ambitious aim of inviting universal inhabitation imbued with a distinct architectural design. Further, the design of the Temple of Golden Trees contributes to the development of a robust evidence-based, trauma-informed multidisciplinary framework for designing children's spaces that support the process of healing through the incorporation of evidence from neuroscience, psychology and environmental design.

This case study contributes to an evidence-based design framework that is grounded in an understanding of, and responsiveness to, an embodied experience of trauma. This case study emphasizes a holistic approach which synthesizes physical, psychological and emotional responses to create opportunities to rebuild a sense of control and empowerment.

6 REFERENCES:

Aspin all, P., Mavros, P., Coyne, R. & Roe, J. (2015). The Urban Brain: Analysing outdoor physical activity with mobile EEG. British Journal of Sports Medicine, 49:272–276.

Bailey, A.W., Allen, G., Herndon, J. & Demastus, C. (2018). Cognitive Benefits of Walking in Natural versus Built Environments. World Leisure Journal, 60:293–305.

Baarendse, P. J. J., Counotte, D.S., O'Donnell, P. & Vanderschuren, L.J.M.J. (2013). Early Social Experience is Critical for the Development of Cognitive Control and Dopamine Modulation of Prefrontal Cortex Function. Neuropsychopharmacology, 38:1485–94.

Berrol, C. (1992). The Neurophysiologic Basis of the Mind-Body Connection in Dance/Movement Therapy. American Journal of Dance Therapy,14(1).

Bell, H.C., Pellis, S.M. & Kolb, B. (2009). Juvenile peer play experience and the development of the orbitofrontal and medial prefrontal cortices. Behavioral Brain Research, 207(1):7-13.

Blaustein, M. E., & Kinniburgh, K. M. (2010). Treating traumatic stress in children and adolescents: How to foster resilience through attachment, self-regulation, and competency. Guilford Press.

Bronstein, I. & Montgomery, P. (2011). Psychological distress in refugee children: a systematic review. Clinical Child and Family Psychology Review,14(1):44-56.

Browder, C. (2018). Recently Resettled Refugee Students Learning English in US High Schools: The impact of students' educational backgrounds. In S. Shapiro, R. Farrelly, & M. J. Curry (Eds.), Educating refugee-background students: Critical issues and dynamic contexts (pp. 17–32). Multilingual Matters.

Brown, J., Miller, J. & Mitchell, J. (2006). Interrupted Schooling and the Acquisition of Literacy: Experience of Sudanese refugee in Victorian secondary schools. Australian Journal of Language and Literacy, 29(2):150-162.

Bürgin, D., Clemens, V., Varghese, N., Eckert, A., Huber, M., Bruttin, E., Boonmann, C., Unternährer, E., O'Donovan, A. & Schmid, M. (2022). Adverse and Traumatic Exposures, Posttraumatic Stress Disorder, Telomere Length, and Hair Cortisol - Exploring associations in a high-risk sample of young adult residential care leavers. Brain Behavior and Immunity - Health, 26.

Byrd, K. R. & Briner, W. E. (1999). Fighting, Nonagonistic Social Behavior and Exploration in Isolation-reared Rats. Aggressive Behavior, 25(3):211–223.

Cantani, C. (2018). Mental Health of Children Living in War Zones: A risk and protection perspective. World psychiatry: Official Journal of the World Psychiatric Association (WPA), 17(1): 104–105.

Chiang, Y.C., Li, D. & Jane, H.A. (2017). Wild or Tended Nature? The Effects of Landscape Location and Vegetation Density on Physiological and Psychological Responses. Landscape and Urban Planning, 167:72–83.

Coburn, A., Kardan, O., Kotabe, H., Steinberg, J., Hout, M.C., Robbins, A., MacDonald, J., Hayn-Leichsenring, G. & Berman, M.G. (2019). Psychological Responses to Natural Patterns in Architecture. Journal of Environmental Psychology, 62:133–145.

Cooley–Quille, M.R., Turner, S.M. & Beidel, D.C. (1995). Emotional Impact of Children's Exposure to Community Violence: A preliminary study. Journal of American Academy of Child and Adolescent Psychiatry, 34:1362–1368.

Cooper, M.A., Grizzell, J.A., Whitten, C.J. & Burghardt, G.M. (2023). Comparing the Ontogeny, Neurobiology, and Function of Social Play in Hamsters and Rats. Neuroscience and Biobehavioral Review,147.

Cruz-Garza, J., Darfler, M., Rounds, J., Gao, E. & Kalantari, S. (2021). EEG-based Investigation of the Impact of Classroom Design on Cognitive Performance of Students. arXiv 2021, arXiv:2102.03629.

https://doi.org/10.52202/082078-0023

da Silva, N. L., Ferreria, V.M.M., de Padua Carobrez, A. & Morato, G.S. (1996). Individual Housing from Rearing Modifies the Performance of Young Rats on Elevated Plus-Maze Apparatus. Physiology & Behavior, 60:1391–96.

Delong, M. (1970). Central Patterning of Movement. Massachusetts Institute of Technology: M.I.T. Press, 10-30.

Fazel, M. & Stein, A. (2002). The Mental Health of Refugee Children. Archives of Disease in Childhood, 87(5):366-70.

Fich, L.B., Jönsson, P., Kirkegaard, P.H., Wallergård, M., Garde, A.H. & Hansen, Å. (2014). Can Architectural Design Alter the Physiological Reaction to Psychosocial Stress? A virtual TSST experiment. Physiological Behavior, 135:91–97.

Grabowska, S., Holtzkinger, C., Wilson, J., Rossbert, L., Macur, R. & Brisson, D. (2021). Architectural Principles in the Service of Trauma-Informed Design. Denver, CO: Shopworks Architecture, Center for Housing and Homelessness Research at the University of Denver, and Group 14 Engineering.

Graham, H., Minhas, R. & Paxton, G. (2016). Learning Problems in Children of Refugee Background: A systematic review. Pediatrics,137(6).

Hadfield, K., Ostrowski, A. & Ungar, M. (2017). What Can We Expect of the Mental Health and Well-being of Syrian Refugee Children and Adolescents in Canada? Canadian Psychology / Psychologie canadienne, 58(2):194–201.

Hagerhall, C.M., Laike, T., Taylor, R.P., Küller, M., Küller, R. & Martin, T.P. (2008). Investigations of Human EEG Response to Viewing Fractal Patterns. Perception, 37:1488–1494.

Himmler, B.T., Pellis, S.M. & Kolb B. (2013). Juvenile Play Experience Primes Neurons in the Medial Prefrontal Cortex to be More Responsive to Later Experiences. Neuroscience Letters, 556:42-5.

Joung, D., Kim, G., Choi, Y., Lim, H., Park, S., Woo, J.M. & Park, B.J. (2015). The Prefrontal Cortex Activity and Psychological Effects of Viewing Forest Landscapes in Autumn Season. International Journal of Environmental Research and Public Health, 12:7235–7243.

Kalcher-Sommersguter, E., Preuschoft, S., Crailsheim, K. & Franz, C.. (2011). Social Competence of Adult Chimpanzees (Pan troglodytes) with Severe Deprivation History: I. An Individual Approach. Developmental Psychology, 47:77- 90.

Kaplan, I., Stolk, Y., Valibhoy, M., Tucker, A. & Baker, J. (2016). Cognitive Assessment of Refugee Children: Effects of trauma and new language acquisition. Transcultural Psychiatry, 53(1):81-109.

Kempes, M. M., Gulickx, M.C., van Daalen, H.J.C., Louwerese, A.L. & Sterk, E.H.M. (2008). Social Competence Is Reduced in Socially Deprived Rhesus Monkeys (Macaca mulatta). Journal of Comparative Psychology, 122:62–67.

Kolb, B. & Whishaw, I.Q. (1985). Fundamentals of Human Neuropsychology (2nd ed.). New York: W.H. Freeman and Company.

Kyle, S.C., Burghardt, G.M. & Cooper, M.A.(2019). Development of Social Play in Hamsters: Sex Differences and Their Possible Functions. Brain Research, 1712:217-223.

Lee, J., Dhauna, J., Silvers, J. A., Houston, M. H. & Barnert, E. S. (2023). Therapeutic Dance for the Healing of Sexual Trauma: A Systematic Review. Trauma, Violence & Abuse, 24(4):2143-2164.

Levine, B. & Land, H. M. (2015). A meta-synthesis of qualitative findings about Dance/Movement Therapy for individuals with trauma. Qualitative Health Research, 26:330-344.

Lindsey, E.W. & Colwell, M.J. (2013). Pretend and Physical Play: Links to Preschoolers' Affective Social Competence. Merrill-Palmer Quarterly, 59:330–60.

Marsden, C.D. (1986). Movement Disorders and the Basal Ganglia. Trends in Neurosciences, 9(10):512-514.

Martinez, P. & Richters, J.E. (1993). The NIMH Community Violence Project: H. Children's distress symptoms associated with violence exposure. Psychiatry, 56:22-35.

Neale, C., Aspinall, P., Roe, J., Tilley, S., Mavros, P., Cinderby, S., Coyne, R., Thin, N., Bennett, G. & Thompson, C.W. (2017). The Aging Urban Brain: Analyzing Outdoor Physical Activity Using the Emotiv Affectiv Suite in Older People. Journal of Urban Health, 94:869–880.

Osofsky, J.D. (1995). The Effects of Exposure to Violence on Young Children. American Psychologist, 50:782–788. Paillard, J. (1986). Cognitive Versus Sensorimotor Encoding of Spatial Information. In P.

Ellen & C. Thinus-Blanc (Eds.). Cognitive Processes and Spatial Orientation in Animal and Man (pp. 1-35). Dordrecht, The Netherlands: Martinus Nijhoff.

Pallasmaa, J. (2012). The Eyes of the Skin (3rd ed.). Hoboken: John Wiley & Sons.

https://doi.org/10.52202/082078-0023

Park, B.J., Tsunetsugu, Y., Kasetani, T., Hirano, H., Kagawa, T., Sato, M. & Miyazaki, Y. (2017). Physiological Effects of Shinrin-yoku (taking in the atmosphere of the forest)—Using salivary cortisol and cerebral activity as indicators. Journal of Physiological Anthropology, 26:123–128.

Pellegrini, A.D. (1995). Boys' Rough-and-Tumble Play and Social Competence: Contemporaneous and Longitudinal Relations. In Anthony D. Pellegrini, (Ed). The Future of Play Theory: A Multidisciplinary Inquiry into the Contribution of Brian. Sutton-Smith, 107–26.

Potochnick, S. (2018). The Academic Adaptation of Immigrant Students with Interrupted Schooling. American Educational Research Journal, 55(4):859-892.

Rosenbaum, S., Vancampfort, D., Steel, Z., Newby, J., Ward, P. & Stubbs, B. (2015). Physical Activity in the Treatment of Post-traumatic stress disorder: A systematic review and meta-analysis. Psychiatry Research, 230:130-136.

Salingaros, N.A. (2015). Biophilia & Healing Environments; Technical report; Terrapin Bright Green: New York, NY, USA.

Salingaros, N. & Masden, K. (2008). Neuroscience, the Natural Environment, and Building Design. Biophilic Design: The Theory, Science, and Practice of Bringing Buildings to Life, 59–83.

Salingaros, N.A. & Sussman, A. (2020). Biometric Pilot-Studies Reveal the Arrangement and Shape of Windows on a Traditional Façade to be Implicitly "Engaging", Whereas Contemporary Façades are Not. Urban Science, 4(2):26.

Schilder, P. (1950). The Image and Appearance of the Human Body. New York: International Universities Press, Inc.
Shemesh, A., Leisman, G., Bar, M. & Grobman, Y.J. (2021). A Neurocognitive Study of the Emotional Impact of Geometrical Criteria of Architectural Space. Architectural Science Review, 64:1–14.

Steel, Z., Chey, T., Silove, D., Marnane, C., Bryant, R. A., & Van Ommeren, M. (2009). Association of torture and other potentially traumatic events with mental health outcomes among populations exposed to mass conflict and displacement: a systematic review and meta-analysis. Journal of the American Medical Association, 302(5):537-549.

Strategic Initiative (2012). SAMHSA's Working Definition of Trauma and Guidance for Trauma-Informed Approach. Rockville, MD: Substance Abuse and Mental Health Services Administration.
Stuss, D. (1988). What the Frontal Lobes Do. Paper presentation, Conference, Rehabilitation: Coma to Community. San Jose, CA.

Sweigard, L.E. (1974). Human Movement Potential: Its Ideokinetic Facilitation. New York: Dodd, Mead & Company, Inc.

Tomaszewski, C., Belot, R. A., Essadek, A., Onumba-Bessonnet, H., & Clesse, C. (2023). Impact of Dance Therapy on Adults with Psychological Trauma: A systematic review. European Journal of Psychotraumatology, 14(2).

United Nations High Commissioner for Refugees (UNHCR) (2023a) Refugee Statistics: Global Trends At-a-Glance. Downloaded at: https://www.unrefugees.org/refugee-facts/statistics/#:~:text=Global%20Trends%20At%2Da%2DGlance,62.5%20million%20internally%20displaced%20people (Accessed October 29, 2023).

United Nations International Children's Emergency Fund (UNICEF) (2023a). Child Displacement. Retrieved from: https://www.unicef.org/press-releases/number-displaced-children-reaches- new-high-433-million. (Accessed November 12, 2023).

van den Berg, C.L., Hol, T., Van Ree, J.M., Spruijt, B.M., Everts, H. & Koolhaas, J.M. (1999). Play is Indispensable for an Adequate Development of Coping with Social Challenges in the Rat. Developmental Psychobiology, 34:129–138.

Van der Kolk, B.A.(2014). The Body Keeps the Score : Brain, mind, and body in the healing of trauma. New York, New York : Viking.

Vartanian, O., Navarrete, G., Chatterjee, A., Fich, L.B., Gonzalez-Mora, J.L., Leder, H., Modroño, C., Nadal, M., Rostrup, N. & Skov, M. (2015). Architectural Design and the Brain: Effects of ceiling height and perceived enclosure on beauty judgments and approach-avoidance decisions. Journal of Environmental Psychology, 41:10–18.

von Frijtag, J.C., Schot, M., van den Bos, R. & Spruijt, B.M. (2002). Individual Housing During the Play Period Results in Changed Responses to and Consequences of a Psychosocial Stress Situation in Rats. Developmental Psychobiology, 41:58–69.

Yin, J., Yuan, J., Arfaei, N., Catalano, P.J., Allen, J.G. & Spengler, J.D. (2020). Effects of Biophilic Indoor Environment on Stress and Anxiety Recovery: A between-subjects experiment in virtual reality. Environment International, 136.

https://doi.org/10.52202/082078-0023

A RETROSPECTIVE REVIEW OF END-OF-LIFE CARE ENVIRONMENTS FROM EDRA1 TO EDRA55

Track 4: Design for Health and Wellbeing
Knowledge Network: Environment-Gerontology, Health
Methods: Mixed Methods
Sharmin Kader (Ball State University College of Architecture and Planning)

1 ABSTRACT

For over 55 years, EDRA has promoted theory, research, and design focused on environmental gerontology. A considerable amount of research has been conducted on environments for older adults (e.g., nursing homes), particularly addressing the 'Third Stage' or 'Fourth Stage' of life. While some of this work has been published and presented at EDRA conferences, relatively few studies have focused on the 'Last Stage' of life—specifically, end-of-life environments such as hospice care. Furthermore, the COVID-19 pandemic introduced unprecedented scenarios and challenges for hospice facilities (Rogers et al., 2021). It is therefore essential to understand the state-of-the-art knowledge in hospice and end-of-life care environments; a retrospective review offers valuable insights for shaping future research and design directions. This paper presents a literature review conducted using multiple sources: a) academic databases (e.g., PsycINFO, Google Scholar, ProQuest Dissertations & Theses), b) EDRA01–EDRA55 proceedings, c) books, reports, and guidelines published over the last 20 years, and d) white papers, webinars, podcasts, and online panel discussions. Over EDRA's 55-year history, notable advancements have been made, including: the identification of therapeutic goals; development of research-based design guidelines; publication of reference texts on hospice environments; and the creation of environmental assessment tools for hospice settings within both the U.S. healthcare and building design system, and the Canadian system, especially concerning hospice care for homeless populations. Additionally, a small body of literature has emerged in response to the pandemic. This article synthesizes the key findings from this body of work and concludes with recommendations for future research and design in hospice and end-of-life care environments.

1.1 Keywords

End-of-life care
Hospice Design
Dying Patients
Covid-19
Healthcare Architecture.

2 INTRODUCTION

Hospice is a philosophy of care that offers medical, social, and emotional support in one's home or care facility (McGann, 2016). Today, people typically die either at home or in an end-of-life care facility such as a hospice. Hospice serves as a place that accommodates the phenomenon of death. It provides end-of-life care to individuals certified as "terminal" and to their families (Government Printing Office, 2000, as cited in Swenson, 2009, p. 9). Environmental design research on end-of-life care environments is significant because the experience of place has a profound impact on a patient's "quality of life" (Cohen et al., 2001) and the possibility of a "good death," as defined by Tong and colleagues (2003).

To introduce the topic of end-of-life care, this paper has briefly discussed the history of death and dying in contemporary society, hospice as a care philosophy, the hospice movement, and hospice as a place. Lastly, this article identifies future research directions for end-of-life care environments, which are significant for enhancing our ability to design more thoughtful and effective spaces amid a growing aging population and the increasing construction of new hospice facilities.

2.2 History of death as a concept

For thousands of years, death was perceived as a "spiritual passage" or a "religious event of taking a soul"; this has been defined as traditional death by Tony Walker in 1996 (p. 195). Beginning of the twentieth century, death became a "medical event" and it was "rationalized in such a way that people wanted to keep their dying and grieving private, and people were pleased to be controlled by medical professionals (doctors); this has been defined as modern death (Walter, 1996, p. 196).

He died a modern death, in hospital, without his family, attended in his minutes by a nurse, months- indeed, years– after medical science has prolonged his life to a point where the terms on which it was being offered were unimpressive. (Barnes, 2008, as cited in Worpole, 2009, p.4)

Most "modern" individuals welcomed medical experts' control over the dying process, "but there was a price: the exclusion of a dying person" (Walter, 1996, p. 197). A dying person confined to a hospital bed, undergoing medical procedures and medications, and "visited by family at the bureaucratically determined visiting

https://doi.org/10.52202/082078-0024

hours, is hardly a person" (Walter, 1996, p. 197). People fear losing themselves and their identity in death, a concern that has become particularly problematic in Western individualistic societies. In response to these issues, the "happy death movement," as Lyn Lofland (1978) termed it, attempted to personalize death and dying (as cited in Walter, 1996). Tony Walter defined this shift as "postmodern death," a "revival of tradition" in which modern expertise was combined with traditional approaches (Walter, 1996, p. 197). The characteristics of these three types of death, as defined by Walter (1996), are presented in the following chart (Table 1).

By the end of the 1960s, an alternative philosophy of care—palliative care—emerged in response to the postmodern consumer's desire to avoid the isolation and depersonalization of institutionalized dying. This model employed an interdisciplinary team of physicians, nurses, social workers, therapists, and clergy, who in turn encouraged family and friends to participate in the caregiving process (Walter, 1996). These determinants set the stage for the contemporary hospice movement (Verderber & Refuerzo, 2006).

aimed to allow patients to die with peace and dignity (Koff, 1980; Buckingham, 1983). Dr. Saunders's vision became reality in 1967 when the first modern hospice, St. Christopher's Hospice, opened in a residential suburb of London. It served as both a religious and medical foundation for the care of the dying (Corr & Corr, 1983; Seibold, 1992). St. Christopher's was planned to combine the old concept of hospitality and care with the modern skills of a hospital (Buckingham, 1983). She disseminated her philosophy internationally in a series of tours of the United States that began in 1963. The dean of the Yale School of Nursing spent a month working with Dr. Saunders and, in 1971, established the first hospice in the United States based on the principles of modern hospice care (Buckingham, 1983; Corr & Corr, 1983; Koff, 1980).

At the same time, in 1965, Swiss psychiatrist Elisabeth Kübler-Ross also began to examine social responses to terminal illness. Her 1969 best-seller "On Death and Dying" significantly influenced how the medical profession responded to terminally ill patients in hospitals (Reed, 2004). She and proponents of palliative care fostered the "Death with Dignity" movement in the United States. Finally, in 1982, Congress enacted the Medicare Hospice Benefit, establishing four "core" principles: medical, nursing, psychosocial, and spiritual care (Bushfield & DeFord, 2010, p. 5). The hospice movement proved successful, and the number of hospice programs nationwide has continued to grow. According to the 2018 NHPCO's Facts and Figures: Hospice Care in America, more than 6,500 programs had been established in the United States since the first one opened in 1974.

Table 1: The Characteristics of the three types of death (Walker, 1996).

	Traditional Death	Modern Death	Postmodern Death
Authority	Tradition	Professional expertise	Personal choice
Authority figure	Priest	Doctor	The self
Dominant discourse	Theology	Medicine	Psychology
Coping through	Prayer	Silence	Expressing feelings
The traveler	Soul	Body	Personality
Bodily context	Living with death	Death controlled	Living with dying
Social context	Community	Hospital	Family

2.3 'Last Place' and Hospice Movement

In medieval times, a hospice was referred to as a place of shelter and rest for pilgrims or travelers on a difficult journey (Kohut & Kohut, 1984; Stoddard, 1978). Centuries later, hospice institutions were established in 17th-century France, 18th-century Germany, and Ireland. In the early 1900s, a group of Irish women founded St. Joseph's Hospice with thirty beds to care for the dying (Corr & Corr, 1983; Koff, 1980; Stoddard, 1978). Another group of Irish laywomen established a facility to care for the dying in New York City in the early 19th century. Throughout the 1900s, other hospice facilities were established in Pennsylvania, Massachusetts, Georgia, Minnesota, and Ohio, many of which are still in operation (Koff, 1980).

In the late 1940s, Dr. Cicely Saunders, founder of the modern hospice movement, envisioned a place where dying patients could receive pain management, family care, bereavement counseling, and support from multidisciplinary teams. Her vision

2.4 End-of-Life Care Philosophy

In the early 1980s, hospice care began to take root in the United States, with an expansion of hospital- and home-based services. The philosophy of hospice care holds that all human beings are inherently valuable, regardless of their circumstances, and capable of self-reflection and decision-making. Individuals are seen as having the ability to creatively relate within the family system to care about and for each other (Miller & McGown, 1997). The openness of the family system allows hospice caregivers to positively influence the quality of life for both the patient and their family, where learning, personal growth, and change are cornerstone concepts

https://doi.org/10.52202/082078-0024

inherent in the hospice care philosophy (Hall & Kirschling, 1990). According to NHPCO's Facts and Figures, in 2017, approximately 94.8% of Medicare hospice patients were 65 years of age or older, with 64.2% being 80 years or older (NHPCO, 2018).

2.5 Hospice as a Care Facility

End-of-life care is a multidisciplinary care and support (non-curative) system designed to address the physical, emotional, psycho-social, and spiritual concerns of terminal patients (Medical Dictionary, 2012), which makes the hospice care model unlike other healthcare models (Siebold, 1992). The hospice setting has dissimilarity in design themes with other healthcare settings. The history of end-of-life care or hospice as a place has been discussed earlier. In the contemporary design philosophy, the end-of-life care service has five basic types of models (Verderber and Refuerzo, 2006): the in-hospital hospice or palliative care units (PCUs); the home; the nursing institute with dedicated beds for hospice care; the medical-center affiliated free-standing hospice; and the non-hospital affiliated autonomous hospice.

3 LITERATURE ON END-OF-LIFE CARE ENVIRONMENTS

3.1 EDRA01-EDRA54 Proceedings

As mentioned earlier, the end-of-life environment is considered under the wide area of environmental gerontology. Since the 1970s, EDRA has been making a significant contribution to the area of environmental gerontology by developing theories, methodology, research-based design and planning, and by developing environmental assessment tools or protocols. All the proceedings (as many available) were searched by the writer twice. A good number of abstracts of individual presentations, group presentations (workshop or symposium), and full-length papers on environmental gerontology were published in the EDRA Proceedings. For example, In EDRA15, Gary T. Moore from the University of Wisconsin-Milwaukee published a paper titled, "Environmental Gerontology: Toward a Conceptual Perspective and News Lines of Research." In EDRA20, the content of the proceeding started organizing the papers and abstracts in the category of 'Environments of Elderly.' In EDRA 21, these papers and abstracts were put into the 'Special Population (elderly, disabled, prisoners)' category. In EDRA 22, again these publications and presentations were grouped as 'The Elderly'; in EDRA33 as the 'Aging' category.

In EDRA37, a good amount of papers on environmental gerontology were published; for example, a paper by Sixsmith, Andrew, from the University of Liverpool, The Theory and Reality of "Aging in Place" Polices; a paper by Rodiek, Susan from Texas A&M University, Exploration of older People's Immediate Environment and Their well-being. In EDRA 38, an invited paper was published by Margaret P. Calkins from IDEAS Institute, Private Bedrooms in Nursing Homes: Benefits, Disadvantage, and Costs. In the same proceeding under the Environmental Gerontology category, two papers were published. All these years of proceedings contain research on various types of places for an aging population, but none found specifically focused on end-of-life care environment (based on the PDFs available online).

In EDRA44, there was a presentation made by Kader which focused on literature reviews of end-of-life care environments. In EDRA46, a poster was presented on the hospice environment by the researcher. In EDRA48 Conference Proceedings, a paper was published that received the best-paper award (first place) by Kader, titled Environmental Characteristics that Enable Dying Patients to Exercise Choice and Personal Preference. In that same conference, a pre-conference intensive (group presentation) was made on The Built Environment as a Factor in the Engagement Process in Health Care, where an abstract was presented, titled Environmental Dimensions of End-of-Life Care to Support Family Members after a Patient's Death.

In EDRA53, Kader published and presented an abstract, Hospice Care Facility Design in the COVID-19 Era. The study has considered a systematic literature review and has identified six major design-related challenges of hospice care facilities: a) improving infection control, b) accommodating COVID-19 patients along with regular patients, c) accommodating patients' families in the facilities, d) creating a safe and supportive place for healthcare workers, e) providing palliative therapies and spiritual or grief support, and f) increasing in-house surge capacity.

3.2 Literature from other sources

A literature search was made focusing on the physical environment of end-of-life care (hospice or palliative care service); some focused on design issues and some studied physical environment in the context of wider issues such as quality of care, patients' perception, and experience, staff' or family members' perspective about hospice care or setting, and patients' views on a particular concept such as dignity or a 'good death'. Five books, seven guidelines, two reports, five dissertations and a thesis, seventeen journal articles, one book chapter, and two articles in the conference proceedings, focused on designing aspects of the hospice and palliative care environment.

In 2006, Verderber & Refuerzo wrote a major reference book, Innovations in Hospice Architecture, an overview of the historical background of the contemporary hospice and its basic principles of design. In 2020, Verderber & Refuerzo published the 2nd edition of that book. This fully revised, new edition of Innovations in Hospice Architecture responds to the need for an up-to-date, theoretically based reference book summarizing key historical and recent developments concerning this rapidly evolving building type. This Second Edition presents a) an overview of the historical origins of the contemporary hospice, b) the diverse

https://doi.org/10.52202/082078-0024

variations on the basic premise of hospice care, c) a review of the scant architectural literature published on this subject to date, d) a broad series of case studies of exemplary hospices around the world, and e) planning and design concepts for palliative care environments. This book also displays case study projects from Japan, Canada, Europe, Africa, Australia, Indonesia, China, the United States, and South America – a total of thirty-six case studies are individually presented. The book provides both technical and conceptual information, blending narrative, images, and diagrams so that the audience may understand and articulate the complexities of this specialized building type in professional practice contexts.

Another author, Ken Worpole from the UK, has published the 1st edition in 2009, Modern Hospice Design, which elaborates on the architectural design challenges for palliative care. Ken Worpole has published the 2nd edition in 2024, 2nd Edition, titled, Modern Hospice Design: The Architecture of Palliative and Social Care. The new edition has updated its timely advocacy of the need for good quality palliative care, the social history of the care of the elderly and terminally ill, and the new conditions of social care in the 21st century, principally in the UK, Europe and North America. This new book discusses the philosophy of design for providing care for the elderly and the vulnerable, the importance of architectural aesthetics and quality materials, and the integration of indoor and outdoor spaces as part of the overall care environment. This book is for architects and interior designers and their students, healthcare professionals, social care providers, estate and facility managers, and hospital administrators.

In 2016, Sarah McGann has published a book, titled, The Production of Hospice Space Conceptualising the Space of Caring and Dying. This book argues the positive and negative impact that space can have in the pursuit of an ideal such as hospice care has previously been underestimated. Also, the book claims that whether it be a purpose-built hospice, part of a hospital, a nursing home, or within the home, a hospice is anchored by space and spatial practices, and these spatial practices are critical for a holistic approach to dying with dignity. Such spatial practices are understood as part of a broad architectural, social, conceptual and theoretical process. By linking health, social and architectural theory and establishing conceptual principles, this book defines 'hospice' as a philosophy that is underpinned by space and spatial practice.

A series of hospice design guidelines and reports have been published in the USA, UK, and Australia. In 2000, the Hospice Unit Generic Brief was published in Australia. In 2005, the Design Guidelines for Specialist Palliative Care Settings was published by the Department of Health and Children of the Irish government. Another guideline published in the UK, A Place to Die with Dignity: Creating a Supportive Environment (NHS Estates 2005). In the USA, the Hospice Education Institute published the Hospice Design Manual for In-Patient Facilities (2006), written by Timothy Moorhouse. In 2010, the Hospice Friendly Hospitals Programme published Quality Standards for End-of-Life Care in Hospitals. The King's Fund Enhancing the Healing Environment (EHE) program published Improving Environments for Care at End of Life on lessons from eight pilot sites (Waller et al. 2008). Another report was published on a summary of evaluation findings by the King's Fund EHE (Aurthur et al. 2010). In 2017, Eshelman et. al. published a guideline, Designing End-of-Life Care Settings to Enhance Quality of Life: A Cornell University Report. Washington, DC: Facility Guidelines. In 2022, the Facility Guideline Institute (FGI) published a chapter, Specific Requirements for Hospice Facilities, in the book Guidelines for Design and Construction of Residential Health Care and Support Facilities. FGI is an independent, not-for-profit organization dedicated to developing guidance for the planning, design, and construction of hospitals, outpatient facilities, and residential health care, and support facilities.

Among the twenty articles in the journal, book, or conference proceedings, two are literature reviews (Brereton et al. 2011; and Rigby et al. 2010). Anderson (2008) studied patients' room design in palliative care units. Sagha & Eshelman (2019) focused on "Palliative design" meets "palliative medicine." Verderber (2014) focused on evidence-based architectural and landscape design considerations for residential hospice environments. Two studies focused on environmental factors of end-of-life care at intensive care units (Fridh et al. 2007; Fridh et al. 2009). Silver (2004) focused on healing environments in end-of-life care. Tan et al. (2005) researched hospice environments on patient spiritual expression. Pease & Finlay (2002) focused on the preference of single versus shared accommodation. Rowlands and Noble (2008) studied environmental impact and ward design. Kayser-Jones et al. (2003) identified the physical environment as one of the three factors that influence end-of-life care in nursing homes. Among the two articles in conference proceedings, one identified the therapeutic goals of an end-of-life care environment (Kader, 2015), and another published in EDRA48 proceedings (Kader, 2017) will be discussed later.

Niedzielski et. al. (2016) focused on sensory experiences and personalization in an inpatient residential hospice setting. Waldrop and McGinley (2018) wrote an article on environmental Serenity in residential hospice care. In 2019, Zadeh and Eshelman published an article in the HERD journal focusing on a strategic approach to the design, construction, and operation of healthcare facilities to improve quality of life and reduce suffering for patients, families, and

https://doi.org/10.52202/082078-0024

caregivers. In 2021, Ferrante and Villani published an article on environmental, physical, and perceived quality in hospice. In 2022, Coomans et al published an article focusing on the role of the built environment in experiences of hospice care: challenges for hospice design. In 2023, Kader published an article in a peer-reviewed book, titled - Designing the post-pandemic hospice environment: 'The Last Place.'

Among the five dissertations and theses, four focused on environmental design features for end-of-life care. Swenson (2009) examined the designated hospice rooms in nursing home facilities, and Sargent (2012) explored design through relationship-centered end-of-life care. The dissertation by Kader (2016) on developing a post-occupant evaluation tool for a hospice environment will be discussed later. Fronsman (2017) focused on exploring dimensions of nature in hospice design for a master's thesis.

A significant amount of studies focused on quality of care with different goals: quality of life in palliative care units (Cohen et al. 2001); quality of life for cancer patients (Cohen & Leis 2002); non-pharmacological caregiving activities (Lindqvist et al. 2012); quality of spiritual care (Puchalski et al. 2009); views on dignity (Franklin et al. 2006); influencing forces of care (Wilson & Daley 1998). Kayser-Jones et al. (2005) suggested a model hospice unit with three factors: care; community, and compassion.

Seven studies focused on patients' perception and dying experience: Chinese patients' dying experience in a nursing home (Chan & Kayser-Jones 2005); experience of living-dying of black and white older adults (Engle et al. 1998); ways of relating to death (Ternestedt & Franklin 2006); minority and non-minority perspectives on what is a "good death" (Tong et al. 2003); physical and psychosocial suffering in the dying process (Schroepfer 2007); and cancer patients' experience towards death (Larkin et al. 2007; and Ryan 2005).

Family perception of hospice care was examined by eleven studies with different objectives: identification of factors that influence quality of end-of-life care (Heyland et al. 2006; Russell et al. 2008; Stajduhar et al. 2011); end-of-life care in hospital (Hawker et al. 2006; and Spichiger 2008); and end-of-life care in nursing home (Kaarbo 2010; Munn & Zimmerman 2006; Vohra et al. 2004; Vohra et al. 2006; Wilson & Daley 1999). Staff's perspective on hospice care was studied by Brazil et al. (2004) and Evans et al. (2006). A dissertation by Nakashima (2002) investigated the psychosocial and spiritual well-being of older adults at the end of life.

4 MAJOR CONTRIBUTIONS: EDRA01-EDRA55

4.1 Theoretical

One of the significant innovations in theory is the identification of the therapeutic goals (TGs) of a dying person. The term 'therapeutic goal' was suggested by Cohen and Weisman (1991) in their Conceptual Framework for the organization of the person-environment system (shown in Figure-1), where the physical environment is part of the entire care milieu, which also includes personal, social, and organizational dimensions. The organizational component is conceptualized in terms of policies and programs, the social component is represented by residents and family, and the architectural component is defined in terms of the experiential qualities or attributes of the environment (Cohen and Weisman, 1991). The environmental experience of the individual involves continuing mutual reciprocity between the person and the environment. The therapeutic goals serve as unifying intentions, which can direct congruent decision-making in the organizational, social, and physical realms and thereby provide a useful foundation for planning and design (Cohen and Weisman, 1991).

There are several significant implications of TGs in place research, planning, and design. The significance in place theories is mentioned here briefly. One point to mention for clarification about the vocabulary of place experience is leading environmental gerontologists use various terms to explain 'experience of place': Lawton, Fulcomer, & Kleban (1984) of, evaluation criteria; Calkins (1988) of, environment-behavior issues; Cohen & Weisman (1991) of, therapeutic goals; Zeisel, Hyde, & Levkoff (1994) of, environmental characteristics, Diaz Moore et al. (2006) of, attributes of place experience; and Diaz Moore (2014) of, place attributes. The notion of developing one coherent and useful framework or conceptual model of Place has been focused on since the 1970s (Gubrium, 1978) as the core concept of environmental gerontology. Forty years ago,

Lawton and Nahemow (1973) published the primary environmental gerontology theory, Ecological Model of Aging. David Canter (1977; 1983; 1991) also provided a central theory (Figure-2), Theory of Place. In 2006, Diaz Moore and his colleagues developed A Model of Place (Figure-3). In 2014, Diaz Moore again developed the Ecological Framework of Place (Figure-4). All these models have theoretical, methodological, and practical implications. To analyze hospice as place using any of these models, identification of therapeutic goals is significant; a useful point of departure to engage in effective and systemic place making.

https://doi.org/10.52202/082078-0024

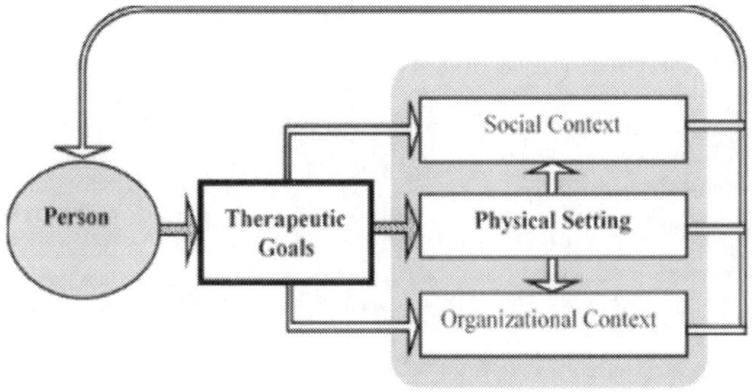

Figure-1. A Conceptual Framework for the organization of the person-environment system (Adapted from Figure 1.2, Cohen & Weisman, 1991).

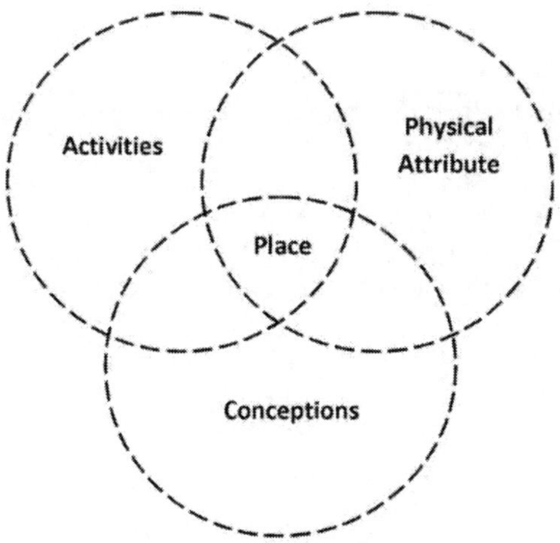

Figure-2. Canter's Theory of Place (Canter, 1977)

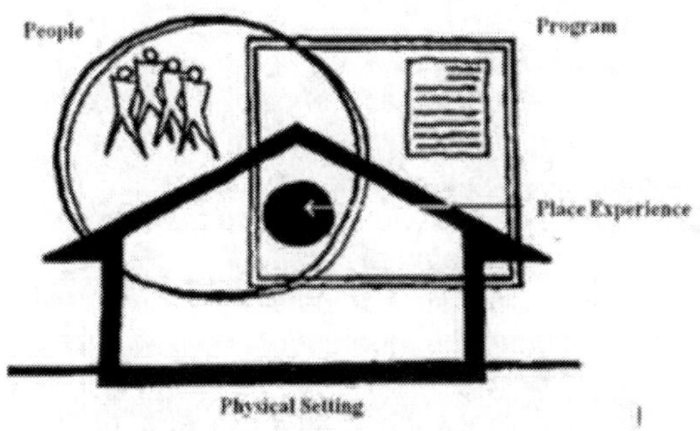

Figure-3. A Model of Place (Diaz Moore et al., 2006)

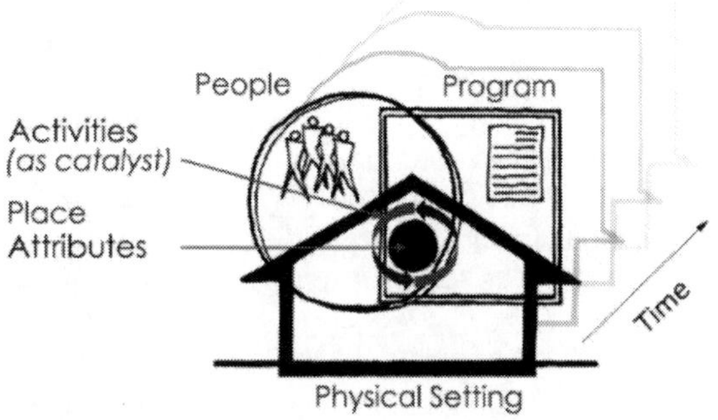

Figure -4. The Ecological Framework of Place for Aging (Diaz Moore, 2014)

https://doi.org/10.52202/082078-0024

Though several researchers have suggested age-specific environmental characteristics since the 1980s (Lawton, 1983; Cohen &Weisman, 1991; Lawton et al., 1992; Calkins, 2001; and Diaz Moore et al., 2006), none of them focused on hospice environments or environments specifically for dying patients.

In 2013 EDRA44, Kader presented the eleven therapeutic goals which were developed focusing on end-of-life care environment and later published in the ARCC conference proceedings (Kader, 2015). The objective of the study was to identify the TGs focusing on patients' experience; "What does the literature suggest are the appropriate dimensions of the 'environment as experienced' and therapeutic goals for the end-of-life care environment?" This study employed a systematic literature review with an approach developed by Hawker and colleagues in 2002 and considered a wide-ranging literature search. The definition of each goal is presented in Table 2;

As this list illustrates, many dimensions are consistent with the aging and environment literature (e.g., safety & security, stimulation), but others emerge as particularly salient in the hospice care environment, such as spiritual care or support after death. Even where the environmental dimensions are consistent, the meanings of those dimensions are considered somewhat differently due to the very specific needs surrounding the dying experience. The first eight dimensions have a direct influence on patients' experience, and the last three goals have an indirect influence on patients' experience (e.g. staff support). These goals reflect two characteristics; each expresses a basic or derived major patient's need, and a potential environmental facilitator for the satisfaction of the need (Lawton et al. 2000).

Table 2: Therapeutic goals of the hospice environment.

Therapeutic Goals	**Definition**
Provide continuity of self	*Environmental characteristics that help preserve or support patients' past activities, preferences and awareness.*
Provision of access to nature	*Environmental characteristics that provide opportunities for visual and physical access to nature.*
Provision of privacy	*Environmental characteristics that facilitate patients' choices in various levels of privacy through regulation of visual and auditory stimuli.*
Facilitate social interaction	*Environmental characteristics that facilitate and enable meaningful interaction between patients with staff, their family and other patients.*
Maximize safety and security	*Environmental characteristics that maximize patient safety and security of self.*
Provision of autonomy	*Environmental characteristics that enable patients to exercise choice and personal preference about their environment and everyday life.*
Regulate stimulation and support sensory therapies	*Environmental characteristics that contribute to an appropriate quantity and quality of sensory experience and support palliative therapies.*
Provision of spiritual care	*Environmental characteristics that facilitate opportunities for patients' spiritual care; religious, philosophical, existential, and personal beliefs, values, practices, and preferences.*
Provide family accommodation	*Environmental characteristics that facilitate patients' family accommodation and support control, functional independence, comfort, privacy, recreation, and spiritual care.*
Provide support after death	*Environmental characteristics that support care and dignity for patients and their families during the moment of death, body removal, bereavement, and remembrance.*
Maximize support for staff	*Environmental characteristics that support staff for better efficiency, communication, observation, satisfaction, and well-being.*

https://doi.org/10.52202/082078-0024

4.2 Environmental evaluation

The theoretical background concerning the person, the environment, and their interrelationship was discussed earlier. To assess an environment, it is essential to first understand the meaning of the term "environment." Cohen and Day (1993) defined the environment as "the interaction of organizational factors (i.e., policy, program, and services), the social environment (e.g., formal and informal caregivers), and the physical setting" (p. 9). Environmental evaluation, or post-occupancy evaluation (POE), of building facilities is essential for providing feedback to architects and building owners to improve design strategies and management practices. According to Lawton et al. (1984), living environments can be evaluated based on three goals: (1) performance criteria, assessing how the environment facilitates or impedes certain behaviors; (2) subjective criteria, involving users' expectations and evaluations measured through their responses; and (3) social criteria, which consider a group's social norms and cultural values.

The history of environmental assessment begins with Kleemeier's 1959 suggestion (as cited in Lawton, 1980) that residential settings could be evaluated based on the dimensions of segregation, congregation, and control. A decade later, Pincus and Wood (1970) introduced additional dimensions: privacy, freedom, resources, integration, and personalization. These concepts formed the foundation for environmental assessment tools developed in the 1980s and 1990s.

4.3 Environmental evaluation tools

In the 1980s, the first multi-dimensional tool was developed, the Multiphasic Environmental Assessment Protocol (MEAP) (Moos & Lemke, 1980) for assessing congregate housing facilities. This was followed by the Therapeutic Environmental Screening Scale (TESS) (Sloane & Mathew, 1990), the Environment-Behavior (E-B) model (Zeisel, Hyde, & Levkoff, 1994), the Nursing Unit Rating Scale (NURS) (Grant, 1996) and the Professional Environmental Assessment Protocol (PEAP) (Norris-Baker et al., 1999). Later, several other instruments were developed, among those the Environmental Quality Assessment for Living (EQUAL) checklists (Cutler et al., 2006) and the Sheffield Care Environment Assessment Matrix (SCEAM) (Parker, et al. 2004) are most prevalent. A few examples are briefly described below:

The Multiphasic Environmental Assessment Protocol (MEAP) was developed by Moos and Lemke (1980) and is the most comprehensive environmental evaluation instrument for use in nursing home facilities. It has five major rating instruments, and each has the subcomponents of Physical and Architectural Features Checklist (PAF), Policy and Program Information Form (POLIF), Resident and Staff Information Form (RESIF), Sheltered Care Environment Scale (SCES) - Form R, and the Rating Scale (RS). The MEAP has been used for two decades in environmental research.

The Therapeutic Environment Screening Scale (TESS) developed by Sloane & Mathew (1990), its revised TESS+ form was developed to evaluate dementia special care units to assess eight domains: general design features; maintenance; inventory of spatial amenities and seating capacity; lighting; noise; amenities; programming; and global environment (Sloane, Mitchell, Long, & Lynn, 1995).

The Environment-Behavior (E-B) model was developed by Zeisel, Hyde, and Levkoff (1994) for SCUs and it assesses eight environment concepts: exit control; wandering paths; individual away places; common space; outdoor freedom; residential scale; autonomy support; and sensory comprehend ability.

The Environmental Quality Assessment for Living (EQUAL) checklists collects data on resident room, nursing unit and overall facility. The room and bathroom checklist have 112-items, the nursing unit checklist has 140-items, and overall facility checklist has 134-item for assessment (Cutler & Kane, 2007).

The Sheffield Care Environment Assessment Matrix (SCEAM), developed by Parker, et al. (2004) in the United Kingdom, provides a comprehensive assessment of the physical environment of residential care facilities for older people and consists of 370 items, each relating to a specific building feature.
The Professional Environmental Assessment Protocol (PEAP) was developed PEAP was developed by a team of environment and aging researchers to provide a standardized method of expert evaluation of special care units (SCUs) for people with dementing illnesses such as Alzheimer's disease (Norris-Baker et al., 1999). PEAP evaluates SCU settings with respect to eight dimensions of therapeutic goals which are developed by Cohen and Weisman (1991): 1) maximize safety and security; 2) maximize awareness and orientation; 3) support functional abilities; 4) facilitate social contact; 5) provide privacy; 6) provide opportunities for personal control; 7) regulate and provide quality of stimulation; and 8) provide continuity of the self.

4.4 Evaluation tool for end-of-life care environment (hospice)

As discussed above, several POE tools are available for assessing the environment of care facilities, mostly nursing homes with an emphasis on dementia special care units. Generally, hospice facilities were assessed by POE tools that have been developed for use in nursing homes, but Swenson (2009) found a significant gap in the usefulness because hospice patients are mostly bed-bound, their physical, social, and spiritual demands are different, and a patient's family accommodation plays a significant role in the patient's dying experience. Considering the need to develop a POE tool for use in hospice facilities to support improvement, and to satisfy the explicit and implicit

https://doi.org/10.52202/082078-0024

needs of dying patients and their families, a doctoral thesis (Kader, 2016) developed the first-ever POE tool for hospice facilities, titled - Development of Hospice Environmental Assessment Protocol (HEAP): A Post Occupancy Evaluation Tool.

HEAP is a built-environment evaluation tool for assessing the hospice environment. It is an Indicative level of the POE tool. This tool was developed following the conceptual framework of PEAP. HEAP assesses only the physical setting and does not include assessment processes of organizational or policy features. One of the major reasons rationalized PEAP as an appropriate choice for developing the HEAP, is the user's experience. End-of-life care patients are mostly bed-bound and most of their cognitive status is similar to the patients of SCUs for dementia care; PEAP was found most relevant for hospice patients. HEAP also provides a standardized method for the expert evaluation of hospice environments in three levels of the physical setting: fixed or structural features, semi-fixed features, and non-fixed features. It provides a global assessment of the quality of hospice care environments on the eleven dimensions deemed to be therapeutic for a dying person (Kader, 2015).
In 2021, a group of Canadian researchers adopted HEAP to develop an environmental audit data collection method using the Aging in the Right Place – Hospice Environmental Assessment Protocol (AIRP-HEAP) and the Secondary Observation (AIRP-HEAP SO) audit tools specifically designed to evaluate spatial and placemaking characteristics of hospice sites that serve OPEH. The AIRP-HEAP tool includes 295 items categorized into 11 therapeutic goals pertaining to the built and natural environment of hospice sites. The AIRP-HEAP SO tool consists of six open-ended questions about the surrounding neighborhoods of hospice sites.

5 COVID-19 and the hospice environment

The COVID-19 pandemic is an unprecedented situation that has created many challenges for healthcare facilities, including hospice or end-of-life care facilities (Rogers et al. 2021). Generally, hospice facilities face unique challenges during times of disaster as their patient populations are particularly vulnerable due to terminally ill conditions, decreased mobility, and dependence on medical technology (Rogers et al. 2021). Additionally, hospice care facilities accommodate patient families and provide spaces for social interaction, and the interdisciplinary care model includes many volunteers and social workers, creating a risk of spreading infection among patients, family members, and staff (Anderson et al. 2020). In addition, restricted family visitations, constraints on social interaction, control over shared activities (such as spiritual care), and the need for staff to wear personal protective equipment all create an extra level of mental stress, increasing depression and patient and staff dissatisfaction (Rogers et al. 2021; and Gergerich et al. 2020).

In 2023, Kader published an article based on a a literature review on post-pandemic hospice environment, and has identified six major design-related challenges of post-pandemic hospice care facilities: a) Improving Infection Control, b) Accommodating COVID-19 patients along with Regular Patients, c) Accommodating Patients' Families in the Facilities, d) Creating a Safe and Supportive Place for Healthcare Workers, e) Providing Palliative Therapies and Spiritual or Grief Support, and f) Increasing In-house Surge Capacity.

6 FUTURE RESEARCH DIRECTIONS

In the post-pandemic era, several emerging issues and unanswered questions warrant further empirical investigation to advance the understanding of end-of-life environments. While prior research has explored hospice care settings within institutional facilities, there remains a critical gap in studying the home environment for individuals nearing the end of life. Specifically, research is needed to define and develop therapeutic goals tailored to the home setting for dying individuals. The eleven therapeutic goals previously identified for hospice environments may serve as preliminary themes; however, each goal, along with its corresponding design considerations, represents a standalone research inquiry deserving in-depth exploration. The Hospice Environmental Assessment Protocol (HEAP) also requires additional field testing to strengthen its reliability and validity. Moreover, the HEAP framework must be expanded to include evaluation criteria related to hospice organizational policies and management practices. Another area for advancement involves developing the second level (indicative) and third level (diagnostic) of Post-Occupancy Evaluation (POE) tools tailored to hospice settings. These enhancements will support a more nuanced and evidence-based approach to assessing the performance of end-of-life care environments. Importantly, the HEAP protocol may also serve as a foundational model for creating an assessment tool specific to end-of-life care in home environments—a domain that remains underexplored. Finally, research must address how hospice settings can enhance resiliency and preparedness in anticipation of future pandemic scenarios, ensuring that care environments remain responsive, adaptable, and supportive under crisis conditions.

7 CONCLUSIONS

Over the past 55 years, EDRA has played a pivotal role in advancing research-based innovation across the full spectrum of environments—from the 'First Place' of childhood to the 'Last Place' of end-of-life care. Through its sustained commitment to interdisciplinary collaboration and knowledge exchange, EDRA has catalyzed meaningful progress in environmental

https://doi.org/10.52202/082078-0024

design. Among the most profound and symbolically rich environments are those supporting individuals at the end of life, where the intersection of vulnerability and spatial experience becomes particularly significant. As the global population continues to age, the need to design and renovate end-of-life care settings—such as hospices—will become increasingly urgent. EDRA remains at the forefront of this emerging frontier, encouraging research that informs more empathetic, effective, and evidence-based design for end-of-life care environments.

8 REFERENCES

Anderson, D. (2008). Palliative care unit design: patient and family preferences. World Health Design, 1(1), 62-67.

Anderson, D. C., Grey, T., Kennelly, S., & O'Neill, D. (2020). Nursing home design and COVID-19: balancing infection control, quality of life, and resilience. Journal of the American Medical Directors Association, 21(11), 1519-1524.

Barnes, J. (2008). Nothing to be Frightened of. London, Cape.

Brazil K, McAiney C, Caron-O'Brien M, et al. (2004). Quality End-of-Life Care in Long-Term Care Facilities: Service Providers' Perspective. Journal of Palliative Care, 20(2): 85-92.

Brereton, L., Gardiner, C., Gott, M., Ingleton, C., Barnes, S., & Carroll, C. (2012). The hospital environment for end-of-life care of older adults and their families: an integrative review. Journal of Advanced Nursing, 68(5), 981-993.

Buckingham, R. W. (1983). The Complete Hospice Guide. New York: Harper and Row Publishers.
Bushfield, S. Y., & DeFord, B. (2010). End-of-life care and addiction. NY: Springer Publishing Company.

Calkins, M. P. (2001). The physical and social environment of the person with Alzheimer's disease. Aging & Mental Health, 5(sup1), 74-78.

Canter, D. (1977). The psychology of place. London: Architectural Press.

Canter, D. (1983). The purposive evaluation of places. Environment and Behavior, 15(6), 659-698.
Canter, D. (1991). Understanding, assessing, a nd acting in places: Is an integrative framework possible? In T. Garling& G. Evans (Eds.), Environment, cognition, and action (pp. 191-209). New York: Oxford University Press.

Chan, J., & Kayser-Jones, J. (2005). The experience of dying for Chinese nursing home residents: cultural considerations. Journal of Gerontological Nursing, 31(8), 26-32.

Cohen, S. R., & Leis, A. (2002). What determines the quality of life of terminally ill cancer patients from their own perspective? Journal of Palliative Care, 18(1), 48.

Cohen, S. R., Boston, P., Mount, B. M., & Porterfield, P. (2001). Changes in quality of life following admission to palliative care units. Palliative medicine, 15(5), 363-371.

Cohen, U. & Day, K. (1993). Contemporary environments for people with dementia. Baltimore, MD: Johns Hopkins University Press.

Cohen, U. & Weisman, G. (1991). Holding onto Home: Designing Environments for People with Dementia. Baltimore: John Hopkins University Press.

Corr, C. A. & Corr, D. M. (1983). Hospice Care Principals and Practice. New York: Springer Publishing Company.

Cutler, L. J., & Kane, R. A. (2007). EQUAL Environment Quality Assessment for Living: Resident Room, Nursing Unit & Overall Facility. Presentation, School of Public Health University of Minnesota. Retrieved on December 2015 from http://www.hpm.umn.edu/ltcresourcecenter/research/environmental_PDFs/GSA%202007%20EQUAL-11-08-07.pdf

Cutler, L. J., Kane, R. A., Degenholtz, H. B., Miller, M. J., & Grant, L. (2006). Assessing and comparing physical environments for nursing home residents: Using new tools for greater research specificity. The Gerontologist, 46(1), 42-51.

Department of Health and Children (2005). Design Guidelines for Specialists Palliative Care Settings. Department of Health and Children, Dublin, Ireland.

Diaz Moore, K. (2004). Interpreting the "hidden program" of a place: An example from dementia day care. Journal of Aging Studies, 18 (3), 297–320.

Diaz Moore, K. (2014). An ecological framework of place: Situating environmental gerontology within a life course perspective. The International Journal of Aging and Human Development, 79(3), 183-209.

Diaz Moore, K., Geboy, L., & Weisman, G.D. (2006). Designing a better day: Guidelines for adult and dementia day services centers. Baltimore, MD: Johns Hopkins University Press.

Engle, V. F., FoxHill, E., & Graney, M. J. (1998). The Experience of LivingDying in a Nursing Home: Self Reports of Black and White Older Adults.Journal of the American Geriatrics Society, 46(9), 1091-1096.

https://doi.org/10.52202/082078-0024

Eshelman, P., Sagha Zadeh, R., Setla, J., & Krieger, A. (2017). Designing End-of-Life Care Setting to Enhance Quality of Life: A Cornell University Report. Washington, DC: Facility Guidelines Institute. https://www.fgiguidelines.org/resource/designing-end-life-care-settings-enhance-quality-life/

Evans, W.G., Cutson, T.M., Steinhauser, K.E. and Tulsky, J.A. (2006). Is There No Place Like Home? Caregivers Recall Reasons for and Experience upon Transfer from Home Hospice to Inpatient Facilities. Journal of Palliative Medicine, 9(1), 100-110.

Ferrante, T., & Villani, T. (2021). Environmental physical and perceived quality in hospice. HERD: Health Environments Research & Design Journal, 14(4), 324-338.

Franklin, L. L., Ternestedt, B. M., & Nordenfelt, L. (2006). Views on dignity of elderly nursing home residents. Nursing ethics, 13(2), 130-146.

Fridh, I., Forsberg, A., & Bergbom, I. (2007). Endof life care in intensive care units–family routines and environmental factors. Scandinavian journal of caring sciences, 21(1), 25-31.

Fridh, I., Forsberg, A., & Bergbom, I. (2009). Close relatives' experiences of caring and of the physical environment when a loved one dies in an ICU. Intensive and Critical Care Nursing, 25(3), 111-119.

Fronsman, A. E. (2017). Exploring Dimensions of Nature in Hospice Design (Master's Thesis, Cornell University).

Gergerich, E., Mallonee, J., Gherardi, S., Kale-Cheever, M., & Duga, F. (2021). Strengths and struggles for families involved in hospice care during the COVID-19 pandemic. Journal of Social Work in End-of-Life & Palliative Care, 17(2-3), 198-217.

Government Printing Office (2000). U.S. Code of Federal Regulations 42CFR418.1. Retrieved from http://www.access.gpo.gov/nara/cfr/waisidx_00/42cfr418_00.html.

Grant, L.A. (1996). Assessing environments in Alzheimer special care units. Research on Aging, 18(3), 275-291.

Gubrium, J. (1978). Notes on the social organization of senility. Urban Life, 7(1), 23-44.

Hall, J. E., & Kirschling, J. M. (1990). A conceptual framework for caring for families of hospice patients. Hospice Journal, 6(21), 1-28.

Hawker, S., Kerr, C., Payne, S., Seamark, D., Davis, C., Roberts, H., ... & Smith, H. (2006). End-of-life care in community hospitals: the perceptions of bereaved family members. Palliative Medicine, 20(5), 541-547.

Hawker, S., Payne, S., Kerr, C., Hardey, M., & Powell, J. (2002). Appraising the evidence: reviewing disparate data systematically. Qualitative Health Research, 12(9), 1284-1299.

Heidegger, M. (1996). Being and time: A translation of Sein und Zeit. SUNY Press.

Heyland, D. K., Dodek, P., Rocker, G., Groll, D., Gafni, A., Pichora, D., ... & Lam, M. (2006). What matters most in end-of-life care: perceptions of seriously ill patients and their family members. Canadian Medical Association Journal, 174(5), 627-633.

Hospice Friendly Hospitals. (2010a). Quality Standards for End-of-Life Care in Hospitals: Making end-of-life care central to hospital care, Dublin. Hospice Friendly Hospitals. Retrieve on April 2013 from http://hospicefoundation.ie/wp-content/uploads/2013/04/Quality_Standards_for_End_of_Life_Care_in_Hospitals.pdf

Hospice Unit Generic Brief. (2000). The Aged, Community and Mental Health Division, Melbourne, Australia: Victoria Government Department of Human Services.

Kaarbo, E. (2011). Endoflife care in two Norwegian nursing homes: family perceptions. Journal of clinical nursing, 20(78), 1125-1132.

Kader, S. (2016). Development of Hospice Environmental Assessment Protocol (HEAP): A post occupancy evaluation tool (Doctoral dissertation, University of Kansas). http://hdl.handle.net/1808/21798

Kader, S. (2017). Environmental Characteristics that Enable Dying Patients to Exercise Choice and Personal Preference. Proceedings of the EDRA48 Conference on Voices of Place. 116-126. https://cdn.ymaws.com/www.edra.org/resource/collection/8F43AFA7-151E-4AE2-8E3F-F71DF0ABF411/EDRA-Proceedings-2017.pdf

Kader, S. (2022). Hospice care facility design in the COVID-19 era. In A. Joseph, L. McCunn, S. Kader, et al. (Eds.), EDRA53: Health in all design—Promoting health, equity, sustainability and resilience through environmental design (Proceedings of the 53rd Annual Conference of the Environmental Design Research Association). Environmental Design Research Association. ISBN 9781713868446

Kader, S. (2023). Designing the post-pandemic hospice environment:"The Last Place". In (Re) designing the Continuum of Care for Older Adults: The Future of Long-Term Care Settings (pp. 237-259). Cham: Springer International Publishing. https://doi.org/10.1007/978-3-031-20970-3_13

https://doi.org/10.52202/082078-0024

Kader, S. and Diaz Moore, K. (2015). Therapeutic Dimensions of Palliative Care Environment. In Proceedings of the ARCC Conference on Future of Architectural Research. 492-499. http://hdl.handle.net/1808/23329

Kayser-Jones, J., Chan, J., & Kris, A. (2005). A model long-term care hospice unit: Care, community, and compassion. Geriatric Nursing, 26(1), 16-64.

Kayser-Jones, J., Schell, E., Lyons, W., Kris, A. E., Chan, J., & Beard, R. L. (2003). Factors that influence end-of-life care in nursing homes: the physical environment, inadequate staffing, and lack of supervision. The Gerontologist, 43(2), 76-84.

Koen Coomans, K., Margo Annemans and Ann Heylighen. The role of the built environment in experiences of hospice care: challenges for hospice design. The Evolving Scholar. 2022. https://doi.org/10.24404/62D015598DF837AEE90961E0

Koff, T. H. (1980). Hospice, a Caring Community. Cambridge, Massachussetts: Winthrop Publishers, Inc.

Kohut, J. M., and Kohut, Sylvester. (1984). Hospice: Caring for the Terminally Ill. Springfield, Illinois: Charles C. Thomas, Publisher.

Kubler-Ross. (1969). On Death and Dying. New York, Macmillan Publishing Company.
Larkin, P. J., Bernadette Dierckx, D. C., & Schotsmans, P. (2007). Transition towards end of life in palliative care: An exploration of its meaning for advanced cancer patients in Europe. Journal of Palliative Care, 23(2), 69-79.

Lawton, M. P, Kleban, M. H., Rajagopal, D., Dean, J. (1992). Dimensions of affective experience in three age groups. Psychology and Aging, 7(2), 171-184.

Lawton, M. P. (1980). Environment and aging. Monterey, CA: Brooks/Cole Publishing Co.

Lawton, M. P. (1983). Environment and other determinants of weil-being in older people. The Gerontologist, 23(4), 349-357.

Lawton, M. P., Fulcomer, M., & Kleban, M. (1984). Architecture for the mentally impaired elderly. Environment and Behavior, 16, 730–757.

Lawton, M. P., Weisman, G. D., Sloane, P. D., Norris-Baker, C., Caulkins, M., & Zimmerman, S. I. (2000). Professional Environment Assessment Procedure for special care units for elders with dementing illness and its relationship to the therapeutic environment schedule. Alzheimer's Disease and Associated Disorders, 14 (1), 23–38.

Lawton, M.P. & Nahemow, L. (1973). Ecology and the aging process. In C. Eisdorfer & M.P. Lawton (Eds.), The psychology of adult development and aging (pp. 619-674). Washington, DC: American Psychological Association.

Lindqvist, O., Tishelman, C., Hagelin, C. L., Clark, J. B., Daud, M. L., Dickman, A., ... & Miccinesi, G. (2012). Complexity in non-pharmacological caregiving activities at the end of life: An international qualitative study.PLoS Med, 9(2), e1001173.

McGann, S. (2016). The Production of Hospice Space: Conceptualising the Space of Caring and Dying (1st ed.). Routledge.

Miller, B., & McGown, A. (1997). Bereavement: Theoretical perspectives and adaptation: Canberra, Australia. American Journal of Hospice and Palliative Medicine, 14(4), 156-177.

Mitroff, I., & Turoff, M. (1975). Philosophical and methodological foundations of Delphi. In H. Linstone, & M. Turoff (Eds), The Delphi Method: Techniques and Applications. (pp. 17-34). Reading, MA: Addison-Wesley.

Moher, D., Liberati, A., Tetzlaff, J., & Altman, D. G. (2009). Preferred reporting items for systematic reviews and meta-analyses: The PRISMA statement. BMJ : British Medical Journal, 339.

Moore, K. D. (2007). Restorative dementia gardens: Exploring how design may ameliorate attention fatigue. Journal of Housing for the Elderly, 21(1-2), 73-88.

Moore, K. D., VanHaitsma, K., Curyto, K., & Saperstein, A. (2003). A pragmatic environmental psychology: A metatheoretical inquiry into the work of M. Powell Lawton. Journal of Environmental Psychology, 23(4), 471-482.

Moorhouse, T. (2006). Hospice Design Manual for in-Patient Facilities. Hospice Education Institute, Maine, USA.

Moos, R.H., & Lemke, S. (1980). The multiphasic environmental assessment procedure (MEAP). Social Ecology Laboratory, VA & Stanford University Medical Center, Palo Alto, CA.

Munn, J. C., & Zimmerman, S. (2006). A good death for residents of long-term care: Family members speak. Journal of social work in end-of-life & palliative care, 2(3), 45-59.

Nakashima, M. (2002). A qualitative inquiry into the psychosocial and spiritual well-being of older adults at the end of life (Doctoral dissertation). Retrieved from ProQuest Dissertations & Theses Global. (Accession Order No. 3071127).

https://doi.org/10.52202/082078-0024

National Hospice and Palliative Care Organization. (2018). NHPCO's Facts and Figures Hospice Care in America. Retrieved on May 2019 from https://39k5cm1a9u1968hg74aj3x51-wpengine.netdna-ssl.com/wp-content/uploads/2019/07/2018_NHPCO_Facts_Figures.pdf

NHS Estates. (2005). Improving the Patient Experience. A place to die with dignity: creating a supportive environment. Leeds: Design Brief Working Group, NHS Estates.

Niedzielski, O.K., Rodin, G., Emmerson, D., et al. (2016). Exploring sensory experiences and personalization in an inpatient residential hospice setting. The American Journal of Hospice and Palliative Care, 33(7), 684–690.

Norris-Baker, C., Weisman, G.D., Lawton, M.P., Sloane, P., & Kaup, M. (1999). Assessing special care units for dementia: The professional environmental assessment protocol. In E. Steinfeld & G.S. Danford (Eds.), Enabling environments: Measuring the impact of environment on disability and rehabilitation (pp. 165-181). New York: Plenum Publishers.

Parker, C., Barnes, S., McKee, K., Morgan, K., Torrington, J., & Tregenza, P. (2004). Quality of life and building design in residential and nursing homes for older people. Ageing and society, 24(6), 941-962.

Pease, N. J., & Finlay, I. G. (2002). Do patients and their relatives prefer single cubicles or shared wards?. Palliative Medicine, 16(5), 445-446.

Pincus, A., & Wood, V. (1970). Methodological issues in measuring the environment in institutions for the aged and its impact on residents. The International Journal of Aging and Human Development, 1(2), 117-126.

Puchalski, C., Ferrell, B., Virani, R., Otis-Green, S., Baird, P., Bull, J., ... & Pugliese, K. (2009). Improving the quality of spiritual care as a dimension of palliative care: the report of the Consensus Conference. Journal of palliative medicine, 12(10), 885-904.

Reed, C. (2004). Elisabeth Kubler-Ross: Psychiatrist who identified five stages of dying - denial, anger, bargaining, depression and acceptance. The Guardian. Retrieved January 2014 from https://www.theguardian.com/society/2004/aug/31/mentalhealth.guardianobituaries

Rigby, J., Payne, S., & Froggatt, K. (2010). Review: what evidence is there about the specific environmental needs of older people who are near the end of life and are cared for in hospices or similar institutions? A literature review. Palliative Medicine, 24(3), 268-285.

Rogers, J. E. B., Constantine, L. A., Thompson, J. M., Mupamombe, C. T., Vanin, J. M., & Navia, R. O. (2021). COVID-19 pandemic impacts on US hospice agencies: A national survey of hospice nurses and physicians. American Journal of Hospice and Palliative Medicine®, 38(5), 521-527.

Rowlands, J., & Noble, S. (2008). How does the environment impact on the quality of life of advanced cancer patients? A qualitative study with implications for ward design. Palliative Medicine, 22(6), 768-774.

Russell, C., Middleton, H., & Shanley, C. (2008). Research: Dying with dementia: The views of family caregivers about quality of life. Australasian Journal on Ageing, 27(2), 89-92.

Ryan PY. (2005) Approaching death: a phenomenologic study of five older adults with advanced cancer. Oncology nursing forum, 32(6), 1101–1108.

Sagha Zadeh, R. & Eshelman, P. (2019). "Palliative design" meets "palliative medicine": A strategic and judicious approach to the design, construction, and operation of healthcare facilities to improve quality of life and minimize suffering for patients, their families, and their caregivers. HERD: Health Environments Research & Design Journal, January.

Sargent, A. C. (2012). Living while dying: An exploration of design through relationship-centered end-of-life care (Doctoral dissertation). Retrieved from ProQuest Dissertations & Theses Global.

Schroepfer, T. A. (2007). Critical events in the dying process: the potential for physical and psychosocial suffering. Journal of palliative medicine, 10(1), 136-147.

Siebold, C. (1992). The Hospice Movement: Easing Death's Pains. New York: Twayne Publishers.
Silver, S. (2004). Optimal healing environments in end-of-life care and beyond. Journal of Alternative & Complementary Medicine, 10(Supplement 1), S-201-S-209.

Sloane, P. D., & Mathew, L. J. (1990). The therapeutic environment screening scale: An observational screening instrument to assess the quality of nursing home environments for residents with dementia. American Journal of Alzheimer's Disease and Other Dementias, 5(6), 22-26.

Sloane, P. D., Mitchell, C. M., Long, K., & Lynn, M. (1995). TESS+ Instrument B: Unit observation checklist (A report on the psychometric properties of individual items and initial recommendations on scaling). Chapel Hill: University of North Carolina.

https://doi.org/10.52202/082078-0024

Spichiger, E. (2008). Living with terminal illness: patient and family experiences of hospital end-of-life care. International journal of palliative nursing, 14(5), 220-228.

Stajduhar, K. I., Funk, L., Cohen, S. R., Williams, A., Bidgood, D., Allan, D., ... & Heyland, D. (2011). Bereaved family members' assessments of the quality of end-of-life care: what is important?. Journal of palliative care, 27(4), 261.

Steinhauser, K. E., Christakis, N. A., Clipp, E.C., McNeilly, M., McIntyre, L., & Tulsky, JA (2000). Factors considered important at the end-of-life by patients, family, physicians, and other care providers. Journal of the American medical association, 284(19), 2476-82.
Stoddard, S. (1978). The hospice movement: A better way of caring for the dying. New York: Vintage Books.

Swenson, D. L. (2009). Designated hospice rooms in nursing homes: A new model of end-of-life care (Doctoral dissertation). Retrieved from ProQuest Dissertations & Theses Global. (Accession Order No. 3455510).

Tan, H. M., Grief, M., Couns, P., Dip, G., Braunack-Mayer, A., Med, B., & Beilby, J. (2005). The Impact of the Hospice Environment on Patient Spiritual Expression. Oncology nursing forum, 32(5), 1049-1055.

Ternestedt, B. M., & Franklin, L. L. (2006). Ways of relating to death: views of older people resident in nursing homes. International Journal of Palliative Nursing, 12(7), 334-340.

Tong, E., McGraw, S.A., Dobihal, E., et al. (2003). What is a good death? Minority and non-minority perspectives. Journal of palliative care, 19(3): 168.

Verderber, S. & Refuerzo, B.J. (2006). Innovations in Hospice Architecture.Taylor & Francis.

Verderber, S. & Refuerzo, B.J. (2019). Innovations in Hospice Architecture (2nd ed.). Taylor & Francis. https://doi.org/10.4324/9780429316104.

Vohra, J. U., Brazil, K., & Szala-Meneok, K. (2006). The last word: Family members' descriptions of end-of-life care in long-term care facilities. Journal of palliative care, 22(1), 33.

Vohra, J. U., Brazil, K., Hanna, S., & Abelson, J. (2004). Family perceptions of end-of-life care in long-term care facilities. Journal of palliative care, 20(4), 297-302.

Waldrop, D.P. and McGinley, J.M. (2018). Magic Happens Here": Environmental Serenity in Residential Hospice Care. Journal of housing for the elderly, 462-478.

Waller, S., Dewar, S., Masterson, A., & Finn, H. (2008). Improving environments for care at end of life. Lessons from eight UK pilot sites. London: King's Fund.

Walter, T. (1994).The revival of death. Psychology Press.

Walter, T. (1996). Facing death without tradition. Contemporary issues in the sociology of death, dying and disposal, 193-204.

Wilson, S. A., & Daley, B. J. (1998). Attachment/detachment: forces influencing care of the dying in long-term care. Journal of palliative medicine, 1(1), 21-34.

Wilson, S. A., & Daley, B. J. (1999). Family perspectives on dying in long-term care settings. Journal of Gerontological Nursing, 25(11), 19-25.

Worpole, K. (2009). Modern hospice design: The architecture of palliative care. Routledge.

Zadeh, R. S., & Eshelman, P. (2019). Palliative design meets palliative medicine: a strategic approach to the design, construction, and operation of healthcare facilities to improve quality of life and reduce suffering for patients, families, and caregivers. HERD: Health Environments Research & Design Journal, 12(3), 179-186.

Zeisel, J. (1975). Sociology and architectural design (Vol. 6). Russell Sage Foundation.

Zeisel, J., Hyde, J., & Levkoff, S. (1994). Best practices: An environment-behavior (EB) model for Alzheimer special care units. American Journal of Alzheimers Disease, 9(2), 4-21.

https://doi.org/10.52202/082078-0024

INDIVIDUAL PRESENTATIONS

Abstracts: Individual Presentations

REWILDING INTERIORS: BIODIVERSITY FROM THE INSIDE OUT

Track 1: Nature and urban ecology
Knowledge Network: Interior Design, Movement in Designed Environments: Nature & Ecology
Methods: Design Methods, Practice-based Research

Meghan Mick (Florida State University); Kendra Ordia (University of Nebraska Lincoln)

1 ABSTRACT

With roots in conservation biology, the term rewilding was first used thirty years ago by conservationist and environmental advocate Dave Forman. Originally intended to describe a method of restoring wilderness areas and networks for keystone species in North America, the concept of rewilding has become broadly used to describe ecological restoration benefiting all living things. It is a life-centric approach for supporting biodiversity that has significantly declined in areas of human-centric development. Biodiversity is critical for healthy ecosystems supporting the well-being of not just flora and fauna, but also human health.

As Ecology teaches us, we inhabit an interconnected web of life, a system that will suffer if siloed and separated both physically and theoretically. The broader use of the term 'ecology' is supported by a variety of disciplines, both in and outside of design. It is also backed by the U.S. Centers for Disease Control's One Health initiative. "One Health ... recognizes that the health of people is closely connected to the health of animals and our shared environment" (2023). This is more of a return to pre-industrial and indigenous ways of living than new or innovative thinking. Still, it will require a rewiring of our cultural aesthetics that currently creates a binary between ecologically healthy, wild places on one end and safe, inhabitable environments on the other.

Rewilding as a design approach has been described as "rebuilding our relationship with the wild in order to receive the plethora of benefits that keep us well" by Sierra Bainbridge, a landscape architect with MASS Design Group (2023). The reconnection between humans and nature needs to occur even within the built environment. The gap is arguably most impacted by the aesthetic perceptions associated with natural vs. developed places. As researcher Joan Nassauer argues, "Perception of human intention may be the difference between nature preserve and dumping ground ... Designing ecosystems so that people will recognize their beauty and maintain it appropriately may depend upon including design cues of human intention" (1995, p.162). Designers of the built environment have an opportunity to help rewild aesthetics in this way. It does happen outdoors, but this aesthetic will need to transcend the shells of our buildings as well. How can concepts of rewilding be translated for interiors? What aspects have the most potential to reframe our relationship with nature, even in controlled environments within buildings? This research examines rewilding in the design of the built environment – in theory and practice - and how it might be applied to interiors to contribute to a symbiosis with the natural world combining ecological knowledge with aesthetic preferences to enhance human health and well-being.

2 REFERENCES

Bainbridge, S. (2023, August 8). A guide to rewilding. Architect. https://www.architectmagazine.com/design/a-guide-to-rewilding_o

Centers for Disease Control and Prevention. (2023, December 1). One Health. Centers for Disease Control and Prevention. https://www.cdc.gov/onehealth/index.html

Nassauer, J. I. (1995). Messy ecosystems, orderly frames. Landscape Journal, 14(2), 161–170. https://doi.org/10.3368/lj.14.2.161

ENHANCING ECOLOGICAL AWARENESS AMONG JUNIOR HIGH SCHOOL STUDENTS THROUGH AESTHETIC EDUCATION

Track 1: Nature and urban ecology
Knowledge Network: Children, Youth & Environment, Student Affairs
Methods: Action/Participatory Research, Grounded Theory

Ya-Wen Cheng (National Cheng Kung University); Li-wen Sung (National Cheng Kung University)

1 MOTIVATION

In response to Nature Deficit Disorder (NDD), introduced by Richard Louv in 2005, this study addresses the diminishing exposure of junior high school students to natural settings and the associated physical and mental health challenges with art and aesthetics education. In practical teaching, we have observed a noticeable decline in students' ability to describe their environment. Therefore, this study employs grounded theory to address these shortcomings. Additionally, multiculturalism became prevalent in art education, with Duncum (2002) proposing Visual Culture Art Education (VCGA), emphasizing the interpretation of images in daily life surroundings. This research further integrates aesthetic education, emphasizing emotional cognition, to guide students in understanding the contemporary societal concern regarding ecological imbalance and alienation.

2 METHOD

As Dewey argues, all teaching occurs through the medium or intermediary of the environment in the school and classroom and should be focused on activity, not learning (Hansen, 2002). In the seventh grade, there is a specific focus on fostering direct interaction between students and the environment. For instance, creative obstacle courses involving dragging or rolling cardboard boxes on the floor are implemented, or picnics on the school lawn.

In the eighth grade, the emphasis gradually shifts towards enhancing cognitive content and expanding the scope to the surrounding school environment. This includes activities like location-based games of plants, and leveraging Google smart lens features to record. Additionally, students analyze plant species in the neighborhood, delving into discussions about their associations with residences, such as medicinal or protective functions.

In the ninth grade, the study delves into architectural techniques related to achieving balance with nature, such as Chinese Feng Shui. Our school districts often feature buildings with diverse ethnic architectural styles such as Japanese, indigenous, or traditional Chinese architecture, allowing students to personally experience how different cultures achieve harmony with nature. With the embodied process, students gain a heightened level of aesthetic perception, contributing to a richer perspective on the world (Berlearnt, 2016).

3 CONCLUSION

The richness and detail of worksheet writing depend on students' expressive abilities, thus serving as a partial reference. Improvement in students' ecological awareness needs to be observed over time by educators through long-term hermeneutic phenomenology of research. Any aesthetic experience is built upon at least three components: the sensory qualities, the associated sets of meanings, and the emotional response (Chatterjee, 2015), so, promoting collaboration, experiential learning, and community involvement, this approach not only enhances ecological awareness but also nurtures a generation of individuals who are inspired to address and mitigate the challenges of ecological imbalance and alienation in their immediate surroundings.

4 REFERENCES

Berleant, A. (2016). Re-thinking aesthetics: Rogue essays on aesthetics and the arts. New York: Routledge.

Chatterjee, A. (2015). The neuropsychology of visual art. Art, aesthetics, and the brain, 341-356. UK: Oxford University Press.

Duncum, P. (2002). Visual Culture Art Education: Why, What and How. International Journal of Art & Design Education, 21(1), 14-23.

Hansen, D. T. (2002). Dewey's Conception of an Environment for Teaching and Learning. Curriculum Inquiry, 32(3), 267–280.

ASSESSING THE RESTORATIVE VALUE OF GREEN CULTURAL HERITAGE IN KOREA USING EEG

Track 1: Nature and urban ecology
Knowledge Network: Cultural Aspects of Design, Historic Preservation
Methods: Experimental/Quasi-Experimental Research, Mixed Methods

Jeong-Yeon Lee; Hyeon-Jeong Hwang (Kyungpook National University); Hyung-Sook Lee (Kyungpook National University)

1 ABSTRACT

The global population grapples with mental health issues like stress, depression, and anxiety. Research has extensively explored restorative environments within urban settings, such as urban parks. Recent studies indicate that exposure to historical and cultural landscapes can significantly benefit well-being, offering considerable restorative potential (Halbwachs, 1950; Scopelliti et al., 2019). Traditional and historical landscapes evoke feelings of nostalgia, familiarity, and attachment.

In particular, green cultural heritage (GCH) refers to landscapes that blend nature, culture, and history, such as traditional gardens that reflect regional lifestyles and cultures over time through their interaction with the natural environment (Xie et al., 2023). GCH holds significant promise for healing and recovery in urban environments.

There has been a growing focus on preserving cultural heritage and recognizing its practical benefits. However, research exploring the psychological stability and emotional recovery linked with cultural heritage is still in its nascent phase. Thus, our study aimed to quantitatively assess the physiological and psychological restorative effects of Korean GCH using Perceived Restorative Scale (PRS) and electroencephalography (EEG). The brain's response to environmental stimuli is influenced by various spatial elements, resulting in physiological and psychological alterations. EEG signals offer a means to gauge an individual's cognitive state within a specific environment. We have assessed the restorative potential of Korean GCH by comparing it with other human-centric environments like cityscapes and urban parks.

In the experimental setup, a group of 34 college students were exposed to videos featuring different environments, namely cityscape, urban park, and GCH in Korea. These videos were captured at various locations across Korea in October 2023. The cityscape video depicted a bustling plaza surrounded by buildings and traffic in a downtown area. The urban park video aimed to represent a typical neighborhood park with diverse vegetation, while the GCH video integrated historical elements with natural features. During the viewing of each video, which lasted for 3 minutes, participants' brain activity was monitored using EEG. Subsequently, participants completed the PRS questionnaire, which encompasses restorative components (being-away, extent, fascination, and compatibility) (Hartig et al., 1997). For analysis, alpha, beta, and theta waves were utilized. Alpha waves are typically associated with feelings of relaxation, focus, and calmness, while beta waves are linked to sensations of nervousness, restlessness, and anxiety (Hashimoto et al., 2003).

As a result, using the PRS, perceptions of being-away, fascination, and extent were notably higher in GCH scenes compared to the other videos. In terms of physiological responses, participants exhibited decreased alpha waves while viewing the GCH video, suggesting heightened engagement. Our findings also revealed an increase in beta wave amplitude in EEG recordings. Beta waves typically prevail during mental tasks or complex cognitive processes. Our observations suggest that GCH scenes stimulate beta activity, thereby enhancing concentration. Additionally, a higher alpha/beta ratio correlated with reduced arousal and vigilance. Specifically, the lower alpha/beta ratio in occipital suggests cognitive processing.
The insights from this study will advance our comprehension of the significance of cultural heritage and offer guidance for creating restorative environments that incorporate historical elements into contemporary landscapes.

A COMPARISON OF CROSS-AGE PERCEPTION OF KOREAN HERITAGE GARDENS USING EYE-TRACKING

Track 1: Nature and urban ecology
Knowledge Network: Cultural Aspects of Design, Historic Preservation
Methods: Mixed Methods, Experimental/Quasi-Experimental Research

Shinyoung Chung; Qian Wang (Kyungpook National University); Hyung-Sook Lee (Kyungpook National University)

1 ABSTRACT

Traditional gardens are essential cultural assets that reflect the unique cultural traits of a nation or cultural area. Korean heritage gardens not only symbolize Korean traditional culture but also encompass both tangible and intangible heritage elements of the community or society. Asian cultures have long acknowledged the importance of tradition and cultural heritage. Nonetheless, recent research suggests a trend where younger generations exhibit declining interest in traditional heritage (Gao, 2021). While there has been considerable attention given to the preservation of heritage gardens, there remains a dearth of studies focusing on the perception of traditional gardens and the variations across different generations in Korea. Previous studies investigating how people perceive gardens have primarily relied on questionnaire surveys, which are limited in capturing data about individuals' unconscious perceptions. Through the utilization of eye-tracking technology, our study had measured where individuals consciously focus their attention and assess the extent to which specific scenes or elements attract their gaze.

The objectives of this study were 1) to examine the perception and preferences on Korean heritage gardens in different age groups; and 2) to explore the relationship of different elements of landscape attractiveness in Korean gardens using eye-tracking technology.

Methods: Volunteers were recruited through poster advertisements displayed on the bulletin boards of the university library and the senior centers in Daegu city, Korea. A total of sixty-four subjects participated in the study, comprising 30 elderly individuals and 34 young adults. Each participant was equipped with an eye-tracking device (Gazepoint GP3 Eye-tracker) and randomly presented with eight photographs showcasing various features of Korean heritage gardens, including historical buildings, traditional patterns, vegetation, water features, hardscapes, etc. These images were sourced from traditional gardens across multiple Korean heritage sites, selected based on a review of previous studies. Following the observation of the eight photographs, participants were interviewed using questionnaires that included items related to their backgrounds, perceptions, and preferences regarding Korean heritage gardens. Results: The majority of elderly participants were in their 60's, while most of the young participants were under 25 years old. The findings revealed that both elderly and young participants displayed preferences for heritage gardens featuring historical buildings and water features. However, elderly participants exhibited a notably stronger preference for gardens with traditional patterns compared to their younger counterparts. The results showed that the appreciation of traditional gardens among young participants is not inferior to those of elderly participants. Eye-tracking results indicated that young participants had the longest fixation duration on water features and vegetation elements, whereas elderly participants focused more on traditional architecture and pavilions. Heatmaps displaying the results suggested that when viewing heritage gardens, young individuals exhibited more visual behaviors and fixation points compared to the elderly group. The analysis suggests that both garden features and individual characteristics should be considered in the design and enhancement of Korean heritage gardens. This study is limited by its use of photographs instead of real gardens due to constraints related to equipment availability and the mobility required to access study sites.

EXPLORING PLACE ATTACHMENT AND CONTRIBUTING FEATURES OF THE PHYSICAL ENVIRONMENT IN FAVORITE LOCAL URBAN PLACES

Track 1: Nature and urban ecology
Knowledge Network: Environmental & Architectural Phenomenology
Methods: Interviews, Photo-based Methods
Tamala Anderson (University of Surrey); Eleanor Ratcliffe (School of Psychology, University of Surrey); Birgitta Gatersleben (School of Psychology, University of Surrey); Clara Weber (School of Life Sciences and Facility Management, Zurich, University of Applied Sciences)

1 BACKGROUND

Place attachment, the cognitive and emotional bond between individuals and a meaningful place, is complex and richly studied. It can influence a broad spectrum of benefits, including positive emotions, a sense of belonging, and individual and social well-being. In addition, the relationship between place attachment and psychologically restorative environments is increasingly recognized.

In place attachment research, physical aspects of place are often studied at general levels, such as place type (like natural environments categorized blue or green; and built environments categorized urban, rural, residential, historic, etc.) and place scale (neighborhood, city, region, country). In this way, the physicality of place often becomes a controlled backdrop, at the expense of examining the physical environment itself. Research investigating the relationship between details of the physical environment and place attachment is infrequent, and little is known about particular physical attributes and features, as experienced by individuals, that support or enable place attachment. Affordance theory describes a relationship between features of the physical environment and an individual's embodied perception of those features, providing a compelling avenue for understanding place-based processes of place attachment.

2 AIM

To explore how attributes and features of the physical environment, experienced by the individual in favorite local urban places, support or enable processes of place attachment.

3 METHODS

Twenty adult residents of Kilburn, a diverse neighborhood undergoing regeneration in northwest London, UK, participated in this exploratory qualitative study of favorite local urban places using resident-employed photo elicitation. Photographs were used to guide in-person interviews; interviews were then transcribed and analyzed using an inductive approach to reflexive thematic analysis.

4 RESULTS

Results presented here focus on themes and subthemes most closely related to the physical environment and place attachment: Relationship with Place, Experiences in Place, and Perceptions of Physical Place. Participants often described favorite places through narratives of meaningful experiences which frequently reflected psychological processes of place attachment: behavior, through social and physical activities; affect through joy and happiness; cognition through memory or imagination. Perceptions of physical place were then related to these meaningful experiences through experiential or sensory descriptors (like sounds, textures), affective appraisals, and descriptions of important physical attributes and features. These physical attributes (like location, spatial definition, greenness, spatial variety) and physical features (like trees, planting, places to sit, points of focus) were perceived to support, or afford, the meaningful experiences underpinning participants' place attachment.

5 CONCLUSION

We found that important attributes and features of the physical environments of participants' favorite places were related to a variety of meaningful experiences in those places. To understand the role the physical environment plays with processes of place attachment, it is then necessary to consider individuals' active, embodied engagement of that environment. With this understanding, environments can be better equipped to support positive place relationships and realize the potential health and well-being benefits such as psychological restoration, sense of belonging, individual and social well-being associated with place attachment.

FINE-GRAINED URBANISM MULTI-SCALER URBAN DESIGN + PHENOMENA AND DATA-DRIVEN DESIGN FOR AFFORDABLE CLT HOUSING IN MARINE DEL PRAT VERMELL: BARCELONA URBAN DESIGN PROGRAM 2022 + 2024

Track 1: Nature and urban ecology
Knowledge Network: Cities & Globalization, Environmental & Architectural Phenomenology
Methods: Design Methods, Mixed Methods
Philip J Speranza (University of Oregon)

1 ABSTRACT

Design of cities is changing. Air pollution, sound pollution, water, energy, and metabolic processes, along with important social inclusion, gentrification, and urban poverty problems, are growing problems. New data-driven urban design methods are critical tools to measure, confirm and provide continued feedback for public and private agencies and the public they serve. The knowledge and expertise to use social media and sensor data has become and will be essential to future urban design professionals. How can mobile phone-based GPS and data-sensors empower participatory design processes with local stakeholders? How can data collection engage the perception of local neighbors, especially around qualities such as air pollution and pedestrian safety, for the adaptive reuse of post-industrial neighborhoods to meet the growing global demand of housing access?

The abstract presentation will share work from two full-term urban design programs set in the Catalan city of Barcelona, Spain: 1) one in summer 2022 that embraced new Superilla street design using both a human-scaled fine-grained urbanism approach and a robust city-scale based GIS approach, and 2) one in spring 2024 that embraced fine-grained urbanism methods to measure ground level qualities working with local planners and neighbors to address the perception of air pollution for street design and adjacent coop housing design. The program will first engage in-situ data acquisition, which allows the integration of both existing and newly acquired datasets using human-scaled data for both analysis and design. Then it will discuss a design approach of qualitative phenomena integrated with quantitative data collection. The phenomena approach will be highlighted by an eight-day workshop at Pritzker Prize RCR Arquitectes in dance, philosophy and photography, a four-day trip to the Alhambra and Great Mosque in Granada and Cordoba and a three-day retreat in the Pyrenees. The Arduino-sensor based method will measure sound, light, heat, humidity, and air quality based on a decade of published research experience from the program director in Barcelona and the United States. A key collaboration will occur with the City of Barcelona Planners for the new Marina del Prat Vermell industrial reuse into residential mixed-use neighborhood, alongside housing COOP and other local neighbor stakeholders. Local instructional collaboration will include past sub-director of Barcelona's pedestrian Superilles project, a local architect and landscape architect firm specializing in ecology and water, and a Cross Laminated Timber expert and factory visit in the Pyrenees.

The objective of this abstract presentation is: 1) to share the design research methods of geospatial social and natural data collection, 2) to understand the challenges and opportunities of the renewed participatory design process as it informs the established Social Interaction Tool, 3) to reflect on the student learning process and 4) to grow a collaborative and continued project with the City of Barcelona Planners to inform innovative design research to support local neighbor health in this new mixed-use and mixed-income neighborhood.

BIOPHILIC DESIGN IN THE REAL WORLD: EMERGING TRENDS IN IMPLEMENTATION AND MEASUREMENT FOR HEALTHY AND EQUITABLE PLACES

Track 1: Nature and urban ecology
Knowledge Network: Interior Design, Work Environments
Methods: Case Study, Meta-Analysis
Angela Loder (Greening of the City)

1 ABSTRACT
Biophilic design continues to be popular at multiple scales: at the building level, the community level, and regional level. Implementing biophilic design faces multiple challenges, however, including uncertainty on the return on investment for building owners and occupiers, evaluation methods that accurately reflect human behavior, performance and health, and conflicts around green gentrification and equity. Despite many decades of research demonstrating the health and cognitive benefits of biophilic design, it is still seen as a nice-to-have amenity versus an evidence-based intervention. This session explores the intersection of evaluation of biophilic design with implementation at multiple scales.

The first module of this session provides a comprehensive and critical overview of emerging research and metrics linking health, performance, and creativity outcomes with biophilia in buildings to build a case for prioritizing them in the workplace. This is particularly relevant with the increased pressure for organizations to provide a welcoming and healthy space for employees and economic challenges in the commercial real estate market, trying to entice them back to the office.

The second module evaluates the implementation of biophilic urban greening at a community and urban level, focusing on case studies of Philadelphia and Chicago. We will apply interdisciplinary evaluation methods to understand tensions and opportunities around cultural, equity, place and health outcomes in these contexts.

The final module focuses on research demonstrating biophilic impacts at a regional level, using geospatial and satellite-based assessments of surrounding greenness and comparing those against educational and health-based outcomes at a population scale. This research will set the foundation for a discussion on governmental and non-profit initiatives that can move the needle on biophilic exposure.

The session concludes with emerging opportunities and challenges of implementing and evaluating biophilic design as part of larger sustainability and placemaking initiatives, and how to link action at multiple scales of intervention.

CREATING AN AUDIT TOOL FOR BIOPHILIC ATTRIBUTES OF COLLEGE CAMPUSES

Track 1: Nature and urban ecology
Knowledge Network: None/Not Tied to Any Network
Methods: Case Study
Gowri Betrabet Gulwadi (University of Northern Iowa); Susana Alves (Sapienza University of Rome)

1 ABSTRACT

Humans have an innate affinitive response towards nature - biophilia (Kellert & Wilson, 1993) that is effortless, immediate, and impactful when they go through cycles of mental fatigue, stress, and a need for deep restoration. Viewing or being in nature has been associated with deeply healing outcomes (Keniger et al., 2013). Physiological effects (e.g., lowered heart rate) and psychological outcomes (e.g., reduced depression) have highlighted a need for meaningful daily interactions with nature. While nature interactions are salutogenic for a variety of people in different cultures and settings, and across the lifespan, on college campuses they have the biophilic potential to provide short- and long-term benefits to teenagers who seek to establish self-care habits, balance mental stimulation with mental fatigue, and explore their identity.

Connections with campus personnel contribute to student satisfaction, retention, and academic success, but enabling ways to connect with campus outdoor nature spaces can support place attachment and a sense of belonging. Everyday sensory nature-related experiences on campuses - e.g. sitting under a tree, listening to birdsong - can foster a student's connectedness to nature on campus, create pleasant settings for connectedness with others, and enable connectedness with oneself through introspective self-regulating experiences.

Well landscaped campus grounds attract and recruit students, but their hidden, untapped biophilic role in health promotion is less recognized. From an existing list of 14 basic patterns containing 70 biophilic attributes provided by Browning et al. (2014) we distilled the ten relevant biophilic patterns that have the most potential to make a difference on college campuses. However, each pattern and its attributes can be best recognized only with the help of detailed observable criteria and mechanisms to observe them on a college campus. This presentation describes the development and refinement process of a multimedia Biophilic Pattern Recognition Audit tool that provides a rubric for documenting the presence of biophilic attributes on college campuses. With this audit tool, universities will be able to recognize biophilic attributes on their own campus and decide if and what needs to be changed to make the campus more health promoting through either design or program interventions.
This project is at the confluence of different branches of research dealing with health and wellness fostered by interactions with nature. The main domain is the study of restorative environments that looks at stress-reduction and protective benefits derived from nature-based activities, and secondary domains are the study of nature-connectedness and its benefits and the emerging discussion of biophilic design attributes on campuses. The most important contribution is translational - how existing research can be applied to the design and management of campuses.

2 REFERENCES

Browning, W.D., Ryan, C.O., Clancy, J.O. (2014).14 Patterns of Biophilic Design. New York: Terrapin Bright Green LLC.

Kellert, S. R., & Wilson, E. O. (1993).The Biophilia Hypothesis. Washington, DC.: Island Press.

Keniger, L. E., Gaston, K. J., Irvine, K. N., & Fuller, R. A. (2013). What are the benefits of interacting with nature?. International journal of environmental research and public health, 10(3), 913-935.

EDIBLE ECOLOGIES: UNDERSTANDING THE DEGREES OF INTERCONNECTIVITY IN OUR FOOD SYSTEM

Track 1: Nature and urban ecology
Knowledge Network: Environmental Design Research Education, Movement in Designed Environments: Nature & Ecology
Methods: Case Study, Ethnography
Zaneta Hong (Cornell University)

1 ABSTRACT

What is palpable in our contemporary food system is that it is always in a state of flux. Food operates concurrently at the biological, ecological, and industrial scales; and with our climate crisis – concomitant with biodiversity loss, deforestation, water availability, soil erosion, pollinator extinction, invasive pests and diseases – the security of our foods will become dire at all scales and for all biotic matter. From shifting seasons, variable precipitation levels, increased global temperatures and greenhouse gas emissions, food production has been impacted by the multitude of anthropocentric disturbances that in turn have impacted seed availability, harvesting yields, supply chain challenges, labor shortages and inequities. When we see the innumerable agents, constituents, and stakeholders represented in our contemporary food system, we must recognize (even respect) the diversity of agendas and modalities that exists. A vision towards a more sustainable and resilient food system requires a multi-faceted, multi-dimensional, and multi-scalar collaborative approach that includes as many people as possible to make change. Acknowledging this network of intersections and entanglements provides us a mutual acceptance that humanity must come together in solving issues on food sustainability, human survival, and in mitigating future weather-related disasters and extreme events.

A systems innovation must go beyond the incremental; it needs to radically reconfigure what has been considered a broken food system. In an advanced design studio at Cornell University, radical change was the enterprise for students, in order to materialize a positive future for our foodscapes. The strengths and responsibilities of our creative design discipline require a nexus approach to global issues, and a means to enact a deep and rapid transformative change with critical leverage points and well-informed decisions that influence future social-ecological trajectories while avoiding any unintended consequences. As landscape practitioners and educators, we need to be increasingly aware that the environmental realities of our practice, including knowing how the design process is interconnected at multiple scales, influenced by multiple species, and operated across multiple locations around the world, is relevant to the local and global scales and for all living organisms. In foregrounding issues, such as food accessibility, land conservation, environmental degradation, and socioeconomic injustices, these dilemmas might seem localized to isolated communities and townships, but they are actually part of a global cause – providing humanity with a much closer degree of interconnectivity that acknowledges what inevitably affects our collective futures.

2 REFERENCES

Metcalfe, Robyn. Food Routes: 2019. Growing Bananas in Iceland and Other Tales from the Logistics of Eating. Cambridge: MIT Press.

Tsing, Anna Lowenhaupt. 2017. The Mushroom at the End of the World: On the Possibility of Life in Capitalist Ruins. Princeton, NJ: Princeton University Press.

Wallace-Wells, David. 2019. The Uninhabitable Earth: Life After Warming. New York: Tim Duggan Books.

ECOSYSTEM SERVICES AND BIODIVERSITY PRESERVATION IN PRAIRIE SUBURBAN AREAS.

Track 1: Nature and urban ecology
Knowledge Network: Health, Sustainable Planning, Design & Behavior
Methods: Experimental/Quasi-Experimental Research, Statistical/Modeling Analysis
Ana Karinna Hidalgo (University of Saskatchewan); Oscar Zapata (University of Saskatchewan)

1 ABSTRACT

The production of ecosystem services is essential for cities to adapt to climate change. Practices that reduce water consumption, control extreme temperatures, protect urban wildlife, and promote physical and mental health will be fundamental to preserving individual and community well-being (Young & Lieberknecht, 2019). Suburban areas have great potential as producers of ecosystem services (Middle et al., 2014), a topic overlooked in the literature (Vargas-Hernández & J. Zdunek-Wielgołaska, 2021).

However, creating markets for ecosystem services faces high transaction costs, including identifying the supply and demand sides of the market, eliciting values and opportunity costs of ecosystem services, and bargaining between producers and users. Even when these markets are created, their long-term operation is still a challenge, especially for urban ecosystem services, where the production of these services has commonly focused on parks and green areas. Our research focuses on the design and sustainability of markets for ecosystem services in suburban areas of prairie provinces in Canada.
We conducted a choice experiment to identify the challenges for producers of suburban ecosystem services (n=240 suburban homeowners; discrete choice regression analysis) and experimental design to elicit preferences and willingness to pay for attributes of native/pollinator gardens in suburban areas (n=180 university students; ANOVA test). With the aid of photo-editing software, we used alternative urban scenarios of suburban front yards with native species.
In this presentation, we discuss our research findings concerning the challenges and opportunities to create a market for suburban ecosystem services in Alberta and Saskatchewan. We will present homeowners' challenges in becoming ecosystem service providers, including monetary and nonmonetary factors and the willingness to pay for the benefits associated with ecosystem services of native/pollinator gardens. Climate-resilient cities are needed to preserve residents' well-being. This research contributes to designing a market for ecosystem services in suburban areas of Prairie cities. It will inform policy and planning in urban areas by determining the feasibility of a market for suburban ecosystem services and identifying the incentives producers and users require to engage in such a market.

2 REFERENCES

Young, R. F. & Lieberknecht, K. (2019). From smart cities to wise cities: ecological wisdom as a basis for sustainable urban development. J. Environ. Plan. Manag., vol. 62, no. 10, pp. 1675–1692, doi: 10.1080/09640568.2018.1484343.

Middle, I., Dzidic, P., Buckley, A., Bennett, D., Tye, M., & Jones, R. (2014). Integrating community gardens into public parks: An innovative approach for providing ecosystem services in urban areas. Urban For. Urban Green., vol. 13, no. 4, pp. 638–645, doi: 10.1016/j.ufug.2014.09.001.

Vargas-Hernández J. G. & Zdunek-Wielgołaska, J. (2021). Urban green infrastructure as a tool for controlling the resilience of urban sprawl. Environ. Dev. Sustain., vol. 23, no. 2, pp. 1335–1354, doi: 10.1007/s10668-020-00623-2.

DESIGNING BORDERS: RIVER-PEOPLE RECIPROCITIES IN THE SALINAS VALLEY

Track 1: Nature and urban ecology
Knowledge Network: Movement in Designed Environments: Nature & Ecology
Methods: Case Study, Mixed Methods

MaFe Gonzalez (BASE Landscape Architecture); Thomas Chastain (University of California, Berkeley)

1 ABSTRACT

Running 175 miles, the Salinas is supporting the whole valley, but is hard to see and acknowledge by human eyes. Its existence has been suppressed by design, hidden behind fences, absorbed by sprinkles, and disconnected from its plains. However, the Salinas River is the irrigation system for the 'salad bowl of the world', and is used to water about 275,000 acres of farmland in the Salinas Valley. Flowing underground for much of the year, the siphoning of its water occurs beneath us, away from our immediate view. Above ground, the river reappears for a few months but is greatly reduced in its extent and quality.

On occasion, the Salinas River still has the power to announce itself, its presence becoming unavoidable and typically unwelcomed. Winter storms can bring floods, sometimes mild, sometimes devastating, as they were in 2023, collapsing levees and taking over hundreds of acres of farmlands and urban areas, renewing attention to the Salinas with calls for further management and control of its water. This is our responsibility. We long ago detached it from the shaping that nature would give it, attempting to control it. The river is now our design problem, and our future actions cannot be more of the same. Designing the Salinas has been a human-centric activity, its outcomes measured in monetary terms that benefit us, not nature. Seeing the river's intrinsic value beyond exploitation requires an adjustment in our thinking. The goal of this work was to design a Floodway Park along the Salinas River, transforming agricultural land into recreational land, giving more room to the river (allowing it to flood, erode, sediment, create habitat), and people to reconnect with nature. We first described the Genius loci of the Salinas River representing different sources of information: Steinbeck's novel "East of Eden", scientific data (Beller et al. 2009, MCWRA 2019), and interviews. Then we defined its erodible corridor (Piégay et al. 2005), described its habitats, and identified access points and potential uses (Hester 2006). With all that information, we proposed management guidelines and a social system for the basin, a schematic design for three sections of the river in Monterey County, and a detailed local design for the floodway park in the city of Salinas.

We blended sciences and design as a holistic approach to deal with environmental and social issues. We think it is possible to design ways of living in which the river and people flourish together, however, the current economic model has been the biggest challenge to this project.

2 REFERENCES

Beller, E., R., Grossinger, and A. Whipple. 2009. Historical Ecology Reconnaissance for the lower Salinas River. San Francisco Estuary Institute. Historical Ecology Program Contribution No. 581.

Hester, R. 2006. Design for Ecological Democracy. MIT Press, Cambridge, Massachusetts.

Monterey County Water Resources Agency (MCWRA). 2019. Salinas River Stream Maintenance Program, Website. https://www.co.monterey.ca.us/government/ government-links/water-resources-agency/programs/salinas-river-stream-maintenance- program-smp

Piégay, H., S. E. Darby, E. Mosselman, and N. Surian. 2005. The erodible corridor concept: Applicability and limitations for river management. River Res. Appl. 21: 773–789.

ARE RAIL-TRAILS ECOLOGICAL CORRIDORS? A CASE STUDY FROM THE CORN BELT

Track 1: Nature and urban ecology
Knowledge Network: Movement in Designed Environments: Nature & Ecology
Methods: Case Study, Mixed Methods
Austin Dunn (Iowa State University); Kat Shiffler (National Park Service)

1 ABSTRACT

Within the Corn Belt region of the USA, large tracts of public land for conservation are sparse. In this context, linear landscape corridors such as rivers and trails are highly consequential in terms of their potential benefit to both human recreation and environmental quality. In recent decades, over 25,000 miles of abandoned railroad corridors have been converted for use as recreational trails in the United States. Rail-trails have been found to have a wide range of benefits to users as well as surrounding communities: encouraging fitness and healthy lifestyles, preserving and creating open spaces, expanding nature access, and strengthening local economies (Moore, 1992). But to what extent do rail-trails also serve as ecological corridors? What metrics and methods allow us to better quantify the environmental benefits of rail-trails? How does the ecological value of a rail trail corridor relate to its context and the composition of the surrounding landscape?

To answer these questions, we consult literature from the fields of landscape ecology, landscape architecture, and road ecology. For Smith and Hellmund (1993), the three most critical variables of a corridor in the landscape are width, connectivity, and quality. These structural elements of a corridor determine how a corridor functions as part of the larger landscape. Because of their previous use as railroads, rail-trail corridors share common structural qualities that bear on their ecological value. However, because railroad alignments tend to transect diverse environments, the relationship between the corridor and its context is varied and dynamic.
We examine the High Trestle Trail in Boone County, Iowa as a case study. Utilizing publicly available GIS data, drone photogrammetry, and spatial analysis, we highlight key structural characteristics of the corridor. We evaluate metrics for analyzing and describing corridor quality, connectivity, and width. The High Trestle Trail transects varied contexts, from low-density suburban development to agricultural, and from upland woodland habitats to riparian lowlands. The relationship between the High Trestle Trail and local drainage patterns is generally transverse—the rail-trail is oriented East-West while local rivers and streams are primarily North-South. Accordingly, the ecological value of the rail-trail corridor may shift as it relates to a changing landscape matrix.

We conclude with recommendations for the design and management of rail-trail corridors, with a focus on rail-trails situated in the predominantly agricultural landscapes of the US Corn Belt region. Drawing from guidance on the larger study of greenway planning and design (Hellmund & Smith, 2013), we find that while some greenway considerations apply, rail-trail corridors are unique landscape types that require a specific set of design and maintenance strategies to maximize their ecological value.

2 REFERENCES

Hellmund, P. C., & Smith, D. (2013). Designing greenways: Sustainable Landscapes for Nature and People, Second Edition. Island Press.

Moore, Roger L. (1992). The impacts of rail-trails: A study of the users and property owners from three trails. Rivers, Trails, and Conservation Assistance Program, National Park Service.

Smith, D. S., & Hellmund, P. C. (1993). Ecology of greenways: Design and Function of Linear Conservation Areas. University of Minnesota Press

ECOLOGY BY WAY OF CULTURE: SEEKING INDIGENOUS KNOWLEDGE IN THE WETLANDS OF VILLARRICA, CHILE

Track 1: Nature and urban ecology
Knowledge Network: Cultural Aspects of Design, Environmental Design Research Education
Methods: Design Methods, Mixed Methods
David de la Pena (University of California, Davis)

1 ABSTRACT
Urban wetlands are indeterminate landscapes that belong both to land and water, that contain cultural meanings and ecological possibilities. Although they provide abundant ecosystem services, these landscapes have been neglected, abused, and often entirely destroyed (Rojas, et al., 2019). This is the case in Villarrica, Chile, part of Wallmapu, the historical territory of the Mapuche people. Facing increasing development as well as a new law protecting urban wetlands, Villarrica enters a critical moment that will define its relationship with its ecology and its history of colonialism. In this paper, I ask whether a culturally inclusive design approach can enhance the ecological prospects for the redevelopment of Villarrica's wetland. I use data collected through a collaboration from 2020-2023, including document analysis, site visits, interviews, secondary texts, and student work.

In academia, including in the design curriculum, ecology remains largely divorced from culture. Historically, a landscape ecology approach has sought to buffer sensitive ecological zones from human disturbance (McHarg, 1969), but indigenous scholars challenge these notions, revalorizing the restorative stewardship practices of first peoples (Ammerman, 2022). Between California and Chile, a longstanding program has sought to apply the scientific expertise and ecological projections of Lake Tahoe to Chile's Lakes District. While some best practices in lacustrine wetland management are readily transferable, cultural context is not. In fact, many of Tahoe's endearing challenges - which Chile should hope not to import - stem from its neglect of cultural contexts while promoting ecology and tourism. Villarrica faces a similar trajectory, in which speculative development is positioned against ecological restoration, but in Villarrica, the lines between what is native and what is transplanted are blurred by legacies of colonialism.

The inclusive design approach demands closer attention to cultural contexts. Simplistically, we might have attempted to superimpose Mapuche stewardship upon the wetlands, but our research suggested that a more complex and nuanced touch was needed. Through this search, we discovered histories that revealed hidden scars, transplanted horticultures, buried monuments and other legacies upon the landscape. We found that Mapuche stewardship of the wetlands was being over-romanticized, and that other localized knowledge was equally important. The most promising design approaches were those that were more tentative, respectful, and revealing of a complex web of social and ecological relationships. Through this project, the wetland itself confirmed its indeterminacy, and its need to reveal its history—from the mounds of original settlements to the scars of political unrest to the now beloved groves of invasive willows—as part of a process of grief, healing and restoration.

2 REFERENCES
Ammerman, Cinthya. (2022). Lessons from Fire: The Displaced Radiata Pine on Mapuche Homelands and the California Roots of Chile's Climate Crisis. Annals of the American Association of Geographers, 112:3, 692-705.

McHarg, I. L., & American Museum of Natural History. (1969). Design with nature. New York: American Museum of Natural History.

Rojas, C., Munizaga, J., Rojas, O., Martínez, C., & Pino, J. (2019). Urban development versus wetland loss in a coastal Latin American city: Lessons for sustainable land use planning. Land use policy, 80, 47-56.

WHAT CAN INTERIORS TELL US ABOUT THE RELATION BETWEEN ENVIRONMENTAL SUSTAINABILITY AND HUMAN-CENTRIC SPACES?

Track 1: Nature and urban ecology
Knowledge Network: Interior Design, Sustainable Planning, Design & Behavior
Methods: Design Methods, Interpretive-Historical Research
William Mangold (Drexel University)

1 ABSTRACT

The relationship between nature and the built environment is a complex and ever-evolving one. In recent years, there has been a growing interest in bringing nature into interior spaces, as evidenced by the popularity of green walls, biophilic design, and other nature-inspired interventions. This trend is driven by a number of factors, including an increasing awareness of the benefits of nature for human health and well-being, a growing demand for sustainable design, and a desire to create more comfortable and inviting living and working environments.

This presentation will consider the connection between built and natural environments, with a specific focus on interior spaces, and explores the complex interplay between human-centric design and environmental sustainability. As a review of recent design and scholarship, this presentation suggests that interactions between the built environment and nature be understood on a dynamic, permeable continuum. Interior spaces, traditionally viewed as distinctly human domains, are increasingly being reimagined as seamless extensions of the natural world. This shift not only redefines architectural aesthetics but also reflects an evolving understanding of human wellbeing. This presentation will look at various types of spaces and ways of bringing nature inside--from personal collections of rocks and leaves to corporate and institutional statement features--and consider both the physical design and the social and psychological conditions of these natural elements.

Historically, the relationship between nature and interiors has been multifaceted, often reflecting prevailing cultural and societal attitudes. (Sparke, 2020) This presentation will briefly trace this trajectory from historical fear and reverence of nature, to industrial efforts to dominate and exploit nature, to the contemporary embrace of biophilic design. These changing attitudes are evident in architectural design and interior features, and suggest how our relation to nature has always been centered on human perception and experience.

This shifts the discussion to the critical balance between human-centric design and environmental sustainability. It raises the question: How can interiors be designed to prioritize human comfort and wellbeing while respecting and sustaining the natural environment? This query leads to the concept of regenerative design--an emerging trend that advocates for designs that actively benefit the environment, going beyond sustainability. (Lyle, 1996) Regenerative design represents a paradigm shift, positioning nature not just as a component to be included in human-centric spaces but as a guiding principle that surpasses human convenience.

The presentation will culminate in a forward-looking perspective, contemplating the future of interior design in an era increasingly defined by environmental consciousness. By examining case studies and emerging trends, we will consider potential pathways for creating spaces that are both nurturing for humans and beneficial to the environment. (Watson, 2019) In conclusion, this presentation aims to illuminate the transformative potential of integrating nature within our built environments, offering a compelling vision where the design of interior spaces not only reflects but also enriches our relationship with the natural world, leading us towards a more sustainable and human-centric future.

2 REFERENCES

Lyle, J. (1996). Regenerative Design for Sustainable Development.

Sparke, P. (2021). Nature Inside

Watson, J. (2019). LoTEK: Design by Radical Indigenism.

MENOMONEE VALLEY: EXAMINING ECO-REVELATORY DESIGN TO CREATE ECOLOGICAL LITERACY

Track 1: Nature and urban ecology
Knowledge Network: Movement in Designed Environments: Nature & Ecology, POE/Programming
Methods: Case Study, Survey
Jennifer Engelke (Virginia Tech)

1 ABSTRACT

Menomonee Valley Community Park case study examined the systematic relationships at a park located in Milwaukee, Wisconsin. This 45-acre area was a brownfield redevelopment designed by Wenk Associates with the intention of promoting eco-revelatory design, a design process that recognizes human-centric environments yet promotes more-than-human elements on the land. The design incorporates the land history of a landscape that was once a wild rice marsh along the Menomonee River. In the late 1800s and early 1900s, the area was filled and turned into the Milwaukee Road Shops and Airline Yards rail site. The site was transformed from a rail yard to a community park through design strategies incorporating ecological restoration, stormwater treatment of nearby neighborhood and highway runoff, and a reconnection of people to the Menomonee River. The site also includes narratives from the local Miller Brewing Company, Milwaukee Stockyards, and sewer pipes. These narratives in the designed site create an interesting juxtaposition in the system of native plant restoration and the postindustrial elements still evident on the site.

This study used a landscape biography methodology to analyze changes in the historical landscape system relationships. Surveys and observations of site users assessed the landscape and ecological literacy people gained from storytelling through design (i.e., site and system history demonstrated through the design [Heatherington, 2011]) and eco-revelatory aspects (i.e., more-than-human elements revealed and made visible to site users through the design [Helphand & Melnick, 1998]). Place attachment (i.e., human connection with the land [Manzo & Devine-Wright, 2019]) was assessed through interviews with the non-profit, community scientists, and project team members.
The study demonstrated that a user's personal experiences of the place and signage elucidated their knowledge of the history and landscape interventions to a greater degree than the design elements did. Ecological literacy was evident from the narratives of general site users, community non-profit participants, and a larger neighborhood community. An adjacent non-profit, the Urban Ecology Center, creates ways for people to learn about and acknowledge the more-than-human elements in the landscape. People are recognizing, appreciating, and seeking out the more-than-human elements. The place attachment formed by users in this landscape contributes to building strong, caring relationships between human and more-than-human elements. Menomonee Valley, once viewed as an eyesore, is emerging as a holistic system in which people look to care for one another and the more-than-human elements in the landscape.

2 REFERENCES

Heatherington, C. (2011). Buried Narratives. In A. Jorgensen & R. Keenan (Eds.), Urban Wildscapes (pp. 171–186). New York: Routledge.

Helphand, K. I., & Melnick, R. Z. (1998). Editor's Introduction. Landscape Journal, 17(Eco-revelatory design: nature constructed/nature revealed).

Manzo, L., & DevineWright, P. (2019). Place Attachment. In L. Steg, A. Van den Berg, & J. de Groot (Eds.), Environmental Psychology: An Introduction (2nd Edition). Chichester, West Sussex: BPS Blackwell.

SEEDING SUSTAINABILITY: TRANSFORMING URBAN ECOLOGY THROUGH COMMUNITY-BASED RESEARCH ON NATIVE PLANT INTEGRATION - THE LATEST REPORT FROM THE SEED PILE PROJECT

Track 1: Nature and urban ecology
Knowledge Network: Movement in Designed Environments: Nature & Ecology, Sustainable Planning, Design & Behavior
Methods: Practice-based Research, Statistical/Modeling Analysis
Sahar Sadeghmalakabadi (University of California, Davis); A. Haven Kiers (University of California, Davis)

1 ABSTRACT

Urban areas are valuable for ecological enrichment and human activity. Human survival depends on our ability to adjust to the surrounding environment, which necessitates acknowledging the interconnectedness of our relationships with wildlife, air, water, and earth. This underscores the need to consciously shape our surroundings to promote the well-being of both humans and the planet (Spirn, 2013). Embracing native plants in the urban landscape provides cities and citizens with resilient green environments, leading to improved human well-being and the creation of urban wildlife habitats (World Health Organization,2016), (Adams, 2020).

The Seed Pile Project is a community science initiative by Miridae Living Labs and UC Davis to better understand the dynamics of native plant seed dispersal in human-dominated landscapes like cities and roadsides. It invites community members to drop small piles of native seeds (such as yarrow, California poppy, and lupine) in urban areas along transportation corridors (alleys, sidewalks, bike paths, medians, etc.) and then monitor the results through monthly observations. By researching the conditions under which certain species of locally native seeds spread, survive, die, or thrive in urban areas, we can gain a better understanding of which native plants to incorporate into the built environment, what benefits they provide, and where to put them for the greatest survivability and resilience. From an ecological perspective, incorporating native flora into urban design can increase biodiversity, support pollinators, and enhance public health by bringing wildlife into the cityscape. Economically, the adoption of native plants in urban areas can reduce costs associated with landscaping maintenance. Most importantly, as a community science initiative, the Seed Pile Project can educate, engage, and empower populations and communities typically left out of the traditional urban design process by providing them with knowledge and resources to help promote healthy ecological development within their own neighborhoods.

This research was initiated in 2020 in response to the question of which native plants can tolerate the urban environment with little to no maintenance. Pilot observations have shown that optimal plant growth occurs within cracks in asphalt and concrete, along the sides of roads, and adjacent to debris.

Manicured locations adjacent to ornamental gardens offer less hospitable conditions for these native species. This paper will discuss the important role of native wildflower seeds in shaping sustainable urban environments, using the project's methodology, most recent data analysis to give a better understanding of the relationship between the germination of native plants in corridors and the enhancement of urban ecology. The implications of these findings can benefit landscape designers, policymakers, and government agencies by helping them make more informed decisions about incorporating native wildflower seeds into urban design projects.

2 KEYWORDS

Urban ecology, green corridors, community engagement, citizen science, native wildflower seeds

RETROFITTING AN URBAN OPEN SPACE NETWORK

Track 1: Nature and urban ecology
Knowledge Network: Movement in Designed Environments: Nature & Ecology, Sustainable Planning, Design & Behavior
Methods: Design Methods, Practice-based Research
Dwaine Carver (University of Idaho- College of Art & Architecture)

1 ABSTRACT

This paper describes contracted professional work for the conceptual design of a retrofitted open space network for Boise, Idaho. A midsize city in the northwest United States, Boise's population is 275,000 within a metropolitan area of 807,000. Projected to grow to 1.4 million by 2060, the present suburban development pattern is typical: six-lane arterials flanked by car-centric retail pad-sites, surface parking, and 'big box' stores. These large-scale roadways are single use, providing only minimal, hostile public space consisting of attached sidewalks, sparse landscape, and primary vehicular access to the adjacent, low-density, single-family neighborhoods.

The city requested design strategies toward enhanced density, ecological performance, connectivity, and landscape conservation. The design team proposes a retrofitted urban open space network as the principal armature able to accomplish these goals. Composed of architects, designers, ecological scientists, and geographers, the design team identifies several existing elements as potential contributors to the city's open space network. Using the USGS/ESRI Green Infrastructure database in concert with local data, components such as disused rail lines, an extensive agricultural water system, surface and storm water infrastructures, existing open space sites, conservation lands, and public paths, are evaluated as potential constituents of a public network able to integrate density, recreation, ecology, and connectivity.
Two particular urban elements, ecological districts and boulevards, are developed for their potential to integrate density, connectivity, and ecological performance. District sites are selected for strategic overlaps with existing infrastructures and their potential to contribute to the open space network. Typical sites are 600' x 1200', consisting of 20 to 80,000 square-foot retail stores combined with 4 to 10,000 square-foot pad sites. Selected district designs take full advantage of recent rezoning, allowing for the addition of housing and reduced parking requirements. Proposed densities range from 35 to 90 units per acre. Public spaces, including small-scale streets, subdivide the site, providing connections to adjacent single-family neighborhoods as well as new public pathways.

The proposed boulevards are retrofits of selected segments of existing 100-foot vehicular right-of-ways flanked by single-use retail zoning. Performance driven, the boulevards integrate carbon sequestration, tree-canopied public walkways, storm- and gray-water captures, and anticipate future modes of transportation.

In its presentation of a retrofitted open space system, the paper argues that a retrofitted open space network, opportunistically developed, can cultivate a potent set of collective assets. Conservation lands can act as discrete network nodes as well as growth boundaries. Agricultural waterways can provide linear, connective public pathways as well as post-agricultural opportunities. Storm water control and retention sites can contribute to ecological remediation as well as public recreation and habitat open spaces. Strategic segments of boulevards and selected districts can engage a multitude of systems at various scales, integrating ecological performance, current and future transportation modes, increased density, community access and health, and new public space typologies.

HU-BE-SIM: SIMULATING OUTDOOR THERMAL COMFORT-BASED HUMAN BEHAVIOR IN PUBLIC SPACES USING AGENT-BASED SIMULATION

Track 1: Nature and urban ecology
Knowledge Network: Movement in Designed Environments: Nature & Ecology, Sustainable Planning, Design & Behavior
Methods: Mixed Methods, Observation
Zahida Khan (Ball State University College of Architecture and Planning)

1 ABSTRACT

Human interaction with the built environment is exhibited through human spatial behavior in outdoor public spaces. It is a significant part of urban ecological settings that serves as a key performance indicator of successful spaces and can be rewarding for urban planning policies. Studies on human behavior in urban environments have contributed substantially towards human-centered sustainable development. Predicting human behavior could prove beneficial to architects and planners in their design process. They seek human behavior simulation tools to help them make informed design decisions through 'what-if' design scenarios. The biggest challenge in developing these tools is modeling the complex nature of human behavior for prediction and simulation. The complexity involves a person's decision-making process to act or adapt while assessing their immediate environment. Such behaviors can be simulated using Artificial Intelligence (AI) such as Agent-Based Models (ABM).

The immediate environments around a person create ecological environmental units, also referred to as behavioral settings. One of the primary aspects of these settings is microclimate, which impacts people's Outdoor Thermal Comfort (OTC) levels and thus their presence in spaces, especially in urban conditions. Modeling microclimate and OTC are very helpful in designing public spaces. They are physics-based models and involve a Computational Fluid Dynamics (CFD) approach for simulations.

This research presents HuBeSIM, a human behavior simulation framework for assessing outdoor public space design. The framework is developed in Rhino-Grasshopper, a popular 3D modeling tool amongst designers and planners. It combines mathematical and physics-based models to simulate OTC-integrated human behavior in urban spaces. Rhino-GH offers numerous libraries that support the coupling of the two models. These include Quelea and Ladybug/Honeybee. Daley Plaza - an urban public plaza in Chicago, was used as a pilot project to test the framework.

This study contributes to the evolving urban ecological settings in cities as a result of the pressing global issues, namely climate change and world population increase. It helps with investigating the relationship between urban space, its microclimate, and human spatial behaviors. The significance of the findings directly impacts the UN's Sustainable Development Goals (SDG- 11 & 13) for climate action and sustainable communities.

RE-GENERATION OF TOWN FACTORIES: TYPOLOGICAL TRANSFORMATION OF COEXISTENCE OF RESIDENCE AND MANUFACTURE IN LOCAL INDUSTRY IN SUMIDA WARD, TOKYO

Track 2: Sustainable Buildings and Infrastructure
Knowledge Network: Environmental & Architectural Phenomenology, International Housing Research
Methods: Practice-based Research, Survey
Akio Yasumori (Chiba University); Yasuaki Morinaka (Kosaka Morinaka Architecture); Tomoya Watanabe (Graduate student, Chiba University); Ryousuke Sugiura (Former Graduate student, Chiba University); Kanako Murayama (Former Graduate student, Chiba University)

1 ABSTRACT

This paper presents a field survey of town factories in the eastern part of Tokyo, revealing a typology of architectural characteristics based on the coexistence of work and residence, including its transformation, and discusses the future vision of these factories with residences for the sustainable neighborhoods.

2 INTRODUCTION

A downtown area in the eastern part of Tokyo, Sumida ward, has long been a manufacturing town where local industries have developed. The density of factories is the highest in Tokyo, and even today there is a concentration of small town factories in diverse industries such as metal, printing, textiles, and leather. Residential and industrial uses coexist in buildings and neighborhoods. There, the number of factories has been decreasing in recent years due to relocation of factories, generational changes in management, and rebuilding to condominiums. At the same time, the creation of new industries and the increase in new residents due to urban residential development are also observed, changing the nature of manufacturing towns from both industrial and residential perspectives. In these town factories, we can see various generations of factories, including those in operation, vacancy, high-rise buildings, and rebuilding. In this study, we consider these changes in local industry as "generations" of town factories, research their typology, and propose a "re-generation" of the architecture and area. As the relationship between factories (work) and residences (housing) changes at both the architectural and urban scales, the project examines the mix of the two and the "opening" of factories to the city. The main subjects of this project are the metal industry in the northern part of the city and the textile and printing industry in the southern part of the city.

3 INPUT AND OUTPUT

With the cooperation of the Sumida Ward Industrial Promotion Division, the Urban Planning Division, local architectural design firms, and town factories, a field survey was conducted in collaboration with the community from 2021 to 2023. Based on the issues interviewed and the direction of each factory, future proposals were made based on an understanding of production and livelihood.

4 TRANSFORMATION OF TYPOLOGY AND GENERATION OF TOWN FACTORIES

We extracted a typology of town factories by classifying the relationship between factories and residences into several types: stacked, flat, and compound. The generations of town factories were divided into the first generation, which consisted of low-rise wooden factories built after the Great Kanto Earthquake and World War II (up to the mid-20th century), the second generation, in which the number of town factories peaked and their scale expanded after the period of rapid economic growth (around 1970s), and the third generation, in which they became high-rise buildings as land prices rose in the city center and they were reconstructed (1990s onward). The fourth generation includes not only coexistence of factories and residences, but also shared offices, stores, welfare facilities, and other uses open to the community (since 2010). Finally, we discuss the potential for architectural and urban sustainability through the mix of industry, residence, and new functions open to the town.

ENHANCING SUSTAINABILITY AND RESILIENCE IN REFUGEE CAMP PLANNING THROUGH CIRCULAR CITY PRINCIPLES: THE CASE STUDY OF JABALIA REFUGEE CAMP

Track 2: Sustainable Buildings and Infrastructure
Knowledge Network: Environmental Design Research Education, Sustainable Planning, Design & Behavior
Methods: Case Study, Observation
Hala Zahalqa (University of Idaho)

1 ABSTRACT

UNHCR reports a continuous increase in the numbers of refugees. It is estimated that forcibly displaced people worldwide have reached 103 million in mid-2022. Moreover, the refugee population reached 32.5 million as of mid-2022 [1]. Palestinian refugees are more than 5.9 million individuals. Their camps have lasted for more than 75 years now with the fifth generation to be born into exile, representing the largest stateless population worldwide, facing protracted situations [2]. Although camps are intended to be temporary habitats for refugees, the history of current refugee camps teaches us otherwise. This temporary yet permanent reality of refugee camps creates inhuman conditions, resulting in a whole population living through persistent challenges. This research focuses on addressing these challenges, which include unsustainable camp design, excessive reliance on aid, and limited economic opportunities. How can we reconcile the intended temporariness of refugee camps and their permanent realities? How can we learn from the current refugee camps' realities for a better resilient future? And what could be the solution?

Actions inspired by circular economy, such as designing circular cities, led this research to investigate the opportunities of refugee camps to become circular cities. A circular city that can redesign, repair, reuse, recycle, recover, and share. It is not just about using waste as a resource, but it's about creating long-term sustainability, resilience, enhanced community engagement, and economic enrichment. This exploration, within the context of protracted refugee camps, will offer a new lens, an opportunity to redefine places of refuge beyond their designated lifespan and reimagine them as sustainable and thriving communities. This will be analyzed in the context of Jabalia Refugee Camp in Occupied Palestine. Jabalia Refugee Camp was chosen for being the largest refugee camp in Palestine. Examining new development approaches, opportunities and challenges in these camps is mainly done according to literature review analysis, previous research, and the author's observations of the camp.

4 REFERENCES

[1] UNHCR. (2023). Mid-Year Trends. https://www.unhcr.org/mid-year-trends

[2] Citino, N., Martin Gil, A., & P. Norman, K. (2023, May 3). Generations of Palestinian Refugees Face Protracted Displacement. Migration Policy Institute. https://www.migrationpolicy.org/article/palestinian-refugees-dispossession

TARGETING URBAN HEAT ISLAND MITIGATION BASED ON NEIGHBORHOOD AND SOCIO-ECONOMIC DATA

Track 2: Sustainable Buildings and Infrastructure
Knowledge Network: Environmental Design Research Education, Sustainable Planning, Design & Behavior
Methods: Interviews, Statistical/Modeling Analysis

Mahdi Afkhamiaghda (Corgan); Kevin Sloan (Corgan Associates); Samantha Sanders Flores (Corgan Associates, Inc.); Weijia Wang (Corgan)

1 ABSTRACT

In the summer of 2023, numerous US counties, notably in southern states such as Texas, experienced record-high temperatures, surpassing 100°F for multiple consecutive days. Continuation of this trend could lead to residents enduring over 70 consecutive days with heat indices exceeding 100°F. Urban heat islands (UHI) arise when urban areas undergo significantly higher temperatures due to factors like urbanization and diminished vegetation, exacerbating extreme temperatures. These areas often exhibit temperatures 15-20°F higher than their greener counterparts.(Del Serrone et al., 2022).

Managing heat in rapidly urbanizing cities with concentrated human activities presents significant challenges. UHI has wide-ranging implications, impacting air quality, public health, etc. in urban areas.

Many studies suggest the correlation between socioeconomic status and UHI, however, they have mostly focused on just the land surface temperature (LST) (Gabbe et al., 2023). This study aimed to understand the relationship between UHI magnitude and multiple socioeconomic factors in various neighborhoods through a mixed methods approach, layering qualitative and quantitative data and insights. The methodology involved using census demographic data to pinpoint areas of study, conducting surveys to establish localized social KPIs through urban heat as well as leveraging Landsat and GIS data to understand the impact of building typologies and vegetation index on UHI. This approach aims to guide resource allocation for effectively addressing the challenges posed by UHI in lower-income communities.

First, census tract data was used to identify low and high-income neighborhoods in the Dallas, TX metropolitan area. Surveys were then conducted among residents to explore patterns and identify potential links between local heat conditions and residents' lifestyles. Five different neighborhoods were chosen based on population density and proximity to downtown, ensuring a comprehensive representation. For comparison, three low-income and two high-income areas were selected. Each neighborhood had 15 participants randomly selected for an in-depth user survey, focusing on lifestyle changes in summer months and neighborhood design features during this period.

The researchers then utilized Google Earth Engine to gather USGS Landsat 8 satellite images of Dallas, TX, from June to September between 2015 and 2022. The researchers derived LST and vegetation index, conducting temporal analysis on the derived values. The results were mapped to GIS parcel data to correlate the quantity and the contribution of different building typologies on the urban heat island. Triangulating this data with the survey findings leads to the development of more data-driven, and accurate, architectural, and green infrastructure design strategies.

This study examines how a neighborhood's socioeconomic status and building types affect urban heat islands. It aims to guide resource allocation for effective strategies to mitigate and adapt to these heat islands. The findings will aid designers and policymakers in reducing the negative impacts of heat islands in affected neighborhoods.

4 REFERENCES

Del Serrone, G., Peluso, P., & Moretti, L. (2022). Evaluation of Microclimate Benefits Due to Cool Pavements and Green Infrastructures on Urban Heat Islands. Atmosphere, 13(10), 1586.

Gabbe, C. J., Mallen, E., & Varni, A. (2023). Housing and urban heat: assessing risk disparities. Housing Policy Debate, 33(5), 1078-1099.

KITCHEN TABLE ISSUES OPEN THE DOOR TO DECARBONIZATION

Track 2: Sustainable Buildings and Infrastructure
Knowledge Network: Health, Sustainable Planning, Design & Behavior
Methods: Action/Participatory Research, Mixed Methods
Adele Houghton (Harvard T.H. Chan School of Public Health)

1 ABSTRACT

Sustainability initiatives like decarbonization are often described from the point of view of the building or infrastructure rather than the value they will bring to society. For example, technologies like solar arrays and heat pumps can help a building achieve decarbonization – but to what end? Unless the building occupants and maintenance team feel invested in the idea of running a net zero building, they are unlikely to support the behavior change that would be necessary to operate the building at the level assumed by the design standards. This presentation will share the results from a mixed-methods action research study that successfully supported development teams, community groups, and local officials in aligning around a vision for three real estate projects (two in New York State and one in Virginia).

The novel approach taken by researchers combined health situation methodology – which uses census tract data for social determinants of health, climate change and health, and community health – with semi-structured focus group sessions held with each stakeholder group separately and also with all of the stakeholder groups together. Results found that starting the conversation by highlighting the link between community members' high priority kitchen table issues and specific climate change hazards can increase community support for sustainability measures such as decarbonization.

EXPLORING EFFECTIVE USES OF CHATGPT IN PEDAGOGY

Track 2: Sustainable Buildings and Infrastructure
Knowledge Network: Interior Design, None/Not Tied to Any Network
Methods: Action/Participatory Research, Practice-based Research
Marwa Abdelmonem (Virginia Commonwealth School of the Arts-Qatar)

1 ABSTRACT

The adoption of ChatGPT has sparked controversy, but its widespread use among students and faculty is undeniable. Currently, the educational community exhibits polarization, where some endorse AI for its accessibility and efficiency, considering it an essential tool, while others express concerns about potential risks to academic integrity and intellectual development (Sağın, Özkaya, Tengiz, Geyik and Geyik, 2023). Nonetheless, it is crucial to take a broader perspective and recognize that ChatGPT constitutes only a fraction of the swiftly advancing AI landscape to which we all actively contribute daily (Calhoun, 2023).

ChatGPT and similar language models offer substantial utility in higher education, serving in writing assistance, language learning, research, and administration. In writing assistance, it aids in generating text, summaries, and outlines, enhancing efficiency and quality. It also performs grammar and style checks for improved clarity. In language learning, ChatGPT produces translations, language-specific summaries, and exercises, aiding skill development. For research, it supports data analysis, summarization, and document creation, enabling quick pattern identification. The tool's multifaceted applications make it a valuable asset in diverse educational contexts, streamlining tasks for both students and educators (Atlas, 2023). Liang, Zou, Xie, and Wang (2023) indicated in their recent study that ChatGPT offers pedagogical benefits in physics education, providing immediate feedback and accessible explanations, supporting scaffolded learning, and engaging learners interactively.

In interdisciplinary domains like interior design, the use of ChatGPT becomes particularly pronounced due to its multifaceted advantages. The model's versatility and capacity for generating diverse content make it a valuable resource in addressing the intricate and varied aspects of the field. Generally, generative AI tools like ChatGPT, Notion, and Quillbot offer students a drafting foundation and support in brainstorming, organizing, and enhancing the structure and flow of their written assignments or projects (Sağın et al, 2023). This exploratory qualitative study aims to evaluate the effectiveness of incorporating ChatGPT into the teaching methodology of the "History of Interior Design," a core course in an interior design program. The research methodology employs content analysis, utilizing data from students' reflective journals. In these journals, students initiate prompts and analyze ChatGPT responses on specific topics, comparing and contrasting this information with content from scholarly sources as part of their reading assignments. Reflection stands out as a powerful learning method. It embodies a form of mental processing or thinking utilized to fulfill specific purposes or achieve expected outcomes. This approach is particularly valuable when dealing with intricate or unstructured concepts that lack clear solutions (Moon, 1999, p. 23).

Preliminary findings suggest that AI support in crafting reflections doesn't necessarily diminish critical thinking; instead, it encourages self-exploration and curiosity by reallocating time from editing to more effective research. This ongoing research aims to further illuminate this phenomenon in the realm of higher education.

HUMAN-CENTRIC NET ZERO SCHOOLS: EMPATHETIC PERFORMANCE OF A NEW EDUCATIONAL BUILDING TYPOLOGY

Track 2: Sustainable Buildings and Infrastructure
Knowledge Network: POE/Programming, Sustainable Planning, Design & Behavior
Methods: Case Study, Mixed Methods

Ihab Elzeyadi (University of Oregon)

1 ABSTRACT

Even though over 200 net-zero energy educational buildings are operating or are in construction in the US and more numbers worldwide, getting to zero in schools can seem an insurmountable goal to many architects and architecture students. The problem is both in the way this building type is conceived as well as the design process and workflow to achieve performance-based design at its highest level. This inquiry is focused on: (1) how and in what ways can we conceptualize a performative-based building typology such as Net-Zero schools (NZS), and (2) based on a comprehensive conceptualization, what are the impacts of NZS that could create a market transformation shift into their adoption. This paper reports on a comprehensive project that took a deeper look at a sample of exemplary net-zero schools built within the last decade. The project compiled a database of verified and emerging NZS in the US and analyzed several building performance metrics to highlight commonalities to successful NZS projects. Out of 41 verified NZS buildings, the study focused on seven case studies representing a wide geographical and economic range of NZS. This analysis included the design process, design strategies, energy performance goals, and indoor environmental quality that affects occupants' comfort, satisfaction, and wellness to develop both a conceptual framework and design process for this building type.

The study conducted a deep dive into the process that design teams followed with school district stakeholders, and the economic analysis to design and deliver these exemplary educational buildings to net zero. The project employed a comparative case study survey design to systematically collect building design and performance data for the studied schools. The paper concludes with the best design strategies and metrics to set as design targets of NZS on six major categories: Design Process, Design Strategies, Site Performance, Building Performance, Envelope Performance, and Indoor Environmental Quality/ Occupant Performance. The findings are summarized in crosscutting best practices, design patterns, and metrics that provide an anatomical guideline of this building typology to both architectural professionals and students.

The collaborative work of architects, engineers, and building scientists has the potential to bring net-zero school buildings into a more prominent place in the building market by measuring and drawing attention to the triple-bottom-line benefits attainable with healthier, comfortable, and high-performing buildings. This work is of a pressing national need. School districts and non-profit organizations are waiting to see evidence that net zero schools are good investments for children's education.

THE ROLE OF MINDFUL AWARENESS IN FOSTERING GREEN BUILDING KNOWLEDGE AND ENVIRONMENTALLY RESPONSIBLE BEHAVIORS IN A GREEN RESIDENCE HALL

Track 2: Sustainable Buildings and Infrastructure
Knowledge Network: Sustainable Planning, Design & Behavior
Methods: Experimental/Quasi-Experimental Research, Survey

Erin M Hamilton (University of Wisconsin-Madison, Design Studies); Laura Brianna Cole (Colorado State University)

1 ABSTRACT

Sustainable buildings offer tangible platforms for environmental education and promotion of environmentally responsible behaviors (ERBs). Previous literature has demonstrated a connection between occupant awareness of green building features and ERBs in the building; yet, effectively communicating sustainable features to the public remains a challenge. Given the myriad ways green buildings may cue ERB, engaging building occupants with these features may seem a matter of capturing their attention. However, solutions that briefly attract building occupant attention may yet fail in a level of engagement that promotes ERBs or knowledge gain. Recognizing that knowledge and behaviors are commonly determined by multiple factors, the current work examines an additional factor: occupant capacity to be actively aware of their surroundings. Contemporary research has demonstrated that mindfulness, the practice of being aware in the present moment, is positively associated with pro-environmental attitudes and behaviors (Wamsler et al., 2018). As awareness of green building features has been shown to support improved ERBs, we hypothesized a predictive relationship between mindful awareness and ERB with awareness of green building features and perceived behavioral control as serial mediating variables.

This study was conducted in a LEED Gold (green) residence hall at a large public Midwestern university where researchers developed and installed a suite of educational green building signage designed to promote awareness of sustainable building features and ERB opportunities (e.g., taking the stairs instead of the elevator, taking shorter showers, refilling reusable water bottles). A pre/post survey was administered to residents of the green residence hall (before/after signage implementation) along with a non-LEED hall next door (n=167). Drawing from several ecological behavior change models (Steg & Vlek, 2009), the survey used validated measures to assess students' green building knowledge (GBK) and ERBs, mindful awareness (Pirson, et al., 2018), perceived behavioral control, and other psychological and demographic variables (e.g., values, environmental concern, academic major, age, and gender). The hypothesized mediational model was analyzed according to the Hayes' Process Model 6 procedure using the PROCESS Macro for SPSS.

The results demonstrate a significant direct relationship between students' Mindful Awareness and ERBs ($p<.001$); and significant mediational relationships between 1) Mindful Awareness and GBK ($p<.05$) and GBK and ERBs ($p<.01$) and 2) Mindful Awareness and PBC ($p<.05$), and PBC and ERBs ($p<.01$). However, the serial mediation pathway between Mindful awareness, GBK, PBC, and ERBs was not found to be significant. Overall, these results suggest that Mindful Awareness supports greater awareness of sustainable features in one's surroundings. These results have implications for organizations seeking to operate sustainable buildings designed to support sustainability education. Policies, practices, and environments that support mindful awareness may also enhance user capacity to connect with their surroundings, have an experience, and notice something new.

COMPARATIVE STUDY ON EVALUATION INDICATORS, ENERGY CONSUMPTION LIMITS, TECHNICAL REQUIREMENTS AND POLICY OF CERTIFICATE SYSTEMS.

Track 2: Sustainable Buildings and Infrastructure
Knowledge Network: Sustainable Planning, Design & Behavior
Methods: Mixed Methods

Xinyi He (Yunnan University); Yifan Zhang (The University of Edinburgh); Pinru Zhu (University of California Berkeley); Hao Zhang* (the University of Bath)

1 ABSTRACT

Formulating certified standards can drive sustainable building practices, reduce energy consumption, promote technological development, foster global low-carbon cooperation, and guide government policies. Since differences persist in the definition and evaluation of certificated standards, this paper is a comparative study of certificate standards among various countries aiming to gain in-depth insights into the origins, characteristics, development trends, similarities, and differences of these standards. This paper includes four standards: BREEAM, LEED, CASBEE, and ESGB. The results indicate significant differences and characteristics from evaluation indicators, energy management, technical requirements, and policy. Considering evaluation indicators, BREEAM focuses on the overall environmental performance of buildings, including ecosystem services and biodiversity; LEED leans towards indoor environmental quality, air, natural lighting, and illumination; CASBEE emphasises life cycle analysis and the actual performance of buildings during the usage phase; and ESGB aligns with Chinese conditions, highlighting water resource management, ecological design, and the application of local sustainable technologies. Regarding energy consumption, BREEAM emphasises the overall energy performance of buildings, requiring low-carbon and highly efficient energy systems; LEED focuses on standardised energy performance indicators; CASBEE emphasises energy consumption during the usage phase; ESGB aligns with Chinese conditions and local standards for energy performance indicators. Certified standards should strengthen the following three aspects: regional differences in the spatial and temporal dimensions, the integrity of system indicators, and the embodiment of new technologies in the standards.

BUILT ENVIRONMENTS & SUSTAINABILITY: MASS TIMBER END-OF-LIFE SCENARIOS

Track 2: Sustainable Buildings and Infrastructure
Knowledge Network: Environmental Design Research Education, Researchers in Professional Practice
Methods: Design Methods, Practice-based Research
Allison Peitz (Lake|Flato Architects)

1 ABSTRACT

Mass timber construction is crucial for combating climate change and supporting the global push for circular economies. This method of construction has been called, "The quintessential carbon negative technology for climate change mitigation," noting that the material's longevity in use drives the magnitude of this impact (Lippke et al., 2021). As an inherently lightweight, durable, and modular system, mass timber construction has great potential for both whole-building adaptive reuse and deconstruction for component reuse. While life-cycle assessment tools demonstrate that mass timber can yield significant carbon benefits through its inherent ability to store carbon, the extent of benefits largely depends on how mass timber is treated at the end of its life and whether it continues its life cycle.

Given that the typical lifespan of a building is 50-100 years, how we design mass timber buildings today will impact their embodied carbon potential by facilitating or impeding our ability to reuse or deconstruct buildings and their components for the next 50-100+ years. Life Cycle Assessment tools indicate that the timing of release is hugely impactful and delaying this release (for the lifetime of a building, for example) has significant benefits in slowing the progression of global warming (Long-Term Biogenic Carbon Storage, n.d.). Current guidance for designing mass timber structures for adaptability and deconstruction is limited and lacks clear precedent. This research is a literature review documenting examples of historic and contemporary wood and mass timber reuse, considers methods for accounting for biogenic carbon as part of a Life Cycle Assessment (LCA), provides recommendations for supplementary tools that can be used to account for reuse scenarios, and culminates in the development of a series of recommendations for design of mass timber structures to optimize potential for adaptability, deconstruction, and reuse.

These recommendations fall into five categories:

1. *Structural Design*
2. *Finishes and Built-Up Assemblies*
3. *Systems Distribution*
4. *Moisture Management*
5. *Documentation*

These guidelines seek to translate research into practice by providing rules of thumb for architects, engineers, contractors, and clients to serve as a roadmap for design for reuse in mass timber structures. Even with these recommendations, the single most effective thing we can do as designers to promote the reuse of existing building stock is to design human-centric, beautiful, high quality, well-functioning buildings that people come to love. In return, people will continue to invest money and effort into maintaining and adapting these spaces over time. Mass timber naturally lends itself to creating warm, inviting, beautiful spaces where people feel comfortable. These recommendations close the loop by ensuring that these building elements can continue to be reconfigured and reused for lifetimes to come.

2 REFERENCES

Link to full whitepaper: https://issuu.com/lakeflato/docs/mt_eol_scenarios_journal_final

Lippke, B., Puettmann, M., Oneil, E., & Dearing Oliver, C. (2021). The Plant a Trillion Trees Campaign to Reduce Global Warming – Fleshing Out the Concept. Journal of Sustainable Forestry, 40(1), 1–31. https://doi.org/10.1080/10549811.2021.1894951

Long-Term Biogenic Carbon Storage. (n.d.). WoodWorks | Wood Products Council. Retrieved March 28, 2023, from https://www.woodworks.org/resources/long-term-biogenic-carbon-storage/

PAVED PARADISE: A VIEW FROM OREGON

Track 2: Sustainable Buildings and Infrastructure
Knowledge Network: Cities & Globalization, Sustainable Planning, Design & Behavior
Methods: Case Study, Interpretive-Historical Research
James Tice (University of Oregon)

1 ABSTRACT

In his 2015 publication, ***The High Cost of Free Parking****, Donald Shoup, identifies parking as one of the most devastating aspects of contemporary American culture. He points out that the automobile and the need to accommodate it through storage—parking—has overtaken our way of life, profoundly affecting our environment. As of 2022, it is estimated that there are eight parking spaces for every registered vehicle in the U.S. Shoup asks why parking is free—requiring acres of public land at enormous cost in resources and deprivation to the environment—while necessities like housing are so expensive? The growing proliferation of electric vehicles promises to have important environmental benefits, but there is no guarantee that this emergent pattern will significantly affect the number of vehicles on the road or their need to be 'housed'.*

This paper proposes to trace the role of parking and its effect on the urban fabric of the traditional American downtown, specifically focusing on Eugene, OR.
Using historic aerial photography and contemporary satellite imaging and GIS inspired visual graphics, I will show the transformation of Eugene through four phases: mapping the city from a compact, walkable city of about 35,000 in 1944 to an auto-dominated and dispersed conurbation of 175,000 in 2023.
Faced with the rise of the regional shopping mall at mid-century with its ease of automobile access and free parking, municipalities like Eugene promoted policies giving precedence to acres of parking at the expense of much else. The Pensacola Syndrome describes the resultant condition whereby parking displaces a city's urban fabric and activities to such an extent that the very reason for going downtown no longer exists. The city falls victim to its own mis-guided policies.

Likewise, a major reason for Eugene's current urban malaise has been the provision for parking seemingly at any cost. It has resulted in compromising public space, parks and the demolition of entire blocks of sound building stock, many of historic value. Today off-street parking in Eugene makes up 29% of the land area of the central city, with 63 acres, which is the equivalent of 8,369 vehicular spaces. I estimate that on any given weekday, surface parking lots are full to only 40% of their capacity.

It is important to realize that these results are consequences of legal codes and policies at every level of government, both local and federal. They have been promoted by planning departments, engineers, design professionals, business leaders, and have been largely approved by a citizenry made to believe in the idea of progress and the infallibility of technology.
Despite these dire circumstances, I will suggest that there is hope. These policies and their deleterious effects have been made through human action and can, therefore, be reversed through human intelligence, public awareness and enlightened policies. One such breath of fresh air are the provisions of the Climate-Friendly and Equitable Communities (CFEC) rules adopted by the Eugene City Council, which has changed its land use code to eliminate off-street parking requirements citywide.

POST-OCCUPANCY EVALUATIONS PRE/POST OCCUPANCY EVALUATION FRAMEWORK: QUANTIFYING GREEN-CERTIFIED BUILDINGS' IMPACT ON OCCUPANTS' HEALTH AND WELL-BEING

Track 2: Sustainable Buildings and Infrastructure
Knowledge Network: POE/Programming, Sustainable Planning, Design & Behavior
Methods: Case Study, Mixed Methods
Ihab Elzeyadi (University of Oregon); Stanley Gatland II (CertainTeed Corporation)

1 ABSTRACT

Employers, developers, and designers are persuaded, in response to an increasing marketing campaign by the green building industry, that LEED-certified buildings affect the health and well-being of occupants in many ways. With staff accounting for 90% of business operating costs, a 1% improvement in productivity can have a major impact on the bottom line and competitiveness of any business. Building developers are also discovering the business value of delivering healthy and green buildings to their markets, which are tied to increasing rent and competitiveness in office building real estate. Despite the favorability of these hypotheses, most recent studies have failed to systemically prove these linkages leading to non-conclusive evidence of green building performance. In addition, the current trend of open-office plans and glazed facades in work environments to achieve LEED certification credits resulted in mixed perceptions of favorability of this design strategy by the building occupants. The specific question this presentation answers is whether a well-planned evaluation study with pre-post occupancy analysis of a retrofitted, green-certified building could lead to more conclusive findings related to the impacts of green buildings, in general, and those with specific high IEQ on occupants' well-being, health, and productivity.

A late 1960s commercial office building was transformed into a state-of-the-art double platinum-rated LEED® corporate headquarters over an extensive 18-month renovation period. Approximately eight-hundred employees work within 270,000 square feet of open-concept office space with an additional 116 meeting rooms available to promote cross-collaboration. The traditional floor plan with perimeter offices and central workstations was reconfigured into an open-concept office space with central workstations, perimeter hallways, and collaborative space. The interior open-plan office space was designed to maximize occupant comfort, as well as improve the health and well-being of the employees.

The presentation will review the current state-of-the-art approaches and studies of building performance evaluation in green-certified buildings. It will follow up with outlining the extensive physical improvements to green the existing building envelope, interior space, and mechanical systems under study. A comprehensive PrePost Occupancy Evaluation protocol (PPOE) was employed to assess and analyze multi-comfort parameters and metrics of the thermal, visual, acoustical, air quality, and spatial quality of the environment over one year, both before and after occupancy. Results of a systematic indoor environmental quality (multi-comfort) assessment and pre-post occupancy evaluation for the building and its occupants pre-and post-retrofit, using continuous and intermittent measurements to quantify indoor thermal, visual, and acoustical comfort, as well as air quality, will be presented and correlated with the occupant's surveyed experience. The presentation will provide evidence-based guidelines for implementing successful design strategies to improve the IEQ of green-certified buildings. Most important, is the application of an integrated design and pre/post-occupancy evaluation framework that would act as a replicable methodology to the human-centric design and operation of the future buildings.

ANALYZING THE UTILIZATION OF SITES AND LEED CERTIFICATION PROGRAMS TO INSTRUCT PERFORMANCE-BASED DESIGN IN A COLLABORATIVE STUDIO

Track 3: Resilient Design for Climate Change
Knowledge Network: Environmental Design Research Education, Sustainable Planning, Design & Behavior
Methods: Design Methods, Survey
Matthew Nicolette (Clemson University); Sallie Hambright-Belue (Clemson University)

1 ABSTRACT

This presentation analyses a strategy for utilizing the SITES and LEED certification programs to teach sustainable design in a collaborative studio with landscape architecture and architecture students. The presentation will address how the certification programs were taught in the course, how the students incorporated them into their designs, how students communicated the performance-based designs to the audience, and how the teaching strategies can be improved.

The Sustainable SITES Initiative (SITES) and Leadership in Energy and Environmental Design (LEED) are professional certification programs that guide development projects toward more energy and ecologically efficient outcomes. The Green Business Certification Inc. (GBCI) was founded to 'provide independent oversight of professional credentialing and project certification under the [LEED] green building rating system' and has since become responsible for additional green building standards programs, including SITES (About GBCI, 2023). These programs have a points system that engages the designer in site staging, materiality, site organization, site program, structure placement, ecosystem services, cultural context, social systems, and economy. These programs can help students learn sustainable design more rigorously and carefully by offering parameters to frame their projects.

Students created collaborative master plans and developed individual site and building designs for the Wind River Trust Historic Arboretum and Nursery in Carson, Washington. The studio contained four teams of landscape architecture and architecture students. The first phase of the course involved analyzing precedent projects, creating a site inventory, and analyzing existing conditions. This phase also introduced students to an overview of SITES and LEED certification programs. In the second phase of the course, the groups developed master plans of the site, which began with a student-led on-site stakeholder engagement and discovery workshop. In the final phase of the course, students developed individual site and building designs for separate zones of the groups' master plans. In this phase, students selected which SITES and LEED parameters to use to guide their design decisions. They chose to use the parameters that best supported their project concept and intended user experience.

After the studio, the students were surveyed to understand how and why they selected specific prerequisites and credits and how the prerequisites and credits influenced their individual designs. We analyzed the effectiveness of the student's communication strategies to educate the audience on their design performance methodology and how these prerequisites and credits helped to defend their design decisions. We also surveyed the students on their perceived benefit of these programs in professional practice and how the instructors can better integrate the programs into the studio course.

SITES and LEED are gaining greater market integration in projects nationwide. Austin, Texas, for example, now requires public projects to be SITES-certified (Su, 2022). Students that enter the profession with an introduction to these programs are well-positioned to be valuable team members. This presentation will analyze one way these programs were utilized to introduce sustainable design in a collaborative studio, understand how students utilized the certification program parameters to create and communicate performance-based designs, and propose how learning outcomes and teaching strategies can be improved.

ADAPTIVE FUTURES: COASTAL DEVELOPMENT FOR SEA-LEVEL RISE IN JAMAICA BAY

Track 3: Resilient Design for Climate Change
Knowledge Network: Environmental Design Research Education, Movement in Designed Environments: Nature & Ecolo
Methods: Case Study, Design Methods

Zhan Chen (Kansas State University)

1 ABSTRACT

Rising sea levels are an increasingly devastating consequence of climate change for populations living in coastal cities worldwide. The warming ocean and melting glaciers are predicted to increase by 12 inches globally by 2050 (Sweet, 2022). With higher sea levels, storms cause more damage due to stronger surges and water reaching further inland. In response to increased flooding, cities continue to employ traditional engineering solutions. Seawalls, levees, piers, and flood channels provide a degree of protection and allow humans to live and work in close proximity to water (Oppenheimer, 2011). However, covering beaches and wetlands with concrete and steel oversimplifies the problem and presents new challenges. In efforts to tame the waterfront, these interventions impair the coastal environments' ability to adapt, creating rigid barriers lacking resilience to absorb systemic shocks.

Living Laboratory is an ongoing design-research studio that examines the urban impacts of sea-level rise and develops new models for coastal development. The proving ground for the project is New York's Jamaica Bay, an estuary covering 25,000 acres. The bay is a prime example of how natural resources and urban communities are entangled and at risk together. In 2012, Hurricane Sandy pushed through Jamaica Bay, flooding dozens of neighborhoods in its watershed and damaging the critical ecosystem of plants, migratory birds, and aquatic species. Over the next 30 years, problems such as tidal flooding, storm surges, coastal erosion, and drowning marshlands are expected to become much worse (Kensinger, 2022).

Working with the Jamaica Bay Conservancy, the Living Laboratory team seeks to uncover the natural and artificial systems at work in the bay. The investigative phase focuses on the far-reaching impacts of sea-level rise and the various ways that species and humans negotiate the dynamic relationship with water. The research builds a scaffold for new ways to design waterfront developments. Rather than thinking about flooding and storm surges as problems to be prevented, how can we design communities to take advantage of these phenomena? Instead of limiting the problem to engineering solutions, how can we leverage and synthesize architecture, landscape, and infrastructure? The rapid rate of sea-level rise requires researchers, designers, and policy makers to question conventional responses and imagine new solutions. Through testing outcomes for Jamaica Bay, Living Laboratory speculates on new futures for how cities can operate and thrive along our volatile coastlines. The project seeks to reframe coastal environments as multifaceted domains that can adapt to multiple scenarios and support diverse communities.

2 REFERENCES

Sweet, W. (2022, February 1). 2022 Sea Level Rise Technical Report. National Ocean Service. Retrieved January 2, 2024, from https://oceanservice.noaa.gov/hazards/sealevelrise/sealevelrise-tech-report.html

Oppenheimer, M. (2011). Climate Change and World Cities. In Rising currents: Projects for New York's Waterfront (pp. 32–39). essay, Museum of Modern Art.

Kensinger, N. (2022). Sea Level Rise and Brooklyn's Jamaica Bay Communities: Storm Surge Barriers and Managed Retreat. Brooklyn Waterfront Research Center, 2. https://academicworks.cuny.edu/ny_pubs/905/

EXPLORING THE ROLE OF BALCONIES ON RESIDENTIAL RESILIENCE

Track 3: Resilient Design for Climate Change
Knowledge Network: Residential Environments, Sustainable Planning, Design & Behavior
Methods: Case Study
Sepideh Masoudinejad (National Research Council of Canada)

1 ABSTRACT

In response to the climate change crisis, population growth, and housing shortage, cities are rapidly densifying and multi-unit residential buildings (MURBs) are becoming a prevalent housing type in many cities. Thus, design strategies that enhance the climate resilience of MURBs should be placed at the forefront of urban housing policies and design. Critical here is to develop integrated solutions addressing various challenges in MURBs, including adaptation to the life-threatening effects of climate change, such as overheating, and improving indoor environmental quality (IEQ) while supporting residents' health and well-being.

This study focuses on the role of balconies as a potential multi-tasking solution for addressing these challenges in MURBs. Despite their potential benefits, balconies are often overlooked in the housing sector due to a lack of knowledge and guidelines for effective design, which has resulted in underused balconies with unwanted consequences. To fill this gap, this study creates a knowledge and technical database with the following objectives:

- *Understand the ways that balconies can play a role in the climate change resilience of MURBs by 1) Enhancing adaptation of dwellings to climate change effects such as overheating; 2) Improving the dwellings' IEQ and residents' comfort; 3) Integrating into nature-based solutions (NBS) on urban façades to enhance resiliency on a local scale;*
- *Investigate existing measures and solutions applicable in design and retrofit of balconies to increase the climate change resilience of MURBs;*
- *Learn how balconies can improve residents' health and well-being;*
- *Explore contributing factors, design parameters, challenges, drivers and barriers that affect the benefits of balconies, their function and acceptance.*

This study conducted a comprehensive literature review based on 175 publications and analyzed over 20 international housing projects with balconies incorporated in either their design or retrofit processes. The results demonstrated how balconies can address multiple challenges in MURBs and dense urban areas by offering a combination of passive sustainable and resilient strategies and NBS with health-promoting effects, as well as potential opportunities for a new wave of technologies for flexible urban surfaces and futuristic ideas. The study showed that the role of balconies in resilient MURBs is a less-explored niche area in research, but it provides a context for a wide variety of interdisciplinary studies from building science and architecture to environmental psychology and behavioral science. The findings alongside the proposed top priority research domains can lead to the development of strategies to support health and resilience in both future and existing MURBs.

DEVELOPING A SUSTAINABLE ADAPTIVE MATERIAL PRACTICAL LEARNING SYSTEM TO FORECAST INTERIOR MATERIAL RESILIENCE

Track 3: Resilient Design for Climate Change
Knowledge Network: Interior Design, Sustainable Planning, Design & Behavior
Methods: Statistical/Modeling Analysis, Mixed Methods
Lisa Sundahl Platt (University of Florida - Florida Institute for the Built Environment Resilience)

1 ABSTRACT

The design of interior environments plays a crucial role in creating settings that support human and ecological resilience. A practical approach to operationalizing building design responsive to human and environmental needs involves understanding the performance capability of the building's material landscape. This session delves into developing a Sustainable Adaptive Material Practical Learning (SAMPL) system for predicting the impact of interior materiality on human and environmental well-being. "Material landscape" refers to the applied surfaces and artifacts within interior environments (Campbell & McDonagh, 2009). Research suggests that the composition and performance characteristics of interior materiality can significantly influence the presence or mitigation of chronic illness (Lu et al., 2022; Maung et al., 2022) and acute health conditions (Harris et al., 2021; Shaughnessy et al., 2016), impact users' emotional and cognitive experiences (Quesada-Garcia et al., 2023), and contribute to or detract from a circular and regenerative economy (Charef et al., 2021).

Understanding the overall performance of materiality involves evaluating the integral effects of discrete yet interconnected finish performance metrics. Interpreting the impact of the material landscape on user safety and environmental sustainability necessitates assessing the overall building performance by analyzing the integration of material specification data (Yazici & Tanacan, 2020). SAMPL fulfills the objective of interpreting safety and sustainability implications based on material performance by employing a resilience engineering taxonomy of material Technical Performance Measures (TPM). These TPMs include composite criteria of robustness, recovery, graceful extensibility, and sustainability (Seager et al., 2017; Woods, 2015) to guide the selection of applied finishes within the material landscape based on the proposed building lifecycle. Conducting this complex systems analysis necessitates creating a cohesive system for mapping material performance data points, calculating outcomes, and developing algorithmic rules to forecast materiality resilience potential over time. SAMPL performance intelligence provides a tunable platform using a validated adaptive neuro fuzzy inference systems (ANFIS) model (Platt & Chen, 2023) for inputting various technical properties of interior materials. This system maps discrete finish data on individual material typologies and predicts their combined environmental performance and adaptability to usage and contextual risks. This approach empowers users to evaluate material capabilities concerning space requirements and ecological impact goals. Designers can then apply the outcomes of material performance analysis to select and specify materials, creating designs responsive to both human and environmental well-being and resilient to risks.

Using SAMPL for material selection, specification, and scheme design provides an empirical evidence basis and a testable logic model to ensure that low-level design performance responses achieve high-level building resilience requirements. This sustainable and healthy material selection and specification process would empower designers to choose interior finishes that are aesthetically pleasing, functional, durable, and have low environmental impacts. Such an approach could afford an increased potential for climate-positive material usage and interior finish performance outcomes, reliability. A long-term aspiration for this effort is that iterative use of SAMPL resources may be instrumental in enhancing collective industry knowledge of how interior finishes may support built environment resilience in the wake of emergent human-induced and natural risks.

ACTIVE LIVING BY DESIGN THE CORRELATION BETWEEN PLACE SATISFACTION AND DISASTER RECOVERY IN COMMUNITIES FACING ENVIRONMENTAL SHOCKS

Track 3: Resilient Design for Climate Change
Knowledge Network: Sustainable Planning, Design & Behavior
Methods: Case Study, Statistical/Modeling Analysis
Saeideh Sobhaninia (Arizona State University); Melika Amirzadeh (University of Art, Tehran)

1 ABSTRACT

The increasing impact of climate change has led to a rise in the frequency and changed patterns of natural disasters, worsened by population growth and urbanization. As a result, the efforts to recover from disasters have become crucial to long-term adaptation to climate change. Previous research has shown that place satisfaction, which is rooted in psychology and urban sociology, significantly affects how communities respond to disasters. This study aimed to explore the link between place satisfaction and disaster recovery in four underserved communities in Puerto Rico following Hurricanes Maria and Fiona. To accomplish this goal, we used an exploratory approach and conducted surveys for data collection. The data collected through questionnaires was then quantitatively analyzed using principal components on JMP. Our findings indicated a positive correlation between place satisfaction and disaster recovery in the selected case studies, demonstrating that the residents' satisfaction with their surroundings may contribute to post-disaster recovery. The framework introduced in this study for examining the relationship between place satisfaction and disaster recovery can assist policymakers, researchers, and urban planners in understanding the connection between these two concepts, and the key findings can help them make informed decisions to develop successful disaster recovery strategies and build more resilient communities in an era of environmental uncertainty.

IMPACT OF OFFICE BUILDING EXTERIOR VENETIAN BLINDS ON ENERGY AND DAYLIGHTING PERFORMANCE DURING CURRENT AND FUTURE WEATHER IN SEVEN US CITIES

Track 3: Resilient Design for Climate Change
Knowledge Network: Sustainable Planning, Design & Behavior, Work Environments
Methods: Statistical/Modeling Analysis
Teresa Moroseos (University of Washington Integrated Design Lab); Chris Meek (University of Washington)

1 ABSTRACT

Window shading systems installed on the exterior of buildings have long been recognized as an effective strategy to improve thermal comfort and reduce cooling loads, particularly in hot climates (Cheng 2015). Exterior Venetian Blinds (EVBs) are a type of exterior shade that consist of dynamic, horizontal slats. These slats lower when direct sunlight is present to minimize glare and heat gains, and they tilt to maximize views to the outside and allow diffused light to enter the building. They can also be raised to maximize daylight and provide completely unobstructed views to the exterior when direct sunlight is not present.

The efficacy of EVBs used in Office buildings for reducing cooling energy, minimizing peak cooling loads, and improving daylighting performance has been studied in warm climates (Lu 2024). Our study adds to the current literature by assessing EVB performance in seven climates in the United States and also assesses the efficacy of EVBs using future climate models. For each location, an analysis of the east, south, and west façade was conducted.
To characterize the performance of EVBs, our team developed a parametric, simulation-based comparative analysis of a series of window strategies for commercial building application. These included manual and motorized interior shades, fixed exterior shades, and two EVB applications: a standard automated EVB and an EVB within and actively-ventilated double skin façade. The analysis documented key performance indicators associated with high-performing buildings for each of the different shading strategies: energy consumption, peak cooling requirements, daylight sufficiency, and the percentage of unobstructed views to the exterior.

By controlling the amount of direct solar radiation entering a building, EVBs reduced the need for air conditioning and electric lighting, leading to significantly reduced overall energy consumption. The advantage of EVBs is particularly pronounced in east and west-facing conditions in climates where cooling energy predominates. In all but two cold-climate south-facing situations, EVBs provided the lowest overall annual energy consumption in all simulation cases. In the cases where EVBs underperformed conventional shading strategies, glare control was prioritized, which resulted in sacrificing some beneficial passive solar gains at south facades to maximize visual comfort.

Where EVBs substantially out-performed other conventional shading strategies was in reducing peak cooling loads in areas near windows. Peak cooling loads are a critical performance indicator in buildings since they drive mechanical system sizing, type, and cost. They are a strong indicator of occupant thermal comfort, which is associated with improved productivity and user satisfaction. In our analysis EVBs typically cut cooling loads by nearly 50%. Reducing peak cooling can reduce energy consumption, allow for downsizing HVAC equipment, and enable super-efficient mechanical system types such as radiant heating and cooling that are not possible with high peak solar loads.

2 REFERENCES

Chen, X., Yang, H., & Lu, L. (2015). A comprehensive review on passive design approaches in green building rating tools. Renewable & Sustainable Energy Reviews, 50, 1425–1436.

Lu, W. (2024). Dynamic Shading and Glazing Technologies: Improve Energy, Visual, and Thermal Performance. Energy and Built Environment, 5(2), 211–229.

FROM SUSTAINABILITY TO RESILIENCE IN URBAN DESIGN

Track 3: Resilient Design for Climate Change
Knowledge Network: Cities & Globalization, Sustainable Planning, Design & Behavior
Methods: Meta-Analysis

Szymon Piotr Nogalski (University of Cincinnati); Vikas Mehta (School of Planning, DAAP, University of Cincinnati)

1 ABSTRACT

The key to continuous existence of a city lies in its sustained capacity to repair, adapt, and regenerate itself. Grand, unifying, and reductive solutions cannot provide this capacity as they don't account for the complex nature of relationships and patterns that form a city. This places a responsibility on the policy maker, planner, urban designer, and architect: to ensure their intervention in the built environment enables and accommodates the diversity of mending and improvement mechanisms to occur.

Most scholarship regarding resilience in the built environment disciplines has been focused in three areas: buildings, landscape, and city. Architects have been researching energy-efficient sustainable solutions through innovating building technologies and systems. Landscape architects and environmental scientists are looking at site-specific and systemic solutions to address issues of water and air quality, management, and more. City planners have been concerned with the city's defensive properties to survive natural events, as well as with topics of social and economic resilience. These scales capture a great deal of what is needed for the built environment to be sustainable and resilient. However, resilience in urban design, as an explicitly defined focus of inquiry, is a rather new field. In parallel with other disciplines, resilience has often been relegated to risk management in urban design, or a dimension of sustainability, with their inherently equilibrium-centered focus. Although much research is presented under the general umbrella of urban design, it is either in the general realm of urban policy or focused on engineering aspects of buildings and landscape. There is limited research on the established tenets and principles of urban design vis-à-vis resilience; few contributions have explored if the long-established fundamental ideas of good urban design add to making places resilient. Additionally, the "resilient cities" moniker often tilts disproportionately toward disaster recovery and biophilia where a detailed understanding and capacity of urban form is missing. This paper analyzes design principles and practices rooted in urban design through the lens of resilience. It focuses on form, function, and flows — bringing urban form to the center. The well-founded tenets of urban design derived from sustainable urban design scholarship are grouped according to broadly defined aspects of human functioning in a city, e.g. shelter, energy production and consumption, movement, access to food, waste, etc. This allows the tenets to be examined through the lens of their interaction with the concepts of resilience in its physical and functional dimension: the built form's capacity to accommodate change, but also its propensity to generate new ways of dealing with change. The framework of resilient urban design emerging from the analysis departs from the stability-centered sustainability principles and focuses on urban design rules that produce urban environments that are capable of absorbing and promoting change and innovation, while preventing the emergence of social, economic, and environmental vulnerabilities (e.g. exploitation, homogeneity, rigidity, standardisation, ...) that can result in irreversible damage of these places.

EXPLORING DETERMINANTS OF RESIDENTS' BEHAVIOR IN GREEN STORMWATER INFRASTRUCTURE ADOPTION: AN EXTENSION OF THE THEORY OF PLANNED BEHAVIOR

Track 3: Resilient Design for Climate Change
Knowledge Network: Communication, Participation
Methods: Statistical/Modeling Analysis, Survey
Mahsa Adib (Penn State University); Hong Wu (Penn State)

1 ABSTRACT

Climate change and increasing imperviousness levels are causing urban areas around the world to experience larger volumes of rainfall runoff. This requires effective management to protect property, and infrastructure, and prevent environmental pollution. The traditional grey stormwater infrastructure is often outdated and incapable of controlling the increased runoff volumes. Green stormwater infrastructure (GSI) can complement grey infrastructure, but public land for its installation is limited. As a result, municipalities often rely on residential properties to install GSI at the lot level. Despite governmental efforts to promote GSI, local residents sometimes lack the motivation to implement individual GSI practices on their properties or contribute to the development of GSI within their communities (Le et al., 2023). Even when provided at no cost or with incentives, resident participation remains low, potentially due to factors such as insufficient information, knowledge gaps, and various social-psychological and demographic influences (Venkataramanan et al., 2020). Currently, there is limited understanding of how these factors influence residents' behavior in GSI adoption. Our study aims to address this gap by identifying the social and psychological determinants impacting residential GSI implementation, contributing to a more comprehensive understanding of this critical area.

The theoretical foundation of the Theory of Planned Behavior (Ajzen, 1991) guides our exploration, positing that attitudes, subjective norms, and perceived behavioral control collectively determine behavioral intentions, ultimately shaping actual behaviors. We extend this framework by incorporating the mediating role of demographic characteristics, as well as knowledge and past experiences. We hypothesize that individuals with greater knowledge and exposure to stormwater issues and GSI practices are more likely to exhibit positive intentions and behavior toward GSI implementation and maintenance. To test the proposed hypotheses, a questionnaire survey was conducted online across three cities in the United States - Pittsburgh, PA, Portland, OR, and Baltimore, MD. To ensure a diverse and representative sample, the survey utilized Dynata, a first-party data and survey platform, for sampling and recruitment purposes. A total of 645 complete responses were collected and analyzed using IBM SPSS and AMOS. The collected data were analyzed using both descriptive statistics and Structural Equation Modeling (SEM) techniques.

Preliminary results indicate that prior knowledge of stormwater issues, past flooding experience, and knowledge and exposure to GSI, play a significant role in individuals' decision of GSI adoption and maintenance. Individuals who hold positive attitudes toward GSI, perceive social approval and support for its adoption (Subjective Norm), and feel they have the necessary control over the process (Perceived Behavioral Control) are more likely to express intentions to implement and maintain GSI features. Demographic characteristics, however, did not seem to affect residents' decisions about GSI installation. We recommend the creation of effective municipal education and outreach programs regarding urban stormwater management that speak to residents' shared responsibility and options for addressing this issue, as well as the creation of financial instruments that provide meaningful subsidies for residential GSI adoption. These findings aim to inform policymakers, urban planners, and educators about the importance of tailored educational programs in promoting sustainable stormwater management practices.

MOTIVATIONS, IMPACTS, AND ETHICAL CONSIDERATIONS IN ENGAGING CITIZEN SCIENTISTS IN CLIMATE CHANGE ACTION- A CASE OF BALTIMORE SOCIAL ENVIRONMENTAL COLLABORATIVE (BSEC)

Track 3: Resilient Design for Climate Change
Knowledge Network: Children, Youth & Environment, Participation
Methods: Action/Participatory Research, Case Study
Mahsa Adib (Penn State University); Derrick Mirindi (Baltimore Social-Environmental Collaborative); Tonya Sanders Thach (Morgan State University); Lisa Iulo (Penn State Hamer Center for Community Design)

1 ABSTRACT

Citizen Science, a community science effort aimed at combating climate change, has grown due to global environmental issues. Citizen Science involves the general public or non-professional scientists (often volunteers) in collecting, analyzing, and sharing climate science data (Peña-Rodriguez et al., 2022). Citizen science projects have been effective in climate action initiatives, particularly by raising awareness about the environment and how climate change affects it (Seibert & Meerveld, 2022). As citizen science continues to grow in popularity, attention is increasingly focused on the ethical issues it raises, including ethical concerns related to data ownership, control of research outputs, and the communication of research findings to participants (Wiggins & Wilbanks, 2019). These challenges underscore the necessity of creating and implementing protocols that tackle essential ethical considerations in climate-focused initiatives.

An examination of the existing literature on citizen science reveals that there is currently no universally accepted standard for ethical considerations. Additionally, despite the extensive climate change science research, there is still a gap between recommendations from the scientific community and current actions taken by policymakers at the national or international level (Diener, 2022). There is also relatively little attention to how citizen science can support society's climate mitigation or adaptation and its community-level impacts on governance for climate change.

This study addresses the identified knowledge gap by utilizing the Baltimore Social-Environmental Collaborative (BSEC) project as a case study. BSEC is a 5-year US Department of Energy-funded Urban Integrated Field Lab that aims to study equitable pathways to climate change mitigation and adaptation in Baltimore City. This project is a prime example of how citizen science can be applied to address specific environmental challenges in an urban context. As the first project of its kind in the United States, BSEC, based in Baltimore, is not only a pioneer in tackling urban environmental issues through citizen science but also in its commitment to ethical research practices. The environmental challenges that this project addresses have shaped a series of Scientific Theme Teams that focus on topics including air quality, atmospheric dynamics, building and energy efficiency, greenhouse gas emissions, sustainable transportation, vegetation and soil, water quality, and community engagement. Considering that BSEC is an ongoing project and community engagement initiatives are not synchronous, this study centers on examining how the Atmospheric Dynamics Scientific Theme Team navigates ethical considerations within the overall BSEC citizen science activities. Lessons learned from this project will be used for the citizen science activities initiated by the other scientific teams. Additionally, the project's approach to ethical considerations has the potential to serve as a foundational model for future research in similar initiatives, filling an important gap in the current discussion on the ethical dimensions of citizen science. By documenting and analyzing the ethical strategies employed by BSEC researchers, including data quality and integrity, data sharing and intellectual property, conflict of interest, and exploitation, this research aims to contribute to the development of a much-needed ethical standard for citizen science projects, especially those focused on environmental justice and community engagement.

INTEGRATING ECOLOGICAL EXPERIMENTATION AND SITE DESIGN STRATEGIES TO MITIGATE VANDALISM AT THE BALD ROCK NATURAL HERITAGE PRESERVE

Track 3: Resilient Design for Climate Change
Knowledge Network: Movement in Designed Environments: Nature & Ecology, Sustainable Planning, Design & Behavior
Methods: Action/Participatory Research, Design Methods
Matthew Nicolette (Clemson University); Paul Russell (Clemson University)

1 ABSTRACT

In the Fall of 2022, a team from the South Carolina Department of Natural Resources, the Friends of Bald Rock, and Clemson's LA-Co-LAB outlined a phased approach to address the current issues of visitor behavior at the Bald Rock Natural Heritage Preserve. The team developed innovative strategies to modify how visitors engage and explore one of South Carolina's cherished natural resources to reduce vandalism and promote ecological health. The team framed the design problem through an order of scales to reach compressive solutions that improve the site experience, restore healthy ecologies, and enhance the site as a cultural resource. The design solutions demonstrate how these diverse scales are interconnected and can be implemented to improve overall site health and experience.

A team of faculty and students from Clemson University's Landscape Architecture program began a comprehensive inventory and analysis of the existing Bald Rock Natural Heritage Preserve. The initial investigations focused on visitor activity and behavior, existing infrastructure and circulation, topography and vegetation, and the site's ownership and land-use history. After inventory and analysis, the consultant team performed a strengths, weaknesses, opportunities, and threats analysis on a series of strategies before testing protocols to mitigate vandalism, allow for ecological restoration, and better engage the site as a cultural resource for the South Carolina Upstate community. The design team executed a series of experiments to address the systemic vandalism issue on the site by promoting expeditious lithosere succession. A series of design strategies also addressed on-site access, site circulation, site experience, and cultural interpretation. These experiments and design strategies were performed concurrently, and the information gleaned was integrated to inform the final design strategies. In this presentation, we will address the results of the physical site design and lithosere succession experiments and how these concurrent processes inform each other to develop a holistic strategy to mitigate vandalism on the Bald Rock Natural Heritage Preserve. We will address strategies that adjust site access, regulate site circulation, promote the site as a cultural resource, and educate site users on the importance and sensitivity of the site. We will also address the protocols tested to promote lithosere succession. The design team explored alternative ways to mass broadcast organic patina and deposit native soil on the rock face, promoting planting zones. We will discuss these experiments in the context of their success in the relationship to re-vegetation, application, scalability, and cost. Finally, we will demonstrate how these strategies inform each other and can be implemented using the Bald Rock Natural Heritage Preserve as a case study.

SLEEPING WHILE POOR: WHY WE NEED TO IMPROVE HOUSING DESIGN FOR LOW INCOME URBAN AREAS

Track 3: Resilient Design for Climate Change
Knowledge Network: Cities & Globalization, Health
Methods: Mixed Methods
Eunhwa Yang (Georgia Institute of Technology); Jennifer DuBose (Georgia Institute of Technology, SimTigrate Design Lab); Arthi Rao (Georgia Institute of Technology)

1 ABSTRACT

Environmental justice research has revealed the disproportionate impact of various challenges faced by low-income communities including proximity/exposure to pollution, limited access to transportation, food deserts, and high crime rates. Importantly, these environmental factors extend to health outcomes, particularly affecting sleep quality. The timing, amount, and quality of sleep are critical for overall health and quality of life and yet, people living in socially disadvantaged communities are also at greater risk of poor sleep. Specifically, sleep disturbances can lead to chronic physical and mental health problems, such as cardiovascular diseases, dementia, and lack of quality of life. The current literature shows the existence of significant sleep disparities in socioeconomically disadvantaged groups. For example, those with lower education and income levels often report very short sleep (less than five hours) compared to those with college degrees and an annual income exceeding $75,000 (Whinnery et al., 2014).

There is growing evidence linking natural, built, social, ambient, and neighborhood environmental characteristics such as lighting, noise, traffic, pollution, and walkability, with sleep outcomes, points to greater difficulty for vulnerable populations with lower SES and racial minorities to sleep (Johnson et al., 2018). Socioeconomically disadvantaged groups are exposed to higher housing insecurity, poorer home environments, neighborhoods, and community-level environments. Climate change stands to make these issues worse in the coming years as urban areas heat up and nighttime temperatures rise during the summer months. Excessive heat has many documented negative health consequences. One that hasn't drawn much attention is the effect that heat has on sleep. Older adults struggle with thermoregulation, and are more likely to live in these at-risk communities. Many in these socioeconomically disadvantaged communities lack access to air conditioning, or even if they have air conditioning units at home, they are non-operational or too expensive to use, leading to uncomfortably hot nights. Existing research confirms that elevated temperatures result in sleep disruption and disproportionately impact the sleep of older and low-income individuals (Obradovich et al., 2017).
In this presentation, we explore how multidimensional environmental factors influence sleep disparities for aging adults in socioeconomically disadvantaged racial minorities (African Americans and Hispanics). We approach this research from a comprehensive view of Bronfenbrenner's ecological systems theory by considering multiple levels of built environments: 1) housing insecurity, 2) house/household physical and social environment, 3) neighborhood, and 4) community-level infrastructure. We discuss built environment strategies and policy solutions that can ameliorate the urban built environments where low-income individuals live so that they can have improved sleep health even as the global climate changes.

2 REFERENCES

Johnson, D. A., Billings, M. E., & Hale, L. (2018). Environmental determinants of insufficient sleep and sleep disorders: implications for population health. Current Epidemiology Reports, 5(2), 61-69.

Obradovich, N., Migliorini, R., Mednick, S. C., & Fowler, J. H. (2017). Nighttime temperature and human sleep loss in a changing climate. Science Advances, 3(5), e1601555.

Whinnery, J., Jackson, N., Rattanaumpawan, P., & Grandner, M. (2014). Short and long sleep duration associated with race/ethnicity, sociodemographics, and socioeconomic position. Sleep, 37(3), 601-611.

BEHAVIORAL AND ENVIRONMENTAL EFFECTS ON BUILT SPACES TEMPERATURE ART INSTALLATIONS: VISUALIZING HEAT

Track 3: Resilient Design for Climate Change
Knowledge Network: Children, Youth & Environment, Participation
Methods: Case Study, Practice-based Research
Carolina Aragon (University of Massachusetts Amherst)

1 ABSTRACT

Among all natural hazards, urban heat is responsible for the largest number of mortalities, especially in lower socioeconomic and minority communities in urban areas (Mitchell et al, 2015). Heat exposure has significant impacts on health, well-being, and labor productivity. In children, exposure to heat decreases learning and, when combined with other factors, can create adverse health conditions that can last a lifetime. While the problem is widespread, with 30% of the world's population already living in areas where heat and humidity exceed the death threshold for humans for 20 days of the year (Mora, et al, 2017), areas with traditionally cooler weather have additional challenges. These include residents' lower ability to acclimatize to warmer temperatures (Hanna et al, 2015), and the lack of infrastructure (such as air conditioning) to mitigate extreme heat events.

This presentation includes 3 case studies of indoor and outdoor public art installations (Heat Trees, Temperature Quilts, and sWARM) designed by the author to engage impacted communities with issues of urban heat in Massachusetts (Boston and Springfield). The artworks demonstrate the potential for using smart materials in the visualization and mitigation of heat in ways that can be made with school-age students and community members. Smart materials are commonly employed for their ability to transform available energy found in the environment into usable forms. They are largely categorized as property-changing– physically transforming because of environmental change; or energy exchanging– transforming input energy into a different form of output energy. The artworks presented here describe the use of thermochromic and smart memory alloys, property-changing materials, to create participatory artworks that can provide visual cues on indoor and outdoor temperatures, serving as record keepers and monitoring devices that enhance awareness of rising temperatures.

Considering the severity of the impact of heat on human health, and the rapidly changing temperatures, these works provide a model by which designers may contribute to education, engagement, and monitoring of rising temperatures. This is particularly important in environmental justice communities, ageing school buildings, and other areas with insufficient infrastructure to mitigate extreme temperatures. Additionally, the artworks point to opportunities for increasing visibility of environmental justice issues through participatory art projects that can serve to connect underserved populations to available mitigation and energy efficiency resources. As our environment rapidly changes, we need new tools for making these changes apparent, and modify behaviors to allow us to better adapt to them.

2 REFERENCES

Hanna, E. G., & Tait, P. W. (2015). Limitations to thermoregulation and acclimatization challenge human adaptation to global warming. International journal of environmental research and public health, 12(7), 8034-8074.

Mitchell, B. C., & Chakraborty, J. (2015). Landscapes of thermal inequity: disproportionate exposure to urban heat in the three largest US cities. Environmental Research Letters, 10(11), 115005.

Mora, C., Dousset, B., Caldwell, I. R., Powell, F. E., Geronimo, R. C., Bielecki, C. R., ... & Trauernicht, C. (2017). Global risk of deadly heat. Nature climate change, 7(7), 501-506.

SAFE, BUT NOT HEALING: NURSE EXPERIENCES OF VALUE CONFLICT IN THE BUILT ENVIRONMENTS OF INPATIENT BEHAVIORAL AND PSYCHIATRIC CARE

Track 4: Design for Health and Wellbeing
Knowledge Network: Health, Work Environments
Methods: Grounded Theory, Interviews
Monica Gripko (Clemson University); Anjali Joseph (Clemson University);
Lucy Easler (Prisma Health Baptist Columbia SC)

1 BACKGROUND

Nursing is anchored in an ethos of personalized, comprehensive patient care that addresses physical and psychosocial dimensions (Fahrenwald et al., 2005). However, the reality of modern healthcare often clashes with this philosophy, leaving nurses feeling drained and disillusioned and contributing to role turnover and professional dropout (Maben et al., 2007). Still, little is known about how nurses' perceptions of their built environments might conflict with their care values.

2 METHOD

This qualitative study uses in-depth interviews and a Grounded Theory methodology to delve into the values upheld by inpatient psychiatric mental health nurses working at two hospitals in the Southeastern United States and their intricate relationships with the built environments of their units.

3 RESULTS

Analysis of thirty nurse interviews revealed a predominant emphasis on safety in inpatient behavioral and psychiatric care environments, overshadowing most other patient and staff needs. While the interviewed nurses acknowledged and supported the importance of patient safety, they grappled with balancing patient safety measures and fulfilling their patients' other needs in line with their values of empathy, patient dignity, and holistic care. The interviews identified several environmental features and characteristics that conflicted with nurses' values for patient care.

4 CONCLUSIONS

The findings underscore inherent conflicts arising from the built environments of inpatient behavioral and psychiatric care units and highlight areas for design improvement to enhance nurses' well-being, performance, and job satisfaction while aligning with their core values..
t

5 REFERENCES

Fahrenwald, N. L., Bassett, S. D., Tschetter, L., Carson, P. P., White, L., & Winterboer, V. J. (2005). Teaching Core Nursing Values. Journal of Professional Nursing, 21(1), 46-51. https://doi.org/10.1016/j.profnurs.2004.11.001

Maben, J., Latter, S., & Clark, J. M. (2007). The Sustainability of Ideals, Values and the Nursing Mandate: Evidence from a Longitudinal Qualitative Study. Nursing Inquiry, 14(2), 99-113. https://doi.org/10.1111/j.1440-1800.2007.00357.x

PHYSICAL ENVIRONMENTAL EVALUATION OF A DEMENTIA VILLAGE IN BRITISH COLUMBIA, CANADA: PRE-RELOCATION FINDINGS

Track 4: Design for Health and Wellbeing
Knowledge Network: Environment-Gerontology
Methods: Mixed Methods
Habib Chaudhury (Simon Fraser University); Ziying Zhang (Simon Fraser University)

1 ABSTRACT

The physical environment in long-term care homes is a key component in supporting quality of care and quality of life for residents with dementia. The first dementia village in the public healthcare system in Canada, currently under construction, will offer a neighborhood environment for residents with familiar places and destinations beyond the households. As part of a broader evaluation project, we are conducting a baseline evaluation of the influence of the physical environment in the existing traditional care home, and planning for a post-relocation evaluation in the new village-inspired care home. Standardized environmental assessments (with Therapeutic Environment Screening Scale for Nursing Homes, Professional Environmental Assessment Protocol, and Person Oriented Environmental Tool), ethnographic observations in the social spaces, and staff focus groups were the research methods utilized to generate data in the existing special care unit to generate findings on care practices and residents' quality of life that complement each other for a comprehensive and in-depth understanding.

The pre-relocation baseline evaluation in the existing unit revealed challenges arising from the physical layout of the unit, scale of the main social space, impacting residents' autonomy and staff interactions with the residents. The built environmental features and staff's practice jointly influenced the use of the spaces. And residents' functioning and wellbeing. The 50-resident special care unit had a large undifferentiated social space and a long intersecting corridor. Lack of access to public washroom and wayfinding cues impeded residents' independent mobility. Residents had difficulty in finding their way as they moved between the main common space (dining and activity space) and the resident rooms. The two long and intersecting corridors as the main circulation paths acted as a barrier in residents' wayfinding abilities and contributed to behaviors to go out of the unit. The residents in the common space also experienced "toileting disempowerment" as they received little support from staff to use the public bathroom or walking back to the private washrooms. Clocks and the activity calendar on the unit's wall provide residents orientation to time, which were standard institutional markers. The unit was decorated with a few artifacts, however, the residents rarely interacted with the objects in the environment, except for the holiday decorations that served as stimuli and reminders for conversations and activities. Secondary social spaces like the "granny room" and the patio provided additional choices for social interactions, albeit constrained by a scarcity of meaningful props, limiting opportunities for spontaneous or self-directed activity engagement. Staff also reported that they did not have time to move residents to adjacent secondary spaces for activities.

STUDENT EXPERIENCE IN THE POST-PANDEMIC CAMPUS: SYSTEMATIC STUDY OF A STUDENT UNION RENOVATION PROJECT AND ITS CONTRIBUTIONS TO STUDENT WELLBEING

Track 4: Design for Health and Wellbeing
Knowledge Network: Health, Student Affairs
Methods: Ethnography, Observation

Caryn Brause (University of Massachusetts); Derek Noble (Shepley Bulfinch)

1 ABSTRACT

Consideration of the impact of the built environment on health and well-being is a priority in the design and construction industry, interrelated with sustainability, resilience, and planetary wellbeing. Meeting these requirements necessitates a commitment from management, operations, and ownership realms and an attendant imperative to educate a broad range of individuals regarding the built environment's role.1 In higher education, institutions are focused on initiatives to enhance individual and community wellbeing.2 The most aspirational and integrative framework employed, the Okanagan Charter, calls for institutions to "Embed health into all aspects of campus culture, across the administration, operations and academic mandates."3 This call acknowledges that institutions have a responsibility to lead by providing transformative education, developing new knowledge and understanding, and advocating to decision-makers for the benefit of society.

This presentation shares a systematic study of a newly renovated student union at an institution that has adopted the Okanagan Charter. Interdisciplinary advanced undergraduate teams worked with campus and design practitioners to frame research questions aimed at understanding the contribution of this space to student wellbeing. Observational and survey methods adapted from the Gehl Public Life Tools and the Toolkit for the Ethnographic Study of Space were employed4; data were collected over two years for a period of three weeks each time.

Initial findings indicate the project is successful in its objective to serve as the campus "living room." Participants rated the building as welcoming and inclusive; they rated it significantly higher than other campus spaces for contributing to their wellbeing. However, granular findings offer recommendations for addressing varied dimensions of wellbeing in students' everyday lives, from furnishings to operating hours. For example, students rated the building as performing better for spending time with others, such as collaborative work, than for studying alone. Analysis suggests this is, in part, due to furniture arrangements - 84% of students mentioned some aspect of seating in their response during the second round of data collection. Further, data analysis offers insights for design and planning of post-pandemic campus environments such as the need for a greater variety of student work spaces throughout campus. Finally, this research considers how teaching built environment topics to non-design students harnesses a broad range of intelligences critical to creating more vibrant, inclusive, and health-promoting campus environments.

2 REFERENCES

Rider, Traci Rose. "How Health Factors into Green Building Rating Systems: WELL Building Standard." American Institute of Architects, October 17, 2017.

Suárez-Reyes, Mónica, Mónica Muñoz Serrano, and Stephan Van den Broucke. "How Do Universities Implement the Health Promoting University Concept?" Health Promotion International 34, no. 5 (October 1, 2019): 1014–24; Travia, Ryan M. et al., "Framing Well-Being in a College Campus Setting," Journal of American College Health 70, no. 3 (2022): 758–72.

"Okanagan Charter: An International Charter for Health Promoting Universities and Colleges," 2015, https://bp-net.ca/program/the-okanagan-charter/.

Gehl Institute. (n.d.). How to Use the Public Life Tools. Gehl. https://gehlpeople.com/tools/; Low, S., Simpson, T., & Scheld, S. (2019). Toolkit for the Ethnographic Study of Space (TESS). The Graduate Center, City University of New York. https://psrg.commons.gc.cuny.edu/the-toolkit-for-the-ethnographic-study-of-space/

HEALTH-CENTERED OUTDOOR PUBLIC PLACES: EXPLORING THE IMPACTS OF CAMPUS DINING EXPERIENCE ON COLLEGE STUDENTS' SENSE OF BELONGING

Track 4: Design for Health and Wellbeing
Knowledge Network: Health, Student Affairs
Methods: Mixed Methods, Survey

Shan Jiang (GBBN Architects); Stephen Mrdjenovich (GBBN Architects); Zahra Daneshvar (GBBN Architects)

1 ABSTRACT

The activities that are associated most with socializing are eating and drinking, and students' dining activities have composed a significant proportion of their social lives on the college campus (Mehta & Bosson, 2010). Students' dining experience can be a critical factor impacting their satisfaction and overall sense of belonging to the university community, leading to long-lasting impacts such as student retention, motivation, and enjoyment (Knight, 2015; Pedler et al., 2022). Understanding the factors influencing students' dining experience extends beyond the servicescape factors, such as food choices, price, and staff service quality (Clark & Wood, 1999). In fact, students' purchasing decisions and satisfaction are also influenced by the sense of place, physical environment, and overall atmosphere of the dining facility. The campus dining facilities' physical environments and architectural attributes could play essential roles in students' satisfaction with their dining experience and mediate their sense of belonging at the university.

The first phase of our study is to conduct a literature review that identifies the common themes in previous research on student satisfaction with campus food services. The literature review spans various dimensions, including student/customer satisfaction, the impact of the built environment, and the intricate relationship between design and student well-being in the context of dining and food services on college campuses. The comprehensive literature review forms the theoretical foundation of the study and led to the development of a survey questionnaire used in the second phase of the study.

Second, our study explores the relationship between a range of variables about dining experience and students' sense of belonging through a survey study. Using the Penn State Berks campus as a case study site, a sample of 132 students answered the questionnaires. Data were analyzed using descriptive statistics, correlation, and regression techniques. Preliminary findings indicated that on- and off-campus students have different behavior patterns at the campus dining facilities. The physical environments of the campus dining facility have substantial impacts on students' overall satisfaction about their dining experience. Noticeable factors include dining furniture for privacy and socialization, cleanliness, lighting, acoustics, and technology support. Aesthetics/design ratings and overall satisfaction with the dining facility jointly predict students' sense of belonging to the university campus.

The survey results also shed light on equity and inclusion when measuring college students' sense of belonging. BIPOC students have experienced a significantly lower sense of belonging compared to their White peers. Different ethnic groups of students have shown notable trends of differences regarding how they perceive and use the dining facilities.

The concept of place attachment emerges as a critical factor in designing public spaces on college campuses, fostering a bond between individuals and their environment (O'Rourke & Baldwin, 2016). This study illuminates the vital role of placemaking in student mental health and well-being. This study filled a significant research gap by examining the relationship between dining facility design, dining experience, and students' sense of belonging, leading to practical design suggestions for future practices.

IDENTIFYING KEY PHYSICAL AND ENVIRONMENT FACTORS THAT IMPACT THE USE OF URBAN OPEN SPACE FOR PHYSICAL ACTIVITY: INSIGHTS FROM EXPERT INTERVIEWS

Track 4: Design for Health and Wellbeing
Knowledge Network: Active Living by Design, Sustainable Planning, Design & Behavior
Methods: Grounded Theory, Interviews
Benjamin Okenwa (University of San Francisco); Marga Munar Bauzá (CARDIFF UNIVERSITY)

1 ABSTRACT

Existing studies on physical activity (PA) in open spaces have examined individual, social, physical, and environmental factors influencing PA engagement (Van Hecke et al., 2016). While various physical and environmental factors influencing open space use for PA have been identified, there needs to be more focus on specifying the key factors that attract users to open spaces for PA. This knowledge gap is significant, particularly amid declining health patterns among young adults due to insufficient PA (Leslie et al., 2001), and rapid urbanization in the 21st century. Consequently, researchers and urban designers are increasingly interested in identifying the key physical and environmental factors that positively impact the use of open spaces for PA.

This study aimed to identify key physical and environmental factors that impact young adults' use of publicly accessible open spaces in urban settings for PA. To investigate the concept outlined above, the study used semi-structured interviews of expert sampling to obtain detailed and context-specific information. Participants (n = 18, including researchers, academics, and practitioners in architecture, landscape architecture, urban design, and public health disciplines) were recruited across seven countries. Participant interviews occurred virtually on Zoom from May to September 2023. All audio-recorded interviews were transcribed and analyzed using NVivo 14 software, and thematic analysis was used to derive categories and subcategories using a grounded theory approach.

The main insights derived from the study validate some of the broader concepts of physical and environmental factors that impact the use of different types of urban open spaces (UOS) for PA but suggest that perceived safety, which emerged as a social and physical factor, had the most significant association with the use of UOS for PA. Also, insights from the study identify attributes both compositional (within) and contextual (outside) as fundamental in attracting users to an UOS to engage in PA. Other important subcategories that supported visiting UOS to perform PA were the presence of trees and vegetation, mixed use of neighborhood, walkable distance to open space, connectivity, and having other destinations around the open space. The research objective focused on understanding physical and environmental factors. Hence, individual, and social factors were not discussed or analyzed, except for safety, which is both physical and social evolving as an emergent theme in the study. Environmental factors such as mixed-use neighborhoods, connectivity, destinations, physical access, and density appeared to be often more influential than physical factors (such as trees and vegetation, place-making elements, views, sidewalks, etc.). However, the combination of physical and environmental factors was found to positively impact young adults' use of UOS for PA.

2 REFERENCES

Leslie, E., Sparling, P. B., & Owen, N. (2001). University campus settings and the promotion of physical activity in young adults: lessons from research in Australia and the USA.Health education, 101(3), 116-125.

Van Hecke, L., Deforche, B., Van Dyck, D., De Bourdeaudhuij, I., Veitch, J., & Van Cauwenberg, J. (2016). Social and physical environmental factors influencing adolescents' physical activity in urban public open spaces: A qualitative study using walk-along interviews.PloS one, 11(5), e0155686.

HEALTHY BY DESIGN: EXPERIENTIAL DESIGN RESEARCH ASPIRATIONS

Track 4: Design for Health and Wellbeing
Knowledge Network: Environmental Design Research Education, Researchers in Professional Practice
Methods: Experimental/Quasi-Experimental Research, Practice-based Research
Gail Brager (University of California at Berkeley)

1 ABSTRACT
Much of experiential design research, practice and standards is more attuned to avoiding the negative rather than enhancing the positive. Sensory neutrality is seen as the ideal, where "success" simply means no one complains or even notices the environment. Researchers and designers need to collaboratively eschew experiential monotony, instead striving for experiential delight (DeKay + Brager, 2023)

What knowledge is missing towards this aspiration? This session—2/3 presentation, 1/3 interaction—collectively explores a research roadmap forefronting positive human experiences, laser-focused on practical findings that bridge to design. It aims to generate inspiring dialogue, from students to emerging professionals/researchers to the more seasoned among us. Below are examples where research under-examines what a knowledge-based designer seeks.

Response complexity. The more complex the human response—from sensation, emotion, awareness, to affinity—the less it can be predicted by scientific methods. How can researchers more deeply explore the nature of experiences in buildings, and investigate states well beyond comfort?

Multi-sensory experience. Most research examines singular stimuli in isolation (e.g., thermal, lighting, air quality), neglecting multi-sensory reality. How can interdisciplinary research better investigate stimuli interactions, link perception with behavioral outcomes, and add more explanatory power to findings (Schweiker et al. 2020)?

Time scales. Research commonly focuses on momentary experience, and less on longer-term experience, particularly relevant to residential and workplace settings. How does cumulative exposure to the rhythms of a place impact human response, and contribute to sustained positive experience and overall life quality?

From physiological to psychological. Physiological response is more predictable than perception and emotion. How can research explore the psychologically sophisticated phenomena of happiness, well-being and awe, and how can hedonic experience findings be translated to building design and operation (Altomonte et al. 2020)? Alliesthesia is the physiological basis for thermal delight; are there analogies for other senses? Dynamic environments. Building inhabitants experience dynamic environments when conditions change in a single space, or when they move between spaces; yet these are less well-studied than steady-state conditions. How can research focus more on designing positive experiences for transient conditions in spatial sequences?

Social justice. Everyone deserves experiential delight regardless of relative prosperity of people or place. Both research and design focus too heavily on average or standard populations, paying little attention to neurodiversity or unique needs of vulnerable communities. What drives dissimilarities in perception and, in being more inclusive, how much range can architecture accommodate?

The missing philia. Science confirms the general value of Nature connections but is particularly silent on the philia of biophilia. Beyond stress reduction and attention restoration, how can we better understand how philia can be engendered by design, and what aspects of responses to indoor Nature connections need more study?

2 REFERENCES
Altomonte, S., Allen, J.,... & Wargocki, P. (2020). Ten questions concerning well-being in the built environment. Building and Environment, 180.
DeKay, M. & Brager, G. (2023). Experiential Design Schemas. ORO Editions.

Schweiker, M., Ampatzi, E.,... & Zhang, S. (2020). Review of multidomain approaches to indoor environmental perception and behaviour. Building and Environment, 176.

DESIGN AND ENVIRONMENTAL FACTORS AFFECTING THE BEHAVIORAL HEALTH PATIENT MANAGEMENT IN THE EMERGENCY DEPARTMENT: A SYSTEMATIC LITERATURE REVIEW

Track 4: Design for Health and Wellbeing
Knowledge Network: Environmental Design Research Education, Health
Methods: Meta-Analysis
Intisar Ameen Tyne (Texas State University - San Marcos); Marzia A Chowdhury (The University of Kansas)

1 BACKGROUND

Minimal research has been carried out to identify the design and environmental factors affecting behavioral health (BH) patients' management in the Emergency Departments (EDs), a critical aspect of a safe and healing care environment for this vulnerable population.

2 OBJECTIVE

This systematic literature review (SLR) explores the impacts of the physical environment of EDs in the management of patients with BH presentations.

3 METHOD

The "Preferred Reporting Items for Systematic Reviews and Meta-analyses (PRISMA)" method was adopted by following the PRISMA flowchart, including identification, eligibility, screening, and publications (Page et al., 2021). PubMed, HERD journal, PsycINFO, and MEDLINE databases were used to identify articles by using several keywords together in various combinations, such as emergency department, emergency room, emergency center, emergency service; environmental design, physical design, interior design, facility design, environment, physical environment, built environment, interior; mental health, mental illness, behavioral health, psychiatric patients, and management. The included articles were peer-reviewed articles written in English, spanning from 2000 to March 2023.

4 RESULTS

2489 articles were listed for title screening after the initial search and the removal of duplicates. A total of 61 articles were shortlisted for full-text evaluation. Finally, this SLR included 21 articles that fulfilled all primary inclusion criteria.

5 FINDINGS

The findings of this SLR suggested that only a handful of articles deal with design and environmental factors in managing BH patients in EDs. Based on the review of selected articles, design, and environmental factors are identified in six domains of ED spaces: triage/assessment area, waiting area, treatment area, toilet, nurse station, and overall physical layout. For the BH patients, design and environmental factors have diverse impacts on safety, security, privacy, patient care, patient experience, communication, and family support. The key findings regarding design factors are as follows:

1. *The proximity, accessibility, and visibility of the treatment area from the nurse station, treatment room size, safe/healing environment, and security camera in the treatment room, as well as the overall effects of the ED area, contribute to the sense of security and safety of patients and staff.*
2. *Physically separating patient spaces from the nurse station and the waiting area increased the sense of security and reduced the number of restraints.*
3. *Reduction of noise in the waiting area to create calm environments for BH patients.*
4. *Creating a balance between safety (constant surveillance, reduction of accessibility, restraint, separation of nurse station) and a therapeutic environment (Visibility, openness, quiet space, noise reduction, proximity of nurse station) for BH patient safety and family support.*

6 CONCLUSION

This SLR summarizes the design strategies for EDs in managing patients with BH by creating a secure, safe, and calming environment.

7 REFERENCES

Page, M. J., McKenzie, J. E., Bossuyt, P. M., Boutron, I., Hoffmann, T. C., Mulrow, C. D., Shamseer, L., Tetzlaff, J. M., Akl, E. A., & Brennan, S. E. (2021). The PRISMA 2020 statement: an updated guideline for reporting systematic reviews. International journal of surgery, 88, 105906.

ORTHOPEDIC CLINIC ENVIRONMENTS: A COMPARATIVE STUDY OF WORKFLOW EFFICIENCY AND SATISFACTIONS

Track 4: Design for Health and Wellbeing
Knowledge Network: Environmental Design Research Education, Health
Methods: Interviews, Survey
Mahshid Jalalian (HGA Architects & Engineers); Danya Hakky (HGA Architects & Engineers); Tanya Chadha (HGA Architects & Engineers); Andrew Zimmer (HGA)

1 ABSTRACT

The increasing demand for Minimally invasive orthopedic procedures and services (Gressler et al, 2022) calls for an in-depth understanding of the built environment in which these services are provided. This study presents a comprehensive examination of two orthopedic clinics within the same healthcare system in Eastern Wisconsin. The primary focus of the study was to assess the functionality of the clinics, as well as understanding staff and patient experience within exam rooms and staff work areas at each of the clinics. The research aimed to identify effective practices and challenges in each clinic, ultimately informing the design of a new orthopedic clinic, which is currently under construction and scheduled to open in Fall 2024.

One of the two clinics understudy has a pod layout, with five staff areas each assigned to a single healthcare provider, while the other has a traditional linear model with two staff work areas, each shared between two providers. The size and spatial configuration of the exam rooms were also slightly different between the two clinics, with the pod layout exam rooms 22% larger in size. The research used a mixed-methods approach, including shadowing of nurses, staff, and patient questionnaire, and staff interviews. Shadowing took place in both clinics, where nurse activities and time spent in different areas were recorded for a total of 13 hours.

Staff and patient questionnaires were also distributed to understand satisfactions with different aspects of the clinic design, which resulted in a total of 54 staff and 74 patient responses. Finally, to further understand how the exam rooms facilitated patient care, 20 interviews were conducted with staff including physicians, nurses, and athletic trainers at both clinics. The use of mixed methods uncovered valuable insights into the factors contributing to the success and/or shortcomings of each clinic.

The study found that exam rooms in the clinic with a pod layout were more favored by staff and patients due to several aspects including having two workstations in the room, one for the physician and one for the scriber which allowed clear communication channels among staff, having a printer in the room which streamlined workflows, access to natural light, and the arrangement of seating. The variations in staff work area layout and design between the two clinics also created differences in staff satisfaction with privacy, noise level, ability to concentrate without distractions, and visual access to exam rooms.

The results of this study served as a foundation for evidence-based recommendations for the design of the new orthopedic clinic. By integrating successful elements and addressing the challenges observed in both Clinics, the proposed design seeks to enhance workflow efficiency, improve patient satisfaction, and create a conducive environment for effective orthopedic care. A post-occupancy evaluation will be conducted after the new clinic is constructed to evaluate the influence of design concepts on the efficiency of workflow and satisfaction.

2 REFERENCES

Gressler LE, Devlin V, Jung M, et al. (2022). Orthopedic Coordinated Registry Network (Ortho-CRN): advanced infrastructure for real-world evidence generation. BMJ Surgery, Interventions, & Health Technologies.

INTEGRATING NATURE IN BUILT ENVIRONMENTS EXPLORING VIRTUAL REALITY AND BIOPHILIC ATTRIBUTES IN THE BUILT ENVIRONMENT TO IMPACT SELF-PERCEIVED PAIN THRESHOLDS

Track 4: Design for Health and Wellbeing
Knowledge Network: Environmental Design Research Education, Interior Design
Methods: Experimental/Quasi-Experimental Research
Luis Mejia-Puig (University of Florida); Ximena Paz (University of Florida)

1 ABSTRACT

Creating built environments that positively influence human behavior is fundamental in architectural design disciplines. Designers manipulate spatial attributes like forms, element placement, material selection, color schemes, and textures to enrich how people engage with a given space. Biophilic design, which integrates elements of the natural world into built environments, has the potential to forge a meaningful connection between the space and the individuals inhabiting it (Kellert & Calabrese, 2015). While biophilic design has traditionally been explored through the lens of sustainability (Zhong et al., 2022), this pilot study takes a more intimate human-centered perspective.

Healthcare interior environments have a profound impact on human behavior from a physical and functional standpoint but also through their psychological implications. An illustrative example of this psychological impact is the correlation between pain and anxiety experienced by individuals in dental settings (Lin et al., 2017). A consistent instrument physicians and researchers use to measure pain is the Numeric Pain Rating Scale (NPRS) (Young et al., 2010). The NPRS is a self-reported measure of pain, where participants rate their pain on a scale of 0 to 10, with 0 corresponding to 'no pain' and 10 implying 'worst possible pain.' This instrument, combined with a physiological pressure algometer, can measure the pain thresholds of individuals under controlled conditions (Mutlu & Ozdincler, 2015). This study explored how biophilic attributes in a simulated interior healthcare environment affected self-perceived pain thresholds.

This study created four distinct virtual reality (VR) spaces, each with varying degrees of biophilic attributes, to influence participants' self-perceived pain. All four rooms were designed as healthcare interior environments representing hospital rooms, where the psychological connections between anxiety and pain can parallel those of a dentistry setting. VR is a potent tool for replicating real-world scenarios while permitting controlled manipulations with varying levels of immersion.

A convenient sample of thirteen participants (n=13) was randomly assigned the order in which they experienced the four VR rooms in a within-subject experimental design. To mitigate potential VR interaction biases, participants underwent a training session in a white lobby area before entering any of the rooms. Pre- and post-experiment questionnaires were administered to collect demographic information and gather spatial perceptions of the environments. Ahead of the VR experience, participants were given a printed handout of the NPRS to acquaint themselves with the pain rating scale. While immersed in VR, a pressure algometer recorded self-perceived pain thresholds both in the lobby area and within the four differently designed rooms.

Statistical analysis of the acquired data affirmed the relationship between the NPRS instrument and measurements obtained from the pressure algometer. Furthermore, the varying levels of biophilic elements across the rooms were found to impact participants' perceived pain levels. Statistically significant differences in self-perceived pain thresholds indicated that biophilic features enabled individuals to endure higher levels of pressure from the algometer.

Beyond the implications for design and the built environment, this study unveils the exciting prospect of leveraging VR and virtual interior design to enhance medical therapy.

RECLAIMING GREEN OPEN SPACES IN BOSTON CHINATOWN

Track 4: Design for Health and Wellbeing
Knowledge Network: Environmental Design Research Education, Health
Methods: Mixed Methods
Lily Song (Northeastern University); Zorana Matic (College of Arts, Media & Design | Northeastern University)

1 ABSTRACT

Boston Chinatown is a vital residential neighborhood, commercial district, and cultural hub that has persevered against racial exclusion, infrastructural injustice, institutional expansion, and gentrification. Situated between the Boston common, downtown commercial district, South Station, and expansive Tufts medical-educational complex, and cut through by heavily-trafficked streets and major highways, Chinatown has the lowest air quality, tree canopy coverage, and green, open space per capita among Boston neighborhoods. It also has an active presence of non-profit and civic organizations that have undertaken several rounds of community master planning to articulate and respond to collective concerns as well as clarify shared priorities to guide the neighborhood's continued changes and growth.

The importance of improving parks and open space has remained a key priority for Chinatown community members and advocates, and more recently catalyzed Chinatown Hope, a collective of eight organizations that are working together to activate and improve neighborhood open spaces. The open space initiative incorporates gardening, arts, cultural programming, and resident leadership development, calls for more city and state investment, and builds urban design collaborations with local professionals and firms. While previous research has focused on the environmental injustices and health risks faced by Chinatown residents, our study investigates the active usage of neighborhood open spaces both by community members and outside visitors, in part to inform and amplify community-led open space initiatives.
The paper examines initial results from the study of seven Chinatown open spaces—Mary Soo Hoo Park, Chin Park on the Greenway, Phillips Square, Reggie Wong Park, Tai Tung Park, Eliot Norton Park, and One Greenway Park. This includes: (1) spatial maps of each site's physical features and qualities (i.e. sidewalks, edges, and paths; trees, shrubs, bushes, grass; seating, trash bins, lighting; outdoor furniture; and topography); (2) area maps of quarter-mile radius surrounding contexts (i.e. land uses, streets, public infrastructures, and major landmarks); (3) behavioral maps of activities and use patterns at the parks (i.e. necessary, optional, social uses; perceived demographics and characteristics of users; active times and spaces within site); and (4) surveys of Chinatown residents and neighborhood visitors on perceptions, experiences, and suggestions related to neighborhood parks, streets, and other open and shared spaces.

The analysis focuses on the relationship between the physical features and qualities of the Chinatown open spaces, surrounding land uses, and usage and behavioral patterns, and considers similarities and differences between the sites. The concluding section discusses the function and significance of Chinatown open spaces to community preservation in the face of gentrification and displacement pressures, growing heat waves, and other climate and public health emergencies, along with implications for community-led activation and redesign.

THE HEALTH IMPACTS OF NATURAL INTERIOR MATERIALS - DEVELOPING AN EVALUATION FRAMEWORK FOR NATURAL MATERIALS IN INTERIOR DESIGN APPLICATIONS: HEALTH, WELLBEING, AND EMBODIED CARBON

Track 4: Design for Health and Wellbeing
Knowledge Network: Environmental Design Research Education, Interior Design
Methods: Interviews, Mixed Methods
Alivia Green (Atelier Ten)

1 ABSTRACT

The built environment impacts the world around us on multiple levels. Of these impacts, concerns include the health of the environment and human health. As definitions of health evolve, we've come to recognize wellbeing as an aspect of overall health, which is also affected by the built environment. Within the built environment, a focus has been placed on materials as a factor that imperatively affects the environment and human health within buildings. Previous evidence has shown that natural materials (NM) can have lower embodied carbon (EC) which impacts the environment through greenhouse gases, and lower indoor air emissions, which can negatively impact the physical health of occupants. However, it is suggested that perception of these materials, through their physical appearance, can also impact mental health and wellbeing through their sensorial properties by way of biophilia. While there is accessible data on the ways in which materials can affect health, it can be challenging for designers and built environment professionals to holistically evaluate materials for their health impacts.

Therefore, this research aims to investigate how NM, within building interiors, can be categorized based on a criteria including human health impacts and embodied carbon content, to develop a framework to assist industry professionals in selecting natural interior materials which positively impact the environment and occupant health. Four material databases were searched for NM which had low EC and low VOCs, to then be analysed for biophilic attributes using a newly developed audit based on biophilic design frameworks. Seven designers and material experts were then interviewed to determine the decision-making process of material selection and to investigate the current understanding of how professionals consider materials for their influences on health. Results from material analysis showed that NM can be scored for their biophilic attributes, and those with minimal processing yielded higher overall scores. Interview findings showed that most participants did not consider biophilia or sensorial properties when selecting materials, and many were not aware of metrics to account for indoor air emissions. All findings were synthesized to create a framework for selecting NM based on their health impacts, which was then represented by a label to graphically communicate a material's biophilic score, and all other factors included in the framework. This study and the proposed framework/label could be the basis for further investigation surrounding NM and their impacts on health, and how these factors can be better integrated into the decision-making process of built environment professionals. Further research is needed on how to account for all material selection drivers within the label, which also requires further testing with clients, building occupants, and designers to evaluate its robustness and practical impact.

ASSESSING THE EFFECTS OF NATURAL VIEWS AND SCENTS ON WORKPLACE PERFORMANCE AND PRODUCTIVITY MEASURES

Track 4: Design for Health and Wellbeing
Knowledge Network: Environmental Design Research Education, Work Environments
Methods: Experimental/Quasi-Experimental Research, Survey
Sarayu Chandramouli (Texas A&M University); Suma Katabattuni (Rekovar Inc.,); Marco A Palma (Texas A&M University); Jay E Maddock (Texas A&M University)

1 ABSTRACT

In the pursuit of "Design for Health and Wellbeing", this study combines architectural concepts from a public health perspective to investigate the dynamic relationship between the built environment and human cognitive wellbeing. Rooted in the Attention Restoration Theory (ART), Stress Reduction Theory (SRT), and the emerging concept of "smellscape," our research explores the combined effects of visual and olfactory elements of nature within controlled office environments. The experiment is based on the well-evidenced fact that time spent in nature improves mental, physical, and psychological wellbeing, combined with the idea that nature in office spaces improves performance.

Although nature is a multisensory experience, the literature emphasizes much on nature views. This paper presents a subset of a study that explores nature views, scents, and the combined effects of views and scents. The incorporation of smellscape theory adds a novel dimension to this investigation, recognizing the significant role of olfactory stimuli in shaping individuals' perceptions and experiences of space. This extended abstract focuses on the combination of views and scents in office workspaces. The hypotheses posit that the integration of views of nature and natural fragrances, as part of the smellscape, yields a synergistic enhancement in cognitive performance required for work performance. Employing a between-subjects 2x2 experimental study design, the experiments were carried out through O-Tree, an open-sourced coding platform, and iMotions 9.3, a neurophysiological eye-tracking software to develop an online environment. The meticulously designed approach delves into uncharted territory, examining how the fusion of sensory nature experiences and smellscape influences stress levels, memory retention, attention, abstract intelligence, risk-taking tendencies, and attitudes toward cheating. The sample consisted of 128 students to resemble the young working adult population, recruited from Texas A&M University System, who were asked to come to the lab during the span of the study.

The results of this paper indicate a significant difference in task performance and positive emotional dimensions in participants with increased nature content, more so with multiple stimuli. The eye tracking data shows increased positive emotions, and positive facial expressions in condition 1 (views) and condition 3 (both). The eye tracking also suggests increased attention and engagement across all three conditions, followed by surprise and frustration. The exception to this positive state was prominent for the Stress test where, despite the condition, contempt was identified multiple times as a leading expression.

By selecting cognitive tasks mirroring real-world office tasks, the study not only deepens the understanding of the intricate relationship between nature, smellscape, and cognition but also provides actionable insights for anyone aiming to craft spaces that holistically support their occupants. On a smaller scale, the findings may also encourage more people to try and incorporate nature into their own personal work areas. This study contributes to understanding how nature views and scents influence cognitive wellbeing, offering insights for architects designing spaces to support occupants' holistic wellbeing and productivity, and users to incorporate nature into their workspaces without too much investment.

2 KEYWORDS

Biophilia, Cognition, Healing Environments, Health outcomes, Nature connectedness, Nature Scents, Nature Views, Offices.

URBAN PLACES FOR HEALTH AND WELL-BEING TOWARDS AN AGE-INTEGRATED LIFELONG LEARNING COMMUNITY ON A UNIVERSITY CAMPUS

Track 4: Design for Health and Wellbeing
Knowledge Network: Active Living by Design, Environment-Gerontology
Methods: Action/Participatory Research, Focus Groups
Brian Schermer (University of Wisconsin-Milwaukee)

1 ABSTRACT

Addressing the future housing needs of the aging population poses a paramount challenge for architects and planners, necessitating a paradigm shift in spatial considerations, building codes, services, and community engagement. This imperative calls for inventive programs and architectural solutions grounded in a profound understanding of the human condition, leveraging the insights and expertise of those aspiring to lead enriched lives.

The Living Well Initiative, a collaborative effort involving Milwaukee-area architects, developers, and care providers, aims to inspire the next generation of architects and designers to envision age-integrated communities that enable individuals to pursue purposeful and meaningful lives as they age.

This session will detail how the studio explored innovative models of university-affiliated senior communities tailored to resonate with the baby boomer generation. The University of Wisconsin-Milwaukee's urban campus served as a dynamic test-bed, allowing students to create age-integrated communities promoting active living and meaningful pursuits for this aging demographic. Integral to the studio learning environment was participatory social research involving lifelong learners. Collaborating with the UWM Osher Center for Lifelong Learning, UWM Retirees Association, and the Shorewood Senior Resource Center, students gained valuable insights into housing preferences through workshops and design review sessions. This inclusive approach facilitated feedback on concepts, forms, relationships, and aesthetics, creating a reciprocal exchange between designers and potential residents.

Design solutions crafted during the studio were submitted to an international housing design competition, focusing on mitigating the isolation experienced by seniors. The students developed designs for three sites on the campus, with each offering unique opportunities to integrate senior living with campus life. The first leveraged proximity to the university library to create an academically focused cohousing hub community. The second leveraged proximity to student residence halls, a campus daycare, and a nearby grade school to create an intergenerational crossroads. The third created a mix of housing that created stronger connections to the surrounding neighborhood. The overarching goal was to develop innovative solutions that integrate seniors into the social fabric of their communities. The session will showcase diverse design solutions and discuss plans for further exploration in future design studios, emphasizing the evolution of current solutions and their potential impact on shaping age-integrated lifelong learning communities.

NEIGHBORHOOD STREET ENVIRONMENTS AND THE DURATION OF OLDER ADULTS ACTIVE TRAVEL - CASE STUDY IN DAOKOU ANCIENT TOWN

Track 4: Design for Health and Wellbeing
Knowledge Network: Active Living by Design, Environment-Gerontology
Methods: Case Study, Observation
Zhe Wang (Henan University); Guoxiang Li (Henan University)

1 INTRODUCTION

Neighborhood streets are convenient places for older adults to engage in behaviors for active living, such as walking around and meeting people (active travel). Older pedestrians' active travel and the built environments along neighborhood streets in ancient towns need investigation. Taking Daokou ancient town in China as an example, this research observed older pedestrians travel on three neighborhood streets and investigated the differences in travel duration by street.

2 METHODS

On-site non-participant observation was conducted for four weekdays with seven 30-minute sections per day. Street environmental factors were measured and classified into four categories in terms of active-travel promotion for older adults: typology, motivators, functionality, and safety. Focusing on the pedestrians with perceived age of 60 years or older, researchers recorded 799 independent active travels on these streets (411 on the D Street, 202 on the SN Street, and 186 on the K street). To identify the difference in travel duration by street, one-way ANOVA tests were conducted and four significant personal or social confounding variables (age, health, travel arrangement, and daypart) were controlled in the tests to clarify environmental influence on behavior.
Results: Older pedestrians' active travel duration significantly varied by neighborhood street. Personal, social, and environmental factors influenced the duration. In the group of healthy pedestrians in their 60s, for both females and males, irrespective of their walking arrangements, active travel durations were significantly longer on the streets with more age-friendly features ($p<=0.01$). However, the differences were no longer statistically significant in the pedestrians aged 70s or older and those without a very good health situation. Older pedestrians who had an accompaniment or engaged in social interactions during walking had significantly longer travel durations than those who walked alone ($p<0.01$).

3 CONCLUSION

Neighborhood street environments may influence older adults' active travel. Highlighting the importance of older adults' active living in ancient towns, our findings can be used to create street affordances for older pedestrians' active travel, including both walking and social interaction, and produce healthy outcomes through the refinement of design and transportation policies and practice on street intervention.

TOWARDS HUMAN-CENTERED URBAN SAFETY: DESIGNING URBAN ENVIRONMENTS TO PREVENT MOTORCYCLE ACCIDENTS

Track 4: Design for Health and Wellbeing
Knowledge Network: Active Living by Design, Health
Methods: Case Study, Statistical/Modeling Analysis

Yujin Kim (KAIST); Lisa Lim (KAIST); Byeongjoon Noh (Soonchunhyang University); Hwasoo Yeo (Korea Advanced Institute of Science and Technology)

1 INTRODUCTION

Recently, the use of online-to-offline (O2O) food delivery services using two-wheeled vehicles has been rapidly increasing in urban environments around the world. While these services provide convenience to city dwellers, the increasing number of two-wheeler accidents poses an increasing threat to citizen safety. According to a report by the National Highway Traffic Safety Administration (NHTSA), the typical fatality rate for motorcycle accidents is approximately 34 times higher than that for automobile accidents (Chung et al., 2014). Therefore, it is urgent to prepare measures to reduce damage caused by motorcycle accidents in order to ensure the safety of city residents. In this study, we focused on ways to reduce damage from the perspective of urban environment design.

This study explores the non-linear relationship between two-wheeler accidents and the urban environment and analyzes how specific urban and demographic factors are closely linked to these accidents. The main questions of the study are: What is the relationship between two-wheeler accidents and the built environment of the urban environment? Under what environmental conditions can two-wheeler accidents in cities be reduced? And how can we design people-friendly roads and urban environments?
This study targeted Seoul metropolitan city in South Korea, adopting land use, demographic information, and traffic environment variables as independent variables, and setting the number of motorcycle accidents as the dependent variable. To investigate the complex interaction between the urban landscape and motorcycle accidents from a pedestrian's perspective, various urban elements were extracted and analyzed through street view. In addition, we explored how the layout and use of urban space, including residential and commercial areas, contribute to the frequency and severity of motorcycle accidents. In terms of demographics, we found that older people (in their 60s) and younger people (teenagers) were particularly vulnerable to accidents.

The study also suggests concrete measures to reduce two-wheeler accidents by strengthening urban design principles and safety measures. It includes road designs to improve interaction between pedestrians and two-wheeler drivers in the city, and road environment proposals for seniors and youth. The results of this study are expected to provide urban planners and policymakers with important information and insights needed to reduce motorcycle accidents in cities and improve the safety of city residents.

In conclusion, this study presents a strategic approach to reduce motorcycle accidents in urban areas, which improves the quality of life of city residents and addresses land use and demographic factors to create a safer and more sustainable urban environment. It also highlights the importance of a well-considered and integrated urban planning and policy strategy. Our study is expected to provide practical guidelines for designing safer urban environments, which will not only improve pedestrian safety for city residents but also contribute to improving the overall quality of urban life.

3 REFERENCES

Chung, Y., Song, T.-J., & Yoon, B.-J. (2014). Injury severity in delivery-motorcycle to vehicle crashes in the Seoul metropolitan area. Accident Analysis & Prevention, 62, 79-86. https://doi.org/https://doi.org/10.1016/j.aap.2013.08.024

ARCHITECTURAL DESIGN GUIDELINES ON COMPREHENSIVE CHILD SUPPORT CENTERS IN JAPAN

Track 4: Design for Health and Wellbeing
Knowledge Network: Children, Youth & Environment, Health
Methods: Mixed Methods, Survey
Ayako Naka (Toyo University)

1 ABSTRACT

Suicide is a significant cause of death among pregnant women and nursing mothers in Japan (Mori, 2017). To combat this, the Japanese government introduced Comprehensive Child Support Centers to extend support through pregnancy, birth, and child-rearing. These centers are modeled after "neuvola" (maternity and child health clinics) in Finland (Yokoyama & Hakulinen, 2018). The centers have been rapidly developed in Japan. According to the latest data, as of 2022, 2,486 centers are open to the public. However, few studies have examined the centers. No specific guidance exists regarding how the centers' facilities should be designed.

This study aimed to examine design guidelines of comprehensive child support centers from an architectural planning point of view.
To clarify the actual condition and requirements for the centers' facilities, we surveyed comprehensive child support centers throughout Japan and obtained data from 611 centers. Our questionnaire is the most extensive survey in this field in sample size and survey area. Statistical analysis and text mining were applied to the data.

The results showed that 62.4% of centers have consultation rooms, 44.0% have playrooms, and 43.7% have breastfeeding rooms. A few centers have nurseries, medical examination rooms, Japanese Tatami (multipurpose) rooms, lunchrooms, kitchens, libraries, workshops, storage areas, and other rooms. The average area of each room is 10.1 square meters (consultation rooms), 51.7 square meters (playrooms), and 5.0 square meters (breastfeeding rooms). Other rooms vary from 2 to 134 square meters.

Based on the text mining, we obtained 142 planning methods and organized them according to four major categories and 18 subcategories. The major categories are 1) site planning, 2) mechanical and electronic facility management, 3) architectural design for each room, and 4) other. Examples of the planning methods are: 1) accessibility is indispensable for users with babies in strollers. At the same time, some users require privacy protection, because they have serious problems and are nervous to be seen by others. 2) Ventilation, air conditioning, lighting, and soundproofing should be carefully considered. 3) Consultation rooms should have windows to make users feel relieved. Restrooms should include small toilets for children. Entrances should be located near the office to monitor visitors. 4) Every corner should be rounded to prevent serious injuries. Semi-outdoor spaces, such as terraces, enable users to appreciate fresh air and nature.

At the end of the design guidelines on the comprehensive child support centers, we provide a diagram to visually illustrate the planning methods.
In conclusion, consultation rooms, playrooms, and breastfeeding rooms are the three principal rooms of the centers that can support mothers, fathers, and babies. Other rooms, such as nurseries, could expand the centers' services. The guidelines we organized could be applied to the design of future comprehensive child support centers. However, further studies are required to evaluate their effectiveness.

2 REFERENCES

Mori R. (2017). Linkage based on Perinatal Medical Data Base. https://mhlw-grants.niph.go.jp/project/26364

Yokoyama Y. & Hakulinen T. (2018). Method of maternal and child health learning from "neuvola (maternity and child health clinics)" in Finland. Ishiyaku Publishers.

DESIGNING INCLUSIVE LIVING ENVIRONMENT - ENHANCING THE BOARDING EXPERIENCE FOR LEFT-BEHIND CHILDREN IN RURAL CHINA

Track 4: Design for Health and Wellbeing
Knowledge Network: Children, Youth & Environment, Residential Environments
Methods: Case Study, Interviews
Tianhong Yan (Deakin University)

1 BACKGROUND

China has one of the largest numbers of left-behind children in the world. Since 2004, the Chinese government has initiated an ambitious plan to use boarding schools as a mechanism to accommodate these young people from rural areas and to provide them with high-quality basic education. Results from the literature review indicate that there is a lack of research focused on understanding the school living environment these young people inhabit. Data from examining the school living environment suggests that planning and designing inclusive left-behind children-oriented accommodation buildings can significantly affect their well-being, academic performance, and overall boarding experience.

2 PURPOSE AND METHOD

The purpose of this study is to understand the physical and cultural environment of the school's boarding accommodation where left-behind children reside, and the life of these adolescents aged between 12 and 15 years old. As part of a broader study on the living environment of left-behind children in rural China, this research was conducted in a rural junior high school in Qichun County, Hubei Province. The qualitative data of this study are derived from visual materials supplied by school teachers (i.e., images), online interviews with 15 teachers, 5 school leaders, and 5 parents of left-behind children. Furthermore, the study has compiled diary entries from 10 child participants. These qualitative data sets support developing an understanding of how boarding school living environments can be designed for adolescent well-being. Subsequently, suggestions for the design of the accommodation buildings for this particular junior high school are proposed, based on analysis of the data, aiming to create an enhanced boarding experience for the left-behind children. This research focuses on the physical elements of this research project.

3 FINDINGS

This study reveals that optimizing the physical structure design of the boarding accommodation (e.g., increasing the area of corridor spaces) and implementing an open spatial planning and redevelopment philosophy (e.g., reducing partitions between different functional areas) can provide satisfactory living and activity spaces for students. These strategies appear to support adolescents in actively utilizing various functions within the accommodation buildings and facilitate communication and interaction with their peers. However, for the left-behind children at this boarding school, there also appears to be a heightened need for regularly scheduled indoor and outdoor environmental design activities (e.g., painting of murals). The research suggests that these design and cultural activities may offer them opportunities to express and reflect on their rural backgrounds.

4 SIGNIFICANCE

The significance of this study is manifested in two aspects. First, school accommodation buildings are designed to support collective living and facilitate the management of these young people. This model, however, often diminishes and obscures opportunities and desires for adolescents to use architecture as a means of expressing their individual backgrounds. Therefore, for left-behind children, this study provided them with a mechanism to reflect on and build in home-related 'things' into their living environment. Moreover, the future design process of accommodation buildings should be more influenced by those children who will live there, rather than solely by school authorities and designers.

IMPACT OF CLASSROOM ILLUMINANCE LEVELS ON THE HEART RATES OF NEUROTYPICAL AND NEURODIVERSE STUDENTS

Track 4: Design for Health and Wellbeing
Knowledge Network: Children, Youth & Environment, Environmental Design Research Education
Methods: Experimental/Quasi-Experimental Research, Mixed Methods
Rudy Dieudonne (University of Wisconsin - Madison)

1 ABSTRACT

The current interior design guidelines for classroom lighting, which specify the minimum and maximum illuminance levels of primary work surfaces, were developed based on the neurological needs of students who are neurotypical. Yet, as the number of children with sensory sensitivities and/or autism increases, this equates to an increase in the population of neurodiverse students. Given that many neurodiverse students have difficulties processing forms of sensory stimulation such as interior lighting (Sadia, 2020). Research is now needed to determine how the current design standards for classroom lighting are impacting this growing population of students.

This study's objective is to examine the influence that the standard classroom illuminance levels have on the heart rates (HR) of neurodiverse students. The HR of an individual is one of many reliable indicators of mental stress and discomfort (Taelman et al., 2009). Ninety-nine children were recruited to participate in this research study. These participants were placed into two groups: children without sensory sensitivities or autism (neurotypical) and children with sensory sensitivities and/or autism (neurodiverse). The standard classroom lighting condition for primary work surfaces is 300-500 lux (Dilaura et al., 2011). The objective of this study was achieved by documenting the HR of the two groups, while they engaged in identical activities under a lighting condition of 250 lux, 300 lux, 400 lux, and 500 lux.

This study proposes two hypotheses. If the standard classroom illuminance levels are impacting neurodiverse students, then the mean HR of this group will elevate once they are exposed to the standard illuminance levels. In addition, if the mean HR of the neurodiverse group elevates when exposed to standard classroom illuminance levels, then when compared to the neurotypical group, the mean HR of neurodiverse students will show the largest increase. To test the first hypothesis, a paired t-test was performed on the mean HR values of the neurodiverse group. The results showed that the mean HR of this group elevated once the group was exposed to the standard classroom illuminance levels; consequently, the first hypothesis was accepted. To test the second hypothesis, an independent t-test was performed on the mean HR values of the two groups. When comparing the mean HR of both groups, the results showed that the mean HR of the neurodiverse group had a larger increase after being exposed to the standard classroom illuminance levels; consequently, the second hypothesis was accepted.

The results of this research could inform the revision of current standards for classroom lighting. In addition, these results are potentially relevant to research within the fields of education, neurology, design, behavior analysis and many more.

2 REFERENCES

Dilaura, D., Houser, K. W., & Mistrick, R. G. (2011). The lighting handbook: reference and application. Illuminating Engineering Society.

Sadia, T. (2020). Exploring the Design Preferences of Neurodivergent Populations for Quiet Spaces. https://doi.org/10.31224/osf.io/fkaqj

Taelman, J., Vandeput, S., Spaepen, A., & Van Huffel, S. (2009). Influence of Mental Stress on Heart Rate and Heart Rate Variability. IFMBE Proceedings, 22, 1366–1369. https://doi.org/10.1007/978-3-540-89208-3_324

DESIGN FOR HEALTH IN RURAL AND POST-PANDEMIC SETTINGS: A NATIONWIDE SURVEY ON EMERGENCY DEPARTMENTS (EDS) PANDEMIC RESPONSES

Track 4: Design for Health and Wellbeing
Knowledge Network: Environmental Design Research Education, Health
Methods: Mixed Methods

Marzia A Chowdhury (The University of Kansas); Hui Cai (University of Kansas School of Architecture & Design); Frank Zilm (The University of Kansas); James J Augustine (ED Benchmarking Alliance)

1 BACKGROUND

The prolonged COVID-19 pandemic created multiple surge demands at various phases of the pandemic. Pandemic responses vary across different EDs. However, there is little research on the connection of the built environment in response to adopted strategies.

2 OBJECTIVES

This study aims to evaluate pandemic preparedness approaches from a facility design perspective for different periods and identify their relationship to EDs' characteristics.

3 METHODS

A qualitative research method has been conducted, including a systematic literature review (SLR) to understand ED design strategies and a nationwide survey to evaluate the strategies during different periods of the pandemic. The survey was distributed through the EDBA (Emergency Department Benchmark Alliance) from September 2022 to October 2022. The survey asked for information on strategies to address pandemic needs in their EDs. A descriptive statistic and Chi-squared test of independence were conducted to evaluate what design strategies various EDs in the United States have implemented to respond to the COVID-19 pandemic.

4 RESULTS

Out of the 130 respondents for the EDBA survey, 57 completed all sections and were included in the analysis. The most adopted strategies are —a) negative pressure room (95%, n=54/57), b) cohort isolation (84%, n=48/57), and c) separate pre-triage or screening area (84%, n=47/57). Almost half (48%) of the strategies were implemented at the onset, whereas only 17% were implemented before the outbreak. The statistical analysis revealed significant differences between types of EDs and the design strategies they have adopted in response to the pandemic. For example, academic EDs are associated with a higher chance of adopting separate triage at twice the rate (1, N = 57) = 7.75, p = 0.004; separate patient flow X2 (1, N = 57) = 5.53, p = 0.019; and separate dedicated donning and doffing areas X2 (1, N = 57) = 3.989, p = 0.046. Hospital size is also significantly associated with the likelihood of adopting separate screening areas, X2 (1, N = 57) = 4.236, p = 0.04.

5 CONCLUSIONS

This study identified key design strategies that could support ED operations during pandemics. It also revealed that few EDs had built-in design features that can support their pandemic responses. Smaller, non-academic EDs face bigger challenges to implementing split flow design strategies to protect patient and staff safety. Future ED designs should consider including more built-in design to allow flexible and adaptive use of spaces within and around the ED.

6 ACKNOWLEDGMENT

EDRA for the contribution of data and expertise. Funding for the design of emergency department pandemic responses from the Foundation for Health Environments Research (Principal Investigator: Hui Cai, Ph.D.), Joseph G. Sprague New Investigator Award (NIA), Griffin/McKahan/Zilm (GMZ) Fellowship.

EFFECT OF ENVIRONMENTAL FACTORS ON HEALTHCARE WORKERS' WELL-BEING AND SAFETY IN THE EMERGENCY DEPARTMENT DURING COVID-19: A QUALITATIVE ANALYSIS

Track 4: Design for Health and Wellbeing
Knowledge Network: Environmental Design Research Education, Health
Methods: Focus Groups
Marzia A Chowdhury (The University of Kansas); Hui Cai (University of Kansas School of Architecture & Design)

1 BACKGROUND

The COVID-19 pandemic has impacted the well-being of healthcare workers (HCWs). It has also raised concerns about infection risk in the Emergency Department (ED) (Sheehan et al., 2021).

2 AIM

This empirical study aims to report on the EDs' environmental features that the HCWs felt were important to work during the COVID-19 pandemic in the ED for their well-being and safety.

3 METHOD

Between 2022 and 2023, an on-site walk-through and semi-structured focus group interviews were conducted at five emergency departments, including academic and non-academic hospitals. The interviews were recorded and transcribed by Microsoft Word. Thematic analysis was conducted using Maxqoda.

4 FINDINGS

The interviews revealed three common environmental components—safety features, respite areas, and work support/amenities that impact the staff's well-being, safety, and support for their work.
Staff identified several design features that would affect safe patient care delivery, staff safety, and their perceived infection risk, which included a) dedicated donning and doffing areas in the ED before entering the COVID-19 hot zone, b) separate patient flow upon patient arrival to departure in separate spaces, c) presence of support spaces (equipment, medication, supply, clean, and soil space, break room) within the hot zone, and d) separate toilet for staff and patient. Staff also endorsed a mix of two break areas—a quick break area close to the nurse station and a respite area or escape room adjacent to the ED with auditory detachment to promote taking necessary breaks. Staff also stated preference for a few features of the respite area: a) access/ view to nature, b) quiet space, and c) away from ED noise.

Snack/eating areas, private/quiet/sleep spaces, lactation spaces, and shower facilities were identified as essential features in work environments.

5 CONCLUSION

For HCWs' well-being, break areas that are proximal to the nurse station and adjacent to the ED with access to nature and private space are important factors in promoting break-taking behavior. Critical safety features such as proper donning and doffing space adjacent to the COVID-19 hot zone with adequate support areas (clean and soiled spaces, separate toilet) for COVID-19 patients upon arrival to departure were important in addressing their concerns about perceived infection risks.

6 ACKNOWLEDGMENT

Funding from Foundation for Health Environments Research (Principal Investigator: Hui Cai, Ph.D.), Joseph G. Sprague New Investigator Award (NIA), Griffin/ McKahan/Zilm (GMZ) Fellowship

7 REFERENCES

Sheehan, O., Sheehan, M., Rau, R. I., Sullivan, I. O., McMahon, G., & Payne, A. (2021). Burnout on the frontline: the impact of COVID-19 on emergency department staff wellbeing. Irish Journal of Medical Science (1971-), 1-9.

Valipoor, S., Bosch, S. J., & Chiu, L. Y. T. (2023). From Stressful to Mindful: Reactions to a Proposed Emergency Department Design for Enhancing Mindfulness and Stress Reduction Among Healthcare Clinical Staff. HERD: Health Environments Research & Design Journal, 19375867231172222.

Ziabari, S. M. Z., Andalib, E., Faghani, M., Roodsari, N. N., Arzhangi, N., Khesht-Masjedi, M. F., & Leyli, E. K. (2023). Evidence-based design in the hospital environment: A staff's burnout study in the COVID-19 era. HERD: Health Environments Research & Design Journal, 16(2), 236-249.

RURAL PARK ACCESSIBILITY AS A DETERMINANT OF RURAL HEALTH

Track 4: Design for Health and Wellbeing
Knowledge Network: Health, Movement in Designed Environments: Nature & Ecology
Methods: Survey
Hans Klein-Hewett (Iowa State University Landscape Architecture Department)

1 BACKGROUND

Health outcomes of rural Americans routinely rank lower than those of their urban counterparts, specifically in the key morbidity indicators of cancer, obesity, and heart disease (Martin et al., 2006). While infrastructural limitations like a lack of healthcare facilities, limited access to broadband internet, and reliable access to healthy foods compound these poor health outcomes, rural parks have been shown to improve health outcomes for rural residents (Graves, et al, 2021). This is because rural parks are often the only low- or no-cost recreational options available in many rural areas, making them critical facilities for rural health and wellbeing (Edwards & Matarrita-Cascante, 2011).

To date, most scholarship on rural parks has been completed by public health scholars who focus on physical activity and the correlating factors of demographics and socioeconomics. However, few researchers are studying the spatial and functional implications of rural parks, and thus are not questioning where rural parks are located, how rural people access those parks, and what qualities in those parks impact physical health.

This presentation aims to fill that gap by discussing the findings from a rural park accessibility survey sent to 1,000 rural Iowans during the summer of 2023. The survey aimed to better understand whether rural residents recreate in public outdoor recreation facilities, and if so, where they recreate, how far they are willing to drive to access those places, what factors cause them to select those locations, and how frequently they visit. The results, which feature responses from 155 rural residents (a 15.5% response rate), will be presented based on the community type of the respondents: unincorporated areas, small towns, small cities, and urban areas. The study suggests that residents in unincorporated areas are willing to and do drive farther to access rural recreation opportunities, though the factors that motivate their decision on where to recreate are different from those who live in smaller towns and cities.

2 REFERENCES

Edwards, M. B., & Matarrita-Cascante, D. (2011). Rurality in leisure research: A review of four major journals. Journal of Leisure Research, 43(4), 447–474.

Graves, Janessa M., Demetrius A. Abshire, Solmaz Amiri, Jessica L. Mackelprang. (2021). Disparities in Technology and Broadband Internet Access across Rurality: Implications for Health and Education. Family Community Health, 44(4) 257-265.

Martin, Sarah Levin, Gregory J. Kirkner, Kelly Mayo, Charles E. Matthews, J. Larry Durstine, and James R. Herbert. (2006). Urban, Rural, and Regional Variations in Physical Activity. The Journal of Rural Health, 21(3), 239-244.

LEARNING FROM OUR BUILDINGS: A MULTI-METHOD FUNCTIONAL PERFORMANCE EVALUATION OF A STATE-OF-THE-ART OUTPATIENT ORTHOPEDIC CENTER OF EXCELLENCE

Track 4: Design for Health and Wellbeing
Knowledge Network: Health, POE/Programming
Methods: Observation, Practice-based Research
Rutali Joshi (HKS Inc.); Deborah Wingler (HKS); Sara Kennedy (Clemson University)

1 OVERVIEW

An evidence-based approach was used to create the ultimate patient and staff experience, as well as improve operational efficiency and flows (people, equipment, and instruments) in the six-story, 180,000-square-foot, outpatient orthopedic facility in Brookhaven, Georgia. The facility housed 117 exam rooms, four ambulatory surgery suites, clinical labs, clinical and research spaces, physical therapy, and imaging together under one roof. The design team conducted a Design Diagnostic of existing clinics within the health system in the pre-design phase. The team also worked closely with an academic partner to conduct a scenario-based simulation, which is a high-fidelity physical mockup that largely informed the design of the operating room (OR). The intent of the design decisions was clearly documented along with desired outcomes during the design process based on the research and best available evidence to set the foundation for a future occupancy evaluation. In collaboration between the architecture firm and an academic partner, a Functional Performance Evaluation (FPE) of the clinics and surgical suite at the facility was conducted in 2022 to understand whether the facility performs as originally intended.

2 METHODS

A 4-day site visit was conducted to collect extensive on-site data through guided tours of the clinics and surgical suite, shadowing of key clinical personnel, photo essays, and semi-structured interviews with staff. In-person observations in the clinic entailed shadowing a clinic nurse and observations in the ORs entailed observing the surgeon, the anesthesiologist, and the scrub nurse. Surgical flow disruptions were observed using a consistent template across six orthopedic surgeries in two mirrored ORs. Surveys rated on a 5-point Likert scale were administered to measure staff perception regarding five constructs: safety, efficiency, experience, access, and flexibility.

3 RESULTS

Findings from the Functional Performance Evaluation suggest that the size and location of key spaces within the clinic help to support increased efficiency and optimized flow of patients and staff, as well as an enhanced patient and staff experience. Lack of physical separation between clinical workspace and patient care areas leaves staff feeling visually exposed and frequently interrupted. Wayfinding and staff safety were identified as opportunities for improvement. Within the surgical suite, the operating rooms were found to support increased efficiency and optimized flow, as well as substantially minimize flow disruptions within the anesthesia work zone and supply area as compared to precedent projects.

4 PRACTICAL IMPLICATIONS

Insights from this evaluation have helped inform corrective measures for quality improvement to existing processes, and as a guide for informing new capital investments. Additionally, findings from FPE contribute to a broader understanding of best practices and lessons learned that can be leveraged to support the continuing development of design guidelines and regulatory processes for the health care industry at large.

WHAT MAKES A PUBLIC GREEN PLACE ACTIVE? IDENTIFYING KEY COMPONENTS OF SUCCESSFUL PUBLIC GREEN SPACES IN RURAL APPALACHIA

Track 4: Design for Health and Wellbeing
Knowledge Network: Active Living by Design, Health
Methods: Mixed Methods
Udday Shankur Datta (West Virginia University); Peter Butler (West Virginia University); Shan Jiang (GBBN Architects); Dr. Sam Zizzi (West Virginia University); Aaron Maxwell (West Virginia University)

1 ABSTRACT

There is a prevailing scarcity of physical activity (PA) opportunities and facilities in rural communities (Carter et al., 2019). Public green spaces such as parks, gardens, and other outdoor recreational areas can provide valuable outdoor activities and social interaction opportunities while promoting physical and mental well-being. The unavailability of public green spaces with appropriate amenities for PA opportunities has been one of the primary causes of the higher prevalence of obesity and physical inactivity-related health issues in rural Appalachian communities (Arcury et al., 2006; Jilcott et al., 2011; Jones et al., 2021). Health discrepancies in this region may become more pronounced if there are not enough convenient and accessible locations to exercise (Jiang et al., 2022). However, access to green space will not reduce obesity if one is inactive, distant from population centers, deemed unsafe, or lacks community programming (Jennings & Gaither, 2015). Intricate planning of public spaces can ensure the active engagement of users in physical activity.

This study aimed to identify critical components of successful rural public green spaces in rural Appalachia. Google API refers to a collection of application programming interfaces (APIs) developed by Google to allow third-party developers to access various Google services and data. These APIs include a wide range of services, including Google Maps. The study used a Python-based tool to search for public green spaces in the most active and inactive rural Appalachian counties. Using the "type" parameter, "park" was used as the place category to specify the locations to search in Google Maps. Once the list of public green spaces was obtained, the Places API was used to get more information about each place, including public reviews. VADER (Valence Aware Dictionary and sEntiment Reasoner) Sentiment-Analysis was used to analyze public reviews for 130 public green spaces in rural Appalachia (Hutto & Gilbert, 2014). The results identified a lack of safety, poor maintenance, and scarcity of kid-friendly activities as major disenabling factors for PA. Outdoor exercise equipment, playgrounds, pet-friendly walking trails, and well-maintained facilities with beautiful natural views were identified as preferred elements that facilitated PA in public spaces. Findings from this study will help planners and policymakers to identify suitable program elements for new public green spaces in rural Appalachia to have positive health outcomes in this region.

2 REFERENCES

Carter, W. M., Morse, W. C., Brock, R. W., & Struempler, B. (2019). Peer Reviewed: Improving physical activity and outdoor recreation in rural Alabama through community coalitions. Preventing chronic disease, 16.

Hutto, C., & Gilbert, E. (2014, May). Vader: A parsimonious rule-based model for sentiment analysis of social media text. In Proceedings of the international AAAI conference on web and social media (Vol. 8, No. 1, pp. 216-225).

Jiang, S., Datta, U., & Jones, C. (2022). Promoting Health and Behavior Change through Evidence-Based Landscape Interventions in Rural Communities: A Pilot Protocol. International Journal of Environmental Research and Public Health, 19(19), 12833.

LIGHTING THE PATH: LESSONS FROM A CLINICAL TRIAL ON IMPROVING SLEEP AND WELL-BEING IN DEMENTIA

Track 4: Design for Health and Wellbeing
Knowledge Network: Environment-Gerontology, Health
Methods: Experimental/Quasi-Experimental Research
Nina Sharp (Arizona State University), Mohammed Nasser Alrahyani (Arizona State University); Mahya Fani (Arizona State University)

1 BACKGROUND

Depression and poor sleep quality are relatively common among older adults with dementia, with all three conditions influencing and amplifying each other in a dynamic manner. Light is a crucial element of the indoor built environment that affects occupants' comfort and health. The importance of light in the indoor environment goes beyond merely enhancing the visual tasks to being the primary stimulus that regulates the occupants' circadian rhythms. The lighting design can deliver a potentially effective non-pharmacological solution to improve sleep, depression, and other associated behavioral problems across different age groups, including older adults with dementia.

2 OBJECTIVES

Beyond presenting the empirical findings, our primary focus is to share valuable insights gleaned from a clinical trial that we conducted to examine the effects of a dynamic lighting intervention on depression, agitation, and sleep quality of older adults with dementia. Emphasis is placed on lessons learned in designing suitable lighting interventions tailored to the unique needs of this population and the selection of appropriate measurement tools.

3 METHODS

The study was conducted as a counterbalanced crossover study design over 7 days of conventional lighting (Baseline) followed by 21 days of room lighting (Placebo, static average room lighting from morning to evening) and then 21 days of dynamic lighting (Treatment). The dynamic lighting intervention provided a high level of circadian stimulation in the morning and circadian-neutral lighting later in the evening. The human-lighting interaction occurred in the common areas of a selected memory care facility. Actigraphy and light trackers were employed on the last 7 days of each condition to monitor sleep quality and daily light exposure. Additionally, depression and agitation were measured using two questionnaires (Cornell Scale for Depression in Dementia (CSDD) and Cohen-Mansfield Agitation Inventory (CMAI)) completed by caregivers. Findings: A total of 10 participants with dementia (Mean age=75.6, SD=9.29) completed the study. The data revealed a significant improvement in sleep duration at night after exposure to both Placebo and dynamic conditions. Exposure to dynamic conditions increased sleep duration at night by 82 minutes. Additionally, a trend of improvement was observed in agitation and depression after exposure to both lighting interventions with better outcomes for the Placebo condition.

4 DISCUSSION AND CONCLUSION

While implementing an effective lighting design exhibited promising effects to improve sleep and mood in older adults with dementia, key insights from this trial underscored the heightened sensitivity of this population to bright ambient lighting, revealing a distinction from responses observed in other age groups. Despite the successful implementation of the initial dynamic lighting intervention in our previous clinical trial with adults, a pragmatic adjustment (50% reduction in morning lighting intensity of the dynamic condition) became imperative, prompted by participants' discomfort experienced on the first day of intervention. Moreover, the study illuminated the challenges associated with employing wearable devices in this population, necessitating exploration of alternative non-invasive measurement methods. Instances of device loss, breakage, and participant agitation prompted reconsideration of wearables' suitability. Ensuring data accuracy mandates meticulous collaboration with caregivers, including comprehensive training in measurement tools and participant observation methods.

NATURAL ENVIRONMENTS' INFLUENCE ON RESPONSIVE BEHAVIOUR, WORK-RELATED STRESS, BURNOUT, AND TURNOVER INTENTION IN NURSING HOMES

Track 4: Design for Health and Wellbeing
Knowledge Network: Environment-Gerontology, Work Environments
Methods: Mixed Methods, Statistical/Modeling Analysis
Peggy Chi (Institute of health policy, management and evaluation, university of toronto); Dr. Whitney Berta (University of Toronto); Dr. Adrian Wagg (University of Alberta - Department of Medicine); Stephen Verderber (University of Toronto); Dr. Olesya Falenchuk (University of Toronto - Ontario Institute for Studies in Education)

1 ABSTRACT

Natural environments in healthcare design are thought to promote health (Joseph et al., 2016; Verderber et al., 2021). However, its empirical research in nursing homes has been limited; the degree of rigor is mixed, focusing on older adults with little research on professional caregivers. This study examined the relationships between the natural environment and the mental health and well-being of older adults living with Alzheimer's disease and related dementias (residents) and of their professional caregivers in nursing homes using a rigorously developed survey to standardize an approach to the evaluation of natural environments in this setting.

A cross-sectional, multiphase mixed method study was conducted that included a scoping review (n=137), a Delphi panel (n=24 experts), cognitive debriefing interviews (n=33), psychometric evaluations (n=83 nursing home areas), and multilevel modeling and regression analyses (n=1599 residents; n=563 professional caregivers).

Based on a conceptual framework of the natural environment in healthcare facilities (Blinded for Review), the Natural Environment Survey for Long-Term Care was developed to evaluate the design of, usage of, and exposure to natural environments among residents and professional caregivers in secure and non-secure home areas. Significant associations were found between various factors of the natural environment and outcomes of responsive behaviour, work-related stress, burnout, and turnover intention.
Practical implications derived from the finding will present opportunities for future research and potential interventions to include in the transformation of nursing homes.

DESIGN FOR WALKABILITY: DIFFERENCES IN URBAN ENVIRONMENTAL CHARACTERISTICS OF FATAL PEDESTRIAN TRAFFIC ACCIDENTS BY AGE GROUP

Track 4: Design for Health and Wellbeing
Knowledge Network: Health, Sustainable Planning, Design & Behavior
Methods: Mixed Methods, Statistical/Modeling Analysis
Dawoon Jeong (Korea advanced institute of science & technology); Jounghyun Lee (Korea Advanced Institution of Science and Technology); Lisa Lim (KAIST)

1 ABSTRACT

Pedestrian accidents pose a significant challenge to the creation of a safe walking environment in urban areas, as they can lead to severe injuries or fatalities. The varying ages of pedestrians further impact the likelihood and severity of traffic accidents, influenced by differences in physical and mental well-being and variations in knowledge and safety awareness levels associated with age.

This study aims to investigate the characteristics of urban environments where pedestrian traffic fatalities occur, specifically emphasizing distinct pedestrian age groups. We used GIS data on pedestrian traffic fatalities in Seoul, South Korea, spanning a decade from 2013 to 2022. The analysis integrated various data, including traffic accident information, demographic factors, urban environmental factors, road information, and regional characteristics. The spatial analysis was conducted using QGIS, depthMapX, and R.

The primary objective of this study is to enhance pedestrian safety and cultivate a secure urban walking environment through a detailed analysis of spaces linked to pedestrian fatalities. The findings can inform targeted urban design interventions in areas frequented by specific age groups or susceptible to traffic accidents. This approach aims to optimize resource utilization and contribute to developing a sustainable urban environment.

DESIGNING EMERGENCY CARE ENVIRONMENTS RESOLVING THE DILEMMA: DESIGNING FOR SAFETY AND DESIGNING FOR HOPE IN THE ED

Track 4: Design for Health and Wellbeing
Knowledge Network: Health
Methods: Case Study, Interviews
Fernanda de Moraes Goulart (Clemson University); Ana Sandoval (Clemson University Architecture + Health); Maithili Awasarikar (Clemson University)

1 PROBLEM SIGNIFICANCE

As mental health concerns among children in the United States have been increasing, children and their families present themselves to the Emergency Department (ED) seeking help. However, this setting is not designed to meet the same standards as inpatient psychiatric facilities, which have rigorous requirements to keep the patients and staff safe (Cox, 2018). To address this problem, the Emergency Department staff must perform the time-consuming task of transitioning regular ED rooms to ligature-minimized rooms. However, these ligature-minimized exam rooms are often dehumanizing and stress-inducing.

2 PURPOSE

The purpose of the project was to develop evidence-based concepts for the design of an ED exam room that facilitates quicker transitions to ligature-minimized rooms while also promoting hope and dignity for patients.

3 METHODS

Data collection consisted of the following sources: (1) a review of literature to identify existing challenges, (2) a review of guidelines to understand the minimum requirements, (3) semi-structured walk-through interviews with four healthcare professionals at a local hospital to identify barriers and facilitators of pediatric ED exam rooms and an inpatient flexible room (4) open-ended questions interview with two former patients using images to trigger positive and negative responses to elements in the ED (5) case study analysis of two existing solutions to understand opportunities for further development. The research team then compiled and analyzed the data to determine the needs that must be addressed when designing Pediatric ED exam rooms that can accommodate MBH patients.

4 FINDINGS

The team identified key challenges in existing exam rooms in Emergency Departments. The most significant challenge is the time-consuming nature of ligature-minimizing these rooms. Due to the number of ligature and tamper points, the amount of equipment, and the lack of storage, transitioning the room can take up to several hours. This issue is further exacerbated when the room needs to be transitioned back, and equipment has been misplaced. Safety is also a significant concern, as ligature and tamper points put patients and staff at risk. This is especially important when the need to transition arises when the patient is already in the room. Finally, interviews with former patients revealed that the environment is not conducive to de-escalation. They described the emergency department as institutional and impersonal, which led them to feel discomfort and exacerbated their condition. They also expressed negative feelings towards decorations catering to much younger children, uncomfortable lighting, and lack of privacy. These challenges were then translated into a set of needs statements that then drove the design of an ED exam room.

5 CONCLUSION

The needs statements will help architects and healthcare leaders design ED exam rooms that enable staff to quickly and safely transition from regular exam rooms to ligature-minimized rooms while also creating a hopeful environment for the patients. Further research is needed to develop a more comprehensive list of needs statements, given that the present study focused on one hospital and a limited number of patients.

6 REFERENCE

Cox, A. M. (2018). Behavioral Design Strategies: Providing a safe and therapeutic ED environment. Health Facilities Management, 34-37.

GREENSPACE MORPHOLOGY ASSOCIATES WITH EMERGENCY DEPARTMENT VISITS DUE TO MENTAL DISORDERS

Track 4: Design for Health and Wellbeing
Knowledge Network: Health
Methods: Statistical/Modeling Analysis
Huaqing Wang (Texas A&M University); Dongying Li (Texas A&M University)

1 ABSTRACT

This study explores the relationship between greenspace morphology and emergency department visits for mental disorders, aiming to elucidate whether greenspace planning and design characteristics, beyond the influence of greenness, contribute to these visits. Utilizing statewide data on mental health-related emergency room visits across all zip code areas in Texas from 2016 to 2019, we employed GIS and R to calculate greenspace morphology metrics, including size, fragmentation, connectedness, aggregation, and shape, based on 10-meter resolution Sentinel-2 land cover data obtained from ESRI.

Conditional autoregressive spatial models with a zero-truncated negative binomial distribution were fitted to assess the associations between each greenspace morphology metric and emergency department visits. The analysis adjusted for urban and social-demographic factors, as well as greenness levels. The results revealed that larger size, less fragmentation, greater connectedness, aggregation, and more complex shapes in greenspace spatial morphology were associated with a lower risk of emergency department visits due to mental disorders.

This research underscores the importance of incorporating spatial morphology considerations into health-promoting greenspace projects during landscape planning and design practices. Beyond merely increasing greenery, our findings suggest that the spatial characteristics of greenspaces play a crucial role in mitigating the risk of emergency department visits for mental health issues.

MICRO OASIS: ENHANCING CARE PROVIDER WELLNESS THROUGH RESPITE POD IN HIGH-STRESS EMERGENCY DEPARTMENT

Track 4: Design for Health and Wellbeing
Knowledge Network: Health
Methods: Design Methods, Interviews
Yueci Jiang; Roshan Jose (Clemson University)

1 ABSTRACT

In emergency departments, healthcare providers face escalating stress and burnout rates, as evidenced by a significant 74.7% of ED physicians experiencing burnout since the pandemic. The ED's heightened workload, training gaps, and social support issues compound the stress on healthcare professionals. Beyond its impact on mental health, stress poses a serious threat to patient care quality, leading to medical errors and diminished patient satisfaction. Furthermore, higher rates of burnout among healthcare professionals might lead to an intent to leave their jobs, which imposes significant financial burdens on healthcare facilities (Shanafelt et al., 2017).

2 PURPOSE

This study aims to address care provider burnout by providing solutions that enable them to obtain respite while also being feasible within the context of their work.

3 METHODS

The study first adopts a comprehensive approach to understanding staff stress in pediatric ED settings. Initiated by a literature review and data analysis from staff interviews at emergency departments within four hospitals in South Carolina, the study explores stress factors, consequences of burnout, and interventions. Furthermore, an investigation at a trauma level 1 emergency department employs qualitative interviews with two nurses, two physicians, and on-site observations. With a profound understanding of the challenges through interviews, the study formulated a preliminary design prototype to address the problem. A full-size physical mock-up simulation is implemented to test and refine the prototype. Six participants engage in 30-minute sessions each to provide feedback on key design elements under different scenarios. The insights gathered from the participants inform the final design solution.

4 FINDINGS

Key challenges include staff shortage and heavy workload due to population growth. Distractions in the working environment and the absence of suitable break spaces contribute to stress. Physicians also face challenges with overtime charting due to limited work area resources and a distracting ED environment. These challenges collectively impact staff well-being and efficiency in the pediatric emergency department. To address the challenges identified, the solution needs to be within close proximity to the ED, away from patient flow, exclusive to its staff, offering quiet and private environments. It should provide multiple functionalities for diverse needs and customizable size options for different ED layouts, ensuring efficiency and flexibility in space utilization. The respite pod idea arises from those needs, offering a quiet, secure space tailored within the working context for different utilization durations from 3 minutes to 2 hours. Mock-up feedback highlighted participants' preferences for the respite pod design features including room size, doors, furniture, monitors, plants, and virtual windows. The final design solution is informed by mock-up feedback, ensuring a human-centered space for ED care providers.

5 CONCLUSION

In conclusion, the respite pod underscores the importance of addressing staff well-being through a human-centered research and design process. By mitigating stress and reducing burnout, the project aims to enhance the overall work experience for healthcare professionals, and the quality of care they provide.

6 REFERENCE

Shanafelt, T., Goh, J., & Sinsky, C. (2017). The Business Case for Investing in Physician Well-being. JAMA Internal Medicine, 177(12), 1826–1832. https://doi.org/10.1001/jamainternmed.2017.4340

PEDIATRIC MENTAL HEALTH: ANALYZING ENVIRONMENTAL BARRIERS AND FACILITATORS IN EMERGENCY DEPARTMENT (CASE STUDY)

Track 4: Design for Health and Wellbeing
Knowledge Network: Health
Methods: Interviews, Observation
Mina Shokrollahi ardekani (Clemson University Center for Health Facilities Design + Testing); Fernanda de Moraes Goulart (Clemson University); Monica Gripko (Clemson University); Devi Soman (Clemson University Center for Health Facilities Design + Testing)

1 BACKGROUND
Over the past decade, there has been an 8% increase in emergency department (ED) visits by children and adolescents with mental and behavioral health (MBH) complications in the U.S., representing approximately 13% of all pediatric ED visits (Bommersbach, McKean et al. 2023). The high prevalence underscores the need for effective resource allocations and accommodations for this population coming to ED, with a particular emphasis on the significant role played by the physical environment. ED environment has been shown to pose barriers impacting both patient and staff safety (Haines et al., 2017; Pati et al., 2016), as well as MBH patient management in the ED (Innes et al., 2014). Therefore, paying attention to the ED environment's physical aspects and spatial qualities is imperative.

2 OBJECTIVES
The primary objective of this paper is to understand the built environment-related barriers and facilitators to providing care to pediatric patients visiting the emergency department with mental and behavioral conditions.

3 METHODS
This study is a qualitative case study research focusing on the barriers of the built environment in providing care for pediatric patients visiting the ED for MBH complaints. Four ED units (two urban, two rural) in the Southeastern United States were part of this study, which included an environmental assessment, onsite observations, and in-depth semi-structured interviews with 55 ED staff (administrators, nurses, physicians, mental health professionals, security staff, and social workers). For the interviews, we conducted content analysis using Atlas.ti software. The emergency departments' floor plans were also analyzed using the Space Syntax method using the Depth map to complement the findings from the interviews and observations related to spatial qualities of the emergency departments' environment.

4 RESULTS
The results show various barriers and facilitators associated with the emergency department's physical environment. These barriers are categorized into two themes: Room-Level barriers and Unit-Level barriers. Concerning patient room design, barriers include limited access to natural light or a view, as well as patient safety issues due to the existing ligature risks. At the Unit-Level, limitations in resources and not having enough rooms to accommodate MBH patients are some of the main barriers leading to the utilization of hallway beds, which could raise additional issues related to patient privacy and patient safety. Nurse stations are also found to be problematic in terms of privacy for staff, visual access to the patients, and workspace availability. The visual access from triage to the waiting room area was another barrier highlighted throughout the interviews, as well as privacy issues for patients in the triage area. Moreover, the physical environment attributes related to the layout of the units and the type of furniture were perceived as facilitators to providing care for pediatric MBH patients.

5 CONCLUSION
EDs around the country are facing significant challenges to providing safe and effective care to children with mental health conditions. The chaotic physical environment of the ED often exacerbates the conditions of this patient population and contributes to negative outcomes. Future ED designs should consider solutions to address those barriers and improve the quality of care for this population.

HEALTHCARE DESIGN INNOVATIONS FINDINGS AND RECOMMENDATIONS FROM A NATIONAL WORKSHOP ON DESIGNING EMERGENCY DEPARTMENTS FOR PEDIATRIC MENTAL HEALTHCARE

Track 4: Design for Health and Wellbeing
Knowledge Network: Health
Methods: Action/Participatory Research, Case Study
Monica Gripko (Clemson University); Anjali Joseph (Clemson University); Deborah Wingler (HKS); Ann Dietrich (Prisma Health); Lucy Easler (Prisma Health Baptist Columbia SC)

1 ABSTRACT
Mental and behavioral health (MBH) conditions – a range of mental, behavioral, and substance abuse disorders – affect one in five children in the United States (OSG, 2021). With a lack of community-based and outpatient MBH resources, especially in rural communities, emergency department (ED) utilization by pediatric MBH patients has risen alarmingly over the last 20 years and is projected to continue increasing. Yet, ED environments are poorly designed, organized, and equipped for safely and effectively caring for these patients. Furthermore, there is a lack of research on how ED work systems should be designed to support pediatric MBH care.

Designing safer, more human-centered work systems in the ED for pediatric MBH care, minimizing unnecessary patient stressors, and improving provider well-being first requires developing a shared, in-depth understanding of the work system facilitators and barriers involved in existing caregiving workflows. As such, our research team organized a full-day workshop and design charette that brought together researchers, designers, clinicians, administrators, and students from across the United States.

2 PURPOSE
The workshop aimed to:

1. *Identify the most urgent issues and challenges that must be addressed to improve the care of children and adolescents with MBH conditions and staff well-being.*
2. *Prioritize issues and challenges for further study.*
3. *Develop goal statements to address critical challenges.*
4. *Brainstorm interventions for solving identified challenges.*

3 METHODS
The workshop began with an introduction to the problem, keynote and panel discussions from industry experts and key stakeholders, and an overview of the research team's findings from a series of case studies of EDs in the southeastern United States. Findings were organized per the Systems Engineering Initiative for Patient Safety 2.0 (SEIPS) model, which considers design based on various work system elements – people, environment, tasks, and organization – and their interactions in the context of processes and workflows Following this introduction, workshop attendees broke out into groups to define the top five challenges they viewed as most critical to the care, safety, and well-being of pediatric MBH patients and providers in EDs. Then, the larger group voted to identify the top six challenges overall. Next, each breakout group took on one of these six challenges and engaged in a design charette to brainstorm possible interventions to address their given challenge.

4 FINDINGS
This presentation will highlight the critical challenges and possible environmental interventions developed through the workshop. The identified challenges pertained to the lack of flexibility, hope, and support for patients, families, and care providers in ED environments and sub-optimal processes. Interventions aimed to reduce stressors, support coping, and increase patient and family engagement, ranging from low-effort changes to high-impact, high-effort redesigns.

6 REFERENCE
Holden, R. J., Carayon, P., Gurses, A. P., Hoonakker, P., Hundt, A. S., Ozok, A. A., & Rivera-Rodriguez, A. J. (2013). Seips 2.0: A Human Factors Framework for Studying and Improving the Work of Healthcare Professionals and Patients. Ergonomics, 56(11), 1669-1686.

Office of the Surgeon General (OSG). (2021). Protecting Youth Mental Health: The U.S. Surgeon General's Advisory [Advisory]. Office of the United States Surgeon General. https://www.hhs.gov/sites/default/files/surgeon-general-youth-mental-health-advisory.pdf

SPACES OF TRANSSPECIES DESIGN IN HEALTHCARE THROUGH MASS TIMBER CONSTRUCTION: MICROBES, PATIENTS, & FORESTS

Track 4: Design for Health and Wellbeing
Knowledge Network: Health
Methods: Experimental/Quasi-Experimental Research
Mark Fretz (University of Oregon); Gwynne Mhuireach (University of Oregon); Jason Stenson (University of Oregon); Dale Northcutt (University of Oregon)

1 ABSTRACT

Spaces designed for healing can be identified in Western society as early as in ancient Greece. Consecrated to the healing god Asclepius, temple-like spaces converged elements of nature to focus on holistic health. In the Middle Ages, healing spaces further developed alongside monastic medicine, which was influenced by the twelfth century abbess, Hildegard von Bingen, who propounded her concept of Veriditas that could be interpreted as the healing power of nature, or green divine. These concepts held significant spatial influence in Western architectural imagination, informing the design of early hospitals through modernist glass temples with massing organized for optimized sunlight, air, and access to adjacent outdoor garden spaces.

However, throughout the twentieth century, existing spatial precedent slowly eroded to the industrialized demands of healing efficiency, which included a preoccupation with environmental control and microbial eradication. Although microbes form the basis of life on earth and most are essential to human existence, health, and survival, current healing spaces wage war on all microbes, which has led to the development of synthetic building materials that visually mimic nature coupled with operational practices intended to support destruction of potential pathogens on these surfaces. The war has produced collateral damage, including an increase in antibiotic resistant genes in areas of greater antimicrobial compounds indoors (Fahimpour, 2018), a decrease in microbial diversity from synthetic material plasticizers (Velazquez, 2019), and depletion of transspecies health due to the embodied impacts of material lifecycle and operation of these spaces.

Today's healthcare organizations are beginning to recognize the impact of healing space design and operations not only on human patient health but also on transspecies and planetary health. Therefore, they are exploring the use of structural mass timber for the creation of these spaces, which has the potential to significantly reduce embodied carbon, reintroduce connection to nature through the visual environment of exposed wood and olfactory environment of terpene emissions, and be sourced from forest restoration activities. Yet, healthcare organizations are empirically driven and risk averse; therefore, they approach issues of infection control cautiously and with data regarding exposed natural wood. The body of scientific evidence is expanding to suggest wood construction materials may exhibit natural antimicrobial activity and transfer fewer microbes from the surface than materials commonly found in healthcare environments, such as concrete, painted drywall (Mhuireach, 2021), plastic, or stainless steel (Munir, 2021).

In this context, the Institute for Health in the Built Environment created microcosm experiments intended to replicate indoor conditions to evaluate cross-laminated timber surface and air microbial communities, with independent variables consisting of wood species, finish, operational relative humidity, and simulated wetting. We analyzed microbial community abundance, diversity, viability, and composition through qPCR and shotgun metagenomics. In a subset of microcosms, we characterized VOCs, analyzing their concentration and emission rate before, during and after wetting of the CLT blocks using gas chromatography-mass spectrometry (GC-MS) paired with microbial community composition from surface swabs. Results indicated that wood may transfer fewer viable microbes compared to non-wood construction materials and should be reconsidered in healthcare facility design and construction.

IMPACTS OF QUALITY OF URBAN STREETSCAPES WITH GREENERY ON MENTAL HEALTH: MEDIATING ROLES OF SOCIAL COHESION ON SOCIAL WELL-BEING

Track 4: Design for Health and Wellbeing
Knowledge Network: Health
Methods: Photo-based Methods, Survey
Nianchu Hou (Auburn University)

1 ABSTRACT

In recent years, more researchers have been discussing the role of natural environments in enhancing people's mental health. Accumulated findings argue that outdoor greenery is entwined with and plays a significant role in an individual's mental health (Chen et al., 2021). There is only one literature found that examines the relationship between the quality of urban streetscapes with greenery and health, and the effect of the quality of urban streetscapes with greenery was significant in potentially mitigating chronic health problems (Liu et al., 2023). However, how the quality of urban streetscapes with greenery affects social well-being remains an open question with limited evidence.

The purpose of this study is to examine whether quality of urban streetscapes with greenery can contribute to users' social well-being. The specific objectives of this study are to examine 1) the relationship between quality of urban streetscapes with greenery and social well-being, 2) the relationship between social cohesion and social well-being, 3) the relationship between quality of urban streetscapes with greenery and social cohesion, and 4) the mediating effect of social cohesion.

This study will use an online survey method to examine whether the quality of urban streetscapes with greenery increases users' social well-being, and specifically to test whether social cohesion can mediate between the perceived quality of urban streetscapes with greenery and social well-being. Participants will complete an image-based questionnaire containing 15 images and will measure their perceived quality of urban streetscapes with greenery of each image. Then, participants will self-report their perceptions of social well-being and social cohesion based on the images. The survey design will use the scales of social well-being and social cohesion developed by Vries et al. (2013).

The target population for this study is residents aged 18-64 years in the United States. Participants will be selected from a database provided by Amazon Mechanical Turk, and representative quota data for gender (male and female) will be selected. This study will use exploratory factor analysis (EFA) to test the convergent validity of each multi-item measure, where principal component analysis and the formation matrix will be utilized. In addition, simple regression and multiple regression will be used to analyze the hypothesis. The research aims to provide insights for urban planners on enhancing residents' life quality through improved urban greenery, acknowledging limitations such as reliance on visual stimuli and self-reported measures.

6 REFERENCES

Chen, K., Zhang, T., Liu, F., Zhang, Y., & Song, Y. (2021). How urban green space impacts residents' mental health: A literature review of mediators. International Journal of Environmental Research and Public Health, 18(22), 11746. https://doi.org/10.3390/ijerph182211746

Liu, Q., Fu, W., van den Bosch, C., Xiao, Y., Zhu, Z., You, D., Zhu, N., Huang, Q., & Lan, S. (2018). Do local landscape elements enhance individuals' place attachment to new environments? A cross-regional comparative study in China. Sustainability, 10(9), 3100. https://doi.org/10.3390/su10093100

Vries, S. de, van Dillen, S. M. E., Groenewegen, P. P., & Spreeuwenberg, P. (2013). Streetscape greenery and health: Stress, social cohesion and physical activity as mediators. Social Science & Medicine, 94, 26–33. https://doi.org/10.1016/j.socscimed.2013.06.030

REPORT FROM TERRA ALTA BIRTH CENTRE, SWITZERLAND: CONSTRUCTING AND PRESENTING A TEAM-BASED LIVE CASE STUDY

Track 4: Design for Health and Wellbeing
Knowledge Network: Health
Methods: Case Study

Davis Harte, PhD (Paradigm Spaces, llcBoston Architectural College); Doreen Balabanoff (Faculty of Design, OCAD University); Alessia Macchi (DIDA - Universit degli Studi di Firenze); Anka Dür (Https://Ankaduer.Com/); Alison Mulvale-Fletcher (OCAD University)

1 ABSTRACT

Recent research reveals that women, infants, supporters and caregivers worldwide are impacted by the absence of a supportive, sensitive environment for birth – one that increases chances of a physiological birth and a salutogenic birth. Today most built birth environments across the globe fail to deliver these desired, valuable and human-centered experiences. (Anonymous). The clinical model of birth space design is not likely to meet the needs and desires of most birthing women and their supporters/companions (Anonymous). However, with the resurgence of midwifery care and the growing research on birth and birth spaces, momentum is building towards a more person-centered, emotionally and culturally sensitive approach. Our work is grounded in these ideas and goals.

The research team's case studies have developed out of a key goal of the 'research-creation' project, optimizing knowledge mobilization in a web-based resource. Case studies, we recognize, are useful not only as design precedents, but as documentary evidence and inspiration for architects, designers, design students, caregivers, managers, policy makers and researchers. We need all of these players to participate if transformational change for birth environment is to become a reality. Finding/using strong 'positive exemplars' that we feel can indeed inform and inspire, we have been conducting 'live' case studies (using on-site visits and interviews with diverse users).

This presentation reports on our first case study - we will show what we have learned from the Terra Alta Birth Centre and its interviewed users and creators, and discuss how our goal of 'optimal knowledge transfer' informed processes of data gathering and analysis. We have asked: What forms of data analysis bring most value for transformational change to birth environment design? What knowledge transfer modes best support efficient provision of knowledge and insight about optimizing birth spaces? With over 1,000 items in our literature review library, we have identified fifty-four that provide significant knowledge and understanding for design of optimal birth spaces. Yet most of these papers lack images, architectural drawings, contextual information, and perspectives of diverse users. Without these elements, the language of architecture and design, visual and sensory, felt and recognized, is missing...To bridge this gap means finding a model of knowledge transfer designed first and foremost for designers and those seeking to create and reimagine birth centers. Efficient and clear contact with relevant and current knowledge about birth and how the design affects all involved is crucial to optimizing childbirth care settings.

The Terra Alta Birth Centre is a private midwifery care birth center in Oberkirch, Switzerland. Here we present/discuss the project and its lessons for designers. We also offer a glimpse of our concerted efforts to create enhanced design knowledge transfer for optimal birth environment design, through highlighting the issues that need addressing – and spatial and architectural concepts made manifest in the forms (constructed forms, spaces and atmospheres) of realized projects. (Oxman, 1994).

2 REFERENCES

Oxman, R. E. (1994). Precedents in design: A computational model for the organization of precedent knowledge. Design Studies, 15(2), 141–157. https://doi.org/10.1016/0142-694X(94)90021-3

INNOVATIONS IN HEALTH ENVIRONMENTS RESEARCH AND DESIGN NAVIGATING THE FUTURE OF WORKPLACE: INSIGHTS INTO FLEXIBILITY AND FREE-ADDRESS DESIGNS

Track 4: Design for Health and Wellbeing
Knowledge Network: Work Environments
Methods: Mixed Methods, Practice-based Research

Hanna Negami (Perkins Eastman Architects); Widya A Ramadhani (Perkins Eastman Architects); Rebecca Milne (Perkins Eastman Architects); Heather Jauregui (Perkins Eastman); Jennifer Askey (Perkins Eastman); Jennica Deely (Perkins Eastman); Connor Glass (Perkins Eastman); Jeff Young (Perkins Eastman)

1 ABSTRACT

The COVID-19 pandemic catalyzed the rapid adoption of remote work (Kniffin et al., 2021). Today, many companies continue to accommodate remote work to various degrees. Terms such as "flex work" and "work from anywhere" are commonplace, reflecting the diverse ways companies provide flexible work arrangements for their employees (Ray & Pana-Cryan, 2021). This flexibility has led to fewer people being in the office simultaneously, prompting many organizations to reduce their office space and introduce unassigned seating. The challenge lies in developing a workplace model that fosters productivity and culture in both physical and virtual settings (Hirsch, 2021). Here, we present a post-occupancy evaluation of one North American office that integrates policy and design to holistically support employees in a hybrid environment.

We conducted a case study of a large architecture and design firm's studio, examining the impact of a flexible work environment that combines policy (i.e., work from anywhere) and design interventions (i.e., free-address office) to support employee well-being, productivity, career growth, and overall organizational performance. This mixed-methods study was conducted in a 60-person office space in Pittsburgh, PA, which opened its newly renovated free-address office in May 2021, when it also introduced a work-from-anywhere policy. To examine the dynamics of this flexible work environment holistically, we employed four data collection strategies: targeted one-on-one interviews (N=10), an office-wide online survey (N=51), behavioral observations, and on-site indoor environmental quality measurements. The study focused on seven areas: look and feel, environmental quality, well-being and stewardship, work from anywhere, career growth, performance, and productivity. In the analysis phase, data types were compared, contrasted, and interpreted to provide nuanced insights into the complex interplay of factors shaping the work environment.

In this session, we present the results of our study, evaluating employee outcomes in a functioning office environment. We demonstrate how the office meets basic needs by providing a healthy workplace, as well as how its spatial design and operational policies come together to provide an enhanced working experience that promotes productivity, well-being, and culture building. Key findings include an overall positive experience with the office. Participants appreciate the free-address office design, which allows them to choose their work locations based on their tasks; however, some feel that the clean-desk policy results in a lack of personalization. Our discussion explores the implications of our findings for workplace design and policy, offering actionable insights for organizations navigating the transition to flexible work arrangements. This study contributes valuable insights to the ongoing discourse on the role of flexibility in modern workplace practices and its impact on employee well-being and productivity.

ASHE: AN EMPIRICALLY-GROUNDED MULTI-AGENT SIMULATOR FOR EVALUATING HEALTHCARE ENVIRONMENT

Track 4: Design for Health and Wellbeing
Knowledge Network: POE/Programming, Sustainable Planning, Design & Behavior
Methods: Experimental/Quasi-Experimental Research, Statistical/Modeling Analysis
Yuanyuan LIU (Southeast University); Yangpeng Xin (Southeast University); Ying Zhou (Southeast University)

1 ABSTRACT

Optimizing through rational spatial layout in healthcare facilities can significantly enhance medical efficiency, safety, employee satisfaction, and operational sustainability, ultimately promoting human-centered health environment design. This study aims to provide a reliable simulation tool for diverse spatial analysis by developing an empirically grounded multi-agent simulator —the ASHE Agent Simulator of the Healthcare Environment.

The ASHE simulator was developed on the multi-agent development platform artisoc4.0, using a Visual Basic-like programming language (Artisoc, Kozo Keikaku Engineering Inc). Its development involved four key stages: empirical data collection, development of behavioral rules, multi-agent modeling, and validation. The development was based on surveillance video observation data from a trauma intensive care unit of a tertiary hospital in East China. Subsequently, we employed the ASHE simulator to conduct a preliminary evaluation and analysis of the layout of award-winning ICUs from the North American SCCM ICU design competition.

The ASHE simulator exhibits the following characteristics: 1) Unlike evacuation simulators, it places a greater emphasis on simulating everyday health environments, thereby capturing the complexity of medical workflows; 2) It can depict the spatiotemporal behaviors of diverse healthcare professionals, including doctors, nurses, and auxiliary staff; 3) Its development is grounded in empirical data, allowing it to serve specific research objectives better. Given the increasing professional collaboration among healthcare practitioners and the growing complexity of the physical environment in healthcare, existing research tools face significant challenges in describing and analyzing diverse occupant behaviors (Verderber et al., 2021). Concurrently, agent-based modeling has emerged as an effective approach for studying complex human-environment systems. The progression from stylized models to empirically grounded complex models has become an inevitable trend in addressing the real world (Sun et al., 2016). The development of this model can be viewed as a beneficial attempt to apply the agent-based approach to the design for health and wellbeing.

2 REFERENCES

Artisoc, Kozo Keikaku Engineering Inc. Available at: https://mas.kke.co.jp/en/artisoc4/

Sun, Z., Lorscheid, I., Millington, J. D., Lauf, S., Magliocca, N. R., Groeneveld, J., Balbi, S., Nolzen, H., Müller, B., & Schulze, J. (2016). Simple or complicated agent-based models? A complicated issue. Environmental Modelling & Software, 86, 56–67.

Verderber, S., Gray, S., Suresh-Kumar, S., Kercz, D., & Parshuram, C. (2021). Intensive care unit built environments: a comprehensive literature review (2005–2020). HERD: Health Environments Research & Design Journal, 14(4), 368–415.

REAL DATA, REAL TIME: LEVERAGING TECHNOLOGY AND PASSIVE ENGAGEMENT METHODS FOR A MORE ENGAGING AND INFORMATIVE MIXED-METHODS POST-OCCUPANCY EVALUATION

Track 4: Design for Health and Wellbeing
Knowledge Network: POE/Programming, Researchers in Professional Practice
Methods: Practice-based Research, Mixed Methods
Rachel Nasland (LPA inc.); Kimari Phillips (LPA, Inc.)

1 ABSTRACT

After moving into a new studio in converted U.S. Navy barracks in San Diego, California, an architecture firm's in-house sustainability and applied research team explored new methods for collecting data on the quality of users' indoor experiences in real-time. A typical post-occupancy evaluation (POE) entails triangulating data collected from surveys, focus groups, site observations and, when feasible, building data1. While this widely-accepted process provides useful information, data is limited to one point in time and can lack critical real-time details regarding location and fluctuations in use, behavior, and external factors, like weather and time of day.

To better understand users' experiences and interactions with their immediate surroundings in the new office, the researchers partnered with their IT and graphics teams to leverage technology to capture experiences corresponding to specific spaces at multiple points in time. Together, they developed a touchscreen tablet application — dubbed the "Human Experience App". A literature review of human factors and workplace research informed content for the app response selections to reflect environmental qualities that influence workplace satisfaction, productivity, and engagement2,3. The app interface allows people to rate their experience of current indoor environmental conditions in up to six categories: lighting, acoustics, air, furniture, thermal comfort, and/or state of being. Researchers hypothesized that the lessons learned in real-time could be used to refine building systems, optimize person-environment interactions, and measure the impact of specific design features on users' experiences.

Four months after occupancy, the tablets were placed in 19 locations throughout the office, along with 9 temperature and humidity sensors, and 3 CO2 sensors to provide quantitative context to the qualitative responses received through the app. During analysis, each user response was linked to a time, location, and surrounding sensor readings. A total of 1,200 real-time ratings were entered into the Human Experience Application over the course of six months, and exploratory analyses were conducted to see correlations between user feedback, location, and environmental data.

This session will discuss successes with the tested real-time data collection methodology and highlight lessons learned and next steps. Several advantages will be shared in comparison to traditional POE methods, such as occupants' ability to provide feedback in real time, unlike surveys which collect only retrospective perceptions of indoor environmental qualities or ratings at only one point in time. The new technology helped shed light on several key areas of the studio design's performance and identify trends to help fine-tune spaces through small-scale design alterations and changes in user interaction. This new research tool allows designers to benchmark a building's performance, test design decisions, rectify issues early after move-in, and inform future designs with lessons learned.

2 REFERENCES

Li, P., Froese, T. M., & Brager, G. (2018). Post-occupancy evaluation: State-of-the-art analysis and state-of-the-practice review. Building and Environment, 133, 187-202.
Heath, O., Ja
ckson, V., & Goode, E. (2018). Creating Positive Spaces Using the WELL Building Standardtm.

Choi, J.-H., Loftness, V., & Aziz, A. (2012). Post-occupancy evaluation of 20 office buildings as a basis for future IEQ standards and guidelines. Energy and Buildings, 46, 167-175.

TRANSFORMING CARE DELIVERY OF A TRANSPLANT CLINIC: EFFECTIVE DESIGN AND TECHNOLOGY STRATEGIES FOR EFFICIENCY AND COLLABORATION

Track 4: Design for Health and Wellbeing
Knowledge Network: POE/Programming, Researchers in Professional Practice
Methods: Mixed Methods, Practice-based Research

Kara N Freihoefer (Flad Architects); Elizabeth Strutz (Flad Architects); Preeti Dalvie (Flad Architects)

1 ABSTRACT

This presentation will provide results from a comprehensive post occupancy evaluation of a new 16,000-square-foot Transplant Clinic that is part of a major academic medical center in the Midwest. The clinic relocated into a compact and repurposed space in a prominent location of the existing hospital in October of 2022. The overall project vision was to "provide the best patient, provider, and staff experience and to remain one of the top transplant centers in the country."

To achieve this, project-based guiding principles were developed that include advancing patient- and family-centered care by streamlining flows with a one-stop-shop clinic model, creating a collaborative care work environment by providing centralized workrooms and team stations, and integrating innovative technology such as electronic-synchronized (e-synch) boards that allow staff to track patients' appointment journeys. The new U-shaped, double-loaded corridor design includes 26 exam rooms with a centralized teaming station and workroom that dissect through the center core for the clinic staff to work. These rooms include a total of 12 workstations with whiteboards and large display monitors for collaboration and communication. In addition, five e-boards were strategically placed throughout the corridor among a grouping of 5-6 exam rooms. These boards display anonymous information about patient visits, and clinic staff frequently update the displays by moving 'tags' along the continuum of patients' appointments.

A multi-method POE was conducted about eight months after opening the clinic to evaluate if the design interventions related to the project guiding principles met a level of success. For example, the study investigated if strategic placement of work areas and e-boards improved collaboration and workflow efficiencies. The POE plan consisted of surveying and observing staff and deploying occupancy sensors in collaboration areas.

A total of 100 staff responded to an online, Likert-type satisfaction questionnaire about elements of the new design and e-boards integration. A total of 122 observations of staff engaging with the e-boards were collected; documenting frequency and duration of use, workflows, and activities performed. Continuous remote sensor monitoring of the workroom and team station was conducted for 18 days to understand utilization, seat fill, and number of people using these spaces during clinic hours.

Results found that staff are highly satisfied with their usage and personal understanding of e-boards (7.83 and 7.82, respectively out of a 10-point scale). However, ease of use was rated slightly lower at 6.40. Observation results found that two e-boards positioned to the front of the clinic received the most usage (54%). An average e-board visit was roughly 20 seconds long, and 79% of the visits entailed touching/scrolling. Seventy-five percent of staff workflow involving e-boards entailed either coming from or going to an exam room, team station, or workroom.

The team station and workroom were 61% and 66% utilized, respectively, during clinic hours. The peak usage was between 9 am and 2 pm. The central location of these rooms makes them ideal to enhance and support staff collaboration and workflow efficiency. Staff indicated that the design supports teamwork with a satisfactory score of 7.14.

OCCUPANTS WELLBEING THE (PATHOGENIC)-CITY: A HISTORY OF URBANIZATION, MINORITY HEALTH, AND RACIAL DISPARITIES IN THE CITY OF BALTIMORE

Track 4: Design for Health and Wellbeing
Knowledge Network: Environment-Gerontology, Health
Methods: Mixed Methods, Observation
Farhana Ferdous (Howard University)

1 ABSTRACT

Since arriving as slaves, the segregation policy and racial disparities among African Americans affected the history of American urbanization, city planning, and the healthcare system. The (pathogenic)-CITY is a chronological history of racial disparities in the American Landscape by focusing on how urbanization and planning movements have transformed the minority's health and well-being from post-industrial society to the present. The study aims are to analyze the connection between the spatial form of city design, and minority health in response to racial disparities and societal changes from the 1840s to the present.

For thousands of years, humans have looked to physical spaces to cure sicknesses; and cities and neighborhoods are redesigned to respond to epidemics. The growth of American cities throughout the 1800s and 1900s shows how a series of issues including racial disparity, lack of sanitation, rapid urbanization, poor waste collection, high levels of pollution, and inadequate housing could create unhealthy environments and cause the spread of infectious or communicable, diseases like polio, cholera, plague, tuberculosis, and smallpox. The mid-20th century changes in city design occurred in response to the adverse health conditions of African Americans. During the industrial era, raw sewage flowed out of the buildings and caused a severe cholera outbreak globally that significantly affected black neighborhoods and eventually led to the sanitary reform movement. In the 1960s, racial tensions were high, and wealthy white citizens quarantined immigrant and minority populations through the Housing Covenants Act that prohibited African Americans from purchasing or renting property. The city commission law was used as an excuse for city development and excluded owners by race because they had higher rates of mortality than white homeowners.

In the late 19th and early 20th c., the morbidity and mortality rate of African Americans with degradable health conditions led many to conclude that the disease-causing factors were racial rather than social or economic. Even some diseases were considered almost exclusively racial and used as a marker for racial identities, such as Syphilis, Sickle-cell, Black Death, Negro Disease, and Polio. The spread of contagious diseases fosters racist ideology by characterizing African Americans as notoriously diseased and ignorant about personal hygiene. But in reality, they were unprivileged with little or no access to hospitals. This combination of discrimination and neglect was responsible for the continued racial disparities, urbanization movements, and the persistent spread of contagious diseases as of today.
Although racial segregation has been a core topic of inquiry in African-American and Social Studies, the notion of this study to critically narrate, analyze, and integrate American urbanization and health disparities is completely absent. With the rapid development of the built and urban environments, the global disease profile is continuously changing and disproportionately affects African-American populations at an astonishing rate, including the recent surge of the COVID-19 pandemic. 'The (pathogenic)-CITY' will contribute to the development of new methodological and theoretical notions for a broader interdisciplinary discourse by combining city design, minority health, and by discussing the role of urban designers, theorists, town planners, health professionals, and architects.

GUIDELINES FOR SENSORY WELLBEING SPACE THROUGH SHARED KNOWLEDGE

Track 4: Design for Health and Wellbeing
Knowledge Network: Interior Design, Participation
Methods: Focus Groups, Practice-based Research

Tina Patel (Kent State University); Lisa Audet (Kent State University); Jennifer Meakins (University of Kentucky - Lexington, KY)

1 ABSTRACT

The concept of neurodiversity encourages brain diversity in the same way that we talk about biodiversity and cultural diversity (Singer, 2017). Integrating neurodiverse students into higher education systems offers a significant opportunity to address and accommodate their diverse sensory needs. It encourages educators to think outside the box and embrace a novel perspective on human diversity. Besides making changes to the curriculum, the learning environment must meet students' sensory needs to allow them to feel physically comfortable, encourage self-regulation, and support academic achievement. A survey of university campuses suggests that classrooms are typically designed to hold the maximum number of students, disregarding individual learning needs. However, the inclusion of a sensory well-being room in an educational setting could alleviate overstimulation by reducing the amount of sensory input and could help both hyper and hypo-sensitive diverse learners re-engage with their surroundings.

This research began by addressing fundamental questions: How can we make our campuses more inclusive for neurodiverse students? Can we create adaptable sensory-supportive well-being hubs on university campuses to cater to the unique needs of diverse learners? Addressing these questions calls for collaboration among designers, researchers, educators, and the wider community to explore interior design as a tool for understanding the educational environment and accommodating all learners (Zingoni, 2019).

This study aimed to bridge the gap through an extensive literature review of architecture, interior design, and special education journals to understand the interior design spatial attributes and sensory needs of neurodiverse students. Participatory research encompasses various methodological approaches that transfer decision-making power from the researcher to participants and end-users (Leavy, 2017). Hence, this research aimed to co-create design guidelines of the sensory supportive, well-being space for neurodiverse college students to improve and enrich their physical, emotional, learning, and social interactions while adding value to the design responses. This process is more end-user-inclusive, providing an agency to the participants so they take charge of the research agenda, methodology, and design. This method deviates from the linear, step-oriented, unidirectional approach of probing, priming, understanding, and generating (Sanders, 2002).

A focus group using universal design for teaching and learning principles was conducted by researchers from interior design and health sciences departments. Nine neurodiverse college students from undergrad through doctoral levels participated. The purpose was to understand their diverse sensory needs and identify interior design attributes and strategies that could best meet their sensory needs and promote engaged learning and well-being. The insights gathered informed the design guidelines for a prototype sensory-supportive, well-being space. Such spaces can positively influence behavior, impacting students' learning, academic performance, and individual and social well-being. Capturing the perspectives and wisdom of neurodiverse students was essential to this research as it empowered them within their own space. This multidisciplinary research study helped the researchers understand the correlation between design factors, sensory needs, and the well-being of neurodiverse students. It further reinforced the belief that these recommendations should be considered in the university-wide design to make campuses more inclusive and promote success for students who are neurodiverse.

THE ROLE OF ARTISTIC EXPRESSION IN THERAPEUTIC DESIGN: ADDRESSING OCCUPATIONAL STRESS IN YOUNG ADULTS

Track 4: Design for Health and Wellbeing
Knowledge Network: Interior Design, Health
Methods: Interviews
Alexis Davies; Kutay Guler (Kansas State University)

1 ABSTRACT

Stress, along with other mental health problems that negatively impact well-being, is uniquely significant for young adults as they are in the process of gaining independence and typically have minimal access to resources or support (Varma et al., 2021). Therefore, young adults are more susceptible to negative coping mechanisms which in turn affect social and workplace success. The COVID-19 pandemic has exacerbated existing stressors specifically in the workplace as a new work model was established resulting in a new "normal". Adapting to this new "normal" while simultaneously coping with traumatic, emotional, and psychological effects of the pandemic leads to a decrease in mental well-being and workplace success, which can culminate in unemployment and financial burden, thus increasing difficulty in receiving needed treatment and support.

Art therapy is a sensory and action based interactive form of psychotherapy with "purposeful application and integration of art, music, dance/movement, dramatic enactment, creative writing, and imaginative play" (Malchiodi, 2023, p. 3). As each artistic outcome is unique to an individual, the practice of art is individually expressive in itself. Art therapy recognizes that everyone expresses themselves differently and, by encompassing a variety of mediums and activities, it allows each individual to communicate in their own way. As art therapy directly relates the study of the mind to the body with "interplay of experiences, emotions, behavior, and physical health," (Masterson et al., 2008, p.21) it can serve as an appropriate method of psychotherapy to address traumatic stress in young adults post-pandemic in addition to overall mental well-being. By implementing sensory experiences, art therapy provides an individual the opportunity to effectively communicate their emotions (Malchiodi, 2023). This not only helps process stressors, but also builds confidence in individual expression, which aids in collaboration and socialization.

Accordingly, this research focuses on the following research question "How can artistic expression within the context of a therapeutic environment improve well-being in young adults experiencing occupational stress?" Data is collected through a series of semi-structured interviews with art therapy experts (n=8+) and working young adults (n=8+). Key discussion points identified in the interview framework include stress, isolation, individuality, expression, belonging, work-life balance, and therapist-patient dynamic. Transcriptions were interpreted in accordance with the Iterative Thematic Analysis method. Based on the findings, a framework for designing therapeutic spaces for artistic expression will be developed. It is expected that the findings will help designers understand the benefits and limitations of artistic expression for addressing stress in young adults, and how artistic expression can be utilized to develop therapeutic spaces.

2 REFERENCES

Malchiodi, C. (Ed.). (2023). Handbook of Expressive Arts Therapy. Guilford Publications.

Masterson, J., King-West, E., Findlay, J., Carr, R., Kaplan, F., Bridgham, T., Christian, D., Hass-Cohen, N., Galbraith, A., & Loya, N. (2008). Art Therapy and Clinical Neuroscience (N. Hass-Cohen & R. Carr, Eds.). Jessica Kingsley Publishers.

Varma, P., Junge, M., Meaklim, H., & Jackson, M. (2021). Younger people are more vulnerable to stress, anxiety and depression during COVID-19 pandemic: A global cross-sectional survey. Progress in Neuro-Psychopharmacology and Biological Psychiatry, 109(1), https://doi.org/10.1016/j.pnpbp.2020.110236

STRESS REDUCTION IN REAL AND VIRTUAL ENVIRONMENTS: DEVELOPMENT OF A SCALE TO ASSESS INDIVIDUALS' ORIENTATIONS TOWARD PHYSICAL AND VIRTUAL ENVIRONMENTS

Track 4: Design for Health and Wellbeing
Knowledge Network: Cyberspace & Digital Environments
Methods: Design Methods, Survey
Ruth Barankevich (University of California Irvine); Daniel Stokols (School of Social Ecology, UC Irvine)

1 ABSTRACT

Over the past few decades, environmental influences on people's behavioral, socioemotional, and health outcomes have changed in a number of respects due to the rapid influx of digital technology and online worlds into everyday life. This study builds on earlier research studies of personality and environment (e.g., Craik, 1977; McKechnie, 1974), and especially research on individuals' dispositional orientations toward their physical (place-based) and virtual (cyber-mediated) settings (e.g., Misra & Stokols, 2012). The present study describes people's shifting patterns of use of online and digital tools over the last decade, and with it a changing understanding and definition of environmental orientations. This presentation articulates the redevelopment and testing of a Physical-Virtual Environmental Orientations (PVEO) scale, meant to gauge an individual's level of use, perceived importance and meaning of, and attachment to physical and virtual places. The scale is comprised of questions that address both physical and digital components, outputting a numerical pair of P-V orientation scores that can be placed on a spectrum, ranging from strongly connected to completely disconnected. We link these results to existing cyber-based, place-based, cyber-placed based, or placeless categories (cf., Misra & Stokols, 2012) and other related constructs. We present the validated measure along with assessments of its validity and reliability. Finally, we identify potential uses for the scale and discuss next steps.

This scale was developed for use in a project funded by the Michael Brill Research Grant.

2 REFERENCES

Craik, K. H. (1977). Multiple scientific paradigms in environmental psychology. International Journal of Psychology, 12(2), 147-157.
McKechnie, G. E. (1974). Environmental response inventory. Consulting Psychologists Press.
Misra, S., & Stokols, D. (2012). A typology of people-environment relationships in the Digital Age. Technology in Society, 34(4), 311-325.

ENHANCING MENTAL WELL-BEING THROUGH SPACE DESIGN: A PILOT STUDY ON ANXIETY REDUCTION IN YOUNG ADULTS

Track 4: Design for Health and Wellbeing
Knowledge Network: Cyberspace & Digital Environments, Health
Methods: Experimental/Quasi-Experimental Research, Statistical/Modeling Analysis

So-Yeon Yoon (Cornell University); Yuexing Hao (Cornell University DEA); Jay Yoon (Cornell University)

1 ABSTRACT

This study examines the potential of creative space design problem-solving as a therapeutic tool for reducing anxiety among young adults, a demographic increasingly facing mental health challenges. The research involved 30 college students (23 females, 7 males, aged 18-30, who undertook a creative task of designing a small home office (12'x10') using an online 3D space configuration tool during finals. The task duration ranged from 8 to 18 minutes.

To assess participants' anxiety levels, we used the State-Trait Anxiety Inventory (Spielberger, 1983) before and after the design task. Their engagement and psychological state during the task were evaluated using the Flow State Scale (Jackson & Marsh, 1996). We also conducted a survey incorporating the Big Five Personality Inventory (John et al., 1991), the Savoring Configuration Inventory (SCI) (Lauzon & Green-Demers, 2020), and the Rational-Experiential Inventory (REI) (Pacini & Epstein, 1999) for individual trait assessments. These were administered separately to avoid additional cognitive load and fatigue for participants.

Data collected from the questionnaires was analyzed using mixed linear regression analysis. This approach was chosen for its ability to handle the complexities of repeated measures and the variability in individual responses. The mixed linear regression model revealed a significant reduction in anxiety levels post-task (p = .049), particularly among participants exhibiting higher neuroticism. This result suggests that creative problem-solving activities like space design can be particularly beneficial for individuals with certain personality characteristics.

Furthermore, the hedonic savoring scores, a subscale of SCI, and the experiential ability scores, a subscale of REI, were found to correlate significantly with the reduction in anxiety. These findings highlight the psychological benefit of engaging in and enjoying creative activities. In addition to these quantitative findings, the study includes qualitative insights derived from observational learning through recorded footage of participants during the design task. This aspect of the study offers a deeper understanding of the participants' behavioral and emotional responses, adding a rich layer of context to the statistical results.

The implications of these findings are particularly relevant given the broadening scope of anxiety-related challenges faced by young adults. While our initial study was focused on a specific demographic – college students during their finals – the results suggest broader applicability. To build upon these findings and address the limitations of the pilot study, such as the predominantly female and environmental design major participant pool, a more inclusive follow-up study will expand the understanding of the therapeutic potential of creative space design in managing anxiety, particularly among a diverse young adult population.

2 REFERENCES

Lauzon, A., & Green-Demers, I. (2020). More of a good thing is even better: Towards a new conceptualization of the nature of savouring experiences. Journal of Happiness Studies: An Interdisciplinary Forum on Subjective Well-Being, 21(4), 1225–1249.

Pacini, R., & Epstein, S. (1999). The relation of rational and experiential information processing styles to personality, basic beliefs, and the ratio-bias phenomenon. Journal of Personality and Social Psychology, 76(6), 972–987.

Richards, R. (Ed.). (2007). Everyday creativity and new views of human nature: Psychological, social, and spiritual perspectives. APA.

HOW DOES TECHNOLOGY ACCEPTANCE OF AUGMENTED REALITY PEDESTRIAN-ORIENTED WAYFINDING APPLICATIONS INFLUENCE THE PERCEIVED WALKABILITY AND TRAVEL BEHAVIOR?

Track 4: Design for Health and Wellbeing
Knowledge Network: Cyberspace & Digital Environments, Health
Methods: Experimental/Quasi-Experimental Research, Survey
Yanhua Lu (North Carolina State University); Celen Pasalar (College of Design, North Carolina State University)

1 ABSTRACT

As one of the key components of physical enticing that aids walkability, pedestrian-oriented wayfinding tools have been widely adopted in complex outdoor environments, including conventional wayfinding aids and digital wayfinding applications.

Conventional pedestrian-oriented wayfinding tools such as signage and printed maps help change people's perception and mindset, by providing information such as walking distance and time, and points of interest that are accessible by walking. To improve users' experience for better navigation and a better understanding of the environment, new technologies such as location-aware GPS (global positioning system) and GNSS (Global Navigation Satellite Systems) technologies have been employed in digital wayfinding applications, which results in 2D or 3D digital maps in various forms. In recent years, Augmented Reality (AR) has been applied in mobile pedestrian wayfinding applications for outdoor navigation, such as Live View on Google Maps. AR mobile wayfinding applications generate a mixed reality that allows people to see the physical environment and the virtual environment simultaneously. While pedestrians see the physical environment, the same scene is also shown on their phone display, with virtual elements added on top of the objects, such as directions and distance.

Theory of Reasoned Action (TRA) and Theory of Planned Behavior (TPB) (Ajzen, 1991) suggest that people's behavior is determined by their intention to perform the behavior, which is influenced by their attitudes towards the behavior, and the subjective norms, and the perception of ease or difficulty to perform the behavior. According to the TAR and TPB theories, whether people use AR wayfinding applications is determined by their intention to use them, which is influenced by their attitudes towards using the AR app, and how important others think of us using the AR app. Literature (Chen et al., 2022) suggests that location-based AR navigation apps may help reduce cognitive load, improve spatial abilities, and provide enjoyment. On the other hand, unfamiliarity with the technology, the interface, and the interaction design may increase the cognition workload, and reduce satisfaction with the tool. Thus, it is important to understand how AR wayfinding applications are accepted by the participants, which can further help interpret their behavior and environmental perception.

To look into this question, this research applies an experiment, a pre-experimental survey, and a post-experimental survey. Each participant is asked to walk on the same route using an assigned pedestrian wayfinding tool, i.e., printed map, 2D Google map, or Live View on Google Maps. Survey data is used for pre- and post-comparison, as well as the cross-comparison between groups of participants that use different types of wayfinding in the experiment. A correlational analysis is conducted between the technology acceptance of the assigned wayfinding, the level of perceived walkability, and travel behavior.

2 REFERENCES

Ajzen, I. (1991). The theory of planned behavior. Organizational Behavior and Human Decision Processes, 50(2), 179-211.

Chen, C., Liu, C., Chiu, T., Lee, Y., & Wu, K. (2022). Role of Perceived Ease of Use for Augmented Reality App Designed to Help Children Navigate Smart Libraries. International Journal of Human–Computer Interaction, 1-18.

IMPACT OF GENERATIVE ART ON MOOD MANAGEMENT, SHOPPING SATISFACTION, AND WELL-BEING: AN EXPERIMENTAL STUDY USING THE METAVERSE RETAIL ENVIRONMENT

Track 4: Design for Health and Wellbeing
Knowledge Network: Cyberspace & Digital Environments, Interior Design
Methods: Design Methods, Experimental/Quasi-Experimental Research
Woo Bin KIM (Cornell University); Jhovanna Perez; So-Yeon Yoon (Cornell University)

1 ABSTRACT

Generative art, a novel form of tech-driven art, utilizes AI algorithms to produce diverse artworks, offering both aesthetic and cost-effective approaches. When integrated into retail or service environments, it serves as a valuable medium, enhancing human well-being beyond the sensory experience (Peralta, 2023). This investigation is guided by the perspectives of retail therapy and mood management theory. Aligned with these theoretical frameworks, individuals can regulate their moods through various stimulating media (Choi et al., 2023), with the retail environment serving as a prominent exemplar (Lee & Lee, 2019). Art is a crucial resource in crafting a distinctive spatial experience, fulfilling a therapeutic function by addressing psychological concerns (Lee & Lee, 2019). Therefore, as visitors traverse retail spaces, exposure to generative art can emerge as a therapeutic experience, effectively nurturing psychological well-being during the shopping process.

Amid COVID-related concerns about physical contact, the metaverse—a virtual environment enabling daily activities and real-time engagement—has captured public interest (Choi et al., 2023). Additionally, the metaverse is recognized as an expandable platform for practitioners, enhancing human-environment interactions beyond realistic time and space constraints (Choi et al., 2023). With profound shifts in individuals' consumption landscape in this digital era, investigating the impact of generative art on users' psychological improvement within the virtual environment becomes both timely and significant.

Specifically, this study explores how generative art experiences in a metaverse retail space impact visitors' mood status, subsequently influencing shopping satisfaction and well-being. For this, 3D retail environments were designed on the Spatial.io metaverse platform. Generative art imagery adorned a large feature wall in a virtual store, with a similar effect achievable through a projection mapping technique in the real-world environment. A between-subject experiment compared a non-art group (n=30) with an art presence group (n=32). After navigating the metaverse environment, participants responded to items related to mood management, shopping satisfaction, and shopping well-being. Using the PROCESS procedure (Models 80 and 84), significant serial mediation paths were identified. When interacting in an art-infused virtual store (vs. non-art), participants experienced reinforced positive mood and alleviated negative mood simultaneously. Elevated mood resulted in increased shopping satisfaction and well-being. Notably, the mediation paths of positive and negative mood were moderated by age, with the therapeutic effect of generative art being stronger for older participants compared to their younger counterparts.

This study initially explored the role of AI-based generative art as a therapeutic stimulus within a digital environment, extendable to real-world contexts. Our findings offer insights for decision-makers, including designers, artists, and technicians interested in developing a human-centric environment, ultimately providing individuals with greater fulfillment of their psychological well-being needs in their daily lives.

2 REFERENCES

Choi, D., Lee, H. K., & Kim, D. Y. (2023). Mood management through the metaverse enhances life satisfaction. International Journal of Consumer Studies, 47(4), 1533-1543.

Lee, J., & Lee, Y. (2019). Does online shopping make consumers feel better? Exploring online retail therapy effects on consumers' attitudes towards online shopping malls. Asia Pacific Journal of Marketing and Logistics, 31(2), 464-479.

Peralta, A. (2023). Generative Art is Helping (re) Humanize the Retail Experience.

CAPTURING THE "WOW" EXPERIENCE IN BUILT FORM: AN EXPLORATORY STUDY

Track 4: Design for Health and Wellbeing
Knowledge Network: Movement in Designed Environments: Nature & Ecology, None/Not Tied to Any Network
Methods: Experimental/Quasi-Experimental Research, Interviews

Newton D'souza (Florida International University); Sam Moshaver (Florida International University); Bimal Balakrishnan (Mississippi State University); Armin Mostafavi (NA)

1 ABSTRACT

When an interior or architectural space elicits a positive emotional response, it creates a memorable and pleasurable experience. This emotional impact, often expressed through people's reactions like "wow," is evident when one experiences evocative architectural spaces such as Unity Chapel or Westminster Abbey. To incorporate this "wow" factor into built form, it is essential to explore methods for measuring and understanding it.

Designers aim to create pleasurable experiences through various spatial strategies. One such manipulation popular in architectural discourse is called "compression/ expansion" (Crouch, 2013). Compression narrows spaces, fostering intimacy or tension, while expansion opens up spaces, instilling freedom and tranquility. Compression is enhanced through low lighting, and expansion is accentuated by bright lighting. In other words, our hypothesis is that the contrast experienced when one moves from a compressed dimly lit space to an expanded brightly lit space makes the experience pleasurable and is expressed as a "wow" factor. Therefore, the study hypothesizes that the relative difference of spatial scale (measured in terms of ceiling height contrast) and brightness (measured in terms of luminance contrast) of a given space will lead to a "wow" factor in design. As an architectural example, the Pantheon in Rome exemplifies this transition, moving from a dark portal to a bright expanded dome characterized by the oculus. In our study, the Pantheon building is simulated by manipulating ceiling height contrast and luminance contrast as the primary independent variables.

The stimulus comprises a two-minute pre-canned animation of the Pantheon space manipulated in the following four conditions) both ceiling height and luminance contrast; 2) only ceiling height contrast with no luminance contrast; 3) no ceiling height contrast but only luminance contrast and 4) no ceiling height contrast nor luminance contrast. The hypothesis is that the 'wow' factor is most evident in condition 1, mirroring the authentic experience of the Pantheon in its current form.

The "wow" factor is measured through emotional responses consisting of two dependent variables, namely arousal and valence (Russel, 1980). Arousal represents the intensity of a stimulus, ranging from dull to stimulating, while valence represents the pleasantness-unpleasantness spectrum, from negative to positive. In phase 1 of the experiment, arousal-valence measures are obtained through a PANAS Likert scale and think-aloud protocols that elicit verbalization of experiences. These measures are then examined for moments of heightened arousal and positive valence in the Pantheon animation. In phase 2, plans are proposed for detecting the 'wow' experience through neural measurements using EEG devices.

While various factors contribute to a "wow" experience in architecture, this study focuses on ceiling height and luminance contrast as primary variables, acknowledging potential gaps in capturing all environmental complexities. The objective is measuring the spatial strategy of "compression-expansion" and its effectiveness of inducing a 'wow" experience in design. The contribution by developing a methodology to quantify the "wow" factor in architecture through emotional and behavioral measures, using spatial and lighting contrasts in a Pantheon-based experiment.

CHILDREN'S HEALTH AND WELL-BEING RANDOM FOREST ANALYSIS OF WALKABILITY: EVALUATING THE IMPACT OF VEGETATION ON URBAN AND RURAL AREAS IN MICHIGAN

Track 4: Design for Health and Wellbeing
Knowledge Network: Health, Sustainable Planning, Design & Behavior
Methods: Meta-Analysis, Statistical/Modeling Analysis
Musab Wedyan (Michigan State University); Fatemeh Saeidi-Rizi (Assistant Professor)

1 ABSTRACT

Walkability is a crucial aspect of both urban and rural planning, influencing public health, and the quality of life for residents. Utilizing multi-source data and machine learning techniques such as random forest, gradient boosting decision tree, and XGBoost, previous studies have investigated the complex nonlinear relationships between the built environment (BE) and mobility behaviors (Zang et al., 2022). However, these studies overlook how vegetation, measured by the Normalized Difference Vegetation Index (NDVI), affects walkability in different settings. Addressing this gap, our study employs the Random Forest algorithm to differentiate the impact of NDVI on walkability between urban and rural areas in Michigan. The primary walkability data was obtained from the United States Environmental Protection Agency (EPA), while the NDVI and Land Use classification data were sourced from the National Environmental Public Health Tracking Network. In 2017, the Environmental Protection Agency (EPA) developed the National Walkability Index, a tool designed to simplify the evaluation of neighborhood walkability (Thomas & Zeller, 2017).

The Index is based on four indicators: intersection density (design), proximity to transit stops (distance), a mix of employment and household types (diversity). Intersection Density relates to the number of intersections affecting walkability. Proximity to Transit Stops measures the distance to the nearest transit stop, influencing walking frequency. Employment Mix refers to the variety of jobs in an area, linked to walking levels. Employment and Household Mix combines job diversity and housing occupancy, also correlating with more walking. The sample size included 2708 census tract cases. NDVI was divided into two categories: the first ranging from 0.04 to 0.5, and the second from greater than 0.5 to 0.8. The results indicate that in rural areas with lower NDVI (0.1-0.5), intersection density emerges as the most important factor (0.6) for walkability, followed by employment mix and employment and household mix, with proximity to transit stops being the least important. As NDVI increases above 0.5, intersection density remains paramount, but proximity to transit stops appears insignificant (0.0). In urban areas, proximity to transit stops is the leading indicator of walkability across all NDVI levels, with its importance slightly diminishing as NDVI increases, yet it remains the strongest predictor. The importance of intersection density in urban areas becomes more pronounced with higher NDVI. To conclude, we suggest that in areas with low NDVI levels, urban policies should focus on boosting green space to enhance walkability, while rural strategies should enhance the connectivity of pathways amidst sparse vegetation. Conversely, in high NDVI regions, urban planning should effectively integrate these green spaces with pedestrian networks, whereas rural policies might focus on maximizing the utility of existing greenery for walkability.

2 REFERENCES

Zang, P., Qiu, H., Xian, F., Yang, L., Qiu, Y., & Guo, H. (2022). Nonlinear effects of the built environment on light physical activity among older adults: the case of Lanzhou, China. International journal of environmental research and public health, 19(14), 8848.

Thomas, J., & Zeller, L. (2017). National walkability index user guide and methodology. Environ. Prot. Agency: Washington, DC, USA.

PLAYING NATURALLY: EXAMINING DIFFERENCES IN SUPPORT FOR DIVERSE OUTDOOR PLAY AND INTERACTION EXPERIENCES IN NATURALIZED VERSUS CONVENTIONAL PLAYSPACES

Track 4: Design for Health and Wellbeing
Knowledge Network: Children, Youth & Environment, Movement in Designed Environments: Nature & Ecology
Methods: Observation
Janet Loebach (Cornell University)

1 ABSTRACT

While research has firmly established the superior benefits of nature-engaged outdoor play, few studies have directly compared play behaviors supported by natural playspaces versus conventional outdoor playspaces, in part because of the dearth of data collection methodologies capable of capturing both children's in-situ play behavior and related environmental supports. Most outdoor playspaces designed for schools, parks, or other public spaces also tend to rely heavily on manufactured play structures; while supportive of physical play activities such as climbing, swinging, and group games, play structures are often unable to provide affordances which support multiple and diverse forms of play, limiting their ability to foster healthy development across all domains. Play structures also tend to be designed for the needs and abilities of younger children (under 10 years), and don't support the differing abilities and interests of older children. Natural playspaces not only tend to provide more varied and complex play opportunities, the additional exposure to and playful engagement with nature can provide additional benefits for children's physical, emotional and social health. Nature-rich outdoor playspaces therefore have the potential to improve both play quality and diversity while providing the myriad benefits of nature engagement. However, there is little scholarship to date which directly demonstrates the increased play and interaction diversity offered by natural outdoor playspaces compared to more conventional playspaces.

This study utilizes the behaviour mapping protocol and outdoor play typology from the Tool for Observing Play Outdoors (Loebach & Cox, 2020) to examine the differences in children's play and social interaction behaviours in outdoor playspaces across the continuum from conventional playgrounds to naturalized playspaces. Observational behaviour mapping was performed at four (4) public outdoor playspaces in San Francisco, USA; study sites included a conventional playground, a highly naturalized playground, and two 'hybrid' natural playspaces which still primarily feature play structures but utilize natural materials in a more nature-rich setting. Observations were conducted at each site across 9 to 12 days (including weekday and weekend days) by 2 to 3 observers between October 2022 and May 2023. This presentation will illustrate findings from the study including examination of the differences in environmental affordances and conditions for play provided across the conventional-naturalized playspace spectrum, as well as differences in outdoor play types and behaviours occurring within each playspace type. Highlighted findings will include impacts of playspace type on diversity of play, peer & adult social interaction, nature & wildlife interaction, and risky play behaviors. The presentation will also include an evaluation of the value and efficacy of observational behaviour mapping for evaluating the play value and affordances offered by outdoor playspaces.

TYPOLOGY AS A TOOL FOR UNDERSTANDING TREATMENT IMPACTS ON ARCHITECTURE: THE EVOLUTION OF PEDIATRIC BEHAVIORAL HEALTH FACILITIES

Track 4: Design for Health and Wellbeing
Knowledge Network: Children, Youth & Environment, Health
Methods: Case Study, Practice-based Research

Lyndsey Deaton (Clemson University); Zahra Ghazanfari (Clemson University); Graham Denton (Clemson University); Shan Jiang (GBBN Architects); Arielle Spencer (Clemson University)

1 ABSTRACT

Urban buildings have always embodied both cultural traditions and practical organization, changing over time in ways that reflect social evolution. The past two centuries have seen a proliferation of new and transformed building types connected with the growth of industrial capitalism and associated changes rendering building types that reflect the means of building production as well as mirroring status quo architectural knowledge (Markus 1993, Pevsner 1979). In the second decade of the new millennium, we are experiencing further changes connected to globalization, the emergence of information technology, climate change, and advances in human health and wellbeing that converge in the design of healthcare facilities. By studying the spatial form of healthcare facilities such as hospitals, behavioral health centers, and ambulatory care facilities and classifying them into distinct typologies, architects stand to gain important insights about the comprehensive impacts of social evolution and scientific advancements on the built environment.

The pediatric behavioral health facility (PBHF) is a healthcare facility that has significantly changed in program and spatial layout as well as increased in demand over the past half-century (Lutterman 2022). Most US state codes similarly define a behavioral health facility as a facility that provides mental health services in an outpatient or residential setting to ameliorate mental, emotional, behavioral, or substance use disorders or issues. Pediatric facilities focus on the wellbeing of children aged 3-17 and as appropriate provide continued care for adults who were diagnosed with chronic disorders as children. This paper specifically focuses on medium to large scale residential PBHF as they starkly highlight social evolution in architect's design intentions—specifically responding to advancements in treatment and rising public health concerns such as mass shootings, anxiety/depression, and obesity. However, models for evaluating the efficiency of PBHFs have been slow to change and still rely on out-dated metrics such as in-patient bed counts and area efficiency ratios. Without a comprehensive understanding of how these facilities have evolved—beyond the scope of a single firm—architects and owners are not able to revise benchmarks and precedents, thus holding back design progress, impeding accurate cost estimates, and limiting programmatic goals.

This paper answers the research question, "What is the PBHF architectural typology commonly used for benchmarking?" by reviewing floor plans and design intentions of 10 PBHFs against a literature review of social and policy changes, this paper cultivates a defined program tightly connected to social evolution during a specific era. The findings were further corroborated through behavior mapping and observations of an iconic PBHF, the Cincinnati Children's Behavioral Health Facility at College Hill, OH, to understand environmental affordance in everyday treatment. The result is an "institutional" PBHF typology that provides a needed foundation for understanding the status quo.

2 REFERENCES

Markus, T. (1993) Buildings and Power: Freedom and Control in the Origin of Modern Building Types. London: Routledge.

Pevsner, N. (1979). A History of Building Types. Princeton: Princeton University Press.

Lutterman, T. (2022) "Trends in Psychiatric Inpatient Capacity..." From Crisis to Care Series. National Association of State Mental Health Program Directors Research Institute.

ANALYSING THE IMPACT OF CAMPUS GREEN SPACES ON STUDENT WELL-BEING: AN AI-ASSISTED APPROACH

Track 4: Design for Health and Wellbeing
Knowledge Network: Children, Youth & Environment, Health
Methods: Mixed Methods, Photo-based Methods

Yu-Chen Yeh (National Taiwan University); Chun-Yen Chang (National Taiwan University); Fatemeh Saeidi-Rizi (Assistant Professor)

1 ABSTRACT

The campus environment plays a crucial role in influencing the well-being of teachers and students. It encompasses physical, psychological, social, and spiritual health aspects. Despite its importance, campus green space, a vital health resource, remains widely overlooked. These spaces offer campus users respite from academic stress, opportunities for physical activity, and venues for social interaction, essential for learning and overall well-being (Foellmer et al., 2021). Previous studies have tended to focus either on land cover or visual elements, but rarely both aspects of environmental structure. This study aims to bridge this gap by using AI technologies to object detection of campus land cover (Aszkowski et al., 2023) and to analyze the percentage of visual elements in landscapes (Ho, 2023), enabling a comprehensive evaluation of environmental factors that contribute to students' social interactions and emotional states.

This study examined the role of green spaces at Michigan State University and their impact on student well-being. Participants (18 -30 years old) are required to provide GPS coordinates of two significant locations pertinent to their daily routines and answer questions related to activities, behavior, and feelings through an online questionnaire. The first location is to be a personal space that the participant frequents individually, while the second location is to be a communal space where the participant engages in activities with others. The focus was on identifying environmental elements and structures that support daily social activities and influence emotional well-being. Subsequently, we analyzed the structures of activity places, including facilities, size, and percentage of landscape features in on-site photos.
In conclusion, this study highlights the effective use of AI and GIS technologies, combined with student feedback, in identifying the emotional impacts of campus environments at Michigan State University. The development of 'emotional maps' has been particularly instrumental in revealing how environmental features affect student well-being. These findings not only facilitate the optimization of campus design to enhance positive experiences but also demonstrate the potential of AI in guiding evidence-based environmental decisions, especially important in settings used by young adults.

2 REFERENCE

Aszkowski, P., Ptak, B., Kraft, M., Pieczyński, D., & Drapikowski, P. (2023). Deepness: Deep neural remote sensing plugin for QGIS. SoftwareX, 23, 101495.

Foellmer, J., Kistemann, T., & Anthonj, C. (2021). Academic Greenspace and Well-Being — Can Campus Landscape be Therapeutic? Evidence from a German University. Wellbeing, Space and Society, 2, 100003.

Ho, L.-C. (2023). LaDeco: A tool to analyze visual landscape elements. Ecological Informatics, 78, 102289.

WHAT MAKES A CITY CHILD-FRIENDLY? DEFINING, MEASURING, AND VISUALIZING CHILD-FRIENDLY INDICATORS

Track 4: Design for Health and Wellbeing
Knowledge Network: Children, Youth & Environment, Cities & Globalization
Methods: Mixed Methods

Muntazar Monsur (Texas Tech University); Nazia Afrin Trina (Texas Tech University);
S Y Andalib (Student at Texas Tech University)

1 ABSTRACT

The concept of a child-friendly city is gaining momentum, yet there's a dearth of research on systematically measuring its aspects. Visualizing a city's child-friendliness through mapped data and attributes would shed light on the intricate processes of defining and gauging such indicators. Sadly, many cities worldwide have become increasingly hostile to children over the past decades, leading to a significant decline in their outdoor time and physical activity, adversely affecting their health. Factors such as excessive car usage, lack of walkable neighborhoods, and limited access to public spaces contribute to this decline. Additionally, rapid urbanization, insufficient infrastructure, and poor urban management exacerbate the situation in some areas.

This research aims to investigate how various aspects of the urban environment and socio-economic factors influence measures of child-friendliness or unfriendliness. Leveraging GIS software and spatial datasets, the goal is to develop a framework for defining, measuring, and visualizing child-friendly indicators across diverse urban contexts. Furthermore, the research aims to offer insights into mapping approaches for cities lacking relevant datasets. By grounding assessments of child-friendliness in data rather than subjective interpretations, this research proposes a novel approach for cities to evaluate their suitability for children.

The study will focus on one or two pertinent locational datasets, such as the locations of childcare centers and parks, and overlay them with demographic and social data in selected cities to illustrate the proposed methodology. Through this case study approach, the primary objective is to grasp the potential and potential pitfalls of this new GIS-based method for understanding, visualizing, and measuring child-friendly indicators in cities.

ENVIRONMENTAL NOVELTY IN DESIGN FOR OLDER ADULTS: INSIGHTS FROM NEUROSCIENCE AND PSYCHOLOGY

Track 4: Design for Health and Wellbeing
Knowledge Network: Environment-Gerontology, Health
Methods: Practice-based Research
Rachael Shields (University of Wisconsin-Madison)

1 ABSTRACT

Environmental novelty is novelty arising from encountering new environments, experiencing reconfigured spaces, or discovering new elements within an environment. Novelty has been proposed as a fundamental psychological need and is recognized for its role in boosting life satisfaction, wellbeing, happiness, and memory (González-Cutre et al., 2020; Schomaker & Meeter, 2015). Older adults, in particular, may benefit from novelty's memory-enhancing effects to combat age-related memory decline. Bonsang, Adam, and Perelman (2012) found a significant drop in cognition approximately a year after an individual retires due to a reduction in mentally stimulating activities—a category in which novelty is underutilized. Furthermore, environmental novelty is recognized for providing a heightened and more enduring cognitive reward than other forms of novelty (Schomaker & Meeter, 2015). Novelty should be intentionally utilized more by designers for its cognitive and psychological benefits. However, the literature in this area may be challenging for designers to navigate due to the abundance of specialized jargon. This study aims to bridge that gap, making the insights more approachable to designers.

This research will employ an integrative review, synthesizing and analyzing literature from neuroscience and psychology to gain a comprehensive understanding of how novelty influences the brain function of older adults, particularly in contexts relevant to environmental design. Articles will be sourced from the PubMed and PsycINFO databases using variants of the keywords "novelty," "exploration," "surprise," "deviance," "older adults," "cognitive impairment," "dementia," "aging," "gerontology," "environment," "environmental factors," "environmental enrichment," "built environment," "design," "neuroscience," "cognition," "cognitive abilities," "memory," and "learning." The inclusion criteria are as follows: peer-reviewed articles, published in the English language, and empirical or relevant review studies published in the last twenty years. ATLAS.ti will be utilized for coding and organizing the data into insights relevant to environmental designers. The expected outcomes of the study encompass recurring themes, methods, results, and discussions, all of which will be synthesized and presented in an integrative review format for environmental designers to utilize.

Environments have a profound impact on how we think and process information, highlighting the necessity of incorporating neuroscientific and psychological insights into design strategies. This study represents a key advancement in environmental design by translating scientific findings into applicable knowledge for designers, emphasizing the potential benefits of novelty for older adults. While additional empirical research is crucial for a deeper understanding of these effects, this work aims to initiate vital discussions on the subject's significance.

2 REFERENCES

Bonsang, E., Adam, S., & Perelman, S. (2012). Does retirement affect cognitive functioning? Journal of health economics, 31(3), 490-501.

González-Cutre, D., Romero-Elías, M., Jiménez-Loaisa, A., Beltrán-Carrillo, V. J., & Hagger, M. S. (2020). Testing the need for novelty as a candidate need in basic psychological needs theory. Motivation and Emotion, 44(2), 295-314.

Schomaker, J., & Meeter, M. (2015). Short-and long-lasting consequences of novelty, deviance and surprise on brain and cognition. Neuroscience & Biobehavioral Reviews, 55, 268-279.

EXPLORING THE IMPACTS OF LIGHTING ON THE SLEEP AND COGNITION OF HEALTHY OLDER ADULTS, A COMPREHENSIVE REVIEW OF METHODOLOGIES

Track 4: Design for Health and Wellbeing
Knowledge Network: Environment-Gerontology, Health
Methods: Interpretive-Historical Research, Meta-Analysis
Mahya Fani (Arizona State University); Nina Sharp (Arizona State University)

1 ABSTRACT
The impacts of lighting conditions on human circadian rhythm, sleep quality, and cognitive performance have been investigated in several studies. Older adults are at a higher risk of developing serious physiological and mental illnesses, such as Alzheimer's or dementia. Research shows that individuals experiencing cognitive decline may have difficulties with memory and executive functions, which can disrupt their sleep quality. Light therapy has emerged as a low-risk intervention to improve sleep quality and cognitive function. However, the optimal methodology for evaluating the efficacy of light therapy in older adults remains unclear. This literature review presents a comprehensive analysis of the methodologies used in previous studies investigating the influence of lighting on sleep and cognitive performance among healthy older adults. The review is structured into six domains: study setting and type, participant characteristics, lighting conditions, study design, sleep quality evaluation methods, and cognitive performance evaluation methods. The findings suggest that current studies lack consistency in study design, lighting conditions, and assessment methods, making it challenging to draw solid conclusions about the efficacy of lighting and light therapy for improving sleep quality and cognitive function in older adults.

2 METHODOLOGY
This review has been conducted according to the Preferred Reporting Items for Systematic Reviews and Meta-Analyses (PRISMA) guidelines. Multiple databases were comprehensively searched for relevant topics in the research area using the relevant keywords. The studies had to investigate the influence of either daylighting or artificial light on healthy older adults' sleep, well-being, and cognitive performances, or a combination of these parameters. After the full screening process —based on the PRISMA guidelines—33 papers were selected and reported in the current literature review.

3 Discussion and Results
We discussed our findings in terms of study setting and design, lighting condition and implementation in practice settings, and evaluation methods of sleep quality and cognitive performance. The review included lab-based and field measurement studies, and the results indicate that most experimental studies were conducted in the lab environment. Although experimental lab-based studies found significant results, they lack the ability to investigate the impact of lighting design in real-world settings. There is a need for further experimental field-based studies that are representative of the everyday environment of older adults. Utilizing a combination of within-subject and between-subject study designs with a crossover approach offers enhanced control over variables. It produces more dependable outcomes when compared to between-subject studies, in which participants are randomly assigned to only one of the experimental lighting conditions. Future studies should control for potential confounding variables and individual differences and ensure that the light exposure conditions are standardized and counterbalanced.

The outcomes of this review furnish a comprehensive analysis of preceding research and recommend that future investigations utilize consistent study designs and lighting conditions and employ a combination of objective and subjective measurement techniques for sleep quality and cognitive performance to attain conclusive findings and establish an evidence-based lighting standard for older adults vulnerable to mental health issues, such as dementia or Alzheimer's.

PLACE'S INFLUENCE ON THE WELL-BEING OF YOUTH IN SAN FRANCISCO

Track 4: Design for Health and Wellbeing
Knowledge Network: Children, Youth & Environment, Cities & Globalization
Methods: Focus Groups
Cory Parker (University of California, Davis); Patsy Eubanks Owens (University of California, Davis)

1 ABSTRACT

For many young people, access to a public realm created for adults leads to inequities, particularly in lower income neighborhoods. Lower income communities and communities of color frequently receive lower investments of funding for public space allocations and improvements (Joassart-Marcelli, 2010). This study examines the role that the physical environment plays in supporting the health and wellbeing of youth in the Tenderloin neighborhood of San Francisco, California. It is part of a larger global effort to understand youth experiences in the urban landscape called Growing Up in Cities. This research is one component of the third generation of the GUIC initiative which builds upon studies in the 1970s (Lynch, 1977) and 1990s (Chawla, 2002).

Youth-engagement methods included focus groups, a photo-framing exercise conducted during youth-led tours, and informal behavior observations, findings which were presented in 2023. Our ongoing research with youth in the Tenderloin has led to additional findings, a youth engagement and art exercise, and a mapping of the neighborhood, called The Adolescent City. Through these engagements, youth offered ideas for making their community a better place to grow up. Additional findings include the continued importance of indoor or semi-indoor facilities at the expense of outdoor space, as well as challenges to youth (or youth worker) burn out while part of the research study. To avoid taxing a few particular youth with research responsibilities, we then recruited adults who grew up in the neighborhood and still work there to describe their experience and changes to the place. The compilation of data into a neighborhood map revealed several particular streets that, with the right design investment, would provide more safe outdoor gathering spaces for youth.

2 REFERENCES

Chawla, L., ed. (2002). Growing Up in an Urbanising World. UNESCO: Paris and Earthscan: London.

Joassart-Marcelli, P. (2010). Leveling the Playing Field? Urban Disparities in Funding for Local Parks and Recreation in the Los Angeles Region. Environment and Planning A: Economy and Space, 42, 5, 1174–1192. https://doi.org/10.1068/a42198

Lynch, K. (1977). Growing Up in Cities. MIT Press: Cambridge, MA

OUTDOOR RECREATION SPACE AND WELLBEING IN SINGAPORE: A SOCIO-TEMPORAL PERSPECTIVE

Track 4: Design for Health and Wellbeing
Knowledge Network: Health, Sustainable Planning, Design & Behavior
Methods: Statistical/Modeling Analysis, Survey
Yohei Kato (Singapore University of Technology and Design);
Belinda Yuen (Singapore University of Technology and Design)

1 ABSTRACT

Outdoor recreation space is a vital element in addressing the social, economic, cultural, and ecological needs of diverse populations and ecosystems within cities (Markevych et al., 2017). Outdoor recreation space refers to public open spaces such as green spaces, outdoor sports facilities, and community spaces. These spaces offer opportunities for people to engage in various outdoor activities like relaxation, socializing, recreational pursuits, and promote wellbeing. As part of Singapore's transformation into a City in Nature, the country has expanded its range of open spaces and recreational options in the last decade (Lee, 2017; Tan et al., 2013). Despite these national efforts, few studies have examined how people interact with and perceive outdoor recreation spaces (Lok & Leng, 2014; Yuen, 1996). There remains a dearth of comprehensive studies on people's perceptions, needs, and experiences across a variety of outdoor settings. This raises questions about how different types and features of outdoor recreation spaces may affect people's perceptions and experiences within these spaces. These factors, in turn, may influence people's satisfaction with these spaces and overall wellbeing.

This study aims to address the literature gap by developing a holistic understanding of Singaporeans' lived experiences within outdoor recreation spaces across diverse settings, socio-demographic groups, and life stages. Guided by the conceptual framework of healthy neighborhood, the person-environment fit theory, and the ecological model of health promotion (Aicher, 1998; Dustin et al., 2009; Kahana et al., 2003), a large-scale, first-of-its-kind outdoor recreation space usage and satisfaction survey was conducted with residents in Singapore (n=1200), from August to October 2023. The survey results identify Singapore residents' outdoor recreation space experience, usage, activity, and satisfaction details. The multi-dimensionality of the constructs—use and satisfaction—will be explored. It will examine: To what extent does satisfaction with outdoor recreation space impact how one uses these spaces? What forms of use may predict satisfaction? What aspects of satisfaction might predict what forms of use of outdoor recreation spaces? Factor and cluster analysis will be used to study lifestyle personas and segmentation models and explain how behavioral, personal, and environmental characteristics influence people's spatial and temporal usage of outdoor recreation spaces. Through moderation and mediation analysis, the study further explores the mechanisms through which distinct usage patterns of outdoor recreation spaces are linked to satisfaction, including accessibility and environmental quality. The findings on person-environment relationships and their implications for planning and design are discussed.

DESIGN FOR HEALTH AND WELL-BEING V ASSESSING NOISE POLLUTION IN SCHOOL BREAK AREAS: MULTIPLE CASE STUDIES IN ORDU, TURKEY

Track 4: Design for Health and Wellbeing
Knowledge Network: Children, Youth & Environment
Methods: Case Study, Interviews

Aybuke Geyik (UW Madison); Halil İbrahim Çakır (Giresun University); Nagihan Baş (Ordu University)

1 ABSTRACT

Studies have shown that high levels of unpredictable noise can result in negative physical and psychological effects, particularly when left uncontrolled (Klatte et al., 2017). The adverse effects are also relevant to the field of education, as high levels of noise could negatively impact students' learning and educational progress (Klatte, Bergstrom, & Lachmann, 2013; Klatte et al., 2017). This study focuses on investigating the effects of noise pollution on students during break times in urban school environments in Ordu, Turkey, incorporating perspectives from teachers and administrators to better understand this environmental stressor.

A number of studies in the literature examine noise levels during school break times, offering valuable insights into the severity of these situations. In their study, Bulunuz et al. (2017) discovered that students perceive noise levels to be notably high during break times. The study found that noise levels during recess were on average 74.56 dBA in a private primary school and 82.18 dBA in a public primary school. These levels were significantly higher than the thresholds established by the Regulation on Assessment and Management of Environmental Noise in Turkey and the European Union Harmonization Laws. Similar studies emphasize the necessity of dealing with noise pollution in schools in order to improve the educational experience and well-being of children.
The study utilized sound level measurements, observations, and interviews to evaluate noise levels, sources of noise, and the perspectives of teachers' and administrators on noise levels in different school environments, including three elementary, three middle, two high schools, a preschool, and a special education school. The findings revealed that noise levels, particularly in outdoor schoolyards, significantly surpassed the World Health Organization's recommended threshold of 55 dB for daytime outdoor residential areas. Three schoolyards exceeded this threshold even in the absence of people, while noise levels during recess averaged 76 dB in schoolyards and reached a maximum of 94 dB in interior areas. These findings emphasize the urgent necessity of addressing noise pollution in schools because of its possible effects on student well-being and academic performance.

In order to expand upon this study, the researchers intend to conduct a survey to gather students' viewpoints on the issue of noise annoyance and impact, thereby providing a more detailed understanding of the wider consequences of noise pollution in educational environments. The researchers also plan to present the participating schools with targeted environmental and school policy suggestions to minimize noise levels and enhance the learning environment for all children.

The significance of this study is in its focused evaluation of noise pollution inside the often neglected setting of school break areas. This research enhances our comprehension of the relationship between environmental elements in educational settings and their influence on student health and learning by examining the distinct problems faced in Ordu, Turkey. The results have the potential to influence policy decisions and school management practices, with the goal of establishing healthier and more favorable learning environments for children, especially in metropolitan areas where the issue of noise pollution is becoming increasingly concerning.

A SYSTEMATIC REVIEW OF LITERATURE ON SURGE CAPACITY IN EMERGENCY DEPARTMENTS

Track 4: Design for Health and Wellbeing
Knowledge Network: Health, Work Environments
Methods: Meta-Analysis, Practice-based Research
Ali Momen-Heravi, Ph.D., EDAC (Page Southerland Page)

1 ABSTRACT

Over the past decade, Emergency Departments (EDs) across the United States experienced more frequent and severe surge events. The combined effect of sudden increases in patient visits, limited resources, potential access blocks, and risks to the infrastructure and staff of EDs during surge events can decrease care delivery performance, negatively impacting the nation's healthcare outcomes.

2 PROBLEM STATEMENT

While healthcare operation professionals have started to optimize the emergency department's operational response to surge events, there is a lack of practical knowledge about the role and potential of the built environment in enhancing emergency department response to overcrowding incidents.

3 METHODS

his study utilized a systematic literature review of peer-reviewed journal articles published between 2012 and 2022 to identify spatial design strategies that enhance surge capacity within the emergency department.

4 RESULTS

The systematic literature review revealed several innovative spatial design features at multiple levels of the environment support clinical workflows, helping to address surge events more effectively. Also, the literature reviews highlighted several spatial need changes due to emerging technologies and best practices.

5 CONCLUSIONS

This study provided new insights into how emergency department spaces can be designed to address surge events and highlighted emerging trends and best practices to enhance surge capacity. The findings from this study can inform healthcare planning professionals and stakeholders to create more resilient emergency departments.

SHAPING TIME: EXPLORING THE IMPACT OF INDOOR BUILT ENVIRONMENT FEATURES ON TEMPORAL PERCEPTION

Track 4: Design for Health and Wellbeing
Knowledge Network: Sustainable Planning, Design & Behavior, Work Environments
Methods: Design Methods, Experimental/Quasi-Experimental Research

Basma Altaf (Stanford University); Sarah Billington (Stanford University)

1 ABSTRACT

Time perception, a key psychological concept, refers to our subjective judgment of the time that has passed within a specific interval. Researchers have explored time perception and its potential impact on shaping human behavior and experiences for an extended period. Numerous studies have established a connection between perceived time availability and positive well-being outcomes, such as increased life satisfaction, reduced stress, better decision making, and increased productivity. On the contrary, time famine, or the perceived unavailability of time, has been linked to many negative consequences, such as depressive symptoms, increased stress, and the risk of hypertension. The way we perceive time can be shaped (Carstensen et al., 1999), making it possible to evoke positive well-being benefits by influencing this perception. Several factors, such as our emotional states, ongoing activities, and the environment we are in, influence how we perceive time at any given moment. Given that humans spend more than 90% of their time indoors, it is important to understand how the built environment affects human well-being.

However, few prior studies have delved into this topic. This exploratory study seeks to fill the gap by investigating how different features of the indoor built environment influence the experience of time. The study comprises two experimental phases, with sample sizes of 270 and 300, respectively. It targeted the US population and was conducted online through the Prolific crowdsourcing platform. Experiment 1 focused on examining the effect of color (blue vs. yellow), while experiment 2 investigated the effect of biophilic features (indoor plants and natural materials vs. no plants and artificial materials) on the experience of time. For convergent validity, we used different environmental contexts for each experiment, i.e., home office for experiment 1, and office for experiment 2. The study was between subjects, with participants being assigned randomly to one of the variable conditions. After exposure to each of the variable conditions through a simulated waiting period, participants were asked to record their experience of time, both subjectively and objectively, using three measures: perceived time availability (Likert scale rating of 1-7), time estimate (quantitative number of seconds), and perceived quickness (Likert scale rating of 1-9). Participants also rated their affective status. We hypothesized that the time perception measures would be sensitive to the manipulations of the built features that the participants experienced during the experiment. Our results confirmed our hypotheses, with participants perceiving time differently in different interior conditions. This study is an initial step to understand how exposure to different indoor features might affect temporal perceptions of occupants. This research can ultimately inform the design of spaces that respond to our temporal needs as well as they do to our physical spatial needs, revolutionizing how we design built structures that enhance the experiences and well-being of occupants.

2 REFERENCES

Carstensen, L. L., Isaacowitz, D. M., & Charles, S. T. (1999). Taking time seriously. A theory of socioemotional selectivity. The American Psychologist, 54(3), 165–181.

INTEGRATING CIRCADIAN LIGHTING DESIGN INTO BEHAVIORAL HEALTH FACILITY

Track 4: Design for Health and Wellbeing
Knowledge Network: Health, Researchers in Professional Practice
Methods: Case Study, Practice-based Research
Mansoureh Jalali

1 ABSTRACT

Research indicates that environmental and psychosocial stressors trigger aggressive behavior in psychiatric facilities, where up to 50% of incidents are violent and over half of nurses experience physical violence annually, yet the impact of architectural design on such behaviors is often overlooked (Browers et al., 2011; Daffern & Howells, 2002). Ulrich et al. (2018) defined an evidence-based conceptual model for designing psychiatric wards to reduce aggression. He introduces ten design features to reduce crowding stress, environmental stress, and positive distractions, and observation. Violence could be reduced if the facility is designed with features that enable patients to control or personalize their space (Ulrich et al., 2018).

Referencing Ulrich et al.'s (2018) study, the creation of stress-reducing environments with positive distractions hinges on factors such as access to daylight, gardens for walking, and nature views. These elements not only enhance patient well-being but also contribute to supporting their circadian rhythm, yielding numerous benefits for both patients and healthcare staff. [Project Name] is a 460,000-square-foot in-patient facility in the state of [State Name], prioritizing safety, flexibility, and dignity. Designed by [Company's Name], the facility accommodates 252 beds, catering to civil and maximum-security needs while considering the well-being of both patients and staff. In this project, the team's approach to electric lighting and daylight design places a deliberate emphasis on circadian rhythm through the incorporation of tunable light sources and sensors. Beyond this, we conducted a detailed room-by-room circadian analysis, extending to common spaces, to ensure equitable access to healthy electric and daylighting design. This inclusive strategy accommodates patients who may be unable to leave their rooms, providing them with the benefits of natural light.

In this project, ALFA facilitated the circadian study conducted on the Rhino 3D model. Electric light sources and materials were meticulously imported from corresponding drawings and schedules to ensure precision. The initial focus was on the living unit—a space where patients congregated in the afternoon—prompted by the client's request for a lighting system to alleviate afternoon drowsiness. Patient rooms were also examined to ensure sufficient daytime stimulus and appropriate evening lighting. Results indicated that, despite adequate daylighting, numerous patient rooms and living units lacked optimal lighting, particularly during overcast winter days, a common occurrence in the project location. These findings prompted recommendations to the lighting designer, highlighting that exceptional LEED-defined daylight design in a project doesn't guarantee adequate lighting in every space to support circadian rhythm. This presentation will delve into the circadian analysis methodology employed, present the findings, and share valuable lessons learned from [project name].

2 REFERENCES

Bowers, L., Stewart, D., Papadopoulos, C., Dack, C., Ross, J., & Khanom, H. (2011). Inpatient violence and aggression: A literature review.

Daffern, M., & Howells, K. (2002). Psychiatric inpatient aggression: A review of structural and functional assessment approaches. Aggression and Violent Behavior, 7(5), 477-497. https://doi.org/10.1016/S1359-1789(01)00073-8.

Ulrich, R. S., Bogren, L., Gardiner, S. K., & Lundin, S. (2018). Psychiatric ward design can reduce aggressive behavior. Journal of Environmental Psychology, 57, 53-66. https://doi.org/10.1016/j.jenvp.2018.05.002.

STUDY ON THE INFLUENCE OF SPATIAL ATTRIBUTES ON PASSENGERS' PATH SELECTION IN FENGTAI HIGH-SPEED RAILWAY STATION: BASED ON EYE TRACKING

Track 4: Design for Health and Wellbeing
Knowledge Network: Environmental & Architectural Phenomenology, Sustainable Planning, Design & Behavior
Methods: Experimental/Quasi-Experimental Research, Statistical/Modeling Analysis

Mo Zhang (Beijing Jiaotong University)

1 ABSTRACT

The daily throughput of passengers in the Beijing high-speed railway station is huge, and the arrangement of different spatial attributes and spatial signs directly affects the transfer efficiency of passengers. In this study, eye-tracking technology in human factors engineering is used to simulate people's path-finding behaviors according to the signage elements in the high-speed railway station, and the following questions are explored: 1, whether the spatial attributes of the transportation space affect the path choice behaviors of the test subjects; 2, the differences in the attention to the environmental elements by the passengers in the transportation space, and the correlation between them and the corresponding spatial attributes.

The experiment selects Fengtai Station, which will be open to traffic in 2022, and constructs a path choice program based on real-life pictures: real-life pictures are used to show different key nodes, and numbers are labeled on the pictures according to the different directions; the subjects find their way according to the signs, and switch to the corresponding spatial pictures after pressing the corresponding numbers, and then they are successful in transferring from the concourse level to the passenger level of Line 10, and then fill in subjective evaluation questionnaires, such as the sense of direction and the validity of the signs. At the end of the test, a subjective evaluation questionnaire was filled out, including the sense of direction and the effectiveness of the signs. Thirty subjects were selected to watch the program on the screen, and the behavioral data including the correct rate of path selection, the number of wrong returns, and the time spent at each intersection were collected simultaneously, and each photo was partitioned by AOI, and the eye movement data of different partitions were collected. The spatial syntax was applied to quantify the spatial attributes of the space represented by each photo and calculate its connectivity and integration. By analyzing the correlation between the spatial attributes of different picture spaces and the subjective data and choice time of the corresponding pictures, the influence of the spatial attributes of different pathfinding spaces on the subjects' path selection can be explored. Analyzing the subjects' eye movement data of different pictures and the spatial attributes of the corresponding space can lead to the difference of visual attention allocation of the subjects in the scenes with different spatial attributes, and according to the questionnaire, we can derive the degree of accuracy of the pathfinding process after the subjects viewed the signs, and analyze the reasonableness of spatial marking.

In this paper, eye-tracking technology is used to carry out wayfinding experiments in Fengtai high-speed railway station to explore the influence of different spatial attributes of transportation space on passengers' subjective behaviors and eye-movement data, to verify the correlation between spatial attributes and the differences in passengers' attention to spatial elements of the environment, and to explore the optimal design of the interchange space of high-speed railway station to improve the efficiency of the high-speed railway station interchange by means of quantitative analysis.

DESIGN, CULTURE & COMMUNICATION BODY CONSCIOUS DESIGN PRINCIPLES

Track 4: Design for Health and Wellbeing
Knowledge Network: Health, Interior Design
Methods: Grounded Theory

Galen Cranz (University of California - Berkeley)

1 ABSTRACT

Over the last twenty-five years, architectural design researchers and educators have written about the importance of the body in the design of objects, furniture, buildings, campuses, and streets, as well as the computer-user interface. This presentation focuses on Cranz (1998), Höök (2018), and Hendren (2020), who advocate body awareness in design for health and well-being, using a somatic perspective. This paper summarizes and analyzes their research in order to clarify the generative principles for Body Conscious Design and the means to achieve them. Designers, design educators, design students, and users of their designs will benefit from having this theory and its methods articulated.

Cranz, in The Chair, explores the cultural significance of the ubiquitous chair, which has shaped the physical and social dimensions of our lives for the last two centuries. Cranz relates the modern era's rampant back pain to increasingly sedentary lifestyle spent in right-angle seating, and goes beyond traditional ergonomic theory to formulate new design guidelines for a healthy lifestyle.

Höök, in Designing with the Body, points out that current design methods emphasize symbolic, language-oriented, and predominantly visual interactions. She proposes a shift to an "experiential, felt, aesthetic stance that focuses on bodies and perception rather than reasoning and intellect" as an alternative to the "aggressive, rapid design processes that dominate commercial interaction design." Soma design allows users to engage in a "smooth, embodied interaction, creating an intimate correspondence between users' actions and system response"—her criteria for successful interactive tools.

Hendren, in What Can a Body Do?, challenges the distinction between able and disabled by arguing that everything humans use from tools to streets is a form of assistive technology, and that all of our bodies change through degrees of ability and disability throughout our lives. She explains that designing for an "average" person helps nobody, because of the significant variation in human bodies.

These three theorists share common principles for Body Conscious Design.
1. Body Conscious Design promotes whole person health and well-being.
2. Body Conscious Design requires somatic education for designers and users.
3. Body Conscious Design encourages bodily movement and a variety of postures.
4. Body Conscious Design includes all of the senses, not only vision.
5. Body Conscious Design accommodates variety in human shapes, sizes, and abilities.
6. Body Conscious Design embraces and insists on choices.
7. Body Conscious Design can be subtle or obvious.
8. Body Conscious Design defies and confronts design and social conventions.
9. Body Conscious Design takes influence from and contributes to both historical and new developments in research, technology, and design.
10. Body Conscious Design accepts and celebrates any new aesthetic consequences of truly accommodating bodies.

2 REFERENCES

Cranz, Galen. (1998). The chair: rethinking culture, body, and design / Galen Cranz. (1st ed.). W.W. Norton.

Höök, Kristina. (2018). Designing with the body: somaesthetic interaction design / Kristina Höök. The MIT Press.

Hendren, S. (2020). What can a body do? : how we meet the built world / Sara Hendren. Riverhead Books.

HEALTHY CHILDREN ENVIRONMENTS DEVELOPING DESIGN GUIDELINES TO SUPPORT HUMAN-CENTERED DESIGN FOR MENTAL AND BEHAVIORAL HEALTH CARE IN THE PEDIATRIC EMERGENCY DEPARTMENT.

Track 4: Design for Health and Wellbeing
Knowledge Network: Health, Researchers in Professional Practice
Methods: Focus Groups, Interviews

Fernanda de Moraes Goulart (Clemson University); Anjali Joseph (Clemson University); Dr. Sahar Mihandoust (Clemson Univeristy); Monica Gripko (Clemson University); Mina Shokrollahi ardekani (Clemson University Center for Health Facilities Design + Testing); Devi Soman (Clemson University Center for Health Facilities Design + Testing); David Allison (Clemson University)

1 ABSTRACT

Over the last ten years, the number of pediatric patients turning to the emergency department (ED) for psychiatric emergency care has increased by 8%, representing circa 13% of all pediatric ED visits (Bommersbach, McKean et al. 2023). Children and adolescents with mental and behavioral health (MBH) complications boarding in the ED are undergoing life-threatening disturbances of behavior, thought, or mood, and oftentimes, the ED environment is not tailored to address to their specifics needs, which may lead to the escalation of disturbances and increased safety risks such as attempts to harm to self, others, or the environment. These circumstances increase the staff's workload, exposing them to physical and psychological risks (Haines, Brown et al. 2017). Considering the increasing concerns about the burden of mental and behavioral health (MBH) problems among youth, it is of importance to develop tools to design safe and efficient ED environments for patients and care providers. The objective of this study was to develop a set of evidence-based design guidelines for designing pediatric Emergency Department units that support patients with MBH conditions, minimizing patient stressors while improving provider well-being. There were three phases of the data collection process: a systematic literature review, followed by in-depth case studies, and finally, a brainstorming workshop with stakeholders. Four ED units were part of the case study analysis, which included an environmental assessment, onsite observations, and qualitative data analysis of 55 interviews with staff (administrators, nurses, physicians, mental health professionals, security staff, and social workers). The workshop facilitated discussions on the current challenges in the care models for MBH children in the ED, bringing together 70 stakeholders, including research team members, technical and clinical advisory committee members, and external experts. Findings are summarized into fourteen evidence-based design guidelines, (1) Provide opportunities for control over environment, sensory stimuli, and belongings. (2) Create opportunities for visual and auditory privacy for secure sharing of patient confidences. (3) Provide opportunities for respite. (4) Maximize visual and situational awareness in the ED through non-invasive surveillance. (5) Optimize movement and flow in the ED. (6) Provide anticipation and clarity of path and steps in healthcare experience. (7) Provide opportunities for quiet, sensory-friendly, age-friendly activity and engagement. (8) Provide opportunities for patient and family engagement. (9) Enhance sense of security for all and reduce opportunities to cause harm. (10) Provide positive distractions. (11) Provide timely access to psychiatric care and reduce unnecessary waiting and delay. (12) Support patient, care partner, and caregiver comfort. (13) Support clear and effective communication between care team members. (14) Provide flexible and adaptable environments to arrange patient and care provider needs. This study highlights many aspects of the pediatric ED environment and improvements that can be incorporated to make these spaces more suitable for MBH patients, care partners, and providers.

2 REFERENCES

Bommersbach, T. J., et al. (2023). National Trends in Mental Health-Related Emergency Department Visits Among Youth, 2011-2020. JAMA 329(17): 1469-1477

Haines, A., et al. (2017). Factors impacting perceived safety among staff working on mental health wards. BJPsych open 3(5): 204-211.

EXPLORING STUDENT AND PARENT PREFERENCES FOR INTERVENTIONS TO INCREASE ACTIVE SCHOOL TRANSPORT

Track 4: Design for Health and Wellbeing
Knowledge Network: Children, Youth & Environment, Health
Methods: Survey

Anna Gannett (Telethon Kids Institute); Paula Hooper (The University of Western Australia); Julie Saunders (The University of Western Australia); Gina Trapp (Telethon Kids Institute)

1 BACKGROUND/AIMS

In recent decades, the number of students who walk or cycle to school has been declining, including within Australia (Aubert et al., 2022; van der Ploeg et al., 2008). Tackling this decline requires targeted interventions to address the key environmental, social and individual factors associated with active school transport (i.e. walking and biking to school). While numerous interventions have been trialled, success has varied across studies. Understanding which interventions students and parents think may be most impactful and if student and parent preferences differ can help prioritise which future interventions are needed. Therefore, this study aimed to identify primary and secondary school student and parent preferences for interventions to increase active school transport participation.

2 METHODS

A cross sectional survey was conducted at primary (n=23) and secondary (n=9) government schools in Perth, Western Australia in 2023. A total of 775 primary school students aged 8-13 years, 453 secondary school students aged 11-18 years, and 885 parents participated. Participants were provided with a list of potential interventions (n=8 were built environment related, n=3 were education-based, n=7 were programs/events, and n=3 were policy-based) and asked to select the top five that would help them/their child walk and/or cycle to school more often.

3 RESULTS

The top five interventions selected by the highest proportion of primary school students were: 'More safe crossings' (60%), 'Paths on both sides of the road' (53%)', 'Dedicated bike paths separated from the roads' (48%), 'More convenient bus/train stops' (42%), and 'Locked bike and scooter storage' (38%). The top five interventions selected by the highest proportion of secondary school students were: 'More safe crossings' (55%), 'More frequent bus/train services' (55%), 'Paths on both sides of the road' (52%)', 'More convenient bus/train stops' (45%), and 'More trees/shade' (44%). The top five interventions selected by the highest proportion of parents were: 'More safe crossings' (59%), 'Paths on both sides of the road' (47%)', 'Dedicated bike paths separated from the roads' (46%), 'Education programs (e.g., cycling education courses, planning safe routes training' (33%), and 'Drop-off zones located at a walkable distance to the school' (29%).

4 CONCLUSIONS

This study found that the majority of parents and students perceived that 'More safe crossings' and 'Paths on both sides of the road' would increase their likelihood of participating in active school transportation the most often. These findings will help inform future interventions to increase active school transport.

5 REFERENCES

Aubert, S., Barnes, J. D., Demchenko, I., Hawthorne, M., Abdeta, C., Nader, P. A., Sala, J. C. A., Aguilar-Farias, N., Aznar, S., Bakalár, P., Bhawra, J., Brazo-Sayavera, J., Bringas, M., Cagas, J. Y., Carlin, A., Chang, C.-K., Chen, B., Christiansen, L. B., Christie, C. J.-A., ... Tremblay, M. S. (2022). Global Matrix 4.0 Physical Activity Report Card Grades for Children and Adolescents: Results and Analyses From 57 Countries. Journal of Physical Activity and Health, 19(11), 700–728. https://doi.org/10.1123/jpah.2022-0456

van der Ploeg, H. P., Merom, D., Corpuz, G., & Bauman, A. E. (2008). Trends in Australian children traveling to school 1971-2003: Burning petrol or carbohydrates? Preventive Medicine, 46(1), 60–62. https://doi.org/10.1016/j.ypmed.2007.06.002

COOL PLAYGROUNDS: INFLUENCE OF PLAYGROUND SHADE ON CHILDREN'S THERMAL COMFORT AND ASSOCIATED PLAY BEHAVIOURS

Track 4: Design for Health and Wellbeing
Knowledge Network: Children, Youth & Environment, Health
Methods: Case Study, Observation

Breann Corcoran (UBC School of Population & Public Health)

1 ABSTRACT

Skin cancer is the most common form of cancer in North America and rates continue to rise. Children are particularly sensitive to the harmful effects of UV radiation and severe sunburn during childhood sets the stage for higher rates of melanoma later in life. Rising temperatures associated with climate change are contributing to increases in skin cancer incidence and raising new concerns regarding heat-related stress and illness. Heat disproportionately affects children, making them more sensitive and vulnerable to hot ambient conditions compared to adults. Many children's play spaces in Canada are not thermally comfortable and existing playground design standards do not adequately consider sun safety and thermal comfort.1 With play being a critical building block for healthy child development, there is a need to ensure children's play settings are thermally comfortable and sun safe, enabling extended outdoor play opportunities.

Provision of playground shade is an effective prevention strategy that has been shown to reduce harmful exposures to UV radiation and promote thermal comfort among children, but its impact on play behaviours has been relatively unstudied. In this mixed-methods case study, we explored the influence of playground shade on children's perceived thermal comfort and associated outdoor play behaviours at a childcare centre pre- and post-shade intervention. The shade intervention involved 2 shade sails and a temporary UV canopy net installed above outdoor play equipment. We partnered with a Vancouver municipality and used purposive sampling to select a childcare centre with limited existing shade and high sun exposure to implement a shade intervention. With parental consent, 12 children aged 2-3 years participated in the study.

With J.J. Gibson's theory of affordances2 as a theoretical framework for our study, we aimed to observe interactions between individual children and their natural environment, focusing on how shade might afford quality outdoor play experiences through enhanced thermal comfort. Using person-centered observational behaviour mapping to capture in-situ behavior, we simultaneously and systematically recorded observed play behaviours, activity patterns and specific environmental location of individual children over a series of 10-minute periods. We collected data over 4 non-consecutive days in July 2023 with the shade intervention and 4 non-consecutive days in August 2023 with the shade removed, totaling 800 minutes of observation. We also conducted short interviews with children asking questions about their thermal comfort and play. A focus group with Early Childcare Educators (ECEs) and a parent survey were also conducted. Using Merriam's qualitative cross-case methodology3, we performed a within-case analysis. Subsequently, we conducted a cross-case analysis to compare these themes, combined with the ECE focus group and parent survey data. This presentation will share key findings from the cross-case analysis including: 1) Improved thermal comfort among children with shade enhancements; 2) Shade enhancements afford quality play behaviours (particularly physical and exploratory play) and extend engagement in these play activities; 3) Shade affords more social play with peers; and 4) Shade enhancements reduce non-play and restorative play behaviours. These results have the potential to influence guidelines and standards for sun safe and thermally comfortable play spaces.

USING PATIENT AND PROVIDER WORKFLOWS TO UNDERSTAND BARRIERS TO PEDIATRIC MENTAL AND BEHAVIORAL HEALTH CARE IN EMERGENCY DEPARTMENTS

Track 4: Design for Health and Wellbeing
Knowledge Network: Children, Youth & Environment, Health
Methods: Case Study, Interviews

Monica Gripko (Clemson University); Devi Soman (Clemson University Center for Health Facilities Design + Testing); Dr. Sahar Mihandoust (Clemson University); Ann Dietrich (Prisma Health); Lucy Easler (Prisma Health Baptist Columbia SC); Danielle Stewart (USC Greenville College of Medicine)

1 BACKGROUND

In the United States, mental and behavioral health (MBH) conditions affecting school-age children are a public health problem, resulting in an increasing number of Emergency Department (ED) visits (Radhakrishnan, 2023). However, ED settings that are well-equipped to handle medical emergencies are unable to meet the needs of children with MBH conditions in a safe and efficient way. This study uses a multiple case-study design approach to understand the barriers to safe and efficient MBH care for children in the ED with the help of patient and staff workflows. Patient and staff flow mapping has been used in previous research as a data collection tool during case study tours (Machry et al., 2021), to understand provider-patient interactions (Holman et al., 2016), and improve the integration of care (Sifaki-Pistolla et al., 2017). Within EDs, patient flow mapping has effectively identified bottlenecks in the care process affecting length of stay, which can be addressed through process and design interventions (Martin et al., 2011).

2 OBJECTIVES

This study uses patient and provider flow mapping to identify systemic barriers to the care received by children with MBH conditions in four EDs in South Carolina.

3 METHODS

Patient and provider workflows were derived through observations and interviews at the ED case study sites. Two researchers independently observed the patient flows for children with MBH conditions at each ED on two different days to map out the patient pathways from their point of entry into the ED, until their transfer or discharge from the ED. Additionally, the researchers interviewed 55 ED staff members from the case study sites, including administrators, physicians, psychiatrists, social workers, nurses, and patient observers, to understand their workflows and interactions with pediatric MBH patients at different points of their care pathway. Based on these interviews, the researchers developed patient flows identifying the different patient pathways for pediatric MBH patients with agitated behaviors or suicidal ideation, showing their transitions through spaces within the ED and their interactions with various care providers at each step. Further, the researchers developed staff workflows for four key care providers, the ED physician, nurse, social worker, and psychiatrist, to understand their interactions with pediatric MBH patients and the patient-related tasks that affect care delivery.

3 RESULTS

The patient and provider flows indicated that slight variations exist in the patient pathways depending on their context (urban or rural) and availability of resources. Factors such as the hospital policy, care processes, availability of in-house psychiatrists, and a dedicated pediatric MBH holding unit affected each hospital's patient and provider workflows.

4 CONCLUSIONS

Patient and provider flow mapping helped understand and represent the complex process of pediatric MBH care in EDs. Additionally, these workflows helped identify barriers at the various steps for an in-depth analysis to design appropriate interventions to overcome these barriers and improve the safety and efficiency of MBH care for pediatric patients in the ED.

DESIGN FOR AUTISM AND NEURODIVERSITY IN THE EMERGENCY DEPARTMENT

Track 4: Design for Health and Wellbeing
Knowledge Network: Health, Children, Youth & Environment
Methods: Case Study, Interviews
Elyssa Bakker (HK); Deborah Wingler (HKS)

1 ABSTRACT

The emergency room (ER) is an overwhelming place that is especially challenging for patients on the autism spectrum, who may have heightened sensory sensitivities, difficulty with communication, and difficulty adapting to unpredictable situations. This population may externalize stress with disruptive behaviors, and often experience excessive sedation (Brown et.al., 2019), forceful restraint, and delayed or foregone care (Kuo et.al., 2021). Current interventions that attempt to improve care for individuals on the autism spectrum are ad-hoc and limited to pediatric spaces in existing facilities. These interventions are often non-standard, restrictive in access, and outside normal staff workflow. This research project aimed to develop feasible design strategies that reduce environmental stress and enhance opportunities for self-regulation, in a way that is more universal and integrated into the emergency care process.

Phase 1 of this research included a literature review, case studies, and expert interviews to understand current needs for neuro-inclusive design and gathered insights into stakeholder narratives and journey models. Stakeholder narratives synthesized a range of neurodivergent profiles to build awareness towards sensory challenges in the hospital. Journey models mapped out top complaints and recommendations across the ER and pointed to key moments of vulnerability and influence where design may have the greatest impact.

Phase 2 involved observational site visits at two emergency rooms (ERs) that had differing approaches to neuro-inclusive design. The first location was a pediatric ER that used a deployable sensory kit within existing exam spaces. The second location was an ER with two highly specialized treatment rooms for patients on the autism spectrum, with one space dedicated to children and one to adults. Site visits included facility tours, photo essays, and semi-structured interviews with staff. The deployable sensory kit, while successful in providing inclusive access to accommodations, declined over time due to its transient nature. The specialized and dedicated sensory rooms, while widely appealing, had challenges with clinical functionality and were similarly underutilized. These two design models have made major strides towards accommodating autism in healthcare design, but they also demonstrated the need for more integrated design strategies that have improved feasibility and impact.

This research culminated into a visual design guide that suggests sensory accommodations based on these guiding principles: inclusive access, clinical functionality, operational ease, individual control, versatile appeal, and support of calm and refuge. These guiding principles stemmed from staff and designer feedback. Future phases would seek patient reactions from individuals on the autism spectrum and beyond. There is high demand for sensory accommodations in the ER that could transform the care experience for, not only people on the autism spectrum, but also the wider population.

2 REFERENCES

Kouo, T., Bharadwaj, N., Kouo, J., Tackett, S., & Ryan, L. (2021). Assessing Ease of Delivering Emergency Care for Patients with Autism Spectrum Disorders. Journal of Development and Behavioral Pediatrics, 42:704–710.

Brown, J. J., G
ray, J. M., Roback, M. G., Sethuraman, U., Farooqi, A., & Kannikeswaran, N. (2019). Procedural sedation in children with autism spectrum disorders in the emergency department. American Journal of Emergency Medicine, 37(8), 1404–1408. https://doi.org/10.1016/j.ajem.2018.10.025

ANALYZING AND SYNTHESIZING THE LINKAGE BETWEEN ENVIRONMENTAL CONTROL THEORIES AND HEALTHCARE DESIGN: A QUALITATIVE INQUIRY

Track 4: Design for Health and Wellbeing
Knowledge Network: Health, Interior Design
Methods: Case Study, Practice-based Research
Suining Ding (Purdue University Fort Wayne)

1 ABSTRACT

Environmental control has been defined as the degree to which people perceive they have control over various characteristics of their environment (Lee and Brand 2005), as well as the ability to regulate exposure to one's surroundings (Evans and Cohen,1987). A significant volume of research has been conducted on the sense of control in the built environment through environment-behavior (EB) studies. According to scientific evidence, people have a strong need for control over their environments and situations. The primary focus of healthcare design research is to develop theories and use scientific research evidence to inform design decisions. Therefore, what is the linkage between environmental control theory and healthcare design? What specific design guidelines can be derived from EB control theories to inform healthcare design decisions? The findings of this study addressed these two issues. This study took a qualitative approach through literature reviews and case studies. The purpose of this study was to examine the linkage between environmental control theories and healthcare design. Another purpose was to synthesize design strategies to be used in healthcare design based on research findings.

Findings show a sense of control significantly influences stress levels and wellness in diverse situations. Research evidence indicates that benefits are associated with controlling the environment, and negative consequences occur when occupants are restricted to these controls. Previous research demonstrated the importance of perceived control over noise, showing that lack of perceived noise control is linked to performance decrements and increased frustration. In healthcare contexts, lack of control is a pervasive problem that increases stress and adversely affects wellness. Many studies have found that lack of control is associated with negative consequences such as depression, passiveness, elevated blood pressure, and reduced immune system functioning. Research findings also show that patients' sense of control can also be reduced by an unsupportive and uncontrollable physical environment, which often is noisy, restricted, and limited in privacy.

Based on the literature reviewed, connections between environmental control theory and healthcare design practice were identified. As a result of research and theory in behavioral sciences, healthcare environments are likely to promote well-being and help users deal with stress if they consider 1) environmental affordances, 2) behavioral competencies, 3) cognition abilities, and 4) motivational factors. Environment-behavior studies related to environmental control provide valuable guidelines for creating healthcare facilities, which were found in case studies. Although some of the design theories are implicitly applied to healthcare design practice, this study established a link between environmental control theory and healthcare design practice and demonstrated how environmental control theory can be applied to healthcare design practice through case studies. Its significance lies in its direct connection to a theoretical framework and healthcare design.

2 REFERENCES

Evans, G. & Cohen, S. (1987). Environmental Stress. In D. Stokols & I. Altman (eds), Handbook of Environmental Psychology (pp 571-610). New York, NY: Wiley

Lee, S. Y. and J. L. Brand (2005). "Effects of control over office workspace on perceptions of the work environment and work outcomes." Journal of Environmental Psychology, 25(3): 323-333.

USING SPACE SYNTAX TO EXAMINE USERS' PERCEPTION OF PRIVACY IN DOCTORS' OFFICE WAITING ROOMS: SEATING VISIBILITY AND SEATING PREFERENCES

Track 4: Design for Health and Wellbeing
Knowledge Network: Health, Interior Design
Methods: Mixed Methods, Practice-based Research
Suining Ding (Purdue University Fort Wayne)

1 ABSTRACT

There has been little research conducted in waiting rooms of doctors' offices. The literature suggests that people prefer end seats and privacy in a waiting room, not middle seats, and furniture arrangements encouraging social interaction may not always be appropriate (Devlin, 2022). Research findings also suggest that seating placement requires more attention. This is because stress and anxiety increase when there is no personal space for people forced to sit next to each other (Becker & Douglass, 2008). In previous studies, online tools or virtual environments have been used to examine waiting room issues. Only a few studies were conducted in real doctors' office waiting rooms to investigate the environment-behavior relationship with a qualitative approach. Since previous research findings showed various results with various interpretations, it is necessary to take a deeper dive and a different approach - a quantitative approach - to validate and/or confirm the previous research findings. This research addressed the following research questions: (1) What are the seating preferences in a doctor's waiting room? (2) Is visibility of the seats more preferred than a more private seat in a doctor's waiting room? These research questions were investigated in this study.

This research used a quantitative research method, Space Syntax, which is a software that applies space syntax theory and methods to examine behavioral-spatial relationships. Using Space Syntax theory and methods, one can understand how spaces are arranged and connected, and measure individual spaces and the layout of an entire building. Isovist is an essential concept in Space Syntax theory, which refers to hypothetical 360-degree visibility from a particular point within a room layout. Since this study aims to examine patients' visibility preferences, Isovist heatmaps and analyses were generated in Space Syntax software. Through this study, findings shed some light on the phenomenon occurring in doctors' office waiting rooms. The findings from this quantitative research confirmed that patients prefer to have privacy in the waiting room and require sufficient personal space when selecting a seat. In the meantime, patients choose more visible seats in waiting rooms instead of the most private ones.

The significance of this study is that it fills a void in current research by exploring seating visibility, and seating preferences in real settings of waiting rooms. By building on the previous qualitative research in waiting room design, this research makes noteworthy contributions to healthcare design. The results of this study come from a mixed research method that increases the rigor of the study. The research findings make valuable recommendations/guidelines for healthcare design, particularly for waiting room design in doctors' offices. Findings from the research can be used to design a better environment to improve the experience of occupants.

2 REFERENCES

Devlin, A. S. (2022). Seating in Doctors' Waiting Rooms: Has COVID-19 Changed Our Choices? HERD: Health Environments Research & Design Journal, 19375867221104248.

Becker, F., & Douglass, S. (2008). The ecology of the patient visit: physical attractiveness, waiting times, and perceived quality of care. The Journal of Ambulatory Care Management, 31(2), 128-141.

TO WHAT EXTENT DOES TECHNOLOGY USAGE IN PRIMARY EDUCATION IMPACT THE COGNITIVE DEVELOPMENT OF CHILDREN IN AUSTRALIA BETWEEN THE AGES OF SIX TO TEN?

Track 4: Design for Health and Wellbeing
Knowledge Network: Children, Youth & Environment, Interior Design
Methods: Ethnography, Interviews

Gowri Maelathil (VCUarts Qatar); Marwa Abdelmonem (Virginia Commonwealth School of the Arts-Qatar); Matthew Holmes Dallimore (VCUQ)

1 INTRODUCTION

Australia is part of the Organization for Economic Co-operation and Development (OECD) and is one of the countries with the greatest technology presence within classrooms (OECD, 2015). Concerns are rising about the negative effects of technology on the cognitive development of children. Cognitive development refers to how children form memories, problem-solve, build connections with their peers, and develop empathy and compassion. These drastically increase between the ages of 6 and 10 in schools ("Cognitive and Social Skills to Expect From 6 to 10 Years", n.d). Around 84% of teachers in Australia expressed that technology is a distraction in the learning environment, 78% stated that students' focus has declined, and 59% said that students' willingness to learn has declined (Graham and Sahlberg, 2021). This research aims to review the correlation between children's cognitive abilities and more technology exposure and integration within educational environments.

2 METHODOLOGY

This content analysis observes two variables, children and technology in schools, that are not strongly correlated in the existing literature. Thus, a thorough literature review was carried out using secondary qualitative and quantitative data to cross-examine topics such as children's technology usage in Australian schools, Australian school curriculums, child psychology, the free play deficit phenomenon, and other relevant themes. To contextualize further research findings, pilot interviews were conducted with Australian child psychologists and school teachers to understand the first-hand experience in schools and child psychology.

3 FINDINGS

The findings show that excessive technology usage in learning environments negatively affects the cognitive development of primary school children in Australia as they become more screen-dependent. In the Journal of Psychiatry and Psychiatric Disorders, the evidence shows that brain development is connected to free play during primary education (Bergen, 2018). To support this statement, research findings and the pilot interviews add that the lack of sensory experience, play-based learning, and free play adversely impact children's behavior and emotional and information processing. Their ability to learn and grow their literacy and numeracy skills, think creatively, and communicate diminishes (Merga, 2019). Moreover, no strong evidence suggests that more technology integration increases children's intelligence, academic scores, and achievements.

4 CONCLUSION

Children are the future and need nurturing to become successful adults. This research proposes to design better learning environments for primary school children in Australia through reduced integration of technology. Creating interactive and sensory spaces for children to learn through play fosters their psychological and physical well-being. No technology in the 21st century is realistically impossible, but less technology can make a significant difference by making these young children's learning experiences enriching.

WHERE SENIORS WOULD STAY IN CITIES: ENVIRONMENT PREFERENCES IN CITIES DURING DAILY TRIPS IN SOUTH KOREA

Track 4: Design for Health and Wellbeing
Knowledge Network: Active Living by Design, Health
Methods: Mixed Methods, Statistical/Modeling Analysis
Jounghyun Lee (Korea Advanced Institute of Science and Technology); Yujin Kim (KAIST); Lisa Lim (KAIST)

1 ABSTRACT

Outdoor activities positively influence the health of older adults, fostering their vibrant daily lives. This study investigates the relationships between the built environment and the staying behaviors of seniors in their everyday lives. The primary objective is to identify seniors' built environmental preferences for different purposes of staying and propose an urban design that promotes the vibrant daily life of the aging population. This analysis explored the association between seniors' duration of staying and the specific attributes of buildings and communities.

For this study, we utilized GPS data and trip diaries from senior individuals living in urban contexts in South Korea. We used GPS data to analyze the dependent variable, quantifying the duration of individuals' stays at specific outdoor locations. For the independent variables, spatial analysis was conducted using space syntax and QGIS. The study illustrates where seniors preferred to stay and reveals variations in their environmental preferences depending on the purpose of their trips, emphasizing the nuanced nature of their preferences. These findings hold significant implications for urban planning and design, facilitating the creation of more age-friendly environments within urban settings.

CONTEXTUAL ENHANCEMENT OF INDOOR ENVIRONMENTAL QUALITY IN JORDAN: A DETAILED ANALYSIS USING THE WELL BUILDING STANDARD

Track 4: Design for Health and Wellbeing
Knowledge Network: Cultural Aspects of Design, Health
Methods: Case Study, Survey
Genell W Ebbini (Purdue University)

1 INTRODUCTION

In pursuing optimal workplace health and productivity, this study focuses on the critical importance of Indoor Environmental Quality (IEQ) within Amman, Jordan's unique environmental and cultural context. Leveraging the WELL v2 Building Standard as an analytical lens, this investigation presents a nuanced approach to examining IEQ enhancements in Jordanian work environments, marking a significant contribution to localized IEQ research in the Middle East.

This research extends the dialogue on sustainable building practices in the Middle East and IEQ assessment (Elnaklah et al., 2020; Shamseldin et al., 2020) to the specificities of Jordan. Diverging with a focused lens on workplaces in Amman, the study shows the intricate dynamics between various IEQ factors and their impact on occupant well-being.

Employing a dual-method approach, the research initiative undertook a comprehensive survey targeting 685 employees across diverse educational and research settings, achieving a notable response rate from 374 participants. This substantial engagement facilitated a rich dataset, offering profound insights into the perceptions and realities of IEQ within Jordanian workspaces. The survey encompassed both quantitative metrics and qualitative feedback, enabling a thorough exploration of acoustic quality, air quality, lighting, workspace configuration, and thermal comfort through statistical analysis and thematic interpretation. This methodology ensures a detailed examination of the WELL Building Standard's criteria and their applicability in Amman's unique architectural and environmental context.

Findings from the study underscore the WELL Building Standard's adaptability to Jordan's workplaces. They identify key areas for IEQ intervention, such as noise management, ventilation, and thermal comfort. These insights reveal the potential of contextually adapted IEQ enhancements to significantly improve workplace environments, emphasizing the necessity of customized approaches that consider Jordan's specific climatic and cultural conditions.

Moreover, the research highlighted a significant gap in IEQ awareness among employees in Jordan, pointing to an urgent need for educational programs to enhance understanding and implementation of IEQ principles. Applying the WELL Building Standard as an evaluative tool and as a basis for education and policy development, this study provides actionable recommendations for stakeholders, including policymakers, designers, and organizational leaders, to champion IEQ enhancements that prioritize well-being through informed, context-aware strategies.

PLACE ANALYSIS IN OLD DELHI: INVESTIGATING SOCIO-SPATIAL INTERACTIONS AND ADAPTIVE UTILIZATION OF BUILT AND UNBUILT FABRIC.

Track 4: Design for Health and Wellbeing
Knowledge Network: Cultural Aspects of Design, Health
Methods: Case Study, Observation

Tasha Singh (Drexel University)

1 ABSTRACT

This presentation discusses an observational study in Old Delhi, a historical area known for its rich culture, intricate streets, and lively community life. Characterized by traditional architecture and bustling markets, Old Delhi's 'Mohalla' neighborhoods offer a distinctive setting for analyzing social interactions. The study focuses on understanding social patterns and space use within these neighborhoods, aiming to derive design insights from these dynamic, human-centered environments.

By closely examining how residents engage with the area's distinct spatial features, from constricted 'Galis' (alleyways) to bustling marketplaces, this study seeks to unravel the intricate ways in which these environmental aspects influence and are influenced by the community's social behaviors (Liddle, S.,2007). The research reveals patterns of social engagement and space utilization that highlight the community's creativity and resilience in adapting to their urban environment. The insights gathered here demonstrate how the physical spaces of Old Delhi facilitate unique forms of social interaction and communal gatherings, reflecting a deep interconnection between the residents and their surroundings.

The methodology utilized here integrates direct observation of spatial environments and interviews with neighborhood inhabitants(Bakos, M.,1979). This approach entails rigorously documenting spatial layouts, usage patterns, and interactions, vital for understanding the tangible aspects of spatial use and spontaneous urban activities. Complementing this approach, structured interviews with residents provided qualitative insights into their experiential narratives, lived experiences, and spatial and cultural significance.

This study includes an analysis of 'Chai Wallahs' (tea stalls) in Delhi, which are more than just tea vendors. They serve as dynamic social hubs, influencing the city's spatial layout by fostering communal spaces. This interaction between social behavior and space utilization is crucial for understanding urban design's impact on social connectivity. The role of Chai Wallahs provides deep insights into small-scale urban interventions and their effects on community engagement. This observation aligns with John Zeisel's assertion in Inquiry by Design that "adaptive traces are significant for designers as they are direct manifestations of design by users" (Zeisel, J., 2006). Understanding these organic conditions adds an essential dimension to the ideas of human-centered design.

This research is particularly pertinent amid global concerns about urban loneliness and social disconnection. Studying how people gather, and the environments that support community, is the basis for designing places that encourage connection. The study elucidates a range of strategies for urban spaces that have led to a 'kit of parts' to cultivate spaces that enhance communal bonds, stimulate social interaction, and foster a sense of belonging.

2 REFERENCES

Zeisel, J. (2006). Inquiry by Design: Environment/ Behavior/Neuroscience in Architecture, Interiors, Landscape, and Planning, Updated and Revised Edition. W. W. Norton.

Whyte, W. H. (1980). The social life of small urban spaces. Washington, D.C.: Conservation Foundation.
Liddle, S. (2017). Chandni Chowk: The Mughal City of Old Delhi. Speaking Tiger.

Michael Bakos, Richard Bozic, David Chapin, Judith Gandrus, & Stephen Kahn. (1979). Privacy, Territory, and Participation: Projects for your Environment [Pamphlet]. Architecture Research Construction Inc.

SUPPORTING COMMUNICATION FOR ADOLESCENTS SEEKING MENTAL AND BEHAVIORAL HEALTHCARE IN EMERGENCY DEPARTMENTS

Track 4: Design for Health and Wellbeing
Knowledge Network: Communication
Methods: Mixed Methods
Zahra Ghazanfari (Clemson University); Rishaad Amirali Amlani (Clemson University)

1 INTRODUCTION

Emergency Departments (EDs) often fail to meet the unique needs of children and adolescents seeking care for mental and behavioral health (MBH) crises. The noise, hectic atmosphere, lack of privacy, long waits, infrequent communication from care providers, and opaqueness of ED processes can exacerbate patients' distress and anxiety, especially for teens in mental health crises. Care providers must build rapport with adolescent MBH patients to diagnose their concerns and develop appropriate care plans. However, adolescents can be reluctant to open up to strangers, hesitant to share information in non-private spaces, prefer self-reliance, and uncomfortable when bombarded with questions. Furthermore, the volume and pace of work in the ED leave staff with little time or ability to engage with these patients in personal, meaningful ways. Often, these teens are left feeling overwhelmed, dismissed, or like their concerns are trivialized.

2 OBJECTIVE

This study seeks to support and engage adolescents seeking crisis care for MBH concerns in the ED through a virtual communication platform.

3 METHODS

This study built off the findings of four ED case studies conducted as part of a larger, grant-funded research project. After identifying the key challenges that emerged from the case studies, the team conducted a literature review and interviews with nurses, psychiatrists, adolescent MBH patients, and their parents to understand the challenges teens face seeking MBH care in the ED and possible interventions to address critical issues.

4 RESULTS

Our team developed "Care Buddy," an app-based platform that facilitates communication between adolescents seeking MBH care in the ED and care providers. While the ED environment can be chaotic, unfamiliar, and uncontrollable, frequently leaving patients feeling exposed and stressed, Care Buddy creates a virtual environment where teens can regain control over their environment and experience. The app, which we propose implementing via kiosks, wall-mounted screens in patient rooms, or iPads (depending on patient acuity, risk level, and ED resources), includes customizable options for patients to create their avatar, see where they are in the care process and what they can expect next, access educational resources, communicate with care staff and pre-approved family or friends, chat with an AI dog, and engage in journaling, art, and mindfulness activities that provide therapeutic experiences and positive distraction. Though not developed in this phase of the project, we also propose that the app could include a staff-facing interface to improve staff workflows and communication with patients.

In this presentation, we will describe the development and features of the Care Buddy app, explore its application in the ED environment, and discuss its implications for care processes, patient and provider experiences, and the physical environment of the ED.

5 CONCLUSIONS

Care Buddy has the potential to improve adolescent MBH patients' care and experiences by creating a comfortable, personal virtual environment for communication, self-expression, and positive distraction – in contrast to the chaotic, uncontrollable, and impersonal ED environment – and facilitates open, honest communication between patients and care providers.

OPTIMIZING WORKPLACE ENVIRONMENTS FOR EMPLOYEE WELLBEING DESIGN FACTORS AFFECTING MILLENNIALS NEW EMPLOYEES' TURNOVER INTENTION IN WORKPLACE: A MIXED METHODS STUDY

Track 4: Design for Health and Wellbeing
Knowledge Network: Interior Design, Work Environments
Methods: Mixed Methods

Huili Wang (Louisiana State University); Michelle Pearson (Texas Tech University); Kristi Gaines (Texas Tech University); Erin M Hamilton (University of Wisconsin-Madison, Design Studies)

1 ABSTRACT

Millennials are the largest generation in the US workforce today and they are thought to have significantly different working preferences compared to previous generations (Pew Research Center, 2018). Previous research has documented increased rates of turnover among millennials, with most turnover occurring in new employees who work less than two years (Kuron, 2015). Moreover, a survey conducted by LinkedIn and CensusWide (2022) revealed that 66% millennials are considering a career change in the next 12 months. Organizations will lose substantial investments if employees quit soon after entering because they invest considerable time and money for recruiting and training new hires. Physical work environment has been identified as an important factor influencing employees' job outcomes. Therefore, this study focuses on investigating what workplace design features will affect millennial new employees' turnover intention.

2 METHODOLOGY

This mixed-methods study consists of two phases. The first phase was qualitative research, which was motivated by the need to deeply understand how millennial new employees experience the new work environment. Two online focus groups were conducted to answer the research question: what the design features millennial new employees need and prefer in an office, and what design features affect their turnover intention. All participants were millennials and identified themselves as IT industry employees. After the completion of two focus groups, the moderator and the assistant moderator agreed that the sample size was large enough to obtain data saturation.

The second phase was quantitative research, which was motivated by the need to explore which design features statistically significantly affect millennial new employees' turnover intention. A survey was developed based on the focus group findings to test the relationship between physical work environment satisfaction and turnover intention. The independent variables are satisfaction with functional space and layout, ambient conditions, personal control, FF&E, flexibility, technology, ambience, privacy, and nature and windows. In total, 252 responses were used for analysis.

3 RESULTS

The focus group participants expressed their consideration to leaving when companies do not satisfy their basic needs to carry out job requirements. Three design features were identified, which affect their decisions to leave jobs. They had insufficient equipment, no windows, and extreme noise.

The quantitative regression analysis indicates that the model was a significant predictor of millennial new employees' turnover intention, $F(9,242) = 3.769$, $p < .001$*. The nine predictors explained 12.3 % of variance in turnover intention. Using the enter method, it was found that satisfaction with FF&E (* $= -.273$, $p = .016$*) significantly decreased turnover intention. Satisfaction with ambiance (* $= -.160$, $p = .081$*) was found marginally significantly negative affect turnover intention.*

4 REFERENCES

Fry, R. (2018, April 11). Millennials are the largest generation in the U.S. labor force. https://www.pewresearch.org/fact-tank/2018/04/11/millennials-largest-generation-us-labor-force/

Kuron, L. K. J., Lyons, S. T., Schweitzer, L., & Ng, E. S. W. (2015). Millennials' work values: differences across the school to work transition. Personnel Review, 44(6), 991–1009. https://doi.org/10.1108/PR-01-2014-0024

Arshadi, N., & Damiri, H. (2013). The relationship of job stress with turnover intention and job performance: Moderating role of OBSE. Procedia-Social and Behavioral Sciences, 84, 706-710.

EXPLORING THEORETICAL CONSTRUCTS AND ENVIRONMENTAL DESIGN SOLUTIONS TO MITIGATE AGGRESSIVE BEHAVIOR

Track 4: Design for Health and Wellbeing
Knowledge Network: Interior Design, Work Environments
Methods: Design Methods

Soheyla MohammadiGorji (San Francisco State University); Sheila Bosch (University of Florida, Interior Design); Lisa Sundahl Platt (University of Florida - Florida Institute for the Built Environment Resilience)

1 INTRODUCTION

This study tackles the issue of workplace assaults on healthcare workers, who, despite being a small fraction of the U.S. workforce, endure nearly 75% of such incidents. Guided by Rational Choice Theory, Lifestyle Exposure Theory, Routine Activity Theory, and Crime Prevention through Environmental Design, in this investigation we delve into understanding the socio-psychological antecedents of aggression and key constructs influencing offensive behaviors in healthcare settings.

Examining the frustration-aggression theory, asserting aggression as a consequence of frustration, we identify triggers in situations like long wait times, unpleasant smells, lack of nourishment, and inadequate amenities. Proposed solutions include strategic space planning and creating a restful ambiance, incorporating features like internal sub-waiting areas and unidirectional traffic flow to prevent setbacks and aggression towards staff (Mohammadigorji et al., 2020; Pati et al., 2016). Building upon criminal opportunity theories, literature suggests that victimization results from the interaction between victims and offenders. Translating this to interior design involves creating features that protect staff against visitors, enhancing concepts like territoriality.

Cohen and Felson's emphasis on the convergence of offenders, suitable targets, and the absence of capable guardians informs an approach to spaces, optimizing visibility and accessibility while identifying potential risk areas.

Aligning with the Routine Activity Theory, we stress the role of supervision in reducing crime opportunities, emphasizing the importance of attributes like visibility and accessibility in interior spaces, drawing insights from relevant space syntax papers (MohammadiGorji et al., 2023).

Acknowledging the need for more empirical research on the quantitative relationship between spatial configuration and violence against healthcare providers, we emphasize the importance of investigating these aspects for a comprehensive understanding. Attention to the physical environment and space planning details allows healthcare facilities to create a positive and less agitating experience, potentially mitigating aggressive behavior. This presentation further delves into translating theory into practice, discussing how design may address the problem. A conceptual framework is proposed, based on the main constructs of the theories discussed, representing offenders' decision-making processes and a cost-benefit analysis for offensive behavior. The framework underscores the impact of environmental design on creating or hindering crime opportunities, providing a comprehensive approach to fostering a conducive atmosphere in healthcare settings.

2 REFERENCES

MohammadiGorji, S., Bosch, S. J., Valipoor, S., & De Portu, G. (2021). Investigating the impact of healthcare environmental design on staff security: A systematic review. HERD: Health Environments Research & Design Journal, 14(1), 251-272.

MohammadiGorji, S., Nubani, L., Bosch, S. J., Valipoor, S., & McElhinney, S. (2023). Using Space Syntax Analysis to Measure Patient Aggression Risk. HERD: Health Environments Research & Design Journal, 16(2), 125-145.

Pati, D., Pati, S., & Harvey Jr, T. E. (2016). Security implications of physical design attributes in the emergency department. HERD: Health Environments Research & Design Journal, 9(4), 50-63.

SUPPORTING ATTENTION THROUGH DESIGN: EXAMINING HOW LIBRARY DESIGN CAN INFLUENCE ATTENTION RESTORATION

Track 4: Design for Health and Wellbeing
Knowledge Network: Interior Design, Student Affairs
Methods: Experimental/Quasi-Experimental Research, Mixed Methods
Evangelina Neri (Corgan & Florida State University); Stephanie Sickler (3066 Waterford Dr.)

1 ABSTRACT
Attention has become increasingly fragmented among college students as the availability of media and information has increased alongside digitally enhanced courses (Copeland et. al. 2021). Attention Restoration Theory is a person-environment theory that posits that experiencing effortless attention – generally through experiencing nature – is critical for restoring directed attention deficits (Kaplan, 2001). Attention Restoration Theory broadly defines ten essential properties for restorative experiences that align with an exposure to nature (Ackerman, 2018). However, the extent of nature exposure needed to see an improvement of attention spans indoors remains unknown, making it difficult to meaningfully apply Attention Restoration properties in interiors. While Attention Restoration Theory presents an evidence-based rationale for incorporating nature into the built environment, there is no existing literature connecting attention restoration theory and interior academic environments. This prompts the questions, to what degree do multisensory biophilic stimuli improve attention, and to what degree do indoor multisensory biophilic stimuli improve attention?

2 METHODS
Twenty-seven college students were invited to participate in a mixed methods study. Participants engaged in a ten-minute break with varying degrees of nature exposure, which comprised of auditory stimuli indoors, multisensory stimuli indoors, or an outdoor walk. Participants took the SART2 attention test as a pre- and post-measure of attention spans. Following the attention test, participants were asked to describe the experience. The results were analyzed for statistical significance and responses were coded in alignment with Attention Restoration Theory's ten properties.

3 CONCLUSION
The auditory and multisensory indoor groups saw a significant improvement in attention scores, whereas the control group and outdoor walk group did not show a significant improvement in scores. Preliminary SART2 results and responses point to patterns that (1) individuals with a mindfulness practice had substantial improvements in attention scores regardless of the degree of nature exposure and (2) acoustic comfort may offer a higher degree of restoration than artificial visual stimuli indoors. While this study points to quality over quantity when designing with nature, it is essential to acknowledge that the small sample size means the critical threshold of nature on attention restoration cannot be clearly concluded by this study.

4 IMPLICATIONS
Attention is a cognitive resource that directs how one consciously interacts with reality. The attentional challenges faced by college students and knowledge workers are precipitating an unprecedented impact on the personal, professional, and social dimensions of daily life. To this discord, designing restorative opportunities in new and existing building programs might encourage dwellers to experience effortless attention. The presentation will highlight how the research was translated to the built environment through a proposal for a restorative library.

5 REFERENCES
Ackerman, C. (2018). What is Kaplan's Attention Restoration Theory (ART)? Benefits + Criticisms. PositivePsychology.Com.

Copeland, W. E., McGinnis, E., Bai, Y., Adams, Z., Nardone, H., Devadanam, V., Rettew, J., & Hudziak, J. J. (2021). Impact of COVID-19 pandemic on college student mental health and wellness. Journal of the American Academy of Child & Adolescent Psychiatry, 60(1), 134-141.e2.

Kaplan, S. (2001). Meditation, Restoration, and the Management of Mental Fatigue. Environment and Behavior, 33(4), 480–506.

USING VIRTUAL REALITY TO CONDUCT PROACTIVE EVALUATION OF THREE ALTERNATIVE ANESTHESIA WORKSPACE LAYOUTS IN THE OPERATING ROOM

Track 4: Design for Health and Wellbeing
Knowledge Network: Health
Methods: Interviews, Mixed Methods

Mina Shokrollahi ardekani (Clemson University Center for Health Facilities Design + Testing); Anjali Joseph (Clemson University); David Neyens (Clemson University); Kenneth Catchpole (Medical University of South Carolina)

1 INTRODUCTION

With the growing need to understand the performance of proposed future designs, teams are increasingly seeking proactive methods to evaluate designs (Dalgleish et al., 2000). This is especially true in critical areas within healthcare settings, such as operating rooms, where design can significantly influence the safety and efficiency outcomes(Taaffe et al., 2021). Within the operating room, the anesthesia provider frequently encounters workflow disruptions due to small, crowded, and poorly organized workspaces (Joseph et al., 2019), potentially impacting medication-related errors (Savoldelli et al., 2010). In response to the need for a better and more efficient anesthesia provider's workspace, a research team studied workflows among anesthesia providers (Mohammadigorji et al., 2024), resulting in two alternative configurations that could support improved workflows and fewer disruptions for anesthesia providers.

2 OBJECTIVES

This study aims to evaluate three alternative proposed anesthesia workspace configurations. The evaluation criteria include the support for anesthesia providers' situational awareness, visual access to the patient, flexibility and the ability to reconfigure the workspace, workflow support in terms of movement and overall workflow, and the efficiency of provided horizontal workspace. In addition, this study assessed the integration of a novel syringe holder design solution for medication organization in the anesthesia workspace instead of kidney trays.

3 METHODS

Three configurations of anesthesia workspace were presented within the same operating room layout for evaluation including two boom-based workspaces with different configurations (A1: having the booms on one side of the surgical table, A2: having the booms on either side of the surgical table), and a typical anesthesia provider's workspace on wheels (Prototype). Due to cost and time constraints, we developed a virtual model of each design to be presented to anesthesia providers for their feedback. We ran the experiment across two academic medical centers, in the Northeastern and Southeastern United States. After exploring the virtual environment of each design, 18 anesthesia providers with an average of 8 years of experience answered multiple open-ended questions and provided their feedback. The responses were analyzed using thematic analysis to provide a nuanced understanding of the participants' perspectives about the presented designs.

4 RESULTS

The results showed that the participants appreciated the VR approach's fidelity and efficiency. In terms of situational awareness, visual access to patient, flexibility, workflow support, and the efficiency of provided horizontal workspace, both configurations with the boom-based design (A1, A2) were preferred compared to the typical anesthesia provider's workspace on wheels (Prototype). Moreover, integration of the syringe holder was recommended among the anesthesia providers compared to typical kidney trays.

5 CONCLUSION

Virtual reality provided an efficient platform to understand the participants' perspectives on the presented anesthesia workspace designs. Even though the providers had a clear opinion about their preferences, they stated that presenting the same design alternatives with specific surgery scenarios would be very important to see whether the proposed designs would support inclusion of different equipment and different team arrangements in the operating room. Future research can consider presenting different scenarios within each design alternative to enhance our understanding of the anesthesia provider's workspace design.

VITALITY IN URBAN ENVIRONMENTS: IDENTIFYING ASSOCIATIONS BETWEEN NEIGHBORHOOD BUILT ENVIRONMENT DESIGN AND SOCIAL CAPITAL

Track 4: Design for Health and Wellbeing
Knowledge Network: Cities & Globalization, Health
Methods: Statistical/Modeling Analysis
Hannah Myers (University of Michigan, Design Science); Jeffrey Morenoff (University of Michigan)

1 ABSTRACT

The U.S. is experiencing an urgent public health crisis of social isolation and loneliness. A 2023 U.S. Surgeon General Advisory recommends efforts to alleviate social isolation and loneliness which include built environment interventions (Holt-Lunstad & Golant, 2023). A growing body of research has connected the design of neighborhood built environments and social capital - a construct that indicates social connection. Walkability, green space access, and mixed-use zoning have been positively associated with social capital, but research is less clear about other neighborhood design elements. Thus, the strategic design of neighborhood built environments might be a promising way to combat loneliness and isolation, but there is some debate about which built environment design elements - if any - contribute to social connection.

Mazumdar and colleagues' systematic review of research investigating the relationship between the built environment and social capital reveals limitations of the current literature (Mazumdar et al., 2018). For example, few studies use standardized measures of built environment elements, which makes it difficult to compare results across studies. To facilitate comparison of these disparate measures, the authors offered a conceptual framework that they call the 4Ds: Density (measures include population density, e.g.), Diversity (e.g., land-use mix), Design (e.g., street network patterns), and Destination (e.g., access to "third places"). To date, no study has included measures from all four of these categories, and only a handful have analyzed two or more categories simultaneously.

2 METHODS

Using the 4D framework, I propose a quantitative study that will contribute to built environment and social capital research by 1) using standardized and previously-validated measures to collect neighborhood built environment data, and 2) examining the relationship between neighborhood built environment and social capital with all 4Ds represented within the same study.

The study site for this inquiry is the city of Detroit, MI. Data on social capital come from the Detroit Metro Area Communities Study, a panel study conducted on a large, representative sample of Detroit households. Social capital is measured using self-report survey items capturing social connection, trust, and participation. The built environment design elements that make up the 4Ds will be quantified using standardized measures, and drawn from secondary sources (e.g., National Neighborhood Data Archive, Census products, GIS resources, and market research data).

3 CONCLUSION

By including many neighborhood built environment design elements in one model, I will be able to compare strength and direction of association with social capital across elements. The results of this study will indicate which elements have the greatest effective relationship with social capital; knowledge that can be directly translated into policy, planning, and development priorities for new and existing neighborhoods.

4 REFERENCES

Holt-Lunstad, J., & Golant, S. (Eds.). (2023). Our Epidemic of Loneliness and Isolation: The U.S. Surgeon General's Advisory on the Healing Effects of Social Connection and Community. Office of the U.S. Surgeon General.

Mazumdar, S., Learnihan, V., Cochrane, T., & Davey, R. (2018). The Built Environment and Social Capital: A Systematic Review. Environment and Behavior, 50(2), 119–158. https://doi.org/10.1177/0013916516687343

URBAN VACANT LANDS, BIODIVERSITY, AND COMMUNITY WELL-BEING

Track 4: Design for Health and Wellbeing
Knowledge Network: Cities & Globalization, Health
Methods: Statistical/Modeling Analysis
Chia-Ching Wu (The University of Illinois at Urbana Champaign);
William C. Sullivan (University of Illinois at Urbana-Champaign)

1 INTRODUCTION

In urban settings, vacant lands are neglected places frequently associated with social and safety concerns, a situation influenced by contextual social factors, untended built environments, and spontaneously growing vegetation. Despite these challenges, this unattended greenery can be beneficial to the local ecosystem by providing undisturbed habitats for wildlife (Anderson & Minor, 2017). Furthermore, studies suggested that enhanced plant diversity correlated with residents' well-being (Methrost et al., 2021).

These underutilized spaces are often found in areas scarce in officially designed green spaces, where vacant spaces can offer unique opportunities for residents to connect nature on a daily basis and for recreational purposes (Kremer et al., 2013). However, the spontaneous and unmanaged state of vegetation in these areas may not align with conventional human aesthetic preferences. In addition, the lack of management can lead to problems such as littering. These concerns not only undermine those potential benefits but also adversely affect the well-being of community members.

The contrasting positive and negative consequences associated with vacant spaces highlight the need to explore the use and impacts of such spaces within communities. This study reviewed 15 pieces of literature that explored both the ecological and social evaluations of vacant lands in urban or suburban areas published in peer-reviewed journals from 2011 to 2024. In doing so, this study will address the following questions:

1. *What are the relationships between vacant land areas, the vegetated land cover quality, and wildlife diversity?*
2. *What are the relationships between vacant land areas, proximity to vacant lands, and residents' well-being?*
3. *What types of transformation have been implemented to vacant lands, and what outcomes have been observed from these changes?*

2 REFERENCES

Anderson, E. C., & Minor, E. S. (2017). Vacant lots: An underexplored resource for ecological and social benefits in cities. Urban Forestry & Urban Greening, 21, 146-152.

Kremer, P., Hamstead, Z. A., & McPhearson, T. (2013). A social-ecological assessment of vacant lots in New York City. Landscape and Urban Planning, 120, 218-233.

Methorst, J., Bonn, A., Marselle, M., Böhning-Gaese, K., & Rehdanz, K. (2021). Species richness is positively related to mental health-a study for Germany. Landscape and Urban Planning, 211, 104084.

EVALUATING URBAN VITALITY WITH BIG DATA: INSIGHTS FROM SOCIAL MEDIA AND GPS DATA ACROSS 10 GLOBAL CITIES

Track 4: Design for Health and Wellbeing
Knowledge Network: Cities & Globalization, Sustainable Planning, Design & Behavior
Methods: Experimental/Quasi-Experimental Research, Mixed Methods

Mingze Chen (UBC); Yongming Huang (Kyoto University of the Arts); Yuqiao Zheng (University College London); Jiani Du (The University of Hong Kong)

1 ABSTRACT

Urban vitality, a crucial indicator of the quality of the built environment (Y. Li et al., 2022), reflects the liveliness and health of urban spaces. This study explores the efficiency of big data and specific machine learning technologies, surpassing traditional methods like on-ground observation and questionnaires, in evaluating urban vitality. Despite the innovative approach, challenges such as bias in data representation, limited demographics, and privacy or ethical concerns inherent in social media and GPS technologies are acknowledged (Ghermandi et al., 2022). Distinguishing itself from previous studies that often concentrated on single regions or datasets (L. Li et al., 2023), this research aims to uncover a universal pattern in leveraging social media and smartphone-based GPS data to assess the spatial and temporal dynamics of human behavior in urban settings. Employing a combination of computer vision, natural language processing, and location service-based technologies, this study comprehensively analyzes citizens' distribution, diversity, and preferences. The comparative analysis, encompassing data from 2018 to 2022 across 10 strategically selected global cities, sheds light on the variations in urban vitality within different cultural, economic, and geographical contexts. The findings reveal that the validity of big data in evaluating urban vitality exhibits significant dynamics in temporal changes (weekday vs. weekend) and across cities with diverse socioeconomic backgrounds. By analyzing a variety of urban environments, this research provides invaluable insights for urban planning and policy-making, highlighting both the universality and uniqueness of urban dynamics and their implications for future urban development.

4 REFERENCES

Ghermandi, A., Depietri, Y., & Sinclair, M. (2022). In the AI of the beholder: A comparative analysis of computer vision-assisted characterizations of human-nature interactions in urban green spaces. Landscape and Urban Planning, 217, 104261. https://doi.org/10.1016/j.landurbplan.2021.104261

Li, L., Du, Q., Ren, F., Huang, L., Voda, M., & Ning, P. (2023). Geolocated social media data for measuring park visitation in Shenzhen, China. Urban Forestry & Urban Greening, 88, 128069. https://doi.org/10.1016/j.ufug.2023.128069

Li, Y., Yabuki, N., & Fukuda, T. (2022). Exploring the association between street built environment and street vitality using deep learning methods. Sustainable Cities and Society, 79, 103656. https://doi.org/10.1016/j.scs.2021.103656

DECISION TREE MODEL-BASED SUBJECTIVE AND OBJECTIVE EVALUATION OF RENEWAL POTENTIAL FOR OUTDOOR RECREATION SPACES IN OLD URBAN COMMUNITIES

Track 4: Design for Health and Wellbeing
Knowledge Network: Residential Environments, Sustainable Planning, Design & Behavior
Methods: Case Study, Mixed Methods
Qi Su (Tianjin University); Yinuo Qin (Tianjin University); Haonan Sun (Tianjin University); Wentian Shi (Tianjin University); Xinyu Cai (Tianjin University); Min Xu (Tianjin University); Jin Zuo (Tianjin University)

1 INTRODUCTION

Amidst the swift pace of urbanization, the United Nations has advocated for the Sustainable Development Goals (SDGs), and the construction of healthy communities is a key topic. The renewal of outdoor recreation spaces within old urban communities to optimize the human settlement and satisfy the inhabitants' leisure and recreational demands emerges as a critical endeavor towards achieving sustainable urban development and fostering the development of healthy communities.

Presently, research on the renewal of outdoor recreation spaces in old urban communities predominantly concentrates on existing spaces, with scant consideration for potential spaces, coupled with an absence of a universally applicable and objective framework for measurement and evaluation to serve as a foundational support. In light of this, the study takes the central urban area of Tianjin as its research scope, selecting the Four Seasons Village area of Tianjin University as a representative study area. It proceeds to craft a detailed database of outdoor recreation spaces, selects multiple indicators from two dimensions—namely, potential feasibility and functional suitability—to formulate a decision tree model, thereby establishing a subjective-objective coupling evaluation system for the renewal potential of outdoor recreation spaces within old urban communities. The results indicate that (1)sunlight exposure, space area, utilizable time, and spatial occupancy significantly influence potential feasibility, whereas sDNA accessibility, space area, distance to residential buildings, and crowd congregation reflect the functional suitability. (2)The outdoor recreational spaces within the study area have high potential feasibility for renewal, with suitability to accommodate regional and linear sports activities, as well as social and solitary leisure. Drawing upon the coupled feasibility and suitability evaluation, the spaces are classified into three levels based on their comprehensive renewal potential. The sum of the areas of the first level spaces is approximately half the total potential spaces, indicating generally high renewal potential. (3)Based on the evaluation results, the study proposes strategies for the renewal and construction of outdoor recreation spaces, including layout optimization, communal sharing and space construction. It also presents the outlook of simulating the renewed environment with the help of digital twin technology and conducting the resident satisfaction assessment.

This study focuses on the systematic, subjective-objective jointness and generalizability of the evaluation system of renewal potential for outdoor recreation spaces in old communities. It offers a new potential evaluation methodology and a whole-process renewal idea for the construction of outdoor recreation spaces in old urban communities, and provides experience and insights for other cities to improve the function of recreation in old communities and realize sustainable development.

EVIDENCE-BASED ENVIRONMENTAL DESIGN: IN FOCUS OR OUT OF FOCUS? INVESTIGATING THE ALIGNMENT OF PERCEPTIONS AND PHYSIOLOGICAL RESPONSES IN GREEN INDOOR ENVIRONMENTS

Track 4: Design for Health and Wellbeing
Knowledge Network: Health, Residential Environments
Methods: Experimental/Quasi-Experimental Research, Mixed Methods
Hossein Saedi (NC State University)

1 ABSTRACT

The fact that human interaction with nature triggers separate processes in the human body is well-established in the fields of environmental psychology. These interactions have proven to have different perceptual and physiological impacts on individuals (Saedi and Rice, 2022; Parasuraman, 1986). However, many researchers have solely focused on human perception to measure such impacts (Calogiuri et al., 2018). What has been examined far less is how the body reacts during those interactions and if the physiological and perceptual reactions are aligned (Lee et al., 2015; Saedi et al., 2021). Therefore, an exploratory experimental research has been conducted to examine whether or not the presence of natural greenery in indoor environments induces restorative physiological responses in individuals. Also, are individuals' perceptions of the restorative qualities of these environments aligned with their physiological responses?

182 residents of a high-rise residential building were randomly assigned to walk through one of the two versions of their building's lobby from the entrance to the elevators, without engaging with the space, in less than a minute. The two versions were modeled in a fully immersive virtual reality environment (FIVRE). One version included natural greenery, and those elements were removed in the other. The participants' eye movement patterns were recorded via VR goggles as indicators of physiological responses, including fixation location, fixation duration, and eye travel distance. At the end of the experiment, participants filled out a Perceived Restorative Scale (PRS) questionnaire (Hartig et al., 1997) to measure the two FIVREs' perceptual impacts.

The results revealed a significant difference in the eye movement pattern between the two groups. Those who walked through the lobby with greenery demonstrated validated patterns related to restorative environments in all three aspects of the recorded eye movement. The members of the group with no greenery did not demonstrate such patterns. However, the members of both groups did not perceive the lobby as restorative. The outcomes of this research suggest that measuring human perception as the indicator of the restorative impact of an environment might not be enough and may lead to premature conclusions.

4 REFERENCES

Hartig, T., Korpela, K., Evans, G., & Garling, T. (1997). A measure of restorative quality in environments. Housing, Theory and Society, 14(4).

Lee, K. E., Williams, K. J. H., Sargent, L. D., Williams, N. S. G., & Johnson, K. A. (2015). 40-second green roof views sustain attention: The role of micro-breaks in attention restoration

Parasuraman, R. (1986). Vigilance, monitoring, and search. (pp. 1-39). Oxford, England: John Wiley & Sons

Saedi, H., & Rice, A. (2022). A Deeper Understanding of the Impact on the Restorative Quality of Green Environments as Related to the Location and Duration of Visual Interaction. Journal of Digital Landscape Architecture, 412-424.

Saedi, H., Einifar, A. R., & Barati, N. (2021). The Impact of Micro Interaction with Natural Green Elements Through Virtual Reality on Attention Restoration (In a High-Rise Residential Building's Lobby). Journal of Architecture and Urban Planning, 13(30), 5-21.

THE POWER OF SENSES: A STUDY ON MULTI-SENSORY ENVIRONMENTS AND THEIR IMPACTS ON SLEEP, MOOD, AND STRESS IN OLDER ADULTS WITH BEHAVIORAL AND PSYCHOLOGICAL SYMPTOMS OF DEMENTIA

Track 4: Design for Health and Wellbeing
Knowledge Network: Environment-Gerontology, Interior Design
Methods: Experimental/Quasi-Experimental Research
Mohammed Nasser Alrahyani (Arizona State University); Nina Sharp (Arizona State University); Dr. Dongwoo (Jason) Yeom (Arizona State University); Mahya Fani (Arizona State University)

1 INTRODUCTION

The physical environment is one of the most influential factors in patients' experience in assisted long-term care facilities. Many studies reported the impact of ambient features of the indoor built environment, such as lighting, aroma, and sound, on occupants' psychological and physiological responses. The multi-sensory environment has gained wide attention as a non-pharmacological intervention that may contribute to delaying the worsening severity of neurocognitive disorders from the mildest to the most severe stages in older adults with dementia and other cognitive impairments. Many studies have examined the impact of the multi-sensory environment on depression, agitation, and functional performance of older adults diagnosed with dementia. However, the type of MSE used in these studies, which is known as Snoezelen room, demands special requirements (e.g., trained therapist, trained MSE facilitator, private room, one-to-one session) and expensive equipment. Furthermore, most of the studies did not investigate the impact of MSE on the sleep quality of Behavioral and Psychological Symptoms of Dementia (BPSD) patients. Thus, we conducted a within-subject field study on 14 older adults (age > 54 years old) diagnosed with BPSD to examine the effectiveness of a novel MSE approach. Our study introduces cost-effective and passive MSEs, Multi-Sensory Stimulating Environment (MSSE), and Multi-Sensory Relaxing Environment (MSRE), using a combination of sensory stimuli such as visual stimuli (lighting and bubble tubes), aroma, and sound.
Over six consecutive weeks of study, participants underwent a baseline week in their existing living environment, followed by three 7-day periods of MSE interventions, each separated by a 7-day washout. Interventions were applied in the common area of our partner memory care facility, which included one hour of morning exposure to MSSE in Week 2, one hour of evening exposure to MSRE in Week 4, and a combination of both (morning and evening) in Week 6. The impact of each intervention on sleep, heart rate variability (HRV), stress, depression, and agitation was assessed through the use of a non-invasive wearable device (Fitbit Sense 2) worn by the participants during the interventions. Additionally, well-validated questionnaires, the Cornell Scale for Depression in Dementia (CSDD) and the Cohen-Mansfield Agitation Inventory (CMAI) were completed by primary caregivers to provide comprehensive insights into participants' responses. Our findings indicate notable improvements in the participants' mood after exposure to all three interventional conditions with better outcomes for the combined environment. The exposures to MSRE and combined interventions positively impacted the sleep quality of participants at night. The results provide designers and practitioners with a profound understanding of the influence of cost-effective sensory stimuli, encompassing lighting, sound, and aroma on the well-being of individuals with BPSD. These insights offer practical guidance for the development of more accessible and universally applicable Multi-Sensory Environments (MSEs), ultimately elevating the quality of life for dementia patients within the constraints of diverse care settings.

EVIDENCE-BASED DESIGN AND PROGRAMMING OF AMBULATORY SURGERY CENTER

Track 4: Design for Health and Wellbeing
Knowledge Network: Health, Interior Design
Methods: Interviews, Mixed Methods

Sharmin Kader (Ball State University College of Architecture and Planning)

1 ABSTRACT

Ambulatory surgery refers to patients who can leave the hospital on the same day as the surgery, without being admitted to an in-patient unit (American Society of Anesthesiologists, 2022). According to the Ambulatory Surgery Center Association (ASCA), outpatient procedures are becoming increasingly common in the United States - approximately 129 million outpatient procedures in 2018, which are expected to grow to 144 million by 2023. Over the years, the number of Ambulatory Surgery Centers (ASCs) has grown in response to fulfilling that demand. In 2008, William E. Lindeman wrote an article in ARON Journal titled, "Design and Construction of an Ambulatory Surgery Center," where he articulated the design goals of ASCs as to provide a high-performance facility with maximum efficiency, high patient satisfaction, and minimal staff turnover, resulting in lower operating costs and correspondingly higher profit potential. He also pointed out that the ASCs should have the ability to meet both initial expectations and longer-term demands for operational flexibility and financial integrity. Since 2008, there has been very little literature published on ACS's design and planning. Meanwhile, the healthcare delivery system has been advancing, such as telehealth, and the unprecedented scenarios of the COVID-19 pandemic have brought new challenges to the ASCs' planning, programming, and design. Innovations and the development of research-based design strategies are evident. This study has focused on developing evidence-based design solutions to address the concurrent issues and demands. The study has considered a mixed-method model: the first step is a graduate-level design studio dedicated to researching and designing a post-pandemic ambulatory surgery center and developing the recommendations, and the second step is conducting interviews with the healthcare planners and designers to validate all those findings. The students were assigned to conduct a series of activities: reviewing relevant literature (design guidelines, journal articles, etc.); analyzing web-based case studies; identifying therapeutic goals for patients, their families, and the staff; and developing functional diagrams and programming. The studio invited several eminent guest speakers (architects, researchers, and interior designers) to share their work and provide feedback on the design goals, strategies, concepts, site analysis, layout, and exterior and interior design. This learning lab model design studio has identified some emerging issues and has led to the development of some innovations. For example, separate patient discharge lobby, private rooms for the post-operative area, and consultancy rooms for telemedicine. Based on the findings generated in the learning lab, an interview was conducted with the industry experts to validate these findings. This session will present a comprehensive list of therapeutic goals, such as increasing infection control, enhancing staff privacy, providing flexibility in space for patient education, and expanding recreational facilities for family support. This presentation will also discuss the programming, space requirements, functional relationships, and diagrams. The findings carry great significance to the decision-makers designing new facilities or remodeling an old ambulatory surgery center.

NATURE'S INFLUENCE: A COMPREHENSIVE STUDY ON SENSATION, PERCEPTION, AND COGNITION IN URBAN ENVIRONMENTS

Track 4: Design for Health and Wellbeing
Knowledge Network: Health, Interior Design
Methods: Experimental/Quasi-Experimental Research, Mixed Methods
Hossein Saedi (NC State University)

1 INTRODUCTION

"Sensation-Perception-Cognition Model of Human Interaction with Their Environment," suggests that human interaction with their environment consists of three phases that happen concurrently (Balasubramanian, 2014). Interaction with an environment may affect each of those phases independently. Thus, to explore the restorative impacts of an environment on individuals, all phases should be considered separately (Saedi et al., 2021). So far, most of the studies that explore the restorative quality of an environment and its impacts on individuals' health and well-being have utilized perception and cognition measurements (Saedi, 2020). Moreover, the results obtained from these measures do not always support each other (Diefenbach et al., 2012). The contradictory outcomes make it harder for researchers to identify an environment as restorative, especially when sensation processes are not considered.

The Attention Restoration Theory and Biophilia suggest that there are physiological traits that are inherent in humans (Kaplan & Kaplan, 1989; van den Berg et al., 2003). Those traits are triggered at the subconscious level when individuals face certain situations (McMahan & Estes, 2015). The advancement of technology makes it possible to quantify those traits and identify certain patterns in them (Graap et al., 2012; Lee & Baek, 2014). The eye movement is among those traits that have proven to demonstrate certain patterns in restorative environments (Valtchanov & Ellard, 2015; Franek et al., 2018).

This research reports on an experiment that measures the restorative impact of micro-interactions with vegetation in a residential building lobby through sensation, perception, and cognition. 182 participants were randomly assigned to walk through one of the two versions of a virtual reality lobby using the VR goggles for less than a minute. One version contained plants and the other version did not. Participants took the SART cognitive test before and after experiencing the lobby. They filled out the Perceived Restorative Questionnaire at the end of the experiment to report their perception of the restorative quality of the lobby. Participants' eye-movement patterns were recorded by the VR goggles throughout the lobby explorations as indicators of participants' physiological responses. The result demonstrated no significant differences between the two groups' perceptions of the restorative quality of the lobby. However, individuals scored significantly higher in the cognitive test after experiencing the vegetated lobby. Moreover, the eye-movement patterns for those who experienced the vegetated lobby were aligned with the patterns related to restorative environments and significantly different than those who experienced the lobby with no vegetation.

This research underlines the fact that perception measures are subjective and may be less reliable for evaluating an environment's restorative quality and that physiological responses were supported by cognitive performances. Current research suggests a more holistic approach to measuring the restorative quality of an environment.

REDESIGNING RECOVERY: APPLYING EVIDENCE-BASED DESIGN TO SUBSTANCE USE DISORDER TREATMENT AND RECOVERY IN THE USA

Track 4: Design for Health and Wellbeing
Knowledge Network: Health, Interior Design
Methods: Ethnography, Observation
Claire Hayes Watson (Self-Employed)

1 ABSTRACT

There is both an unintentional and intentional history of exerting control over vulnerable populations and the spaces they inhabit through dehumanization. One of these vulnerable populations is women struggling with substance use disorder (SUD), often referred to as addiction. Sometimes, they are dehumanized by institutions and government services simply because they are not designers and some do not consult designers in decisions because there's such a high demand for housing and health services in our county. When designers are hired, unless they have a holistic understanding of SUD and its spatial implications, their efforts are often misguided.

For this paper, I have chosen to specifically focus on houseless women with substance use disorder because SUD is prevalent among the houseless population and can be both a cause and effect of being houseless (Polcin, 2015), and women are under-represented in SUD treatment services as a result of historical trauma and victimization, shame, stigma, and other gender-specific barriers (Rizzo et al., 2022). For the women who overcome these barriers and access treatment, the services rendered and the design of the environment in which they are offered, must be evidence-based, addressing the unique needs of women and the psychological traumas and spatial needs of being houseless.

A literature review was conducted to learn about the neurological effects of substance use and houselessness and how they impact spatial perception. Observational studies were used to understand how women interact with the surrounding environment on the street during active substance use, and in clinical treatment and recovery settings. The populations observed were houseless populations in Florida and Massachusetts, women in medically-assisted treatment (MAT) programs, which have spatial inequities of their own, and women in a Florida recovery residence. Women living in the recovery residence were also interviewed to understand their past and current spatial experiences. The research provides an understanding of the internal and external experiences of each population, which can be translated into new approaches and redesigns.

This paper aims to provide a framework of evidence-based design that accompanies evidence-based treatment and recovery services. This includes gender-specific and trauma-informed design guidelines that can be used to inform the design of facilities from medical detoxification facilities to long-term residential recovery environments.

2 REFERENCES

Polcin, D. L. (2015). Co-occurring substance abuse and mental health problems among homeless persons: Suggestions for research and practice. Journal of Social Distress and the Homeless, 25(1), 1–10. https://doi.org/10.1179/1573658x15y.0000000004

Rizzo, D., Mu, T., Cotroneo, S., & Arunogiri, S. (2022). Barriers to accessing addiction treatment for women at risk of homelessness. Frontiers in Global Women's Health, 3. https://doi.org/10.3389/fgwh.2022.795532

POST-OCCUPANCY EVALUATIONS FOR ENHANCING THE HEALTH AND WELLBEING IN THE UAE RESIDENTIAL SECTOR

Track 4: Design for Health and Wellbeing
Knowledge Network: Health, Residential Environments
Methods: Case Study
Sara Adel Abdulmaksoud (American University of Sharjah); Salwa Beheiry (Dr)

1 INTRODUCTION

Human-centric design, and particularly the design for health and wellbeing, continue to be a prominent focus of research in the Architecture, Engineering, Construction and Operations (AECO) industry. Building design attributes guide the construction and ultimately the performance of residential buildings. Building performance, in turn, influences health and wellbeing, by shielding occupants from environmental stresses such as noise and extreme temperatures (Foster et al., 2022). One approach to monitor and record the functional performance of residential buildings is Post Occupancy Evaluations (POE) (Almalki, 2021). POEs involve all stakeholders in a systemic evaluation process of buildings, after a period of occupancy, to inform operation, retrofitting, and future design decisions that support productivity and wellbeing. While POEs can play a significant role in improving design techniques by considering the end users' experience and highlighting the important design attributes (Hassanain & Mahroos, 2023), they are rarely conducted in countries characterized by hot and arid climates (Kim et al., 2022). The limited adoption of POEs in harsher climates may be attributed to more complicated occupancy patterns and other building performance factors. Thus, the aim of this study is to examine the relationship between a set of design attributes and their impact on the satisfaction and wellbeing of occupants in the UAE residential sector using multiple holistic case studies. The objectives include an analysis of the POEs' methodologies, levels of investigation, and strategies, a review of the tangible design attributes that relate to occupants' health and wellbeing, synthesis of the findings to highlight the applicable criteria within the context of the UAE residential sector and the validation of the findings using an expert panel.

2 REFERENCES

Almalki, A. (2021). Postoccupancy Evaluation for Student Housing with a Focus on Maintenance Planning, Design Process, and Client Requirement. Journal of Performance of Constructed Facilities, 36(1). https://doi.org/10.1061/(ASCE)

Foster, S., Hooper, P., Turrell, G., Maitland, C., Giles-Corti, B., & Kleeman, A. (2022). Grand designs for design policy: Associations between apartment policy standards, perceptions of good design and mental wellbeing. SSM - Population Health, 20. https://api.semanticscholar.org/CorpusID:254012324

Hassanain, M. A., & Mahroos, M. S. (2023). A preliminary post-occupancy evaluation of the built-environment in office buildings: a case study from Saudi Arabia. Property Management, 41(4), 564–581. https://doi.org/10.1108/PM-11-2022-0089

Kim, Y. K., Abdou, Y., Abdou, A., & Altan, H. (2022). Indoor Environmental Quality Assessment and Occupant Satisfaction: A Post-Occupancy Evaluation of a UAE University Office Building. Buildings, 12(7). https://doi.org/10.3390/buildings12070986

SUSTAINABLE DWELLINGS: NARRATING SENSE OF PLACE AND AGENCY IN AFFORDABLE HOUSING COMMUNITIES IN JORDAN

Track 4: Design for Health and Wellbeing
Knowledge Network: International Housing Research, Sustainable Planning, Design & Behavior
Methods: Case Study, Interviews
Genell W Ebbini (Purdue University); Sahera Bleibleh (United Arab Emirates University)

1 ABSTRACT

This study focuses on the nuanced intricacies of sustainable living within affordable housing communities of Jordan, specifically in Ajloun and Qasr Hallabat. Employing a grounded theory approach centered on analyzing inhabitants' narratives and success stories, the research recognizes the significant role of individual and community sense of place and agency in shaping sustainable urban environments. It also seeks to provide a nuanced understanding rooted in the initiatives and experiences of the residents themselves.

Through a grounded qualitative methodology including observations, open-ended interviews, and in-depth thematic analysis of personal narratives, the study aims to develop a grounded framework. This framework unveils how residents' mechanisms, driven by their sense of place and agency, shape aspects of sustainable housing contexts. By focusing on successful stories within these communities, the research seeks to identify patterns, opportunities, and challenges that contribute to sustainable well-being. This narrative-driven approach allows for a holistic exploration of the lived experiences of residents, shedding light on the multifaceted factors that operationalize their sense of belonging and active engagement in community development. Through the lens of grounded theory, the study aspires to offer rich insights into the social, cultural, and environmental dynamics that underpin successful sustainable practices within Ajloun and Qasr Hallabat affordable housing settings.

The outcomes of this research are poised to not only contribute to the academic understanding of sustainable urban development but also provide practical guidance for policymakers, urban planners, and community stakeholders. Grounded in the stories of those who call these communities home, this study aims to bridge the gap between theory and lived reality, ultimately fostering a more inclusive, resilient, and sustainable future for affordable housing in Jordan, particularly Ajloun and Qasr Hallabat, and across the Middle East region.

WINDOW VIEWS OF TREES, SHRUBS, OR FLOWERS: WHICH ONE MATTERS MORE TO WELLBEING?

Track 4: Design for Health and Wellbeing
Knowledge Network: Health, Residential Environments
Methods: Statistical/Modeling Analysis, Survey
Sara Hadavi (Kansas State University)

1 INTRODUCTION

It has been well documented that green views from the window are important for wellbeing (Li & Sullivan, 2016; Kaplan, 2001). There is growing evidence that it also affects neighborhood satisfaction, however, there are still many unknown aspects of these associations in residential settings. Which specific green elements visible from the window may exhibit stronger connections with particular aspects of wellbeing and neighborhood satisfaction? Some studies show that proximity to certain types of green elements may have positive associations with wellbeing mediated by neighborhood satisfaction. How about the window view of such plants? Do various green elements exert indirect effects on wellbeing through the mediation of neighborhood satisfaction? This study aims to address these questions using the survey data (n=434) collected in parts of four community areas in Chicago, IL, including Logan Square, Avondale, Humboldt Park, and West Town, that consisted of moderately dense residential blocks with multi-story houses that did not provide private green backyards to residents. The independent variable was 'view from window' which included view of plants in general, tree canopies, shrubs and bushes, flowerbeds, and houses or apartment buildings. Outcome variable was 'mental wellbeing' measured by 31 items including the Attentional Functioning Index (Cimprich, Visovatti, & Ronis, 2011), Positive and Negative Affect Scale (Watson, Clark, & Tellegen, 1988), and Social Connectedness Scale (Lee & Robbins, 1995). These three aspects of wellbeing were averaged to measure overall wellbeing with a high reliability test result. We hypothesized a mediating role for neighborhood satisfaction in the association between window view and mental wellbeing. Mediating variables included satisfaction with the amount of green features and satisfaction with the quality of public spaces. Pearson correlations, standard linear regression modellings, and mediation analyses were conducted in SPSS to examine associations. Neither views of tree canopies nor views of flowerbeds exhibited a significant association with overall wellbeing, while the view of shrubs and bushes from the window was found to be positively associated with overall wellbeing. Digging deeper into different aspects of wellbeing, we found that different types of plants have different associations with each aspect of wellbeing. For example, the view of tree canopies was found to be the predictor of attentional functioning in residents, while the view of shrubs was the predictor of positive and negative affect. The results of mediation analyses indicated that satisfaction with the quality of public space partially mediates the associations between the view of plants from the window (particularly, the view of shrubs) and overall wellbeing. More specifically, the prediction of attentional functioning by the view of tree canopies was found to be partially mediated by satisfaction with the quality of public spaces. These findings offer valuable insights for the planning and design of outdoor spaces in residential areas and other campuses, tailored to the needs of the building occupants. Further, this study contributes to the body of literature on the window view and wellbeing in residential neighborhoods.

HEALTH-CENTERED HOUSING DESIGN IN-LAB TESTING OF ASSISTIVE TECHNOLOGY FOR DEMENTIA HOMECARE: SEQUENCING AND DELIVERY CUES FOR TASKS AND NAVIGATION

Track 4: Design for Health and Wellbeing
Knowledge Network: Environment-Gerontology
Methods: Mixed Methods, Observation
Emily Roberts (Oklahoma State University); Guoliang Fan (Oklahoma State University); Xiaowei Chen (Oklahoma State University)

1 ABSTRACT

This research supports caregivers and care recipients via the continued development of prompting cognitive assistive technology (P/CAT) for multistep tasks in dementia homecare. Globally, approximately 50 million people are affected by dementia, and it is predicted that this number will rise to 152 million by 2050 (Patterson, 2018). For people living with dementia, information-processing capabilities and previously acquired knowledge are generally limited. Daily tasks such as making coffee or brushing may become too difficult to complete independently. People with normal cognitive functioning learn about the environment as they move through it with spatial information obtained sequentially, requiring mental activity to process successive inputs into a comprehensive understanding of a task. Without this ability, individuals with dementia in many cases then rely on repetitive cueing for tasks from their family member or caregiver, adding significant stress and mental exhaustion in the caregiving role. While other assistive technology systems are being developed to support single, pre-determined tasks, there is a persistence of structural limitations to successful adoption due to a lack of clinical validation and insufficient focus on end-user needs (Harris, 2021). There is often a mismatch between the user's cognitive profile and the technology developed, which may be the consequence of a top-down approach to technology design (Lee-Cheong, Amanullah, & Jardine, 2022). A bottom-up user-centered design (UCD) approach is needed to develop technologies that are useful and valuable. This research includes the ongoing development of a wearable APP-based audio-visual (AV) prompting system with a first-person camera which recognizes the user's location and the intended task for that location and provides necessary AV cues to assist the user accomplish the task successfully. In addition to the ongoing P/CAT testing, testing will also take place with a fully customizable companion Robot as a Service solution (RaaS) which provides a mobile AI-powered autonomous platform, with smart assistant and cloud-based services. All testing of the P/Cat and RaaS will take place in the Human Environmental Factors Lab (HEFL) located in a 2-bedroom campus apartment. Using an iterative user-centered design process, we propose to evaluate the tested outcomes of the P/CAT and RaaS system with 2 user populations to determine the similarities and differences in optimization and flexibility of the two systems. This will include 10 college-age students and 10 older adults with normal cognition. This testing in the HEFL lab prior to in-home system testing with individuals living with dementia will improve understanding of the prompting trigger, sequencing of information, and the delivery mode which may impact opportunities and barriers in household task completion.

2 REFERENCE

Harris, N. et al. (2021). A preliminary evaluation of a client-centered prompting tool for supporting everyday activities in individuals with mild to moderate levels of cognitive impairment due to dementia. Dementia. 20(3):867-83.

Lee-Cheong, S., Amanullah, S. and Jardine, M. (2022). New assistive technologies in dementia and mild cognitive impairment care: Asian Journal of Psychiatry. 22:103135.

Patterson, C. (2018). World Alzheimer's Report Alzheimer's Disease International.

AGING IN PLACE KITCHEN – EXPLORING TASK ZONE HEIGHTS FOR OLDER ADULTS

Track 4: Design for Health and Wellbeing
Knowledge Network: Environment-Gerontology, Residential Environments
Methods: Mixed Methods
Shauna Mallory-Hill (University of Manitoba); Jacquie Ripat (University of Manitoba); Sarah Bohunicky (Acadia University); Reyhane Aliakabari (University of Manitoba); Cristine Perdomo Delgado (University of Manitoba); Taravat Nourian (University of Manitoba)

1 INTRODUCTION

Across all provinces and territories, Canada's population is projected to increase in the proportion of people aged 65 and older to as much as 29.8% in the coming decades (Statistics Canada, 2023). A 2021 March of Dimes Canada-sponsored study found that while over 75% of Canadians wish to age in their current homes, only 26% reported they could do so. One of the barriers to aging in place is the failure of the built environment to meet the physical and cognitive needs of adults as they age. Home modifications, particularly for high utilization areas, like kitchen and bath areas, are required to age-in-place safely and independently. According to Fernihough & McGovern (2015), one of the significant changes over the course of one's life is a decline in physical stature, 0.08 -0.10% for males and 0.12%-0.14% for females, which is equivalent to approximately 2-4 cm. This means that work surface heights, such as kitchen counters and shelves, may no longer provide appropriate ergonomic support for completing various tasks. This presentation describes the research design of a mixed-methods study exploring the relationship of work surface heights in five kitchen task zones to functionality and risk in older adults. The study involved 13 older adult females (mean age 74 years) who undertook a standardized kitchen activity in their homes and in a fully adjustable "smart kitchen." Investigators collected anthropometric data, field observations, evaluated exposure to ergonomic risk of injury, and conducted semi-structured interviews. Data was analyzed using a paired-sample t-test for ergonomic risk and qualitative thematic analysis for interview data. Quantitative analysis revealed significant risk differences in two kitchen zones (cleaning and preparation). Qualitative analysis revealed four key themes: lifestyle, social connections, safety and accessibility. Design considerations emerging from the investigation include the importance of adjustable surface heights, storage, lighting, layout and incorporation of assistive devices.

ASSESSING THE ROLE OF BUILDING FEATURES ON THE STRESS AND SENSE OF BELONGING OF RESIDENTS OF PERMANENT SUPPORTIVE HOUSING TO SUPPORT TRAUMA-INFORMED CARE

Track 4: Design for Health and Wellbeing
Knowledge Network: Residential Environments
Methods: Interviews

Eva Bianchi (Stanford University); Sarah Billington (Stanford University)

1 ABSTRACT

Permanent supportive housing (PSH) is a type of housing designed to house and support formerly unhoused or low-income individuals. Residents of PSH buildings are likely to have experienced traumatic events as a result of being unhoused, thus prompting the need to adopt a trauma-informed approach when designing residences for such individuals. While research on the impact of building features on well-being has become increasingly prevalent over the past few decades (e.g., Allen and Macomber, 2020), very few studies have focused on settings occupied by trauma-survivors. Assessing how built features affect the well-being of these understudied populations can inform future decision-making and ensure that the built environment does not hinder their healing process. Our work focuses on assessing the impact of built features, particularly features that promote greater connection to nature (e.g., views through windows and natural materials), on the stress and sense of belonging of PSH residents. The outcome variables of stress and sense of belonging are chosen based on their relevance to most of the principles of trauma-informed care: safety, trustworthiness, peer support, collaboration, and empowerment (CDC, 2020). The focus on connection to nature stems from prior work on the importance of biophilia for human well-being (Kellert et al., 2011).

We partnered with a non-profit organization and selected two study buildings in San Francisco based on their differences in building design. In each building, thirteen to fifteen residents participated in semi-structured interviews and answered questions about their well-being and the building in which they resided. The interview transcripts were qualitatively coded in pairs using an inductive approach in Dedoose by two co-coders, resulting in an average inter-rater reliability of 0.81. We found that the themes of safety and relationships with building staff and other residents were the most important factors affecting the stress and sense of belonging of participants in both buildings, thus echoing the principles of trauma-informed care. Our analysis also showed that participants' perceived agency, an individual factor also related to the principles of trauma-informed care, interacted with participants' connection to nature through multiple pathways (e.g., feeling increased agency when bringing natural elements into their home or when spending time in public green spaces in their neighborhood). We discuss how designers and practitioners can implement our findings into PSH development and management to promote trauma-informed care via building design.

2 REFERENCE

Allen, J.G. and Macomber, J.D. (2020) Healthy Buildings: How Indoor Spaces Drive Performance and Productivity. Harvard University Press.

Centers for Disease Control and Prevention (2020) '6 Guiding Principles To A Trauma-Informed Approach'. CDC. Available at: https://www.cdc.gov/cpr/infographics/6_principles_trauma_info.htm.

Kellert, S.R., Heerwagen, J. and Mador, M. (2011) Biophilic Design: The Theory, Science and Practice of Bringing Buildings to Life. John Wiley & Sons.

HEALTHY DESIGN IN OLDER ADULTS ENVIRONMENTS INTERGENERATIONAL HOUSING: FRAMEWORK, PROSPECTS, CHALLENGES, AND FUTURE RESEARCH DIRECTIONS

Track 4: Design for Health and Wellbeing
Knowledge Network: Environment-Gerontology, Residential Environments
Methods: Meta-Analysis

Dowon Lee (University of Wisconsin Madison); Jung-hye Shin (University of Wisconsin-Madison); Uchita Vaid (University of Wisconsin - Madison)

1 INTRODUCTION

The concept of intergenerational housing has gained prominence as a potential solution to various societal challenges, including population aging, health and social care crises affecting both older adults and youth, rising housing costs, and the social isolation exacerbated by the COVID-19 pandemic. Nevertheless, persistent issues surround the unclear conceptual definition of intergenerational housing and the untested nature of its presumed benefits, especially considering the recent marketing movement targeting the baby boomer generation rejecting age-segregated senior housing. Through a systematic literature review using scholarly databases, organizational reports, guidelines, and other grey literature, the study aimed to elucidate its conceptual boundaries, examine benefits, identify valuable design features, delineate challenges, and chart future research directions.

We define intergenerational housing as an intentional community where multiple generations cohabit, fostering integrated interactions between individuals aged 18 and younger and those aged 65 and older, along with their adult caregivers. Examples such as senior housing with on-site daycare or senior housing in close proximity to college campuses are excluded. Our literature review revealed that intergenerational housing benefits seniors by reducing loneliness and depression, fostering trust and social capital, promoting interaction with children, and delaying declines in cognitive and physical functions, ultimately delaying institutionalization. Children and youth benefit from improved academic performance, communication and empathy skills, social skills, reduced anxiety, improved self-worth, positive perceptions of the elderly, and positive attitudes toward aging. Primary caregivers also experience lower levels of loneliness, supervision intensity, and caregiver burnout. However, evidence supporting these benefits often remains anecdotal or is based on small sample sizes or a limited number of case studies, possibly intertwined with validity issues such as self-selection and confounding factors.

For promoting intergenerational contact, key features in the built environment included spatial layouts facilitating activity layering, visibility, and flexibility. Successful design decisions prioritized elements that offered choice, comfort, and accessibility for individuals of all ages. Incorporating aspects like prospect and refuge, as well as ensuring privacy alongside opportunities for social interaction, were identified as useful design strategies. Numerous challenges persist in realizing and expanding intergenerational housing. Many approaches predominantly focus on aging-friendly strategies, neglecting the younger generation's role in creating mutual benefits. Funding models fail to cover additional costs for promoting new ways of living and sustaining intergenerational interactions for operating organizations. At the individual level, challenges include fluctuating community composition and participation among residents, as well as potential friction between existing residents (founding members) and newcomers.

Future research should investigate the long-term effects of intentional intergenerational community living on residents' health and well-being, considering validity concerns such as self-selection and distinguishing individual characteristics from residential setting influences. More information is also needed on how these communities can facilitate prolonged independent living, particularly for older adults and people with disabilities. Simultaneously, balanced attention should be given to younger generations across diverse domains of human development, examining the synergistic impact of the intergenerational living setting. Scrutinizing specific aspects of community design, programming, and services contributing to these outcomes will help identify the pathways through which purported health benefits are achieved.

A STUDY ON THE IMPACT OF WORKPLACE SPATIAL FACTORS ON PSYCHOLOGICAL WELL-BEING: ASSESSING STRESS THROUGH VARIOUS PHYSIOLOGICAL SIGNALS

Track 4: Design for Health and Wellbeing
Knowledge Network: Health, Interior Design
Methods: Experimental/Quasi-Experimental Research, Interviews
Daehwa Baek (KAIST); Lisa Lim (KAIST)

1 ABSTRACT

In recent times, many individuals working in various workplaces have been experiencing low levels of happiness and high levels of stress. In response, some of these workplaces are being transformed into stress-reducing spaces, aimed at promoting well-being and work efficiency. Amidst these changes, diverse research has been conducted to understand the impact of physical environments, such as office spaces, on stress. However, there is still a lack of studies that involve both subjective and objective measures for understanding the stress levels of the users, and that conduct experiments using actual physical environments for testing the effects of the design elements on stress levels.

Addressing this gap, specific research conducted at the Institute in South Korea provides profound insights into the simultaneous impact of environmental factors on workplace well-being. This study utilized design elements like windows, artificial windows, and high ceilings to create test beds for experiments. It involved 37 participants and employed a multifaceted evaluation approach, including the Depression Anxiety Stress Scales (DASS). For a comprehensive analysis of stress, the study gathered extensive biometric data, including EEG, ECG, PPG, and GSR, meticulously collected using the Biopac160 system. This approach was aimed at analyzing stress through various measurement techniques. Furthermore, participants' assessments of various photographs provided qualitative insights, complementing the quantitative data.

The research utilized a robust statistical analysis toolkit, comprising Repeated Measures ANOVA, Friedman's Test, Paired Sample T-Test, and the Wilcoxon Signed Rank Test, for a detailed examination of time-frequency, event-related potential (ERP), and the interrelations between variables.

This study underscores the vital role of incorporating natural elements and thoughtful architectural design in the creation of workspaces. It highlights the significance of natural light as a key element in such designs. The study supports design principles that emphasize visibility of external environments in reducing employee stress.

Acknowledging the constraints inherent in short-term studies, this research advocates for more extensive longitudinal studies to fully comprehend the long-term effects of these environmental conditions. The findings offer strategic guidance for organizations seeking to develop work environments conducive to improved well-being and heightened innovative capabilities.

FALL RISK BEHAVIORAL MARKERS OF OLDER ADULTS: OBSERVING INDOOR ENVIRONMENTS AND BEHAVIORS BASED ON INDIVIDUAL HEALTH CONDITIONS

Track 4: Design for Health and Wellbeing
Knowledge Network: Environment-Gerontology, Health
Methods: Case Study, Observation
Jeongyeop Baek (KAIST); Lisa Lim (KAIST)

1 INTRODUCTION

The incidence of falls among older adults is increasing and severe injuries and complications lead to fatalities. Approximately one-third of adults aged 65 and above in the United States experience falls annually, and more than half of the physical injuries are attributed to falls. To prevent fall incidents, risk behavioral markers and high-risk groups should be identified through observational studies. Previous studies have identified that fall risk factors are generally divided into personal and environmental factors. Personal factors include static information such as age, gender, and types of medication, as well as time-series factors such as sequential physical activity changes, which is crucial in older adults due to their age-related health conditions. Environmental factors include climate-related information such as temperature and humidity.

This study attempts to identify the differences in personal health conditions and observed indoor environmental and behavioral patterns between older adults who have experienced fall emergencies and those who haven't. In this study, lifelog data collected from the residential settings of older adults were utilized. The lifelog data were measured through activity detectors, radar sensors, door sensors, and environmental sensors. Additionally, the study focuses on the analysis of the lifelog data of individuals who experienced fall emergencies and were transferred to the hospital between January 2022 and May 2023. The study samples consist of 418 participants, with 209 having experienced fall emergencies and an equal number selected as controls, matching personal factors such as age, gender, and health conditions. Variables in the study are categorized into static (medication types, house height; number of rooms) and time-series (stay time and movement in different rooms, indoor and outdoor climates, and time spent outdoors). The time-series variables were collected over 30 days right before the occurrence of a fall emergency. Chi-square tests evaluate differences in the frequency of fall emergencies based on static information, while T-tests analyze differences in recorded time-series data between the two groups.

This study identified personalized fall risk behavioral markers based on individual health conditions, providing insights for fall prevention strategies. The results not only propose personalized preventive measures but also indicate the potential for enhancing the quality of life in older adults through improvements in indoor environments and behavioral patterns. Moreover, the identified fall risk behavioral markers aid in the early diagnosis of fall emergencies and enable efficient healthcare service implementation through extended connected healthcare networks.

IMPROVING SLEEP FOR BLACK OLDER ADULTS WITH MILD COGNITIVE IMPAIRMENT: A BUILT ENVIRONMENT CASE STUDY

Track 4: Design for Health and Wellbeing
Knowledge Network: Environment-Gerontology
Methods: Case Study, Mixed Methods

Andrea Green (Stanford University); Ibrahim A Bilau (Georgia Institute of Technology); Eunhwa Yang (Georgia Institute of Technology); Ece Erdogmus (Georgia Institute of Technology); Felicia Goldstein (Emory University); Julie Blaskewicz Boron (University of Nebraska Omaha); Aziza Siddiqui (University of Nebraska Omaha)

1 ABSTRACT

Improving sleep quality can slow down the advancement of Mild Cognitive Impairment (MCI) in older adults, as insufficient sleep has been associated with the progression of MCI. While health interventions have been studied to improve sleep, built environment interventions may be a viable option as lighting, noise, and temperature can disrupt sleep. To best support sleep, bedroom lighting/darkness and daytime lighting should be <1 lux and ~250 lux, respectively (Brown et al., 2022). The World Health Organization recommends that average sound exposure during sleep should be less than 30 dB (Fietze et al., 2016). The recommended bedroom temperature for older adults ranges from 68°F to 77°F (Baniassadi et al., 2023).

Our study objectives were to identify the perceptions of socially disadvantaged Black older adults with MCI regarding their built environment and its impact on their sleep, validate these perceptions through in-situ measurements of the built-environment factors, and propose potential interventions to improve the sleep environment. We used validated measures to assess perceptions of sleep quality and environment and sensors to measure the indoor environmental conditions in two participants' homes.

The findings revealed that the average daytime lighting for both participants and the sleep-time lighting of one participant did not align with the recommendations. Because the lighting experienced by the participants does not support a healthy circadian rhythm, the exposure may potentially affect sleep quality, thereby affecting the progression of their MCI. Based on our findings, we propose three interventions tailored to the participants' financial capabilities: 1) replacing current bulbs with cool, bright alternatives, 2) encouraging opening the blinds during the daytime, and 3) providing recommendations for optimal built environment lighting.

2 REFERENCES

Baniassadi, A., Manor, B., Yu, W., Travison, T., & Lipsitz, L. (2023a). Nighttime ambient temperature and sleep in community-dwelling older adults. Science of The Total Environment, 899, 165623. https://doi.org/10.1016/j.scitotenv.2023.165623

Brown, T. M., Brainard, G. C., Cajochen, C., Czeisler, C. A., Hanifin, J. P., Lockley, S. W., Lucas, R. J., Münch, M., O'Hagan, J. B., Peirson, S. N., Price, L. L., Roenneberg, T., Schlangen, L. J., Skene, D. J., Spitschan, M., Vetter, C., Zee, P. C., & Wright, K. P. (2022). Recommendations for daytime, evening, and nighttime indoor light exposure to best support physiology, sleep, and wakefulness in healthy adults. PLOS Biology, 20(3). https://doi.org/10.1371/journal.pbio.3001571

Fietze, I., Barthe, C., Hölzl, M., Glos, M., Zimmermann, S., Bauer-Diefenbach, R., & Penzel, T. (2016). The effect of room acoustics on the sleep quality of healthy sleepers. Noise and Health, 18(84), 240. https://doi.org/10.4103/1463-1741.192480

HEALTHY WORKPLACE DESIGN: FACILITATING FLOW STATE AND RESTORATION IN POST-PANDEMIC CREATIVE WORKSPACES

Track 4: Design for Health and Wellbeing
Knowledge Network: Researchers in Professional Practice, Work Environments
Methods: Case Study, Mixed Methods

Hongfei Li (University of Washington)

1 INTRODUCTION

As we continue to move through the pandemic, the transition to hybrid work models and greater reliance on digital tools necessitates a redefinition of traditional office spaces. In the creative industry, the organic and fluid interactions among colleagues remain crucial for dynamic, team-based creativity. An increasing number of firms require a return to in-person work. This transition highlights the need for workspaces that blend the opportunity for deep-focused work and restorative breaks necessary for productive creativity and well-being. When office design does not meet these needs, employees experience more fatigue and burnout. As a common ground, how can the workplace support well-being and facilitate individual and collective human flourishing? A key element for human flourishing and well-being in the work context is achieving a flow state, a state of deep immersion and heightened performance and creativity (Csikszentmihalyi, 1990).

This study proposes a new framework based on the dynamic stages of flow (Benson, 2003), focusing on the cycles that include focus, struggle, release, flow, and recovery. This study explores the interplay of these stages with environmental factors, seeking to understand how transitions between stages of flow can be facilitated. This research aims to understand (1) the experience of flow, its dynamic stages, and how environmental factors impact them in the creative workplace; (2) what restorative transition helps employees achieve and sustain the flow state. Regarding this restorative aspect, the study considers mindfulness-based interventions (Menardo et al., 2022) and biophilic design as critical elements in promoting workplace well-being. This research hypothesizes that design strategies informed by mindfulness and biophilia can enable the critical shift necessary for restorative stages, thereby enhancing the flow state. The goal is to provide actionable design insights for facilitating mental shifts and restorations in workplace environments.

The presentation will report on a pilot study conducted in a Pacific Northwest design firm. The study has three phases. The first phase establishes the framework for understanding flow, its dynamic stages, and the environmental factors that support it. The second phase collects data through observations, interviews, and surveys to understand how employees experience flow, creativity, struggle, and fatigue and how office space impacts them. The last phase consists of constructing and testing a physical prototype in the office setting, informed by mindfulness-based intervention and biophilic design. This prototype helps to determine what kind of environmental factors can help to restore and facilitate the shift to flow. Lastly, this research aims to pioneer new paradigms in workspace design, transforming post-pandemic work environments into catalysts for well-being and creative excellence.

2 REFERENCES

Csikszentmihalyi, M. (1990). Flow: The psychology of optimal experience
Benson, H., & Proctor, W. (2003). The break-out principle: How to activate the natural trigger that maximizes creativity, athletic performance, productivity, and personal well-being. Scribner.
Menardo, E. et al (2022). Nature and Mindfulness to Cope with Work-Related Stress: A Narrative Review. International Journal of Environmental Research and Public Health, 19(10), 5948.

HOW DO ACTIVITY-BASED WORKPLACES IMPACT EMPLOYEE STRESS LEVELS? A QUALITATIVE INQUIRY TO EXPLORE HOW TO CREATE SUPPORTIVE ACTIVITY-BASED WORKPLACES

Track 4: Design for Health and Wellbeing
Knowledge Network: Researchers in Professional Practice, Work Environments
Methods: Interviews, Practice-based Research
Hande Burcu Deniz (University of Wisconsin - Madison)

1 ABSTRACT

Work-related stress is a global challenge that costs U.S. industries approximately $300 billion annually in reduced productivity, absenteeism, and accidents. It also has expensive costs to individuals by causing health issues, such as unhealthy eating behaviors, failure in the immune system, and cardiovascular diseases. The goal of this study is to understand how the physical features of workplaces impact employees' stress levels.

This study was informed by Ulrich's stress reduction theory (Ulrich, 1991) and the Job Demands-Requirements model (Bakker & Demerouti, 2006). The majority of existing studies in the literature focused on the impact of singular environmental factors on stress levels or the influence of stress on singular outcomes. This study aims to fill this gap by adopting a more holistic approach to produce actionable knowledge to support employees' well-being in activity-based workplaces (ABW). ABW is a work style that allows employees to choose where they work within the office environment based on the task at hand. ABW showed a strong potential to make a positive impact on employee stress levels in previous research (Engelen, et. al., 2018) and has attracted attention from designers and researchers.

The present study used an online survey and semi-structured interviews to collect data with the target sample of 15 participants who work at technology companies that adopted ABW. This population was chosen because the work environment in the technology field is fast-paced and employees usually report high stress levels. The survey included questions about demographics, the nature of their job, and open-ended questions about the impact of the physical qualities of their work environment on their stress level and served as a screening tool for inclusion criteria. The semi-structured interviews explored how the work environment impacted employees' stress levels, as well as their experiences with ABW and what features influenced their stress levels. The interviews were held on Zoom to reach a larger population.

Thematic analysis was conducted to analyze interview transcripts. The first theme derived from the analysis was programmatic requirements. Some participants shared that when their needs to perform their jobs are not fulfilled, they experience more stress. The second theme was choices, which involved individual preferences, interpersonal differences, and the level of agency employees have over their work environment, all of which are mentioned by participants to play a significant role in reducing stress for employees in an ABW context. The third theme was social dynamics, which concerns social relationships between employees, such as organizational hierarchy, teamwork, and how they perceive each relationship to impact their work and performance. The last theme derived from the analysis was distractions. This theme uncovers how the environment can create positive or negative distractions for employees that may in turn impact their stress levels.

The findings of this study will inform future empirical research about workplace design by identifying environmental stressors and de-stressors. This study will help produce actionable knowledge and allow for leveraging design to improve the mental health and well-being of employees in the workplace.

MEASURING WORKPLACE DESIGN SATISFACTION AND UNDERSTANDING THE HYBRID WORK EXPERIENCE: A PRE- AND POST-PANDEMIC CASE STUDY

Track 4: Design for Health and Wellbeing
Knowledge Network: Researchers in Professional Practice, Work Environments
Methods: Practice-based Research, Survey
Oshana Reich (HGA); Heather Kampa (HGA); Renae Rich (HGA)

1 INTRODUCTION

Since the early months of the COVID-19 pandemic, hybrid work has become common across workplaces. However, many leaders are still trying to determine the right balance between in-office and remote work to optimize performance (Trevor & Holweg, 2022) and the appropriate mix of space types and technology needed to support both in-office and hybrid work. For those working in an office, there is growing evidence that interior office space design (including layout, furniture, light, greenery, controls, and noise elements) impacts users' physical, psychological, and social well-being (Colenberg, et al., 2021). In addition to relying on existing evidence to inform design, understanding employees' specific experiences and behaviors in a workplace provides the means to gauge current functionality of the design and receive feedback about ways to improve their productivity and satisfaction while at work. Occupant evaluations can provide valuable information by identifying the activities they engage in at work, preferred spaces for tasks, and perceived opportunities and barriers for hybrid work.

In this session, a study that engaged employees to inform and evaluate workplace design will be presented. A workplace study was conducted in 2018 to inform design solutions for an organization focused on facilitating collaboration among area governments. The resulting renovation consisted of workspaces, support areas, and a flexible technology-driven board room. Following the completed renovation, a 2023 post-occupancy study provided an occupant-experience assessment of the new workplace. Through an online employee questionnaire, feedback was gathered about experiences working in the office, functionality of the spaces, and work patterns. Several measures from the pre-occupancy survey were repeated to assess changes in satisfaction from the old to new office design. Due to the impact of COVID-19 on in-person work that happened between study periods, additional insights were gathered regarding how hybrid work has changed employee perspective and use of in-person workplace environments.
Results demonstrated the success of the design to meet project goals and will help shape the organization's hybrid work policy going forward. Overall, the new office experience was positive, and satisfaction significantly improved with furniture, storage, lighting, and views outside. Respondents' top reasons for working in-office highlighted the office as a place for person-to-person connection: opportunities for informal or impromptu collaboration, attending scheduled in-person meetings, and to be more visible and/or accessible to others. Post-renovation experience with collaborative work areas was very positive, however, the addition of virtual meetings in the same space made noise a concern. Commute, productivity, and work-life balance were the main factors when choosing to work remotely, however respondents indicated that they would like to spend more time in-office than they currently are. Solutions to support more in-office work and productivity in a hybrid work environment will be discussed in this session.

2 REFERENCES

Colenberg, S., Jylhä, T., & Arkesteijn, M. (2021). The relationship between interior office space and employee health and well-being—a literature review. Building Research & Information, 49(3), 352-366.

Trevor, J., & Holweg, M. (2022). Managing the New Tensions of Hybrid Work. MIT Sloan Management Review.

POST-OCCUPANCY FOR RESEARCH-INFORMED PROTOTYPING AND ITERATIVE DESIGN IMPROVEMENT

Track 4: Design for Health and Wellbeing
Knowledge Network: Researchers in Professional Practice
Methods: Mixed Methods, Survey
Helen Ho (DLR Group)

1 ABSTRACT

Post-Occupancy Evaluations (POEs) compile knowledge for future designs, generating iterative improvement, and are used to benchmark building performance for future comparisons (Roberts et al, 2019). This case study provides an in-depth look at a POE for a health facility prototype, ultimately highlighting the knowledge gains POEs generate for designers, clients, and facilities operations alike.

For this case study, secondary data sources and efficiencies with AI-supported data sorting kept the analysis within a manageable budget, ultimately justifying the value of POEs in the industry. The presentation will discuss the collaborative development of the POE process with the client, how the Center for Health Design Post-Occupancy Toolkit was used to justify the process, and ultimately how the evaluation process was structured around data the client was comfortable providing. The presentation will also show how secondary data sources (in the form of employee feedback to HR, Press Ganey patient surveys, and utilities data) were combined with primary data (in the form of change orders and other project documents), to create a holistic view of the facility's day-to-day performance and operations.

At the end of the process, POE data validated new communication pathways for cross-functional teams to increase the efficiency of internal processes and identified areas for improved training resources for facilities managers. Findings triangulated from multiple data sources highlighted priority changes to support patient-centered care, from programming adjustments and staffed waiting rooms for a more human touch, to increased privacy and thermal comfort. A big decision to move to modular interior walls was justified by multiple points of analysis: modular walls not only support future-focused flexibility needs on the client side (as evidenced by the rate of change orders), but were also validated by questionnaire responses pointing to the need for acoustical privacy and comfort. The modular system also promised construction efficiencies at future sites.

Ultimately, this case study shows how a holistic approach to a post-occupancy evaluation can generate improved team efficiencies, support materials specifications and design at future sites, solidify a strong relationship with a key client, and justify the value of research and researchers in industry practice.

2 REFERENCES

Roberts, C. J., Edwards, D. J., Hosseini, M. R., Mateo-Garcia, M., & Owusu-Manu, D. G. (2019). Post-occupancy evaluation: a review of literature. Engineering, Construction and Architectural Management, 26(9), 2084-2106

BEHAVIORAL CHARACTERISTICS OF ZOO VISITORS' ROUTE SELECTION USING BLUETOOTH COMMUNICATION

Track 4: Design for Health and Wellbeing
Knowledge Network: Health, POE/Programming
Methods: Experimental/Quasi-Experimental Research
Takayuki Kumazawa (Ibaraki University)

1 INTRODUCTION

It is difficult to establish a suitable tour route in a zoo because visitors tend to select a route based on their preferences and visual information at sites. However, the behavioral characteristics of zoo visitors during viewing have not been quantitatively verified. In previous research, although there has been some focus on driving characteristics in safari parks (Matthew 2021), very few studies have been conducted on zoo viewing. In this study, Bluetooth communication was used to record human flow and extract visitor behavior data. The impact of the relationship between a zoo's tour route and visitors' route on selection behavior was then clarified. The Kamine Zoo in Hitachi City, Ibaraki Prefecture was used as a case study.

Bluetooth devices that record human flow were placed at 10 junctions along the visitor path in the zoo. The experimenter gave each visitor a sensor to detect the Bluetooth device and recorded the route and time of the visitor's passage. Specifically, the experimenter used the "Raspberry Pi3 A+" microcomputer to create a sensor and receiver that detected Bluetooth devices. When a person holding the sensor passes by the Bluetooth device, the location and time are recorded. The experimenter extracted the number of times and duration of the subjects' passage at each point. The subjects were the 446 groups that visited the zoo in October 2023.

The study's findings were:
(1) Infrequent visitors tend to follow the tour route, while frequent visitors tend to choose their own route.
(2) Visitors tend to choose the route with no slope.
(3) Visitors tend to choose the paths facing the high exhibits.
(4) Visitors tend to choose the paths with a panoramic view.
Based on the above-stated findings, it is necessary to design the exhibition space according to the behavioral characteristics of zoo visitors' routes.

2 REFERENCES

Matthew Lloyd, Naomi Davies Walsh: Investigating Visitor Activity on a Safari Drive, Journal of Zoological and Botanical Gardens, pp.576-585, 2021.2

BIM SYSTEM DEVELOPMENT FOR BUILDING PERFORMANCE EVALUATION OF HOSPITAL WARDS

Track 4: Design for Health and Wellbeing
Knowledge Network: Health, POE/Programming
Methods: Mixed Methods
Akikazu Kato (Mie University)

1 ABSTRACT

The challenge of using Building Information Modeling (BIM) systems is becoming an important issue in the building industries of the USA and Japan, as well as in other countries. BIM enables front-loading decision-making in planning, programming, and design phases, meaning more is discussed and decided in the basic design stage, rather than in the succeeding detailed design stage. This shift of workload increases liability issues for the architects and engineers, however, the process enhances the quality of each phase in the project's development.

The use of BIM provides a holistic picture for the general contractors, where, in the conventional construction process, the information is scattered throughout architectural, structural, mechanical, and electrical drawings. Thus, the system will increase productivity in the construction phase. When the information needed in the management and operation of the building is suitably extracted before the actual completion of the building, as in the virtual hand over stage, the use of BIM can support efficiency in the facility management stage. Lastly, BIM enables simulations by use of evidence uncovered through research, benefitting the building performance assessment stage.

This presentation explains these advantages and discusses the development of meta-BIM as a major player in the current innovations to the building delivery cycle. The BIM system that enables effective facility planning, programming, design, and management of hospital wards. A circular corridor floor plan type was focused and management types regarding patient placement and nursing travel distance were considered.

The nursing travel survey was conducted in four wards of two hospitals. Simulations were carried out to show optimal and inappropriate versions of patient placements using a genetic algorithm. Various fits between the floor plan type and management type were clarified, including the difference in nursing travel distances, which was small in the circular corridor floor plan type.

INVESTIGATING THE RELATIONSHIP BETWEEN PHYSIOLOGICAL STRESS AND OFFICE WORKSTATION DESIGN: A SYSTEMATIC LITERATURE REVIEW

Track 4: Design for Health and Wellbeing
Knowledge Network: Health, Work Environments
Methods: Meta-Analysis
Sophie Schuller (TUe)

1 PURPOSE

This study aims to explore the link between workstation design and physiological stress among office workers through a systematic literature review. Theoretical frameworks such as Neuroergonomics (Dehais et al., 2020), the James-Lange Theory of Emotion (Fehr & Stern., 1970) and Embodiment theories (Magnon et al., 2018) offer valuable perspectives to evaluate the relationship between workstation design and physiological stress, potentially mediated by workstation-associated physiological motor processes, posture and ergonomics.

2 METHODS

For this systematic literature review, the PRISMA guidelines were used. Reviewed papers were searched in Scopus, PubMed, PsycINFO, and Web of Science databases, using keywords associated with 'office, workstation design and physiological stress'. Physiological stress measurements are based on widely accepted bio and clinical markers, such as heart rate variability (HRV), cortisol secretion and blood pressure. Papers were limited to primary research, published between 1980 and 2024 and written in English.

3 FINDINGS

Research indicates poor ergonomic workstation design, typified by un-adjustable furniture or poor ergonomic-employee fit, is linked to an increase in muscle tension and poor posture associated with increased activity of the sympathetic nervous system and heightened cardiovascular reactivity, which are also shared biomarkers of stress. Conversely, ergonomic workstations with adjustable features, such as sit-stand desks and workstation designs that promote movement, show cardiovascular, ambulatory and cortisol reactivity improvements. In addition, limited but exciting findings suggest that the angular positioning of hips, knees and torso in relation to one another may mediate physiological stress effects through simulating reclining and restoration stances. These findings suggest that workstations encouraging physical activity and movement may reduce physiological stress, potentially strengthening the nervous and cardiovascular systems and building subsequent resilience.

4 ORIGINALITY

The originality of this paper lies in being the first systematic literature review to explore the relationship between workstation design and physiological stress. The significance of these findings is that they have the potential to enhance our understanding of the impact of workplace environmental design on the aetiology and development of physiological stress.

SENSORY ENVIRONMENTS AND HEALTH EXPERIENTIAL DESIGN TEACHING

Track 4: Design for Health and Wellbeing
Knowledge Network: Environmental & Architectural Phenomenology, Environmental Design Research Education
Methods: Design Methods
Gail Brager (University of California at Berkeley)

1 ABSTRACT

Architecture education has increasingly steered away from centering human experience in the design process. At the same time, students seem to be hungry for meaningful perspectives and direct experiences that counter our digitally saturated world. Architecture pedagogy grounded in a people-based approach—where subtle environmental variations alleviate experiential monotony and bring inhabitants' back to the center of design intentions—can be pivotal in enhancing occupant well-being while simultaneously minimizing energy use, shore resilience, and supporting environmental stewardship.

This presentation provides lessons learned from two recent offerings of a class about how to create rich multisensory experiences in buildings, connecting people to nature, and embracing the value of environmental variability and delight. The class advocates for designers to embrace a broader view of experiential aesthetics, choreographing visual, thermal, acoustic and olfactory elements of the sensescape to contribute to the beauty and memorability of space. It provides new frameworks and nuanced language for understanding the spatially and temporally varied sensory world around us, and communicating experiential-based design intentions. It uses research-based evidence to forefront individual agency, and provide for neurodiversity, personal control and choice. It showcases ways to provide sensory delight in a range of scales (materials, rooms, whole buildings, ecosystems) and settings (from residential to institutional, from monumental projects to everyday spaces). Every element of the class bridges interdisciplinary research and design, integrating the science of sensory experience with generative schemas for creating dynamic, healthy, human-centric architecture.

The seminar-format class is structured around lectures, in-class interactive exercises, and both individual and group assignments to encourage collaborative thinking. It guides students through a series of increasingly complex observational activities, case studies and design projects that build on each other, yet can also be used by themselves in modular ways, easily modified for various class formats. As examples, the topics and assignments explore concepts of sensory/experiential awareness, mapping / representing spatial and temporal variability, zoning for sensory diversity, meaning and application of sensory equity, experiential programming, and retrofit design of existing places. Extensive reading assignments include DeKay + Brager (2023), Erwine (2017), Kellert et al (2011), and many more.

A goal for 2024 is to package the instructional materials and make them freely available online, for people to use and revise however it suits their purpose. This session will thus serve to educate and inspire on its own, as well as be an introduction to these eventual resources.

2 REFERENCES

DeKay, M. & Brager, G. (2023). Experiential Design Schemas. ORO Editions.

Erwine, B. (2017). Creating sensory spaces: the architecture of the invisible. Routledge, Taylor & Francis Group.

Kellert, S. R., Heerwagen, J., & Mador, M. (2011). Biophilic design: the theory, science and practice of bringing buildings to life. John Wiley & Sons.

SENSORY PROCESSING IN THE WORKPLACE – NEUROTYPE WORKPLACE SURVEY

Track 4: Design for Health and Wellbeing
Knowledge Network: Environmental & Architectural Phenomenology, Interior Design
Methods: Action/Participatory Research, Survey
Sarah Oppenhuizen (HOK); Kay Sargent (HOK)

1 ABSTRACT

Over the past 8 years our firm has undertaken continuous research on the topic of designing for neurodiversity and expanded that to include sensory processing and cognitive well-being.

Our workplace team and global practice leaders have conducted surveys, focus groups, and research to develop the initial business case, call to action, and initial design strategies for consideration.

In a continuation of our research, we recently undertook an assessment in our San Francisco office that included a survey of a mixed neurotype population to assess their sensitivities to sensory stimulation. Those findings were then compared to the findings from the previous research done with a neurodivergent sample group. The goal is to identify similarities and differences between the groupings.

ENHANCING TRAVELER EXPERIENCES: USING DATA-DRIVEN CUSTOMER JOURNEY MAPS TO INFORM AIRPORT DESIGN

Track 4: Design for Health and Wellbeing
Knowledge Network: Environmental & Architectural Phenomenology, Researchers in Professional Practice
Methods: Mixed Methods, Practice-based Research

Melissa Hoelting (Corgan); Samantha Sanders Flores (Corgan Associates, Inc.); Mahdi Afkhamiaghda (Corgan); Arezoo Zeinali Kalkhoran (University of Florida)

1 ABSTRACT

In the realm of airport design, understanding and optimizing the passenger experience is vital for long-term success. These complex facilities require seamless navigation, comfort, and efficiency while also accommodating diverse and individual user needs, all of which design plays a key factor in accomplishing. At a California airport, the client's goal was to complete a study that defines their current customer experience, analyzes passenger touchpoints, and proposes strategies for future development. A data-driven approach was taken to develop a Customer Journey Map that provides a clear and comprehensive understanding of a passenger's journey throughout the airport to elevate their overall experience.

This study took a mixed-methods approach to comprehensively investigate user behaviors, needs, and preferences and how the overall design and quality of space influences passenger satisfaction. The initial phase, secondary data collection, utilized existing data sources to provide insights into current passenger demographics, flow patterns, documented passenger sentiment through social media, and emerging trends and technologies.

Once the study team had gained insight into who the airport's unique set of passengers were, how the airport was designed to operate, and current satisfaction levels, primary data collection occurred to dive deeper into the physical experiential aspects within the environment. First, a sense of place study was conducted with 30 locals and non-locals to understand how the character of the airport staff, space, operations, and design contribute to a localized experience. Next, on-site data collection occurred to observe and measure throughout the airport. This consisted of interviews with over 50 passengers and staff, 60 documented behavior maps, 31 arriving and departing passenger shadow paths, and 21 areas of study with environmental quality sensors. In addition, 17 experiments with sourced participants who had never been to the airport while wearing retina-scanning glasses were conducted, triangulating eye movements, location in space, and task completion to assess level of difficulty, identify distractors, and determine where interventions might be needed.

The data was then analyzed to identify pain points, areas of success and opportunities for improvements at key touchpoints throughout the airport for a defined set of unique passenger profiles. After a gap analysis to identify the difference between what is currently being provided and what is desired or needed by all users of the space, insights were developed into how future design, policy change, and technology integration can enhance the experience throughout the journey. These insights were consolidated to build a comprehensive journey map to design an intuitive and enjoyable experience for a variety of passengers.

From this project, this airport applies captured data and resulting insights to make informed decisions, prioritize investments, optimize services, and ultimately enhance customer satisfaction. Resulting design strategies from the research were prioritized and organized from immediate, low-hanging fruit opportunities through innovative long-term solutions. Findings, recommendations, and phased implementation strategies were documented in comprehensive journey maps and a guidebook that serve as a resource for future enhancements. This holistic approach deepens the airport's understanding of passengers' needs, fostering innovation in service and infrastructure improvements to deliver a world-class airport experience.

A STUDY ON THE INFLUENCE OF TRANSPORTATION SPATIAL ATTRIBUTES AND REPRESENTATION ON PASSENGER PATH SELECTION BASED ON VIRTUAL REALITY SCENARIOS

Track 4: Design for Health and Wellbeing
Knowledge Network: Environmental & Architectural Phenomenology, Sustainable Planning, Design & Behavior
Methods: Experimental/Quasi-Experimental Research, Statistical/Modeling Analysis
Mo Zhang (Beijing Jiaotong University)

1 ABSTRACT

Beijing's rail transit is complex and has a large daily throughput, so the wayfinding signs in the stations determine the efficiency of passengers' self-orientation and spatial guidance, and passengers' behavior is different when they are wayfinding in different spatial attributes of the traffic space, which directly affects the efficiency of passengers' transfer. In this study, we used virtual reality to simulate people's wayfinding behaviors according to the signage elements in a complex transportation space, and explored the following questions: 1) whether the spatial attributes of the transportation space affect the subjects' path choice behaviors; 2) the differences in passengers' attention to the environmental elements in the transportation space, and the correlation with the corresponding spatial attributes.

The experiment was conducted at Xizhimen subway station in Beijing, where three subway lines (Line 2/4/13) intersect and the paths of pathfinding are more complicated, and a detailed three-dimensional model was constructed. Thirty subjects wearing virtual reality headsets were selected to start from the train level of Line 13, find their way according to the spatial attributes of the environment and signs, and arrive at the station hall level of Line 2 to complete the transfer. The subjects' wayfinding behaviors were segmented according to the important turn markings, and the spatial syntax was applied to quantify the spatial attributes of the transportation spaces in different segments, and calculate the spatial attributes such as the degree of connection and integration. Behavioral data including turn, stop, turnback, head-up, etc. of different segments of the subjects are collected synchronously. Route choice data includes walking trajectory and walking time, and the calculation of the two can obtain the average walking speed of the subjects in different segments. Collect eye movement data, i.e. eye movement data on signs, advertisements, decorations, and other elements in the wayfinding space. After completing and filling out the questionnaire to collect subjective evaluation data, including sense of direction, sense of disorientation, sign validity, etc., in different segments. By analyzing the correlation between spatial attributes and behavioral data in different segments, the influence of spatial attributes on subjects' path choice can be explored. By analyzing the eye movement data and corresponding spatial attributes of different segments of space, the difference in attention allocation of subjects under different spaces can be derived, and the sign validity during pathfinding in different segments can be derived according to the questionnaire to analyze the reasonableness of spatial labeling.

This paper uses virtual reality technology to carry out wayfinding experiments on high-speed rail station to Beijing Xizhimen subway station to explore the influence of different spatial attributes of transportation space on passengers' subjective behavior data, verify the correlation between spatial attributes and passengers' attention to the differences in the spatial elements of the environment, and explore the rationalization of the design for optimizing the interchange space of high-speed rail station with the support of the data to promote the transformation, upgrading, and retention of high-speed rail station.

COMMUNITY MOBILIZATION AND PLACEKEEPING EFFORTS IN RESPONSE TO DISPLACEMENT IN SEATTLE'S HISTORICALLY BLACK NEIGHBORHOOD

Track 5: Spatial Justice, Equity, and Inclusion by Design
Knowledge Network: None/Not Tied to Any Network
Methods: Action/Participatory Research
Lynne Manzo (University of Washington); Kristin McCowan (Wa Na Wari); Inye Wokoma (Wa Na Wari)

1 ABSTRACT

Black urban communities are particularly vulnerable to displacement as their neighborhoods are revalorized and targeted for development and "revitalization" (Hern, 2016). Fullilove (2016) describes the experience of displacement as "root shock" -- "traumatic stress reaction to the destruction of all or part of one's emotional ecosystem" (p. 11). It is a rupture of both the geographic and the social. As such, it can be considered a collective moral injury (Barbot, 2020).

Seattle's Central District is one urban neighborhood where residents are experiencing displacement and root shock. The Central District has been a vital African-American community for over 75 years, but it has seen a dramatic decline in the Black population due to gentrification. In 1970, Blacks comprised 73% of the population, and that number has shrunk consistently in subsequent decades. If this trajectory continues, the Black population will be less than 10% by 2025 (U.S. Census Bureau).

This presentation describes the work of an ongoing community-driven action research project called "CACE 21" – Central Area Cultural Ecosystem for the 21st Century. The research takes a case study approach and draws on oral histories and interviews to understand residents' experience of neighborhood change, their efforts to maintain a foothold in the neighborhood and the ongoing project of placekeeping (Dempsey et al., 2014). Here, placekeeping is considered a practice of care, a grounding intervention that enables people to interrupt harm and engage in the process of collective healing through the maintenance of social and geographic ties.

This presentation also documents and reflects on community-mobilization and placekeeping efforts as a form of activism and resistance that is a joyful celebration of the history, legacy and beauty of the Black community in the Central District. In this way, it is a form of "milieu therapy – a "working through the creation [and maintenance] of healing places for an environment to lift the spirit" through opportunities for relatedness (Fullilove, 2016, p. 6). Implications for community-driven projects in partnership with university researchers in other communities of color will also be discussed.

2 REFERENCES

Barbot, O. (2020). George Floyd and our collective moral injury. American Journal of Public Health, 110(9), 1253-1253.
Fullilove, M. T. (2016). Root shock: How tearing up city neighborhoods hurts America, and what we can do about it. New Village Press.
Hern, M. (2016). What a city is for: Remaking the politics of displacement. MIT Press.

DE-URBANISM AND WAR: FRAGMENTED COMMUNITIES BETWEEN RESILIENCE AND SURVIVAL

Track 5: Spatial Justice, Equity, and Inclusion by Design
Knowledge Network: Cities & Globalization, Residential Environments
Methods: Case Study, Ethnography
Sahera Bleibleh (United Arab Emirates University)

1 ABSTRACT

In a war-torn context, the repercussions of spatial justice and urban warfare extend beyond military strategies, wavering a complex array of contradictory and ambivalent social inequality. In occupied Palestine, the profound insecurities permeate all aspects of everyday life, fostering a persistent struggle for renewal that not only shapes social bonds but also reinforces a profound sense of place. Against the backdrop of Israeli colonial strategies, individuals' attachment to home becomes inseparable from their unwavering steadfast resistance and sumoud. Despite the cumulative threats imposed by ongoing attacks during the Israeli occupation, home, and its remaining after being demolished, paradoxically assumes the dual role of a "forbidden space" of sanctuary and a source of peril.

This study intricately delves into the interplay of home, de-urbanism, and war, examining how military attacks disrupt the social dynamics and overall resilience of urban communities. Embracing the themes of urbicide, fragmentation, displacement, as well as resilience and survival, the research explores the coping mechanisms and adaptive strategies employed by communities to endure, resist, survive, and rebuild amidst the harsh realities of de-urbanization during warfare.

Through a multidisciplinary ethnographic approach that seamlessly integrates urban studies, sociology, and urban violence analysis, this research contributes valuable insights into the dynamics of hope within post-war urban landscapes. Ultimately, the findings aim to inform policies and interventions that foster community recovery, emphasizing the importance of resilience as a catalyst for rebuilding shattered urban environments.

WHAT MAKES A PLACE UNIQUE: CONTENTS OF PLACE IDENTITY AMONG MIGRANTS AND NATIVES

Track 5: Spatial Justice, Equity, and Inclusion by Design
Knowledge Network: Cities & Globalization, Sustainable Planning, Design & Behavior
Methods: Case Study, Focus Groups
Tiago Dionisio (Gent University); Fátima Bernardo (Instituto Superior Técnico); Isabel Loupa Ramos (IST/University of Lisbon, Portugal); Seppe De Wit (Ghent University); Veerle Van Eetvelde (Ghent University, Department of Geography)

1 ABSTRACT

People attribute different meanings to features from the landscape to build their own identity with their place based on their experience (Loupa-Ramos et al., 2019). The same place can be perceived differently by different groups, which can lead to conflicts of ownership of a place (Storz et al., 2020). This conflict is relevant to understand in the context of migration as this event brings challenges for both natives and migrants. On one hand, migrants need to re-adjust their identity to the place due to its new social and physical characteristics, while natives might feel their place identity threatened due to the presence of these new groups. Thus, different meanings can be attributed to the same place feature as each group can characterize their place based on different or similar elements. Overall, studies have focused on this relationship without acknowledging objective features of the landscape and the perceived value to migrants and natives. Thus, this study focuses on comprehending which characteristics of the place are relevant for the place identity of both migrants and natives.

To comprehend this phenomenon, four case studies were studied: an urban and a rural case study each in Portugal and Belgium with high migrant populations. For each case study, a focus group was made with only migrants and another with natives. Participants were asked to identify and discuss what characterizes and makes their place unique.

For each of these landscape meaning dimensions, both migrants and natives reveal different levels of a shared identity. In some case studies, people distinguish different groups based on their perception of diversity. Plus, migrants tend to describe elements that are not specific to the neighborhood and emphasize the importance of green areas, services, and lifestyle quality. Finally, natives characterize their place more based on their prestige and history of the place. These results suggest that the understanding of the meaning migrants and natives attribute to the different elements of the landscape can reflect how they experience the landscape, with more similarities or differences between them. Depending on the case study, migrants and natives can share a similar experience of the landscape or a more group-exclusive view of the place they live in.

In sum, this study emphasizes the importance of understanding the meaning of the characteristics of a place by both migrants and natives. Comprehending how these groups relate to the landscape can support future planning and actions to improve territorial cohesion by acting on elements of the place that both groups use as part of their identity. Thus, this knowledge can help identify elements that can create bridges between both groups through their place.

2 REFERENCES

Loupa-Ramos, I., Bianchi, P., Bernardo, F., & Van Eetvelde, V. (2019). What matters to people? Exploring contents of landscape identity at the local scale. Landscape Research, 44(3), 320–336.

Storz, N., Martinovic, B., Verkuyten, M., Žeželj, I., Psaltis, C., & Roccas, S. (2020). Collective psychological ownership and reconciliation in territorial conflicts. Journal of Social and Political Psychology, 8(1), 404–425.

SPACES IN BETWEEN: THE GENDERED SPATIAL TEXTURES PRODUCED BY THE DOMESTIC 'MAIDS' OF KOLKATA, INDIA

Track 5: Spatial Justice, Equity, and Inclusion by Design
Knowledge Network: Communication, Cultural Aspects of Design
Methods: Action/Participatory Research, Ethnography
Aparajita Santra (University of Illinois at Urbana Champaign)

1 ABSTRACT

This study explicates the spatial textures in the urban fabric produced and altered by the domestic 'maids' of Kolkata. The culture of domestic servitude dates back to the colonial period in Kolkata and continues to be a part of the fabric of daily life in Kolkata (Ray & Qayum, 2009). Historically, domestic helps or the 'maids' who have engaged in so-called menial activities have been constructed in binary opposition to the respectable middle- or upper-class women. Terms like common women, public women, or chhotolok (in Bengali that translates to lower caste women) have often been used to refer to these subaltern women who have engaged in spatial practices that have defied the established norms of morally, socially/ sexually appropriate behavior in urban spaces of the city. These women negotiate multiple spatial precarities as they lead their ordinary lives and move around the city for their work and other everyday needs. As these 'maids' of Kolkata negotiate a wide range of structural and systemic constraints that are embedded in urban space, they produce liminal spaces within the urban fabric that I call spaces in between. These liminal spaces are produced through informal support systems and kinship affiliations. To understand the production of these liminal spaces, this study draws on the concept of kitchen-table knowledges (Collins, 1999) building on Black feminist thought. Collins (1999) discusses kitchen-table knowledge as being the "collective secret knowledge generated by groups on either side of power that are shared in private when the other side's surveillance seems absent." These spaces of informal collective knowledges produced by the working-class women communicate a spatial language to other women who are part of this knowledge system. This study uses a feminist ethnography methodologically. Lacking access to different forms of social, material and cultural capital (Bourdieu, 1986), ordinary leftover spaces in the city like parking lots, street pavements, terraces and balconies become meaningful spaces for these women as they occupy them for their various needs. The study examines the various psychological processes of these women that affect their spatial behavior in urban space like their sense of territorial control and personal space, their perception of environmental cues, their mental constructions and cognitive imagery that they overlay on physical space, as well as the ways in which their behaviors are adopted and adapted constantly during the person-place transactions. The spatial histories of these subaltern women have been systemically obscured in the configuration of urban spaces of Kolkata, especially because of their marginalized caste and class. Thus, the exploration of these spatial textures produced by the 'maids' of Kolkata is significant as it foregrounds spatial knowledges that have been historically rendered unintelligible due to gendered, classed, racialized and sexualized dispossession.

2 REFERENCES

Bourdieu, P. (1986). The forms of Capital. In Handbook of Theory and Research for the Sociology of Education. Greenwood.

Collins, P. H. (1999). Reflections on the Outsider Within. Journal of Career Development, 26(1), 85-88.

Ray, R., & Qayum, S. (2009). Cultures of Servitude: Modernity, Domesticity, and Class in India. Stanford University Press

URBAN DESIGN AND SPATIAL ANALYSIS OF CANADIAN INDUSTRIAL DISTRICTS: HEALTH AND EQUITY

Track 5: Spatial Justice, Equity, and Inclusion by Design
Knowledge Network: Health, Sustainable Planning, Design & Behavior
Methods: Mixed Methods, Statistical/Modeling Analysis
Ana Karinna Hidalgo (University of Saskatchewan)

1 ABSTRACT

Industrial districts are physically recognizable for their unique urban landscape, featuring large parcels and massive warehouses. In addition, these sites are identifiable in North America due to the scant or nonexistent urban design features such as sidewalks, bike paths, public transportation, and other amenities. However, every day, thousands of workers use these spaces. In 2022, 40% of persons in the labour force were employed in the goods-producing sector (Statistics-Canada, 2023). Furthermore, 25% to 40% of the workforce is exclusively located in industrial areas in North American cities (Hatuka & Ben-Joseph, 2022). Moreover, one in five people in Canada experiences mental health issues each year and half of the population experiences mental illness by age 40.

In addition to the urban design conditions, other environmental conditions also affect human health. Air quality, surface temperatures and green coverage could determine people's well-being in cities. Our study assesses Canadian industrial districts by considering these indicators and using evidence-based urban design focusing on health and equity. We selected fifteen districts, including two districts per major province, one for each province and one from the territories. Using ArcGIS, Google Earth, and Google Maps, we determine each industrial site's state of the art concerning environmental indicators and urban design.

Based on an interdisciplinary literature review, we determined an analytical and conceptual framework to conduct our analysis. Evidence shows that public spaces containing biodiversity (Konijnendijk, 2023) and other urban design features promote communities' physical, mental and social health. Specific design theories and concepts were addressed to understand the urban environment's qualities concerning health and equity. Our framework consists of three pillars: 1. Biophilic design, urban nature and therapeutic landscapes; 2. Indigenous values and sustainability; and 3. Prospect refuge theory and quality urban design. Using this framework, we looked at landscape features, the district's environmental and urban history, main roads and connectors, greenery, public transportation, walkability and bikeability, accessibility, legibility, water management, building setbacks and buffers.

Our analyses confirmed that industrial districts are not designed for health, the environment, or people's well-being. The analysis of environmental indicators in industrial areas compared with other areas in the city, and in contrast with the literature, demonstrates that these areas lack quality spaces and are linked to health-related issues in the built environment.

These differences in city areas discourage equity and inclusion. We are convinced that urban industrial areas should promote mental, physical, and psychological restoration to contribute positively to all individuals and the city. The benefits of healthy urban environments go beyond the area's physical boundaries. This study constitutes the first phase of a long-term research agenda in Canada's healthy and environmentally friendly industrial districts. A healthy place for daily activities and work will benefit everyone.

2 REFERENCES

Hatuka, T., & Ben-Joseph, E. (2022). New Industrial Urbanism. Designing Places for Production. Routledge: New York and London.

Konijnendijk, C. (2023). "Evidence-based guidelines for greener, healthier, more resilient neighbourhoods: Introducing the 3-30-300 rule." Journal of Forestry Research 34: 821–830.

Statistics Canada. (2023). Table 14-10-0023-01. Labour force characteristics by industry, annual (x 1,000).

CLASSROOMS FOR WHOM: DESIGN ELEMENTS SHAPING INTERACTIONS BETWEEN PERCEPTIONS OF GENDER AND SENSE OF BELONGING

Track 5: Spatial Justice, Equity, and Inclusion by Design
Knowledge Network: Interior Design, Researchers in Professional Practice
Methods: Experimental/Quasi-Experimental Research, Survey
Julia Pascutto (Lemay); Michael Ralph (Multistudio)

1 ABSTRACT

The built environment embodies and communicates the beliefs, values, and expectations of the designers. In the interior design of classrooms, research has shown the physical environment plays an active role in influencing the experience of the occupants of those spaces as it relates to their gender identities and gendered societal experiences (Cheryan et al., 2009). Removing marginalizing pressures and barriers to participation for women and gender non-binary people is a persistent need in higher education (Ballenger, 2010; Teague, 2015). Given this well-documented need to improve representation of women and gender non-binary people in many fields (including the higher education workforce itself), a key lever in change-making is the improvement of career preparation pathways. Said directly, institutions should improve access to positive, productive experiences for undergraduate students to improve retention, graduate, and matriculation rates into the academic workforce. However, the mechanisms of gendered classroom design in higher education remain understudied.

We investigated how design elements in classroom design are perceived by two key groups: the end users (undergraduate students) and the designers (design industry professionals). We collected survey data from 97 undergraduate students and subsequently 88 design firm professionals to study their perceptions of femininity and masculinity in a randomized sequence of 4 rendered classroom environments. We also collected data on their expectations of belonging in each space. Our results showed a positive correlation between femininity and masculinity in both samples, but a dramatically different relationship between femininity and expectation of belonging. Among the undergraduate students there was a medium-strong positive correlation, but among the industry professionals there was a medium-strong negative correlation. There were no consistent differences between gender subgroups. The difference between groups' perceptions of femininity is potentially attributable to multiple design elements in the rendered environments.

We identified 11 design elements that we manipulated between environments. Our presentation will examine the potential impact of these elements on perceptions. Examples include differences in an accent wall finish, lighting arrangement, and seating configuration. Future study may further refine the measurement strategy and manipulation of design elements. These studies also have clear implications for both research and practice themselves, such as the importance of reflexivity for designers who may have different personal perceptions of their designs than those of their intended users.

2 REFERENCES

Ballenger, J. (2010). Women's Access to Higher Education Leadership: Cultural and Structural Barriers. In Forum on Public Policy Online (Vol. 2010, No. 5). Oxford Round Table. 406 West Florida Avenue, Urbana, IL 61801.

Cheryan, S., Plaut, V. C., Davies, P. G., & Steele, C. M. (2009). Ambient belonging: How stereotypical cues impact gender participation in computer science. Journal of Personality and Social Psychology, 97(6), 1045–1060. https://doi.org/10.1037/a0016239

Teague, L. J. (2015). Higher Education Plays Critical Role in Society: More Women Leaders Can Make a Difference. In Forum on Public Policy Online (Vol. 2015, No. 2). Oxford Round Table. 406 West Florida Avenue, Urbana, IL 61801.

INCLUSIONARY HOUSING IN URBAN CHINA: A SILVER BULLET TO REDUCE HOUSING INEQUALITY AND FOSTER SOCIAL INCLUSION?

Track 5: Spatial Justice, Equity, and Inclusion by Design
Knowledge Network: International Housing Research, Residential Environments
Methods: Case Study, Mixed Methods

Weican Zuo (University of Michigan, Taubman College of Architecture & Urban Planning)

1 ABSTRACT

Over the past three decades, China has launched several affordable housing programs to combat the housing affordability crisis and help low- and moderate-income families own or rent a decent home in the city (Deng et al., 2011) To increase affordable housing provision without overburdening local governments and to avoid concentrated poverty, the State Council of China introduced an inclusionary housing scheme in 2007 and required that affordable rental housing units should be mainly produced through inclusionary housing developments (Huang, 2015). This research responds to a present lack of studies that systematically examine how inclusionary housing works in China and assess its accomplishments, problems, and challenges. Specifically, this research aims to answer the question: To what extent does inclusionary housing serve as a social development tool that reduces housing inequality and fosters social inclusion in Chinese cities? This research employed a mixed methods case study design to investigate inclusionary housing in urban China, examining policy design and implementation and, most importantly, exploring policy outcomes from the residents' perspectives.

This research selected the city of Zhengzhou as the case study. At the city level, the study examined the development process of inclusionary housing projects and reflected on how the inclusionary housing scheme transforms the socio-spatial patterns of the city through a review of government documents and interviews with government officials and developers. At the neighborhood and individual level, the study focused on two inclusionary housing neighborhoods as case studies, using survey instruments combined with qualitative interviews to investigate policy outcomes from the residents' perspectives. Specifically, the study inquired about the extent to which the housing situation of migrants and low-income families has been improved by inclusionary housing and whether the residents' perceived level of social inclusion has been increased by the mixed-living patterns.

This study found that the inclusionary housing scheme has helped reduce housing inequality in urban China by making affordable housing more accessible for migrants and low-income households and by distributing affordable housing more evenly across the city, which allows disadvantaged families to live in better neighborhoods and have better access to public services. Nonetheless, the study also found that inclusionary housing falls short of fostering social inclusion on a neighborhood scale. Despite different social groups living near each other in the same neighborhood, there were limited social interactions and integrations across lines of income or tenure. Inequities embedded in neighborhood planning and design have further deepened the issue of stereotyping and affordable housing residents' feelings of being stigmatized and marginalized. This study concluded with a discussion of how governments, developers, and designers could work together to make inclusionary housing reach its social goals and make mixed-income neighborhoods more livable, sustainable, and inclusive. In doing so, this study contributes to the global knowledge of inclusionary housing and mixed-income development practices.

2 REFERENCES

Deng, L., Shen, Q., & Wang, L. (2011). The emerging housing policy framework in China. Journal of Planning Literature, 26(2), 168–183. https://doi.org/10.1177/0885412210390220

Huang, Y. (2015). Bolstering Inclusionary Housing in Chinese Cities. The Paulson Institute/Paulson Policy Memorandum.

WORTH THE WAIT? EXAMINING THE INFLUENCE OF WAITING ENVIRONMENTS ON MEDICAL TRUST

Track 5: Spatial Justice, Equity, and Inclusion by Design
Knowledge Network: Interior Design, Researchers in Professional Practice
Methods: Experimental/Quasi-Experimental Research, Survey

Amy M. Huber (Florida State University, Dept. of Interior Architecture and Design);
Rachel Bailey (Florida State University)

1 ABSTRACT

Women's longer life expectancies make them more vulnerable to a host of health issues, but the primary healthcare system in the U.S. is failing to meet their needs, and young adults are increasingly less likely to use primary care physicians. This shift may be especially deleterious to those from minority, underrepresented groups, who are less likely to seek healthcare services and visit a usual healthcare provider (Levine et al., 2020), but are more likely to report higher levels of waiting room anxiety (Abdou & Fingerhut, 2014) and physician distrust (Rosenthal & Lobel, 2020). Consequently, the experiences of minority women waiting for care appointments may vary from those of other patients. Amidst these contexts, building medical trust in primary care is crucial. Still, short visit durations can hinder trust development between women and healthcare providers, which necessitates an exploration of the healthcare environments' impact on medical trust perceptions. Yet, research on the waiting room's physical environment has focused on universal perceptions, and its impact on medical trust is unclear.

2 PURPOSE

This mixed methods research used surveys and semi-structured interviews to identify which waiting room factors influence medical trust for African American, Hispanic/Latina, and White, non-Hispanic women aged 18-35. These participants represent the three largest demographic groups of women in the U.S., at an age when they are most likely to bear children.

3 FINDINGS

1,114 women responded to a Qualtrics survey. They rated calming, followed by welcoming environments as the highest scoring trust-eliciting characteristics (Tables 1 & 2). Pairwise comparisons revealed that each ethnic/race cohort scored these factors statistically significantly higher than provider credentials or images, local photos/art, ethnically similar patients, and images of ethnically/racially similar providers. Yet, a Tukey post hoc test revealed that AA women scored patient ethnic/racial similarity, provider images, and images of ethnically/racially similar providers as statistically more important than the other participants. Quota sampling was used to recruit 25 interviewees representing varying healthcare trust levels across the three demographic cohorts (Tables 3 & 4). Trustworthy environments were characterized by having clear signage, informative materials, provider images, well-maintained spaces, positive distractions, and minimal marketing. Conversely, interviewees suggested that untrustworthy environments exhibit poor maintenance, inappropriate marketing, and lack clear information, positive distractions, or diverse representation in art and materials.

4 IMPLICATIONS

By prioritizing potential determinants of medical trust in waiting room environments, these findings help to disentangle innate environmental preferences from those that may be more culturally constructed amongst women in the three largest demographic groups in the U.S. While similarities emerged regarding how medical trust perceptions may be fostered within waiting rooms, the differences revealed between participant cohorts underscore the importance of representation amongst minority groups – a value which can be imbued into the built environment during the design process and its outcomes. Accordingly, this study provides valuable insights for those who are involved in designing healthcare environments. By making informed design decisions, the physical environment of medical waiting areas can better cultivate trust. These enhancements may positively impact women's healthcare-seeking behaviors and ultimately improve their health outcomes.

SPATIAL JUSTICE, EQUITY, AND INCLUSION BY DESIGN: LEARNING ENVIRONMENTS FOR NEURODIVERSE USERS: HOW DESIGN CAN SUPPORT INCLUSIVE HIGHER EDUCATION CLASSROOMS

Track 5: Spatial Justice, Equity, and Inclusion by Design
Knowledge Network: Health, Interior Design
Methods: Interviews, Photo-based Methods

Elif Tural (Virginia Tech); David Kniola (Virginia Tech); Andrew Kim (Steelcase)

1 ABSTRACT

Neurodiverse, neurodivergent or neuroatypical are terms used to describe students with differences in brain function and behaviors, or have atypical intellectual and cognitive abilities. Generally, these terms are used to describe students with Autism Spectrum Disorders (ASD) but may include other developmental conditions such as Asperger's Syndrome and ADHD. The college years are seen as an important period of transition to adulthood and identity development. Little is known about how neurodiverse students experience this transition or what contributes to successes in and out of the college classroom.

There is growing attention in the academic and professional design practice literature on how the built environment can support the lives of children and adults with autism spectrum disorders (ASD). The underlying theme among those studies is an approach to neurodiversity and ability as a spectrum, with an understanding that individual needs and priorities are different and may shift at different times (see for example, Gaines, et al, 2018; Mostafa, 2018; 2021). The evidence-based design recommendations usually align with the universal design approach, i.e., a position to propose fully inclusive design solutions for all user groups. Yet, the use of the terms "neurodiverse" or "neurotypical" in the built environment literature is very scarce, and most design recommendations target the user groups with ASD.

The purpose of this project was to investigate design decisions that enhance learning spaces for neuroatypical students in higher education. The research project had two main phases. (1) Systematic literature review, addressing the salient literature in education and architecture to define and frame current knowledge and understanding of neurodiverse student learners and the existing evidence on how the built environment may support this user group. (2) Engagement of student learners. We first interviewed with traditional-age college students (n = 8) who identified as neuroatypical/neurodiverse. These interviews were designed to elicit descriptions of classroom experiences and beneficial augmentation to learning spaces. The data were analyzed using a thematic content analysis approach to identify the common themes or patterns to guide the following design guideline development phase. We then created 3D conceptual models representing the proposed guidelines in classroom environments. Neuroatypical students (n = 7) were recruited to experience those models in VR and provide feedback on the proposals. In line with the existing literature, the findings of the study highlight the significance of providing choice in seat location in various sizes of classrooms. The participants also emphasized proximity to natural light and windows as a significant factor in choosing their seat. The inconsistent yet frequent comments on furniture types in relation to ergonomics, geometry, colors, textures, and materiality underscore the significance of making informed, evidence-based design decisions for more inclusive interior learning environments.

2 REFERENCES

Gaines, K., Bourne, A., Pearson, M., & Kleibrink, M. (2018). Designing for autism spectrum disorders. Routledge, Taylor & Francis Group.

Mostafa, M. (2018). Designing for autism: an aspectss™ post-occupancy evaluation of learning environments. Archnet-Ijar, 12(3), 308–326.

Mostafa, M. (2021). The autism friendly university design guide.

EMPOWERING MARGINALIZED COMMUNITIES: A COLLABORATIVE APPROACH FOR CRIME REDUCTION IN DETROIT

Track 5: Spatial Justice, Equity, and Inclusion by Design
Knowledge Network: Health, Participation
Methods: Action/Participatory Research, Survey
Linda Nubani (Michigan State University); Julie M Krupa (Michigan State University);
Alaina De Biasi (Wayne State University)

1 ABSTRACT

Geographic Information System (GIS) and spatial analysis applications have served as effective techniques in understanding hotspots and criminal activity in neighborhoods. However, citizen participation in community crime prevention has in recent years become a strategy for addressing crime (Hipple & Saunders, 2020). Little is known about the best strategies to build trust and maximize citizens' participation in marginalized and underserved communities to address crime. This abstract presents ongoing research stemming from a collaborative effort with the Detroit Police Department to address violent crime in Detroit's Tenth Precinct. The area has an estimated population of 56,515 residents (ASC 2010-2014) with 89.4% Black or African American, 4.4% White, and 3.9% mixed races. It holds 36,583 housing units, with a 45.75% poverty level and a 38.53% long-term vacancy rate based on the 2010 Census. Additionally, the area contains more than 2,900 land-bank-owned vacant homes and commercial structures.

Over the first year of a three-year grant funded by the U.S. Department of Justice, the focus was centered on fostering citizen participation and involvement in creating an action plan to address crime. First, the research team analyzed historical crime data (2013-2023) to identify five crime hotspots in the 10th Precinct. Further analysis was conducted to analyze crime around vacant properties. Second, the team conducted three impactful community engagement events and administered a comprehensive community-wide survey to elicit resident input. Engagement activities involved a series of exercises designed to empower community members in identifying the root causes of crime within their community. The outcome of these efforts guided the formulation of an action plan. The community demonstrated consensus on five priority areas critical to their neighborhood's enhancement, such as challenges associated with problematic gas stations and businesses, loitering, and speeding, through the implementation of code enforcement and Crime Prevention Through Environmental Design (CPTED) strategies. Furthermore, the community expressed the need for improved community policing initiatives, enhanced police presence, and addressing blighted properties through creative avenues like art, community gardens, and rehabilitation programs. While the results of community input aligned with the findings of the historical hotspot analysis, the qualitative input provided unique insights into the role of environmental design in shaping crime and quality of life in their community. The implementation of the environmental design changes and enhancing code enforcement are planned during years two and three of the grant.

Findings from this presentation are useful for planners, policymakers, and policing agencies in developing targeted policing strategies, addressing environmental design issues, and allocating resources effectively. The presentation delves into an in-depth analysis of the most effective strategies employed to bolster citizen participation in co-creating comprehensive and impactful measures aimed at mitigating violence and enhancing the overall quality of life within the community. This includes strategies that tackle challenges in recruitment and the formation of partnerships among key stakeholders in the area.

2 REFERENCE

Hipple, N. K., & Saunders, J. (2020). Evaluation of the Innovations in Community-Based Crime Reduction (CBCR) Program: Executive Summary and Final Report. Final Report submitted to the National Institute of Justice (Grant No. 2016-BJ-BX-0069).

A STUDY ON THE IMPROVEMENT OF THE MAIN GATE AREA OF ELEMENTARY SCHOOLS: A COMPARATIVE ANALYSIS BETWEEN RESIDENTIAL AND COMMERCIAL AREAS

Track 5: Spatial Justice, Equity, and Inclusion by Design
Knowledge Network: Children, Youth & Environment, Interior Design
Methods: Interviews, Observation
Seonju You (Yonsei University); Suk-kyung Kim (School of Planning, Design, & Construction, Michigan State University)

1 ABSTRACT

Many elementary school main gates have been discovered to be poorly designed, especially posing a safety hazard due to their proximity to roads. According to news reports, numerous pedestrian accidents surrounding elementary school students occur during school hours. This is particularly hazardous due to the heavy traffic, as numerous cars converge during both work and school hours.

Each local government and education office has its own school design guidelines. However, many of these guidelines either recommend positioning the main gate adjacent to the road for vehicular access, or they lack specific design guidelines for positioning the main gate. This study aims to propose a main gate area design that ensures the safety of elementary school students by drawing on insights from pattern language based on safety design elements, new urbanism principles, and transitional space. To achieve the purpose of the study, two types of schools with different main gate environmental conditions were selected.
The research methods include site-based observations, behavioral mapping, video recording, and interviews. The researcher visited 'A' school located in a residential area and 'S' school, located in a commercial area, in Seoul, Korea, to assess the physical and environmental characteristics of each main gate and to observe the behavior of elementary school students during their commutes. Each case is compared and analyzed according to the traffic volume checklists, which include the number of cars passing per minute and a place-centered map. The observation was conducted at both arrival time 8:00 a.m.-9:00 a.m. and dismissal time 1:45 p.m.-2:45 p.m. of the school hours and the results were also compared and analyzed accordingly. The primary findings are as follows: First, 'A' elementary school is located in an apartment complex and is adjacent to a wide road and operates its main gate exclusively during arrival time and closes it during dismissal time. At arrival time, approximately 315 cars passed by in an hour, and during dismissal time, approximately 315 cars passed by in an hour, and all students left by using the rear gate.
Second, 'S' elementary school is located in a commercial area and is adjacent to a narrow road. There was no rear gate in this school. At arrival time, approximately 375 cars passed by in an hour and at dismissal time, approximately 120 cars passed by in an hour, many parents picked up their children by car, leading to significant congestion in front of the main gate and several near-accidents being witnessed. This study suggests a design plan to prevent children from going directly onto the road. This will be achieved by creating a transitional space that acts as a buffer zone between the school building and the main gate. The design of this transitional space incorporates elements from the safety pattern language and embraces the concept of New Urbanism.

2 REFERENCES

Kim, Seul-Ki, Lee, You-Mi, Kim, Seo-Young. (2014). A Study on Students' Satisfaction for Pedestrian Environment around Elementary School in Detached

Housing Area - Focused on Elementary Schools in Ilsan New Town -. Journal of the Regional Association of Architectural Institute of Korea, 16(3), 35-42.

PERSPECTIVES OF MUNICIPAL PROFESSIONALS ON ADOPTING A DEMENTIA-FRIENDLY AND INCLUSIVE APPROACH IN URBAN PLANNING AND DESIGN IN BRITISH COLUMBIA, CANADA

Track 5: Spatial Justice, Equity, and Inclusion by Design
Knowledge Network: Active Living by Design, Environment-Gerontology
Methods: Interviews
Kishore Seetharaman (Simon Fraser University); Habib Chaudhury (Simon Fraser University)

1 ABSTRACT

There are around 50 million people living with dementia worldwide and this number is estimated to increase to over 150 million by 2050 [1]. The recognition of the role of community-based supports in enhancing the quality of life of people living with and affected by dementia is evidenced by the prominence of the dementia-friendly and inclusive communities (DFCs) framework in policy and practice, centred on the creation of spaces of inclusion, empowerment, respect, and support through individual, environmental, societal, and cultural change [2]. Creating an accessible and navigable neighbourhood physical environment is a critical part of DFCs and facilitating outdoor mobility and social participation. Municipalities in British Columbia (B.C.), Canada have committed to the vision of DFCs with action plans outlining priorities in planning, design, engineering, and services. City planners and designers play an instrumental role in implementing these plans. This study aims to understand their needs and challenges in adopting a dementia-friendly and inclusive approach in planning and design. Semi-structured interviews (45-60 min via Zoom and phone) were conducted with 16 planners and designers in five Metro Vancouver municipalities in B.C. Thematic analysis of data generated two themes:

- *Augmenting DFC Knowledge and Awareness: Understanding the access needs, challenges, and requirements of people living with dementia was seen as central to DFC planning and design. Participating in municipal DFC education and awareness activities had a positive impact on participants' knowledge of dementia-friendly design attributes, principles, and strategies. Gaps in understanding the lens challenged planners' ability to determine dementia-friendliness and inclusivity of general planning and design strategies, raising the issue of paucity of evidence, guidelines, and best practices. Participants' recommendations to improve DFC knowledge and awareness included having: 1) easy-to-apply, practical, and evidence-based guidelines, 2) ongoing staff training and education, 3) a staff role for DFC advocacy and coordination of DFC action plan, 4) regular consultation with DFC and access advisory committees, and 5) community walkabouts with people with lived experience.*
- *Integrating DFC Approach in General Planning and Design: Participants cited synergies between DFC and other municipal plans, emphasizing the need to contextualize DFC as part of broader access and inclusion work. Recommendations for the integration of DFC in planning and design included: 1) integrating plans with shared goals and objectives and creating municipal transportation strategy covering all aspects of mobility, wayfinding, and transportation in the city, and a municipal accessibility strategy covering a wide range of access needs and requirements in the public realm, including those related to DFC, 2) integrating dementia-friendly design standards with general planning and design standards, and 3) creating accessibility standards at regional and provincial levels.*

- *Future research should seek to augment the body of knowledge required to support planners and designers in differentiating and balancing the access needs of different groups of people with disabilities. An evidence-based approach that offers perspectives on diverse access needs and requirements could contribute towards more inclusivity in accessibility.*

2 REFERENCES

World Health Organization. (2020). Dementia. https://www.who.int/news-room/fact-sheets/detail/dementia

Alzheimer's Disease International. (2016). Dementia Friendly Communities: Key principles. https://www.alz.co.uk/adi/pdf/dfc-principles.pdf

METRIC BASIS OF CONFIGURATION ANALYSIS: EXPLORATIONS ON WALKABILITY AND NEIGHBORHOOD DESIGN

Track 5: Spatial Justice, Equity, and Inclusion by Design
Knowledge Network: Movement in Designed Environments: Nature & Ecology
Methods: Case Study, Mixed Methods
Saif Haq (Texas Tech University)

1 ABSTRACT

Neighborhood design strategies owe a great deal to Clarence Perry, who in 1929 proposed that (among other things) convenience regarding walkability to schools, parks, and local shops was an important factor. He also suggested that in single family neighborhoods, a half-mile radius should be the controlling dimension because it approximated the pedestrian walk range. While this distance idea for neighborhood design has been ubiquitous, the process of measuring the walking distances was rudimentary. For Perry and subsequent many, a circle drawn with a radius of a quarter mile, superimposed on a plan, sufficed. Unfortunately, this could not deal with the actual path configurations that existed on the ground.

Space Syntax theory, on the other hand, takes the actual layout as its base of analysis, and considers topological, metric, and geometric relationships between paths to measure configurational values. More importantly, it permits for the analyses to be set at pre-defined distances allowing investigation of different distance measures, (such as half mile, quarter mile, etc.). Numerous empirical evidence and theoretical developments now indicate that larger level city functions such as connections between various city-parts are captured by geometric analyses, but at a neighborhood scale, distance-based analyses are useful. While a lot of research has been done regarding the former, it is not the case for neighborhood investigations. This is the focus of this paper.

The concerns to be investigated are: (1) is Space Syntax metric analysis a useful tool to investigate the formation of neighborhood centers, and if so, what distance of analysis is it revealed? (2) Do neighborhood layout patterns and socio-cultural factors play a role in walking distances, and (3) do neighborhood walking distances vary across and within cities, and why? This research will study both large and small towns in Texas to explore these questions. The presentation will explain the differences between topological, metric and geometrical analyses of Space Syntax, the concept of analyzing at a predefined distance and their usefulness, and explore the selected cities to shed some light on distance modelling of neighborhoods to understand walkability and its associated social dimension.

A PEOPLE'S GUIDE TO BOSTON CHINATOWN

Track 5: Spatial Justice, Equity, and Inclusion by Design
Knowledge Network: Participation
Methods: Action/Participatory Research
Lily Song (Northeastern University)

1 ABSTRACT

The design studio is a foundational pedagogical model in architecture and planning education for teaching practice through open-ended problem setting, skills synthesis, real or constructed clients, and teamwork (Long 2012). While invaluable as pedagogical practices, material spaces, and creative design practices, design studios have traditionally struggled to engage current contexts of students, refocus from solutions-driven design process to exploring, feeling, and empathizing as integral parts of design process, and overcome disconnection from real-world problem scenarios (Corazzoa 2019; Dutton 1987). This paper examines a teaching and learning practice undertaken by an architectural research studio that aims to retool design pedagogy and methods to support community-led anti-displacement planning and design initiatives.

Over fall 2023 semester, the anti-displacement studio at Northeastern University worked with the Asian Community Development Corporation (ACDC) in Boston's Chinatown on a People's Guide to Chinatown. Twelve undergraduate and graduate architecture students in the studio gained guidance and feedback from ACDC staff as they conducted research on neighborhood sites and buildings as spaces of community-led resistance, organizing, and transformation in the face of systemic oppression and injustice and created visual summaries of their findings. In addition to supporting the partners' youth leadership development and coalition building activities, the people's guide offers an experimental approach to teaching and learning place-based, community-engaged architectural research at the intersection of migration, race/ethnicity, and economic inequality that is enriched by lived experiences.

The paper examines the creative process for the People's Guide to Chinatown, including: (1) learning from books, articles, and multimedia resources on how local residents and community members have persevered against racial exclusion, highway projects, urban renewal, public dis/mis-investment, gentrification, and other spatial injustices in Boston Chinatown; (2) participating in a guided tour of Boston Chinatown with ACDC staff; (3) conducting site-specific research and connecting it with social histories; (4) crafting each entry—comprising a narrative summary, timeline, photos, plans/maps, and diagrams; (5) a collaborative workshop with ACDC staff; and (6) final round of revision based on received feedback. The discussion focuses on tensions and challenges that emerged as we crossed university-community boundaries and sought to ground our work in the real-world context and concludes with implications for design teaching and learning on topics of spatial justice, equity, and inclusion.

2 REFERENCES

Corazzoa, J. (2019). Materialising the Studio. A systematic review of the role of the material space of the studio in Art, Design and Architecture Education. The Design Journal, 22(sup1), 1249-1265.

Dutton, T. A. (1987). Design and studio pedagogy. Journal of architectural education, 41(1), 16-25.

Long, J. G. (2012). State of the studio: Revisiting the potential of studio pedagogy in US-based planning programs. Journal of Planning Education and Research, 32(4), 431-448.

EXAMINING COMMUNITY ENGAGEMENT THROUGH NETWORK ANALYSIS: A MIXED-METHODS STUDY OF A PLACEKEEPING INITIATIVE IN JAPAN

Track 5: Spatial Justice, Equity, and Inclusion by Design
Knowledge Network: Participation, Sustainable Planning, Design & Behavior
Methods: Mixed Methods, Statistical/Modeling Analysis
Yohei Kato (Singapore University of Technology and Design)

1 ABSTRACT

Community engagement plays a crucial role in urban planning and design, contributing to the creation of inclusive and equitable environments. It not only allows experts to grasp community needs and insights but also empowers community members to impact decision-making in their lives (Melcher, 2013; Sanoff, 2006). This process fosters interaction and collaboration among community members such as residents and various stakeholders, underscoring the importance of relationship building in community engagement. Previous research has primarily explored community engagement from individual viewpoints using case studies and participatory action research (Francis, 2001; Rios, 2006). These methods have yielded valuable insights, but they fall short in addressing the dynamic and complex relational aspects of community engagement (Neal & Christens, 2014; Shin, 2022). Specifically, there is limited understanding of how individual engagement behaviors evolve over time, influenced by personal characteristics, perceptions, other community members' behaviors, and network structure (e.g., shapes, patterns, density).

Addressing this, the study investigates community engagement from both temporal and network perspectives, employing Social Network Analysis (SNA) to examine the engagement of community members in a long-term park management initiative in Japan from 2021 to 2022. The participants, ranging from 19 to 22 individuals, have been involved in managing an urban park for over seven years, taking on various roles and activities. The study uses SNA and Exponential Random Graph Models (ERGMs) to understand how participants collaborate and support each other, and to identify factors that influence the sustainability of the initiative. Qualitative interviews and quantitative surveys provide a comprehensive view of engagement, incorporating multidimensional measures (breadth, duration, frequency, intensity) and using "name generators" to gather information about ties that facilitate discussion and opinion sharing.

The findings, illustrated through network graphs and descriptive analysis, reveal an evolving network of community members influenced by various internal and external factors. Statistical analysis shows that those with higher engagement levels are more likely to form trustful relationships and support other members. This study enhances the understanding of community engagement by demonstrating the benefits of SNA and ERGMs in analyzing interpersonal relations within community engagement initiatives. It sheds light on the dynamic process of relationship building in community engagement and offers insights for sustaining engagement over time. Advantages, limitations, and potential applications of network approach to community engagement are discussed.

THE ART OF COMMUNICATING ARCHITECTURE USING UNIVERSAL DESIGN FOR LEARNING

Track 5: Spatial Justice, Equity, and Inclusion by Design
Knowledge Network: Communication, Participation
Methods: Case Study, Observation

Rachel Updegrove (HERA laboratory planners); Jinhee Lee (HERA Laboratory Planners)

1 ABSTRACT

When writing a paper or presentation, we are asked to "know our audience" and write with them in mind. In architectural education, pedagogy explicitly focuses on how design students need to communicate to design professionals (i.e., via technical drawing skills and construction terminology), as the student's "audience" comprises architectural professionals who are hiring. However, when designers join the professional industry after graduation, their "audience" not only includes a wider range of built environment professionals (i.e., architects, interior designers, landscape architects, construction managers, engineers), but now includes someone new and who is likely not a design professional -- the client. Now, the student-turned-designer must adapt, communicate, and learn how to educate, listen, and foster understanding with this melting-pot "audience", much like how in higher education, general education courses are designed for all majors and all minds. With the use of Universal Design for LEARNING (UDL) principles used in the daily activities of architectural professional/educational practice, everyone is included, communication is clearer, and everyone learns something new, resulting in a human-centered designed process and spatial solution.

Universal Design is a widely used term in the architectural world that means quite the opposite from Universal Design for LEARNING (UDL). Universal Design looks for a design solution that will fit the needs of the collective without adaptation1. Universal Design for LEARNING (UDL) is typically used by K-12 educators, to provide options for methods in which students can learn and communicate their understanding on a topic2. We believe that as designers, we must communicate with our students, clients, and peers in various ways to meet the heterogeneous – not universal – needs of all. Across and within each industry, communication preferences vary between audiences and disciplines because they are made up of individual people who have intersectional layers, impacting why they prefer to communicate in one way over another.

For example, one of the presenters is a designer, educator, and an autistic, neurodivergent woman, who prefers explicit instructions in text form because she struggles with executive functioning due to her autism and ADHD. The other presenter is an architect, mentor, and Korean woman, who prefers visual communication instead of large bodies of text, due to English being her second language. When both presenters collaborate, communication strategies of verbal and visual thinkers are represented, removing barriers to participation, rather than if one method of communication was privileged.

This presentation is a dialogue between real-life case studies of lessons learned in higher education and industry practice, exploring how different design activities (technical drawings, user surveys, post-occupancy evaluations, workshops) can be designed to be human-centered, through the lens of Universal Design for LEARNING. The presenters will use interactive polling to engage audience members to participate and recall their own preferences of communication and reflect on how they accommodated individuals' different styles of communication.

2 REFERENCES

The Center for Universal Design (1997). The Principles of Universal Design, Version 2.0. Raleigh, NC: North Carolina State University.

CAST (2018). Universal Design for Learning Guidelines version 2.2. http://udlguidelines.cast.org

PLACE IDENTITY VARIATION ACROSS AGES IN A HYPER-SEGREGATED NEIGHBORHOOD

Track 5: Spatial Justice, Equity, and Inclusion by Design
Knowledge Network: Cultural Aspects of Design, Environmental & Architectural Phenomenology
Methods: Action/Participatory Research, Interviews
Aakanchcha Raj (The University of Kansas)

1 ABSTRACT

This study explores and comprehends the intricacies of place identity within Wendel Phillips, a Kansas City, Missouri neighborhood established in unjust socio-economic conditions and racial segregation. Place identity embodies the subjective and emotive ties individuals or communities forge with a specific geographic location, encompassing personal meanings, emotions, memories, and symbolic affiliations. These elements significantly shape one's sense of belonging, attachment, and identification with a particular environment, intertwining with the construction of a distinct identity closely linked to the attributes of that place. The research methodology involved qualitative methods, including unstructured interviews, and mapping exercises, engaging 20 residents from diverse age groups (above 18 years old), genders, and residency durations within Wendel Phillips. This comprehensive approach aimed to straighten out how demographic variances influence the perception of identity within a shared environment characterized by similar challenges and aspirations. By amplifying the voices of the community, this study seeks to spotlight multifaceted perspectives, diverse familiarities, and future visions, ultimately proposing inclusive changes.

Place identity plays a pivotal role in an individual's self-concept and their connection with the broader world, fostering a profound sense of belonging to a specific place or community. By scrutinizing a spectrum of personal perceptions across different ages and genders within the neighborhood, this research endeavors to reshape the prevailing narrative from suppression to optimism for the future. It emphasizes the profound impact of stories on shaping present perceptions and empowering the potential trajectory of the neighborhood. Recognizing the evolutionary nature of identity with perceptions, this study challenges the subjective essence of place identity. It sheds light on residents' proposed future changes, offering a glimpse into transformative pathways within the community. By embracing diverse narratives, this research underlines the dynamic nature of identity within a shared space and the implications for future development.

In essence, this study not only unties the complexities of place identity within a marginalized community but also advocates for a more inclusive and empowered future. By acknowledging and amplifying diverse perspectives, it paves the way for understanding the nuanced interplay between individual identities and the environment, envisioning a neighborhood that thrives on collective narratives and inclusive transformations.

THE U.S. MOSQUE: A CRITICAL INTEGRATIVE LITERATURE REVIEW

Track 5: Spatial Justice, Equity, and Inclusion by Design
Knowledge Network: Cultural Aspects of Design, Interior Design
Methods: Interviews, Meta-Analysis
Hassnaa Mohammed (University of Wisconsin Madison)

1 ABSTRACT

Over the last twenty years, U.S. mosques have gained a lot of attention, focusing on the contested grounds on which they stand (e.g. Bagby, 2020; Numrich, 1997; Prickett, 2019). While literature on U.S. mosques has been published in different fields including architecture, sociology, religious studies, and political science, no work has been done to assess the state of research and how mosques might develop as they continue to accommodate a largely increasing U.S Muslim population. Given the 31% increase in building mosques between 2010 and 2020 (Bagby, 2020), synthesizing existing work may lead to a renewed perspective on ways to understand these complex institutes and sustainably design them in the future.

2 METHODS

This study sets forth a systematic critical integrative review of literature published on mosques in the U.S. in peer-reviewed journals. To identify publications that examine mosques in peer-reviewed journals, I searched the leading databases for research in humanities and social sciences, namely, "EBSCOhost", "Sage", and "Scopus" databases. Focusing on studies published before December 2022, I searched for the following: keywords "mosque", "American", "United States", "USA", and "US". After multiple rounds of screening relevant publications, the final sample size included 68 articles that were thematically analyzed.*

3 FINDINGS

First, the study presents overarching trends in publications related to mosques before December of 2022, such as type of publications, year published, and number of publications per identified themes. Second, the study identifies seven main themes in existing literature related to mosques structures and factors that impact the development of mosques. These themes are perception of Muslims and challenges facing the construction of mosques in the United States; The role of mosques in Muslims' integration, political engagement, and civic involvement; Race, ethnicity, and Socio-economic Diversity in U.S. mosques; Mosques architectural development; Women experiences in mosques; Mosques in the U.S. urban context; Socio-emotional support of mosque communities. Each theme is presented as a synthesis of findings from relevant publications with analytical statements identifying the gaps.

4 CONCLUSION

In conclusion, the study reflects on the strengths and weaknesses of current literature, providing insights to guide future research on U.S. mosques and the role they play in Muslim placemaking.

2 REFERENCES

Bagby, I. (2020). The American Mosque 2020: Growing and Evolving. ISPU. https://www.ispu.org/report-1-mosque-survey-2020/

Numrich, P. D. (1997). Recent immigrant religions in a restructuring metropolis: New religious landscapes in Chicago. Journal of Cultural Geography, 17(1), 55–76. https://doi.org/10.1080/08873639709478330

Prickett, P. (2019). Changing constructions of religious visibility: The case of an African American Muslim community in South Central LA. Social Compass, 66(1), 49–61. https://doi.org/10.1177/0037768618813765

A CASE STUDY OF INCLUSIVE DESIGN TO IMPROVE SENSE OF BELONGING IN HIGHER EDUCATION

Track 5: Spatial Justice, Equity, and Inclusion by Design
Knowledge Network: Interior Design
Methods: Case Study, Focus Groups

Gabriela Fonseca Pereira (Illinois State University); Danielle Gruber (Illinois State University); Ana Rebeca Bicudo (Illinois State University)

1 ABSTRACT

Inclusive design considers the population's capabilities and needs to reduce differences in the use of the designed object, providing all with comfort and dignity, and giving them the confidence to be independent and participate equally in activities (Patrick & Hollenback, 2021). With the activist movement and protests across the country in 2020, diversity and inclusion committees were created in higher educational institutions to foster a culture of inclusion and belonging. These are relevant topics when it comes to interior spaces, especially educational buildings that serve as second places to many undergraduate and graduate students, faculty, and staff. Therefore, the purpose of this research study was (1) to investigate the perceptions of the inclusive design and sense of belonging in a specific building of a state university located in the Midwestern United States, and (2) to develop a design proposal to make the physical space of this building more inclusive to promote belonging.

The data for this project was collected through eight focus groups with a total of 27 participants—21 students from different majors, four faculty from various departments, and two Dean's office staff members. Participants were asked how long they have been using the building, their perceptions of accessibility and sense of belonging, and suggested improvements for the entrances, hallways, classrooms, and offices.

Some barriers were pointed out; stairs and exits not ADA-compliant, lack of signage coming from other campus buildings, hard-to-identify hallways due to their cross shape and similarity, outdated student lounge that lacks seating space, and dysfunctional and tight bathrooms that are difficult to maneuver in a wheelchair. Classrooms were described as accessible to the general population, but not for wheelchair users, and the furniture is small and uncomfortable for most body types. The professors' offices were described as hard to find, lacking storage, and extremely small, making students feel that they do not have an inviting environment to reach out to professors.

As a result, some entrances to the building were modified, creating their own identity, thus making it easier for new users to identify classrooms and offices. Hallways were adapted with more seating and study space areas, and professors' offices were moved to the second floor and grouped according to their majors. Bathrooms were modified, giving their users more privacy and a non-gender one was also added to increase the sense of belonging of those who do not identify themselves with any gender. Signage, wayfinding, and a directory were added in every hallway to easily conduct new users. Some classrooms and computer laboratory rooms were also moved and added to the second floor, with height-adjustable furniture, and increased circulation space. This study shows the relevance of the application of Inclusive Design to promote a sense of belonging in college buildings.

2 REFERENCE

Patrick, V. M., & Hollenbeck, C. R. (2021). Designing for All: Consumer Response to Inclusive Design. Journal of Consumer Technology, 31(2), 360-381. doi: 10.1002/jcpy.1225

SOCIAL JUSTICE AND STUDENT ENGAGEMENT IN HOUSING DESIGN

Track 5: Spatial Justice, Equity, and Inclusion by Design
Knowledge Network: Residential Environments, Sustainable Planning, Design & Behavior
Methods: Design Methods, Practice-based Research
Jordan Kanter (UMASS Amherst Department of Architecture); Erika Zekos (UMass-Amherst)

1 ABSTRACT

In the context of the deeply interrelated and seemingly intractable issues of housing insecurity, racial and income inequality, and mental, physical, and social health disparity, there is an urgent need to rethink our approach to architectural design that takes a more critical, human-centric approach (Rios, 2011). At the heart of this shift is the need to re-center social justice as crucial to the design and planning process. While there is rich discourse in higher education around social justice pedagogy (Adams et al., 2023) and social justice in architectural practice and scholarship (Cheng et al., 2020), there are still opportunities to integrate social justice pedagogy in architectural design education. This presentation outlines recent efforts at UNIVERSITY to close this gap and develop social justice and community engagement-informed pedagogy as part of the core curriculum of a Master of Architecture program.

For the past two years, a studio, co-taught by the authors, has engaged CITY, STATE, a small post-industrial city, as a site for mixed-use affordable housing. Working with city and local non-profit development agencies, this studio engages students in an active dialogue regarding the future of its housing. It begins with a series of qualitative and quantitative mapping exercises, introducing students to different ways of making sense of the neighborhood. It then moves into a research phase that includes engagement with community members and city leadership, with the goal of identifying development strategies for mixed-use affordable housing. This phase invites students to identify a client population, enumerate spaces that will support occupants and the community at-large, and identify approaches, policies, and funding strategies to enable affordability.

Throughout this phase, the studio foregrounds programming as a vital part of the design process. This phase also emphasizes group work, both to facilitate cohort-building, and as a method to open the design process to multiple, at times competing perspectives. Building on the outcome of these first two phases, students embark on the design of their studio project. Our teaching practice unites social justice education frameworks with the traditional demands of the architecture studio. Students utilize team-based asset-mapping to understand a built environment affected by generational poverty; meaningfully engage with people and spaces; and learn to advocate for neighborhood needs through the development of housing and public programming initiatives. Building on this diverse set of practices, our approach supports students to become "effective dilettantes", capable of deriving and applying knowledge gained from a wide range of sources, including community engagement, policy analysis, social research, as well as conventional architectural methods. This emphasis on research-as-design allows for a deepening understanding of the complex human and built environments in the formulation of an architectural design process oriented towards social justice.

2 REFERENCES

Rios, J. (2011). Reconsidering the Margin. In Service Learning in Design and Planning: Educating at the Boundaries. (T. Angotti, et al., Eds.). New Village Press.

Adams, M., et al. (Eds.). (2023). Teaching for diversity and social justice (4th ed.). Routledge.

Cheng, I., et al. (Eds.). (2020). Race and modern architecture. University of Pittsburgh Press.

EVALUATING CHARACTERISTICS OF URBAN STREETS OF AMASYA BY USING BUILT ENVIRONMENT ASSESSMENT TOOL AND UNIVERSAL DESIGN CRITERIA

Track 5: Spatial Justice, Equity, and Inclusion by Design
Knowledge Network: Residential Environments, Sustainable Planning, Design & Behavior
Methods: Case Study, Photo-based Methods

Ozlem Demir (Amasya University); Duygu Kalkan Açıkkapı (Amasya University)

1 ABSTRACT

Creating a human-centric built environment requires effective, efficient, well-connected, accessible, comfortable, and all other human-oriented conditions. This study aims to define and evaluate the characteristics of four urban streets in Amasya. The city has a rich historical background, marked by various civilizations such as the Hittites, Persians, Pontus Kingdom, Roman, Byzantine, Seljuks, and more. Despite its historical legacy, Amasya is experiencing rapid change due to an increasing permanent and temporary population, including college students and tourists. This demographic shift is driving morphological and demographic transformations in the city, evident in its linear expansion and rising built density within existing settlements.

As these changes unfold dramatically, this study questions whether the evolving built environment in Amasya possesses potential human-centric criteria. The primary objective is to define and evaluate the physical characteristics of four selected streets in Amasya. This case study includes two streets of historical significance and two contemporary streets. The evaluation is based on universal design criteria, utilizing a built environment assessment tool.

The intended audience for this study comprises practitioners, designers, planners, and policymakers seeking a deeper understanding of the specific characteristics of urban streets strongly correlated with environmental quality. The study sheds light on design features that promote environmental quality in various urban settings within a growing city, striving to bring equity to the streets amid rapid change. It emphasizes the importance of designing streets with equity purposes to maintain cultural, social, and economic homogeneity in societies.

STUDY ON THE EDGE AREA OF THE UNIVERSITY CAMPUS CAUSED BY FOOD PICK-UP LOCKERS——A CASE STUDY IN CHINA

Track 5: Spatial Justice, Equity, and Inclusion by Design
Knowledge Network: Student Affairs
Methods: Case Study

Yilin Song (Tianjin University); Yanlin Wu (Ms.); Lingfei Li (Tianjin University); Qianhui Qu (Tianjin University); Qianya Yu (Tianjin University)

1 ABSTRACT

The Image of the City, authored by Kevin Lynch, posits that cities are composed of five fundamental elements: paths, edges, districts, nodes, and landmarks. He mentioned in Site Planning that the imperative for universities to prioritize openness to the surrounding community and foster interaction with the external environment.

Chinese university campuses are akin to miniature cities, encompassing diverse functions such as study, work, residence, and daily life activities. The edge area of a campus serves as a pivotal transitional zone, shared by individuals both within and beyond its confines, potentially functioning as a nexus for various activities. In response to the COVID-19 pandemic, most universities in China have enforced stringent management policies, restricting campus access exclusively to students and staff. Consequently, although food delivery remains a prevalent choice among university students, delivery personnel are now prohibited from entering campuses. Instead, food pick-up lockers have been introduced at the campus edge to supplant the previous food delivery paradigm. These lockers not only facilitate food retrieval but also emerge as novel functional units seamlessly integrated into the campus edge, influencing student activities in the vicinity.

Based on the theory from the Image of the City, this study employs Tianjin University as a case study to examine the perception and utilization of campus food pick-up lockers and their impact on the surrounding edge area. Utilizing questionnaire surveys, field visits, and interviews, the study collected 165 valid responses and conducted 7 interviews. Analysis of the data revealed that 90.96% of respondents had previously availed food delivery services. Among the four primary food pick-up lockers at Tianjin University, 66.89% of respondents favored those situated at Pengxiang Dormitory, exhibiting a significantly higher utilization rate than others. Moreover, 67.33% of respondents made purchases from off-campus vendors, contributing to a distinct edge business ecosystem. Based on the variations in food pick-up locker usage highlighted in the survey data, this study categorizes campus edges into four types: hard, flexible, permeable, and melting edges. An in-depth examination of the melting edge, represented by the campus edge around Pengxiang Dormitory's pick-up lockers, identifies its emergence as an edge business primarily driven by advantageous geographic positioning, student demand, vendor convenience, the promotional impact of takeaway lockers, and the demand for resource sharing between the city and the university.

This study presents planning and design strategies for activating and utilizing the campus edge area surrounding food pick-up lockers, offering a novel perspective to comprehend the potential impact of settings, campus environment, and behavioral models.

CULTURAL ASPECTS OF DESIGN IDEAL VS. REALITY: STATE-DEVELOPED HOUSING AND MIXED-INCOME INFORMAL SETTLEMENTS IN THE GLOBAL SOUTH

Track 5: Spatial Justice, Equity, and Inclusion by Design
Knowledge Network: Cultural Aspects of Design, International Housing Research
Methods: Interviews, Mixed Methods
Sana Ahrar (Pennsylvania State University); Alexandra Staub (Pennsylvania State University)

1 ABSTRACT

The prevailing planning discourse on urbanization and housing in Global South cities has portrayed formal, state-planned housing as the 'ideal' urban growth model. Meanwhile, informal, 'unplanned' developments have been seen as a 'challenge' to urban planning. However, this narrative overlooks the significant role that informal housing plays in providing accommodation to a vast segment of low-income populations and, notably, less acknowledged middle-class demographic. This research aimed to delve into the nuanced experiences of middle-class residents living in state-developed housing versus those in mixed-income informal settings, a demographic often overlooked in urban planning narratives. It sought to investigate the perceived advantages and disadvantages associated with both informal and formal housing from the middle-class residents' perspective. The methodology included a mixed-methods approach that integrates discourse analysis of academic literature, governmental reports, and media publications with in-depth interviews with middle-class residents in both informal and formal housing sectors in Delhi. The study findings reveal that formal housing is generally perceived to offer greater investment security, planned infrastructure, and efficient services. However, it is also criticized for its rigid layouts, poor aesthetics, and a lack of diversity in housing options. In contrast, the primary findings show that housing in the informal sector offers more flexibility, faster delivery, and more agency to the middle-class community. These insights are critical for understanding the middle class's housing preferences and perspectives, thereby informing more inclusive and effective urban housing policies. This study contributes to a more nuanced understanding of urban living, advocating for policies that consider the realities and aspirations of a significant yet underrepresented segment of the urban population.

PARTICIPATORY ACTION RESEARCH TO EMPOWER VULNERABLE COMMUNITIES IN KENYA

Track 5: Spatial Justice, Equity, and Inclusion by Design
Knowledge Network: Cultural Aspects of Design, Participation
Methods: Action/Participatory Research, Design Methods

Donghwan Moon (NC State University)

1 ABSTRACT

This project focuses on employing a participatory action research methodology to explore the potential of community engagement and participatory mapping in articulating community knowledge, narratives, and guiding the trajectory of design interventions for vulnerable communities in Kenya. The primary goal of this engaged research is to foster inclusive and equitable built environments, with a particular emphasis on addressing the specific needs of these communities.

Upon investigating rural Kenyan communities, significant social, economic, and environmental challenges were revealed, primarily due to their remote locations and governmental neglect. The absence of development plans and support leaves these areas without the resources to address critical needs, such as healthcare and education. Additionally, the lack of formal land ownership rights and unclear property borders exacerbate the situation, making it increasingly challenging for people to invest in their community, thus creating an uncertain future.

To address these issues, this research creates comprehensive solutions to protect community land, establish ownership rights, and reinforce local culture. Participatory mapping and inclusive design methods are employed, involving workshops, interviews, and ongoing monitoring with community members and stakeholders. This collaborative approach, grounded in the core concepts of participatory action research, identifies challenges and cultivates opportunities for collective spatial use and sustainable development. The resulting projects, co-created by the research team and the communities, include the creation of village maps, inclusive public spaces, house prototypes utilizing local resources, and community facilities offering essential programs. The insights gained from these projects act as a catalyst for raising awareness among community leaders, local government officials, and NGOs, with the hope of inspiring similar initiatives in rural contexts worldwide.

INDIGENOUS INTELLIGENCE: CONTINUITIES AMID DISRUPTIONS

Track 5: Spatial Justice, Equity, and Inclusion by Design
Knowledge Network: Cultural Aspects of Design, Sustainable Planning, Design & Behavior
Methods: Case Study, Grounded Theory
James Patrick Miller (Western Washington University); Howard Davis (University of Oregon)

1 ABSTRACT

The publication of the Handbook of Contemporary Indigenous Architecture in 2018 along with the Canadian exhibition, "Unceded: Voices of the Land," at the 2018 Venice Architecture Biennale activated a renewed engagement with Indigenous knowledge in architecture and design leading to a revolution: a rebirth of Indigenous architecture. The contemporary wave is significant as it is led by Indigenous architects, scholars, and designers as opposed to the documentation of Indigenous material culture and interpretation by non-indigenous people who are primarily Westerners. The revolution is significant as it centers Indigenous design epistemologies within the design process, allowing for a multitude of material manifestations. Indigenous architecture, designed and built by Indigenous people, emphasizes process over artifact, giving a renewed meaning to the vernacular, and showing how it transmits architectural knowledge in ways that support cultural and political identities.

Through a critical examination of several case studies in the U.S. and Canada, we seek to disrupt the idea of the architect as leader, placing the emphasis on the process from conception through occupancy. The shift away from the architect to the process highlights the significance of continuity within building practices, a factor in cultural continuity. Particularly in the consideration of disruptions across time through colonization, migrations, and globalization, the continuity of vernacular tradition challenges these disruptions, and now the resurgence in these practices demonstrates the importance of deep patterns. The idea that people in the community need to be doing the work to make meaningful place-based architectures is evident in the case studies. There is an important distinction between the use of vernacular ideas by a designer and the application of Indigenous intelligence, which is inclusive of community in the building production, where tradition becomes embedded in the process. Through the generative nature of processes, the form that arises is unpredictable, but predictable patterns are employed.

The case studies include a mix of built projects and projects currently underway. They are: s'gi gi altx: House of Welcome by John Paul Jones (Choctaw) at Evergreen State College in Washington; Skokomish Community Center by Daniel Glenn (Crow) on the Skokomish Indian Tribe; Ho'oulu 'ina in Honolulu, Hawaii; and 'Ike o Keahialaka by Metaamo studio in Puna, Hawai'i. The case study methodology includes: analysis of archival materials and news articles, interviews with designers, and a critical analysis of the project's overall through Indigenous architecture principles established in the literature (Wanda Dalla Costa, Hirini Matunga, Albert Refiti). These case studies provide a wide array of Indigenous architecture that demonstrate the continuity of traditions through process.

Until recently, Indigenous architecture has been largely disregarded within architectural disciplines, appearing on the periphery as traditional and vernacular practices. With the renewed interest of social justice and decolonization, Indigenous architecture has resurfaced, led by Indigenous practitioners. We see the rise in Indigenous Architecture as an opportune moment to recenter the importance of Indigenous intelligence. As climate change necessitates a closer examination of the human-environment relationship, Indigeneity provides real solutions.

INVESTIGATING INTERACTIONS WITH NATURE WITH VULNERABLE POPULATIONS

Track 5: Spatial Justice, Equity, and Inclusion by Design
Knowledge Network: Children, Youth & Environment, Environmental Design Research Education
Methods: Action/Participatory Research, Mixed Methods
Julie Stevens (Iowa State University); Tricia Neppl (Iowa State University); Jeenkyoung Lee (Iowa State University)

1 ABSTRACT

Existing evidence demonstrates that for vulnerable youth, interaction with nature has positive psychosocial and behavioral outcomes, with enduring post-treatment effects (Overbey et al., 2021). However, traditional research methods are not typically tailored to this unique and vulnerable population and fail to uncover their specific needs. Our community partner, a crisis stabilization and addiction treatment shelter in the Midwest region of the U.S., is moving from an urban setting to a new nature-based campus, intentionally designed with a trauma-informed lens. Our aim is to systematically study the effect of the outdoor environment on psychological and behavioral well-being of youth, and staff, in residential facilities as they make this shift from urban to nature-based.

This study has developed a mixed method research strategy to better engage with youth who are living with the effects of trauma and who are currently living within a residential treatment setting. In addition to traditional survey instruments, the research team used two innovative methods: interaction with nature response cards and a biophilic assessment. The brief "before" and "after" response cards, available to youth at their daily counseling session, ask individuals to select their level of interaction with nature followed by an emotional self-assessment using a range of emoji images. The biophilic assessment evaluates the nature-based qualities of the shelter and treatment houses by recording which spaces contain any of the six principles and 72 attributes of biophilic design as outlined by Stephen Kellert.

The outcome of the proposed research project will be a better understanding of the effect of the nature-based campus on the well-being of youths living in shelter and treatment facilities. Our research has a broad significance since the findings from the intervention effect of the nature-based campus and activities can be used to assess and improve the environment, programs, and activities at youth services agencies. The knowledge gained with this research will guide future designs and interventions of youth shelters and residential treatment facilities to better promote the psychological, behavioral, and social well-being of youths and the staff who care for them. Additionally, improved environments have the potential to assist youths in developing coping and relational skills that are useful upon release.

"AGE-INCLUSIVE URBAN DESIGN": NAVIGATING MOBILITY CHALLENGES AND OPPORTUNITIES FOR PERSONS WITH MOBILITY, SENSORY AND COGNITIVE DISABILITIES ACROSS THE LIFESPAN THROUGH THE SWAN TOOL

Track 5: Spatial Justice, Equity, and Inclusion by Design
Knowledge Network: Active Living by Design, Environment-Gerontology
Methods: Action/Participatory Research, Mixed Methods
Atiya Mahmood (Simon Fraser University); Rojan Nasiri (Student); Niloofar Hedayati (Simon Fraser University); Sogol Hosseini (Student); Ben Mortenson (Associate Professor, University of British Columbia)

1 ABSTRACT

In the pursuit of fostering inclusive urban environments, concerning spatial justice and equitable design, it is vital to explore the intricate dynamics of mobility disability experiences across different life stages. This research investigates how age influences the accessibility of neighborhoods for individuals with disabilities, shedding light on the challenges and opportunities faced by both young and older adults. Employing a community engagement approach and utilizing the Stakeholders' Walkability/Wheelability Audit in Neighborhoods (SWAN) tool, this study delves into the objective and subjective aspects of neighborhood street segments.

2 RESEARCH DESIGN AND METHODS

This study adopts a community-engaged design to comprehensively explore mobility disability experiences across diverse age groups. At its core, the Stakeholders' Walkability/Wheelability Audit in Neighborhoods (SWAN) tool, a micro-scale assessment tool, evaluates functionality, safety, land use, appearance and maintenance, and social aspects within neighborhoods. This quantitative approach provides objective insights into the physical attributes of the built environment, forming a foundational understanding of urban accessibility. In addition to the quantitative foundation, the study integrates a qualitative approach to delve into the subjective dimensions of participants' experiences. Aimed at capturing detailed narratives, the interviews explore participants' interactions with their environment, the social dynamics they navigate, and the adaptive strategies they employ. Also focusing specifically on young and older adults (aged 55+) in Metro Vancouver, the study seeks to draw nuanced comparisons.

3 RESULTS

Insights from the study reveal a divergence in the experiences of younger and older adults with disabilities, with the former perceiving higher levels of support in various domains. This emphasizes age as a crucial factor influencing individuals' experiences in a neighborhood, particularly regarding accessibility. The findings contribute to a nuanced understanding of how age-specific dynamics shape the landscape of spatial justice for individuals experiencing disabilities.

4 DISCUSSION AND IMPLICATIONS

This research underscores age-specific challenges in neighborhood accessibility for individuals with disabilities. Utilizing the SWAN tool and qualitative interviews, the study uncovers nuanced differences between older adults and younger individuals, informing targeted interventions for improved accessibility and social participation. The implications extend to design professionals and academics, offering insights into effective approaches and fresh perspectives on fostering inclusive urban environments. The study contributes to the broader discourse on creating environments that address spatial justice and equity without explicitly using the conference focus language.

5 CONCLUSION

In navigating the complexities of inclusive design, this research underscores the importance of understanding diverse disability experiences across the lifespan. Insights contribute valuable perspectives for creating equitable urban environments. Uncovering nuanced differences between older and younger individuals informs targeted interventions for enhanced accessibility and social participation. The study offers valuable insights for design professionals and academics, contributing to the broader discourse on creating spatially just environments. Justice and equity, emphasizing the importance of age-specific considerations in enhancing the overall quality of urban spaces.

CARVING OUT SPACE FOR BIKE POLO: LESSONS FROM THE PANDEMIC

Track 5: Spatial Justice, Equity, and Inclusion by Design
Knowledge Network: Active Living by Design, Health
Methods: Interviews, Photo-based Methods
Carey Clouse (University of Massachusetts, Amherst)

1 ABSTRACT

The COVID-19 pandemic caused a reshuffling of recreational activities to allow for social distancing in public outdoor spaces. In the years that followed, the informal transitioning of recreational landscapes in public parks to new types of uses helped to underscore the ways in which landscape architecture and planning processes underserve particular populations. Literature supporting informal use of public space includes insurgent urbanism, guerrilla urbanism, DIY urbanism, and right to the city (Douglas 2018; Harvey 2012; Hou 2020, Marinic & Meninato 2022).

This paper examines one example of unofficial new land use, in which a public tennis court supported weekly bike polo club practices in Eugene, OR. Through participant surveys, stakeholder interviews and spatial research methods, a series of recommendations for the continued support of multifunctional recreational uses in public park landscapes emerged. These findings address the ways in which the more proactive engagement of design professionals can facilitate a re-imagining of public outdoor space, and in so doing, represent the needs of multiple stakeholder groups, build more diverse representation in parks, and ease tensions inherent to multifunctionality.

2 REFERENCES

Douglas, G. (2018). The Help-Yourself City: Legitimacy and Inequality in DIY Urbanism. New York: Oxford University Press.

Harvey, D. (2012). Rebel Cities: From the Right to the City to the Urban Revolution. New York: Verso.

Hou, J. (2020). Guerilla urbanism: urban design and the practices of resistance. Urban Design International(25), 117-125. DOI 10.1057/s41289-020-00118-6

Marinic, G. and P. Meninato. (2022). Informality and the City: Theories, Actions and Interventions. Cham, Switzerland: Springer. DOI: 10.1007/978-3-030-99926-1

RURAL EDUCATION IN PAKISTAN: DEVELOPING A FRAMEWORK FOR AN INCLUSIVE LEARNING ENVIRONMENT

Track 5: Spatial Justice, Equity, and Inclusion by Design
Knowledge Network: Children, Youth & Environment, Cultural Aspects of Design
Methods: Interviews, Ethnography

Huda Muazzam (VCU); Marwa Abdelmonem (Virginia Commonwealth School of the Arts-Qatar); Matthew Holmes Dallimore (VCUQ)

1 ABSTRACT

In 2012, Malala Yousafzai survived a gunshot to the head by a Tehrik-e-Taliban gunman while en route from school, sparking a focus on Pakistan's rural education landscape (Blumberg, 2018). Currently, 22.7 million children in these areas remain out of school, primarily in regions grappling with environmental degradation and limited access to fundamental needs (Powering Better Education in Pakistan, n.d.). Neglect of rural schools is evident in challenging physical conditions where the existing infrastructure has a significant impact on learning (Sattar, 2007). This is further exacerbated by families earning a per capita income of $443 (Iram & Butt, 2004). The overarching issue of child labor and restricted education access intertwines with complex economic and social dynamics, prioritizing immediate sustenance over educational investments and perpetuating a challenging poverty cycle (Ray, 2002). Entrenched gender roles that place men above women continue to influence whether a child stays in school. (Hawke, 2015). Furthermore, resulting in the offshoot of a weak, polarized, and non-unified education system. (Ahmad et al., 2014)

In addition, the study delves into the person-environment fit theory, aiming to identify key constructs within a framework that elucidates the impact of the learning environment on community strength (Azizan et al., 2021). The main aim of this study is to shed light on the factors contributing to educational disparities and propose a framework for adopting a holistic learning approach in rural settings. By leveraging these insights, educators can create tailored programs, such as peer mentoring initiatives, to boost student engagement in remote Pakistani schools. Additionally, flexible learning spaces and sustainable energy sources can be implemented to address specific rural education challenges.

In this exploratory qualitative research, a triangulation approach was embraced, incorporating diverse datasets such as scholarly sources, interviews, and non-scholarly media to address the research question: What factors contribute to educational disparities in rural areas of Pakistan? Which theoretical framework is conducive to creating an inclusive environment? The primary methodology employed is content analysis, entailing a thorough review of existing literature and examining two case studies. Furthermore, pilot semi-structured interviews were conducted with local educators and administrators using convenience sampling to validate identified themes, offering guidance for the research and facilitating timely refinement.

The findings of this study indicate that building their cognitive and social skills through community connectedness remains the cornerstone of building a conducive environment. Emotional embeddedness, feelings of safety, and belongingness associated with an environment, in turn, affect the child's behavior in a classroom (Brown & Perkins, 1992). Children in close-knit communities struggle with motivation when disconnected, emphasizing the key role of creating a space where belonging is central. Contributors to educational disparities include the socio-economic conditions of rural communities and environmental degradation, resulting in student dropout rates and hindering educational progress (Shaheen, 2013). This study aims to enhance the understanding of how the rural learning environment influences educational outcomes, informing evidence-based interventions for greater inclusivity and effectiveness.

The proposed framework may serve as a foundation for future designers to establish inclusive learning environments that foster belonging and identity.

SPATIAL ENVIRONMENTAL JUSTICE: AN ANALYSIS OF CONDITIONS AFFECTING MARGINALIZED COMMUNITIES

Track 5: Spatial Justice, Equity, and Inclusion by Design
Knowledge Network: Participation, Sustainable Planning, Design & Behavior
Methods: Action/Participatory Research, Statistical/Modeling Analysis
Daniel Cronan (State University of New York); Iman Fayyad (Syracuse University); Eliana Abu-Hamdi (Syracuse University)

1 ABSTRACT

This research addresses the intricate relationship between the built environment and the overall wellbeing of marginalized communities. Our approach hinges on the examination of three pivotal variables: a) physical elements linked to income inequality, b) potential sites for intervention, and c) agency for various community partners. Within this research, we examine our constructed database of variables affecting wellbeing within Syracuse, New York communities to be disseminated through platforms of engagement throughout the project. Our project demonstrates a digital analysis through a digital 'hub' and its development, which we intend to serve for the project, the communities, and potentially for future researchers to adjust and replicate for similar research initiatives. The data hub development, an enriched geospatial dataset to examine a gradient of variables, demonstrates gathering of data and the construction of an integrated tool for addressing challenges in infrastructure, but potentially suitability for site-scale interventions to address said challenges. The hub serves as a dynamic platform that provides a holistic view of the physical environment's impact on community wellbeing, therefore we intend for this platform to act as a transferable framework for future projects.

The methodology imparts a geographically weighted regression (GWR) framework to address the issue of wellbeing in marginalized communities through a holistic, multi-variable approach for spatial analysis (Brunsdon, 1998; Leong and Yue, 2017). We intend for this analysis, illustrated through a digital platform, to inform a set of structures to create a proof of concept of co-production of knowledge (Norstrom, 2020) informing community-based design. The research intends to seed future projects and work as a catalyst for connections with other research groups involved with similar projects in the future. The research seeks to provide communities with tools for fostering meaningful change and imparting a geospatial blueprint for inclusive community development through a process utilizing co-production of knowledge.

The results intend to provide an analysis of data gathered extended to contiguous and adjacent urban areas within Syracuse. The hub technology incorporates building information modeling (BIM), which was developed to mimic decision-making trends, generating scenarios of alternative future conditions within the built environment. Through scenario generation, policymakers, planners, and community members will be provided with a design framework to understand the effects of past policies, present conditions, and applicable designs at the site scale to impact change. In this way, community groups and policy-makers can, through our produced technology, better understand the conditions that generated current developmental problems and the potential interventions that could reverse detrimental effects through design.

2 REFERENCES

Norström, A. V., Cvitanovic, C., Löf, M., West, S., Wyborn, C., Balvanera, P., Bednarek, A. T., Bennett, E. M., Biggs, R., de Bremond, A., Campbell, B. M., Österblom, H. (2020). Principles for knowledge co-production in sustainability research. Nature Sustainability, 3(3), 182–190. https://doi.org/10.1038/s41893-019-0448-2

Brunsdon, C., Fotheringham, S., & Charlton, M. (1998). Geographically Weighted Regression-Modelling Spatial Non-Stationarity. Journal of the Royal Statistical Society. Series D (The Statistician), 47(3), 431–443. https://www.jstor.org/stable/2988625

Leong, Y.-Y., & Yue, J. (2017). A modification to geographically weighted regression. International Journal of Health Geographics, 16(1), 11. https://doi.org/10.1186/s12942-017-0085-9

PROGRAMMING AS AN INTEGRATIVE PLATFORM FOR SOCIAL ISSUES IN DESIGN

Track 5: Spatial Justice, Equity, and Inclusion by Design
Knowledge Network: POE/Programming
Methods: Meta-Analysis

Lubomir Popov (Bowling Green State University); Franklin Goza (University of Wisconsin Whitewater)

1 ABSTRACT

Currently, in environmental design research and related fields, there are a lot of social issues studied in separate research domains, like environment and aging, gender, race, justice, inequality, equity, and so forth. There are attempts to integrate research domains in order to provide a holistic view, but these attempts are mostly "mechanical," discussing several different issues in one paper. This leads to emerging "competition" for attention and resources among stakeholder groups, as well as among different social issues, goals, and priorities. This leads to new kinds of imbalances. The problem is how to integrate all these issues and goals into a holistic model for discussing them not separately, but in a system of stakeholders and available resources. The goal of this presentation is to propose a foundation for creating an integrative platform for social issues in design. Our study is theoretical.

Because we are looking at the integration of social issues/problems and their consequences for design, we adopt a sociological perspective, including Anthony Giddens' structuration theory as a methodological lens. In addition, we use the theory of design to identify the phases of the project delivery process where the social issues mentioned above are discussed and integrated most intensively. The processes of architectural programming are devoted to researching building users and stakeholders, as well as their needs, problems, and priorities. It is during this phase that the most important decisions are made about optimizing user needs and relating them to available resources. We propose that the programming process is reinvented as an integration platform that will facilitate stakeholder research, discussions, negotiations, conflict management, and agreements. This type of integration platform will benefit greatly the underrepresented and underprivileged groups and the balancing of the social issues mentioned above.

BRINGING EQUITY TO THIRD PLACES: A LIBRARY CASE STUDY IN A HIGHLY DIVERSE CANADIAN COMMUNITY

Track 5: Spatial Justice, Equity, and Inclusion by Design
Knowledge Network: Cities & Globalization, Cultural Aspects of Design
Methods: Case Study, Mixed Methods

Francesqca Jimenez (HDR Architecture, Inc.); Melissa Templeton (HDR); Jeri Brittin (HDR Inc)

1 ABSTRACT

As important information hubs and gathering spaces within a community, today's public libraries resonate as what Ray Oldenburg describes as a "third place"—a space that is not home, not work, but another meaningful place supporting social connections and community wellbeing. Recent scholarship highlights how a shifting paradigm for library services and the library's relationship to its users strengthens and expands this link. Wood (2021) describes libraries as "a rare cross-section of the public sphere and third place, satiating community-based and political needs" (Wood 2021, pp 144). The library, as a place or community center, contributes to the overall provision of community empowerment and development (Sørensen, 2020). Libraries may also be leveraged to conduct health-related work that advances public health in promoting equitable community-scale health outcomes (Philbin et al., 2019).

As part of a larger community needs assessment, our research team measured library utilization as well as barriers to use to understand the role the library plays in modern community life. The study was situated in Richmond, BC, a vibrant city near Vancouver experiencing rapid urbanization and substantial population growth, and an influx of recent immigrants. The city is distinctly ethnically and linguistically diverse. Like many North American cities, Richmond's population is aging, and rising costs threaten housing security. Across the city's geography, Richmond's library system has four branches, two book vending machines and a regular pop-up shop. Additionally, in 2020, library staff completed training on unhoused populations with goals to improve communications with individuals experiencing homelessness and to deescalate tense interactions.

The study comprised a mixed methods approach including a survey of a demographically and geographically representative sample of 1000 community members, and in-depth focus groups with an additional 52 Richmond residents. Among other metrics, library utilization was assessed based on users' demographics, activities, motivations, access, and experience. A main goal of the study was to understand library users' needs and to determine how those responsible for planning city services and facilities can increase library access and ensure the library system continues to support all residents of the community. Contextualizing the findings from this study in Richmond with prevalent social complexities and public health topics, we consider the importance of the library as an accessible and linguistically diverse third place. Building on the theme of this year's conference, "Human-Centric Design," and specifically on the "Spatial Justice, Equity, and Inclusion by Design" track, this research examines how libraries can be better designed as a third place that provides local need-based solutions to support diversity and health in a community.

2 REFERENCES

Philbin, M. M., Parker, C. M., Flaherty, M. G., & Hirsch, J. S. (2019). Public libraries: A community-level resource to advance population health. Journal of community health, 44, 192-199.

Sørensen, K. M. (2020). The values of public libraries: A systematic review of empirical studies of stakeholder perceptions. Journal of Documentation, 76(4), 909–927. https://doi.org/10.1108/JD-10-2019-0201

Wood, E. (2021). Libraries Full Circle: The Cross Section of Community, the Public Sphere, and Third Place. Public Library Quarterly, 40(2), 144–166. https://doi.org/10.1080/01616846.2020.1737491

TRANSPLANTED MODERNITY: A CRITICAL HISTORY OF ARCHITECTURE ALONG THE TAZARA RAILWAY

Track 5: Spatial Justice, Equity, and Inclusion by Design
Knowledge Network: Cities & Globalization, Cultural Aspects of Design
Methods: Case Study, Design Methods

Yunyun Zhu ; Zhongzhong Zeng (Beijing Jiaotong University); Bo Zhang (Oklahoma State University)

1 ABSTRACT

This study delves into the understudied architectural landscape of the Tazara Railway, a 1,860 km lifeline constructed between 1970 and 1975 as a cornerstone of China's foreign aid initiatives in Africa. Beyond its strategic function – connecting Zambia's Copperbelt to the Tanzanian port of Dar es Salaam and bypassing colonial routes – the Tazara embodied powerful ideologies. Its moniker, the "Great Freedom Railway," resonated with the Pan-African aspirations of the era.

Focusing on thirteen railway stations, as well as ancillary structures like schools and warehouses, this research offers a nuanced fragment of architectural history. It navigates the intricate interplay of artistic modernism, political post-colonialism, material locality, and intercultural exchange. By situating these projects within the Cold War and the broader trajectory of modern architecture, the study seeks to answer several key questions: (1)How did Chinese architects and policymakers adapt modern architectural principles to the cultural and functional contexts of East Africa? (2) In what ways did this architecture evolve as a symbol of post-colonial identity over time? (3)To what extent did the railway stations contribute to local economic development and urbanization in Tanzania and Zambia? (4)How do these structures reflect the design vocabulary and construction techniques of their time? To address these questions, the study employs a multifaceted methodological approach. Archival materials such as architectural drawings, historical photographs, and newspaper/TV reports are combined with in-depth interviews with designers and their descendants. This triangulation of data sheds light on the conception, construction, and reception of this unique architectural corpus.

This research contributes to a deeper understanding of: (1)Global modern architecture: By examining the adaptation and reinterpretation of Western architectural principles in a non-Western context, this study adds a valuable chapter to the narrative of global modernism. (2) Regional architecture: The analysis of the Tazara's architectural language elucidates the development of distinct regional identities within the broader currents of modern architecture. (3) Postcolonialism: The symbolic and functional roles of these structures offer insights into the complex processes of postcolonial nation-building and identity formation. (4)Tectonic theories: The study delves into the specific construction techniques employed in the Tazara's buildings, enriching our understanding of regional building traditions and their interaction with international currents.

ENCOURAGING SOCIO-CULTURAL AWARENESS BETWEEN SUB-SAHARAN AFRICANS AND MOROCCANS IN MOROCCO

Track 5: Spatial Justice, Equity, and Inclusion by Design
Knowledge Network: Cultural Aspects of Design, Interior Design
Methods: Interviews, Mixed Methods

Yara Faraj (VCUarts Qatar); Marwa Abdelmonem (Virginia Commonwealth School of the Arts-Qatar); Matthew Holmes Dallimore (VCUQ)

1 ABSTRACT

Due to its geographically advantageous position, Morocco has become a common transit point for Sub-Saharan African migrants seeking to reach Europe to better their social, educational, or economic standing (Mourad, 2022). Nevertheless, the aspirations of these individuals often falter due to the stringent border controls imposed by the European Union (Papadopoulou-Kourkoula, 1970). Consequently, this phenomenon results in their prolonged confinement inside a nation previously only deemed a transitional passage to reach their desired destination. Based on a study conducted by the Migration Policy Institute in 2019, approximately 700,000 irregular migrants are presently in Morocco with thousands more migrating yearly. This estimate discounts those without appropriate documentation and residing unlawfully in the subject country (El Ghazouani, 2019).

It is also worth noting that these irregular migrants, mostly occupying Rabat and Casablanca, are exposed to significant mistreatment by both governmental authorities and residents, displaying a lack of regard for their human rights and individual identities (Mourad, 2022). The maltreatment experienced by migrants leads to various mental and emotional health challenges, as they are subjected to repetitive acts of mistreatment daily. Hence, considering the significant presence of Sub-Saharan African migrants in Morocco, the imperative of their socio-cultural integration arises as a pivotal aspect in mitigating the psychological adversities they encounter in the receiving nation. In order to address the research question: "How can the cultural integration of Sub-Saharan African migrants within the Moroccan community eliminate discrimination and racism?" a triangulation of qualitative methods has been performed. The methodology includes an extensive literature review that analyzes the challenges and barriers to integrating Sub-Saharan African migrants in Morocco and the impact of cultural integration in mitigating discrimination and racism. To fill the research gaps identified in the literature review, semi-structured pilot interviews were conducted with individuals from Sub-Saharan African organizations in Morocco, who closely interact with the subject migrants addressing their needs and demands, and integrating them within the Moroccan context.

The findings indicate that economic and socio-cultural dimensions are required to achieve reciprocal and efficient interaction between the involved parties. Nevertheless, the existing administrative obstacles faced by Sub-Saharan African migrants in Morocco hinder their socio-cultural assimilation with the local community, ultimately leading to their marginalization. Despite the limitations in identifying effective solutions at the administrative level, involving the subject migrant population in Morocco may be achieved through innovative socio-cultural activities and workshops that promote socio-cultural interchange and linguistic courses to enhance communication between the two communities.

In summary, it is vital to create an inclusive environment that tackles the lack of equity and socio-cultural awareness among these populations. This segment of the ongoing study serves as the foundation for further efforts focusing on design considerations in cultural centers, which will be undertaken in the spring of 2024.

CHILDREN, YOUTH & ENVIRONMENT EQUITY AND INCLUSION IN ENVIRONMENTAL DESIGN STUDIOS: DETRIMENTAL EFFECTS OF GRADES TO ENGAGED LEARNING OF DIVERSE STUDENTS

Track 5: Spatial Justice, Equity, and Inclusion by Design
Knowledge Network: Environmental Design Research Education, Interior Design
Methods: Case Study, Focus Groups
Nisha A. Fernando (University of Kansas)

1 ABSTRACT

Teaching and learning in design studios include identifying critical problems to solve, incisive design thinking, understanding and applying research evidence, and subsequently creating human-centric design solutions. This general design process calls for a variety of skills, knowledge and deep engagement by students. Successful pedagogical environments of design studios must allow equity where individual students can apply their own design approaches using their unique learning styles and skills. While this creative and intellectual freedom is an inherent component of inclusion and diversity in design studios, evaluation of student learning has been limited to a rather conventional and monopolized approach: the traditional grading system with points and letter grades with a GPA as the end goal. Recent progressive pedagogical discourses point out these conventional systems are not only outdated now, but also devalue the purpose of engaged and immersed learning (Blum, 2020; Kohn, 2011).

This paper presents an evidence-based, methodical, alternative learning evaluation approach used in an undergraduate interior architecture studio in place of letter grades. The particular method was chosen among many other alternative examples (Blum, 2020) that avoid conventional grading systems, and after a semester-long participation in on-campus teaching and learning discussion groups. First, the paper discusses various negative effects of conventional grading systems that impair the levels of engagement and learning among diverse students with diverse learning styles, especially in interactive design studio learning contexts. Secondly, the paper discusses at length the details of the alternative evaluation approach used – how students were made in charge of evaluating their own learning and the particulars of regular in-class, in-person and two-way discussions and feedback. Instead of the convention that the instructor is in charge of evaluation, students measured their own learning, and this allowed for various learning styles to be used without any worry or prohibition in the studio, creating an inclusive learning environment. Discussions and feedback approaches were methodically documented by the author. Points and letter grades were not used at any time, while each student performance was evaluated through deliberate verbal and interactive discussions. Using case studies of interior architecture student performances and feedback (n=33), the paper presents research evidence on (a) how students' stress levels significantly decreased in studio, (b) levels to which students became more engaged and enthusiastic about learning, creativity, and critical thinking in design instead of focusing on 'working for a grade,' and most importantly, (c) how students with diverse learning styles improved their studio performance, thus increasing equity in studio learning. The presentation also includes detailed student feedback collected at the end of the semester illustrating the success levels of the alternative evaluation processes used in the studio.

2 REFERENCES

Blum, S. (Ed). (2020). Ungrading: Why Rating Students Undermines Learning (and What To Do Instead). West Virginia University Press.
Kohn, A. (2011). The Case Against Grades. Educational Leadership, 69(3),58-62.

DESIGN, EQUITY & INCLUSION AUTISM, NEURODIVERSITY AND SENSORY DECOLONIZATION: "ESCAPESCAPE" AS ACUPUNCTURES FOR SENSORY EQUITY IN OUR OVERSTIMULATED CITYSCAPES

Track 5: Spatial Justice, Equity, and Inclusion by Design
Knowledge Network: Cities & Globalization, Health
Methods: Experimental/Quasi-Experimental Research, Survey

Magda Mostafa (The American University in Cairo and StudioTM); Sherif Goubran (The American University in Cairo); Mahinour Yasser Awad (The American University in Cairo)

1 ABSTRACT

Our cities have increasingly become multisensory canvases of spatial commodification, commercialization, and exclusion. Like exclusionary practices of gentrification, the sensory environment has become a growing threat to equitable access to cities and their public space to our most vulnerable communities, but particularly to the autistic and neurodivergent citizens seeking to be included. (Mostafa et al, 2023) As cities strive to be Human-Centered and more aware of the diverse needs of their citizens, we are confronted with the notion that the profile of "human" at the center of this process is often very narrowly defined. The predominant profile has been, and largely remains that of the able-bodied, cis-gendered, straight, white, English-speaking, financially stable, seeing, hearing, ambulatory, neurotypical male.

This paper presents the spatial construct of "EscapeScapes" which was developed through the lens of the autistic spatial experience based on the Autism ASPECTSSTM Design framework. (Mostafa, 2021) Specifically, it explores how" EscapeScapes" can be mobilized in the city to hybridize the notion of prospect and refuge- present "refuge as prospect" and "prospect of refuge"- moments of sensory reprieve and escape that allow at once retreat and safety but also positive restoration and resilience.

This is illustrated through the results of a six-month experimental spatial installation at the Venice Architecture Biennale hosted at the European Cultural Center and part of an exhibit titled "A Case for Sensory Decolonisation: Autistic Escape". The experiment asked multiple questions: To what extent can acoustically modified, built environments mitigate sensory overload and create a respite for neurodiverse and autistic individuals? What is the perceived impact of such environments on focus and anxiety levels in neurodiverse and autistic individuals? Do these benefits extend and generalize to the neurotypical population? The installation consisted of two notions: a simulated sensory overstimulation using crowd-sourced visual voices from the autistic community; and an interactive spatial installation, testing the performance of sensory escape opportunities by proposing the introduction of "EscapeScapes." The installation allowed users to step away from the noise of the simulated cityscape and experience the value of sensory refuge.

The results of the survey conducted at the exhibit indicate that the acoustically modified spatial acupuncture afforded by the proposed "EscapeScapes" provides a positive reduction in perceived anxiety and stress across autistic and neurodivergent users surveyed. Interestingly, the responses indicated that these benefits carry over to the neurotypical users surveyed. These preliminary results indicate the specific benefits of "EscapeScapes" as urban strategies for improving access for autistic and neurodivergent users to public spaces and that such benefits could be amplified when used in a suite of connected interventions. The results also support the value of such interventions to the larger typical population, supporting the proposition that when we design for autism and neurodiversity, we design better for all.

2 REFERENCES

Mostafa, M., et al, (2023, July). Multi-Sensory Wayfinding: Lessons from the Margins Towards the Design of Equitable and Healthy Spaces. Design for Inclusivity (pp. 731-741). Springer International Publishing.

Mostafa, M. (2021). The Autism Friendly University Design Guide. Dublin City University. https://issuu.com/magdamostafa/docs/the_autism_friendly_design_guide.

REDEFINING INCLUSIVITY THROUGH THE LENS OF HERITAGE CONSERVATION

Track 5: Spatial Justice, Equity, and Inclusion by Design
Knowledge Network: Cities & Globalization, Historic Preservation
Methods: Action/Participatory Research, Experimental/Quasi-Experimental Research
Claudine Déom (University of Montreal)

1 ABSTRACT

Redefining inclusivity through the lens of heritage conservation

This proposal for an individual presentation is based on research conducted in 2022-2023 and funded by the Social Sciences and Humanities Research Council of Canada. It is about the relationship between people and the places they deem important in city neighbourhoods. It aims to contribute to the reflection on the role that heritage can play in the integration of ethnocultural communities in Montreal, one of the three most cosmopolitan cities in Canada. The research also aims to position heritage conservation as a means to achieve social cohesion, well-being and inclusivity in the context of superdiversity and some of the issues it raises such as racial tensions. (Avrami, 2020)
Once associated exclusively with the idea of monument and with a moment in history, heritage gradually evolved from the 1990s in response to globalization, to the increased awareness of climate change and to the emergence of the sustainable development discourse. UNESCO's Universal Declaration on Cultural Diversity (UNESCO, 2002) represents a turning point in the evolution of the concept of heritage, which is increasingly focused on human values. This shift in heritage emphasizes the importance of including the voices of different stakeholders - including those of the marginalized - in the identification of heritage and conservation processes. It no longer confines heritage to exceptional places, but can include those of everyday life such as daycares, places of worship, local retail shops and parks. Heritage is therefore no longer exclusively old and beautiful. As such, it can play a determining role in policies aiming for inclusion and social cohesion for its ability to create and sustain a sense of belonging and well-being. This paradigm shift allows heritage to open up to other disciplines invested in the study of people and places such as sociology and urban geography. (Oldenburg, 1999)

This presentation will summarise this exploratory research and some of the results it yielded. The different phases of the study, from the initial contact with community leaders to the organization of mapping workshops and the publication of exploratory maps of places, will be explained. The social and cultural values of the places identified by members of the Haitian, Moroccan and Algerian communities reveal the importance of preserving them for the future. Achieving this goal will, however, require transdisciplinary methods to develop the tools that will go beyond those currently used by the heritage sector.

2 REFERENCES

Avrami, Erica (2020). Preservation and Social Inclusion. Columbia Books.
UNESCO (2003). Convention for the Safeguarding of the Intangible Cultural Heritage.

Oldenburg, R. (1999). The great good place: cafés, coffee shops, bookstores, bars, hair salons, and other hangouts at the heart of a community. (2nd ed.) Hachette Books.

WHO IS AT THE CENTER OF HUMAN CENTERED DESIGN?: LEARNING FROM ASPECTS AND DEAFSPACE, A COMPARATIVE ANALYSIS

Track 5: Spatial Justice, Equity, and Inclusion by Design
Knowledge Network: Cultural Aspects of Design, Participation
Methods: Design Methods, Practice-based Research
Magda Mostafa (The American University in Cairo and StudioTM); Hansel Bauman (HB a+p)

1 ABSTRACT

In current discourse around Human-Centered Design and policies of diversity, equity and inclusion (DEI) one often finds broad differences in the identities included as well as a perpetuation of siloed perspectives from previous paradigms. We find ourselves asking "Who is at the center of Human-Centered Design? and Who do we mean when we say Diversity, Equity and Inclusion?". Responses all too often are culturally specific, politically informed, and institutionally centered, creating as much diversity in the definition of Human-Centered and DEI themselves, as the spectrum of the human condition they set out to capture.

Case in point is the call for papers to which this abstract responds. In its specific call for Spatial Justice, Equity and Inclusion by Design it plans to include work that "celebrate(s) environmental design research, scholarship, projects, and advocacy at the intersections of gender, race, ethnicity, indigeneity, sexuality and economic inequality". As noble a call this is, rather conspicuously absent is the identity of disability.

This paper looks at this exclusion, but also expands to a second level of exclusion, that of the sensory and neurodivergent from the definition of disability. It expands on the call for work to include that which "celebrate(s) environmental design research, scholarship, projects, and advocacy at the intersections of disability, neurodiversity and the sensory" through the presentation and comparative analysis of the Autism ASPECTSS Design Index and Gallaudet University DeafSpace Design Guidelines. ASPECTSS is an evidence-based set of 15 design constructs, developed from the perspective of the autistic lived experience and demonstrated to support access to architecture for autism. (Mostafa, 2014 & 2018) DeafSpace Guidelines are a catalog of over 100 architectural strategies for creating spaces attuned to visual language, multi-sensory spatial awareness and Deaf culture. (Bauman, Murray 2014)
It aims to explore the practicalities of intersectional design across the deaf and autistic experiences by mapping the connections, overlaps and conflicts between and across ASPECTSS and DeafSpace. Using theoretical framing, language mapping, and comparative analysis, it looks at both the conceptual language of the two approaches as well as the practical spatial devices proposed in both.

The objective here is threefold: to propose a method to conduct similar comparative analyses between other previously siloed identities and sets of user needs specific to those identities; to illustrate the opportunities, gaps and win-wins created by designing with the broad sensory needs of the autistic and deaf experience; and to demonstrate the broader impact on typical populations when adopting these approaches. In short, it sets out to explore the questions "When we design for neurodivergence, autism and deaf experiences, do we exclude? Amplify inclusion? or even design better for everyone?"

2 REFERENCES

Bauman, H. D. L., & Murray, J. J. (Eds.). (2014). Deaf gain: Raising the stakes for human diversity. U of Minnesota Press.

Mostafa, M. (2014). Architecture for autism: Autism ASPECTSS™ in school design. International Journal of Architectural Research: ArchNet-IJAR, 8(1), 143-158.

Mostafa, M. (2018). Designing for autism: an ASPECTSS™ post-occupancy evaluation of learning environments. Archnet-IJAR: International Journal of Architectural Research, 12(3), 308.

POST-OCCUPANCY EVALUATIONS OF PERMANENT SUPPORTIVE HOUSING

Track 5: Spatial Justice, Equity, and Inclusion by Design
Knowledge Network: POE/Programming
Methods: Practice-based Research, Survey
Corey Schnobrich (Leddy Maytum Stacy Architects)

1 ABSTRACT

Permanent supportive housing, which combines stable shelter with social services, is the most successful model for the health and housing stability of the formerly homeless (Fitzpatrick-Lewis, 2011, p. 10). Designing and programming for this type of housing faces unique challenges, however. Future residents may be difficult to reliably reach or may not even be homeless during the programming phase. They may also be wary of participating in research; social workers described their clients as an "over-surveyed but under-focused group" (Nolan, 2018, p. 16).

The presentation will discuss a means of navigating these challenges through post-occupancy evaluations (POEs) at four permanent supportive housing facilities in northern California. The POEs included both quantitative and qualitative questions. The quantitative question responses provided easy comparison across building spaces and between projects, while the qualitative question responses provided an opportunity for detailed and nuanced feedback.

The presentation will focus on the qualitative responses and resident perceptions of their new homes. The responses suggest the favorite features of the building include shared courtyards and community rooms, while the least favorite features include acoustics, laundry rooms, and lack of storage. Among residents' wish list for amenities, gym spaces are the most frequent request. In addition, building management policies are closely tied to resident grievances. The presentation will conclude with reflections on POE revisions and implications for future projects.

2 REFERENCES

Fitzpatrick-Lewis, D., Ganann, R., Krishnaratne, S., Ciliska, D., Kouyoumdjian, F., & Hwang, S. W. (2011).

Effectiveness of interventions to improve the health and housing status of homeless people: A rapid systematic review. BMC Public Health, 11(1), 638. https://doi.org/10.1186/1471-2458-11-638.

Nolan, J. (2018). Post-Occupancy Evaluation: Findings and Recommendations (p. 68). San Francisco, CA: Community Housing Partnership.

WHO IS THAT STADIUM FOR? ATHLETIC FACILITIES, STUDENT SUCCESS, AND ECONOMIC EQUALITY

Track 5: Spatial Justice, Equity, and Inclusion by Design
Knowledge Network: POE/Programming, Student Affairs
Methods: Statistical/Modeling Analysis

Michael Darfler (Brightspot Strategy)

1 ABSTRACT

The relationship between university facilities and student success has been well studied, especially the relationship between athletics and graduation (McCollum, 2018; Norton, 2023). These studies have shown a strong positive relationship between access to and usage of athletic facilities and retention rates and graduation rates. This would seem to indicate that investment in recreational facilities by a university or college would be a valuable investment. However, this might not be the case for all students. This study examines the relationship between access to athletic facilities and graduation rates and its interaction with socio-economic status. The results confirmed that the total square footage of athletic facilities per FTE student was positively correlated with graduation rates. However, this was not true for all students. The data also showed that there was a significant interaction effect between socio-economic status and athletics area and that for Pell and Stafford Grant Recipients, more athletic facilities were negatively correlated with graduation rates. While the cause of this relationship requires further investigation, recent studies have shown that student athletes tend to come from high-income families (Chetty et al., 2023). Therefore, it may be the case that low-income students may not be as likely to take advantage of athletic facilities and that for low-income students, institutional resources could be better spent elsewhere.

2 REFERENCES

Chetty, R., Deming, D., & Friedman, J. (2023). Diversifying Society's Leaders? The Determinants and Causal Effects of Admission to Highly Selective Private Colleges (w31492; p. w31492). National Bureau of Economic Research. https://doi.org/10.3386/w31492

McCollum, G. (2018). Collegiate Recreation Participation and Student Retention, Progression, and Graduation. https://digitalcommons.georgiasouthern.edu/etd/1854/

Norton, T. (2023). The Influence of Campus Facilities on the Satisfaction and Retention of Students in a Higher Education Setting. Culminating Projects in Higher Education Administration. https://repository.stcloudstate.edu/hied_etds/70

SPATIAL JUSTICE IN THE YOUNG POPULATION STUDENT PROTOTYPES OF GENDER-INCLUSIVE RESTROOM DESIGN AND GUIDELINES

Track 5: Spatial Justice, Equity, and Inclusion by Design
Knowledge Network: Cultural Aspects of Design, Interior Design
Methods: Case Study, Design Methods
Casey Gail Franklin (University of Kansas)

1 ABSTRACT

We have a bathroom problem; public restrooms are usually sex-segregated spaces, which is leaving a growing group of trans- and gender-diverse (TGD) users without restrooms they feel comfortable using. This lack of appropriate facilities can impact physical and mental health as TGD users may try to not use the restroom at all, may travel long distances to use a safe restroom, may experience workplace discrimination, or may face harassment when using the restroom. All-gender or gender-inclusive restrooms seek to solve this problem by providing spaces that are not designated male or female but are instead open to users of all gender identities. Multi-user all-gender restrooms are becoming more commonplace but are still a new architectural typology with its own set of unique design needs.

This presentation will showcase student design explorations of multi-user gender-inclusive restrooms including research, design guidelines, and design prototypes. The goal of this graduate-level architectural assignment was to help students understand the need for inclusive restrooms, to identify what design guidelines should be used for this space type, and to explore various forms these spaces could take on. Students conducted secondary and primary research by reviewing literature, analyzing case studies of built gender-inclusive restrooms, and assessing designated on-campus gender-inclusive restrooms. Students used this information to create a list of design guidelines and created a gender-inclusive restroom prototype embodying those guidelines. Despite similar design recommendations, such as zero sightline stall doors, student work demonstrated a variety of potential outcomes that challenge traditional restroom design norms.

EXAMINING SPATIAL JUSTICE IN LOS ANGELES' CULTURAL PLANNING: INSIGHTS FROM OPEN-SOURCE AND CROWDSOURCED DATA

Track 5: Spatial Justice, Equity, and Inclusion by Design
Knowledge Network: Cities & Globalization, Cultural Aspects of Design
Methods: Case Study, Statistical/Modeling Analysis
Norhan Abdelgawad (Virginia Tech University); Shalini Misra (Virginia Tech)

1 ABSTRACT

This research uses crowdsourced and open-source data to analyze spatial justice in arts and cultural planning in Los Angeles. Using Lefebvre's right to the city (1967) and Giddens' structuration theory (2004) as our dual theoretical lenses, we develop a set of dimensions and associated indicators to evaluate spatial justice in arts and cultural planning.

This study offers several learning outcomes, where first, through using the lens of arts and culture as a tool for building healthy communities, we can identify critical gaps to address inequity, such as diminished civic power, economic disinvestment, and/or social exclusion. Second, incorporating a spatial dimension in our analysis allows us to better understand the distribution of cultural resources in relation to demographic characteristics. Third, the development of spatially just cultural indicators and dimensions enable us to establish consistent metrics for assessing spatial justice in the city. Fourth, by leveraging two different types of data, this research offers an opportunity to examine and understand the distinctions in how relevant cultural resources are categorized and classified from both the city and community perspectives.

The research is implemented through four main phases. In the first phase, we review the literature to identify relevant cultural planning paradigms and approaches. In the second phase, we synthesize prior work through developing a set of spatial justice dimensions such as economic impact, arts' supply, participation, education, social impact, and public support. The study develops a set of indicators for each of these dimensions, such as the number of available arts programs, and the number of resident artists, among others. The third phase uses the Los Angeles County open data portal and user-generated blogs, including the Los Angeles Mural Art Project, to visualize art as a network infrastructure using GIS and urban network analyses. These visualizations and empirical analysis provide insights regarding the allocation of resources for art in the city, identifying any existing policy gaps and areas with a 'locational disadvantage'. In the fourth phase, we use the insights from the GIS and network analyses to evaluate spatial justice in arts and cultural planning in Los Angeles, operationalizing the theoretical dimensions and indicators we developed.

We hypothesize that the community's role in creating and sustaining art would vary depending on the allocation of resources. That is, urban areas lacking formal arts and cultural resources would develop informal ones. We consider how these dimensions and indicators as well as the big data approaches could be used to inform spatially just arts and cultural planning.

2 REFERENCES

Giddens, A. (2004). The constitution of society: Outline of the theory of structuration: Elements of the theory of structuration. In Practicing History (pp. 121-142). Routledge.

Lefebvre, H. (1967). Le droit à la ville. L'Homme et la société, 6(1), 29-35.

EQUITY ISSUES: SLUM REDEVELOPMENT IN INDIA FOR CHILDREN'S MOBILITY AND PLAY

Track 5: Spatial Justice, Equity, and Inclusion by Design
Knowledge Network: Children, Youth & Environment, International Housing Research
Methods: Statistical/Modeling Analysis, Survey
Uchita Vaid (University of Wisconsin - Madison)

1 ABSTRACT

Urban informal housing communities in various cities in India are undergoing significant social and environmental transformations due to the implementation of slum redevelopment policies. A widespread approach to housing improvement in India includes in-situ redevelopment practices, which construct new apartment-style housing at original slum sites. While much scholarly attention has focused on potential socio-economic, and health impacts of such policies, the consequences for children's critical developmental aspects of play and mobility have been understudied. There is extensive research on informal settlements and their impact on children's physical health and development1, which are critical without doubt, but investigation of opportunities for children's mobility and play in these communities is scarce. Having opportunities for outdoor play (OP) and independent mobility (IM) play a critical role in children's development as they enable socializing with peers and neighbors, developing independence; self-identity, and taking an active part in community life2. Despite these documented benefits, the consequences of slum redevelopment policies for children's IM and OP have been understudied. Furthermore, there is even less attention on differential impacts of such policies for boys and girls, creating questions for equitable redevelopment policies.

In addressing this gap, the study investigated the level of OP and IM for children (6-12 years old) living in two informal settlements at different stages of redevelopment and further includes a sex sensitivity analysis of consequences. This is a cross-sectional comparison between two groups; one group of children living in urban slums, waiting to move into redeveloped housing, and the second group that has already undergone redevelopment. The study answers two research questions: Is the amount of children's OP and IM different in slum settlements and redeveloped settlements? Does housing redevelopment have a differential impact on OP and IM for girls in comparison to boys? Outdoor play3 and independent mobility4 were measured using validated scales filled out by parents. The results indicated that children in slum settlement were reported to have more IM licenses and at an earlier age, and a broader IM range than children in redeveloped settlement. Sex differences are also evident, girls in the redeveloped settlement received licenses at a later age in comparison to girls in the slum settlement. For OP, girls in redeveloped settlement engaged in less OP than those in slum settlement, and this was converse for boys. These findings underscore the inequities resulting from slum redevelopment, especially detrimental for girls' development.

2 REFERENCES

Unger, A. (2013). Children's health in slum settings. Archives of Disease in Childhood, 98(10), 799-805.

Fyhri, A., Hjorthol, R., Mackett, R. L., Fotel, T. N., & Kyttä, M. (2011). Children's active travel and independent mobility in four countries: Development, social contributing trends and measures. Transport policy, 18(5), 703-710.

Aarts, M. J., et. al. (2012). Outdoor play among children in relation to neighborhood characteristics: a cross-sectional neighborhood observation study. International Journal of Behavioral Nutrition and Physical Activity, 9(1), 1-11.

Larouche, R., et. al. (2017). Test-retest reliability and convergent validity of measures of children's travel behaviours and independent mobility. Journal of Transport & Health, 6, 105-118.

RELICS TO REALITIES: PRESERVING NATIVE AMERICAN HERITAGE USING 3D SCANNING

Track 5: Spatial Justice, Equity, and Inclusion by Design
Knowledge Network: Cyberspace & Digital Environments, Historic Preservation
Methods: Case Study, Mixed Methods

Tilanka Chandrasekera (Oklahoma State University); Kat Gardner-Vandy (Oklahoma State University); Nishan Wijetunge (Oklahoma State University); Kasun Gayantha Dodanduwa Waduge (Oklahoma State University)

1 ABSTRACT

The preservation of cultural heritage artifacts is a critical scholarly and societal endeavor. They serve as invaluable repositories of human history, social structures, and cultural identities (Smith, 2006). Far from being mere historical objects, these artifacts are archives of cultural narratives and collective memories, often forming the cornerstone of community identities (Brown, 2015). In this context, the role of museums and cultural institutions is interpretive and representational, extending beyond mere custodianship (Hooper-Greenhill, 1992).

Preserving Native American museums and artifacts is vital for safeguarding cultural identity, enriching educational programs, and facilitating scholarly research. The scattered locations of these cultural assets and the resource-intensive nature of physical preservation, vulnerable to natural and man-made disasters, underscore the need for digital methods like 3D scanning. These digital archives offer a resilient solution, democratizing global access and serving as a form of reparative justice.

Traditional preservation methods, while effective, have limitations, including exposure to environmental risks like light, temperature, and physical handling (Michalski, 1994). Every interaction, even with precautions, contributes to incremental degradation (Caple, 2000). Digital technologies, particularly 3D scanning, have transformed cultural heritage preservation, enabling non-intrusive, high-resolution documentation of artifacts (Bentkowska-Kafel, 2014; Proctor, 2011).

In this study, we launched a comprehensive project with well-defined short-term and long-term objectives focused on the digital preservation of Native American cultural heritage. In the preliminary phase, the project employed volumetric scanning techniques to create a high-fidelity digital representation of a historically significant Native American museum at Spiro Mounds, along with 3D scanning of the artifacts housed within. A supplementary aim of this initiative was to rigorously evaluate the impact of Culturally Relevant Pedagogy (CRP) on STEM education outcomes through the use of the digital repositories created in the first phase. In the project's subsequent phase, we incorporated the high-resolution 3D scans of artifacts and spatial data from the museum into an existing STEM educational module designed specifically for Native American middle school students. This module, overseen by the research team, leveraged immersive Virtual Reality experiences to enhance place-based education within after-school programs focused on STEM subjects for Native American communities. To assess the module's effectiveness, we employed a multi-faceted evaluation strategy. This included observational metrics to gauge student engagement in STEM activities related to the digital artifacts, as well as pre- and post-module assessments focused on career-life preparedness. Through this comprehensive approach, we aimed to quantify the transformative potential of culturally relevant educational resources in enhancing STEM efficacy among Native American youth.

FOSTERING SOCIAL CONNECTIVITY IN LONG-TERM CARE FACILITIES THROUGH ENVIRONMENTAL DESIGN

Track 5: Spatial Justice, Equity, and Inclusion by Design
Knowledge Network: Environment-Gerontology, Environmental & Architectural Phenomenology
Methods: Case Study
Niloofar Hedayati (Simon Fraser University); Habib Chaudhury (Simon Fraser University)

1 ABSTRACT
As the global demographic landscape undergoes a significant shift with an increasing aging population, this research explores the critical intersection of environmental design and its profound impact on social interaction within long-term care facilities Emphasizing the demand for supportive living environments, the study underscores the vital link between the physical environment and the well-being of older individuals.

2 OBJECTIVE
The research aims to bridge the gap in social integration within long-term care communities, acknowledging the pivotal role of social interaction in shaping the quality of life for older adults. Lawton and Nahemow's press-competence model is adopted, conceptualizing aging as an interplay between physical and social environmental constraints and individual coping abilities. The primary objective is to identify specific physical environment characteristics within long-term care facilities that encourage social interaction, thereby enhancing the well-being of older adults.

3 METHODS
Through a meticulous review of 25 studies, the methodology synthesizes major themes and subthemes, exploring the nuanced relationship between the physical environment and social dynamics within long-term care settings. Critical aspects such as spatial layout, accessibility, facility size and scale, visual connectivity, availability of social hubs, and wayfinding are analyzed. The comprehensive analysis aligns with the multifaceted nature of environmental design research, transcending disciplinary boundaries. Results: The exploration of environmental design features influencing social interaction draws insights from 25 studies, categorizing major themes and revealing complexities. Critical components such as spatial layout, visual connectivity, the availability of social hubs, and wayfinding emerge as influential factors in shaping social dynamics within long-term care facilities. Major issues identified include gaps in the literature, limited exploration of cultural influences, neglect of technological interventions, and the absence of long-term impact assessments.

4 DISCUSSION AND IMPLICATIONS
The study advocates for inclusive design practices that consider the diverse needs of older adults in care settings. Emphasizing participatory design and stakeholder integration, the research underscores the complex interplay between the physical environment and social interactions. These implications extend beyond immediate concerns, fostering spatial justice and equity through informed design practices that celebrate diversity and prioritize the well-being of marginalized communities.

5 CONCEPTUAL FRAMEWORK
The nuanced conceptual framework focuses on creating a "Supportive Place for Fostering Social Engagement." Comprising Spatial Layout, Quality of Environment, Residential Security and Individualized Comfort, and Natural Environment, the framework highlights the interconnectedness of these elements, emphasizing the need for a holistic and integrated approach to design.

6 CONCLUSION
The research not only contributes valuable insights into the impact of environmental design on social interaction among older adults but also advocates for transformative design practices that prioritize spatial justice, equity, and inclusion. Addressing gaps in the literature, the work aims to shape the future of long-term care facilities, promoting positive social interactions and enhancing overall well-being across diverse contexts.

ENVIRONMENTAL ACCOMMODATIONS TO SUPPORT AGING-IN-PLACE: INSIGHTS FROM A MULTI-SITED ETHNOGRAPHY OF THE ELDERLY IN INDIA

Track 5: Spatial Justice, Equity, and Inclusion by Design
Knowledge Network: Environment-Gerontology, Residential Environments
Methods: Ethnography, Observation

Gourab Kar (Indian Institute of Technology Delhi); Sanket Subhash Naik (Indian Institute of Technology Delhi); Sidharth Sonowal (Indian Institute of Technology Delhi); Aswin S (Indian Institute of Technology Delhi); Sabyasachi Paldas (Indian Institute of Technology Delhi)

1 ABSTRACT

India is aging rapidly. The population of people aged 60 years and above is estimated to rise from 149 million in 2022 to 347 million in 2050. While India is rapidly transitioning to an older society, there is a lack of research data on how the built-environment of home enables or hinders activities-of-daily-living for the elderly. Therefore, this qualitative research evaluates the degree of person-environment fit for the elderly to inform the design of environmental accommodations that support aging-in-place.

A multi-sited ethnography of elderly aging-in-place was conducted in three cities - Dibrugarh, Panaji, and New Delhi. Sixty-four elderly, aged 55 - 85 years, were observed performing activities-of-daily-living at home. Using environmental traces, participant observation and user interviews, researchers documented human-environment interactions of the elderly. Informed consent was obtained prior to the study; participants were not compensated for participation. Research data included sketches, floor plans, field notes, video recordings, photographs, and interview transcripts. Qualitative data analysis tools were used to translate research data into analytical memos that were coded and categorized.

Four key insights emerged from the qualitative data analysis:

1. *Postural transitions – such as getting up from bed, rising up from a chair/toilet seat – were difficult for the elderly because of changes in strength, balance, and coordination. Change in posture may be accompanied by feeling dizzy, leading to increased risks of falls. To aid the elderly, environmental accommodations such as guide rails on walls, grab-bars in toilets and bathrooms, chairs with armrests, and beds with supports to aid ingress and egress - are needed.*
2. *Functional reach for the elderly was significantly reduced with progression in age. Since the elderly found it difficult to bend down or reach out for things, work surfaces and storage zones at home should be designed to be in their zone-of-comfortable-reach.*
3. *Safe and independent mobility at home was hindered by stairs. The elderly were afraid of using stairs due to a fear of falls. Staircases rarely had handrails, tread and riser dimensions were not optimized for elderly use, illumination levels were inadequate and staircase configurations could not be retrofitted with a stair-lift. Environmental accommodations to aid staircase use by elderly are needed to enhance safe and independent mobility.*
4. *Culturally-rooted behaviours such as floor-seated activities and use of squat-toilets are difficult for the elderly due to their reduced functional capabilities. A few elderly expressed regret in their inability to sit on the floor, especially in the prayer room. Environmental accommodations are needed to enable the elderly to perform activities while floor-seated.*

This study used multi-sited ethnography to investigate the degree of person-environment fit for sixty-four elderly residing at home in three cities in India. Investigations revealed how the built-environment enabled or hindered the elderly in performing activities-of-daily-living. Findings suggest that environmental accommodations to support aging-in-place should respond to lack of person-environment fit experienced by elderly in postural transitions, functional reach, independent mobility and culturally-rooted practices. With India in early stages of a transition to an older society, there is a critical need for research to inform the design of inclusive home environments that support aging-in-place for the elderly.

HOMEMAKING AFTER HOMELESSNESS

Track 5: Spatial Justice, Equity, and Inclusion by Design
Knowledge Network: Interior Design, Residential Environments
Methods: Observation, Photo-based Methods
Christina Bollo (University of Oregon)

1 ABSTRACT

Permanent Supportive Housing provides dwellings and on-site assistance for people who have experienced long-term homelessness and its related trauma. A safe and stable home enables people to transcend survival mode and to work through past trauma in a supportive environment (Henwood, et al, 2014). Recent scholarship shows that living in full apartments correlated with the trauma-informed principles of choice, control, stability, and security (Watson et al., 2019). Gary Evans and colleagues found, "floorplan configurations that afford greater opportunity for being alone also attenuate the negative impacts of residential density on psychological wellbeing" (Evans et al., 2002, 222).

Maximizing choice is a key principle of trauma-informed design. The research question for this study is whether the residents believe they have ample choices in their domestic sphere, particularly in how they arrange their belongings in their small apartments. The data collection for this study employed a videovoice approach as documented by Catalani and colleagues in their research with Hurricane Katrina survivors (2012). Four floorplan configurations were represented across two buildings. Each of the 17 residents wore a chest-mounted camera and narrated a tour of their apartment. The imagery and narration of each video were analysed using Thematic Analysis (Braun & Clarke, 2006) first with inductive coding and then with a deductive codebook that included theoretical concepts related to trauma-informed design such as choice, empowerment, and safety (Hopper, K.E et al, 2010). The results from this study show that despite the small size of apartments, residents create a sense of home through the acquisition and placement of belongings and that this ability increases with time of tenancy. Residents exhibited a strong sense of control over their environment, especially entering the dwelling, and in making choices over which outsiders are invited in. Residents valued a continuum of choices of furniture placement and demonstrated an interest in flexible and ample storage.

While PSH has received attention from scholars in nursing and social work, this is the first study to provide evidence to support design decisions related to the apartment layout. Designing apartments that work upon move-in and for many years beyond, can increase the potential for trauma recovery and ensure that the dwellings can indeed be permanent homes.

2 REFERENCES

Braun, V., & Clarke, V. (2006). Using thematic analysis in psychology. Qualitative research in psychology, 3(2), 77.

Catalani, C. E., et al (2012). Videovoice: community assessment in post-Katrina New Orleans. Health Promotion Practice, 13(1), 18-28.

Evans, G. W., et al (2002). Crowding and children's mental health: the role of house type. Journal of environmental psychology, 22(3), 221-231.

Henwood, B. F., et al (2014). Quality of life after housing first for adults with serious mental illness who have experienced chronic homelessness. Psychiatry research, 220(1-2), 549-555.

Hopper, K.E., et al (2010). Shelter from the storm: Trauma-informed care in homelessness services settings. The Open Health Services and Policy Journal, 3(1).

Watson, J., et al (2019). A home, but how to connect with others? Health & social care in the community, 27(3), 546-564.

DIGITAL ENVIRONMENTS: HOW REAL IS ENOUGH? AN ANALYTICAL REVIEW ON VR REALISM IN BUILT ENVIRONMENT RESEARCH

Track 6: Transdisciplinary Design and Collaboration/Open Track
Knowledge Network: Cyberspace & Digital Environments, Interior Design
Methods: Grounded Theory
Khanh Hoa Thi Vo (Indiana University Bloomington)

1 ABSTRACT

The realism of Virtual Reality (VR) determines its effectiveness in researching the built environment (Tseng & Do, 2022). But how real is enough? While high realism in VR enhances presence and induces positive emotions, especially in natural sceneries, Newman et al. (2022) observed no significant differences between high and low realism in VR for built environments. So how can we strike a balance in VR realism? A balance that avoids the excessive costs associated with advanced and complex hardware and software systems while upholding the depth and quality of environmental research.

This presentation features an analytical review of VR realism in built environment research, focusing on three platforms: Spatial.io, Twinmotion, and NVIDIA Omniverse. Framed upon the Model of Realism Score for Immersive VR Systems (Perroud et al., 2019), the review explores different realism levels as research modalities with implications for both cost and effectiveness. Further, the platform choices strategically respond to Newman et al.'s note (2022), attributing the minimal differences between high versus low realism in VR for built environments to the lack of restorative qualities, such as the dynamic visuals and sounds associated with a crowd. Hence, this review examines Spatial.io with multi-user interactions, Twinmotion with soundscape elements, and NVIDIA Omniverse with physical simulations. However, each platform either compromises visual (e.g., graphical quality) or immersion cues (e.g., interaction), two measures of the Realism Score Model (Perroud et al., 2019). The intention is to compare the trade-offs between realism and interactivity on presence and emotion (e.g., means of self-reported measures) from the existing literature.

Results from Google Scholar with the search term "VR realism + [name of the platforms] + presence + affects (emotions)" included 172, 131, and 229 publications since 2019 for Spatial.io, Twinmotion, and NVIDIA Omniverse respectively. A snapshot of these results indicated that emotional differences exist across VR realism levels. For instance, Sptail.io is high in immersion yet low in usability, and less rewarding. Twinmotion achieves eminent realism and induced emotions with the use of animated digital humans but requires adequate hardware. NVIDIA Omniverse excels in realism, immersion, and induced emotions with personalized avatars, easy locomotion, and compatibility across Building Information Modeling (BIM) applications at the cost of expensive hardware.

With added restorative qualities, VR realism of built environments leads to differences in induced emotions. Higher scores on the Realism Score Model (Perroud et al., 2019) mean greater immersion and induced emotions, but at higher costs. Striking a balance between specific constructs of interest and VR realism is a case-by-case decision.

2 REFERENCES

Newman, M., Gatersleben, B., Wyles, K., & Ratcliffe, E. (2022). The use of virtual reality in environment experiences and the importance of realism. Journal of Environmental Psychology, 79, 101733.

Perroud, B., Régnier, S., Kemeny, A., & Mérienne, F. (2019). Model of realism score for immersive VR systems. Transportation research part F: traffic psychology and behaviour, 61, 238-251.

Tseng, K. C., & Do, N. G. T. (2022). A feasibility study of using virtual reality as a pre-occupancy evaluation tool for the elderly. Automation in Construction, 134, 104037-104037.

ARTIFICIAL INTELLIGENCE IN DESIGN EDUCATION

Track 6: Transdisciplinary Design and Collaboration/Open Track
Knowledge Network: Cyberspace & Digital Environments, Interior Design
Methods: Design Methods, Survey

Tilanka Chandrasekera (Oklahoma State University); Zahrasadat Hosseini (Oklahoma State University);
Anna Bazhaw-Hyscher (Oklahoma State University); Wellabadage Ubhaya Upendra Perera (Oklahoma State University)

1 ABSTRACT

The integration of Artificial Intelligence (AI) is transforming design education, automating processes, and reshaping the educational landscape. Understanding emerging technology in its formative stages is important to the comprehension of foundational concepts. In addition, early exposure to AI allows students a competitive edge in critical thinking and problem-solving skills.

Students can use AI as inspiration, in space planning, and as a means of improving workflow. Due to the consequent reduction in cognitive load, mental processes may be redirected from minor tasks to innovation. The VARK model, which categorizes learning styles into Visual, Auditory, Reading/Writing, and Kinesthetic, can significantly influence how interior design students engage with generative AI tools. Visual learners, who prefer learning through images and spatial understanding, might find generative AI tools particularly beneficial, as these tools often provide visual representations like 3D models and floor plans, enhancing their comprehension and design process. Auditory learners, on the other hand, might prefer AI tools that offer verbal explanations or interactive discussions about design concepts. Reading/writing learners could benefit from AI that provides extensive textual information, such as design specifications or historical context, while kinesthetic learners, who learn best through hands-on experience, might prefer AI tools that allow for interactive manipulation of design elements in a virtual space. Understanding these learning preferences is crucial for educators in tailoring AI tools to suit different learning styles, thereby maximizing the educational benefits for interior design students. This approach can lead to more effective learning outcomes and a deeper engagement with the subject matter.

To understand how generative AI tools can be an asset in education, a study was developed to examine the potential benefits of incorporating AI into an interior design studio course. 33 sophomore interior design students participated in this study in two phases: first, they were invited to experiment with several generative AI planning tools intended for residential interior design. The participants used one of these tools to help automate a basic floor plan for a home. Using these as a guide, they were able to use another generative AI tool to create inspirational 3D images to variously depict the design style of the home. With the AI image generator, the students developed several conceptual images.

In the second phase of the study, a survey was conducted among the students to gather information on both their style of learning (visual, aural, reading/writing, kinesthetic) and level of technology acceptance. The results suggested that the visual learners easily adopt technology that incorporates visual elements. Understanding the correlation of learning styles and the acceptance of technology helps in informing the implementation of AI in a classroom setting.

As artificial intelligence continues to evolve, the acceptance of the technology will transform the educational landscape. Students that prefer learning styles compatible with the use of AI in design education will be able to produce creative solutions to design problems, as well as have the opportunity to contribute to the responsible development of AI.

MULTI-SITE ETHNOGRAPHY AND DIGITAL MEDIA IN ENVIRONMENTAL PLANNING: INSIGHTS FROM THE SUSTAINABLE EXUMA PROJECT

Track 6: Transdisciplinary Design and Collaboration/Open Track
Knowledge Network: Cyberspace & Digital Environments, International Connections
Methods: Design Methods, Ethnography
Marcus Owens (Washington State University)

1 ABSTRACT

The Sustainable Exuma project, an innovative initiative aimed at sustainable development in the Exuma archipelago, Bahamas, serves as a case study for exploring the integration of multi-site ethnography and digital media in environmental planning. This paper delves into the project's methodology, characterized by its multidisciplinary and participatory approach, employing both traditional and innovative ethnographic techniques to deeply understand the local context. Central to this exploration is the role of digital media in 2013, a period marked by significant transformations in the social media landscape across the globe, albeit with regionally specific impacts. The project leveraged various digital platforms for communication, data collection, and community engagement, bringing new perspectives into the planning process.

The empirical portion of the paper situates the development of Exumatopics, a digital platform created during the project, within the context of the emerging Web 2.0 social media landscape. It compares ethnographic observations of media technologies in the field, such as marine radio, WhatsApp, satellite internet, and Facebook, highlighting the nuances of digital communication in environmental planning.

The conclusion situates the analysis of this case within subsequent developments in participatory digital media as a site of ethnographic inquiry, offering insights into the evolving landscape of digital ethnography and its implications for environmental planning.

This paper contributes to a broader understanding of applied multi-site ethnography in environmental planning, demonstrating how digital media can enhance the capacity of planners and researchers to engage with complex environmental challenges in a rapidly changing world. The insights gained from the Sustainable Exuma project provide valuable lessons for future environmental planning efforts, particularly in settings characterized by ecological diversity and dispersed communities.

THE INFLUENCE OF BRIGHTNESS ON PERCEPTIONS AND BEHAVIORAL INTENT IN A CAFÉ ENVIRONMENT

Track 6: Transdisciplinary Design and Collaboration/Open Track
Knowledge Network: Cyberspace & Digital Environments, Sustainable Planning, Design & Behavior
Methods: Experimental/Quasi-Experimental Research, Statistical/Modeling Analysis
Kyuho Ahn (University of Oregon); Siobhan Rockcastle (University of Oregon)

1 ABSTRACT

TLighting design characteristics such as brightness, color, and distribution can impact building performance and occupant experience. Light intensity can influence the duration of time consumers spend at merchandise displays (Summers & Herbert, 2001), and ambient lighting conditions can contribute to the perceived brand/merchandise quality (Sharma & Stafford, 2000). This project presents findings from an online survey that compares the influence of brightness, correlated color temperature, and luminaire layout on behavioral responses in an immersive café setting. Immersive virtual representations have been found to provide a reasonable surrogate to physical spaces for research on lighting perception (Heydarian et al., 2015; Chamilothori et al., 2018). Our survey used navigable panoramic renderings to provide an immersive viewing experience in a web-based format (Rockcastle et al., 2023). This paper will disseminate our findings on the effect of brightness on perceived arousal, comfort, pleasure, and shopping intent (desire-to-stay-longer, return, purchase, talk-to-others, willingness-to-tell-friends, perceived friendly-staff, aesthetic-quality, and professional-staff).

An IRB approved Qualtrics online survey was used to collect subjective responses to a series of five-point Likert questions. The lighting conditions included brightness (dim, medium, and bright), color temperature (cool vs. warm), and luminaire layout (orthogonal vs. non-orthogonal). Participants were recruited within the university community and compensated with a $5 gift card. A total of 210 samples from 256 responses were used for data analysis. Freeman's tests were used to compare perceived arousal, comfort, pleasure, and behavioral intents under different brightness conditions: dim, medium, and bright.

The results found that when light intensity increases, perceived arousal (measured by wide-awake and excitement) also increases. This suggests that brightness can be modulated to shape customer behavior. The medium brightness condition produced the highest ratings of comfort and pleasure, followed by the bright and dim conditions. For shopping intent, the dim condition resulted in less favorable behavioral intents than the medium/bright conditions with no significant differences between medium and bright conditions on all behavioral intents ($p>.05$) It was notable that Cronbach's Alpha (reliability test) of shopping intents was increased to .886 when "aesthetic quality" was removed from the clustered list of shopping intents. When "aesthetic quality" is added to the cluster of ratings related to pleasure (like, satisfied, feel good, and aesthetics), the Cronbach's Alpha of pleasure showed a slight increase.

In conclusion, an intermediate brightness level can increase positive impressions in a café environment than dimmer or comparably brighter lighting levels. While increasing light intensity can increase ratings of being awake, it does not guarantee positive consumer responses. More research is needed to determine the effects of arousal in relation to pleasure and shopping intent. It may suggest that a mid-level of arousal would optimize positive behaviors over low or highly arousing environments. It is notable that both the medium and bright lighting conditions elicited positive shopping behavior over the dimly lit environment, even though all three conditions were deemed acceptable by IES standards for dining spaces (50 lux, 150 lux, 300 lux).

COMMUNITY-ENGAGED DESIGN: TIPPING THE NARRATIVE IN RURAL WEST VIRGINIA COMMUNITIES WITH THE FULCRUM PROJECT.

Track 6: Transdisciplinary Design and Collaboration/Open Track
Knowledge Network: Participation
Methods: Action/Participatory Research, Survey

Udday Shankur Datta (West Virginia University); Peter Butler (West Virginia University)

1 ABSTRACT

The Fulcrum Project is a collaboration between the Davis College of Agriculture, Natural Resources and Design, and the West Virginia Community Development Hub to establish connections with current and former Hub Communities. The Fulcrum Project is supported through funding from the Claude Worthington Benedum Foundation. The Fulcrum Project seeks to connect communities to the academy's resources through partnership development. It connects students, communities, and expert faculty to solve the challenges of rural communities, building capacities towards positive change. There are many challenges facing rural West Virginia communities: environmental, social, and economic. The Fulcrum project's asset-based planning approach focuses on leveraging existing strengths, community connections, and identified community needs to pair groups of interdisciplinary students with motivated and innovative faculty to envision positive community change and empower local leaders toward progress. This work builds on the need to work with disenfranchised and underserved populations (Hester, 2006; Thering, 2007; Sanoff, 2010) through service-learning projects (Butler, 2019; Angotti, Doble & Horrigan, 2012; Bose, Horrigan, Doble, & Shipp, 2014) designed to increase citizen control, delegate power, and form partnerships (Arnstein's "Ladder of Citizen Participation," 1969). Through partnerships for capacity-building, eligible communities worked with the West Virginia Community Development Hub and its community coaching programs, such as Cultivate, HubCap, and Blueprint. Through capacity-building initiatives, the Fulcrum Project integrated design and planning with well-established key stakeholder groups in communities. Since its inception, The Fulcrum Project has benefitted 30 community projects in rural West Virginia. Ninety-five students from various backgrounds have been involved with the Fulcrum Project.

Community design projects completed through the Fulcrum Project primarily included environmental design within marginal landscapes: floodplains, brownfields, empty downtown lots, and buildings. Communities partnered with faculty and students from landscape architecture, interior architecture, environmental and community planning, and other disciplines to create solutions to enhance community health, equity, sustainability, and resilience. The designs have led to investment in communities that turn liabilities into assets. Seed grants allow communities to move forward with professional services toward implementation. Evaluation of the Fulcrum Project includes student learning outcomes tied to citizen leadership and transformative learning (Mezirow, 2011). Community capital (Flora and Flora, 2018) outcomes are measured through pre- and post-engagement surveys and interviews to measure the long-term viability and sustainability of implemented projects: Are local leaders empowered? How do communities steward projects toward implementation? Findings point to the project's success in achieving outcomes in student growth and community capacity development. A well-established Institutional Research Board protocol made it possible to gather long-term data, improve the community engagement process to have increased impacts, and create scholarship on community engagement.

2 ACKNOWLEDGEMENTS

Project funded by the Claude Worthington Benedum Foundation With support from: USDA NIFA Hatch Project NE 1962: Outdoor Recreation, Parks and Other Green Environments: Understanding Human and Community Benefits and Mechanisms & USDA NIFA Hatch Project WVA00709: Trans-Disciplinary Approach to Community Planning and Development for Heritage and Recreation Tourism

BRIDGING DESIGN THINKING AND DESIGN ANTHROPOLOGY

Track 6: Transdisciplinary Design and Collaboration/Open Track
Knowledge Network: Participation, Researchers in Professional Practice
Methods: Design Methods, Ethnography

Amin Mojtahedi (HGA Architects & Engineers); Jacob Culbertson (HGA); Bryanna Benicia (HGA); Adaheid Mestad (HGA Architects & Engineers)

1 ABSTRACT

Bridging Design Thinking and Design Anthropology

This proposal sets out to explore the intersection of Design Thinking and Design Anthropology, underlining their combined potential to foster fresh insights and creative solutions in the context of complex architectural projects. Design Thinking, known for its emphasis on human-centered discovery, ideation, and iteration processes, has been used in addressing and resolving complex design challenges (Plattner, Meinel, & Leifer, 2013). Simultaneously, Design Anthropology is defined by its methodological commitment to "knowing from the inside" (Ingold 2013)—that is, to understand spaces and systems through participating in the lives of users and stakeholders rather than relying on how they report their experiences and preferences—and to situating design projects (and design as a discipline, cf. Suchman 2011) in the cultural, historical, and ecological contexts in which they take place (Gregory 2018).

Despite their individual strengths, integrating these disciplines presents a unique opportunity to transcend traditional limitations. This paper advocates for a transdisciplinary strategy that blends the in-depth cultural insights of design anthropologists with problem-solving capabilities of design thinkers. Such a collaborative approach encourages a process of "reflecting in action," where practitioners revisit and refine their theories and methodologies. This not only enhances the relevance of their work but also uncovers hidden aspects and paves the way for groundbreaking solutions (Yaneva, 2010).

The presentation will detail case studies showcasing the collaboration between the two disciplines in architectural projects across different sectors of healthcare, workplace, higher education, and community. These examples will highlight reframing of design topics such as community health, students' sense of belonging, workplace social dynamics, employees' sense of anxiety, museum visitors' confidence, and spatial distribution of emotions at the scale of campuses. Through these case studies, the effort aims to highlight the distinct advantages and challenges of this synergistic collaboration, demonstrating its capacity to generate innovative and effective design solutions.

2 REFERENCES

Gregory, S. 2018. Design Anthropology as Social Design Process. Journal of Business Anthropology, 7(2): 210-234.

Ingold, Tim. 2013. Making: Anthropology, Archaeology, Art, and Architecture. New York: Routledge.

Leifer, Larry, Hasso Plattner, and Christoph Meinel (eds). 2013. Design Thinking Research: Building Innovation Eco-Systems. New York: Springer.

Suchman, Lucy. 2011. Anthropological Relocations and the Limits of Design. Annual Review of Anthropology, 40: 1-18.

Yaneva, A. 2011. From Reflecting-in-Action Towards Mapping of the Real. In I. Doucet and N. Janssens (eds). Transdisciplinary Knowledge Production in Architecture and Urbanism: Towards Hybrid Modes of Inquiry (pp. 117-128). New York: Springer Publishing

ASSESSING WALKABILITY AT TEXAS TECH UNIVERSITY: INTEGRATING PARTICIPATORY MAPPING AND CAMPUS COMMUNITY PERCEPTIONS OF THE ENVIRONMENT

Track 6: Transdisciplinary Design and Collaboration/Open Track
Knowledge Network: Participation and Sustainable Planning, Design & Behavior
Methods: Action/Participatory Research, Mixed Methods
Gisou Salkhi Khasraghi, Ph.D Candidate Texas Tech University

Abstract Text:
Enhancing Campus Well-Being: A Comprehensive Evaluation of Walkability at Texas Tech University through Community Surveys and Student Participatory Mapping

1 ABSTRACT
University campuses play a crucial role in shaping the lifestyles of entering students, impacting their adaptation to healthy living. The built environment on campuses has the potential to encourage walking, recognized as a health-promoting form of transportation. Despite numerous health initiatives, walking consistently emerges as a key factor in fostering physical activity among students who face frequent stressors during their education and may find it challenging to allocate dedicated time for daily exercise.

This research presents a comprehensive approach to assess walkability at Texas Tech University. A campus-wide survey is administered to the diverse TTU community, including students, professors, and staff, aiming to capture a broad spectrum of perspectives and experiences related to walking on campus. Concurrently, geography students participate in focus groups, engaging in student participatory mapping activities. Through detailed walking diaries and trajectories, participants mark locations on campus, expressing preferences for walking and identifying areas perceived as less conducive to pedestrian activities.

The interdisciplinary methodology integrates quantitative data from the community survey with qualitative insights obtained through student participatory mapping. This holistic approach aims to provide a nuanced understanding of walkability, considering both spatial aspects and individual perspectives in navigating the campus environment. Recognizing the significance of the campus environment in promoting healthy living, this research addresses the need to enhance walkability for the well-being of the TTU community. The findings contribute valuable insights to inform campus planning strategies, promoting physical activity and creating an environment conducive to healthy lifestyles.

2 KEYWORDS
Campus Walkability, Community Survey, Participatory Mapping and Perception

3 LEARNING OBJECTIVES

- *Grasp the complexities involved in assessing walkability through a comprehensive approach that integrates community surveys and student participatory mapping.*
- *Gain insights into the impact of walkability on the well-being of the Texas Tech University community, recognizing the role of the campus environment in promoting healthy lifestyles.*
- *Appreciate the integration of quantitative and qualitative methodologies, understanding how combining community surveys and student participatory mapping provides a nuanced understanding of walkability.*
- *Acquire practical knowledge to apply the research findings in informing campus planning strategies aimed at enhancing walkability, promoting physical activity, and creating an environment conducive to healthy living for the university community.*

NEW RESEARCH METHODS FOR FACILITIES PROGRAMMING: AGILE ETHNOGRAPHY

Track 6: Transdisciplinary Design and Collaboration/Open Track
Knowledge Network: POE/Programming
Methods: Ethnography
Lubomir Popov (Bowling Green State University); Franklin Goza (University of Wisconsin Whitewater)

1 ABSTRACT

The "ethnographic turn" is all over the corporate world and professional practices. Although it is spurred by the blue-chip companies and management consultancies, it is important to consider it for the needs of environmental design and in particular, for architectural/facilities programming. Conventional ethnography is an extremely time-consuming methodology that only academic researchers can afford. There are also several other burdens and restrictions that constrain its use in practice, including programming. Researchers in the areas mentioned above are working on new methods that can provide reasonably good information in a short period of time and within a small budget. The big question is about the quality of information and the epistemological compromises that can be made without giving up research integrity. Our goals are to inform environment and behavior researchers about this new method, as well as to initiate a discussion and a collaborative project on the epistemological considerations regarding the use of this method in programming. Our methodological plan for this project starts with an analysis of the nature of programming practices, their requirements, and constraints. In the next step, we compare programming process conditions and the epistemological requirements of conventional ethnography in order to find out the possible epistemological compromises programmers can make without sacrificing the integrity of programming field research. Our findings show that the agility will come at an epistemological cost, but even so, it is possible to provide a reasonably good trustworthy information. In our presentation, we will inform about several techniques that can be used to control the research process for the trustworthiness of data and findings. At the same time, we believe that agile ethnography methodology needs further development and adaptation to the typical facilities programming situations, considering time, budget, politics, and many additional constraints of programming projects.

FACILITIES PLANNING AS A WAY TO BRIDGE THE RESEARCH—(DESIGN) PRACTICE GAP

Track 6: Transdisciplinary Design and Collaboration/Open Track
Knowledge Network: None/Not Tied to Any Network
Methods: Meta-Analysis
Lubomir Popov (Bowling Green State University); Franklin Goza (University of Wisconsin Whitewater)

1 ABSTRACT

Designers often complain that Environment and Behavior Studies (EBS) knowledge is abstract and of little help. Researchers have to comply with academic norms and standards, as well as the requirements for tenure and promotion. Bridging the research—(design) practice gap has been a major problem in the first decades of Environment and Behavior Studies (EBS) and continues even now. There have been multiple ideas ranging from research utilization, application, and usability up to evidence-based design and translational research, but the pursuit is the same: How to make EBS knowledge usable and practical? Our goal is to search for feasible ways to bridge the gap. We have built a methodological platform and an analytical framework based on philosophy of science and design theory. We have analyzed the specifics of knowledge production and the nature of knowledge used in design. From the design "end," we have explored the project delivery process with the intent to identify the most information-intensive phases. Our findings show that the "core" design phase engages mostly with organizing materiality in order to fulfil programmatic requirements. Designers need project-specific information for decision making. In this line of thinking, we have found that the predesign phase is the realm where most research activities take place and most of the information is produced. We focus our attention on this phase and delimit our current presentation to the programming (sub)phase. Programming involves project-specific field research, use of EBS theories and knowledge, knowledge from other fields, building type knowledge, information from building evaluation studies, and so forth. Programmers reorganize, transform, and synthesize different types of knowledge into what designers understand best—user requirements and guidelines. Although this process might have shortcomings and limitations, it is the best possible option for bridging the research—practice gap.

ENHANCING EXISTING HEALTH AND SAFETY PROCESSES FOR MULTICULTURAL WORKERS IN CONSTRUCTION PROJECTS

Track 6: Transdisciplinary Design and Collaboration/Open Track
Knowledge Network: Work Environments, Cyberspace & Digital Environments
Methods: Mixed Methods, Survey

Aditya Jayadas (Oklahoma State University); Wellabadage Ubhaya Upendra Perera (Oklahoma State University); Tilanka Chandrasekera (Oklahoma State University); Nishan Wijetunge (Oklahoma State University); Kasun Gayantha Dodanduwa Waduge (Oklahoma State University)

1 ABSTRACT

In the contemporary global landscape, the unprecedented mobility of workers across borders and continents has become a prevalent phenomenon, driven by diverse motives such as employment prospects, educational pursuits, and familial reunification. Amid this dynamic environment, the construction industry stands out as one of the most perilous occupations globally, marked by elevated rates of fatalities, injuries, and health risks (Occupational Safety and Health Administration, 2018). Contributing to this peril are various inherent factors intrinsic to construction work. Notably, recent years have witnessed a notable surge in casualties related to maintenance work, as reported by the US Department of Labor (US Bureau of Labor Statistics, 2023), with over 811 fatal work injuries recorded in 2022 for building and grounds cleaning and maintenance occupations.

Mitigating the high incidence of injuries and fatalities within the construction industry necessitates a focus on effective training programs. Such programs are designed to impart knowledge about potential hazards, safe work practices, and proper equipment usage. Occupational Safety and Health Administration (OSHA) emphasizes that safety training equips workers with the skills to identify and avert hazards, thereby reducing workplace accidents (OSHA, n.d.). Furthermore, Dong et al. (2004) substantiated the impact of safety training, highlighting its significant role in decreasing the risk of injury, particularly among younger or less experienced workers.

However, the unique challenges faced by the United States' multilingual construction workforce, coupled with the industry's high fatality and injury rates, necessitate innovative approaches to safety training (Wang et al., 2018). The current modes of safety training, including text-based manuals, pictorial guides, lectures, and videos, fell short of addressing the linguistic and cultural diversity within the workforce. Therefore, there was a critical need for culturally and linguistically tailored safety training programs that emphasize participation, interactivity, cultural relevance, technological proficiency, and practical significance.

This study searches into the essential factors influencing health and safety issues within the construction industry and proposes more comprehensive safety training processes for a multicultural workforce. Through a survey questionnaire distributed to construction companies engaged in public and private sector projects with diverse workforces, along with input from safety managers actively monitoring the industry, the research aims to identify vital factors affecting safety issues, particularly in the learning process. The study's data could prove instrumental for safety agencies in developing innovative, customized safety training methods, ultimately minimizing health and safety hazards for the multicultural construction workforce. Recognizing the significance of ensuring the safety of these workers, given the heightened risk of workplace hazards and their crucial role in contributing to the nation's economy, this ongoing study seeks to provide valuable insights. The preliminary results are anticipated to shed light on the current state of health and safety training within the construction industry, paving the way for future advancements in safeguarding the well-being of the diverse and dynamic workforce.

2 REFERENCES

Occupational Safety and Health Administration (OSHA), 2018. Construction Industry.
US Bureau of Labor Statistics (BLS), 2023. Census of Fatal Occupational Injuries Summary-2022.

Occupational Safety and Health Administration (OSHA). (n.d.). Training Requirements in OSHA Standards.

AN ARCHITECTURE + HORTICULTURE COLLABORATIVE DESIGN STUDIO TO PROMOTE BUILDING-INTEGRATED AGRICULTURE

Track 6: Transdisciplinary Design and Collaboration/Open Track
Knowledge Network: Health, Movement in Designed Environments: Nature & Ecology
Methods: Design Methods

Gwynne Mhuireach (University of Oregon); Gail Langellotto (Horticulture Department, Oregon State University); Brad Gaolach (Washington State University)

1 ABSTRACT

With an uncertain climate future ahead, urban agriculture is increasingly becoming critical to ensure nutritional security, community resilience, and equitable access to healthy foods for growing urban populations. However, the lack of open space and surging land prices present a constraint to the establishment of new urban farms and gardens.

A possible solution and a way to establish new agricultural sites is to produce food on and in buildings using methods like rooftop farming, green facades, and controlled environment agriculture. Further co-benefits can be gained by creating a closed-loop system where building outputs are recycled as inputs to the agricultural system, and agricultural outputs ultimately provide inputs to building systems and users. Despite the many potential benefits of such building-integrated agriculture (BIA), examples of commercial BIA food production remain scarce, due to a lack of qualified experts and a research base since BIA requires multi-disciplinary expertise, as well as other obstacles.

University of Oregon, in partnership with OSU and WSU, offered a multi-disciplinary design studio to engage architecture, landscape architecture, and horticulture students in identifying and proposing solutions to critical challenges, barriers, benefits, and opportunities. The project was situated as an adaptive re-use of the barge fabrication building at Zidell Yards in Portland, Oregon. The site has a century-long history as an industrial ship-breaking facility, thus this last remaining large parcel fronting the Willamette River suffers from soil contamination, which makes it unsuitable for traditional ground-based urban agriculture.

We brought in Extension professionals, practicing architects, horticultural experts, and marketing/production specialists to provide guest lectures and critiques of students' work. Ultimately, the student teams generated BIA designs that challenged our perceptions of what was possible to accomplish and offered key social goods and ecosystem services, while also optimizing building (and food system) carbon footprints.

A STUDY OF SPATIAL COGNITION IN LIBRARIES FOCUSES ON USERS' WALKING SPEED ADJUSTMENT AND VISUAL PERCEPTION DURING BOOK-FINDING

Track 6: Transdisciplinary Design and Collaboration/Open Track
Knowledge Network: Environmental & Architectural Phenomenology
Methods: Experimental/Quasi-Experimental Research
Natsuho Hanawa (Tohoku university); Yasuaki Onoda (Tohoku University); Momoka Hirai (Tohoku University)

1 ABSTRACT

The role of libraries is changing in an era dominated by e-books and the Internet. To address this shift, libraries are varying their architectures with diverse shelf layouts and spatial configurations, trying to provide opportunities for "unexpected" encounters with information. This research aims to clarify how users receive information while adjusting their body movements during book-finding in different libraries by analyzing focal and ambient vision, coupled with walking speed.

Spatial cognition has been studied by examining the visual perception and kinetics. Vision is categorized into ambient and focal vision, which interact to gather necessary information, and a description for measuring ambient visual information is studied. (Ohno, 1991) Gibson (1979) mentions that obtaining space information means extracting the invariant from the ambient optic array, emphasizing the crucial role of movement. Besides, people change their walking speed based on objectives. (Rajat, 2011)

In this study, a behavioral experiment utilizing an eye-tracking recorder was conducted in each library with two to five participants, given a book genre before walking. Four libraries —three in Japan and one in Denmark— were chosen for their unique characteristics: conventional plan as a benchmark, commercial operation, and architects' design. For analysis, the average speed per three seconds was classified into three levels and gazing distance into five. Shelf signages and book covers were tracked for ambient vision analysis.

Results reveal a pattern in the book-finding process:
. Look around the entire space at the entrance to determine the walking direction. More than 45% of gaze was at long distances of 3.6 m or more and users walked slowly.
. Refer to floor maps to locate the target shelf.
. Walk fast toward the approximate shelf while gazing at spatial information. Most gaze was at a distance of 1.0m to 3.6m.
. Searching for the target shelf by reading shelf signages and book titles at a slower pace.
. Inspect each book on the target shelf while walking at a slower speed or stopping.

Comparing libraries indicates variations in these steps based on spatial configuration, book layout, and bookcase composition. Also, the four libraries are categorized into two types based on ambient vision: those with few shelf signages visible at once and those where multiple front covers are in view. Users adapt their search methods accordingly, relying on shelf signages and spine titles in the former type, or cover images and titles in the latter.

In conclusion, it was found that users adjust their search strategies based on their objectives, spatial information, and visual perception, developing their library search activities by modulating walking speed in response to changing gazing targets.

2 REFERENCES

Ohno, R. (1991). Ambient Vision of the Environmental Perception: Describing Ambient Visual Information. EDRAZZ: Proceedings of the 22nd Annual Conference of the Environmental Design Research Association. 237-246

Gibson, J. J. (1979). The Ecological Approach to Visual Perception. Houghton Mifflin.

Rastogi, R., & Chandra, S. (2011). Design Implications of Walking Speed for Pedestrian Facilities. Journal of Transportation Engineering, 137(10), 687-696.

COMPREHENSIVE EVALUATION OF PEDESTRIAN FLOW ORGANIZATION OF PEDESTRIAN BRIDGES IN THE INFLUENCE AREA OF BEIJING SOUTH RAILWAY STATION

Track 6: Transdisciplinary Design and Collaboration/Open Track
Knowledge Network: Environmental & Architectural Phenomenology
Methods: Mixed Methods, Statistical/Modeling Analysis
Zuyue Liao; Zhongzhong Zeng (Beijing Jiaotong University); Bo Zhang (Oklahoma State University)

1 ABSTRACT

Driven by the TOD development model, the slow travel system combined with rail transit has become one of the main travel models for citizens. It is not only an independent travel mode in short solo travel, but also an important part of solving the "last mile" of station travel. Pedestrian bridges are an integral part of the slow traffic system. Therefore, the reasonable setting of pedestrian bridges is a decisive factor that affects the applicability of slow-moving systems around rail stations. The organization of pedestrian flow in pedestrian bridges improves the efficiency of the use of pedestrian bridges. On the one hand, it can avoid the chaotic organization of pedestrian flow lines and non-haul vehicle flow lines. On the other hand, it can also promote the overall interests of surrounding areas.

This article takes Beijing South Railway Station as an example, and selects 16 pedestrian overpasses within one kilometer of influence with Beijing South Railway Station as the center as samples to conduct research on the optimization design of slow-travel space comfort at high-speed rail stations. During the investigation, there were many imperfections in the pedestrian facilities around Beijing South Railway Station. Among them, three overpasses have been suspended, respectively at the intersection of Beijing South Station Road and Majiabao East Road, and west of Dahuangzhuang Bridge on the Chaoyang Road station, the subway biomedical base station. In order to improve the comfort of the target group, the pedestrian bridge in the slow traffic space of Beijing South Railway Station will be scored and evaluated. This article will use an evaluation method that combines software simulation and questionnaire surveys. The scores will be divided into comfort, safety, and detours. Based on the three major themes of traffic flow, some quantitative factors are intuitively scored through software simulation and some qualitative factors are scored through questionnaires and expert evaluation standards. Finally, the pedestrian overpass pedestrian flow organization is evaluated by combining the scoring values of qualitative and quantitative indicators.

Overall, the pedestrian system planning in the area affected by the high-speed rail station should be designed to facilitate pedestrians to quickly enter and exit the core area of the high-speed rail station. At the same time, a diversified street system should be used to enhance the vitality of the neighborhood, attract more passengers to various surrounding service facilities and promote neighborhood development. Through a three-dimensional and people-oriented pedestrian system, the high-speed rail station, streets, passenger flow, and surrounding neighborhoods are closely integrated. While the high-speed rail station promotes the prosperity of the surrounding neighborhoods, the neighborhoods can then feed back the development of the high-speed rail station. This is the true meaning of station. Urban integration is also a key link in sustainable urban development.

INVESTIGATING SPACE-TIME RELATIONSHIP IN ARCHITECTURAL DESIGN THROUGH SPACE SYNTAX ANALYSIS

Track 6: Transdisciplinary Design and Collaboration/Open Track
Knowledge Network: Environmental & Architectural Phenomenology, Interior Design
Methods: Case Study, Survey
Pariya Sheykhmaleki (Texas Tech University); Saif Haq (Texas Tech University - College of Architecture)

1 ABSTRACT

In 1989, Hiller and Hanson expanded on a fundamental idea -- spatial order can impact social life, and social life can influence spatial order. To explain further, they developed the Space Syntax theory, arguing for connections of space as basic. The associated methodology considered connections to immediate and faraway spaces to analyze spatial arrangements, and this was then used to investigate certain social behaviors with remarkable success. By analyzing the spatial arrangement, its impact on social life could be analyzed; however, the effect of social behavior on spatial analysis has not been proposed. This is the intention of this paper. Here, we suggest an analytical technique to add social behavior to the calculations of configuration.

The original Space Syntax methodology considered topological connections between spaces to calculate values of 'closeness' or 'betweenness'. Real-world testing contributed to the evolution of both the methodology and the theories. Recently, metric and geometric relationships have been argued for and implemented in the computerized toolset –leading to new ways of urban analysis. Thus, Space Syntax theory and analytical techniques have expanded substantially. This paper introduces 'time' as an important additional factor in the relationship between different spaces of a layout. From an Einsteinian point of view, space and time are intertwined with each other such that neither can be defined separately. From a cognitive point of view, the time spent in different spaces may affect recall and behavior. Much like the theory of General Relativity speculating spacetime deformation by gravity, this paper proposes that Space Syntax configuration will also deform with the inclusion of time. Traditional Space Syntax methodology is based on a binary understanding of connections between spaces. The proposed methodology adds the portion of time that a spatial connection supports in the analysis. Therefore, the ratio of the time during which the connection between spaces is active, influences the steps between these spaces. Instead of having a value of one between two spaces (as in traditional Space Syntax calculation), it will be based on the portion of time that that connection supports. This paper will briefly explain Space Syntax, introduce the new concept of time, and propose an extension to the methodology by including a time factor in the relationship between spaces in proportion to their anticipated use. It will also explain the normalizing procedures of the new variables tentatively called Time Values or TVa to Time Weighted Closeness (TWC) and Time Weighted Betweenness (TWB) values. These will then be tested with data being collected by a nurse survey. In general, these new values will integrate social behavior with layout values and are expected to be useful for studies of interior environments where human movement and time spent in various spaces are of special interest.

EFFICIENT CELL THERAPY MANUFACTURING FACILITY DESIGN: A PROCESS-LED SIMULATION FOR PLANNING AND BOTTLENECK ANALYSIS

Track 6: Transdisciplinary Design and Collaboration/Open Track
Knowledge Network: Researchers in Professional Practice
Methods: Case Study, Statistical/Modeling Analysis
Maryam Hosseini (EwingCole); Adrienne Erdman (EwingCole)

1 ABSTRACT

The rapidly developing field of cell therapy manufacturing promises disruptive, personalized medicine for a wide variety of conditions; however, the processes themselves present complex challenges in supply chain, facility layout, and performance of the system. It necessitates a flexible facility design to balance standardized manufacturing processes and scalable supply chain strategies with temperature-sensitive, just-in-time, and aseptic requirements. Subsequent validation of the design is important with any facility but is of critical importance in rapidly growing fields where time is of the essence to provide disruptive services and for start-ups who typically have limited funds for investment.

This presentation will discuss a case study in designing for the unique cell and gene therapy environment. Using iterative process mapping, simulation modeling, and co-design meetings, this work will demonstrate methods to effectively identify and address bottlenecks, enhancing supply chain configurations and optimizing resource allocation. By involving stakeholders in continuous feedback loops, the research efforts prioritize practical and scalable solutions, highlighting the approach's potential in advancing cell therapy manufacturing processes and aiding stakeholders in making informed decisions about production scaling and quantification.
This case study specifically covers a two-phase research effort with objectives of defining manufacturing operations, configuring supply chain parameters, and optimizing workflows in order to meet a start-up client's production targets – all without increasing the cost or footprint of the in-progress facility design. Phase 1 scope included the day-one facility design with only a fraction of the planned manufacturing machines. The phase was used to identify bottlenecks, configure the initial supply chain, analyze key performance indicators (KPIs), and determine the resources needed for scaling up production by 1600%. Phase 2 built upon these findings, implementing iterative adjustments in supply chain strategies, processes, and resource allocation to refine the manufacturing process for optimal performance and scalability. Each phase involved designing various scenarios and continuous stakeholder engagement to validate and enhance the system configuration. This approach provided a cost- and time-efficient proof of concept for the manufacturer to not only validate their facility design but also define their long-term operational strategy.

This presentation will cover in-depth the processes and metrics that can be used to ensure timely and efficient product manufacturing while considering uncertainty factors and fluctuating demand. Additionally, the presentation will discuss environmental factors that are flexible and conducive to operational scalability.
Ideal attendees include process engineers, manufacturing facility designers and owners, supply chain managers and data analysts who are responsible for defining and analyzing facilities and associated manufacturing operations. In particular, practitioners in highly controlled environments, such as pharmaceutical manufacturing, will benefit from the presentation.

REDEFINING MOVEMENT IN ARCHITECTURE: A NEW AGENT-BASED SIMULATION APPROACH INFORMED BY HUMAN NAVIGATION HEURISTICS

Track 6: Transdisciplinary Design and Collaboration/Open Track
Knowledge Network: Sustainable Planning, Design & Behavior
Methods: Mixed Methods
Yeinn G Oh (Georgia Institute of Technology); Gonzalo L Vegas Olcese (Georgia Institute of Technology); Elliot Jeongwon Park (Marble Fairbanks Architects)

1 ABSTRACT

This study advances the field of Space Syntax by proposing a novel approach to Agent-Based Simulation (ABS) in architectural spaces, aiming to better replicate human movement and behavior within buildings. Traditional ABS in Space Syntax, as exemplified by Turner & Penn (2002), involves agents selecting random destinations within their field of vision and moving accordingly, thereby generating generic movement patterns. However, we posit that human movement within a building is often driven by specific purposes or tasks, necessitating a more sophisticated algorithm.

To develop this new model, we engaged participants in navigating a virtual environment using both Virtual Reality (VR) and 2D screens. This approach enabled us to gather detailed insights into the heuristics and strategies of human navigation. Heuristics in this context refer to the intuitive strategies or mental shortcuts that humans employ to make decisions and navigate complex environments, effectively simplifying the cognitive process of wayfinding. These findings informed the development of an advanced agent-based simulation within Rhino Grasshopper.
Our study involved a two-fold comparison: firstly, replicating the Turner & Penn model to understand its mechanism, and secondly, testing our newly developed model. We then evaluated the paths taken by both models, comparing them with the empirical movement data we collected. Our findings suggest that ABS, with our proposed model, transcends generic movement simulation, becoming a potent tool for testing behavioral theories and embedded heuristics. This approach not only offers a more realistic representation of human navigation in buildings but also provides a deeper understanding of the underlying cognitive processes guiding movement in architectural spaces. Furthermore, our study contributes a novel perspective by testing a simple human behavior algorithm, incorporating heuristics that closely mimic realistic human wayfinding behavior.

2 REFERENCES

Turner, A., & Penn, A. (2002). Encoding natural movement as an agent-based system: an investigation into human pedestrian behaviour in the built environment. Environment and planning B: Planning and Design, 29(4), 473-490.

KPI-DRIVEN DESIGN TOOL: INFORMING POST-COVID WORKPLACE DESIGN TO IMPROVE KEY PERFORMANCE INDICATORS

Track 6: Transdisciplinary Design and Collaboration/Open Track
Knowledge Network: Work Environments
Methods: Design Methods
Helia Taheri (Arcadis Inc.); Shreya Somani (Arcadis Inc.)

1 ABSTRACT

While remote work will remain a key part of business operations due to the COVID pandemic, physical offices will still play a crucial role in many industries and cities. As companies reassess their workplace strategies, it is important to understand how building design impacts key performance indicators (KPIs). On the other hand, researchers in academia have done a noticeable number of peer-reviewed studies in the past three years to understand the post-COVID workplace KPIs and design approaches to improve them. However, despite the credibility and validity of these studies, there is often a gap between the academic research and industry implications, which is caused by the lack of accessibility and communication of the results to the practitioners who can benefit from them. Therefore, there is a need for tools that can bridge this gap by synthesizing the academic findings in an easy-to-understand language for the designers to inform their design strategies. In this presentation, we will share an interactive tool based on a review of 26 peer-reviewed papers published in the past three years, which studied post-COVID workplace KPIs related to employee performance and the building design strategies.

The identified KPIs are divided into two categories: (1) subjective KPIs, which reflect employees' perceptions of their workplace experience, and (2) objective KPIs, which illustrate employee performance through cognitive tests and sensors. The study also finds seven design approaches to improve KPIs, including acoustic design, lighting design, office layout, office dimension, workstation, space control, and biophilia design. The tool's user-friendly interface allows designers to easily access information about the relationship between KPIs and architectural design in their proposals and client meetings. The tool also helps designers to back up their design decisions by evidence and data, which can enhance their credibility and confidence in achieving desired design outcomes. The tool is intended to support designers during the early stages of the design process, such as proposals and client meetings.

A BIKE RIDE AND A LANDSCAPE: 50 TRANSECTS ACROSS IOWA

Track 6: Transdisciplinary Design and Collaboration/Open Track
Knowledge Network: Active Living by Design, Movement in Designed Environments: Nature & Ecology
Methods: Mixed Methods, Statistical/Modeling Analysis
Austin Dunn (Iowa State University)

1 ABSTRACT

In 2023, a bicycle touring event named RAGBRAI celebrated its 50th anniversary. Each July since 1973, thousands of RAGBRAI cyclists have traversed the entire State of Iowa from west to east. Each year has taken a single and unique route across the state. These routes represent 50 transects across the Iowa landscape. We analyze 50 RAGBRAI routes, across five decades, to understand landscape change in Iowa--from the perspective of a person riding a bicycle.

According to Liu. et al. (2021), cycling offers a mode of travel that offers a distinct impression and understanding of the environment. The experience of cycling is distinct from pedestrian or vehicular travel. The cycling experience can be conceptualized as a social, sensory, and spatial phenomenon (Liu et al., 2021). On RAGBRAI, in Iowa, these phenomena are tied to the characteristics of the landscape. Iowa's geological history, its robust agricultural economy, its extensive network of secondary roads, and patterns of rural to urban migration—all these bear on the experience of a cyclist riding across Iowa in July. In this study, we employ spatiotemporal GIS (geographic information systems) analysis to identify changes in land use and socioeconomic patterns in Iowa during the last half century. We contextualize RAGBRAI routes with spatial & temporal data related to agriculture, energy, and demographics. We ask, "what can a bicycle ride reveal about a landscape?" The bicycle as a mode of transportation and a mode of experiencing offers a human-scale frame to discuss larger issues in the landscape.

Fifty annual bicycle-transects across Iowa reveal the diversity and morphology of the landscape – a landscape that continues to change, and a landscape that can be uniquely interpreted at the speed of the bicycle. Innovative cartographic strategies (Kraak, 2020) allow for visual representation of landscape characteristics across space & time.

2 REFERENCES

Kraak, M. J., & Ormeling, F. (2020). Cartography: visualization of geospatial data. CRC Press.

Liu, G., Krishnamurthy, S., & van Wesemael, P. (2021). Conceptualizing cycling experience in urban design research: a systematic literature review. Applied Mobilities, 6(1), 92-108.

BUILDING ECOLOGY: GEOGRAPHICAL INSIGHTS INTO SPACE-PLACE DYNAMICS IN A LIVING-LEARNING COMMUNITY

Track 6: Transdisciplinary Design and Collaboration/Open Track
Knowledge Network: Interior Design, POE/Programming
Methods: Mixed Methods

Elif Tural (Virginia Tech); Tim Baird (Virginia Tech); Thomas Pingel (Binghamton University); David Kniola (Virginia Tech); Nicole Abaid (Virginia Tech); Tanner Upthegrove (Virginia Tech); David Franusich (Virginia Tech); Addison Flack (Virginia Tech); Shashank Karki (Virginia Tech)

1 ABSTRACT

A growing body of literature underscores the significance of meaningful social interactions and a sense of belonging for college students' educational experiences (Strayhorn, 2018; MacCabe, 2016). Living-learning communities (LLCs) provide opportunities for students to foster more explicit social relationships with peers and faculty. Previous studies suggest that students find meaning in their LLCs with regards to obtaining social and academic support, and engaging in multicultural experiences (Spanierman et al., 2013). Most research on living-learning communities seeks to understand the outcomes of students' participation in terms of their interaction with faculty, peer interaction, and social interaction (Wawrzynski et al., 2009). However, less attention has been paid to understanding how the design of the physical environments influences these outcomes.

This multidisciplinary study examines building interiors and how space is transformed into "place" by studying a living learning community on a university campus. While many disciplines have examined the relationships between humans and indoor environments, geography's contributions have been limited. And yet the theories and constructs from the field of geography, such as "disturbances" and "productive collusions" as well as the data collection tools are well suited to study place attachment and meaning.

Adopting a mixed-methods, concurrent triangulation research design, this longitudinal project relies on remotely sensed lidar data, student surveys and focus groups, and interviews with building occupants, designers, and campus administrators as well as anonymized audio recordings extracted from soundfield microphones placed throughout the main public gathering space of the LLC. The information will then be integrated and analyzed to create a mathematical model that reveals patterns in how place emerges, gets disrupted, and is rebuilt over time, and how various design factors support and hinder these processes.

This presentation will discuss our strategies to address this broad research question and focus on the research design and initial findings of this mixed methods study. We will also share the lessons from our novel research approaches and the interdisciplinary research process where we build on theories and research methods from geographical sciences, environmental design, cultural anthropology, complex systems science, computer vision, and mathematics.

2 REFERENCES

McCabe, J. M. (2016). Connecting in college. University of Chicago Press.

Spanierman, L. B., Soble, J. R., Mayfield, J. B., Neville, H. A., Aber, M., Khuri, L., & De La Rosa, B. (2013).

Living learning communities and students' sense of community and belonging. Journal of Student Affairs Research and Practice, 50(3), 308-325.

Strayhorn, T. L. (2018). College students' sense of belonging: A key to educational success for all students. Routledge.

Wawrzynski, M. R., Jessup-Anger, J., Stolz, K., Helman, C., & Beaulieu, J. (2009). Exploring students' perceptions of academically based living-learning communities. College Student Affairs Journal.

PHOTOGRAPH LIKE SOMEONE ELSE: LEARNING THROUGH ANOTHER PERSON'S EYES

Track 6: Transdisciplinary Design and Collaboration/Open Track
Knowledge Network: Cultural Aspects of Design, Participation
Methods: Experimental/Quasi-Experimental Research, Photo-based Methods
Anne C Godfrey (SUNY College of Environmental Science and Forestry); Anne C Godfrey (SUNY ESF)

1 ABSTRACT

We use photography as an investigative tool across the fields of environmental design. Photographs are made easily and photographing remains an accessible and democratic form of landscape representation. Photographs, though, are not visual copies of a place, instead they are a complex representation of the photographer's interest, desires, values and knowledge. Thus, each photograph made is a unique act of creation.

This presentation is interested in working with this condition by presenting an engagement method called Photograph Like Someone Else. Similar to the notion of standing in another person's shoes, photographing like someone else is a way to see the world from another person's eyes. This method supports our growing interest in cultivating broad and more inclusive understandings of any place we affect with environmental design.

The method, Photograph Like Someone Else, asks one designer and one community member to pair together. Instead of making their photograph, each person describes to the other person how they would photograph a particular place or subject, then the other person makes their photograph. This process lets both people "see through the other person's eyes" and instigates dialogue about a place.

The presentation will articulate the steps of the method, present examples of work created through this method, and reflections by participants. Examples include photographs made by students in workshops and design studios, and examples from community engagement processes concerned with green infrastructure proposals.

This method also confronts the notion of what makes a "good photograph." Photographing Like Someone Else can become a portal to accepting more types of photography beyond compositional norms such as the picturesque and the beautiful, which are situated in cultures from Western Europe and the post-colonized United States (Burke 1757, Gilpin 1794). The author will discuss some engagement issues to pay attention to, such as "improvement" and "correcting" in order to fit the other person's photograph (consciously or unconsciously) into one of these norms. Through this discussion, we can also assess our own tendencies and biases. Learning from others through photographing like someone else is a process of learning from how another person sees and evaluates the world – in this case through photography. The action of photography, always physical in nature, becomes an action of embodying another person's experience through the physical process of photographing how the other person wants to portray a place or subject.

Finally, sharing and listening are key components to both of these methods (Sanders 2022). The author will discuss some engagement issues to pay attention to, such as "improvement" and "correcting." Further, this explores how the accessibility of photography can help designers transcend these tendencies, confront aesthetic norms, and accept broader types of photography.

2 REFERENCES

Edmund Burke. 1757. A Philosophical Enquiry into the Origin of Our Ideas of the Sublime and Beautiful.

Gilpin, William. 1794. Three Essays on Picturesque Beauty, on Picturesque Travel, and on Sketching Landscape. London, England.

Sanders, E. (April 25, 2002). From user-centered to participatory design approaches. 1-8.
Design and the Social Sciences. J.Frascara (Ed.), Taylor & Francis Books Limited.

HERITAGE AND HEALTH IN RESIDENTIAL INTERIORS

Track 6: Transdisciplinary Design and Collaboration/Open Track
Knowledge Network: Health, Historic Preservation
Methods: Case Study
Bryan Orthel (Indiana University)

1 ABSTRACT

Historic buildings may be healthy buildings. We know of specific health-related concerns related to older residential buildings (e.g., lead paint, asbestos), but we do not know how the forms, materials, operations, and relationships of these buildings may improve human health. To date, research on how historic buildings support human health has been limited. This presentation uses a case study of a historic house to describe how and why we should rigorously consider how historic buildings interact with physiological, cognitive, and social aspects of health.

Historic residential interiors provide a meaningful scale of study because of the long duration of time we spend in such spaces, the close personal identities and cultural practices based in these spaces, and the rapid pace of changes humans make to interior spaces. Each of these characteristics amplifies the potential health benefits or concerns associated with historic building materials, forms, and uses. Long-term exposure may increase risks (e.g., materials and indoor air quality); continuation of practices and uses may protect health (e.g., forms and materials); building forms may amplify health issues.

The presentation uses a heritage systems model to analyze the interior of a 1910s-era midwestern US house. Systems models reveal how "events accumulate into dynamic patterns of behavior" (Meadows 2008, p. 88). Systematically evaluating heritage requires us to consider the myriad, obscured ways heritage intersects with social, economic, ecological, and other structures. Heritage professionals, designers, and everyday individuals must understand these systematic interactions to be preprepared for the decisions they make about how to use, evaluate, and alter historic spaces, behaviors, and values (Hutchings & Casser, 2006; Fouseki & Bobrova 2018). Specifically for heritage and interiors, these decisions have profound, compounding implications for our health at multiple scales.

The heritage systems model guides topical discussion with focus on influences and supports, tangible and intangible aspects of heritage, individual and community experiences, human implications, and the flows and actions that underpin change. The analysis identifies responsibilities and consequences to consider in decision making. The analysis relies on interdisciplinary scholarship on topics as varied as microbiomes, indoor air quality, energy policy, material culture, and social customs.

Outcomes are presented as a taxonomy of questions to guide decisions about heritage, health, and interiors: How does this interior protect me? How does this interior hurt me? How does this interior reflect me? And, how must I adapt to this interior?

Interior environments embody personal identity and social position. Interior environments should also provide health, safety, wellbeing, and ecological support for human living. This presentation reveals core questions to guide future heritage work with interior spaces. The systems model reveals structural relationships that may be otherwise obscured by disciplinary frames and personal experience, as well as prompting reconsideration and recognition of new values. Asking how a historic building can be a healthy building reframes heritage and preservation with new issues that must be balanced. These questions expand the value of heritage and historic places for individuals and society.

EXPLORING STUDENT AGENCY: VARIATIONS IN ENGAGEMENT AND COLLABORATION WITHIN AN ENGINEERING DESIGN ACTIVITY

Track 6: Transdisciplinary Design and Collaboration/Open Track
Knowledge Network: Children, Youth & Environment, Sustainable Planning, Design & Behavior
Methods: Case Study

Sepideh Fallahhosseini (University of Missouri - Columbia); Laura Brianna Cole (Colorado State University); Laura Zangori (University of Missouri); Suzanne Otto (University of Missouri)

1 ABSTRACT

Sustainable infrastructure, such as green buildings and urban spaces, requires a future workforce positioned to collaboratively solve sustainability challenges related to the built environment. This study examines an educational intervention in the middle school (MD) setting that nurtured student goal-driven agency (GDA) in the engineering design (ED) process, which focused on energy-efficient building design. GDA refers to the human capacity to set internal goals and make choices that direct behavior towards achieving those goals. Recognizing students as active agents in the design realm empowers them to engage in collaborative problem-solving, instilling a sense of ownership (Zimmerman, 2000). Understanding and studying GDA sheds light on how students navigate teamwork, communicate ideas, and contribute to collective goals. By emphasizing students' agency, educators and practitioners can tailor educational experiences that nurture teamwork, creativity, and adaptability, laying the foundation for future engineers capable of sustaining productive collaborations that drive innovation and success in the design professions.

The infusion of STEAM (Science, Technology, Engineering, Arts, and Mathematics) into MS classrooms via ED holds the potential to inspire the next generation of professionals dedicated to human-centric design principles. Our interdisciplinary team across architectural studies and science education creates standards-aligned green building curricula for MS classrooms. This qualitative case study focuses specifically on the implementation of an ED lesson within one Midwestern public elementary school (19 students in 6 teams and 1 teacher). The unit laid the foundation for energy literacy broadly with a focus on energy-efficient building design. In the culminating unit activity, students conceptualized and constructed an energy-efficient one-room schoolhouse using the ED process. Student teams made decisions pertaining to roofs, walls, insulation, windows, building orientation, and renewable energy sources. This study provides valuable insights into how students exercise agency in the ED process when immersed in real-world applications associated with sustainable practices. In this investigation, our focus is narrowed to examine the active engagement and collaboration of students in the ED process. The dataset comprises audio recordings, student worksheets, interviews, and researcher observations. Our analysis of the qualitative data adopts a holistic approach, utilizing GDA as our analytical framework based on social cognitive theory (Bandura, 1989), scrutinizing students' self-efficacy, self-regulation, and self-reflection (Zimmerman & Campillo, 2003). This examination unveiled the natural utilization of agency by students in the design process, highlighting the ways that unique learners navigated the ED process. The patterns discovered in this analysis led to the identification of diverse learner profiles. In this presentation, we share our ED curriculum along with the successes and challenges revealed by this study. A more profound comprehension of this phenomenon aids in the development of an innovative green building curriculum, inspiring youth to become future professionals in STEAM fields. These individuals will promote human-centric design across diverse scales and environments.

2 REFERENCES

Bandura, A. (1989). Human agency in social cognitive theory. American psychologist, 44(9), 1175.

Zimmerman, B. J. (2000). Self-efficacy: An essential motive to learn. Contemporary educational psychology, 25(1), 82-91.

Zimmerman, B. J., & Campillo, M. (2003). Motivating self-regulated problem solvers. The psychology of problem solving, 233262.

FOSTERING CREATIVITY IN HIGHER EDUCATION ARCHITECTURAL STUDIO SPACES

Track 6: Transdisciplinary Design and Collaboration/Open Track
Knowledge Network: Interior Design, Work Environments
Methods: Mixed Methods, Observation
Casey Gail Franklin (University of Kansas); Hui Cai (University of Kansas School of Architecture & Design)

1 ABSTRACT

Fostering Creativity is one of the primary goals of architecture education. Creativity is largely due to personal attributes, but the physical environment plays an important role in supporting creativity (J. H. Lee & S. Lee, 2023). Architecture studio spaces impact how students work individually and collectively, faculty teaching methods, and how students learn from faculty and peers. With most architecture design firms shifting to large open office spaces, many architecture schools' studio spaces have also changed from the traditional dedicated space for one studio to large open shared spaces that collocate multiple studio sections. A common perception is that shared spaces can enhance creativity through increased social interactions. However, limited research has been conducted to evaluate the impacts of studio spaces on students' creativity. This pilot study used mixed methods to explore the impact of studio spaces on architectural student perceptions of creativity. The study used a natural experiment opportunity that assigned students at the same year level in two different studio environments: individual versus shared spaces. GIS-based behavioral mapping and observations, interviews with professors, and a student survey (n=16) were used to identify differences in academic and social interactions and student perceptions of creativity. Initial findings indicated that students in the shared space rated themselves as feeling less creative, noting that environmental factors like noise limited creativity.

Structured behavioral mapping was used for academic and social interactions in studio spaces. Interactions were logged every 15 minutes during classes for a week resulting in 460 total observation points. Of 219 points in shared studio space, 58.9% were social interactions, 28.9% were academic interactions, and 37.4% were individual. In the individual studio space, 241 observation points were made, with 7.8% being social, 21.2% being academic and 70.9% being individual. Data for this period indicates the shared studio space fostered far more social interactions. Interestingly though, in the survey students' ratings of perceived creativity in the shared environment were overall lower (M=2.88) compared to the individual environment (M=3.17). Explanations included information that the shared studio space was loud and distracting during concurrent studios, making it difficult to concentrate. Interviews with three studio professors supported these findings, each professor noted a preference for individual studio space. However, more than one professor noted that they liked the idea of a shared or visually shared space if it had quieter support spaces for activities like presentations. This pilot study demonstrated that fostering creativity in studio spaces is more complex than simply combining studios in one area. When designing future shared studio spaces, it is important to balance the increased natural encounter and interaction with careful considerations of the acoustical and functional needs of activities like presentations, tutorials, and both group and individual student work. This study has built a foundation for a large-scale follow-up study to investigate environmental features of studio spaces on creativity.

2 REFERENCES

Lee, J. H., & Lee, S. (2023). Relationships between physical environments and creativity: A scoping review. Thinking Skills and Creativity, 48, 101276. Retrieved from https://www.sciencedirect.com/science/article/pii/S1871187123000469. doi:https://doi.org/10.1016/j.tsc.2023.101276

RESEARCHERS IN PROFESSIONAL PRACTICE ORCHESTRATING AEC TEAMWORK SUCCESS: COMPOSING, ENGAGING, AND MANAGING

Track 6: Transdisciplinary Design and Collaboration/Open Track
Knowledge Network: Communication, Environmental Design Research Education
Methods: Action/Participatory Research, Mixed Methods

Nancy Yen-Wen Cheng (University of Oregon)

1 ABSTRACT

A co-taught Architecture-Engineering-Construction collaboration course challenges students, in two universities separated by 50 miles, to work in small groups to design, detail and prototype structures. This paper analyzes seven iterations of the class to extract lessons on what makes remote interdisciplinary collaboration successful. Emphasis will be on the latest efforts to design pavilions with reusable timber reciprocal frame parts.

Questions: How does the team's mix of backgrounds influence the quality of collaboration? What can motivate more equitable participation?
To understand the relationship between the group structure, the collaboration process, and the design product, the study looks at the mix of student backgrounds, markers of collaboration, and overall project grades. The vertical class brought together students with different majors, educational level, and backgrounds (gender, work experience, U.S. vs. international origin). The small group collaboration's success was measured by examining the students' reflective writing and project reports for markers of support, coordinated work, knowledge transfer, conflict negotiation, data sharing, equity of participation among team members, and satisfaction reported. Student descriptions of team dynamics and project outcomes reveal:

- *Heterogeneous groups that mix seniority and disciplines have a greater chance of success. Overlaps in expertise and lack of hierarchy require more interpersonal negotiation or third-party mediation/intervention.*
- *Activities such as collaborative free play with building components, team guidelines and process reflections can streamline teamwork.*
- *Defining and assigning specific roles can help keep all members engaged, especially as a project moves towards production and delivery. Interested and motivated participants benefit from clear plans and role assignments to understand how their abilities can contribute to the bigger picture.*
- *Team members appreciate role options: they can choose familiar focus areas to flex existing skills or choose less familiar areas to explore and grow.*
- *Tradeoffs will occur between individual voices shining and the group working as a whole. For example, one student was able to individually cut all the pieces for his team's assembly in his home workshop. While this circumvented teamwork for this step, his teammates were happy to assemble the results for a strong final presentation.*

2 REFERENCES

Caroline T.W. Chan and Sher, W. (2014), "Exploring AEC education through collaborative learning", Engineering, Construction and Architectural Management, Vol. 21 No. 5, pp. 532-550. https://doi.org/10.1108/ECAM-04-2013-0036

Maii Emam, Dina Taha and Zeyad ElSayad (2019). Collaborative pedagogy in architectural design studio: A case study in applying collaborative design. Alexandria Engineering Journal, 58(1), 163-170. https://doi.org/10.1016/j.aej.2018.03.005

D.W. Johnson and R.T. Johnson, R.T. (1998) "Cooperative Learning and Social Interdependence Theory" in Tindale, L., Heath, J., Edwards, E.J., et. al. Eds.; Theory and Research on Small Groups, Plenum Press: New York, NY, USA, 1998; Volume 4, pp. 9–36.

Riggio, M., & Cheng, N. (2021). Computation and Learning Partnerships: Lessons from Wood Architecture, Engineering, and Construction Integration. Education Sciences, 11(3), p.124. https://doi.org/10.3390/educsci11030124

Tran, V.D. (2013). Theoretical Perspectives Underlying the Application of Cooperative Learning in Classrooms. International Journal of Higher Education. 2(3), 102-115. http://doi:10.5430/ijhe.v2n4p101

Richard Tucker & John Rollo (2006) Teaching and Learning in Collaborative Group Design Projects, in Architectural Engineering and Design Management, Vol 2, no. 1-2, pp. 19-30. https://doi.org/10.1080/17452007.2006.9684602

RESEARCH ON OPTIMIZATION OF COMMERCIAL SPACE IN WAITING HALLS OF LARGE HIGH-SPEED RAILWAY STATIONS BASED ON BIG DATA TOOLS --TAKING SIX LARGE HIGH-SPEED RAILWAY STATIONS AS AN EXAMPLE

Track 6: Transdisciplinary Design and Collaboration/Open Track
Knowledge Network: Cyberspace & Digital Environments
Methods: Design Methods, Mixed Methods
Yuzhu Wang; Zhongzhong Zeng (Beijing Jiaotong University); Bo Zhang (Oklahoma State Univeristy)

1 ABSTRACT

As a part of the design of large-scale high-speed railway station second only to the function of transportation interchange, the design of commercial space is not only related to the use of high-speed railway station, but also to the formation of urban sub-center centered on high-speed railway station. Optimized design based on consumer experience is especially important for the future development of commercial space. Network big data brings a new perspective to the study of commercial space.

Compared with the traditional questionnaire and interview survey method, the network comment data has the characteristics of large sample size, real-time and dynamics, which can quickly reflect the public user's evaluation of the spatial environment feeling in the Internet era, and at the same time, summarize the significant issues of public usage reflected on the Internet. Big data will provide an important means for the cognition, association and discipline discovery of space and other related information in the process of commercial space analysis.

By analyzing the commercial facility layout of waiting space of large high-speed railway stations that rank among the top 15 in China in terms of both the number of undertaking train arrivals and departures and the number of platform strands, this study has selected Zhengzhou East Station, Shanghai Hongqiao Station, Nanjing South Station, Guangzhou South Station, Changsha South Station, and Xi'an North Station as the research objects, and these six high-speed railway stations can represent the six modes of the commercial facility layout of the waiting halls.

This study uses python technology to collect all the valid evaluations for the six high-speed railway stations from 2019-2024, and after refine words and frequency classification, visualization and analysis are carried out using gephi software, so as to compare and analyze the commercial space layout of the waiting halls of the six stations at three levels: monthly analysis, word frequency analysis and word linkage analysis. Through the above analysis, we can conclude: 1. what kind of commercial layout mode users pay more attention to 2. what kind of commercial mode has the highest user satisfaction 3. the usage characteristics of waiting hall commercial space under different commercial modes 4. the future optimization direction of waiting hall commercial space in large high-speed railway stations.

Through the processing and comparative analysis of network comment data, this study basically understands the passengers' concerns and needs for the commercial space of waiting lounges in large high-speed railway stations, which provides data support and new analytical ideas for the subsequent optimization of the actual use and functions of similar commercial spaces. In the future, we can introduce the evaluation and analysis from the commercial space managers and operators, as well as expand the access to big data and optimization analysis tools, so as to coordinate the relationship between multiple parties through the data, and make the optimization method of commercial space in waiting halls of high-speed railway stations have a higher feasibility.

BIOPHILIC DESIGN INDIRECT ATTRIBUTES IN HAWAII HOTEL INTERIOR DESIGN: A NLP-ENHANCED COMPARATIVE STUDY

Track 6: Transdisciplinary Design and Collaboration/Open Track
Knowledge Network: Cyberspace & Digital Environments, Interior Design
Methods: Photo-based Methods, Statistical/Modeling Analysis
Chunya Wu (University of Hawaii at Mānoa)

1 ABSTRACT

According to Kellert's biophilic design theory, incorporating natural attributes in architecture and interior design can enhance well-being and satisfaction (Kellert & Calabrese, 2015). Studies also show that direct biophilic attributes, such as water and sunlight, are associated with positive consumer intentions (Lee, 2019) in hotel design.

2 RESEARCH GAP

However, the ability of different hotels to introduce direct natural attributes and landscapes may vary and is not entirely under the control of designers. Furthermore, research on the biophilic design indirect attributes (BDIA), which are design elements that offer indirect natural experiences like natural materials, colors, and patterns, is also limited. Therefore, this study focuses on the impact of BDIA in interior design on hotel evaluations.

3 CASE STUDY

This research selects 26 representative four-star hotels in Waikiki beach, Hawaii, with consistent geographic locations and facilities, and similar local interior design style, which has rich and varied BDIAs, to explore their impact on customer experience and booking intentions.

4 METHODOLOGY

The study applies a transdisciplinary comparative approach to analyze and quantify the impact of these interior design elements on customer experience. It collected over 35,000 open-sourced customer online reviews, using Natural Language Processing (NLP) technology to quantify customers' emotional tendencies and design quality evaluations of hotel interiors. Additionally, referencing the construction method of the BID-M quantification model (McGee et al., 2019), BDIA-Index was developed by four rounds of feature engineering, to quantitatively assess BDIA quality in three hotel interior spaces: guest rooms, lobbies, and lounges. By comparing and analyzing data from different resources using the random forest machine learning algorithm, a predictive model for real hotel online rating score based on BDIA-index was established.

5 RESULTS

Hawaii local identity and vintage style is a less used BDIA (34%) related to higher Booking scores, higher Quantitative Feeling And Evaluation, and more positive consumption decisions. Some widely applied(>50%) BDIAs, such as Nature Image and Geometries, have shown significant correlation with positive results in multiple analyses, and are strong predictors. Natural colors and materials related to BDIA cannot have a significant impact on quality perception and positive sentiments. Testing by other Hawaii hotels shows that BDIA-Index can predict Hawaii hotels Booking.com rating with an explanatory power of 50.4%. Pathway analysis shows that BDIA-index is indirectly related to willingness to pay premiums.

6 SIGNIFICANCE

The results provide insights into the effectiveness of biophilic interior design in hotels, offering practical suggestions for hotel design and marketing strategies. The study also supports sustainable development in the local economy, providing empirical evidence for life-affirming and nature-inclusive design, promoting sustainability, well-being, and prosperity in human-centric design.

2 REFERENCES

Kellert, S., & Calabrese, E. (2015). The practice of biophilic design. London: Terrapin Bright LLC, 3, 21-46.

Lee, S. H. (2019). Effects of biophilic design on consumer responses in the lodging industry. International Journal of Hospitality Management, 83, 141–150. https://doi.org/10.1016/j.ijhm.2019.05.006

McGee, B., Park, N., Portillo, M., Bosch, S. J., & Swisher, M. E. (2019). DIY Biophilia: Development of the Biophilic Interior Design Matrix as a design tool. Journal of Interior Design, 44(4), 201–221. https://doi.org/10.1111/joid.12159

CREATING A FUTURE FOR CORAL REEFS BY REIMAGINING HUMAN ACTIVITY: LIFE-INSPIRED DESIGN FOR THE HEALTH OF CORALS

Track 6: Transdisciplinary Design and Collaboration/Open Track
Knowledge Network: Environmental Design Research Education, Sustainable Planning, Design & Behavior
Methods: Case Study, Design Methods
Luis Angarita (Arizona State University); Michelle Fehler (ASU, The Design School);
Hazal Gumus Ciftci (Arizona State University); Judith Klein-Seetharaman (Arizona State University);
Darren Petrucci (Arizona State University)

1 ABSTRACT

Corals are the architects who built highly productive tropical coastal reef ecosystems over the past 25 million years. They are now on the brink of extinction due to a litany of factors linked to human influence, including rising ocean temperatures, acidification, pollution, overfishing, and ecological imbalances (NOAA, 2019). If we do nothing, the coral reef surface area is predicted to vanish by 2050, with devastating consequences for biodiversity, food security, coastal protection, tourism, trade, and cultural heritage. As the causal agents in coral reefs decline, we must design novel solutions to halt this predicted outcome. Here, we propose a highly interdisciplinary team at the intersection of art, design, science, and technology to partner with nature conservation agencies working closely with native Hawaiians to save the coral reefs. Our initiative embodies a life-centered design approach within the framework of a multidisciplinary graduate studio course to assist in organismal resilience - the capacity for survival in the face of an ever-changing environment, and the coral reefs were our first organism.

As Malina et al (2016) wrote in their paper, we established a widely multi-disciplinary student, faculty, and researcher body for this project. The journey of our graduate students began in August 2023, following exhaustive research and a series of research presentations, the students embarked on a field trip to Oahu, Hawaii, where they uncovered insights through immersive experiences, dialogues, and observations. The multidisciplinary design studio comprises graduate students specializing in Architecture, Industrial Design, and Visual Communication Design, complemented by faculty experts in collaboration, Natural History, and Molecular Science. Our project is a pioneering exploration of the ecotone between rim-to-reef watershed communities, fishponds, and coral reefs, with a mission to devise regenerative solutions benefiting all life forms.

The urgency of our mission cannot be overstated. Coral reefs, these magnificent ecosystems that harbor an extraordinary array of marine life, are teetering on the brink of devastation due to human-induced pressures (Lin et al, 2023). Our dedication lies in crafting innovative designs that alleviate human activity's adverse impacts on coral reefs and actively contribute to their revitalization and enduring resilience.

This paper investigates our first hybrid creative innovation and our project which seeks to transcend conventional problem-solving boundaries, ushering in a new era of sustainable design thinking. We envision a future where tangential design thinking and scientific research harmoniously converge to address critical environmental challenges, yielding tangible solutions that safeguard our planet's invaluable coral reefs. The studio projects reimagined human activity to create a better future for coral reefs and also for a better world.

2 REFERENCES

Lin, B., Zeng, Y., Asner, G. P., & Wilcove, D. S. (2023). Coral reefs and coastal tourism in Hawaii. Nature Sustainability, 6(3), 254-258.

Malina, Roger F., Strohecker, C., LaFayette, C. Steps to an Ecology of Networked Knowledge and Innovation: Enabling New Forms of Collaboration among Sciences, Engineering, Arts, and Design. (2016). Art Journal, 75(3), 65–. College Art Association, Inc.

NOAA (2019, February 1). Coral reef ecosystems. National Oceanic and Atmospheric Administration. Retrieved October 4, 2023, from Coral Reef Ecosystems.

FAST TRACK TRANSFORMATION: REDEFINING HOSPITAL LOBBY DYNAMICS FOR EFFICIENCY, SAFETY, AND PATIENT EXPERIENCE

Track 6: Transdisciplinary Design and Collaboration/Open Track
Knowledge Network: Health, Movement in Designed Environments: Nature & Ecology
Methods: Case Study, Statistical/Modeling Analysis
Maryam Hosseini (EwingCole); Suzy Genzler (EwingCole); LarriJo Boone (136 Mt. Vernon Court)

1 ABSTRACT
Enhancing hospital efficiency has been a long-standing goal of most clinical institutions; however, with increased instances of violence and the recent pandemic, the goal has evolved to include safety, without sacrificing efficiency. Achieving this goal involves careful consideration of processes and flow management to ultimately decrease patient waiting times and operational costs while optimizing resource utilization and minimizing safety risks. This presentation will discuss the opportunity to use advanced decision-making tools, such as discrete event simulation (DES) and multi-criteria decision-making (MCDM), for evaluating design options to meet this goal and build consensus among a diverse set of stakeholders.

The presentation will share a case study of a hospital lobby redesign aimed at increasing operational efficiency, optimizing resource utilization, and integrating new security protocols. Case study objectives included decreasing check-in times and addressing safety and security concerns amplified by challenges like the COVID-19 pandemic. Additionally, the project goal was to demonstrate the rapid identification and implementation of proposed solutions, underscoring the importance of evidence-based decision-making using real data and scenario modeling to guide the healthcare facility's redesign and operational strategies.

Through a combination of DES and value-focused thinking, the team identified various patient inflow and outflow types, quantifying volumes and assessing the impact of alternative employee flow paths. The team's access to the client site was restricted due to the pandemic, so instead of using observation to inform the model, the team analyzed security footage of the lobby to identify traffic volumes and times to complete lobby-based processes such as registration. MCDM was then employed to evaluate potential lobby design options, prioritizing patient experience, flow efficiency, and risk reduction in the context of pandemic safety measures. The optimal lobby layout achieved through DES significantly reduced waiting times and congestion, while improving staff and patient safety through the implementation of self-service kiosks and turnstiles.

The case study, completed in an expedited six-week period, demonstrates the effectiveness of using real data and scenario modeling for rapid healthcare flow redesign, with a notable improvement in patient check-in times and stakeholder engagement. The presentation will provide attendees with the framework and lessons learned for conducting their own decision-making process to quickly build consensus among stakeholders with diverse perspectives—hospital management, staff, and patients—to work towards optimal healthcare processes, layouts, and experiences

PLACEMAKING: REVITALIZING ANDERSON HALL TO REVEAL TRADITION AND ENHANCE THE PLACE IDENTITY OF KANSAS STATE UNIVERSITY

Track 6: Transdisciplinary Design and Collaboration/Open Track
Knowledge Network: Historic Preservation, Interior Design
Methods: Survey, Observation

Karis Sandin (Kansas State University Architecture); Mariana Garcia Junqueira (Kansas State University - APDesign)

1 ABSTRACT

At K-State, in Manhattan, KS, family and community are rooted deep in their core. By revitalizing an iconic site where students' and the university's identity meet, the celebration of community on campus can be enhanced and become more recognizable to visitors. "The interweaving of place identity and human identity is clearly a broadly encountered phenomenon, which perhaps helps to explain why the term 'identity' is used, in everyday speech, equally in relation both to places and to people" (Butina-Watson & Bentley, 2007, p.4). With the intention of designing a gathering place that best represents its unique site, the reveal of its traditions and values benefits the overall development. For visitors experiencing a campus tour, a large component of their route is highlighting iconic locations to establish a sense of belonging at the university.

This qualitative inquiry investigates how tradition has influenced K-State's identity and values as a university. Data has been collected via online survey from community members with an affiliation to K-State, resulting in 102 responses with ages ranging from 18-60+. Flyers advertising the survey were posted in buildings across campus and the link was open for 8 weeks. The most important takeaways from this process include the reveal of Anderson Hall as the best architectural representative, uncovering the tradition considered by the majority as the favorite, and identifying the preferred hang out environment. Ultimately, the community's input is key to designing a space that meets the needs of the people and encourages longevity in usage.

Currently, campus leadership recognizes that there is no primary location to celebrate the culture of the university. Valuable informal learning experiences are shown through generous amounts of public and common space in the design of teaching and research facilities to foster a sense of place and sense of community (Mallon, 2015, p.8). A historical site on campus that is adjacent to a variety of public spaces and regarded with reverence is Anderson Hall. Clearly marked by tradition, this location allows for an experience of pride and engagement for faculty, students, and visitors alike. Placemaking is a "key purpose in spatial planning [that] aims to create, reproduce or mold the identities of places through the manipulation of various activities, feelings, meanings and fabric that combine into place identity" (Strydom, Puren, & Drewes, 2018, p.166).

Using the learned information, this placemaking effort takes the shape of a design intervention in Anderson Hall and its exterior surroundings. The survey findings informed the project design guidelines, including encouraging the tradition of community by incorporating a multi-use conversation lounge pit as well as adding value to campus by revitalizing an iconic building, making it more relevant with new usage and circulation. These guidelines aid in the recruitment of potential new students by creating meaningful and memorable experiences. Consequently, a strong desire to be a part of K-State is cultivated through the adaptive reuse of Anderson Hall plus the addition of a gathering place and showcases the deep traditions that have led to the success of the university.

PARK VITALITY AS A FRAMEWORK FOR VIBRANT URBAN PARKS: A SCOPING REVIEW

Track 6: Transdisciplinary Design and Collaboration/Open Track
Knowledge Network: Movement in Designed Environments: Nature & Ecology, Sustainable Planning, Design & Behavior
Methods: Interpretive-Historical Research, Mixed Methods
Asim Khanal (University of British Columbia); Keunhyun Park (University of British Columbia)

1 ABSTRACT

Park vitality as a framework for vibrant urban parks: a scoping review

The concept of urban vitality initially put forward by Jane Jacobs in the 60s (Jacobs, 1961), and then theorized further by notable scholars like Kevin Lynch in the 80s and John Montgomery in the 90s among others, has also been applied to urban parks (Banchiero et al., 2020). This paper identifies conceptual and practical gaps in the park vitality concept, and also reviews the methodologies used to determine user volume and user diversity in the context of urban parks. Academic databases like Scopus and Web of Science have been utilized, and only peer-reviewed journal articles are included. The scoping review follows the JBI guidelines and PRISMA-ScR checklist (Tricco et al., 2018).

In the backdrop of an ever-increasing urban population, growing inequity, and racial/ethnic tensions around the world, this review takes into account studies that determine or discuss user volumes and user diversity in urban green spaces and how internal park attributes, external neighborhood characteristics and broader socio-cultural context can affect usage of these spaces. This review also discusses the need to apply the park vitality framework in park planning, design, and operation in order to achieve vibrant urban parks. Preliminary insights highlight the relatively recent popularity of this topic (e.g., 73.7% of shortlisted papers are less than ten years old) although the first articles that discuss park vitality were published in the 70s. Findings also indicate that the user volume is influenced by park amenities, surrounding neighborhoods' built environment characteristics, demographics of users and residents living nearby, and personal preferences including perception. User diversity on the other hand is more complicated, and is influenced by broader socio-political phenomena, government policies, space design, and management approaches.

1.2 KEYWORDS

park vitality, urban parks, urban green spaces.

2 REFERENCES

Banchiero, F., Blečić, I., Saiu, V., & Trunfio, G. A. (2020). Neighbourhood park vitality potential: From Jane Jacobs's theory to evaluation model. Sustainability (Switzerland), 12(15). Scopus. https://doi.org/10.3390/su12155881

Jacobs, J. (1961). Jane Jacobs. The Death and Life of Great American Cities, 21(1), 13–25.

Tricco, A. C., Lillie, E., Zarin, W., O'Brien, K. K., Colquhoun, H., Levac, D., Moher, D., Peters, M. D. J., Horsley, T., Weeks, L., Hempel, S., Akl, E. A., Chang, C., McGowan, J., Stewart, L., Hartling, L., Aldcroft, A., Wilson, M. G., Garritty, C., ... Straus, S. E. (2018).

PRISMA Extension for Scoping Reviews (PRISMA-ScR): Checklist and Explanation. Annals of Internal Medicine, 169(7), 467–473. https://doi.org/10.7326/M18-0850

"GETTING IT RIGHT." A NETWORK-BASED APPROACH TO SITE SELECTION FOR COLLABORATION

Track 6: Transdisciplinary Design and Collaboration/Open Track
Knowledge Network: None/Not Tied to Any Network
Methods: Case Study, Statistical/Modeling Analysis
Felichism Kabo (CannonDesign)

1 ABSTRACT

Site selection is a complex, multicriteria process and a key, but poorly understood contributor to the success of new scientific facilities. A list of factors that are deemed important during the site selection process would include things like the budget, sustainability, accessibility, utility and infrastructure costs, and environmental factors such as air quality. Typically omitted from the list are the organizational imperatives that drive the construction of new scientific facilities. For higher education institutions, this includes stimulating interdisciplinary scientific collaboration within their built environments. Traditional site selection processes treat these factors reductively and are ill-equipped to account for the complex effects of the built environment's topology on interdisciplinary collaboration. Systems approaches, like network analysis, enable the simultaneous examination of a system (e.g., built environment) and its parts. In this paper, we discuss the findings from a study that applied spatiosocial network analysis to examine site selection with respect to collaboration for BLDX, a new building for the plant and environmental sciences at a major research university.

We used electronic floor plans to generate and, subsequently, analyze the spatial networks consisting of the potential paths of movement and navigation at two potential sites for BLDX, A and B. This enabled us to identify which of the two sites is better configured to foster potential encounters and interactions. These play a key role in birthing the connections within the research community slated to move into BLDX that would lead to more interdisciplinary scientific collaboration. The future BLDX research community spans multiple scientific disciplines. While it is mainly comprised of the plant sciences, it also includes disciplines such as biochemistry, chemistry, biomedical sciences, and physical sciences. The members of this research community are currently collocated across multiple buildings in a larger neighborhood that encompasses sites A and B. The spatial network for each site represents all possible connections between spaces or rooms, floors, and the buildings housing future members of the BLDX research community. We used the networks to generate and visualize whole network measures on a platform that is specific to spatial network analysis, and on two other platforms that are typically used for social network analysis. These measures enabled us to compare sites A and B with respect to their current (before the addition of BLDX) capacity for supporting potential encounters and interactions.

Site A's spatial network has both a smaller diameter and average distance, indicating lower costs at site A when individuals must physically seek out potential collaborators that are on-site. The results also indicate that site A's paths on average are not only less deep than site B's, but they also have more connections between other pairs of paths. Further, site A's spatial network has both higher cohesion and compactness, and lower fragmentation and clustering. The findings indicate site A is better configured to support potential encounters and interactions, and its higher spatial legibility translates to better cognitive maps for occupants. The study shows that network analysis enables site selection to factor in key organizational imperatives like interdisciplinary collaboration.

ENVIRONMENTS FOR DESIGN PEDAGOGIES: ASSESSING THE DESIGN STUDIO SPACE

Track 6: Transdisciplinary Design and Collaboration/Open Track
Knowledge Network: Interior Design, POE/Programming
Methods: Case Study, Focus Groups
Roberto Castillo (American University of Sharjah)

1 ABSTRACT

The design studio is the most relevant component in the education of architects and designers. Current literature explains the relevance of the studio model and the diverse pedagogical approaches to it (Schön,1988; Boyer & Mitgang, 1996; Salama & Wilkinson, 2007), yet existing research on the physical environment and features of the studio and its impact or influence in the design pedagogies is scarce. This study attempts to contribute to the discussion by interrogating how the physical environment of the design studio can contribute to or hinder the development of pedagogical agendas. In architecture and interior design schools, the learning environment not only provides students and faculty with the required spaces and fittings but can also become subject matter and proof of concept of design ideas, potentially enhancing the educational experience.

The study focuses on a new studio space recently renovated by Interior Design students in the framework of a design-build experience at the Department of Architecture in an American university located in the Middle East's Gulf region. This design studio project attempts to present a model for future studio space in the institution. Current studio spaces in this college are not different from other architecture and interior design schools, primarily focused on providing individual workspaces for students to use inside and outside studio schedules and for one-on-one critique sessions. In this new classroom, the design direction attempts to balance individual and collective activities by granting a variety of work and critique/presentation settings; it also redefines the personal workspace using a different type and arrangement of furniture and provides amenities, including a lounge area for resting and socializing. After almost two years of implementation, this study uses a post-occupancy evaluation setting to assess user experience and perception of this renovated space through observation, focus groups, interviews, and surveys. In this process, the research asks respondents to evaluate and compare the new space with the existing studios and propose changes or additions. Preliminary findings include feedback about specific components of the setting and discussions on topics linking the implementation of the new studio space and the pedagogical direction of the institution.

2 REFERENCES

Boyer, E. L.; Mitgang, L. D. (1996). Building community: a new future for architecture education and practice: a special report. Carnegie Foundation for the Advancement of Teaching.

Salama, A. M. A., & Wilkinson, N. (2007). Design studio pedagogy: horizons for the future. Urban International Press.

Schön, D. A. (1988). Toward a marriage of artistry & applied science in the architectural design studio. Journal of Architectural Education (1984-), 41(4), 4–10.

FUTURE LABORATORIES: SPACES FOR CONVERGENCE AND SCIENTIFIC COLLABORATION

Track 6: Transdisciplinary Design and Collaboration/Open Track
Knowledge Network: Cultural Aspects of Design, Researchers in Professional Practice
Methods: Mixed Methods, Observation
Somayeh Mousazadeh (HDR); Abigail Heithoff (HDR Architecture)

1 ABSTRACT

The Covid-19 pandemic and remote work experience changed our understanding of a singular 'fixed' workplace. Some companies, most famously Shopify (Roy, 2022), have since moved away from the physical office. Meanwhile, inspired by the success of multidisciplinary collaboration in Covid-19 vaccine development, academic and research institutions are embracing an emerging building type: 'convergence' facilities for multidisciplinary collaboration (Mäkinen, 2020), challenging the notion that physical research space will become obsolete. This leads us to an important question: What if addressing some of the world's biggest scientific challenges, like developing the next pandemic vaccine, treating cancer, or combating global warming, depends on the physical research environment and the way we design them?

The latest in science architecture, convergence facilities, pose two questions that challenge how we design typical disciplinary buildings: How should we design the research space to embody multidisciplinary collaboration? And what are the spatial and infrastructure implications of bringing multiple disciplines under one roof? Our year-long fellowship research journey, adopting an anthropological approach, delved into these questions. We found that, just like in other places, 'static' buildings are irrelevant. Learning from timeless science architecture of the past, like the Salk Institute — a result of close collaboration between scientist Jonas Salk and architect Louis Kahn — we discovered that only a 'living' building can adapt to the fast-evolving science needs. The paper relies on hours of interviews with researchers in these facilities to reimagine science architecture in a world where the concept of space is transforming, yet its significance in the research environment remains strong as ever.

BUILDING A BRIDGE BETWEEN RESEARCH AND PRACTICE: ONE FIRM'S APPROACH TO DESIGN RESEARCH FOR MORE HUMAN-CENTRIC ENVIRONMENTS

Track 6: Transdisciplinary Design and Collaboration/Open Track
Knowledge Network: Researchers in Professional Practice
Methods: Mixed Methods, Practice-based Research
Rutali Joshi (HKS Inc.); Renae Mantooth (HKS, Inc.)

1 ABSTRACT

The 2019-2020 Architectural Research Agenda published by the American Institute for Architects (AIA) stated that "research available for study of architecture and buildings is disproportionate to its impact on societies and economies" and proposed three goals: (1) prioritizing research within the architect core competencies and firm culture; (2) increasing investment in research through new and expanded funding, and (3) expanding dissemination of research and promote exchange of findings, as well as methodologies and failures. Despite the growing emphasis, only a handful of architecture firms recognize the value of investing in research capabilities and are dedicated to rethinking their practice to integrate an evidence-based approach. Leveraging research-oriented methods, mindsets, and tools into the typical design process requires a different model for architectural practice, creating opportunities for design researchers and alternative career pathways. Practice-based research requires a more agile approach, adapting to the pragmatic and varied needs of real-world architecture projects. This requires a different mindset than academic research, but the process-oriented and rigorous research methods unite the two ways of working.

The purpose of research in practice is to communicate scientific evidence, explore alternatives outside of pre-established norms or empirical considerations, and create a framework of knowledge to ultimately inform design decisions in pursuit of better people, planet, and business outcomes. Graduate education in the field of design research undeniably trains students in the hard skills required to conduct rigorous research, however, it is not necessarily translational or industry oriented. Seldom do we discuss not only the challenges of conducting research in practice but how to apply the unique skillset gained through academic research projects to an applied project. To advance the industry forward, it is imperative that we not only discuss the what and the why of practice-based design research, but also the how.

In this session, speakers will discuss their transition from academia to researchers in practice, while bridging the two. Session attendees will learn how an international architecture firm not only invests in design researchers but invests in a holistic evidence-based practice. The speakers will share glimpses into ongoing applied research efforts within the Healthcare and Education market sectors, along with tools for discovery and diagnostics, prototyping and testing, and occupancy evaluations through the design process. In addition to employing research methods on applied projects, they conduct independent scholarly research studies that serve as a catalyst for innovation in design. They are also involved in firmwide initiatives such as Living Labs, research incubators, and knowledge sharing, collaboratively pursuing various lines of inquiry in building technology, human experience, material performance, and resilience. By remaining engaged with various academic pursuits including teaching and grant-funded research, in addition to their roles in industry, the speakers will discuss current and future opportunities for pracademics.

As society faces expanding complex challenges and innovation is taking many different forms, more diverse, iterative, and cyclical design research processes have opened new areas of exploration and experimentation in architecture. A bridge between practice and academia is required, with consideration of pragmatic challenges, to advance human-centric environments.

DESIGN STRATEGIES TO PROMOTE TRANSDISCIPLINARY COLLABORATION IN HIGHER EDUCATION SCIENTIFIC RESEARCH FACILITIES

Track 6: Transdisciplinary Design and Collaboration/Open Track
Knowledge Network: POE/Programming, Researchers in Professional Practice
Methods: Mixed Methods, Practice-based Research
Renae Rich (HGA); Oshana Reich (HGA); Chun-Ling "Vanya" Hsu (HGA); Mahshid Jalalian (HGA Architects & Engineers)

1 ABSTRACT

The integration of multiple perspectives, diverse knowledge, and varied specialties is essential for understanding and solving complex problems that face society today. For this to happen, successful collaborations between research disciplines must be formed and sustained. Even 60 years ago, faculty members from the University of Ghent in Belgium sought to establish a "Center of Interdisciplinary Synthesis" to combine the power of research and teaching across their departments. While ultimately unsuccessful at that time, more recent initiatives to create interdisciplinary environments for teaching, learning, and research on higher education campuses are intended to educate a new generation to think beyond the confines of one subject and to tackle difficult challenges together (Budwig and Alexander, 2020).

In-person interaction has been shown to be an important factor in the creation of new social and professional connections, with a higher likelihood of collaborations, or receipt of joint grant funding in the case of university faculty, among those in close physical proximity and traveling similar paths (Hall et al., 2018). A study on the MIT campus found an increase in coauthored publications between researchers relocated to the same building between three to five years after the move, with an even larger effect among those from similar fields across different departments (Miranda and Claudel, 2021).

Two universities in the U.S. have embraced the concept of transdisciplinary research through construction of scientific research buildings that are not owned by a single department but house faculty and research across several disciplines, primarily in engineering and health, with the intention to bring people and their research together. Both incorporate a range of dry to wet lab environments, faculty offices, graduate student workspace, and public amenities, but in very different configurations. One building features a multi-story innovation makerspace, while the other has connected research labs and open student work areas. Evaluations of both buildings were conducted in 2023 to understand how spaces were being used, including collaborative interactions and the frequency and locations in which they were occurring. Opportunities and spaces for collaboration were rated highly in both facilities. Interactions between students occurred most frequently in the research labs, while exchanges with or between faculty members happened most in their offices. However, occupants of both buildings experienced challenges with finding spaces to be alone for individual-focused work. More details about the design features of each building, their impact on fostering and supporting collaboration in research and education, and considerations and recommendations for designing transdisciplinary buildings will be shared in this presentation.

2 REFERENCES

Budwig, N., & Alexander, A. J. (2020). A transdisciplinary approach to student learning and development in university settings. Frontiers in Psychology, 11, 576250.

Hall, K. L., Vogel, A. L., Huang, G. C., Serrano, K. J., Rice, E. L., Tsakraklides, S. P., & Fiore, S. M. (2018). The science of team science: A review of the empirical evidence and research gaps on collaboration in science. American psychologist, 73(4), 532.

Salazar Miranda, A., & Claudel, M. (2021). Spatial proximity matters: A study on collaboration. PLOS One, 16(12), e0259965.

"BALANCING SERVICE AND LEARNING: ASSESSING DATA GATHERING TECHNIQUES FOR COMMUNITY ENGAGEMENT RESEARCH PROJECTS IN ARCHITECTURE AND URBAN PLANNING COURSES"

Track 6: Transdisciplinary Design and Collaboration/Open Track
Knowledge Network: Environmental Design Research Education, Participation
Methods: Mixed Methods
Susanne Cowan; Sarah P Church (Montana State University)

1 ABSTRACT

At its best, service learning can provide useful research for the community, and introduce students to new professional skills. Literature on service learning shows it can enhance students' reflective thinking and engaged citizenship. However, service learning can also face hurdles in aligning projected learning outcomes with community objectives, coordinating semester timelines with the evolution of the longer project, and in ensuring accuracy of student-collected data and analysis. Some service-learning advocates emphasize that the benefit to the students and the community should be balanced; however, most of the literature on service learning focuses on assessment of student learning outcomes rather than measuring community benefits.

This paper will explore how to balance the benefits to the community, specifically accurate, applicable, and publishable research, with student learning. The paper focuses on a case study of a two-year service-learning project by the School of Architecture and the Department of Earth Sciences at Montana State University (MSU), working with a neighborhood association and city staff. This project has documented the existing character of the Northeast neighborhood of Bozeman, MT, and the residents' perceptions of the social, economic, and architectural changes that have occurred in the face of growth and gentrification. Faculty and students have helped to conduct and analyze a physical inventory of over 750 homes, a survey of 143 stakeholders, and qualitative analysis of interviews with nine residents. The final product, a 176-page professional report, included GIS maps, charts, and quotes, which provide data for future planning in the Northeast Neighborhood, including revisions of the downtown business zoning district and revisions to the Neighborhood Conservation Overlay District. During this service-learning research, the faculty learned several lessons about how to integrate long-term city planning research investigations into a semester-long class. Breaking the project into small chunks that could be accomplished with different student groups allowed us to accomplish this project over four semesters in four different courses, while giving each student group a sense of the big picture of community issues and a clear outcome of which they could be proud. While we kept the student projects aligned with community goals, we also did not perfectly align student class products with the final product for the community; this allowed students to explore creative directions with more freedom, while still having quality control over the final research report. Streamlining data collection with Survey 123, a GIS compatible phone application, and spot checking was key to balancing efficient student data collection with accuracy. Hiring four upper division and graduate students to complete data collection and analysis was also key to ensuring robust outcomes worthy of publication and application to city planning projects. Overall, this project was successful at balancing student learning outcomes with community objectives; a few tweaks and lessons learned from literature on the best practices in service learning can make this project a model for future service-learning projects addressing city planning issues.

CREATING HUMAN-CENTERED SPACES: INTEGRATING RESEARCH INTO INTERIOR DESIGN SENIOR CAPSTONE STUDIO COURSE FOR SERVICE-LEARNING PROJECTS

Track 6: Transdisciplinary Design and Collaboration/Open Track
Knowledge Network: Environmental Design Research Education, Interior Design
Methods: Case Study
Suining Ding (Purdue University Fort Wayne)

1 ABSTRACT

Environmental design has undergone rapid changes when the design discipline expands its scope and confronts increasingly complex social and ecological challenges. Additionally, the design discipline has been intertwined with research, science, and technology, and the design approach has become more human-centered. Since the design discipline is both grounded in research and practice and the advent of evidence-based design, design research has been an integral component in the design process. Although the fundamental design principles remain unchanged, the accelerated pace of discoveries in interdisciplinary research has made rethinking design education necessary. This accelerated change demands designers to swiftly respond to the drastic shift in the design process and methodology and create design solutions that are evidence-based and research-informed.

According to the literature review, service-learning is a course-based, credit-bearing educational experience in which students participate in an organized service activity that meets identified community needs. In a service-learning course, students use knowledge and skills directly related to the course content. Students also reflect on the service activity in such a way as to gain a further understanding of course subjects, a broader appreciation of the discipline, and an enhanced sense of personal values and civic responsibility (Bringle & Hatcher, 2000). Additionally, research indicates that students with field experience demonstrate a greater understanding of the complexity of real-world problems than those without (Eisman, J. E. Zlotkowski, & Tsang, 2000). Therefore, service-learning as a methodology for teaching provides irrefutable benefits. How to integrate research into a senior capstone course with service-learning projects? How to create human-centered spaces for community partners? This presentation addresses these two teaching issues.

This study analyzes service-learning projects in senior capstone studio courses in the past ten years at one institution. The findings show that when research topics and components are clearly identified, students gain a better understanding of evidence-based design. Each student identified a specific design feature and integrated the research components, such as sustainability, human factors, universal design, and privacy needs. As part of the project programming phase, students interacted with community partners, gathered information, identified research evidence through literature reviews, and incorporated research evidence into their designs. This presentation demonstrates how research is integrated into a service-learning senior capstone course. Course modules, students' learning outcomes, and students' projects that take different approaches to integrating research evidence are presented. This presentation aims to provide a curriculum map that articulates the course modules addressing different research theories and components. Similarly, these course modules correspond directly to student learning outcomes based on the accreditation standards set by the Council for Interior Design Accreditation (CIDA).

2 REFERENCES

Bringle, R. G., & Hatcher, J. A. (2000). Institutionalization of service learning in higher education. The journal of higher education, 71(3), 273-290.

Eisman, G. S., J. E. Zlotkowski, & Tsang, E. (2000). What I never learned in class: Lessons from community-based learning. Projects that matter: Concepts models for service-learning in engineering, Service-Learning in Disciplines, Washington, D.C.: American Association for Higher Education, 19, 13-25.

MAKING A DIFFERENCE THROUGH COMMUNITY VISIONING: TEACHING HUMAN-CENTERED DESIGN

Track 6: Transdisciplinary Design and Collaboration/Open Track
Knowledge Network: Environmental Design Research Education
Methods: Case Study
Elizabeth Payne Tofte (South Dakota State University)

1 ABSTRACT

As urban communities grapple with environmental design issues such as climate change, urbanization, and social inequality, we recognize students in the arts and built environment disciplines have a part to play. They are well positioned to lead and to contribute in creating positive, transformative, and resilient change. This presentation describes how university students in an interdisciplinary design course participated in a community visioning project with local stakeholders in Moline Acres, Mo. The community was facing decisions to repurpose a dilapidated 1960s-era housing complex.

2INTRODUCTION

The presentation includes a description of the federally funded project, goals of the funding agency (EPA/TAB) and learning objectives for students taking the design course. This presentation also examines the means and practices used to engage stakeholders in a public process to realize a healthy solution to the problem. Course instructors believe robust pedagogy requires students visit the site and successfully interact with community stakeholders. These are what Steve Revington refers to as authentic learning opportunities. Students' authentic learning experiences may include technical skills, discipline-specific knowledge, and behaviors relevant to conducting community visioning in the real-world. Authentic learning "is about developing a set of culminating skills sets, within a realistic timeline, using self-motivated, natural inquiry methods to create a useful product to be shared with a specific, meaningful audience... It's a quality outcome or product of significance that is intended for community consumption or betterment" (Steve Revington). Faculty utilized Revington's hierarchy of instructional skill development to conduct systematic inquiry of student learning during a semester-long community visioning project.

3 METHODS

This case study is based on students' written reflections, photo documentation, critical thinking and enquiry, as well as detailed sketches they generated to illustrate ways the dilapidated housing tower might be repurposed. These were included in the final (professional) work product funded by the Environmental Protection Agency's Technical Assistance to Brownfields Program. Students were interviewed one final time and asked to reflect on the significance of this project on their current output using Revington's Learning Artifact Comparison Chart.

4 RESULTS

Eight students experienced immersion into the world of community visioning, adaptive reuse of a brownfield site, placemaking, and compiling a final (professional) work product. Students reported being more motivated to take leadership roles in the design studio and consider issues such as climate change, urbanization, and social inequality when generating initial ideas for new studio projects.

5 CONCLUSION

Shaping more sustainable, equitable, and resilient communities through externally funded visioning projects is an effective way to assist communities in preparing an adaptive reuse plan while providing the next generation of design professions an opportunity to gain real-world skills and knowledge. Students reported being more motivated to take leadership positions, share skills and knowledge with classmates, and understand the nuanced details involved in preparing community visioning reports as compared to project-based learning or reading documented case studies.

6 REFERENCE

Authentic Pedagogy: Proposition for Education Dr. N. D. Trivedi, et. al. IJRAR- International Journal of Research and Analytical Reviews IJRAR- Research Paper, E ISSN 2348 –1269E, ISSN 2348 –1269

INSTRUCTOR-LED BLENDED PRACTICES FOR ACCESSIBLE AND FLEXIBLE STUDIO TEACHING AND LEARNING: INSIGHTS FROM US ARCHITECTURAL DESIGN STUDIO INSTRUCTORS

Track 6: Transdisciplinary Design and Collaboration/Open Track
Knowledge Network: Cyberspace & Digital Environments, Environmental Design Research Education
Methods: Interviews

Manju Adikesavan (CUNY Graduate Center)

1 ABSTRACT

This presentation reflects on the increasingly blended (online and in-person) nature of the post-pandemic campus-based design studio and its implications for facilitating accessible, inclusive, and flexible studio learning and teaching experiences. Despite the widespread digitalization of learning and architectural design and construction practices and the post-COVID intensification of remote and hybrid operations, the architectural design studio is taught in predominantly in-person or campus-based mode. Post-COVID interviews of undergraduate and graduate students (N=27) and instructors (N=32) of US architecture programs indicate a continuing preference for in-person or campus-based studio teaching and learning to maximize tactile, socio-constructivist, and vicarious learning experiences. However, the instructors also favored small but meaningful blending of online and in-person practices for office hours, lectures, group critique sessions, final juries, etc., to make studios more accessible to students with competing work, caregiving, parenting, and related commitments. Such instructor-led blended practices offer alternate options for students and instructors to be proximate and present with each other where and when it is most needed, thus helping enable responsive and engaged studio learning experiences. Thematic analysis of the interviews suggests that instructor-led blended practices in campus-based programs foster relational proximity between instructors and students by facilitating instructor-student interactions when and where needed. Instructor-led blended practices in campus-based practices suggest the emergence of an alternate ethos of student-instructor interaction and relatedness in post-pandemic campus-based studios.

RESEARCH ON THE COMMERCIAL FORM CHARACTERISTICS OF URBAN RAILWAY STATIONS-TAKING THE ZHEJIANG METROPOLITAN AREA AS AN EXAMPLE

Track 6: Transdisciplinary Design and Collaboration/Open Track
Knowledge Network: Environmental & Architectural Phenomenology
Methods: Mixed Methods, Statistical/Modeling Analysis
Yingying Huang (Benzhi Studio, BJTU); Zhongzhong Zeng (Beijing Jiaotong University); Bo Zhang (Oklahoma State University)

1 ABSTRACT

The space-time compression effect of urban railway construction promotes the integrated development of urban agglomerations. However, different railway station areas within the metropolitan area have very different commercial development characteristics, and there is a lack of systematic quantitative research on the influencing factors. The construction of urban railways is in full swing in Zhejiang, China. In the future, a total of 11 urban railways in Zhejiang's four major metropolitan areas of Hangzhou, Ningbo, Wentai and Jinyi will operate in their own networks. Therefore, this article selects 4 intercity railways in the Hangzhou metropolitan area that have been opened to traffic and a total of 45 railway stations for research. The aim is to use a radius of 3,000 m as the research area of the station area to create a circular space construction model. The open map platform is used to obtain planar morphological indicators such as the number, density and distribution of POIs in the site area; combined with GIS data to generate indicators related to development intensity such as land completion rate, building coverage rate, height floor area ratio, etc. Then, the typology analysis method was used to summarise the commercial form characteristics of the station area under the guidance of urban railways, and SPSS was used to conduct a correlation analysis on the commercial form indicators and potential influencing factors of the station area. The research shows that commercial density has a significant positive correlation with the annual GDP of the district and county and the proportion of tertiary industry, a significant negative correlation with the distance between the site and the administrative centre, and a significant positive correlation with the number of years the site has been built. Further analysis revealed that the impact of GDP on commercial distribution patterns is much stronger than other factors, and that a longer station completion age is a necessary condition for high intensity development in the outer circle of the station area.

PLACES OF HEALTH DISPARITIES: A TRANSDISCIPLINARY APPROACH TO THE INTEGRATION OF HUMAN-CENTRIC DESIGN IN BUILT AND NATURAL ENVIRONMENTS

Track 6: Transdisciplinary Design and Collaboration/Open Track
Knowledge Network: Children, Youth & Environment, Health
Methods: Case St Sara Clement (Kennesaw State University); Pegah Zamani (Kennesaw State University); Joshua Robinson (Firm)

1 ABSTRACT

In what ways can the design of the built environment enhance spatial efficiency without compromising the spatial experience of its diverse occupants? How can such design interventions be mapped out, measured, and monitored? This transdisciplinary research focuses on the possibilities of designing optimized built environments while advancing their inhabitants' well-being. The study centers on a pivotal spatial setting: healthcare emergency facilities, with a particular emphasis on a safety net hospital serving uninsured patients from marginalized communities - particularly with physical and/or mental health disorders. In the context of emergency care delivery, as a case, we highlight the potential oversight of human-centered experience within the internalized and efficiency-driven nature of emergency departments (EDs) that impact inhabitants, especially in moments of extreme stress. We analyze visible and invisible, clinical and environmental factors, along with human-centered design interventions associated with efficient space planning that fosters connection between clinicians, patients, and visitors. Striking a delicate balance between efficient space planning, streamlined staff workflows, and a healing environment is crucial in critical healthcare settings. The research employs a multi-faceted transdisciplinary methodology, incorporating an extensive literature review and case studies, to identify innovative practices that improve the overall experience for all stakeholders. The research utilizes evidence-based design as a catalyst to formulate a set of parameters for developing design strategies that aid in the patient's healing process while responding to the needs of a diverse range of users as well as future needs. Methodologically, the study also fosters a systematic inquiry into the development and integration of disciplinary and interdisciplinary research methods. The research addresses the complexities of optimizing occupants' well-being through design across a variety of built environment settings from the waiting area to the clinical spaces. By examining these critical spaces, we aim to identify ways in which designers can spearhead creating more effective sustainable, equitable, and healthier environments for diverse populations.

2 REFERENCES

Winkel G. H., Holahan C. J. (1985). The environmental psychology of the hospital: Is the cure worse than the illness? Journal of Prevention & Intervention in the Community, 4(1–2), 11–33.

Ulrich R. S. (1991). Effects of interior design on wellness. Theory and recent scientific research. Journal of Health Care Interior Design, 3, 97–109.

TRANSFORMATIVE [INTER]PLAY: INTERIOR DESIGN, EDUCATION, AND COMMUNITY COLLABORATION IN CRAFTING INNOVATIVE LEARNING ENVIRONMENTS

Track 6: Transdisciplinary Design and Collaboration/Open Track
Knowledge Network: Children, Youth & Environment, Interior Design
Methods: Case Study, Practice-based Research
Rana Abudayyeh (University of Tennessee, Knoxville)

1 ABSTRACT

"Education is a natural process carried out by the child and is not acquired by listening to words but by experiences in the environment." These words of Maria Montessori formulate the charge for this project, which is led in collaboration with a local elementary school and aims to expand the understanding of the connection between interior design and educational environments. How can interior design activate and empower experiences in educational settings that are integral to the natural learning processes and nurture curiosity and community in elementary-aged children? As our society, and especially our educational communities, recover from the effects of the COVID-19 pandemic, rethinking learning environments through the lens of interior design is essential. Such investigations are the catalyst for pioneering new approaches to designing holistic learning settings that support mental and physical health and well-being. Transformative [Inter]play addresses the study and development of an exterior classroom located in an internal courtyard of a public elementary school (name redacted for anonymous review). As with many public schools in the area, our partner school grew in size over the years to accommodate the influx of students. Several additions were made to the original structure, which dates back to 1952. The school's central courtyard unites the original school building with newer additions and emerges at the resulting intersection. While it is a liminal exterior space, it is interior and collective in nature where activities converge. Our research team will work with the school's faculty, staff, students, and larger community to reinvent the space as an exterior classroom/learning common that nurtures exploration and establishes a robust learning collaborative. We will engage in community-based workshops and town-hall style dialogue, where all interested parties can discuss the vision for the learning commons. The first stage of this endeavor was the research and design development phases of the project. These phases are the foundation of the project and conclude and paved the way for developing a comprehensive set of design guidelines that can be utilized by design professionals, educators, students, the larger community, and other stakeholders. These design strategies ground design approaches for innovative interactive learning spaces compatible with new teaching methods and align with the engagement priorities and shared governance of the learning commons. Furthermore, they actively demonstrate the public benefits of interior design and highlight the advantages of community problem-solving and co-design to empower responsible and resilient placemaking. Not only is this work essential to the school's future, but it also serves as a case study for addressing the changes in spatial narratives and needs of learning environments and their development in a rapidly changing educational landscape. The project's commitment to advancing the school's mission through Interior Design's ethos of empathy, service, and engagement. It is fueled by a partnership that undoubtedly brings together diverse constituencies and will be fruitful on many fronts, positively impacting generations of young learners and committed educators for years to come.

HARMONIZING INTERIOR DESIGN EDUCATION: A CROSS-CULTURAL PEDAGOGICAL EXPLORATION OF CO-INSTRUCTION ACROSS CONTINENTS IN STUDIO PRACTICE

Track 6: Transdisciplinary Design and Collaboration/Open Track
Knowledge Network: Interior Design, International Connections
Methods: Action/Participatory Research, Mixed Methods
Haroon Sattar (University of Central Missouri); Mehrab Iftekhar (University of Asia Pacific);
Tahmina Rahman (University of Asia Pacific)

1 ABSTRACT

Teaching in international institutions holds a crucial role in shaping students for the globally interconnected landscape of the 21st century. This study explores the dynamic experience of teaching Interior Design Studio in two diverse locations during the 2022 academic year, employing a distinctive co-instruction model. The lead instructor facilitated the studio remotely through online platforms, while a local instructor conducted in-person sessions. The curriculum integrated various teaching methods, including lectures, presentations, and design projects, engaging students with diverse materials such as reading assignments, short papers, and evidence-based design research.

The course's outset emphasized clear learning objectives, incorporating reading assignments, lectures, and papers to establish a foundational knowledge base. A structured approach guided studio design projects, commencing with research to understand project objectives, goals, and client needs, aiming for a comprehensive grasp of interior design facets. Initial reflective assignments prompted students to contemplate their perspectives on the profession and professional aspirations. Core principles of design thinking were solidified through a combination of reading assignments and hands-on exercises encompassing interior spaces, furniture design, and detailing. The semester concluded with a challenging design project, requiring students to develop a branding concept incorporating cultural aspects in interior space planning, followed by self-evaluations. The presentation will delve into highlighting the students' learning experiences, the challenges of offering such a hybrid course, and the wealth of experiences the faculty gathered in such a unique teaching scenario.

The presentation will highlight the difficulties associated with implementing a collaborative co-instructional approach within a transdisciplinary framework. The challenges encountered include the limitation of offering only one studio in an Interior Design context within an architecture program. Prior exposure to an Interior Design lecture course before the studio added an extra layer to the curriculum. The architecture program's overarching mission to instill an understanding of interior design among students aiming for an architectural degree further shaped the course objectives. Additionally, managing time differences between the two locations posed logistical challenges. Effective communication between instructors and students utilizing various online platforms and in-person participation in a studio setting were additional challenges. Encouraging students to expand critical thinking beyond their local context emerged as a pivotal challenge.

Despite challenges, teaching Interior Design Studio across diverse locations, utilizing co-instruction and varied learning methods, offers a rich educational experience. Inherent to international education complexity, the study emphasizes learning outcomes focused on cultural sensitivity, global market navigation skills, innovation, adaptability, and a refined design critique process. The paper aims to contribute to effective teaching strategies in the realm of international interior design education, recognizing its pivotal role in preparing students for a globally connected future.

ABSTRACTS CONTINUED

Group Presentations: Practitioner/Researcher Storytelling

DECARBONIZATION OF HEALTHCARE BUILDINGS

Track 2: Sustainable Buildings and Infrastructure
Knowledge Network: Health, Sustainable Planning, Design & Behavior
Methods: Case Study, Mixed Methods

Heather Burpee (University of Washington, Integrated Design Lab); Mara Baum (DIALOG Design); Walt Vernon (Mazzetti)

1 ABSTRACT

This practitioner/researcher-led presentation will outline current research, methods, and case studies for decarbonizing new hospital buildings. Rapid decarbonization is imperative to ameliorate the environmental impacts of climate change. Buildings, representing a large portion of overall greenhouse gas (GHG) emissions have a large part to play in realizing overall emission reductions. Healthcare buildings have disproportionately high GHG emissions – rather than healing the environment, they are rapidly contributing to harm of the people and planet. Several reports outline the global GHG impact of healthcare. These reports estimate that healthcare represents 4.4 4.6% of total global emissions (HCWH 2019), (Eckelman et al. 2020). To put this in perspective, according to Healthcare without Harm, "if the global health care sector were a country, it would be the fifth-largest greenhouse gas emitter on the planet (HCWH 2019)." To better understand healthcare's global impact on GHG emissions, it is helpful to know where those emissions arise and what methods to use to mitigate those impacts.

This session will focus on evidence that shows healthcare's climate impacts, and solutions for decarbonizing hospitals. These impacts and solutions are based on the forthcoming ASHE/ASHRAE "Decarbonize Hospital Buildings Guide." The authors of that guide will share in this session a peek "under the hood" of the guidebook with technologies and methods that are key to decarbonizing our healthcare system. You will learn best practices to decarbonize that can be applicable beyond the healthcare setting, and review case studies from healthcare decarbonization projects.

2 REFERENCES

Eckelman, M. J., Huang, K., Lagasse, R., Senay, E., Dubrow, R., & Sherman, J. D. (2020). Health care pollution and public health damage in the United States: An update. Health Affairs, 39(12), 2071–2079. https://doi.org/10.1377/hlthaff.2020.01247

Health Care Without Harm (HCWH). (2019).

Health care's climate footprint: How the health sector contributes to the global climate crisis and opportunities for action. https://noharm-global.org/sites/default/files/documents-files/5961/HealthCaresClimateFootprint_092319.pdf

CLIMATE-CONTEXT-CULTURE-BASED DESIGN

Track 2: Sustainable Buildings and Infrastructure
Knowledge Network: Researchers in Professional Practice, Sustainable Planning, Design & Behavior
Methods: Mixed Methods, Practice-Based Research
Syed Abu Sufian Kushol (Bangladesh University of Engineering and Technology);
Tasneem Tariq (Pennsylvania State University)

1 ABSTRACT

This presentation aims to foster awareness and understanding of the pivotal role that climate, context, and culture play in shaping architectural design. Through the exploration of three distinct projects, this session will highlight the integration of environmental parameters, local materials, techniques, and community culture to create a cohesive design vocabulary.

The first project, a water pavilion in Bangladesh, draws inspiration from the region's climatic conditions. The design features generous overhangs that provide shaded areas conducive to social interactions. Encompassing a swimming pool, gym facilities, and a recreational pavilion, the Urban Oasis integrates local architectural forms while prioritizing framed views, privacy, and user comfort.

The second project is a community learning center located in northern Bangladesh. Taking a multidisciplinary approach, this project facilitates cultural and artistic exchange between artists and local communities. The pavilion-like structure, inspired by vernacular architecture, incorporates sustainable materials such as Greek cement sheets for roofing and repurposed old doors and window shutters.
The third project, an award-winning multi-story green office building in Bangladesh's capital, exemplifies energy efficiency and environmental responsibility. The design prioritizes daylight access, natural ventilation, and open spaces, while also incorporating water bodies, rainwater harvesting, and solar power. Material selection minimizes environmental impact, adhering to the principles of reduce, reuse, and recycle, fostering a seamless connection between design and natural elements.

By examining these projects, this presentation will guide design professionals in initiating projects with a comprehensive consideration of site-specific environmental forces, available techniques, and opportunities for reusing materials. This holistic approach enhances environmental consciousness and inspires impactful designs that positively contribute to the community.

TRAUMA-INFORMED DESIGN FOR PERMANENT SUPPORTIVE HOUSING: FIVE YEARS LATER, THE BUILT RESULTS OF A PRACTITIONER/ RESEARCHER COLLABORATION

Track 4: Design for Health and Wellbeing
Knowledge Network: Researchers in Professional Practice, Residential Environments
Methods: Practice-Based Research
Christina Bollo (University of Oregon); Amanda Donofrio (BDA Architecture & Planning)

1 ABSTRACT

Following a presentation at EDRA50 (2019), we report on the results of a practitioner/researcher collaboration that has resulted in 96 trauma-informed apartments in two new-construction permanent supportive housing (PSH) projects in Oregon. The research presented at EDRA50 was later developed into a paper published in Housing & Society (Bollo & Donofrio, 2021), outlining actionable patterns for the design of trauma-informed PSH. BDA Architecture and Planning applied these patterns and principles in designing The Commons on MLK, completed in January 2021, and The Nel, completed in July 2022. This presentation describes the transition from incorporating trauma-informed design (TID) principles in the first building to refining them in the second. The evolving practice/research process led to the development of replicable TID patterns, each combining at least two core principles. These patterns for PSH include:

- *Connecting common area spaces with interior windows.*
- *Providing common areas that maximize resident choice.*
- *Supplying a visible, central third stair to encourage engagement.*
- *Including two exits from spaces that host one-on-one interactions.*
- *Creating wide, orienting circulation.*
- *Designing future opportunities for resident personalization and identity expression.*
- *Clearly defining the entry with an in-swinging door and nearby shelf or closet.*
- *Reducing concealed spaces inside dwelling units while allowing for optional covering.*

Illustrations from completed buildings will highlight these patterns. We will also discuss specific lessons learned from The Commons on MLK and how they were applied to The Nel. For example, after The Commons was occupied, staff reported that residents preferred to avoid conflicts at the front desk but were required to pass through this zone to exit. In The Nel, the design includes two exit doors at the lobby to provide an alternative route. Similarly, while The Commons' reception area was intentionally designed as an open, welcoming space, staffing changes left the desk frequently empty, creating challenges. In The Nel, a transaction window was used to balance openness with security and flexibility.

The collaboration began in 2018 when the architect was hired to design the first PSH project for her community and discovered that, at the time, no professional resources addressed PSH design comprehensively. Since then, in addition to completing two buildings, the team has given over 50 presentations to designers, nonprofit developers, and public agencies on trauma-informed design in PSH. Feedback from these presentations has contributed to further refining TID patterns.

In the next phase of our collaboration, we will conduct post-occupancy evaluations of both buildings. We look forward to engaging with the EDRA community for potential research questions and methodologies.

2 REFERENCES

Bollo, C., & Donofrio, A. (2021). Trauma-informed design for permanent supportive housing: Design patterns for dignity and stability. Housing & Society, 48(1), 43-65. https://doi.org/10.1080/08882746.2021.1916269

CREATING A VIDEO SERIES TO PROMOTE HANDS-ON GARDENING IN CHILDCARE CENTERS

Track 4: Design for Health and Wellbeing
Knowledge Network: Children, Youth & Environment
Methods: Mixed Methods

Muntazar Monsur (Texas Tech University); S. Y. Andalib (Texas Tech University); Nazia Afrin Trina (Texas Tech University); Robin Moore (Natural Learning Initiative, College of Design, NC State University); Nilda Cosco (North Carolina State University)

1 ABSTRACT

The Extended Reality (XR) Lab at Texas Tech University and the Natural Learning Initiative (NLI) at NC State University collaborated to create a video series promoting hands-on gardening in childcare centers. The goal of the videos is to translate research findings on the benefits of hands-on gardening for young children into an engaging visual format, encouraging childcare professionals to integrate gardening into their programs.

The health benefits of hands-on gardening in early childhood have gained increasing attention, particularly for its potential to improve physical activity and healthy eating behaviors. Research indicates that habits related to physical activity and food intake are established early in life (Wethington, 2005). Introducing young children to gardening can positively shape their long-term health trajectories (Soltero et al., 2021). Additionally, hands-on gardening enhances cognitive and social development, improving language and communication skills, math and science literacy, and emotional regulation (Miller, 2007).

Despite these benefits, hands-on gardening remains rare in most childcare centers due to barriers such as teacher workload, limited outdoor time, and lack of resources or expertise. To address these challenges, the video series provides a step-by-step guide to starting and maintaining a garden, demonstrating its integration into early childhood education. However, translating academic research into compelling visual content for non-academic audiences presents unique challenges.

This presentation will discuss how the two teams collaborated on research, storyboarding, scripting, voiceover, revisions, and final production to create an accessible, research-based video series. It will also explore how the videos were received by childcare professionals and their potential to advance a culture of hands-on gardening in early childhood education.

2 REFERENCES

Miller, D. L. (2007). The seeds of learning: Young children develop important skills through their gardening activities at a Midwestern early education program. Applied Environmental Education & Communication, 6(1), 49–66. https://doi.org/10.1080/15330150701318828

Soltero, E. G., Parker, N. H., Mama, D., Scherezade, K., Ledoux, T. A., & Lee, R. E. (2021). Lessons learned from implementing a garden education program in early child care. Health Promotion Practice, 22(2), 266–274. https://doi.org/10.1177/1524839919868215

Wethington, E. (2005). An overview of the life course perspective: Implications for health and nutrition. Journal of Nutrition Education and Behavior, 37(3), 115–120. https://doi.org/10.1016/S1499-4046(06)60265-0

IMPACT OF FLUORESCENT LIGHTING COLOR TEMPERATURE ON AUTISTIC COGNITIVE PROCESSES: A COLLABORATIVE EXPLORATION BETWEEN DESIGN, NEUROSCIENCE, AND BEHAVIORAL SCIENCES

Track 4: Design for Health and Wellbeing
Knowledge Network: Environmental Design Research Education, Interior Design
Methods: Interviews, Statistical/Modeling Analysis

Cherif Amor (Virginia Commonwealth University); Ghizlane Bendriss (Weill Cornell); Hiathem El-Hammali (Virginia Commonwealth University); Ryad Ghanam (VCUarts Qatar); Edward Boone (VCU Richmond)

1 ABSTRACT

Behavioral evidence suggests that fluorescent lighting, among various indoor environmental variables (e.g., noise, ambient temperature, and air quality), plays a critical role in either facilitating or hindering daily activities (Amor, 2019; Rashid & Zimring, 2008). Specifically, autistic individuals often experience increased distraction under fluorescent lighting, which can lead to agitation, hyperactivity, stress, and impaired cognitive skills, ultimately affecting their overall well-being and performance (Gaines et al., 2014).

Despite ongoing debates in environment-behavior research regarding the impact of fluorescent lighting on cognitive, behavioral, and psychosocial outcomes, little is known about the correlation between brain neural activity and the color temperature of fluorescent lighting in autistic individuals. Neuroscience research indicates that seeing color activates the ventral occipital cortex, including the fusiform and lingual gyri (Hsu et al., 2012; Morita et al., 2004). However, the impact of different lighting color spectra—referred to as correlated color temperature (CCT)—on autistic cognitive processes remains largely unexplored. This funded pilot study aims to investigate the behavioral and neural brain wave responses of three autistic individuals when exposed to four lighting conditions:

- *Three chromatic lighting colors:*
- *Warm White (WW) with a 2700K CCT*
- *Cool White (CW) with a 4100K CCT*
- *Daylight (DX) with a 5500K CCT*
- *One achromatic color condition (black, white, and gray)*

Each lighting condition will be evaluated across three test applications: numerical reasoning, reading comprehension, and problem-solving. Using an immersive virtual environment (HTC VIVE Focus V3) and an EMOTIV EPOC+ (14-channel wireless EEG headset), neural brain wave data will be recorded. Additionally, behavioral data will be collected through a post-experiment satisfaction survey.

This research aims to establish behavioral and neural benchmarks for lighting color temperatures that either facilitate or inhibit cognitive processes in autistic individuals. Preliminary observations suggest that higher-color-temperature fluorescent lighting may enhance prefrontal cortex activation, which is responsible for cognitive processing. However, further large-scale studies are needed to substantiate these initial findings.

OPTIMIZING ED EFFICIENCY THROUGH MULTIDISCIPLINARY COLLABORATION: EXAMINING THE IMPACT OF THE BUILT ENVIRONMENT ON EFFICIENCY & THROUGHPUT

Track 4: Design for Health and Wellbeing
Knowledge Network: Health, Researchers in Professional Practice
Methods: Mixed Methods, Practice-Based Research

Lesa Lorusso (Gresham Smith); Tatiana Orozco (Gresham Smith); Dagmar Rittenbacher (Gresham Smith); Doug Spies (AIA, AHCA

1 ABSTRACT

The Emergency Department (ED), known as the gateway into the healthcare system, has been paradoxically described as both a safety net and a war zone (Chang et al., 2021). This is especially the case in a community-based regional hospital providing care to more than 300,000 residents. This presentation provides insights into a multidisciplinary collaboration between practitioners, design researchers, and medical professionals who partnered to understand ways to optimize efficiency in a community-based ED.

2 OBJECTIVES

The team sought to identify areas of improvement for operational challenges plaguing the ED post-pandemic. Although staffing within the ED was close to pre-pandemic levels through a staff-sharing strategy integrating inpatient floor staff into the emergency department, annual volume and average daily census had not yet returned to full capacity. Despite this, throughput times continued to increase, leading executive leadership to seek assistance in understanding the causes of persistent operational inefficiencies.

3 METHODS

A mixed-methods approach included analysis of ED throughput data, a design thinking workshop, staff survey, and on-site behavior mapping. ED throughput data included patient volume, length of stay, emergency severity index level, staffing, arrival times, and diagnostic turnaround times. The design thinking workshop aimed to identify opportunities for improvement and future needs. An online staff survey was conducted on workplace culture, safety, quality and efficiency, technology integration, and adaptability and resiliency. Accelerometers tracked staff steps and activity levels (Zhu et al., 2022), and spaghetti maps of shift routes were recorded. Accelerometer data were analyzed by step counts and time spent in different activity levels ("light" and "moderate"). Longitudinal graphs of step counts over time per participant were also generated.

4 OUTCOMES

Improvements were identified in four categories: person, place, process, and product.
Person: Staff culture was critically low and required improvement through amenities and additional environmental services. Ambulance volumes also needed to be reduced to divert overflow patients elsewhere.

- **Place:** *Recommendations included developing a psychiatric crisis center off-unit, reducing hallway beds, and providing staff respite spaces.*
- **Process:** *Identified improvements included categorizing avoidable ED admissions, implementing a wheelchair storage solution, and standardizing medication dispensers and storage.*
- **Product:** *Suggested changes included streamlining the charting process, improving communication on room releases, installing telemetry in provider spaces, and adding performance monitors and lab printers within the ED.*

2 REFERENCES

Chang, Y. C., Nkambule, N. S., Xiao, X., Ngerng, R. Y., & Monrouxe, L. V. (2021). Safety net, gateway, market, sport, and war: Exploring how emergency physicians conceptualize and ascribe meaning to emergency care. Social Science & Medicine, 287, 114338.

Zhu, X., Ory, M. G., Xu, M., Towne Jr, S. D., Lu, Z., Hammond, T., & Lee, C. (2022). Physical activity impacts of an activity-friendly community: A natural experiment study protocol. Frontiers in Public Health, 10, 929331

PLANTING DESIGN FOR EARLY CHILDHOOD ENVIRONMENTS: A STORYTELLING APPROACH TO PROMOTE PLAY AND LEARNING VALUES OF PLANTS IN CHILDCARE CENTERS

Track 4: Design for Health and Wellbeing
Knowledge Network: Children, Youth & Environment
Methods: Mixed Methods
Muntazar Monsur (Texas Tech University); Nazia Afrin Trina (Texas Tech University); Umme Haque (Texas Tech University); Jared Horsford (Texas Tech University)

1 ABSTRACT

Licensed childcare centers and family care homes are critical early childhood environments where nearly 21 million children aged 0 to 5 years spend the majority of their waking hours in the U.S. (Cui & Natzke, 2021). The quality of these environments significantly impacts children's health and well-being (Smith et al., 2016). As young children spontaneously learn from their surroundings, the built environment plays an important role in advancing informal learning opportunities. Nature, as a constant source of environmental learning, has gained wide attention in early childhood education.

Plants, one of the most important natural elements, play a vital role in childcare centers by fostering a nurturing environment essential for holistic development. They provide shade, creating comfortable outdoor spaces that encourage physical activity and imaginative play. The presence of plants promotes curiosity, stimulating cognitive development as children explore the natural world. Beyond aesthetic appeal, plants offer educational opportunities, teaching children about growth cycles, biodiversity, and environmental stewardship.

Interacting with plants fosters sensory development, as children touch, smell, and observe their surroundings (Trina, 2022). Greenery also has a calming effect, positively influencing emotional well-being and reducing stress—benefits that extend to overworked childcare professionals. Teaching environmental responsibility through plant care instills values of sustainability and prepares children to become conscientious global citizens.

Despite having so many benefits to children and teachers, such play and learning values of plants are often ignored in childcare planting design and species selection. Innovative approach and content creation is needed to help childcare professionals and landscape architects understand the delicate planting design approach needed in childcare centers to compliment the developmental needs of children 0 to 5 years of age. A group of faculty and graduate students at the Department of Landscape Architecture at Texas Tech University adopted a storytelling approach to create content that will meaningfully translate the research and justification behind a unique approach of planting design in childcare centers necessary to advance children's health, well-being, play, and learning.

2 REFERENCES

Cui, J., & Natzke, L. (2021). Early childhood program participation: 2019. National Center for Education Statistics, U.S. Department of Education.

Smith, W. R., Moore, R., Cosco, N., Wesoloski, J., Danninger, T., Ward, D. S., Trost, S. G., & Ries, N. (2016). Increasing physical activity in childcare outdoor learning environments: The effect of setting adjacency relative to other built environment and social factors. Environment and Behavior, 48(4), 550–578.

Trina, N. A. (2022). The play value of plants: Altering parts of plants as loose parts play materials in early childhood learning landscapes (Doctoral dissertation).

PLAYFUL PLACEMAKING: METHODS AND TOOLS TO ENHANCE COMMUNITY VOICE IN DESIGN

Track 4: Design for Health and Wellbeing
Knowledge Network: Participation, Sustainable Planning, Design & Behavior
Methods: Focus Groups, Action/Participatory Research
Cheryl Salazar; Joy Kuebler (PLAYCE Studio Ltd.); Lisa Iulo (Penn State Hamer Center for Community Design); Sara Hadavi (Kansas State University); Claire Schneider (Kansas State University)

1 ABSTRACT

Social science and play research show that every human plays, regardless of age, ability, culture, or language. Play is universal and serves as the ultimate inclusive language. It provides a compelling entry point for community engagement, as individuals naturally become collaborative problem solvers through play. When we move from a language of drawing and presentation to a language of play, the community, as well as the design team, CAN imagine situations they have never encountered before and share what they see as possible. Even when the project is intended to improve their community, members often voice opposition citing risk avoidance and fear of failure as major motivations. However, through play they can try on possibility for themselves and are willing to embrace risks and come to consensus even if it is not unanimous. Play is universal and is the ultimate inclusive language.

By taking the scientific benefits of social science and play and aligning them with this unique placemaking approach, the design team can elevate community perspectives, empower communities, and generate successful public spaces that align with community values.

This session will present the PLAYCE methodology and its application within the context of architecture and landscape architecture classes at three universities - Kansas State, Penn State and U.Buffalo.

At Kansas State, the PLAYCE approach to community engagement was introduced to final year graduate students through an intensive workshop in their Master's Project Studio. The workshop inspired one student to apply the PLAYCE approach in her master's project research on place attachment. The PLAYCE approach was woven into the qualitative data collection method by Claire Schneider and will be presented here.

Collaborative Community-Engaged Design is a Directed Research Studio (CCED-DRS) for advanced professional architecture students (4th and 5th year B.Arch and 3rd year M.Arch) at the Pennsylvania State University. CCED-DRS is a participatory action research studio "where design research, design participation, and design development are in inextricably linked activities."1 Students pursue urban and architectural interventions informed by place and co-designed with a community.2 Students were introduced to PLAYCE community engagement methods through a series of in-class workshops and asked to adapt or apply what they learned throughout the semester-long project. PLAYCE Studio's methods and tools enhanced the students' learning experience and outreach capacity. This presentation will reflect on the process and methods of the CCED-DRS studio and consider roles of academia and community towards future project realization.

Arc 606- Engage and Empower was a graduate level Architecture Studio at SUNY Buffalo School of Architecture and Planning (UB). Together professors and students crafted the content for four community meetings. As each event unfolded, relationships were forged, and vulnerabilities were shared. The community asked for a renewed sense of shared responsibility and for forgiveness within the community to be possible. Students were then tasked with designing for those possibilities to be realized. Outcomes will be shared during the presentation.

PORTLAND MUSEUM OF ART: COLLABORATION IN COMMUNITY-CENTERED DESIGN

Track 5: Spatial Justice, Equity, and Inclusion by Design
Knowledge Network: Cultural Aspects of Design, Participation
Methods: Mixed Methods, Practice-Based Research

Raissa Xie (Openbox); Karen Braitmayer (Studio Pacifica); Mahmoud Hassan (Maine Language Connect); Christopher Newell (Akomawt Educational Initiative); Chandra Robinson (LEVER Architecture)

1 ABSTRACT

How might we implement a community-centered process in the design of a cultural space? This panel brings together collaborators from different parts of the building process through a case study and conversation on the Portland Museum of Art's (PMA) ongoing campus expansion and unification project to discuss equity in the built environment.

When Maine's most prominent art institution began planning a major campus expansion and unification, it sought to ensure that the corresponding new visitor experience would embody its "Art for All" mandate. Through an international design competition, a design team was entrusted with the task of welcoming and engaging historically underserved communities. Architectural design projects often emphasize aesthetic and structural aspects, but understanding both the visible and invisible impact of a project on communities is crucial. Meaningful community integration in the design process requires intentional collaboration between designers, clients, and community partners. The PMA team sought to achieve this by embedding community-centered research at the project's outset, allowing community voices to shape the foundation for design decisions.

This panel brings together diverse perspectives, including representatives from the client, community, and design collaborators, each contributing unique lived experiences and expertise. The session underscores the importance of inclusivity—not just in the final product but throughout the entire process—ultimately working toward engagement that is equitable rather than performative.

The presentation will begin with a high-level project overview (20 min), followed by panelists introducing themselves, their perspectives, and their roles in the project (20 min). A facilitated discussion (40 min) will explore themes such as collaboration and partnership, building trust, implementing community feedback, and the future of community and client partnerships in the built environment. The session will conclude with an audience Q&A (20 min).

By examining the PMA project, this session aims to provide insights into how early community involvement, intentional collaboration, and inclusive research methods contribute to spatial justice, equity, and inclusion in the built environment. Panelists will share practical strategies that attendees can apply in their own work, fostering a broader dialogue on transforming design processes toward a more inclusive future.

ARCHITECTURAL SPECULATIONS: REIMAGINING THE BUILT ENVIRONMENT IN A PAGAN PARADIGM

Track 5: Spatial Justice, Equity, and Inclusion by Design
Knowledge Network: Cultural Aspects of Design, Environmental & Architectural Phenomenology
Methods: Interpretive-Historical Research
Madalyn Asker (University of Idaho); Randall Teal (University of Idaho)

1 ABSTRACT
The architectural landscape serves as a palimpsest, bearing the imprints of ideological shifts, power dynamics, and cultural transformations. This paper presents a speculative architectural history, propelled by a fundamental inquiry: In a world of nascent Christianity, what if Pagan Roman Emperor Julian the Apostate had a 30-year reign as opposed to the 5 years he was given?

We begin this inquiry by drawing on the insights of Bernard Steigler, who confronts the inseparability of humanity from technics, elucidating how the nexus of power and material world makes manifest societal structures. In particular, we are interested in Stiegler epiphylogenesis, a phenomena he describes wherein we inherit memories of things we have never lived through. The implication of this is that we are coming into the world potentially indoctrinated by the world we are brought into. Technics in general is the means of this indoctrination but what we are interested in here is how architecture and the constructed environment function as this site of tertiary memory.

Because we want to point to the generative force of these material assemblages we will refer to cultural-architectural formations as machines. The first machine is the Christian-machine, structured by the typologies of Christian architecture. The second machine is the Pagan-Machine, powered by a speculative pagan architectural vision.

Being born into the Christian-machine, one already possesses the practices and narratives of Christianity, which are then reinforced via the specificity of cathedrals, monasteries, baptisteries, reliquaries, pilgrimage churches, marteria, catacombs, etc. Through these architectures different types of subjects – believers, adherents, copyists, scholars, the saved, the devout, the crusader, the proselytizer– are produced. All of this is underpinned by a relationship with time that requires constant striving for the timeless eternal afterlife. In such a world, one's place is clear, one's possibilities are circumscribed, and expectations for right and wrong, good and evil, sacred and profane, as well as notions about civility, duty, and agency are pre-formed and given a priori.

A pagan machine would generate awareness of nature and "what is;" it would mark time in both its flow and cycles with particular emphasis on breaking down the boundaries between our bodies and the world around us; it would reveal and amplify our fundamental ontological emmeshments, open up attunements to our own potentialities, and erase self-limiting definitions of what it means to be human. The pagan machine would slow one down; the pagan machine would be ecological; the pagan machine would point to the imminent sacredness of the world.

We believe this thought experiment is worth conducting and exploring, not simply for the fantastical interest that it might generate, but more importantly, to help us envision alternate futures for our built environment and the communities they enable. We imagine an architecture not made to dominate, dictate or control, but rather one that accommodates, interrogates, and reminds us to give ourselves over the mystical realm of immanent phenomena.

THE HOME PHYSICAL ENVIRONMENT ATTRIBUTES ASSOCIATED WITH CHILD SELF-REGULATION AND FAMILY INCOME LEVELS

Track 5: Spatial Justice, Equity, and Inclusion by Design
Knowledge Network: Children, Youth & Environment, Interior Design
Methods: Interviews

Reem Bagais (Illinois State University); Debajyoti Pati (Texas Tech University);
Erin Hamilton (University of Wisconsin-Madison, Design Studies)

1 ABSTRACT

Research indicates that early exposure to poverty can significantly impact children's physiological, emotional, social, and cognitive development. Children from low-income families are less likely to break the cycle of poverty and are more likely to experience life patterns aligned with the Family Stress Model (Baker & Brooks-Gunn, 2019). This model describes how financial hardship can lead to economic pressures that, in turn, increase parental emotional distress, affecting children's development.

A child's home is part of a complex microsystem referred to as the immediate environment, consisting of different factors such as family income, parenting responsivity, and a chaotic home environment. The home environment is an important external source that affects a child's cognitive, behavioral, psychological, and social development. Literature has established the association between physical environments and child self-regulation through the accumulation of different stressors. This study aimed to fill the gap and explore design elements in homes of families with different income levels and contribute to the body of literature that supports diversity, equity, and inclusivity among different socioeconomics classes from the lens of interior design and family sciences.

2 METHOD

This research qualitatively explored the home physical environment in relation to child self-regulation. The current research asked two questions: (1) What physical environment factors in the home environment generate stress/calming moments for children, and (2) what are the differences and similarities in physical home environment factors identified in Question One among families of different income levels? Data were collected through surveys and interviews and included participants from three different income levels (upper =2, middle =2, and lower =6) with diverse ethnicities. Outcomes: Thirty-nine environmental attributes were identified from parents/caregivers' perspectives. The key findings of the study include (1) the more the affordances in the environment, the better for child's self-regulation; (2) spatial attributes of the environment influence the child's self-regulation negatively and positively; and (3) there are few variations between income levels in relation to home environmental attributes that impact child self-regulation.

3 SIGNIFICANCE

This research will benefit children aged 3-5 years old in low-income homes, interior designers, and stakeholders. By determining physical environmental attributes, children 3-5 can be better supported in an environment that helps create a calming and less stressful experience. Specifically, by identifying and addressing the impact of the physical environment on children's self-regulation, this study seeks to develop cost-affordable design solutions, fostering equity by making these solutions accessible regardless of financial status. The failure to investigate the importance of the physical environment continues to limit the impact of interior design research concerning self-regulatory behavior. Studies in the field of psychology on self-regulatory behavior have explored limited aspects of home physical design. This research is intended to help shift the design community's focus to this issue. Furthermore, by increasing awareness among stakeholders and policymakers about the significance of interior design and the built environment in enhancing self-regulation, this research intends to advocate for more inclusive practices in low-income housing and preschools.

AMPLIFYING DIVERSE OCCUPANT VOICES DURING PREDESIGN TO INFORM DESIGN DECISION-MAKING

Track 6: Transdisciplinary Design and Collaboration/Open Track
Knowledge Network: Participation, Researchers in Professional Practice
Methods: Case Study, Practice-Based Research

Danya Hakky (HGA Architects & Engineers); Mahshid Jalalian (HGA Architects & Engineers); Luke Nichols (HGA); Dominique Thomas (HGA); Oshana Reich (HGA)

1 ABSTRACT

This presentation aims to share participatory design research insights from project case studies of varying scales, owners, team compositions, and types while creating a space for knowledge exchange.
Historically, for many design projects, a traditional engagement model consisting of a design team, owner, and contractor was sufficient to meet project needs. However, project teams—including planners, architects, landscape architects, and interior designers—are increasingly asked to tackle more complex, intersecting problems that are difficult to solve.
To address these challenges, design processes are evolving to include collaborative approaches with extended predesign phases. This crucial predesign phase defines project goals and explores opportunities and solutions. It is also when designers and researchers collaborate to collect and interpret insights.

Complex problems require interdisciplinary teams of experts and stakeholders, including future occupants, who are considered "experts of their experience" (Sanders & Stappers, 2016). Overlooking occupant inclusion may decrease user buy-in, overlook underlying issues, and perpetuate historical injustices. Involving users in the design process requires thoughtful planning for "mutual learning" (Luck, 2018) between designers and occupants. Researchers can develop tools and facilitate processes to enable this exchange so users ultimately contribute to design solutions.

Literature across planning and design fields—including the public realm—has defined participatory design and shared examples. This session will contribute to this growing knowledge by demonstrating the value of participatory processes in contemporary design practice and highlighting their meaningful and effective applications. Led by a cross-disciplinary group of designers, researchers, and allied experts, the session will also introduce generative design research, a practice within participatory design that brings people into the design process and enables them to share their dreams and ideas for the future (Sanders & Stappers, 2016). The evolving and overlapping roles of researchers, designers, and allied experts will be discussed as the future of design.
Presenters will share case studies of varying scales and types, identifying team composition, design problems/questions, data collection methods, and outcomes for each project. Through case study discussions, presenters will explore:

- *Process: Recognizing challenges and opportunities of cross-disciplinary collaboration and occupant engagement.*
- *Engagement Tools: Developing and selecting the most effective tools and exercises.*
- *Equity: Identifying the right team members and empowering diverse voices.*
- *Translation: Converting findings into design solutions.*
- *Storytelling: Communicating benefits and return on investment to owners and the design industry.*

Finally, lessons learned will be shared, along with reflections on the future of participatory practice and generative design research in the built and natural environments through a guided conversation with attendees.

2 REFERENCES

Luck, R. (2018). Participatory design in architectural practice: Changing practices in future making in uncertain times. Design Studies, 59, 139–157. https://doi.org/10.1016/j.destud.2018.10.003

Sanders, E., & Stappers, P. (2016). Convivial toolbox: Generative research for the front end of design (3rd ed.). BIS Publishers.

TRAUMA-INFORMED COLOR DESIGN IN SPACES

Track 6: Transdisciplinary Design and Collaboration/Open Track
Knowledge Network: Interior Design
Methods: Mixed Methods
Rene Berndt (Mahlum Architects Portland); Esther Hagenlocher (University of Oregon)

1 ABSTRACT
Addressing mental health has become a focus area for institutions educating young people and beyond. The proposed workshop project examines the impact of trauma-informed design strategies on mental health in the context of residential campus life. The questions asked were if and how the framework of biophilic design could reduce instances that trigger trauma in the built environment. Color and material play an integral part in this. During a collaborative workshop, students examined design strategies and color and material concepts beyond the known field - which not only leads to the design proposal and considers the perception and performance of color reflectivity in daylit spaces and shapes our aesthetic value.

ABSTRACTS CONTINUED

Group Presentations: Workshops

COMMUNITY, CULTURE AND PLACE BETWEEN RESILIENCE AND RESISTANCE

Track 1: Nature and Urban Ecology
Knowledge Network: Environmental Design Research Education, Cultural Aspects of Design
Methods: Ethnography, Mixed Methods

Rula Awwad-Rafferty, (University of Idaho- College of Art & Architecture); Sahera Bleibleh, (United Arab Emirates University); Rula Awwad-Rafferty, (University of Idaho- College of Art & Architecture);
Sahera Bleibleh, (United Arab Emirates University)

1 ABSTRACT

It is not difficult to sense the liveliness while walking in the vaulted stone alleyways, the low tunnels, and the labyrinths of narrow lanes of the old town of Nablus. It is a sensual, interactive, and mixed experience of optics, acoustics, and aromatics. ... Since the invasion, people cautiously start every day with the fear and uncertainty that they might be unable to enjoy what used to be a normal day (Bleibleh, 2012).

Places are both heritage and lived experiences for those who call them home, encompassing geography, history, culture, flora and fauna, stories, and decisions made over time. They are remembered and imprinted into the collective consciousness, shaping group identity and individual well-being. Indigenous communities worldwide emphasize the intrinsic healing properties of their surroundings. They speak of a medicinal quality inherent in their places, where "food is medicine," "ritual is medicine," and "stories are medicine." Essentially, the land is medicine, providing a foundation for cultural resilience and representation. This interactive workshop focuses on what happens when the relationship between people and their homeplace is disrupted and on the transformations of the narratives that maintain a connection to home. It seeks to nurture community empathy toward the "everyday fundamentals" of heritage landscape practices, planting a seed of appreciation for the authentic, inherited wisdom embedded within these traditions and fostering a stronger bond with indigenous cultures and their profound connection to the land. Methods include icebreaker activities, foundational premises, contextual Palestinian narratives, audience lived-experience reflections, and discussions.

As places become intertwined with people's lives, a legacy of inherited habits emerges—ways that connect land, plants, seasons, and memories, along with acceptable "habits/rituals/behaviors" that promote stewardship, land-people relationships, and harmony with the place. These attributes manifest in seasonal behaviors, rituals, and economies, preserving the authenticity and integrity of lived experiences and ensuring the continuity of culture and relational practices between community and place. Participants are prompted to explore and connect with the indigenous experiences of "making-do" and their everyday practices, deepening their understanding of the intersection between place, culture, and resilience. In a time of massive change, global movement, migration, gentrification, infringements, and connectivity, the community (people)–Culture (ritual)–Place (land and its treasures) continuum is disrupted, targeted, and undermined. In many cases, access to the homeplace is restricted or denied. What was once a source of identity, resilience, and continuity now evokes pain, dissonance, and resistance.

Healing and reconnection take many forms. Stories and sensorial experiences trigger memories of place and lived experience, despite the passage of time and geography, bringing us home to another time. In "We Are Water Protectors," inspired by Indigenous environmental justice and resistance, the story begins with the quote, "Water is the first medicine, Nokomis told me...It affects and connects us all" (Lindstrom & Goade, 2020). This interactive storytelling workshop emphasizes the significance of place, identity, and community in stewarding integrative cultures of resilience, responsibility, and authenticity in the face of massive change, exploring ways that honor people-environment relationships amidst an increasingly complex and ever-changing world.

ADAPTING CULTURAL LANDSCAPES IN SHIFTING URBAN CONTEXTS

Track 1: Nature and Urban Ecology
Knowledge Network: Cultural Aspects of Design, Historic Preservation
Methods: Case Study, Design Methods
C Ungureanu; C Ungureanu; Nate Cormier, (RIOS); Jenn Low, (Openbox)

1 ABSTRACT
What happens when cultural landscapes are no longer serving their cities? How do we, as designers of human-centric environments, engage in efforts to adapt cultural landscapes physically, programmatically, and experientially to meet the evolving needs of our communities? We'll explore these questions through the lens of repositioning efforts at three sites in the American West - Balboa Park (San Diego, CA), Skyline Park (Denver, CO), and Ghirardelli Square (San Francisco, CA).

In the Balboa Park Cultural District—one of San Diego's most significant historic parks—the Balboa Park Cultural Partnership sought to create a more distinctive and inclusive experience for its visitors. The effort was to reimagine the Cultural District's interstitial spaces to center the experiences of visitors both inside and outside the region, including communities historically underserved by the Cultural District: San Diego and Tijuana's Spanish-speaking border communities, Kumeyaay Indigenous groups, Black and Asian families, among others.

In Skyline Park—originally designed by Lawrence Halprin and located in the heart of downtown Denver—the City and its partners witnessed gradual decline and sought to reimagine this three-block landscape with more flexible and interactive spaces in mind. This effort involved multiple design iterations and an extensive citywide public outreach process.

At Ghirardelli Square in San Francisco, adaptation meant improving public access for day, night, and year-round use, and updating landscape materials, plantings, and lighting to comply with evolving sustainability and maintenance needs.

In conversation with one another and the audience, we'll explore the common challenges these spaces have faced since their inception—spatial, economic, programmatic, and perceptual. We'll discuss how urban researchers and design practitioners have begun to revive these cultural landscapes while balancing multiple goals: preserving historic legacy, sustaining these spaces economically, engaging multifaceted audiences in the redesign process, and welcoming an evolving demographic.

Although necessary, the evolution of cultural landscapes is not without controversy. Decisions to upgrade or alter their physical form are deeply contested, and through this workshop, we'll explore those debates—particularly how the public was empowered to weigh in on those tradeoffs.

STUDYING THE IMPACTS OF HEALTHY & SUSTAINABLE STUDENT HOUSING ON RESIDENT WELL-BEING

Track 2: Sustainable Buildings & Infrastructure Ecology
Knowledge Network: POE/Programming, Sustainable Planning, Design & Behavior
Methods: Case Study, Survey
Michael Ralph, (Multistudio); Teresa Jan, (Multistudio)

1 ABSTRACT

Post Occupancy Evaluations (POEs) are increasingly adopted by industry researchers to study the success of architectural projects in achieving their goals. Efforts to improve and formalize POE methods include work that identifies a need to refine and align definitions of performance across studies. Hassanain (2008) operationalized performance in the context of student housing as "technical performance" and "functional performance." Technical performance includes thermal comfort and indoor air quality, while functional elements encompass engagement with facility programming and amenities. Although Hassanain's study provides a precedent for operationalizing performance, few POEs include an empirical examination of these components. Our workshop will present POE methods for student housing via a recently completed POE of Manzanita Square, a project that prioritizes sustainability and resident wellness. The POE aims to identify strategies to continue reducing both operational and behavioral carbon, feeding the positive cycle of building a sustainable and wellness-focused collegiate residence culture.

Manzanita Square is a student residential complex designed to be affordable, sustainable, and to serve students as an academic and community resource. Following project completion, our team conducted a multi-phase POE in collaboration with the developer and university partners to collect performance data in both categories using utility billing and waste data, physical environmental measurements, and student and operator surveys. Manzanita Square prioritized passive design features such as super-insulation, natural ventilation, and the elimination of active cooling. It also emphasizes healthy resident movement by promoting stair use and public transit. Recent research shows that transportation can be an effective lever for improving climate stewardship (Cherry et al., 2018).

Our workshop is organized into three sections: research context, findings for technical performance, and findings for functional performance. It will emphasize a blended format that integrates active learning through a cycle of presentation, discussion, and application, allowing participants to apply their learning to their own projects or research. Section 1 will present the current state of research in student housing and POEs. We will share details on Manzanita Square, which will lead to a POE planning exercise where participant groups ideate on how they would plan and collect data for their own POE of the project. Section 2 will provide an overview of technical performance data, including prompts to analyze real project data and our team's actual analysis. In Section 3, we will share the results from multiple student and operator surveys, including lessons learned from the first phase of survey delivery and insights from the subsequent second round of data collection. The workshop will close with a full group discussion of potential directions for subsequent study at and beyond Manzanita Square.

2 REFERENCES

Boissonneault, A., & Peters, T. (2022). Concepts of performance in post-occupancy evaluation post-probe: a literature review. Building Research & Information, 1-23. https://doi.org/10.1080/09613218.2022.2132906

Cherry, C. R., Riggs, W., Appleyard, B., Dhakal, N., Frost, A., & Jeffers, S. T. (2018). New and unique aspects of university campus transportation data to improve planning methods. Transportation research record, 2672(8), 742-753. https://doi.org/10.1177/0361198118781659

Hassanain, M. A. (2008). On the performance evaluation of sustainable student housing facilities. Journal of facilities management, 6(3), 212-225. https://doi.org/10.1108/14725960810885989

SMART KIDS AND COOL SENIORS: COMMUNITY ENGAGED RESEARCH FOR CLIMATE RESILIENCE

Track 3: Resilient Design for Climate Change
Knowledge Network: Environmental Design Research Education, Participation
Methods: Action/Participatory Research, Mixed Methods

Jennifer Senick, (Rutgers Center for Green Building); Jennifer Senick, (Rutgers Center for Green Building); Gediminas Mainelis, (Department of Environmental Sciences, Rutgers, The State University of New Jersey); Abigail Andrews, (Rutgers University); Sunyoung Kim, (School of Communication and Information at Rutgers, the State University of NJ); Jackie Park Albaum, (Groundwork Elizabeth, Elizabeth, New Jersey)

1 ABSTRACT

Elizabeth, NJ, is an environmental justice community in the heart of the densely populated Northeast Corridor that endures air pollution emanating from oil refineries, major highways, a large port, and an international airport. In addition to air pollution, residents increasingly suffer from heat stress during the summer months. The compounded effects of environmental hazards and poverty especially threaten the health and well-being of seniors living in public housing. A collaborative led by Groundwork Elizabeth, the Housing Authority of the City of Elizabeth (HACE), and Rutgers University seeks to mitigate these problems with a bold project that brings together youth and seniors to characterize indoor and outdoor air quality and thermal conditions and develop solutions through a National Science Foundation (NSF) Civic Innovation Challenge (CIVIC) Grant. The NSF Civic Grant is a $1 million grant for a year-long research and action implementation pilot that accelerates research and technology through community leadership and engagement. Over the past year, the research team has deployed a network of fixed outdoor and indoor air pollution and temperature sensors, in addition to environmental sensor-equipped backpacks for youth to carry along routes used by seniors to measure personal exposure. The research team developed community engagement strategies, identified personal exposure risks, created a public-facing data hub, and built a scenarioized digital twin model of individual apartments as well as whole complexes. These presentations will show how state-of-the-art technologies and simulation tools can advance community engagement, citizen science, and design for well-being.

The presentations within this symposium aim to equip planners, community members, and designers with information on environmental exposure risks. They will address the following questions and challenges:

- *Can the same behavioral models apply to people as they move between and perform actions in both indoor and outdoor environments?*
- *How are system dynamics affected when both the people and the building technologies are smart and have agency?*
- *Do more adaptive possibilities emerge as we add levels to our multi-level system?*
- *Is it possible to design interfaces that improve the odds of successful adaptation for poor urban seniors?*

Presentations:

- *Tailoring scientific information for lay audiences to improve resilience to climate change and resilient design (Jennifer Senick, Sunyoung Kim, Jackie Park Albaum)*
- *An integrated approach to monitor and model air pollutants, environmental stressors, and occupant exposure to determine mitigation strategies (Abigail Andrews, Gediminas Mainelis, Jie Gong, Clinton Andrews)*

COMMUNITY ENGAGEMENT & DESIGN RESEARCH PARTNERSHIPS IN A UNIVERSITY-TRIBE COLLABORATION: NOTES FROM A RAPIDLY ERODING SHORE

Track 3: Resilient Design for Climate Change
Knowledge Network: Cultural Aspects of Design
Methods: Action/Participatory Research

Lynne Manzo, (University of Washington); Julie Kriegh, (Kriegh Architecture Studio);
Daniel Abramson, (University of Washington)

1 ABSTRACT

Climate change-driven sea level rise and greater environmental volatility are increasingly putting coastal communities at risk worldwide (Kousky & Berger, 2021). Among these "frontline communities" are many indigenous tribes. In the US, the Shoalwater Bay Indian Tribe is one such community. This Tribe lives on the most rapidly eroding portion of the US Pacific coastline (Talebi et al., 2017). In response to increased environmental risk, the Tribe has purchased former timberlands on higher elevations adjacent to their reservation to expand their community onto safer ground and to maintain self-reliance (Judkins, 2020).

The purpose of this group presentation/workshop is to share the work and outcomes of an ongoing university-community collaboration among the Shoalwater Bay Tribe, the University of Washington, and design practitioners. This collaboration involves research and design to support climate change resilience and cultural continuity. In particular, this session will reflect on lessons learned from the collaboration for community-engaged and culturally responsive design and planning. It will consider how university partners—both faculty and students, who largely have white settler identities—can respectfully engage with Tribal communities and support larger decolonizing efforts (Greenwood, 2012; Tuck & Yang, 2012). The session will also address strategies for melding traditional ecological knowledge and indigenous ways of living and building on the land with technological innovations for sustainability.

Additionally, the session will discuss implications for broader practitioner-researcher collaborations as well as pedagogical strategies for community engagement. One key lesson is that research-design collaborations can be effective when community-driven work places the priorities, assets, and needs of the community first. A second lesson is that community-driven design research projects involving students as researchers and designers can result in substantial, change-making action at the community level. However, achieving this requires a range of pedagogical temporal scales or "paces" of community engagement—from short, intensive workshops and charrettes to studio courses, course sequences, and multi-term theses and capstone projects. The discussion will also focus on the meaningful engagement of design practitioners with university researchers/instructors and the Tribe. Ultimately, the goal of this session is to describe and discuss translatable approaches and methods of collaboration that can be applied to other community-driven design and research projects.

2 REFERENCES

Greenwood, D. (2012). A critical theory of place-conscious education. In R. Stevenson, M.Brody, J. Dillon & A. Wals. (eds.). International handbook of research in education (pps. 93-100). NY: Routledge.

Judkins, J. (2020), for "Health and Coastal Perils." In Hutchison, R. and Abramson, D. eds., American roundtable: Dynamic landscapes, South Beach, Washington. Architectural League of New York. https://archleague.org/article/southbeach-washington-health/

Kousky, C. Fleming, B. and Berger, A. (2021). A Blueprint for coastal adaptation: Uniting Design, Economics and Policy. Washington: Island Press

Talebi, B., Kaminsky, G.M., Ruggiero, P., Levkowitz, M., McGrath, J., Serafin, K., McCandless, D. (2017). "Assessment of Coastal Erosion and Future Projections for North Cove, Pacific County," State of Washington Department of Ecology Publication 17-06-010.

ENVIRONMENTAL PSYCHOLOGY, EDRA AND FIVE CAREERS: A ROUNDTABLE

Track 4: Design for Health and Wellbing
Knowledge Network: Environmental Design Research Education, Researchers in Professional Practice
Methods: Case Study

Karen Franck, (New Jersey Institute of Technology); Lynn Paxson, (Iowa State University); Ruth Rae, (Tandon School of Engineering of New York University); David Stea, (Geography and International Studies at Texas State University); Richard Wener, (NYU)

1 ABSTRACT

The emergence and development of the field of environmental psychology, the long-time existence of the Environmental Design Research Association, and the annual conferences it has held have played a significant role in the careers of many of its members. Many times, EDRA has provided collegial connections for scholars and practitioners who are otherwise the lone E&B person in their organization. This session will provide an opportunity to share these professional histories. Participants on this panel will tell the stories of their education and careers to illustrate this point. These stories should be of interest both to older EDRA members, who have followed similar (or different) career trajectories, and to younger members who are at the beginning of their careers. After the four participants briefly describe their careers, people attending the session will be invited to tell their career stories, ask questions, and raise additional points.

WHAT MAKES A SPACE HUMAN? A WORKSHOP ON EMPATHETIC DESIGN

Track 4: Design for Health and Wellbeing
Knowledge Network: Environmental Design Research Education, Sustainable Planning, Design & Behavior
Methods: Action/Participatory Research, Mixed Methods

Connie Lin, (Perkins&Will); Kalpana Kuttaiah, (Perkins&Will); Connie Lin, (Perkins&Will); Shailagh McCormack, (Perkins&Will); Saran Megha Parimi, (Perkins&Will)

1 ABSTRACT

The objective of this workshop is to influence a paradigm shift in the world of healthcare design that places empathy at the forefront of design decision-making. With a growing emphasis on health and wellness in the built environment, designers and practitioners alike are doubling down on the occupant experience. However, there are no set standards or guidelines for an empathetic approach. Therefore, through a series of participatory design exercises and group discussions, this workshop seeks not only to educate and inform participants on the environmental influences on human emotions but also to provide a kit of parts for applying empathetic design.

An interdisciplinary approach incorporates insights from design professionals, researchers, students, and practitioners. Interdisciplinary design is significant because collaborative teamwork—across several disciplines—is essential to producing innovative work and finding new solutions (Jutraž & Zupančič, 2014). Architects, designers, contractors, and healthcare practitioners will participate in a collaborative effort to generate design principles and interventions that are rooted in empathy. Fittingly, the speakers are exemplary hosts for this workshop due to their interdisciplinary expertise in both healthcare design and research. To kick off the workshop, they will deliver a 15-minute presentation discussing their collective expertise on mental health topics such as depression, burnout, and patient autonomy in healthcare settings.

The first 15-minute breakout exercise will be modeled after the community mapping method—a participatory method used to provoke ideas, inform participants on the subject, and encourage collaboration (Community Places, 2022). The question posed will be, "How do we design to mitigate negative human emotions and stimulate positive ones?" Each team will be given a specific healthcare space (waiting room, care team area, patient exam room, etc.) to discuss how the environment can foster both positive and negative effects. The second 15-minute breakout exercise will be modeled after the art and creativity method, intended to foster hands-on interaction. Participants will utilize the spaces discussed in the previous exercise and develop an experiential procession focused on the transitions between spaces and emotions. Regrouping for 20 minutes, each team will also develop and present their own "empathy guidelines" to be shared with the overall workshop group. For the final 25 minutes of the workshop, the discussion will focus on comparing and contrasting the "empathy guidelines" with the key takeaways from the workshop.

The outcome of this workshop will be a comprehensive set of design guidelines that provide participants with the necessary tools to foster a positive human experience in the design of the built environment. Additionally, participants will be equipped to understand the correlation between space and emotions. By leveraging an interdisciplinary approach and utilizing participatory engagement, this symposium aspires to stimulate empathy within the healthcare design profession and create spaces that nurture a sense of comfort while enhancing overall well-being.

2 REFERENCES

.Community Places. (2022). Guide to Engagement Methods (p. 22).

Jutraz, A., & Zupancic, T. (2014). The Role of Architect in Interdisciplinary Collaborative Design Studios. Igra Ustvarjalnosti - Creativity Game, 2014, 034–042. https://doi.org/10.15292/IU-CG.2014.02.034-042

CO-DESIGNING A SOLIDARITY LANDSCAPE FOR ECOLOGICAL, SOCIAL, AND INDIVIDUAL WELL-BEING AND CAPABILITIES

Track 4: Design for Health and Wellbeing
Knowledge Network: Children, Youth & Environment, Participation
Methods: Action/Participatory Research

Deni Ruggeri, (University of Maryland, College Park); Mary Christensen, (University of Maryland, College Park); Deni Ruggeri, (University of Maryland, College Park); Mary Christensen, (University of Maryland, College Park)

1 ABSTRACT

In his travels to the United States in the early 1800s, Alexis de Tocqueville observed early American society's unique sense of community and mutual help (Putnam, 1995). With the advent of industrialization and globalization, this aspect of American society has been lost, replaced by individualism and self-interest. This trend is also evident in Europe and was apparent during the COVID-19 crisis, when responses to the pandemic revealed a lack of coordination and resource sharing, accompanied by increased competition for medical supplies (Gray, 2020; Pornschlegel, 2021). Sociologist Robert Putnam (1995) has observed a decline in solidarity, trust, and mutualism as a result of a preference for bonding over bridging social networks.

Living and working environments, access to health care, education, job opportunities, and social and family contexts are all social determinants of health. Inequities in these determinants profoundly affect health and well-being and are often experienced in underserved and fragile communities. Communities that embody the shared value of solidarity "tend to promote equality, social rights, well-being, and friendship" (Arban, 2021), leading to an increased quality of life and enhanced well-being.

This workshop is devoted to developing a shared definition of solidarity as it relates to well-being in underserved and fragile communities. It will examine the relational health benefits of solidarity and how it can affect community, ecological, social, and individual well-being and capabilities.

Villaggio Solidale is a co-housing community in northern Italy that serves children, individuals, and families experiencing health, social, and economic fragilities. The community views fragility as a shared human condition and focuses on empowering individuals through education, safe and healthy housing, improved social capital, and access to care. Villaggio Solidale embodies the concept of solidarity as its members share their fragilities, weaknesses, and strengths while working toward their goals of independence and well-being.

This workshop will explore themes of solidarity and well-being in a fragile community while drawing on research methods, findings, and experiences from an ongoing participatory action research project with Villaggio Solidale. The project has employed several methods, including participatory action research, observation, and community engagement. Over the past two years, researchers have collaborated with the community in goal setting, drawing, co-design, mapping, listening, collaging, site visits, and informal interviews during two eight-day participatory workshops. The content of the drawings, co-designs, group discussions, and interviews conducted during these workshops has been analyzed to identify and enumerate recurring themes and concepts regarding landscapes important to the community's collective well-being. The goals and objectives of this participatory action research project include co-designing a solidarity landscape and developing a training curriculum to integrate nature-based interventions, biophilia, agroecology, and well-being into the larger ecosystem and community surrounding Villaggio Solidale.

2 REFERENCES

Arban, E. (2021). The Principle of Equality in Diverse States. In (Ed.), "Chapter 4 Italy: The Principle of Solidarity as a Principle of Equality."

Pornschlegel, S. (2021). Solidarity in the EU: More hype than substance? EPC Issue Paper 28/07/2021.

Putnam, R. D., & Goss, K. A. (1995). Robert D. Putnam. Bowling alone: America's declining social capital. Journal of Democracy, 6(1), 65-78.

HOSTILE GROUNDS – HOW CAN THEY HOLD PEOPLE CENTRIC PLACES?

Track: 5 Spatial justice, equity, and inclusion by design
Knowledge Network: Cultural Aspects of Design, Participation
Methods: Mixed Methods

Lynn Paxson, (Iowa State University); Lynn Paxson, (Iowa State University); Rula Awwad-Rafferty, (University of Idaho-College of Art & Architecture); Geraldene Blackgoat, (., Indigenous Design Studio + Architecture, LLC.,); Gerard Rodriguez, (PLACE)

1 ABSTRACT

Our world is full of hostile—sometimes actively contested—grounds: lands and places where narratives collide. Colonization and power dynamics over the last centuries, and even today, have brought about exclusions and entitlements that are "woven into the narrative of dominant ideologies." While the topic of this session is global in scope, affecting many, this workshop will explore it primarily for Indigenous Peoples of America. Many Indigenous peoples were removed and suffered genocide during colonization, while many others and their descendants were moved off their traditional lands and ways of being that they had known since time immemorial. A significant source of "hostile terrain" is the construct of borders and boundaries. Even today, many Indigenous/First Nations groups' lands cross borders—national, state, municipal, and even individual ownership/property borders. The border is not simply a mark in the sand; it is imbued with policies and hostilities targeting sovereignty and identity by design.

Indigenous peoples have demonstrated resilience and struggled as communities while resisting various forms of hostility—colonization, assimilation, adoption, or displacement—since the earliest contact with settlers. They moved off their traditional lands and made "place" in the new spaces to which they were forced. A traditional value among Indigenous peoples is to understand the world as a "relative," not a "resource," and to treat it accordingly. Indigenous peoples see themselves as part of place, just one part among many communities or "peoples" that populate the earth. This perspective recognizes not only the diversity among human groups but also the peoples of nature—the four-legged, the winged, the plants, etc. You cannot create environmentally, ecologically, or socially just places without a balance and recognition of all these multiple peoples and groups. Hostility toward one is hostility toward all.

Can Indigenous place-making design and community-based design processes support the creation of ever-changing cultures and places while resisting hostile ground? Perhaps the first step is to broadly acknowledge that these areas are, in fact, hostile and contested grounds.

Presenters will discuss borders and identities, the historical and present impacts of border policies, and attempts at decolonization with examples from the Tohono O'odham, Dine (Navajo), the Shoshone Bannock, the Nez Perce (Nimiipuu), Coeur d'Alene (Schitsu'umsh), Dine, and the Siletz, Umatilla, Yakama, and Warm Springs Nations. The session will also examine Indigenous design and the challenges Indigenous practitioners face when working for or with primarily mainstream firms and organizations—even when those firms work with Indigenous clients and purport to engage in Indigenous design. Additionally, the discussion will address how service-learning studios can help non-Indigenous students better understand a less hostile way to work with Indigenous peoples and places, while also addressing issues of hostility in place. Using concrete examples, the panel will engage workshop attendees in a lively and productive discussion about how people-centric places might be transformed through action-oriented, holistic environmental behavior research and design.

HOUSING FOR AGING - WHAT TO PRIORITIZE? A FAST AND RELIABLE WAY TO ESTABLISH THE RELATIVE VALUE OF MULTIPLE ELEMENTS

Track 6: Transdisciplinary design and collaboration/open track
Knowledge Network: Environment-Gerontology, Interior Design
Methods: Mixed Methods

Nic Campbell, (Ohio University); Nic Campbell, (Ohio University); Alexis Earley, (University of South Florida)

1 ABSTRACT

For the architecture and design fields, the primary value in environmental and behavior research is its usefulness in providing Evidence-Based Design (EBD) recommendations for application in practice. While research provides evidence indicating which factors are important, what is often left undone is indicating how important these factors are compared to other recommended factors. This matters because design recommendations may be too numerous to include within a particular project's constraints, or recommendations may be collected from a number of studies.

To help guide practitioners, researchers must establish both what is important and how important each item is compared to the other priorities. While simple ranking can order factors, optimal decision-making tools are generally preferred because they offer more precise details for establishing which elements have the greatest value relative to one another. The application of these decision-making tools produces numerical weights that can be used to put different kinds of factors on equal footing for "apples-to-apples" comparison when establishing the relative importance of each element.

Tools such as Multi Attribute Utility Theory (MAUT) analysis are commonly applied in fields such as economics, water and energy management, and actuarial sciences (Velasquez & Hester, 2013). In environment and behavior research, these optimal decision-making tools were first used by Hunt and Ross (1990), who utilized MAUT analysis to establish the relative importance of senior-friendly design characteristics in naturally occurring retirement communities. Subsequent applications have been sporadic. Despite the value of these tools, they have not become commonplace in architecture and design research.

Modeled after the Hunt and Ross study, a recent study utilized MAUT analysis to understand how to create neighborhoods and their social spaces (e.g., cafes) to optimally meet older adults' needs. For this purpose and to explore the utility of MAUT as an environment and behavior research tool, the study's goal was to understand the relative importance of 40 factors identified in prior research as aiding seniors aging in place. In this study, MAUT was applied to interview data from 63 seniors living within the case study community to provide mathematical insight into each factor's relative value, thereby identifying design priorities to create neighborhoods and social centers that optimally support aging in place.

The primary objectives of this workshop are to introduce, within the context of this study, optimal decision-making methods as an underused but valuable research tool and to demonstrate and critique a new tool of this type called Intuitive Comparative Value Assessment (ICVA), which was created for applications in architecture and design research and practice. The workshop agenda includes:

- *Describing the usefulness and the mechanics of using optimal decision-making tools such as MAUT analysis to establish the relative value of factors.*
- *Comparing and demonstrating a new evolution of this tool, ICVA, developed specifically to provide a more convenient, accessible option for design researchers and practitioners.*
- *Utilizing conference attendee feedback to further critique and refine ICVA to improve its utility and ease of use in future research and architecture/design practice.*

BEYOND DIASPORA: INTENSIVE HUMAN CENTRIC COLLABORATIONS

Track 6: Transdisciplinary design and collaboration/open track
Knowledge Network: Cultural Aspects of Design, Participation
Methods: Action/Participatory Research, Design Methods
Rula Awwad-Rafferty, (University of Idaho- College of Art & Architecture); Rula Awwad-Rafferty, (University of Idaho-College of Art & Architecture); Shauna Corry-Hernandez, (University of Idaho); Kiana Carey, (Architects West)

1 ABSTRACT

Migration, immigration, and diaspora—iconically scattering people across the world away from their home base due to war, famine, climate, and work—have been dominant themes in human history. In a roundtable with hands-on participation from the audience, this workshop shares the story of a specific diaspora vision and mission, research and design practice, a design and teaching collaboration, and service-learning engagement that transcends disciplinary boundaries.

Any work with complex problems such as diaspora calls for a deeper level of engagement to unearth and expose worldviews, narratives, and experiences beyond a sanitized version of history. This process forces emotional and intellectual engagement that goes beyond typically professionally defined terminologies, roles, and processes (Brubaker, 2005). The presenters, through a choreographed storytelling and inquiry-based approach, share their processes and explore the potential of human-centric collaborations that build a more resilient, humane, and purposeful space to educate about diasporic people, their journeys, and their stories, thereby opening a future for cultural engagement and enrichment. In doing so, the ultimate vision is to build a more humane, hospitable, and enriched world.

"Engaging with migration helps us stay relevant. It allows us not just to be a museum about past immigration. It allows us to remind and educate the public that immigration and migration are still very much modern phenomena that are constantly happening" (Yates, Curator, Swedish American Museum–Erdogan-O'Connor, 2023).

This partnership, even in the most complex contexts for higher education, afforded us intensive learning and action opportunities that are core to environmental design research and practice. The session investigates the community-university-profession partnership and ongoing dialogues, complexity, and context, followed by the main narrative that began the exploration—which holds both personal and collective forms of trauma and narrative; the approach and changes in pedagogical engagement; and outcomes.

2 BRIEF AGENDA:

Two Domains: The World and the University
- *Leadership, partnership, and coordinating buy-ins*
- *Service learning and co-authored processes*
- *Keeping the light*

Human-Centric Stories
- *What is Diaspora? And why does it matter? The world in Idaho*
- *Stories within stories: Each person carries a story—the power to see*

Human-Centric Collaboration
- *In the times of COVID and social reckoning, time and place impact us*
- *Crossing disciplinary boundaries*
- *Building authenticity of voice*
- *Case studies and pitching a vision*

Professional Capital
- *Investing in the project*
- *Overcoming barriers*
- *Professional voice of knowing*
- *Evolutionary understanding*
- *Expanding the circle*

Negotiating the Landscape
- *Designing across lines of difference*
- *· Space and the story's integrity*
- *· Listening with knowing*

A Movement and a Mission
- *· Final critiques and reflections*
- *· Diversity and professional expectations*
- *· Prologue and epilogue*

Lessons and Takeaways
- *· Saying yes*
- *· Pushing forward; learning anew*
- *· Communication, respect, managing stress*
- *· Keeping an eye on the initial request*
- *· Now what?*

Audience Participation:
- *Audience participation is planned throughout the session. At the conclusion, the audience will be invited to create a tangible web connecting issues, productive scenarios, tools, and experiences to empower our environmental-behavior toolkits for change and resilient action.*

ABSTRACTS CONTINUED

Group Presentations: Intensives

A DEEP DIVE INTO CITY INTELLIGENT SERVICE MODULES WITH THE "STEP" THEORETICAL MODEL IN URBAN ENVIRONMENTS

Track 6: Transdisciplinary design and collaboration/open track
Knowledge Network: Participation
Methods: Case Study, Design Methods

1 ABSTRACT

This intensive session aims to delve deeply into the strategic planning and incubation strategies for city intelligent service modules based on the STEP theoretical model. The course will focus on the latest developments in smart city services, analyze the impact of social, technological, entrepreneurial, and policy dimensions on intelligent service modules, and reveal the key success factors through case studies. The learning objectives include: (1) grasping the fundamental concepts and trends of city intelligent service modules; (2) understanding the application of the STEP theoretical model in planning, designing, and promoting intelligent services; (3) learning how to assess the actual effectiveness of intelligent services through case studies; and (4) exploring the significance of cross-sectoral collaboration and policy support in the development of smart city services.

Agenda (4-hour length):

Part 1:
15 minutes: Introducing the course background and objectives
30 minutes: Explaining the concepts, categories, and characteristics of city intelligent service modules
15 minutes: Interactive discussion and Q&A

Part 2:
15 minutes: Elucidating the STEP theoretical model and its application in city intelligent services
30 minutes: Case study analysis of a bike-sharing system, emphasizing the importance of government planning and management
15 minutes: Interactive discussion and Q&A

Part 3:
15 minutes: Introducing the case of the Huo-Yan project, demonstrating the practical effectiveness of the STEP theoretical model driven by big data
30 minutes: Group discussions analyzing the success factors and challenges in the case
15 minutes: Group sharing and plenary discussion

Part 4:
20 minutes: Exploring the roles of cross-sectoral collaboration and policy support in the development of smart city services
10 minutes: Tea break and networking
20 minutes: Summary and outlook, sharing learning insights and practical experiences
10 minutes: Closing remarks and acknowledgments

Unlocking Urban Intelligence: A Comprehensive Exploration of Concepts, Categories, and Characteristics in City Intelligent Service Modules

Yunsheng Su (Professor)

Goal:
Enhance participants' understanding and analytical skills in the realm of city intelligent service modules through an intensive and interactive session.

Objectives:
Analyze Concepts: Enable participants to critically analyze the fundamental concepts associated with city intelligent service modules.

Learn Categories:
Facilitate learning about diverse categories and classifications pertinent to city intelligent service modules.

Understand Characteristics:
Foster an understanding of the key characteristics that define and distinguish various city intelligent service modules.
Interactive Discussion: Encourage active participation through an interactive discussion and Q&A, consolidating learned concepts and addressing participant queries.

Agenda:
Introduction (15 minutes): Provide a comprehensive background of the course, outlining its significance and setting the stage for the ensuing content.
Concepts, Categories, and Characteristics (30 minutes): Delve into the core concepts, elaborate on diverse categories, and expound on the characteristics defining city intelligent service modules.
Interactive Discussion and Q&A (15 minutes): Engage participants in a dynamic exchange, encouraging them to share insights, ask questions, and foster collaborative learning.

Strategic Theoretical Enhancement Process (STEP) Essentials

Xihan Cheng (Tongji University)

Goal:
This course aims not only to provide essential knowledge but also to ensure participants can seamlessly integrate the STEP model into their professional settings, enhancing their ability to address challenges and drive innovation.

Objectives:
1. *Learn: Gain a foundational understanding of the STEP theoretical model, encompassing Social Needs Analysis, Technological Innovation Design, Entrepreneurial Business Strategies, and Policy Design Support.*
2. *Analyze: Develop basic analytical skills to assess societal demands, identify technological opportunities, formulate preliminary business strategies, and recognize initial policy implications within urban contexts.*
3. *Understand: Familiarize yourself with the interconnected nature of social, technological, entrepreneurial, and policy dimensions, laying the groundwork for further exploration.*
4. *Application in Work Contexts: Acquire practical knowledge on applying the STEP model to diverse work scenarios, enabling participants to integrate theoretical insights into their respective professional environments.*

Agenda:
1. *Introduction to STEP (15 minutes)*
2. *Brief overview and historical context*
3. *Significance in post-pandemic urban challenges*
4. *Social Needs Analysis (S) (15 minutes)*
5. *Rapid understanding of societal demands*
6. *Introduction to needs assessment methods*
7. *Technological Innovation Design (T) (15 minutes)*
8. *Quick exploration of technological opportunities*
9. *Overview of design thinking in urban innovation*
10. *Entrepreneurial Business Strategies (E) (10 minutes)*
11. *Basic principles for formulating business strategies*
12. *Brief insight into the startup ecosystem*
13. *Policy Design Support (P) (10 minutes)*
14. *Quick understanding of the role of policy in urban innovation*
15. *Overview of regulatory frameworks and initial policy considerations*
16. *Application Workshop (15 minutes)*
17. *Practical exercises on applying STEP in participants' work contexts*
18. *Discussion and sharing of real-world applications*

Unlocking Practical Insights with the Huo-Yan Project

Mingjun Zhong (Designer)

Goal:
Enhance participants' ability to analyze, learn, and understand the practical effectiveness of the STEP theoretical model, driven by big data, through a focused exploration of the Huo-Yan project.

Objectives:
Analyze Practical Effectiveness: Enable participants to critically analyze the practical application and effectiveness of the STEP theoretical model within the context of the Huo-Yan project.

Learn Success Factors:
Facilitate learning about the key success factors that contribute to the effectiveness of the STEP model in the real-world scenario of the Huo-Yan project.

Understand Challenges:
Foster an understanding of the challenges encountered in implementing the STEP theoretical model, particularly in big data-driven initiatives.

Group Collaboration:
Encourage collaborative learning through group discussions, providing a platform for participants to analyze, learn, and understand the nuances of the Huo-Yan project.

Agenda:

1. *Introduction to the Huo-Yan Project (15 minutes): Provide an overview of the Huo-Yan project, showcasing practical applications and demonstrating the effectiveness of the STEP theoretical model in a big data-driven environment.*
2. *Group Discussions on Success Factors and Challenges (30 minutes): Participants engage in group discussions to analyze and identify the success factors and challenges inherent in the Huo-Yan project, applying the principles of the STEP model.*
3. *Group Sharing and Plenary Discussion (15 minutes): Groups share their insights, followed by a plenary discussion, allowing participants to collectively analyze, learn, and understand the practical implications of the STEP model in the Huo-Yan project.*
4. *Tea break and networking (10 minutes)*
5. *Summary and outlook, sharing learning insights and practical experiences (20 minutes)*
6. *Closing remarks and acknowledgments (10 minutes)*

ANTHROPOCENTRISM, HUMANISM, AND SOCIAL ENGINEERING, DESIGNING FOR THE AUTHENTIC PERSON IN AN ERA OF POLY-CRISIS

Track 6: Transdisciplinary design and collaboration/open track
Knowledge Network: Cultural Aspects of Design
Methods: Mixed Methods

Session Abstract:
Humans are unique in the world as the only animals believed to have domesticated themselves, detaching from the natural environment. We wear clothing to protect ourselves from temperature extremes and build houses to keep the weather at bay. To reinforce these constructs—at least in the Euro-centric worldview pushed by colonialism and conquest—we have built an entire infrastructure to support them. We farm, raise animals, build economies, specialize in trades, and create vast cities and political structures to ensure it all works as intended. With each step, we remove ourselves more from the natural world that, despite our attempts to ignore, we remain a part of.

This intensive explores the idea of the "authentic" person—connected to others and to the natural environment—and examines strategies to reconnect to a wider world where humans are only one of many elements that must co-exist. We begin by defining the problem of the poly-crisis in "To Thine Own Self Be True," then lay out principles for designing for the authentic person in "This Above All." Next, we consider how late-stage capitalism co-opts authenticity, and how Utopia inevitably becomes dystopic in "Places for Creatures of Habit - on the Limits of Spatialized Social Engineering in Late Capitalism." We then question the very concept of humanism rooted in a Eurocentric worldview in "Is 'Humancentric' a Desirable Goal?" After dissecting fundamental issues around humanocentrism, we examine a wider range of human experiences in "Not Like Everyone Else." Finally, we present an alternative approach via "The Fosbury Case: De-Anthropocentrism in Community Identity."

These papers interrogate the built environment and authenticity in an era of Commodification, Social Engineering, Globalism, and Climate Catastrophe, forming the basis of a collaborative discussion in the intensive session.

To Thine Own Self be True

Bambi Yost (Iowa State University)

What is it that makes us think that we will eventually live in a world full of people who believe in resilient and regenerative design; inclusive and diverse cultures; and an inherent love for living things (biophilia) with a need to care for the planet, when in fact, the majority of our societies, cities, and peoples have done everything but believe in these things for thousands of years? If we truly were an altruistic species, determined to save our natural world above all else, we would have killed ourselves off long ago much like the Shakers. But that is not our nature.

Human beings are an anomaly. We are the only big, fierce creatures to live in such large numbers. We are the only species that is capable of adapting our environments so drastically, and so rapidly, to serve our own needs. Evolutionary biologists have written about our abnormal place in the world for over 25 years now, and yet, we still think that we belong and that we are part of "Nature." We possess a special kind of arrogance and ignorance thinking that we can simply design more "sustainably" while destroying natural systems to create these new sustainable and regenerative design solutions.

When we lie to ourselves, we do far more damage than good. It would be more honest to acknowledge that we are, in fact, conquerors, capitalists, and crusaders. While there are many who strive to do less harm and to do more good than bad, that is not enough. We are the problem. We take more than we can give to a planet that we are actively populating and destroying. The naïve ideal that somehow, we can design our way out of this dilemma is narcissistic, short-sighted, and anything but altruistic.

To thine own self be true.

This Above All: To Thine Own Self be True!

Joseph Juhasz (University of Colorado)

I shall begin by stating a number of highly debatable propositions—on which (at least for argument's sake) I take one side—hoping that we can have a vigorous debate about and over them.

To design for the authentic person:
1: The designer(s) must be authentic.

2: The person(s) for whom the design is meant must be authentic.

3: To be authentic is to be true to one's self.

4: The human self is paradoxical: our selves are deceptive and self-deceptive, and in this respect, we are a unique species and a unique life form.

5: We are an outcome of a process that has made us unique, and as such we can never fully grasp that process.

6: Thus, to be authentic to one's self one must submit to the reality of deception: to comb one's hair, to cover, to be domestic, in the sense of living in a domus, a human shelter fundamentally different from that of any other species or life forms, though continuous with them.

7: For all that—what makes us truly distinct—is that we alone on this planet know the difference between right and wrong—good and evil—and that we have the capacity to adhere to the right or to do evil.

8: It is therefore possible to design that which is good, not in the utilitarian sense, but in the moral sense of the word.

9: We are social beings—in part because, or as a result of, the fact that we are utterly helpless for a greater portion of our lifespan than any other species. We are utterly dependent on one another, and thus there really are no authentic "individuals". The authentic act of creation is communal.

Amen

Places for Creatures of Habit – On the Limits of Spatialized Social Engineering in Late Capitalism

Joern Langhorst (University of Colorado Denver)

While philosophy (and humanity) is still trying to figure out what the good life actually is, environmental designers are habitually arrogant (or clueless) enough to not only think they know the answer to one of the most fundamental questions, but how to design the places where it happens. Authenticity has become – or better, has been abused as – a buzzword that connotes a value system that pretends to be an alternative to the Western-Eurocentric model of design. This paper will disentangle the dimensions and aspects of authenticity in the processes of intentional environmental change and frame an urgently needed critical debate not just on this concept, but on the actors, processes/agendas involved in environmental design in general.

A crucible of environmental design is that it involves ideas about a better world (or some part of it) – and it is inevitable that it proposes some sort of utopia – and historically, those have without fail or exception morphed into dystopias. Better worlds are codependent on better people to inhabit them, and it is precisely here – in the social engineering inherent to any and all environmental design – that authenticity might become a useful category for a critical discourse. Deeply entangled in notions of authority, validity, and originality it is mostly applied to the "source" (the author(s) or designer(s), or the product (the design) – but rarely to the people who live with, in, and enact such design. How can they assert an influence on, or participate in the processes of, intentional environmental change. How can we know and ascertain that their positions and actions are "authentic"? Are there processual and procedural steps that can be taken, and if so, which?

Or – is authenticity a red herring that serves to camouflage capitalist and colonialist actors and agendas?

Is "Humancentric" a Desirable Goal?

Lynn Paxson (Iowa State University)

The topic or term 'humancentric' for this year's EDRA conference has some issues embedded in it for me. As a long-time "EDRoid" I'm all for places that support humans. However, a label that is solely focused on 'humans' as inhabitants, or prioritizes them over other inhabitants is, I think, problematic. It underscores or represents a 'western', or Eurocentric, or colonizer, understanding of the world we live in. This view extracts and puts humans above everything (or everybody else), it sees the rest of the world as a set of "resources" to be extracted and or used to support humans, some may believe humans are "stewards" of these "resources." The Indigenous human communities see themselves as an embedded part of the world, and understand they are no more important to the system than other "peoples" (earth, water, plants, animals, rocks, etc. etc.) You could describe the Indigenous peoples' vision as seeing the rest of the world as a "relative" vs a resource, and it requires respecting all those others. This presentation of the workshop will look at examine these issues. I'm not talking about 'appropriating' Indigenous architecture. What if we tried to design and inhabit the world in this more inclusive, more respectful way? Would we be less likely to marginalize certain human groups? Have less climate and environmental issues? Create places that are more "all-centric" and therefore also 'humancentric"?

Not like Everyone Else – the Impact of Designing for "the Average"

Scott Crisman Sworts
(Oxford Brookes University)

In architecture, we often think of design for the majority, for the typical, for the "normal." However, embedded in this is a complete dismissal of the fringes – the people who by birth, or by experiences, who lie outside the mainstream. Environments that are perfectly accessible and welcoming for the majority can be stressful or even disabling for groups that lie outside of the mainstream. When we design for the "average" we actually design for something that isn't actually real. Like the Utopic visions that are actually no-place, universal design is actually for no one.

This problem begins with the ergonomic norms that actually only "fit" a vanishingly small segment of the population and are actually either too large/tall or too small/short for almost everyone, and yet are held up as the ideal proportion. This becomes even more problematic when we begin to look at things like the differences in perception from neuro-divergence, generational trauma, and structural racism.

The problem is exacerbated by the imperative of designing for intersectionality, where people identify with not just one marginalized group, but into several, creating an overlapping, and sometimes even conflicting, set of barriers. As architects who must design spaces for buildings that are to be used by many, how do we navigate the waters of these conflicts? How can we create space that removes barriers rather than creates them? And, most importantly, how can we put ourselves in the mindsets of people whose lived experiences differ so dramatically from our own?

The Fosbury Case: De-Anthropocentrism in Community Identity Towards a Genuine Harmony between Culture & Nature

Francesco Proto (Oxford Brookes University)

An architectural practice called Fosbury plays a crucial role in building up a sense of identity and belonging in communities by challenging the traditional notion that humans are at the center of the universe. Instead, the interconnectedness of all living beings is emphasized. By prioritizing sustainability, Fosbury architects create spaces that harmonize with their natural surroundings.

They incorporate renewable materials, energy-efficient designs, and green spaces into their projects. This not only minimizes the negative impact on the environment but also fosters a sense of belonging within communities. When individuals feel connected to their surroundings, they develop a deeper appreciation for nature and are more likely to engage in sustainable practices.

Furthermore, Fosbury's architectural practice encourages community involvement and participation. They prioritize local materials and craftsmanship, which not only supports local economies but also strengthens community ties. By involving residents in the design process, Fosbury Architects empower individuals to take ownership of their built environment. This collaborative approach builds a collective sense of identity and belonging within communities.

This paper addresses Fosbury architects, who represented Italy at the recent Venice Architecture Biennale 2023. It highlights their creative approach to de-anthropocentrism in architecture and presents an alternative view of architecture as a step back from the rampant narcissism that it has recently been associated with. Through the application of an archaeological method designed mainly to infuse poetic meaning into otherwise barren and worn-out paths, de-anthropocentrism here takes the form of a transdisciplinary network in which architecture relinquishes its historic authority to function as a bridge between culture and nature.

FOUR PARADIGMS FOR BUILDING EVALUATION - HOW THEY ADDRESS USERS' NEEDS

Track 4: Design for health and wellbeing
Knowledge Network: POE/Programming
Methods: Case Study, Mixed Methods

Session Abstract:
Although owners have undoubtedly been evaluating buildings for as long as humans have been constructing, the last 60 years have seen a number of models emerge to structure the process of building evaluation. Four of the major approaches to emerge include post occupancy evaluation (POE), building performance evaluation (BPE), facility condition assessments (FCAs), and building forensic investigations. Each has a constituency and one or more specific uses, but they also overlap one another in ways that suggest, if not a continuum, at least a constellation of services in which the four are bound and overlapping.

This half-day intensive session focuses on transdisciplinary design and collaboration to explore the relationships and boundaries between these four approaches to building evaluations and how they relate to a more comprehensive context of building evaluation. The session will consist of a brief introduction of the agenda for the session, the featured speakers, and participants lead by a moderator. The presentations by the speakers will begin with conceptual content followed by case studies: • an overview of the four models and their characteristics followed by a case study of the use of the Serviceability Tools and Methods in an evaluation of a design proposal for a U.S. Passport Agency office; • a discussion of five issues whose resolution can guide the future teaching and practice of BPE; • a case study of the use of BPE to assess the implementation of United Nations sustainable development goals in the facility management of hospitals; • BPE evaluations of the indoor air quality performance of six LEED-rated buildings; • POEs of four prototype residences for unhoused First Nation youth to inform the following year's building design and delivery process; • a case study of using experience mapping to evaluate the social, intellectual, restorative, and symbolic associations one building has within the context of its site and surrounding buildings.

*The presentations will be followed by participants' questions and discussion of a duration similar to each speaker's presentation, led by a moderator and followed by a summary of observations. Participants who wish to receive HSW CEUs will be given the opportunity to participate in a 10-minute feedback activity at the end of the session.
The session is sponsored by the Programming POE Knowledge Network.*

Overview of Four Paradigms for Building Evaluation and Case Study of Design Evaluation

Greg Allen Barker (Greg Allen Barker, AIA)

Building evaluation has become increasingly formalized over the last 60 years, with additional breadth as well as specialization along the way. Post Occupancy Evaluation, began with the idea of evaluating an individual building after a full year's operating cycle. Building Performance Evaluation extended the idea of evaluating buildings through an entire design, construction, occupancy, and decommissioning cycle. Specialized disciplines emerged independently: facility condition assessments emphasize the deferred and ongoing maintenance needs of building from a facility management perspective while building forensic investigations examine very specific building failures with an eye towards remediation and/or informing standards and practices to prevent future failures. An overview of these four paradigms sets the context for the conceptual and case study presentations to follow.

The overview is followed by the first case study, the use of the Serviceability Tools and Methods (ST&M) to evaluate construction documents for a design's ability to meet a profile of user's needs. A team experienced in the use of ST&M worked with employees of the U.S. Passport Agency in New York to develop a user requirements profile for their office space, then evaluated the construction documents for new agency offices for the features needed to support that requirements profile. This served as a demonstration of an approach to facility procurement that would be fully responsive to user needs. For the purposes of the intensive session, the case study illustrates an application of BPE at the point of design and construction documentation within a facility's life cycle.

Teaching Building Performance Evaluation for Sustainable Buildings: Chances and Challenges

Ulrich Schramm (Hochschule Bielefeld, University of Applied Sciences and Arts)

With its consideration of the entire building life cycle, the Building Performance Evaluation (BPE) process model constituted a challenge for planners in the building sector when it was introduced in the 1990s: Traditionally, the building delivery process has been understood as a linear course of activities that ends with the opening of the building. At that time, with the new understanding of the building life as a cyclic sequence of phases – including occupancy – and internal review loops, the establishment of performance criteria and their evaluation from a human-centric perspective became crucial. This basic idea of the BPE process model has been verified with the emergence of similar concepts in the building industries, such as certification systems for sustainable buildings (like LEED in the U.S. or DGNB, BNB and QNG in Germany). Thus, universities connected to the building sector have to take these changes and developments into consideration and adapt their existing degree programs – or even develop new ones! But what is the main content and who are the addressees for BPE? Some of the unresolved challenges and questions are: • Which are the pivotal processes, phases, and review loops to focus on in the context of BPE? • What are the relevant topics, methodologies, and instruments to be taught? • Does BPE fit into programs for designers, construction project managers, or facility managers? • Should BPE become part of other transdisciplinary programs for new specialists to appear yet in the building sector? • How can the collaboration with building users be strengthened throughout the BPE process in order to meet environmental, economic, and social challenges in general and to optimize user satisfaction in particular?

The presentation will discuss these issues based on the author's professional background as an architect, his teaching experience, and his research activities in the area of BPE.

Facility Management Approach to Apply UN SDGs and JCI Standards for Building Performance Evaluation of Hospitals

Akikazu Kato (Mie University)

The 2030 Agenda for Sustainable Development adopted by all United Nations Member States in 2015 provides a shared blueprint for peace and prosperity for people and the planet. More hospitals are advocating 17 Sustainable Development Goals (SDGs) to challenge the change of their services for the better. However, those reports of advocacy tend to be anecdotal and do not clarify the depth of their commitment. The lack of suitable Key Performance Indicator KPI may be the cause. Although 17 SDGs and supported by 169 Targets showing numerical values of endeavor, some Targets do not fully meet the management conditions of Japanese hospitals and those in other industrial countries. It should be noted that SDGs for hospital management require a comprehensive approach to promote health of citizens. J

CI Joint Commission International is an independent not-for-profit organization to identify, measure and share best practices in quality and patient safety around the world. Currently, JCI Standards for Hospitals, seventh edition is used in number of hospitals outside USA. In Japan about thirty hospitals are accredited by JCI. The standard consists of over three hundred clauses requiring over 1,200 reports on measurable elements or evidence for rigorous assessment.

Thus, the application of both approaches should provide for the comprehensive, scientific and evidence-based evaluation. The presentation will discuss the extracted suitable Key Performance Indicator KPI for the building performance evaluation. Lastly, the issue of Facility Management is included in both approaches, and the presentation will focus on the impact of physical environment.

Building Performance Evaluation and Indoor Environmental Quality Research – Evidence-Based Design Guidelines for Sustainable Buildings

Ihab Elzeyadi (Dept. of Architecture)

How can we design a better indoor environmental quality (IEQ) in sustainable buildings? The current approach for achieving IEQ in high-performance buildings relies on segregated metrics that predict user-independent comfort attributes--such as visual, thermal, and acoustical comfort--with failed accuracy. In real settings, especially those designed with sustainable objectives, the occupant interacts with the overall ambiance of the environment combining space performance, indoor comfort metrics, user experience, as well as spatial functionality and aesthetics. This presentation reports on a series of comparative Building Performance Evaluation (BPE) projects of LEED™ rated buildings that provide insights into the complex nature of IEQ findings and how BPE protocols can help clarify the interrelationships between multi-comfort phenomena and building performance from the user's perspective. For these studies, we employed a systemic BPE protocol for assessing IEQ parameters, building performance, and occupants' multi-comfort and satisfaction of ten LEED-rated buildings. A comparative analysis of the design strategies and spatial performance of the buildings and patterns of indoor experiences are outlined and compared. In addition, the presentation will focus on six different sampled spaces --of different spatial configurations and IEQ strategies—to develop a framework for the relationship between the various IEQ parameters from the occupant's perspective. The session will present the patterns and LEED credits that contribute to a better IEQ as perceived by the occupants and measured through validated metrics. Methodological innovations for conducting comprehensive BPE studies as a research paradigm will be summarized. The hope is to provide a decision support process and methods for environmental researchers, practitioners, occupants, and owners that would help them design and evaluate green buildings and IEQ comprehensively.

One House Many Nations: Community-Led Sustainable Housing Design for Houseless First Nation Youth

Shauna Mallory-Hill (University of Manitoba)

One House Many Nations (OHMN) is an initiative created by the Idle No More movement of Treaty People in Manitoba, Saskatchewan and Alberta, Canada, to address the poor housing conditions of First Nation (FN) and other Indigenous communities. The initiative seeks to use land-based and community-based processes to design and build sustainable housing using local skills and resources to increase indigenous autonomy, sovereignty, and self-sufficiency. This presentation reports on the findings of OHMN's five-year-long community-led project to explore solutions to the critical housing needs of youth living in FN communities who are at risk of, or are experiencing, houselessness. The interdisciplinary and diverse project team comprises scholars, architects, engineers, tradespeople, land-based users, knowledge keepers and community members. Participatory design sessions with FN youth determined housing design requirements and site selections. The homes were constructed in partnership with a special collegiate program, which helps students aged 18-20 get grade 12 credits as they build. After each building was delivered, investigators undertook a post-occupancy evaluation (POE) process to inform the following year's building design and delivery process. To date, four tiny houses have been constructed and delivered, each a single case study that makes up the overall project. The project's findings reveal the opportunities and ongoing challenges to achieving affordable and sustainable housing for youth living on First Nation reserves in Canada.

Student Experience Mapping

Brian Schermer (University of Wisconsin-Milwaukee)

In conventional building evaluations, the focus is often on individual structures as the primary unit of analysis. However, within the university campus milieu, there are scenarios where it becomes imperative to shift the unit of analysis, considering the building within the broader context it inhabits. This paradigm shift forms the core of Student Experience Mapping, an innovative approach to building evaluation.

Student Experience Mapping utilizes an online, participatory survey to assess social, intellectual, restorative, and symbolic engagement with the physical campus. By soliciting input from students, it aims to evaluate the significance of a specific building relative to other key locations on and near campus that hold personal importance to them. The assessment comprises four distinct types of 'campus capital': • Social: Spaces facilitating social bonding and the creation of new connections. • Intellectual: Environments for thinking, studying, collaborating, and creating. • Restorative: Areas designated for relaxation, rest, and re-energizing. • Symbolic: Places contributing to campus pride and a sense of belonging.

Supplemented by demographic data, the survey sheds light on the role of the physical campus in fostering community, inclusion, and belonging. Going beyond mere evaluation, Student Experience Mapping serves as a strategic tool for guiding planning decisions and future capital investments. It establishes a baseline understanding of how specific buildings contribute to student experiences and provides benchmarks against which the success of future physical changes can be measured.

With participation from over 5,000 students across nine campuses, insights derived from Student Experience Mapping have played a pivotal role in shaping student center planning and design. This impact extends to both ongoing projects and those in the planning stages, showcasing its practical significance in enhancing the overall student experience within university environments.

LESS BAD IS NOT GOOD ENOUGH: THE CIRCULAR ECONOMY, WORKPLACE AND WELLBEING

Track 3: Resilient design for climate change
Knowledge Network: Work Environments
Methods: Practice-based Research

Session Abstract:
Following the Work Environments Network tradition, in this Intensive attendees will explore circular economy principles, the organizations supporting this economic model and methodologies for implementation. Sessions in this Intensive integrate insights gleaned from both research and practice to create a framework of resource efficiency, cost savings, innovation, environmental impact reduction, and embodied carbon implications. Individual sessions will appraise aspects of economic frameworks, stakeholder expectations, wellbeing, notes from the field and hands on talks and tours from Portland area circular economy leaders.

The Circular Economy (CE) is defined as the methodology whereby "products and materials are kept in circulation through processes like maintenance, reuse, refurbishment, remanufacture, recycling, and composting." (MacArthur, 2023). The circular economy is different from sustainability in that the "CE is not about reducing waste, it is about designing a world where waste is not created in the first place," (Circular Economy Institute, 2021). The main goal is "not doing less harm; the main goal is doing good to our ecosystems," (McDonough, 2014). The eight pillars of the CE are Circular Ecodesign; Resource Management and Circular Procurement; Extension of Life Cycle; Responsible Use; Reverse Logistics; Industrial Symbiosis; New Life of Products and Materials; Functional Approaches (Circular Economy Institute, 2021).

A field of interest "almost exclusively developed and written by practitioners," (Corvellec et al., 2021), with rising material extraction, the Circularity Metric stands at 7.2% which accounts for the number of secondary materials which are cycled back into the global economy, (Circular Economy, 2023). This translates into more than 90% of materials that are not cycled back into the economy after the end of their useful life. Organizations are moving to adopt circularity practices in their workplaces to satisfy economic, environmental, and social factors. Thus, the questions of this session will explore opportunities and challenges in research and application; and how practitioners, researchers and academics can understand and apply CE principles in workplace design, implementation and operations.

The session goals are twofold: (1) to explore the circular economy research through the lens of human prosperity, wellbeing, and the environment, assessing the gaps in research and practice; (2) to equip Intensive attendees with tools to integrate circular economy principles into their work.

The agenda for the intensive is as follows:
Welcome & Introductions (15min);
The Circular Economy: Stakeholders, Frameworks and Impacts (30min);
The Circular Economy Impacts on Wellness and Why it Matters (30min);
The Circular Economy: Notes from the Field (30min);
Travel to tour site (20min);
Talk/Tour: How the Circular Economy impacts the Built Environment (45min);
Travel to tour site (20min);
Talk/Tour: How the Circular Economy impacts the Built Environment (45min);

References
Corvellec, H., Stowell, A.F., & Johansson, N. (2021). Critiques of the circular economy. Journal of Industrial Ecology, 26(2), 421–432. https://doi.org/10.1111/JIEC.13187
Circle Economy. (2023). The circularity gap report. Retrieved from: https://www.circularity-gap.world/2023
Circular Economy Institute. (2021). Circular economy assessment. Retrieved from https://ceinstitute.org/
MacArthur, E. Foundation. (2023). What is a Circular Economy. Retrieved from: https://www.ellenmacarthurfoundation.org/topics/circular-economy-introduction/overview#:~:text=The%20circular%20economy%20is%20a,remanufacture%2C%20recycling%2C%20and%20composting.
McDonough, W. (2014). VIDEO: Why design is key to the circular economy. Retrieved from: https://mcdonough.com/video-why-design-is-key-to-the-circular-economy

The Circular Economy: Stakeholders, Frameworks and Impacts

Cynthia L Milota

This session will kick off the Work Environments Network Intensive by outlining the key aspects of the Circular Economy (CE) business economic model, address the actors and stakeholders involved, the fundamental principles and frameworks, and the motivations behind and impacts on adopting a CE mindset and practice.

Key Actors and Stakeholders

Perspectives for the CE model include business and industry, consumers, governments and policymakers, innovations and entrepreneurs. In addition, third party certification of spaces designed within a circular economic model, credentialing of professionals and organizations devoted to promoting and educating about the CE will be explored. This segment will outline, define and suggest a method of adoption for attendees' specific organizational requirements.

Frameworks and Principles

Two CE frameworks will be appraised: the Ellen MacArthur Butterfly Diagram (MacArthur, 2023), and the Harvard Circular Business Model (Atasu et al, 2021). Attendees will explore and understand the difference between the linear (make, take, waste) model and the linear (share, lease, re-use, repair, refurbish) models.

Motivations and Impacts

Beyond the reporting of ESG (environment, social, governance) goals, individuals, organizations and governments are adopting CE principles as a means for resource conservation, waste reduction, climate impact mitigation, economic opportunities and social benefits. Informed by anti-waste and circular economy laws in Europe, and recent research in the US, this segment will examine motivations and impacts on the implementation of CE principles.

Embracing the CE model by understanding the perspective of stakeholders, utilizing frameworks, and clearly outlining outcomes will establish a methodology to implement the Circular Economy principles.

References

MacArthur Foundation. (n.d.). The butterfly diagram: Visualizing the circular economy. Retrieved from: https://www.ellenmacarthurfoundation.org/circular-economy-diagram

Atasu, A., Dumas, C., & VanWassenhove, L. (2021). The circular business model: Pick a strategy that fits your resources and capabilities. Retrieved from: https://hbr.org/2021/07/the-circular-business-model

The Circular Economy: Impacts on Wellbeing and Why It Matters

Deb Bucci (Live Well Strategies)

The Circular Economy is often described in general terms as a progression from a linear model of 'take, make, use, dispose' into a circular model by adding in components encompassing restorative and regenerative elements. It is argued that this resource-oriented definition focuses on the importance of closed loops for material management. However, other definitions of CE go beyond material management to include dimensions of health, well-being, and the workplace, which drive human behavior toward consumption (World Health Organization 2018). Incorporating the concept of well-being into the circular economy can play a crucial role in enhancing personal, social, and mental growth and sustainable consumption. This discussion will focus on the peer-to-peer sharing model (PSS) and the product service model (PSM) (Ho, M. X., & Yanagisawa, H. 2023).

Peer-to-Peer Sharing:

1. *Community Building and Social Capital: o Fosters community by encouraging interactions and trust between individuals*
2. *Access to Goods and Services: o Gain access without the need for ownership o Reduced financial stress*
3. *Environmental Awareness and Sense of Purpose: o Increases environmental awareness o A sense of purpose in contributing to sustainable practices*

Product-Service Model:

1. *Predictable Expenses and Financial Stability: o Pay for services rather than owning products*
2. *Access to Upgraded Technology: o Provide access to latest and most efficient technologies without need for constant product replacement*
3. *Business Opportunities and Innovation: o Encourages innovation in service delivery and product design.*
4. *By adopting circular economy principles, societies have the potential to create a more sustainable and resilient future while improving the well-being of individuals and communities.*

References

Ho, M. X., & Yanagisawa, H. (2023). Design for Well-Being and Sustainability: A Conceptual Framework of the Peer-to-Peer Sharing and Reuse Platform in the Circular Economy. Sustainability, 15(11), 8852.

World Health Organization. (2018). Circular economy and health: opportunities and risks. Retrieved from: https://iris.who.int/bitstream/handle/10665/342218/9789289053341-eng.pdf?sequence=1

The Circular Economy: Notes from the Field

Cynthia L Milota, Janika McFeely (JLL); Grace Manuel (Haworth)

A Circular Economy (CE) framework is based on three principles: "eliminate waste and pollution; circulate products and materials; regenerate nature," (MacArthur, 2023). This segment explores how the adoption of CE principles impacts employee engagement/performance. Utilizing the upcoming ISO-59.010 International Standards on Circularity Pillars I-VIII, (ISO/DIS 59010, 2023), the discussion will draw upon real world examples and lessons learned. Panelists will assess the utilization of circular principles to measurably improve the work environment.

The ISO Pillars will be applied from three perspectives, specifically discussing strategies organizations have implemented and how it worked out:
Employee Engagement • Employee mission, vision and values alignment with organization • Climate literacy, recognition, motivation, empowerment, change management
Environmental Stewardship • Resource efficiency and waste reduction • Responsibility for products through their lifecycle • Decarbonization strategies in retrofits and new construction
Corporate Initiatives • Resource management and circular procurement • Technological advancements • ESG (environment, social, governance) and SDG (sustainable development goals) objectives
Analysis of challenges and opportunities are faced by individuals making sustainable choices and by organizations delivering an optimal workplace mindful of circularity. Through a comprehensive examination of assessments, initiatives and change management, this session will highlight key findings, lessons learned, and potential implications for those involved in the design, development and management of workplaces developed with a circular economy mindset. The insights will contribute to a deeper participant understanding of the implementation of circularity. In addition, panelists will offer recommendations on the lifecycle of performance for employees and buildings based on circular economy principles.

References

MacArthur Foundation. (2023). The circular economy in detail. Retrieved from: https://www.ellenmacarthurfoundation.org/the-circular-economy-in-detail-deep-dive
ISO-59-010. Circular economy-guidance on the transition of business models and value networks. Retrieved from: https://www.iso.org/obp/ui/en/#iso:std:iso:59010:dis:ed-1:v1:en

The Circular Economy's Impact on the Built Environment: Talks and Tours

Elizabeth Baudler (Le Cercle Brands); Tracy Danyluk (Reclaim NW); Jackie Kirouac-Fram (ReBuilding Center)

Managing how we design, construct, maintain and operate our workplaces matters. The built environment uses "almost half the materials extracted globally every year and is a significant contributor to greenhouse gas emissions," (MacArthur Foundation, 2023). This segment will explore the circular economy's impact on the Portland built environment. Stakeholder perspectives will include private enterprises, community outreach organizations, service providers and governmental bodies.

Organized as two tours, the presenters, local to the Portland area, will explore a wide range of hands-on topics including opportunities and challenges for implementing circular economy initiatives in the work environment during initial design phases through construction and daily operations. Appraising the 2016 Portland Residential Deconstruction requirements (Portland.gov, 2016), via specific examples of integrating re-use into the process from the outset, will inform concepts for future circular economy principles in commercial applications.

Additional factors for discussion include how adopting material reuse and deconstruction methods can aid in meeting metrics assessments for green building certifications such as Living Building Challenge or LEED and criteria/reporting to meet organizational ESG (Environment, Social, Governance) goals.

The WEN Intensive group will travel as a group to visit RECLAIM NW and Rebuilding Center, both warehouses and resource centers for reclaimed building materials. After a presentation by each organization's executive director, there will be a guided tour and discussion on their business operations, training classes, donation programs, and local volunteer opportunities. Presenters will share their experience with circular economy initiatives and measured outcomes, providing actionable strategies for attendees to integrate into their practice, research and classroom.

References

City of Portland. (2016). Deconstruction Requirements. Retrieved from: https://www.portland.gov/bps/climate-action/decon/deconstruction-requirements
MacArthur Foundation. (n.d.). Reimagining our buildings and spaces for a circular economy. Retrieved from: https://www.ellenmacarthurfoundation.org/topics/built-environment/overview

PLANNING DECISION MECHANISMS FOR DATA-DRIVEN DECISION PLATFORMS FOR HYDROGEN ENERGY MANAGEMENT

Track 6: Transdisciplinary design and collaboration/open track
Knowledge Network: Movement in Designed Environments: Nature & Ecology
Methods: Case Study, Practice-based Research

Session Abstract:
In the wake of the global energy transition, hydrogen energy stands out as a promising sustainable alternative. To harness its full potential, this intensive course proposes an in-depth exploration of a data-driven hydrogen energy management decision-making platform. The primary objective is to equip participants with the knowledge and skills necessary to understand and implement advanced planning mechanisms that integrate the Internet of Things and big data analytics for optimal hydrogen infrastructure planning.
The course aims to achieve the following specific goals: (1) introduce participants to the concept of dynamic planning in hydrogen energy management, (2) demonstrate the application of spatio-temporal supply-demand mapping tools, (3) discuss the challenges and opportunities in breaking traditional static planning barriers, and (4) showcase the real-world effectiveness of the platform through a case study of the Jinshan Hydrogen Valley Town in Shanghai.

The four-hour agenda is structured as follows: The first hour will provide an overview of hydrogen energy and its role in sustainable development, along with an introduction to data-driven decision-making. The second hour will delve into the core components of the proposed platform, including the integration of IoT and big data technologies. Hour three will feature a hands-on session on dynamic planning techniques and spatio-temporal analysis using real-world data sets. Finally, the fourth hour will be dedicated to a case study analysis, group discussions, and a Q&A session, allowing participants to apply their learnings to a practical scenario and gain insights from experts in the field.

By the end of this intensive course, participants will have a comprehensive understanding of the proposed platform and its potential to revolutionize hydrogen energy planning, leading to more efficient, sustainable, and resilient energy systems.

1. *Learn the fundamental principles and concepts of data-driven hydrogen energy management, including the role of the Internet of Things (IoT) and big data analytics in optimizing hydrogen infrastructure.*

2. *Understand the challenges associated with traditional static planning approaches and how dynamic planning concepts can be applied to achieve better spatial and temporal allocation of hydrogen resources.*

3. *Analyze real-world case studies, such as the Jinshan Hydrogen Valley Town, to grasp the practical applications of the proposed platform and its impact on sustainable energy development.*

4. *Develop the ability to critically evaluate and discuss the potential benefits, limitations, and future prospects of data-driven hydrogen energy management systems in the context of global energy transitions.*

By achieving these learning objectives, participants will gain a comprehensive understanding of the proposed platform and its potential to revolutionize hydrogen energy planning, leading to more efficient, sustainable, and resilient energy systems.

Course Learning Summary: Hydrogen Energy, Sustainability, and Data-Driven Decision-Making

Yunsheng Su (Professor)

With the growing global economy, population, and living standards, the depletion of fossil fuels and environmental concerns are becoming pressing issues. China is shifting towards clean hydrogen production, anticipating it to surpass fossil fuels by 2030. However, the hydrogen industry faces challenges in planning and development. Traditional static planning methods fail to address the dynamic relationship between stations and vehicles. Therefore, innovative approaches like dynamic planning and spatio-temporal supply-demand mapping are crucial for optimizing hydrogen infrastructure. This course highlighted hydrogen's role in sustainability and emphasized the importance of data-driven decision-making for efficient and reliable energy systems.

1. *Learn about the role of hydrogen energy in promoting sustainability and the challenges faced by the hydrogen industry in planning and development.*
2. *Understand the importance of data-driven decision-making in optimizing hydrogen infrastructure and its impact on creating efficient and reliable energy systems.*
3. *Analyze the dynamic relationship between hydrogen stations and vehicles, and how innovative approaches like dynamic planning and spatio-temporal supply-demand mapping can address this complexity.*
4. *Apply knowledge gained from the course to evaluate the feasibility and potential benefits of hydrogen energy in China's transition to clean energy sources by 2030.*

Course Learning Summary: Methods for Hydrogen Energy Planning and Decision-Making

Xihan Cheng (Tongji University)

In the context of growing global energy demands and environmental concerns, hydrogen energy emerges as a promising sustainable solution. This course delved into the methodologies for effective hydrogen energy planning and decision-making. Emphasizing data-driven approaches, we explored dynamic planning techniques, incorporating historical and real-time data to optimize infrastructure planning. Key Influencing Parameters (KIPs) such as transport distances and traffic flow were critical in determining the spatial layout of hydrogen refueling stations. The course highlighted the importance of adapting to different development stages and leveraging technological advancements to enhance planning accuracy and efficiency. Overall, this course provided valuable insights into the evolving field of hydrogen energy planning, equipping learners with the tools and knowledge to contribute to a sustainable energy future.

1. *Learn the methodologies for effective hydrogen energy planning, including data-driven approaches and dynamic planning techniques.*
2. *Understand how Key Influencing Parameters (KIPs) such as transport distances and traffic flow impact the spatial layout of hydrogen refueling stations.*
3. *Analyze the importance of adapting to different development stages and leveraging technological advancements in enhancing the accuracy and efficiency of hydrogen energy planning.*
4. *Apply the knowledge gained from the course to contribute to the evolving field of hydrogen energy planning, with a focus on promoting a sustainable energy future.*

Dynamic Planning and Decision-Making for Hydrogen Energy Infrastructure: Lessons from the Hydrogen Source Carbon Valley in Shanghai Jinshan

Mingjun Zhong (Designer)

In this study, we delved into the planning and decision-making mechanisms for hydrogen energy infrastructure, employing innovative approaches such as dynamic planning and spatio-temporal supply-demand mapping. The Hydrogen Source Carbon Valley town in Shanghai Jinshan was used as a case study to demonstrate the effectiveness and feasibility of these methods. The results highlighted the advantages of dynamic planning in optimizing the capacity and spatial location of hydrogen infrastructure, considering key influencing parameters (KIPs). However, the current reliance on static data poses limitations, emphasizing the need for improved data sources and refined KIP analysis. Looking ahead, we anticipate the platform's enhanced maturity and precision, offering reliable support for hydrogen energy management.

1. *Learn about the innovative approaches employed in hydrogen energy infrastructure planning, such as dynamic planning and spatio-temporal supply-demand mapping.*
2. *Understand the advantages of dynamic planning in optimizing the capacity and spatial location of hydrogen infrastructure, considering key influencing parameters (KIPs).*
3. *Analyze the limitations posed by the current reliance on static data and the need for improved data sources and refined KIP analysis in hydrogen energy planning.*
4. *Evaluate the potential for enhanced maturity and precision of planning platforms to offer reliable support for hydrogen energy management, based on the insights gained from the case study of Hydrogen Source Carbon Valley town in Shanghai Jinshan.*

PLAYFUL PLACEMAKING: SOCIAL SCIENCE OF PLAY APPLIED TO COMMUNITY ENGAGEMENT METHODS

Track 5: Spatial justice, equity, and inclusion by design
Knowledge Network: Cultural Aspects of Design
Methods: Design Methods, Mixed Methods

Session Abstract:
Session Abstract: Engagement is crucial to design to having strong communities and neighborhoods. Organizations and researchers involved in the built environment have seen the interest in engagement grow tremendously in recent years, as the commitment to justice, equity, diversity and inclusion continues to rise. Designers, municipal leaders, and community leaders however, have not been trained as facilitators, making this crucial component sometimes challenging, and often leaving the community feeling disconnected.

The long held methods for outreach have been transactional, outbound information only, with an understood requirement of the community to freely GIVE themselves to outsiders. Many have seen the negative impacts of this, namely gentrification, and are creating alternative models for engaging, rather than just outreach. Having participated in team building exercises around corporate strategic plans, we saw value in the methodology and sought to adapt that work to the built environment.

In 2021, PLAYCE authored a white paper called Playful Placemaking that linked human play-science to the intentions of meaningful community engagement. The universal human experience of play allows a space for empowered listening and the lifting of voices that no other process can match. Through play the community can voice their concerns and discuss potential outcomes and solutions without a contentious atmosphere. Play allows for participating community members to feel heard and create a relatedness to others involved in the process, including project leaders.

Through this collaborative co-design process, the project leaders guide the community through creating a vision to use as a benchmark in realizing their vision. All projects and initiatives must measure up or the existing vision must evolve to include the new initiatives. In this technique the project leader is also a player, shifting the role of "designer as expert" to designer as part of community voice. When using this technique, project leaders can leverage the power of play to build trust within a group, create consensus around the community's vision, and develop strategies to move those ideas forward. Allowing designers to then use their training and expertise to bring community voice into existence. This playful co-design process assists all involved stakeholders to have a voice and contribution to the project generating community champions in the restoration of the urban fabric, recovery of surrounding ecosystems, promoting the stewardship of the environment and incorporating the contributions of community members.

As this session is about playful community engagement, we will of course be playing. Panelists will introduce themselves and present an overview of engagement limitations, the science of play and the background of this playful placemaking methodology.

The session is a dynamic mixture of presentation, group activities and facilitated discussion. After each game, panelists will lead group sharing of what participants experienced, as well as the applicability of the various activities to classroom studio, research and design practice.

The final facilitated discussion explores where each participant sees this methodology could be helpful in their own work, practical implementation tactics, and invites participants to try on facilitating an activity themselves before the session concludes.

Social Science of Play

Cheryl Salazar (PLAYCE Studio Ltd.)

Social science and play research shows that every human plays, regardless of age, ability, culture, or language. Play is universal and is the ultimate inclusive language. Play is a very interesting beginning place for community engagement, because when one is playing, one is naturally involved as a collaborative problem solver.

When we move from a language of drawing and presentation to a language of play, the community, as well as the design team, CAN imagine situations they have never encountered before and share what they see as possible. Even when the project is intended to improve their community, community members often voice opposition to the project citing risk avoidance and fear of failure as major motivations. However, through play they can try on possibility for themselves and are willing to embrace risks. The community is willing to come to consensus even if it is not unanimous.

Play creates major breakthroughs in how they see themselves and how they imagine the project would impact their lives. Play builds trust within a group, creates consensus around an idea, and supports the community and designer to develop strategies that move that idea forward. And most importantly, it is FUN! By taking the scientific benefits of social science and play and aligning them with this unique placemaking approach, the design team can elevate community perspectives, empower communities, and generate successful public spaces that align with community values.

This session will provide an overview of the science of play, how it directly aligns with playful placemaking, a unique playful placemaking methodology, and provide opportunities to experience various games used in public participation meetings and events.

Playful Placemaking

Joy Kuebler (PLAYCE Studio, Ltd)

Joy opened her landscape architecture studio in 2003, and in 2014 started to develop and implement "playful planning" strategies. After working with a professional from the organizational development world for her own team, Joy saw the direct application of play and team building as a compelling methodology for design and planning professionals. Since then she has developed a full spectrum of games and activities for site analysis and project visioning, project scoping, initiative prioritization, and phasing. Additionally she has developed games and activities to support champion building, stewardship planning, and citizen action initiatives. Her design firm has led or participated in projects that range from comprehensive plans and brownfield redevelopment plans, to large streetscapes, and urban parks, as well as school yards and even for spaces honoring those impacted by violent acts such as domestic violence and mass shootings. She has found this work to be highly valuable at every scale and scope of working with people and the built environment. Play is a unique tool, our earliest language, and the highest offering of equitable participation; every human on the planet plays regardless of language, culture, physical or cognitive ability. Over and over she has seen community members and the project leaders empower one another in creating the vision for, and ultimately delivering highly meaningful projects that add real value to the community's quality of life.

In 2019, she created an organization separate from her design practice, with the intention of changing the global conversation around community engagement and design. By sharing this work, empowering others to add play to their engagement process in order to deliver truly impactful projects, the opportunity for architecture and design to positively impact the quality of life for all people can become a reality.

EDUCATION PHILOSOPHY, PROGRAM, AND PEDAGOGY FOR DESIGNING HEALTH AND HEALING ENVIRONMENTS

Sharmin Kader
Ball State University College of Architecture and Planning

Session Abstract

Background

The first program of Architecture for Health was established in 1966 at Texas A&M University (HCD, 2005). Since then, many universities in the United States have developed curricula integrating efforts and philosophy to create future generations with the knowledge of health, healing, and healthcare environments. Pedagogy and curriculum development should be aligned with the fundamental and state-of-the-art research and socioeconomic needs. In the post-pandemic era, all these programs and philosophies need a reassessment to identify the gaps, and may need to ask the question, "Do we want to create a designer without fundamental knowledge of health? "Hasn't the matter of health become crucial in building environments?" However, it is significant to introduce the principles, processes, tools, and techniques for designing health and healing environments.

Having cutting-edge knowledge and competencies strengthens designers' problem-solving ability and provides a seamless transition from academic life to professional practice.

Vision

This intensive wants to identify the future directions of educational programs and pedagogy of preparing upcoming and current professionals to integrate health or healing aspects in their design and research. With this vision, the discourse focuses on a variety of approaches and intends to identify the strengths and weaknesses of these programs, pedagogies, and resources. Preparing design professionals is challenging in many ways, for example, the accreditation criteria, the industry needs versus industry wants, lack of expertise, and unique pedagogies.

Structure

This intensive is divided into two sessions: the first one focuses on preparing professionals, and the second one focuses on training post-professionals. The first session will discuss various curricula of accredited and non-accredited programs, programs related to architecture, interior design, and landscape architecture, various pedagogy considering health, healing, or healthcare environments, and feedback from the students. The session focuses on post-professional education, training, certifications, and resources available by various organizations, such as the Center for Health Design (CHD), the American Institute of Architects (AIA), and the Nursing Institute for Healthcare Design (NIHD). At last, this intensive will end with a workshop using the Six Thinking Hats Technique.

Expected Outcome

This session intends to develop a holistic approach that equips the future generation with appropriate skills and enriches the student's experience through the long process of development and learning. Each speaker will develop an article based on the feedback they have received during this intensive session and will submit the article to the organizer after the conference. The organizer intends to develop a report or book compiling all the abstracts, articles, and summaries of the charrette.

Question and Answer

This session intent to develop a holistic approach that equip the future generation with appropriate skills and enrich the students' experience through the long process of development and learning. This intensive session is composed with ten speakers and is divided into two sessions: first one focuses on presentation topics related to preparing design professionals, and the second one focuses on training post-professionals. The first session will discuss about various curriculum of accredited and non-accredited programs, programs related to architecture, interior design, and landscape architecture, various pedagogy considering health, healing, or healthcare environments, and feedback from the students. The session focuses on post-professional education, training, certifications, and resources available for by various organizations, such as, Center for Health Design (CHD). At last, this intensive will end with a workshop using Six Thinking Hats Technique.

After the EDRA55 conference, each speaker will develop an article based on the feedback they have received during this full-day intensive session, and will submit the article after the conference, by the end-of-summer to the organizer. The organizer wants to develop a report or book out of this intensive session compiling all the abstracts, articles and summary of the charrette

Incorporating Evidence-Based Design in Healthcare Design Studio: Course Planning and Pedagogy

Sharmin Kader
Ball State University College of Architecture and Planning

Evidence-based design (EBD) creates a physical environment based on research findings to achieve the best possible outcomes. EBD is a methodology that emphasizes the use of data acquired to influence the design process in healthcare settings, such as measuring the physical and psychological effects of the built environment on its users. The major goals of EBD are to 1) reduce stress for patients, families, and caregivers, 2) reduce errors in patient care, 3) provide a safe and efficient environment for all occupants, and 4) most importantly, provide a healing environment for patients. Teaching healthcare design studios is challenging; students need to learn about the healthcare-delivery model, population demography, codes and FGI guidelines, cutting-edge knowledge or research findings, and ensure efficiency, safety, and a healing environment.

Nonetheless, the classification of the students, senior level or grad level, acquiring all that knowledge and implementing those in their projects is overwhelming. Developing a studio model or technique is necessary to teach the EBD process in the healthcare design studio. With this aim, this session shares examples of three design studios and the course planning, pedagogy, and outcome of those studios. The examples are from the University of Kansas (architecture students), Kent State University healthcare design program (mixed, architecture and interior design, grad and undergrad students), and Ball State University (all interior design students). After analyzing the outcome and students' feedback, this study has developed the following key recommendations a) to allocate a reasonable amount of time for literature reviews, b) to develop EBD goals and design strategies to achieve those goals, and c) to distribute the research findings development among all students so that there is no repetition of work, and the entire class ends up having an enriched pre-design report to start their design development.

The Genesis and Development of the 'Design for Health' Program, Huckabee College of Architecture, Texas Tech University

Saif Haq
Texas Tech University

This presentation will describe the DESIGN for HEALTH program at Huckabee College of Architecture, Texas Tech University. It will describe its founding vision, explain the opportunities available at Texas Tech University and Texas Tech University Health Sciences Center, the chain of events that has led to both the formation of the program and its development in the last decade, and its status at this time.

A special focus of this presentation are the efforts made to understand societal awareness and needs for environments at different scales that support being and getting well, professional requirements for certain knowledge and skills, opportunities available locally, statewide, and nationally, and the negotiations that have shaped this program.

The presentation will conclude by commenting on the program itself, and on lessons learned for creation and/or development of similar academic endeavors.

Industry-Academia Collaborative Studios in Healthcare Design Education: Lessons Learned

Roxana Jafarifiroozabadi
Texas A&M University

The healthcare design education at the School of Architecture, Texas A&M University, has benefitted from collaborations with industry professionals in recent years to expose students to real-world healthcare projects in diverse contexts prior to their graduation. This initiative has offered students immersive, hands-on experiences within the realm of healthcare design profession as well as numerous networking opportunities and mentorship programs. In addition to the educational benefits and employment opportunities, such initiatives have provided future employers with insights into the students' specific skills and capabilities prior to the recruitment process. Nevertheless, such collaborations can involve challenges and unforeseen complications that might impact the learning outcomes of the architectural studios.

This report will investigate an undergraduate healthcare design studio that engaged senior-level students in the design process of a 250-bed mental and behavioral healthcare facility in Texas, U.S. in collaboration with an industry partner during the spring 2024 semester at Texas A&M University. The report will use an ethno-methodological approach to explore the nature of learning within the collaborative healthcare design studio, focusing on the relationship between the students, professionals, and the contexts of learning. The student-professional and student-user interactions will be recorded using field-notes, observations, and photographs, and additional data collections will take place using semi-structured interviews and questionnaires. The goal is to evaluate the overall strategies involved in forming student-mentoring sessions with industry professionals in the architectural studio, assess methods employed to involve the professionals as well as end-users during the design process, and develop a framework to guide future collaborative healthcare design studios in the undergraduate architecture education.

Architectural Pedagogy and Practice: How to Design for Individuals with Disabilities

Neda Norouzi
UTSA

This study responds to the dynamic changes in pedagogy and curricula development post-pandemic, focusing on the imperative of creating inclusive environments for individuals with disabilities. Initiated through a collaborative effort between a non-profit organization and a healthcare design-focused architecture professor, our research commenced with a comprehensive survey conducted by the non-profit to discern the specific needs of individuals with disabilities within the built environment. The results highlighted a compelling demand for environments exceeding the minimum accessibility standards of the Americans with Disabilities Act (ADA).

Traditionally, architectural studios have revolved around hypothetical design problems that challenge students to apply creative thinking. While the classroom-to-career model has positively impacted architectural pedagogy by integrating academic experiences with professional practices, a significant disparity remains between students and the ultimate users of built environments. To address this gap, we present the development of an innovative architectural design curriculum for architecture studios. The approach incorporates three key components: (1) fostering evidence-based design principles, (2) collaborating with the non-profit organization as the project owner, and (3) conducting workshops and presentations for community members with disabilities as the end users.

Course evaluations reveal a profound impact on students, characterizing the studio as challenging yet rewarding. Emphasis was placed on the amalgamation of research and architecture, with insights on the significance of evidence-based design in healthcare environments. Particularly impactful were the reflections on the distinction between presenting to the project owner and directly engaging with the end users.

This transformative pedagogical approach equips students with practical skills while instilling heightened awareness of ethical and social responsibilities in architectural practice. As we refine and expand this curriculum, our research highlights the potential for academic institutions to play a pivotal role in fostering inclusivity and advancing architecture towards a more empathetic, user-centered paradigm.

Harmonizing Healing Spaces and Integrating Wellness: A Comprehensive Exploration of Holistic Healthcare Design in Interior Design Studio Courses

Haroon Sattar
University of Central Missouri

Integrating healthcare design into interior design programs is of paramount importance, stressing health, safety, and welfare in built environments. This preparation is pivotal for future designers to create spaces that meet aesthetic standards and significantly enhance the well-being of occupants and healthcare practitioners alike. To align with Council of Interior Design Accreditation's (CIDA) Standard 7: Human Centered Design- "that graduates understand theories of human-centered design, and identify, analyze, and apply information from a variety of stakeholders and sources to develop a successful response to user needs and to promote health and wellbeing." (CIDA Professional Standards 2022). The emphasis is for students to understand: a) impact of the built environment on human dynamics; b) designed environments' impact on human experience, wellbeing, behavior, and performance."
A comprehensive approach to healthcare design in interior design curricula integrates human-centric philosophies and pedagogies. Students learn to prioritize health, safety, and welfare in spaces through a blend of evidence-based design, empathetic understanding of user needs, and biophilic principles. Emphasizing a humanistic perspective, which foster physical and emotional well-being, prioritizing safety and embracing a compassionate, sustainable approach. Healthcare design pedagogy in interior design merges evidence-based principles, emphasizing health and safety, with a focus on healing elements. Studio projects include comprehensive case studies and site visits, exposure to the intricate balance between aesthetics and the creation of therapeutic environments. The curriculum also underscores the importance of understanding building and environmental systems integral to healthcare design, necessitating the review of white papers, presentation of precedent studies, and the development of design goals rooted in evidence-based research outcomes.

The holistic philosophy and teaching pedagogy emphasizes evidence-based, human-centric approaches, ensuring interior design students prioritize health, safety, and welfare. This comprehensive education cultivates students to harmonize aesthetics with creating health-promoting spaces, fostering career growth and meaningful societal contributions.

Teaching and Fostering Design Empathy

Naomi A. Sachs
University of Maryland

Landscape architects and other designers work for everyone, including individuals and communities far more diverse than our current professional demographic. Teaching and cultivating empathy in design students helps them become better designers at the beginning of their careers by inculcating a deeper cognitive, affective, and somatic understanding of people and their environment. Research also suggests that teaching empathy can increase students' mental health and creativity. We are at a turning point in history when disparity is increasing and our very life on earth–and the earth itself–is threatened. In addition to evidence-based design, which is critical for vulnerable populations, empathy is essential. As we gain knowledge of how to design for people with varying abilities, designers must do more than follow the minimum standard guidelines. Empathy is the key. One method of teaching and fostering empathy is with "design empathy equipment"–wheelchairs, walkers, and other assistive devices–that enable students to experience the physical world on a somatic level. The researcher has used these "design empathy exercises" in undergrad and graduate studios for the past 3 years. Additional strategies to address differences and invisible disabilities include reading and discussion, interviews and focus groups, storytelling, and even the use of VR and AR.

This presentation explores tools for teaching and cultivating empathy in landscape architecture and other environmental design programs. The presenter will cover the literature and theory related to "design empathy"; the design empathy exercises used for the past three years; other potential tools that can be explored; and potential future avenues of research in this realm.

Designing for Health and Healing

Laurie T Waggener
Page

Young professionals as well as established design professionals, working in private architectural practice and interior design, are on a perpetual learning curve. New challenges emerge frequently within the healthcare design and construction industry. Case studies and lessons learned must be documented and shared for continuing education to occur relative to meeting the ever-changing and complex needs of the occupants in our built environments. Additionally, post-professional training is a key, critical ingredient in training future design and thought leaders. Professionals specializing in design and construction of healthcare settings have multiple constraints to consider in the design process. Post-professional training and education occur using multiple means to address these constraints. Ease of access, ease of documentation and affordability are top priorities articulated by design professionals. How can we make post-professional training more inclusive?

Post-professional training is offered in two models: 1) Low cost, live presentations/ virtual webinars as well as local conferences and 2) Expensive in-person attendance at long-distance national conferences. The most affordable options include opportunities offered by an architecture firm. Live lunch and learn webinars offered by senior staff, building product specialists, as well as the firm's membership to national organizations offer affordable and accessible learning. The second model requires expensive tuition and travel.

Dedicating employees for staff education has proven to be integral to staff professional development and staff satisfaction. Leaders in architectural practice must keep a keen eye on identifying the knowledge gap in their practice as well as in the industry at large. Proactive planning for education to meet the needs of all ages, all cultures and all end users is something that firms must plan for strategically. This presentation will discuss what has worked and what must be our focus for a future in inclusive learning and inclusive design.

A Case Study Focusing on an Interior Design Curricular Agenda to Aid in the Study of Healthy and Healing Environments

Sarah Angne Alfaro
Ball State University

This presentation focuses on a four-year undergraduate interior design program and investigates how students' knowledge of restoration unfolds within the curriculum. Restoration takes on many meanings. From a historic lens, restoration differs from preservation in that rather than focusing on the maintenance and repair of existing historic materials, it depicts a property at a particular period of time in its history, while removing evidence of other periods(Condo, 2016). From a sustainable lens, restoration is exhibited in biophilia, as an approach that attempts to connect people with nature (Kellert, 2008; Browning et al., 2014); a form of psychological restoration that focuses on the renewal of individual depleted resources caused by adaptation of demands, such as stressors and challenging tasks (Honold et al., 2016). From a spiritual lens, restoration personifies the mind, body and spirit and can aid in restoring directed attention as well as fostering meditation and reflection (Herzog et al, 2010).

As students enter the four-year design program, they are immersed in a design plan where the entire design process unfolds and at each level restoration is emphasized. Thread throughout, a tri-part of environmental awareness, social vitality and economic conservation affords students a well-rounded education focused on inclusivity as emphasized in design that accommodates as many users as possible. By focusing on universal design, evidence-based design, environmental responsibility, and environment and behavior tenants, students are prepared to solve design problems for a variety of spaces within environments built for human habitation. Students encounter immersive learning opportunities in their courses that provide opportunities to design for real clients instilling tangible meaning as they focus on healthy and healing environments. These experiences, which emphasize restorative solutions, position the students to have design sensibility and intellectual inquiry as they accept a career in an interior design professional setting.

Bad Design Kills: How Architecture Schools Propagate Design for Human Harm

Erin Peavey
HKS Inc.

Our form-centric focus on architecture misses the need to first understand the humans we are designing for. Da Vinci recognized human proportions as foundational to his architecture, we in the modern era have ignored the humans in strained and highly technical focus on the form of the building itself. But without a foundational understanding of the principles of psychology, sociology, biology, and the basics of what it means to be human and well, we are creating spaces that are unintended and ignorant propagate harm to the very people we claim to serve.

For decades we've been sold the narrative that healthy buildings have to be expensive. The false dichotomy between being offered ugly buildings or no buildings at all. These narratives have taken on a fast-food mentality, cheap, easy, quick – in both of these cases, the long-term impact is detrimental to our health.

To transform the practice of architecture, we must change the way we educate future architects. Currently, there is little if any requirement for architectural education to include psychology, sociology, and a fundamental understanding of what makes us sick or well. Without that understanding, we are doomed to repeat the flaws of modernist architects who with good intentions, strong opinions, and poor science who propagated a world where buildings are often sources of stress, sickness, and exacerbation of chronic illness.

This session will suggest concrete changes that can make way for a future that embraces the role of architecture as interconnected with human health and well-being. This includes changes to National Architecture Accreditation Board educational requirements, a transformation in how architecture is viewed by the public, and a code of ethics that is inherent in everyday practice.

From Academia to Action: Twenty Years of Research Evolution in a Global Design Firm

Emily Chmielewski
Perkins Eastman

Over the last several decades, the design industry has witnessed a profound shift: Rigorous academic research, once confined to scholarly journals, now sits at the heart of many design firms, shaping their strategies and solutions. This presentation explores this transformation through the lens of one design researcher's twenty-year journey, a testament to the challenges and triumphs of integrating research with practice.
Our presenter began her career steeped in the methodologies of academia, wielding social science research with finesse. Crafting dense reports was second nature, but translating these findings into actionable insights for a dynamic design environment proved a crucial hurdle. This presentation delves into her evolution, detailing how she honed her skills to become an agile research navigator, adept at bridging the gap between theoretical rigor and practical implementation. The audience will see firsthand how her experience transformed, witnessing the shift from academic-style research and reporting to a more practice-based approach. Through her story, the audience will also gain insights into the diverse tapestry of research projects undertaken within her design firm.

But research isn't a solitary endeavor. Our presenter will also share her strategies for fostering effective collaboration – between designers and researchers, clients and creatives, and even the broader design and research communities. Through workshops, presentations, and collaborative initiatives, research becomes a shared language, driving innovation, deepening client relationships, and resulting in greater positive outcomes and stakeholder experiences from design.

Witness the metamorphosis of research in action: from dense dissertations to succinct, impactful client briefings, from ethnographic expeditions to trend-forecasting white papers. Join us as we explore the dynamic convergence of research and practice, and leave an understanding how designers and researchers can work together to leverage the power of human-centered insights.

TRAUMA-INFORMED DESIGN ODYSSEY: CRAFTING SPACES FOR WELL-BEING AND EMPATHY

Davis Harte, PhD
Paradigm Spaces, llcBoston Architectural College

Session Abstract
Background

This full-day intensive offers an exploration of Trauma-informed Design (TiD), delving into the psychological and physiological dynamics of trauma and its complex impacts. The morning will include presentations and a panel discussion on foundational trauma responses, design requirements, societal impacts, and research principles. Attendees will then deepen learning with the DrawnOut! Mental Health sketching activity. We will visit examples of TiD in practice, such as Flora Health (Holst Architecture) and PHFS Family Village (Carlton Hart Architects), pending confirmation.
The day will emphasize creating environments that prioritize safety, empowerment, and comfort, including:

- *Trauma and societal impacts, establishing a foundational understanding of the multifaceted ways trauma manifests, physiologically, psychologically, and sociologically. We will introduce key frameworks from SAMHSA1 and the landmark ACEs study2. Understanding the nuanced landscape of TiD, we aim to underscore its role in fostering environments that transcend trauma. This session aligns TiD with principles of justice and inclusivity, focusing on how it can contribute to positive societal change. Addressing local challenges such as homelessness and strained police relations, TiD becomes a catalyst for constructive transformations.*

- *Conducting Trauma-Informed Research. Delving into trauma-informed research intricacies, we'll emphasize sensitivity at every design stage and the significance of researching specific populations. Our guidance covers planning inclusive logistics, accommodating diverse perspectives in interviews, focus groups, and workshops. We'll explore strategies framing findings with consideration for the emotional toll on both participants and researchers, ensuring a respectful research process.*

- *Panel Discussion Exploring Design Process and User Needs: Innovative design strategies are only effective if they align with the organization's operation and models of care. We will explore TiD implementation in project contexts. From policy considerations to clinical insights, we'll delve into the holistic inputs needed for TiD, addressing needs-driven design, care models, and systemic issues. Panelists will share insights on designing trauma-informed applications, emphasizing the crucial step of validating designs before implementation for seamless alignment.*

- *DrawnOut! Mental Health Sketching: To help process and apply the morning's information we'll participate in a guided collaborative sketching activity. This interactive experience aims to demonstrate how creative approaches can contribute to TiD, fostering a deeper connection between design and mental well-being.*

- *Lunch and Tours to Portland locations (TBC) (industry sponsorship in process): The final segment will comprise immersive tours, offering encounters with Trauma-informed Designed spaces.*

We aim to demystify Trauma-informed Design, positioning it as a potent force for creating inclusive and supportive environments. Together, let's navigate this transformative journey toward holistic well-being.

References

1 Center for Substance Abuse Treatment (US). Trauma-Informed Care in Behavioral Health Services. Rockville (MD): Substance Abuse and Mental Health Services Administration (US); 2014. (Treatment Improvement Protocol (TIP) Series, No. 57.). Available from: https://www.ncbi.nlm.nih.gov/books/NBK207201/

2 Felitti, Vincent J., Robert F. Anda, Dale Nordenberg, David F. Williamson, Alison M. Spitz, Valerie Edwards, and James S. Marks. "Relationship of childhood abuse and household dysfunction to many of the leading causes of death in adults: The Adverse Childhood Experiences (ACE) Study." American Journal of Preventive Medicine 14, no. 4 (1998): 245-258.

Trauma-Informed Design Approach: Navigating Design, Policy, Operations, and Research with Sensitivity

Christine Cowart
Cowart Trauma Informed Partnership

Embarking on an evidence-based, Trauma-informed Design journey first requires a foundational understanding of trauma and stress responses, in alignment with organizational operations, care models, and research practices1. We will host two presentations, a panel discussion with local designers, community engagement, and clinicians, and a sketching activity informed by Trauma-informed Design principles. Breaks include an opportunity for guided mindfulness sessions.

By integrating trauma understanding, SAMHSA principles, adverse childhood experiences (ACEs), and stress regulation with design principles and research practices, the morning session aims to empower participants with a detailed yet holistic approach, inspiring empathy in design and developing foundational understanding of trauma and its varied implications.

The second presentation will spotlight trauma-informed research practices, to accommodate emotional impacts on participants and researchers in sensitive topics2. The panel discussion that follows will ask Portland-based experts questions such as "Is it appropriate for designers to suggest a different model of care?" Panelists will then be available to accompany us on the afternoon tours (confirmation pending).

We offer a roadmap for Trauma-informed Design practices, including: 1) cultural sensitivity: understanding local culture, practices, and guidelines to tailor assessment tools to the unique context of the target setting; 2) Interdisciplinary collaboration: engage diverse professionals in relevant fields for a perspective enriched with multifaceted insights; 3) visual communication, such as journey mapping, to identify key user experience moments to ensure compassionate comprehensive over time; 4) comparative analysis: identifying differences between the new and initial contexts to perceive transferable content.

References

1 Owen, Ceridwen, and James Crane. "Trauma-informed design of supported housing: A scoping review through the lens of neuroscience." International Journal of Environmental Research and Public Health 19, no. 21 (2022): 14279.

2 Dietkus, Rachael. "The Call for TraumaInformed Design Research and Practice." Design Management Review 33, no. 2 (2022): 26-31.

Trauma-Informed Design Approach: Navigating Design, Policy, Operations, and Research with Sensitivity 2

Molly Pierce
Molly Pierce OT Consulting

Clinical Occupational Therapy contribution to the morning sessions.

Trauma-Informed Design Approach: Navigating Design, Policy, Operations, and Research with Sensitivity 3

Adrienne Erdman
EwingCole

Focus on the trauma-informed research process within the morning sessions

Drawnout! Mental Health Sketch Workshop

Stephen Parker
Behavioral + Mental Health Planner at Stantec

Mental wellbeing and health has been rising in public awareness, catalyzed by the pandemic. When socially isolated, our built environment can have an even more profound impact upon our mental health. This has led to a surge in demand for mental health services, with the limited number of psychiatric inpatient facilities, outpatient clinics and crisis stabilization centers often overwhelmed. This is also an opportunity to learn more deeply about this growing field of practice and craft an inspired response, collectively.
With this interactive workshop-within-the-workshop,, participants will learn from a nationally-recognized mental health planner regarding sensory-sensitive approaches to mental health design.

A design primer will set the stage for the audience to roll up their sleeves, team up and sketch up ideas for space in a mental health facility - the visitation suite. This 75-minute workshop will have participants gather in 2 person teams to rapidly explore the persona of visiting family member or friend to a mental health facility, ideate new insights, inspire us through empathy and design with dignity.

The sketches will be gathered and on display as a living mural and will be added to as an evolving exhibit, with contributions throughout the conference and beyond. The hashtag #MentalHealthSketch will help engage a greater audience.

Trauma-Informed Design Showcase: A Journey through Award-Winning Projects Transforming Lives

Janet Roche
The Boston Architectural College

Spotlight on the big picture of Trauma-informed Design in the context of the sites visited in the afternoon.

Trauma-Informed Design Showcase: A Journey through Award-Winning Projects Transforming Lives 1

Evon Rose Calabrese
Clark / Kjos Architects

The afternoon session will give attendees the opportunity to tour three local award-winning projects that employ the Trauma-informed Design process: a day shelter; an overnight family shelter; and a substance abuse and treatment headquarters. All three projects underwent extensive community engagement to develop innovative design strategies specific to users' lived experiences. While touring the buildings, attendees will hear from the facility operators to learn about their experience as the client/user group and any adjustments they've made to their space since opening. Confirmation is pending.

Rose Haven is the only day shelter and community center in Portland dedicated to serving women, children, and gender non-conforming folks. In 2022 they moved from their original basement space into a new 10,500 SF facility that was designed by Gensler. The design team transformed a historic warehouse into a state-of-the-art facility while maintaining its industrial and historic charm. The community engagement process led to an innovative approach to their clinical space that has been highly successful.

The Path Home - Family Village is the first shelter of its kind in the state of Oregon, serving families experiencing housing insecurity. The design team, Carlton Hart Architects and Jessica Helgerson Interior Design converted an old church into a transitional residence featuring shared and private spaces for families. The program includes a children's library, play space, classrooms for life-skills, as well as apartments with private bedrooms for families.

Fora Health's new headquarters unites its adult substance use treatment programs with administrative offices. Designed by Holst, the treatment facility features stabilization, outpatient, and residential programs. The design team took a trauma-informed approach when determining the configuration of the lobby, wayfinding throughout the building, and material palette. Holst's early work consisted of exceptionally designed market-rate housing, and they've maintained that mindset, as is evident in the spa-like interior.

ABSTRACTS CONTINUED

Group Presentations: Symposia

SURVIVAL THROUGH DESIGN IN 2024: NEW RESEARCH DIRECTIONS

Track 4: Design for Health and Wellbeing
Knowledge Network: None/Not Tied to Any Network
Methods: Experimental/Quasi-Experimental Research, Statistical/Modeling Analysis

Session Abstract:
The great Modernist architect Richard Neutra was mostly known for the style of his buildings. But behind all of them was a deep philosophical and psychological rationale that he most famously expressed in his 1954 book, Survival Through Design. This year, that book is being republished with extensive annotations and illustrations, and a new foreword by Dr. Raymond Neutra, the last surviving son. In recognition of this event, this symposium session explores some of the themes present in that book and shows how Richard Neutra's ideas and designs are still relevant today, 70 years later.

The session begins with a POE of Neutra's and Alexander's famous UCLA Lab School, recently performed by Perkins-Eastman. The questions there were how well its design has held up to the introduction of new teaching methods, new computer technologies, and the increased temperatures and air quality changes that accompany climate change impacts in the LA area.

This is followed by a study of one of Neutra's most famous innovations, the "Spider-Leg," which became a hallmark feature of many of his homes. It was meant to increase the sense of interior spaciousness and connection to nature for viewers from indoors. But does it? And if so, how does that work, neurally and perceptually? This study attempts to find some answers by utilizing EEG and eye-tracking methodologies. The results also suggest some interesting new uses for "spider-leg"-like design enhancements for isolated and confined environments, like lunar and Mars bases. Our third presentation describes a study performed in a closed psychiatric ward for schizophrenic patients in Melbourne, Australia. This utilized specially designed nature tree form murals, using a fractal strategy. Most fractal research focuses on the fractal dimension. Here, the researcher has made all murals adhere to the same most preferred fractal dimensionality and investigated other scalar properties of the images that produce very different viewing experiences. Neutra observed that "No one laments the monotony of a tree." This study shows why. It also shows that Neutra's thoughts can still inspire cutting-edge design research.

The session concludes with attendees' discussion and a closing statement by Dr. Raymond Neutra.

A POE of Richard Neutra's UCLA Lab School: How Does It Meet Its Original Intent?

Authors: Heather Jauregui, Emily Chmielewski
Affiliation: Perkins Eastman

Abstract:
The UCLA Lab School was designed and built by architects Robert Alexander and Richard Neutra, beginning in 1950. It was designed as an open-air school with easy access to an open, treed central space. The school's pedagogy was built into the design. But much has changed since then. In collaboration with the Neutra Institute, Perkins Eastman undertook a POE of the Lab School to ascertain how its design features and rationale have held up over the last 70 years. This presentation by Dr. Chmielewski will highlight the findings and suggest how educational facilities designed for one climate and pedagogical intent can meet the challenges created by the advent of new technology and the impacts of climate change.

HUMAN-CENTRIC LIGHTING: RECENT RESEARCH AND DESIGN APPLICATIONS

Track: Design for Health and Wellbeing
Knowledge Network: Health, Researchers in Professional Practice
Methods: Experimental/Quasi-Experimental Research

Session Abstract:
This symposium discusses the human-centric lighting approach and highlights new aspects to be considered in architectural and lighting design guidelines and practices.

The first presentation by Dr. Elzeyadi will propose an evidence-based framework of human-centric lighting. This framework addresses aspects not often discussed in green building certification systems, such as LEED and WELL, including occupants' ergonomics, mood, status, empathy, spatial equity, and aesthetics. With the proposed framework in mind, the second and third presentations provide examples of some of the aspects to be considered in a human-centric lighting design framework.

In the second presentation, Dr. Golshany shares findings from recent studies that she led, where she investigated the impact of spatial and spectral light patterns on the impression, visual preference, and cognitive performance of elderly participants. The findings support the inclusion of these light attributes, as well as outcomes, in human-centric lighting frameworks.

The third presentation by Dr. Abboushi focuses on the facial geometry of occupants and investigates whether facial geometry plays a role in informing the design of overhead light sources (e.g., common ceiling-mounted luminaires). Findings suggest that facial geometry—specifically whether the pupil is shaded or unshaded from an overhead light source—explains variations in glare ratings among individuals. A human-centric framework could consider occupants who tend to be more sensitive to glare from overhead sources.

Architectural Lighting for Occupants' Wellbeing: A Human-Centric Design Framework

Author: Ihab Elzeyadi
Affiliation: University of Oregon

Abstract:
How can we design spaces with better lighting quality (LQ) from the occupant's perspective? The current approach for providing lighting design for different space types relies on an engineering perspective that quantifies occupant comfort and requirements based on a segregated set of metrics designed to satisfy building codes, standards, and LEED and WELL certifications. This prescriptive method predicts occupants' comfort based on lighting's physical attributes—illumination levels, glare indices, and light sources' temporal and spectral distribution.

In real settings, especially those designed with an engaged human-centric objective, the occupant interacts with the overall ambiance of the environment through both visual and non-visual receptors, impacting their physical, physiological, psychological, and social environments. These receptors are further influenced by both objective and subjective environmental factors affecting well-being.

Despite the rise of the term "human-centric lighting" in both design and product marketing media, its impact on human outcomes has been debatable, and its predictive performance marred by errors in conceptualization. This presentation critiques the current framework's lack of inclusion of parameters related to occupants' ergonomics, mood, status, empathy, spatial equity, and aesthetics. It proposes a more inclusive framework that acknowledges both the systems and outcomes of the lighting environment from the occupant's perspective. The presentation will summarize evidence from the literature supporting this conceptual framework and its application in research and practice for more informed lighting design that positively impacts occupant well-being.

Enlightening Well-Being: Unraveling the Influence of Human-Centric Lighting on Older Adults' Perception and Cognitive Brilliance

Author: Nasrin Golshany
Affiliation: California State University

Abstract:
The aging population presents a demographic challenge, increasing the risk of sleep issues, depression, and cognitive decline among older adults. Safety concerns at home heighten, prompting a shift to assisted living facilities. Research emphasizes the impact of healthcare environmental design, particularly architectural lighting, on residents' well-being. To improve the health of older adults, it's crucial to consider both visual (task performance, visual acuity) and non-visual effects (circadian rhythm, sleep quality, mood, cognitive performance) of architectural lighting. This underscores the importance of a holistic approach to human-centric lighting in indoor spaces. Current gerontology studies have primarily emphasized light-level attributes, neglecting exploration into spectral and spatial pattern parameters in indoor lighting. This research addresses this gap by delving into both quantitative and qualitative aspects of lighting design, exploring spatial layout characteristics (uniformity, direction, centrality), and spectral attributes such as correlated color temperature (CCT). The study specifically investigates the impact of these factors on the visual perception, preference, mood, cognitive performance, and overall well-being of older adults in assisted living facilities.

Employing a multi-method approach across two research phases, Phase I utilizes a Q-sort survey with 60 participants to assess the impact of various spatial light patterns on visual perception and preference. Phase II delves into the relationship between spatial and spectral light patterns and cognitive performance through virtual reality testing with 32 participants.

The study's results show that different spatial light patterns significantly influenced older adults' impressions, affecting visual preference, stress levels, and cognitive performance. Preferences leaned toward uniform and indirect lighting, with no notable differences between peripheral and central light arrangements. Non-uniform lighting induced a relaxed impression, while uniform lighting increased perceived stress. These findings highlight the substantial impact of spatial and spectral light patterns on older adults' perception and cognitive performance in assisted living facilities.

Facial Geometry Mediates the Impact of an Overhead Glare Source

Author: Belal Abboushi
Affiliation: PNNL (Pacific Northwest National Laboratory)

Abstract:
Discomfort from glare is the sensation of discomfort or annoyance due to unsuitable luminance distribution or range, or extreme luminance contrasts. Many environmental and observer-related factors can affect glare perception. One observer-related factor that contributes to variation in perceived glare across individuals is the portion of the face exposed to direct illumination. The larger the exposed area, the higher the perceived glare. This suggests that facial geometry may affect perceived glare; we hypothesized that individuals with deeply recessed pupils (within the eye socket) may tolerate higher luminance for an overhead glare source because their pupils could be shaded.

To test this hypothesis, an experiment was conducted where the facial geometry of 29 participants was collected using a 3D scanner. Participants were then seated at a spherical apparatus and asked to set the luminance of a light source positioned 60° above the line of sight to the border between comfort and discomfort (BCD). The facial models were used to measure the pupil obscuration angle—the angle between a line normal to the pupil and another line restricted by eyebrow ridges. This quantifies how recessed the pupil is from the facial plane.
Findings showed that the pupil obscuration angle ranged between 23° and 76° (mean = 57°). Participants with a shaded pupil set the BCD luminance to a significantly higher value compared to those with an unshaded pupil (pupil obscuration angle > 60°). This further supports findings from an earlier study and suggests that individuals with a larger pupil obscuration angle (less shading on the pupil) are likely more sensitive to overhead sources. This can inform the design and application of overhead sources, considering individuals who are likely more sensitive to overhead glare.

Examining the Neutra Spider-Leg Design with Eye-Tracking Technology

Author: Kris Mun
Affiliation: Sensory Architectures

Abstract:
The prominent Austrian-American architect Richard Neutra played a pivotal role in modern architecture. A notable characteristic of many of his houses was the "Spider-Leg" system. This design feature, marked by slender beams and columns extending beyond the building envelope, appears to extend both interior visual space and a sense of connection to the landscape outside.

How does this work? In this study, we investigate the neural and perceptual engagement with viewing the spider leg from indoors and compare this with altered versions of it. This was done using eye-tracking and EEG technologies. We were interested in gaze attention and visual exploration of this kind of distinctive design element.

Results will be presented that describe shifts in spatial perception and directed visual attention. These suggest that spider-leg-like strategies may be used to enhance Isolated and Confined Environment (ICE) designs, such as lunar and Mars bases.

Testing "the Monotony of a Tree" in a Closed Residential Psychiatric Ward

Author: Marc Jurblum
Affiliation: University of Melbourne

Abstract:
Closed residential wards for schizophrenic patients are deliberately designed to be very plain and not arousing. This study tested the placement of specially designed tree-form murals placed in a dayroom for short periods of time in two otherwise identical wards. The murals were designed using specific fractal strategies that kept all four different murals in the most preferred "fractal water hole" range, well documented in previous studies. But they differed in manipulations of scale properties that created very different viewing experiences.

This presentation describes the most significant results and suggests that, though "No one laments the monotony of a tree," as Richard Neutra observed, there may be ways of improving on it.

POTENTIAL CHANNELS FOR EMERGING PROFESSIONALS INTERESTED IN RESEARCH-INFORMED DESIGN OF INPATIENT HEALTHCARE FACILITIES

Track 4: Design for Health and Wellbeing
Knowledge Network: Environmental Design Research Education, Researchers in Professional Practice
Methods: Mixed Methods

Unpacking the Milieu: A Comparative Analysis of Environmental Affordances on Spatial Autonomy in Behavioral Health Facilities

Author: Connie Lin
Affiliation: Corgan

Abstract:
According to a report by the World Health Organization, patients who feel empowered are more likely to participate in their care, which can, in turn, improve the quality of care. However, the lack of patient empowerment can have negative consequences. Physical violence and aggression in behavioral health facilities have become increasingly intensified. In addition, loneliness felt by a mental health inpatient has been shown to exacerbate their mental health condition. Seventy-one percent of patients receiving psychiatric rehabilitation have reported feelings of loneliness and isolation.
In light of these cumulative issues, previous studies have closely examined design interventions to stimulate autonomy and well-being in "defensible spaces" or the private patient room. However, there is little insight as to how milieu spaces—or social spaces—can foster patient autonomy, control, ownership, and recovery. Therefore, the purpose of this study is to investigate the effects of design interventions in milieu spaces associated with instigating patient autonomy on the patient's recovery outcomes in behavioral health inpatients.

It is hypothesized that environmental affordances that provide greater stimulation of patient autonomy would also be associated with faster recovery outcomes. This study utilizes a triangulated methodology through a pilot site assessment tool, behavioral observations within milieu spaces, and an analysis of published medical data. Outcomes from this study will include an inventory of specific design features that are acuity-adaptable for implementation. Best practice design recommendations to stimulate patient autonomy in milieu spaces will also be provided.

Reimagining Care Environments: Architectural Strategies to Alleviate Depression in Long-Term Nursing Patients

Author: Shailagh McCormack
Affiliation: Perkins&Will

Abstract:
In the United States, a staggering 48.8% of patients in nursing facilities were diagnosed with depression, accounting for nearly 645,000 individuals in 2018. Nursing facilities present a unique challenge when it comes to the emotional well-being of their residents. While these spaces are primarily designed to provide medical care, they often lack the empathetic design qualities necessary to address the emotional needs of patients. To tackle the psychological distress that plagues these environments, architectural interventions are explored as a solution.

This research is dedicated to the exploration of architectural interventions to enhance the emotional well-being of long-term patients in skilled nursing facilities. It begins with a narrative literature review focusing on clinical patient needs and care delivery methods, aiming to shed light on the crucial role of spatial conditions and existing programmatic elements in these settings. Additionally, the study conducts a descriptive evaluation of current facility standards and workflow through comparative case studies of skilled nursing facilities.The culmination of this research creates a comprehensive design framework that aligns with the established design goals. This framework not only represents a significant step toward a more holistic approach to design within skilled nursing facilities but also holds the potential to benefit a broader spectrum of healthcare settings.

This research endeavor serves as a vital resource for skilled nursing facilities and extends its utility to various healthcare contexts. By prioritizing the emotional well-being of patients through thoughtful architectural interventions, it strives to transform these spaces into environments that foster healing, comfort, and an improved overall quality of life.

Staff Experience and Well-Being in Behavioral Health Settings: A Measure to Assess the Qualities and Impacts of the Built Environment for Caregivers

Author: Saran Megha Parimi
Affiliation: Perkins&Will

Abstract:
According to recent statistics from the Journal of Internal Medicine, the percentage of burnout among nurses is 56%, clinical staff is 54%, and non-clinical staff is 46%. These numbers have increased after the COVID-19 pandemic, significantly impacting the physical and psychological well-being of caregivers. A striking 78% of psychiatrists self-reported burnout, with 16% of them showing signs of major depression through screening. This has resulted in decreased staff retention and an increased demand for mental health workers, projected to reach approximately 250,000 by 2025, as predicted by the U.S. Department of Health and Human Services.

The proposed study aims to address the research gap by focusing on stress reduction and the enhancement of the psychological and physiological well-being of staff in behavioral health settings. This research utilizes a mixed-methods approach, initially concentrating on spatial conditions, workflow, and organizational strategies in the work environment. Additionally, various tools such as a stress-audit tool, staff experience survey, and staff interviews are employed for data collection and analysis of case studies. These approaches are intended to understand both quantitative and qualitative aspects of design strategies within the built environment and their impacts on caregivers.

The primary objective of this study is to develop a measurement tool for evaluating and analyzing staff experiences in behavioral health facilities. The insights gained from this research will inform future practices and help improve facility design for better staff experiences and outcomes.

URBAN MORPHOLOGY AND HEAT STRESS: STRATEGIES FOR IDENTIFYING AND MITIGATING MORBIDITY IN VULNERABLE GROUPS

Track 4: Design for Health and Wellbeing
Knowledge Network: Residential Environments, Work Environments
Methods: Statistical/Modeling Analysis

Session Abstract:
Scope of Symposium
Climate change is expected to cause an increase in extreme environmental conditions like heat waves, storms, and floods. However, it is becoming evident that specific populations are more vulnerable to the effects of climate change. These vulnerable populations include the elderly, children, athletes, people with disabilities and chronic health conditions, as well as outdoor and indoor workers. This symposium will focus on addressing the challenge of heat stress in urban settings, specifically for two vulnerable populations — the elderly and outdoor workers. The aim is to present recent research work on this critical issue.

Goals and Objectives:
The symposium focuses on urban environments and human resilience, with an emphasis on environmental design. Its goal is to promote collaborations, advance interdisciplinary knowledge, and foster innovative solutions for mitigating heat-related risks in urban settings. Specifically, it aims to:

1. *Examine the impact of urban morphology on external heat stress, particularly under prolonged exposure to extreme heat conditions.*
2. *Investigate the effects of heat stress on vulnerable populations, such as the elderly and construction workers, in various urban contexts.*
3. *Develop a comprehensive understanding of human tolerance to heat stress through a cumulative heat stress index that applies to diverse environmental conditions.*

Symposium Agenda:
Introduction:
- *Symposium Organizer: Dr. Adil Sharag-Eldin*

Session 1:
- *Investigating the Effect of Urban Morphological Features on External Heat Stress During Heat Waves*
- *Speaker: Ms. Fahimeh Gholami*

Session 2:
- *The Effect of Exposure of Construction Workers to Heat Stress in Different Urban Settings Under Different Work Intensities and Times of Exposure*
- *Speaker: Ms. Mehraneh Aladini*

Session 3:
- *Development of a Cumulative Heat Stress Index to Estimate Tolerances to Exposure to Heat Stress Under Various Environmental Conditions and Settings*
- *Speaker: Dr. Adil Sharag-Eldin*

Conclusion:
The symposium will conclude with a panel discussion, integrating insights from all sessions to formulate comprehensive strategies for mitigating heat stress in urban environments. This will include policy recommendations, design guidelines, and research agendas for future work in this crucial area.

Expected Outcomes:
- *Enhanced understanding of the interactions between urban design and heat stress*
- *Practical guidelines and strategies for reducing occupational heat stress risks*
- *A validated heat stress index applicable across various environmental conditions*
- *Strengthened interdisciplinary collaboration and knowledge exchange*

Target Audience:
The symposium targets researchers, practitioners, policymakers, and students in urban planning, architecture, environmental science, public health, and related disciplines.

Sharing research outcomes on Environmental Design and Human Resilience: Addressing Heat Stress in Urban Settings may foster a holistic understanding of urban heat stress and inspire innovative solutions for a more heat-resilient future.

Investigating the Effect of Urban Morphological Features on External Heat Stress During Heat Waves

Author: Fahimeh Gholami
Affiliation: Kent State University

Abstract:
Background: *This study critically examines the 2021 Northwest heat wave in Portland, addressing the intricate ways in which urban design contributes to microclimatic variations. It emphasizes how these design choices can intensify the impacts of heat waves, focusing on the urban heat island effect and its implications for city dwellers.*

Justification: *Understanding the relationship between urban morphology and localized climate responses is crucial in mitigating heat stress conditions exacerbated by climate change. This is especially important in extreme weather events like heat waves. The research aims to bridge the gap in knowledge regarding how urban designs can impact these conditions.*

Objectives: *The research analyzes heat stress in Portland's urban areas using UTCI (Universal Thermal Climate Index) to evaluate the impact of extreme heat events on microclimatic conditions. It studies how distinct urban morphologies influence these conditions.*

Research Methods: *The methodological approach includes selecting five distinct sites in Portland, each representing different urban morphologies. These sites are chosen based on elements believed to mitigate the Urban Heat Island (UHI) effect, as suggested in existing literature. Using the ENVI-met simulation program, the study estimates environmental conditions across these urban structures and environments, specifically during the ten days of the heat wave.*

Findings: *The study uncovers significant variations in heat stress levels across the selected urban areas. These variations are closely linked to specific urban design features, such as the presence of vegetation and the density of buildings. This direct correlation underscores the impact of urban planning on microclimate regulation.*

Significance: *Urban planning and design play a pivotal role in enhancing public health and safety. Insights from this study on how urban designs influence heat stress during extreme weather events can guide policymakers and urban planners to incorporate heat-mitigating features in urban design, better preparing us for increasing challenges from heat waves and climate change.*

The Effect of Exposure of Construction Workers to Heat Stress in Different Urban Settings Under Different Work Intensities and Times of Exposure

Author: Mehraneh Aladini
Affiliation: Kent State University

Abstract:
Background: *This presentation investigates the Urban Heat Island (UHI) phenomenon, explaining how urban environments—characterized by dense infrastructure and reduced greenery—lead to higher temperatures than surrounding areas. It emphasizes the elevated risk this poses to construction workers exposed to these conditions, exacerbated by ongoing climate change trends.*

Justification: *The research expands on the urgent need for effective strategies to protect construction workers from increased health risks associated with UHI. It argues for the development of targeted prevention strategies and specialized guidelines, considering the unique challenges of heat stress in urban settings.*

Objectives: *This investigation aims to comprehensively understand how urban design elements—(1) building density, (2) pavement materials, (3) lack of vegetation—contribute to workers' heat stress, and (4) how the duration of exposure under extreme heat conditions affects health. The research also seeks to evaluate and propose effective strategies to mitigate these impacts.*

Research Method: *The methodological approach involves using ENVI-met, an environmental simulation tool, to model the urban microclimate of various sites in New York City during work hours in one of the hottest Julys in 2011. The Universal Thermal Climate Index (UTCI) is applied to assess heat stress in different urban scenarios and under varying metabolic rates.*

Findings: *The research highlights significant variations in heat stress levels across different urban morphologies. It underscores how specific design features can exacerbate or alleviate thermal stress in construction workers.*

Significance and Conclusions: *The presentation concludes with a discussion of the implications of the research. It emphasizes how the findings inform urban planning and design strategies aimed at protecting the health and safety of construction workers and enhancing overall livability in urban environments in the face of climate change.*

POTENTIAL CHANNELS FOR EMERGING PROFESSIONALS INTERESTED IN RESEARCH-INFORMED DESIGN OF INPATIENT HEALTHCARE FACILITIES

Track 4: Design for Health and Wellbeing
Knowledge Network: Environmental Design Research Education, Researchers in Professional Practice
Methods: Mixed Methods

Session Abstract:
Mental health in healthcare facilities today presents a complex landscape characterized by stigma, workplace stress, resource constraints, and a critical need for systemic change. Patients often hesitate to seek help due to the enduring stigma surrounding mental illnesses, and healthcare workers are susceptible to burnout. Insufficient resources in facilities result in prolonged wait times and compromises in the quality of care delivered. The intersection of mental health, patient safety, and individuality in healthcare settings requires careful consideration. Addressing these various issues demands a comprehensive and holistic approach that places empathy and support at its core.

With a growing emphasis on health and wellness in the built environment in both institutional research and professional practice, the field of evidence-based healthcare design has attracted the interest of many emerging professionals. The integration of research into the design process has enhanced the patient/provider experience, engaged stakeholders, and improved user health. As the field continues to expand, architects, designers, landscape architects, and urban planners have many career avenues through which to pursue healthcare design from a research-integrated perspective. The role of research in design is crucial to a holistic design approach.

One outlet for research-integrated design is the E. Todd Wheeler Health Fellowship, which is dedicated to fostering emerging healthcare design and planning leaders. Named after the late visionary architect E. Todd Wheeler, this fellowship is awarded to recent graduates who share a passion for innovating healthcare design. Wheeler's commitment to advancing healthcare design inspires critical questions in practice today, including how the built environment affects well-being, the humanization of healthcare settings, and the potential for these settings to exhibit resilience.

The E. Todd Wheeler Class of 2024 Fellows — Connie Lin, Shailagh McCormack, and Saran Megha Parimi — are based in Los Angeles, Boston, and Houston, respectively. Despite their geographic distance, their areas of inquiry show overlap, as each focuses on mental health in inpatient facilities. During this presentation, the fellows will discuss their experiences and research topics — depression, burnout, and patient autonomy within healthcare settings — and how working together on tangential mental health topics has benefited the research process.

The goal of this session is to increase awareness about potential outlets for integrated design-research opportunities in the architectural and design industries centered around the mental health of inpatient care facility users. It aims to demonstrate and discuss pathways for emerging researchers and designers to engage in translational design and contribute to a greater understanding of mental health in design.

Development of a Cumulative Heat Stress Index to Estimate Tolerances to Exposure to Heat Stress Under Various Environmental Conditions and Settings

Author: Adil Sharag-Eldin
Affiliation: Kent State University

Abstract:

Background: *Heat waves pose a significant risk to urban populations, with vulnerable groups like the elderly facing severe health challenges. Traditional indices for measuring heat stress may not fully capture the cumulative effects of prolonged exposure to high temperatures..*

Justification: *There is a need for a more comprehensive tool to assess the impact of heat stress, especially in urban environments where heat islands exacerbate conditions. This research seeks to address this gap by developing a new index.*

Thesis: *This study posits that a Cumulative Heat Stress Index (CHSI), incorporating Universal Thermal Climate Index (UTCI) data, can more accurately estimate human tolerance to heat stress under various environmental conditions. The index is expected to reflect the compounded effects of continuous heat exposure on human health.*

Research Method: *The research employs a simulation-based approach using historical data from the 1995 Chicago, IL, and 2021 Portland, OR heat waves. The ENVI-met simulation program calculates the UTCI, focusing on temperatures above 28°C. Data are then cumulatively assessed to gauge the heat stress on building occupants in selected residential areas. These areas correspond to actual sites with reported fatalities during the heat waves..*

Findings: *The study demonstrates that the cumulative index can predict stress levels beyond which the elderly population studied could no longer tolerate heat stress. This finding is crucial in identifying high-risk periods and areas during heat waves.*

Significance and Conclusions: *This research addresses the urgent need for better assessment tools to predict heat stress risks in urban environments. The development of the CHSI offers a novel approach to understanding and mitigating the impact of heat waves, particularly on the most vulnerable urban populations. The index's ability to incorporate cumulative exposure effects makes it a valuable tool for urban planners, health officials, and environmental researchers in developing strategies to protect at-risk communities.*

INDIGENOUS KNOWLEDGE AND PLACE MAKING

Track 1: Nature and Urban Ecology
Knowledge Network: Active Living by Design
Methods: Action/Participatory Research

Session Abstract:

This symposium focuses on interdisciplinary efforts to rethink approaches to research, design, student engagement, and placemaking through a unique collaboration between Portland State University's Indigenous Nations Studies Department and the School of Architecture on a landscape and building project. Since 2011, Judy BlueHorse Skelton has worked on the project site known as the PSU Oak Savanna with Indigenous Studies students in efforts to learn about and restore native habitat to the PSU campus. The site is dedicated to the relationship between the First People and the native vegetation of oak savannas. A vision for the site developed to create a minimally managed outdoor classroom and community gathering building that shares the natural history of the Willamette Valley, offers space for reflection and healing, creates wildlife habitat, and demonstrates Indigenous land practices for food, medicine, art, and ceremony.

In 2019, Professor Sergio Palleroni from the PSU School of Architecture partnered with Judy BlueHorse Skelton on a disaster resilience project that considered the Oak Savanna as a hub for support in the event of a natural disaster. This initiated a partnership which later manifested as designs by architecture students for an ITECK Center on the Oak Savanna. ITECK stands for Indigenous Traditional Ecological and Cultural Knowledge, and is the framework within which Judy BlueHorse Skelton, Emma Johnson, and their colleagues support the ITECK Certificate program and major/minor students.

In 2021, Professor Palleroni invited architecture faculty and students to engage with Indigenous Studies faculty and students to develop ideas for an ITECK Center on the Oak Savanna through design studios, seminars, and fieldwork opportunities. One project led by Professor Travis Bell in 2023 resulted in the design of a stage for the Pickathon music festival that was meant to be deconstructed after the festival and rearranged as a pavilion on the Oak Savanna. This pavilion marks the first step toward activation of the site through a collaboration with the School of Architecture and Center for Public Interest Design. Collectively, these design projects helped to support a dialogue about project aspirations and imagery for a new space that has resulted in the allocation of $2,000,000 toward the construction of a new ITECK center.

Building upon the collaborations to date, the PSU School of Architecture and Center for Public Interest Design are using this opportunity to examine the role of the architect in centering Indigenous knowledge toward more equitable and resilient communities and ecologies, and to reimagine an architecture pedagogy informed and transformed by Indigenous Traditional Ecological and Cultural Knowledge. As part of these efforts, this project is providing the groundwork for international dialogue around learning from Indigenous knowledge within the architecture community through the Design for the Common Good Network.

This symposium will explore lessons learned to date through this effort, concepts driving this work—such as un-building and culturally specific design—and challenges in overcoming the dominant narratives of contemporary education and professional practice.

The Oak Savanna: Design Build as Cultural Imperative

Author: Travis Bell
Affiliation: Portland State University (PSU)

Abstract:
Design-build projects provide an ideal context for students to understand the relationship between a site, community, culture, building materials, and ecosystem. Since 2012, the PSU School of Architecture has hosted the "diversion design-build" studio in which students design a temporary stage for the Pickathon music festival, with the materials later transformed and put to use rather than discarded.
The 2023 project resulted in composite timber arches to be repurposed into a new outdoor classroom structure at Portland State's Oak Savanna. This natural space on the PSU campus is a lightly forested grassland that was restored in 2012 by a group of students for use by the Indigenous Nations Studies program. The repurposed stage will become a permanent installation in the Oak Savanna, to be used as part of the new Indigenous Traditional Ecological and Cultural Knowledge (ITECK) initiative.

This presentation will cover the research and process of the diversion design-build project and discuss new layers of design considerations that emerged as a result of applying a new lens through collaboration with partners in Indigenous Nations Studies.

The Oak Savanna: Creating a Space for Gathering and Cultural Practices

Author: Judy Shelton BlueHorse
Affiliation: Portland State University (PSU)

Abstract:
Creating a space for gathering and cultural practices is crucial for building community and collective knowledge. The journey to reclaim an Oak Savanna in the Portland State University urban landscape—for food, medicine, and ceremony—began more than a decade ago with students seeking sanctuary on campus, a place to come together.
Guided by student voices and their desire to experience and understand the cultural significance and value of connection with the land to heal and build community health and resilience, the Oak Savanna site was incorporated into Indigenous Nations Studies courses centered in Indigenous Traditional Ecological and Cultural Knowledge (ITECK).

Over the years, the Oak Savanna site has been the focus of student research and projects, as well as seasonal gatherings and work groups. We recognized that reclaiming the urban forest for food, medicine, ceremony, and healthy lifeways had found a timely synergy with agencies and community partners working in culturally responsible restoration, education, reclamation, and protection practices.

Indigenous students and community members visited the site and imagined the possibilities of a building to host diverse gathering spaces for people, plants, and items of cultural significance for learning, understanding, and healing. Community and students envisioned indoor spaces: a kitchen, Indigenous library, a classroom, and seating areas transitioning seamlessly into the outdoors, featuring culturally significant native plants, a storytelling circle, and an outdoor kitchen for traditional salmon bakes and camas roasting.

This presentation will cover the site's evolution and current efforts to share Indigenous knowledge through the development of the site.

The Oak Savanna: Applied Research in Food Sovereignty

Author: Emma Johnson
Affiliation: Portland State University

Abstract:
Food sovereignty is the right of peoples to access culturally appropriate, nutritious food as defined through their own cultivation, tending, and harvesting practices. The Western worldview places humans separate from Mother Earth, and because of this, many people often have an immense disconnect from their surrounding environments—making it even more challenging for them to understand the importance of food sovereignty.

It is because of this that we constantly overlook some of our greatest teachers—our plant relatives—many of which have been here since time immemorial. Indigenous Traditional Ecological and Cultural Knowledge (ITECK), paired with place-based and community-based learning, offers people a chance to better connect with the landscape they call home and develop a sense of belonging.

This presentation provides an overview of projects, species, and places where a food sovereignty framework is weaving community together. Only by building a relationship with our surrounding environments do we begin to understand how important it is for our personal health to truly know how to care for the land—which, without question, cares for us day after day.

The Oak Savanna: Applied Research Toward Indigenous-Informed Placemaking

Author: Eric Giovannetti
Affiliation: Portland State University

Abstract:
Working on the Oak Savanna Project has generated new and exciting opportunities for us as students to embrace the positive synergies and negotiate the challenges of cross-cultural design. Our participation in these design studios has added new forms of learning beyond the typical design studio education. We have learned significantly from immersion in Participatory Action Research, incorporating ITECK values into architecture, and considering the diverse stakeholders at an urban university. As we enter into the professional field, we see the growing importance of this knowledge and process in environmentally responsible practice. We have both learned ways for architecture to begin embodying these Indigenous values and for communicating those values with different stakeholders. This presentation examines the student experience from two perspectives: Athena will present her experience as an Indigenous architecture student in this design-build context. Eric will discuss participating in an Oak Savanna design studio, and the learning outcomes of collaboration between ITECK and architecture students.

SOCIALLY ENGAGED PROCESSES TO [RE] CONTEXTUALIZE HUMAN-CENTERED DESIGN

Track 5: Spatial Justice, Equity, and Inclusion by Design
Knowledge Network: Environmental Design Research Education, Interior Design
Methods: Case Study, Design Methods

Session Abstract:
Human-centered design is grounded in the belief that seemingly insurmountable problems, such as poverty, systemic inequalities, and ecological crises, are solvable (Norman, 2021). This underscores the importance of equipping students with tools that empower them to become agents of change, addressing crucial issues like racial equity, culture, social justice, and belonging. However, design pedagogy has a longstanding history of privileging formal ideas over lived experience (Mangold, 2014). This has resulted in a built environment that is often alienating and inaccessible to many people. In this panel discussion, we explore a range of critical perspectives and models for interior design education that place people first. Drawing on principles from feminism and anti-racism, this discussion will help us define and contextualize human-centered design within our discipline. It proposes ways for interior designers to learn how to create spaces that are inclusive, accessible, and comfortable for all.

Traditional design pedagogy often concentrates on teaching students how to craft beautiful and functional spaces. However, under this approach, students frequently find themselves immersed in resolving functional design challenges while struggling to construct impactful spatial narratives that address users' emotional and psychological needs, especially in unfamiliar design scenarios. Confronted with accelerating global crises—ranging from climate calamities to unequal wealth distribution, political inefficiencies, and inadequate systems—there is a compelling need for designers to respond with innovation and empathy. This symposia aims to explore how interior design education can surpass traditional boundaries to effectively tackle emerging issues of spatial justice, equity, and inclusion. Through conversation, we will examine effective approaches, advocate for new perspectives, and explore avenues for empowering marginalized and underserved communities in interior environments. For instance, feminist design pedagogy challenges the traditional notion of the "universal user" and the idea that there is a "typical" person to design for (Ahmed, 2017). It seeks to create spaces that are inclusive of all people, regardless of their gender, race, ability, or other identity. Similarly, anti-racism pedagogies challenge traditional assumptions, specifically recognizing that racism is a systemic problem that creates barriers to access and participation for people of color (Harris, 2023). Anti-racism design pedagogy aims to create spaces that are equitable and inclusive for people of color. These critical and socially engaged perspectives offer new and empathetic ways of thinking about human-centered spaces. They focus on the lived experiences of people and how these experiences can be used to transform the built environment and create a more just and equitable society.

We will explore how we comprehend and deploy these critical, socially oriented pedagogies within our discipline. Can these tools enhance design innovation while responding to our increasingly complex social, cultural, and political realities? How do we evaluate these engagements in design education? These are the questions that shape and initiate this panel discussion. Panelists will contemplate the methods and pedagogies crafted and tested in studios, seminars, and courses to nurture human-centered design processes. The discussion will culminate by outlining strategies to elevate the interior design discourse, inspiring the next generation of empathetic design thinkers and practitioners.

Empathy, Space, and the Body

Author: Jennifer Meakins
Affiliation: University of Kentucky – Lexington, KY

Abstract:
Through the lens of feminist and anti-racist theory and pedagogy, this presentation will discuss the goals and outcomes of an undergraduate architecture design studio centered around empathy, space, and the body. Moving beyond empathy for others who experience the harmful effects of laws and regulations, the studio sought to "inspire modes of affective perspective-taking and affective practices that call subjects (e.g., teachers, students, parents) into account for their own complicity in perpetuating coloniality" and inequity through pedagogical methods highlighting the use of fiction and narrative, power structures, and the hidden curriculum, for example (Zemblas, 2014; Johnson, et al., 2013; Gore, 1998). This approach reflects the need to "know," to recognize the "pain" of others, while simultaneously acknowledging the role of one's privilege in the inability to truly understand the impact of anti-abortion and anti-transgender laws and regulations across the United States (Ahmed, 2014). Space, defined by physical elements, is ultimately produced and defined by the body or the gathering of bodies, and changes depending on the experience, social, and cultural perceptions of the user (and those around them) (Lefebvre, 1991). Through the three projects, focused on different scales—body, space, and architecture, the studio moved between research, analysis, diagramming, making, and design across three different sites. First, the body—reading and analyzing dystopian fiction dealing with bodily autonomy, in order to explore the link between dystopian fiction, reality, and the spatial and architectural implications. Second, space takes from dystopian fiction, realities, and personal experiences, to construct a personal wearable object to define, enforce, or challenge social space. Finally, pulling from the previous investigations, the students conducted an analysis and design of a site intervention at a place of power, the Ohio Statehouse. The presentation will conclude with reflection and critical analysis of the process and outcomes looking toward future courses.

People First: Critical Pedagogies from Interior Design

Author: William Mangold
Affiliation: Drexel University

Abstract:
Socially responsible design is characterized by attitudes that value justice, equity, participation, sharing, sustainability, and practices that intentionally engage social issues and recognize the consequences of decisions and actions. It is important to consider how these attitudes can be cultivated through the teaching and practice of interior design. In particular, recent years have seen a significant increase in awareness and efforts towards diversity, equity, and inclusion (DEI) that are informing pedagogy and design. This presentation will briefly review current literature and thinking about DEI in architecture and interior design, including approaches such as JEDI and decolonizing design (Tunstall, 2022), and discuss efforts to enhance DEI through teaching design. The presentation will look specifically at classroom practices intended to develop theoretical understanding, and opportunities for design students to engage DEI issues in real environments. Students are often familiar with identity issues, and can benefit from chances to study and discuss these ideas in ways that are both safe and rigorous. Centering discussion on issues of place identity and place attachment (Altman & Low, 2012) allows student-designers from a variety of backgrounds to engage with DEI questions on equal footing, enabling them to draw upon personal experience, while beginning to look at those experiences critically. Students unpack things they've encountered as they learn how place and experience are intersectional and layered with dynamic social and cultural interactions. Student designers are interested in engaging questions of belonging and inclusive community (Wise, 2022), as well as understanding how issues of power and subjectivity can create social and spatial inequalities. Designing for real world places is another way that student designers learn to actively engage DEI issues. This presentation will discuss how class projects have been developed, utilizing typical campus spaces, to provide students with opportunities to design for diversity, equity, and inclusion.

Mindfulness Techniques in Architectural Pedagogy

Author: Kristen Mimms Scavnicky
Affiliation: Kent State University

Abstract:
Design disciplines must recognize mental wellness for our present and future practitioners. Turning a blind eye to the ever-present human afflictions of stress and burnout can only continue to deteriorate the quality of our design eyes and honed skills as a collective. This pedagogical philosophy promotes students' mental well-being through studio policies, intertwining mental health awareness with the design curriculum. By advocating for an enhanced architectural pedagogy that integrates mindfulness practices, specifically meditation, meditative drawing, and focused breathing, instructors can equip students not only with design skills but also with strategies to navigate the demanding and often stressful nature of design practice.

Integration of mindfulness techniques in architectural pedagogy can be transformative. Norman (2005), a cognitive scientist, notes, "Design profoundly influences our well-being," emphasizing the impact of the built environment on mental states. Therefore, design curriculum, being the precursor to design practice, should actively engage in nurturing the well-being of its students. Meditation is an accessible tool of self-care that has proven benefits in stress reduction and enhanced focus. It offers students a means to cultivate mindfulness as an essential step through their design process and to navigate the creative process. Incorporating meditative drawing practices encourages a deep connection between creativity and introspection, fostering a more balanced and self-aware approach to design prompts and developing an empathetic design response to the users and the project contexts. Mindful practices, as emphasized by Davidson and Kabat-Zinn (2003), empower students to accomplish focused tasks while simultaneously heightening their empathy for the evolving complexities within our social, cultural, and political spheres. This, in turn, facilitates a more human-centric approach to addressing intricate studio projects.

Civic Engagement and Agency

Author: Kimberly Montes
Affiliation: Youth Design Philly

Abstract:
Interior design, often perceived as a luxury service for the affluent, has the potential to be inclusive and accessible to all. This presentation looks at ways of bridging the accessibility gap in interior design, emphasizing community engagement and creativity as catalysts. In particular, this work engages interior design students in teaching tools and methods to youth. The goal is to enhance their environments and expose them to creative opportunities, opening doors to potential careers in design and architecture.

The work is focused on residential spaces in Philadelphia—the homes and rooms inhabited by youth, mainly low-income, who participate in the learning process facilitated by Youth Design Philly. According to the Pew Charitable Trust (2019), the poverty rate in certain areas within Philadelphia surpasses 45%. City schools lack funding for creative initiatives, so this program fills the gap, nurturing creativity and introducing interior design. Teaching students how to repurpose and refurbish items to enhance their surroundings can instill a sense of sustainability—an essential perspective for the future. In the context of impoverished circumstances, where students may lack the resources to purchase or renovate their surroundings, creating awareness of available opportunities becomes pivotal. Simple acts like refurbishing or utilizing donated items can result in well-designed spaces, and students learn creativity and agency.

Involving college students in this process of sharing design knowledge is a form of civic engagement that further cultivates a sense of community. For interior design students it centers their learning on the whole-life experiences of the people they work with. It teaches them empathy, and ways to think and approach design differently, especially when working in situations with limited resources. This presentation advocates for making design accessible to youth in inner-city communities and explores what design students learn by working directly in these circumstances.

CREATING NEIGHBORHOODS TO SUPPORT OUTDOOR MOBILITY AND PARTICIPATION OF PEOPLE LIVING WITH DEMENTIA

Track 5: Spatial Justice, Equity, and Inclusion by Design
Knowledge Network: Active Living by Design, Environment-Gerontology
Methods: Focus Groups, Interviews

Session Abstract:
The neighborhood environment plays a key role in promoting and maintaining brain health in later life, both through physical and social affordances. For people aging with cognitive impairment in the community, the neighborhood environment offers meaning, value, and purpose, as they navigate through the growing demands of their condition. Thus, a supportive and inclusive neighborhood is instrumental to maintaining and enhancing mobility and social participation of people living with dementia in the community. Ensuring the supportiveness of neighborhoods for people living with dementia strongly aligns with the Dementia-Friendly and Inclusive Communities (DFC) movement. A key objective of DFCs is to create an accessible neighborhood built environment that allows people to easily and safely move around, take part in everyday activity, and access destinations in the community. To achieve this objective requires a clear understanding of how the condition of dementia shapes people's use of streets and outdoor spaces, and how neighborhood built environmental features influence movement and activity of people living with dementia. This understanding is fairly limited in current research and practice, owing to the paucity of empirical evidence on the role of the built environment on mobility, spatial navigation, and community participation of people living with dementia, as well as evidence-based tools and resources that can guide decision-making and planning to create a dementia-friendly and inclusive neighborhood built environment.

These research gaps are addressed in a multi-site, mixed-methods study in British Columbia, Canada, titled "Dementia-inclusive Spaces for Community Access, Participation, and Engagement (DemSCAPE)." The aim of this study is to explore the outdoor mobility patterns and experiences of people living with dementia in the community, and examine the role of the neighborhood built environment in shaping their mobility and participation in outdoor activities. Further, the objective of this study is to develop evidence-based resources that can support local government agencies and community-based organizations to implement dementia-friendly and inclusive planning and design of the neighborhood built environment. This symposium will present findings from the knowledge generation and knowledge mobilization phases of the DemSCAPE study. Presentations focused on knowledge generation in the study offer perspectives from qualitative and quantitative analyses of data collected with participants living with dementia and care partners in British Columbia. The latter portion of the symposium focuses on knowledge mobilization activities involving three key resources developed in our project: documentary video, photo exhibit, and dementia-inclusive planning and design guide. Together, these four presentations offer insights from both the evidence base, as well as practical tools and resources that can help relevant stakeholder groups identify features of the neighborhood built environment that are integral to the outdoor mobility and participation of people living with dementia and apply this knowledge towards the creation of dementia-friendly and inclusive outdoor spaces. Understanding how to create a neighborhood built environment that supports safety, mobility, engagement, and social participation can foster functioning and well-being of people living with dementia.

"It's So Hard for Me to Walk": A Qualitative Study of the Role of Neighborhood Environments in Outdoor Walking Experiences of People Living with Dementia

Author: Habib Chaudhury
Affiliation: Simon Fraser University

Abstract:
This presentation is based on a study that extends the scope of previous research on design and dementia from identifying properties, attributes, and features of the neighborhood built environment to exploring the lived experience of the neighborhood environment by people with dementia in their everyday life. The experientially grounded approach explored ways in which people living with dementia can inform the process of dementia-friendly and inclusive planning and design. The focus on lived experience can also help planners and designers gain insights into the actions and strategies that people living with dementia adopt to cope with challenges in mobility and navigation in the neighborhood, thus gaining useful knowledge to generate contextually relevant design solutions. Four-part sequential sit-down and video-documented walk-along interviews were conducted with 32 participants living with mild to moderate dementia or mild cognitive impairment in British Columbia, Canada. Thematic analysis of interview transcripts and videos helped identify themes relevant to the ways in which the neighborhood environment shaped participants' outdoor walking experiences. Findings underscored both supportive and constraining aspects of the neighborhood environment as it pertains to outdoor walking. Awareness of environmental barriers and stressors in the context of their physical and cognitive challenges informed participants' planning and decision-making processes in relation to outdoor walks, as well as coping strategies to enhance familiarity, safety, comfort, and independence in the neighbourhood environment. Planning and designing supportive built environments is key to facilitating the outdoor mobility of people living with dementia and should be a focus of research, policy, and practice related to building dementia-inclusive communities.

Exploring Factors Influencing Walking Activity in Individuals with Dementia: A Quantitative Spatial Analysis in Metro Vancouver

Author: Mohammadjavad Nouri
Affiliation: Simon Fraser University

Abstract:
Promoting outdoor walking activity is a fundamental goal of creating dementia-friendly neighborhoods, facilitating social engagement and physical well-being of people living with dementia. While previous research has delved into the qualitative aspects of neighborhood environments' impact on the walking behavior of people living with dementia, there is a notable lack of quantitative studies in this area. Leveraging the capabilities of GPS technology, our study presents a unique quantitative approach to analyze the mobility patterns of this vulnerable population. This study utilizes GPS and Geographic Information System (GIS) technology to conduct a comprehensive investigation into the complex interplay between individual and neighborhood environmental factors and their influence on the duration and distance of walking activity among individuals living with dementia. We focus on Metro Vancouver, Canada, engaging 25 participants. This quantitative spatial analysis explores a myriad of variables, including sociodemographic attributes, land use patterns, public transportation accessibility, and the intricate street network within walking routes and their corresponding 20-minute walkshed areas. The findings underscore the significance of several key factors that significantly impact walking activity in people living with dementia, including physical health limitations, the diversity of micro-scale land use, proximity to green spaces, accessibility to bus stops and routes, street slope, street hierarchy, sidewalk connectivity, the presence of street furniture, the quality of intersections, and the subtle curvilinearity of the street network. This in-depth analysis provides a nuanced understanding of how these elements interact to influence the walking activity of individuals with dementia. By shedding light on these critical associations, this study contributes useful insights to the development of dementia-friendly environments. The results offer practical guidance for urban planners, policymakers, and healthcare professionals seeking to create spaces that facilitate and enhance outdoor walking activity for those living with dementia, ultimately promoting their mobility and social inclusion in the community.

Developing Knowledge Mobilization Strategies to Create Dementia-Inclusive Neighborhoods for Community Access, Participation, and Engagement (DemSCAPE): An Educational and Awareness-Raising Video

Author: Caralyn Randa-Beaulieu
Affiliation: Simon Fraser University

Abstract:
To promote community engagement and increase awareness of dementia-inclusive neighborhood built environment, we have co-created with people with lived experience (PLWE) multiple knowledge mobilization (KM) resources. These resources include an educational and advocacy video and photo exhibit showcasing the project findings to stakeholders and the general public in British Columbia, Canada and beyond. A seven-minute documentary video was produced to illustrate the lived experiences of three people living with dementia in Metro Vancouver and Prince George. This video is one of several public engagement tools for dialogue and challenging stigma surrounding living with dementia in the community that will be mobilized during this phase of the project. This presentation will share the process of community engagement based on the World Café method in two municipalities in Metro Vancouver, Canada. The World Café method encourages conversations, collaborative learning, and generation of new ideas with the video and photo exhibit. This KM project aims to increase awareness and understanding of the features of a dementia-inclusive neighborhood built environment and advocate for positive actions for a broad range of stakeholders, including advocacy groups, community-based organizations, municipalities, and the general public. In partnership with the City of Burnaby and the City of Richmond in Metro Vancouver, these KM activities maximize the impact of the DemSCAPE study's findings and model various evidence-based approaches to achieving the vision and objectives for developing dementia-inclusive communities.

Knowledge Mobilization of a Dementia-Inclusive and Friendly Planning and Design Guide: Engagement with Municipal Professionals in British Columbia, Canada

Author: Kishore Seetharaman
Affiliation: Simon Fraser University

Abstract:
Municipal dementia-friendly and inclusive community (DFC) action plans in British Columbia (BC), Canada acknowledge the importance of the built environment and identify corresponding strategies and actions under the purview of different departments of local government, such as planning, design, engineering, transportation, parks, and community services to meet the goals of DFC. However, there is a lack of evidence-based design guidelines tailored to the context of British Columbia to support the implementation of these plans. To address this gap, a DFC planning and design guide was generated as part of the DemSCAPE study, consisting of planning and design principles and guidelines to provide support for wayfinding, safety, social interaction and access to amenities for people living with dementia. A series of knowledge mobilization (KM) activities will explore and augment the relevance, uptake, and implementation of the DFC planning and design guide. Two focus groups (60–90 min; N = 5–7 per session) will be conducted with municipal professionals working in two municipalities (i.e., City of Burnaby and City of Richmond) in Metro Vancouver of BC. The existing municipal action plans emphasize the need for collaboration and dialogue across different departments of local government, e.g., social planners and seniors services coordinators (staff who are more closely connected with and help coordinate these plans) and planners, designers, and engineers. The focus groups are intended to trigger inter-departmental dialogue centered on DFC planning and design and examine DFC strategies and actions from various stakeholders' perspectives within the scope of municipal planning. This KM project is expected to make meaningful contributions to the DFC initiatives of local government in BC by providing guidelines to municipal planners on how to apply a dementia-friendly and inclusive lens in planning and design.

CREATING HEALTHY AND EQUITABLE NEIGHBORHOODS: UNCOVERING AND ACTIVATING COMMUNITY LOCAL KNOWLEDGE AND RESOURCES

Track 5: Spatial Justice, Equity, and Inclusion by Design
Knowledge Network: Health, Participation
Methods: Action/Participatory Research, Mixed Methods

Session Abstract:
The conditions of the places where we live, work, and spend time affect our health and well-being, and there can be significant inequities in the conditions of neighborhoods based on multiple factors, including historic and contemporary decisions, actions, and demographic characteristics (Gullon & Lovasi, 2018). These include high levels of crime and poorly maintained properties and infrastructure that can exacerbate physical and mental health problems as well as decrease social cohesion within neighborhoods (Holliday, et al., 2021; HUD, 2016).

This symposium examines processes and methods that give voice to local community knowledge about the effects of neighborhood conditions on their health and well-being and uses those findings as a foundation on which to build individual and collective capacity to create and sustain equitable environmental, social, and economic change that support better health outcomes. It centers on "Adapting in Place: Exploring Community and Environmental Wellness in a Time of Climate Change," a research project funded by AmeriCorps and organized by a multidisciplinary team of academic researchers and community advisors that are engaging residents of five different neighborhoods in Syracuse, NY. Anticipated as a three-year project, first-year questions focused on residents' definition of health and well-being, their identification of characteristics of a healthy neighborhood, and describing conditions in their neighborhoods that support or detract from those values. Mixed methods in this participatory research have included questionnaires, mapping, brainstorming, photovoice, and Q-Methodology, applied over a series of neighborhood focus groups and a gathering event involving all of the neighborhoods. Findings from analysis of community responses identified several priority areas for change and clusters of several health-related topics around which there is consensus and others around which there is difference.

The presenters are among the research team members who come from disciplines in landscape architecture, environmental geography, and environmental sociology. Community participants' and advisory group members' voices will be included in one or more of the presentations, each of which will be 15 minutes long. There will be 30 to 40 minutes for questions and discussion in a moderated panel or facilitated group format, with notes taken and distributed to attendees. The primary goal of the session is to share the approach and innovative methods and to spark discussion and contributions from attendees on creative application of participatory methods and urban environmental education as pathways to impactful community involvement in decision-making and taking action.

Urban Environmental Education to Support Civic Action for Healthy Neighborhoods

Author: Caralyn Randa-Beaulieu
Affiliation: Simon Fraser University

Abstract:
In urban communities, exploration and understanding of the impacts of built, economic, social, and cultural characteristics and systems, in addition to those that are defined as "natural," are relevant topics for environmental education that is based in history, current lived experience, and desired futures. (Bellino and Adams, 2017)

Opportunities for public participation in planning and community development are proliferating. Due to the infusion of COVID and infrastructure funding, community members are being asked to take part in multiple project meetings; they answer questions, provide comments, and contribute ideas. Through findings from our ongoing Neighborhood and Health project, we have heard from participants that they rarely hear back from the organizers of the public participation on significant redevelopment projects, or if they do, they don't see their ideas or values represented or even acknowledged. While this results in "meeting fatigue" and lack of trust, there is the expressed desire by participants for education and information, so that they can take part in decision-making and that they can become the educators in their own communities (Nabatchi and Leighninger, 2015).

Through literature review of historic and contemporary research practices, investigation of recent municipal and other agency public participation processes, and interviews with organizers and participants of recent community participatory research and education events, this paper and presentation (1) provides an overview of characteristics of effective urban environmental education programs with focus on environmental and social justice, that prepare community members to take action, and (2) describes how these characteristics can be integrated into municipal and other participatory planning and research processes around community and neighborhood redevelopment.

Defining Community Health and Well-Being in Syracuse, NY: A Q-Methodology Study of Community Perspectives

Authors: Elinor Hanjian, Joshua J. Cousins
Affiliation: SUNY-ESF; State University of New York, Environmental Science and Forestry

Abstract:
This paper presents results from a co-productive effort in Syracuse, NY to spur transformative change in relation to community health and well-being in under-resourced communities. Syracuse is a city highly segregated by race and class and that is mirrored in the disinvestment in public infrastructure and housing in the built environment.

Drawing on a series of focus groups across five distinct areas of the city, we sought to understand how community members define health and well-being in their neighborhood. From these focus groups, we used Q-methodology to understand the different assets communities value and how their diverse perspectives converge and diverge around different aspects of community health and well-being among and between individuals and their neighborhoods.

Our results indicate that community residents agree on reforming issues around housing and improving walkability but diverge around the role of policing and green space in supporting community health. We discuss what these differences mean for supporting transformative change and addressing inequality within and among Syracuse neighborhoods.

FROM SHELTER TO HOME

Track 5: Spatial Justice, Equity, and Inclusion by Design
Knowledge Network: Cities & Globalization, Cultural Aspects of Design
Methods: Design Methods, Practice-Based Research

Session Abstract:
This symposium will discuss the experience of unhoused individuals in the public realm and the role that designers (our methods of inquiry, potential impact, and cultural and political orientation) have to untangle one of the most critical issues of today.

The growing houseless crisis has created an urgent demand for as many beds as possible, as quickly as possible. In response, a flurry of design toolkits, guidelines, and pallet houses have been rolled out. Our work has uncovered how communities have "hacked" these formulaic spaces—making them more familiar, more like a home. Consequently, houseless communities have become spaces of experimentation, creativity, and resilience. Learning from these innovations not only improves the quality of life for transient communities through a dignified, respectful, participatory design/build process, it also daylights lessons of resilience for everyone.

The symposium will be moderated by Dr. Kory Rusel, chair of the CBSA (Container Based Sanitation Alliance) and co-founder of Landscape for Humanity. The presenters (Brice Maryman, Jared McKnight, Jean Yang, and Todd Ferry) are all designers who have worked with, for, and on behalf of houseless communities. They have created award-winning designs that balance community needs, regulatory processes, and safety requirements. This includes partnerships with diverse houseless communities and neighborhoods (Skid Row, Seattle's "Jungle," Everyone Village, Opportunity Village, South Eugene, Echo Park), city agencies (NYDOT, City of LA Mayor's Office of City Homeless Initiatives, City of Eugene Community Development Division), non-profits and supportive service providers (Seattle Street Sink Clean Hands Collective, LA Community Action Network, Homeless Healthcare LA ReFresh Spot, Downtown Women's Center, LA LGBT Center, LA Poverty Department, Square One, St Vincent de Paul, and Community Supported Shelters), and many others.

By discussing projects currently underway in Los Angeles, New York, Oregon, Washington and beyond, the symposium will explore how inclusive and equitable impact in the public realm is realized through community-driven, climate-responsive, strategic partnerships, and collaborative design. How can we invite unhoused individuals to stake a more dignified claim in a landscape and build spaces that reflect their unique needs and interests?

Making Space for Homelessness

Author: Brice Maryman
Affiliation: MxM Landscape Architecture LLC

Abstract:
To better respond to increased homelessness in the city's streets, parks and other public spaces, Brice applied for and received the 2017–2018 Fellowship for Innovation and Leadership from the Landscape Architecture Foundation. Through his fellowship project and podcast HomeLandLab, he explored the complexity of homelessness, collecting diverse perspectives to develop empathetic responses that designers can employ when working in the public realm. As the pandemic struck, this work continued with the development and deployment of the Seattle Street Sink, an innovative, open-source solution to persistently deteriorating hygiene conditions among the area's homeless population as parks and libraries closed. In his work designing the Seattle region's public spaces, his work has taken on a trauma-informed design perspective, centering equity and humanity in strengthening our civic infrastructure.

Criminalized for Their Very Existence: The Spatial Politics of Homelessness

Author: Jared McKnight
Affiliation: WRT (Wallace Roberts & Todd, LLC)

Abstract:
Jared's research explores the explicit and implicit nature of policy in design, looking at the systems of rules, codes, and ordinances that govern human behaviors in our civic spaces in Los Angeles. Through a deep investigation into the Los Angeles Municipal Code, and the complexity of regulations found across numerous chapters, articles, and sections of the code that target individuals for the simple fact that they might not have another option for shelter, his work re-organized the categories of regulations to better understand their spatial implications, finding that almost every instance of one's daily life and routine is punishable while unhoused.

This talk will focus on Los Angeles's infamous Skid Row, exploring the democratization of our civic spaces to propose alternate ways that the landscape can be a material for social advancement. Based on dozens of interviews at the LA LGBT Center with young adults experiencing homelessness, Jared will discuss how codified landscapes can govern one's ability to have a dignified existence on the street: the effect that "looking the other way" has on the unhoused individual's psyche; the physicality of hate crimes that shadow and threaten life on the streets; the increased policing, citations, and incarcerations; lack of access to public restrooms; hostile landscapes that criminalize one's existence; and the enforced transiency that occurs in the process of a Street Sweep.

Much of the daily experience tells people what they can't do, but what if our landscapes began asserting what you can? What if our landscapes learned from these narratives, and invited us to hope? How can we welcome emotions at a human scale through ideas that lessen the stress and trauma that already exist in these spaces and give someone an opportunity to do the things they might not think they can—pavilions to recharge your humanity?

Alternative Shelter

Author: Todd Ferry
Affiliation: Portland State University

Abstract:
The village model is an increasingly popular form of alternative shelter being explored by organizations, activists, and municipalities around the country. Portland's Dignity Village is the country's first and longest-running village, serving as a touchstone for community dialogue in Oregon around the subject of supporting people experiencing homelessness since 2000. More recently, the region has seen the rapid increase in alternative shelters informed by or following the village model, sparked by a state of emergency declaration on housing and homelessness in Portland in 2015, and further accelerated by the COVID-19 pandemic.

As the model continues to grow and morph with each iteration, the research team at Portland State University's Homelessness Research and Action Collaborative (HRAC) wanted to know if the village model was working as intended, whom it was serving, and what lessons could be learned to improve future village efforts based on insights from those with personal experience living in, working at, or designing for a village.

This multi-year study resulted in extensive interviews with 42 villagers, 9 village support staff, 7 village designers/architects, 6 village creators/builders, 16 neighbors of villages, as well as a survey about homelessness and villages that included 2,065 Portlanders, 436 of whom reported living near a village. The results provide a framework for best practices for the creation of alternative shelter, highlight shortcomings within the model, and suggest pathways for new and unexplored opportunities within this realm.

Utilizing Geospatial Tools to Validate Stakeholder Perspectives: A Triangulation of Lived Experiences, Consensus, and Existing Data

Authors: Daniel Cronan, Akosua Pomaa Asabere
Affiliation: State University of New York; SUNY ESF

Abstract:
Community-informed research often lacks practical outputs for decision-making (Anthony, 2023). This study assesses geospatial, community-driven tools' effectiveness in validating multifaceted stakeholder perspectives. The research methodology encompasses three phases to address: "Does a self-reported, web-based tool effectively support and validate lived experiences, consensus, and existing data from stakeholders?" (Anthony, 2023).

Phase 1 focuses on exploring lived experiences through a focus group session where participants describe and map neighborhood conditions impacting health using mixed participatory methods including focus group discussions and Q-Sort methodology (Millar et al., 2022).

Phase 2 integrates photographs submitted by focus group participants, enhancing narratives through photovoice methodology, a tool employed in Community-Based Participatory Research (CBPR) (Hergenrather et al., 2009). These photographs form the foundation for a web-based application using ArcGIS Online (ESRI, 2023) to explore stakeholder perspectives. The application intends to disseminate findings to focus group participants and allow ongoing data input. Phase 3 involves surveys and interviews to evaluate application usage and effectiveness in validating stakeholder perspectives. Anticipated findings aim to validate stakeholder perspectives and emphasize geospatial tools' potential in enhancing authenticity, consensus-building, and validating existing data. Employing qualitative and quantitative data, along with geospatial visualization, presents a replicable methodology (Anthony, 2023).

The study highlights geospatial tools' role in amplifying stakeholder narratives and validating community and public datasets. This research carries implications for multiple fields, offering a framework for using geospatial tools to analyze and validate stakeholder opinions. It aims to refine participatory research methods, confirming lived experiences and existing data, facilitating a thorough understanding of societal challenges within communities (Anthony, 2023).

FOREGROUNDING HEALTH IN THE DESIGN PROCESS THROUGH COMMUNITY ENGAGEMENT & DATA

Track: 5 Spatial Justice, Equity, and Inclusion by Design
Knowledge Network: Health, Researchers in Professional Practice
Methods: Mixed Methods, Practice-Based Research

Session Abstract:
Without consideration of occupant well-being, our built environment is just a collection of energy-efficient sculptures. In the worst case, it harms and exacerbates injustices in our communities—from chemicals of concern to inadequate public transit. Yet, traditional design limits community engagement and inconsistently embeds public health principles into the process. At every level of decision-making (individual, organizational, state, national), opportunities exist to address past systemic inequities and advance environmental restorative justice.

This Oregon-centered symposium shares the challenges and opportunities to advance health equity through funding, engagement, on-the-ground efforts, research, and data. Representatives include leaders from the Oregon Health Authority and AfroVillage PDX, as well as a local transportation planner and environmental health researcher. Together, we will talk about practical barriers and potential solutions for prioritizing health and equity in our built environment, especially for priority populations. We will highlight a unique funding model in Oregon—the Health Equity fund—which directly pays partners through centralized state funding and amplifies the voice of smaller organizations. We will share the perspective of both the funder and recipient. We will zoom out to think about how health information and data can prepare designers to have more robust health equity conversations with clients and community. We will feature a new tool, PRECEDE, the Public Repository to Engage Community and Enhance Design Equity.

Our symposium goal is to advance the audience's understanding of effective community engagement and public health-informed design. It invites and immerses participants into a session that models community engagement strategies that build trust, transparency, and respect. The symposium will be organized into three key parts. First, each panelist will provide a foundational presentation to share their unique perspectives with the audience (~30 minutes total). Second, the audience will be invited to participate in a hands-on exercise to strengthen and tailor their community engagement approaches (~20 minutes), followed by a short bio-break (~5 minutes). Lastly, there will be a moderated panel discussion with structured prompts and time to reflect on audience participation. We will welcome Q&A from audience members (~35 minutes).

Integrating Public Health Information in the Design Process

Author: Erika Eitland
Affiliation: Perkins & Will

Abstract:
The built environment can exacerbate or alleviate health conditions by modifying environmental exposures and social behaviors. It is critical for designers to prioritize and correctly identify the most pressing health challenges for their community of interest regardless of project scale (interiors, architecture, urban planning). Budget or client constraints can further inhibit the integration of all features of a healthy built environment. Variations in users, building types, geographies, and social and environmental health contexts highlight the limitations of a one-size-fits-all "healthy building" approach to design. But public health integration requires unique training, methodology, and vocabulary, which traditionally is not included in design education. Therefore, transdisciplinary collaboration and authentic community engagement is critical for health-driven design.

To address these challenges, the Public Repository to Engage Community & Enhance Design Equity (PRECEDE) tool was created to educate designers on key health factors (e.g., extreme heat, air pollution), quantify health priorities using publicly accessible geospatial databases (e.g., EPA EJ Screen, CDC PLACES), validate these findings through community engagement, and identify potential health-based design solutions. This 15-minute presentation will provide an overview of the PRECEDE tool with emphasis on their community engagement resource. It will highlight case studies where PRECEDE identified health priorities early in the design process and created opportunities for more thoughtful discussion and design outcomes. By the end of this presentation, participants will know (1) why data alone is not enough to promote healthy design, and (2) what tools exist to encourage more holistic thinking throughout the design process.

Resilient Design for Climate Change

Author: Josillia Johnson
Affiliation: Oregon Health Authority

Abstract:
Racism, settler colonialism, and historic and contemporary injustices have created policies and programs that have led to unfair and unjust health inequities over time. Oregon Health Authority's (OHA's) Public Health Division took responsibility and recognized the essential role of community-based organizations in creating community-driven, culturally, and linguistically responsive strategies for addressing public health issues. In centering community strengths and wisdom for health, the Public Health Equity fund was created. This grant opportunity aims to shift power, increase engagement and shared leadership with community-based organizations, and center equity in communities of color, Tribal communities, disability communities, immigrant and refugee communities, undocumented communities, migrant and seasonal farmworkers, LGBTQIA+ communities, faith communities, older adults, houseless communities, and other priority populations.

The Public Health Equity funding for modernizing our state's public health system is momentous for OHA, as it reflects a coordination of eight different OHA programs coming together to center health equity and community priorities in one centralized funding opportunity. OHA's commitment to eliminating health inequities by 2030 relies upon building trusted relationships with organizations from every county and supporting their efforts to uplift community health priorities that are grounded in equity and accessibility. This presentation will be led by an OHA Community Engagement Coordinator who helped award $31 million to 157 organizations. By the end of this presentation, participants will be able to (1) evaluate and summarize key opportunities and challenges in creating and maintaining meaningful engagement with community partners; (2) demonstrate how shifting power and increasing engagement and shared leadership with community builds trust and creates a stronger public health system; (3) describe tools for engagement that center community; and (4) learn how impactful engagement with partners at every level—state, local, and community—can build community-driven solutions.

AfroVillage PDX: Community-Grounded, Action-Driven, and Racial Equity-Focused

Author: LaQuida Landford
Affiliation: AfroVillage PDX

Abstract:
AfroVillage PDX is a Portland-based organization that began as a grassroots movement to build community, reduce social isolation, and support the needs of Black women, people of the African Diaspora, and others impacted by systems of oppression and who have been excluded from existing resources through systemic racism.

AfroVillage's approach is action-driven, community-grounded, and racial equity-focused. We provide a variety of basic services and integrated outreach efforts that address the physical, mental, and emotional needs of marginalized and Black community members across the city. We provide services for those who are housing insecure, in need of substance use supports or harm reduction resources, or a place to gather and share a meal, and to take care of basic hygiene needs. We believe that conversation, listening, and community engagement are key to fully understanding the needs of the populations we serve. One of our aims is to create dignified and healing places for the community.

Participants will be able to: (1) Identify how equitable engagement, direct care, and community-driven strategies build community trust and promote safe, inclusive, and healing spaces that elevate the human experience of Black and Brown communities; (2) evaluate and summarize key opportunities and challenges in creating, managing, planning, and designing healing, inclusive, and equitable spaces for our Black and Brown communities; and (3) learn about a new initiative that responds to historical harms and injustices by centering healing called the AfroFuturism Oasis. This strategic initiative explores reimagining and repurposing retired MAX train cars to benefit communities by creating spaces that integrate nature, culture, art, and technology—spaces that are community-designed, safe, inclusive, and healing. The AfroFuturism Oasis will be designed as a healing oasis for Black and Brown communities to rest, collaborate, innovate, and access pathways to economic empowerment while bridging the past and the future.

ACCREDITATION IN THE BUILT ENVIRONMENT FIELDS: PANEL DISCUSSION WITH LEADERS FROM NAAB, LAAB, PAB AND CIDA

Track 6: Transdisciplinary Design and Collaboration
Knowledge Network: None / Not Tied to Any Network
Methods: Mixed Methods

Session Abstract:
This panel discussion with leaders of NAAB, LAAB, PAB, and CIDA will focus on issues of accreditation in built environment fields. Discussion will include:

1. *Similarities and differences in the conditions and procedures of the boards*
2. *Opportunities and threats to accreditation, higher education, and professions*
3. *Support from relevant professional and academic associations*
4. *Globalization*

The panel will generate actionable ideas for academics and professionals in our respective fields, a learning opportunity for EDRA and the accrediting boards.

National Architectural Accrediting Board: Current Issues

Author: Stephen Schreiber
Affiliation: UMass

Abstract:
This panel discussion with leaders of NAAB, LAAB, PAB, and CIDA will focus on issues of accreditation in built environment fields. Discussion will include:
1. *Similarities and differences in the conditions and procedures of the boards*
2. *Opportunities and threats to accreditation, higher education, and professions*
3. *Support from relevant professional and academic associations*
4. *Globalization*

The panel will generate actionable ideas for academics and professionals in our respective fields, a learning opportunity for EDRA and the accrediting boards.

Council for Interior Design Accreditation: Current Issues

Author: Felice Silverman
Affiliation: CIDA

Abstract:
This panel discussion with leaders of NAAB, LAAB, PAB, and CIDA will focus on issues of accreditation in built environment fields. Discussion will include:
1. *Similarities and differences in the conditions and procedures of the boards*
2. *Opportunities and threats to accreditation, higher education, and professions*
3. *Support from relevant professional and academic associations*
4. *Globalization*

The panel will generate actionable ideas for academics and professionals in our respective fields, a learning opportunity for EDRA and the accrediting boards.

Planning Accreditation Board: Current Issues

Author: Weiping Wu
Affiliation: PAB

Abstract:
This panel discussion with leaders of NAAB, LAAB, PAB, and CIDA will focus on issues of accreditation in built environment fields. Discussion will include:
1. *Similarities and differences in the conditions and procedures of the boards*
2. *Opportunities and threats to accreditation, higher education, and professions*
3. *Support from relevant professional and academic associations*
4. *Globalization*

The panel will generate actionable ideas for academics and professionals in our respective fields, a learning opportunity for EDRA and the accrediting boards.

Landscape Architecture Accrediting Board: Current Issue

Author: Roxi Thoren
Affiliation: LAAB

Abstract:
This panel discussion with leaders of NAAB, LAAB, PAB, and CIDA will focus on issues of accreditation in built environment fields. Discussion will include:
1. *Similarities and differences in the conditions and procedures of the boards*
2. *Opportunities and threats to accreditation, higher education, and professions*
3. *Support from relevant professional and academic associations*
4. *Globalization*

The panel will generate actionable ideas for academics and professionals in our respective fields, a learning opportunity for EDRA and the accrediting boards.

K-12 PRE/POST-OCCUPANCY EVALUATIONS: HOW A DESIGN FIRM IS USING RESEARCH TO INFORM PRACTICE

Track 6: Transdisciplinary Design and Collaboration / Open Track
Knowledge Network: Children, Youth & Environment; Sustainable Planning, Design & Behavior
Methods: Mixed Methods, Practice-Based Research

Session Abstract:
How does the built environment impact learning? A growing body of knowledge demonstrates that cognitive, social/emotional, and physical development is related to the built environment. But how do designers take this knowledge and bring it to their real-world practice? How can an investment in research lead to an investment in educational environments through school modernization that achieves Net Positive Education™?

This symposium will present three research studies that were conducted by a global design firm to delve into the profound relationship between the built environment and the learning experience, thereby introducing the innovative concept of Net Positive Education™. Net Positive Education™ is a forward-thinking approach that leverages the pursuit of Net Zero Energy to get to the real goal—improving education—through the use of strategies that impact energy and well-being, thereby promoting education.

The first study to be presented, "Addressing a Multi-Billion Dollar Challenge: Advancing Knowledge of How High-Quality School Environments Can Positively Affect Educational Outcomes," will share how a multidisciplinary research team unraveled the complexities of school modernization to understand the best investment of the limited funds available to school districts for capital improvements. The study engaged 28 schools in two cities, comparing modernized to non-modernized facilities across three key variables—Indoor Environmental Quality, Educational Adequacy, and Community Connectivity—to assess the impact of school modernization on educational outcomes, occupants' well-being, and neighborhood relationships.

The second presentation will share the study, "Above and Beyond: School Modernization to Improve Building, Human, and Educational Performance," which explored the interconnected elements of school design that create high-performing buildings and healthy learning environments. The pre/post-occupancy evaluation comparative study focused on the impact of modernization in two schools, where design went above and beyond conventional approaches to achieve Net Zero Energy and Net Positive Education™.

The last study to be shared, "Biorealism Today: Modern(ist) Lessons Learned from Richard Neutra's UCLA Lab School," explored how the UCLA Lab School, an important artifact of visionary architect Richard Neutra's legacy, can continue to be a thriving educational institution by adapting to meet the pressing demands of climate change and pedagogical shifts, and how schools across the nation can learn from this case study. Findings include the impact of biorealistic design on the educational experiences and the health and well-being of learners and educators.

All three studies are meaningful resources for researchers, educators, and designers, adding to the body of knowledge about how built and natural environments directly impact teaching and learning. The audience will gain insights into how research is being used to understand the intersection of education, architecture, and sustainability to envision a future where schools positively impact educational outcomes and contribute to the health and well-being of individuals and communities.

Addressing a Multi-Billion Dollar Challenge: Advancing Knowledge of How High-Quality School Environments Can Positively Affect Educational Outcomes

Authors: Heather Jauregui, Emily Chmielewski
Affiliation: Perkins Eastman

Abstract:
An estimated 53% of schools in the United States are said to need renovations to be considered in good overall condition, the cost of which would total $197 billion (Alexander & Lewis, 2014). Addressing the nation's inadequate school facilities is an important and necessary step in supporting equitable educational and community improvement.

To address the multi-billion dollar challenge of modernizing the nation's schools, this study's objective was to develop new knowledge that would be of value to both school district leaders and design firms working on modernization plans for new and renovated schools, as well as provide thoughtful and convincing information for decision-making and advocacy for school modernization on a national scale.
To achieve this goal, the study focused on the variables of Indoor Environmental Quality (IEQ), Educational Adequacy (EA), and Community Connectivity (CC). Conducted by a multi-disciplinary team of researchers, the study used an explanatory sequential mixed methods design to collect data at 28 schools in two cities, comparing modernized to non-modernized facilities. The data collected included: on-site IEQ measurements, visual assessments of school environments, occupant questionnaires, stakeholder interviews, and analysis of community and school/district archival data.

Our findings demonstrated that as school districts strive to enhance educational outcomes, investment in the school's built environment plays a critical role in achieving that goal. School modernization had a significant, positive impact on key educational indicators, including test scores and enrollment over time; modernized learning environments improved their occupants' well-being and satisfaction; and the modernization process enhanced community engagement and connectivity. We also identified where planning and design strategies focused on improving areas within the variables of IEQ, EA, and CC can derive the most value from continued and enhanced investment in modern and modernized learning environments, thereby helping to prepare children for success in the 21st century.

Above and Beyond: School Modernization to Improve Building, Human, and Educational Performance

Authors: Heather Jauregui, Emily Chmielewski
Affiliation: Perkins Eastman

Abstract:
This study explores two pre/post-occupancy evaluations (POEs) conducted at Net Zero Energy schools that went above and beyond industry norms. Rather than simply complying with building codes or baseline energy efficiency goals, these schools intentionally integrated high-performance design strategies, healthy materials, and educational programming into the modernization process to achieve what the research team calls Net Positive Education™.

The POEs focused on three core areas: building performance (including indoor environmental quality and energy data), human experience (students' and teachers' perceptions of their school environments), and educational outcomes (curriculum alignment, school culture, and student engagement). The study incorporated qualitative and quantitative data from students, teachers, administrators, and facilities personnel, combined with third-party energy use analysis.

Findings indicate that integrated design strategies—especially when paired with curriculum—enhanced students' sense of agency, environmental literacy, and emotional well-being. Teachers reported greater satisfaction with lighting, acoustics, and access to daylight and views. Students expressed pride in their schools, describing them as beautiful, comfortable, and "made for us."

These findings underscore the value of designing beyond compliance, and how thoughtful school design can become a lever for equity, learning, and health

Biorealism Today: Modern(ist) Lessons Learned from Richard Neutra's UCLA Lab School

Authors: Heather Jauregui, Emily Chmielewski
Affiliation: Perkins Eastman

Abstract:
This case study explores Richard Neutra's legacy at the UCLA Lab School and how biorealistic principles still inform contemporary educational design. Neutra believed that architecture could support human survival through design by responding to physiological and psychological needs. His design for the Lab School embodied these values by offering access to nature, transparency, natural ventilation, and a strong connection between indoor and outdoor environments. Perkins Eastman partnered with the Neutra Institute to conduct a post-occupancy evaluation of the school. The team evaluated how Neutra's original principles have held up over time and how the school has adapted to modern pedagogical and environmental demands.

The study included a literature review of Neutra's writings, interviews with educators and alumni, and on-site evaluation of environmental conditions and spatial layout. Findings revealed that while many of Neutra's biorealistic intentions remain intact, updates are needed to support 21st-century education, environmental resilience, and digital learning.

The presentation will reflect on how designers can carry forward lessons from Neutra's work—particularly his belief that "no one laments the monotony of a tree"—to create modern learning environments that support whole-child development and ecological well-being.

STANDARDIZING K-12 POST OCCUPANCY EVALUATIONS THROUGH A MULTI-FIRM RESEARCH COLLABORATION

Track 6: Transdisciplinary Design and Collaboration / Open Track
Knowledge Network: POE/Programming, Researchers in Professional Practice
Methods: Experimental/Quasi-Experimental Research, Survey

Session Abstract:
Our K-12 schools are both public resources and significant investments. Research has shown the physical characteristics of the learning environment influence student and teacher outcomes, including academic performance, well-being, and behavior (Maxwell, 2018). Despite this history of research, connections between design and other relevant domains like psychology and human health remain piecemeal. Seminal studies have shown design research can connect to measures like academic performance and sense of belonging (Barrett et al., 2015; Cheryan et al., 2014). These examples notwithstanding, there remains a lack of uptake of this methodology in industry research.

This symposium represents a multi-firm team coming together to advocate for improving the rigor and generalizability of post-occupancy evaluations (POEs). We will foreground problems with existing POEs in K-12 settings, with representation from varied fields relevant to design. We will also discuss the practical barriers facing efforts to standardize POEs, including those from the client side of projects. Finally, we will present our vision for a Coalition for the Advanced Understanding of School Environments (CAUSE). CAUSE may offer a solution to common barriers by leveraging the strengths of open-source science and the potential for a multi-firm collaboration to overcome issues that have stymied past efforts from outside industry.

Our symposium goal is to promote awareness and initial interest in participating in CAUSE on the research and evaluation side of POEs. In the first phase of the symposium, each presenter will share relatively brief (approximately 10 minutes) structured remarks summarizing the opportunity for improvements in K-12 POE practice through their particular lens of expertise: health & wellness, clients & facilities operation, school design, and educational outcomes. In the second phase, all presenters will conduct a panel discussion of K-12 POEs comprehensively. The facilitator will provide structured prompts and provocations, and presenters will engage the prompts and each other in an authentic dialogue. In the final phase, the facilitator will engage the audience in a robust Q&A with the full panel of presenters. This Q&A will be a substantial component of the session and is essential to achieving the symposium goal of promoting interest and participation in advancing CAUSE in the coming year.

2 REFERENCES

Barrett, P., Zhang, Y., Moffat, J., & Kobbacy, K. (2015). A holistic, multi-level analysis identifying the impact of classroom design on pupils' learning. Building and Environment, 89, 118–133. https://doi.org/10.1016/j.buildenv.2015.02.013
Cheryan, S., Ziegler, S. A., Plaut, V. C., & Meltzoff, A. N. (2014). Designing classrooms to maximize student achievement. Policy Insights from the Behavioral and Brain Sciences, 1(1), 4–12. https://doi.org/10.1177/2372732214548677
Maxwell, L. E. (2018). School building condition, social climate, student attendance and academic achievement: A mediation model. Journal of Environmental Psychology, 55, 21–30. https://doi.org/10.1016/j.jenvp.2017.11.003

POEs & School Communities – Schools That Support Teacher Change

Author: Raechel French
Affiliation: University of Melbourne / DLR Group

Abstract:
This presentation connects the theory and practice of post-occupancy evaluations (POEs) through Raechel's work as Director of Planning at Austin ISD and her current PhD research at the University of Melbourne. Her work focuses on the impact of innovative learning environments on teachers and teaching practices. Drawing from large-scale research with the ILETC project and district-led initiatives in Austin, Raechel emphasizes the limitations of traditional POEs and the need to incorporate pedagogical and operational practices into evaluation models. She argues that POEs should not only measure environmental satisfaction but also reflect teaching and learning experiences. Her presentation identifies gaps in current approaches, explores the impact of physical environments on teacher identity and efficacy, and calls for more nuanced, collaborative tools to support school communities during and after occupancy.

POEs & Human Health – Schools That Support Student Wellness

Author: Erika Eitland
Affiliation: Perkins & Will

Abstract:
Drawing on her expertise in public health and building science, Erika Eitland presents a post-occupancy evaluation model that places student health and wellness at the forefront of school design assessment. She shares findings and strategies from her work with AIA Massachusetts and the Massachusetts School Building Authority to develop and test scalable tools for assessing indoor environmental quality (IEQ) in K–12 settings. Erika emphasizes the importance of including human health metrics such as air quality, light, acoustics, and thermal comfort in POEs—especially when layered with behavioral, demographic, and geographic data. She proposes a path forward for POEs that is inclusive, health-centered, and scalable, while also advancing the broader goals of health equity and evidence-based school design.

POEs & Education – Schools That Improve Learning and Attitudes

Author: Michael Ralph
Affiliation: Multistudio

Abstract:
Michael Ralph presents a vision for making K–12 POEs more directly responsive to learning goals and educational metrics. His background in scientific research and education design informs a unique perspective on the intersection of physical space and pedagogy. He focuses on opportunities to evaluate education-aligned outcomes such as academic performance, engagement, and student belonging. Michael advocates for the development of open-source instruments, data-sharing initiatives, and interdisciplinary research partnerships that will allow POE results to speak more directly to priorities familiar to district stakeholders. He explores how linking POEs to academic and psychosocial outcomes can support stronger design arguments and enable cumulative learning across projects and firms.

POST-OCCUPANCY EVALUATION: A FIRMWIDE MIXED-METHODS EXPLORATORY RESEARCH AND A CASE STUDY IN EDUCATION

Track 6: Transdisciplinary Design and Collaboration / Open Track
Knowledge Network: POE/Programming, Researchers in Professional Practice
Methods: Case Study, Mixed Methods

Session Abstract:
Post-occupancy evaluation (POE) is a standardized and customized data-driven process to enhance design quality and workflow by evaluating the alignment between the design intents and outcomes from occupant experience and building performance perspectives. However, implementing POE across architectural practice is still rare, even though architects understand the value POE brings into their design decision-making process. POE also often refers to a wide variety of services with unknown associated time and cost, which makes it even more intimidating for designers to include in their workflow.

In the past two decades, several studies have been conducted to understand the benefits and obstacles of implementing POE, but there are few focused on cross-market conversations and a framework for identifying easy-to-implement practice areas and criteria to select suitable projects.

This symposium addresses that gap by presenting an internal exploratory research project about POE and an example of POE conducted in an educational setting. The first part of the symposium shares findings from a firmwide mixed-methods exploratory research initiative—part of an internal research program at a global design and consultancy firm—to understand the advantages and obstacles of conducting POE across practice areas and regions and to identify sectors that are easiest for POE implementation. The study also presents recommended tools and deliverables to make POE more feasible with practical outcomes to improve design.

The second part of the symposium presents a case study of a POE process in an educational setting used to assess educational adequacy, where results helped a school district shape and inform capital improvement construction bonds and school design.

The session concludes with an interdisciplinary discussion with the audience to address questions and comments.

Post-Occupancy Evaluation: A Firmwide Mixed-Methods Exploratory Research

Author: John Naranjo
Affiliation: Arcadis Inc.

Abstract:
Despite the benefits of learning from past projects, architects rarely get the opportunity to return to their buildings and compare actual occupant experiences to the predicted outcomes intended in the design phase. While the value of POE in the design process is acknowledged in literature, few studies focus on the cross-market and cross-regional benefits, challenges, and strategies to overcome those challenges.
This paper presents findings from a mixed-methods exploratory research initiative conducted within a global firm (800+ employees) aimed at understanding the benefits, obstacles, and recommendations for implementing POE across different practice areas and regions. Methods included (1) a firmwide survey with over 160 responses to gauge existing POE knowledge and interest, (2) eight focus groups with 36 employees, and (3) two interviews exploring diverse perspectives. The findings indicate POE improves client relationships, advances marketing, and enhances design quality. Recommendations for overcoming barriers include developing a benchmark database, identifying "right-fit" clients and projects, and providing POE education to design teams. Healthcare, workplace, and education projects were found to be the most viable sectors due to their long-term planning horizon and access to repeat users.

Post-Occupancy Evaluation Case Study: Educational Adequacy Assessment

Author: Rebecca Stuecker
Affiliation: Arcadis Inc.

Abstract:
An educational adequacy assessment evaluates how well the physical aspects of a building and site support teaching, learning, and the daily needs of students and staff. In a post-pandemic world, the pressure on schools to meet psychological, social, and emotional needs has grown exponentially—and the built environment plays a key role in their ability to do so.

This POE focused on how the facility supported current and future instructional methods and health-related initiatives. The methodology and field tools used by Arcadis will be presented, along with a case study in which findings from this POE were used by a school community to inform both capital improvement construction bonds and the design of new schools.

DESIGNING FOR RESILIENCE: A TRAUMA-INFORMED APPROACH

Track 6: Transdisciplinary Design and Collaboration / Open Track
Knowledge Network: Environmental Design Research Education, Researchers in Professional Practice
Methods: Case Study, Mixed Methods

Session Abstract:
Our clients' experiences and well-being are shaped by the spaces that surround them—the spaces we design. Clients may feel safe or threatened, welcome or rejected, cared for or neglected in some spaces. These considerations become even more significant when working with trauma survivors who may be seeking safety and care for themselves and their loved ones. Environments designed to center the unique needs of trauma survivors can provide much-needed safety and healing. In contrast, environmental design, if done poorly, may exacerbate their trauma responses. It is imperative that spaces in which they receive services—e.g., justice centers, shelters, or community resource centers—are designed to consider their unique needs and experiences. There are myriad events and experiences that elicit a trauma response—e.g., violent victimization, child abuse, sexual assault, natural disasters, to name a few. When faced with a traumagenic event, the body and brain respond in physiologically wired ways to facilitate survival. The result is myriad physiological, emotional, relational, and spiritual impacts that shape the survivor's life at the time of the event and moving forward. There is considerable knowledge about these impacts, the resulting needs of survivors to build resiliency and to heal, and the implications for working with trauma survivors. Less is known about the ways in which environmental design can exacerbate or mitigate the trauma response and support healing and resiliency. This gap in knowledge presents an opportunity to explore and create trauma-informed and healing-centered design that prioritizes the needs of trauma survivors and attends to diverse trauma responses. Such a design framework draws on holistic, evidence-based design principles shown to facilitate healing and resiliency and compassionately engages survivors in the design process.

This session introduces evidence-based design principles for environments that facilitate healing and resiliency, offers a way to assess existing environments for their healing characteristics, and presents thoughtful, nimble, and relational participatory design processes well suited for clients who have experienced trauma. Participants will learn from the presenters' real-world experiences in three different contexts: a residential treatment center providing shelter care and substance use treatment for young people; a Safe Healing Center residence for youth as an alternative to juvenile halls and detention camps; and a space for families who have lost loved ones to criminal fatalities as their case makes its way through the justice system. The session also calls attention to the need for a shift from a pathological, deficit model of design to one that identifies and fosters strengths and promotes individual, family, and community resilience. Doing so highlights the interconnectedness of contexts and the need for collaboration across systems (e.g., criminal justice, child welfare and health, education, housing, etc.) to design environments that promote healing and resilience across the life span and in diverse contexts.

Trauma-Informed Housing for Youth Involved in the Justice System

Author: Garrett Jacobs
Affiliation: Designing Justice Designing Spaces

Abstract:
This presentation reviews a conceptual design that has been recently completed for a youth residential program and building type intended to replace juvenile halls and detention camps throughout LA County. These facilities, called Safe Healing Centers, would be unlocked and therefore youth would be free to leave. To establish the initial design and programming criteria for the buildings, DJDS used its trauma-informed design process to first listen to and engage those closest to the problem of youth incarceration—youth, formerly incarcerated individuals, and systems-impacted youth. By engaging individuals in these communities, their experiences were not only acknowledged but honored as a valuable first step in a healing process to address the impacts of punitive institutionalization.

The design firm conducted literature reviews, site visits to a similar project in New York called Close to Home, and interactive workshops to determine the trauma-informed design criteria and emergent themes for a Safe Healing Center environment. This presentation shares the high-level programming criteria, activities that emerged, and the environmental characteristics that will guide initial pilot projects later this year.

Nature-Based Healing Environments to Improve Critical Protective Factors for Vulnerable Youth

Author: Julie Stevens
Affiliation: Iowa State University

Abstract:
All children deserve opportunities to grow and thrive. However, more than two-thirds of children are victims of at least one traumatic event. Trauma-informed design and exposure to nature can help buffer the impacts of trauma and alleviate its systemic effects. This presentation explores these issues through a campus design project with a local residential treatment center providing shelter care and substance use treatment for young people. Through a university-community partnership, we created developmentally appropriate, sensitively designed therapeutic landscapes to alleviate the enduring effects of trauma and toxic stress.

This partnership demonstrates the ways in which environmental design for youth shelters aligns with trauma-informed care practices to create nature-based environments essential for optimal health and well-being. It also explores how such design can reduce risk factors, prevent re-traumatization, and promote healing. Working with vulnerable populations requires design teams to spend more time in research and participatory phases than traditional contracts often allow. This complexity demands expertise beyond design, such as in social work and human development. Taking time to understand these needs can positively impact residents and support long-term recovery and well-being.

A Healing Haven: Trauma-Informed Space for Families of Criminal Fatality Victims

Author: Barb Toews
Affiliation: University of Washington Tacoma

Abstract:
Each year, millions of people experience violent victimization, which greatly impacts their physical, emotional, and social well-being. When a person is arrested and charged, survivors may find themselves participating in court proceedings that can re-traumatize them due to the lack of consideration for their needs or the proximity to the person who harmed them.

This presentation explores the design of a day-use facility created to support survivors during court-related processes. This organization offers two spaces—Healing Haven and Heart Spaces—designed to provide trauma-informed staff support, services, and a refuge away from the courthouse.

Home-like in design, these spaces include an eat-in kitchen, living and dining rooms, and private nooks to foster calm and empowerment. In addition to providing physical space, support services such as court accompaniment, smudging, food hospitality, and therapeutic activities deepen the healing experience. The presentation shares lessons learned from the space's design, staffing model, and a recent program evaluation to demonstrate the effectiveness of trauma-informed environments on survivors, staff, and support professionals.

TOWARDS EMERGING POSSIBILITIES FOR BUILT SPACES

Track 6: Transdisciplinary Design and Collaboration / Open Track
Knowledge Network: Environmental & Architectural Phenomenology, Sustainable Planning, Design & Behavior
Methods: Case Study, Mixed Methods

Session Abstract:
The built environment is responsible for significant impacts including climate, health, and social justice issues. The lived experiences of people and places globally, due to local decision-making, necessitate a dramatic shift in the way our places are imagined, designed, and constructed.

This 90-minute participatory symposium actively explores how practices of engagement and an expansion of who and what we consider in decision-making can reveal new and emerging possibilities that challenge our own assumptions related to the design, policy, and pedagogy that influence how our built environments are developed and shape our experiences.

The 3-person panel includes a participant/facilitator and brings together expertise and perspectives from each of the three principal conference themes: prosperity, well-being, and global environment. The panelists use Oregon as a case study to develop a holistic look at the conference themes, bringing experience in design and architecture, a sense of place and diverse lived realities, along with planetary boundaries and the implications for well-being, adaptability, and resiliency toward equitable and just prosperity.

The symposium includes:

- *An introduction bridging the three themes to set the dialogue within "the big picture" (10 minutes)*
- *A synopsis of an extended engagement with communities across Oregon (10 minutes)*
- *Reflective exercises with the audience on human-centeredness and placemaking, exploring ideas of scale, prosperity, and well-being—both planetary and societal (45–50 minutes)*
- *A closing discussion (15–20 minutes)*

Issues explored include: assumptions of human-centeredness; alternative decision-making frameworks; the built environment's scale and impacts; well-being as a counterpoint to ill-being; and the environment beyond abstractions like pollution. Reflection exercises may use poetry, art, imagery, and more. The panel facilitates the discussion toward the emerging, with an eye toward transdisciplinary collaboration while holding space for the unknown or unexpected.

Learning Objectives:

1. *Understand the interconnectivity of global material flows*
2. *Explore potential rebound effects*
3. *Consider scale as a continuum affecting planetary functioning*
4. *Deepen understanding of the entanglements among prosperity, well-being, and environment*

Towards Emerging Possibilities in Built Spaces

Author: Amanda Ingmire

Abstract:
Amanda Ingmire's role includes the initial design of this proposal (with Minal Mistry), and subsequent concept development, narrative flow and exercises, and visual aesthetic and cohesion. During the symposium's delivery, dialogue, and engagements, Amanda's role is to spotlight key areas of concern in the built environment including climate, toxics and other health issues, and equity issues such as economic opportunity, housing affordability, displacement, and forced labor—and to tie these to the lived experience of people and places both locally (through community engagement insights) and globally (through research).

She will also speak to project-level and systems-level changes needed to shift from a built environment underlaid with harm to one centered on responsibility, care, and reciprocity—toward prosperity and well-being for people and places. Amanda will participate throughout the symposium.

Towards Emerging Possibilities in Built Spaces

Author: Adam Davis
Affiliation: Oregon Humanities

Abstract:
Adam Davis's role as a presenter is to provoke session attendees to reflect on the relationship between a sense of place and a sense of belonging. He draws on experience facilitating community conversations across Oregon and the U.S. on the conditions that shape our collective well-being.

In this session, Adam will guide participants through a participatory, reflective dialogue aligned with the symposium's themes—inviting attendees to interrogate how built environments support or suppress inclusion, identity, and social equity in local and planetary contexts.

ABSTRACTS CONTINUED

Individual Presentations: Digital Media Shorts

DECODING THE STREET: A DATA-DRIVEN APPROACH FOR UNRAVELING URBAN PERCEPTION BASED ON THE 5DS FRAMEWORK IN BEIJING'S OLD CITY

Track 5: Sustainable Buildings and Infrastructure
Mingze Chen (UBC), Hao He (Beijing Forestry University), Wenchen Jian (Beijing Forestry University), Yuxin Yan (Beijing Forestry University), Boya Wang (Beijing Forestry University)

Session Abstract:
Streets serve as a crucial interface in shaping a city's image. Understanding the structural relationship between the physical environment and human perception is fundamental to human-centric street design. Traditional survey-based investigations provide detailed assessments, but their time-consuming nature limits applicability to larger scales. Recent studies have employed big-data approaches at a city-wide scale; however, these studies primarily focus on visual traits, with limited exploration of other environmental factors. This study proposes a hybrid approach to fully understand urban perception in Beijing's Old City (BOC).

To quantify the physical environment, we developed a GIS model for BOC using multi-source data. Additionally, we established an indicator system based on the 5Ds framework (Cervero et al., 2009), encompassing Density (population and building density), Diversity (image and POI entropy), Design (imageability, enclosure, greenness, openness, walkability, safety, facility, and culture), Destination accessibility, and Distance to transit. Some indicators were derived from GIS data, while others were extracted from panoramic Street View Images (SVIs) collected in BOC. The extraction process utilized a pre-trained deep-learning image-segmentation model, DeepLabV3+.

For public perception assessment, we conducted a localized and cost-effective evaluation using the human-machine adversarial scoring framework proposed by Yao et al. (2019). We examined street perception in four major aspects: safety, vitality, aesthetics, and restoration. By training a deep learning model, we achieved over 90% accuracy in predicting public perception of the street within BOC from over 100,000 first-person perspective SVIs.

Cluster analysis was employed to categorize streets based on their 5Ds, and we subsequently investigated the spatial heterogeneity of the four perception aspects. Statistical analyses identified environmental factors contributing to specific street perceptions. Findings include: (1) Density, indicative of urbanization, is positively related to safety and vitality but negatively associated with restoration; (2) Design and Diversity indicators strongly correlate (positively or negatively) with all four perceptual aspects; and (3) Destination accessibility and Distance to transit have less impact on human perception. A correlation network visually illustrates these intricate relationships.

This study explored how people perceive different streets, offering valuable insights into the nuanced qualities of the street environment. The findings provide a practical model for portraying detailed streetscape perception, examining complex urban environments, and informing human-centric city planning and street renewal.

References

Cervero, R., Sarmiento, O. L., Jacoby, E., Gomez, L. F., & Neiman, A. (2009). Influences of built environments on walking and cycling: Lessons from Bogotá. International Journal of Sustainable Transportation, 3(4), 203–226. https://doi.org/10.1080/15568310802178314

Yao, Y., Liang, Z., Yuan, Z., Liu, P., Bie, Y., Zhang, J., Wang, R., Wang, J., & Guan, Q. (2019). A human-machine adversarial scoring framework for urban perception assessment using street-view images. International Journal of Geographical Information Science, 33(12), 2363–2384. https://doi.org/10.1080/13658816.2019.1643024

TANGATA WHENUA (PEOPLE OF THE LAND): THE ENDEMIC INFLUENCE OF MĀORI CULTURE ON VISITOR EXPERIENCE IN AOTEAROA NEW ZEALAND

Track 5: Spatial Justice, Equity, and Inclusion by Design
Knowledge Network: Cultural Aspects of Design, International Connections
Methods: Observation, Photo-Based Methods
Meghan Mick (Florida State University)

Session Abstract:
Indigenous peoples around the world—over 475 million, according to Amnesty International—suffer injustice and discrimination. The centuries-old imperialist ideology that led to the invasion and colonization of native lands persists in the mistreatment, removal, and continued marginalization of Indigenous populations. Although reparations are being made in some places, historic injustices linger, reflected in society and the built environment, where public art and location names often honor recent settlers rather than presenting a more thorough and layered heritage, including natural, cultural, and political histories.

A recent trip abroad illustrated a sharp contrast between the U.S. and Aotearoa New Zealand in the representation and celebration of native peoples. Despite enduring similar struggles to other Indigenous populations, Māori heritage has not been erased. Instead, it is interwoven into the environment, shaping the experiences of residents and visitors alike. While the U.S. attempts to acknowledge original land inhabitants through statements, New Zealand celebrates Indigenous culture as part of the everyday experience.

Since 1974, the Māori have been legally referred to as tangata whenua (people of the land) in New Zealand. In 1987, Te Reo Māori was officially declared a national language under the Māori Language Act. These governmental policies acknowledge the original inhabitants of Aotearoa New Zealand, but legislation alone does not always translate into societal and environmental representation. The recent inclusion of the Māori language in all public signage enhances cultural awareness. Other visible influences of Māori heritage appear in symbols, advertising, rituals, food, and environmental protection laws, which preserve one-third of the country's natural landscapes (The World Bank). Travelers are encouraged to take the Tiaki Promise, an ethical oath rooted in Māori principles of Manaakitanga (hospitality) and Kaitiakitanga (respect and guardianship), which asks visitors to: "care for land, sea, and nature," "travel safely, showing consideration for all," and "respect culture with an open heart and mind." This photographic study captures how Māori culture is not only acknowledged but empowered as a fundamental component of New Zealand's identity.

References
Amnesty International. (n.d.). Indigenous peoples. https://www.amnesty.org/en/what-we-do/indigenous-peoples/

New Zealand Immigration. (n.d.). Māori culture. https://www.live-work.immigration.govt.nz/live-in-new-zealand/tips-for-settling-in/maori-culture

The World Bank. (n.d.). Terrestrial protected areas (% of total land area). https://data.worldbank.org/indicator/ER.LND.PTLD.ZS

HACKING FOR WELLNESS: USING BIG BOX SYSTEMS FOR PROTOTYPING

Track 4: Design for Health and Well-being
Knowledge Network: Health
Methods: Design Methods, Mixed Methods
Jennifer Webb (University of Arkansas), Angela Carpenter (University of Arkansas)

Session Abstract:
Wellness environments, self-care strategies, and wellness products have been part of design discourse for decades. Today, they dominate popular media, industry-specific websites, and professional design communications. Across environmental scales, from aromatherapy and sensory interventions to entire rooms and outdoor spaces, the Global Wellness Institute estimated the global wellness economy to be $5.6 trillion in 2022. This industry is entrenched in the luxury market, targeting those with disposable income and time, as well as organizations with resources to invest. However, respite, sensory control, privacy, and engagement with nature are essential for all people and organizations, regardless of financial capacity.

This video presentation illustrates the prototyping of a wellness pod that is healthful, flexible, affordable, and adaptable to diverse settings. The project prioritized low-VOC materials and modular construction, ensuring it could be assembled with minimal tools and construction skills. These priorities led to the use of IKEA-hacking, a strategy in which the company's flat-packed systems are reconfigured for new purposes (Mars, 2014).

IKEA products were selected following site visits and online research. Affordable shipping, material quality, and modular add-on parts were key strengths. The IKEA Bror System, featuring unfinished pine plywood and powder-coated galvanized steel, was chosen for its minimal off-gassing and biophilic benefits. Multiple corner pieces, wooden and metal shelves, and a polycarbonate sheet were ordered for experimentation. A basic tool kit from Harbor Freight was also purchased, though the Bror System included its own Allen wrenches.

Time-lapse photography documented the prototyping process. Units were arranged iteratively, each composition suggesting new ideas. One, two, three, and four-unit configurations were explored with varying degrees of enclosure. Shelves were oriented vertically to create enclosure and allow openings at different heights. A single unit was tested as an open and enclosed bench for sensory elements, though stability issues were noted when sitting or rising. The manufacturer recommends anchoring units to a wall.

Two-unit configurations provided three different spatial arrangements, with a back-to-back orientation accommodating diverse sitting styles and enclosure needs. Three-unit configurations maximized flexibility in enclosure and seating arrangements. Four-unit configurations expanded this potential further, supporting both individual and group use.

While the exploration confirmed the flexibility of a modular unit to meet dynamic user needs, weaknesses were identified. These include the stability of assembled units, the maximum weight they can support, and potential user misuse. The wellness pod system will be documented in different settings for preliminary evaluation. Future iterations will investigate sensory elements such as felt padding, nature boxes, and lighting. This video presentation will highlight the full development of this wellness pod system.

References
Mars, R. (2014). The new and improved IKEA hacker: IKEA projects that are as unexpected as they are clever. Skyhorse Publishing.

ENVISIONING BELONGING: AN EXPLORATION OF INTERIOR SPACES AND THE SENSE OF BELONGING AMONG BIPOC AND FIRST-GENERATION COLLEGE STUDENTS

Track 5: Spatial Justice, Equity, and Inclusion by Design
Knowledge Network: Cultural Aspects of Design, Student Affairs
Methods: Focus Groups, Photo-Based Methods
Brodie Bain (NAC Architecture), Alana Pulay (Washington State University), Dustin R. Saalman (NAC Architecture)

Session Abstract:
A significant proportion of post-secondary students in the United States report a lack of connection to their academic institutions. A sense of belonging in higher education is defined by feelings of respect, value, acceptance, care, inclusion, and significance. Research shows that fostering belonging is crucial, as it correlates with enhanced persistence, engagement, and mental health, prompting students to use campus resources and improve academic achievement (Johnson, 2022). Numerous studies demonstrate the positive association between belonging and various aspects of the higher education experience, including social and psychological functioning (Hagerty et al., 1996), support and caring (Freeman et al., 2007), college activities (Hurtado & Carter, 1997), self-confidence, academic success (Pittman & Richmond, 2007), and social recognition and acceptance (Freeman et al., 2007). Several studies have identified belonging as a key factor in student retention (Thomas, 2012; Tinto, 1975, 1996; Astin, 1999, as cited by Ahn & Davis, 2020), emphasizing its importance for administrators.

Architects and designers make intentional decisions to create campus spaces that foster a welcoming atmosphere and enhance students' sense of belonging. Elements such as space design, furniture, artwork, flooring, and wall treatments are selected with the expectation that they will positively impact students. However, predicting how diverse user populations—particularly racially and ethnically minoritized (REM) and first-generation students—perceive these spaces remains a challenge. Given the limited exploration of this issue, identifying specific interior design variables that contribute to a student's sense of belonging is critical. Despite a space appearing aesthetically pleasing from a design perspective, it may not necessarily create a welcoming environment, particularly for REM and first-generation students. Uncovering the built environment factors that contribute to a sense of belonging is essential. This understanding will empower designers and decision-makers to create campuses that foster a shared sense of belonging among all students, potentially increasing retention rates and academic success. This research will specifically address underrepresented populations and first-generation students to ensure that campus design is inclusive and supportive of all.

DESIGNING FOR PANDEMIC PREPAREDNESS: WHAT WE LEARNED FROM EBOLA AND COVID-19 OUTBREAKS ABOUT FLEXIBILITY AND HUMAN-CENTERED DESIGN

Track 4: Design for Health and Well-being
Knowledge Network: Health
Methods: Observation, Survey
Herminia Machry (Georgia Institute of Technology), Zorana Matić (Northeastern University)

Session Abstract:
In the last decade, healthcare facilities worldwide have faced extraordinary challenges due to outbreaks of infectious diseases such as Ebola and COVID-19. Existing building layouts had to be rapidly adapted to safely manage new complex workflows and high patient volumes while maintaining rigorous infection prevention protocols. In the U.S., the Ebola and COVID-19 pandemics also highlighted a series of safety challenges and environmental stressors for healthcare staff.

Adequate doffing of personal protective equipment (PPE) is one of the most critical behaviors for protecting staff from virus exposure. Studies show that environmental design can help reduce risks and cognitive load for healthcare workers during PPE doffing tasks (DuBose et al., 2018). The design of spaces where PPE doffing occurs plays a key role in reducing contamination risks by ensuring proper placement of hand hygiene stations, storage, and door configurations (Machry et al., 2022). To understand how the built environment impacts risky behaviors and occupational health during PPE doffing, our team conducted extensive research in inpatient care settings. Between 2017 and 2019, in the context of Ebola preparedness, we observed staff behaviors in real and simulated isolation spaces, mapping PPE-related behaviors and administering surveys. During the COVID-19 pandemic in 2021, we performed behavioral mapping and surveys at four inpatient nursing units treating COVID-19-positive patients. These studies recorded staff behaviors that led to increased risks of self- and cross-contamination, ultimately informing design guidelines for PPE doffing spaces.

Our presentation will discuss findings compiled in our upcoming book chapter, demonstrating how the built environment can support human-centered outcomes such as staff safety and occupational health during pandemic-like situations. We will share strategies for rapidly redesigning and operationalizing hospital spaces for PPE doffing and highlight key actions for designers and health systems related to flexibility. Furthermore, we will discuss the urgent need to equip healthcare facilities with adaptable infrastructure to better respond to future pandemics while maintaining a human-centered approach to caring for patients with serious communicable diseases.

References
DuBose, J. R., Matić, Z., Sala, M. F. W., Mumma, J. M., Kraft, C. S., Casanova, L. M., Erukunuakpor, K., Durso, F. T., Walsh, V. L., Shah, P., Zimring, C. M., & Jacob, J. T. (2018). Design strategies to improve healthcare worker safety in biocontainment units: Learning from Ebola preparedness. Infection Control & Hospital Epidemiology, 1-7.

Machry, H., Matić, Z., Oh, Y., DuBose, J. R., Morgan, J. S., Love, K. L., Jacob, J. T., & Zimring, C. M. (2022). Healthcare design to improve safe doffing of personal protective equipment for care of patients with COVID-19. Infection Control & Hospital Epidemiology, 1-10.

LEVERAGING MACHINE LEARNING AND REMOTE SENSING TO EVALUATE WINDOW VIEW QUALITY DURING THE DESIGN PHASE FOR COMFORT ANALYSIS THROUGH VIWIN – AN ONGOING ACCEPTABILITY STUDY

Track 4: Design for Health and Well-being
Knowledge Network: Building Process Alliance, Interior Design
Methods: Design Methods, Interviews
Naghmeh Pak (University of Wisconsin-Madison)

Session Abstract:
Given the extensive time people spend indoors, prioritizing human well-being and comfort is increasingly critical in architectural design. This ongoing research project, currently in its proof-of-concept stage, focuses on a novel technology for building designers: ViWin (View and Window Assessment). ViWin leverages satellite imagery, remote sensing, and computer vision and machine learning (CVML) technologies to offer a new method for assessing windows during the design phase to improve occupants' visual comfort.

Research on window view quality is hindered by a lack of unified standards and comprehensive studies (Cadena et al., 2022; Kim et al., 2022). Previous work has partially addressed the relationships between window characteristics, external conditions, and comfort (Ko et al., 2022). Established metrics for visual comfort exist (Konstantzos et al., 2015), but the use of advanced tools like CVML with satellite data remains underexplored. ViWin aims to bridge these gaps by applying CVML to satellite data for precise land feature detection, enhancing window view quality analysis accuracy, and enabling three-dimensional assessments. This allows for a comprehensive understanding of a building's visual environment and surroundings.

The methodology integrates BIM data with satellite imagery, involving site coordinate extraction, building design analysis for window specifications, satellite image acquisition, and machine learning model training for landscape detection. The final product generates a comfort index for each window. ViWin is designed for seamless integration with architectural design platforms like Revit, allowing designers to create 3D visualizations of comfort indices within their models.

ViWin is currently undergoing acceptability and appropriateness testing to ensure easy implementation in architectural practice. Feedback from architects is informing further development.

In conclusion, this research represents a crucial step toward creating a user-centric tool that enables designers to make informed decisions about window design. By integrating technology with architectural planning, ViWin has the potential to improve occupants' visual comfort and overall well-being.

References

Cadena, J. D. B., Poli, T., Koir, M., Lobaccaro, G., Mainini, A. G., & Speroni, A. (2022). Current trajectories and new challenges for visual comfort assessment in building design and operation: A critical review. Applied Sciences.

Kim, J., Kent, M., Kral, K., & Dogan, T. (2022). Seemo: A new tool for early design window view satisfaction evaluation in residential buildings. Building and Environment, 214, 108909. https://doi.org/10.1016/j.buildenv.2022.108909

Ko, W. H., Schiavon, S., Altomonte, S., Andersen, M., Batool, A., Browning, W., Burrell, G., Chamilothori, K., Chan, Y.-C., Chinazzo, G., Christoffersen, J., Clanton, N., Connock, C., Dogan, T., Faircloth, B., Fernandes, L., Heschong, L., Houser, K. W., Inanici, M., & Jakubiec, A. (2022). Window view quality: Why it matters and what we should do. LEUKOS, 18(3), 259–267.

Konstantzos, I., Chan, Y.-C., Seibold, J. C., Tzempelikos, A., Proctor, R. W., & Protzman, J. B. (2015). View clarity index: A new metric to evaluate clarity of view through window shades. Building and Environment, 90, 206–214.

THE AUGMENTED REALITY HOME ASSESSMENT TOOL

Track 6: Spatial Justice, Equity, and Inclusion by Design
Knowledge Network: Environment-Gerontology, Residential Environments
Methods: Design Methods

Jung-hye Shin (University of Wisconsin-Madison), Jenny Lee (University of Wisconsin-Madison), Jade Ho (University of Wisconsin-Madison), Rachael Shields (University of Wisconsin-Madison), Ross Tredinnick (University of Wisconsin-Madison), Hyun-seok Chang (University of Wisconsin-Madison), Beth Fields (University of Wisconsin-Madison), Kevin Ponto (University of Wisconsin-Madison)

Session Abstract:
Traditional home assessments conducted by occupational therapists evaluate the suitability of an individual's living environment for their functional capacities, serving as a precursor to home modification. This comprehensive process, similar to a medical diagnosis, involves measuring and examining various aspects of the home, such as doorways and countertops, to identify potential challenges in daily activities. While home assessments are essential for successful home modifications, securing these services is challenging due to their complexity, cost, and a shortage of professionals, particularly in rural areas.

To address these challenges, the Augmented Reality Home Assessment Tool (ARHAT) was developed to streamline the home assessment process into a user-friendly mobile application. The app provides step-by-step guidance on where and how to measure using a series of questions and automated AR tools, allowing users to store measurements within the platform. The questions are based on the Housing Enabler (Iwarsson, 1999) and a simplified version of ADA guidelines. ARHAT utilizes the LiDAR scanner on an iPhone, giving users the option to take measurements using augmented reality (AR). This eliminates the need for traditional measurement tools, enhancing accessibility and efficiency.

Within the app, users begin by selecting any functional limitations they may have from a list of 14 different categories, such as visual impairment, poor balance, or reliance on a wheelchair. They can then choose specific areas of their home for measurement, such as the entrance, kitchen, or bathroom. These inputs filter questions and measurements, focusing the assessment only on relevant aspects of the home. The final report generated by the app includes measurements, identified barriers, relevant guidelines, and suggested solutions. This report can be shared with an occupational therapist for further analysis or with a building professional to facilitate modifications.

Conceptualized as disruptive technology, ARHAT democratizes home assessments and redefines the home modification process. This innovation enhances convenience and accessibility, empowering users to ensure their living spaces are tailored to meet their specific needs, potentially contributing to an improved quality of life.

ABSTRACTS CONTINUED:

Posters

POST OCCUPANCY EVALUATION OF THE COUNTRY'S FIRST CENTER FOR RESTORATIVE JUSTICE AND RESTORATIVE ECONOMICS

Track 5: Spatial Justice, Equity, and Inclusion by Design
Knowledge Network: Interior Design, POE/Programming
Methods: Interviews, Mixed Methods
Garrett Jacobs (Designing Justice Designing Spaces)

Backgrounds:
The practice of restorative justice is becoming more common, but there exists a void of dedicated spaces for restorative justice practice that accurately reflect its values. In 2019, Restore Oakland opened to the public as the country's first center for restorative justice and restorative economics. It is home to the Ella Baker Center for Human Rights, Restaurant Opportunity Center, Causa Justa: Just Cause, and the newly formed Restore Oakland Inc. The building contains a dedicated space for restorative justice practice, to be used by the owners, community, and partnering organizations. Throughout 2024 the original designers of the space conducted a post-occupancy evaluation (POE) of the Restore Oakland building with a particular focus on the community who uses the dedicated restorative justice space.

Methods
The designers conducted a POE of the Restore Oakland building through 2023. The evaluation methodology was designed with the building operators and consisted of: a digital survey distributed to all occupant organizations and their employees; two building walkthroughs with the building owners and operators; and a focus group discussion following the walkthrough. A participatory process was then conducted with a focus on the dedicated space for restorative justice. A participant from Restore Oakland's healing justice program was hired as a community researcher, they then hosted three healing circles within their existing weekly community building circle format which focused on space and then designed and conducted eleven interviews with program participants and staff. The field notes, along with the interview and focus group transcriptions were all coded and analyzed to determine the trends.

In developing the process and surveys the authors conducted research into other POE surveys, including the American Institute of Architecture's Courthouse POE and the BUSS POE template with reference to the Evidence-Based Design Accreditation surveys and multiple Sustainable, Green, Active and Wellbeing Building certifications in the field. The questions posed were also influenced by original research by the authors focused on creating spaces for survivors of violence. The authors then collaborated with the community researcher to edit the survey and interview questions to ensure their cultural relevance to those in the building.

Results
Outcomes of this POE include gaining an understanding of:

- *The impact of the physical building on occupants, and their ability to serve their constituents, and opportunities for improvement;*
- *The success of the building in meeting the initial needs of occupants and owners;*
- *The relevance of the building to the occupants and their organizations' missions;*
- *The importance of dedicated restorative justice - or healing justice space;*
- *The importance of "spiritual spaces" in healing and transformation work;*
- *The importance of building circulation and entry sequencing;*
- *How signage and wayfinding impact building occupants and visitors.*

Conclusions
This evaluation is a baseline for the designer's future project evaluations and initiates a database for the evaluation of restorative justice spaces. This evaluation has supported the building owners in prioritizing future changes, established that dedicated spaces for transformative justice work are of vital importance, and there are many spatial elements which contribute to spiritual spaces for healing.

READING UNDERVALUED LANDSCAPES: SHARING CURRICULUM FOR EXPANDING PRECEDENT STUDIES

Track 1: Nature and urban ecology
Knowledge Network: Cultural Aspects of Design, Sustainable Planning, Design & Behavior
Methods: Design Methods, Observation
Carey Clouse (University of Massachusetts, Amherst)

Backgrounds:

This poster explains a pedagogical structure for developing a better understanding of the physical environment through the reading of undervalued landscapes in the US. The class is called "Hidden Landscapes," and is open to undergraduate honors college students (typically non-majors) as a discussion-based elective course. Together, we spend the semester looking at case studies of under-represented landscapes in our own region. Each student identifies a research topic and then works with observational, mapping and interview methods to develop and improve documentation about their landscape "phenomenon." By centering under-represented landscapes, we also critically discuss the state of the canon in landscape architecture, broadly evaluating access to material and drawing conclusions about perceptions of value in the design disciplines.

This poster shares the process used by students to learn about their own unsung landscapes, and then demonstrates how that research approach may enhance other types of inquiries as well. A list of case studies is shared, as well as the reading material used in the class. Matthew Gandy's articulation of "unintentional landscapes," Gilles Clement's "Third Landscapes" and de Solà-Morales Rubió's conceptualization of "terrain vague" form the basis of our shared reference material. Then, students engage in deeper reading and research about their own landscape project, seeking out people to interview and new ways of seeing those landscapes. The shift in awareness that emerges through this class–about sites of meaning or value that may previously have been under-examined–helps students to see landscapes differently, and also critically examine their own preconceived notions about the design disciplines.

References

Clément, G. (2005). The Third Landscape. Accessed 12.06.2023.
De Solà-Morales Rubió. (1995). Terrain Vague, in Anyplace, ed. Cynthia Davidson. Cambridge: MIT Press.
Gandy, M. (2016). Unintentional landscapes. Landscape Research 41(4), 433-440. DOI: 10.1080/01426397.2016.1156069

EXAMINING ECOVILAGES: THE DESIGN AND IMPLEMENTATION OF EQUITABLE AND ACCESSIBLE SUSTAINABLE COMMUNITIES

Track 5: Spatial Justice, Equity, and Inclusion by Design
Knowledge Network: Participation, Sustainable Planning, Design & Behavior
Methods: Case Study, Mixed Methods
Sophie Kramer (University of Massachusetts Amherst)

Backgrounds:

The benefits of living in an Ecovillage are innumerable, but how can those benefits be accessed if the Ecovillage itself is inaccessible?

As the Earth is increasingly devastated by the impacts of climate change, evidence continues to show that specific communities are more affected. These are the very communities that have historically experienced inequity: including low-income households, communities of color, people with disabilities, and immigrant populations (Gamble 2016). One community-based approach to addressing climate change has been the founding and development of ecovillages-sustainable cohousing communities with shared values and resources centering the environment (Walker 2005). There are more than 10,000 of these ecovillages globally, with a concentration of the communities in the Global North. The reality for many of these communities is that these ecovillages, "despite the frequently expressed interest in diversity, have a homogenous profile that is mainly middle or upper middle class, ethnically 'white' with higher education levels" (Dias et al. 2017).

This research addresses the barriers to DEIA (diversity, equity, inclusion, and accessibility) in sustainable community design. Cohousing communities, including ecovillages, throughout the Eastern United States, served as precedents for this work, and interviews as well as participatory design methodologies, created a framework for developing recommendations. This poster identifies strategies for making inroads into DEIA in ecovillage communities. These suggestions address the needs of local inhabitants, foster the creation of a space that is welcoming to diverse populations, and support inclusive strategies for groups of people that have typically not been well represented in cohousing neighborhoods. Only once ecovillages are welcoming to all can they fully tackle both environmental and social sustainability.

References:

Dias, M. A., Loureiro, C. F. B., Chevitarese, L., & Souza, C. D. M. E. (2017). The Meaning and Relevance of Ecovillages for the Construction of Sustainable Societal Alternatives. Ambiente & Sociedade, 20, 79–96. https://doi.org/10.1590/1809-4422ASOC0083V2032017

Gamble, J. L., Balbus, J., Berger, M., Bouye, K., Campbell, V., Chief, K., Conlon, K., Crimmins, A., Flanagan, B., Gonzalez-Maddux, C., Hallisey, E., Hutchins, S., Jantarasami, L., Khoury, S., Kiefer, M., Kolling, J., Lynn, K., Manangan, A., McDonald, M., ... Wolkin, A. F. (2016). Ch. 9: Populations of Concern. The Impacts of Climate Change on Human Health in the United States: A Scientific Assessment. U.S. Global Change Research Program. https://doi.org/10.7930/J0Q81B0T

Walker, L. (2005). Ecovillage at Ithaca: Pioneering a Sustainable Culture. Gabriola, BC: New Society Publishers.

ENHANCING SUSTAINABILITY IN A DENTAL SCHOOL BUILDING CONSIDERING LEED-BASED DESIGN STRATEGIES

Track 2: Sustainable Buildings and Infrastructure
Knowledge Network: Health, Sustainable Planning, Design & Behavior
Methods: Case Study

Sepideh Niknia (Texas Tech university); Pariya Sheykhmaleki (Texas Tech University); Debajyoti Pati (Texas Tech University); Armin Piriyaei (Texas Tech University)

Abstract:

An increasing danger to the environment is caused by the built environment impacting on user's function and well-being and earth by significant carbon emissions and resource usage. In order to solve this problem, the green building movement has developed as a paradigm change focused on reducing environmental effects and improving every aspect of building. Significant environmental, economic, financial, and social gains are anticipated as a result of this transformation. Central to this transformation is the implementation of green building rating systems, such as the LEED-based Green Building Rating System, which serves as a comprehensive framework and tool for project teams striving to achieve sustainable development goals.

The presented methodology in this poster introduces an innovative approach, elucidating how integrated design strategies, guided by the LEED-based Rating System, can significantly enhance building performance. The Dental School Building serves as a meticulously designed case study, showcasing the practical application of forward-thinking design strategies. These include Algae panels adorning the building envelope, Solar Panels harnessing renewable energy, a Roof Garden providing green space, Rainwater collection systems for sustainable water management, and a Wind Turbine contributing to renewable energy generation.

Crucially, the design doesn't stop at the structural and environmental aspects. The well-being and functionality of the users, encompassing both students and patients, are carefully considered. This involves the incorporation of more respite spaces, an increase in social engagement areas, and the seamless integration of green outdoor perspectives to balance potentially stressful indoor spaces, such as dental units.

The collective impact of these integrated strategies is multifaceted. They not only contribute to increased renewable energy production and effective rainwater management but also play a role in mitigating heat island effects, optimizing overall energy performance, and addressing user well-being concerns. By showcasing the successful integration of these strategies, the presentation underscores the immense potential for substantial environmental gains within the construction industry, laying the groundwork for a future characterized by sustainable and eco-conscious buildings.

Moreover, the adoption of the Triple Bottom Line Concept has further shaped the sustainability of the Dental School Building. This holistic approach emphasizes a reduction in energy consumption, the enhancement of energy efficiency, the promotion of self-sufficiency, and the fostering of social interaction, all while minimizing stress burdens. This comprehensive strategy not only aligns with environmental goals but also reflects a broader commitment to creating spaces that are economically viable, socially inclusive, and environmentally responsible.

References:

Campagnol, G., & Shepley, M. M. (2014). Positive distraction and the rehabilitation hospitals of João Filgueiras Lima.HERD: Health Environments Research & Design Journal, 8(1), 199-227.
Cottrell, M. (2014). Guide to the LEED green associate V4 exam. John Wiley & Sons.
Setyowati, E., Hardiman, G., & Budihardjo, M. A. (2019). On the role of acoustical improvement and surface morphology of seashell composite panel for interior applications in buildings. Buildings, 9(3), 71.
Habib, F., Alborzi, F., & Etessam, I. (2013). Light Processing in Iranian Houses, manifestation of meanings and concepts.

NIGHT MIXED USE

Track 2: Sustainable Buildings and Infrastructure
Knowledge Network: Residential Environments, Sustainable Planning, Design & Behavior
Methods: Mixed Methods
Noura Philip Bishay (NPB Design Studio)

Background:
The emergence of Night Mixed-Use Buildings marks a departure from traditional structures that remain dormant after daylight hours. These innovative spaces challenge the notion of buildings operating solely during the day by repurposing them to serve the homeless population or low-income community during nighttime. Not only does this approach offer a solution to homelessness, but it also maximizes the use of infrastructure, transforming these spaces into multifunctional hubs. By accommodating the homeless at night and utilizing the building for other purposes during the day, it reflects a contemporary understanding of continuous, versatile building occupancy to address societal needs effectively. Ideal candidates of spaces are city owned libraries, schools, parks, parking lots and structures.

References:
Urban Land Institute. (2019). Understanding Mixed Use and Multi Use. https://knowledge.uli.org/-/media/files/reading-list/reading-list-pdfs/readinglist_mixeduse

THE RELATIONSHIP BETWEEN THE INDOOR COURTYARD GARDENS IN NURSING HOMES AND SENIORS' HEALTH

Track 4: Design for Health and Wellbeing
Knowledge Network: Environment-Gerontology, Health
Methods: Case Study
Shireen Kanakri (Ball State University); Adrienne Barron (Interior Designer)

Background:
The garden setting that is available throughout all seasons will promote and stimulate exercise, social interaction, and mental acuity with anyone that interacts with this type of environment. The presence of live plants has been associated with decreasing stress and stabilizing sleep wake cycles among seniors. The interaction of working with plants will improve the physical activity, social activity, and health of anyone, especially seniors in nursing homes. Given that the elderly has declining health, it is essential that this type of environment is installed within all nursing homes. The goal of this research is to find the relation between the health and wellbeing of the elderly living in nursing homes by using indoor courtyards. The hypothesis will be testing if there is an impact between indoor year-round courtyards gardens in nursing homes on residents emotional, social, and physical wellbeing, compared to outdoor gardens that can be utilized only a few months out of the year. Case study method used to conduct this research. Aa strong relationship found between the indoor year-round courtyards gardens use and seniors improved health.

Keywords: *indoor garden setting; social interaction; mental acuity; physical activity; live plants; nursing homes*

NATURAL CURIOSITY: INVESTIGATING HOW PLAYFUL NATURE INTERACTIONS INFLUENCE INFORMAL STEAM LEARNING DURING PRESCHOOL AGE

Track 1:Nature and Urban Ecology
Knowledge Network: Children, Youth & Environment, Environmental Design Research Education
Methods: Experimental/Quasi-Experimental Research, Mixed Methods

Nazia Afrin Trina (Texas Tech University)

Abstract:
The early years of a child's life are crucial for developing positive attitudes towards science, technology, engineering, arts, and mathematics (STEAM), as these perceptions are often formed before elementary school and can be challenging to alter later on. Research indicates that informal science activities foster long-term interest and learning in STEM fields. Nature-based STEAM learning is particularly beneficial for children's development and academic success, as it aligns with their innate tendencies to explore, question, and understand the world around them.

These natural inclinations are the foundation of informal STEAM learning, which is supported by the Experiential Learning Theory (ELT). ELT emphasizes the importance of direct experiences in learning, particularly in outdoor settings where the environment itself can be designed to facilitate STEAM education. The physical aspects of outdoor learning spaces are, therefore, critical, as they can provide opportunities for experiential learning that are essential for cognitive development in young children. However, assessing and evaluating the cognitive achievements of young children in STEAM is complex due to the significant variations in personal, developmental, and environmental factors that influence their behavior. There is a growing concern about the misuse of testing and the interpretation of results in early childhood education.

As a result, there is a pressing need for valid and reliable indicators to assess the quality of STEAM learning, especially concerning outdoor environments. Despite the increased focus and funding for early STEAM education, there is a notable gap in assessment guidelines for STEAM-based outdoor learning. This research aims to fill this void by identifying suitable indicators and methods for evaluating preschool children's learning progress during outdoor STEAM activities and determining how outdoor learning landscapes can support STEAM education. This is essential for ensuring that educational support in the early years improves school readiness and achievement in STEAM subjects.

EXPLORING DIFFERENCES IN FOREST LANDOWNERS' PROPENSITY TO PERFORM FUELS REDUCTION UNDER DIFFERENT INCENTIVE REGIMES

Track 3: Resilient Design for Climate Change
Knowledge Network: Sustainable Planning, Design & Behavior
Methods: Photo-based Methods, Statistical/Modeling Analysis
Robert Ribe (University of Oregon); Bart R. Johnson (U. Oregon); Max Nielson-Pincus (Portland State University)

Background:

As wildfires' extent and severity grow, forest fuels reduction is widely promoted to mitigate catastrophic risks. Fuels reduction is a common pool resource. More projects across ownerships produce increasing gains in everyone's protection (Charnley et al., 2020). Incentives are likely required to get sufficient participation. Efficient approaches would target incentives to avoid unnecessarily paying willing owners while bringing in recalcitrant owners. Is this possible (Hamilton et al., 2019)?

Research employed a mail survey of 361 landowners in the foothills of the Willamette Valley, Oregon. Respondents identified the forest types they owned and were offered corresponding alternative fuels reduction projects that included tree removal. Photo-illustrations, estimated cost ranges, and expected changes in wildfire risk, habitat values, and property rights risks were provided. They rated the probability of performing each project across four accumulating incentive sets: (1) bear all costs, (2) no costs due to subsidies or market conditions, (3) additional guarantee of a forest-type-appropriate minimum profit/acre, (4) additional offer of no lost property rights. Respondents answered questions about ownership motivations, their property, political attitudes about wildfire risks and climate change, perception of forests' values and aesthetics, and demographics.

For each of the incentive sets, a best regression model was estimated across all the rated projects. The dependent variable for the first 'bear all costs' model was the respondents' corresponding rated probabilities. The dependent variables for the other models were the gain or loss in rated probabilities each incentive set produced against the first 'bear all costs' model. Independent variables were selected from all the other question responses or project information items in the survey. Comparing the models in the table below suggests how landowners might respond to different incentives: Unincentivized landowners are prone to reduce fuels if wildfire risks can be substantially reduced, if costs are reasonable, and the post-project forest is acceptably beautiful. It helps if they are more nature-centric, have experienced adverse wildfire impacts, and do not feel hopelessly vulnerable.

Landowners recruitable by no-cost incentives have larger parcels, more education, and higher incomes, more so if corporate managers, not farmers. It helps if owners are optimistic about post-project habitats, favorably predisposed to ground fuels reduction, more concerned about costs, and female. Owners averse to no-cost incentives tend to be conservatives and unimpressed by available wildfire risk reductions with small financial gains.

All the above factors, except gender, apply again to assured profit incentives. Owners averse to such incentives don't like the beauty of post-project forests, feel they are already protected, favor leaving nature alone, are anthropocentric, feel hopelessly vulnerable, are concerned about lost property rights, and wish to avoid the hassle.

Added property rights protections tend to be favored by owners with more education and larger parcels but not higher incomes. Recruited owners here tend to be financially stressed, prioritize land investment, concerned about keeping property rights, active land managers, predisposed toward ground fuels reduction, and concerned about costs. Owners averse to responding to property rights guarantees are conservative and disinterested in post-project wildfire risk reduction and habitat values.

RESPONSIBLE AND COLLECTIVE DESIGN PROCESS: A PATHWAY TO MITIGATE CLIMATE CHANGE

Track 3: Resilient Design for Climate Change
Knowledge Network: Interior Design, Sustainable Planning, Design & Behavior
Methods: Design Methods, Practice-based Research

Andrea Sosa Fontaine (Kent State University)

Abstract:
Design professionals increasingly acknowledge the severity of climate change, and their professional social responsibility to mitigate continued contributions of the built environment to the climate crisis. However, there are accepted and normative practices within architecture and interiors that have yet to be modified to respond to this ethical and professional responsibility. If we are to design with resiliency, perhaps we need to reconsider our foundational design practices.

Methodology:
Analysis of the typical professional design process for interior design and architecture, within North America, reveals that design process overwhelmingly follows a standard phased practice, including variations on the phases of design; predesign, programming, schematic design, design development, bidding, and contract administration. A similar analysis of interior design and architecture pedagogy, accreditation standards (CIDA, 2022) reveals that the same phased design process continues to be the default practice integrated in design pedagogy. Additionally, this process is tested for on licensing examinations and appears in standard contracts for design services, such as in the AIA A201 contract, which notes that obligations of design services are complete at the phase of building inspection or occupancy (AIA, 2017).

Implications:
This research and analysis reveal that although the typical phased process used by interior designers and architects, were implemented prior to the prevalence of sustainable design practices, minimal modifications have been made in response to new circumstances. Specifically, since the typical design process ends at occupancy, there are minimal barriers for the interior designer and architect to free themselves of moral and social responsibility for the impact of the space on the climate crisis. Currently ,both contractually, and as a standard of professional practice, the relationship between design professional and occupants of the space has no obligation to continue after construction is complete.

Certainly, there has been some movement toward climate responsibility by design professionals, through enhanced certification of both individual and construction project. At the same time, there has also been some movement in enhanced responsibility of design professionals though the work of The Architecture Lobby, in advocating for the designation of design professionals as climate resilient works within the Green New Deal(The Architecture Lobby, 2023).

Responsible design practices for resilient futures in the built environment, includes a collective, human centered approach, where each stakeholder, whether it be owner, occupant, designer, architect, or politician, holds a collective and equal responsibility for the climate impact of the space, even after construction is complete. Furthermore, we need to include discussions of these moral obligations in design pedagogy, as a precursor to foster change in professional practice. When we acknowledge our collective responsibility in building resilient futures amid the climate crisis, we stand a better change to sustain the future of human inhabitation on this earth.

References:
American Institute of Architects(AIA) (2017). Contract A201-2017 General Conditions of the Contract for Construction. https://www.aiacontracts.com/contract-documents/22076-general-conditions-of-the-contract-for-construction
The Architecture Lobby (2023). "The Climate Resilience Workforce Should Include Architects." Op-Ed, in The Architect's Newspaper. November 28th, 2023. https://www.archpaper.com/2023/11/climate-resilience-workforce-should-include-architects/
Council for Interior Design Accreditation (2022). Professional Standards 2022. https://www.accredit-id.org/professional-standards

THE TRANSFORMATION OF VACANT LOTS INTO URBAN GREENWAYS IN DETROIT NEIGHBORHOODS

Track 1: Nature and Urban Ecology
Knowledge Network: Environmental Design Research Education,
Movement in Designed Environments: Nature & Ecology
Methods: Case Study

Shno Karimi (Michigan State University); Fatemeh Saeidi-Rizi (Michigan State University);
Jun-Hyun Kim (Michigan State University)

Background:
In the context of contemporary American cities, there exists a discernible pattern of unequal population redistribution dynamics. Over time, neighborhoods within a given urban area exhibit fluctuations, including population growth, decline, relative constancy, and demographic shifts. (Schilling & Logan, 2008) Post-industrial cities like Detroit have experienced substantial decentralization, witnessing significant population loss, economic shifts, and industrial decline from 1950 to 2010 (Hall, 2010). Approximately 165,000 industrial jobs have been lost, as well as 147,000 housing units in Detroit, an emblematic city in this trend. Since 2010, Detroit has seen a population decline of over 50%. According to the World Population Review of 2023, the current population stands at 621,193, an indication of continued decline.

It is estimated that 32% of Detroit's landmass is vacant, resulting in urban voids or negative spaces called 'urban shrapnel' (Newman & Kim, 2017). The physical characteristics of these vacant parcels, such as size, shape, and location, contribute to long-term vacancy, perpetuating future vacant land accumulation (Lee, Newman, & Park, 2018). Surpluses in vacant land pose challenges for financially distressed cities, leading to urban decline triggered by factors like depopulation, ineffective regeneration policies, economic downturns, and decentralization (Zhang, Yuan, & Li, 2015; Audirac, 2007; Németh & Langhorst, 2014).

The presence of long-term vacant land exacerbates issues like crime, reduced quality of life, visual blight, and safety concerns (Kivell, 1993). These factors contribute to declining property values, further depopulation, and erosion of the tax base, establishing a self-reinforcing cycle where vacant land perpetuates its own persistence (Immergluck, 2016).

There are an estimated 24 to 40 square miles of vacant and abandoned land in the 139-square-mile city of Detroit, which has faced a high number of vacant and abandoned lots for decades (Detroit Free Press). Complex and multifaceted reasons contribute to this abundance, including the decline of the auto industry, once the city's economic backbone, leading to a population exodus and surplus housing and commercial properties.

This study explores creative solutions to transform these vacant lots into valuable community assets. Recognizing the challenges posed by excessive vacant land, we propose a novel approach focusing on converting vacant lots into urban greenways. This approach aligns with the principles of sustainable urban development, aiming to enhance community well-being, address safety concerns, and revitalize neighborhoods.

In this study, we examine the feasibility and potential benefits of repurposing vacant lots into green spaces, taking into account environmental, social, and economic factors. Our goal is to develop a comprehensive strategy that addresses the negative externalities associated with vacant land while also contributing to the renewal of the city by engaging local communities, policymakers, and urban planners. This innovative approach seeks to redefine vacant lots as opportunities for positive transformation, fostering a more sustainable and resilient urban landscape in the face of ongoing demographic and economic challenges.

Keywords: *Urban Greenways, Vacant Lots, Sustainable Urban Development, Vacant Land Transformation, Post-Industrial Cities*

A COLLABORATIVE DESIGN BUILD EFFORT TO REVITALIZE A ONCE VIBRANT URBAN COMMUNITY

Track 5: Spatial Justice, Equity, and Inclusion by Design
Knowledge Network: Environmental Design Research Education, Interior Design
Methods: Action/Participatory Research, Practice-based Research
Denise McAllister (Texas State University)

Abstract:
An exciting project is underway in a midsize, midwestern town to provide low-income persons who might not otherwise be able to purchase homes the opportunity to be first time homeowners. Students and faculty from three different departments of a local university are collaborating on the project to bring resources and expertise in the areas of Historic Preservation, Construction Management, and Interior Design.

A Purpose-Built Community (PBC) model (Buck, 2022) is being used to resolve challenges facing a previously vibrant south side neighborhood. The project, known as The Porch Initiative (People Organized to Revitalize Community Healing) is providing opportunities for home ownership for low-income residents. Diversity is honored in this historic community where families of different traditions are building lives together. At the foundation of the project is collective eco-friendly stewardship with the intent being to reuse exciting homes with as little new construction as possible. The historic community is a rich source of homes, churches, schools, and mixed-use buildings which originally served as retail spaces with owners living above. Bakeries, grocery stores, small mercantile and the like supported residents of the once vibrant area. From the remnants of this once thriving neighborhood, a new community is beginning to evolve. Participation of historic preservationists provide an opportunity to educate students and community members regarding the history and future potential of this once vital community. These efforts are informing a new generation of homeowners, for whom the existence such a walkable community of homes, multi-use commercial buildings, churches and even schools are not even a distance memory.

Recent data is indicating that similar revitalization efforts globally are proven to increase property values exponentially (Agnieszka Bieda, 2023)This anticipated increase in values will support the initiative's primary goal of giving residents a hand up out of poverty.
Interior design students are providing space planning services to maximize the available square footage of the houses which were typically built in the 193's through the 1950's. The homes are relatively small, usually two bedrooms with one bath. While in disrepair, with many currently abandoned, the homes retain much of their original structural integrity and inherent charm. Arched doorways, expansive double hung windows and charming porches still cling to these soon to be restored homes. The designers are modernizing the kitchen and bath layouts, while specifying environmentally friendly cabinetry, fixtures, and appliances. They are also working to create layouts and product specifications for gender neutral bedroom design to provide opportunities for sharing space while maximizing privacy and storage. Off season storage will be planned for basement and cellar areas where available. Implementing a design-build project delivery method, the design team is working together with the builders to insure unity and collaboration throughout the process (Design Build Project Delivery, 2024) Representatives from Construction Management are participating in the cost estimating and scheduling while Historic Preservation is educating the entire construction and design team regarding the wealth of cultural history in the area being impacted.

THIRD PLACE FOR THE ELDERLY: A STUDY ON THE CONCENTRATION PATTERN OF ELDERLY POPULATION IN THE NAKWON ARCADE AREA IN SEOUL AND IMPROVEMENT STRATEGIES FOR THE REGION

Track 1: Nature and Urban Ecology
Knowledge Network: Cultural Aspects of Design, Sustainable Planning, Design & Behavior
Methods: Mixed Methods, Observation
Soyeon Kim (Yonsei University); Suk-kyung Kim (School of Planning, Design, & Construction, Michigan State University); Seho Park (Yonsei University)

Abstract:
Modern architectural buildings in Korea were constructed to fit the urban systems prior to industrialization. While they hold important historical meaning, they face issues of maintenance due to economic challenges and building deterioration in old downtown areas. This has led to the decline of local communities, which has been exacerbated by low birth rates and the aging population.

The Nakwon Arcade in Seoul, Korea has been known for having a lot of large musical instrument stores and representative modern commercial buildings in Seoul. However, the activities from older adults who have been killing their times in the area have contributed to the slumification around the Nakwon area. If left unattended, this high elderly density could become detrimental to local development through environmental issues and landscape degradation.

This study aims to propose environmental design solutions for the slumification of the Nakwon Arcade area in Jongro district in Seoul caused by the elderly population. It adapted the 'Third Place' theory as the research frame. The concept of 'Third Place' refers to important places outside of home and work, as introduced in Ray Oldenburg's "The Great Good Place" (1989). Through a literature review of previous studies, this research identified six characteristics of 'Third Place': Openness, Interactivity, Accessibility, Comfort, Playfulness, and Neutrality.

The research method employed observations and behavioral mapping. The Nakwon area is divided into three locations: (1) Nakheuidong, (2) Tapgol Park Exterior, (3) Tapgol Park Interior. They are located in the middle of old downtown in Seoul. Each location was visited six times on weekdays and weekends at 10am, 12pm, and 7pm. Two researchers recorded user distribution patterns through photographs taken at different time slots. Through observations, this study identified the physical elements that provide 'Third Place' characteristics to the elderly at each location. The results from the analysis of behavioral maps showed the density of the elderly population across different time periods, behavioral patterns, and activity types. For behavioral mapping, this study referred to previous studies that identified types of users' behaviors in urban open spaces.

Primary results are as follows: First, it identified physical elements that support 'Third Place' for the elderly. People tended to gather around typical furniture arrangements to engage in activities. Second, it identified the areas with high elderly density and also the types of behaviors exhibited there during different time periods.

This study finally suggested the improvement plan for the densely-populated area with many older adults in Nakwon Arcade in Seoul. Main elements to be considered are territoriality and sense of ownership of the spaces by the elderly. Aligned with this direction, the study suggests installing tangible design elements, such as customized seating areas, partitions, and activity zones. This can lead to improving spatial usage and enhancing place attachment for elderly residents.

References
1. Lee,H.(2015). Cultural and Creative Space for Revitalizing idle Space-1. Focused on Busan City-. Journal of the Korean Society Design Culture, 21(4), 549-559.

2. Oldenburg, R.,& Brissett,D.(1982). The third place. Qualitative sociology, 5(4), 265-284.

APPLYING CULTURALLY SENSITIVE DESIGN IN INTERIOR SPACES FOR NATIVE AMERICANS

Track 5: Spatial Justice, Equity, and Inclusion by Design
Knowledge Network: Cultural Aspects of Design, Interior Design
Methods: Focus Groups, Survey

Jenna Nicole Warneke (Oklahoma State University); Tilanka Chandrasekera (Oklahoma State University); Paulette Robert Hebert (Oklahoma State University); Aditya Jayadas (Oklahoma State Univeristy)

Background:

The United States Census Bureau reported that in 2023, out of the total 4,053,824 individuals who reside in Oklahoma, 9.5% of the population in Oklahoma belong to one of the 39 different Native American tribal nations (U.S. Census Bureau, 2023). Native Americans who reside on tribal lands and other rural areas often face unique housing challenges (Findling et al., 2022). Oftentimes, spaces are designed for Native Americans without applying culturally sensitive design principles (Young, 2022). Purpose: This study aimed to understand better the design preferences of interior spaces for Native Americans.

Methodology:

The study was carried out in two phases: a) a focus group study with open-ended questions on design preferences, and b) a survey on preference of design attributes for interior spaces. Eight participants were recruited for the focus group study. Four of the eight participants identified as Native American and were enrolled with a tribal nation (18-25), and four identified as Non-Native American (18-52). The participants were asked nine open-ended questions on preferences for interior spaces. The focus group session was audio recorded. The data was then transcribed, and content analysis was used to determine themes and sub-themes. Next, nine Native Americans (age: 18-39) and 24 Non-Native Americans (age: 18-52) were recruited for the survey. The survey was conducted using 20 Likert scale questions on perceived preferences for the design of interior spaces. The survey data was analyzed using a two-sample unequal variance t-test with a significance level of 0.05. Results: The qualitative data from the focus group showed that while Native Americans valued artifacts such as tribal artwork and blankets, Non-Native Americans preferred lots of color and bigger spaces. The other big difference was the Native Americans preferred plants and natural elements, while Non-Native Americans preferred hardwood and marble. As for the quantitative data, there was a statistically significant difference in the preferences of historical items ($p=0.03$), with Native Americans reporting a higher score than Non-Native Americans. There was a statistically significant difference for a non-visual attribute, the smell of sage and cedar in interior spaces ($p=0.02$), with Native Americans preferring this more when compared to Non-Native Americans.

Discussion:

The Native American participants in the study preferred to incorporate culture, tradition, and religion to design interior spaces, while Non-Native American participants preferred materialistic options such as lots of color choices and larger spaces. Implications: This study has important implications for designing interior spaces for Native and Non-Native Americans using culturally sensitive design.

USING ABM TO SIMULATE STUDENT BEHAVIOR TO STUDY THE SPACE CONFIGURATION OF 'HOMEBASE' IN HIGH SCHOOL.-FOCUSING ON HIGH SCHOOL CREDIT SYSTEM IN SOUTH KOREA-

Track 2: Sustainable Buildings and Infrastructure
Knowledge Network: Cyberspace & Digital Environments, Sustainable Planning, Design & Behavior
Methods: Design Methods, Statistical/Modeling Analysis
Yunjeong Hwang (University of Yonsei); Seohyeon Lee (epartment of Interior Archictecture and Built Environment, University of Yonsei); Gidong Byun (YONSEI UNIVERSITY); MiKyoung Ha (Yonsei University)

Abstract:
In response to 'digital transformation', 'climate change', and 'declining school-age population due to low birthrate', South Korea has changed its educational curriculum overall to strengthen learner-customized education and improve competencies required for the future society. At a time when a fundamental change in the education paradigm is required due to the evolution of the times, the 'High School Credit System' announced by the Ministry of Education called for the reform of existing policies as well as the creation of new school spaces.

The importance of making the Homebase environment suitable for students was emphasized, especially in terms of supporting a variety of flexible functions as well as community among students during breaks and between classes. It is notable that the interaction with fellow students that occurs in the Homebase is an ideal space to develop the interpersonal and social skills required in adolescence, at a time when students are required to move to their individual classrooms for every class. However, currently, the focus of the Homebase is on simply providing a space rather than considering the needs of the students in depth.

Therefore, considering the long-term social and economic effects, there is a need for detailed planning of various space configurations for Homebase. For this reason, this study is basic research on homebase as a space that supports interaction between students, and aims to understand and comparison the effects of different types of space configurations by simulating students' behavior during break time using ABM simulation. ABM facilitates the analysis of complex social phenomena in that it can derive and explain macroscopic outcomes from the behavior of micro-level actors. In this study, 'Seoul Gonghang High School' was selected as the target site because it has been in operation for a long time and has secured representativeness among the schools implementing the high school credit system. In the simulation, when it is time for break, agents (students) perform various resting behaviors according to their decision-making stages (behavior tree). When each agent (student) enters Homebase and completes a given task, it is recorded as a dot, which means the number of people who visited Homebase during the break. Finally, the simulation ends when all agents have finished their decision-making process. When explaining the appropriateness of the designated space, the number of visitors entering the space has descriptive capacity, as it is the most primary factor that can support opportunities for communication and interaction between students.

The results of this study are different from previous studies in that it attempts to verify the utilization and feasibility of the space by processing architectural and non-architectural elements of the space at once through simulation, but it is limited in that the personal characteristics of the agents were extracted and applied from previous studies. In future studies, it may be helpful to find sustainable learning community spaces if actor characteristics are specified in more detail based on empirical data.

PLANNING INDICATORS FOR WORKPLACE OF "HEALTH CENTERS SEAMLESSLY LINKING DAILY LIFE AND DISASTER SITUATIONS" -HEALTH CENTER ARCHITECTURE WITH RESILIENT CAPABILITIES-

Track 3: Resilient Design for Climate Change
Knowledge Network: Researchers in Professional Practice, Sustainable Planning, Design & Behavior
Methods: Design Methods, Survey
Ayako Kita (Kochi University of Technology); Kenji Kobayashi (Setsunan University); Junko Ikeuchi (Setsunan University)

Background:

On New Year's Day this year, the news of an unexpectedly large earthquake at Noto Peninsula left all over Japan speechless, and the aftermath is still unfathomable. This has given us an opportunity to rethink what it means to be prepared for disasters. The Great Hanshin-Awaji Earthquake (1995) prompted Japan's disaster medical system to focus on the establishment of disaster base hospital facilities and the development of Disaster Medical Assistance Team (DMAT) personnel. Subsequently, after the Great East Japan Earthquake (2011) and other disasters, the need for a function to manage overall health and medical care was pointed out, and in 2017, the Ministry of Health, Labor and Welfare presented a model for health and medical activities in affected areas during natural disasters, with "health centers" serving as bases for supporting Disaster Health Emergency Assistance Team (DHEAT) activities, which was adopted by the Ministry of Health, Labour and Welfare in 2017. However, there was no accumulated discussion of "health centers" from the perspective of architecture, and we thought that their structure and functions might not be sufficient to serve as base buildings in times of disaster. Furthermore, local governments across Japan are reviewing and reorganizing public facilities, leading us to consider what a phase-free health center should be in the future.

In this study, we assume that, in considering the redevelopment of health centers, it is necessary to take a viewpoint that does not consider daily life and disasters separately, but rather seamlessly (phase-free), and consider planning indicators for the health center's main workplace. The research method consisted of analyzing room composition and area size from administrative data to ascertain the actual conditions of health centers, and conducting interviews regarding the use of workplaces. The average site area of the health centers studied was 2615 m2 (maximum: 3801 m2, minimum: 1598 m2). The average total floor space is 2399 m2 (maximum: 4633 m2, minimum: 1395 m2). Following this, based on the results, planning conditions for a health center that seamlessly links daily life and times of disaster were presented to three architectural design firms, and a discussion session was held to exchange opinions on their technical proposals. As a conclusion, the following five types of locations were identified as important for a health center that seamlessly links daily life with the increased number of personnel, supplies, and activities that occur only during disasters: 1. large areas such as lobbies, offices, and auditoriums that can be converted into workplaces; 2. areas not equipped with specialized equipment; 3. areas facing the exterior walls; 4. ground-level areas including outdoor areas; 5. Ground level locations, including outdoors; 5. Semi-outdoor locations, such as under eaves. These five points are applicable not only to future health center rebuilds, but also to considerations for upgrading existing health centers.

EXPLORING EFFECTIVE APPROACHES TO REDUCE EDUCATIONAL DIFFICULTIES FACED BY NOMADIC PASTORALIST CHILDREN IN LEARNING ENVIRONMENT: A CASE STUDY IN MONGOLIAN SCHOOLS

Track 2: Sustainable Buildings and Infrastructure
Knowledge Network: Children, Youth & Environment, Cultural Aspects of Design
Methods: Interviews, Observation
Munkhsuuri Amartuvshin (Tokyo institute of technology); Naoko Saio (Tokyo Institute of Technology)

Abstract:
Nomadic pastoralist children in various countries have faced challenges in reconciling formal primary schooling with the traditions of pastoralism, despite the latter's importance in promoting diversity. In Mongolia, this issue is particularly pronounced for elementary school children from nomadic pastoralist families who must attend school in settled areas far from their mobile homes and nomadic lifestyle due to the long commute1. This separation from their parents and way of life can negatively impact their academic interests and lead to social problems such as homesickness. Moreover, inadequate infrastructure in elementary schools has led to overcrowded classrooms and a lack of functional space for children to spend outside-of-class hours, which can affect their personal and educational development.

Previous research mentions that distance learning can provide children with digital resources, eliminating long commutes and circumventing the constraints of inadequate school facilities2. However, there is a lack of research on learning environments combined with distance learning for Mongolian nomadic children in the post-COVID-19 pandemic era. This study aims to explore the current situation of nomadic children's schooling and identify potential ways to create sustainable learning environments that combine distance (DL) and classroom learning (CL). The research question is: To what extent can the learning environment combined with distance learning address the educational challenges faced by nomadic pastoralist children during their schooling? This study comprises three steps. Firstly, it compares the current schooling practices and challenges faced by nomadic pastoralist children in different countries. Secondly, it conducts a case study of three schools in Mongolia through spatial observation and interviews with 25 teachers, 24 parents, and guardians, examining the feasibility of combined learning (CL and DL) approaches and the needs of school facilities for nomadic children. Lastly, the study identifies influencing factors for a sustainable learning environment combined with learning approaches.

The research reveals that nomadic children in many countries face social difficulties with regard to school attendance, living arrangements, and cultural practices. However, the current school environment in Mongolia does not mitigate the social issues, nor does it meet the needs of academic learning and extracurricular activities due to the lack of facilities. To address these challenges effectively, the study suggests incorporating CL+DL approaches in school planning to establish sustainable infrastructure. The research identified that the hybrid learning approach could help alleviate social problems among nomadic children as they study with their parents in their main home in DL. It can also help promote face-to-face communication in CL and facilitate school facilities for both academic and extracurricular activities. The research findings offer insights into creating sustainable learning environments for rural school planning, particularly in sparsely populated countries. These insights can be applied in culturally and educationally similar contexts, with special consideration for the unique needs of ethnic minorities, such as nomadic pastoralist children.

References:
1. Ahearn, A., & Bumochir, D. (2016). Contradictions in Schooling Children among Mongolian pastoralists. Human Organization, 75(1)

2. Sadeghi, M. (2019). A Shift from Classroom to Distance Learning: Advantages and Limitations. International Journal of Research in English Education, 4 (1)

A CASE STUDY OF URBAN REGENERATION CULTURE AND ART SPACE FROM THE PERSPECTIVE OF AMENITY

Track 5: Spatial Justice, Equity, and Inclusion by Design
Knowledge Network: Interior Design
Methods: Case Study

Seung Hee NAM (Hanyang University); Youlim Hwang (Hanyang University); YeonSook Hwang (Hanyang University)

Background:

Recently, Korea's declining urban space has been regenerated as a space for the creation of culture and arts in consideration of amenities. Amenity does not mean simple convenience facilities or specific facilities, but a comprehensive living environment that integrates harmony, comfort, and beauty with the surrounding environment. Space regeneration using amenities provides a friendly and comfortable environment for people to enjoy, and promotes new urban cultural experiences.

The purpose of this study is to analyze cultural spaces using urban regeneration in Seoul from the perspective of amenities and to suggest ways to revitalize cultural spaces. To analyze the urban regeneration culture space from the perspective of amenity, an analysis framework was based on the characteristics of environment, aesthetics, convenience, and culture. An on-site survey was conducted by seven case studies that were reopened as urban regeneration spaces belonging to the Seoul Cultural Foundation.

The results of this study are as follows. First, from the perspective of amenity, it can be classified into environmental, aesthetic, convenient, and cultural characteristics of urban regeneration culture spaces. Environmental characteristic provides people with a sense of comfort and psychological stability by providing a nature-friendly environment and minimizing costs and environmental pollution. Aesthetic characteristic provides the comfort of facilities such as natural landscapes and public sculptures in harmony with space. Convenient characteristic provides easily accessible to users by forming an online network that can quickly obtain a diversity of information and public transportation. Cultural characteristic promotes convergence and cooperation of local communities by user's participation in cultural activities and education. Second, urban regeneration culture spaces are composed of exhibition, creation, education, rest, sales, and community spaces. Third, among the characteristics of urban regeneration culture, aesthetic characteristic was the best and environmental characteristic was the most insufficient. In order to promote amenities, the environment that can provide users with not only comfort, but also various possibilities are needed by creating a community and a rest area that emphasizes eco-friendly landscaping.

For a sustainable environment of cultural spaces, efforts are needed to improve the comfort of urban regeneration cultural spaces by cooperating with many citizens and artists to develop shared programs and spaces.

Keywords:
Amenity Perspective, Culture and Arts Space, Urban Regeneration

References:

Kim,S,H, & Kwon Y.J. & Kim, J. (2018). A study on the characteristics of interior space design in Seoul subway stations from the perspective of cultural and artistic amenities. Journal of the Korean Society of Space Design.

Jung J,E. (2021). Research on revitalizing the operation of cultural and artistic spaces as cultural urban regeneration: focusing on the case analysis of the Incheon Art Platform. Culture Industry Research.

Kim K.S & Kim M.S. (2015). A study on ways to improve urban amenities through case analysis of activation of idle space on the lower part of an overpass. Research in digital design.

INCLUSIVE BY DESIGN: EXPLORING LACTATION ROOM DESIGN AND ADHERENCE TO SUPPORT SPATIAL EQUITY AND NURSING MOTHERS' NEEDS

Track 5: Spatial Justice, Equity, and Inclusion by Design
Knowledge Network: Interior Design
Methods: Case Study, Photo-based Methods
Molly Cai (Indiana University); Miyoung Hong (Indiana University Bloomington)

Objective:
To evaluate lactation room spaces in Midwest U.S. public sector buildings in for compliance with interior design recommendations and their ability to foster a supportive environment for nursing mothers.

Background
As paid maternity leave time in the U.S. lags behind other developed countries, the development of lactation rooms for nursing mothers' is a physical representation of spatial equity by design.

Existing literature for lactation rooms emphasize emotional design elements to create a supportive and nurturing atmosphere (AIA, 2019), accessibility of location and abundance of spaces (Hoffman & Hoffman, 2023), and comfort through biophilic design (WELL, 2020). However, implementation discrepancies persist due to the lack of universal regulations. This situation prompts the research questions, including: 1. How do lactation spaces differ and compare to current interior design recommendations (AIA, 2019; WELL, 2020)? 2. How well do these recommendations promote designed equity for nursing mothers? 3. How might lactation room spaces be utilized to support spatial equity?

Justification
In the height of gender equity movements, lactation rooms are spaces to show organizational support of mothers returning to the workplace and public spaces, therefore analyzing adherence and effectiveness of interior design recommendations is crucial to furthering designed equity.

Method
A total of 14 lactation rooms were analyzed by using design recommendations by AIA (2019), and WELL (2020). These features included location, room size, function, acoustical and visual privacy, materiality, furnishings, lighting and additional amenities.

We narrowed down the data to 8 lactation room cases through quantitative and qualitative data from online public websites, comprehensive photo studies, and in-person observations. These encompassed diverse public sectors, including healthcare, academic, and recreational spaces situated in larger cities in the Midwest.

Discussion
While all analyzed lactation spaces met the legal requirement of being a private and dedicated space other than a restroom, they varied widely in their quality of restorative design features. Concerns arose regarding visual and acoustical privacy due to being in high-traffic areas such as near restrooms, public lounges, kitchenettes, and elevators.

7 of the 8 cases satisfied the size recommendation of being 7' x 7' or larger. The lactation rooms fell into three major typologies: independent rooms, multipurpose rooms, and rooms accessed through women's restrooms. Most rooms were accommodations that retrofitted existing spaces particularly restrooms. And while biophilic features referencing nature such as indoor plants, artwork, and material patterns were present, they were negligently added.

Implications
Design practitioners in public sector buildings, especially in higher education, healthcare, and recreation, should prioritize the strategic placement and design of lactation rooms. Emphasizing accessibility while maintaining privacy and incorporating elements like task lighting and enhanced biophilic features is crucial for creating supportive environments for nursing mothers.

Conclusion
Lactation rooms counteract maternal leave inequities, promoting spatial equity. In supporting nursing mothers across the United States, the next crucial step lies in the thoughtful implementation of design features to ensure that lactation rooms consistently provide the privacy and comfort that every mother deserves, transforming these spaces into sanctuaries of empowerment and care.

RESTAURANT DESIGN FEATURES IMPACTING THE EXPERIENCE OF HARD-OF-HEARING INDIVIDUALS

Track 5: Spatial Justice, Equity, and Inclusion by Design
Knowledge Network: Interior Design
Methods: Observation

Lorin Kmieciak (Illinois State University); Olivia Ross (Illinois State University); Alaina Zajac (Illinois State); Reem Bagais (Illinois State University)

Background:

Restaurants are popular social gathering spaces that need to be equipped to handle a wide variety of people on a daily basis. According to US Foods, 63% of people care most about atmosphere and experience when dining out (The Dining Dispatch: 2023 American Dining Habits, 2023). Additionally, the National Institute of Deafness and Other Communication Disorders stated that about 15% of American adults, approximately 37.5 million people, ages 18 and older report having some trouble hearing (Quick statistics about hearing, 2021). Hearing related disabilities often go unnoticed and are typically invisible, eliminating the opportunity for impromptu accommodations.

Therefore, addressing the increasing need for inclusivity in restaurant design can be best met through the enhancement of the pre-existing space. The purpose of this research is to highlight the deficiencies of restaurant design for hard-of-hearing individuals and provide design guidelines in that regard. This research allows for diverse individuals to co-exist within commercial spaces more effortlessly.

Method:

This research aims to address how existing restaurants can adapt new design considerations to a diverse and inclusive range of users, including those with hearing difficulties. The study used two methods: (1) Observing human behavior in an upscale, bi-level restaurant setting and (2) identifying design criteria related to people that are hard-of-hearing established in the literature. The research question is: What design guidelines can be easily implemented in an existing upscale, bi-level restaurant to fit the needs of people who are hard-of-hearing? Data was gathered from on-site observations based on environmental conditions, path and wayfinding, behavioral settings, and ergonomics of restaurant design.

Findings:

The research findings underscored accessible design guidelines and solutions specifically tailored for individuals with hearing impairments in upscale, bi-level restaurants. These insights were derived from both direct observations and empirical findings. The observation provided insights into areas of strength, exemplified by light fixture selection, ceiling design, and separation of spaces, and areas that require improvement, such as noise level, material selection, and wayfinding. The empirical findings provided evidence of design solutions related to the population of interest, including acoustical solutions, signage, and choreographed lighting design.

Significance:

The incorporation of inclusive design elements into an existing space offers the improvement of function, accessibility, and experience for all users. This research will provide an understanding of the importance of inclusive restaurant design and offer design guidelines for the hospitality industry, restaurant owners, and designers. The creation of design guidelines allows all users, including the hard-of-hearing users, to comfortably recognize and utilize restaurant spaces.

References:

Levins, J. (2013, June). A holistic approach to room design for hearing impaired populations. In Proceedings of Meetings on Acoustics (Vol. 19, No. 1). AIP Publishing.

U.S. Department of Health and Human Services. (2021, March). Quick statistics about hearing. National Institute of Deafness and Other Communication Disorders. https://www.nidcd.nih.gov/health/statistics/quick-statistics-hearing#:~:text=Quick%20Statistics%20About%20Hearing%201%20About%202%20to,over%20report%20some%20trouble%20hearing.%203%20More%20items

US Foods. (2023). The Diner Dispatch: 2023 American Dining Habits. Food Supplier & Distributor | Restaurant Supply. https://www.usfoods.com/our-services/business-trends/american-dining-out-habits-2023.html

ESTABLISHING THE NEED FOR FEMINIST FRAMEWORK IN ENVIRONMENT BEHAVIOR STUDY (EBS) TO EXPLORE WOMEN'S PLACE-MAKING

Track 5: Spatial Justice, Equity, and Inclusion by Design
Knowledge Network: Cities & Globalization, Cultural Aspects of Design
Methods: Case Study, Interpretive-Historical Research
Ankita Karmakar (Pennsylvania State University); Mallika Bose (Pennsylvania State University)

Objective:
Research has shown that the Environment Behavior (EB) framework is underrepresented in women's place-making. EBS, or Environmental Behavior Study, associated with architecture and urban design, explores how the socio-cultural environment influences human behavior and vice versa. It examines the complexities of cultural values associated with the diverse needs, requirements, and latent activities of different groups, affecting physical forms at multiple scales. The culture-environment relationship, place-based attachment, and embodying places–are three existing lenses in the EB framework that can help us understand place-making.

The feminist framework articulated by Low (2014), Massey (1994, 2005), Butler (1990), Ahmed (2006), and McDowell (2014) on the other hand, addresses women's place-making as 'embodying spatial culture,' influenced by power factors associated with constraints and resistance stemming from intersectional identities, socio-physical settings, and scales of everyday ordinary livelihood. Drawing from spatial geography and psychogeography, 'Embodying spatial culture' emphasizes the significance of the body and emotions in shaping space utilization and its meaning. This concept encompasses metaphors, ideologies, and language that influence behaviors, habits, skills, and spatial orientations within specific local contexts. It aims for human-centric development, prioritizing the needs and experiences of individuals and groups to create places characterized by attachment, emotion, and ethical considerations.

Feminist scholars such as Dutta (2020, 2022), Dutta and Odendaal (2019), and Desai (2022) have criticized the narrow practice of exploring women's place-making based on safety, accessibility, and economic empowerment. A failure to acknowledge the complex socio-spatial embodied relationship between women and the environment in the production of everyday places has consistently conflated the physical environment of different groups of women. Addressing this gap, this poster will explore how the EB perspective can be a possible analytical framework for understanding women's placemaking informed by the feminist principles of 'embodying spatial culture'.

The objectives are twofold – 1) to integrate the feminist principles in EBS as an analytical framework for women-friendly environments, and 2) to devise the feminist framework in EBS as a tool to understand women's place-making in human-centric development.

This research conducted a systematic literature review (n=80), incorporating various place-based theories, case studies, historical research on the human and built environment, and women's place-making. It involved (n=15) literature searches focused on culture, the built environment, and place-making, with an emphasis on Rapport's work, as well as (n=25) advanced theories related to place attachment, and meaning. Additionally, (n=40) literature reviews explored women's place-making and feminism's role. Based on the theoretical understanding and existing knowledge gap in the feminist framework, the poster highlights the EBS framework and feminist principles to understand the physical environment, compares traditional EB approaches with feminist perspectives, and proposes a theoretical framework integrating EBS with feminism for examining women's place-making. Findings cover spatial embodiment in understanding everyday life, public-private dynamics, home, place identity, democracy, stigma, attachment, and appropriation. The poster concludes by discussing the effectiveness of the feminist framework in EBS to guide women-friendly environments in design and planning practices for human-centric development.

Keywords:
Environment Behavior Study (EBS), Feminist framework, Women's Place-Making, Embodied Place

RECLAIMING AGENCY: A COMMUNITY CENTERED APPROACH TO ZOMBIE MINE RECLAMATION IN CENTRAL APPALACHIA

Track 5: Spatial Justice, Equity, and Inclusion by Design
Knowledge Network: Health, Participation
Methods: Interpretive-Historical Research, Mixed Methods

Claire Henkel (North Carolina State University)

Background:

Mountaintop removal (MTR) is a form of surface coal mining that was popularized in Appalachia as an unintended consequence of amendments to the Clean Air Act (Hendryx & Holland, 2017). Despite this origin, MTR practices are less labor intensive, less labor protected, and more environmentally destructive than traditional underground mining methods (Boyles et al., 2017). Decades of this exploitative industry in the region have led to poor public infrastructure (Stewart Burns, 2005), high rates of depopulation (Shwartzman, 2021), ecosystem degradation, biodiversity loss (Price et al., 2015), critical public health crises (Hendryx et al., 2019) and high vulnerability to disasters such as floods and landslides (Reed & Kite, 2020). As coal extraction declines, many companies are filing for bankruptcy, abandoning operations, and leaving mines in "zombie mode". Both the literature review and policy analysis indicate that from federal oversight to state bonding, the systems built to protect local communities from corporate greed have remained complicit or, at times, co-conspiratorial in the destruction of Central Appalachian ecologies, economies and the wellness of its communities. Geospatial analysis revealed an overlap between MTR practices and key factors within the social determinants of health, such as economic stability, cancer rates, and environmental health. As recent climate and environmental justice funding increases, the present moment is one ripe with opportunity to reimagine what the reclamation process can look like. As a response to these findings, the author developed the Reclaiming Agency framework built upon principles of social permaculture, environmental planning, and community-first engagement to advance a community-centered approach to reclamation. The proposed framework is aligned with local, regional, and federal funding sources that make it adaptable and actionable with the goal of creating tangible benefits for those most impacted by MTR practices.

References:

Boyles, A. L., Blain, R. B., Rochester, J. R., Avanasi, R., Goldhaber, S. B., McComb, S., Holmgren, S. D., Masten, S. A., & Thayer, K. A. (2017). Systematic review of community health impacts of mountaintop removal mining. Environment International, 107, 163–172. https://doi.org/10.1016/j.envint.2017.07.002

Hendryx, M., & Holland, B. (2016). Unintended consequences of the Clean Air Act: Mortality rates in Appalachian coal mining communities. Environmental Science & Policy, 63, 1-6.

Hendryx, M., Yonts, S. D., Li, Y., & Luo, J. (2019). Mountaintop removal mining and multiple illness symptoms: A latent class analysis. Science of The Total Environment, 657, 764-769.

Price, S. J., Muncy, B. L., Bonner, S. J., Drayer, A. N., & Barton, C. D. (2015). Effects of mountaintop removal mining and valley filling on the occupancy and abundance of stream salamanders. Journal of Applied Ecology, 53(2), 459-468. doi:10.1111/1365-2664.12585

Reed, M., & Kite, S. (2020). Peripheral gully and landslide erosion on an extreme anthropogenic landscape produced by Mountaintop Removal Coal Mining. Earth Surface Processes and Landforms, 45(9), 2078–2090. https://doi.org/10.1002/esp.4867

Shwartzman, G. (2021). POWER and the Future of Appalachia: Discursive Framings of an Economic Transition. Southeastern Geographer, 61(4), 343–356. https://www.jstor.org/stable/27094639

Stewart Burns, S. L. (2005). Bringing down the Mountains: The Impact of Mountaintop Removal Surface Coal Mining on Southern West Virginia Communities, 1970--2004. https://doi.org/10.33915/etd.2310

EXPLORING THE SPATIAL DIMENSIONS OF POLICE BRUTALITY: CASE STUDY OF INDIANAPOLIS (2014-2023)

Track 5: Spatial Justice, Equity, and Inclusion by Design
Knowledge Network: None/Not Tied to Any Network
Methods: Observation, Photo-based Methods

Dorna Eshrati (Ball State University); Paria Taheri (Master of Urban Design Student); Sawyer Knuteson (Master of Urban Design Student)

Objective:
Police violence is a pressing issue in the United States, requiring extensive investigation into its underlying factors. While the occurrence of police brutality is widespread, certain regions experience higher incidence rates. This research delves into the physical attributes of locations where such incidents transpire, with the primary aim of identifying any correlation between spatial characteristics and the occurrence of police misconduct. Unlike most existing literature, this research does not center on the racial identities of officers or victims or the reasons for police involvement at the scene. Instead, we adopt a design-oriented approach to discern commonalities among spaces where police brutality has occurred. To begin our research, we gather data on 40 cases of police brutality in Indianapolis, IN, between 2014 and 2023 from various sources, including media reports and Google Earth imagery.

We will compile this information into a spreadsheet or database, noting details such as the date, location, type of incident, zoning, degree of enclosure, accessibility to transportation, lighting, density, and other relevant factors. We will cross-check the data from multiple sources to ensure its accuracy, which may require site visits or interviews with individuals who have firsthand knowledge of the incidents. Using charts and graphs, we will identify patterns, trends, and correlations between different variables. Finally, we analyze and compare the data to draw conclusions about the relationship between physical attributes of locations and the occurrence of police brutality. This research can inform city planners and policymakers on how to design safer and more equitable communities by identifying potential solutions to prevent future incidents.

MAPPING THERMAL RESILIENCE: RESIDENTIAL BUILDING TYPOLOGIES IN AFRICAN CITIES

Track 3: Resilient Design for Climate Change
Knowledge Network: Environmental Design Research Education, Residential Environments
Methods: Case Study, Statistical/Modeling Analysis

Alpha Yacob Arsano (Northeastern University); Elisabeth Molnar (Northeastern University); Kwan To Ryan Chan (Northeastern University)

Background:

Hotter temperatures associated with climate change will disproportionately affect vulnerable communities with less access to air conditioning units, especially those living in hot climates. The building construction boom in Africa's largest cities, driven by urbanization on an unprecedented scale, is marked by new residential building typologies that give prime importance to increasing residential units rather than the environmental performance and well-being of occupants. Unlike the typical residential typologies built before the turn of the millennium, there are now large numbers of social and affordable condominiums that are not designed to take advantage of passive building strategies to maintain the comfort and health of occupants. Projections expect urban extreme heat exposure in Africa of above 40 C to reach 45-95 billion person-days per year in the 2060s up from 2 billion person-days per year during 1985-2005, with West Africa experiencing the greatest exposure ("Africa," 2023). The research presented in this poster reviews existing literature on some of the residential building typologies and construction techniques in African cities, lays out a methodology for evaluating their thermal resilience, and proposes a mapping approach to identify vulnerable communities that are at high health risk due to high temperature under global warming.

The study provides summaries of residential building typologies and their performance analysis that can be used in further research to identify design strategies to mitigate extreme temperatures associated with climate change in African urban dwellings. A scientific methodology packaged as a digital toolkit called climabox (Arsano, 2022; Arsano & Reinhart, 2019) will be used to formulate combinations of sustainable building strategies (CSBS) for the building typologies customized to the vulnerable communities' climatic and physical contexts. This innovative scientific approach focuses on human health and carbon emission reduction while considering the socio-economic dynamics of communities at the location of interest. Ten different residential typology groups have been identified thus far in different regions of Africa categorized primarily based on building form and envelope construction. These are based on available literature and additional studies will be conducted in the future using different data sources. In future work, a computational framework using satellite maps and street view images, and onsite data collection using surveys will be implemented to further study typologies in regions that have not been documented in this study.

Reference:

Africa. (2023). In Climate Change 2022 – Impacts, Adaptation and Vulnerability (pp. 1285–1456). Cambridge University Press. https://doi.org/10.1017/9781009325844.011

Arsano, A. Y. (2022). CLIMATE-CARBON-EQUITY Making Sustainable Design Concepts Accessible for All. In B.S in Architecture. Massachusetts Institute of Technology.

Arsano, A. Y., & Reinhart, C. (2019). Early-Design Optimization of Target Ventilation Rates for Hybrid Buildings Using Single-Node Analytical Model. Proceedings of BS2019: 16th Conference of International Building Performance Simulation Association.

TURNING THE LIGHTS ON: INTERVENTIONS TO INCREASE ELECTRICITY ACCESS FOR VULNERABLE GROUPS IN KINSHASA

Track 2: Sustainable Buildings and Infrastructure
Knowledge Network: Sustainable Planning, Design & Behavior
Methods: Case Study

Terrance Dumoulin (Northeastern University); Alexandra Conrad (Northeastern University); Alpha Yacob Arsano (Northeastern University)

Objective:

Kinshasa, the capital city of the Democratic Republic of the Congo, is projected to be the largest population center in Africa by 2050 (Hoornweg & Pope, 2017). The city is experiencing rapid population growth and expansion of physical boundaries, ultimately resulting in informal and atomistic settlement typologies bordering traditional urban zones. A significant percentage of the population lives with limited or nonexistent electrical access. For those with access, current electrical transmission and distribution losses countrywide sit around 45%, compared to approximately 6% losses in the United States (Electric Power Transmission and Distribution Losses (% of Output) - Congo, Dem. Rep. | Data, n.d.). This presentation addresses the expansion of energy infrastructure to prepare for projected population density and increased electricity demand in 2050.

The interventions explored in the project included completion of unfinished hydroelectric dam projects along the Congo River and microgrid stations strategically placed across informal settlements, as well as a baseline case. Each scenario focused on either improving overall infrastructure or developing separate distribution networks. Emphasis was placed on proposing renewable energy solutions to comply with United Nations Sustainable Development Goal 7 (Goal 7 | Department of Economic and Social Affairs, n.d.).

Five analysis categories—electricity access, time, benefiting population percentage, potential for expansion, and public perception—concentrated recommendations around human-centric aspects of the research outcomes. The areas were weighted according to qualitative analysis in the literature review process, enabling the comparison of scenarios via a single score for simplifying recommendations.

Additional equations were developed to synthesize historical and projected population density, household size, electricity demand, and delivered energy per year under each case. The data streams produced concise values for comparison in each measure of efficacy. The scoring process resulted in grades for microgrid, hydroelectric power, and the baseline scenarios of 7/10, 9/10, and 4/10, respectively. Collected data and existing literature supported the scoring outcome, but the holistic analysis process encouraged further consideration of human factors affected by infrastructural changes. Results had significant limitations given availability of data and a subjective process of assigning weights to each analysis area. However, there was still value in the recommendations as a theoretical exercise in reframing urban planning decisions from the perspective of a region's most vulnerable stakeholder groups.

Infrastructure to aid in transitioning informal settlements towards integrated urban fabric is essential to improving inhabitants' living situations. Electricity access serves as an important step in providing utilities and connecting individuals with educational and workforce opportunities beyond the bounds of the immediate community. The goal of the project was met through exploring scenarios which would form a foundation of sustainable development beginning on a human scale.

References:

Electric power transmission and distribution losses (% of output) - Congo, Dem. Rep. | Data. (n.d.). Retrieved November 21, 2022, from https://data.worldbank.org/indicator/EG.ELC.LOSS.ZS?locations=CD
Goal 7 | Department of Economic and Social Affairs. (n.d.). Retrieved January 28, 2024, from https://sdgs.un.org/goals/goal7#progress_and_info
Hoornweg, D., & Pope, K. (2017). Population predictions for the world's largest cities in the 21st century. Environment and Urbanization, 29(1), 195–216. https://doi.org/10.1177/0956247816663557/ASSET/IMAGES/LARGE/10.1177_0956247816663557-FIG4.JPEG

RAINWATER CONSERVATION THROUGH ECO-DESIGN

Track 2: Sustainable Buildings and Infrastructure
Knowledge Network: Cities & Globalization, Environmental Design Research Education
Methods: Case Study, Mixed Methods
Fisiha I Likke (Lesley University/Northeastern University)

Abstract:

Freshwater, a cornerstone of life, development, and environmental sustainability, is increasingly becoming a vulnerable resource on a global scale. With the challenges of climate change, rapid urbanization, water risk management, and urban drinking water scarcity/stress looming, particularly in underdeveloped cities and underserved communities, the urban built environment emerges as a significant influencer of water consumption patterns. As cities, especially in the Sub-Saharan countries, continue to expand at an alarming rate, integrating rainwater conservation principles as alternative water resources to the city-piped water into urban policy, planning, and design becomes imperative. This research concept explores the nexus between an urban built environment, rooftop rainwater collection, storage mechanism, filtration, ground infiltration, surface design principles, green coverage, and urban water management operationalized as an ecosystem.

The research study spotlights Addis Ababa, Ethiopia, to accentuate the pivotal role of deliberate, systematic construction materials, form-based architecture, rooftop typologies, and green coverage for bio-filtration and ground infiltration systems. With a high precipitation rate annually compared to other cities in the Horn of Africa or the sub-Saharan African countries, its dismal freshwater accessibility due to underdeveloped infrastructure and lack of urban design approaches calls for eco-design interventions. By zeroing in on low-income or underdeveloped water infrastructure communities, the research aspires to shed light on strategies that could be taken progressively as an alternative direction for a sustainable future and equitable water distribution and uses.

A PLACE IN PUBLIC SPACE: DESIGNING AN AUTISM-FRIENDLY EXPERIENCE FOR ADULTS IN PALMER PARK

Track 5: Spatial Justice, Equity, and Inclusion by Design
Knowledge Network: Health, Participation
Methods: Focus Groups, Interviews
Micaela Ada (the Neighborhood Design Center)

Objective:

In 2023, the CDC reported that approximately 1 in 36 children in the U.S. is diagnosed with autism spectrum disorder (ASD) (Center for Disease Control and Prevention, 2023). When children with ASD enter adulthood, many face new milestones and challenges such as working, living away from the family home, and navigating the world with more autonomy.

Design professions —including architecture, interior design, and landscape architecture— have progressively incorporated considerations for ASD into the built environment. Adult-specific, autism-friendly design guidelines existing today have honed in on improving workplace, home, and therapeutic environments. There is a noticeable gap in addressing landscapes made for the social and leisure aspects of life.

The purpose of this research is to begin addressing this gap by investigating design considerations for creating public park settings that promote positive outdoor recreation experiences for adults with ASD.

Along with a literature review of established universal and ASD-specific principles across design professions, this research utilizes a mixed-methods approach including a review of relevant case studies; interviews with design professionals; a focus group with adults with ASD. Once distilled, the results were six themes to consider when designing a public park space for adults with ASD: sensory flexibility, navigability, clarity, predictability, variety, and safety.

A concept design and program for a neighborhood public park was developed in conjunction to illustrate the application of these themes.

Overlap can be seen between these design themes for ASD-friendly public parks and design recommendations for other groups highlighted in the American Society of Landscape Architects (ASLA) Universal Design Principles. These groups include people with limited mobility, blindness and low vision, deafness and hardness of hearing, neurocognitive disorders, neurodevelopmental and/ or intellectual disabilities (ASLA Publishes Guide to Universal Design, 2020; Dillon & Green, 2019).

The Autism Community makes up a considerable portion of the population, many of whom have reached or are close to adulthood. Knowing that, this research argues that the Autism Community, adults in particular, are stakeholders in public park space. Utilizing these themes in design can result in more inclusive landscapes for adults with ASD. Overall, these themes also add to the growing body of research promoting landscape equity across populations of various abilities and support needs.

References:

ASLA Publishes Guide to Universal Design. American Society of Landscape Architects. (2020, August 16). Retrieved May 6, 2023, from https://www.asla.org/NewsReleaseDetails.aspx?id=62173

Dillon, I., & Green, J. (2019, August 20). Universal Design. American Society of Landscape Architects. Retrieved May 6, 2023, from https://www.asla.org/universaldesign.aspx

Centers for Disease Control and Prevention. (2023, March 22). Autism prevalence higher, according to data from 11 ADDM communities. Centers for Disease Control and Prevention. https://www.cdc.gov/media/releases/2023/p0323-autism.html#:~:text=One%20in%2036%20(2.8%25),1%20in%2044%20(2.3%25).

RESEARCH ON COMPONENTS OF UNDERGROUND WALKING SPACE

Track 2: Sustainable Buildings and Infrastructure
Knowledge Network: Cities & Globalization
Methods: Action/Participatory Research, Case Study

Go Honjo (Tokyo Denki University); Hiroshi Tsuchida (Tokyo Denki University); Chiaki Konishi (Tokyo Denki University); Yuka Honda (Tokyo Denki University)

Background:
In Japan, many people use railroads as a means of transportation. Subways, in particular, are often found in central Tokyo, and redevelopment projects in close proximity to stations are striving to create comfortable spaces for daily users by directly connecting underground walking spaces. However, the underground space is buried in underground facilities and basements, making the underground walking space difficult for users to understand and recognize the space. Furthermore, in the event of a disaster such as an unusual fire or earthquake, underground spaces are required to allow users to escape to the surface as quickly as possible. Therefore, there is a need to create safer, more comfortable, and easier-to-understand underground walking spaces in the Tokyo central subway system, which is used by many people.

Objective
The purpose of this study is to clarify the factors that contribute to increasingly complex underground walking spaces in order to gain insight into the creation of safe and comfortable walking spaces in future redevelopment projects near subway stations. Focusing on stations where there are many subway lines, this study will focus on the flow lines from platform to platform at stations, and understanding the components may lead to clarifying the factors that cause more comfortable spaces.

Surveyed area
Assuming that one of the factors contributing to the complexity is due to the large number of subway station lines and levels, the top six stations of all Tokyo Metro and Toei Subway stations are targeted.

Spatial Characteristics of Underground Walking Space
There are various patterns of conduits, and the underground walking space is composed of several components. Therefore, focusing on five components, we defined five spatial change points: "stairs and escalators" that move up and down, "spread-out spaces" with multiple directions of travel, "single paths" with only one direction of travel, "turnings" such as curves and U-turns, and "ticket gates" for transfers outside latches.

Research Methods
A plan view of all underground walking spaces in the study site was created to classify and understand the spatial characteristics in the conduits in order to understand the components of the underground walking space. Two patterns of conduits were observed: in-latch and out-of-latch. Since there are a myriad of above-ground exits in the conductor outside the latch, we will focus on four stations with conductors only inside the latch as a foundational view.

Discussion
I believe that the number of spatial change points, changes in direction of travel for curves, U-turns, and other turnarounds, and the increase in the number of spreading space options are factors that increase complexity, and that fewer of these components would lead to the creation of more comfortable and easier-to-understand underground walking spaces. What do you think should be done to create a more comfortable and understandable underground walking space?

1) "List of Higashi-Kyoto Stations" Tokyo Metro and Toei Subway
2) "Subway Station-related Redevelopment Projects" Tokyo Metropolitan Government Bureau of Urban Development

WHAT IS A STABLE URBAN STRUCTURE?- FOCUSING ON THE URBAN AREAS ALONG THE NEW AND OLD ROUTES-

Track 2: Sustainable Buildings and Infrastructure
Knowledge Network: Cities & Globalization
Methods: Statistical/Modeling Analysis, Survey

Yuto Yakita (Tokyo denki univercity of master); Hiroshi Tsuchida (Tokyo Denki University); Chiaki Konishi (Tokyo Denki University); Yuka Honda (Tokyo Denki University)

Background:
In Japan, the population and urban functions are concentrated in Tokyo, disrupting the balance of urban structure between urban and rural areas. Therefore, we believe that the development of a self-sustaining urban structure by suburban cities located around Tokyo will lead to the formation of a sustainable city by transforming the metropolitan area into a multipolar decentralized type.

Objective
The purpose of this paper is to obtain knowledge on future urban structure from urban area formation process of suburban city located around Tokyo. The new urban area was formed when Tsukuba Express opened in 2005. The Joban Line, with a history of more than 100 years, is located near and parallel to it. These two lines radiate from the Tokyo metropolitan area. It is considered that the analysis of the urban area along these two lines will lead to the knowledge for changing the metropolitan area to the multipolar dispersion type.

Research Subject
Targeting urban areas along the new Tsukuba Express line and the old Joban Line. It will also focus on Kashiwa City, Chiba Prefecture, where two lines pass through, and will target five stations: Kashiwanoha Campus Station, Kashiwa-tanaka Station, Minami-Kashiwa Station, Kashiwa Station and Kita-Kashiwa Station.

Methods
This paper analyzes the urban transformation of two routes before and after the opening of Tsukuba Express. The analysis is carried out from the viewpoint of the number of passengers, population dynamics, and land use transformation.

Analysis 1
The new urban area along the Tsukuba Express line and the old urban area along the Joban line showed differences in the areas where the population is concentrated around the stations. In the new urban area, the population density increases as one gets closer to the station, while in the old urban area, rapid population growth in recent years was observed within a radius of 500 m to 1,000 m.

Analysis 2
In terms of land use, the area of residential land in relation to the population around the station indicates that in the new urban area, high-rise condominiums are being developed in high-density residential areas. In contrast, the old urban area is being developed mainly with detached houses.

Result
One of the factors that have contributed to the continued growth of the station-area population in the old urban areas along the Joban Line, even after more than 100 years, is the formation of urban areas centered on detached houses. Detached houses are easy to demolish, leading to stable land prices for the entire town. As a result, the gradual population growth is expected to continue.

Future new urban areas should be planned not only for high-density residential land development, such as condominium development, but also for development of detached housing.

References:
Department of Transportation National Land Numerical Information Download Service (January 11, 2023)
Institute of Transport Policy Studies (2023). FY 2020 Annual Report on Urban and Regional Transportation, Page 316
Japan Statistics Bureau, Statistical window for e-Stat government statistics (January 11, 2023)

WHAT DO YOU NEED FOR PEDESTRIAN SPACE?

Track 2: Sustainable Buildings and Infrastructure
Knowledge Network: Cities & Globalization
Methods: Photo-based Methods, Survey

Nene Wakamatsu (Tokyo denki university of master); Keito Kawahira (Tokyo denki univercity of master); Hiroshi Tsuchida (Tokyo Denki University); Chiaki Konishi (Tokyo Denki University); Yuka Honda (Tokyo Denki University)

Background:

In Japan's major regional cities, housing is becoming more suburbanized, but the corresponding public transportation infrastructure is lagging behind. This has led to the formation of automobile-dependent urban structures. If this trend continues, the combination of a declining population, falling birthrates, and an aging population will result in the formation of low-density cities. Therefore, it is necessary to develop the cities for urban intensification. For these purposes, we will focus on pedestrian space development, since it is important to break away from automobile dependence and to develop a pedestrian-centered city in the center of the city.

Purpose

There are many sustainable urban plans around the world to secure pedestrian spaces that can serve as a reference for solving Japan's problems.
Therefore, by comparing the plans of Europe, which was developed as an existing urban structure, with those of the United States, which was developed as a new development, and by extracting the elements to form a pedestrian space, I believe that this will serve as a reference for sustainable city development in Japan in the future.

Target districts and survey methods

This paper compares and analyzes two urban structures with an emphasis on pedestrian space: the European Traffic Zone System (TZS) and the American New Urbanism (NU) (Katz, 1993). Europe organized and analyzed traffic regulations and major facility functions on a photographic basis while capturing the entire picture on a map. The United States based its reading on the figures in the book (Katz, 1993). Two districts each considered particularly noteworthy for comparison in the U.S. and European plans are included in this study.

Discussion

Viewpoint
Method of pedestrian maintenance
functional distribution

Europe

Pedestrian space is secured in a 500 m area, and a landmark church and plaza exist in the center.
Since the degree of freedom of passage is high, the center is used uniformly. However, a peripheral road is needed to cover the zone where ordinary cars cannot pass. Historic architecture is used, and buildings of major uses are distributed. It is necessary to consider safety when moving, but cities are used uniformly.

America

It was zoned and maintained with a basic unit of about 100 acres, and there is an open space in the center. Although the degree of freedom of walking is reduced by improving the sidewalk at the back of the building, it is made to be continuous linearly. However, moving between areas of large districts requires a car. Commercial, residential and other functions are arranged in an integrated manner. It is efficient because people concentrate but travel is reduced.

Result

Although the backgrounds varied from city to city, the continuity of central plazas and pedestrian spaces was common. In the central plaza, there were landmarks with churches in Europe and public facilities and malls in America.

References

Peter Katz (1993). The New Urbanism:Toward an Architecture of Community,McGray-Hill Education.

Europe is survey-based, so there is no reference.

A STUDY ON FACILITY LOCATION AND USER ACCESSIBILITY OF SERVICES THAT SUPPORT THE LIVES OF CHILDREN AND PERSONS WITH DISABILITIES

Track 5: Spatial Justice, Equity, and Inclusion by Design
Knowledge Network: Active Living by Design, Sustainable Planning, Design & Behavior
Methods: Case Study, Statistical/Modeling Analysis

Eiji Satoh (Utsunomiya University); Sorano Tanaka (Utsunomiya University); Kurumi Takezawa (Utsunomiya University); Asuka Yamada (Tokyo Denki University); Yoshimune Hiratsuka (Juntendo University)

Background:

The living environment for people with disabilities has been discussed worldwide on topics such as social coexistence, equality issues, etc. In Japan, the government has established a law to support the lives of children and persons with disabilities by planning and developing services within the scope of "Disability Health and Welfare Areas" consisting of a single municipality or multiple municipalities. The government established the new policy will come into effect in 2024.

In response to the government's policy to expand support, local cities are also developing small-scale community-based living centers such as group homes. However, many challenges remain in building a support system for persons with disabilities, such as the lack of places to receive severely disabled persons and the difficulty in developing a network of care for persons with disabilities in the community. In addition, Japan is currently in a period of population decline, and the problem of human resources to provide services has been pointed out. Therefore, the purpose of this study is to measure the accessibility of users and to quantitatively clarify the current service provision system in the region using big data in order to obtain knowledge for designing an appropriate provision system for welfare services for persons with disabilities. This study of appropriate service locations and the accessibility of services show the status of livelihood support for children and persons with disabilities in local cities in Japan.

Method

Tochigi Prefecture is one of Japan's regional cities. Tochigi Prefecture is located about 100 km north of Tokyo, with an area of 6,408 km^2 and a population of about 2 million. Tochigi Prefecture has six "disability health and welfare zones" and plans services. Tables 1 and 2 show the number of population and the number of service centers for each service. The following procedure was used to measure the accessibility of users.
1) Create a 500m mesh data of the population distribution of the entire Tochigi Prefecture from the national statistical data.
2) Based on the national data, we assume that approximately 6% of the population has some kind of disability, and multiply the population mesh of Tochigi Prefecture by 6% to obtain a hypothetical distribution of disabled persons.
3) Calculate the latitude and longitude from the address data of each facility and plot them on an electronic map.
4) Using a geographic information system (GIS), measure the road distance to the nearest service establishment, with the starting point as the virtual residence and the ending point as the service establishment, using the mesh center point as the starting point.
5) Create a life style model for children and people with disabilities, and evaluate it by combining services based on this model.

Discussion

Based on these results, we discuss the external environment in which they live while using the services. Figure 2 in the sample visualizes the distance zones in which hearing-impaired children can live if their behavior is modeled as attending school and using social services after school.

3D PRINTED CERAMIC AIR COOLER

Track 3: Resilient Design for Climate Change
Knowledge Network: Environment-Gerontology, Residential Environments
Methods: Design Methods
Magda Rojas (UC Davis)

Background:
Due to climate change, using HVAC systems (Heating, Ventilation, and Air Conditioning) to enhance indoor comfort is increasingly necessary. However, the reliance on mechanical appliances presents a paradox. While they effectively control temperatures, they are both energy-intensive and a source of carbon emissions (Abdullah, 2022). Additionally, only 8% of the world's population can access HVAC systems. (Burrows, 2022)

This project aims to address the challenges of hot and dry climates by exploring a 3D-printed ceramic Air Cooler prototype inspired by traditional terracotta jars. These jars have been historically used to cool down drinking water by utilizing the porosity of the clay surface, which absorbs water and allows heat to dissipate through contact with the air, thus keeping the water cool (Ortega-Casanova et al., 2021). This same principle can be applied to reduce indoor temperatures when combined with a source of airflow.

The prototype incorporates additive manufacturing technology that uses Sio-2 3D printing clay to create a water jar in the shape of a concentric cylinder. The design, made using CAD Rhinoceros 3D, is inspired by the continuous spiral movements of the Potterbot 3D printer extruder. The outer cylinder is a water reservoir that has a textured surface to enhance the interaction between clay and water. The inner cylinder functions as an air conduit that houses a solar-powered fan at its base. This fan generates cool air that speeds up the evaporation cooling process. The 9x12 inch prototype was printed using a 3mm nozzle to ensure stability and detailed texture. This project aims to evaluate the performance and ease of adoption of the Ceramic Air Cooler, by measuring different parameters under controlled conditions, such as water temperature, evaporation levels, humidity, and surrounding temperature changes. User feedback will also be collected through surveys to determine perceived temperature differences, increases in thermal comfort, user satisfaction, and ease of installation.

References
Abdullah, A. (2022). Experimental study of natural materials for an evaporative cooling design in hot-arid climate. Building and Environment, 207, 108564.
Burrows, L. (2022, July 20). In a hotter world, air conditioning isn't a luxury, it's a lifesaver. Harvard John A. Paulson School of Engineering and Applied Sciences. https://seas.harvard.edu/news/2022/07/hotter-world-air-conditioning-isnt-luxury-its-lifesaver
Cain, A., Afshar, F., Norton, J., & Daraie, M. R. (1976). Traditional cooling systems in the Third World. The Ecologist, 6(2), 60–64.
Ortega-Casanova, J., Ortega-Cortes, M., Cortes-Carretero, P., & Ortega-Cortes, J. (2021). A primitive method for cooling water: does the shape matter? Case Studies in Thermal Engineering, 26, 101022.

A MULTIDIMENSIONAL APPROACH TO ENHANCING THE OVERALL PRODUCTIVITY OF THE CORPORATE WORKFORCE IN GHANA THROUGH EVIDENCE-BASED DESIGN STRATEGIES

Track 4: Design for Health and Wellbeing
Knowledge Network: Active Living by Design, Health
Methods: Case Study, Experimental/Quasi-Experimental Research

Daphne A.U. Chilaka (Texas Tech University); Nishan Khatiwada (Texas Tech University); Debajyoti Pati (Texas Tech University)

Background:

Qavah Retreat presents a comprehensive framework for corporate wellness, grounded in evidence-based design principles, and informed by the Hebrew concept of "Qavah," signifying expectant hope in waiting. The abstract highlights the key elements of the research, focusing on integrating physical, mental-emotional, and social-community health strategies within the retreat's framework. The redesign of Qavah's base layout emphasizes community-based design, incorporating pathways for walking, friendship benches, and informal care lodge spaces. By leveraging panoramic views of nature and maximizing natural light, the retreat promotes stress reduction and well-being among its guests. Furthermore, strategic considerations address challenges such as communicable diseases and environmental hazards, ensuring a safe and conducive environment for wellness activities. The retreat's physical health strategies prioritize promoting physical activity and mindful eating through accessible lodges, coupled pathways, and an indoor botanical restaurant. Medicinal gardens featuring insect-repellent plants enhance the physical experience, contributing to overall well-being. Mental-emotional health strategies focus on providing mental breaks through sensory design, tranquil soundscapes, and aromatherapy-zoned spaces.

The seamless integration of indoor and outdoor environments fosters a peaceful atmosphere that reduces anxiety and restores mental health. Social-community health strategies aim to facilitate social interaction and informal care through strategically placed spaces like the botanical kitchen restaurant and friendship benches. Outdoor group activities and community-focused initiatives extend the retreat's impact beyond its borders, creating opportunities for job creation and community engagement.

In conclusion, Qavah Retreat exemplifies a multidimensional approach to corporate wellness, incorporating evidence-based design strategies to address the diverse needs of the workforce in Ghana. By fostering resilience and hope through thoughtful design interventions, the retreat aims to empower individuals to thrive in both their personal and professional lives. This abstract serves as a concise overview of the research conducted, highlighting the significance of the Qavah framework in promoting holistic well-being within corporate settings.

A STUDY ON THE SEMI-BASEMENT AS AN ALTERNATIVE SPACE FOR START-UPS

Track 1: Nature and Urban Ecology
Knowledge Network: Environmental & Architectural Phenomenology, Interior Design
Methods: Case Study, Design Methods
John Mut Lee (Chungnam University)

Background:
Young people flock to cities to get a chance to start a business, but they are not getting a chance to fail due to high opportunity costs. The government has created childcare spaces such as co-working spaces to solve this problem, but due to the limitations of government buildings, start-up items must be specific and it is still difficult to demonstrate the creativity of space utilization that early entrepreneurs must have. In addition, as the operator of the co-working space in the private sector of the government is striving to attract start-ups with high possibility of success, the opportunity for the ultra-low start-up team, which needs actual incubation, to try various businesses is decreasing. (Low-cost) start-up childcare spaces or (high-cost) private co-working spaces are effective when start-up items are clearly determined. Therefore, for the very early prospective start-ups who have to think about start-up items, it is also a threshold to overcome the co-working space.

For young people who cannot dare to start semi-underground on their own, concrete research on construction methods and lifestyles that can be used by fixing semi-underground spaces with low capital is needed, and there is a need to build an extremely low start-up space through the spread of programs such as DIY, DIT. The preceding studies and background of this study are as follows.

ARTFUL ALLEYWAYS: TOWARDS BEAUTIFICATION AND WELLBEING IN OTHERWISE STAGNANT SPACES

Track 4: Design for Health and Wellbeing
Knowledge Network: Residential Environments, Sustainable Planning, Design & Behavior
Methods: Interviews, Observation
Nicholas Wise (Arizona State University)

Background:

Public art is deeply rooted in the cultural and historical contexts of society, and thus a humanities perspective can provide insights into the significance of public art within a community. This study focuses on the humanistic connections people make with public art and how the humanities is a lens for developing new insights for public health and safety. Alleyways in urban areas can be spaces associated with stigma or crime because they are secluded and can encourage anti-social behaviors. While alleyways are spaces that people use, albeit not regularly, some see artwork as a way to beautify such spaces to give a safer impression. Public art as an expression of place and culture in residential settings. One way to better understand connections people make with art in public settings is through how people engage with art. There is a need to further understand connections with leisurely activity as people in neighborhoods often walk and encounter such art daily. But we take for granted art that is available that we pass on a daily basis and place less consideration on art that is emerging is spaces that are less traveled through. Alleyways are primarily used for garbage removal and do not always have lighting.

There is a need to look at how art can give presence to stagnant areas to make them feel safer. City management offices fund art and culture projects for these reasons, and a main consideration is to connect people. Understanding how art in stagnant spaces impacts how people interact with such spaces helps us address how art can shirt negative associations and if the beautification of such spaces offers insight into positive well-being as people pass of traverse these spaces. Undertaking phenomenological research challenges us as researchers to gain perspectives of time and space, relations with others, and objects in our surroundings. Such an approach will help us understand how public art reflects the values, beliefs, and concerns of a community, as the presence of art and the use of the humanities in public spaces impact public health and safety. This work can offer new implications for public policy as well as theoretical implications surrounding how we engage with public art in community settings. The city of Phoenix, Arizona is currently undertaking a public art plan that runs from 2023 to 2028.

BEYOND AMERICANS WITH DISABILITIES ACT (ADA) COMPLIANCE IN RETAIL SPACES: INTERVIEWS WITH INCLUSIVE DESIGN EXPERTS ON STRATEGIES AND CHALLENGES IN ARCHITECTURAL PRACTICE.

Track 5: Spatial Justice, Equity, and Inclusion by Design
Knowledge Network: Researchers in Professional Practice
Methods: Interviews
Mariángel Meza Santiviago

Background:
The Americans with Disabilities Act (ADA) standards establish a regulatory framework for accessibility that prohibits discrimination against individuals with disabilities in various areas and is the only mandatory standard under federal law that architects and designers must comply with regarding inclusion in physical spaces. However, previous research underscores the shortcomings of ADA compliance in architectural practice, particularly in achieving true inclusivity in physical settings, which is especially aggravated in retail settings. This study addresses the identified gap in understanding the practical adoption of Inclusive Design (ID) beyond ADA regulations, with a specific focus on the architectural dynamics of the retail sector. Through a qualitative approach, in-depth, semi-structured interviews were conducted with ID experts in the U.S. Thematic analysis was employed to comprehensively explore the interconnection between ADA, ID, the retail sector, and architectural practice.

The primary goal is to identify common design strategies, approaches, and methods, along with prevalent challenges experts face when striving to surpass regulatory compliance and implement ID in the retail store design process. The findings reveal critical insights across three clusters: advocating for innovative design practices, critiquing ADA standards, and highlighting challenges in implementing ID in retail spaces. These insights contribute to a broader understanding of ID considerations in the retail sector, guiding future research and promoting a holistic approach to the design process. Additionally, the implications advocate for a shift in perspectives and practices, especially within the dynamic and competitive landscape of retail design. Overall, this study contributes valuable knowledge to the field, offering practical guidance for practitioners and influencing the evolution of inclusive architectural practices, particularly in retail design.*

Keywords:
Inclusive Design, Architectural Practice, ADA Compliance, Retail Design, Interviews, Qualitative Research.
**** Pilot / Preliminary Findings. Interviews are still being conducted. Theoretical Saturation has not been reached yet.***

CREATING NEURODIVERSE-FRIENDLY SPACES IN INDIA: EXPLORING THE IMPACT AND CHALLENGES OF SENSORY CALMING ROOMS

Track 4: Design for Health and Wellbeing
Knowledge Network: Environmental & Architectural Phenomenology, Health
Methods: Ethnography, Mixed Methods

Shivani Shah (Accessible Designs LLP)

Background:

Neurodiversity encompasses a spectrum of neurological differences, including but not limited to autism, ADHD, dyslexia, and sensory processing disorders. While the concept of accessibility has gained significant traction in recent years, there remains a critical gap in understanding and addressing the unique needs of individuals with neurodivergence and their caregivers, particularly from an Indian perspective. This study delves into the multifaceted dimensions of creating neurodiverse-friendly spaces, aiming to unravel the intricacies of accessibility in public spaces such as festivals and cultural events.

This study focuses on sensory processing challenges and investigates the creation of sensory regulation and calming opportunities in public spaces through a pilot study conducted at a festival in India. Sensory and Calming Rooms with engaging activities were implemented for six days, attracting over 500 participants of various demographics, including different ages, genders, and disability types.

Findings revealed significant positive impacts of these sensory calming rooms on the overall festival experience. Participants utilized these spaces for socializing, participating in activities, relaxation, exploration, and enjoyment. However, challenges were encountered in setting up these rooms, particularly in creating affordable, high-quality temporary setups. Resource constraints, including material availability and labor, posed challenges in installation and operation. Each room required a team of six individuals, consisting of three student-trained volunteers, two special educators, one therapist, and two caretakers. The study recognizes the need for innovative approaches to reduce costs associated with installation, operations, logistics, and human resources required to run such rooms.

Moving forward, the study aims to develop innovative strategies to address these challenges and enhance the feasibility and sustainability of such interventions in public spaces. By exploring cost-effective solutions and streamlining operational processes, the aim is to make sensory regulation and engagement opportunities more accessible and inclusive for individuals across diverse settings. Through continued research and collaboration, this study contributes to advancing neurodiverse accessibility and inclusion in public spaces, ultimately fostering a more inclusive society for all.

References

Mostafa, M. M. (2019). Sensory design: An approach to creating a positive sensory experience in the built environment. Frontiers of Architectural Research, 8(1), 87-101. Ward, J., & Schmied, D. (2018). The impact of sensory rooms on mental health recovery in acute inpatient psychiatry: A pilot study. International Journal of Mental Health Nursing, 27(1), 218-226. Lutzenberger, K., & Klein, E. M. (2020). Exploring the effectiveness of sensory rooms in public spaces: A systematic review. Journal of Environmental Psychology, 71, 101499.

EVALUATING OCCUPANT INTERACTION WITH COGNITIVE-ENABLED BUILDING TECHNOLOGIES: SCALE DEVELOPMENT TO MEASURE EQUITY AND PRIVACY

Track 5: Spatial Justice, Equity, and Inclusion by Design
Knowledge Network: POE/Programming, Work Environments
Methods: Survey

Angelica Ulinara (Cornell University); Ying Hua (Cornell University)

Background:

In today's rapidly evolving digital landscape, the integration of artificial intelligence, machine learning, and other cognitive technologies into work environments is enhancing operational efficiency and user experience, leading to the emergence of Cognitive-Enabled Building (CEB). Amidst these advancements, concerns about the impact on workplace equity and individual privacy are increasing, underlining the importance of these factors in Human-Building Interaction (Becerik-Gerber et al., 2022). This research aims to develop and validate an instrument for assessing how CEB technologies influence workplace equity and privacy. It seeks to measure the effects of cognitive system—such as automated environmental controls, data-driven space optimization, and AI-assisted security—on employees' perceptions of privacy and equitable access to technological resources. In order to build the scale for measurement, a multidisciplinary approach is used to create a framework that could capture the breadth of the theoretical concept of occupant interaction with CEB technologies. The process involves semi-structured interviews, and instrument-testing that targets a diverse population to represent the broad spectrum of potential occupants and stakeholders in the design and operation decision-making processes of CEBs. Data analysis methods are outlined, including thematic analysis for the semi-structured interview transcripts to identify patterns and identifying correlations between variables using exploratory factor analysis (EFA) for initial scale evaluation. By using a robust method, this scale instrument will be a tool to bridge gaps in current knowledge about balancing technological advancement and individual rights in the workplace, potentially influencing future technology design and policy formulation. Insights from this research are expected to contribute to the development of guidelines and policies that ensure these technologies are implemented in a manner that respects and promotes workplace equity and privacy.

References:

Becerik-Gerber, B., Lucas, G., Aryal, A., Awada, M., Bergés, M., Billington, S. L., Boric-Lubecke, O., Ghahramani, A., Heydarian, A., Jazizadeh, F., Liu, R., Zhu, R., Marks, F., Roll, S., Seyedrezaei, M., Taylor, J. E., Höelscher, C., Khan, A., Langevin, J., ... Zhao, J. (2022). Ten questions concerning human-building interaction research for improving the quality of life. Building and Environment, 226, 109681. https://doi.org/10.1016/j.buildenv.2022.109681

Becerik-Gerber, B., Lucas, G., Aryal, A., Awada, M., Bergés, M., Billington, S., Boric-Lubecke, O., Ghahramani, A., Heydarian, A., Höelscher, C., Jazizadeh, F., Khan, A., Langevin, J., Liu, R., Marks, F., Mauriello, M. L., Murnane, E., Noh, H., Pritoni, M., ... Zhu, R. (2022). The field of human building interaction for convergent research and innovation for intelligent built environments. Scientific Reports, 12(1), 22092. https://doi.org/10.1038/s41598-022-25047-y

Carpenter, S. (2018). Ten Steps in Scale Development and Reporting: A Guide for Researchers. Communication Methods and Measures, 12(1), 25–44. https://doi.org/10.1080/19312458.2017.1396583

EXPANDING DAYLIGHT DESIGN BEYOND CONVENTIONAL METRICS: SIMULATION MODELING FOR ENHANCED DAYLIGHTING ON HIGHER EDUCATION CAMPUSES

Track 4: Design for Health and Wellbeing
Knowledge Network: Researchers in Professional Practice, Sustainable Planning, Design & Behavior
Methods: Design Methods, Practice-based Research
Jiewei Li (Goody Clancy); Elaine Hoffman (Goody Clancy)

Background:
This research poster investigates the optimization of daylight utilization in academic campus buildings, particularly focusing on simulation of existing buildings retrofits and major additions. Beyond conventional metrics like Spatial Daylight Autonomy (sDA) and Annual Solar Exposure (ASE), the study delves into the nuanced aspects of glare and explores the most effective daylight qualities for various space typologies prevalent in academic environments.

In the pursuit of creating environments conducive to learning, creativity, and well-being, the research examines distinct space types including laboratories, art studios, residence hall student rooms, gathering spaces, and teaching areas. By employing advanced simulation techniques and project case studies, the investigation aims to discern optimal daylighting strategies tailored to the specific functional requirements and user experiences inherent to each space type.

To refine our understanding of daylight's impact on occupant comfort and wellbeing, this study broadens the spectrum of daylight metrics by incorporating measures such as Useful Daylight Illuminance (UDI) and Daylight Glare Probability (DGP), and further innovates with the introduction of a novel index aimed at quantifying the nuanced behaviors of direct sunlight. This new index, distinguishing from traditional metrics like ASE, seeks to embrace and accurately measure the presence of direct sunlight, highlighting its desirable aspects in specific indoor scenarios. By doing so, the study acknowledges the complexity of daylight's 'interest' and dynamic qualities in educational spaces, ensuring that both the quantity and quality of light are meticulously evaluated. Parameters such as intensity, distribution, color rendering, and now, the nuanced appreciation of direct sunlight, are integrated into tailored daylight simulation strategies for diverse space typologies across academic campuses, fostering a balanced interaction with natural light. Ultimately, this research endeavors to contribute to the creation of vibrant, sustainable, and human-centric academic environments where daylight serves as a catalyst for enhancing the overall quality of life and educational experiences for students, faculty, and staff alike. By setting new benchmarks for daylighting excellence in educational settings, the study fosters an enhanced relationship between academic spaces and their natural surroundings, aiming to support the evolving needs of its occupants.

RESEARCH PROSPECTUS: EXPLORING THE IMPACT OF INCLUSIVE RETAIL ARCHITECTURE BEYOND AMERICANS WITH DISABILITIES ACT (ADA) STANDARDS COMPLIANCE ON CUSTOMER INDEPENDENCE, COMFORT AND SATISFACTION.

Track 5: Spatial Justice, Equity, and Inclusion by Design
Knowledge Network: Interior Design, Researchers in Professional Practice
Methods: Interviews, Mixed Methods
Mariángel Meza Santiviago

Background:
In architectural practice, Inclusive Design (ID) can be implemented not only to address the needs of individuals with disabilities but to positively impact all users' experiences. Nonetheless, a considerable gap exists in how contemporary architects and designers incorporate ID in practice, especially in Retail Spaces. Often, their efforts are limited to complying with Americans with Disabilities Act (ADA) standards, which offer a basic regulatory foundation for accessibility but fall short of achieving genuine inclusivity. This challenge demands a paradigm shift in architectural practice, encouraging professionals to embrace inclusion, and presents an opportunity to use ID to elevate the overall consumer experience within the competitive retail landscape. This research study aims to provide companies and retailers with compelling ways to embrace ID in their spaces, as well as practical guidance for architects and designers to accomplish it. It is grounded in the frameworks of ID, the Social Model of Disability, and the principles of user-centered design, ensuring designs are tailored to the needs and preferences of users. Moreover, it incorporates insights from environmental psychology, shedding light on its impact on customer satisfaction, while also considering the implications of ADA legislation. Structured into two stages, combining qualitative and correlational research, the first objective was to identify inclusive parameters and strategies within architecture that could be effectively implemented in the design of retail spaces through interviews with ID experts. In the subsequent stage, coffee shops were selected as the focal point to empirically evaluate the impact of these previously identified strategies and parameters on the independence, comfort, and satisfaction of customers. This evaluation involved surveys, observations, and physical assessments.

The project reveals strategies to create inclusive retail spaces, specifically focusing on coffee shops. Empirical analysis demonstrates the positive impact of inclusive architectural approaches on customer independence, comfort, and satisfaction. The findings provide practical recommendations for architects and designers to ensure coffee shops are designed with diverse needs in mind, advancing inclusive architecture in the retail sector. This approach not only caters to the requirements of individuals with disabilities but also underscores how ID in a retail space significantly enhances the overall satisfaction of a broader range of customers.*

Keywords:
Inclusive Design, ADA Compliance, Architectural Design, Retail Spaces, Customer Satisfaction.
**Work in progress.*

REVEALING UNEXPLORED DATA: ARCHITECTURAL DESIGN INSIGHTS THROUGH CROWD BEHAVIOR SIMULATIONS

Track 6: Transdisciplinary Design and Collaboration/Open Track
Knowledge Network: Environmental Design Research Education, Sustainable Planning, Design & Behavior
Methods: Interviews, Mixed Methods

Mariángel Meza Santiviago

Background:
A Case Study from Professionals' Perspectives.

Understanding the movement, interaction, and behavior of end-users within designed spaces is crucial in architectural design. Pedestrian and Crowd Simulations are computational tools that make this level of insight possible. However, there is a gap in addressing the insights derived from simulations, as well as in explaining the differences observed when employing this tool during the design process compared to when it is not utilized. Additionally, the theoretical frameworks of current research predominantly center on the foundational principles and intersections of computer science, often overlooking the involvement of practicing architects as active participants in research endeavors and as the primary audience for the studies.

This research explores the potential of Pedestrian and Crowd Simulations within the architectural practice from the viewpoint of current design professionals of an architectural firm in Minneapolis, MN. The study targets an existing building, 'Campbell Hall Institute of Child Development', designed by the main participants of the study, for a retrospective analysis as a case study. Grounded in decision-making, simulation driven design, architectural design processes, and grounded theory frameworks, the study employs semi-structured interviews, task-oriented interviews, and simulations as methods with a case study approach.

The findings allow for the identification of two key aspects: (1) various insights derived from simulation data that architects can utilize to inform their design decisions and (2) types of data that remained unexplored during the building's design phase but could have been unveiled through simulations. This underscores the potential impact of these insights on design decisions and aims to inform architects about the extractable insights from simulations. Consequently, this may enhance the effectiveness of design considerations in the initial stages of future projects when utilizing this type of software as a support tool.

Keywords:
Crowd behavior simulations, Architectural Design Process, Architectural decision-making.

**Work in Progress.*

STREETWISE SAFE HAVENS

Track 4: Design for Health and Wellbeing
Knowledge Network: Health, Interior Design
Methods: Ethnography, Mixed Methods
Claire Hayes Watson (Self-Employed)

Background:
Substance Use Disorder (SUD) and houselessness have many overlaps in terms of internal and external experiences. The two often go hand-in-hand - with substance use disorder as both a cause and effect of being houseless (Polcin, 2015). For this project, I have chosen to focus on houseless women with SUD because women are under-represented in SUD treatment services as a result of historical trauma and victimization, shame, stigma, and other gender-specific barriers (Rizzo et al., 2022). Using observation and theoretical review, the research aims to uncover how houseless individuals create a sense of shelter for themselves when living on the street and how that design language can be translated into the design of treatment and recovery environments.

References:
National Institute of Mental Health. (2023, March). Substance use and co-occurring mental disorders. https://www.nimh.nih.gov/health/topics/substance-use-and-mental-health
Polcin, D. L. (2015). Co-occurring substance abuse and mental health problems among homeless persons: Suggestions for research and practice. Journal of Social Distress and the Homeless, 25(1), 1–10. https://doi.org/10.1179/1573658x15y.0000000004
Rizzo, D., Mu, T., Cotroneo, S., & Arunogiri, S. (2022). Barriers to accessing addiction treatment for women at risk of homelessness. Frontiers in Global Women's Health, 3. https://doi.org/10.3389/fgwh.2022.795532

STROLL CITY

Track 2: Sustainable buildings and infrastructure
Knowledge Network: Cities & Globalization
Methods: Action/Participatory Research, Design Methods
Kweonneung Cho (Ji bang.co)

Background:
We are creating a bar street that experiences various bars from all over the world under the theme of 'alcohol' in the city center of Gunsan. The reason for building the streets with the core themes of alcohol and bar is that when the image of Gunsan was recalled, it was mixed in the neighborhood of Japanese journalism and the Chinese who were stationed in Gunsan and the Chinese port after liberation. And the various foods and residential cultures they made in perfume and residential culture also affected the local lifestyle of Gunsan, such as Japanese confectionery, sake, champon, bath culture, American club, etc. I would like to say that BAR culture, which is easy to accept new things and not obsessed with meeting and breaking up, is the locality of Gunsan.

UNVEILING INSIGHTS INTO CITY INTELLIGENT SERVICE MODULES

Yunsheng Su (Professor)

Goal:
Conduct a concise academic symposium to introduce, analyze, and discuss key concepts, categories, and characteristics of city intelligent service modules, fostering a deeper understanding among participants.

Objectives:
Analyze Course Context: Participants will critically analyze the background and objectives of the course, gaining insights into the broader context of city intelligent service modules.

Learn Essential Concepts:
Attendees will learn the fundamental concepts, categories, and characteristics shaping city intelligent service modules through a comprehensive presentation.

Understand Categorizations:
Scholars will deepen their understanding of the categorizations inherent in city intelligent service modules, exploring how these contribute to urban development.

Promote Interactive Discourse:
Encourage interactive discourse through a brief session of discussions and Q&A, providing a platform for participants to engage, share perspectives, and address queries.

Agenda:
Introduction to Course Background and Objectives (2 minutes):
Provide a succinct overview of the course, emphasizing its objectives and contextual relevance within the landscape of city intelligent service modules.
Explanation of Concepts, Categories, and Characteristics (3 minutes):

Delve into the core concepts, categories, and characteristics that define city intelligent service modules, facilitating an in-depth learning experience.
Interactive Discussion and Q&A (5 minutes):

Engage participants in a brief interactive session, encouraging them to analyze and discuss key takeaways while addressing any queries.

STEP UNLEASHED - CROSS-INDUSTRY DIALOGUES FOR URBAN INNOVATION

Xihan Cheng (Tongji University)

Goal:

This succinct symposium aims to transform the STEP theoretical model into a catalyst for cross-industry dialogues, encouraging participants to collectively learn, analyze, and understand urban innovation challenges in a condensed yet impactful session.

Objectives:

1. *Learn: Facilitate a collective learning experience for participants, providing a foundational understanding of the STEP theoretical model, including Social Needs Analysis, Technological Innovation Design, Entrepreneurial Business Strategies, and Policy Design Support.*
2. *Analyze: Encourage participants from diverse industry backgrounds to analyze urban challenges through the lens of STEP, fostering unique perspectives and insights.*
3. *Understand: Create an environment that promotes mutual understanding among symposium participants about the interconnected nature of social, technological, entrepreneurial, and policy dimensions within urban contexts.*

Agenda:

This dynamic symposium, condensed to a focused 20 minutes, is designed to ignite vibrant discussions:
Introduction to STEP (5 minutes)
Concise overview and historical context
Significance in addressing post-pandemic urban challenges

1. *Cross-Industry Panel (7 minutes)*
2. *Experts from varied sectors share condensed experiences applying STEP*
   Rapid-fire Q&A session to extract key insights
3. *Interactive Breakout Sessions (4 minutes)*
   Brief small-group exploration of urban challenges using the STEP model
4. *Highlighting unique perspectives and common challenges*
5. *Knowledge Synthesis (2 minutes)*
   Group representatives share succinct key findings
   Facilitated discussion on commonalities and diverse perspectives
   Application Workshop (2 minutes)
6. *Rapid-fire collaborative exercises applying STEP in diverse industry contexts*
   Quick sharing of real-time applications, challenges, and success stories

UNVEILING INSIGHTS FROM THE HUO-YAN PROJECT

Mingjun Zhong (Designer)

Goal:
Conduct an academic symposium to delve into the practical efficacy of the STEP theoretical model in the Huo-Yan project, driven by big data, fostering scholarly discussions and insights.

Objectives:
Analyze Practical Application:
Scholars will critically analyze the practical application of the STEP theoretical model in the Huo-Yan project, with a focus on its effectiveness within the context of big data.

Learn Success Factors:
Attendees will gain knowledge of the key success factors contributing to the effective implementation of the STEP model in the Huo-Yan project.

Understand Challenges:
Scholars will deepen their understanding of the challenges encountered in applying the STEP model, particularly in the domain of big data-driven initiatives.

Promote Scholarly Discourse:
Encourage scholarly discourse through group discussions, providing a platform for attendees to share their analyses and contribute to a collective understanding of the Huo-Yan project's intricacies.

Agenda:
Introduction to the Huo-Yan Project (15 minutes):

Overview of the Huo-Yan project, its academic significance, and the practical application of the STEP model driven by big data.
Group Discussions on Success Factors and Challenges (30 minutes):

Scholars engage in focused group discussions to analyze and discuss success factors and challenges in applying the STEP model within the Huo-Yan project.
Group Sharing and Plenary Discussion (15 minutes):

Groups share their scholarly insights, followed by a plenary discussion to deepen the academic understanding and address emerging research questions.

NEEDS AND DESIGN FOR AGING-FRIENDLY TOILETS IN RURAL AREAS

Track 4: Design for Health and Wellbeing
Knowledge Network: Environment-Gerontology, Residential Environments
Methods: Interviews, Observation
Zhe Wang (Henan University); Jingang Zhen (Henan University)

Background:
Typical toilets in traditional rural residence in China are aqua privies in order to avoid smell and located far away from living spaces. This type of layout design is unfriendly to older residents and even poses safety risks.

Aims:
Focusing on rural residence, this research aims to 1) analyze the factors influencing toilet uses in older residents and the needs for age-friendly toilets, 2) clarify the differences in toilet use among older residents by the spatial layout of their residence.

Methods:
Taking Daokou ancient town as an example, data on the spatial layout of rural residential environments and older residents' toilet use were collected through observation and interview. Specific measures include the size of the courtyard, location of the toilet(s) and bedroom(s), and the distances between them. Five residents and their residents were recruited in research. Quantitative and qualitative data analyses were conducted to identify factors affecting toilet use in older residents.

Results:
The main factors affecting older residents' toilet use included the availability and quality of toilet lighting, floor surface evenness and materials, the distance from a bedroom to the toilet, the availability of a roof in toilet cubicles, and the presence of handrails in the toilet.

Conclusion:
The safety of toilet environments is of top importance if promoting age-friendly rural residence. Safe toilet use can reduce the risk of falling and associated dangers, and age-friendly design can improve the experience of safe toilet use, especially in rural areas.

MINIMIZING GLARE AND IMPROVING VIEWS IN NASA'S SUSTAINABILITY BASE

Track 3: Sustainable Buildings and Infrastructure
Knowledge Network: Health, Work Environments
Methods: Case Study, Design Methods
Phillip Mead (University of Idaho)

Goal:

William McDonough's design of NASA's "Sustainability Base" research facility in Mountain View, California incorporates daylighting as one of its main strategies for conserving energy and improving the health of its occupants. However, this investigation found that glare is a common problem throughout the building, as it is in many daylit buildings, which can trigger headaches (Terman) and unneeded stress. This causes some users to override the automatic shading systems, compromising energy efficiency. (Inkarojrit) Using standard light meters, this research found the glare index for the buildings was 1 to 50, significantly higher than the recommended 1 to 10 light ratio which the design recommendations strive to implement.

Using the Vital Signs Case Study methodology developed by the Society of Building Science Educators, on-site light meter readings and analysis of views were taken in key areas of the building. Design recommendations were given to the NASA staff on how to improve views from parking lots to views of plant walls and to reduce glare, making work more comfortable and less stressful for the majority of the users. Simple recommendations were given from re-location of program workspaces to the placement of horizontal plant walls to block glary views of parked cars and improve views of nature.

Additionally, it was recommended that the insertion of simple solar tubular skylights on the interior side of outer walls could reduce glare to more tolerable levels near the 1 to 10 ratio. To conclude, if NASA was able to make the improvements, then further follow-up studies could be made for glare reduction and stress reduction.

References:

Inkarojrit, Vorapat. "Balancing Comfort: Occupants' Control of Window Blinds in Private Offices," Doctoral Dissertation, UC Berkeley. 2005
Terman, Michael, and Jiuan, "Light Therapy for Seasonal and Nonseasonal Depression: Efficacy, Protocol, Safety, and Side Effects." CNS Spectrums, Number 8, Volume 10, August 2005

A STUDY ON THE NUMBER OF REVIEWS AND RATING DIFFERENCES OF HOTELS IN THE SUBURBS OF BEIJING BASED ON CTRIP.COM DATA--TAKING 10KM AROUND YANQI LAKE AS AN EXAMPLE

Track 3: Sustainable Buildings and Infrastructure
Knowledge Network: Cities & Globalization, Sustainable Planning, Design & Behavior
Methods: Case Study, Statistical/Modeling Analysis
Mo Zhang (Beijing Jiaotong University)

Background:

With the rapid economic and social development, the living standard of residents has been greatly improved, suburban leisure and vacation tourism has become the main vacation choice for urban residents on weekends and holidays. Tourists are not only concerned about the quality of tourism programs, but also concerned about the environmental quality of hotels in the tourist suburbs. Good suburban hotels can fully meet the needs of tourists' tourism activities, and at the same time, they are strongly supported by government policies, which in turn can realize the comprehensive economic, social and ecological benefits of tourism resources, and ultimately promote the sustainable development of the tourism industry.

This study takes the suburban hotels in the 10km area around Yanqi Lake in Huairou District, Beijing, which is representative of suburban scenic spots, including 3A, 4A and 5A scenic spots within 10km, and utilizes new data such as government public data, POI location data and Ctrip travel data to explore the influence of factors such as transportation, service facilities, price and internal supporting facilities of suburban hotels on the development of scenic suburbs by means of spatial syntactic analysis and statistical analysis. The internal supporting facilities of the hotel influence the number of reviews and review star rating of suburban hotels around scenic spots, so as to explore the reasons that lead to the differences in the number of reviews and ratings of different suburban hotels in tourist scenic spots, and to explore a new path of recognizing the influencing elements of suburban hotels in tourist scenic spots, so as to provide certain reference for the site selection of new suburban hotels.

The main conclusions are: the crowd's choice of suburban hotels is positively related to the degree of choice of dining, business and sports and 3000-meter radius of suburban hotel facilities, of which the number of dining functions has the most significant effect on the number of reviews of suburban hotels. There was a significant positive correlation with the number of suburban hotel rooms, star rating and distance to the city center, and a regression equation was developed to predict the number of reviews: the top 25% of suburban hotels with the highest number of reviews were selected, and the two parameters of CHOICE_R3000 and the number of suburban hotels in a 1000-meter radius were counted as requirements to be fulfilled by new suburban hotels.

REGENERATING THE URBAN VOID: DESIGN STRATEGIES FOR SUSTAINABLE & INCLUSIVE COMMUNITIES

Track 3: Sustainable Buildings and Infrastructure
Knowledge Network: Sustainable Planning, Design & Behavior
Methods: Case Study, Photo-based Methods
Tyler Young (UMASS Amherst); Caryn Brause (University of Massachusetts)

Abstract:
Central to sustainable urban renewal is the aspiration to cultivate spaces that evoke creativity, joy, and promote community engagement. However, amidst careful design lies a stark reality - spaces neglected, misused, underused, or simply wasted (Hwang & Lee, 2018). An estimated 30% of urban land worldwide is vacant or underused (UH-Habitat 2016). In the face of the impending climate crisis, the demand for livable and sustainable cities has never been greater. Within this context, the potential to leverage urban voids in service to urban regeneration holds tremendous potential for transformation.

This thesis employs a design research methodology to propose strategies for urban regeneration in Springfield, Massachusetts. This thesis responds to the needs of the site while fostering Transit-Oriented Development and the revitalization of a historic theater. Design proposals at the urban, block, and human scale aim to create built environments that actively contribute to catalytic investment in order to link Main Street to an adjacent river while restoring and renewing natural systems (Mang et al., 2016). These initiatives, integrated with broader urban renewal efforts, aim to enhance community connectivity and preserve cultural heritage, embodying a pivotal paradigm shift that perceives neglected urban spaces not as liabilities but as untapped potential.

References:
1. *Hwang, S. W., & Lee, S. J. (2018). Unused, underused, and misused: An examination on urban void spaces. Urban Research & Practice.*
2. *Mang, P., & Haggard, B. (2016). Regenerative Development and Design: A Framework for Evolving Sustainability. Wiley.*
3. *UN-Habitat. (2016). Urbanization and Development: Emerging Futures. United Nations Settlements Programme.*

RESEARCH ON ENVIRONMENTAL PLANNING TECHNIQUES TO REDUCE CRIME RATE: FOCUSING ON CASE STUDIES IN LOW-RISE RESIDENTIAL AREAS NEAR UNIVERSITIES OF SOUTH KOREA

Track 4: Design for Health and Wellbeing
Knowledge Network: POE/Programming, Residential Environments
Methods: Case Study, Observation
Sungeun Shin (Yonsei university); Soohyun Lee (University of Yonsei); MiKyoung Ha (Yonsei University)

Abstract:
In Korea, despite a low crime victimization rate of 1.49%, fear of crime is high at 23.1%. Seoul's Dongjak District, particularly 6th Street, a high-crime, low-rise residential area near universities with many students and women, necessitated a design project in 2020 for crime reduction. This study reviewed the project's design process, impact on residents' perceptions, and assessed daytime and nighttime CPTED (Crime Prevention Through Environmental Design) applications in various spaces.

VERNACULAR ARCHITECTURE AS HUMAN-CENTRIC ENVIRONMENT TO PROMOTE SUSTAINABLE DESIGN APPROACH IN ARCHITECTURAL EDUCATION IN TUNISIA

Track 3: Sustainable Buildings and Infrastructure
Knowledge Network: Cultural Aspects of Design, Environmental Design Research Education
Methods: Case Study, Observation
Khansa Dhaouadi (University of Liège)

Background:
This paper investigates principles of vernacular architecture within the framework of a Human-Centric Environment, aiming to advance a Sustainable Design Approach in Architectural Education in Tunisia. The principles embedded in this heritage not only emphasize the human experience, but also foster a deep commitment and connection to one's surroundings.

Employing an empirical approach, our research study combines two observations conducted within the context of Tunisian architectural schools.

The first case study involves a workshop for a sustainable design project for Djerba Island, as a Summer School for students from different levels. The other case study focuses on the Second-year design studio within a contemporary architectural and urban intervention in Nefta.

These two environments are fascinating microcosms of popular vernacular resilience in Tunisia in response to human needs, identities, and socio-cultural influences. The findings highlight the potential of vernacular architecture as an effective strategy for the enhancement of sustainable design education from a human-centric perspective. Through an analysis of pedagogical practices and pedagogies, opportunities for integrating these principles into the curriculum emerge.

The focus of the results is on the lessons learned from this holistic approach to promoting a comprehensive understanding of sustainability, while also being contextualized.

These opportunities may be accompanied by the recognition of potential challenges, including change resistance, resource constraints, knowledge, and teaching methodology needs.

It is imperative to address and overcome these challenges to ensure the integration of vernacular architectural principles as we strive for a paradigm shift towards a more human-centered approach to architectural education.

A STUDY ON THE REGENERATION SPACE UTILIZING CLOSED SCHOOL APPLYING THE PLANNING ELEMENTS OF PLACEMAKING

Track 4: Design for Health and Wellbeing
Knowledge Network: Interior Design, Sustainable Planning, Design & Behavior
Methods: Case Study, Survey
Yujin Kim (Dept.of Interior Architecture Design, Hanyang University,Seoul, Republic of Korea);
YeonSook Hwang (Hanyang University)

Background:
Recently, the school-age population has continued to decrease due to the low birth rate, and as a result, schools are closing and becoming neglected. Although the closed school has lost its educational function, it is considered an important resource for promoting regional revitalization from a geographical and emotional perspective. In this respect, it is closely related to placemaking that improves the quality of life of local residents through the values and culture unique to the community based on a people-centered approach. The purpose of this study is to derive planning elements of placemaking and analyze regeneration spaces utilizing closed school. Based on previous research and the guidelines for the use of closed school, a checklist was derived that reorganized planning elements of placemaking applicable to regeneration space utilizing closed school. An on-site survey was conducted by eight case studies that were reopened as cultural and experiential education spaces located in metropolitan cities in Korea. The results of this study are as follows.

First, this study derived 'Enhancement of Publicness', 'Forming a Bond', 'Utilization of Resources', 'Convenience of Access', and 'Image Formation' as planning elements of placemaking. Second, the regeneration space utilizing closed school is created through the methods of 'Remodeling' and 'New Construction'. Both remodeling and new construction methods preserve and utilize existing school, which are important resources for the community, it is necessary to consider ways to directly or indirectly express the preservation value of space. Accordingly, rather than simply organizing an exhibition space related to the historical elements of existing school, the space itself, in the non-physical aspect of placemaking, provides a variety of experiences to users through emotional connections such as 'memories, experiences, and emotions' related to the existing place. Third, applying the 'Image formation' element was the most insufficient. Regarding image formation, it is necessary to consider the reflection of a design that visually expresses the new purpose of a closed school through regeneration. Visual effects can be created by symbolic images in the main entrance and lobby space, where the most movement and inflow of users occurs. Fourth, in order to enhance of publicness, the number of users should be limited through reservations in consideration of congestion in experiential spaces, but other convenience spaces should be free of time restrictions so that local residents can use them at any time. Regeneration spaces utilizing closed school should be used as community spaces for local residents in both physical and emotional aspects by more actively applying the elements of placemaking.

Keywords:
Placemaking, Closed School, Regeneration Space Utilizing Closed School

References:
Jung, A. R. (2009). 'Place Making' As a Cultural City Strategy [Unpublished master's thesis]. Pusan National University.
Local Culture Promotion Agency. (2023). 2022 Project to Revitalize Cultural Regeneration in Idle Space: A Study on Cultural Regeneration Using Closed Schools.
Shin, S. G. (2011). A Study on the Efficient management methods of the Property of Closed School [Unpublished master's thesis]. Sang-Ji University.

ADDRESSING THE CHILDHOOD OBESITY CRISIS

Track 6: Transdisciplinary Design and Collaboration/Open Track
Knowledge Network: Health
Methods: Mixed Methods

Shahd Alkaabi (VCUarts Qatar); Matthew Holmes Dallimore (VCUQ);
Marwa Abdelmonem (Virginia Commonwealth University)

Abstract:
Qatar's high frequency of physical inactivity, combined with other variables such as an inadequate diet, has contributed to the rise in childhood obesity (Ibrahim et al., 2018). Lack of sufficient sports facilities, time, social support, and motivation were the most generally stated impediments to physical exercise in the region, followed by gender and cultural norms, undereducation, and arid climate (Chaabna et al., 2022). In Qatar, over half of all school-aged children are overweight or obese. Childhood obesity is linked to adult health problems such as hypertension, diabetes, metabolic abnormalities, cardiovascular risks, and psychological difficulties (M;). To further explore this issue, a qualitative approach was adopted for the present study. An extensive review of existing literature was conducted to examine Qatar's children's obesity problem, including its causes, impacts, and preventative measures. Addressing gaps in the literature regarding the causes and preventive strategies for childhood obesity in the region, the study conducted three pilot semi-structured interviews. These interviews involved professionals, including a physician, a nutritionist, and a physical therapist. The research findings show that obesity in Qatar can be attributed to a genetic abnormality but the increased prevalence results from societal environmental factors. The society in Qatar is affected and at more risk due to the infrastructure, female barriers, and living in extended families that all encourage obesity. Urbanization has also drastically contributed to the increase in obesity in the country because more people are reliant on cars, and the culture often involves high-calorie foods associated with socioeconomic statuses. A problem caused by obesity is poor performance in school. Obesity and poor performance in school have a complex relationship that affects both physical and academic well-being. Furthermore, children who suffer from obesity are less physically active, which reduces peer interaction and their capacity to play with other children, thus affecting social bonding negatively. By addressing childhood obesity, an action should be taken to tackle this problem. A positive change in children's lifestyles and eating habits is encouraged. The goal of the design is to foster children to healthy habits without feeling the force to battle it. An edutainment way of attracting the children for a better healthy lifestyle.

References

Ibrahim, I., Hammadi, E. A., Sayegh, S., Zimmo, L., Neama, J. A., Rezeq, H., Al-Mohannadi, A. S., & Al-Thani, A. A. (2018, January 2). Results from Qatar's 2018 report card on physical activity for children and Youth. Human Kinetics. https://journals.humankinetics.com/view/journals/jpah/15/s2/article-pS400.xml

Chaabna, K., Mamtani, R., Abraham, A., Maisonneuve, P., Lowenfels, A. B., & Cheema, S. (2022, June 16). Physical activity and its barriers and facilitators among university students in Qatar: A cross-sectional study. MDPI. https://www.mdpi.com/1660-4601/19/12/7369

M;, A. (n.d.). Obesity among children and adolescents in Qatar: Protocol for a scoping review. BMJ open. https://pubmed.ncbi.nlm.nih.gov/35173008/

AN INVESTIGATION OF THE ROLE OF BRIDGES IN VULNERABILITY ASSESSMENTS: THE CASE OF CEDAR KEY

Track 6: Transdisciplinary Design and Collaboration/Open Track
Knowledge Network: Sustainable Planning, Design & Behavior
Methods: Case Study, Observation
Ana Orosco (University of Florida); Jeffrey Carney (University of Florida); Christian Calle (University of Florida)

Background:
Bridges are a familiar component of the built coastal environment. They are an essential feature of paths that maintain connectivity across rivers, bays, canals, or gulfs. Although easily recognizable as a benchmark, the dynamic relationship between bridges and water might go unnoticed by the casual observer beyond the functional crossing the structure provides. When it comes to vulnerability assessments and the role of bridges through these same landscapes, the challenges are inherently transdisciplinary. Hydrologic features are in constant flux, requiring bridge engineers to follow strict design guidelines and monitor frequently. City officials and planners are concerned with various policy and funding matters that affect evacuation routes, maintenance, and public perception. In addition, an abundance of ecological issues intersect with all of these matters. This paper studies how bridges are treated in a vulnerability assessment (VA) to support the decision-making process of adaptation planning. In particular, the unique conditions in Cedar Key, Florida (FL), will be analyzed as a compelling case study for understanding the possibilities of broadening the framework for bridge data collection as issued by the state's Standardized Vulnerability Assessment (Standardized Vulnerability Assessment: Scope of Work Guidance, 2022). This research was carried out in conjunction with the ongoing adaptation plan for the city.

Through an investigative approach, a combination of Florida Department of Transportation (FDOT) bridge inspection reports, historical hurricane data, topographic Lidar survey, sea-level rise (SLR) projections, and flood risk mapping is used to analyze existing processes and identify potential gaps. The study found that the minimum requirements in the FL VA may not be sufficient to flag bridges on evacuation routes as a high priority, while a more in-depth exploration of additional markers suggests otherwise. In conclusion, the paper underscores the significance of a comprehensive and transdisciplinary approach to vulnerability assessments, particularly in assessing bridges on evacuation routes. It proposes further investigations into past vulnerability assessments across municipalities and the inclusion of additional criteria for bridges in future guidelines. The lessons learned from Cedar Key offer valuable insights into refining vulnerability assessment practices and addressing the complex intersection of sea-level rise, flooding, and infrastructure vulnerabilities. This deeper insight into the role of bridges in a VA is meant to encourage more significant cross-pollination between the different fields that overlap where a bridge stands.

ANALYZING USER INTERACTIONS IN VIRTUAL REALITY CLASSROOMS: INSIGHTS FROM VIDEO DATA ANALYSIS FOR ENHANCING EDUCATIONAL OUTCOMES IN LANDSCAPE ARCHITECTURE

Track 6: Transdisciplinary Design and Collaboration/Open Track
Knowledge Network: Communication, Cyberspace & Digital Environments
Methods: Design Methods, Mixed Methods
S Y Andalib (Student at Texas Tech University); Muntazar Monsur (Texas Tech University)

Abstract:
This study delves into the utilization of VR in landscape architecture education, emphasizing the analysis of video data to enhance educational outcomes. XR technologies, including VR, have been increasingly integrated into educational settings, showing promise in boosting learning outcomes, student engagement, and practical skill development (Shankar et al., 2023). VR, in particular, has been linked to enhanced social interactions in higher education contexts (Slavova & Mu, 2018), suggesting its potential to improve student performance (Alfarsi et al., 2021).

This research focuses on the critical role of video data analysis in understanding user interactions within VR classrooms. By examining behavioral patterns and engagement levels, the study aims to assess the effectiveness of VR as an educational tool. Prior research has indicated that a deeper understanding of user interactions in VR can lead to more effective VR applications for education, highlighting the importance of real-time data analysis for user behavior and interaction design (Kose et al., 2019; Hedrick et al., 2022). Rossi et al. (2020) further stress the need for comprehensive data analysis to grasp user navigation and interaction patterns, essential for creating user-centric VR systems. Employing the Video Data Analysis (VDA) framework outlined by Nassauer and Legewie (2021), this study analyzes screen capture data from sixteen participants engaged in a VR session on brick bonding techniques. The goal is to identify and address the limitations of the VR session, proposing improvements to enhance its educational value.

In summary, this research underscores the importance of video analysis in dissecting user interactions within VR classrooms, aiming to inform the development of more effective and engaging VR-based educational platforms.

References
Alfarsi, G. M., Yusof, A. B. M., Romli, A. M., Tawafak, R. M., Malik, S. I., Jabbar, J., & Rsuli, M. E. B. (2021). A Review of Virtual Reality Applications in an Educational Domain. Int. J. Interact. Mob. Technol.
Hedrick, E., Harper, M., Oliver, E., & Hatch, D. (2022). Teaching & learning in virtual reality: Metaverse classroom exploration. 2022 Intermountain Engineering, Technology and Computing (IETC),
Kose, A., Tepljakov, A., Abel, M., & Petlenkov, E. (2019). Towards assessment of behavioral patterns in a virtual reality environment. Augmented Reality, Virtual Reality, and Computer Graphics: 6th International Conference, AVR 2019, Santa Maria al Bagno, Italy, June 24–27, 2019, Proceedings, Part I 6,
Nassauer, A., & Legewie, N. M. (2021). Video data analysis: A methodological frame for a novel research trend. Sociological methods & research, 50(1), 135-174.
Rossi, S., Ozcinar, C., Smolic, A., & Toni, L. (2020). Do users behave similarly in VR? Investigation of the user influence on the system design. ACM Transactions on Multimedia Computing, Communications, and Applications (TOMM), 16(2).
Shankar, D. A. U., Tewari, D. V., Rahman, M., Mishra, D. A., & Bajaj, M. K. K. (2023). Impact of Virtual Reality (Vr) and Augmented Reality (Ar) in Education. Tuijin Jishu/ Journal of Propulsion Technology.
Slavova, Y., & Mu, M. (2018). A Comparative Study of the Learning Outcomes and Experience of VR in Education. 2018 IEEE Conference on Virtual Reality and 3D User Interfaces (VR), 685-686.

ARCHITECTURAL PSYCHOLOGY OF STRESS: ANALYZING THE IMPACT OF MATERIAL, FORMAL, AND AESTHETIC FEATURES ON STRESS REACTIONS USING BIOPHILIC PATTERNS

Track 4: Design for Health and Wellbeing
Knowledge Network: Health, Sustainable Planning, Design & Behavior
Methods: Experimental/Quasi-Experimental Research, Mixed Methods
Clarice Meffert (Human Health & Performance Lab - Carnegie Mellon University)

Background:
Stress, by the definition set forth by Cohen et al. 1995, is inherently environmental. Yet, somehow, in a world where we control many aspects of our environment, stress is becoming an increasingly prevalent and severe concern. The modern environment – unnaturally dull, and lifeless – is not only unstimulating but also lacks the biophilic connection to nature that relieves psychological (Yin et al., 2020, Demattè et al., 2018; Zhang et al., 2017) and physiological stress (Zhang et al., 2016; Tsunetsugu et al., 2007).

The impact of environmental materiality on stress is largely overlooked despite its broad human mental health implications. Existing research in environmental materiality focuses on traditional woods and overlooks alternative materials despite their sustainability and availability. Additionally, biomorphic shaping of finish material, though beautiful, lacks empirical evidence regarding its possible benefits for stress management. This study investigates the architectural psychology of stress in an independent working environment using three distinct finish materials: conventional plaster, bare biophilic concrete, and bamboo paneling. We developed a within-subject study where 25 participants experienced each environment in independent 30-minute sessions over three days. A gamified Stroop task was used for psychological stress induction, while stress reduction interventions included mindfulness or control 20-minute audio recordings. Biometric data collection involved continuous pulse rate and blood pressure measurements, utilizing a clinical-grade GE DinaMap Carescape™. Self-report measures gauged stress levels, mood, and somatic symptoms. Data analysis employed IBM-SPSS, revealing significant findings regarding intervention efficacy and environmental conditions. Bamboo exhibited comparable stress mitigation to conventional woods. Bamboo environments showed the most pronounced, significant, physiological stress reduction of blood pressure throughout the session, while biomorphic texture neutralized physiological stress induction. Although environmental design significantly decreased physiological stress, it did not significantly impact self-reported psychological stress. Self-reported somatic stress was significantly decreased and mood improved by the mindfulness audio, which also positively shifted environmental perceptions. These findings offer insights into the role of architectural design and programmatic interventions in stress management, highlighting the potential of biophilic elements and mindfulness practices in promoting psychological well-being.

References;
Cohen, S., Kessler, R. C., & Gordon, L. U. (Eds.). (1995). Measuring stress: A guide for health and social scientists. Oxford University Press.
Demattè, M.L., Zucco, G.M., Roncato, S. et al. New insights into the psychological dimension of wood–human interaction. European Journal of Wood and Wood Products. 76, 1093–1100 (2018). https://doi.org/10.1007/s00107-018-1315-y
Tsunetsugu, Y., Miyazaki, Y., & Sato, H. (2007). Physiological effects in humans induced by the visual stimulation of room interiors with different wood quantities - journal of wood science. SpringerOpen. https://jwoodscience.springeropen.com/articles/10.1007/s10086-006-0812-5
Yin, J., Yuan, J., Arfaei, N., Catalano, P. J., Allen, J. P., & Spengler, J. D. (2020). Effects of biophilic indoor environment on stress and anxiety recovery: A between-subjects experiment in virtual reality. Environment International, 136, 105427. https://doi.org/10.1016/j.envint.2019.105427
Zhang, X., Lian, Z., & Ding, Q. (2016). Investigation variance in human psychological responses to wooden indoor environments. Building and Environment, 109, 58–67. https://doi.org/10.1016/j.buildenv.2016.09.014
Zhang, X., Lian, Z., & Wu, Y. (2017). Human physiological responses to wooden indoor environment. Physiology & Behavior, 174, 27–34. https://doi.org/10.1016/j.physbeh.2017.02.043

ASSESSMENT OF INTERIOR DESIGN ATTRIBUTES FOR OLDER WOMEN WHO LIVE IN RURAL AREAS TO AGE-IN-PLACE

Track 4: Design for Health and Wellbeing
Knowledge Network: Health, Interior Design
Methods: Focus Groups, Survey

Aditya Jayadas (Oklahoma State University); Rose Small (Oklahoma State University); Tilanka Chandrasekera (Oklahoma State University)

Background:

The United States (US) is aging, and the US Census Bureau (n.d.) reported that 16.8% of all the individuals who currently reside in the US are over the age of 65 years. The Center for Disease Control (CDC) reports that there is a difference in life expectancy at 65 years between men and women, with women expected to live 2.7 years longer (CDC, 2022). Farage, Miller, Ajayi and Hutchins (2012) reported changes in vision, hearing, mobility, reflexes, flexibility, coordination and balance as you age. These changes can vary across gender. For instance, it has been reported that men have higher visual spatial ability when compared to women (Pauls, Pettermann & Lepach, 2013). In addition to gender differences, living in a rural or urban area can pose unique challenges for older adults. Paul, Fonseca, Martin & Amado (2003) reported that older adults who live in urban areas can have a greater sense of autonomy while also having a greater feeling of anxiety when compared to those who live in rural areas. Thus, there is a critical need to not only better understand differences in physical and cognitive challenges in older adults as they age-in-place, but also understand differences in preferences between male and female older adults, and older adults who live in urban or rural areas. The purpose of this study is to determine if there are differences in physical and cognitive challenges that older men and women face in their home as they age-in-place with regard to their environment along with their preferences.

Methodology:

*The study will be carried out in two phases. For Phase I, 24 older adults will be recruited for a focus group study via Zoom. Specifically, two groups of 12 older adults (six older women and six men) who live in rural and urban areas will be recruited. The participants will be asked open-ended questions on physical and cognitive challenges they face in their home environment along with preferences for different design attributes within their home. Audio recording will be used to capture this data, and content analysis will be used to analyze the data. For Phase II, 60 older adults, in two groups of 30 older adults (15 older women and 15 men) who live in rural and urban areas will be recruited. A Qualtrics survey with Likert scale questions on physical and cognitive challenges that older adults experience in their home environment along with preferences for different design attributes in their home. For analysis, a 2*2 analysis of variance (ANOVA) will be used with two levels of gender (male vs female) and location (rural vs urban). A significance level of 0.05 will be used.*

Results:

The data collection is ongoing. The preliminary findings indicate that there are differences between older male and females living in rural and urban areas.

Implications:

The successful completion of this study will result in a positive impact on the physical and mental health of older adults, especially older women who live in rural areas.

BIOPHILIC DESIGN AND THE COVID-19 PANDEMIC: LITERATURE REVIEW AND NOVEL TOOLS TO ASSESS MENTAL HEALTH AND WELL-BEING

Track 4: Design for Health and Wellbeing
Knowledge Network: Health, Residential Environments
Methods: Mixed Methods

Erin M. Driver (University of South Carolina); Altaf Engineer (College of Architecture, Planning and Landscape Architecture, University of Arizona); Wooyoung Jung (Department of Civil and Architectural Engineering and Mechanics, University of Arizona)

Background:
The COVID-19 pandemic brought concerns about mental health and well-being to the forefront of public health discussions due to closures of community institutions (e.g., restaurants, gyms), the transition from in-person to remote work and education, and at-home isolation practices due to the risk of morbidity and mortality. Subsequent research into the relationship between greenspace access and mental health showed improved health associated with increased access to greenspace. The authors were interested in biophilic design and its relationship to health and well-being during the COVID-19 pandemic, as well as novel methods for assessing behaviors related to biophilia during the same period. In this study, a literature review was performed using the search criteria: biophilic design or biophilia, COVID-19, and mental health or stress. Results revealed only 20% of peer-reviewed publications explored this relationship using experimental research during the target period (2019-2023), with <10% explicitly focused on building design. With these limited results, a subsequent assessment was performed on publicly available information from the US housing market. Housing information was collected beginning November 2023 using the Zillow real estate marketplace where data were sorted by for sale, for rent, and recently sold categories for all units and stratified by city, mountain, park or water view.

These analyses were conducted at the state-level including the District of Columbia and at the county-level covering 10% of US counties. Available data from the Substance Abuse and Mental Health Services Administration, Mental Health American, and the US Department of Agriculture were included in the analyses. At the state-level, results showed a consistently lower percentage of rental units with a water, park, or mountain view (1.4%), as compared to those listed for sale or previously sold (12.9% and 10.1%, respectively). Preliminary assessments using data comparisons show a statistically significant inverse relationship between state-level poverty data and the percentage of units with a biophilic view (water, park, mountain) for rentals, and those units recently sold or for sale. Additional analyses will be performed using state- and county-level data. The goal of these assessments is to determine if these publicly available housing data are correlated with health-related datasets and if there is utility in using these real-time housing data as a surrogate for less frequently collected health metrics.

CHILD'S PLAY: A CASE STUDY OF AN EVIDENCED-BASED HEALING SPACE FOR CHILDREN IN A REFUGEE CAMP

Track 4: Design for Health and Wellbeing
Knowledge Network: Children, Youth & Environment
Methods: Case Study
Renee Desneige Christensen (Extensions)

Background:
The purpose of this case study was to identify the impact of applying an evidence-based approach which incorporates neuroscience to designing therapeutic environments for children in refugee camps. This case study included a transdisciplinary literature review to establish a preliminary framework for an outdoor play space designed for children at a refugee camp in Cyprus. The project was selected to evaluate the design-based strategies as an essential modality for trauma treatment for children. This project achieves the ambitious aim of translating concepts from neuroscience into trauma-informed design strategies. The outcome of this case study supports the development of a theoretical framework for the design of children's play spaces as an important factor in trauma-based interventions in refugee camps.

CLASSIFICATION OF AIR CLEANING TECHNOLOGIES IN THE BUILDING SECTOR: A FOCUS ON HEALTH AND SAFETY

Track 4: Design for Health and Wellbeing
Knowledge Network: Health
Methods: Statistical/Modeling Analysis
Elham Hasani Aalvy (TTU)

Background:
As a result of the global COVID-19 pandemic, there has been an emerging market for air cleaning development to reduce the risks of airborne pathogen transmission within indoor spaces. Most of these technologies, however, are under questions regarding their safety, and health concerns. In this context, this research emphasizes how important it is to select technology carefully within this bombarded air cleaning market, tailoring the choice based on the environment and task involved. The purpose of this study is to use a straightforward scoring system to systematically compare various air cleaners to provide an objective comparison. We assign numerical values to key metrics, including the removal efficiency, the energy consumption, the noise level, and the clean air delivery rate. Using a methodical approach that identifies criteria, assigns weights, and evaluates technologies, we offer an objective assessment of each technology's effectiveness. From a human-centered perspective, the paper will highlight the potential harm and ineffectiveness associated with certain technologies that produce unhealthy byproducts.

DESIGN COMMUNITY HOSPITALS FOR LOCAL OLDER ADULTS' MEDICAL NEEDS

Track 4: Design for Health and Wellbeing
Knowledge Network: Environment-Gerontology, Health
Methods: Observation, Survey
Zhe Wang (Henan University); ChenChen Chi (Henan University)

Background:
The number of older Chinese (65+) continues to rise and will reach up to 366 million by 2050. Many of them suffer from one or more chronic diseases that require regular hospital visits. In the hierarchical healthcare systems in China, community hospitals are at the first level and of high-level importance providing medical consultation and treatment for local residents. Community hospitals are challenged by their needs for geriatric care. The needs and the appropriate design of a community hospital to address the needs should be investigated in the local contexts.

Aims:
The purposes are 1) to clarify the needs of community older adults for medical care and 2) to investigate the real-world challenges that community hospitals meet in order to address the needs.

Methods:
The research site is in the central part of City of Zhengzhou, China. Based on the distribution of older populations, we selected seven community hospitals as research examples and focused on the five residential communities assigned to these hospitals for healthcare services. Questionnaire surveys will be conducted to collect first-hand data on the medical needs of older community residents. Field observations and staff interviews have been applied in the community hospitals, with an emphasis on the type and amount of healthcare services provided on-site (e.g., clinical departments, preventive healthcare, medical technologies), building layouts and functional flows in the hospital, and staff's experience providing geriatric care. Data will be analyzed by using quantitative and qualitative methods, including statistical investigations and content analysis.

Results:
Regarding the needs for geriatric care, findings will be summarized based on the data from the 'real' world. Regarding the challenges to community hospital design, the expected findings include insufficient floor areas, unappropriated functional flows and layouts of clinical or inpatient departments for geriatric care. The positioning of a community hospital, the overall functions of a hospital building, and specific medical care services required by local older adults will be investigated in the local contexts.

Conclusion:
From the perspective of environmental design and planning, the current situations of community hospitals in China need investigation and the associated environmental design and planning should be refined with empirical evidence. Our findings can be used as a theoretical reference for community hospitals' renovation, future design and construction.

A TRAUMA-INFORMED APPROACH TO ACADEMIC SOCIAL SPACES

Track 4: Design for Health and Wellbeing
Knowledge Network: Environmental Design Research Education, Interior Design
Methods: Mixed Methods

Anna Ellis

Background:

Trauma can be minor, short-lived, or long-term and can have any combination of physical, emotional, psychological, or cognitive effects on individuals. Generation Z ages 13 to 25 are facing higher rates of anxiety, depression, and distress than any other generation, resulting in an increased demand for mental health services (American Psychological Association, 2022). This prevalence of trauma-related challenges emphasizes the impact of trauma as a significant public health concern.

In addition to affecting one's life experiences, trauma influences how one experiences the environment (SAMHSA, 2014). Specifically, the design of the built environment can impact users' perceptions of those spaces and affect their wellbeing. Because of the pervasive effects that trauma can have on people, users who have experienced trauma may not always feel comfortable or motivated to engage with public spaces. This research study concerns trauma-informed design approaches and their potential to provide spaces that address the unique needs of students who have experienced trauma. Such an approach may effectively mitigate stress levels and, in turn, positively impact users by enhancing their overall wellbeing.
Previous research suggests that trauma-informed design approaches in homeless shelters and supportive housing may have a positive effect. However, there is a gap in inquiry on how a trauma-informed design framework can be applied to other building typologies, such as academic environments, for the betterment of trauma-experienced students. University administrations, architects, and designers seek to design academic spaces that would better meet student needs; however, existing university campus facilities may not support the needs of students who have experienced trauma (American Psychological Association, 2022).

This Master of Fine Arts thesis research study uses a mixed methods approach to answer the primary research question, "What architectural qualities influence the academic social space preferences of college students who have experienced trauma?" Surveys, including image-based questions to gauge architectural preferences and reflection questions to learn more about user experiences, were distributed online to currently enrolled college students who have experienced trauma, yielding a total of 100 completed surveys. Ten follow-up interviews took place to gain a deeper understanding of participants' choices. Participants took into consideration these built environment qualities based on the researcher's informed expertise and literature review: scale and proportion, access to views, curvilinearity and rectilinearity, biophilic elements, lighting, color, and materiality. Participants were further asked how these design qualities affect their emotional wellbeing.
This study intends to assist designers in making meaningful design decisions that will have a positive impact on college students who have experienced trauma and mental health issues. Data analysis is currently in progress. However, early findings suggest that certain spatial qualities may positively contribute to students' sense of wellbeing, including features such as higher ceilings, access to views, curvilinear elements, biophilic elements, layered lighting, prospect and refuge, lighter colors, and natural materials. Final results will be shared in the presentation if selected.

References:

American Psychological Association. (2022, November 15). Increased need for mental health care strains capacity. https://www.apa.org/news/press/releases/2022/11/mental-health-care-strains
Substance Abuse and Mental Health Services Administration. (2014). Trauma-Informed Care in Behavioral Health Services. https://store.samhsa.gov/sites/default/files/d7/priv/sma14-4816.pdf

EFFICIENT PATIENT CARE: UNDERSTANDING THE ROLE OF THE ENVIRONMENTAL DESIGN

Track 4: Design for Health and Wellbeing
Knowledge Network: Health, Interior Design
Methods: Observation

Elaheh Malek Zadeh (Kent State University); Sara Bayramzadeh (Kent State University); Parnia Azini (Kent State University); Haya Mehar (Kent State University)

Background:
The critical environment of the Level I trauma room must be prepared to stabilize patients with life-threatening injuries at all times. Therefore, efficient patient care in the form of uninterrupted workflow must be provided. Inefficiencies, such as avoidable interruptions, are identified as one of the causes of deaths among patients. This ongoing study aims to identify design factors that lead to inefficiencies in trauma rooms during patient resuscitation.

Using an observational approach, this study video recorded cases of patient treatment in trauma rooms in a Level I trauma center. The study obtained approval from the Institutional Review Board. A total number of 10 videos are analyzed and more videos are expected to be recorded and analyzed. The videos have been analyzed both quantitatively and qualitatively. The videos were first reviewed to identify design factors that seemed to hinder or facilitate workflow. Then, categories of themes were extracted to represent such factors more systematically. The video analysis was conducted by two researchers independently, after an initial training and screening period where discrepancies of reviews were discussed and resolved. This was to ensure the intercoder reliability for rigorous analysis.

The results found various design elements that can play a role as facilitators for or barriers to workflow in Level I trauma rooms. Some of these design factors include but are not limited to adjacencies, distances, proximities as well as sightlines, lighting, and acoustical issues. Examples include difficulty reaching instruments, searching for supplies that are not visible, inappropriate lighting for specific tasks, and physical components obstructing movement among others.

Understanding design factors related to workflow such as facilitators and barriers is key in improving desired efficiencies, especially in a complex environment such as Level I trauma rooms. Such knowledge can be indispensable to architects, designers, and healthcare professionals who work together to create safer and more efficient environments of care.

EXPLORING BEHAVIOR MAPPING: HISTORY AND METHODS

Track 4: Design for Health and Wellbeing
Knowledge Network: Health, POE/Programming
Methods: Mixed Methods
Tatiana Orozco (Gresham Smith); Dagmar Rittenbacher (Gresham Smith); Lesa N. Lorusso (Gresham Smith)

Background:
Behavior mapping refers to an array of methods for observing behaviors in a particular physical space and time (Ng, 2016). Professionals in Architecture, Design, and Engineering can use behavior mapping in conjunction with other data or information to better understand how people are navigating through and interacting with the built environment. For example, combining step count data with route diagrams can reveal workflow efficiencies or inefficiencies within a workspace for people whose job functions are not stationary (such as nurses in a hospital). Collecting heart rate data tagged to geolocation while navigating a route or path can help to reveal specific locations or built environment features which may be associated with stress. Even counts of human presence in a certain location over a specified time can reveal popular features or "hotspots" within the built environment. In healthcare settings particularly, this information can provide insights for producing facilities which, for example, facilitate both staff and patient safety, increase efficiency, or generally enhance the human experience.

Methods in behavior mapping have adapted over time to incorporate advances in technology, ranging from an observer taking notes on types and frequencies of specific behaviors, to using wearable technology equipped with real-time location services. This poster will thus provide a brief background of behavior mapping and its use in evidence-based design. This poster will then provide examples of several different behavior mapping methods, provide an overview of how to implement them, and explain the type of information that can be gleaned from them. Finally, this poster will present an example case study with preliminary findings.

Behavior mapping carries great potential to contribute to the body of evidence on optimal healthcare architecture and design, thereby empowering architectural and design professionals to produce high-quality healthcare facilities with the best possible experience for staff, patients, and their families.

References:
Ng, C. F. (2016). Behavioral Mapping and Tracking. In Research Methods for Environmental Psychology (pp. 29–51). John Wiley & Sons, Ltd. https://doi.org/10.1002/9781119162124.ch3

GENERATIVE AI SUPPORTED INTERIOR CONCEPT DESIGN: ASSESSMENT OF END PRODUCT CREATIVITY BETWEEN DESIGNERS AND NON-DESIGNERS

Track 6: Transdisciplinary Design and Collaboration/Open Track
Knowledge Network: Interior Design
Methods: Practice-based Research, Survey
Tilanka Chandrasekera (Oklahoma State University); Kasun Gayantha Dodanduwa Waduge (Oklahoma State University); Hebatalla Nazmy (Oklahoma State University); Mallikarachchige Nishan Rasanga Wijetunge (Oklahoma State University); Wellabadage Ubhaya Upendra Perera (Oklahoma State University)

Background:

The conceptual phase in interior design is a pivotal moment where ideas transform into tangible forms, serving as a roadmap for the entire design process. This phase is crucial as it establishes a coherent theme, ensuring that all elements within a space are harmoniously integrated (Ching & Binggeli, 2014). The development of a design concept involves a deep understanding of the client's needs, the functionality of the space, and the aesthetic aspirations. It is during this phase that designers engage in creative exploration, drawing inspiration from various sources, including art, culture, and environment, to create a unique and meaningful design narrative (Piotrowski, 2014). The concept not only guides the selection of colors, materials, and textures but also influences the spatial planning and the interaction of users with the space (Gibson, 2014). A well-articulated design concept ensures that the final outcome is not only aesthetically pleasing but also resonates with the users on a deeper, more personal level, enhancing their experience of the space (Lawson, 2005). Therefore, the importance of concept development in interior design cannot be overstated, as it is the foundation upon which successful and impactful designs are built.

The design process and design thinking methodologies differ significantly between trained designers and non-designers, particularly in the realm of concept development. Designers, with their formal education and professional experience, typically approach design problems through a structured process that involves research, ideation, prototyping, and iterative refinement (Cross, 2011). This process is underpinned by design thinking, a solution-focused approach that encourages deep understanding of the user, creative problem-solving, and the exploration of a wide range of possibilities before converging on a solution (Brown, 2009). In contrast, non-designers often adopt a more intuitive and less formalized approach. Lacking formal training, they might rely more on personal experience and subjective judgment, potentially leading to solutions that are less innovative or not fully optimized for the user's needs (Lawson, 2006). While non-designers can be creative and resourceful, their process may lack the systematic exploration of alternatives and the rigorous testing of ideas that characterize professional design practice. The contrast in approaches highlights the value of design education and experience in developing effective, user-centered design solutions and underscores the unique skill set that designers bring to the table in concept development and problem-solving.

Key words:

Generative AI, Interior design, Conceptual design, Creative product Semantic scale

Ching, F. D., & Binggeli, C. (2014). Interior Design Illustrated. John Wiley & Sons.
Piotrowski, C. (2014). Designing Commercial Interiors. John Wiley & Sons.
Gibson, J. J. (2014). The Ecological Approach to Visual Perception. Psychology Press.

INTELLIGENT INTERIOR DESIGN FRAMEWORK | THE GRAND SPHERIC CENTER

Track 4: Design for Health and Wellbeing
Knowledge Network: Health, Interior Design
Methods: Design Methods, Experimental/Quasi-Experimental Research
Abid Garcia Garcia (CSU Fresno); Uris Giron (California State University Fresno)

Background:
The Grand Spheric Center has been a progressive innovative project with an extensive list of algorithms and toolkits from the book by Holly M Sowles "Intelligent Interior Design Framework: Pedagogy and Practice", in hopes of developing a healing space that enhances their wellbeing. The Grand Spheric Center has been the center of attention in technological architectural boundaries with enclosed nature.

The Grand Spheric Center is a commercial center located in the center openings of a forest. It is a grand combination of green design and innovative design. The Center has three domes, which are utilized as restaurants, shopping centers, etc. The domes are carefully crafted with the smart geometry toolkit, which showcases parametric design within the structural panels which can dynamically adapt to the needs of the space as well as the functionality of the environment. The center also includes a responsive design with features that allow tracking and interaction with the numerous visitors who visit the area.

Another state-of-the-art component that was built into the space was the ambient intelligence toolkit, the ambient intelligence toolkit is seamlessly integrated with a projection mapping wall that is preserved in the pool area. This interactive wall can captivate clients by responding to their awareness and preferences for their commercial taste. It also includes customizable user interfaces for each client group to create a unique experience.

The Grand Spheric Center also includes the information modeling toolkit, whereas this entire project was developed through the essence of Revit, a program that allowed precision and efficiency throughout the innovation process of this architectural functional establishment. Another way in which this project used the information modeling toolkit was through the incorporation of dynamic and kinetic panels, these panels react to the atmospheric movement and through light, mimicking the weather pattern or closing when there is a beam of sunlight or simply for the regards of privacy. Information Modeling also allowed the creator to explore different solar studies deriving from an extensive amount of seasons. Last but not least the individual reading this, can explore this three-dimensional model throughout the enscape web for a further detailed exploration of the Grand Spheric Center as part of the information modeling toolkit.

Therefore all of these toolkits combined allowed a different approach to further enhance the atmosphere of innovative and smart design.

LEARNED LESSONS IN DESIGNING NICS UNIT USING EVIDENCE BASED DESIGN: STUDENT DESIGN PROJECTS

Track 4: Design for Health and Wellbeing
Knowledge Network: Children, Youth & Environment
Methods: Action/Participatory Research, Case Study
Shireen Kanakri (Ball State University); Danielle M Hembrock (Ball state university)

Background:
Neonatal Intensive Care Units are a vital part of early life care. For any baby who is sick or not strong enough to go home when they are born the NICU provides a space and care for both the infant and the family. The aim of this design is to create a space using user centered design. In doing research on NICU design some key points to focus on were single parent rooms as opposed to cluster units (Smith, 2018), private family spaces in the unit (O'Callaghan, 2019), and workspaces prioritizing the staff (O'Callaghan, 2019). This design includes 24 single parent rooms, two twin rooms, and one triplet room. Having private rooms for a each parent is shown to have positive results not just for the infant but for staff and family as well. To prioritize staff heath in the work space several of the staff only areas offer direct access to daylight and have a added layer of privacy with staff only hallways. Offering a calming and caring environment for parent, family and staff is top priority.

NAVIGATING ONLINE DESIGN CLASSES WITH VIRTUAL "DESK" CRITIQUE

Track 5: Spatial Justice, Equity, and Inclusion by Design
Knowledge Network: Communication
Methods: Observation, Survey
Qing Lana Luo (Oklahoma State University)

Background:
Desk critique is one of the key instruction methods that benefits student critical thinking in design classes. It involves the professor sitting down at each student's desk and reviewing their work progress displayed on the desk. An effective desk review consists of giving verbal feedback and, more importantly, marking on students' design to offer ways of improving and exploring further on the design problems.

For an online design class, professors could not conduct the traditional close-distanced desk. Marking and sketching over student work during a class would have become a challenge without technology's assistance. With the advancement of video conferencing technology, the Zoom Annotate tool was used in an online design class in landscape architecture program to be in place of in-person critique to provide an equivalent level review for students. This accomplished a live digital review with the ability to mark on student work, which the traditional desk critique can offer.

Twenty two students were given a short questionnaire based on their perceptions of their likes, dislikes, relation to Zoom critique, and potential to facilitate course learning. Student perceptions of Zoom critique and the Annotate tool as a pandemic solution for desk crit were mostly positive. Students have reported the benefits of having undisrupted valuable feedback time via the screen without needing to sit down side-by-side with the professor. A few negatives include students longing for in-person and close distance interactions.

Keywords:
virtual critique, online class, technology, interactive

PAPER TO PIXELS: A STUDY ON THE IMPACT OF DIGITAL DRAFTING ON INTERIOR DESIGN PEDAGOGY

Track 6: Transdisciplinary Design and Collaboration/Open Track
Knowledge Network: Interior Design, Student Affairs
Methods: Mixed Methods, Survey
Zahrasadat Hosseini (Oklahoma State University); Tilanka Chandrasekera (Oklahoma State University);
Aditya Jayadas (Oklahoma State University); Wellabadage Ubhaya Upendra Perera (Oklahoma State University)

Background:

In contemporary design firms, manual drafting has become obsolete, replaced by digital applications for several years. However, the educational landscape still largely adheres to traditional manual methods, notably in universities. Addressing this disparity, the interior design department of a Midwestern university pioneered a transformative approach by shifting its Graphics for Interior Design class from manual drawing techniques to a fully digital format. This transition involved leveraging digital design and illustration applications on iPads, primarily utilizing Procreate as the illustrative app. This strategic move marked a significant milestone, illustrating the department's commitment to aligning with industry trends and providing students with relevant and practical learning experiences. To assess the impact of this digital transition, researchers conducted a comprehensive survey inspired by the Technology Acceptance Model (TAM) questionnaire (Davis, 1989). The survey, administered to both design students and industry professionals, aimed to compare their technology acceptance levels and experiences. Specifically, it evaluated user acceptance through perceived usefulness and perceived ease of use factors, supplemented by open-ended questions to capture participants' insights and feedback. The study comprised two phases, with the initial phase focusing on a needs assessment conducted among industry professionals.

The survey sample for the first phase consisted of 62 designers within the industry, while the second phase involved 45 design students utilizing Procreate. By delving into attitudes, perceptions, and experiences concerning the incorporation of digital technology into the design process, the research team aimed to gain insights into the divergent perspectives of designers in both academic and professional settings. Additionally, the team aimed to gather insights into any challenges faced during the transition in the academic setting. Anticipated outcomes include the recognition of the benefits of using Procreate over traditional manual drafting techniques and materials. Procreate not only aligns education with industry trends but also expedites the production of design work. Moreover, the shift from manual drafting to digital applications offers economic advantages by consolidating multiple drawing materials into a single iPad. The adoption of Procreate applications and iPads ensured that students experience a more dynamic and contemporary learning environment. This transformation not only better prepares students to meet the evolving demands of the interior design profession but also fosters resourcefulness and adaptability. In conclusion, by integrating digital design and illustration applications on iPads, this initiative harmonized the curriculum with industry trends, enriches the learning experience, and encourages resourceful, sustainable practices. Lastly, the comparison of designers' attitudes in educational and industrial settings provided valuable insights into the broader implications of this shift.

Keywords:

Technology acceptance model, Digital Technology, Perceived Usefulness, Perceived ease of use, design professionals, design students

References:

Davis, F. D. (1989). Technology acceptance model: TAM. Al-Suqri, MN, Al-Aufi, AS: Information Seeking Behavior and Technology Adoption, 205-219.

PLACEMAKING STRATEGIES FOR CO-DESIGN OF PUBLIC SPACES: CASE OF RIYADH NEIGHBOURHOODS' PARKS

Track 6: Transdisciplinary Design and Collaboration/Open Track
Knowledge Network: Participation, Sustainable Planning, Design & Behavior
Methods: Action/Participatory Research, Mixed Methods
Lobna Abdel Aziz Mostafa (AMAKEN Placemaking Org.)

Background:

Since the early rise of (The right to the city) notion first proposed in 1968 by Henri Lefebvre in his book Le Droit à la Ville, Lefebvre described it as an attempt to 'reclaim' the city as a co-created space' [1], this idea has been widely embraced recently by thinkers, social movements, progressive local authorities, and urbanists as a call for action to reclaim the city as a place for life. As a result, the co-design participatory approach has gained more attention over the past decade as a tool to engage the community in the decision-making process to build ownership and trust with communities. [2] It has also proven its effectiveness to involve the government's professionals in public co-design workshops and charette instead of taking decisions behind their disks. [3] Recently, it has well observed the growing evolution of Placemaking practices in Saudi Arabia. Though few participatory experiences have been recorded over the past few years, with the active application of the Saudi Quality of Life Program launched in 2020, both the enhancement of open spaces and community engagement became the focus of attention of the municipalities that some of them accredited (Placemaking) as a tool to achieve both goals.

This research demonstrates the placemaking co-design process as a tool that brings all the stakeholders together in the whole design process of public spaces and facilitates community engagement and participation in designing open spaces.

The research documents the placemaking strategy in the "Al-Wadi Park" Project in Riyadh, Saudi Arabia. The research follows an empirical qualitative methodology, where a descriptive-analytical study of the placemaking strategy will be presented.

Given the practical and hands-on experience gained during the placemaking co-design process, the research proposes an applied framework for facilitating a top-down participatory approach through the placemaking practice to ensure a comprehensive, inclusive, and community-based design that allows democratic participation in the decision-making processes. The research findings further reveal the critical role of stakeholders in initiating the design of neighborhood parks through the placemaking participatory strategy. The research emphasizes the importance of replacing target-based decision-making with impact-based one, positively affecting both users' satisfaction and quality of life.

References:

[1] Purcell, M. Excavating Lefebvre: The right to the city and its urban politics of the inhabitant. GeoJournal. 58, 99–108 (2002).
[2] Pólvora A., Nascimento S., Sanders E.B.N., Grael M., Co-Designing Participatory Approaches for Communities, EUR 29091 EN, Publications Office of the European Union, Luxembourg, 2018, ISBN 978-92-79-79334-9 doi:10.2760/448511, JRC110752
[3] Verloo, N. 2019. Captured by bureaucracy. In M. Raco, M. & F. Savini (Eds.), Planning and knowledge: How new forms of technology are shaping contemporary cities (pp.75–89). Bristol: Policy Press.

PRESENCE OF PEOPLE AND PERCEPTIONS OF URBAN PARKS: A MIXED METHODS APPROACH USING EYE TRACKING AND VIRTUAL REALITY

Track 4: Design for Health and Wellbeing
Knowledge Network: Health
Methods: Experimental/Quasi-Experimental Research, Mixed Methods

Shuai Yuan (Clemson University); Matthew Browning (Clemson University); Matthew Nicolette (Clemson University); Andrew Duchowski (School of Computing, Clemson University); Jeffrey Hallo (Clemson University)

Background:

Urban parks are vital for enhancing health and well-being. However, the impacts of other people on these benefits are not well understood, with studies yielding both positive and negative results (e.g., Jorgensen et al., 2013; Lis & Iwankowski, 2021). Also, urban design theories highlight the importance of social interaction, while park management tends to focus on the challenges of crowding. We introduced a novel method combining 360-degree videos in virtual reality (VR), eye tracking, and structured interviews. 360° videos can deliver human movements and emotional cues that capture involuntary attention. Eye tracking can reveal automatic attention responses to detailed environmental features. Our research questions include: 1) How does the presence of others affect perceptions of urban parks? 2) How does the presence of others influence where people look in different park settings? 3) How is visual attention linked to perceptions?

We employed a mixed methods approach with a convergent design to study visual attention and perceptions in urban park scenes with varied visitor densities. Forty university students watched eight 360° videos representing visitor density levels (low vs. medium) and two naturalness levels. A 2 X 2 within-subject design was used. Visual attention was indicated by fixation durations on areas of people and other environmental features in video frames, identified through semantic segmentation. Perceptions of urban parks were assessed through self-reported measures following White (2020) (i.e., willingness to visit, perceived safety, perceived restorativeness) and psychophysiological response (i.e., blink rate, skin conductance). Participants received a follow-up interview about past experiences in urban parks (not used) and explanations of the survey responses.

We expect that more visitors will lead to greater attention to other visitors, reduced attention to natural and built features, and increased arousal. However, we do not hypothesize specific directions of the influence of visitor density. We also expect visitor density to interact with naturalness, visitor motivations, and nature relatedness, affecting perceptions both positively and negatively. The qualitative findings will be used to provide contexts and interpretations of the findings. We will discuss the transferability of the findings in urban public spaces and green spaces and provide potential insights into the controversial roles of others. The results will help identify challenges and improvements in stimulating environments with human presence. Furthermore, we will discuss the implications for environmental design, focusing on strategies to enhance positive visitor impacts.

References:

1. Jorgensen, L. J., Ellis, G. D., & Ruddell, E. (2013). Fear Perceptions in Public Parks: Interactions of Environmental Concealment, the Presence of People Recreating, and Gender. Environment and Behavior, 45(7), 803–820.
2. Lis, A., & Iwankowski, P. (2021). Where do we want to see other people while relaxing in a city park? Visual relationships with park users and their impact on preferences, safety and privacy. Journal of Environmental Psychology, 73, 101532.
3. White, M., Smith, A., Humphryes, K., Pahl, S., Snelling, D., & Depledge, M. (2010). Blue space: The importance of water for preference, affect, and restorativeness ratings of natural and built scenes. Journal of Environmental Psychology, 30(4), 482–493.

PROMOTE AGE-FRIENDLY COMMUNITY PUBLIC SPACES IN URBAN CHINA

Track 4: Design for Health and Wellbeing
Knowledge Network: Environment-Gerontology, Residential Environments
Methods: Meta-Analysis, Mixed Methods
Zhe Wang (Henan University); Xbh Xia (Henan University)

Background:
Aging in place has been a traditional approach for elderly care in China. As the number of older populations increase fast in urban areas, there's a pressing need for age-friendly design of residential communities. However, community spaces are frequently seen falling short in meeting the diverse needs of older adults for daily living, leisure, and social interaction. Many studies have been done in this field and the findings need to be reviewed for age-friendly community design.

Objectives:
The aim is to summarize the current empirical findings on community public spaces in urban China for high-quality age-friendly design.

Methods:
A combination of database searches (Social Sciences Citation Index, CNKI) and manual searches was utilized. Keywords used in the searches included "urban areas," "community senior care," "older adults," "age friendly," and "public spaces." The literature and relevant studies were categorized into two groups: social or physical environments, for a comprehensive understanding. From an initial 275 articles, 92 were selected for abstract review, and 43 articles were collected after full-text review. Seventeen articles with empirical data collection and analysis were identified as appropriated to this research. Content analyses were conducted to summarize their findings.

Result:
Findings of the aforementioned articles were majorly regarding older adults' daily activities, physical environment design of community spaces, and the simultaneous analysis of both aspects. It has been found that urbanization and high population density in China have resulted in a deficiency of community public spaces that fail to meet the diverse needs of the growing older populations. Social environment studies often lack attention to community diversity, while physical environment studies lack quantitative measurements, compromising scientific validity.

These findings highlight the necessity of focusing on diverse activity needs and the importance of implementing multi-level designs. Future research should explore social behaviors across different cultural backgrounds. Standardized physical environment measurements are crucial for scientific and comparable research findings, enabling comprehensive analysis across studies.

Conclusion:
Design of community public spaces significantly influenced the quality of environments for older adults. Based on a systematic review of the existing literature, we categorized and analyzed the findings and identified gaps in research. It provides directions and recommendations for future research and practice, with the ultimate goal of creating safe, comfortable, and barrier-free environments for age-friendly communities.

QUANTIFYING NEUROPHYSIOLOGICAL AND COGNITIVE IMPACTS OF INTERIOR DESIGN ON HUMAN WELLNESS THROUGH EEG, VR, AND COGNITIVE TESTS

Track 4: Design for Health and Wellbeing
Knowledge Network: Health, Interior Design
Methods: Experimental/Quasi-Experimental Research, Statistical/Modeling Analysis

Sieun Kim (Yonsei University)

Background:
This study explored the impact of interior design on human wellness based on the Stimulus-Organism-Response model from environmental psychology, acknowledging that the environment influences physical perceptions as well as emotional, physiological, and behavioral responses. Noting the scarcity of quantitative research on interior design's impact on emotional and physical wellness, this study employs virtual reality (VR), electroencephalography (EEG), and cognitive tests to quantitatively evaluate the neurophysiological and cognitive responses of occupants to interior design elements. Through this approach, the study aimed to provide empirical evidence of the interactions between the environment and humans.

An experiment with 40 mixed-gender university students was conducted in a virtual reality simulation of a university's indoor student lounge, featuring five design elements as variables: a control model, closed space, curvature, wood finishing, and natural view. Neurophysiological responses to these designs were measured using EEG. The EEG data were analyzed by calculating power spectral density using time-frequency analysis, evaluating the relaxation state via alpha-theta band ratio (RAT) and the arousal state via beta-alpha band ratios (RBA). Additionally, subjective spatial perception and emotional responses were assessed through surveys, and cognitive responses were measured in terms of attention, using spelling error detection and Stroop tests. The experimental results were statistically analyzed using repeated measures ANOVA to compare the control model and design elements variables. Analyzing 32 valid samples, key findings include: three design elements (curvature, wood finishing, and natural view) exhibited higher RAT than the control model in EEG, indicating their positive influence on the occupants' relaxation, while the closed space showed lower RAT. Conversely, the closed space had higher RBA, associated with increased arousal and attention, in contrast to wood finishing and natural view. Survey results echoed these trends. Cognitive tests revealed that natural view and wood finishing scored lower than the control model, whereas closed space scored the highest, suggesting its efficacy in enhancing attention. Curvature, as indicated by EEG, was associated with relaxation but did not significantly correlate with attention.

This study demonstrates that wood finishing, natural view, and open-plan in interior designs positively influence users' emotional and physical wellness. The consistent results from both EEG and cognitive tests, particularly for the wood finishing, natural view, and closed-space variables, provide evidence for a correlation between emotional, physiological, and cognitive responses to interior design.

The findings indicate that spaces with wood finishing and natural views are associated with relaxed emotional states and reduced attention, leading to lower cognitive capabilities. Conversely, closed spaces are linked to heightened arousal and attention, enhancing task performance abilities. These findings suggest that the design of indoor environments significantly affects users' psychological and physiological responses, as well as their cognitive performance. Furthermore, the results highlight that spaces conducive to relaxation and attention may differ, thereby providing important evidence of the interaction between environment and users. This underscores the importance of quantifying user responses in the development of human-centered design. Such insights can serve as foundational data for future development of interior designs optimized for users' psychological, physiological, and cognitive responses.

REEVALUATING POST-DISASTER HOUSING: A LONGITUDINAL STUDY ON USER SATISFACTION AND ADAPTIVE DESIGN STRATEGIES IN POST DISASTER HOUSING

Track 4: Design for Health and Wellbeing
Knowledge Network: International Housing Research, Residential Environments
Methods: Interviews, Survey

Tilanka Chandrasekera (Oklahoma State University); Wellabadage Ubhaya Upendra Perera (Oklahoma State University); Mallikarachchige Nishan Rasanga Wijetunge (Oklahoma State University); Kasun Gayantha Dodanduwa Waduge (Oklahoma State University)

Background:
Post-disaster housing settlements, especially after natural disasters like tsunamis and tornadoes, require an understanding of the socio-psychological and emotional needs of affected communities. The design and development of these settlements must go beyond architectural and structural solutions to include a deep comprehension of the human aspects of disaster recovery. Lizarralde et al. (2009) highlight the importance of emotional and psychological well-being, alongside physical safety, in post-disaster scenarios. This approach is crucial for ensuring that the rebuilt environment supports the mental health and social cohesion of its inhabitants (Rodríguez et al, 2007). Involving affected communities in the design process, as argued by Haas et al. (1977), enhances the effectiveness of housing solutions, fostering a sense of ownership and emotional connection. This is particularly relevant in culturally and socially diverse contexts. Therefore, post-disaster housing design must integrate a holistic approach that addresses the complex emotional and social dynamics of disaster-affected populations.

Following the 2004 tsunami, which claimed over 30,000 lives, Sri Lanka undertook numerous resettlement projects to shelter victims who lost approximately 65,000 homes. By 2005, 485 resettlement projects were planned and finalized. However, these projects primarily focused on achieving housing numbers, often overlooking the socio-psychological and emotional aspects crucial for the affected communities. This oversight led to diminished long-term user satisfaction, a trend more pronounced in Sri Lanka than in Thailand, which faced similar challenges. User satisfaction in residential environments encompasses happiness and satisfaction with the living environment, influenced by demographic features, physical residential environment, and the gap between current living conditions and residents' expectations and needs. Long-term user satisfaction is essential for the sustainable existence of post-disaster resettlements.
To understand the changes in long-term user satisfaction, a study was conducted on a post-tsunami resettlement in Midigama, Sri Lanka. This two-stage assessment, first in 2014 and then in 2023, investigated user satisfaction regarding the physical attributes of houses, future expectations, and demographic features of the households. Data were collected from a random sample of 30 homes using the same questionnaire in both stages. The study aimed to assess stage-wise user satisfaction and identify differences in long-term satisfaction under various attributes.

The findings revealed a decline in long-term user satisfaction, particularly in aspects like house plan, living room size, maintenance, and utility consumption. This decline underscores the need for flexible, adaptable design strategies in post-disaster housing to accommodate changing needs and ensure sustained satisfaction. These insights are vital for informing future post-disaster housing policies and practices globally, highlighting the necessity for a dynamic, user-centered approach in designing and developing such communities to ensure their long-term viability and effectiveness.

Key words:
Post tsunami resettlements, sequential study, house attributes, long term residential user satisfaction,

References

Lizarralde, G., Johnson, C., & Davidson, C. (Eds.). (2009). Rebuilding after disasters: From emergency to sustainability. Routledge.
Rodríguez, H., Quarantelli, E. L., Dynes, R. R., & Fordham, M. (2007). Disaster and development research and practice: A necessary eclecticism?. Handbook of disaster research, 335-346.
Haas, J. E., Kates, R. W., & Bowden, M. J. (1977). Reconstruction following disaster. MIT Press.

RESEARCH ON FAIR AND EQUITY URBAN PLANNING FOR WELFARE, BASED ON ANALYSIS OF ACTUAL UTILIZATION OF MEDICAL AND LONG-TERM CARE SERVICES

Track 4: Design for Health and Wellbeing
Knowledge Network: Sustainable Planning, Design & Behavior, Active Living by Design
Methods: Survey, Statistical/Modeling Analysis

Kurumi Takezawa (Utsunomiya University); Eiji Satoh (Utsunomiya University); Takayuki Miyake (Assistant Professor)

Background:

Japan is rapidly progressing into a super-aging society, and the demand for medical and long-term care services is expected to increase more. Therefore, the government is aiming to realize the establishment of a "Community-based Integrated Care System" that provides necessary support integrally within the community. Especially, coordination between medical and long-term care is important, but there are three major issues: first, medical and long-term care have different insurers; second, different insurers have different responsibilities for establishing plans for medical and long-term care; and third, they have different planning units.

In addition, Japan is a country with a universal health insurance system, which means all citizens are enrolled in some form of public medical insurance. Long-term care services for the elderly are also managed by this system. This system allows citizens to freely choose medical institutions and long-term care facilities and to receive necessary services equitably. However, in regional cities, resources such as facilities and human resources to support services are unevenly distributed, and there are issues in ensuring fairness in accessibility to medical and long-term care services. It's anticipated that welfare plans should be revised to be more human-centered and reflect the actual situation.

In this study, we analyze the actual utilization of medical and long-term care services focusing on the combination of service types and accessibility. The purpose of this study is to gain knowledge on how to plan sustainable, fair and equitable welfare planning for the elderly that reflects the actual situation, and on the concept of human-friendly urban planning.

Research Methods

Using medical and long-term care receipt data, identify the actual utilization of services in each " Sphere of Daily Life," which is the planning area of the Community-based Integrated Care System. Receipt data contains actual service provision performance data, such as who used what services, at what facilities, and in what quantities. By using a geographic information system, geospatial information such as road network data, land use data, and population mesh data are integrated with the care utilization data extracted from receipt data. This enables visualization and analysis of road network distances between user residences and facility locations, as well as actual service utilization inside and outside of the planning area. The following three analyses will be conducted in this study.

1) Analysis of the geographical characteristics of each planning area using population data, medical and long-term care facility data, land use data, etc.

2) Analysis and evaluation of accessibility disparities in these services using road network distances

3) Analysis of the relationship between the geographical characteristics and accessibility disparities of each planning area and the actual utilization of these services.

Discussion

Based on the analysis results of this study, we'd like to discuss the following points

1) How to balance the reflection of user needs and actual conditions of service use with the rationality of planning area establishment

2) The possibility of urban consolidation to ensure sustainable, fair, and equity service provision in a society with a declining and a super-aging population

ROOTED IN WELLNESS: RURAL CRITICAL ACCESS HOSPITAL DESIGN ANCHORED IN CONTEXTUAL UNDERSTANDING

Track 4: Design for Health and Wellbeing
Knowledge Network: Environmental Design Research Education, Health
Methods: Case Study
Matthew Maldonado (University of Texas at San Antonio); Luis Victoria (University of Texas at San Antonio); Neda Norouzi (UTSA)

Background:
Critical Access Hospitals (CAHs) play a vital role in delivering accessible healthcare to rural communities by addressing immediate healthcare needs and promoting well-being. However, current CAH designs often neglect the importance of understanding and emphasizing cultural, environmental, and architectural identity within the community context. It is crucial for designers to prioritize creating spaces that not only meet medical requirements but also resonate with the site's context and the natural identity of the community. This project focuses on designing a CAH in south Texas, extending its impact to neighboring counties. The design is focused on patient-centered care, staff efficiency, and connecting with the town's identity.

Method:
Data was collected through a comprehensive literature review conducted to learn about critical access hospitals, understand the societal and interpersonal attributes of these facilities within the community, and determine how spatial and programmatic interventions contribute to fostering these relationships and responding to the needs of the served communities.

Results:
The results integrated in the design of this project include (1) biophilic design to improve patients' well being and decreased hospital stays, (2) wayfinding as a pivotal element in augmenting user experience, (3) Ulrich's theory of supportive design to offer positive distractions and social support decreasing stress and improving patients' mood.

Design Solution:
The site visit informed a design concept that harmonizes with the town's identity while also modernizing its essence. Research focused on hospital design, prioritizing comfort for patients, visitors, and staff. The form of the building was intentionally chosen to allow for future expansion while the modular design supports scalability and replication, ensuring flexibility over time. Integrating local stone in the building connects to the town's history while the modern glazing creates a path to the future for a therapeutic environment with a central courtyard and green roofs. Optimized wayfinding, defined hallways, and supportive design elements contribute to a comforting atmosphere. The overall hospital campus was designed to be adaptable to evolving community needs, technological advances, and medical practices, maintaining resilience in the healthcare landscape.

Objectives:
1. *Understand the critical role of CAHs in addressing healthcare needs in rural communities.*
2. *Explore the integration of biophilic design and its positive impact on patient well-being.*
3. *Investigate the principles of supportive design and their application in creating comforting healthcare environments.*

References
Ali Khan, M., Amin, N., Khan, A., Imtiaz, M., Khan, F., Ahmad, I., Ali, A., & Islam, B. (2016). Plant therapy: A nonpharmacological and noninvasive treatment approach medically beneficial to the wellbeing of hospital patients. Gesunde Pflanzen, 68 (4), 191–200. https://doi.org/10.1007/s10343-016-0377-1
Amato, C., McCanne, L., Yang, C., Ostler, D., Ratib, O., Wilhelm, D., & Bernhard, L. (2022). The hospital of the future: Rethinking architectural design to enable new patient-centered treatment concepts. International Journal for Computer Assisted Radiology and Surgery, 17 (6), 1177-1187. https://doi.org/10.1007/11548-021-02540-
Mackett, R. L. (2021). Mental health and wayfinding. Transportation Research Part F: Traffic Psychology and Behaviour, 81 , 342–354. https://doi.org/10.1016/j.trf.2021.06.014

THE EFFECTS OF LIGHTING COLOR TEMPERATURE ON TEACHER SELF-EFFICACY

Track 4: Design for Health and Wellbeing
Knowledge Network: Health, Interior Design
Methods: Survey

Alyssa Oates (Oklahoma State University); Tilanka Chandrasekera (Oklahoma State University); Aditya Jayadas (Oklahoma State University)

Background:
Early childhood classrooms play an important role in the development of students as they spend the majority of their early years in these classrooms. An important attribute in early childhood classrooms is lighting, specifically the type of lighting (LED vs. fluorescent), and what effect it has on student engagement (Pulay & Williamson, 2019). Color temperature of lighting has been briefly explored, but results are not in-depth and are not specifically applied to teacher self-efficacy. Students are not the only users being affected by the classroom environment; educators also spend the majority of their day within one classroom. Teaching can be stress inducing, and student involvement and motivation can suffer because of teacher burnout (Ghanizadeh & Jahedizadeh, 2015). The educator's ability to effectively teach materials to students is imperative, especially during the early childhood phase of life due to the fact that children's brains are rapidly developing during this time. Educators' stressors can come from a multitude of variables, including student behavior problems, a feeling of isolation or inferiority, and a lack of control (Richards, 2012). Often in preschool or early childhood education settings, teachers and students spend the majority of their day in one classroom. Because of the extensive time spent in these settings,.it is important for the space to provide functional and aesthetic value in order to maximize positive relations for all end-users. By introducing various lighting solutions into an early childhood education classroom, we can improve the positive engagement in these spaces between students and their peers, teachers, and tasks.

Purpose:
The primary objective of this study is to assess if the integration of cool tone lighting in a virtual simulation of an early childhood classroom improves positive emotions and cognitive measurements in educators.

Methodology:
30 pre-service teachers were recruited for the study. The participants were required to experience two different, but similar virtual classroom environments to determine the effects of cool and warm tone lighting on their self-efficacy, cognitive load, and physiological measurements. The classroom design was modeled after an existing classroom in a child development laboratory on the campus of a mid-western university. A one-tailed paired t-test was used for the analysis, with a significance level of 0.05.

Results:
The results showed that the participants expressed more positive emotions (happy - $p < .001$; calm - $p < .001$; excited - $p < .001$), felt like they had more control ($p < 0.001$), autonomy ($p < .001$), and also felt more satisfied ($p < .001$) when they were immersed in the virtual classroom environment with cool tone lighting as opposed to the warm tone lighting. In addition, the participants also felt like the mental demand was less ($p < .001$), and the ability to motivate students was higher ($p < .001$) when they were in the cool tone lighting virtual classroom environment.

Implications:
This study has important implications for the design of early childhood classroom environments, especially with regard to lighting, in order to improve positive emotions and cognitive measures in early childhood educators.

THE HEALTH IMPACTS OF NATURAL INTERIOR MATERIALS: DEVELOPING AN EVALUATION FRAMEWORK FOR NATURAL INTERIOR MATERIALS: HEALTH, WELLBEING, AND EMBODIED CARBON

Track 4: Design for Health and Wellbeing
Knowledge Network: Health, Interior Design
Methods: Interviews, Mixed Methods
Alivia Green (Atelier Ten)

Background:
The built environment impacts the world around us on multiple levels. Of these impacts, concerns include the health of the environment and human health. As definitions of health evolve, we've come to recognize well being as an aspect of overall health, which is also affected by the built environment. Within the built environment, a focus has been placed on materials as a factor that imperatively affects the environment and human health within buildings. Previous evidence has shown that natural materials (NM) can have lower embodied carbon (EC) which impacts the environment through greenhouse gases, and lower indoor air emissions, which can negatively impact the physical health of occupants. However, it is suggested that perception of these materials, through their physical appearance can also impact mental health and wellbeing through their sensorial properties by way of biophilia. While there is accessible data on the ways in which materials can affect health, it can be challenging for designers and built environment professionals to holistically evaluate materials for their health impacts. Therefore, this research aims to investigate how NM, within building interiors, can be categorized based on a criteria including human health impacts and embodied carbon content, to develop a framework to assist designers in selecting natural interior materials which positively impact the environment and occupant health.

Four material databases were searched for NM which had low EC and low VOCs, to then be analysed for biophilic attributes using a newly developed audit based on biophilic design frameworks. Seven designers and material experts were then interviewed to determine the decision-making process of material selection and to investigate the current understanding of how professionals consider materials for their influences on health. Results from material analysis showed that NM can be scored for their biophilic attributes, and those with minimal processing yielded higher overall scores. Interview findings showed that most participants did not consider biophilia or sensorial properties when selecting materials, and many were not aware of metrics to account for indoor air emissions. All findings were synthesized to create a framework for selecting NM based on their health impacts, which was then represented by a label to graphically communicate a material's biophilic score, and all other factors included in the framework. This study and the proposed framework/label could be the basis for further investigation surrounding NM and their impacts on health, and how these factors can be better integrated into the decision-making process of built environment professionals. Further research is needed on how to account for all material selection drivers within the label, which also requires further testing with clients, building occupants, and designers to evaluate its robustness and practical impact.

THE IMPACT OF NATURE-BASED OUTDOOR PLAY AREA ON PLAY AND LEARNING BEHAVIORS OF CHILDREN WITH AUTISM SPECTRUM DISORDER

Track 4: Design for Health and Wellbeing
Knowledge Network: Children, Youth & Environment
Methods: Experimental/Quasi-Experimental Research, Mixed Methods
Umme Tasnima Haque (Texas Tech University); Muntazar Monsur (Texas Tech University)

Background:
Nature plays a vital role in the development of children. Unfortunately, not all children follow the same growth curve. A 2023 WHO report found that 1 in 100 children worldwide has Autism Spectrum Disorder (ASD), totaling about 20 million globally. This ratio is significantly higher in the US at approximately 1 in 44 children (2.3%), totaling nearly 1.7 million. These children often face challenges in sensory integration, impacting communication, learning, social skills, motor development, and stress-related issues. However, many guidelines and design practices discuss only the indoor environment and design guidelines for children with ASD but rarely show any evidence that nature-based outdoor play areas help to minimize the difficulties experienced by children with ASD.

This thesis investigates whether the nature-based outdoor play can help children with ASD to minimize their learning difficulties through transforming the existing playground of Burkhart Center for Autism Education and Research at Texas Tech University by implementing nature-based learning criterion and analyzing data by conducting behavior mapping with ArcGIS Field Map, and surveys among parents and caregivers using Qualtrics to measure the area's effectiveness. This research aims to enhance learning and behavioral development in a nature-based play area and aid future research and design efforts for outdoor play environments for children with ASD.

HOW LIVED EXPERIENCES WITH WATER GUIDE WATER CONSERVATION BEHAVIOUR ON HIGHER EDUCATION CAMPUSES

Track 2: Sustainable Buildings and Infrastructure
Knowledge Network: Participation, Sustainable Planning, Design & Behavior
Methods: Action/Participatory Research, Mixed Methods
Kulsum Fatima (University of Calgary); Zainab Fatima (CLRS)

Abstract:
The study utilizes institutional conditions to assess underlying reasons for water conservation behaviour at higher educational campuses in Canada. This study investigates the social component which would help understand "what people think of the existing water sustainability features on university campus" and "how people relate their behaviour towards these water features". For this purpose, this ongoing study includes a series of experiential learning workshops to evaluate campus community daily experiences with water on campus and help them reflect what motivates their behaviour towards water use and actual water savings. This proposed presentation at EDRA55 will present a comparative of results obtained from experiential learning workshops conducted at three different canadian university campuses. This is an ongoing investigation and therefore more university campuses are expected to be included in the near future.

The results obtained from these experiential learning workshops are used to propose a set of water literacy strategies to promote water conservation behaviour among educational campus communities. These strategies not only help in sustaining water resources in general but are also instrumental in enhancing the achievement of United Nations Sustainable Development (UN SDG) Goals 6 "clean water and sanitation for all" in specific. The findings of this study will highlight the underlying reasons that motivates water sustainable behavioural practices for Canadian higher education campus communities. These findings are instrumental in defining the sustainable character of higher educational campuses and help them lead their communities and peers towards best practices for water sustainable behavioural norm.

THE RELATION BETWEEN PERCEIVED SENSORY DIMENSIONS AND PERCEIVED RESTORATIVENESS IN URBAN ROOFTOP GARDENS

Track 4: Design for Health and Wellbeing
Knowledge Network: Cyberspace & Digital Environments, Health
Methods: Mixed Methods, Photo-based Methods

Nianchu Hou (Auburn University); Melanie Duffey (Auburn University)

Background:
The paper highlights the importance of green spaces in urban areas for promoting mental health and well-being. The authors discuss the various dimensions of perceived sensory dimensions (PSD), including nature, social, space, serene, prospect, rich in species, culture, and refuge. In previous studies, researchers examine the inconsistent findings on the relationship between PSD and perceived restorativeness (Akpınar, 2021; Malekinezhad et al., 2020; Stigsdotter et al., 2017).This paper proposes a research study to investigate the factors in PSD that influence users' perceived restorativeness in urban rooftop gardens. The study aims to provide a comprehensive evaluation of the factors that impact PSD and inform city planners and municipalities about the potential benefits of urban rooftop gardens for mental health. In addition, this paper provides an overview of the theoretical frameworks of stress reduction theory (SRT) and attention recovery theory (ART) that explain the positive effects of exposure to nature on well-being. The proposed research design involves online surveys to collect quantitative and qualitative data on participants' perceptions of restorativeness in urban rooftop gardens. The findings will be analyzed using descriptive statistics, confirmatory factor analysis, linear regression, and thematic coding. Overall, this study emphasizes the importance of understanding the relationship between PSD and perceived restorativeness in urban rooftop gardens for promoting mental health in urban environments.

References

Akpınar, A. (2021). How perceived sensory dimensions of urban green spaces are associated with teenagers' perceived restoration, stress, and mental health? Landscape and Urban Planning, 214, 104185. https://doi.org/10.1016/j.landurbplan.2021.104185

Malekinezhad, F., Courtney, P., bin Lamit, H., & Vigani, M. (2020). Investigating the mental health impacts of university campus green space through perceived sensory dimensions and the mediation effects of perceived restorativeness on restoration experience. Frontiers in Public Health, 8. https://doi.org/10.3389/fpubh.2020.578241

Stigsdotter, U. K., Corazon, S. S., Sidenius, U., Refshauge, A. D., & Grahn, P. (2017). Forest design for mental health promotion—using perceived sensory dimensions to elicit restorative responses. Landscape and Urban Planning, 160, 1–15. https://doi.org/10.1016/j.landurbplan.2016.11.012

UNIVERSITY CAMPUS HEALTHSCAPES: USING THE UNIVERSITY AS AN URBAN LAB FOR DESIGNING HEALTH-PROMOTING BUILT-ENVIRONMENTS.

Track 4: Design for Health and Wellbeing
Knowledge Network: Health, Sustainable Planning, Design & Behavior
Methods: Case Study, Mixed Methods
Rana Taher F. Abdelkader (TU Dresden - IOER - DLGS)

Background:
Approximately 23% of global deaths could be avoided by creating healthier environments (WHO, 2019). The fact that human health is impacted by the built-environment's design just as much as individual choices and genetic factors highlights the importance of studying 'how' to design healthy built-environments that not only promote the quality of inhabitants' lives, but also provide a layer of resilience against future shifts. Despite the existence of multiple Building and Occupant Rating systems such as LEED, WELL, Fitwel,..etc.; understanding 'how' to design healthy built-environments is still questionable and understudied. They mostly focus on assessing the buildings' health status based on checklists. Yet, they do not explicitly suggest design guidelines, strategies, or an answer to the 'how' question.

This research focuses on answering such questions by employing a comprehensive and holistic approach. It aims to identify the architectural and urban design elements that should be integrated into design practices to promote human health and well-being. This is achieved by utilizing the university campus as the study setting, which functions like a small society, replete with diverse amenities. It is also often referred to in urban planning as a micro-city or an experimental setting. Accordingly, the urban form, spatial composition, and scale of a campus enable it to be used in this research as an urban lab, for cultivating healthy built-environments.
Cairo University, Egypt and TU Dresden, Germany were examined and objectively analysed in terms of their urban form and spatial composition. After extracting the main spatial components that constitute the campus built-environment, architectural and urban design elements that impact human health were further investigated. Space syntax, combined with manual mapping, were used to conduct spatial analysis on the studied campuses to precisely indicate and break-down the spatial components into design elements and features. Finally, systematic literature review and focus group discussions with experts in the fields of Architecture, urban design and public health were conducted to deduce the impact of the investigated design elements on human health and well-being, and compile the findings into a set of design strategies for health-promoting environments.

The findings showed that the university campus built-environment consists of main spatial component families that enable it to function as an urban lab, including buildings, meeting spaces (nodes), pathways/roads, focal points, and thresholds (edges). Each category is broken down into a set of types and design elements. Hierarchical correlational levels between these elements and users' well-being were also revealed. There was significant correlation with users' mental health and physical activity; which are strongly related to developing non-communicable diseases.

The research calls for transforming built-environments into catalysts to improve human health. It underscores the concept of using university campuses as urban labs to foster healthy built-environments. This involves identifying architectural and urban design elements within a campus that affect human health and well-being. These elements are to inform design strategies guiding architects and urban designers in creating human-centric, healthy environments.

Reference:
WHO. (2019). Healthy environments for healthier populations: Why do they matter, and what can we do? World Health Organization.

USE OF THE ASSEMBLY HALL AND RESIDENTS' INTERACTIONS IN THE DISASTER PUBLIC HOUSINGS IN MABI-CHO, JAPAN: A STUDY ON THE PREVENTION OF THE SOCIAL ISOLATION IN THE DISASTER PUBLIC HOUSINGS

Track 4: Design for Health and Wellbeing
Knowledge Network: Environment-Gerontology, Residential Environments
Methods: Interviews, Survey
Mitoko Nakashima (KAGAWA UNIVERSITY)

Background:
Throughout recovery processes from natural disasters, some elderly people have lost their social relationships and been facing the risk of social isolation because of the repeated relocations from their original residence to evacuation shelters, temporary housings, and disaster public housings (hereinafter DPHs). Existing studies on the post-disaster residences in Japan have pointed out the necessity of utilizing existing local resources such as neighborhood relationships in communities to prevent the social isolation of the elderly (e.g. Maeda et al., 2020). However, there is a limit to utilizing existing local resources for elderly people whose existing social relationships have been severed or dismantled through their repeated relocations (Nakashima 2018).

Therefore, rather than relying solely on the existing local resources, new social relationships must be established to prevent the social isolation of the elderly in the post-disaster residences. Through a case study of DPHs in Mabi-cho, Kurashiki City, one of the areas affected by the Western Japan floods in 2018, this paper aims to explore clues for creating new social relationships in the DPHs with focusing on the use of the assembly hall and the interactions among the residents.

On the basis of data obtained from questionnaire survey of 75/113 (66.4%) households and in-depth interviews with 25 elderly persons conducted at four DPHs in Mabi-cho in 2022, three (A, B, C) of which have not maintained the original community and one (D) of which has maintained it, this paper examines 1) actual conditions of interactions among residents in the DPHs, 2) the actual usage of the assembly hall in the DPHs, and 3) the relationships between the assembly hall use and the establishment of social interactions.

As a result, we found that 1) regarding the interactions among residents, 71% of the total respondents answered "no" or "few", while 63% in the DPH D with maintaining original community answered "frequent" or "a little", 2) regarding the use of the assembly hall, 91% of the respondents in the DPH D answered "no", while 35% in the DPH B without maintaining the original community answered "yes", and 3) the relationships between the assembly hall use and social interactions observed in concrete cases are not unidirectional but multidirectional. It can be concluded that the relationship between the assembly hall use and the social interactions in the DPHs varies according to each context, including cases where the use of existing local resources has not led to the assembly hall use, and cases where the assembly hall use has led to new social interactions even in the absence of existing local resources.

References
Maeda, M., Tsukuda, H., Onoda, Y., Takada, M., Amakusa, H. & Nakamura, K. (2020). A study on house and life restoration with household separation and reorganization in group resettlement. Journal of Architecture and Planning: Transactions of AIJ, 85(770), 793-803.
Nakashima, M. (2018) Study on the Prevention of Social Isolation of the Elderly in the Collective Relocation Site for Disaster Prevention of Ofunato City, Japan. AIJ Proceedings of the Housing Studies Symposium 13: 99-106.

WALKING THE WALK: THE AFFECT-ALTERING EFFECTS OF ACTUAL VERSUS VIRTUAL WALKS ACROSS SETTINGS THAT VARY ALONG AN URBAN TO NATURE CONTINUUM

Track 4: Design for Health and Wellbeing
Knowledge Network: Cyberspace & Digital Environments, Health
Methods: Experimental/Quasi-Experimental Research
James Harris (University of Luxembourg)

Introduction:
Historically, built environments have been employed as a comparator in nature exposure research, demonstrating the affective benefits of greenspaces and cautioning against the deleterious effects of rapid urbanization on mental wellbeing. Recent trends involve comparing natural environments to their digital representations, reflecting increased digital immersion and indoor lifestyles (Hitchings, 2021). Our study aimed to investigate the relative impact of the environment (natural or built) and exposure type (actual versus virtual) on mood regulation, with implications for understanding societal developments and strategies to foster mental wellbeing.

Methods:
Conducted in two phases (1a & 1b) across different university sites, the study examined the mood-altering effects of actual and virtual walking in varied urban-nature settings. Employing a randomised within-subject design, students (between the age of 18 to 32) experienced all four conditions:
1. *Actual walk in a natural environment*
2. *Virtual walk (i.e. sitting and watching a first-person filmed video on a laptop) of the same natural environment.*
3. *Actual walk in the urban campus*
4. *Virtual walk (i.e. sitting and watching a first-person filmed video on a laptop) of the same urban campus.*

Results:
Both Study 1a (n=80) and 1b (n=50) revealed significant effects for both environment and exposure type on mood outcomes over time. Post-hoc analyses indicated that actual walks, including urban campus walks, had a significantly larger improvement effect than virtual walks. The park and forest settings, across the same exposure type, yielded a larger positive effect on mood than campus settings. Observing consistent results across differing population samples, locations and time-points supports the ecological validity and robustness of our findings, while further substantiating existing evidence.

Conclusion:
Our findings indicate young adults who briefly step away from their contemporary digitally immersive and sedentary lifestyle, through a leisurely walk outdoors, may experience significant mood-enhancing benefits. Prioritizing the integration of (real) nature and (light) physical activity features in urban planning aligns with a human-centric perspective, and serves to improve mental well-being by increasing the enjoyment and time spent in outdoor experiences. This is especially crucial in the context of emerging technologically saturated built environments.

References:
Browning, M. H. E. M., et al. (2020). An Actual Natural Setting Improves Mood Better Than Its Virtual Counterpart: A Meta-Analysis of Experimental Data [Systematic Review]. Frontiers in Psychology, 11.
Hitchings, R. (2021). The unsettling outdoors: Environmental estrangement in everyday life. John Wiley & Sons.

EXPLORING HUMAN BEHAVIOR IN UMD'S ACADEMIC SPACES: AN ENVIRONMENTAL PSYCHOLOGY INVESTIGATION

Track 4: Design for Health and Wellbeing
Knowledge Network: None/Not Tied to Any Network
Methods: Case Study, Mixed Methods
Ayden Caviglia Harris

Background:
Bridging insights from environmental psychology and architecture/design thinking, this study investigates how design choices of non-instructional indoor spaces in academic buildings influence self-reported student wellbeing. This study aims to contribute to the body of knowledge on student wellbeing and behaviors within academic buildings. There is a research gap in understanding this specific category of space: non-instructional indoor areas, such as lobbies, within academic buildings on higher education campuses.

Framework
The configuration of design features affects social implications in physical spaces (Peponis and Wineman, 2002). Siramkaya and Aydin find that indoor space with "high value of movement and accessibility in spatial configuration increases the level of social interaction"(2018). This study will explore how design features in non-instructional indoor spaces in academic buildings impact student behaviors. McCoy categorizes design features into resources, views, ambient conditions, spatial organization and architectural details (2002). This framework assesses how physical properties influence behavior and therefore self-reported wellbeing. Views are defined as sights observed through a window. Resources are the facilities, equipment, and information that enable users to fulfill objectives. Ambient conditions are measurable environmental factors including temperature, light level, and air quality. Spatial organization is the arrangement of internal components (e.g. furniture), creating spaces. Architectonic details are style-defining elements like materials and finishes.

Research Questions and Methods
The study addresses the question "How do design features of non-instructional indoor spaces in academic buildings impact self-reported student wellbeing?". By observing behavior and tying it to design features, and asking about wellbeing and tying it to design features, this study will describe how design in case studies impact student wellbeing.

I will investigate three case studies on a state university campus to perform a descriptive analysis and identify common themes across cases. Employing a mixed methods approach, the study collects qualitative survey data and quantitative observational data.

The process will begin with architectural site analysis to preliminarily identify design features and behaviors exhibited in the field. Students will be intercepted in the field and asked to participate in a survey inquiring about their perceptions of wellbeing and design features. Subsequently, observational data via behavior mapping will record student behaviors like studying, resting, socializing, and passing through on floor plans of the site. Concurrently, students will be surveyed in the field to assess their state of wellbeing at the moment.

Expected Findings
The study's findings will generate an inventory of design features within non-instructional academic spaces that can contribute to wellbeing.

References
Büyükşahin SiramkayaÖ. Ü., & Aydin, D. (2018). The Effect Of Spatial Configuration On Social Interaction: Assessment Of Social Interactional Spaces Spatial Qualities In A Faculty Building. Journal of Social and Humanities Sciences. https://jshsr.org/index.php/pub/article/view/2109/2026
McCoy, J. M. (2002). Work environments. In R. B. Bechtel & A. Churchman (Eds.), Handbook of environmental psychology (pp. 443-460). John Wiley & Sons, Inc..
Peponis and Wineman (2002). Spatial structure of environment and behavior. In R. B. Bechtel & A. Churchman (Eds.), Handbook of environmental psychology (pp. 271-291). John Wiley & Sons, Inc.

INTEGRATING BIOMIMICRY AND SUSTAINABLE DESIGN PRINCIPLES FOR ECOSYSTEM RESTORATION AND COMMUNITY HEALTH AND WELL-BEING

Track 2: Sustainable buildings and infrastructure
Knowledge Network: Environmental Design Research Education, Sustainable Planning, Design & Behavior
Methods: Case Study, Practice-based Research
Daphne A.U. Chilaka (Texas Tech University); Nishan Khatiwada (Texas Tech University);
Debajyoti Pati (Texas Tech University)

Background:
Zoe is derived from the Greek word for life. It encapsulates the essence of the project's mission, which is to restore life to a land reclaimed from the sea and transform an existing event center into a thriving ecosystem that harmonizes with its surroundings. Situated on Victoria Island in Lagos, Nigeria, the project responds to the urgent need for sustainable development in a rapidly urbanizing landscape. The site, characterized by flat, low-lying topography and flood-prone swampy soil, presents both challenges and opportunities for ecological restoration. By leveraging biomimicry strategies and sustainable design principles, Zoe seeks to mitigate the environmental impact of human activity while enhancing biodiversity and promoting human well-being. The existing event center, with its nine floors and various amenities including event halls, office spaces, and rooftop recreation areas, serves as the foundation for the redesign. Strengths such as ease of access, solar energy generation, and opportunities for rainwater harvesting are identified, while weaknesses such as lack of native vegetation and susceptibility to flooding are addressed through innovative design interventions. Key sustainability concepts employed in Zoe include rain gardens and bioswales for stormwater management, coconut palm trees and fountains to enrich the ecosystem and mitigate pollution, and biomimetic technologies inspired by nature's solutions to environmental challenges.

These include humidity-control walls, water-reactive window shutters, and solar panels designed to mimic the efficiency of natural systems. Priority is given to sustainable site development, energy efficiency, water conservation, and materials optimization. By incorporating features such as pedestrian pathways, green roofs, and passive cooling strategies, Zoe aims to mitigate the urban heat island effect and promote a healthier, more resilient urban environment. The design ensures the restoration of natural habitats, the creation of open and inviting spaces for social interaction, and the integration of biophilic elements to enhance human well-being.

Through careful consideration of local climate conditions, cultural context, and ecological principles, Zoe represents a holistic approach to sustainable design that seeks to restore life to the built environment. In conclusion, Zoe offers a model for sustainable development in urban areas, demonstrating how thoughtful design interventions can revitalize degraded landscapes, mitigate environmental impact, and create spaces that promote the health and well-being of both people and ecosystems. This abstract serves as a concise overview of the research conducted, highlighting the innovative approaches and interdisciplinary collaboration essential to addressing the complex challenges of contemporary urbanization.

INTEGRATING COMMUNITY ENGAGEMENT IN URBAN DESIGN DECISION-MAKING: A FRAMEWORK FOR BLUE-GREEN INFRASTRUCTURE IMPLEMENTATION FOR MITIGATING URBAN HEAT ISLAND EFFECT IN BALTIMORE

Track 1: Nature and Urban Ecology
Knowledge Network: Environmental Design Research Education, Sustainable Planning, Design & Behavior
Methods: Action/Participatory Research, Mixed Methods

Tasneem Tariq (Pennsylvania State University)

Abstract:

Rapid urbanization and the consequential Urban Heat Island (UHI) effect pose profound challenges to the livability of cities, impacting outdoor thermal comfort conditions. This paper explores the critical role of community engagement in urban design, specifically focusing on the integration of Blue-Green Infrastructures (BGIs) to address the UHI effect in selected Baltimore neighborhoods. Drawing from a comprehensive literature review, this study synthesizes key objectives, levels, and methods of community engagement, emphasizing its significance for sustainable and community-driven development. Community engagement is defined as a purposeful process fostering collaboration between communities, organizations, and public and private entities to identify and act upon community needs and aspirations (Scotland et al., 2007). The study builds upon Glass's five stages of public participation and the International Association for Public Participation's spectrum of community engagement, ranging from informing to empowering citizens (Glass, 1979; IAP2, 2000). A review of literature reveals twelve distinct purposes and thirty-four methods of community engagement, stressing the importance of tailored approaches based on project requirements and contextual factors (Geekiyanage et al., 2021).

The research question of this study is: How can community engagement be effectively integrated into the urban decision-making process to collect feedback and preferences on BGI integration in Baltimore? Utilizing a human-centric perspective, the study aims to analyze the seamless integration of natural environments into cities to mitigate UHI effect. Acknowledging the challenge of combining community opinions with urban environmental simulation studies, the research outlines a framework for transdisciplinary collaborations to address human needs for natural elements within urban neighborhoods. Balancing community insights with scientific studies is crucial for a holistic understanding of the urban environment. The methodology involves an in-depth exploration of decision science, tools, and techniques for community engagement, leveraging ongoing literature studies. Successful case studies on community engagement in urban design decision-making will be analyzed, providing lessons to inform the development of a Baltimore-specific framework. The study employs a mixed-methods approach, combining qualitative analysis of community opinions with quantitative data from environmental simulations. Expected outcomes include policy guidelines for designers, urban planners, and policymakers, derived from collaborative efforts between community opinions and scientific insights. By synthesizing insights from literature studies and exploring transdisciplinary collaborations, this research aims to offer practical guidance for integrating community engagement into urban design processes. Ultimately, the study contributes to the knowledge base on incorporating natural elements in urban design, providing a blueprint for sustainable and community-driven urban development. The implications extend to cities globally facing similar environmental challenges, emphasizing the importance of community-driven solutions for enhanced environmental quality and community well-being.

References:

Geekiyanage, D., Fernando, T., & Keraminiyage, K. (2021). Mapping Participatory Methods in the Urban Development Process: A Systematic Review and Case-Based Evidence Analysis. Sustainability, 13(16), Article 16.

Glass, J. J. (1979). Citizen Participation in Planning: The Relationship Between Objectives and Techniques. Journal of the American Planning Association.

IAP2. (2000). Core Values, Ethics, Spectrum – The 3 Pillars of Public Participation—International Association for Public Participation.

Scotland, C., House, T., & Terrace, H. (2007). The National Standards for Community Engagement.

INVESTIGATING TRAUMA-INFORMED DESIGN PRINCIPLES FOR CHILDREN SUFFERING FROM PTSD

Track 5: Spatial Justice, Equity, and Inclusion by Design
Knowledge Network: Children, Youth & Environment, Residential Environments
Methods: Case Study, Meta-Analysis
Daphne A.U. Chilaka (Texas Tech University); Lori Guerrero (Texas Tech University); Nishan Khatiwada (Texas Tech University)

Background:
Post-Traumatic Stress Disorder (PTSD) is an increasingly common challenge that children under the age of seventeen experience after being exposed to traumatic events. The principles of trauma-informed design perform a critical role in ensuring that physical environments responsive to the healing and recovery needs of individuals with PTSD are created. Therefore, the goal of this study was to examine current research on trauma-informed design and to focus on identifying the design principles that can be effectively incorporated into the creation of living spaces for children suffering from PTSD. To achieve this goal, some relevant literature on PTSD in children, trauma-informed design for children, and case studies on their application in different physical space settings for children were reviewed. The data retrieved were compared and analyzed using qualitative analysis, and the results revealed that key trauma-informed design principles that alleviate the symptoms children with PTSD face are safety, trust, privacy, choice, empowerment, connection, and access to nature. This implies that designers of the built environment should pay adequate attention to the incorporation of design opportunities integral to these trauma-informed design principles when creating physical spaces that promote the healing and recovery of children with PTSD.

CASE STUDIES RESEARCH: FINDING THE IMPACT OF NATURAL LIGHTING ON CHILDREN IN EARLY CHILDHOOD SCHOOLS

Track 2: Sustainable Buildings and Infrastructure
Knowledge Network: Children, Youth & Environment, Sustainable Planning, Design & Behavior
Methods: Case Study
Shireen Kanakri (Ball State University); Alondra Rojas (Texas A&M University)

Background:
This study gathers the past known literature and studies revolving the significance of natural light's impact on student's performance as it relates to their behaviors in an educational environment. There have been several studies regarding the impact of natural light in office settings as it relates to energy consumption, with some studies overlapping with artificial light's impact on adult performance in office settings. This study, however, identifies the factors in the design criteria that impact student behaviors that lead to improvement when using natural light rather than artificial light. The methods used in this research study, both qualitative and quantitative, were chosen to reveal the correlation between the use of natural light in classrooms and the performance calculated by analytically comparing three case studies. The literature presented reveals natural light's ability to assist in developing healthier behaviors and overall performance during students' time in educational spaces by reducing stress, enhancing moods, promoting healthy sleep, and increasing alertness and attention.

A SELF-PLACE IDENTIFIED NETWORK DESCRIBING HOW RESIDENTS CREATE AND MAKE SENSE OF THEIR LOCAL COMMUNITY THROUGH PLACE-BASED LIFE HISTORIES - CASE STUDY IN DISTRICT

Track 4: Design for Health and Wellbeing
Knowledge Network: Environmental & Architectural Phenomenology, Health
Methods: Interviews, Survey

Rina Mekata (Tokyo Denki University); Asuka Yamada (Tokyo Denki University)

Abstract:

According to third-place theory, modern urban dwellers need a third place, in addition to their home, workplace, school, etc., to balance self-establishment and social existence (Oldenburg, 1991, chap. 1). People experience home, school, and third places during temporal transitions, and they change and grow through relationships with these places. In other words, people are constructed by self-place or the place where "I" am created. This poster describes place-based life history, with reference to research on life history and place. The characterization of a local community through a sense of "my/our place" forms a self-place identified network (SPiN) (Fig. 1). Based on data from place operators and users, we describe the meaning of place in local community as a community hub that provides a foundation for individuals' quality of life as well as support for consumers, while making people who they are.

A survey was conducted in District A (a municipal administrative district), where community activities aimed at creating places for residents and fostering community are held in vacant houses. The survey targeted operators and users of activity centers run by private organizations that provide places to support children and child-rearing and alleviate poverty and social isolation. The survey comprised open-ended interview questions that encouraged free narration; guideline questions were also prepared to understand the narrators' place-based life histories (Table 1).

The following is an example of how the narrators' place-based life histories were organized for analysis (Fig. 2). One operator had the idea of opening a café 20 years ago, and through interaction with the operator of a child-rearing support class that he attended with his own children, he developed an interest in creating a place to connect people with each other. That she now operates a "third place base for local child-rearing support" shows the self-change that occurred due to her experiences in the place.

From the narrators' place-based life histories, we attempted to describe the SPiN comprising relationships between people through places (Fig. 3). For example, a café operator attempted to conduct poverty alleviation activities for children but realized that offering low-priced or free food and beverages would negatively impact the café's profit-making business. Therefore, instead of providing direct support, the café leases the space to a non-profit organization that conducts support activities, showing that a location can provide indirect support by attracting support groups. Additionally, the surveyed center operators formed an imaginary shopping district with other merchants who shared their philosophy, demonstrating community reformation through idea sharing, independent of geographical ties.

The SPiN, as described by place-based life histories, visualizes a regional society created by places and relationships between people and society. Constructing this concept will help people share place and local community's meanings.

Reference

Oldenburg, Ray. (1991). The Great Good Place. New York: Marlowe & Company

HEALTHY BUILDINGS FOR EVERYONE: TAPPING INTO BIOLOGY TO GROW THE NEXT GENERATION OF BUILDINGS

Track 4: Design for Health and Wellbeing
Knowledge Network: Environmental Design Research Education, Researchers in Professional Practice
Methods: Case Study, Practice-based Research
Eric Corey Freed (CannonDesign)

Abstract:
While every healthcare provider is wondering about how technology will shape the future of care, they are only focusing on the digital.

Biology, chemistry, and symbiosis are also a type of technology. With over 3.8 billion years of research and development, Nature has already figured out the solutions to all of the problems facing humanity. Nature found a way to produce oxygen, sequester carbon, and develop complex structural forms for free without emissions or waste.

Based on our research in designing healthcare facilities, we'll explore a new approach to achieving high performance facilities using a symbiotic mindset. We'll explore the six categories of symbiosis: Systems, Senses, Signals, Cycles, Biomes, and Genomes, to uncover new approaches to boost patient recovery times, boost immune response, and lower pain medication needs.

While we all have been working with general sustainability in our projects, what if we fast-forwarded sustainability to where it is headed? As we move from the Digital Age to the Biological Age, the line between the natural and the manmade will start to disappear. In this talk, we'll explore how to embrace biology as the ultimate technology to develop a new approach to healthcare design that triggers specific physiological responses in patients using human systems, human senses, circadian rhythm, neurotransmitters, and microbiomes to transform how we provide care.

DID YOU KNOW?: You're more microbe than human. Human cells make up only 43% of your body's total cell count. The rest are microscopic colonists (bacteria, parasites, microorganisms). We'll explore how to design probiotic spaces to trigger beneficial physiological responses in patients using biophilia, healthy materials, and microbiomes.

IMPACT OF NATURAL AND ARTIFICIAL LIGHT ON COLLEGE STUDENTS PERFORMANCE

Track 4: Design for Health and Wellbeing
Knowledge Network: Environmental Design Research Education
Methods: Survey
Shireen Kanakri (Ball State University); Jenna Schroader (Ball state university)

Abstract:
The importance of this research is to broaden the research field of interior design elements and their effects on people with autism. This can help many public and private spaces be more accommodating to those who are affected by certain lighting types.

The goal of this project is to determine the type of lighting that people with autism find to be the most comforting in both a bedroom and a classroom setting.

It is hypothesized that the relationship between people with autism and natural light will be the most positive. A survey was given to Ball State University students 18 years of age and older who have been diagnosed with autism or another similar sensory sensitivity.

HERITAGE PRACTICE FOR HUMAN-CENTRIC DESIGN: RELEVANCE OF 'PEOPLE CENTRIC APPROACH' TO HERITAGE AND 'LIVING HERITAGE'

Track 6: Transdisciplinary Design and Collaboration/Open Track
Knowledge Network: Cultural Aspects of Design, Historic Environments
Methods: Interpretive-Historical Research, Practice-based Research
Neel Kamal Chapagain (Ahmedabad University, Centre for Heritage Management)

Background:
Heritage discourses and practices have come a long way in their evolution and relevance, i.e. from a broader shift from a 'monument-centric intervention' to a 'site and landscapes based intervention' as well as recent decades' acknowledgement of people as the core of heritage. Discourses like 'people-centric approach', 'critical heritage studies', 'living heritage' are some of the ways in which this shift of focus from monuments and materials to people and sustainable environment have been captured. Such shifts have also been reflected in the heritage practices such as 'cultural landscapes', 'historic urban landscapes', 'safeguarding of intangible cultural heritage or living heritage', and so on. However, the impacts and outcomes of these shifts have not been perceived in the same ways. This paper critically examines the evolution of these discourses and practice frameworks, and assesses their demonstrated efficacies.

The paper primarily will look into heritage frameworks related to the built environment espoused by the key players in the current global heritage system, viz. UNESCO, ICCROM, and ICOMOS. UNESCO's 1972 convention concerning the Protection of the World Cultural and Natural Heritage (popularly known as the World Heritage convention) and its 2003 convention for the safeguarding of Intangible Cultural Heritage (ICH) are two key conventions from UNESCO that have shaped the heritage discourse and practice across the world today. Within the World Heritage system, the frameworks of cultural landscapes and historic urban landscapes (HUL) have advanced the heritage conservation in relation to people and landscapes. ICOMOS' Nara document on authenticity and several other charters, Getty Conservation Institute's Values Based approach, and ICCROM's 'People Centric Approach' as well as the programme on 'Living Heritage' contributed in refining the heritage conservation practice at the global scale. However, how these global frameworks have been perceived and used by governments and local agencies including communities themselves is a big question to ponder upon. On the one hand, dedicated heritage professionals would like to project and refer to these heritage frameworks as the panacea, whereas in reality these end up being just a pariah. This paper attempts to explore these confusions and nuances of heritage concepts as applied in the built environment. A critical understanding of heritage discourse and practices could help take appropriate measures to pursue heritage practice as one way to work towards 'human-centric design' of sustainable environments.

STAFF, FAMILY AND PATIENT CARE IN SWEET LIKE HONEY: NEONATAL INTENSIVE CARE UNIT

Track 4: Design for Health and Wellbeing
Knowledge Network: Children, Youth & Environment, Health
Methods: Action/Participatory Research, Case Study
Shireen Kanakri (Ball State University); Aysiah Hernandez (Ball state university)

Background:
Neonatal Intensive Care Units are important entities to the healthcare environment and design. Through careful consideration the design of these spaces can positively influence and

impact not only the patient but also the staff and families as well. The goal of Sweet like Honey NICU was to design a space that embraces tranquil atmosphere through soft lighting and natural elements, promotes family-centered care with dedicated spaces for parental involvement, integrates innovative technologies for real-time monitoring and care, ensures safety and sanitation with easy-to-clean surfaces, and fosters holistic healing through serene environments and counseling support, creating a comprehensive and nurturing haven for our tiniest patients and their families. Thorough research was conducted for optimal understanding of NICUs and their impact on staff, families, and patients. The following key takeaways were taken into consideration while designing Sweet like Honey. First, there is the pressing issue of mental and behavioral health disorders faced by staff and families. Enhanced mental health therapy and care in these critical environments should be implemented. Secondly, Interior design of Single-Family Rooms significantly influences family engagement in NICUs. Family engagement in NICUs is crucial for infant development. Parents transition from passive to active caregivers during NICU stay. Lastly, emphasis on supporting staff with materials such as underfoot cushioning is essential for healthcare staff who experience long and physically demanding shifts.

Staff support spaces include a large break room, locker room, on-call rooms, and respite room. Natural light and noise-reducing materials have been applied to the respite room to ensure staff privacy and care. Flooring is specified with underfoot cushioning to support staff's physical needs. Family support spaces include a waiting room and meditation room. Natural light has been utilized in both family support rooms. There is plenty of space for families to support each other during their time at the NICU. There are 22 Single Family Rooms, 3 Multi Patient Rooms, and 2 Isolation Rooms available. Natural light has been utilized in all patient rooms as this can be beneficial to the patient. Family space is included in each room to promote patient and family engagement.

PERCEPTION OF "CITY-LAKE" LANDSCAPE SYSTEM IN TRADITIONAL CITY: FORMATION PATH FROM HUMAN CENTRIC INSIGHT

Track 1: Nature and Urban Ecology
Knowledge Network: Environmental & Architectural Phenomenology, Historic Environments
Methods: Case Study, Statistical/Modeling Analysis
Xiyao Zhao (Beijing Forestry University)

Background:
City-lake System is a compound system and a model of artificial transformation of natural land to benefit human development. Previous studies focused on the aesthetic ideas and hydraulic engineering techniques but ignored the real motivation behind. This study attempts to explore the history of environmental construction in this system, and to describe the changes in landscape development activities by extracting from local history the needs arising from the social and cultural environment in which local communities lived in different historical stages. The study uses geographic information technology to highlight and normalize the spatial and type distribution of landscape development activities in different historical stages. The results show that there is heterogeneity and historical inheritance in the process of multi millennial scale landscape development. This reflects the integrated mapping of the environment to the economic, political, and cultural elements of the historical process of the region, as well as the dynamic adaptation to the needs of local communities. The study highlights the value of the lessons learned from this ancient example of human-centered design for contemporary cities, demonstrates the importance of the concept of regional system in preserving and perpetuating urban cultural landscapes, and provides a way to study historical cultural landscape systems beyond the site scale.

THE EVOLUTION OF WEDDING VENUE DESIGN IN THE AGE OF SOCIAL MEDIA: A CASE STUDY OF VIRGINIA, USA

Track 6: Transdisciplinary Design and Collaboration/Open Track
Knowledge Network: Interior Design
Methods: Interpretive-Historical Research, Photo-based Methods

Bryn Ghio (Farmstead Design Studio); Dorna Eshrati (Ball State University)

Abstract:
Weddings hold a cherished place in American culture, symbolizing personal expression and love. Their significance has grown over time, and today, wedding venue design stands as a crucial element of modern weddings, culturally and economically. In the pre-internet era, printed magazines primarily influenced wedding design. The transformative emergence of social media, blogs, Pinterest, and Instagram in 2010 significantly shaped wedding venue design, reflecting the evolving preferences of couples planning their special day. These platforms facilitated the dissemination of information and inspiration while also playing a pivotal role in shaping the landscape of venue design. This paper explores the interdisciplinary realm of wedding venue design, encompassing architecture, landscape architecture, and interior design, despite its relative underexposure within these fields. Focusing on Virginia, USA, as a case study, we investigate the characteristics of wedding venue design that have evolved since the rise of blogs and social media platforms. Our analysis aims to comprehend how social media and generational shifts have influenced various aspects of venue design, such as location, design characteristics, and their representation on social media platforms. Our study examined twelve wedding venues in Virginia established post-2010 found through Internet, Pinterest, and Instagram searches. Data analysis involved document analysis of published photographs and text of the twelve venues in venue websites, recommendations, reviews, news articles, Instagram posts, blog entries, and Pinterest pins. Our findings highlight the reception space as the primary defining factor, with three distinct design divisions: historical, barn, and modern. These venues were further categorized into Historical House, Historic Building, Historical Barn/Farm, and Modern Rustic. Historic venues showcased a range of design aesthetics, from simple to lavish, with a continuum of historical elegance interpretations emphasizing historical features. The Historical House, Historical Barn/Farm, and Modern Rustic categories employed design aesthetics ranging from country-style rustic to luxurious interpretations. The natural countryside landscape was pivotal in these venue designs and featured prominently on social media. In contrast, historic building venues, primarily situated in downtown cities, focus on showcasing historic facades and design elements within the venue itself. In conclusion, the undeniable synergy between wedding venue design and the environment design field significantly impacts the creation of memorable celebrations and lasting memories. This dynamic interplay offers future research and design opportunities for interdisciplinary endeavors of architecture, landscape architecture, and interior design to continue shaping the evolution of wedding venue design in an era marked by digital connectivity and creative innovation.

ENHANCING HEALTHCARE DESIGN DECISION-MAKING: A COMPARATIVE ANALYSIS OF AI-IMAGE GENERATORS AND PHYSICAL MOCK-UPS

Track 6: Transdisciplinary Design and Collaboration/Open Track
Knowledge Network: Health
Methods: Interviews, Survey
Turki Ibrahim (Texas A&M University); Zhipeng Lu (Department of Architecture); Stephen Parker (Stantec)

Background:
Healthcare design is a complex process that necessitates collaboration among clinicians, patients, and architects. However, involving non-architects in the design process poses challenges due to potential unfamiliarity with design media, concepts, and language (Lin et al., 2018). Furthermore, there is a lack of communication between clinicians and designers because of not only the different languages used in their own disciplines but also the limitations of the ability for clinicians to visualize a space as it is built (Jouppila & Tiainen, 2021). Designers have used different tools to enhance the effectiveness of clinicians' participation in the design process, such as 3D modeling programs, physical mock-ups, and building information models (BIM) (Jouppila & Tiainen, 2021). These tools are more effective than conventional design visualization methods such as floor plans and sections. (Wingler et al., n.d.). Recognizing this limitation, architects have recently integrated artificial intelligence (AI) technologies, such as the AI-image generator, a text-to-image-based AI technology, into the design, representing a substantial leap in expanding design imagination and offering multiple high-quality visual design alternatives (Beyan & Rossy, 2023). This study investigates individuals' perceptions of utilizing physical mock-ups and AI-image generators to assess design options in healthcare projects to enhance decision-making processes.

Method:
The survey is the primary research instrument to gather quantitative data. The data were gathered from 39 participants in a workshop that provided experiences of physical mock-ups and an AI-image generator in simulating mental health facilities. The participants were divided into two groups, assigned to evaluate either an adult inpatient room for low-acuity eating disorders or an acute pediatric psychiatric inpatient room, each with its own front desk. Throughout mock-up development, teams contributed to designs, with AI accelerating creative problem-solving by introducing inspirational imagery. This approach process involved photographing each mock-up iteration and incorporating AI-driven design options to mitigate biases and explore new opportunities effectively.

Finding:
Preliminary findings suggest participants appreciated the dynamic nature of the iterative design process facilitated by AI, especially when combining photographs of physical mock-ups. The study highlights the contribution of AI-image generators and mock-ups to a comprehensive understanding of design options, empowering stakeholders, particularly non-designers, in the evaluation process to make informed decisions.

References:
Beyan, E. V. P., & Rossy, A. G. C. (2023). A Review of AI Image Generator: Influences, Challenges, and Future Prospects for Architectural Field. Journal of Artificial Intelligence in Architecture, 2(1), 53–65. https://ojs.uajy.ac.id/index.php/JARINA/article/view/6662
Jouppila, T., & Tiainen, T. (2021). Nurses' Participation in the Design of an Intensive Care Unit: The Use of Virtual Mock-Ups. HERD: Health Environments Research & Design Journal, 14(2), 301–312. https://doi.org/10.1177/1937586720935407
Lin, Y.-C., Chen, Y.-P., Yien, H.-W., Huang, C.-Y., & Su, Y.-C. (2018). Integrated BIM, game engine and VR technologies for healthcare design: A case study in cancer hospital. Advanced Engineering Informatics, 36, 130–145. https://doi.org/10.1016/j.aei.2018.03.005
Wingler, D., Machry, H., Bayramzadeh, S., Joseph, A., & Allison, D. (n.d.). Comparing the Effectiveness of Four Different Design Media in Communicating Desired Performance Outcomes With Clinical End Users. HERD: Health Environments Research & Design Journal, 12(2), 87–99. https://doi.org/10.1177/1937586718796626

DATA-DRIVEN DECISION-MAKING AND IOT BIG DATA IN HYDROGEN ENERGY PLANNING AND MANAGEMENT: UNLOCKING SUSTAINABLE GROWTH THROUGH OPERATIONAL EFFICIENCY, SAFETY, AND INTELLIGENCE

Yunsheng Su (Professor)

Abstract:
This presentation focused on the use of data-driven decision-making and IoT big data in hydrogen energy planning and management. We highlighted the advantages of data-driven approaches, such as real-time responsiveness and accuracy, while acknowledging challenges like data quality. We also discussed the role of IoT and big data in infrastructure monitoring and supply-demand analysis for hydrogen energy. These technologies enhance operational efficiency, safety, and intelligence in management, driving the industry's sustainable growth. Overall, data-driven decision-making and IoT big data show great promise in hydrogen energy, contributing to its development as technology advances.

1. *Learning Objective 1: Learn the fundamental principles and advantages of data-driven decision-making in the context of hydrogen energy planning.*
2. *Learning Objective 2: Analyze the limitations and challenges associated with data quality and source diversity in data-driven decision-making for hydrogen energy.*
3. *Learning Objective 3: Understand the key roles and potential applications of the Internet of Things (IoT) and big data technology in managing hydrogen energy infrastructure.*
4. *Learning Objective 4: Evaluate the prospects and future trends of integrating data-driven decision-making and IoT big data solutions in the sustainable development of the hydrogen energy industry.*

DATA-DRIVEN DECISION-MAKING PLATFORM FOR HYDROGEN ENERGY MANAGEMENT: INTEGRATING IOT AND BIG DATA FOR DYNAMIC PLANNING AND SPATIAL OPTIMIZATION

Xihan Cheng (Tongji University)

Abstract:
This presentation focused on the planning mechanism of a data-driven hydrogen energy management decision-making platform, which deeply integrates IoT and big data technology. By applying dynamic planning concepts and spatio-temporal supply-demand mapping tools, the mechanism accurately determines the planning capacity and spatial layout of hydrogen infrastructure, achieving dynamic synergy between hydrogen refueling stations and vehicles. Taking Shanghai Jinshan Hydrogen Valley as a case study, the effectiveness and practicality of this platform were comprehensively validated in real-world scenarios like logistics and public transportation. The implementation of dynamic planning in hydrogen infrastructure layout faces challenges such as policy environment, market acceptance, and technological updates. Lessons learned from the Jinshan case can guide the promotion of this approach in other regions and domains

1. *Learn and apply dynamic planning concepts in the context of hydrogen energy infrastructure layout to optimize resource allocation and utilization.*
2. *Analyze the challenges encountered in implementing dynamic planning for hydrogen energy infrastructure, including policy environment, market acceptance, and technological updates.*
3. *Understand how to integrate dynamic planning with other planning methods to develop a more comprehensive and holistic strategy for hydrogen energy infrastructure planning.*
4. *Analyze the unique features of the Shanghai Jinshan Hydrogen Valley case study and learn how its data-driven decision-making mechanism can be effectively applied in real-world scenarios, such as logistics and public transportation, while identifying areas for improvement and potential challenges in its wider adoption.*

EXAMINING HYDROGEN ENERGY INFRASTRUCTURE PLANNING: INNOVATIVE APPROACHES, CASE STUDIES, AND FUTURE PERSPECTIVES

Mingjun Zhong (Designer)

Abstract:
In this study, we examined the intricacies of planning and decision-making for hydrogen energy infrastructure, utilizing cutting-edge techniques like dynamic planning and spatio-temporal supply-demand mapping. To demonstrate the practicality and efficacy of these methods, we selected the Hydrogen Source Carbon Valley town in Shanghai Jinshan as a case study. Our findings underscored the advantages of dynamic planning in optimizing the capacity and spatial placement of hydrogen infrastructure, considering key influencing parameters (KIPs). Nevertheless, the current reliance on static data poses constraints, emphasizing the requirement for improved data sources and meticulous KIP analysis. Moving forward, we envision an enhanced platform with greater maturity and precision, offering dependable support for hydrogen energy management

1. *To Learn about Innovative Approaches in Hydrogen Energy Infrastructure Planning:*
2. *Participants will gain knowledge of dynamic planning and spatio-temporal supply-demand mapping, two innovative methods used in the planning of hydrogen energy infrastructure.*
3. *To Analyze the Advantages of Dynamic Planning in Optimizing Capacity and Spatial Location:*
4. *Participants will be able to analyze the advantages of dynamic planning in terms of optimizing the capacity and spatial location of hydrogen infrastructure, considering key influencing parameters (KIPs).*
5. *To Understand the Limitations of Static Data in Hydrogen Energy Planning:*
6. *Participants will understand the current reliance on static data in hydrogen energy planning and the associated limitations, emphasizing the need for improved data sources and refined KIP analysis.*
7. *To Evaluate the Potential of Enhanced Maturity and Precision of Planning Platforms:*
8. *Participants will evaluate the potential for enhanced maturity and precision of planning platforms to offer reliable support for hydrogen energy management, based on insights gained from the case study of the Hydrogen Source Carbon Valley town in Shanghai Jinshan.*

University of Luxembourg

Walking the Walk: The Affective Impact Of Actual And Virtual Leisurely Walks Along-An-Urban To Nature Continuum

Authors: James Harris, Damien Brevers, Gunnthora Olafsdottir, Claus Vögle

RESEARCH AIMS

1. To evaluate the size of the influence of the type of exposure (Actual versus Virtual) on affective outcomes
2. To investigate the size of the influence of environmental variations i.e. everyday urban campus setting to local greenspace)
3. To evaluate the efficacy of each type of walk in improving the emotional health of young adults.
4. To identify any significant individual differences across our results

INTRODUCTION

Access to greenspace is increasingly recognised to be a vital component of urban health, yet it remains insufficient for many residents. According to WHO guidelines, only 40% of European city residents adequate access to greenspace (Barboza et al., 2021). Regular visits to these areas are further hindered by barriers such as extreme weather conditions, technological distractions, and a pervasive sense of time scarcity (Hitchings, 2021).

Our study aimed to evaluate the affective benefits of leisurely walking through actual greenspaces compared to more frequented everyday urban environments and accessible digital greenspaces. While previous research has explored the differences between urban and natural settings, as well as actual and virtual experiences, these comparisons have not been comprehensively assessed together. By examining these various walking experiences, we gain a deeper understanding of how variations in environment and type of exposure contribute to the psychological benefits of walking through space.

STUDY 1A

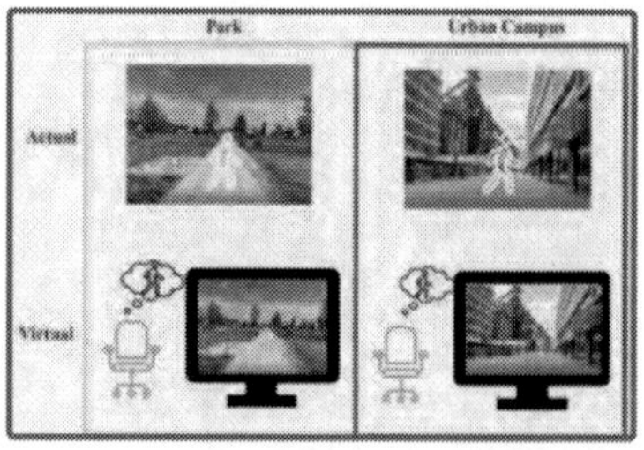

METHODS

- Within-Subject Design with 4 levels: Each participant completed 4 distinct 15-minute walking experiences across different days: 1. Actual Greenspace Walk 2. Virtual Greenspace Walk (Sitting down) 3. Actual Urban Campus Walk 4. Virtual Urban Campus Walk (Sitting down).
- Affective states were measured prior to the walk and immediately after the walk, using several self-report questionnaires: PANAVA-KS, POMS, STAIT-S. Demographic variables and currently well-being levels were collected before the first walking session.
- Eligibility criteria – healthy and 18-32 years old. Exclusion: Any health issue that would adversely impact actual or virtual walking experience.
- Data collection: Winter - Spring 2022 at the University of Luxembourg, Luxembourg.

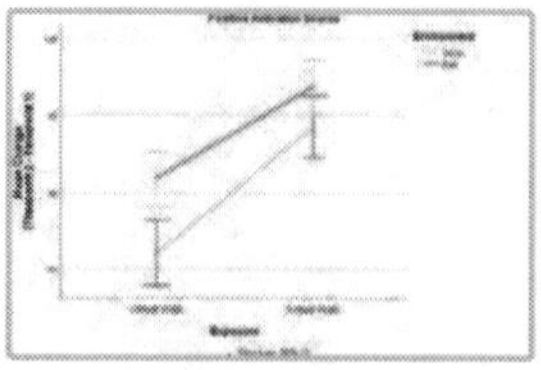

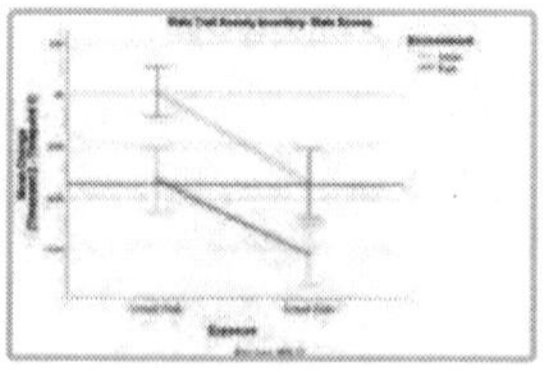

RESULTS

- 80 participants completed the study (43 f, 37 m)
- MANOVA – Found a significant two-way interaction for Time x Environment (p<.001, partial η2 = .33, observed power = 1.00) and Time x Exposure (p<.001, partial η2 = .49, observed power = 1.00).
- Follow-up tests: The actual park walk had a large improvement effect (Cohen's d: 0.95 to 1.13), while the urban walk mostly had a moderate influence (Cohen's d: 0.38 – 0.79) across affective states..
- Neither virtual sessions significantly influenced positive affect or valence (p>.05).
- The virtual park walk led to a significantly moderate improvement in state anxiety, negative affect and total mood disturbance (Cohen's d: 0.32 to 0.59).
- A significant main effect for sex was found – (females generally had larger improvements than males)

STUDY 1B

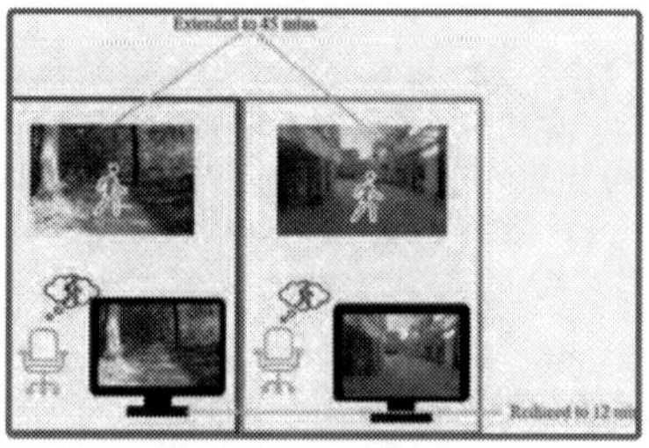

METHODS

- Replicated Study1A design with slight modifications to optimise walking experiences and reduce participant burden based on feedback.
- Focused solely on State Anxiety (STAI-S).
- Increase actual walks to ~45 mins & reduced virtual walk to 12 mins.
- Assessed current physical activity levels as a covariate.
- Data collection: Winter 2023 at Universite Catholique de Louvain, Belgium.

INCREASING YOUNG ADULTS' VISITS TO GREENSPACES IS THE MOST COMPREHENSIVE AND EFFICACIOUS WAY OF IMPROVING THEIR EMOTIONAL HEALTH.

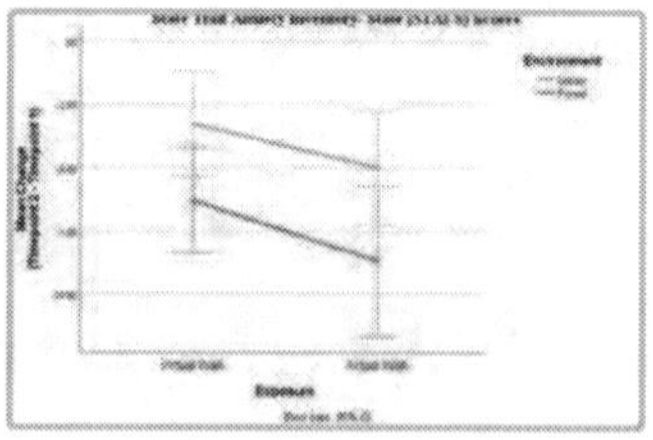

- 46 participants completed the study (30 f, 16m)
- A three-way ANOVA revealed that Time x Environment (partial η2 = .23, p < .001, observed power = 0.95) and Time x Exposure (partial η2 = .23, p < .001, observed power = 0.62) had a moderate effect on state anxiety.
- Follow-up tests revealed that all walking sessions significantly reduced state anxiety.
- The actual forest walk had a large effect (1.00), followed by the virtual forest walk (0.69) and the actual urban walk (0.54), with the virtual urban walk having the smallest effect (0.36).
- No significant differences were found across physical activity levels.
- A significant main effect for Sex was found with females on average showing larger improvements than males.

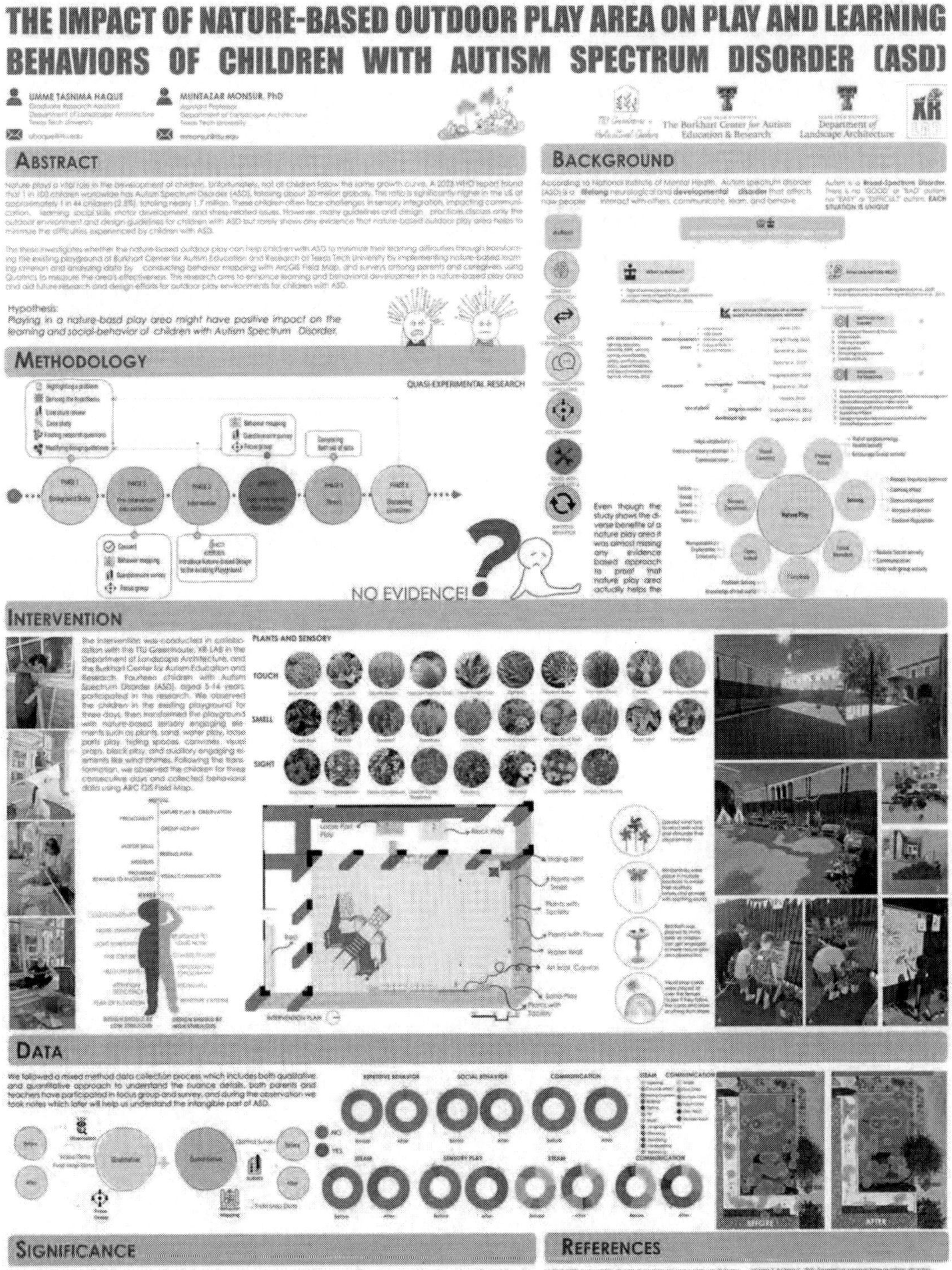

THE IMPACT OF NATURE-BASED OUTDOOR PLAY AREA ON PLAY AND LEARNING BEHAVIORS OF CHILDREN WITH AUTISM SPECTRUM DISORDER (ASD)
UMME TASNIMA HAQUE
MUNTAZAR MONSUR, PhD
The Burkhart Center for Autism Education & Research
Department of Landscape Architecture
ABSTRACT
Nature plays a vital role in the development of children. Unfortunately, not all children follow the same growth curve. A 2023 WHO report found that 1 in 100 children worldwide has Autism Spectrum Disorder (ASD), totaling about 20 million globally. This ratio is significantly higher in the US at approximately 1 in 44 children (2.8%), totaling nearly 1.7 million. These children often face challenges in sensory integration, impacting communication, learning, social skills, motor development, and stress-related issues. However, many guidelines and design practices discuss only the outdoor environment and design guidelines for children with ASD but rarely shows any evidence that nature-based outdoor play area helps to minimize the difficulties experienced by children with ASD.
This thesis investigates whether the nature-based outdoor play can help children with ASD to minimize their learning difficulties through transforming the existing playground of Burkhart Center for Autism Education and Research at Texas Tech University by implementing nature-based learning criterion and analyzing data by conducting behavior mapping with ArcGIS Field Map, and surveys among parents and caregivers using Qualtrics to measure the area's effectiveness. The research aims to enhance learning and behavioral development in a nature-based play area and aid future research and design efforts for outdoor play environments for children with ASD.
Hypothesis:
Playing in a nature-basd play area might have positive impact on the learning and social-behavior of children with Autism Spectrum Disorder.
BACKGROUND
METHODOLOGY
QUASI-EXPERIMENTAL RESEARCH
NO EVIDENCE!
INTERVENTION
PLANTS AND SENSORY
TOUCH
SMELL
SIGHT
INTERVENTION PLAN
DATA
We followed a mixed method data collection process which includes both qualitative and quantitative approach to understand the nuance details. both parents and teachers have participated in focus group and survey, and during the observation we took notes which later will help us understand the intangible part of ASD.
NO
YES
BEFORE
AFTER
SIGNIFICANCE
This research serves as an excellent case study for evidence-based approaches in designing nature-based outdoor play areas for children with Autism Spectrum Disorder (ASD). It has the potential to contribute significantly to the development of future guidelines for outdoor learning areas. Additionally, this study provides valuable information for landscape designers to enhance the quality of nature-based outdoor spaces. Furthermore, it offers insights for parents, teachers, and caregivers about the diverse aspects of nature-based outdoor play, helping children with ASD to thrive in a more meaningful way.
REFERENCES

Navigating Online Design Classes with Virtual "Desk" Critique

Qing Lana Luo, PLA, MLA, ASLA, Associate Professor of Landscape Architecture
Oklahoma State University, Stillwater, OK, USA

Background

The Desk critique is one of the key instruction methods that benefits student critical thinking in design classes (Goldschmidt et al., 2010). It involves the professor sitting down at each student's desk and reviewing their work progress displayed on the desk. An effective desk review consists of giving verbal feedback and, more importantly, marking students' designs to offer ways of improving and exploring further the design problems (Al-Qawasmi et al., 2000).

Qing Lana Luo is helping a student to improve design work by drawing on the student's drawing.

Challenges for Online Design Courses

Professors could not conduct the traditional close-distanced in-person desk critique for online design classes. Although student work can be marked and sketched over by the instructor and then sent over to student for them to look at. But the asynchronized process lacks of real-time, immediate feedback, and non-verbal communication, such as gestures, facial expressions, and body language, which often provide additional context and clarity. All the benefits of desk critique in a design class could have lost without technology's assistance.

Methods & Solutions

With the advancement of video conferencing technology, such as Zoom Annotate and other creative digital drawing tools were used in design classes instead of in-person critique to provide an equivalent level review for students. This accomplished a live digital review with the ability to mark student work, which the traditional desk critique can offer.

Twenty two students were given a short questionnaire based on their perceptions of their likes, dislikes, relation to Zoom live critique, and potential to facilitate course learning.

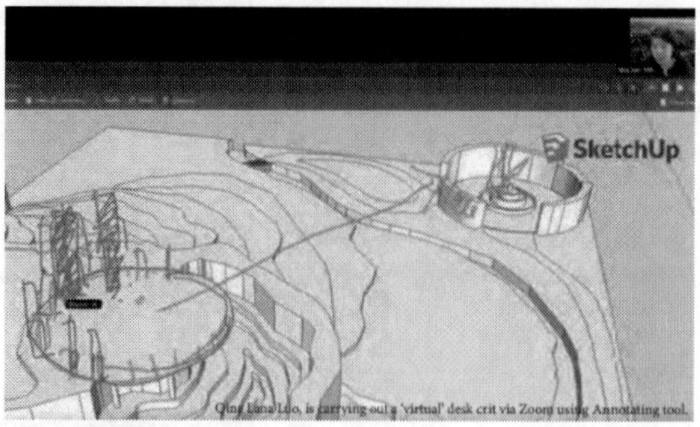

Qing Lana Luo, is carrying out a 'virtual' desk crit via Zoom using Annotating tool.

Function Analysis

Just like a painting board, instructor can change annotation pen colors as using real markers.

- Zoom & its Annotate tool (and other digital annotation tools) have made desk crit a close alternative
- Annotate tool can be used by both meeting host(s) & participants
- The only requirement is to share screen
- Sharing screen can be initiated by a host or participnt

Traditional Desk Critique	*Live Virtual Critique*
Close-distance in person	*Face-to-face virtually*
Professor marks on paper	*Professor marks on screen*
Student need to print out work before hand	*Student only need to share screen; no need to print*
-------	*Student can use any software*

Results & Discussions

1. The virtual crit via digital tools saves time:
   - No need to require the students to submit the files to you before annotation (streamlined process to save time)
   - Easy to initiate:
   - Just share screen (host or participant)

2. Virtual crit process is highly interactive:
   - Both host and participants can annotate interactively
   - No fuss changing editing software for faculty, such as Photoshop, Adobe Acrobat, SketchUp, DrawBoard, AutoCAD, etc.

3. Students have reported the benefits of having valuable visual feedback time via a computer screen at any location. The only limitation is potential internet speed issues.

Conclusions

The implications for using digital crit to teach online design class include 1) providing a more confident and effective tool for online design classes, 2) offering flexibility for both synchronous and non-synchronous crit due to the recording and all-in-one function, and 3) allowing faculty to expand a valued and effective paradigm to technology-enhanced learning for design.

References

Goldschmidt, Hochman, H., & Dafni, I. (2010). The design studio "crit": Teacher–student communication. Artificial Intelligence for Engineering Design, Analysis and Manufacturing, 24(3), 285–302. https://doi.org/10.1017/S089006041000020X

Al-Qawasmi, & Clayton, M. J. (2000). The Effectiveness of Digital Desk Crits in Design Education. Computing in Civil and Building Engineering, 365–372. https://doi.org/10.1061/40513(279)47

Key Words: desk critique, online, technology, interactive, virtual

APPLYING CULTURALLY SENSITIVE DESIGN IN INTERIOR SPACES FOR NATIVE AMERICANS

Jenna Warneke, Tilanka Chandrasekera, Ph.D., & Aditya Jayadas, Ph.D.

Introduction

In 2020, an estimated 3.7 million U.S. residents identified as American Indian and Alaska Native alone, accounting for 1.1% of the US population, while an additional 5.9 million U.S. residents identified as American Indian and Alaska Native and another race group (Office of Minority Health, 2020). One-third (32%) of Native households on Tribal lands live in poverty, compared to 18% of households nationwide, and the annual median household income among Native Americans and Alaska Natives was $37,408, or 33% less than the national median income (Native American Housing, 2017). In addition, often time spaces are designed for Native Americans without applying culturally sensitive design principles (Young, 2022).

A study conducted by Findling in 2022 highlighted the numerous housing challenges faced by rural adult Native Americans in their daily lives. The study revealed that 24% of Native American adults reported experiencing discrimination when attempting to rent housing, 48% faced significant housing problems, 20% encountered issues paying for housing, and 19% noted homelessness issues within their community. These statistics indicate that nearly half of the Native American adults surveyed reported at least one serious housing issue. Despite these ongoing challenges, housing for Native Americans continues to be designed without their specific needs and cultural considerations in mind. Homes are often built by non-tribal members who adhere to Western standards of housing, which can limit Native Americans' ability to participate in cultural practices, separate them from their tribal communities, and result in poorly designed, low-quality housing.

Extensive research shows that Western culture's special housing needs are addressed with substantial support and innovation. For example, holistic care design for older American adults in the United States, as pointed out by Rantz & Ersek, 2023), has gained significant backing. Facilities are now recognizing the importance of redesigning nursing homes to enhance the quality of life for older adults. An influential article by Mercer in 1996 discussed transformative changes in a nursing home within the Navajo Nation. These changes included providing older adults with access to traditional meals, prayer spaces, bedding, and furniture, as well as options for Western furniture, meals, and clothing. This inclusive approach allowed individuals, whether raised traditionally or not, to choose what made them feel most comfortable and connected to their culture.

Purpose

The purpose of this study was to better understand design preferences of interior spaces for Native Americans.

Methods

The study was carried out in two phases: a) a focus group study with open ended questions on design preferences, and b) a survey on preference of design attributes for interior spaces. Eight participants were recruited for the focus group study. Four of the eight participants identified as Native American and were enrolled with a tribal nation (age: 18-25), and four participants identified as Non-Native American (age: 18-52). The participants were asked nine open-ended questions on preferences of interior spaces. The focus group session was audio recorded. The data was then transcribed, and content analysis was used to determine themes and sub-themes. Next, nine Native Americans (age: 18-39) and 24 Non-Native Americans (age: 18-52) were recruited for the survey. The survey was conducted using 20 Likert scale questions on perceived preferences for design of interior spaces.

Results & Discussion

The qualitative data from the focus group showed that while Native Americans valued artifacts such as tribal artwork and blankets, Non-Native Americans preferred lots of color and bigger spaces. The other big difference was the Native Americans preferred plants and natural elements, while Non-Native Americans preferred hardwood and marble. The themes, sub-themes and quotes from the Native American and Non-Native American participants are summarized in Table1 & Table 2 respectively. As for the quantitative data, there was no statistically significant differences between Native American and Non-Native American participants for natural elements or natural lighting. Similarly, there was no statistically difference between the two groups for preference of having beaded decorations and displaying jewelry in interior spaces. However, there was a statistically significant difference for the preferences of historical items ($p=0.03$) with Native American reporting a higher score than Non-Native Americans. There was a statistically significant difference for a non-visual attribute, the smell of sage and cedar in interior spaces ($p=0.02$) with Native Americans preferring this more when compared to Non-Native Americans. The quantitative data is summarized in Figure 1-5.

Table 1. Themes, sub-themes and quotes from Native American participants

Themes	Subthemes	Quotes
Culture Value	Spirituality	"I pray everyday, but try to smudge pray every two weeks"
	Family	"I want an open space for my family…it keeps to my tribal values"
Material Culture	Tribal Artwork	"Displaying my Native American heritage would be great, through native artwork, pottery, blankets"
	Tribal Blankets	"I always had for blankets on all of my bedding"
	Eagle Feathers/Sage	"I pray everyday, but try to smudge pray every two weeks" "I highly value and protect my grandfather's eagle feather"
Nature/ Natural Elements	Plants	"Presence of plants would be nice"
	Natural Lighting	"I like high ceilings, natural lighting and big windows"
	Living Sustainability	"Culturally living off the land and being sustainable, natural things regarding the environment"
Floorplan	Open Space	"I want open space for my family"
	Privacy	"I want privacy space for a safe place for me to pray and keep my special belongings"

Table 2. Themes, sub-themes and quotes from Non-Native American participants

Themes	Subthemes	Quotes
Home Elements	Lighting	"Lots of Natural Lighting" "Change my yellow color lighting to white color lighting"
	Kitchens	"I want a bigger Kitchen" "I want a bigger Kitchen" "I want a more open Kitchen"
Culture	Color	"I like bold Prints and bright colors" "I like Mid Century… and lots of colors, everything you can think of" "Changing the color of my lights, and changing the color of my baseboards to a darker wood color"
Floorplan	Expanding home	"I wouldn't be living on the acreage I live on right now, I want to expand" "I want to live in a bigger house, with a bigger kitchen" "I want a bigger shower"
	Adding personal Usage Rooms	"I want better workspace" "I definitely want a sewing room"
Materials	Hardwood	"changing the color of my baseboards to a darker wood color"
	Marble	"I want marble in my kitchen"

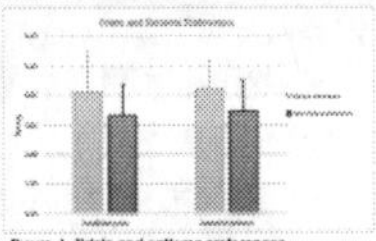

Figure 1. Prints and patterns preferences

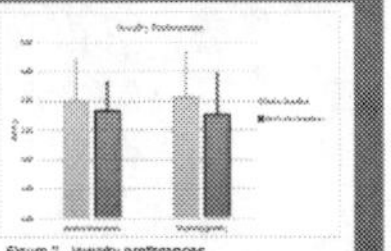

Figure 2. Jewelry preferences

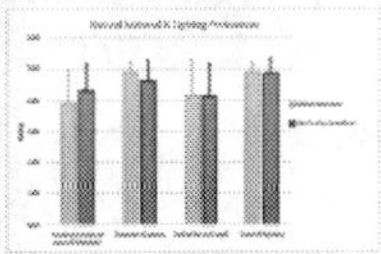

Figure 3. Natural material/lighting preferences

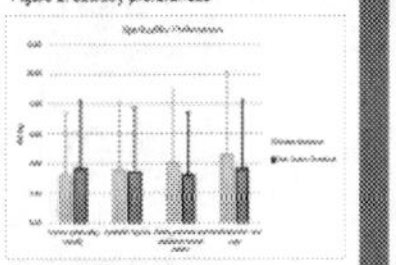

Figure 4. Spirituality preferences

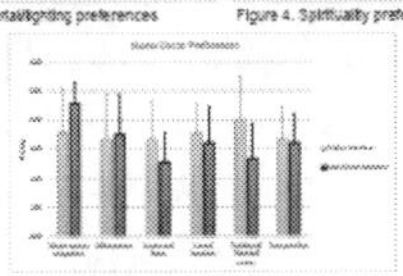

Figure 5. Home decor preferences

Conclusion

The Native American participants in the study preferred to incorporate culture, tradition, and religion into the design of interior spaces. In contrast, Non-Native American participants favored more materialistic options such as a wide range of color choices and larger spaces. This study highlights important implications for the design of interior spaces for both Native Americans and Non-Native Americans, emphasizing the need for culturally sensitive design.

Culturally sensitive design acknowledges and respects the cultural practices, traditions, and preferences of diverse groups. The implications of this study extend to architects, interior designers, and policymakers. For architects and interior designers, it is crucial to engage with community members and stakeholders to understand their unique needs and preferences. This collaborative approach can lead to the creation of living spaces that are not only functional but also culturally enriching and personally meaningful.

References

1. Findling, M. T., Blendon, R. J., Benson, J. M., & Miller, C. (2022). The unseen picture: Issues with healthcare, discrimination, police and safety, and housing experienced by Native American populations in rural America. *The Journal of Rural Health*, 38(1), 180-186.
2. Mercer, S. O. (1996). Navajo elderly people in a reservation nursing home: Admission predictors and culture care practices. Social work, 41(2), 181-189.
3. Native American Housing (2017). Housing needs on Native American and Alaska native tribal lands. Retrieved from https://nlihc.org/explore-issues/policy-priorities/native-american-housing
4. Office of Minority Health (2020). American Indian/Alaska Native Health. Retrieved from https://minorityhealth.hhs.gov/american-indianalaska-native-health
5. Rantz, M., & Ersek, M. (2023). Care delivery, quality measurement, and quality improvement in nursing homes: Issues and recommendations from the National Academies' Report on the Quality of Care in Nursing Homes. *Public Policy & Aging Report*, 33(Supplement_1), S11-S15.
6. Young, B. (2022). Building Cultural Competency in Interior Design: When Good Intentions go Awry. *International Journal of Designs for Learning*, 13(1), 128-136.

Design & Merchandising Dept.

The Relation Between Perceived Sensory Dimensions and Perceived Restorativeness in Urban Rooftop Gardens

Nianchu Hou, Melanie Duffey

Purpose

The study's main objective is to investigate the relationship between PSD and users' perceived restorativeness in urban rooftop gardens and determine the factors in PSD that influence users' response to urban rooftop gardens.

Key Terms

Perceived Restorativeness: subject's perception of the restorative of the green environment (Scopelliti & Vittoria Giuliani, 2004; Van den Berg et al., 2014).

Theoretical Framework

- **Stress Reduction Theory (SRT)** (Ulrich, 1983; Ulrich et al., 1991)
- **Attention Recovery Theory (ART)** (Kaplan and Kaplan, 1989).
- Both theories propose that exposure to natural environments can alleviate stress and restore cognitive functioning.

Significance

This research aims to better inform city planners and municipalities about how urban rooftop gardens may promote better mental health and what dimensions both social and physical are important to consider.

Research Question

RQ1: What factors in PSD effect the high rating in response?
RQ2: What factors in PSD effect the low rating in response?
RQ3: In high ratings, what perceived sensory dimension (PSD) is associated with users' perceived restorativeness?

Methodology

- Online survey
- Explanatory sequential design
- Measurement: Perceived Restorative-

Data Analysis

- Confirmatory Factor Analysis (CFA) to check the construct validity of scale items.
- Partial least squares structural equation modeling (PLS-SEM)
- Heat Map & Eye tracking

What is Perceived Sensory Dimensions (PSD)?

A tool to assess the quality of the landscape, containing eight dimensions namely nature, serene, space, rich in species, social, prospect, culture, and refuge (Grahn & Stigsdotter, 2010).

Eight PSDs	Factors
Space	- Areas not crossed by paths - Lots of trees - Sunny and shady places - Places where people can gather
Refuge	- Many bushes and Tables and benches - Watching people being active - Play equipment
Nature	- Nature quality - Wild and untouched - Free growing lawns
Serene	- Clean and well maintained - No traffic noises
Social	- Entertainment - Restaurant
Culture	- Fountains - Statues - Flowers
Prospect	- Plane, well-cut grass
Rich in Species	- Natural plant and animal populations - Many native plants to study

Research Design

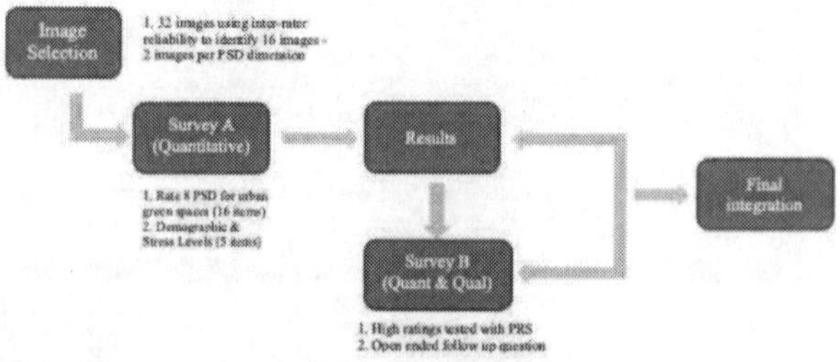

AUBURN UNIVERSITY
College of Human Sciences
Department of Consumer and Design Sciences

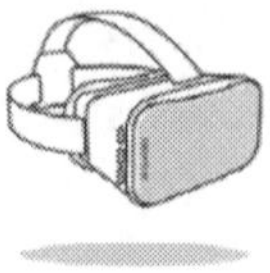

MAPPING THERMAL RESILIENCE:
AFRICAN RESIDENTIAL BUILDING TYPOLOGIES.

Hotter temperatures associated with climate change will disproportionately affect lower income communities with less access to air conditioning units, especially those living in hot climates whose homes do not take advantage of passive cooling techniques. Cities in particular which experience the heat island effect will be particularly vulnerable to risks associated with higher temperatures. Projections expect urban extreme heat exposure in Africa of above 40 C to reach 45-95 billion person-days per year in the 2060s up from 2 billion person-days per year during 1985-2005, with West Africa experiencing the greatest exposure ("Africa," 2023). This paper reviews existing literature on some of the residential building typologies and construction techniques in African cities. It aims to provide summaries of typologies and their compositions that can be used in further research to identify design strategies that can mitigate thermal discomfort and extreme temperatures associated with climate change in African urban dwellings.

Image: Coral rag house (AWL Images)

Image: Karshif house in Egypt

Image: Corrugated metal house

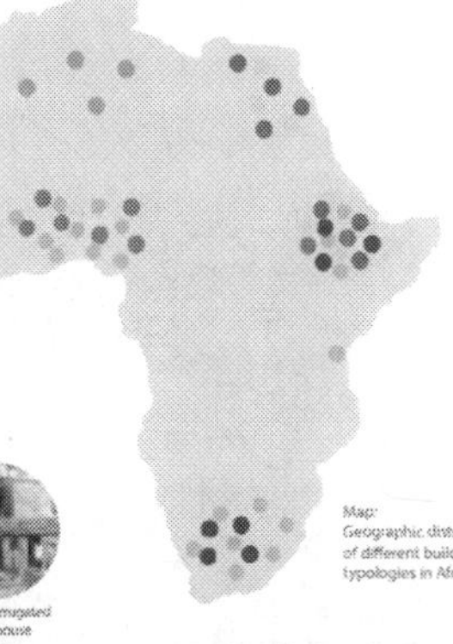

Map: Geographic distribution of different building typologies in Africa.

The Single-Story Brick Masonry House is roughly 40 square meters with 4-meter-high ceilings. They usually contain 3-4 small windows and 2 doors. This typology was identified in South Africa and Ethiopia.

Image: Red Brick House Typology in Nigeria

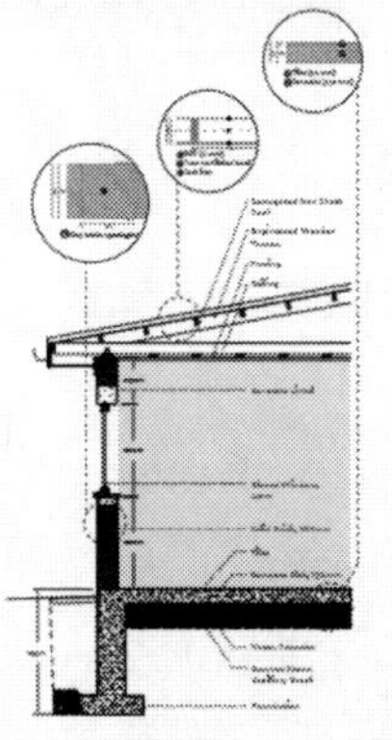

Red Brick House

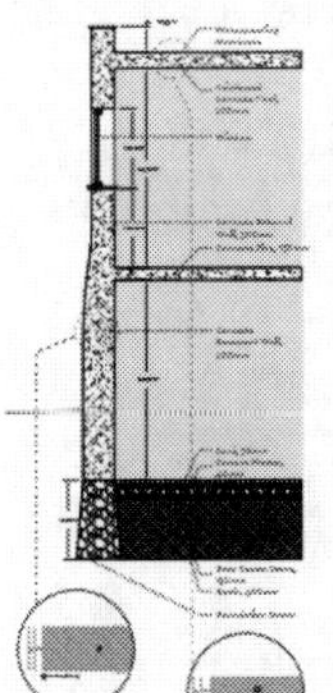

The Ksour Dar typology has 3 to 4 rooms built around an open court. The houses share walls with other buildings with only one wall facing the outdoors.

Image: Ksour House Typology in Ghardaia City, Algeria

Ksour House

Image: Concrete Block House Typology in Namibia

The most abundant typology with walls of reinforced concrete blocks and a metal roof.

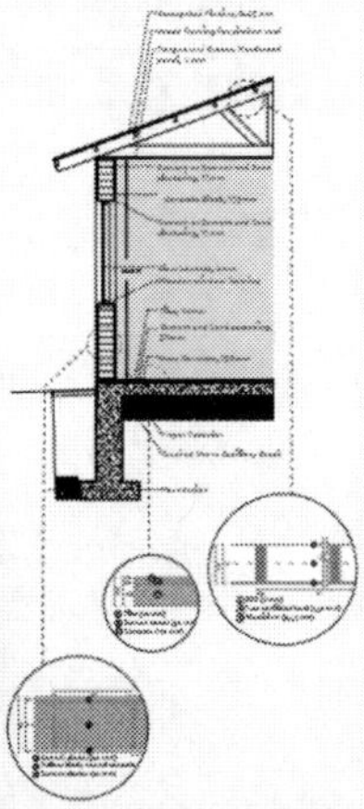

Concrete Block

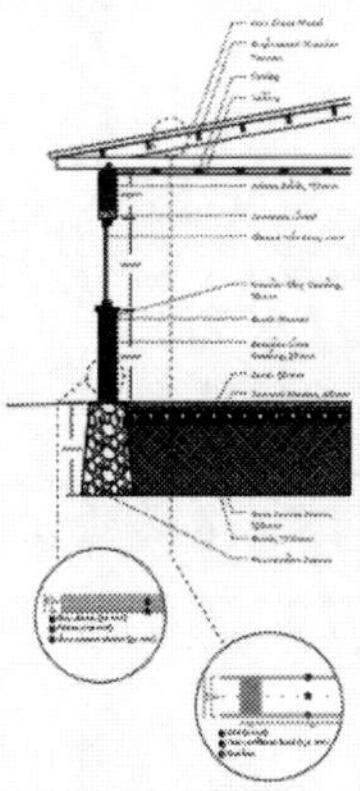

Image: Mud House in Ouagadougou, Burkina Faso

The mud house is a vernacular style primarily made of adobe. The earth used must be 10-40% clay and is mixed with water and sometimes straw and the walls are loadbearing.

Mud House

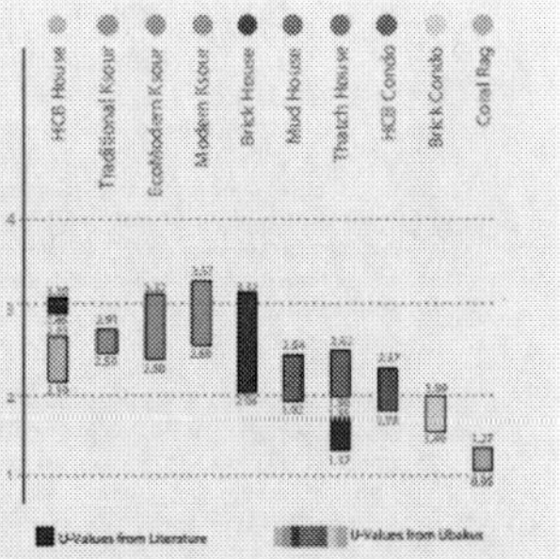

Image: Predicted ranges of external wall U-values for ten selected typologies

For all the typologies, we evaluated thermal performance for the external wall, the roof, and the floor using data gathered from literature and additional assessement conducted using the web-app Ubakus. We highlight the variations we found between numbers in the literature and our analysis.

Distribution of Building Typologies

Twelve different residential typology groups have been identified during this research project. These typologies vary by home and region but retain core features with the categorization scheme primarily based on building form and envelope construction. The typologies in this paper are only a few housing types in Africa with many more to be identified, and they likely also exist in regions that have not been noted in this study. The majority are found in low-income housing as low-income housing appears to be the most studied group in existing literature, but these typologies are often expanded upon with some variation in housing of other income classes.

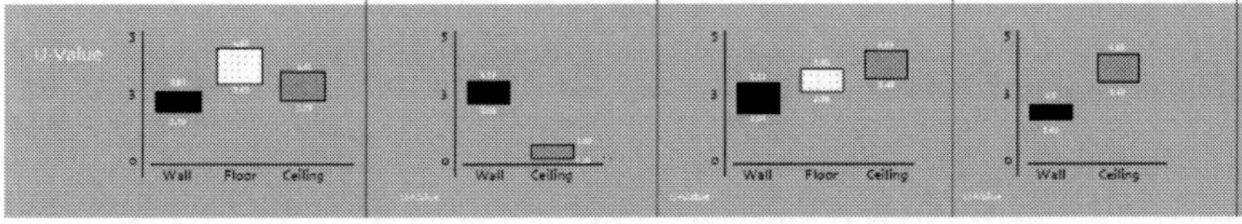

These figures show selected four building typologies from different parts of Africa with different envelope performances. For example, the roof or ceiling construction used in the Ksour house has relatively the highest performance, mainly because it is made out of adobe bricks while the other three typologies we showed in the sectional drawings have sheet metal roofing.

Summary

Proposed Workflow and Next Steps

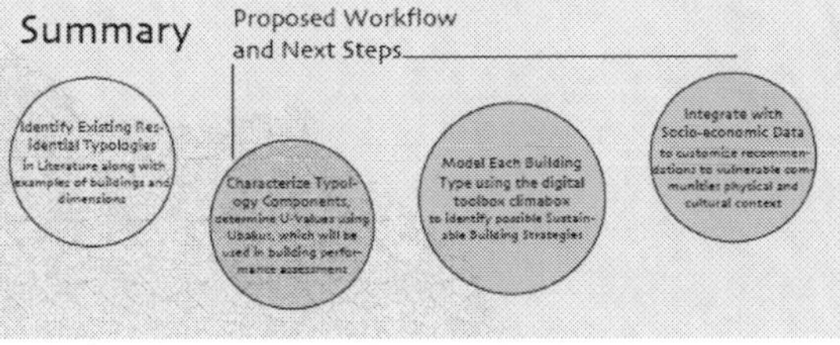

References:

Africa. (2023). In Climate Change 2022 – Impacts, Adaptation and Vulnerability (pp. 1285–1455). Cambridge University Press. https://doi.org/10.1017/9781009325844.011
Arsano, A. Y. (2022).

CLIMATE-CARBON-EQUITY Making Sustainable Design Concepts Accessible for All. In S.S in Architecture. Massachusetts Institute of Technology

Authors:
Alpha Arsano, Assistant Professor
Elisabeth Molnar, MArch
Kwan To Ryan, BS in Architecture

EDRA55 2024, Portland

Northeastern University
College of Art, Media and Design

Assessment of Interior Design Attributes for Older Women who Live in Rural Areas to Age-In-Place

Rose Small, Tilanka Chandrasekera, Ph.D., and Aditya Jayadas, Ph.D.

Introduction

The United States (US) is aging, and the US Census Bureau (n.d.) reported that 16.8% of all the individuals who currently reside in the US are over the age of 65 years. The Center for Disease Control (CDC) reports that there is a difference in life expectancy at 65 years between men and women, with women expected to live 2.7 years longer (CDC, 2022). The differences in life expectancy are often a result of differences in sensory and memory decline, reduced strength and motor functions between older men and women. Farage, Miller, Ajayi and Hutchins (2012) reported changes in vision, hearing, mobility, reflexes, flexibility, coordination and balance as you age. These changes can vary across gender. For instance, it has been reported that men have higher visual spatial ability when compared to women (Pauls, Pettermann & Lepach, 2013). In addition to gender differences, living in a rural or urban area can pose unique challenges for older adults. Paul, Fonseca, Martin & Amado (2003) reported that older adults who live in urban areas can have a greater sense of autonomy while also having a greater feeling of anxiety when compared to those who live in rural areas. Thus, there is a ***critical need*** to not only better understand differences in physical and cognitive challenges in older adults as they age-in-place, but also understand differences in preferences between male and female older adults, and older adults who live in urban or rural areas.

Objectives

The ***primary objective*** is to carry out a needs assessment to determine if there are differences in cognitive and physical challenges that older men and women face in their home as they age-in-place with regard to their environment along with their preferences. The ***secondary objective*** is to better determine if there are differences in challenges that older adults who live in rural areas face along with their preferences for design of their home environment to age-in-place when compared to older adults who live in urban areas.

Specific Aims

1. Determine if the cognitive challenges within their home environment experienced by older women are different from older men as they age-in-place; 2. Determine if the physical challenges within their home environment experienced by older women are different from older men as they age-in-place; 3. Determine if the cognitive challenges experienced by older adults who live in rural areas are different from older adults who live in urban areas as they age-in-place; 4. Determine if the physical challenges experienced by older adults who live in rural areas are different from older adults who live in urban areas as they age-in-place.

Methods

The study will be carried out in two phases using a desktop computer in the Mixed Reality laboratory. For Phase I, 20 older adults (age: 65-93yrs) were recruited for a focus group study via Zoom. Specifically, two groups of 10 older adults (five older women and five men) who live in rural and urban areas will be recruited. The older adults will be asked open-ended questions on physical and cognitive challenges they face in their home environment along with preferences for different design attributes within their home. Audio and video recording will be used to capture this data, and content analysis will be used to analyze the qualitative data gathered. For Phase II, 34 older adults (age: 65-87yrs) were recruited. Specifically, there were 8 female and 6 male participants from rural areas, and 16 female and 4 male participants from urban areas. A Qualtrics survey was used with Likert scale questions on cognitive and physical challenges that older adults experience in their home environment along with preferences for different design attributes in their home. For analysis of the quantitative data, a 2*2 analysis of variance (ANOVA) was used with two levels of gender (male vs female) and location (rural vs urban). A significance level of 0.05 was used.

Results & Discussion

A summary of the ***findings*** from the qualitative data analysis is summarized in *Table 1* below. There were differences in likes, dislikes and preferences between older females and males, residing in rural and urban areas.

Table 1. Themes and sub-themes from the qualitative data analysis

	[illegible]	[illegible]	[illegible]
[illegible]	[illegible]	[illegible]	[illegible]
		[illegible]	[illegible]
	[illegible]	[illegible]	[illegible]
		[illegible]	[illegible]
[illegible]	[illegible]	[illegible]	[illegible]
		[illegible]	[illegible]
	[illegible]	[illegible]	[illegible]
		[illegible]	[illegible]
[illegible]	[illegible]	[illegible]	[illegible]
		[illegible]	[illegible]
	[illegible]	[illegible]	[illegible]
		[illegible]	[illegible]
[illegible]	[illegible]	[illegible]	[illegible]
		[illegible]	[illegible]
	[illegible]	[illegible]	[illegible]
		[illegible]	[illegible]

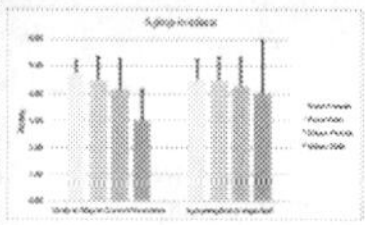

Figure 1. Aging-in-place perceptions across different groups

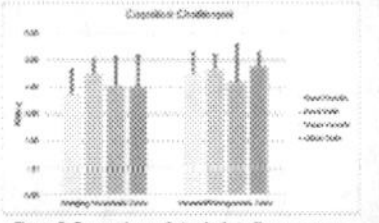

Figure 2. Perceptions of physical challenges across different groups

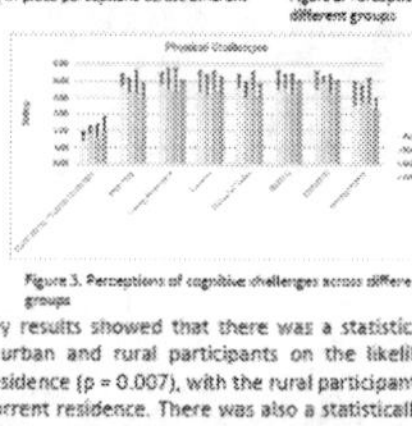

Figure 3. Perceptions of cognitive challenges across different groups

The survey results showed that there was a statistically significant difference between urban and rural participants on the likelihood of staying in their current residence ($p = 0.007$), with the rural participants wanting to age-in-place in their current residence. There was also a statistically significant difference in physical challenges with personal tasks between male and female participants ($p = 0.006$). None of the other dependent variables showed statistically significant differences, for main effects and interactions. Lastly, there was no statistically significant difference for the cognitive challenges between male and female participants, and those living in rural and urban areas. The survey findings are summarized in *Figures 1-3*.

Conclusion

Future studies should include more older adults from different geographical locations across the country. This study has important implications for older adults to age-in-place in rural and urban areas, while improving their physical and cognitive health.

References

1. CDC (2022). Mortality in the United States. Retrieved from https://www.cdc.gov/nchs/data/databriefs/db456.pdf
2. Farage, M. A., Miller, K. W., Ajayi, F., & Hutchins, D. (2012). Design principles to accommodate older adults. *Global Journal of Health Science*, 4(2).
3. Paul, C., Fonseca, A. M., Martin, I., & Amado, J. (2003). Psychosocial profile of rural and urban elders in Portugal. *European Psychologist*, 8(3), 160–167.
4. Pauls, F., Petermann, F., & Lepach, A. C. (2013). Gender differences in episodic memory and visual working memory including the effects of age. *Memory*, 21(7), 857-874.
5. US. Census Bureau (n.d.). Quick Facts. Retrieved from https://www.census.gov/quickfacts/fact/table/US/LFE046221

| Design & Merchandising Dept.

Exploring Behavior Mapping: History and Methods

Tatiana Orozco, PhD; Dagmar Rittenbacher, MID, CKBD; Lesa Lorusso, PhD, RID, NCIDQ
Gresham Smith, Orlando, FL

Definition & History

Behavior Mapping is a set of methods for observing behaviors in a particular physical space and time (Ng, 2016). Findings include descriptions of behaviors, actors/participants, and relationships between behavior and physical environment.

Place-Centered: assess usage of a space.

Person-Centered: learn about a person or group's activities and movement in relation to space and time.

Observer Effects: Actors may behave differently when they know they are being observed (Adair, 1984; Bechtel, 1967; Fisher, 1993). Thus, observing inconspicuously became a key concern.

Technology and Machine Observers: Bechtel (1967) was one of the earliest to use "machine observers," using a network of mats across a museum exhibit floor with sensors to detect location and number of visitor footsteps. Other potential machine observers include proximity sensors, RFID, camera footage with facial recognition software, etc.

Case Study: Background & Methods

A hospital in the southeastern U.S. renovated two floors of inpatient units. A **person-centered behavior mapping study** was conducted, comparing employee activity and movement from two renovated and two non-renovated floors.

Materials & Procedures:

Step-tracking devices: GT9X Link

Hanging file-folder bins: storage for devices and floor plans between shifts

Printed floor plans: for 1 week, participants draw travel routes, write shift and activity notes

Follow-up interviews to review drawings and notes.

Case Study: Findings

Route Diagrams by Shift Time

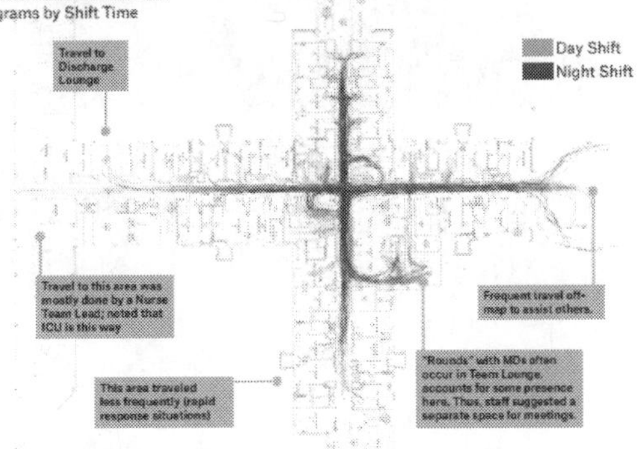

Steps by Renovation Status

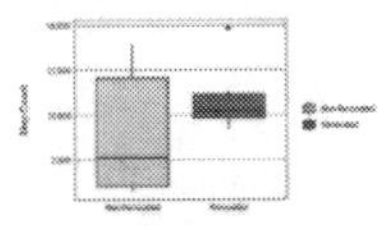

Steps Summary

No significant difference in average or median steps between renovated and non-renovated floors ($p > .05$).

The interquartile range (IQR; spread of step counts) was much larger for non-renovated than renovated floor shifts; there was more step count consistency across shifts for post-renovation.

Shift-Level Steps Summary

Study Floors	Shifts: N (%)	Step Counts: Median (IQR)
Non-Renovated	14 (50%)	7,806 (4,985 – 12,080)
Renovated	14 (50%)	10,254 (9,807 – 11, 141)

Discussion

Behavior mapping has great potential to contribute to the body of evidence on optimal healthcare architecture and design. Findings can empower architectural and design professionals to produce high-quality healthcare facilities with the best possible experience for staff, patients, and their families.

References

Adair, J. G. (1984). The Hawthorne effect: A reconsideration of the methodological artifact. Journal of Applied Psychology, [illegible]
Bechtel, R. (1967). Human Movement and Architecture. [illegible]
Fisher, R. J. (1993). Social Desirability Bias and the Validity of Indirect Questioning. Journal of Consumer Research, [illegible]
Ng, C. F. (2016). Behavioral mapping and tracking. Research methods for environmental psychology, [illegible]

DESIGNING AN AUTISM-FRIENDLY EXPERIENCE FOR ADULTS IN PALMER PARK

Micaela Ada | Advised by Dr. Naomi Sachs | University of Maryland

EDRA 2024 Poster Session

1. Introduction

As of 2020, **1 in 36 children** in the US are estimated to have Autism Spectrum Disorder (ASD). Youth with ASD and their families are eligible for support programs under the *Individuals with Disabilities Education Act*, including health-related services, personal counseling, personal assistance, and case management. At the age of 21, autistic individuals lose access to such services, and must transition to arranging their own support services. **A dominant concern for autistic individuals and their families during this transition into adulthood is quality of life.**

Creating an ASD-friendly built environment is a burgeoning focus across design professions, **largely improving workplace, home, and therapeutic environments.**

Addressing this gap has potential to create better landscapes of leisure for adults with ASD. Further, expanding the current body of research around ASD-friendly environmental design further aligns the landscape architecture practice with the American Society of Landscape Architecture's Universal Design Principles and, broadly, the United Nation's Sustainable Development Goals 10 and 11.

UN Sustainable Development Goals 2015

Aim and Objectives

Distinguish design considerations a public park settings that create positive recreation experiences for autistic adult.

Identify barriers and promoters to public park use through community engagement with autistic adults and practitioners in the ASD community

2. Literature Review

ASD in Adulthood

For adults with ASD, quality of life is frequently challenged by external circumstances including: **lack of awareness about the needs of people with ASD, and fewer opportunities and environments that accommodate ASD-related needs.**

Adults with ASD experience disparities in health outcomes including high rates of cardiovascular problems and mental health conditions. It is common for conditions to be co-occurring.

Quality of life in adulthood is heavily influenced by **self-determination, social inclusion and relationships, personal well-being, and personal development and autonomy.**

Nature and Well-Being

Time spent in nature is associated with improved mental and physical health

Outdoor environments can encourage social cohesion and integration

Outdoor environments can be a platform to strengthen connection to one's community

Designing for ASD

Transitions between activity areas

Multimodal communication

Shelter from elements

Non-aggravating materials and stimuli

Distinct areas of varied sensory stimulation levels

Safe routes to and within parks

3. Methodology

Voices from the autism community were foundational to the design research. Community members were invited to participate in a focus group and interviews.

Focus Group with Self-Advocates
Participants recruited from University of Maryland's Social Interaction Group Network for All
Participants:
3 Autistic Self-Advocates
Male-identifying | Ages 18-29
Facilitation:
14 Questions | 1 Hour Zoom Session

Interviews
Three 1-hour interview sessions were conducted with three different professionals from various perspectives on the relationship between individuals with ASD and the landscape

Participants:

- An autism parent and landscape designer
- A landscape architect who designed a garden for autistic adults with high support needs
- A caretaker for autistic adults

Case study analysis acted as the final component

Case Study
Analysis of 3 existing landscapes designed through a lens of ASD
Landscape Typologies: trail system, therapeutic garden, residential campus

Autism Nature Trail | FWLA

Morgan's Garden | Virginia Burt Designs

Sweetwater Spectrum | Roche and Roche

4. Findings

Across the research methods, there was a common desire from autistic adults, and their communities of support, for park spaces to encourage personal autonomy and a sense of safety, especially how these may show up in four significant moments of park use:

- **Journey planning and preparedness**
- **Getting to and around a park**
- **Spontaneous situation control in the park**
- **Particpating in park experiences**

6 considerations of public park space distilled as the key qualitites for better design for adults with ASD.

Sensory Flexibility
Having areas of various levels stimulation to accommodate an individual's sensory needs as they arise

Navigability
Providing ample tools to create legible journeys to different destinations

Clarity
Making design choices that distinguish the boundaries of each activity area

Predictability
Using layout, signage, and built-in transitions that aide the ability to know and prepare for what is coming next in a journey

Variety
Providing interesting, unique experiences that encourage exploration, especially experiences that are inviting to adults

Safety
Promote safety of a visitor and accompanying support with sightlines, robust materiality, communication infrastructure, and routes to distance from a situation

5. Conclusion

For individuals with ASD, some needs related to adjusting to sensory stimuli and environment flexibility carry into adulthood. Based on the findings, distinguishable needs for lower support autistic adults center around independence, including means to easily get to and around parks, spaces that complement self-regulation, and points of interest that encourage use. Though not the scope, the needs of higher support autistic adults also come up in the research. These center around safety and highlighted caretakers and travel companions as stakeholders, emphasizing robust park materials, legible landscape navigation, and communication lines.

6. Recommendations

Scan QR code for full bibliography

STREETWISE SAFE HAVENS | Applying Observation + Theoretical Review to Designing Environments for Houseless Women with Substance Use Disorder

CLAIRE H. WATSON, MIA, ALLIED ASID, ASSOC. AIA
8TH SENSE DESIGN

ABSTRACT
Substance Use Disorder (SUD) and houselessness have many overlaps in terms of internal and external experiences. The two often go hand-in-hand - with substance use disorder as both a cause and effect of being houseless (Polcin, 2015). For this project, I have chosen to focus on houseless women with SUD because women are under-represented in SUD treatment services as a result of historical trauma and victimization, shame, stigma, and other gender-specific barriers (Rizzo et al., 2022). Using observation and theoretical review, the research aims to uncover how houseless individuals create a sense of shelter for themselves when living on the street and how that design language can be translated into the design of treatment and recovery environments.

RESEARCH QUESTIONS

- What are the overlaps in houselessness and substance use disorder?
- How can existing theories and frameworks inform the design of recovery and treatment environments for houseless women with substance use disorder?
- How do houseless individuals create a sense of shelter when living on the streets?
- What are the characteristics of an evidence-based designed recovery or treatment facility? What does it look like? What are the desired psychological effects for the end users?

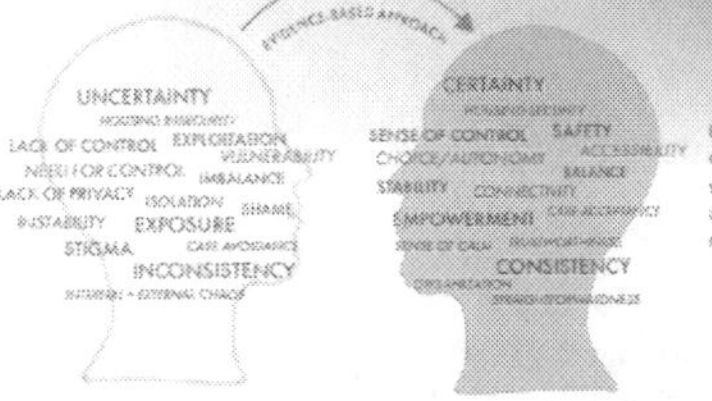

Understanding the overlaps in the internal and external experiences of houselessness and substance use disorder for women and creating an evidence-based design framework that responds to the needs of this population.

DIAGRAM | OBSERVATIONS OF HOUSELESSNESS
THROUGH THE LENS OF JAY APPLETON'S PROSPECT-REFUGE THEORY (1975)

Naturalistic observation was used as a research method to understand how the houseless population interacts with the environment around them without interference. The population was observed in how they create a sense of shelter for themselves in a vulnerable setting (on the street, in parks, etc.). The elements on the top level provide protection from the ground, while the elements below the line makes contact with the ground, but provide enclosure from other sides. The elements closer to the right side of the horizontal line are closer to being a "personal safe haven," which means they provide a sense of security and privacy on various levels. Benches in bus shelters or having their back against the wall provides a layer of protection (from the back), without obstructing a view out - high prospect-refuge.
These findings related to Jay Appleton's Prospect-Refuge Theory, which is supports the tendency for people to prefer environments with unobstructed views (prospect) and areas of concealment and retreat (refuge) (Appleton, 1996).

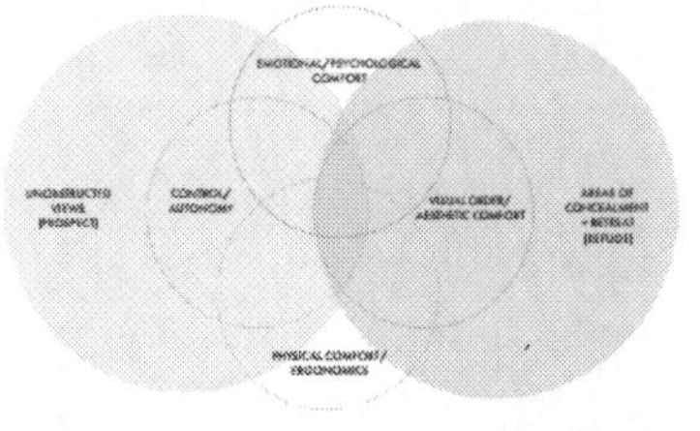

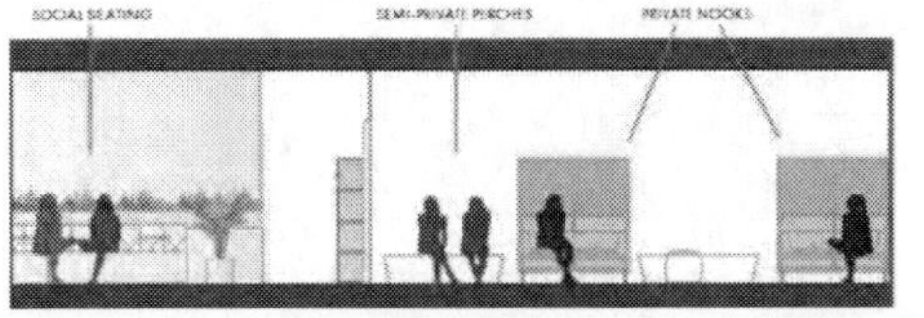

INTAKE/WAITING AREA - PROSPECT/REFUGE | BEAUTY, CHOICE/AUTONOMY, POSITIVE DISTRACTIONS

CLIENT BEDROOMS - REFUGE | PRIVACY, PERSONAL SPACE/STORAGE, RETREAT, SENSE OF CALM

GROUP ROOMS - PROSPECT/REFUGE | CHOICE/AUTONOMY, SEATING OPTIONS, BIOPHILIA, VISUAL BEAUTY

REFERENCES

Appleton, J. (1996). The Experience of Landscape (Revised). John Wiley & Sons.

National Institute of Mental Health. (2023, March). Substance use and co-occurring mental disorders. https://www.nimh.nih.gov/health/topics/substance-use-and-mental-health

Polcin, D. L. (2015). Co-occurring substance abuse and mental health problems among homeless persons: Suggestions for research and practice. Journal of Social Distress and the Homeless, 25(1), 1–10. https://doi.org/10.1179/1573658x15y.0000000004

Rizzo, D., Mu, T., Cotroneo, S., & Arunogiri, S. (2022). Barriers to accessing addiction treatment for women at risk of homelessness. Frontiers in Global Women's Health, 3. https://doi.org/10.3389/fgwh.2022.795532

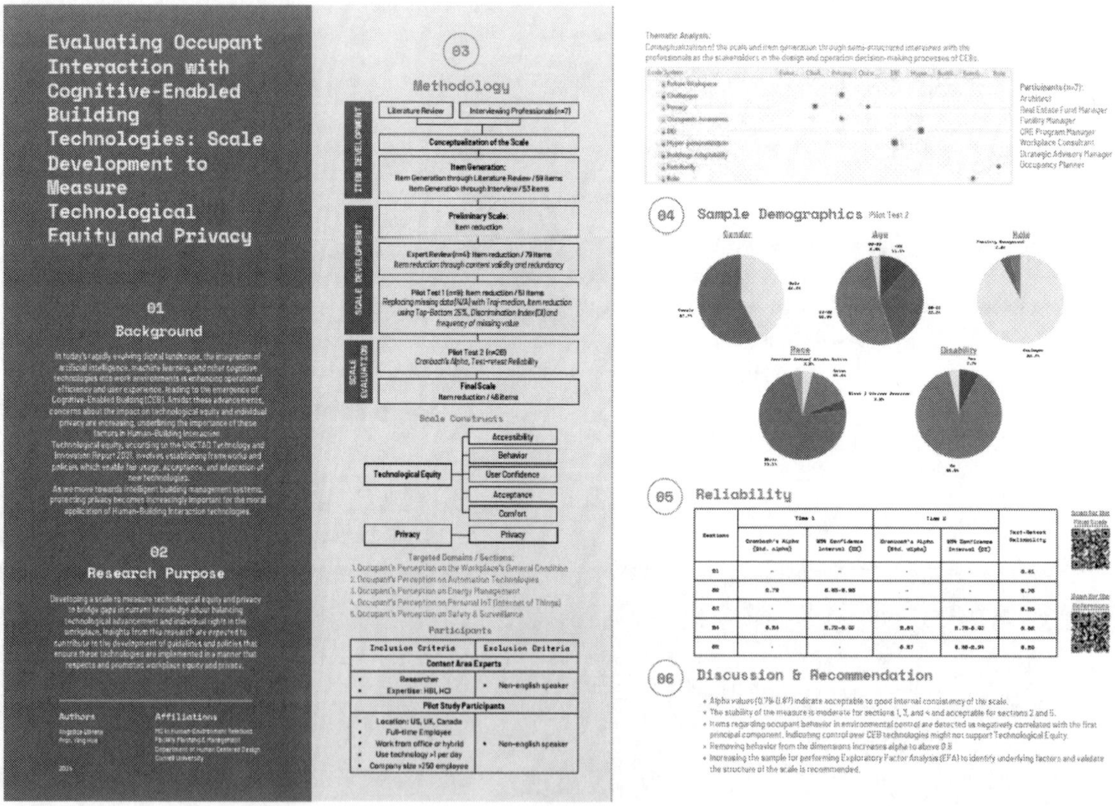
Evaluating Occupant Interaction with Cognitive-Enabled Building Technologies: Scale Development to Measure Technological Equity and Privacy
01
Background
02
Research Purpose
Authors
Affiliations
03
Methodology
ITEM DEVELOPMENT
SCALE DEVELOPMENT
SCALE EVALUATION
Literature Review
Interviewing Professionals(n=7)
Conceptualization of the Scale
Item Generation:
Item Generation through Literature Review / 59 items
Item Generation through Interview / 53 items
Preliminary Scale:
Item reduction
Expert Review(n=4): Item reduction / 79 items
Item reduction through content validity and redundancy
Pilot Test 1 (n=9): Item reduction / 51 items
Replacing missing data(N/A) with Trai-median, Item reduction using Top-Bottom 25%, Discrimination Index(DI) and frequency of missing value
Pilot Test 2 (n=26)
Cronbach's Alpha, Test-retest Reliability
Final Scale
Item reduction / 48 items
Scale Constructs
Technological Equity
Accessibility
Behavior
User Confidence
Acceptance
Comfort
Privacy
Privacy
Targeted Domains / Sections:
1. Occupant's Perception on the Workplace's General Condition
2. Occupant's Perception on Automation Technologies
3. Occupant's Perception on Energy Management
4. Occupant's Perception on Personal IoT (Internet of Things)
5. Occupant's Perception on Safety & Surveillance
Participants
Inclusion Criteria
Exclusion Criteria
Content Area Experts
Researcher
Expertise: HBI, HCI
Non-english speaker
Pilot Study Participants
Location: US, UK, Canada
Full-time Employee
Work from office or hybrid
Use technology >1 per day
Company size >250 employees
Non-english speaker
Thematic Analysis:
Conceptualization of the scale and item generation through semi-structured interviews with the professionals as the stakeholders in the design and operation decision-making processes of CEBs.
Participants (n=7):
Architect
Real Estate Fund Manager
Facility Manager
CRE Program Manager
Workplace Consultant
Strategic Advisory Manager
Occupancy Planner
04
Sample Demographics
Pilot Test 2
05
Reliability
06
Discussion & Recommendation
Alpha values (0.79-0.87) indicate acceptable to good internal consistency of the scale.
The stability of the measure is moderate for sections 1, 3, and 4 and acceptable for sections 2 and 5.
Items regarding occupant behavior in environmental control are detected as negatively correlated with the first principal component. Indicating control over CEB technologies might not support Technological Equity.
Removing behavior from the dimensions increases alpha to above 0.8
Increasing the sample for performing Exploratory Factor Analysis (EFA) to identify underlying factors and validate the structure of the scale is recommended.

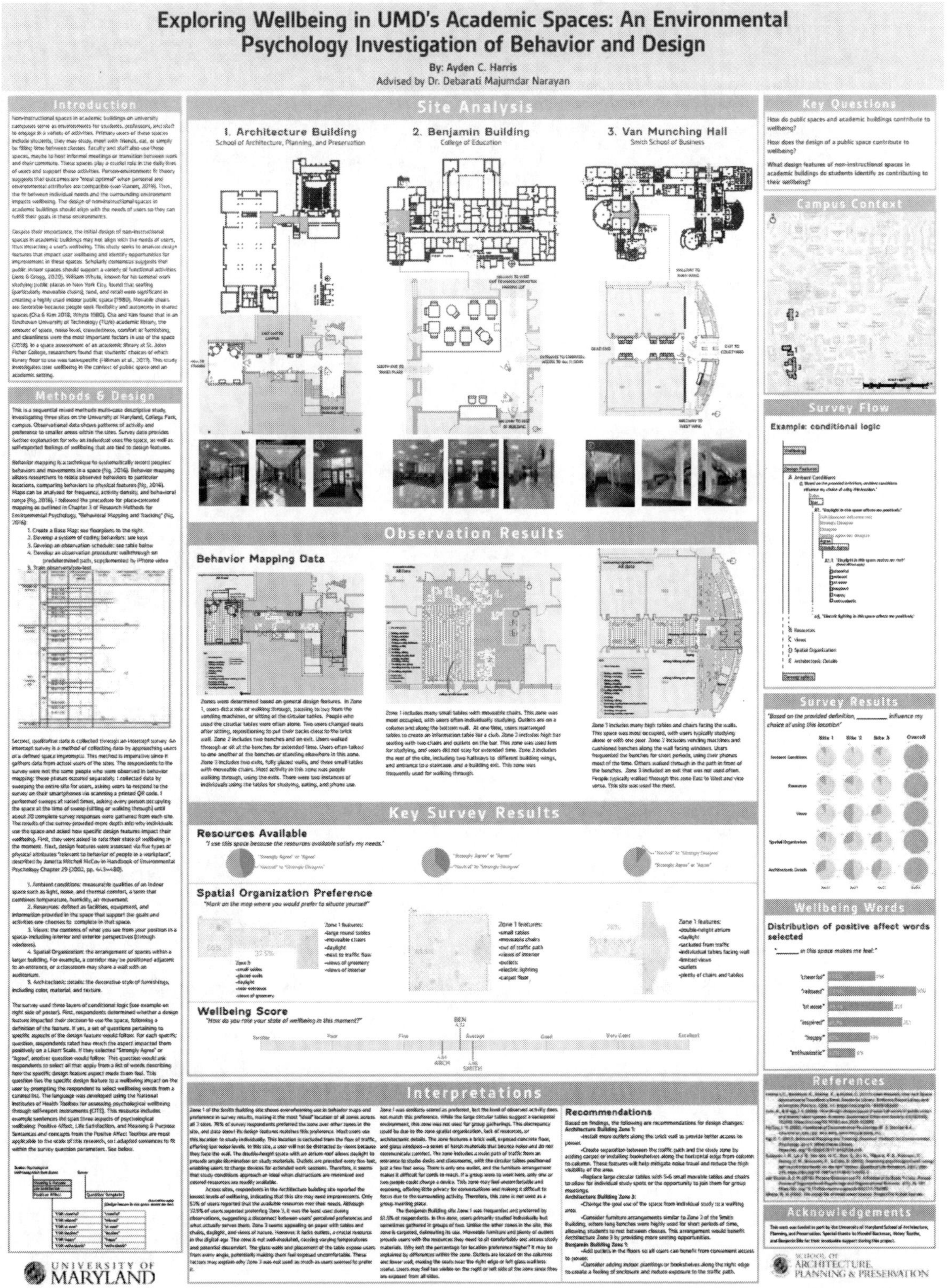
Exploring Wellbeing in UMD's Academic Spaces: An Environmental Psychology Investigation of Behavior and Design
By: Ayden C. Harris
Advised by Dr. Debarati Majumdar Narayan
Introduction
Site Analysis
1. Architecture Building
School of Architecture, Planning, and Preservation
2. Benjamin Building
College of Education
3. Van Munching Hall
Smith School of Business
Key Questions
Campus Context
Methods & Design
Observation Results
Behavior Mapping Data
Survey Flow
Example: conditional logic
Survey Results
Key Survey Results
Resources Available
Spatial Organization Preference
Wellbeing Score
Wellbeing Words
Distribution of positive affect words selected
Interpretations
Recommendations
References
Acknowledgements
UNIVERSITY OF MARYLAND
SCHOOL OF ARCHITECTURE, PLANNING & PRESERVATION

Architectural Psychology of Stress:
Analyzing the Impact of Material, Formal, and Aesthetic Features on Stress Reactions Using Biophilic Patterns

By Clarice Meffert
Advised by: Dana Cupkova, Dr. Azadeh Sawyer, PhD, Dr. David Creswell, PhD, Sinan Goral, Dr. Kushagra Varma, PhD

Introduction

Stress, by the definition set forth by Cohen et al. 1995, is inherently environmental. Yet, somehow, in a world where we control many aspects of our environment, stress is becoming an increasingly prevalent and severe concern. The modern environment – unnaturally dull, and lifeless – is not only unstimulating but also lacks the biophilic connection to nature that relieves psychological (Yin et al., 2020, Dematté et al., 2018; Zhang et al., 2017) and physiological stress (Zhang et al., 2016; Tsunetsugu et al., 2007).

The impact of environmental materiality on stress is largely overlooked despite its broad human mental health implications.Biophilic design has gained popularity for its cognitive and physiological benefits, as evidenced by studies from Yin et al. (2018), Castro et al. (2022), and Choe et al. (2020). These studies report that biophilic environments reduce stress and negative emotions while enhancing memory, positive emotions, and mindfulness. However, many studies combine multiple biophilic elements, complicating the isolation of specific effects. Wood is one of the few finish materials that has been well-studied in isolation and generally found to be more beneficial for mental health than conventional materials like plaster, influencing stress levels, blood pressure, and overall comfort. Although promising, alternative sustainable materials like bamboo need further research. Concrete, while sustainable, often receives negative perceptions unless enhanced with biophilic textures. To optimize interior design for stress reduction, it's essential to understand how specific materials and textures may impact stress. This study aims to isolate and examine the effects of different interior finish materials and textures to identify environments most conducive to stress reduction

Methods

As determined by G*Power, 25 participants were recruited to participate in three 30-minute sessions over three days. Carnegie Mellon University's SONA system was used to recruit, offering course credit or a $25 gift card. Flyers were also posted on campus.

Most participants were undergraduates (84%) with an average age of 20.26 years. 64% idenitied as female, 28% male, 4% non-binary/third gender, and 4% undisclosed. Ethnically, 44% were Asian, 28% white, 16% mixed race, 4% Black or African American, and 4% undisclosed.

The study involved three environments: conventional plaster (control), biophilic-textured concrete, and bamboo, covering 50% of participants' visual field. Bamboo panels from CALI and TAKTL's 'Reeds' pattern textured concrete were used.

Biometric data collection employed the GE DinaMap Carescape™ monitor to measure pulse rate and blood pressure at two-minute intervals. Stress was induced using a gamified Stroop task. For stress reduction, participants listened to a 20-minute mindfulness meditation or control audio, rating their focus at chime intervals.

Self-report data was gathered via Qualtrics surveys at baseline, pre-session, mid-intervention, and post-session, including the Perceived Stress Scale, Mindful Attention and Awareness Scale, Nature Relatedness Scale, PANAS, POMS, and STICSA.

Data analysis was conducted with IBM-SPSS 28, using MRANOVA, MANOVA, and ANOVA. Significant interactions were further analyzed with Tukey's HSD Post Hoc test and pairwise comparisons with Bonferroni correction.

Figures (top left to bottom right)
Figures 1-2 Conventional environment
Figures 3-4 TAKTL Reed biophilic patterned concrete environment
Figures 5-6 Bamboo environment

Hypotheses & Results

Material Results
H1: Bamboo is comparable to conventional woods.
Accepted. Bamboo maintained systolic blood pressure at 113-118 mmHg, similar to Zhang et al.'s findings (114-116 mmHg) with 50% wood rooms. Bamboo also resulted in more positive ($M = 15.07$, $SD = 0.99$) than negative emotions ($M = 7.71$, $SD = 0.44$) on PANAS.

Intervention Results
H2: Mindfulness decreases stress more than control.
Accepted for psychological stress, not physiological. Mindfulness participants reported better mood and affect ($F_{1,73} = 15.09$, $p < .001$; PANAS, $F_{1,73} = 18.02$, $p < .001$), lower mood disturbance ($M = -1.05$, $SD = .625$, $F_{1,73} = 5.33$, $p = .024$), and reduced negative affect ($M = 8.42$ to 7.40, Wilks' Lambda = .78, $F_{1,73} = 21.13$, $p < .001$, $\eta^2 = .22$) and somatic distress ($M = 14.66$ to 12.71, Wilks' Lambda = .79, $F_{1,73} = 20.00$, $p < .001$, $\eta^2 = .22$).

Environmental Condition and Interaction Results
H4: Biophilic designs improve Stroop task performance.
Rejected. No significant differences in Stroop task performance were found.

H5: Biophilic environments reduce initial stress induction.
Accepted. Conventional environments significantly increased systolic blood pressure ($M = 118.93$ to 122.04, Wilks' Lambda = .85, $F_{3,66} = 3.93$, $p = .01$, $\eta^2 = .15$) compared to biophilic environments.

H6: Biophilic environments enhance stress reduction during mindfulness.
Rejected. No significant differences in physiological or psychological stress measures

H7: Bamboo reduces initial stress induction the most.
Rejected. No significant differences between biophilic environments were found.

H8: Bamboo is the best primer for stress reduction.
Accepted. Bamboo significantly decreased systolic blood pressure from session start ($M = 117.53$ to 114.29) and stress induction ($M = 120.61$ to 114.29, Wilks' Lambda = .66, $F_{3,66} = 2.44$, $p < .001$, $\eta^2 = .34$).

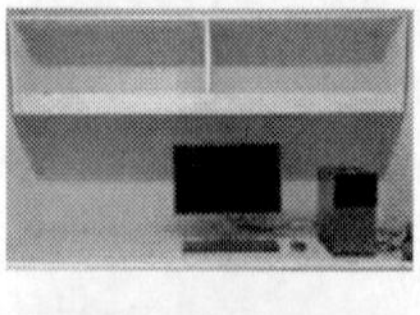

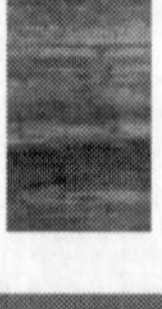

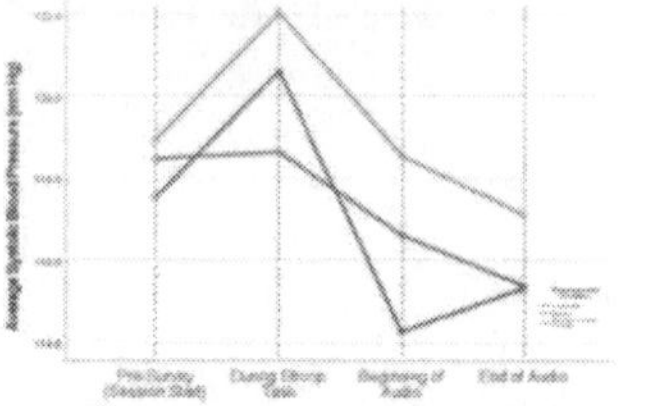

Figure 7 Systolic blood pressure by environment over the course of a session

Discussion

This study underscores the importance of environmental materiality in design and highlights the reciprocal relationship between individuals and their surroundings. Findings reveal that environmental conditions notably affect physiological stress responses. For instance, while the bamboo environment initially allowed blood pressure elevation, it then rapidly reduced systolic blood pressure, indicating stress reduction capabilities. Similarly, biophilic concrete environments moderated stress reactivity, preventing significant elevations in blood pressure to begin with.The study suggests that bamboo environments may be ideal for quick stress reduction, while mildly biophilic concrete environments are better for stress moderation.

Listening to mindfulness meditation enhances mood, decreases somatic distress, and improves environmental perception. Participants engaged in mindfulness meditation perceived their environment as more pleasant, natural, protected, and complex over time. These improvements suggest that negatively perceived environments can be ameliorated through meditation and improved mood.

Exploratory analyses looked at the impact of environmental conditions and personal traits on environmental perception. The biophilic concrete room was perceived as warmer, while the bamboo environment was seen as more natural. Individuals with low nature relatedness perceived biophilic environments as less familiar and the conventional environment as more familiar. High nature relatedness participants found biophilic environments more natural. Thus, environmental perceptions are influenced not only by the environment itself but also by occupant traits.

Overall, while modern environments differ from ancestral landscapes, considering environmental materiality is crucial to prevent unintentionally stressful surroundings. This study emphasizes the complex interplay between environment, personal traits, and stress responses, advocating for consideration of individual differences and intended program for an overall more mindful material selection for environmental design.

Limitations and Future Directions

This study highlighted the impact of the environment on stress levels while also suggesting areas for future research. Limitations included the inability to control room temperature and the need for larger sample sizes to explore the influence of dispositional mindfulness and nature relatedness.

Future studies could investigate the effects of different biophilic textures on stress and evaluate various stressors' impacts. Exploring how internal states and meditation affect environmental perception could inform design strategies to enhance occupants' well-being.

References

Reading Undervalued Landscapes

Sharing curriculum for expanding precedent studies

Carey Clouse
Professor of Architecture and Landscape Architecture
University of Massachusetts, Amherst

This poster shares a pedagogical structure for examining under-studied and under-valued landscapes in North America. These include historical and contemporary examples of invisible, unsanctioned, and under-appreciated types of landscapes, and interrogate the reasons for their relative distance from the public eye. Many of these spaces have developed at an intentional remove from normative design and planning processes, and yet they serve enormously important functions for groups and individuals. Examples of such types of places include foraging landscapes, wayfinding trees, and the Jewish sabbath markers called eruv. These counter-hegemonic design interventions can be studied to teach students about diverse in-terests, to make a case for the preservation of local sites and approaches, and per-haps even inform prevailing planning practices.

The class is called Hidden Landscapes, and it serves as a one-credit optional sur-vey course in the Commonwealth Honors College for non-design majors who are interested in learning more about a new discipline. Because of this audience, I typ-ically work with students who are admittedly non-experts, and as a result, lack so many of the preconceived notions that come with extensive design training. In this course, disciplinary blindness is an asset, allowing students to encounter material that is, perhaps, at odds with disciplinary norms, or the canon, with fresh and open eyes. And because the class is small and discussion based, the students have the time and space for lively discourse about the course content.

The class is structured around a weekly presentation of a hidden landscape, which is paired with a reading on the topic. The precedents change every semester, to bring in different voices and expand the work presented. Students research, re-flect, and present on these topics weekly. They begin by encountering a landscape feature or phenomenon that is new or different, perhaps barely visible, and then try to make sense of it, through research, interviews, and observation. Many of the features that my students have studied over the years are not actually invisible, but they are tucked away, perhaps only understood by one subset of people, and per-haps not widely understood, in the context of landscape architecture education. These are the markers found in everyday spaces, for instance, the types of features that hold meaning for certain groups of people, usually not glorified, but im-portant all the same.

Over the years, students have worked to pick apart hidden landscapes in an effort to center otherwise marginalized human experiences, because these landscapes have much to reveal about peoples' lives, interests and values. As such, they be-come important indicators of use, of memory, or shared experience.

I have worked with students in this course to intentionally develop or enlarge the types of landscapes that are featured. As such, this class could be offered as a companion credit to a more typical landscape history class: one that centers obser-vation and awareness, that encourages students to look for meaning embedded within every day, landscapes, and that calls on students to highlight the DIY or low budget design interventions in their own backyards. This type of class is not meant to displace history/theory curriculum, but could be seen as complementary to the more dominant or popular narratives about landscapes that currently make up the canon.

The course is divided up into several different sections organized by topics such as race, religion, indigenous practices and environmentalism. Whenever possible, I invite an external champion representing each landscape to attend the class. This enables an expert to introduce the topic and serve as an interviewee for the stu-dents. Each landscape is then studied, interviewed and mapped or drawn. We also frame this inquiry with theory, beginning with Matthew Gandy's articulation of "unintentional landscapes," Gilles Clement's "Third Landscapes" and de Solà-Morales Rubió's conceptualization of "terrain vague." The students select a single case study to investigate, and read about that particular landscape phenomenon in detail.

In foregrounding the experience and values that comprise under-represented land-scapes, we are able to see more clearly that those processes, approaches and inter-ests operate at the margins. By looking to invisible landscapes, we see what is un-der-reported, not dominant, nonhegemonic. These stories shine a light on actors and groups who might not otherwise be considered in landscape architecture dis-course.

Throughout this course about reading hidden landscapes, students are asked to think critically about what constitutes a landscape with meaning or value. The un-earthing of unsung stories or experiences or histories helps to shape my students reading of these new landscapes. In this sense the students move beyond an aes-thetic or performative reading of landscape to see value in cultural, social and re-ligious sites. They are asked to understand spaces from another racial, ethnic or socioeconomic experience than their own, and they are asked to consider, for themselves, what matters in a landscape. These questions and exercises help them to move beyond their initial preconceptions about the discipline of landscape ar-chitecture. Hopefully, it gives students experience in applying research, observa-tion, and storytelling or interviews to their understanding of new material.

Do you know of a hidden landscape? Please share yours below!

Race	Religion	Environmentalism	Indigenous Practices	Food Security	Other
Nation of Walls	Defiant Gardens	Lava Parasites	Wayfinding Trees	Black Forager	The Grey Ghost
Mapping MLK streets across the US	The eruv	Underwater reef production	Species Conservation on Reservations	Opensource Foraging Maps	Tennis Shoes on powerlines
Mapping Black statues in DC				The Guerilla Grafters	Yarn Bombing
					Hobo signage

Native American Wayfinding Trees (Photo credit: Winnetka Historical Society and Timothy Schuler)

Remnant wall (Photo credit: Author)

Remnant wall (Photo credit: Author)

Hobo Moniker from 1927. (Photo credit: Charlie Wray)

Greenpoint. (Photo credit: Tim Daoust)

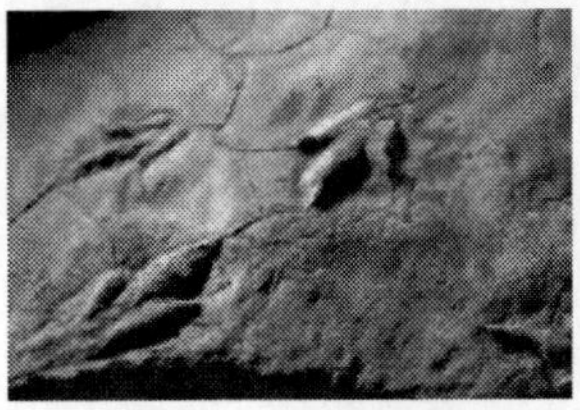

Dinosaur tracks. (Photo credit: Beneski Museum)

Grafted apple (Photo credit: Thomas Levy)

Segregation wall (Photo credit: Chat Travieso)

Screenshot of Fallingfruit.org (website by Ethan Welty and Caleb Phillips)

Piper Bernbaum holds a piece of eruv (Photo credit: University of Waterloo)

Integrating Community Engagement in Urban Design Decision-Making:

A Framework for Blue-Green Infrastructure Implementation for Mitigating Urban Heat Island Effect in Baltimore

PhD Candidate

Doctoral Committee Members

Tasneem Tariq, Penn State

Lisa Domenica Iulo (Co-advisor), Penn State

Ute Poerschke (Co-advisor), Penn State

Hong Wu Penn State

Travis Flohr Penn State

Antonia Hadjimichael, Penn State

ABSTRACT

Urban Heat Island (UHI) effect is an alarming phenomenon where the ambient temperature in cities becomes warmer than in the surrounding rural areas. Baltimore is an eastern American city that faces several environmental and equity challenges including the UHI effect and energy burden. There is a lack of data-driven urban decision-making approaches that can give directives for urban planners to create a comfortable urban microclimate for the community. To address the gap, this research aims to develop a data-driven urban decision-making approach to integrate Blue-Green Infrastructure (BGI) within a city to create a comfortable urban microclimate, which will be accepted by the community. A combination of field survey, computational studies and community engagement workshops will be conducted in two of the most adversely affected neighborhoods of Baltimore by extreme heat. While applied to Baltimore, the intention is to create a framework that is applicable to other cities with similar environmental and social contexts.

***Figure 01:** Urban context of Baltimore with insignificant natural elements (Sources: Author, 2023)*

OBJECTIVE

To provide a framework to support the BGI interventions in Baltimore that will benefit both

1. **Environment** impact on environmental conditions
2. **Community** impact on outdoor thermal comfort and community acceptance

STATEMENT OF INTEREST

- **Co-design** urban design methodology
- **Policy and BGI design support** environmentally conscious urban design
- **Comprehensive framework** applicable to other cities

RESEARCH QUESTIONS

What are the environmental impacts of Blue-Green Infrastructure (BGI) implementation in Baltimore? Which BGI interventions are suggested by computational modeling and embraced through community engagement?

SUB-QUESTION 01 Can the BGI interventions mitigate the UHI effect and improve the outdoor thermal comfort conditions of Baltimore?

SUB-QUESTION 02 How do BGI interventions affect Baltimore's outdoor environment, including air temperature, mean radiant temperature, relative humidity, and air velocity?

SUB-QUESTION 03 How can knowledge co-production enhance decision-making by integrating simulation optimization and community engagement for BGI implementation in Baltimore?

SIGNIFICANCE OF THE RESEARCH

1. Context-based and **Interdisciplinary** BGI interventions
2. Addressing the **Limitations** of Traditional Design
3. **Active Community Participation** in Decision-Making
4. Informed through **Digital Tools**
5. **Local** Contextual Knowledge
6. Integration of **Participatory Design**

METHODOLOGY

Context-Research Problem: Need of data-driven urban decision-making; Need of Community Engagement (CE)

Research Questions: Impacts of BGI integration in Baltimore? Suggested by both computational modeling and community engagement?

Research Activities:

Preparation	CE Stage 01	Simulation	CE Stage 02	Decision Science
Study area → Climatic data	Sharing potential BGIs → BGI preferences	3D model → Test BGIs	Sharing simulation results → BGI acceptance	Identify effective and acceptable BGIs

Research Outcome:

- **BGI design directives** generation for Baltimore
- **Policy formulation** on selection of effective BGI
- **Framework generation** for BGI integration

FIGURE: RESEARCH DIAGRAM OUTLINING THE PROPOSED RESEARCH QUESTIONS AND PLANNED PROGRAM ACTIVITIES. (SOURCE: AUTHOR, 2024)

EXPECTED OUTCOMES

1. **Decision-making** approaches
2. **Comprehensive** framework
3. **Urban design policymaking** guidelines
4. **Community acceptance**
5. **BGI Design decision support**
6. **Environment conscious** urban design

ACKNOWLEDGEMENTS

This work is supported by DOE Award DE-SC0023217, the Baltimore Social Environmental Collaborative Urban Integrated Field Lab. Support for this research is provided by the Department of Energy IFL program, Hamer Center for Community Design, and Penn State Department of Architecture. Envi-met software license supplied by Ecology+Design (E+D), a research lab at Penn State Stuckeman School of Architecture and Landscape Architecture.

This poster presentation is supported by the Department of Architecture, College of Art and Architecture, School of Architecture and Landscape Architecture (SALA), Pennsylvania State University, and the Nana Kirk Travel Scholarship from EDRA.

APPLYING CULTURALLY SENSITIVE DESIGN IN INTERIOR SPACES FOR NATIVE AMERICANS

Jenna Warneke, Tilanka Chandrasekera, Ph.D., & Aditya Jayadas, Ph.D.

Introduction

In 2020, an estimated 3.7 million U.S. residents identified as American Indian and Alaska Native alone, accounting for 1.1% of the US population, while an additional 5.9 million U.S. residents identified as American Indian and Alaska Native and another race group (Office of Minority Health, 2020). One-third (32%) of Native households on Tribal lands live in poverty, compared to 18% of households nationwide, and the annual median household income among Native Americans and Alaska Natives was $37,408, or 33% less than the national median income (Native American Housing, 2017). In addition, often time spaces are designed for Native Americans without applying culturally sensitive design principles (Young, 2022).

A study conducted by Findling in 2022 highlighted the numerous housing challenges faced by rural adult Native Americans in their daily lives. The study revealed that 24% of Native American adults reported experiencing discrimination when attempting to rent housing, 48% faced significant housing problems, 20% encountered issues paying for housing, and 19% noted homelessness issues within their community. These statistics indicate that nearly half of the Native American adults surveyed reported at least one serious housing issue. Despite these ongoing challenges, housing for Native Americans continues to be designed without their specific needs and cultural considerations in mind. Homes are often built by non-tribal members who adhere to Western standards of housing, which can limit Native Americans' ability to participate in cultural practices, separate them from their tribal communities, and result in poorly designed, low-quality housing.

Extensive research shows that Western culture's special housing needs are addressed with substantial support and innovation. For example, holistic care design for older American adults in the United States, as pointed out by Rantz & Ersek, 2023), has gained significant backing. Facilities are now recognizing the importance of redesigning nursing homes to enhance the quality of life for older adults. An influential article by Mercer in 1996 discussed transformative changes in a nursing home within the Navajo Nation. These changes included providing older adults with access to traditional meals, prayer spaces, bedding, and furniture, as well as options for Western furniture, meals, and clothing. This inclusive approach allowed individuals, whether raised traditionally or not, to choose what made them feel most comfortable and connected to their culture.

Purpose

The purpose of this study was to better understand design preferences of interior spaces for Native Americans.

Methods

The study was carried out in two phases: a) a focus group study with open ended questions on design preferences, and b) a survey on preference of design attributes for interior spaces. Eight participants were recruited for the focus group study. Four of the eight participants identified as Native American and were enrolled with a tribal nation (age: 18-25), and four participants identified as Non-Native American (age: 18-52). The participants were asked nine open-ended questions on preferences of interior spaces. The focus group session was audio recorded. The data was then transcribed, and content analysis was used to determine themes and sub-themes. Next, nine Native Americans (age: 18-39) and 24 Non-Native Americans (age: 18-52) were recruited for the survey. The survey was conducted using 20 Likert scale questions on perceived preferences for design of interior spaces.

Results & Discussion

The qualitative data from the focus group showed that while Native Americans valued artifacts such as tribal artwork and blankets, Non-Native Americans preferred lots of color and bigger spaces. The other big difference was the Native Americans preferred plants and natural elements, while Non-Native Americans preferred hardwood and marble. The themes, sub-themes and quotes from the Native American and Non-Native American participants are summarized in Table1 & Table 2 respectively. As for the quantitative data, there was no statistically significant differences between Native American and Non-Native American participants for natural elements or natural lighting. Similarly, there was no statistically difference between the two groups for preference of having beaded decorations and displaying jewelry in interior spaces. However, there was a statistically significant difference for the preferences of historical items (p=0.03) with Native American reporting a higher score than Non-Native Americans. There was a statistically significant difference for a non-visual attribute, the smell of sage and cedar in interior spaces (p=0.02) with Native Americans preferring this more when compared to Non-Native Americans. The quantitative data is summarized in Figure 1-5.

Table 1. Themes, sub-themes and quotes from Native American participants

Themes	Subthemes	Quotes
Cultural Value	Spirituality	"I pray everyday, but try to smudge pray every two weeks"
	Family	"I want an open space for my family… it keeps to my tribal values"
Material Culture	Tribal Artwork	"Displaying my Native American heritage would be great, through native artwork, pottery, blankets"
	Tribal Blankets	"I always had for blankets on all of my bedding"
	Eagle Feather/Sage	"I pray everyday, but try to smudge pray every two weeks" "I highly value and protect my grandfather's eagle feather"
Natural/Natural Elements	Plants	"Presence of plants would be nice"
	Natural Lighting	"I like high ceilings, natural lighting and big windows"
	Living Sustainability	"Culturally living off the land and being sustainable, natural things regarding the environment"
Floorplan	Open Space	"I want open space for my family"
	Privacy	"I want privacy space for a safe place for me to pray and keep my special belongings"

Table 2. Themes, sub-themes and quotes from Non-Native American participants

Themes	Subthemes	Quotes
Home Elements	Lighting	"Lots of Natural Lighting" "Change my yellow color lighting to white color lighting"
	Kitchens	"I want a bigger Kitchen" "I want a bigger kitchen" "I want a more open Kitchen"
Culture	Color	"I like loud Prints and bright colors" "I like Mid Century… and lots of colors, everything you can think of" "Changing the color of my lights, and changing the color of my baseboards to a darker wood color"
Floorplan	Expanding home	"I wouldn't be living on the amount I live on right now, I want to expand" "I want to live in a bigger house, with a bigger kitchen" "I want a bigger shower"
	Adding personal Usage Rooms	"I want better workspace" "I definitely want a sewing room"
Materials	Hardwood	"changing the color of my baseboards to a darker wood color"
	Marble	"I want marble in my kitchen"

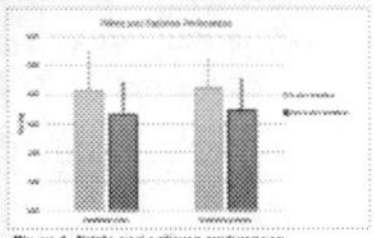

Figure 1. Prints and patterns preferences

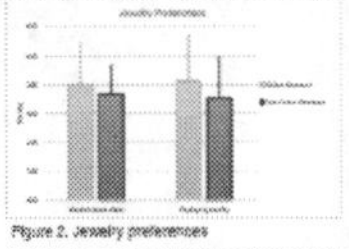

Figure 2. Jewelry preferences

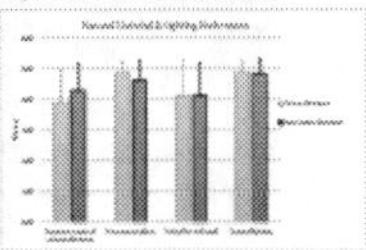

Figure 3. Natural materials/lighting preferences

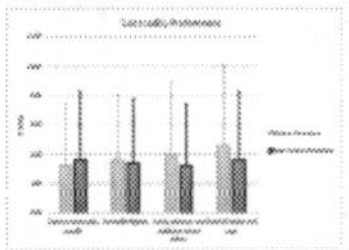

Figure 4. Spirituality preferences

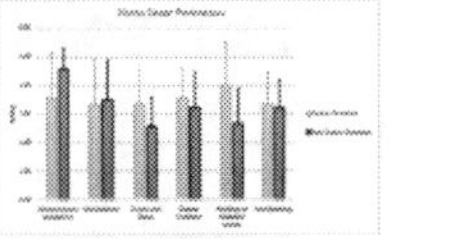

Figure 5. Home decor preferences

Conclusion

The Native American participants in the study preferred to incorporate culture, tradition, and religion into the design of interior spaces. In contrast, Non-Native American participants favored more materialistic options such as a wide range of color choices and larger spaces. This study highlights important implications for the design of interior spaces for both Native Americans and Non-Native Americans, emphasizing the need for culturally sensitive design.

Culturally sensitive design acknowledges and respects the cultural practices, traditions, and preferences of diverse groups. The implications of this study extend to architects, interior designers, and policymakers. For architects and interior designers, it is crucial to engage with community members and stakeholders to understand their unique needs and preferences. This collaborative approach can lead to the creation of living spaces that are not only functional but also culturally enriching and personally meaningful.

References

1. Findling, M. T., Blendon, R. J., Benson, J. M., & Miller, C. (2022). The unseen picture: Issues with healthcare, discrimination, police and safety, and housing experienced by Native American populations in rural America. *The Journal of Rural Health*, 38(1), 180-186.
2. Mercer, S. O. (1996). Navajo elderly people in a reservation nursing home: Admission predictors and culture care practices. *Social work*, 41(2), 181-189.
3. Native American Housing (2017). Housing needs on Native American and Alaska native tribal lands. Retrieved from https://nlihc.org/explore-issues/policy-priorities/native-american-housing
4. Office of Minority Health (2020). American Indian/Alaska Native Health. Retrieved from https://minorityhealth.hhs.gov/american-indianalaska-native-health
5. Rantz, M., & Ersek, M. (2023). Care delivery, quality measurement, and quality improvement in nursing homes: Issues and recommendations from the National Academies' Report on the Quality of Care in Nursing Homes. *Public Policy & Aging Report*, 33(Supplement_1), S11-S15.
6. Young, B. (2022). Building Cultural Competency in Interior Design: When Good Intentions go Awry. *International Journal of Designs for Learning*, 13(1), 129-136.

Design & Merchandising Dept.

Minimizing Glare and Improving Views in NASA's Sustainability Base

William McDonough's design of NASA's "Sustainability Base" research facility in Mountain View, California incorporates daylighting as one of its main strategies for conserving energy and improving the health of its occupants. However, it was found that glare is a common problem throughout the building, as it is in many daylit buildings, which can trigger headaches (Terman) and unneeded stress. This causes some users to override the automatic shading systems, compromising energy efficiency. (Inkarojrit) Using light meters, this research found the glare index for the buildings was 1 to 50, significantly higher than the recommended 1 to 10 light ratio.

Using the *Vital Signs* Case Study methodology developed by the Society of Building Science Educators, on-site light meter readings and analysis of views were taken in key areas of the building. Design recommendations were given to the NASA staff on how to improve views of glary parking lots into non-glary plant walls, making work more comfortable and less stressful. Design recommendations were also given by switching of program workspaces on the second floor that required less daylight to the 1st floor where less daylight was available. Additionally, it was recommended that the insertion of simple skylight tubes on the interior side of outer walls to throw daylight on walls next to windows to reduce glare contrast to more tolerable levels near the 1 to 10 ratio.

Inkarojrit, Vorapat, "Balancing Comfort: Occupants' Control of Window Blinds in Private Offices," Doctoral Dissertation, UC Berkeley. 2005

Terman, Michael, and Jiuan, "Light Therapy for Seasonal and Nonseasonal Depression: Efficacy, Protocol, Safety, and Side Effects," CNS Spectrums, Number 8, Volume 10, August 2005

GLARE and VIEW Conditions

Existing

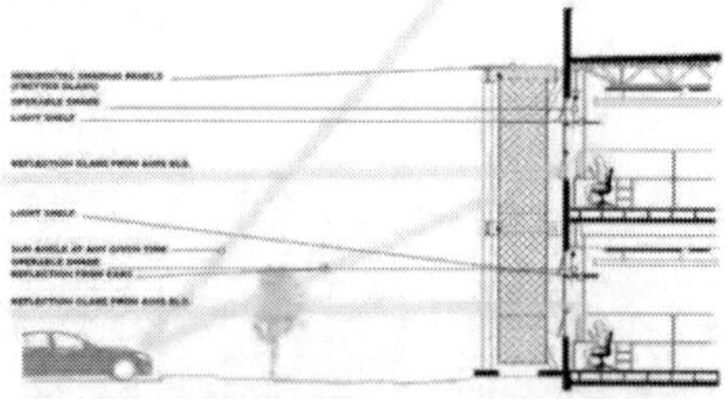

Outside Wall Glare Conditions

Existing Glary Day and Night Parking Lot Views

Proposed Designs

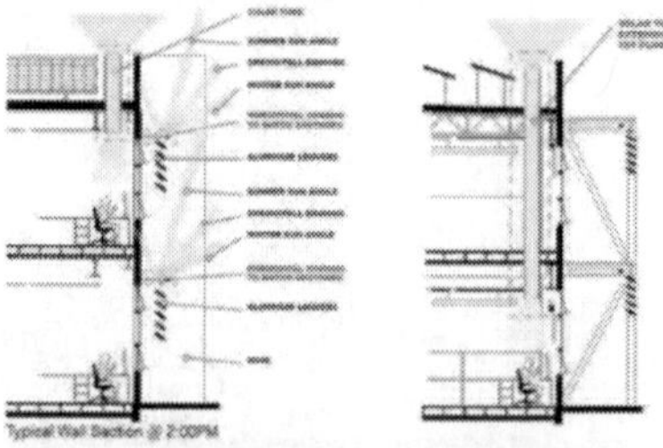

Glare Reduction from use of Solar Tubes Placed Next to Wall

Proposed Plant Wall Outside Window to Block View of Glary Parking Lot

DAYLIGHT and PROGRAM Conditions

Existing Program Configuration

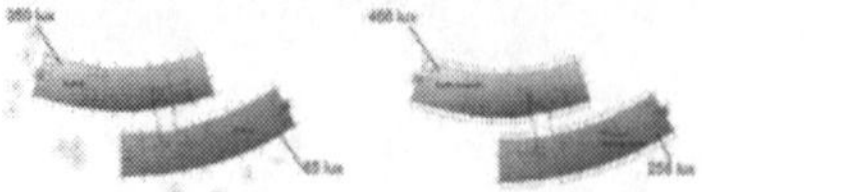

Spaces that need less light are currently placed in areas with more light

Program Re-Configuration

Spaces that need less light are moved to darker areas

Conclusion, Follow up studies can be made for glare and stress reduction if NASA makes the improvements.

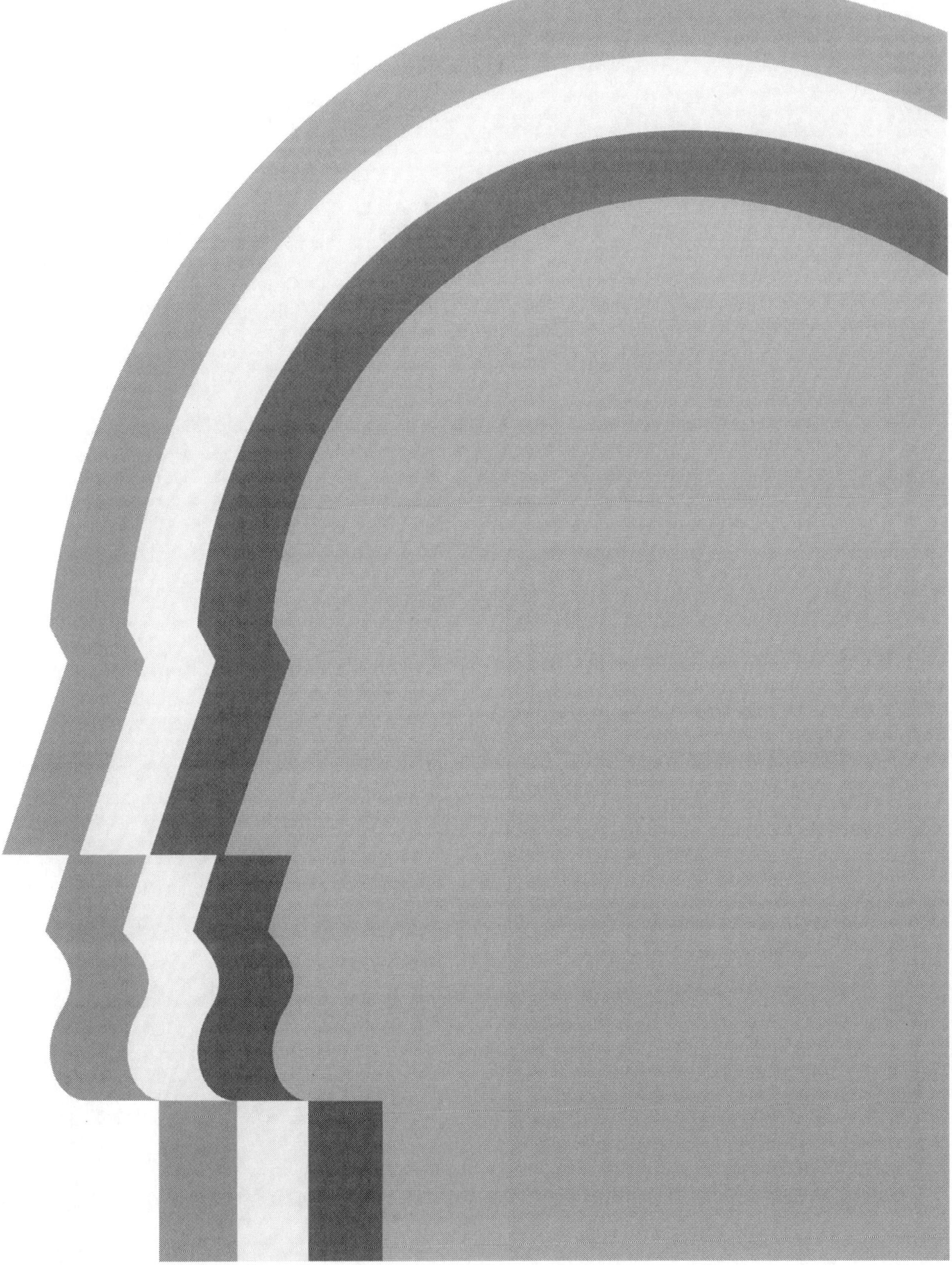

AKNOWLEDGEMENTS

The EDRA55 committee extends its sincere thanks to all our wonderful reviewers, mentors, and volunteers for their assistance with the planning, preparation, and execution of this year's conference!

AKNOWLEDGEMENTS

REVIEWERS

Norhan Abdelgawad
Mahsa Adib
Mahdi Afkhamiaghda
Sana Ahrar
Enkela Alimadhi
Rula Awwad-Rafferty
Caryn Brause
Daniela Casanello Frisius
Colin Caso
Sarayu Chandramouli
Tilanka Chandrasekera
Neel Kamal Chapagain
Nancy Cheng
Ting-ting Cheng
Xihan Cheng
Yongyeon Cho
Marzia Chowdhury
Renee Christensen
Carey Clouse
Jacob Culbertson
Irene Curulli
Michael Darfler
Udday Datta
Fernanda de Moraes Goulart
Lynne Dearborn
Khansa Dhaouadi
Rudy Dieudonne
Suining Ding
Jiaying Dong
Sabeen Durrani
Jill E Bambury
Genell Ebbini
Ihab Elzeyadi
Adrienne Erdman
Kimia Erfani
Carolina Espinoza-Sanhueza
Nisha Fernando
Gabriela Fonseca Pereira
Forough Foroutan Jahromi
Kara Freihoefer
Mariana Garcia Junqueira
Yasaman Gholami
Nasrin Golshany
Maria Gonzalez
Alivia Green
Monica Gripko
Gowri Betrabet Gulwadi
Hazal Gumus Ciftci
Sara Hadavi
Erin Hamilton
Davis Harte
Peter Hecht
Ana Hidalgo
Melissa Hoelting
Zahrasadat Hosseini
Nianchu Hou
Adele Houghton
Hao Huang
Julia Jaffe
Shreya Jaiswal
Mansoureh Jalali
Aditya Jayadas
Sharmin Kader
Gourab Kar
Tulay Karakas
Shno Karimi
Yohei Kato
Ladan Khalvati
Asim Khanal
Mahdi Khansefid
Hans Klein-Hewett
Sophie Kramer
Dowon Lee
Fu LI
Hongfei Li
Connie Lin
Chunyao Liu
Yuanyuan LIU
Zhongyi Liu
Mathilde Loiselle
Quanlin Lu
Yanhua Lu
Qing Luo
James Macard
Caroline Machado
Herminia Machry
Debarati Majumdar Narayan
William Mangold
Lorraine Maxwell
Denise McAllister
Haleh Mehdipour
Luis Mejia-Puig
Derrick Mirindi
Donghwan Moon
Lobna Mostafa
Claudia Berenice Muciño Garcia
Matthew Nicolette
Sepideh Niknia
Benjamin Okenwa
Bryan Orthel
Naghmeh Pak
Cory Parker
Tina Patel
Lynn Paxson
Bo Peng
Wellabadage Ubhaya Upendra Perera
Lubomir Popov
Robin Puttock
Xiaobo Quan
Aakanchcha Raj
Michael Ralph
Widya Ramadhani
Robert Ribe
Chandra Robinson
Kim Rollings
Deni Ruggeri
Hossein Saedi
B Sanborn
Aparajita Santra
Haroon Sattar
Sophie Schuller
Leah Scolere
Kishore Seetharaman
Mina Shokrollahi Ardekani
Devi Soman
Philip Speranza
Julie Stevens
Jodi Sturge
Jing Tian
Elizabeth Tofte

AKNOWLEDGEMENTS

REVIEWERS (CONTINUED)

Elif Tural
Uchita Vaid
HOA VO
Richa Vuppuluri
Huili Wang
Xinxin Wang
Yuzhu Wang
Zhe Wang
Claire Watson
Musab Wedyan
Richard Wener
Nicholas Wise
Chia-Ching Wu
Chunya Wu
Tianhong Yan
Haoyue Yang
Fatemeh Yazdandoust
Mingming Zhao
Xuanyu Zhou
Parisa Ziaesaeidi
Weican Zuo

Mentors

Tatiana Berger
Roxana Jafarifiroozabadi
Hannah Myers
Tom Albright
Elisabetta Caneppa
Kim Rollings
Kati Peditto
Erin Peavey
Davis Harte
Colin Ellard
Arkar Hein
David Borkenhagen
Sara Codarin
Gladis Maestre
Angela Spangler
Payam Tabrizian
Heia Tahri
Widya Ramadhani
Michael Ralph

Floating Mentors (Board Members)

Debarati Majumdar Narayan
Elif Tural
Sharmin Kader

Faculty Volunteer

Prof. Christina Bollo

Student Volunteers

Sherif Khaled A. Abdelhamid
Quincey Dunlap
David Israel
Natalie Price
Chi Rocha
Nicole So

AUTHOR INDEX

— A —

— B —

— C —

— L —

— M —

— N —

edra55

PORTLAND, OREGON

A publication from the Environmental Design Research Association.